Molecular Vaccines

From Prophylaxis to Therapy

分子疫苗

马蒂亚斯·吉斯（Matthias Giese） 编著
马维民 何继军 吴国华 译著

中国农业科学技术出版社

备案号：卷 1（01-2019-7566），卷 2（01-2019-7999）

图书在版编目（CIP）数据

分子疫苗：Molecular Vaccines：From Prophylaxis to Therapy / 马维民，何继军，吴国华译著；（德）马蒂亚斯·吉斯编著．--北京：中国农业科学技术出版社，2021.10

ISBN 978-7-5116-5452-6

Ⅰ．①分…　Ⅱ．①马…②何…③吴…④马…　Ⅲ．①疫苗　Ⅳ．①R979.9

中国版本图书馆 CIP 数据核字（2021）第 163943 号

责任编辑　朱　绯　姚　欢
责任校对　马广洋
责任印制　姜义伟　王思文

出 版 者　中国农业科学技术出版社
　　　　　北京市中关村南大街 12 号　邮编：100081
电　　话　（010）82106632（编辑室）（010）82109702（发行部）
　　　　　（010）82109709（读者服务部）
传　　真　（010）82106631
网　　址　http://www.castp.cn
经 销 者　各地新华书店
印 刷 者　北京建宏印刷有限公司
开　　本　210 mm×285 mm
印　　张　52.25
字　　数　1 542 千字
版　　次　2021 年 10 月第 1 版　2021 年 10 月第 1 次印刷
定　　价　398.00 元

本书由

中国农业科学院兰州兽医研究所

家畜疫病病原生物学国家重点实验室

边境地区外来动物疫病阻断及防控体系研究（No. 2017YFD0501800）

种畜场口蹄疫净化技术集成与示范（No. 2016YFD0501500）

兽用生物制品及检测试剂质量评价标准研究（No. 2017YFF0208600）

资助出版

《分子疫苗》译者名单

主译：马维民　何继军　吴国华

译者　（按姓氏笔画排序）：

马维民　中国农业科学院兰州兽医研究所

吕　律　中国农业科学院兰州兽医研究所

朱学亮　中国农业科学院兰州兽医研究所

杨亚民　中国农业科学院兰州兽医研究所

杨欣燐　兰州市第一人民医院

吴国华　中国农业科学院兰州兽医研究所

何继军　中国农业科学院兰州兽医研究所

张　杰　中国农业科学院兰州兽医研究所

张　强　中国农业科学院兰州兽医研究所

郑海学　中国农业科学院兰州兽医研究所

赵志荀　中国农业科学院兰州兽医研究所

高顺平　中国农业科学院兰州兽医研究所

郭建宏　中国农业科学院兰州兽医研究所

傅松波　兰州大学第一附属医院

靳　野　中国农业科学院兰州兽医研究所

目　　录

第一部分　分子疫苗：从预防到治疗

第二部分　疫苗免疫学

第三部分 传染病疫苗

第四部分 癌症疫苗

第五部分　非传染性疾病疫苗和非癌症疫苗

第六部分　佐剂和纳米技术

第七部分　计算机和输送系统

第八部分　疫苗的专利申请、制造和注册

第一部分

分子疫苗：从预防到治疗

第1章　从巴斯德到个性化疫苗

Matthias Giese①

1.1　1885年：第一个治疗疫苗

1881年，有望成为当时顶尖的疫苗学家路易斯·巴斯德（Louis Pasteur）与其竞争对手兽医亨利·图桑（Henry Toussaint）都在研发首个人工减毒疫苗，最终巴斯德首先研制成功。此前一年巴斯德研制了一种预防禽霍乱的疫苗，他分离出了人畜共患病病原体——多杀性巴氏杆菌（一种革兰氏阴性、非运动性球杆菌）。禽霍乱造成的经济损失无论是当时还是现在都是巨大的。

霍乱疫苗　巴斯德的助手 Emile Roux 提出了一个想法，用不同时间的巴氏杆菌菌液做了一系列免疫实验：（a）用新鲜培养的巴氏杆菌菌液对12只鸡进行免疫；（b）用陈旧巴氏杆菌菌液对12只鸡进行免疫；（c）最后用更旧的巴氏杆菌菌液进行免疫。8天后，A组12只鸡全部死亡（12/12），但是B组4只鸡存活了下来（4/12）；而令人惊讶的是，C组11只鸡存活了下来（11/12）。在接下来的攻毒实验中，用新鲜巴氏杆菌菌液注射11只存活下来的鸡，有8只免疫鸡存活了下来（8/11）。这就证实了用一种古老巴氏杆菌菌液作为疫苗可以预防禽霍乱。众所周知，现在证实这种疫苗会产生严重的副作用，不能减少排毒且免疫持续时间非常短。

巴斯德和氧气　巴斯德得出结论，根据培养条件，巴氏杆菌毒性是可以控制的。培养过程中断或完全停止，巴氏杆菌依然存活，但毒力已经减弱。巴斯德认为，只有细菌暴露在氧气中才是细菌毒力减弱的原因。

琴纳（Jenner）的天花疫苗和巴斯德霍乱疫苗之间最大差异是病原体致弱形式。众所周知，微生物例如天花，可以以有毒和无毒的自然形式存在。琴纳应用了无毒的自然形态，而巴斯德通过特殊的实验室条件分离并致弱了这种病原体，他首次研制出一种人工减毒传染病疫苗。

此外亨利·图桑也在钻研疫苗，但与巴斯德描述的方法不同。巴斯德认为只有活疫苗才能对免疫系统产生适当刺激，而图桑用防腐剂酚灭活炭疽疫苗，后来用重铬酸钾灭活霍乱疫苗。

谁是第一个引进人工生产疫苗概念、验证并证明其可行性和在该领域有应用潜力的科学家？

1881年的炭疽疫苗　1881年6月，巴斯德在小村庄 Pouilly-le-Fort 应邀首次公开演示了实验室生产的羊炭疽疫苗。A组25只羊用炭疽疫苗免疫，B组25只羊作为对照组。随后用强炭疽菌液注射两组羊。结果令人印象深刻：24只（24/25）接种了疫苗的羊存活了下来，但23只未接种疫苗的羊死亡（23/25）。由于巴斯德拒绝公开实验方案，拒绝展示这种疫苗开发的所有细节——这是有充分理由的，真正的诽谤立即接踵而至。

图桑方案　巴斯德采用图桑灭活方案，即加热并结合重铬酸钾来处理病原菌，而不是基于他自己的氧气程序；公众和科学界在这个时候是如何被说服的？巴斯德自己的研究没有成功，因为氧气无法杀灭炭疽芽孢。巴斯德赢得了与图桑的比赛，但他从来没有把成功归功于图桑，而图桑才是这种失活原理的真正发明者。巴斯德赢得了荣耀和知名度，而图桑却被深深湮没，赢家通吃……医学历史学家 Gerald L. Geison 将巴斯德的复制粘贴行为解释为公平的科学欺诈（Gerald L. Geison：《路易斯·巴斯德的私人科学》，普林斯顿大学出版社，1997年，美国）。

1885年的狂犬病疫苗　1885年，巴斯德的研究在全世界范围内取得了突破。巴斯德再次受

---

①　M. Giese，博士（德国海德堡 IMV 分子疫苗研究所，E-mail：info@ imv-heidelberg. com）

到Emile Roux狂犬病实验的启发，研制了一种由受感染兔子的脊髓经风干、乳化而制成的灭活疫苗。当时人们还不知道是否有病毒存在。众所周知的是这种疾病的人畜共患特征和神经症状，因此医生们据此推测患者神经组织受到了感染。

Joseph Meister在9岁的时候被一只患狂犬病的狗咬伤，巴斯德逐步使用具有不同毒性的新鲜分离神经组织作为疫苗给这名男孩注射了13次（图1.1）。1885年10月26日，巴斯德宣布Joseph Meister痊愈，巴斯德因此被誉为英雄。这一系列免疫接种是否是治愈的原因尚未得到证实，因为据估计这种暴露后感染狂犬病的风险在10%左右。值得注意的是，作为预防性治疗，巴斯德只在11只狗身上测试了狂犬病疫苗，在Joseph Meister接种疫苗之前，只对1只感染的兔子进行了1次治疗性免疫接种。此外在Joseph Meister之前不久，对感染狂犬病的11岁女孩Antoinette Poughon第一次治疗失败了。女孩和兔子都死了。

这是世界上第一次有文献记载的人类医学治疗性疫苗接种（Hervé Bazin：L' Histoire des vaccinations，John Libby Eurotext，2008，France；Pasteur Vallery-Radot（ed）：Œuvres de Pasteur. Volume 6：Maladies virulentes，virus-vaccins et prophylaxie de la rage-Méthode pour prévenir la rage après morsure，Masson，Paris 1933，France）。

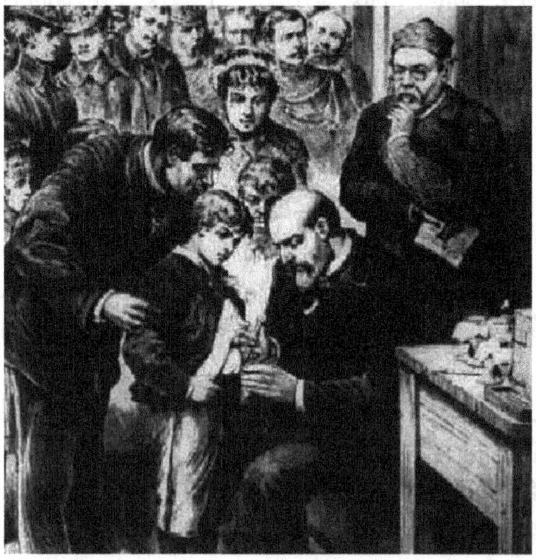

图1.1　1885年巴斯德给9岁的Joseph Meister注射了第一支灭活狂犬病疫苗

1.2　系统疫苗学

病原体分离、灭活或致弱以及注射是任何疫苗研制的基本规则。百年来巴斯德疫苗一直伴随着我们。据估计，全球所有许可的动物疫苗中95%仍然是巴斯德疫苗，这种疫苗开发既便宜又简单，但人类疫苗的情况则完全不同。

毫无疑问，这些经典疫苗极大地提高了我们的生活质量。分子科学的巨大进步几乎以同样的方式完全改变了疫苗的研究和开发，但大多数都没有被公众注意到。

乙型肝炎表面抗原（HBsAg）疫苗　首个注册的人类重组疫苗是预防乙肝病毒感染的。1984年将表面抗原（HBsAg）的cDNA克隆到转染酵母细胞的载体中作为生产系统[1-3]。这种重组产品取代了以前的乙型肝炎疫苗，即所谓的血浆疫苗：乙型肝炎表面抗原粒子从慢性感染患者的血液中分离、纯化并灭活。临床前和临床研究证明了第一种重组疫苗的安全性、有效性和经济性。分子生物学的严格增强和一致应用被证明是药物和疫苗开发中生物医学研究的正确方式。

反向遗传学　建立于 20 世纪 90 年代的反向遗传学首先通过基因测序开始基因功能分析。一旦解码，这个序列被很好地定向（靶向）并通过使用各种技术改变以研究对生物体的影响。因此，靶向诱变和产生的新表型之间因果关系是严格的、精确的和可重复的。相反的方向即由表及里是由经典正向遗传学来解决的问题。

从给定的表型开始，研究该表型的遗传基础。通过反向遗传学技术改变的第一种病原是流感病毒[4,5]。同时，位点特异性突变可以对具有特定生物学特性的病毒进行工程改造。

更复杂的有机体如细菌也可以通过反向遗传学来研究疫苗开发。

非洲脑膜炎地带　脑膜炎奈瑟氏球菌是一种革兰氏阴性双球菌，通常被称为脑膜炎球菌，是一种有荚膜的细菌，寄居于人鼻咽部，偶尔会引起侵袭性疾病。它只感染人类，目前还不清楚其宿主动物。根据细菌荚膜多糖抗原不同，将脑膜炎球菌分为 13 个血清群，但是只有 6 个群，即 A、B、C、W-135、X 和 Y 与严重侵袭性疾病有关并可能导致流行[6]。世卫组织报告，脑膜炎主要发生在第三世界，每年有 4 亿生活在“非洲脑膜炎地带”（从塞内加尔到埃塞俄比亚的 21 个国家）的人患有由奈瑟氏菌引起的细菌性脑膜炎，2010 年病死率为 10.6%（GHO：流行性脑脊髓膜炎，http：//www.who.int/gho/epidemic_diseases/meningitis/en/ index.html，WHO 2013）。

脑膜炎球菌　流行性脑脊髓膜炎是一个全球性健康问题，在世界各地零星发生。脑膜炎球菌 A 群、W-135 群和 Y 群是非洲主要流行病原体，而在工业化国家主要是脑膜炎球菌 B 群、C 群及 Y 群（美国）占主导地位。市场上已经有几种商品化疫苗可以应用。2010 年世卫组织预审了专为非洲开发的第一种 A 群脑膜炎球菌结合疫苗（PsA-TT）（MenAfriVac，印度血清研究所有限公司）。同时一些针对血清群 A、C、W-135 和 Y 的 4 价脑膜炎球菌结合疫苗在欧洲、北美和亚洲获得许可，可以对 2~55 岁的人群进行预防接种。这些结合疫苗取代了以前的胶囊多糖疫苗。胶囊多糖疫苗在婴儿中免疫原性很差，由于这种疫苗不依赖于 T 细胞，所以没有可测的记忆反应，且诱导的免疫很短暂（参见第 2 章）。

反向疫苗学　其瓶颈是脑膜炎球菌 B 群（MenB）。脑膜炎球菌 B 群引起全球 50%的流行性脑脊髓膜炎。主要基于胶囊多糖的疫苗开发受到阻碍，这种细菌糖类分子抗原结构与胎儿脑细胞黏附分子相同[7]。此外大多数表面蛋白序列高度可变。但是对自身免疫的高度关注导致使用非囊状结构开发疫苗。按照在病毒上展示的反向遗传学路线，以及细菌整个基因组首次公开[8]，对 MenB 强毒株 MC58 基因组进行完全测序，以鉴定候选疫苗株[9]。后来这种方法被称为反向疫苗学，包括免疫学和生化研究[10]。

rMenB　脑膜炎奈瑟球菌血清 B 群中 MC58 株染色体是单倍体染色体，含有 2.2 百万个碱基对，2 272 360个核苷酸，2 225个基因（ 2 063个蛋白质）。已对所有基因进行了测序，共鉴定出 570 个用于基因预测的开放阅读框（ORF），其中 350 个在 *E. coli* 中成功表达。从这些潜在候选疫苗蛋白（包括脂蛋白、外膜和内膜蛋白以及跨膜蛋白和未知蛋白）中，发现 28 种蛋白质在小鼠体内能产生抗 MenB 的中和抗体，显示出杀菌活性。在几个试验中只有 7 个重组蛋白显示是阳性并被选用进一步研究。最后将 5 种候选疫苗蛋白组合成一种疫苗制剂，命名为 rMenB（诺华公司：瑞士制药公司）[11]。这种疫苗正在进行临床试验。

反向疫苗学的唯一目标是发现新抗原并预测 B 细胞和 T 细胞应答表位，其关键策略为比较基因组测序与基因鉴定、蛋白质抗原预测与表位定位、候选抗原重组表达、体内外免疫研究。这种方法侧重于微生物有机体的分离部分，不使用系统生物学。在干实验中，整个过程大致可以分为 4 个主要活动，前两步是在电脑上进行的。

DNA 微阵列　序列分析、基因发现、与相关病原比较分析以及通过电脑运用生物信息学方法实时预测数千种蛋白质（另见第 42 章）。DNA 芯片（微阵列）用于研究基因表达模式，可以识别基因簇以及上调和下调基因，检测生物学途径和网络；可以描述已知的和新的免疫相关基因和途径[12,13]。

电脑模拟　首要挑战是运用电脑定位表位来预测蛋白质，也称为干实验，区别于在实验室做的湿实验。并不是病原体的每一种蛋白质都适合疫苗开发，选择标准和准确预测算法必不可少，如交叉反应性和自身抗原、表面表达免疫识别、与序列变异区相对的保守区域、膜蛋白以及螺旋数量等。严肃的蛋白质预测可以节省大量时间和金钱。蛋白质精准定位有助于确定潜在的抗原特性，例如N末端信号肽（前导序列）提供了细胞内的“邮政编码”，为向细胞表面输出提供了途径。由于细胞表面蛋白质是暴露于抗体的，如B细胞和T细胞，所以这样的蛋白质是首选。蛋白质的物理和化学复杂性会影响蛋白质在异源系统（如*E. coli*）中的表达。*E. coli*是工业和制药蛋白质生产中最常用的表达系统。并非每一种异源蛋白质都可以被克隆、表达和纯化。

T细胞表位预测　抗原表位（抗原决定簇）是抗原的结合部分，预测抗原决定簇有助于确定可能的疫苗靶点。如果抗原是蛋白质，则其表位是短肽。MHC Ⅰ类分子长度为8~10个氨基酸的肽，能够被$CD8^+$T细胞识别，引起细胞反应。在电脑模拟中，T细胞表位预测的阳性率为90%~95%。现有多个生物信息学数据库[14-20]。

B细胞表位预测　B细胞表位的预测要复杂得多[21]。B细胞表位由连续和不连续的结构组成。连续表位由抗原与抗体相互作用的线性初级氨基酸序列来确定。不连续表位由蛋白质构象、特征性三维形态来确定。不连续表位产生绝大多数抗体结合表位。但目前表位预测主要针对命中率低的线性表位[22,23]。蛋白质二级结构几乎无法预测更高级蛋白质结构。一旦在大肠杆菌中表达，每个重组候选抗原都会经历一系列生化和免疫实验。为了研究蛋白质活性、功能和相互作用[24]开发的高通量筛选蛋白质芯片，可以检测已知的和新的免疫相关蛋白和途径。

概念验证　基因组学和蛋白质组学结合是鉴定新候选疫苗的有力方法，计算机有助于从成千上万种蛋白质中选择假定抗原[25]。但是在每次电脑实验结束时，必须在湿实验、细胞培养和动物模型免疫研究中进行概念验证。这是Lackmus测试。图1.2给出了嵌入式系统疫苗学的反向遗传学示意图。

生物信息学　与此同时，反向疫苗学已经应用于大量病原菌。尽管生物信息学取得了快速发展，而且开发出了更好的工具来管理海量信息，但我们仍然面临着巨大的问题。仍未满足的最重要需求是艾滋病、结核病和疟疾（WHO）。目前唯一获得许可的结核病疫苗卡介苗已经使用了90年，但效果并不理想。迫切需要一种更好、更先进的现代结核病疫苗。此外，目前还没有开发出预防西尼罗河病毒、登革热病毒、丙肝病毒、埃博拉病毒等感染的疫苗。甚至每年研制一种标准流感疫苗也跟不上流感病毒变异（第52章）。高度可变病原（如艾滋病病毒或丙肝病毒）会导致持续性和潜伏性感染吗？目前既没有预防性疫苗，也没有治疗性疫苗。这些病原基因已被识别，为什么从基因组到疫苗的生物信息学方法并不奏效？

局限性　与体内复杂的免疫反应相比，新抗原的鉴定相对简单，如登革热疫苗抗体依赖性增强作用可导致感染性增加（第7章）。此外通过抗炎细胞因子建立局部免疫抑制环境（第2章）会降低疫苗接种效力。我们对引起影响B细胞和T细胞功能的组织微环境因素的认识非常有限（第2章）。如果一方面结核病疫苗需要强有力的T细胞表位[26]，另一方面艾滋病病毒以T细胞为靶目标，那么如何设计一种结核病疫苗呢？因为在第三世界国家，结核病和艾滋病经常混合感染（第13章）。尽管从1997年开始已经完成了对HRSV（人类呼吸道合胞病毒）基因组测序和蛋白质检测，但到目前为止还没有开发出HRSV疫苗。HRSV是一种重要呼吸道病毒，尤其对新生儿而言，新生儿不成熟的免疫系统和Th2主导的免疫反应导致疫苗增强的疾病是主要障碍[27]。

整体疫苗学　我们必须更深入地了解，生物系统是由一系列或多或少相互依赖的子系统组成的，如免疫系统、神经系统或消化系统，它们通过动态功能相互影响。在此基础上，我们必须对疫苗学有一个整体认识。有必要更多地了解（免疫）基因调控。人类基因组编码22 000个基因。

microRNA　迄今为止，超过1 500个microRNA（miRNA）被确定为基因活性调节剂

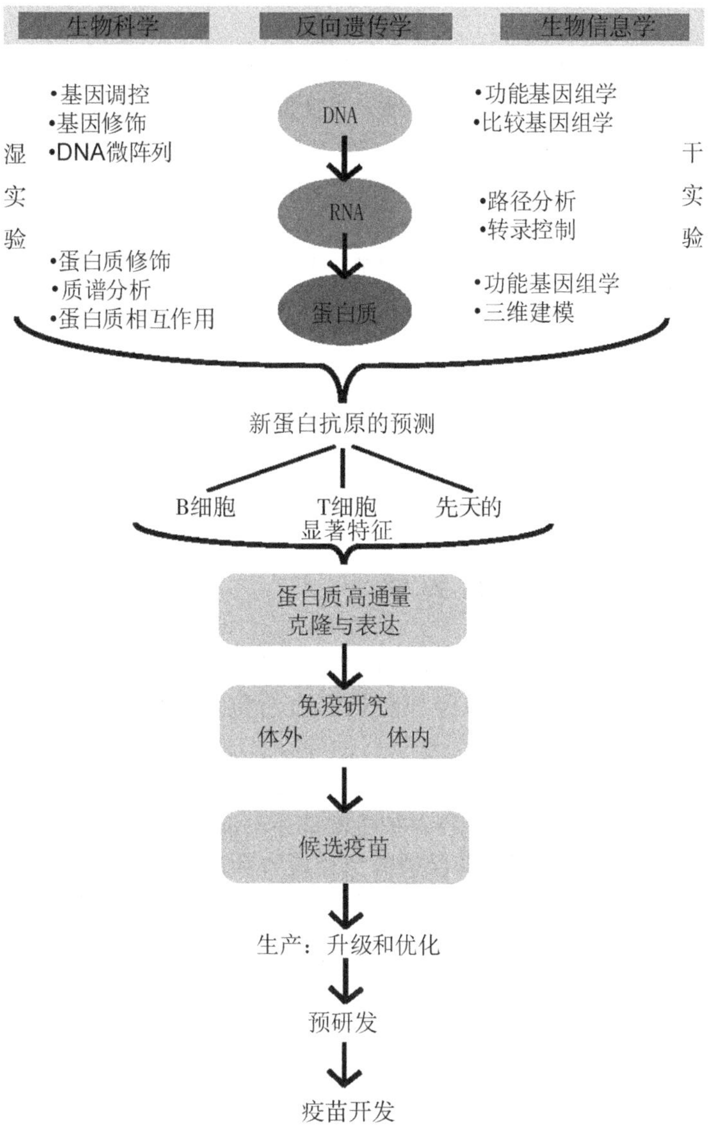

图1.2　与系统疫苗学结合的反向遗传学示意图

定义新抗原的方法从干实验（计算机）开始，生物信息学研究比较DNA水平上的基因组数据。在RNA和蛋白质水平上分析转录控制、途径和网络。B细胞和T细胞的选择标准和精确的预测算法以及天然免疫应答对于预测新蛋白抗原至关重要。蛋白质的精确定位有助于确定潜在的抗原特性。在湿实验中，预测的蛋白质被构建、交叉检查（DNA微芯片、质谱分析、蛋白质网络），并由分子遗传工具定制。高通量克隆和表达系统提供了数千种抗原。由生物信息学数据支持的体外和体内免疫研究提供候选疫苗。蛋白质的物理和化学复杂性会影响异源系统（如*E. coli*）中的表达，*E. coli*是工业和医药蛋白质生产中最常用的生物系统。并非所有的异源蛋白质都能被克隆、表达和纯化。对选定的预成型候选疫苗的生产流程进行升级和优化是进入临床生产过程前的最后一步。

（http：//mirbase. org/index. shtml）。miRNAs启动或关闭基因。这种小RNA分子对B细胞和T细胞发育非常重要并影响免疫应答[28-31]。miRNAs的具体信号是什么，在哪个时间序列中？未解决的基本问题还包括一些病原不能在细胞培养物中培养；或者对于一些病原体，不存在合适的动物模型（第5章）。我们正在为第三世界开发廉价疫苗，但严重贫困是导致许多感染的真正原因。慢性饥饿可能通过T细胞库调节免疫系统，但仍未发现其他免疫功能。早期生活中长期饥饿会显著影响幼年胸腺发育，这是一个非常关键的阶段，对T细胞发育有着长期的影响（第2章）。

“组学”症结 也许我们生活在“组学无限”时代是正确的[32]。基因组学和蛋白质组学已经确立并充分描述了这些疫苗的主要活性领域。与此同时，“组学”也出现了膨胀，如过敏原组学、文献组学、细胞组学、基因组学、时间组学、交叉组学、诊断组学、碎片组学、功能组学、操作组学或最近的疫苗组学。要理解这种激增的“组学废话”，你需要一个庞大词汇表和另一个数据库的合作支持——疫苗本体论（VO；http：//www. violi net. org/vacine ontology/），2012年疫苗本体论中组学和术语就已经超过了5 000项，且呈上升趋势。这些组学和术语是关键的问题，统计分析也是如此。临床试验中统计数据越多，就越需要批判性思维，这关系到疫苗是否有效。

“我们生活在一个‘组学’的时代，我们不再认为自己是谷歌。”（Matthias Giese）

数据库和统计不是圣杯；对复杂结构，根据传统进行假设驱动研究和实验验证仍然是迫切需要的。“想象力比知识更重要，因为知识只限于我们现在知道和理解的一切，而想象力涵盖了整个世界，所有的一切都将会被认识和理解。”（阿尔伯特·爱因斯坦）

1.3 生物标志：保护的相关性

几乎所有获得许可的疫苗都是通过诱导抗体起作用的，这些抗体可以中和或抑制病原体，使细菌或细胞易于被吞噬。对于许多经典疫苗，血清抗体IgG滴度与疫苗诱导的免疫是相关的[33]。

差异 有种差异被观察到，即在大多数疾病中所测量到的免疫应答与疫苗诱导的保护无关。在许多情况下我们评估一种保护却连免疫保护机制都还不清楚。在临床描述中，术语“相关物”和“替代物”常用于免疫应答，即负责保护的免疫应答和通过替代物测量未知参与者的免疫应答[34]。在疫苗药效试验中，评估免疫相关物或替代物以预测个体和群体水平、跨人群、性别和年龄层对感染或疾病的保护[35-37]。

保护相关性 临床前研究阶段是临床研究成功的关键。基因组学和蛋白质组学提供了各种工具来开发生物标志，以确定疫苗效力的免疫相关性，也用于已获得许可和新疫苗开发，特别是用于慢性感染或癌症治疗性疫苗。美国食品和药物管理局（FDA）将“保护相关性”定义为实验室参数，该参数已被证明与预防临床疾病有关联（FDA Guidance for Industry，1997，report no. 97N-0029，IV. C. Efficacy）。与此同时，大量关于“保护相关性”的文献如此令人困惑，所以本章只关注实验室参数。

系统生物学 抗体为机体提供了第一道防线，可以预防早期感染并与保护相关。在细胞内有病原体存在情况下，抗体既不能阻止病原复制，也起不到保护作用，这是T细胞发挥作用的地方，T细胞免疫应答的测量更准确地反映了功效相关性。

为了了解疫苗接种的免疫机制，现在可以详细研究抗原摄取、处理和递呈给B细胞和T细胞的时间顺序（第2章和第50章）。使用单一基因组和单一蛋白质组数据，我们越来越能够了解天然免疫和获得性免疫的复杂生物网络。我们正处于系统生物学研究的中间阶段，把生物学作为一个整体来研究，包括所有免疫细胞、细胞因子以及与环境的相互作用[38]。当然，我们谈论的是免疫系统，不应该忘记整体复杂性。

治疗性马动脉炎病毒（EAV）疫苗 我们正在实验室使用系统生物学来开发一种治疗马动脉炎病毒（EAV）持续感染的疫苗，作为开发人类疫苗的模型系统。EAV是动脉炎病毒科一种小ssRNA病毒，可引起呼吸道疾病和流产。疫苗研发已经开始倒退。首先，我们通过DNA芯片研究了体外假定的逃避策略（图1.3），发现EAV上调了12个干扰素相关基因，下调了2个干扰素相关基因。这种下调基因之一是IFN-γ诱导蛋白（Ifl30，acc. no. NM-023065），这是EAV

在体外持续存在的重要生物标志。

在另外的实验中，我们测量到细胞内肽转运系统 TAP（与肽相关的转运体，也是与抗原处理相关的转运体）在感染 EAV 之后表达下调，MHC Ⅰ类分子也一样。综上所述，我们假设 EAV 引发一系列细胞内和抗原处理相关事件以逃避免疫应答（图 1.4）。

T 细胞疫苗　根据 EAV 能降低细胞免疫力这一认识，我们开发了一种基于两种包膜蛋白和胞内核衣壳蛋白多价 DNA T 细胞表位疫苗[39]。在几次治疗性疫苗接种试验中，我们发现这种疫苗能够通过依赖于病毒载量的强 $CD8^+$T 细胞活性（在自体细胞毒性 T 细胞试验中测量）治愈持续性 EAV 感染的马。由于病毒太多，因此疫苗治疗失败。疫苗效力测定是依据血清中的 IFN-γ 水平。从病毒生物学角度来看，IFN-γ 与体外病毒载量精确相关。体内病毒载量相关性有待进一步研究。然而 IFN-γ 可以被确认为马 EAV 感染等级的生物标志，痘苗病毒可以被确认为治疗性疫苗接种效果标志。病毒越多，IFN-γ 就越少，疫苗治疗效果越差；病毒越少，IFN-γ 就越多，疫苗治疗效果就越好。

IFN-γ 是由 T 细胞和 NK 细胞产生的一种免疫调节剂，不仅影响 MHC Ⅰ类分子和 MHC Ⅱ类分子的表达，而且影响 B 细胞和 T 细胞分化。

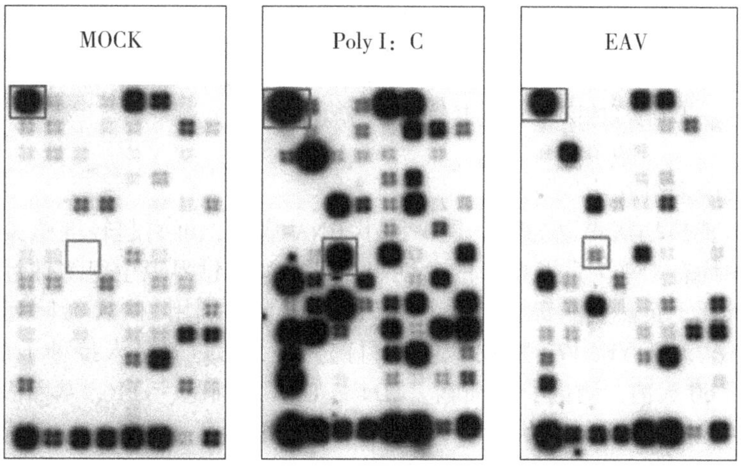

图 1.3　C2C12 细胞 RNA 的寡核苷酸序列

未经处理的 MOCK 对照；Poly I：C 处理；EAV 控制基因感染，GAPDH（红点）；IFN-β（绿点）EAV 上调和下调基因可直接与 MOCK 和 Poly I：C 对照进行比较检测。（Anett Heinrich Schulz，博士论文 2010，Matthias Giese）

唯一获得许可的结核病疫苗是卡介苗（BCG）。对这种疫苗的保护机制知之甚少，然而 IFN-γ 可能是评价卡介苗效力的一个生物标志[40,41]。

生物标志　生物标志本身并不是蛋白质："用生物标志的特性作为一种指示剂来客观测量和评估干预治疗的正常生物过程、致病过程或药理反应……"[42]（生物标志定义工作组，2001）。

利用分子工具，可以实时跟踪疫苗接种痕迹，检查细胞、蛋白质、DNA 和 RNA 水平上的分子特征。可以测量携带抗原的 DC 类型，也可以检测 TLR 和 NF-κB 激活基因的类型，并且可以预测所有免疫应答效应机制方向。利用 cDNA 微阵列可以描述宿主免疫基因的应答，通过疫苗与非疫苗对比分析可以分离出与保护相关的基因。在血液样本中可以检测作为疫苗接种的大量假定生物标志。通过疫苗接种控制疾病还应该研究染色体区域，染色体区域是疫苗接种引起遗传变异的基础。

仅仅一个生物标志是不够的，还需要一组用于天然和获得性免疫因子的各种标志，从基因到蛋白质，再到免疫细胞，能够描述宿主对疫苗接种的反应特性。一个强大的小组将由真正相

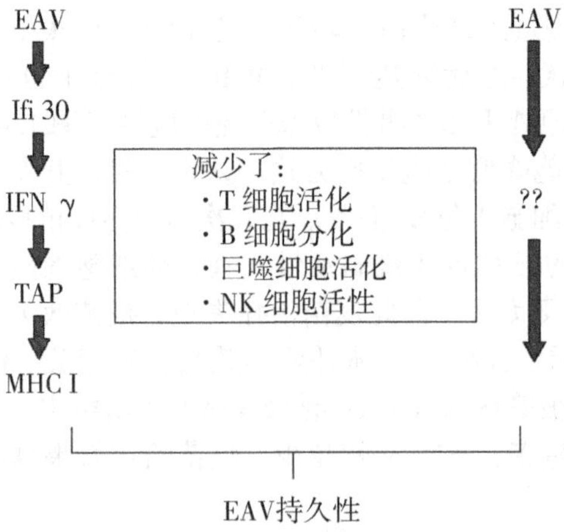

图 1.4　EAV 持久性的发展

进入靶细胞后，EAV 抑制其他基因中干扰素诱导基因 30、Ifi 30。Ifi 30 编码的蛋白质在 MHC 呈递中起着重要作用。缺乏 IFN-γ 导致 TAP 的低表达，导致通过 MHC Ⅰ类分子的抗原递呈减少，EAV 持续存在。这种下调似乎是 EAV 规避策略的一部分。需要进一步研究以揭示 EAV 的其他逃逸机制。

关因素和必要替代因素组成。

黄热病病毒　黄热病病毒（YFV）是一种小 RNA 病毒，可引起轻度感染，伴有发热、头痛，在所有病例中约有 15%为危及生命的急性出血性疾病。世界卫生组织估计，未接种疫苗人群每年有 30 000人死于黄热病病毒感染，主要发生在非洲。利用 cDNA 微阵列和生物信息学方法，可以评价黄热病疫苗 YF-17B 疗效[43]。到目前为止，超过 6 亿人接种了这种减毒活疫苗。在接种前及接种后 60 天内，测量了 B 细胞、$CD4^+$、$CD8^+$T 细胞、TLR、基因表达、转录因子和细胞因子等大量实验室参数。本研究旨在寻找与 $CD8^+$T 细胞高表达量和抗体滴度相关的分子信号。令人惊讶的是，相关生物标志是天然应激反应系统中形成 $CD8^+$T 细胞应答的一些基因[44]。应激被认为是一种有效免疫调节剂。

1.4　个性化疫苗：从基础研究到临床应用

个性化疫苗［P（personalized）-vaccines］使用患者的抗原、组织或细胞来制作疫苗，在体外操作这些抗原、组织或细胞，作为自体疫苗（主要用于肿瘤）再次注射给同一患者。1964 年首次发表了实验性自体癌症疫苗科学报告[45]。

1994 年　30 年后即 1994 年 12 月，一种抗黑色素瘤疫苗首个商业方法在维也纳开始了一项多中心临床试验[46]。

黑色素瘤细胞从一名患者身上分离出来并被送到我们的实验室。首先癌细胞在滚筒瓶中大量扩增，然后用^{60}Co 辐射灭活。选择腺病毒 5（Ad5）作为白细胞介素 2（IL-2）的后续基因转移载体。IL-2 通过复合链霉亲和素-生物素桥与 Ad5 相连。

整个实验室过程花费了至少 3 周时间，并且还存在一些问题，例如成功分离出足够数量的癌细胞用于繁殖、细胞生长缓慢或没有细胞生长以及转染率波动等。经基因修饰的放射癌细胞再次注入患者体内。这种自体细胞疫苗理论概念是向免疫系统提供一组黑色素瘤抗原，结合 IL-2 作为免疫增强剂。实际上这种疫苗不起作用，在 PhⅡ之后研究结束了。

这种失败至少有两个主要原因：①在第二次应用后，所有患者的抗载体抗体都检测出来了，疫苗被中和。②但最基本的错误是在临床前研究中，使用了一种简单的非转移性（局部包膜肿瘤）小鼠模型，而不是更真实的 B-16 黑色素瘤模型。IL-2 表达与非转移小鼠模型疗效之间的相关性在临床试验中无法得到证实。

2010 年　在首次自体肿瘤疫苗实验研究近 50 年后，FDA 批准了第一个树突状细胞（DC）治疗癌症疫苗[47]。自 20 世纪 90 年代首次体外操纵树突状细胞临床试验以来，又启动了约 50 项临床试验，但直到今天基于 DC 的疫苗基本失败了[48]。

医学分析　个性化疫苗主要是为癌症患者开发的；然而也适用于患有慢性感染、代谢紊乱和自身免疫疾病的患者。个体特征分析的目的是了解患者疾病和免疫系统之间的特异性相互作用，并根据测量到的个体生物学状况定制合适的疫苗。定制个性化疫苗的前提是收集年龄、性别和家族史等个体数据，分析基因组流行病学研究结果，并结合生理状态定期监测以确定一般遗传变异。在一般情况下，患者在生病后会先去看医生，因此会缺乏基因和蛋白质数据健康基线，也没有患者健康状况的纵向轮廓，这就限制了基因组学和蛋白质组学使用。

指纹图谱　医学分析从疾病相关指纹图谱开始，这是一种使用 DNA 微阵列基因分析技术。为了简化上调或下调基因复杂性，将根据相关基因通路进行分组分析。随着不同基因收集，活性通路识别更具有预测性[49]。基因损耗或增益功能可以通过差异显示和消减杂交来研究[50]。使用 DNA/RNA 和蛋白质阵列技术对功能基因和蛋白质进行分析以检测疫苗诱导的免疫反应。

外周血单个核细胞（PBMC）可在体外轻松分离、刺激，基因表达谱可常规测定。在对黑色素瘤患者进行 IFN-α 治疗临床研究中，发现体外 IFN-α 刺激 PBMCs 基因表达谱与体内 IFN-α 治疗后 PBMCs 基因表达谱相似。这说明将体外单个 PBMC 基因分析作为免疫应答的预测因子具有很大益处[51]。

MicroRNAs　microRNAs（miRNAs）对医学领域影响越来越大。如前所述，由短的（~20nt）非编码 RNA 在转录和转录后水平上调控基因表达。miRNAs 通过与 mRNA 分子内互补序列碱基配对导致基因沉默。体液中稳定 miRNAs 可用作诊断、预后和预测性生物标志。此外由 miRNAs 引起的靶向基因沉默可以用于治疗性研究[52]。许多癌症特异性 miRNAs 生物学功能已经得到阐明[53]。单个 PBMCs 或更普通的血液和血浆样品中，miRNAs 图谱可以用于预测个体抗癌治疗（免疫）应答[54,55]。同样在急性或慢性病毒性疾病（如甲型肝炎或乙型肝炎）中，预后和预测性循环 miRNA 生物标志可用于治疗性个性化疫苗开发和监测[56-58]。

PBMCs 指纹图谱技术在个性化医学中发挥重要作用。

ABPP　利用蛋白质功能分析而不仅仅是对各种蛋白质记录，为监测活性蛋白质提供了一个平台，称为基于活性的蛋白质图谱（ABPP）[59]。由反应基（RG）和标签组成的化学试剂共价结合到酶活性位点，而非活性酶不能结合反应基（RG）。标签是一个报告物如荧光基团或亲和标签（如生物素），用于测量这种化学蛋白质复合物。ABPP 可以通过分析丝氨酸水解酶来研究病毒-宿主相互作用[60]，并开发对慢性病毒感染的单独、个性化疫苗。丝氨酸水解酶参与多种生理和病理过程，也参与病毒性感染[61]。

生物信息学工具对从活性通路中翻译数据或者解释蛋白质-蛋白质相互作用和网络是必不可少的，用以过滤海量的数据[62,63]。

一个基因，不同的 mRNAs　总之，个性化疫苗使用来自系统生物学中的个人生物信息数据设计针对特定患者的癌症、慢性感染以及非癌症-非感染（NINC）失调症的个体疫苗（图 1.5）。疾病预测、治疗管理或研发个体疫苗的个人基因图谱应该由经过专门临床基因服务认证的实验室掌握，而不是由直接面向客户提供 DNA 图谱的私营公司掌握。仅凭 DNA 指纹图谱是不够的，必须在系统生物学的背景下解释。

在分析的 DNA 和转录 mRNA 以及翻译的蛋白质之间没有一一对应的关系：细胞可以使用不同剪接产生不同的 mRNA，并且蛋白质会经历翻译后修饰。因此，一个基因可以转录生成不同的 mRNAs，最终产生具有不同功能的蛋白质，并与其他蛋白质有不同的相互作用。此外，一个严

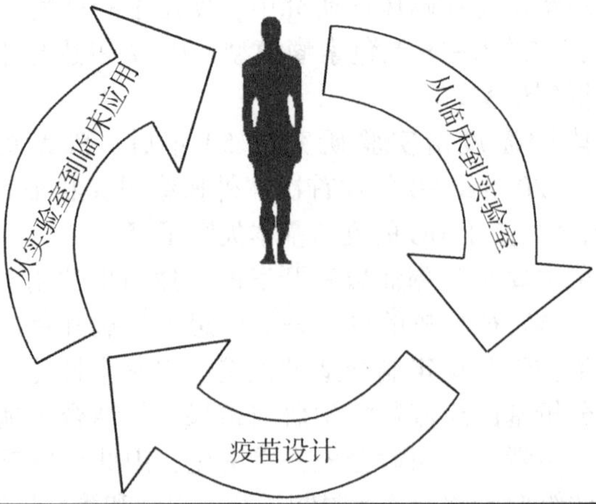

1.个人医疗分析: •疾病导向指纹 • B细胞、T细胞、天然免疫的预测特征 •炎症特征：如Treg。T细胞衰竭，功能紊乱 • miRNAs免疫调节 •治疗窗口和风险评估:有效和安全治疗的前提条件	2.疫苗设计 •多表位的合成生物学*和炎症抑制剂 •免疫应答的重建 •作为基于载体/重组疫苗的应用(加强免疫)
3. 制造工艺: •GLP/GMP控制	*合成生物学: • 从自然界没有发现的分子或遗传回路中产生可预测的生物结果

图 1.5　个性化疫苗的研制

（1）医学分析使用 DNA 微阵列进行疾病导向指纹。基因的丢失或获得功能可以通过差异显示和消减杂交来研究。利用 DNA/RNA 和蛋白质阵列技术，通过功能基因和蛋白质分析监测疫苗诱导的免疫反应。（2）在风险分析和治疗窗口规划之后，可以设计单独的疫苗。其目的必须是重建免疫反应，同时控制炎症。人工载体或特殊纳米颗粒等人工佐剂可以提高疫苗的疗效。（3）另外 P-vaccines 的生产和安全问题必须在认可的实验室按照国际标准处理。

谨的基因图谱总是与临床遗传学家的建议联系在一起。当然，提供个性化药物并不能保证患者从严重疾病中康复。许多测试只有有限的信息价值，因此不能提供准确的个人预测。每个测试结果都只是一个类似复杂生物系统的分形应用，又形成了一个无限谜题。

1.5　治疗窗口

由于免疫衰老，免疫系统效率也会降低（第 2 章）。与年龄相关的免疫变化发生在天然和获得性免疫系统中，不仅影响淋巴细胞，还影响骨髓细胞，导致前炎症细胞因子变化。此外，某些病原引起的慢性感染（如巨细胞病毒和艾滋病病毒），会重塑免疫系统，使其适应老化 T 细胞。免疫能力丧失也是许多癌症疾病的主要风险[64]。免疫衰老和疾病降低了治疗性疫苗效力。

作为医学分析的一部分，治疗窗口（TW）应该基于各种参数的风险分析，例如细胞样本、DNA 和 RNA 阵列以及蛋白质分析（图 1.6）。在免疫系统失衡变得不可逆转之前，治疗窗口是成功激活大多数免疫功能的最后机会。治疗窗口无法与有效剂量和毒性剂量之间的药理学窗口（药物的明确范围）相比。

疫苗诱导长期效应需要足够的天然和获得性免疫功能，特别是在医学分析中可以通过基因组和蛋白质组学方法来测量多种 T 细胞。新抗原主要被幼稚 T 细胞识别。T 细胞库多样性丧失与免疫应答受损相关。治疗性疫苗也可能面对功能衰竭 T 细胞，没有对治疗性疫苗做出适当反应。因此，治疗性疫苗接种应该将阻断抑制途径和激活 $CD8^+T$ 细胞的刺激信号结合起来，$CD8^+T$ 细胞是对抗癌症和慢性病毒感染的主要效应细胞。

平衡　描述与免疫疾病相关失衡的含义似乎比解释健康患者的免疫稳态更容易。免疫系统是模糊的、保护性的和有害的；一方面是免疫，另一方面是计算的免疫病理学，被冗余调节机制很好地平衡[65,66]。维持免疫稳态能够精确调节体液应答和细胞应答的持续时间和强度，这是健康的基础，也是抵御许多肿瘤早发的基础。

免疫系统至少能部分控制肿瘤生长。1863年Rudolf Virchow在肿瘤组织中发现了活跃的白细胞。1909年，Paul Ehrlich提出了抗肿瘤细胞的“身体自身保护系统”假说。“免疫监视假说”诞生于20世纪五六十年代：T细胞介导的免疫逐步进化为对抗癌细胞的特异性防御措施，T细胞经常巡视身体寻找异常体细胞[67]。直到今天仍有许多临床数据表明免疫系统/监测与肿瘤发展之间的相关性：结肠癌自发性缓解和急性髓细胞白血病或由肺癌引起的肺转移和肝转移缓解。只要肿瘤负荷由免疫系统控制，只要疾病与天然和获得性免疫之间的这种平衡起作用，生命就不会受到威胁，传染病也一样。

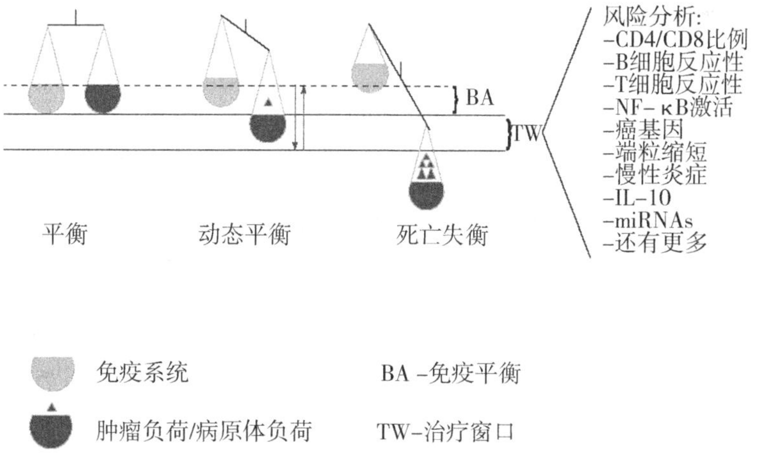

图1.6　治疗窗口

每名患者/每种疾病/每种疫苗接种的治疗窗口应基于各种实验室参数的风险分析来定义，从促炎细胞因子的产生到B细胞和T细胞的反应性、癌基因的过度表达↑↑、miRNAs的低表达↓↓、淋巴细胞端粒的缩短或作为基因调控的一部分寻找特定的miRNAs。这种治疗窗口是在免疫系统失衡变得不可逆转之前，在动态平衡期间成功重新激活大多数免疫功能的最后机会。

动态平衡　同时描述了免疫功能障碍与肿瘤发展的相关性，如T细胞修饰、T细胞失能、信号通路分子表达减少和细胞因子分泌减少。免疫风险表型（IRP）定义为CD4/CD8比值。在健康状况下$CD4^+$T细胞的数量比$CD8^+$CD28-Treg细胞（以前称为T抑制细胞）数量高，CD4/CD8比值很少低于1.0。在进行性疾病中，这个比例可能会下降到0.1。在HIV感染情况下，$CD4^+$T细胞下降与病毒载量成正比；病毒滴度越高，$CD4^+$T细胞下降越快。正如前面我们针对马动脉炎病毒（EAV）开发的治疗性疫苗中所描述的，效力很大程度上取决于病毒载量。病毒太多从而导致疫苗治疗失败。Treg细胞可以抑制抗肿瘤T细胞反应；因此，Treg细胞数量可以用作预后因子[68,69]。从细胞因子到B细胞和T细胞的反应活性，有许多宿主因素可以作为治疗性疫苗应答的决定因素。

动态平衡被定义为一种免疫状态，在这种状态下疾病已经部分破坏了一些效应器和调节功能，但是免疫系统由于其冗余性可以恢复并对疫苗接种做出应答。这是治疗窗口，必须为每种疾病和个性化疫苗定义。

死亡失衡 回到1994年的临床试验以及抗黑色素瘤疫苗，所有患者肿瘤载荷非常大，以至于一些患者在治疗过程中死亡。T细胞亚群平衡及其在肿瘤组织中的位置是影响免疫疗效的一个预后因素。肿瘤部位Treg细胞积累与功能性细胞毒性T细胞（CTLs）减少相结合导致预后不良。疫苗接种太少且为时已晚。这种死亡失衡实验室参数可能是转录因子如NF-κB永久激活导致癌基因异常过度表达、白细胞端粒致命缩短、慢性促炎细胞因子产生、T细胞反应性缺失以及与晚期癌症相关miRNAs下调等[70]，必须对这些参数进行风险分析。

参考文献

[1] McAleer, W. J., et al.: Human hepatitis B vaccine from recombinant yeast. Nature307, 178-180 (1984).

[2] Hollinger, F. B., Troisi, C. L., Pepe, P. E.: Anti-HBs responses to vaccination with a human hepatitis B vaccine made by recombinant DNA technology in yeast. J. Infect. Dis. 153, 156-159 (1986).

[3] Stephenne, J.: Production in yeast versus mammalian cells of the first recombinant DNA human vaccine and its proved safety, efficacy, and economy: hepatitis B vaccine. Adv. Biotechnol. Processes14, 279-299 (1990).

[4] Luytjes, W., Krystal, M., Enami, M., Parvin, J. D., Palese, P.: Amplification, expression, and packaging of foreign gene by influenza virus. Cell59, 1107-1113 (1989).

[5] Enami, M., Luytjes, W., Krystal, M., Palese, P.: Introduction of site-specific mutations into the genome of influenza virus. Proc. Natl. Acad. Sci. U. S. A. 87, 3802-3805 (1990).

[6] Tan, L. K., Carlone, G. M., Borrow, R.: Advances in the development of vaccines against Neisseria meningitidis. N. Engl. J. Med. 362, 1511-1520 (2010). doi: 10. 1056/NEJMra0906357.

[7] Finne, J., Leinonen, M., Makela, P. H.: Antigenic similarities between brain components and bacteria causing meningitis. Implications for vaccine development and pathogenesis. Lancet2, 355-357 (1983).

[8] Fleischmann, R. D., et al.: Whole-genome random sequencing and assembly of Haemophilus influenzae Rd. Science 269, 496-512 (1995).

[9] Pizza, M., et al.: Identification of vaccine candidates against serogroup B meningococcus by whole-genome sequencing. Science 287, 1816-1820 (2000).

[10] Sette, A., Rappuoli, R.: Reverse vaccinology: developing vaccines in the era of genomics. Immunity 33, 530-541 (2010). doi: 10. 1016/j. immuni. 2010. 09. 017.

[11] Giuliani, M. M., et al.: A universal vaccine for serogroup B meningococcus. Proc. Natl. Acad. Sci. U. S. A. 103, 10834-10839 (2006). doi: 10. 1073/ pnas. 0603940103.

[12] Gay, C. G., et al.: Genomics and vaccine development. Rev. Sci. Tech. 26, 49-67 (2007).

[13] B ambini, S., Rappuoli, R.: The use of genomics in microbial vaccine development. Drug Discov. Today 14, 252-260 (2009). doi: 10. 1016/j. drudis. 2008. 12. 007.

[14] Helmberg, W.: Bioinformatic databases and resources in the public domain to aid HLA research. Tissue Antigens80, 295-304 (2012). doi: 10. 1111/tan. 12000.

[15] De Groot, A. S., Berzofsky, J. A.: From genome to vaccine-new immunoinformatics tools for vaccine design. Methods34, 425-428 (2004). doi: 10. 1016/j. ymeth. 2004. 06. 004.

[16] Wang, P., et al.: A systematic assessment of MHC class Ⅱ peptide binding predictions and evaluation of a consensus approach. PLoSComput. Biol. 4, e1000048 (2008). doi: 10. 1371/journal. pcbi. 1000048.

[17] He, Y., Xiang, Z., Mobley, H. L.: Vaxign: the first web-based vaccine design program for reverse vaccinology and applications for vaccine development. J. Biomed. Biotechnol. 2010, 297505 (2010). doi: 10. 1155/2010/297505.

[18] Schafer, J. R., Jesdale, B. M., George, J. A., Kouttab, N. M., De Groot, A. S.: Prediction of well-conserved HIV-1 ligands using a matrix-based algorithm, EpiMatrix. Vaccine16, 1880-1884 (1998).

[19] Rammensee, H., Bachmann, J., Emmerich, N. P., Bachor, O. A., Stevanovic, S.: SYFPEITHI: database for MHC ligands and peptide motifs. Immunogenetics50, 213-219 (1999).

[20] Ehrenmann, F., Kaas, Q., Lefranc, M. P.: IMGT/3Dstructure-DB and IMGT/DomainGapAlign: a da-

tabase and a tool for immunoglobulins or antibodies, T cell receptors, MHC, IgSF and MhcSF. Nucleic Acids Res. 38, D301-D307 (2010). doi: 10. 1093/nar/gkp946.

[21] Gowthaman, U., Agrewala, J. N.: In silico tools for predicting peptides binding to HLA - class II molecules: more confusion than conclusion. J. Proteome Res. 7, 154 - 163 (2008). doi: 10. 1021/pr070527b.

[22] Blythe, M. J., Flower, D. R.: Benchmarking B cell epitope prediction: underperformance of existing methods. Protein Sci. 14, 246-248 (2005). doi: 10. 1110/ ps. 041059505.

[23] Barlow, D. J., Edwards, M. S., Thornton, J. M.: Continuous and discontinuous protein antigenic determinants. Nature322, 747-748 (1986). doi: 10. 1038/322747a0.

[24] Sharma, A., et al.: Identification of potential universal vaccine candidates against group A Streptococcus by using high throughput in silico and proteomics approach. J. Proteome Res. 12, 336-346 (2013). doi: 10. 1021/pr3005265.

[25] Flower, D. R., Macdonald, I. K., Ramakrishnan, K., Davies, M. N., Doytchinova, I. A.: Computer aided selection of candidate vaccine antigens. Immunome Res. 6 (Suppl 2), S1 (2010). doi: 1 0. 1186/ 1745-7580-6-S2-S1.

[26] Hanekom, W. A.: The immune response to BCG vaccination of newborns. Ann. N. Y. Acad. Sci. 1062, 69-78 (2005). doi: 10. 1196/annals. 1358. 010.

[27] Kim, J. Y., Chang, J.: Need for a safe vaccine against respiratory syncytial virus infection. Korean J. Pediatr. 55, 309-315 (2012). doi: 10. 3345/kjp. 2012. 55. 9. 309.

[28] Fernando, T. R., Rodriguez-Malave, N. I., Rao, D. S.: MicroRNAs in B cell development and malignancy. J. Hematol. Oncol. 5, 7 (2012). doi: 10. 1186/1756-8722-5-7.

[29] Okada, H., Kohanbash, G., Lotze, M. T.: MicroRNAs in immune regulation-opportunities for cancer immunotherapy. Int. J. Biochem. Cell Biol. 42, 1256-1261 (2010). doi: 10. 1016/j. biocel. 2010. 02. 002.

[30] Malan-Muller, S., Hemmings, S. M., Seedat, S.: Big effects of small RNAs: a review of microRNAs in anxiety. Mol. Neurobiol. (2012). d oi: 1 0. 1007/ s12035-012-8374-6.

[31] Asirvatham, A. J., Magner, W. J., Tomasi, T. B.: miRNA regulation of cytokine genes. Cytokine45, 58-69 (2009). doi: 10. 1016/j. cyto. 2008. 11. 010.

[32] Kandpal, R., Saviola, B., Felton, J.: The era of-omics unlimited. Biotechniques46, 351-352 (2009). doi: 10. 2144/000113137. 354-355.

[33] Plotkin, S. A.: Vaccines: correlates of vaccine - induced immunity. Clin. Infect. Dis. 47, 401 - 409 (2008). doi: 10. 1086/589862.

[34] Plotkin, S. A.: Correlates of protection induced by vaccination. Clin. Vaccine Immunol. 17, 1055 - 1065 (2010). doi: 10. 1128/CVI. 00131-10.

[35] Qin, L., Gilbert, P. B., Corey, L., McElrath, M. J., Self, S. G.: A framework for assessing immunological correlates of protection in vaccine trials. J. Infect. Dis. 196, 1304 - 1312 (2007). doi: 10. 1086/522428.

[36] Le Polain de Waroux, O., Maguire, H., Moren, A.: The case-cohort design in outbreak investigations. Euro. Surveill. 17, 1-6 (2012).

[37] Halloran, M. E., Longini Jr., I. M., Struchiner, C. J.: Design and interpretation of vaccine field studies. Epidemiol. Rev. 21, 73-88 (1999).

[38] Wang, K., Lee, I., Carlson, G., Hood, L., Galas, D.: Systems biology and the discovery of diagnostic biomarkers. Dis. Markers28, 199-207 (2010). doi: 10. 3233/DMA-2010-0697.

[39] Giese, M., et al.: Stable and long-lasting immune response in horses after DNA vaccination against equine arteritis virus. Virus Genes25, 159-167 (2002).

[40] Weir, R. E., et al.: Comparison of IFN-gamma responses to mycobacterial antigens as markers of response to BCG vaccination. Tuberculosis (Edinb.) 88, 31-38 (2008).

[41] Abebe, F.: Is interferon-gamma the right marker for bacille Calmette-Guerin-induced immune protection? The missing link in our understanding of tuberculosis immunology. Clin. Exp. Immunol. 169, 213 - 219 (2012). doi: 10. 1111/j. 1365-2249. 2012. 04614. x.

[42] Atkinson, A. J., Colburn, W. A., DeGruttola, V. G., DeMets, D. L., Downing, G. J., Hoth, D. F., Oates, J. A., Peck, C. C., Schooley, R. T., Spilker, B. A., Woodcock, J., and Zeger, S. L.: Biomarkers Definitions Working Group: Biomarkers and surrogate endpoints: preferred definitions and conceptual framework. Clin. Pharmacol. Ther. 69, 89–95 (2001). doi: 1 0. 1067/mcp. 2001. 113989.

[43] Gaucher, D., et al.: Yellow fever vaccine induces inte-grated multilineage and polyfunctionalimmune responses. J. Exp. Med. 205, 3119–3131 (2008). doi: 10. 1084/jem. 20082292.

[44] Pulendran, B., Li, S., Nakaya, H. I.: Systems vaccinology. Immunity33, 516–529 (2010). doi: 10. 1016/j. immuni. 2010. 10. 006.

[45] Aswaq, M., Richards, V., McFadden, S.: Immunologic response to autologous cancer vaccine. Arch. Surg. 89, 485–487 (1964).

[46] Stingl, G., et al.: Phase I study to the immunotherapy of metastatic malignant melanoma by a cancer vaccine consisting of autologous cancer cells transfected with the human IL–2 gene. Hum. Gene Ther. 7, 551–563 (1996). doi: 10. 1089/hum. 1996. 7. 4–551.

[47] Hovden, A. O., Appel, S.: The first dendritic cell–based therapeutic cancer vaccine is approved by the FDA. Scand. J. Immunol. 72, 554 (2010). doi: 10. 1111/j. 1365–3083. 2010. 02464. x.

[48] Lesterhuis, W. J., et al.: Dendritic cell vaccines in melanoma: from promise to proof? Crit. Rev. Oncol. Hematol. 66, 118–134 (2008). doi: 10. 1016/j. critrevonc. 2007. 12. 007.

[49] K hatri, P., Sirota, M., Butte, A. J.: Ten years of pathway analysis: current approaches and outstanding challenges. PLoSComput. Biol. 8, e1002375 (2012). doi: 10. 1371/journal. pcbi. 1002375.

[50] Munir, S., Singh, S., Kaur, K., Kapur, V.: Suppression subtractive hybridization coupled with microarray analysis to examine differential expression of genes in virus infected cells. Biol. Proced. Online. 6, 94–104 (2004). doi: 10. 1251/bpo77.

[51] Zimmerer, J. M., et al.: Gene expression profiling reveals similarities between the in vitro and in vivo responses of immuneeffector cells to IFN–alpha. Clin. Cancer Res. 14, 5900–5906 (2008). doi: 10. 1158/1078–0432. CCR–08–0846.

[52] Sibley, C. R., Seow, Y., Wood, M. J.: Novel RNA–based strategies for therapeutic gene silencing. Mol. Ther. 18, 466–476 (2010). doi: 1 0. 1038/ mt. 2009. 306.

[53] Chen, P. S., Su, J. L., Hung, M. C.: Dysregulation of microRNAs in cancer. J. Biomed. Sci. 19, 90 (2012). doi: 10. 1186/1423–0127–19–90.

[54] Gamez–Pozo, A., et al.: MicroRNA expression profiling of peripheral blood samples predicts resistance to first–line sunitinib in advanced renal cell carcinoma patients. Neoplasia14, 1144–1152 (2012).

[55] Ng, E. K., et al.: Circulating microRNAs as specific biomarkers for breast cancer detection. PLoSOne8, e53141 (2013). doi: 10. 1371/journal. pone. 0053141.

[56] Elfimova, N., et al.: Circulating microRNAs: promising candidates serving as novel biomarkers of acute hepatitis. Front. Physiol. 3, 476 (2012). doi: 10. 3389/ fphys. 2012. 00476.

[57] Cortez, M. A., Calin, G. A.: MicroRNA identification in plasma and serum: a new tool to diagnose and monitor diseases. Expert Opin. Biol. Ther. 9, 703–711 (2009). doi: 10. 1517/14712590902932889.

[58] Ajit, S. K.: Circulating microRNAs as biomarkers, therapeutic targets, and signaling molecules. Sensors (Basel) 12, 3359–3369 (2012). doi: 10. 3390/s120303359.

[59] Liu, Y., Patricelli, M. P., Cravatt, B. F.: Activity–based protein profiling: the serine hydrolases. Proc. Natl. Acad. Sci. U. S. A. 96, 14694–14699 (1999).

[60] Shahiduzzaman, M., Coombs, K. M.: Activity based protein profiling to detect serine hydrolase alterations in virus infected cells. Front. Microbiol. 3, 308 (2012). doi: 10. 3389/fmicb. 2012. 00308.

[61] Steuber, H., Hilgenfeld, R.: Recent advances in targeting viral proteases for the discovery of novel antivirals. Curr. Top. Med. Chem. 10, 323–345 (2010).

[62] Bindea, G., Galon, J., Mlecnik, B.: CluePedia Cytoscape plugin: pathway insights using integrated experimental and in silico data. Bioinformatics (2013). doi: 10. 1093/bioinformatics/btt019.

[63] Henderson–Maclennan, N. K., Papp, J. C., Talbot Jr., C. C., McCabe, E. R., Presson, A. P.: Pathway analysis software: annotation errors and solutions. Mol. Genet. Metab. 101, 134–140 (2010). doi: 10.

1016/j. ymgme. 2010. 06. 005.

[64] Pawelec，G.，Derhovanessian，E.，Larbi，A.：Immunosenescence and cancer. Crit. Rev. Oncol. Hematol. 75，165-172（2010）. doi：10. 1016/j. critrevonc. 2010. 06. 012.

[65] Barnaba，V.，Paroli，M.，Piconese，S.：The ambiguity in immunology. Front. Immunol. 3，18（2012）. doi：10. 3389/fi mmu. 2012. 00018.

[66] Germain，R. N.：Maintaining system homeostasis：the third law of Newtonian immunology. Nat. Immunol. 13，902-906（2012）. doi：10. 1038/ni. 2404.

[67] Ostrand-Rosenberg，S.：Immune surveillance：a balance between protumor and antitumor immunity. Curr. Opin. Genet. Dev. 18，11-18（2008）. doi：10. 1016/j. gde. 2007. 12. 007.

[68] Wilke，C. M.，Wu，K.，Zhao，E.，Wang，G.，Zou，W.：Prognostic significance of regulatory T cells in tumor. Int. J. Cancer127，748-758（2010）. doi：10. 1002/ijc. 25464.

[69] Ogino，S.，Galon，J.，Fuchs，C. S.，Dranoff，G.：Cancer immunology-analysis of host and tumor factors for personalized medicine. Nat. Rev. Clin. Oncol. 8，711-719（2011）. doi：10. 1038/nrclinonc. 2011. 122.

[70] Huang，L.，et al.：Downregulation of six microRNAs is associated with advanced stage，lymph node metastasis and poor prognosis in small cell carcinoma of the cervix. PLoSOne7，e33762（2012）. doi：10. 1371/journal. pone. 0033762.

（马维民译）

[illegible] 2010.06. [illegible]

[64] [illegible]

[65] [illegible] The ambiguity [illegible]

[66] [illegible] Mitochondrial [illegible]

[67] [illegible]

[68] [illegible] T cell [illegible]

[69] [illegible] meta-analysis [illegible]

[70] [illegible] associated with [illegible] PLoS One [illegible]

第二部分

疫苗免疫学

概　述

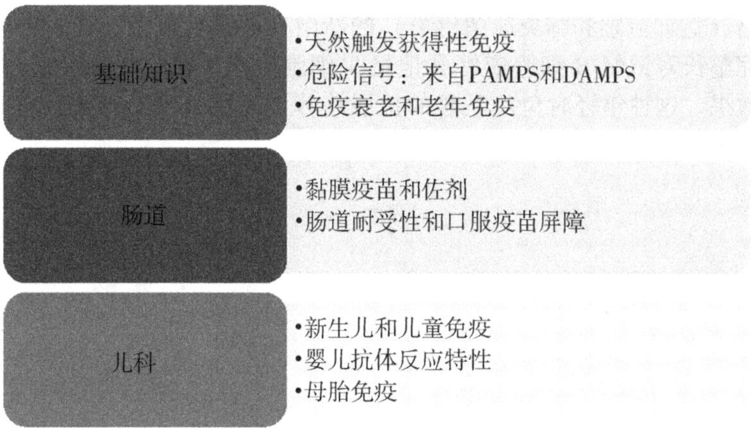

长期以来，先天性免疫被低估，被认为是远古时代的遗存，这对昆虫来说是足够了，但对哺乳动物是否足够？这种无知反映在经典疫苗的开发中，认为血清抗体 IgG 滴度是与保护相关的唯一关键参数。抗体提供了第一道防线，可以预防早期感染，并与保护相关。在细胞内病原体存在的情况下，抗体既不能阻止复制，也起不到保护作用，这是 T 细胞发挥作用的地方，与功效更准确相关的是 T 细胞免疫应答的测量。

用于产生具有高效应能力的强 $CD8^+$T 细胞应答的疫苗应该以适当的 $CD8^+$T 细胞应答的质量、数量和持续时间解决这些基本的关键问题。减毒活载体疫苗（即复制子和 DNA 疫苗），能够刺激 $CD8^+$T 细胞，但不能杀死 $CD8^+$T 细胞或常规的肽/基于蛋白质的疫苗。T 细胞疫苗的研究集中在精确、保守的 T 细胞表位上。慢性感染或癌症的治疗性 T 细胞疫苗开发应该解决持续性 $CD8^+$T 细胞活化不良反应问题。这些 T 细胞疫苗具有产生免疫病理的潜力，如衰竭（功能丧失，从 IL-2 丧失开始）和炎症。PD-1 过度表达是 T 细胞耗尽的一种生物标志。

现代疫苗学的目标之一是开发疫苗，使之成为降低发展中国家 5 岁以下儿童肠道疾病造成的毁灭性发病率和死亡率的工具。了解口服抗原诱导免疫应答所涉及的过程、肠道免疫激活和耐受之间的微妙平衡以及黏膜免疫和全身免疫之间的联系，对于帮助开发和评估候选疫苗至关重要。

肠道获得性免疫的主要效应部位是上皮组织和固有层。这些组织中存在大量抗原特异性抗体分泌细胞（ASC）和活化的 T 细胞，为抵抗肠道病原体提供了第一道防线。口服疫苗的一个标志是诱导长寿的抗原特异性黏膜分泌 IgA 的浆细胞。

营养不良、T 细胞过早衰老和易受感染之间有一个界面。慢性饥饿可通过 T 细胞库调节免疫系统。饥饿还会影响 $CD8^+$T 细胞的端粒长度，导致端粒缩短。早期的慢性饥饿会显著影响幼年胸腺的发育，这是一个非常关键的阶段，对 T 细胞的发育有长期影响。锌是免疫系统必需的矿物质，缺锌会导致胸腺萎缩，这表明营养障碍使这一人群易受感染。营养不良是第三世界国家人群口服疫苗失效的一个重要原因。

全世界 5 岁以下儿童中有一半以上死于传染病。这些疾病中有许多可以用疫苗预防。世卫组织估计，2008 年约有 150 万 5 岁以下儿童死于此类疾病：肺炎链球菌（肺炎球菌）感染和轮状病毒感染是主要原因，其次是新生儿期流感嗜血杆菌B（HIB）、百日咳博德氏菌（百日咳）、麻疹病毒和破伤风梭菌引起的感染（新生儿破伤风）。

新生儿免疫系统的一个主要缺点是大多数B细胞和T细胞都是幼稚的，与物种无关。幼儿在抗原呈递细胞的成熟和这些细胞对细菌和病毒的反应能力方面都受到损害，但TLR8是个重要的例外。使用TLR8配体作为疫苗佐剂可能是一个机会之窗。

虽然非常年幼的儿童不能对免疫做出充分应答，但老年患者已经失去了对抗原作出充分应答的关键免疫功能。针对老年人的最佳免疫方法是在衰老到来之前开始接种疫苗：开发具有广谱抗原性、多价谱（例如流感和肺炎球菌感染）的新型（偶联）疫苗，通过产生强大的CTL活性，诱导强烈的完整长寿记忆B细胞应答和记忆T细胞应答；在成年期，通过反复接种疫苗，还可以达到最佳效果。这种年轻时免疫系统的扩展培养是记忆的坚实基础，也是在老年时身体对抗新抗原的强大免疫背景。

第 2 章　疫苗免疫学基础

Matthias Giese[①]

摘要

本章描述了疫苗接种后免疫反应最突出的方面。但是详细描述所有免疫功能超出了本章的篇幅。我们着重于对疫苗免疫的一般描述。

2.1　抗原的化学性质

病毒、细菌、寄生虫和真菌是我们生活中永久的自然威胁。此外，对花粉、霉菌、动物毛发或室内尘螨的不同类型过敏能够引发抗原诱导的抗体应答。此外，环境过敏原，天然或工业生产的超细碳颗粒物质（PM），会引起过敏并引发哮喘。PM 沉积在肺泡中导致炎症反应。

因此，抗原世界看起来是多种多样的，根据其化学性质，所有已知的抗原（工业 PM 除外）可分为五类：蛋白质、糖蛋白、脂质（脂蛋白）、糖类和核酸（图 2.1）。抗原能够通过表

抗原	化学结构	抗原识别位点
蛋白质	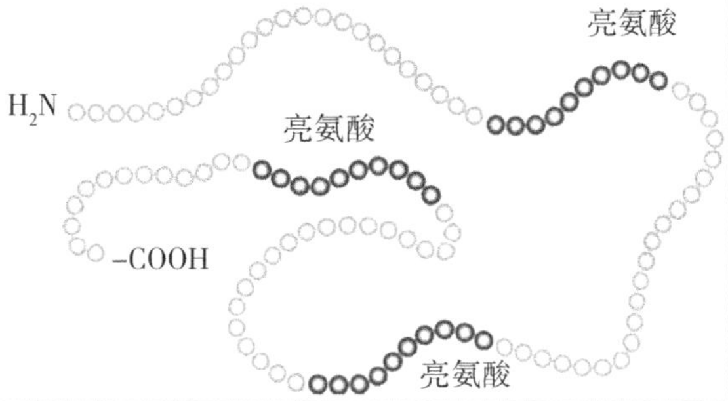	富含亮氨酸重复结构域(LRR)-由2~45个 LL 重复序列组成，每个重复序列长20~30个残基
糖蛋白	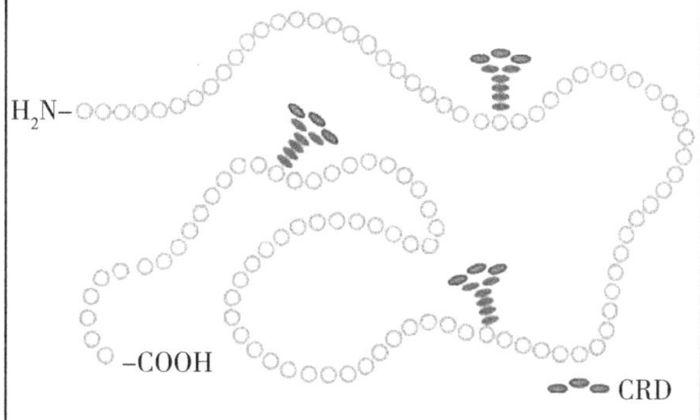	复合糖，如甘露糖、葡萄糖或半乳糖，通过它们的碳水化合物识别域(CRD)识别抗原结构

① M. Giese，博士（德国海德堡 IMV 分子疫苗研究所，E-mail：info@ imv-heidelberg. com）

脂类	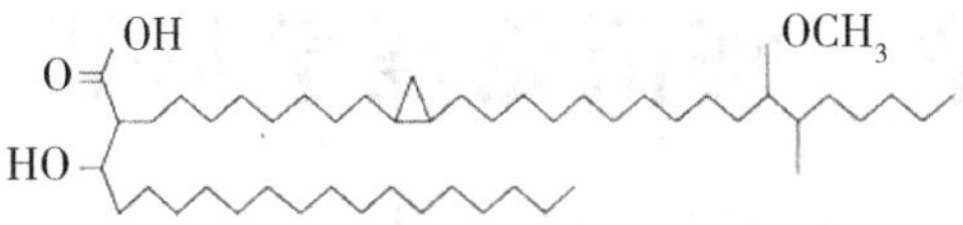 游离分枝菌酸 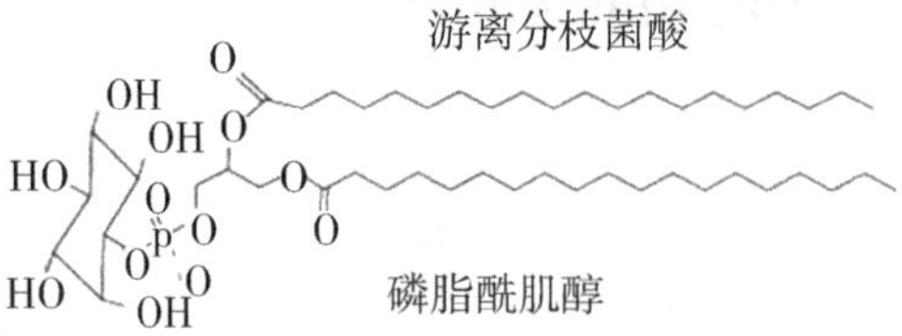 磷脂酰肌醇	链长是由多种多样的脂类、糖脂类和脂肽组成
糖类	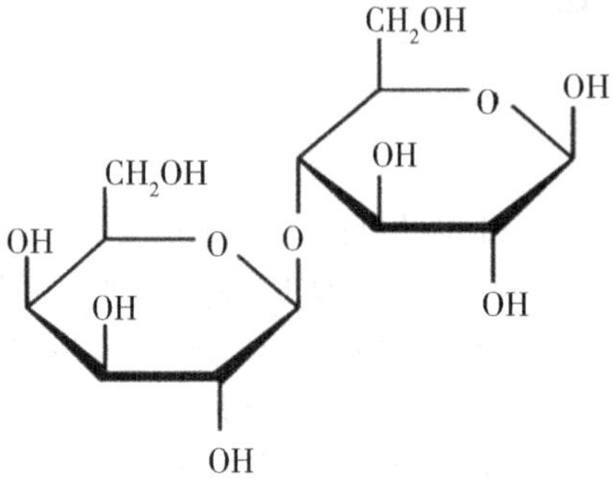	通过C-凝集素样受体识别甘露糖、岩藻糖、半乳糖、葡聚糖寡糖
核酸	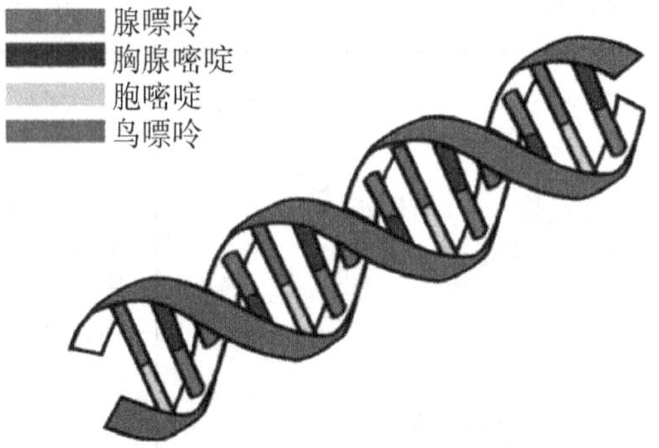	DNA：通过与脊椎动物DNA相比的低甲基化（CPG）识别细菌DNA RNA：通过一级核苷酸序列、二级和三级结构或细胞内定位识别

图 2.1　抗原的化学性质

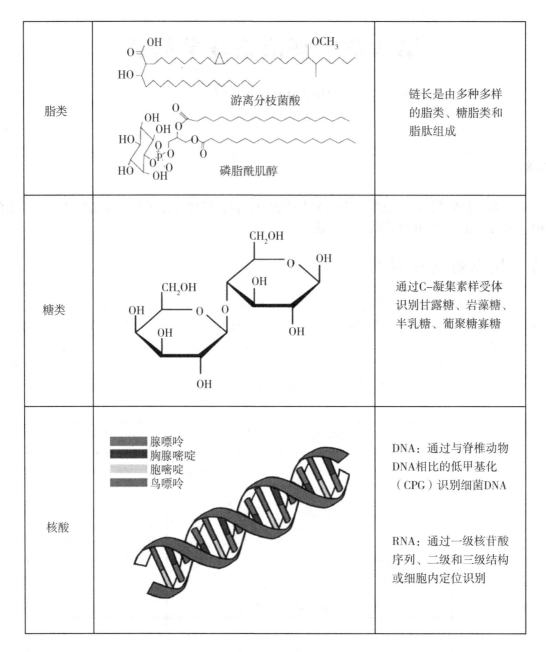

每种蛋白质都由基因编码，并由 1 条或多条由肽键连接的氨基酸链（一级序列）构成。蛋白质的基本功能由三级结构决定，这是一种稳定的三维结构。许多蛋白质与不同的复合糖如甘露糖、葡萄糖，或被脂质（糖蛋白或脂蛋白）修饰。糖蛋白通过其碳水化合物识别域（CRD）识别抗原结构。脂质可以通过它们不同的链长来识别。脂质抗原结合 CD1 分子，其中它们的极性脂质头基暴露于 CD1 限制性 T 细胞。在与 TCR 结合后，这些 T 细胞与巨噬细胞、树突状细胞、自然杀伤细胞、B 细胞和 T 细胞相互作用，从而促进先天和后天免疫反应。各种没有蛋白质结合的糖类通过 C 凝集素样受体被识别。与脊椎动物的 DNA 相比，微生物 DNA（细菌、病毒、真菌）甲基化程度较低，可以被 Toll 样受体 9 检测到。与 DNA 不同，外源 RNA 的抗原性是特别基于结构差异和细胞内定位。

面表位与抗体结合。抗原表位或抗原决定簇是抗原的独特分子特征。根据抗原性质和大小，几个不同的表位可以修饰1个抗原，并可以与各种抗体结合。特殊的细胞表面、细胞内和分泌的受体统称为模式识别受体（PRRs），广泛分布在体内，识别各种抗原结构，从而启动先天和后天免疫应答。

PRRs通过其独特的、高度特异性的、高度保守的PAMPs（病原体相关分子模式）来识别入侵者。所有传感器都使用PAMPs来区分“自我”和“非自我”。这种冗余PRR系统保证通过不同的传感器并行检测病原体。如果一个传感器失灵，另一个PRR会主动跟踪该抗原。这种高度冗余可能是健康患者中估计98%的感染被免疫系统检测和治愈的原因之一。PRRs和PAMPs将在后面讨论。

树突状细胞（DCs）向T细胞递呈自身抗原（细胞碎片、宿主编码的前体蛋白）而不存在任何导致T细胞凋亡的共刺激分子。免疫系统通常能耐受自身抗原，但如果免疫控制失败，自身抗原就会变得危险。

不同抗原在免疫应答的定量和定性上存在差异，在持续时间和疗效上也存在差异。抗原的免疫原性由几个因素决定：

- 蛋白质比脂质和糖类更具免疫原性。
- 抗原和自身抗原的化学和结构差异。
- 抗原的大小和复杂性：更多的表位用于结合抗体。
- 颗粒抗原比可溶性抗原更具免疫原性。

高度复杂的外源蛋白质，其自身结构不存在类似抗原，会被认为是入侵者，并引起强烈的免疫应答。

2.2　抗原呈递细胞

疫苗只能诱导完全的免疫应答，包括先天免疫和后天免疫以及体液和细胞活动。因此B细胞和T细胞、单核细胞、巨噬细胞、自然杀伤细胞（NK）、粒细胞或抗原呈递细胞（APCs），例如树突细胞，由各种细胞因子触发，与补体因子和抗体一起协同作用。

APCs在每次疫苗接种和抗原靶细胞中都处于早期前沿。APCs结合不同的抗原并连接先天免疫和后天免疫。APCs使用先天的、进化的、高度保守的模式来识别各种抗原，从而激活B细胞和T细胞。因此后天免疫由先天免疫成分调节。

抗原识别是先天系统的一部分，由两个相应的元素控制。元素一是模式识别受体（PRR）[1]。元素二是病原相关分子模式（PAMPs），仅在微生物上表达。属于一个类别中的微生物，共享相同的PAMPs[2]。

特异性抗原（Ag）呈递的主要驱动力是3种类型的专职APCs，即单核细胞[3]/巨噬细胞[4]、B细胞[5]和属于白细胞的树突状细胞（表2.1）。所有APCs都是由骨髓中的造血干细胞产生的。

表2.1　专职抗原呈递细胞（APCs）

细胞类型	位置	功能	标记	参考文献
单核细胞：	血液	吞噬作用，抗原呈递		[3]
1. 经典的			$CD14^{+}$，$CD16^{-}$	
2. 中间的			$CD14^{+}$，$CD16^{+}$	
3. 非经典的			$CD14^{+}$，$CD16^{++}$	
巨噬细胞	组织	吞噬作用，抗原呈递	$CD14^{+}$，$CD40^{+}$，$CD64^{+}$	[4]

（续表）

细胞类型	位置	功能	标记	参考文献
B 细胞	血液，淋巴器官	抗原递呈，抗体，调节	$CD19^+$，$CD20^+$，$CD21^+$	[5]
树突状细胞	组织	抗原递呈	见表 2.2	

树突状细胞 单核细胞/巨噬细胞激活幼稚 T 细胞的能力有限，B 细胞的抗原递呈主要是抗体产生的自动递呈。特别是树突状细胞（DCs）是真正的专职细胞，用于向 T 细胞递呈抗原。DCs 是可移动的，可以在血液中随着未成熟的细胞移动，一旦被激活，就会从血液中迁移到不同的组织，再迁移到次级淋巴器官。DCs 是免疫系统的主要守门人。DCs 拥有完整的分子设备来摄取抗原，处理和递呈抗原以诱导免疫应答。

20 年来，DCs 一直是疫苗开发的热点，这并不奇怪。DCs 存在于包括大脑[6]在内的大量淋巴组织和非淋巴组织中，主要存在于外部和内部环境贯穿的组织中：皮肤、肺和胃肠道。因此，DCs 对于免疫耐受和通过 MHC Ⅰ和 MHC Ⅱ与协同刺激分子（如 B7）和促炎细胞因子结合诱导保护性反应都是必不可少的。DCs 是由不同的子集组成的网络，这些子集具有表型和功能上的区别。

未成熟 DCs 作为未成熟细胞，DCs 始终处于运动状态，随时在体内巡逻并捕获抗原。它们是免疫系统的前哨细胞。在这个阶段，MHC 分子（人类的 HLA）数量极微。此外共刺激信号也不存在，这是 T 细胞完全活化所必需的。

DCs 不能被任何独有的细胞标志检测和区分。相反地，标志组合（存在和不存在）被用来识别 DCs。此外，由于单核细胞、巨噬细胞和 DCs 有共同的巨噬细胞 DC 祖细胞，单核细胞/巨噬细胞在受到特定刺激后发育成 DCs，所以分化过程比较复杂。表 2.2 显示了 DCs 的功能子集[7-11]。

根据其个体起源不同，DCs 分为髓系（也称为常规或经典）和淋巴系（非常规）[12]。髓样树突状细胞（mDCs）与单核细胞最相似，可分为几个子集，分为迁移性树突状细胞和淋巴组织驻留性树突状细胞。髓样树突状细胞（mDCs）的表面标志是 $CD11c^+$、$CD11b^+$、$CD1a^+$，有时还有 $CD103^+$。这些外周树突状细胞位于表皮、皮肤真皮、黏膜和间质，称为朗格罕细胞。除朗格罕细胞外，所有 mDCs 都产生大量不同的 Th2 细胞因子，如 IL-4 和 IL-10。在抗原捕获和处理之后，髓样树突状细胞（mDCs）迁移到局部淋巴结去递呈肽抗原，并激活 $CD4^+$和 $CD8^+$T 细胞。

表 2.2 外周血树突状细胞（DCs）的功能子集

DCs 子集	定位	标志和标记	参考文献
髓样（m）DCs：		Th2 细胞因子	
真皮（d）DCs	真皮	$CD1a^+$，$CD14^+$	[7]
朗格罕细胞	上皮	$CD1a^+$，$CD1b^-$，$CD1c^-$	[8]
脑（bDCs）	脑	$CD11b^+$，$CD103^+$	[9]
间质（iDCs）	间质	$CD11c^+$，$CD11b^+$，$CD1a^+$	[10]
浆细胞样（p）DCs	血液，淋巴器官	Th1 细胞因子，INF -α，$CD11c^-$，$CD11b^-$，$CD1a^+$	[11]

淋巴 DCs 被称为浆细胞样树突状细胞（pDCs），看起来像浆细胞。淋巴 DCs 在血流中循环，其特征是能够产生大量 I 型干扰素[13]和其他 Th1 细胞因子。为此，pDCs 必须被病毒激活。pDCs

嵌入淋巴组织和外周组织，形成淋巴组织驻留树突状细胞群。表面标志为 $CD11c^-$，$CD11b^-$ 和 $CD1a^+$。DC 子集也可以通过寿命来区分，特别是通过解剖定位来区分。因此，DC 的特定功能被调整到其定位[14]。

2.2.1　朗格罕细胞对抗原的捕获

朗格罕细胞（LCs）是皮肤树突状细胞的一个子集，耐辐射，也存在于套索、口腔或生殖道的黏膜中，在第一道防线中起关键作用。尤其是表皮基部和基部上层的 LCs，在捕获疫苗抗原、处理抗原和迁移到引流淋巴结方面具有战略优势。几乎 2%的表皮细胞是 LCs。抗原捕获后，LCs 通过真皮迁移到淋巴管。两种不同的 LCs 表面受体通过内吞作用（胞饮作用和吞噬作用）捕获糖蛋白，一种是 C 型凝集素受体 DEC-205（CD205），另一种是 langerin 受体（CD207）。在成熟过程中，LCs 表达 MHC Ⅰ类和 MHC Ⅱ类分子。

凝集素样受体　免疫系统的许多细胞，如树突状细胞、单核细胞和巨噬细胞，都具有凝集素或凝集素样受体（LLR）。这些 LLR 是模式识别受体，其功能和稍后讨论的 Toll 样受体（TLR）家族相同。凝集素样受体 DEC-205（CD205）是巨噬细胞甘露糖受体（MMR）家族的一部分，通过糖识别结构域（CRD）识别微生物和真菌的末端单糖残基。langerin 受体（CD207）是Ⅱ型受体家族成员，识别并处理微生物糖蛋白。DEC-205 与 CD207 的区别在于对末端糖识别不同。体内糖类的形态如此多样，LLR 的糖识别结构域也是如此的多样[15]。与野生型相比，即使糖基化模式的最低差异也激活了与其特异性糖类抗原具有高亲和力的合适 LLR。这意味着 LLR 结合域的高度可变性。LLR 抵抗外来糖类抗原的另一个技巧是 LLR 的多聚化。作为对给定抗原的重复糖类单元的反应，LLR 可以形成并因此增强结合能力。

C 型凝集素和 langerin 受体通常用于鉴定朗格罕细胞：LC（$CD\ 207^+/CD205^+$）。皮内应用的抗原在几分钟内通过这些受体被 LCs 吸收，甚至像抗体这样的大分子也能穿过将表皮与真皮分离的基底膜，并被 LCs 树突吸收[16]。

同时这些树突可以向相反方向扩展，通过紧密连接穿透上皮角质层而不受损伤，并吸收抗原[8]。因此 LCs 控制硬皮空间以及真皮内部空间。LCs 是可移动的，是免疫系统忠实的前哨细胞。然而 LCs 主要捕获细菌和真菌抗原（包括过敏原），但不捕获病毒抗原[17]。

一旦吸收抗原，LCs 通过交叉启动 $CD8^+T$ 细胞递呈处理过的抗原，诱导强烈的细胞免疫应答。在这个活化活过程中，LCs 产生大量 IL-15、IL-6 和 IL-8。这些必需细胞因子支持细胞毒性 T 淋巴细胞（CTLs）成熟[18]。

交叉启动　交叉启动最早在 1976 年被提出来[19]，这个概念代表由 MHC Ⅰ类分子通过抗原呈递细胞 APCs（如 LCs）将外源性摄取的抗原（如凋亡或坏死细胞的碎片）递呈给 $CD8^+T$ 细胞。与直接（经典）启动不同，交叉启动是来自细胞内病原内源性合成抗原的递呈。交叉启动也可能是免疫系统的另一个冗余机制，以确保针对这种病毒产生 CTL 应答，这种病毒不能感染 APCs 或感染 APCs，并使 MCMV 这样的处理和递呈机制失效[20]。交叉启动在自然病毒感染中的作用仍有争议[21]。疫苗接种需要交叉启动。

效应机制　一旦被启动和激活，$CD8^+$细胞毒性 T 淋巴细胞（CTLs）具有大量破坏靶细胞的效应机制：①细胞毒性蛋白，例如穿孔蛋白和颗粒酶 A 和颗粒酶 B[22]。在这个过程中，T 细胞受体（TCR）和 MHC Ⅰ的细胞间直接接触是必不可少的。②CTLs 通过靶细胞上的 Fas 受体直接与 Fas 配体（CD95L）结合。这激活了半胱天冬酶级联反应，导致细胞凋亡[23]。③细胞毒性 T 淋巴细胞（CTLs）分泌大量 TNF 和 IFN-γ。TNF 与靶细胞上的受体结合诱导凋亡，而 IFN-γ 增强 MHC Ⅰ和 Fas 配体表达，导致靶细胞凋亡[24]。图 2.2 显示了 $CD8^+CTLs$ 的交叉启动和直接启动，以及一些针对靶细胞的效应机制。

2.2.2　树突状细胞对抗原的捕获

皮肤的不同层被构造成一个方阵。如果病原体能成功侵入表皮，下一个前线就是真皮树突状细胞（dDCs），这是髓系树突状细胞家族成员。这些细胞也是可移动的，表达高水平 MCH Ⅱ类分子，并且可以交叉递呈不同于 LCs 的病毒抗原。可以描述为两个不同的群体：dDCs $CD1a^+$ 和 dDCs $CD14^+$。这两个群体在表型和功能上都有很大的不同。

dDCs $CD1a^+$群体表现出具有典型 DC 标志的树突状形态，如 CD1、CD83 或 CD208。dDCs $CD14^+$特征是抗原的摄取以及 T 细胞的活化处理和转移到淋巴结，其他亚群呈现巨噬细胞形态，具有典型的巨噬细胞标志，如 CD14、CD68 或 CD209。这些真皮树突状细胞（dDCs）还具有典型的巨噬细胞功能即吞噬细菌和病毒。但是在这两个亚群中也有杂交种，具有树突形态的吞噬细胞 dDCs[7]。共同的个体发育有助于解释树突状细胞和巨噬细胞的混合物。树突状细胞可以在特定刺激后由巨噬细胞发育而来。GM-CSF 诱导 CD1，LPS 诱导 CD14 树突状细胞。

细胞因子 LCs 除了产生 IL-15 外，几乎不产生细胞因子。与 LCs 相比，dDCs 在 CD40 刺激后分泌大量不同的细胞因子：IL-1α、IL-1β、IL-6、IL-7、IL-8、IL-10、IL-12、GM-CSF、TNF-α 和 TGF-β[25,26]。大多数细胞因子都是由两个 dDCs 亚群产生的，但 IL-10 只由 CD14-DCs 分泌。促炎和抗炎细胞因子的模式指示免疫应答的方向。

2.3 炎症与细胞因子

炎症是由微生物感染引起的，如细菌、病毒、真菌或其他病原体。炎症的进一步诱因是创伤或毒素、反应性 T 细胞等。炎症反应的功能是通过产生促炎细胞因子（如 IL-1、IL-6、IL-8 和 TNF-α），在早期预防感染。其他细胞因子具有多效性，同时起到促炎和抗炎的作用[29]。

促炎细胞因子 IL-1 和 TNF 以协同方式运作，并上调基因级联，例如Ⅱ型磷脂酶 A2（PLA2）和环氧化酶-2（COX-2）基因，或诱导型 NO 合成酶（iNO）产生趋化酶。这些酶将免疫细胞（如巨噬细胞、NK 细胞或中性粒细胞）吸引到感染部位。同时血管扩张并变得更加通透。其他细胞因子如 IL-4、TGF-β，尤其是 IL-10 控制和抑制炎症过程，从而支持愈合。这些细胞因子起到抗炎作用，并负责关键的免疫平衡。因此必须非常仔细地计算免疫稳态的每一种治疗干预。

抗炎细胞因子 IL-10 在免疫抑制中起关键作用。它是一种主要的细胞因子，能有效监测感染后的免疫病理环境。IL-10 在正常情况下是调节炎症所必需的。不同的 T 细胞子集产生 IL-10，如 Th1、Th2 和 Th17，也产生嗜酸性粒细胞和中性粒细胞[30]、NK 细胞、DCs 和 B 细胞[31,32]。这种抗炎细胞因子产生的细胞变异性和冗余性在多个水平上引起多重免疫调节，其唯一目的是抑制由 IFN-γ、IL-2 或 TNF 和其他促炎细胞因子引起的炎症反应。

由于 IFN-γ 和 IL-2 对于建立细胞应答很重要，受 IL-10 影响导致以 Th1 应答为主向 Th2 应答为主转变。这种细胞因子不仅影响 T 细胞介导的免疫功能，而且影响树突状细胞成熟和功能，随着 IL-10 分泌，细胞因子的总体格局也发生变化。IL-10 直接作用于 APC 细胞以降低 MHC Ⅰ和 MHC Ⅱ类分子所必需的辅助分子（如 B7）。另外，IL-10 刺激 $CD4^+$T 细胞增殖，从而增强体液应答。IL-10 是连接先天免疫应答和后天免疫应答的关键细胞因子，一方面由巨噬细胞产生，另一方面由 T 细胞在特异性刺激后产生。

IL-10 作为生物标志 IL-10 在多种病原体引起的细胞内感染中具有特殊作用。一些细菌、真菌、病毒或寄生虫通过与其特异性模式识别受体（PRRs）结合并激活基因表达来诱导 IL-10。因此，IL-10 也可以作为生物标志，作为某些感染的替代标志。IL-10 产生水平似乎取决于 PRRs 刺激强度。TNF 将立即下调，整个炎症应答因此下降。由于缺乏 TNF（有可能是主要的诱导因子），细胞凋亡也会下降。

必须着重考虑 IL-10 对慢性感染的作用。在这些特殊情况下，IL-10 起到了魔鬼分子的作用，因为它可以阻止炎症反应，从而避免病原体被清除[33]。显然，各种微生物通过制定靶向诱导 IL-10 的逃逸策略来维持其感染持久性。

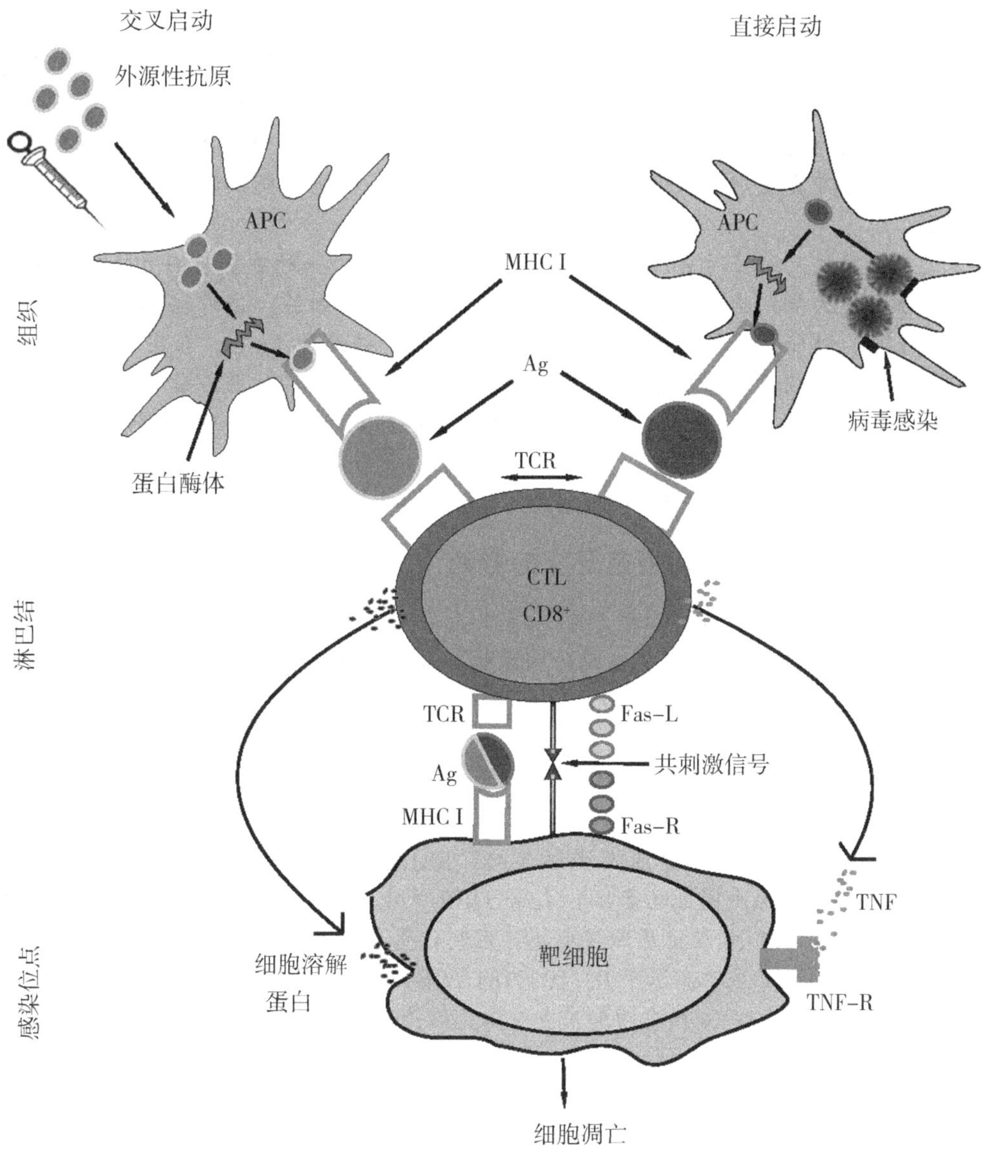

图 2.2 交叉启动与直接启动 CTLs 比较

交叉启动是指 APCs 对凋亡细胞或坏死细胞碎片等抗原的外源性摄取。DNA 疫苗也使用交叉启动激活细胞免疫应答。相反地，直接（经典）启动是指由细胞内病原体产生的内源性合成抗原。在细胞内处理后，抗原通过 MHC Ⅰ呈现给 $CD8^+$细胞毒性 T 淋巴细胞。活化的 CTL 通过 T 细胞受体（TCR）识别目标细胞表面的同源抗原。如前所述，几个效应器机制可以摧毁目标。Ag 抗原、Fas-LFas 配体、Fas-RFas 受体、TNF 肿瘤坏死因子、TNF-R 肿瘤坏死因子受体。

知识点 2.1 APCs 对疫苗开发的影响

人体最大的器官是皮肤。根据体型大小，皮肤表面积在 1.5~2.0 米^2（16.1~21.5 英尺^2）。皮肤是组织损伤和病原体的解剖屏障，是最大的免疫器官。

经皮免疫（TCI）目标是 LCs 和其他 DCs。局部应用将疫苗递送到表皮的 LCs。无针和无创方法是含有生物可降解纳米颗粒的水凝胶贴剂，如作为载体的壳聚糖、脂质体或其他聚合物如聚乙二醇凝胶（PEG）。亲水性大分子（如肽/蛋白质）不能穿透皮肤，需要一个载体。这种应用方法的一个局限性可能是只有可溶性抗原和无颗粒抗原能够诱导免疫应答[27]。载体粒径和亲水性/疏水性可能影响吞噬作用的吸收。

一种低侵入性方法是微针贴片。这种贴片能够携带不同种类的抗原，从DNA到肽/蛋白质，或者灭活或活病毒。与传统皮内注射相比，需要更低的疫苗剂量。针头足够结实，可以穿透皮肤，并且可以无痛地将疫苗递送给LCs。这种简单、安全、有效、低成本装置适用于自我注射。

电穿孔靶抗原在真皮树突状细胞中的应用。电脉冲可逆地改变细胞膜通透性，因此抗原可以穿透细胞。

除了通过皮肤免疫途径对APCs进行简单的体内靶向外，一种更复杂的直接靶向树突状细胞的体外方法被开发为全细胞疫苗。从患者血液中分离树突状细胞，体外培养，并用抗原如肿瘤相关抗原（TAA）攻击。通过摄取TAAs激活DCs成熟，并处理TAAs。在实验室进行了3~5天的复杂程序后，病人被注射了自体TAA负载的树突状细胞，这种细胞在体内会提醒T细胞和B细胞。2010年，美国食品和药物管理局批准了Sipuleucel-T，这是第一个治疗晚期前列腺癌个性化细胞自体免疫疗法[28]。

要更深入了解疫苗的各种递送方法，请参见本书第七部分。

知识点 2.2 IL-10 对疫苗开发的影响

利什曼病是由利什曼属专性细胞内和人畜共患寄生虫引起的，它是内脏（VL）、皮肤或黏膜/黏膜皮肤利什曼病等一系列相关疾病的病原体，严重者可以引起致命的全身感染。

据估计，全球流行病例为1 200万，年死亡率为6万，在4个大陆的88个国家流行，其中72个国家是发展中国家。根据世卫组织预计，每年将新增病例200万。

内脏利什曼病（VL）被认为是与艾滋病病毒（HIV）相关的一种重要的机会性混合感染病，它具有特殊的情况，这是一种被低估的组合。两种病原体相互增强。目前还没有有效疫苗。一个原因是由IL-10引起的利什曼原虫逃逸策略。BALB/C小鼠用DNA疫苗接种，编码LACK（激活的C-激酶利什曼原虫同源物受体）或者TRIP（胰蛋白酶过氧化物酶），并用经过改良的Ankara痘苗病毒（MVA）相应抗原加强免疫，用利什曼原虫进行攻毒。由于调节性T细胞作用，接种LACK疫苗的小鼠产生了高水平IL-10。Th1应答同时下调，导致IFN-γ和TNF产量不足。但是IL-10受体阻断使IFN-γ产量增加了3.5倍。结论：DNA/MVA-LACK疫苗不起作用。不同于用TRYP接种，IL-10水平较低，而IFN-γ水平较高。用利什曼原虫进行攻毒后，接种TRYP的小鼠得到完全保护[34]。

这些研究证实利什曼原虫能够利用宿主细胞因子IL-10存活和持续感染。此外，与IFN-γ水平相关的IL-10滴度可作为疫苗接种成功的预测性生物标志。选择合适的抗原将影响免疫应答的方向，即朝向Th1应答或Th2应答以及是朝向细胞应答还是体液应答。

为了监测内脏利什曼病感染临床过程，测量IL-10滴度等指标。与寄生虫载量的相关性非常显著。同样，IFN-γ相应滴度也很显著：低IFN-γ滴度伴随着高IL-10滴度。因此，IL-10滴度是反映这种寄生虫感染严重程度的间接生物标志[35]。其他病原体如克氏锥虫、肺炎克雷伯氏菌或白色念珠菌也会劫持IL-10以防止免疫攻击。

IL-10作为免疫调节剂的影响也在某些病毒中有所表现。这种病毒编码自身IL-10同源物（vIL-10），因此可以直接影响对这些病毒自身的免疫应答。1990年首次报道了用于埃-巴二氏病毒（Epstein-Barr virus，EBV）的vIL-10[36]。其与huIL-10在蛋白水平上同源性为70%。

还有一些病毒不具有编码这种病毒因子的自身基因。这些病毒通过表达一种能上调宿主IL-10产量的特殊病毒蛋白来避免缺乏vIL-10[37]。

同时，IL-10也可作为某些癌症疾病的预后因子。高IL-10表达与黑色素瘤和鳞状细胞癌的侵袭性临床表现显著相关[38,39]。

2.4 治疗性疫苗与免疫抑制环境

促炎细胞因子如白细胞介素-1β（IL-1β）、白细胞介素-6（IL-6）和肿瘤坏死因子-α（TNF-α）负责引发炎症以应对感染、组织损伤或癌症发展[40]。这激活了细胞介导的免疫应答。抗炎细胞因子也被释放来微调，以平衡这种免疫反应，并限制持续的炎症反应。

免疫抑制 如前所述，无论在微生物感染还是在癌症疾病中，IL-10在监督炎症反应中的中心作用是明显的。其他细胞因子也参与炎症控制，如TGF-β，通过IL-1受体类似物（IL-1ra）阻断单核细胞/巨噬细胞的IL-1α/β控制炎症；或IL-4，通过阻断促炎细胞因子与其特异性受体的绑定控制炎症；或IL-6，通过抑制TNF控制炎症；或IL-1或IL-11等其他细胞因子通过促进Th2应答控制炎症。必须指出，除了IL-1ra和可溶性细胞因子受体外，所有抗炎细胞因子具有双重性，也能够执行促炎活性[41]。

在所有免疫抑制事件中，抗炎细胞因子旨在下调T细胞效应器功能，导致局部免疫抑制环境。只有在这种前提下，病原体才能持续存在，肿瘤细胞才能增殖。免疫调节网络已经从对病原体的保护转变为对嵌套病原体的严格保护。一些微生物病原体编码自己的IL-10作为逃逸策略的一部分。一些肿瘤分泌IL-10或TGF-β作为逃逸策略的一部分，增强抗炎分子的局部优势和建立免疫抑制环境以逃避保护性免疫应答。此外，肿瘤相关巨噬细胞分泌许多促进免疫抑制的细胞因子[42]。

任何治疗性疫苗接种都必须克服这种强大的抗疫苗活动堡垒，通过DCs或MHC Ⅰ和MHC Ⅱ的抗原递呈减少或取消，CTL活性减少或停止。这种疫苗失败是不能有效地恢复T细胞对持续病原体免疫，进而维持这种针对免疫抑制状态的炎症反应恢复状态。细菌、病毒和寄生虫使用病原体编码或宿主编码的蛋白质在宿主体内存活，这一事实本身就表明，作为抵御入侵微生物和癌症疾病的防御策略，促炎细胞因子和相关的基本细胞活动具有很高的优先性。

基于这一知识，治疗性疫苗必须至少具备阻断抗炎过程、促炎环境重建、反复暴露多种抗原以重建T细胞免疫3项基本功能。

2.5 病原识别

抗原特异性免疫反应长级联始于APCs通过其模式识别受体识别抗原，从而产生B淋巴细胞和T淋巴细胞免疫反应。

这些细菌编码的PRR是先天免疫系统的一部分，并且使先天免疫与后天免疫交联，后天免疫性依赖于B细胞和T细胞的克隆扩增。先天免疫和后天免疫之间存在着一种具有明确层次的永久性串扰。PRRs构成了一个高效和冗余的受体网络基础，用于永久监测病原体和危险的自身抗原。PRRs在细胞表面表达或位于细胞核内体或细胞质中。这种冗余系统保证了病原体可以被不同的传感器并行检测到。如果一个传感器失灵，另一个PRR就被激活以追踪该抗原。这种高水平冗余可能是健康患者中98%的感染被免疫系统检测出来和治愈的原因之一。

模式识别受体（PRRs） Charles Janeway提出了病原体识别机制的理论基础。1989年，他假设每个APC能表达识别病原体的模式识别受体（PRRs），识别病原体的受体分子结构的特定保守模式称为“病原体相关分子模式”（PAMPs）[43]。

2.5.1 非Toll样受体

除了Toll样受体（TLRs），还描述了与TLRs在化学和结构上不同的其他类型的先天性PRRs。相对于TLRs，对大多数非TLRs的精确分子机制知之甚少。

（1）*NLRs-核苷酸结合和寡聚结构域（NOD）样受体（NLRs）* 迄今为止，已知人类NLR

家族有 22 个成员，其中最突出的是 NOD1 和 NOD2，它们控制细菌感染和炎症[44]。

（2）*RLRs*-视黄酸诱导型基因 I（RIG-I）样受体（RLRs） 这个家族由视黄酸诱导基因 I（RIG-I）、黑色素瘤分化相关基因 5（MDA5）和遗传生理学实验室 2（LGP2）3 个成员组成，在检测病毒 RNA 中起着关键作用[45]。

（3）*CLR*-C 型凝集素受体 这个由 17 个亚科组成的多样化家族成员起到与可溶性 PRR 或膜结合的作用。Dectin-1 和 Dectin-2 受体，即主要检测甘露糖和葡聚糖的结构，但是除了这些主要的碳水化合物外，还有蛋白质和脂类。CLRs 尤其与抗真菌免疫有关[46,47]。

（4）*DNA* 传感器 即 DAI、IFN 调节因子的 DNA 依赖性激活剂，或 AIM2（在黑色素瘤 2 中不存在），或 RNA 聚合酶Ⅲ对 dsDNA 的间接检测，RNA 聚合酶Ⅲ是一种将富含 AT 的 dsDNA 转录为 dsRNA 的聚合酶，现在能够激活 RIG-I 样受体[48]。这些 DNA 传感器识别细胞内 DNA[49]。

Toll 样 1996 年，在果蝇中描述了第一个 Toll PRR 的特征[50]。仅 1 年后，第一个"Toll-样"受体（TLR）在人类中被发现[51]。1988 年从果蝇中分离到 *Toll* 基因，该蛋白为具有细胞质和大细胞质外结构域的完整膜蛋白[52]。

与大多数主要是细胞外病原体传感器的 TLR 相比，非 TLR 位于细胞内[53]。表 2.3 概述了不同类别的 PRRs。

细菌系编码的 PRRs 在动物界大量存在和分布突出了其对病原体免疫防御的巨大意义。从节肢动物到脊椎动物，从昆虫到哺乳动物，整个进化过程保存在这一受体家族中。2007 年的一项研究统计了大约 500 个 PRRs，其中 177 个 TLRs 在 77 种不同生物体中检测到[54]。昆虫并不具有特定的获得性免疫力，而是像植物一样只具有先天性免疫力。此外，进化上古老 PRRs 的发现，强调了先天免疫对哺乳动物获得性免疫的主导作用。免疫应答级联的起搏器是先天系统，尽管先天免疫和后天免疫是协同工作的。表 2.4 简要介绍了完整免疫系统发展关键步骤的进化，将说明先天免疫的古老作用。

不同 PRRs 共同工作来检测单个病原体，根据其在感染过程中显示的演化毒性结构来调整免疫防御。

表 2.3 不同 PRRs、配体和细胞分布的分类

受体	配体	细胞分布
Toll 样受体 TLR2/1	三酰脂肽，可溶性脂蛋白	B 细胞、NK 细胞、DCs、巨噬细胞、成纤维细胞、上皮细胞和内皮细胞
TLR2	肽聚糖、脂磷壁酸、脂阿拉伯甘露聚糖、酵母多糖	B 细胞、NK 细胞、DCs、巨噬细胞、成纤维细胞、上皮细胞和内皮细胞
TLR4	脂多糖，黄体酮	B 细胞、NK 细胞、DCs、巨噬细胞、成纤维细胞、上皮细胞和内皮细胞
TLR5	鞭毛蛋白	B 细胞、NK 细胞、DCs、巨噬细胞、成纤维细胞、上皮细胞和内皮细胞
TLR6/2	二酰脂肽	B 细胞、NK 细胞、DCs、巨噬细胞、成纤维细胞、上皮细胞和内皮细胞
TLR3	dsRNA	B 细胞、NK 细胞、DCs、巨噬细胞、成纤维细胞、上皮细胞和内皮细胞
TLR7	富含鸟苷或尿苷的 SSRNA	B 细胞、NK 细胞、DCs、巨噬细胞、成纤维细胞、上皮细胞和内皮细胞
TLR8	富含鸟苷或尿苷的 SSRNA	B 细胞、NK 细胞、DCs、巨噬细胞、成纤维细胞、上皮细胞和内皮细胞

（续表）

受体	配体	细胞分布
TLR9	未甲基化 CpGs，寄生虫中的疟原虫色素	B 细胞、NK 细胞、DCs、巨噬细胞、成纤维细胞、上皮细胞和内皮细胞
NOD 样受体 NOD1	γ-D-谷氨酰-*m*-二氨基二甲酸	例如 DCs
NOD2	胞壁酰二肽（MDP）	例如 DCs
NLRP1	被致命毒素激活	例如 DCs
NLRP3	病原体源性毒素，如流感 M2	例如 DCs
NLRP7	酰化脂肽	例如 DCs
RIG 样受体 RIG-1	钝末端 dsRNA，5 端具有三磷酸的 ssRNA，RNA 柄状结构	除 pDCs 外的所有细胞
MDA5	长链合成的单磷酸端 dsRNA RNase l 裂解自 RNAs	除 pDCs 外的所有细胞
LGP2	负/正，RIG-1/MDA5 调节器	除 pDCs 外的所有细胞
C 型凝集素受体 Dectin-1	β-1，3-葡聚糖（真菌病原体）	吞噬免疫细胞
Dectin-2	高甘露糖结构	主要在组织巨噬细胞、树突状细胞和炎性单核细胞上表达
巨噬细胞诱导型 C 型凝集素	多种内源性和外源性配体，例如 SAP130、α-甘露聚糖、海藻糖二霉菌素、分枝杆菌索因子	主要由活化巨噬细胞表达
DNA 传感器 AIM2	细胞溶质 DNA	巨噬细胞
p202 IFI16 IFIX DDX41	细胞溶质 DNA	

表 2.4 数百万年前从天然免疫到获得性免疫发展中一些主要进化事件的短时间表

年代	MYA	生命的进化	免疫进化
新生代	1.8	智人	完整的先天和后天免疫系统
		空气：O_2=20，95%	
中生代	100	蜜蜂	开花植物（被子植物）
	144	第一哺乳动物	
古生代	350	空气：O_2=26%	获得性免疫：MHC 和抗体
	450	颚鱼（七鳃鳗） 无颚鱼（盲鳗）	

（续表）

年代	MYA	生命的进化	免疫进化
前寒武纪	600	水螅（第一种动物?） 后生动物	先天免疫：抗菌活性和 TLRs
	1.200	第一种多细胞生物（藻类）	性和减数分裂!
	3.500	单细胞生物	
	4.500		地球起源

注：百万年以前（MYA）。

根据其功能，PRRs 可分为两组，①清除剂受体：在不传递细胞内信号的情况下负责微生物的吞噬作用，例如甘露糖受体。

②信号受体：通过传递细胞内信号来触发先天性和获得性免疫，例如 TLRs。

2.5.2 Toll 样受体

TLRs 是 PRRs 异质组中最具特征的传感器，可识别外源性和内源性病原体相关分子模式（PAMPs），但也可识别内源性损伤相关分子模式（DAMPs）。DAMPs 产生于癌细胞、坏死细胞、组织损伤或降解过程。

目前在人类中发现了 10 个 TLRs，在小鼠中发现了 13 个。TLR-10 仅在人类中发现。

分布　TLRs 存在于 DCs 细胞、巨噬细胞和单核细胞、T 细胞、B 细胞和 NK 细胞中。此外，TLRs 还在正常组织细胞中表达，如内皮细胞、成纤维细胞、肌肉细胞、软骨和骨软骨组织，即软骨细胞、成骨细胞和破骨细胞。最近在脂肪组织、脂肪细胞[55]、肝脏、肝细胞[56]、脾脏[57]中也检测到 TLR 的表达。TLRs 在各种细胞、不同组织和不同器官中的广泛分布突出了天然免疫及其炎症应答和获得性免疫发展的系统性意义[58]。

TLRs 和细胞表面　TLRs 1、2、4、5、6 和 10 位于细胞表面。它们是 1 型 α-螺旋跨膜糖蛋白：富含亮氨酸的胞外域包括配体结合位点、单层膜固定 α-螺旋和细胞质 Toll/IL-1 受体（TIR）结构域，负责下游信号转导。这些细胞表面 TLRs 专门识别和结合革兰氏阴性和革兰氏阳性细菌的结构。

TLR1 识别可溶性脂蛋白，例如脑膜炎奈瑟菌。TLR2 识别脂蛋白以及糖脂，例如支原体和嗜麦芽密螺旋体。TLR2 与 TLR1、TLR6 或 TLR10 一起充当异二聚体。TLR2 还专门用于真菌配体，如酵母多糖。

脂多糖　脂多糖（LPS）是研究最多的病原结构之一。TLR4 通过与髓细胞分化蛋白 MD-2 相互作用与脂多糖结合，形成串联 TLR4/MD-2。这个串联是脂类 A（脂多糖生物活性成分）结合的分子基础（脂多糖的示意图见第 33 章）。在这个过程中，辅助受体 MD-2 负责通过结合酰基与脂肪酸结合[59]。

除类脂 A 外，TLR4 其他非细菌配体有 MMT 病毒的 gp52 包膜蛋白、RS 病毒的融合蛋白或内源性配体，例如热休克蛋白 HSP60 和 HSP70。

TLR5 是唯一与纯蛋白结合的 TL 受体。TLR5 识别细菌鞭毛蛋白的高度保守区域，因此该受体能够识别并结合广谱微生物。

除了 TLR10 和 TLR1 能形成异二聚体之外，对 TLR10 知之甚少。直到今天，TLR10 的天然配体还是未知的。

总之，TLR 1、2、4、5、6 和 10 是专门用于结合细菌和其他致病性脂蛋白、糖脂和脂多糖的，超出了组织损伤或其他降解过程（如热休克蛋白）描述的病毒包膜蛋白或内源性结构。

TLRs 和核酸　TLRs 3、7、8 和 9 位于核内体，专门识别细菌、病毒的核酸或宿主衍生的 DNA 和 RNA。TLR3 结合 dsRNA，也可以被合成 dsRNA polyI：C 激活。此外，该受体也结合宿

主源 mRNA 和 tRNA。

通过 I 型干扰素（IFNs）诱导以及与 TLR3 结合后的天然抗病毒活性，证明了 TLR3 作为抗病毒受体的特殊特性。天然 dsRNA 通常在病毒复制过程中产生。然而天然衍生的 dsRNA 在体外只是 TLR3 的弱激活剂。TLR7 优先在 pDCs 上表达，如 TLR8 和 TLR9，结合 ssRNA，特别是被引起慢性感染的病毒激活。这些受体的激活也诱导干扰素。与其他核酸结合受体相比，TLR9 识别未甲基化的细菌 ssDNA、CpGs，也识别病毒 DNA，例如 HSV-1 和 HSV-2[60,61]。

TLR7 和病毒 灭活病毒疫苗使用 TLR7。这可以通过灭活整个甲型流感病毒非常清楚地证明。病毒 RNA 与 pDCs 上的 TLR7 结合，导致产生大量 IFN-α。需要注意的是，病毒灭活过程不影响结合特性，也不影响核内体受体的能力。这与仅由蛋白质组成的重组流感疫苗或裂解疫苗形成对比：病毒被去污剂破坏，病毒 RNA 被纯化程序去除。重组疫苗或裂解疫苗不会激活 TLR7[62,63]。

TLR9 和 CpG DNA 疫苗本身的天然细菌骨架能够激活 TLR9。但是这种天然佐剂效果非常弱，可能是因为疫苗 DNA 主要是 dsDNA，而不是 ssDNA。为了改善这种弱佐剂效应，可以在同一质粒或额外的质粒上将几种合成物种特异性 CpGs 结合到 DNA 载体中[60]。

2.6 PAMPs：病原相关分子模式

TLRs 以及其他 PRRs 通过其独特、高度特异性和高度保守的病原体相关分子模式 PAMPs 来识别细菌、病毒、真菌、原生动物和寄生虫。所有传感器都使用 PAMPs 来区分“自我”和“非自我”。并非所有配体识别机制都被完全理解。主要的 PAMP 配体和一些已知的结构特征如下。

(1) *脂质* 可以识别链长，例如脂多糖的三酰基（TLR1）、二酰基（TLR2）和六酰基（TLR4）。LPS（脂多糖）保守模式是脂质 A。用作佐剂的免疫刺激剂单磷酸脂质 A（MLP）仅包含 5 条脂质链。与天然 LPS 相比，炎症活动约减少了 100，是先天性免疫应答弱激活剂[64]。

(2) *蛋白质* 富含亮氨酸的重复序列（LRR）结构域由 2~45 个 LL 重复序列组成，每个重复序列 20~30 个残基，是 PRRs 的靶 PAMP。这些 PRRs 本身包含 LL 结构域，产生的受体-配体复合物由蛋白质-蛋白质相互作用组成[65]。

(3) *糖类* 蛋白质和糖蛋白或脂质和糖脂形成的复合糖。一些 TLRs 和 CLRs 通过其糖识别结构域（CRD）识别复合糖，如甘露糖、葡萄糖或半乳糖，CRD 是凝集素多结构域蛋白中的一个结构域[66]。

(4) *RNA* 许多 PRRs 在病理条件下不仅识别非自身 RNA，而且识别自身 RNA。因此，一些 RNA 识别特征对于“非自身”RNA 的鉴定是必不可少的，例如初级核苷酸序列、二级和三级结构或细胞内定位[67]。

(5) *DNA* 与脊椎动物 DNA 相比，细菌 DNA PRR 识别的分子模式是严格的低甲基化。微生物 DNA 含有高频率的非甲基化胞嘧啶-磷酸-鸟嘌呤（CpG）二核苷酸[68]。

2.7 DAMPs：损伤相关分子模式

在这里谈论免疫系统对抗外来病原能力是很常见的。但是也许免疫系统不会对非自我做出明确的反应，而是只对任何“危险”做出反应，不管其来源如何。Matzinger 在 1994 年中首次提出了这一危险假设[69]：“免疫系统的主要目标是区分自我和非自我，这一观点多年来一直为免疫学家服务。我相信……它的主要驱动力是需要检测和防范危险……”

损伤相关分子模式（DAMPs）或警报素 现在危险信号被定义为类似于 PAMPs 的内源性宿主编码蛋白，在完全没有病原体的情况下，它们只在细胞应激、组织损伤和坏死细胞破坏后才释放。这些分子统称为损伤相关分子模式（DAMPs），有时也称为警报素。许多 DAMPs 是细胞核或胞浆蛋白，也有细胞外基质蛋白（ECM），它们在组织损伤后上调或释放。这些平常隐藏的蛋白发出信号“危险”，导致无菌炎症。在此过程中，从受损细胞释放的物质从细胞内还原变成细胞外氧化环境，引起细胞外氧化还原环境急剧变化。

氧化应激 这种紊乱氧化还原环境的一个后果可能是慢性炎症。这些物质是不同的所谓无领导分泌蛋白（LSP），如高迁移率基团 B1 分子（HMGB1）或 IL-18。这些蛋白是炎症强介质，参与氧化应激反应[70,71]。像 PAMPs 这样的 DAMPs 由同一组 PRR 绑定，描述了 TLR 1、2、3、4、7、8 和 9 的 DAMP 配体[72]，还描述了内源性 DAMPs 与外源性微生物 PAMPs 之间的协同作用，导致共同的先天免疫应答[73]。除了细胞内外蛋白、脂蛋白、糖蛋白或基因组 DNA 和 RNA 分子等非蛋白外，一些能够激活免疫系统的主要 DAMP 配体如下。

（1）线粒体 DAMPs，MTDs 来自线粒体多肽和 DNA，负责提供细胞能量的细胞内细胞器，受到创伤后细胞破裂释放[74]。

（2）高迁移率族 B1 分子，HMGB1 一种普遍存在的核染色质相关蛋白，充当转录因子，如从坏死但不凋亡的癌细胞中释放，或者在早期炎症期由巨噬细胞主动分泌[75]。

（3）热休克蛋白，HSPs 一个高度保守普遍存在的蛋白质家族，充当伴侣，负责蛋白质的正确折叠和运输，如 Hsp90 在 DCs 的抗原递呈和成熟中起重要作用，并且由坏死细胞释放[76]。

（4）S100 蛋白质 钙结合蛋白家族至少有 21 个成员。除了细胞生长、分化中的其他功能外，这些蛋白质也参与炎症反应。一些 S100 蛋白可作为肿瘤标志，细胞内 S100A8 或 S100A9 释放到细胞外介质中是一个强烈的危险信号[77]。

（5）透明质酸片段，HA 一种碳水化合物聚合物，是非硫酸化糖聚糖以及细胞外基质（ECM）的一部分。炎症部位组织损伤后，可迅速发现 HA 的降解产物[78]。

2.8 TLR 细胞内信号级联与免疫应答

抗原与 TLR 或另一种 PRR 结合的唯一功能是诱导针对该捕获抗原的免疫应答，而不管其来源是 PAMP 还是 DAMP。这种反应在 Th1 和 Th2 作用之间必须是合理的和平衡的，更多的 Th2 或更多的 Th1 应答，或 Th1 和 Th2 之间的转换。这种应答也应该是有效的，而不是针对幻觉，如诱饵表位。基于这些假设，免疫反应必须是有时间限制的、协调的、可控的、适当的，并且能够导致记忆效应。

为了满足所有这些假设，APC 内通常发生针对个别情况精确调整的信号级联。最终结果是选择性免疫相关基因特异性激活。因此大约 150 个基因可以被激活和表达。为了诱导保护性而非病理性的免疫应答，信号级联中的串扰是必不可少的。一组（丝裂原-活化蛋白）MAP 激酶控制信号级联的激活。MAP 特征是 3 层经典通路，3 个激酶串联连接。与经典 MAPKs 相比，针对非典型 MAPKs 我们描述了一种双层系统。

LR 依赖的信号级联有以下 5 个主要步骤。

（1）衔接分子的结合 配体与 TLRs 富含亮氨酸的 N 端胞外结构域结合，使胞外结构域二聚化，形成马蹄形结构。这导致细胞质 TIR 胞内结构域的构象改变，形成 TIR-TIR 同源二聚体。TIR 结构域（Toll/IL-1 受体）是连接免疫刺激和适当免疫应答的关键分子，在所有其他 TLRs 中高度保守。TIR 二聚体随后从细胞质中招募 1 个衔接子。有 5 种不同的细胞质衔接子可用，每种分子都能够结合 TIR[79,80]：

MyD88 髓样分化初级应答基因 88

TRIF 含 TIR 结构域的衔接子诱导 IFN-β

TIRAP 含 Toll-白细胞介素 1 型受体结构域的衔接蛋白

TRAM TRIF 相关的衔接分子

SARM 含无菌 α 和犰狳基序的蛋白质

与不同类型的细胞一起，这些不同的衔接分子解释了基因表达的巨大变异性，因此解释了不同的免疫活动。

主要的衔接分子是 MyD88 和 TRIF。两种途径是不同的，一种是 MyD88 依赖性的，另一种是 MyD88 非依赖性的。

大多数 TLRs 都使用 MyD88。TLR3 使用 TRIF，TLR4 使用 MyD88 和 TRIF。TLRs 2、4 和 6 位于细胞膜内，TRIF 使用衔接分子 TIRAP 代替 MyD88 和 TRAM。

（2）IKK 复合物活化 MyD88 的 N 末端，即所谓的死亡结构域（DD，在所有凋亡蛋白中发现的一束特征性的 6 个 α 螺旋），参与 IRAK（IL-1 受体相关激酶，丝氨酸/苏氨酸激酶家族）的结合。IRAK 通过自磷酸化自激活，并在解离后与 TRAF-6 结合（TRAF-6，TNF 受体相关因子 6）。激活的 TRAF-6 再激活 IKK 复合物，该复合物由 IKK-γ 也称为 NEMO（κ-B 激酶抑制剂）和两个催化亚单位 IKK-α（或 IKK1）和 IKK-β（或 IKK2）组成。IKK 的激活是转录因子释放的关键步骤，也是 NF-κB（活化 B 细胞核因子 κ-轻链增强子）途径的标志。

（3）转录因子的释放 TRAF-6 与催化聚泛素链形成的酶相互作用。这些泛素是 IKK 激活所必需的，并以两种不同的途径结束。泛素附着于 IκB 蛋白上，IκB 蛋白是细胞质 NF-κB-IκB 复合物的一部分，也是 NF-κB 的抑制剂。IκB 因此被标记为蛋白水解降解。IκB 失活启动了 NF-κB 通路，NF-κB 正在进入细胞核的通路上。NF-κB 是一个大群体中最显著的转录因子。另一个途径也从泛素化的 IκB 蛋白开始，涉及激活 AP-1（激活蛋白 1）的 c-Jun N-末端激酶（JNK）。AP-1 也像 IFR3（另一种转录因子）一样转移到细胞核中。

（4）通过 NF-κB 激活基因 NF-κB 通常被称作“免疫反应的中枢介质”。超过 150 个基因由该转录因子控制[81]（表 2.5）。几乎所有这些基因都直接参与免疫应答。基于诱导型基因调控，生物体能够对负面影响如感染、（氧化）应激或组织损伤做出非常迅速和有针对性的反应。NF-κB 的失调有助于病理免疫，也被讨论在癌症疾病中扮演重要角色，因为 NF-κB 也控制着负责细胞增殖的基因[82,83]。

NF-κB 不是单一的蛋白质，而是一些具有独特特征的蛋白质复合物，由 300 个氨基酸（aa）组成的 Rel 同源结构域（RHD）。

RHD 也是其他转录因子的一部分，负责蛋白质-DNA 的连接。NF-κB 复合物形成同源和异源二聚体，通过与 IκB 结合，NF-κB 在细胞质中安静下来。IkB 解离激活 NF-κB，然后转移到细胞核并结合到 1 个 10 bp 的 DNA 位点，称为 κB。哺乳动物细胞中最广为人知的复合物由 p50 和 RelA 蛋白组成，也称为 NF-κB。

NF-κB 二聚体不具有内在的酶活性，仅介导靶基因的转录控制，因为二聚体结构不允许与核小体 DNA 直接结合。该二聚体结合 1 个协同调节因子，如组蛋白乙酰转移酶（HAT）和组蛋白去乙酰酶（HDACs）。这些协同调节物具有酶活性并转移乙酰基。DNA 表达受乙酰化和去乙酰化调节。通过组蛋白乙酰化打开染色质结构后组装预起始复合物（PIC），并帮助 RNA 聚合酶 II 通过结合到启动子的 TATA 盒（任何基因的转录起始位点）来激活转录[84]。

表 2.5 TLR 信号通过 NF-κB 作用于免疫和非免疫基因

细胞因子	受体	APPs	Reg P	GF	其他因子
IFN-β	MHC Ⅰ	Tissue F1	TAP1	GM-CSF	p53
IFN-γ	B7.1	CRP	ICAM-1	G-CSF	IRF-1
IL-1α/β	CD48	CF C4	ELAM-1	M-CSF	IRF-2
IL-2	IgG heavy c.	CF B	FAS ligand	PDGF Bc	IkBα
IL-6	IgE heavy c.		IAPs		c-myc
IL-8	Ig light c.				c-myb

（续表）

细胞因子	受体	APPs	Reg P	GF	其他因子
IL-9	TNFr				junB
IL-12	CD23				
IL-15	CD69				
TNF-α/β					

缩略词：APPs 急性期蛋白，Reg P 调节蛋白，GF 生长因子，IFN 干扰素，IL 白细胞介素，TNF 肿瘤坏死因子，MHC 主要组织相容性复合物，B7.1 T 细胞共刺激因子，CD48 淋巴细胞抗原，Ig 免疫球蛋白重/轻链，TNFr 肿瘤坏死因子受体，CD23 细胞表面分子，CD69 C 型凝集素蛋白，Tissue F1 组织因子 1，CRP C-反应蛋白，CF C4 补体因子 C4，CFB 补体因子 B，TAP1 转运蛋白 ER，ICAM-1 细胞内黏附分子，ELAM-1 内皮细胞白细胞黏附分子，FAS 配体细胞凋亡诱导剂，IAPs 细胞凋亡抑制剂，CSF 集落刺激因子/粒细胞/单核细胞/巨噬细胞，PDGF Bc 血小板衍生生长因子 B 链，p53 肿瘤抑制因子，IRF 干扰素调节因子，IkB Rel/NF-κB 的抑制剂，c-myc/c-myb/junB，原癌基因。

（5）*活化免疫分子释放* NF-κB 或其他转录因子对基因表达的瞬时改变是快速的，并且使免疫系统能够对由任何 PRRs（如 TLRs）安排的危险、PAMP 或 DAMP 作出迅速和强健应答。

TLRs 在 DCs、单核细胞和巨噬细胞、T 细胞、B 细胞和 NK 细胞上表达，此外还在内皮细胞、成纤维细胞、肌肉细胞、软骨细胞、成骨细胞和破骨细胞、脂肪细胞、肝细胞和脾细胞上表达。这种无处不在的分布有助于无处不在的免疫应答。

除了 PAMPS 和 DAMPs 对 TLRs 的不同激活之外，细胞类型对免疫基因的选择也有很大影响，并最终影响细胞因子、受体、急性期蛋白（APPs）、调节蛋白（Reg P）或生长因子（GF）的产生和分泌。

因此免疫活性多样性很大程度上取决于各种刺激、不同的 PRRs/TLRs 以及受刺激细胞内的细胞类型特异性转录谱。

如果病毒激活 NF-κB 作为其逃避策略的一部分来调节复制，免疫病理反应将会成功[85,86]。在这种情况下，病毒编码的基序通过绕过正常的信号通路与 κB 位点结合。一种定向细胞因子基因激活以支持完整的、不受干扰的病毒复制。如前所述，IL-10 再次充当免疫抑制原病毒细胞因子。

乙型肝炎病毒表面抗原（HBsAg）覆盖了乙型肝炎病毒（HBV）表面的 90%，是预防性疫苗的一部分，并与 NF-κB 结合[87]。这些结合研究的发现和所观察到的病毒对靶免疫基因的操作，可能有助于开发治疗性抗病毒疫苗。

迄今为止，还没有一种获得许可的药物可以直接阻断 NF-κB 的活性，这不仅对肿瘤学来说是有意义的。NF-κB 在受控的实验室条件下易于操作和阻断。然而经过 25 年 NF-κB 深入研究之后，人们普遍担心治疗性抑制可能会导致不可预测和无法控制的基因抑制严重后果。

知识点 2.3 TLRs 在疫苗开发中的影响

识别病原体和细胞内不同结构（PAMPS 和 DAMPs）上的危险信号。此外许多 TLRs 与疾病有关，如系统性红斑狼疮（SLE）中的 TLRs 2、7 和 9 以及自身抗体的产生；细菌和病毒感染中的 TLR2；过敏或类风湿性关节炎以及各种感染疾病，哮喘、心脏病和肝病中的 TLR4；恶性黑色素瘤中的 TLRs 3、4、7 和 9。

TLRs 桥接先天和后天免疫。同样免疫应答的类型（Th1 或 Th2）也是由 TLRs 触发的。对于某些依赖于疾病的 TLRs，开发配体的机会将促进细胞活性，例如对于病毒感染、癌症，或者比体液应答（如在过敏性哮喘中）更能促进细胞活性。TLRs 是治疗性靶点，可由激动剂和拮抗剂操纵[88]。

结合某些TLRs的激动剂被开发为疫苗佐剂。已经获得许可的Fendrix®是乙型肝炎疫苗佐剂：HBsAg是由AS04C作为佐剂的，含有单磷酸脂A（MPL）。MPL在结构上与LPS相关（参见33章和34章），MPL也是TLR4的配体。据报道，两次免疫后可实现完全保护，而传统以明矾作为佐剂的HBsAg疫苗需要3次免疫。

2012年11月，美国食品和药物管理局（FDA）疫苗和相关生物产品咨询委员会（VRBPAC）批准了Heplisav®，又一种基于HBsAg的疫苗，Heplisav®与合成的CpG寡脱氧核苷酸（CpG-ODN）、免疫刺激序列（ISS）和TLR9配体结合。VRBPAC还审查和评估了有关使用这些CpG佐剂的疫苗有可能导致自身免疫疾病的理论风险相关数据。

Krieg于1995年首次描述了CpG-ODN的免疫刺激能力[89]。他发现未甲基化的CpG基序是重复的胞嘧啶-磷酸-鸟嘌呤，广泛分布在微生物中，但在脊椎动物中绝对罕见。因此认为未甲基化的ssCpG-ODN是以病原相关分子模式（PAMPs）触发免疫应答。TLR9仅在B细胞和浆细胞样树突状细胞中组成性表达。目前还不能完全理解CpG的特定序列、内部回文或磷酸盐修饰是否与免疫刺激或分子两端的环等特定结构有关。因此合成的CpGs长度为18~28个核苷酸，是非常不均匀的。

除了协同作用的CpG佐剂外，CpGs还作为化疗后的辅助免疫治疗在一些癌症疾病中得到临床测试[90]。没有肿瘤相关抗原的CpGs作为单一免疫刺激剂的效果确实令人失望。到目前为止，由于无法显著延长患者生存时间，因此还没有一种基于CpGs的癌症患者免疫治疗方案获得批准。

TLRs拮抗剂处于临床前和临床研究阶段。LPS与TLR4结合并参与脓毒症。脂质A衍生物和其他化合物阻断TLR4的LPS结合位点，从而抑制信号级联。TLR2和TLR4参与心血管疾病，与动脉硬化进程有关。拮抗剂降低两种受体的表达显示出抗动脉粥样硬化作用。

2.9　T细胞和启动

天然免疫基于高度保守的受体结构，固定在种系内。后天免疫特征是B细胞和T细胞的克隆扩张。这些细胞携带抗原特异性受体，其多样性依赖于基因重排。

未成熟DCs（最有效的APC）通过TLRs/PRRs捕获抗原后，DCs通过血液或淋巴管从非淋巴组织迁移到淋巴结，在淋巴结中成熟、加工并通过MHC将捕获的抗原递呈给幼稚T细胞受体（TCR）。

启动　T细胞需要3种不同的信号才能完全激活，称为启动。

除了抗原递呈（信号1），协同刺激分子（信号2）对于完整的T细胞启动是必需的之外，促炎分子（信号3）也是必不可少的。

信号　不完全启动导致T细胞失能，一种功能性失活，但仍然存活，处于耐受状态。T细胞的主要协同刺激信号是CD28，可能通过增强细胞内TCR信号使T细胞扩增和分化[91]。CD28与DCs的B7分子（B7.1=CD80，B7.2=CD86）结合。同时T细胞和DCs中（用括号表示）的额外信号分子都被上调，如4-1BB（4-1BBL）、CD27（CD70）和CD40（CD40L）。在这个过程中T细胞和DCs之间发生了非常强烈的串扰。DCs分泌特异性促炎细胞因子（信号3）以支持T细胞成熟、启动和存活，如IL-1α、IL-6、IL-7、IL-12、IL-15、IL-18、IL-27、IL-33、IFN、IFN-γ和TNF-α[25]。这是一种快速的1型免疫应答。

T细胞家族　完成启动后，T细胞分化为各种效应细胞：$CD4^+$T辅助细胞（T_H），传统的Th1、Th2子集；最近发现IL-17生产Th17子集[92]；T滤泡辅助细胞（TFh细胞专门调节B细胞应答）[93]和调节性T细胞[94]（Treg，原名抑制性T细胞）以及$CD8^+$T细胞毒性T细胞（CTLs）。所有这些T细胞随后主要从血液迁移到感染或危险的部位。因此T细胞表达迁移受体，使其能够渗透到外周组织。大T细胞家族的共同特征是独特的T细胞受体（TCR）。

$CD4^+$Th1 细胞产生细胞因子，如 IL-2、IL-12 和 IFN-γ，并支持由巨噬细胞和 $CD8^+$细胞毒性 T 淋巴细胞（CTLs）安排的细胞免疫。这些免疫细胞通过颗粒酶 B 和穿孔素杀死感染细胞并分泌 IFN-γ 和 TNF-α。$CD4^+$Th2 细胞产生 IL-4、IL-5、IL-6、IL-10 和 IL-13 来支持 B 细胞激活从而产生体液免疫。从未成熟的 T 效应细胞到成熟和启动的 T 效应细胞发育需要 3~5 天。

T 细胞-B 细胞连接　B 细胞是另一种 APCs，与 MHC Ⅱ类分子结合、内化和加工抗原并递呈给 $CD4^+$T 滤泡辅助细胞（TFh 细胞），用同源抗原一起启动。TFh 细胞迁移到次级淋巴器官的 B 细胞滤泡中，在那里它们通过 TCRs 识别这种 MHC Ⅱ抗原复合物并激活 B 细胞产生抗体。

循环结束　抗原呈递细胞广泛分布于全身，通过无处不在的 TLR/PRR 结合危险信号，无论是病原体还是宿主衍生的。在处理和递呈后，APC 内最多表达 150 个基因，准备诱导免疫应答。疫苗抗原模仿危险信号，与病原具有相同的免疫激活过程。有关概述请参见图 2.3。

2.10　免疫记忆

记忆性免疫记忆依赖于后天免疫系统的细胞，如 B 细胞和 T 细胞。此时免疫记忆的持续时间取决于疾病、免疫状态、患者年龄以及疫苗类型。

基本上，自然感染可以导致终生免疫，而活疫苗产生的免疫记忆时间比重组疫苗长。负责维持特异记忆免疫的是特异性记忆 B 细胞和 T 细胞，这是疫苗接种的概念基础。人类 T 细胞记忆检测可长达 75 年[95]。

2.10.1　记忆 T 细胞

记忆 $CD4^+$T 细胞　在免疫应答结束时，大多数 $CD4^+$T 细胞发生凋亡。只有一小部分效应记忆 $CD4^+$TEM 细胞存活并继续产生细胞因子以支持正在进行的免疫活动。TEM 细胞大致分为两组：分化程度较低的记忆性 $CD4^+$T 细胞和分化程度较高的多功能 T 细胞。这些高度分化的细胞可以通过不同模式的细胞因子来区分，如 IL-2、IFN-γ 和 TNF-α，从而产生不同程度的保护[96]。记忆性 $CD4^+$T 细胞保持其通过外周组织的迁移能力，$CD4^+$T 细胞在外周组织中占优势，例如皮肤局部真皮层中都是记忆性 T 细胞。$CD4^+$T 细胞产生 IL-2 帮助 B 细胞分化为抗体，生成浆细胞，并导致记忆 $CD8^+$T 细胞的二次扩增[97,98]。

记忆 $CD8^+$T 细胞　大多数 T 效应细胞经历凋亡，并在病原体清除后立即从循环中消失。只有少数（高达 10%）记忆 T 细胞能够非常迅速地对同源抗原或同源疫苗抗原重新感染做出应答。这些长期存活的记忆 T 细胞的维持与抗原无关，主要通过 IL-7（也称为 T 细胞存活细胞因子）和 IL-15 来维持。

效应 T 细胞另一个不同之处在于记忆 T 细胞增殖能力增强和寿命延长。包括抑制分子如程序性细胞死亡蛋白（PD-1）的表达增加、效应功能丧失，如颗粒酶、IFN-γ 或 IL-12 的产生丧失。

此外 NF-κB 依赖性（细胞因子）基因的转录也将会因 NF-κB 抑制剂而在记忆 T 细胞中降低。

子集　根据记忆 T 细胞的解剖定位和功能状态[99-102]，可以鉴定出记忆 T 细胞的 3 个主要子集（一些文献中只提到两个子集）：中央记忆性 T 细胞（TCM）、效应记忆性 T 细胞（TEM）和组织驻留记忆性 T 细胞（TRM）。

TCM 细胞表达淋巴结归巢受体 CD62L 和 CCR7，位于次级淋巴器官，具有高增殖力，但效应潜能低[103]。TCM 细胞能够在血液中迁移和循环，因此被认为是其他记忆 T 细胞的循环和相互转化池。

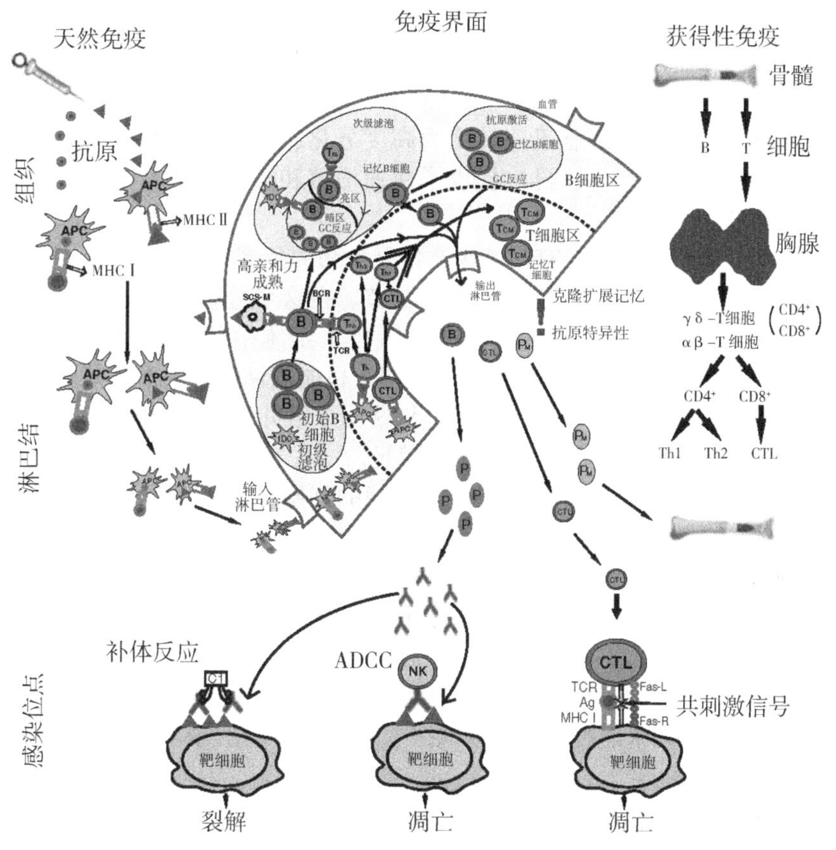

图2.3　天然免疫与获得性免疫在疫苗接种中的相互作用

所有淋巴细胞［B细胞、浆细胞、T细胞或自然杀伤细胞（NK）］都起源于一种共同的淋巴祖细胞，这种细胞由骨髓中的造血干细胞发育而来。T细胞。T细胞迁移到胸腺皮质在无抗原的环境中成熟。成熟后有几种形式的T细胞（MHC限制性αβ-T细胞、Treg细胞以及非MHC限制性γδ-T细胞），可以迁移到外周淋巴组织。B细胞。B淋巴增殖只发生在骨髓中。成熟后B细胞离开骨髓迁移到外周淋巴组织。淋巴结。淋巴结（LN）是天然免疫和获得性免疫的接口。在肌内注射后，例如疫苗抗原被抗原呈递细胞（APC/DCs）捕获并通过MCH Ⅱ/Ⅰ类分子转运到局部淋巴结，在那里它们被递呈到B细胞和T细胞。抗原也可以不经APCs进入淋巴结并被残留被膜下淋巴窦（SCS）巨噬细胞捕获并呈递。淋巴结（LN）中B细胞的发育受T滤泡辅助细胞（TFh）的调节：（1）APC启动的TFh细胞遇到淋巴结原始B细胞的初级B细胞滤泡，处理SCS巨噬细胞转递的同源抗原。这种通过CD27/CD70的T细胞-B细胞相互作用激发了B细胞快速克隆扩增和分化为短命血浆细胞并产生了最早的免疫球蛋白IgM。（2）随着扩增和免疫球蛋白类别向IgA、IgE及IgG转化，淋巴结次级滤泡中的生发中心（GC）反应开始，这是记忆B细胞产生的关键基础，而在细胞外滤泡中淋巴结短命浆细胞分泌抗体。（3）生发中心（GC）反应在不同步骤的循环中进行。B细胞扫描毛囊DCs寻找次级毛囊暗区的抗原并致力于B细胞记忆。TFh细胞在光带中支持这些B细胞，并驱动抗原特异性B细胞高亲和力成熟为非分泌记忆B细胞和长寿分泌性浆细胞以在感染组织远端发挥功能。然而，这些抗原特异性记忆B细胞的一部分不离开次级滤泡，而是重新进入GC循环并经历另一轮亲和成熟。GC循环反应可以连续产生稳定的长寿命记忆B细胞。同源抗原召回后记忆B细胞被记忆TFh重新激活，分泌记忆性浆细胞的抗体快速增殖并诱导另一GC反应。记忆浆细胞（PM）在骨髓中生存并持续提供血清抗体。T细胞发展为效应性细胞和中央记忆性细胞（Tcm）。中央记忆性T细胞（Tcm）能够在血液中迁移和循环，因此被认为是其他记忆性T细胞的循环和转化池。获得性免疫的独特特征是克隆扩增、记忆效应和抗原特异性。经过培养后B细胞和T细胞离开淋巴结并迁移到感染部位。细胞毒性T淋巴细胞（CTLs）在细胞间直接接触并释放如溶细胞蛋白后杀死靶细胞。抗体与细胞表面抗原结合导致抗体依赖性细胞介导的细胞毒性（ADCC），例如由NK细胞介导的。经典的补体途径（通过C1复合物与抗原结合的抗体分子结合）被激活以裂解靶细胞。

TEM细胞产生细胞毒性蛋白，驻留在非淋巴组织中，增殖能力低，也在血管中循环，对全身感染提供有效快速的保护，对局部感染的保护作用有限。TEM细胞池随着时间推移逐渐减少，逐渐向TCM细胞转化。

TRM细胞不会迁移和循环。TRM细胞更喜欢生活在一线防御病原体的组织中，如皮肤、肠道（肺和肠）和阴道黏膜组织，或者在对病原体高度敏感的组织中如脑组织。这些细胞发挥效应功能，产生不同的细胞毒性分子，如颗粒酶 B。表 2.6 概述了记忆 T 细胞。

维持记忆 T 细胞不需要持续抗原[104]，但组织特异性微环境影响迁移受体如 $CD103^+$ 表达。非特异性局部炎性细胞因子足以武装 TRM 细胞[105]。这种通过在外周位置嵌入 TRM 细胞的保护可以持续数年。

面食细胞 从休眠的初始 T 细胞迅速转变为高度分化、活跃和迁移的 T 效应细胞，随着效应细胞迅速减少和记忆 T 细胞产生，再次提供了快速增殖和按需快速减少的能力，这预示了淋巴细胞稳态的精确调节。

任何一种失调都会导致严重的、有时甚至会危及生命的免疫疾病。

控制这种稳态的一个关键机制是新陈代谢的调节。在这个过程中，休眠 T 细胞从柠檬酸循环和呼吸链中葡萄糖的脂肪酸氧化和脱羧转变为活化 T 细胞中的有氧糖酵解（细胞内存在氧气）产生能量[106]。快速生长的肿瘤细胞在氧气作用下也能从糖酵解中产生能量。

Warburg 很早就描述了这一发现，并于 1931 年获得了诺贝尔生理学奖[107]。活化的 T 细胞和癌细胞对能量有很高的需求，只能通过糖酵解来满足。

当氧化脱羧位于线粒体中时，糖酵解发生在细胞的细胞质中。这个位置对活化 T 细胞的好处是快速提供 ATP 能量。新陈代谢的巨大调节使 T 细胞能够有效地对抗病原体，抵御危险，从而 T 细胞可以快速生长、增殖、分化、迁移和分泌细胞因子。

外在因素控制着新陈代谢的增加。T 细胞受体（TCR）起着关键作用。TCR 调控葡萄糖转运蛋白（Glut1）的表达。如果没有 TCR 信号，Glut1 表达会减少，导致葡萄糖吸收也减少，进而导致营养应激和细胞凋亡[108,109]。IL-7 在代谢转换中起着额外的关键作用。IL-7 通过激活 Akt 激酶促进葡萄糖摄取[110]。如果这种代谢转换缺失，例如由于酶的缺陷或葡萄糖缺乏，T 细胞就会失去功能。

完成任务后效应 T 细胞转变成休眠 T 细胞，新陈代谢又转变回线粒体脂肪氧化和氧化脱羧：面食细胞就像马拉松运动员比赛前夜在必备的面食聚会上吃糖以便很快激活这个简单生化能量库。

表 2.6　人类记忆 CD+T 细胞子集

细胞类型	位置	标志	功能
TCM	血液，脾脏，淋巴结	$CD62L^{+++}CCR7^+$ $CD45RA^-$	外周保护 循环 T 细胞池，IL-2
TEM	血液，脾脏，肝脏，外周器官	$CD62L^{\pm}$，$CCR7^-$， $CD45RA^+$	IFN-γ 快速和强健的外周保护，细胞毒性蛋白，IFN-γ
TRM	皮肤，黏膜（肺，肠，阴道），唾液腺，大脑	$CD62L^-$，$CCR7^-$ $CD103^+$，$CD69^+$	快速和强健的局部保护，细胞毒性蛋白

知识点 2.4　T 细胞对疫苗开发的影响

理想的疫苗应能诱导完整体液免疫应答，也能诱导细胞免疫应答。在急性或慢性感染中，自然 T 细胞应答 $CD4^+$ 和 $CD8^+$ 受到多种因素影响，例如病原体及其宿主靶组织的性质。

产生具有高效应能力的强 $CD8^+T$ 细胞应答的疫苗应该以合适的 $CD8^+T$ 细胞应答质量、数量和持续时间来解决这些基本的关键问题。减毒活疫苗或载体疫苗、复制子和 DNA 疫苗能够刺激 $CD8^+T$ 细胞，但不能杀死或刺激常规的基于肽/蛋白质疫苗。T 细胞疫苗研究集中在精确、保守的 T 细胞表位上。

表位是免疫显性肽，与 MHC 等位基因结合是表位向特异性 T 细胞递呈的需要。MHC Ⅰ和 MHC Ⅱ肽长度不同，这取决于等位基因。MHC Ⅰ在 8 和 11 残基之间，MHC Ⅱ在 13 至 25 个残基之间。通过利用一组合成的重叠肽序列（肽库）绘制一系列表位图可以鉴定和表征TCR结合位点。新T细胞表位预测和候选疫苗筛选也用生物信息学方法[111]（参见第42章）。然而，湿实验必须遵循、确认或删除已识别的表位。

自2007年以来，免疫表位数据库和分析资源（IEDB）提供了T细胞和B细胞表位目录以及预测新表位的工具[112,113]。目前197 005个T细胞、221 898个MHC结合和7 621个MHC配体洗脱分析的数据列在本目录中。访问IEDB网站是免费的：http://www.iedb.org。

激活$CD8^+$T细胞最大胆也是最困难的方法是分离直接与MHC Ⅰ类分子结合的抗原[114]。尽管付出了很多努力，但是基于蛋白质的T细胞疫苗还没有成功的临床试验。最近一项研究表明，体外新鲜血液细胞的肽致敏可以通过诱导$CD8^+$T细胞来帮助控制HIV感染[115]。更广泛的方法旨在通过确定的TLR选择抗原直接靶向DCs。然而DCs有许多子集，因此重要的是要了解哪种类型的DCs是靶向的，这些DCs位于何处，TLR库是什么，以及哪些细胞因子参与（佐剂开发）。

记忆性T细胞的维持与同源抗原无关。TEM为全身感染提供有效保护，TRM则负责保护局部感染。非特异性局部炎性细胞因子似乎足以武装TRM细胞[105]。成功的T细胞发育应该包括一种高度特异性的佐剂以迫使这些必要的局部炎症细胞因子与适当的传递系统相结合，例如皮肤传递，以激活HSV-1感染中的皮肤驻留TRM细胞[116]。

针对慢性感染或癌症疾病的治疗性T细胞疫苗的开发应该解决由持续$CD8^+$T细胞活化引起的副作用问题。这些T细胞疫苗有可能产生免疫病理问题，如衰竭（功能丧失，从IL-2产生丧失开始）和炎症。PD-1高表达是T细胞耗尽的生物标志之一。数学模型有助于预测T细胞疫苗对慢性病毒感染的潜在免疫不良反应[117]。

2.10.2 记忆B细胞

对疫苗应答抗体的测量被假定为保护水平和持续时间。这种疫苗免疫原性只是一种可能保护的指示，但肯定不等于实际保护。还有一些报告称，患者接种疫苗后出现抗体，但实际上并没有受到保护。最后但最重要的是一些病原体会诱导初级免疫应答，但是清除病原体后记忆应答就消失了，人体也无法抵御同一病原体再次感染。例如呼吸道合胞病毒（RSV）就是这种情况。RSV是婴儿呼吸道疾病的病原体，也是儿童和成人发病的原因[118,119]。

和T细胞一样，B细胞在初次接触抗原后也会产生长寿记忆细胞。和记忆T细胞一样，记忆B细胞能够在没有新抗原接触情况下持续存在[120,121]。

记忆细胞亚型 B细胞分为两个亚型[122]：记忆B细胞和记忆浆细胞。记忆B细胞存在于次级淋巴器官中，而记忆浆细胞存在于骨髓中，并持续向身体提供血清抗体（表2.7）。

在1915年或之前出生的人群中发现了人类记忆浆细胞的最长存活时间（大约是10年）。1918年H1N1流感病毒大流行的幸存者被纳入一项研究，测试对重组1918年流感血凝素（HA）蛋白的抗体应答。所有幸存者都有这种H1N1大流行性病毒的中和抗体，这种抗体也能与遗传相似的1930年流感感染的HA发生交叉反应[123]。

表2.7 人类B细胞记忆细胞子集

细胞类型	位置	标志	功能
记忆B细胞	次级淋巴器官（淋巴结，扁桃体，派伊尔结）	$CD19^+$，$CD20^+CD27^+$，$CD38^-$	记忆浆细胞快速生产池；次级生发中心反应
记忆浆细胞	骨髓	$CD19^+$，$CD20^-CD27^+$，$CD38^{+++}$	刺激后的高亲和力抗体

亲和力成熟 未成熟B细胞在次级淋巴器官中发展为成熟B细胞。在蛋白疫苗接种后淋巴

结（LN）中的B细胞发育受TFh细胞调节，按3个阶段进行[124]：①DC致敏的TFh细胞在LN幼稚B细胞的初级B细胞滤泡中相遇，处理由被膜下淋巴窦（SCS）巨噬细胞传递的同源抗原[125]。这种通过CD27/CD70的T细胞-B细胞相互作用激发了B细胞快速克隆扩增和分化为短命浆细胞，产生最早的免疫球蛋白IgM。②随着扩增和免疫球蛋白类别向IgA、IgE和IgG转化，LN次级滤泡中的生发中心（GC）反应开始，这是记忆B细胞产生的重要基础。而在细胞外滤泡中LN短命浆细胞分泌抗体。③GC反应按不同步骤循环进行。B细胞扫描滤泡DCs以寻找次级滤泡暗区的抗原。TFh细胞在光带中支持这些B细胞，并驱动抗原特异性B细胞亲和力成熟为非分泌记忆B细胞和长寿分泌浆细胞，以在感染组织远端发挥功能。然而这些抗原特异性记忆性B细胞的一部分不离开次级滤泡，而是重新进入GC循环，并经历另一轮亲和成熟。GC循环反应在蛋白疫苗接种后持续可长达8个月[126]，持续产生稳定的长寿记忆B细胞。同源抗原再次刺激后，记忆性B细胞被记忆性TFh激活，分泌记忆浆细胞抗体迅速增殖，并诱导另一GC反应。但是存活的细胞因子，如针对T细胞描述的IL-7，对于B细胞而言并不存在。

2.11 免疫衰老和疫苗接种

疫苗在老年人（>65岁）中疗效下降通常归因于免疫衰老[127-129]。因此老年人更容易感染传染病。与年龄有关的免疫变化发生在自然和获得性免疫系统中，不仅影响淋巴细胞，而且影响骨髓细胞，促炎细胞因子也发生了变化。

值得注意的是，免疫衰老不仅限于衰老。在慢性感染过程中，某些病原体如巨细胞病毒（CMV）和艾滋病病毒（HIV），重塑了针对老年T细胞的免疫系统[130,131]。

营养不良　营养不良、T细胞过早衰老和易患感染之间也有关系[132]。慢性饥饿可能通过T细胞库调节免疫系统。饥饿也影响$CD8^+$T细胞的端粒长度，导致端粒缩短[132]。早期生活中慢性饥饿显著影响了幼年胸腺的发育。这是一个非常关键的阶段，对T细胞发育有长期影响。锌是免疫系统必不可少的矿物质，缺锌导致胸腺萎缩[133]。有迹象表明，营养不良使这一人群易受感染。营养不良是第三世界国家人群口服疫苗失效的一个重要原因（另见第3章）。

T细胞　从青少年早期青春期开始，胸腺在循环性激素指导下退化：胸腺上皮细胞（TEC）（淋巴细胞发育生态位）主要被脂肪细胞取代，成年期每年有3%胸腺上皮细胞被取代[134]，胸腺大小和功能急剧下降，幼稚T细胞生成和输出减少[135]。70岁时胸腺上皮组织萎缩至10%。胸腺发育所必需的重要细胞因子丢失，也由这些上皮细胞产生，如IL-1、IL-3、IL-6、IL-7和转化生长因子（TGF-β）。IL-7是T细胞发育和稳态的生存细胞因子[136]。随着年龄增长，幼稚T细胞免疫力逐渐减弱，老年人对新抗原的反应能力也逐渐减弱。随着幼稚T细胞产量下降，TCR信号（NF-κB，MAPK）应答也减弱，导致抗原刺激后T细胞增殖减缓[137]，T细胞分化减少[138]。这与早期诱导的$CD4^+$和$CD8^+$细胞记忆池形成鲜明对比，后者可以持续个体一生[139]。不仅如此，幼稚T细胞丢失导致记忆T细胞克隆扩增，使得年轻人和老年人的T细胞总数相当[140]。记忆T细胞增加可由老年人长期抗原接触来解释。然而这种扩增与年龄依赖的TCR库剧减有关[141]。另一个缺点是衰老$CD8^+$T细胞端粒缩短，称为复制性衰老。它不仅影响复制潜能，而且能显著显示抗病毒效应功能，如IFN-γ产生或抗原特异性细胞毒性[142,143]。衰老T细胞缺乏重要的协同刺激分子，如CD28和CD27，但能表达衰老标志：PD-1、CD57和KLRG-1。这种计数对$CD4^+$T细胞的扩展程度较低，对$CD8^+$T细胞的扩展程度尤其低，其细胞具有最严重的功能缺陷。

因此，T细胞生命中与年龄相关的变化包括数量和质量变化，导致免疫反应性降低。

B细胞　大多数关于B淋巴细胞生成与年龄相关变化的结果来自小鼠模型。但是最近对人类的研究发现有一些共同机械特征和相似之处。小鼠造血衰老是人类造血衰老的合理模型[144]。

高亲和力抗体下降部分是由于T细胞缺陷所致。除了影响B细胞功能的外部因素/微环境因素外，衰老B细胞还会发生重要内在变化[145]。

外周血中人类B细胞数量（血液、脾脏、淋巴结）随着年龄的增长而轻度下降[146,147]。老年人外周血B细胞库多样性较低，这与健康状况不佳相关[148]。B淋巴细胞生成受限于初始滤泡B细胞的丢失。所有血细胞谱系，如淋巴细胞（T和B细胞）、髓细胞（多形核细胞，单核细胞/巨噬细胞）都是由骨髓中的造血干细胞（HSCs）不断产生的。造血干细胞（HSCs）能够自我更新，是维持细胞稳态的关键。HSC区中一个显著变化是：以淋巴系祖细胞为代价向髓系祖细胞生产转变[149-151]。基因转录水平及表观遗传改变是这些改变的原因[152]。总的来说，随着功能改变，老年人造血干细胞经历了数量和质量的变化。

小鼠和人B细胞在体外显示出固有变化。老的B细胞表现出免疫球蛋白类转换重组（CSR）能力受损，这是通过抗体产量下降来衡量的。活化诱导的胞苷脱氨酶（AID）是免疫球蛋白（Ig）基因类转换重组（CSR）和体细胞超突变（SHM）所必需的。由于转录因子E47受损，老年B细胞中的胞苷脱氨酶（AID）表达与青年相比显著降低[145]。

DCs和TLRs　DCs是最有效的APCs，也是细胞疫苗的目标。皮肤中郎格罕细胞（LCs）随着年龄增长而减少，导致免疫监视能力下降[153]。小鼠老化LCs具有与年轻LCs相同迁移到LNs的能力；然而它们逐步丧失了刺激T细胞的能力[154]，这在老年人应用疫苗贴剂时应当考虑。

老年人外周血和次级淋巴器官中浆细胞样树突状细胞（pDCs）在数量和功能上发生了变化，而髓样树突状细胞的数量没有变化。受损DCs向$CD4^+$和$CD8^+$T细胞递呈抗原能力降低，$CD8^+$细胞中缺乏细胞毒性诱导。此外，年老pDCs分泌IFN比年轻pDCs分泌得少，导致抗病毒活性丧失[156]。树突状细胞对自身抗原的反应性增加，并可导致自身免疫[157,158]。最近有研究表明，不仅某些TLRs在树突状细胞上表达减少，尤其是老年人的TLR信号通路表达也减少，这与疫苗反应性减弱有关[159,160]。相反地，在小鼠和人类中年轻和年老的巨噬细胞之间没有观察到TLR表达的显著差异[161]。但是观察到TLRs的主要信号衔接子水平（如MAPK、NF-κB或MyD88）随着年龄增长而降低[162,163]。

炎症　老年人体内某些细胞因子如IFN-α、TNF-α、IL-6和CRP的上调统称为炎症，可能是由抗原和应激持续超载引起的[164]。其他研究解释了炎症部分不依赖于这种免疫刺激剂，而是依赖于微环境，例如组织特异性细胞因子产生细胞之间的串扰减少[165]。各种细胞类型参与了促炎细胞因子的持续过量生产（如T细胞、树突状细胞、巨噬细胞和内皮细胞），提高了患年龄相关疾病的风险[166]。表2.8总结了免疫衰老的一些关键特征。

表2.8　免疫衰老的一些关键特征

老化细胞类型	功能障碍	后果
$CD4^+$T细胞	IL-2产量降低	对B细胞的支持减少
	同源辅助活性降低少	
$CD8^+$T细胞	次优信号通路	CTL活性降低
	复制性衰老	
	降低分化	
	TCR库丧失	

（续表）

老化细胞类型	功能障碍	后果
B 细胞	初始滤泡 B 细胞丧失	抗体应答多样性降低
	GCs 数量减少	
	B 细胞分化减少	
	Ig 类别转换减少	
浆细胞	在骨髓中积累抗原经验的老细胞	抗体产量减少和高亲和力丧失
树突状细胞	受损的 TLR 信号	降低 T 细胞免疫力
	自我反应性增强	自身耐受能力丧失
	细胞因子的过度生产	慢性炎症
炎症	不同细胞类型（IFN－α、TNF－α、IL-6、CRP）引起促炎细胞因子分泌增加	与年龄相关疾病（如动脉粥样硬化）有关的动脉粥样硬化特征

知识点 2.5　免疫老化对老年人疫苗的影响

到 2050 年，世界 60 岁以上人口比例将增加到 22%，即 60 岁以上人口的绝对数量将达到 20 亿（据世卫组织估计，2012 年 3 月）。建议老年人接种疫苗，特别是预防感染流感病毒或肺炎球菌。由于老年人流动性较强，建议旅行者接种预防（亚热带病原体）的疫苗，如伤寒或黄热病。

免疫衰老是自然和获得性免疫系统中一个多细胞、多因素的过程，是主要免疫功能的多重衰退。可以确定几个内在缺陷，描述了一些明显外部因素。但是对于组织特异性微环境对免疫细胞深层影响人们知之甚少。每一个体外实验设计都无法模拟体内这种特定自然生态位。

是否存在同步所有其他更改的早期关键事件？衰老的 HSCs 是了解免疫衰老过程的开门器吗？或者胸腺萎缩是领头人吗？

一些针对老年人的疫苗开发集中在增加抗原剂量[167]、新佐剂策略[168]、缩短间隔、载体驱动的疫苗、改变应用途径[169]，或生长激素和细胞因子使胸腺恢复活力[170,171]。这些方法能避免造血干细胞（HSCs）补充滤泡 B 细胞区域的能力降低、免疫球蛋白类别转换受损、高亲和力抗体丧失、新抗原能力降低、幼稚 T 细胞丧失、复制性衰老、TCR 库丧失、TLR 信号受损、转录因子下降或白细胞端粒长度缩短吗？这几乎是办不到的事！

为老年人准备的最好的疫苗在老化前就开始了：开发具有广泛抗原性、多价谱的新（偶联）疫苗，例如流感、肺炎球菌感染，通过产生强有力的 CTL 活性，诱导强烈的、完整的、长寿的记忆性 B 细胞和记忆性 T 细胞应答。成年期反复加强疫苗接种初免–加强机制。这种年轻时对免疫系统的培育是记忆的坚实基础，也为晚年对抗新抗原提供了强大的免疫学背景。

2.12　增强剂和疫苗接种策略

抗原　在本章开头，我们介绍了各种不同的抗原：糖蛋白、脂蛋白、糖、糖脂和核酸。

抗原化学性质不同，诱导的免疫应答也不同。不同的疫苗类型符合病原体的性质，如减毒活病毒/细菌疫苗、灭活病毒/细菌疫苗、结合多糖疫苗、重组蛋白疫苗以及 DNA/RNA 疫苗。

抗原的性质，如化学结构（越复杂，免疫原性越强）、大小（分子越大，免疫原性越强）和物理形式（颗粒抗原比可溶性抗原免疫原性更强）决定了免疫应答的类型。

蛋白质和核酸暴露于 T 细胞上相应的 PPRs/TLRs 导致获得性、T 细胞依赖的免疫应答。这方面的一个关键特征是记忆应答的发展。建立长寿免疫记忆对于再次接种和加强免疫至关重要。一旦由接种（或自然感染）引发，增强剂引起的二次应答通常就增加了抗体和 CTL 对同源抗原的应答。

脂类抗原　细菌多糖（肺炎链球菌、伤寒链球菌、流感嗜血杆菌）和脂质在不涉及 $CD4^+T$ 细胞的情况下激活先天免疫系统。以不依赖于 MHC 的方式将脂-CD1 复合物递呈给不变 NKT 细胞（iNKT）后，CD1 分子（如朗格罕细胞）可以识别脂质抗原[172,173]。iNKT 细胞是自然杀伤（NK）细胞的一个子集，NK 细胞是 T 细胞的一个子集。

多糖抗原　已经描述了两类多糖抗原：胸腺无关的 1 型抗原（TI-1）通过与 B 细胞受体（BCR）结合直接激活 B 细胞，作为第一个信号，并与 B 细胞上的 TLR 结合作为第二个信号；TI-2 抗原与 BCRs 结合作为第一个信号，导致聚集作为第二个信号。这种 $CD4^+T$ 细胞非依懒性激活导致抗体产生。脂多糖通过与 TLR4 结合引起 TI-1 反应。然而 B 细胞应答是短暂的，不产生记忆 B 细胞（仅在 T 细胞依赖性激活时产生），因此增强剂不能增强抗体应答。此外，最近研究表明，使用纯多糖疫苗增强剂可导致免疫耐受状态或由记忆 B 细胞凋亡引起的反应性低下[174,175]。

结合疫苗即糖类与蛋白质载体结合的疫苗，最初是为新生儿和婴儿开发的。由于 B 细胞受体不成熟，新生儿和婴儿无法产生针对重复多糖抗原的抗体。结合疫苗诱导完全的 T 细胞和 B 细胞应答（另见第 4 章）。结合疫苗现在也用于成人和老年人，其效果明显比纯多糖免疫好[176]。

初免-加强免疫　传统的再接种是通过同源增强剂进行的。多样化初免-加强免疫策略涉及针对同一抗原的两个不同载体平台，目前正在进行（前期）临床研究，特别是针对艾滋病病毒、疟疾和结核等慢性感染病。用于初免-加强免疫的重组（rec）载体组合可能非常不同可用用重组病毒初免-用重组病毒 B 加强[177]、用 DNA 疫苗初免-用重组病毒加强[178]、用 DNA 疫苗初免-用重组蛋白质加强[179]、用病毒样颗粒（VLP）初免-用 DNA 加强[180]、用重组蛋白质初免-用重组病毒加强[181]等。但在所有的初免-加强免疫机制中，基本原则是针对相同抗原应用不同的疫苗结构。

同样在第一次强化免疫接种 20 年后，其背后的机制仍不清楚。更清楚的结果是，随着溶解能力的增加，$CD8^+T$ 细胞数量增加，抗体特异性扩大，抗体质量也提高了。这些结果反映出不同平台呈现的抗原可以结合不同的 TLRs，从而拓宽了 B 细胞和 T 细胞的反应。此外，使用质粒 DNA 作为主要成分为免疫系统提供了原始和正确的折叠蛋白。这些 DNA 编码的蛋白质共享显性和亚显性表位，导致更广泛的应答。相比之下，重组蛋白大多共享免疫显性表位导致狭窄的应答范围。

缺点　然而，DNA 疫苗的瓶颈是免疫原性低。纯裸 DNA 不起作用，必须辅以强佐剂（Giese，未发表的结果）。

任何病毒载体（作为初免或加强作用）的缺点是对载体的抗体反应。反复接种病毒载体疫苗会自动诱导中和抗体，从而显著降低该疫苗效力。腺病毒血清 5 型（Ad5）是最常用的病毒载体之一，血清阳性率非常高，而先前存在的免疫力也降低了该载体的效力[182]。为了克服这一障碍，其他人类，也包括猿类腺病毒血清型（Ad26 和 Ad35）正在研究中[183]。它们在先天特征和免疫潜能方面存在差异。并不是所有被测试的腺病毒都能强烈刺激 $CD4^+$和 $CD8^+$细胞，对于痘病毒载体也是如此；它们仅限于刺激 $CD8^+T$ 细胞，但对 $CD8^+T$ 细胞有强大的促进作用。从这些实验可了解适当的时间顺序。

总结这种疫苗策略，即与 DNA、蛋白质或载体疫苗相比，初免-加强免疫策略在激发强大的完整免疫应答方面更有优势。一些单一疫苗本身会失败，但联合使用会引起免疫应答。尽管如此，我们离把初免-加强免疫策略结合在一起的奇迹还很远。用加强免疫策略在泰国进行的临床Ⅲ期 HIV 疫苗 RV-144 试验仅获得 30%的保护[184]。初免-加强免疫策略还需要进一步研究。

参考文献

[1] Kratky, W., Reis e Sousa, C., Oxenius, A., Sporri, R.: Direct activation of antigen-presenting cells is required for CD8$^+$T-cell priming and tumor vaccination. Proc. Natl. Acad. Sci. U. S. A. 108, 17414-17419 (2011). doi: 10. 1073/pnas. 1108945108.

[2] Takeda, K., Akira, S.: Toll-like receptors in innate immunity. Int. Immunol. 17, 1-14 (2005). doi: 10. 1093/intimm/dxh186.

[3] Ziegler-Heitbrock, L., et al.: Nomenclature of mono-cytes and dendritic cells in blood. Blood116, e74-e80 (2010). doi: 10. 1182/blood-2010-02-258558.

[4] Khazen, W., et al.: Expression of macrophage-selective markers in human and rodent adipocytes. FEBS Lett. 579, 5631-5634 (2005). doi: 10. 1016/j. febslet. 2005. 09. 032.

[5] Ding, Q., et al.: Regulatory B cells are identified by expression of TIM-1 and can be induced through TIM-1 ligation to promote tolerance in mice. J. Clin. Invest. 121, 3645-3656 (2011). doi: 10. 1172/JCI46274.

[6] Gottfried-Blackmore, A., et al.: Acute in vivo exposure to interferon-gamma enables resident brain dendritic cells to become effective antigen presenting cells. Proc. Natl. Acad. Sci. U. S. A. 106, 20918-20923 (2009). doi: 10. 1073/pnas. 0911509106.

[7] Ochoa, M. T., Loncaric, A., Krutzik, S. R., Becker, T. C., Modlin, R. L.: "Dermal dendritic cells" comprise two distinct populations: CD1+dendritic cells and CD209+macrophages. J. Invest. Dermatol. 128, 2225-2231 (2008). doi: 10. 1038/jid. 2008. 56.

[8] Ouchi, T., et al.: Langerhans cell antigen capture through tight junctions confers preemptive immunity in experimental staphylococcal scalded skin syndrome. J. Exp. Med. 208, 2607-2613 (2011). doi: 10. 1084/jem. 20111718.

[9] Zozulya, A. L., et al.: Intracerebral dendritic cells critically modulate encephalitogenic versus regulatory immune responses in the CNS. J. Neurosci. 29, 140-152 (2009). doi: 10. 1523/JNEUROSCI. 2199-08. 2009.

[10] Barchet, W., Cella, M., Colonna, M.: Plasmacytoid dendritic cells-virus experts of innate immunity. Semin. Immunol. 17, 253-261 (2005). doi: 10. 1016/j. smim. 2005. 05. 008.

[11] Hart, D. N., McKenzie, J. L.: Interstitial dendritic cells. Int. Rev. Immunol. 6, 127-138 (1990).

[12] Kushwah, R., Hu, J.: Complexity of dendritic cell subsets and their function in the host immune system. Immunology 133, 409-419 (2011). doi: 10. 1111/j. 1365-2567. 2011. 03457. x.

[13] Vremec, D., et al.: Production of interferons by dendritic cells, plasmacytoid cells, natural killer cells, and interferon-producing killer dendritic cells. Blood 109, 1165-1173 (2007). doi: 10. 1182/blood-2006-05-015354.

[14] Chopin, M., Allan, R. S., Belz, G. T.: Transcriptional regulation of dendritic cell diversity. Front. Immunol. 3, 26 (2012). doi: 10. 3389/fi mmu. 2012. 00026.

[15] Cambi, A., Figdor, C. G.: Dual function of C-type lectin-like receptors in the immune system. Curr. Opin. Cell Biol. 15, 539-546 (2003).

[16] Flacher, V., et al.: Epidermal Langerhans cells rapidly capture and present antigens from C-type lectin-targeting antibodies deposited in the dermis. J. Invest. Dermatol. 130, 755-762 (2010). doi: 10. 1038/jid. 2009. 343.

[17] Allan, R. S., et al.: Epidermal viral immunity induced by CD8alpha+dendritic cells but not by Langerhans cells. Science 301, 1925-1928 (2003). doi: 10. 1126/ science. 1087576.

[18] Oh, S., Perera, L. P., Burke, D. S., Waldmann, T. A., Berzofsky, J. A.: IL-15/IL-15Ralpha-me-

diated avidity maturation of memory CD8$^+$ T cells. Proc. Natl. Acad. Sci. U. S. A. 101, 15154-15159 (2004). doi: 10. 1073/pnas. 0406649101.

[19] Bevan, M. J.: Cross-priming for a secondary cytotoxic response to minor H antigens with H-2 congenic cells which do not cross-react in the cytotoxic assay. J. Exp. Med. 143, 1283-1288 (1976).

[20] Snyder, C. M., Allan, J. E., Bonnett, E. L., Doom, C. M., Hill, A. B.: Cross-presentation of a spread-defective MCMV is sufficient to prime the majority of virus-specific CD8$^+$T cells. PLoS One 5, e9681 (2010). doi: 10. 1371/journal. pone. 0009681.

[21] Freigang, S., Egger, D., Bienz, K., Hengartner, H., Zinkernagel, R. M.: Endogenous neosynthesis vs. cross-presentation of viral antigens for cytotoxic T cell priming. Proc. Natl. Acad. Sci. U. S. A. 100, 13477-13482 (2003). doi: 10. 1073/pnas. 1835685100.

[22] Henkart, P. A.: CTL effector functions. Semin. Immunol. 9, 85-86 (1997).

[23] Nagata, S.: Fas-mediated apoptosis. Adv. Exp. Med. Biol. 406, 119-124 (1996).

[24] Andersen, M. H., Schrama, D., Thor Straten, P., Becker, J. C.: Cytotoxic T cells. J. Invest. Dermatol. 126, 32-41 (2006). doi: 10. 1038/sj. jid. 5700001.

[25] de Saint-Vis, B., et al.: The cytokine profile expressed by human dendritic cells is dependent on cell subtype and mode of activation. J. Immunol. 160, 1666-1676 (1998).

[26] Ueno, H., et al.: Harnessing human dendritic cell sub-sets for medicine. Immunol. Rev. 234, 199-212 (2010). doi: 10. 1111/j. 0105-2896. 2009. 00884. x.

[27] Matsuo, K., et al.: A low-invasive and effective trans-cutaneous immunization system using a novel dissolving microneedle array for soluble and particulate antigens. J. Control. Release 161, 10-17 (2012). doi: 10. 1016/j. jconrel. 2012. 01. 033.

[28] Sheikh, N. A., et al.: Sipuleucel-T immune parameters correlate with survival: an analysis of the randomized phase 3 clinical trials in men with castration-resistant prostate cancer. Cancer Immunol. Immunother. (2012). doi: 10. 1007/s00262-012-1317-2.

[29] Dinarello, C. A.: Proinflammatory cytokines. Chest 118, 503-508 (2000).

[30] Cyktor, J. C., Turner, J.: Interleukin-10 and immunity against prokaryotic and eukaryotic intracellular pathogens. Infect. Immun. 79, 2964-2973 (2011). doi: 10. 1128/IAI. 00047-11.

[31] Maynard, C. L., Weaver, C. T.: Diversity in the contribution of interleukin-10 to T-cell-mediated immune regulation. Immunol. Rev. 226, 219-233 (2008). doi: 10. 1111/j. 1600-065X. 2008. 00711. x.

[32] Brooks, D. G., McGavern, D. B., Oldstone, M. B.: Reprogramming of antiviral T cells prevents inactivation and restores T cell activity during persistent viral infection. J. Clin. Invest. 116, 1675-1685 (2006). doi: 10. 1172/JCI26856.

[33] Netea, M. G., et al.: Toll-like receptor 2 suppresses immunity against Candida albicans through induction of IL-10 and regulatory T cells. J. Immunol. 172, 3712-3718 (2004).

[34] Stober, C. B., Lange, U. G., Roberts, M. T., Alcami, A., Blackwell, J. M.: IL-10 from regulatory T cells determines vaccine efficacy in murine Leishmania major infection. J. Immunol. 175, 2517-2524 (2005).

[35] Verma, S., et al.: Quantification of parasite load in clinical samples of leishmaniasis patients: IL-10 level correlates with parasite load in visceral leishmaniasis. PLoS One 5, e10107 (2010). doi: 10. 1371/journal. pone. 0010107.

[36] Moore, K. W., et al.: Homology of cytokine synthesis inhibitory factor (IL-10) to the Epstein-Barr virus gene BCRFI. Science 248, 1230-1234 (1990).

[37] Dolganiuc, A., et al.: Hepatitis C virus core and nonstructural protein 3 proteins induce pro-and anti-inflammatory cytokines and inhibit dendritic cell differentiation. J. Immunol. 170, 5615-5624 (2003).

[38] Mahipal, A., et al.: Tumor-derived interleukin-10 as a prognostic factor in stage III patients undergoing adjuvant treatment with an autologous melanoma cell vaccine. Cancer Immunol. Immunother. 60, 1039-1045 (2011). doi: 10. 1007/s00262-011-1019-1.

[39] Chen, C. J., et al.: High expression of interleukin 10 might predict poor prognosis in early stage oral squamous cell carcinoma patients. Clin. Chim. Acta415, 25-30 (2013). doi: 10. 1016/j. cca. 2012. 09. 009.

[40] Lee, S. J., et al.: Identifi cation of Pro-inflammatory cytokines associated with muscle invasive bladder cancer; the roles of IL-5, IL-20, and IL-28A. PLoS One 7, e40267 (2012). doi: 10. 1371/journal. pone. 0040267.

[41] Opal, S. M., DePalo, V. A.: Anti-inflammatory cytokines. Chest 117, 1162-1172 (2000).

[42] Hao, N. B., et al.: Macrophages in tumor microenvironments and the progression of tumors. Clin. Dev. Immunol. 2012, 948098 (2012). doi: 10. 1155/2012/948098.

[43] Janeway Jr., C. A.: Approaching the asymptote? Evolution and revolution in immunology. Cold Spring Harb. Symp. Quant. Biol. 54 Pt 1, 1-13 (1989).

[44] Krishnaswamy, J. K., Chu, T., Eisenbarth, S. C.: Beyond pattern recognition: NOD-like receptors in dendritic cells. Trends Immunol. (2013). doi: 10. 1016/ j. it. 2012. 12. 003.

[45] Ng, C. S., Kato, H., Fujita, T.: Recognition of viruses in the cytoplasm by RLRs and other helicases-how conformational changes, mitochondrial dynamics and ubiquitination control innate immune responses. Int. Immunol. 24, 739-749 (2012). doi: 10. 1093/intimm/ dxs099.

[46] Palm, N. W., Medzhitov, R.: Pattern recognition receptors and control of adaptive immunity. Immunol. Rev. 227, 221-233 (2009). doi: 10. 1111/j. 1600-065X. 2008. 00731. x.

[47] Akira, S.: Innate immunity and adjuvants. Philos. Trans. R. Soc. Lond. B Biol. Sci. 366, 2748-2755 (2011). doi: 10. 1098/rstb. 2011. 0106.

[48] Ablasser, A., et al.: RIG-I-dependent sensing of poly (dA: dT) through the induction of an RNA polymerase III-transcribed RNA intermediate. Nat. Immunol. 10, 1065-1072 (2009). doi: 10. 1038/ni. 1779.

[49] Cridland, J. A., et al.: The mammalian PYHIN gene family: phylogeny, evolution and expression. BMC Evol. Biol. 12, 140 (2012). doi: 10. 1186/1471-2148-12-140.

[50] Lemaitre, B., Nicolas, E., Michaut, L., Reichhart, J. M., Hoffmann, J. A.: The dorsoventral regulatory gene cassette spatzle/toll/cactus controls the potent antifungal response in Drosophila adults. Cell 86, 973-983 (1996).

[51] Medzhitov, R., Preston-Hurlburt, P., Janeway Jr., C. A.: A human homologue of the Drosophila toll protein signals activation of adaptive immunity. Nature 388, 394-397 (1997). doi: 10. 1038/41131.

[52] Hashimoto, C., Hudson, K. L., Anderson, K. V.: The toll gene of Drosophila, required for dorsal-ventral embryonic polarity, appears to encode a transmembrane protein. Cell 52, 269-279 (1988).

[53] Meylan, E., Tschopp, J., Karin, M.: Intracellular pattern recognition receptors in the host response. Nature 442, 39-44 (2006). doi: 10. 1038/nature04946.

[54] Lata, S., Raghava, G. P.: PRRDB: a comprehensive database of pattern-recognition receptors and their ligands. BMC Genomics 9, 180 (2008). doi: 10. 1186/1471-2164-9-180.

[55] Fortis, A., Garcia-Macedo, R., Maldonado-Bernal, C., Alarcon-Aguilar, F., Cruz, M.: The role of innate immunity in obesity. SaludPublica Mex. 54, 171-177 (2012).

[56] Benias, P. C., Gopal, K., Bodenheimer Jr., H., Theise, N. D.: Hepatic expression of toll-like receptors 3, 4, and 9 in primary biliary cirrhosis and chronic hepatitis C. Clin. Res. Hepatol. Gastroenterol. (2012). doi: 10. 1016/j. clinre. 2012. 07. 001.

[57] Z hu, W., et al.: Overexpression of toll-like receptor 3 in spleen is associated with experimental arthritis in rats. Scand. J. Immunol. 76, 263-270 (2012). doi: 10. 1111/j. 1365-3083. 2012. 02724. x.

[58] Medzhitov, R.: Toll-like receptors and innate immunity. Nat. Rev. Immunol. 1, 135-145 (2001). doi: 10. 1038/35100529.

[59] Ohto, U., Fukase, K., Miyake, K., Satow, Y.: Crystal structures of human MD-2 and its complex with antiendotoxiclipid IVa. Science 316, 1632-1634 (2007). doi: 10. 1126/science. 1139111.

[60] Bauer, S., et al.: Human TLR9 confers responsiveness to bacterial DNA via species-specific CpG motif recognition. Proc. Natl. Acad. Sci. U. S. A. 98, 9237-9242 (2001). doi: 10. 1073/pnas. 161293498.

[61] Boivin, N., Menasria, R., Piret, J., Boivin, G.: Modulation of TLR9 response in a mouse model of herpes simplex virus encephalitis. Antiviral Res. 96, 414-421 (2012). doi: 10. 1016/j. antiviral. 2012. 09. 022.

[62] Geeraedts, F., et al.: Superior immunogenicity of inactivated whole virus H5N1 influenza vaccine is primarily controlled by toll-like receptor signalling. PLoSPathog. 4, e1000138 (2008). doi: 10. 1371/journal. ppat. 1000138.

[63] Koyama, S., et al.: Plasmacytoid dendritic cells delineate immunogenicity of influenza vaccine subtypes. Sci. Transl. Med. 2, 25ra24 (2010). doi: 10. 1126/ scitranslmed. 3000759.

[64] Park, B. S., et al.: The structural basis of lipopolysaccharide recognition by the TLR4-MD-2 complex. Nature 458, 1191-1195 (2009). doi: 10. 1038/ nature07830.

[65] Ng, A., Xavier, R. J.: Leucine-rich repeat (LRR) proteins: integrators of pattern recognition and signaling inimmunity. Autophagy 7, 1082-1084 (2011).

[66] Lloyd, D. H., Viac, J., Werling, D., Reme, C. A., Gatto, H.: Role of sugars in surface microbe-host interactions and immune reaction modulation. Vet. Dermatol. 18, 197-204 (2007). doi: 10. 1111/j. 1365-3164. 2007. 00594. x.

[67] Dalpke, A., Helm, M.: RNA mediated toll-like receptor stimulation in health and disease. RNA Biol. 9, 828-842 (2012). doi: 10. 4161/rna. 20206.

[68] Heeg, K., Dalpke, A., Peter, M., Zimmermann, S.: Structural requirements for uptake and recognition of CpG oligonucleotides. Int. J. Med. Microbiol. 298, 33-38 (2008). doi: 10. 1016/j. ijmm. 2007. 07. 007.

[69] Matzinger, P.: Tolerance, danger, and the extended family. Annu. Rev. Immunol. 12, 991-1045 (1994). doi: 10. 1146/annurev. iy. 12. 040194. 005015.

[70] Rubartelli, A., Lotze, M. T.: Inside, outside, upside down: damage-associated molecular-pattern molecules (DAMPs) and redox. Trends Immunol. 28, 429-436 (2007). doi: 10. 1016/j. it. 2007. 08. 004.

[71] Khanna, A.: Interleukin-18, a potential mediator of inflammation, oxidative stress, and allograft dysfunction. Transplantation 91, 590-591 (2011). doi: 10. 1097/TP. 0b013e31820d3b82.

[72] Piccinini, A. M., Midwood, K. S.: DAMPeninginflammation by modulatingTLRsignalling. Mediators Inflamm. 2010, 21 (2010). doi: 10. 1155/2010/672395.

[73] Yanai, H., et al.: HMGB proteins function as universal sentinels for nucleic-acid-mediated innate immune responses. Nature462, 99-103 (2009). doi: 10. 1038/ nature08512.

[74] Zhang, Q., et al.: Circulating mitochondrial DAMPs cause inflammatory responses to injury. Nature 464, 104-107 (2010). doi: 10. 1038/nature08780.

[75] Hirsiger, S., Simmen, H. P., Werner, C. M., Wanner, G. A., Rittirsch, D.: Danger signals activating the immune response after trauma. Mediators Inflamm. 2012, 315941 (2012). doi: 10. 1155/ 2012/315941.

[76] Basu, S., Binder, R. J., Suto, R., Anderson, K. M., Srivastava, P. K.: Necrotic but not apoptotic cell death releases heat shock proteins, which deliver a partial maturation signal to dendritic cells and activate the NF-kappa B pathway. Int. Immunol. 12, 1539-1546 (2000).

[77] Ichikawa, M., Williams, R., Wang, L., Vogl, T., Srikrishna, G.: S100A8/A9 activate key genes and pathways in colon tumor progression. Mol. Cancer Res. 9, 133-148 (2011). doi: 10. 1158/1541-7786. MCR-10-0394.

[78] Termeer, C., et al.: Oligosaccharides of Hyaluronan activate dendritic cells via toll-like receptor 4. J. Exp. Med. 195, 99-111 (2002).

[79] O'Neill, L. A., Fitzgerald, K. A., Bowie, A. G.: The Toll-IL-1 receptor adaptor family grows to five members. Trends Immunol. 24, 286-290 (2003).

[80] Zhang, Q., Zmasek, C. M., Cai, X., Godzik, A.: TIR domain-containing adaptor SARM is a late addition to the ongoing microbe-host dialog. Dev. Comp. Immunol. 35, 461-468 (2011). doi: 10. 1016/j. dci. 2010. 11. 013.

[81] Pahl, H. L.: Activators and target genes of Rel/NF-kappaB transcription factors. Oncogene 18, 6853-6866 (1999). doi: 10. 1038/sj. onc. 1203239.

[82] Baker, R. G., Hayden, M. S., Ghosh, S.: NF-kappaB, inflammation, and metabolic disease. Cell Metab. 13, 11-22 (2011). doi: 10. 1016/j. cmet. 2010. 12. 008.

[83] Ben-Neriah, Y., Karin, M.: Inflammation meets cancer, with NF-kappaB as the matchmaker. Nat. Immunol. 12, 715-723 (2011). doi: 10. 1038/ni. 2060.

[84] Hayden, M. S., Ghosh, S.: NF-kappaB, the first quarter-century: remarkable progress and outstanding questions. Genes Dev. 26, 203-234 (2012). doi: 10. 1101/gad. 183434. 111.

[85] Griffin, G. E., Leung, K., Folks, T. M., Kunkel, S., Nabel, G. J.: Activation of HIV gene expression during monocyte differentiation by induction of NF-kappa B. Nature 339, 70-73 (1989). doi: 10. 1038/339070a0.

[86] Rong, B. L., et al.: HSV-1-inducible proteins bind to NF-kappa B-like sites in the HSV-1 genome. Virology 189, 750-756 (1992).

[87] J an, R. H., et al.: Hepatitis B virus surface antigen can activate human monocyte-derived dendritic cells by nuclear factor kappa B and p38 mitogen-activated protein kinase mediated signaling. Microbiol. Immunol. 56, 719-727 (2012). doi: 10. 1111/j. 1348-0421. 2012. 00496. x.

[88] O'Neill, L. A., Bryant, C. E., Doyle, S. L.: Therapeutic targeting of toll-like receptors for infectious and inflammatory diseases and cancer. Pharmacol. Rev. 61, 177-197 (2009). doi: 10. 1124/pr. 109. 001073.

[89] Krieg, A. M., et al.: CpG motifs in bacterial DNA trig-ger direct B-cell activation. Nature 374, 546-549 (1995). doi: 10. 1038/374546a0.

[90] Jahrsdorfer, B., Weiner, G. J.: CpG oligodeoxynucleotides as immunotherapy in cancer. Update Cancer Ther. 3, 27-32 (2008). doi: 10. 1016/j. uct. 2007. 11. 003.

[91] Butte, M. J., et al.: CD28 costimulation regulates genome-wide effects on alternative splicing. PLoS One 7, e40032 (2012). doi: 1 0. 1371/journal. pone. 0040032.

[92] Ouyang, W., Kolls, J. K., Zheng, Y.: The biological functions of T helper 17 cell effector cytokines in inflammation. Immunity 28, 454-467 (2008). doi: 10. 1016/j. immuni. 2008. 03. 004.

[93] Deenick, E. K., Ma, C. S.: The regulation and role of T follicular helper cells in immunity. Immunology 134, 361-367 (2011). doi: 10. 1111/j. 1365-2567. 2011. 03487. x.

[94] Sakaguchi, S.: Regulatory T cells: key controllers of immunologic self-tolerance. Cell 101, 455-458 (2000).

[95] Hammarlund, E., et al.: Duration of antiviral immunity after smallpox vaccination. Nat. Med. 9, 1131-1137 (2003). doi: 10. 1038/nm917.

[96] Rudolph, M., Hebel, K., Miyamura, Y., Maverakis, E., Brunner-Weinzierl, M. C.: Blockade of CTLA-4 decreases the generation of multifunctional memory $CD4^+$T cells in vivo. J. Immunol. 186, 5580-5589

(2011). doi: 10. 4049/jimmunol. 1003381.

[97] Litjens, N. H., et al.: IL-2 producing memory $CD4^+$ T lymphocytes are closely associated with the generation of IgG-secreting plasma cells. J. Immunol. 181, 3665-3673 (2008).

[98] Williams, M. A., Tyznik, A. J., Bevan, M. J.: Interleukin-2 signals during priming are required for secondaryexpansion of $CD8^+$memory T cells. Nature 441, 890-893 (2006). doi: 10. 1038/ nature04790.

[99] Kaech, S. M., Cui, W.: Transcriptional control of effector and memory CD8 (+) T cell differentiation. Nat. Rev. Immunol. 12, 749-761 (2012). doi: 10. 1038/nri3307.

[100] Sallusto, F., Langenkamp, A., Geginat, J., Lanzavecchia, A.: Functional subsets of memory T cells identified by CCR7 expression. Curr. Top. Microbiol. Immunol. 251, 167-171 (2000).

[101] Okada, R., Kondo, T., Matsuki, F., Takata, H., Takiguchi, M.: Phenotypic classification of human $CD4^+$T cell subsets and their differentiation. Int. Immunol. 20, 1189-1199 (2008). doi: 10. 1093/ intimm/dxn075.

[102] KeshavarzValian, H., et al.: CCR7 (+) Central and CCR7 (-) Effector Memory CD4 (+) T Cells in Human Cutaneous Leishmaniasis. J. Clin. Immunol. 33, 220-234 (2013). doi: 10. 1007/s10875-012-9788-7.

[103] Vazquez-Cintron, E. J., et al.: Protocadherin-18 is a novel differentiation marker and an inhibitory signaling receptor for $CD8^+$ effector memory T cells. PLoS One 7, e36101 (2012). doi: 10. 1371/journal. pone. 0036101.

[104] Lau, L. L., Jamieson, B. D., Somasundaram, T., Ahmed, R.: Cytotoxic T-cell memory without antigen. Nature 369, 648-652 (1994). doi: 10. 1038/369648a0.

[105] Mackay, L. K., et al.: Long-lived epithelial immunity by tissue-resident memory T (Trm) cells in the absence of persisting local antigen presentation. Proc. Natl. Acad. Sci. U. S. A. 109, 7037-7042 (2012). doi: 10. 1073/pnas. 1202288109.

[106] Michalek, R. D., Rathmell, J. C.: The metabolic life and times of a T-cell. Immunol. Rev. 236, 190-202 (2010). doi: 10. 1111/j. 1600-065X. 2010. 00911. x.

[107] Bensinger, S. J., Christofk, H. R.: New aspects of the Warburg effect in cancer cell biology. Semin. Cell Dev. Biol. 23, 352-361 (2012). doi: 10. 1016/j. semcdb. 2012. 02. 003.

[108] Maciver, N. J., et al.: Glucose metabolism in lymphocytes is a regulated process with significant effects on immune cell function and survival. J. Leukoc. Biol. 84, 949-957 (2008). doi: 10. 1189/ jlb. 0108024.

[109] Jacobs, S. R., et al.: Glucose uptake is limiting in T cell activation and requires CD28-mediated Akt-dependent and independent pathways. J. Immunol. 180, 4476-4486 (2008).

[110] Finlay, D. K.: Regulation of glucose metabolism in T cells: new insight into the role of Phosphoinositide 3-kinases. Front. Immunol. 3, 247 (2012). doi: 10. 3389/fi mmu. 2012. 00247.

[111] Khan, J. M., Kumar, G., Ranganathan, S.: In silico prediction of immunogenic T cell epitopes for HLA-DQ8. Immunome Res. 8, 14 (2012).

[112] Beaver, J. E., Bourne, P. E., Ponomarenko, J. V.: EpitopeViewer: a java application for the visualization and analysis of immune epitopes in the Immune Epitope Database and Analysis Resource (IEDB). Immunome Res3, 3 (2007). doi: 10. 1186/1745-7580-3-3.

[113] Kim, Y., Sette, A., Peters, B.: Applications for T-cell epitope queries and tools in the Immune Epitope Database and Analysis Resource. J. Immunol. Methods 374, 62-69 (2011). doi: 10. 1016/j. jim. 2010. 10. 010.

[114] Herd, K. A., et al.: Cytotoxic T-lymphocyte epitope vaccination protects against human metapneumovirus infection and disease in mice. J. Virol. 80, 2034-2044 (2006). doi: 10. 1128/JVI. 80. 4. 2034-2044. 2006.

[115] De Rose, R., et al.: Control of viremia and prevention of AIDS following immunotherapy of SIV-infected macaques with peptide-pulsed blood. PLoS Pathog. 4, e1000055 (2008). doi: 10. 1371/journal. ppat. 1000055.

[116] Torti, N., Oxenius, A.: T cell memory in the context of persistent herpes viral infections. Viruses 4, 1116-1143 (2012). doi: 10. 3390/v4071116.

[117] Johnson, P. L., et al.: Vaccination alters the balance between protective immunity, exhaustion, escape, and death in chronic infections. J. Virol. 85, 5565-5570 (2011). doi: 10. 1128/JVI. 00166-11.

[118] Walsh, E. E.: Respiratory syncytial virus infection in adults. Semin. Respir. Crit. Care Med. 32, 423-432 (2011). doi: 10. 1055/s-0031-1283282.

[119] Graham, B. S.: Biological challenges and technologi-cal opportunities for respiratory syncytial virus vaccine development. Immunol. Rev. 239, 149-166 (2011). doi: 10. 1111/j. 1600-065X. 2010. 00972. x.

[120] Zinkernagel, R. M.: On natural and artificial vaccinations. Annu. Rev. Immunol. 21, 515-546 (2003). doi: 10. 1146/annurev. immunol. 21. 120601. 141045.

[121] Maruyama, M., Lam, K. P., Rajewsky, K.: Memory B-cell persistence is independent of persisting immunizing antigen. Nature 407, 636-642 (2000). doi: 10. 1038/35036600.

[122] Caraux, A., et al.: Circulating human B and plasma cells. Age-associated changes in counts and detailed characterization of circulating normal CD138-and CD138+plasma cells. Haematologica 95, 1016-1020 (2010). doi: 10. 3324/haematol. 2009. 018689.

[123] Yu, X., et al.: Neutralizing antibodies derived from the B cells of, influenza pandemic survivors. Nature 455, 532-536 (1918). doi: 10. 1038/nature07231 (2008).

[124] McHeyzer-Williams, M., Okitsu, S., Wang, N., McHeyzer-Williams, L.: Molecular programming of B cell memory. Nat. Rev. Immunol. 12, 24-34 (2012). doi: 10. 1038/nri3128.

[125] Phan, T. G., Gray, E. E., Cyster, J. G.: The microanatomy of B cell activation. Curr. Opin. Immunol. 21, 258-265 (2009). doi: 10. 1016/j. coi. 2009. 05. 006.

[126] Dogan, I., et al.: Multiple layers of B cell memory with different effector functions. Nat. Immunol. 10, 1292-1299 (2009). doi: 10. 1038/ni. 1814.

[127] Wong, K. L., et al.: Gene expression profiling reveals the defining features of the classical, intermediate, and nonclassical human monocyte subsets. Blood 118, e16-e31 (2011). doi: 10. 1182/blood-2010-12-326355.

[128] McElhaney, J. E., Effros, R. B.: Immunosenescence: what does it mean to health outcomes in older adults? Curr. Opin. Immunol. 21, 418-424 (2009). doi: 10. 1016/j. coi. 2009. 05. 023.

[129] Cicin-Sain, L., et al.: Loss of naive T cells and reper-toire constriction predict poor response to vaccination in old primates. J. Immunol. 184, 6739-6745 (2010). doi: 10. 4049/jimmunol. 0904193.

[130] Appay, V., et al.: Old age and anti-cytomegalovirus immunity are associated with altered T-cell reconstitution in HIV-1-infected patients. AIDS 25, 1813-1822 (2011). doi: 10. 1097/QAD. 0b013e32834640e6.

[131] Miles, D. J., et al.: Cytomegalovirus infection in Gambian infants leads to profound CD8 T-cell differentiation. J. Virol. 81, 5766-5776 (2007). doi: 10. 1128/JVI. 00052-07.

[132] Ngom, P. T., et al.: Thymic function and T cell parameters in a natural human experimental model of seasonal infectious diseases and nutritional burden. J. Biomed. Sci. 18, 41 (2011). doi: 10. 1186/1423-0127-18-41.

[133] Jones, K. D., Berkley, J. A., Warner, J. O.: Perinatal nutrition and immunity to infection. Pediatr. Allergy Immunol. 21, 564-576 (2010). doi: 10. 1111/j. 1399-3038. 2010. 01002. x.

[134] Taub, D. D., Longo, D. L.: Insights into thymic aging and regeneration. Immunol. Rev. 205, 72-93

(2005). doi: 10. 1111/j. 0105-2896. 2005. 00275. x.

[135] Simpson, J. G., Gray, E. S., Beck, J. S.: Age involution in the normal human adult thymus. Clin. Exp. Immunol. 19, 261-265 (1975).

[136] Plum, J., De Smedt, M., Leclercq, G., Verhasselt, B., Vandekerckhove, B.: Interleukin-7 is a critical growth factor in early human T-cell development. Blood 88, 4239-4245 (1996).

[137] Linton, P. J., Dorshkind, K.: Age-related changes in lymphocyte development and function. Nat. Immunol. 5, 133-139 (2004). doi: 10. 1038/ni1033.

[138] Gillis, S., Kozak, R., Durante, M., Weksler, M. E.: Immunological studies of aging. Decreased production of and response to T cell growth factor by lymphocytes from aged humans. J. Clin. Invest. 67, 937-942 (1981).

[139] Kedzierska, K., Valkenburg, S. A., Doherty, P. C., Davenport, M. P., Venturi, V.: Use it or lose it: establishment and persistence of T cell memory. Front. Immunol. 3, 357 (2012). doi: 10. 3389/ fimmu. 2012. 00357.

[140] Posnett, D. N., Sinha, R., Kabak, S., Russo, C.: Clonal populations of T cells in normal elderly humans: the T cell equivalent to "benign monoclonal gammapathy". J. Exp. Med. 179, 609-618 (1994).

[141] Messaoudi, I., Lemaoult, J., Guevara-Patino, J. A., Metzner, B. M., Nikolich-Zugich, J.: Age-related CD8 T cell clonal expansions constrict CD8 T cell repertoire and have the potential to impair immune defense. J. Exp. Med. 200, 1347-1358 (2004). doi: 10. 1084/jem. 20040437.

[142] Effros, R. B., Dagarag, M., Spaulding, C., Man, J.: The role of $CD8^+$T-cell replicative senescence in human aging. Immunol. Rev. 205, 147-157 (2005). doi: 10. 1111/j. 0105-2896. 2005. 00259. x.

[143] Dock, J. N., Effros, R. B.: Role of CD8 T cell replica-tive senescence in human aging and in HIV-mediated immunosenescence. Aging Dis. 2, 382-397 (2011).

[144] Pang, W. W., et al.: Human bone marrow hematopoietic stem cells are increased in frequency and myeloid-biased with age. Proc. Natl. Acad. Sci. U. S. A. 108, 20012-20017 (2011). doi: 10. 1073/ pnas. 1116110108.

[145] Frasca, D., Diaz, A., Romero, M., Landin, A. M., Blomberg, B. B.: Age effects on B cells and humoral immunity in humans. Ageing Res. Rev. 10, 330-335 (2011). doi: 10. 1016/j. arr. 2010. 08. 004.

[146] McKenna, R. W., Washington, L. T., Aquino, D. B., Picker, L. J., Kroft, S. H.: Immunophenotypic analysis of hematogones (B-lymphocyte precursors) in 662 consecutive bone marrow specimens by 4-color flow cytometry. Blood 98, 2498-2507 (2001).

[147] Rossi, M. I., et al.: B lymphopoiesis is active throughout human life, but there are developmental age-related changes. Blood 101, 576-584 (2003). doi: 10. 1182/blood-2002-03-0896.

[148] Gibson, K. L., et al.: B-cell diversity decreases in old age and is correlated with poor health status. Aging Cell 8, 18-25 (2009). doi: 10. 1111/j. 1474-9726. 2008. 00443. x.

[149] Cho, R. H., Sieburg, H. B., Muller-Sieburg, C. E.: A new mechanism for the aging of hematopoietic stem cells: aging changes the clonal composition of the stem cell compartment but not individual stem cells. Blood 111, 5553-5561 (2008). doi: 10. 1182/ blood-2007-11-123547.

[150] Rossi, D. J., et al.: Cell intrinsic alterations underlie hematopoietic stem cell aging. Proc. Natl. Acad. Sci. U. S. A. 102, 9194-9199 (2005). doi: 10. 1073/ pnas. 0503280102.

[151] Kuranda, K., et al.: Age-related changes in human hematopoietic stem/progenitor cells. Aging Cell 10, 542-546 (2011). doi: 10. 1111/j. 1474-9726. 2011. 00675. x.

[152] Schroeder, T.: Hematopoietic stem cell heterogeneity: subtypes, not unpredictable behavior. Cell Stem Cell 6, 203-207 (2010). doi: 10. 1016/j. stem. 2010. 02. 006.

[153] Sunderkotter, C., Kalden, H., Luger, T. A.: Aging and the skin immune system. Arch. Dermatol. 133,

1256-1262 (1997).

[154] Xu, Y. P., et al.: Aging affects epidermal Langerhans cell development and function and alters their miRNA gene expression profile. Aging (Albany NY) 4, 742-754 (2012).

[155] Jing, Y., et al.: Aging is associated with a numerical and functional decline in plasmacytoid dendritic cells, whereas myeloid dendritic cells are relatively unaltered in human peripheral blood. Hum. Immunol. 70, 777-784 (2009). doi: 10. 1016/j. humimm. 2009. 07. 005.

[156] Sridharan, A., et al.: Age-associated impaired plasmacytoid dendritic cell functions lead to decreased CD4 and CD8 T cell immunity. Age (Dordr.) 33, 363-376 (2011). doi: 10. 1007/s11357-010-9191-3.

[157] Agrawal, A., Sridharan, A., Prakash, S., Agrawal, H.: Dendritic cells and aging: consequences for autoimmunity. Expert Rev. Clin. Immunol. 8, 73-80 (2012). doi: 10. 1586/eci. 11. 77.

[158] Agrawal, A., Tay, J., Ton, S., Agrawal, S., Gupta, S.: Increased reactivity of dendritic cells from aged subjects to self-antigen, the human DNA. J. Immunol. 182, 1138-1145 (2009).

[159] Qian, F., et al.: Impaired interferon signaling in den-dritic cells from older donors infected in vitro with West Nile virus. J. Infect. Dis. 203, 1415-1424 (2011). doi: 10. 1093/infdis/jir048.

[160] Renshaw, M., et al.: Cutting edge: impaired Toll-like receptor expression and function in aging. J. Immunol. 169, 4697-4701 (2002).

[161] Hajishengallis, G.: Too old to fight? Aging and its toll on innate immunity. Mol. Oral Microbiol. 25, 25-37 (2010). doi: 10. 1111/j. 2041-1014. 2009. 00562. x.

[162] Medzhitov, R.: Recognition of microorganisms and activation of the immune response. Nature 449, 819-826 (2007). doi: 10. 1038/nature06246.

[163] Gomez, C. R., Nomellini, V., Faunce, D. E., Kovacs, E. J.: Innate immunity and aging. Exp. Gerontol. 43, 718-728 (2008). doi: 10. 1016/j. exger. 2008. 05. 016.

[164] Franceschi, C., et al.: Inflamm-aging. An evolutionary perspective on immunosenescence. Ann. N. Y. Acad. Sci. 908, 244-254 (2000).

[165] Salvioli, S., et al.: Immune system, cell senescence, aging and longevity-Inflamm-aging reappraised. Curr. Pharm. Des. 19 (19), 1675-1679 (2013).

[166] Bruunsgaard, H., Pedersen, M., Pedersen, B. K.: Aging and proinflammatory cytokines. Curr. Opin. Hematol. 8, 131-136 (2001).

[167] Keitel, W. A., et al.: Safety of high doses of influenza vaccine and effect on antibody responses in elderly persons. Arch. Intern. Med. 166, 1121-1127 (2006). doi: 10. 1001/archinte. 166. 10. 1121.

[168] Vandepapeliere, P., et al.: Vaccine adjuvant systems containing monophosphoryl lipid A and QS21 induce strong and persistent humoral and T cell responses against hepatitis B surface antigen in healthy adult volunteers. Vaccine 26, 1375-1386 (2008). doi: 10. 1016/j. vaccine. 2007. 12. 038.

[169] Hutt, H. J., Bennerscheidt, P., Thiel, B., Arand, M.: Immunosenescence and vaccinations in the elderly. Med. Klin. (Munich) 105, 802-807 (2010). doi: 10. 1007/s00063-010-1137-0.

[170] Taub, D. D., Murphy, W. J., Longo, D. L.: Rejuvenation of the aging thymus: growth hormone-mediated and ghrelin-mediated signaling pathways. Curr. Opin. Pharmacol. 10, 408-424 (2010). doi: 10. 1016/j. coph. 2010. 04. 015.

[171] Aspinall, R.: T cell development, ageing and Interleukin-7. Mech. Ageing Dev. 127, 572-578 (2006). doi: 10. 1016/j. mad. 2006. 01. 016.

[172] Juno, J. A., Keynan, Y., Fowke, K. R.: Invariant NKT cells: regulation and function during viral infection. PLoS Pathog. 8, e1002838 (2012). doi: 10. 1371/journal. ppat. 1002838.

[173] Pena-Cruz, V., Ito, S., Dascher, C. C., Brenner, M. B., Sugita, M.: Epidermal Langerhans cells efficiently mediate CD1a-dependent presentation of microbial lipid antigens to T cells. J. Invest. Dermatol.

121, 517–521 (2003). doi: 1 0. 1046/j. 1523–1747. 2003. 12429. x.

[174] O'Brien, K. L., Hochman, M., Goldblatt, D.: Combined schedules of pneumococcal conjugate and polysaccharide vaccines: ishyporesponsiveness an issue? Lancet Infect. Dis. 7, 597–606 (2007). doi: 10. 1016/S1473–3099 (07) 70210–4.

[175] Brynjolfsson, S. F., et al.: Hyporesponsivenessfol–lowing booster immunization with bacterial polysaccharides is caused by apoptosis of memory B cells. J. Infect. Dis. 205, 422–430 (2012). doi: 10. 1093/ infdis/jir750.

[176] Paradiso, P. R.: Pneumococcal conjugate vaccine for adults: a new paradigm. Clin. Infect. Dis. 55, 259–264 (2012). doi: 10. 1093/cid/cis359.

[177] Li, S., et al.: Priming with recombinant influenza virus followed by administration of recombinant vaccinia virus induces $CD8^+$T–cell–mediated protective immunity against malaria. Proc. Natl. Acad. Sci. U. S. A. 90, 5214–5218 (1993).

[178] Estcourt, M. J., et al.: Prime – boost immunization generates a high frequency, high – avidity CD8 (+) cytotoxic T lymphocyte population. Int. Immunol. 14, 31–37 (2002).

[179] Zhang, M., et al.: DNA prime–protein boost using subtype consensus Env was effective in eliciting neutralizing antibody responses against subtype BC HIV–1 viruses circulating in China. Hum. Vaccin. Immunother. 8, 1630–1637 (2012).

[180] Rattanasena, P., et al.: Prime–boost vaccinations using recombinant fl avivirus replicon and vaccinia virus vaccines: an ELISPOT analysis. Immunol. Cell Biol. 89, 426 – 436 (2011). doi: 1 0. 1038/ icb. 2010. 99.

[181] Chen, H., et al.: Optimisation of prime–boost immunization in mice using novel protein–based and recombinant vaccinia (Tiantan) –based HBV vaccine. PLoS One 7, e43730 (2012). doi: 10. 1371/journal. pone. 0043730.

[182] Mast, T. C., et al.: International epidemiology of human pre–existing adenovirus (Ad) type–5, type–6, type–26 and type–36 neutralizing antibodies: correlates of high Ad5 titers and implications for potential HIV vaccine trials. Vaccine28, 950–957 (2010). doi: 10. 1016/j. vaccine. 2009. 10. 145.

[183] O'Hara, G. A., et al.: Clinical assessment of a recombinant simian adenovirus ChAd63: a potent new vaccine vector. J. Infect. Dis. 205, 772–781 (2012). doi: 10. 1093/infdis/jir850.

[184] Rerks–Ngarm, S., et al.: Vaccination with ALVAC and AIDSVAX to prevent HIV–1 infection in Thailand. N. Engl. J. Med. 361, 2209–2220 (2009). doi: 10. 1056/NEJMoa0908492.

（马维民译）

第3章　肠道免疫与口服疫苗接种

Sharon M. Tennant，Khitam Muhsen，
和 Marcela F. Pasetti①

摘要

口服免疫是降低发展中国家5岁以下儿童肠道疾病造成毁灭性发病率和死亡率的一种切实可行方法。获得许可的预防脊髓灰质炎、轮状病毒、霍乱和伤寒口服疫苗对减轻全球疾病负担和死亡率产生了重大影响。尚待解决的一个难题是，与生活在工业化国家的人相比，这些疫苗对生活在世界欠发达地区受试者的免疫原性和效力都降低了。社会经济地位低、恶劣生活条件、营养不良和影响生活在贫困国家人们的自然障碍是疫苗性能的主要决定因素。目前正在探索新的保护性抗原、佐剂和免疫方法来克服这些障碍。了解口服抗原诱导的黏膜免疫和全身免疫所涉及的过程、多种竞争元素的影响以及肠道免疫激活和耐受之间的微妙平衡，对于帮助开发和评估候选疫苗至关重要。现代技术（例如基因组学、蛋白质组学、使用非侵入性样本高通量免疫分析以及数学建模分析保护的相关性）对这项任务至关重要。最终，需要在高收入和低收入国家进行良好的临床试验，以确定候选疫苗的安全性和有效性——这些候选疫苗在动物模型中似乎很有希望。本章综述了胃肠道口服抗原引发免疫反应的基础，以及随后在人体上的效应应答。本章还概述了现有口服疫苗、影响发展中国家口服疫苗效力的因素以及目前为开发更有效候选疫苗所做的努力。

3.1　引言

口服疫苗是预防肠道病原体引起疾病的一种实用有效方法。口服疫苗具有诱导局部免疫应答的潜力，在黏膜表面提供了一线防御以及全身免疫。与胃肠外接种相比，所有年龄口服疫苗接种依从性更高，而且可以很容易地实施，以造福广大人群。现代疫苗学目标之一是开发疫苗，作为降低发展中国家5岁以下儿童肠道疾病造成的毁灭性发病率和死亡率的工具。获得许可的口服脊髓灰质炎、轮状病毒、霍乱和伤寒疫苗对减少全球疾病负担和死亡率有重大影响。尚待解决的一个难题是，与生活在工业化国家的人相比，这些疫苗对生活在世界欠发达地区的人的免疫原性和效力降低了。社会经济地位低、恶劣的生活条件、营养不良以及影响生活在贫困国家人们的自然障碍现在被认为是疫苗性能的主要决定因素。目前正在探索新的保护性抗原、佐剂和免疫方法来克服这些障碍。了解口服抗原诱导免疫应答所涉及的过程、肠道中免疫激活和耐受之间微妙平衡以及黏膜和全身免疫之间的联系，对于帮助开发和评价候选疫苗至关重要。本章回顾了口服抗原在胃肠道启动免疫反应的基础，以及随后在人体中的效应反应。还概述了获得许可的现有口服疫苗，以及为开发更有效候选疫苗和利用口服疫苗成功进行免疫而取得的成功、遇到的障碍和正在进行的努力。

---

① S. M. Tennant，博士 · K. Muhsen，博士（美国马里兰州巴尔的摩市马里兰大学医学院疫苗开发中心医学系）· M. F. Pasetti，博士（（✉）马里兰大学医学院疫苗开发中心儿科，美国马里兰州巴尔的摩市西巴尔的摩街685号480室，MD 21201，E-mail：mpasetti@ medicine. umaryland. edu）

3.2 肠黏膜疫苗吸收和免疫启动

了解口服抗原和佐剂取样、加工和刺激免疫细胞产生保护性免疫的分子和细胞机制对于设计更有效口服疫苗和免疫战略至关重要。当针对特定群体（如婴幼儿）时，仔细分析影响这些过程的宿主因素也很重要。抗原到达肠黏膜诱导位点后，被滤泡相关上皮细胞（FAE）内的微褶（M）细胞捕获，并被主动转运至FAE下的树突状细胞（DC）和其他抗原呈递细胞（APC，如巨噬细胞）。上皮细胞产生的趋化因子在黏膜诱导位点[1]招募和激活DC、B细胞和T细胞，或抗原可以直接由固有层中DC携带，这些固有层树突延伸至整个肠道上皮。

固有层DC与诱导全身性IgG口服抗原有关，而派伊尔结（PP）中的DC主要与肠T细胞依赖性IgA产生有关[2]。抗原负载的DC迁移到淋巴滤泡的滤泡间T细胞区或局部/区域淋巴结，在那里它们递呈抗原并刺激初始T细胞。这些细胞将扩增并分化为$CD4^+$和$CD8^+$效应细胞和记忆T细胞（Th17或T调节细胞），并通过传出淋巴管和血液迁移到黏膜效应部位固有层[3]。肠黏膜DC（即常规的$CD11c^{hi}$、浆细胞样DC和CD103 DC）通过确定T细胞（即效应细胞、记忆性细胞、调节性细胞）和B细胞（即IgA、IgG）应答的性质，在抗原决定的淋巴细胞上诱导和印记特异性归巢受体（即α4/β7，CCR9和CCR10），而在黏膜应答效果中发挥关键作用[4]。由疫苗抗原激活的B细胞将聚集在派伊尔结或肠系膜淋巴结（MLN）中，形成生发中心（GC），这是一种促进B细胞生长和分化的特殊微环境。在生发中心B细胞将经历IgA类别转换和亲和力成熟，成为IgA^+浆母细胞。

GC反应取决于同源B细胞-T细胞、CD40-CD40L相互作用和IgA诱导信号，包括TGF-β、IL-4、IL-10和维甲酸（RA）[1]。在GC中，B细胞也与称为滤泡树突状细胞（FDC）的基质细胞相互作用，并从滤泡辅助T（T_{FH}）细胞接收信号和协同刺激，从而进一步促进GC形成和IgA类别转换（如TGF-β1）[5]。GC中产生的IgA^+浆母细胞将迁移到固有层，在固有层中最终分化为聚合的IgA分泌浆细胞，其过程涉及$CD4^+$T辅助细胞（如IL-2、IL-5、TGF-β和IL-10）、DC［如RA、IL-10、TGF-β、IL-6、B-细胞激活因子（BAFF）和增殖诱导配体（APRIL）］和肠上皮细胞（如TGF-β，IL-6）产生的细胞因子[4,6,7]。在小鼠和人类中已经描述了B细胞活化和通过T细胞独立机制产生IgA的过程[8,9]。

这一过程似乎发生在派伊尔结中，但也发生在分离的淋巴滤泡和固有层中。与派伊尔结（PP）不同，固有层缺少分离的T细胞区。它涉及B细胞募集和激活，这些细胞对自身受体（包括“天然受体”）识别的抗原（或微生物产物）或DC递呈的抗原，以及由DC不同亚群、局部基质细胞和上皮细胞产生的分子，如TGF-β、BAFF、APRIL、IL-6、IL-10和RA做出反应[5]。T细胞非依赖性、低亲和力的“天然”IgA应答是针对共生生物而产生的，而“经典的”高亲和力T细胞依赖性IgA似乎是由致病菌、毒素和病毒产生的[10]。然而，应该注意的是，虽然大量关于肠道免疫学的机理信息已经从小鼠研究中获得，但它们与人类的相关性和适用性仍有待确定。图3.1描述了肠道中疫苗吸收和免疫启动过程。

3.3 口服疫苗接种诱导的免疫效应物

肠道获得性免疫的主要效应部位是上皮和固有层。大量抗原特异性抗体分泌细胞（ASC）和活化T细胞存在于这些组织中，为抵抗肠道病原体提供了第一道防线。口服疫苗接种的一个标志是诱导长寿抗原特异性黏膜IgA分泌浆细胞；这一过程主要发生在派伊尔结（PP）和分离淋巴滤泡中。这些细胞以一种看似健壮和自给自足方式产生二聚性IgA，通过上皮屏障经聚合的Ig受体（pIgR）运输到肠腔中[5]。在管腔中分泌性IgA（sIgA），与微生物表面抗原结合，并阻止它们附着在黏膜界面上。通过这种“免疫排斥”机制，IgA可以阻止潜在病原体入侵[11,12]。

黏膜 sIgA 还可以中和上皮细胞内或固有层内的毒素、病毒和微生物抗原，并将它们带到管腔。除了阻止病原体进入外，IgA 还有助于维持协调共生体与宿主免疫系统之间的相互作用[5]。人类黏膜分泌物中存在两种 IgA 亚类：IgA1 首先在唾液和近端小肠（这也是血清中发现的亚类）中发现的；IgA2 主要存在于末端小肠和结肠中[3,13]。IgA1 主要识别蛋白质，而 IgA2 识别多糖和脂多糖（LPS）[9]。IgG 也由黏膜浆细胞产生，并经新生的 Fc 受体（FcRn）跨上皮屏障运输[14]。人们还认为，血清 IgG 通过细胞旁路渗漏跨上皮屏障扩散[15]，这可以解释肠外疫苗对某些肠道病原体（如志贺氏菌属）的保护作用[16]。IgA 和 IgG 都能通过上皮细胞将抗原从肠腔运输到固有层，或从基底外侧细胞表面输送回肠腔[17]。

在小鼠中抗体介导的微生物通过 M 细胞运输，这与炎症减轻和黏膜破坏有关[18]。可以想象，类似的方法使疫苗抗原穿梭于肠道上皮细胞，这可能提供一种改进口服疫苗的方法，或可以通过载体和特定配体将抗原靶向 M 细胞、肠上皮细胞和 DC 细胞，从而增强疫苗吸收[2,19]。

肠黏膜中由疫苗抗原引发的 B 淋巴细胞和 T 淋巴细胞迁移到肠系膜淋巴结，在那里进一步分化。它们通过胸导管离开淋巴结进入血流，并全身扩散到达黏膜效应部位。循环 IgA ASC 的存在是肠道感染或口服疫苗后肠道免疫启动的典型指标[20]。这些细胞在迁移到效应位点或其他淋巴组织时，会在循环中被短暂检测到（抗原暴露后 7~10 天）。实际上口服免疫诱导的所有 IgA 和一些 IgG ASC 表达整合素 α4/β7，整合素 α4/β7 结合固有层高内皮微静脉，表达黏膜血管递质素细胞黏附分子-1（MAdCAM-1）。许多肠道和所有的结肠 IgA ASC 也表达 CCR10，这使得它们能够对黏膜上皮趋化因子 CCL28 做出应答[21]。

这些口服接种启动 ASC 表达的外周淋巴结归巢受体 CD62L（L-选择素）比例较小，主要是在全身免疫后诱导的[22]。现在人们普遍认为，决定活化淋巴细胞归巢特性的是局部淋巴环境，而不是抗原性质。来自肠道的 DC 和 MLN（如 $CD103^+DC$）刺激 IgA 产生，并将肠道归巢分子印在黏膜启动的 B 细胞上，这一过程需要 RA、IL-6 或 IL-5[21]。针对霍乱弧菌[23]、伤寒弧菌[24,25]、志贺氏菌和轮状病毒[26]的感染或口服疫苗接种后，人体中已证明存在表达 IgA 和 IgG 特异性 ASC 肠道归巢受体。

在理想情况下，口服疫苗也会产生包括抗原特异性上皮内和固有层 T 淋巴细胞在内的强细胞介导免疫。存在于固有层（小鼠和人类）中的大多数 T 淋巴细胞是准备快速防御的效应记忆 $CD4^+T$ 细胞[27]，包括经典亚群（Th1、Th2 和 Th17）、Treg 细胞和自然杀伤性 T（NK T）细胞[8]。通过与固有层和 MLN 中 $CD103^+DC$ 相互作用，诱导 T 细胞表达 CCR9 和 α4β7 而获得迁移到黏膜效应部位的能力。在空肠和回肠中，由腺窝上皮细胞表达的 CCL25 介导含 CCR9 的记忆 $\alpha4\beta7^{hi}$ $CD4^+$和 $CD8^+T$ 淋巴细胞向固有层的趋化性[28]。

黏膜组织中的细胞毒性 $CD8^+T$ 淋巴细胞，通过分泌能激活吞噬细胞杀伤的 IFN-γ 和 TNF-α，来介导感染细胞溶解并促进炎症环境。$CD8^+$ T 细胞也参与免疫调节；具有抑制活性的 $CD8^+$ T 细胞显示在与上皮肠细胞相互作用时扩大，并通过细胞间接触介导免疫抑制[29]。$Foxp3^+$ $CD4^+$ Treg 细胞在维持肠道内免疫平衡和对共生生物和食物抗原的耐受方面也发挥了关键作用[30,31]。$CD4^+T$ H17 细胞是 Th17 产生的一个亚群，存在于小肠和大肠固有层[32]。根据来自肠道共生体的信号，发现 $CD4^+Th17$ 细胞在增加[33]，并且还与防止幽门螺旋杆菌和其他细菌感染有关[34]。Th17 相关的细胞因子，如 IL-22、IL-17A、IL-17F 和粒细胞-巨噬细胞集落刺激因子（GM-CSF）对细胞外病原体的保护很重要，这些细胞因子也与炎症性肠病（IBD）等病理疾病发展有关，炎症性肠病（IBD）可被 $FoxP3^+Treg$ 细胞抵消[35]。也有人认为，通常归因于 IL-12 和 Th1 细胞，肠道炎症可能是由于 Th17 细胞和 IL-23 引起的[36]。接受口服伤寒疫苗 Ty21a 的成年志愿者在接种疫苗后，产生了分泌 IL17A 和其他细胞因子的 $CD8^+T$ 细胞[37]。

口服疫苗的另一个理想属性是诱导免疫记忆能力。黏膜引发的肠道黏膜抗原特异性 B 细胞可分化为长寿浆细胞或记忆 B（B_M）细胞。当 B_M细胞转移到生态位时，在循环中被检测到；这

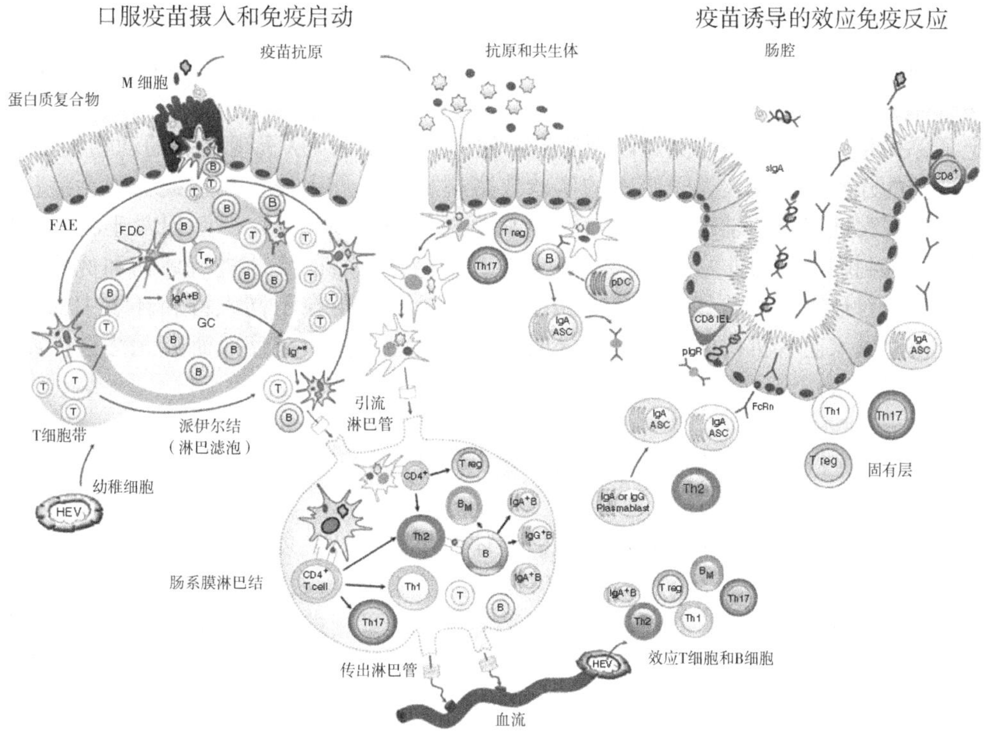

图 3.1　滤泡相关上皮（FAE）含有微皱褶（M）细胞

这些细胞用于细胞内吞作用和完整抗原快速转运。疫苗抗原由 M 细胞穿过上皮屏障转运，并被上皮下未成熟树突状细胞（DC）吸收。肠固有层上皮层内 DC 也可通过其延伸的树突从管腔中捕获抗原；其中一些 DC（CD103$^+$，CD11b$^+$）参与诱导对共生体的耐受。未成熟 DC 以及幼稚和记忆性淋巴细胞通过高内皮小静脉（HEV）进入黏膜；这些细胞被滤泡相关上皮（FAE）肠细胞和其他介质产生的趋化因子（CCL20、CCL23 和 CXCL16）吸引。携带疫苗抗原的 DC 迁移到滤泡间 T 细胞区或引流淋巴结，以向幼稚 CD4$^+$T 细胞递呈抗原。B 细胞在淋巴滤泡外通过与 DC 细胞和 T 细胞相互作用被抗原激活；这些激活 B 淋巴细胞重新进入滤泡，与滤泡树突状细胞（FDC）相互作用后，这些滤泡树突状细胞（FDC）捕获迁移抗原和 Treg 衍生的 T 滤泡 B 辅助细胞（TFH），成为生发中心（GC）细胞。在 GC 中，B 细胞经历 IgA 类别转换和体细胞超突变，分化为 IgA$^+$浆母细胞（或浆细胞）和潜在记忆性 B（B_M）细胞。滤泡间 T 细胞区域启动的效应 CD4$^+$Th 细胞也可以进入 B 细胞滤泡，在那里它们激活 B 细胞（通过 CD40-CD40L 相互作用），并释放诱导 GC 反应的细胞因子（即 IL-4、IL-5、IL-10、TGF-β），导致 B 细胞初始成熟和 IgA 同种型转换。IgA 浆细胞也可以通过 T 细胞非依赖性机制诱导，包括与 DC 直接接触的 B 细胞活化以及由不同的 DC 亚群（如 FDC 和浆细胞样 DC）和上皮细胞产生的 B 细胞活化和生长因子（如 BAFF、APRIL、TGF-β 和 NO）。在派伊尔结（PP）中诱导 T 细胞非依赖性（低亲和力，“天然”）IgA 应答，但也可在分离的淋巴滤泡和固有层中产生。在 T 细胞和 B 细胞活化的同时，黏膜 DC 产生的维甲酸（*RA*）将肠道归巢受体（即 α4β7、CCR9、CCR10）印在抗原特异性 Th 细胞和 IgA 决定的 B 细胞上。由疫苗抗原激活的肠黏膜 B 细胞和 T 细胞迁移到区域（肠系膜）淋巴结，在那里它们进一步分化为不同的效应细胞群（例如 Th1、Th2、Treg、Th17 和 B_M细胞）。疫苗特异性 T 细胞和 B 细胞通过胸导管排出淋巴结并进入血流；它们通过循环传播并重新进入黏膜效应部位。在固有层中黏膜 B 细胞最终分化为浆细胞，其中大多数生产 IgA，通过上皮细胞输出并作为分泌 IgA（sIgA）。黏膜 IgG 由局部血浆细胞产生或可能从血液中渗出。这里所描述的是简化的过程；还应注意的是，在口服免疫后诱导适应性（激活或调节）免疫应答的机制只有部分被了解。

些细胞不分泌抗体，但在抗原暴露后迅速分化成 ASC[38]。在人类中已证实，B_M细胞存在是对轮状病毒[39]、霍乱弧菌[40]、志贺氏菌[41]和沙门氏菌[42,43]感染或口服疫苗接种的反应。口服疫苗可诱导的主要免疫效应物概要如图 3.1 所示。

根据每种病原体致病机制，需要不同类型的疫苗来产生特异性免疫效应物，这些效应物可以针对不同的毒力因子阻断发病机制。黏膜疫苗可能足以预防非侵入性生物体，如霍乱弧菌和产肠毒素大肠杆菌（ETEC）引起的疾病，而黏膜-胃肠外联合途径可能更适合抵御引起肠道炎症和细胞破坏的病原体，如志贺氏菌属和轮状病毒以及诱发全身性疾病的病毒，如斑疹伤寒和脊髓灰质炎病毒。除了宿主相关因素之外，所选抗原、呈递方法和是否包含佐剂将决定诱导应答的特征和程度。仔细评估这种应答，对于确定根据保护的已知免疫相关性（或替代物）保护特定病原体的可能性很重要[44]。

确定了保护相关性，免疫终点可以预测疫苗效力。它们可以从重现疾病的动物模型中获得；人体随机可控的疗效研究；人类志愿者的实验攻毒模型；以及比较自然感染者和无症状暴露者的免疫状态的观察性流行病学现场研究（包括病例对照研究）[45]。这一知识对于将候选疫苗通过调控途径转移到临床试验非常有价值。还允许对获得许可后的疫苗进行评估和优化，比较配方、免疫计划或对最终产品的其他修改。相关性可以预测疫苗在不同人群和环境中的性能，并比较候选疫苗。国家监管机构通常接受定义明确的免疫终点作为在疗效研究无法进行或在没有得到保证的情况下获得许可的依据。

3.4 肠耐受性和疫苗接种有关的免疫调节

在促进肠道黏膜免疫系统激活以对潜在病原体产生获得性免疫应答，同时保持耐受性以预防免疫紊乱［如食物过敏、炎症性肠病（IBD）和腹腔疾病］之间有一个微妙平衡。可以说，考虑到肠道的普遍耐受性环境，在不干扰免疫平衡和生理平衡的情况下期望口服抗原产生强烈的免疫应答是违反直觉的。抗原性质和输送方式（如剂量）可决定最终结果（免疫激活或耐受）。最成功的口服疫苗包括主动复制（如轮状病毒）或强免疫原性刺激（如活的减毒菌株、免疫调节分子）。

大颗粒通常被 M 细胞捕获，并可能经历抗原递呈和免疫启动的正常过程，而可溶性抗原（单次大剂量或多次小剂量口服）可能被 APC 在肠道中优先摄取和递呈，以通过克隆无反应/缺失或抑制效应细胞诱导耐受性[2,46]。几个与维持口腔耐受性和肠道内环境稳定有关的 T 细胞亚群被描述如下：$Foxp3^{+}$ $CD25^{+}CD4^{+}$nTreg 细胞，$CD25^{-}$ $Foxp3^{-}$ IL-10 分泌 $CD4^{+}$Tr1 细胞，TGF-β 分泌 Th3 T 细胞[47]。nTreg 细胞在胸腺中发育，被认为能调节系统内环境稳定，预防自身免疫疾病。肠道免疫的抗原特异性涉及维持免疫抑制环境的调节性细胞因子；抗原非特异性调节机制产生抗原特异性失能或 Treg 细胞，可以维持系统免疫静止[47]。

通过紧密连接扩散到固有层或通过跨细胞运输转移到固有层的可溶性抗原，可被具有独特耐受表型的特殊固有层 $CD103^{+}$ DC 吸收[48]。这些细胞迁移到 MLN，通过产生 RA 促进 $CD4^{+}$ T 细胞分化为 $FoxP3^{+}$ $CD25^{+}$ $CD4^{+}$Treg 细胞，RA 是 TGF-β 介导的 T 细胞转化为 Treg 细胞的辅因子[49,50]。驱动和维持黏膜耐受性的一个重要成分是 IL-10，它主要由 Tr1 和 nTreg 细胞产生，也由肠道中的髓细胞产生。IL-10 和 TGF-β（由 Tr1、Th3 和 nTreg 产生）都具有抑制活性 T 细胞产生细胞因子、APC 上协同刺激分子表达以及抗体产生的抑制功能。众所周知，微生物群也会影响 $Foxp3^{+}$Treg 细胞的诱导，从而影响耐受状态。PP 中一些 $Foxp3^{+}$Treg 细胞分化成 T_{FH} 细胞参与 GC 形成，并支持 IgA 应答[1]；这与 IgA 支持（不受干扰的）黏膜界面的看家功能是一致的。

3.5 许可口服疫苗及其在不同人群中的表现

少数口服疫苗已获得许可证，可用于预防脊髓灰质炎、轮状病毒、伤寒和霍乱弧菌 O1（表 3.1）。包括通过组织培养传代（Rotarix© 和脊髓灰质炎病毒）获得的减毒活生物体[57]或重组技术（RotaTeq©）[58]、毒性因子（霍乱）基因缺失[59]或化学诱变（伤寒）[55]以及结合免疫调节重组毒素亚单位的灭活生物体（霍乱）[60]。这些疫苗成功地预防了特定目标人群的疾病：工业化国家的婴儿（轮状病毒疫苗）、发展中国家的婴儿（脊髓灰质炎疫苗）和发展中国家的学龄儿童（伤寒和霍乱疫苗）。然而，尽管这些疫苗在工业化国家取得了成功，但越来越多的证据表明，它们在世界欠发达地区某些亚群中降低了免疫原性能力和效力[61-64]。

表3.1　获得批准的疫苗

目标病原	许可的疫苗	疫苗细节	参考文献
脊髓灰质炎病毒	Sabin's tOPV 减毒活病毒株	1型、2型和3型（三价脊髓灰质炎病毒疫苗）	[51]
轮状病毒	RotaTeq®（默克公司）	五价减毒重组轮状病毒疫苗	[52]
	Rotarix®（葛兰素史克生物制品）	减毒单价轮状病毒活疫苗	[53, 54]
沙门氏菌伤寒	Vivotif®（Berna 生物科技有限公司）	减毒伤寒活疫苗 Ty21a	[55]
霍乱弧菌	Dukoral®（Crucell）	与重组霍乱毒素B亚单位（CTB）混合的全杀霍乱弧菌O1 Inaba 和 Ogawa（ElTor 和经典生物型）	[56]
	Shanchol®（Shantha Biotechnics - Sanofi Pasteur）/ mORCVAX®（VaBiotech）	全杀霍乱弧菌O1株和O139株（无CTB）	[56]

3.5.1　口服脊髓灰质炎疫苗

脊髓灰质炎病毒有3种血清型：1型、2型和3型。这种单链RNA病毒在胃肠道上皮下的淋巴组织中复制，通常产生轻微症状（发热和喉咙痛）或无症状。然而在<1%的病例中，病毒传播到运动神经元并导致弛缓性麻痹。Albert Sabin 开发了一种三价口服脊髓灰质炎病毒疫苗（tOPV），该疫苗含有来自血清型1、2和3的活菌株，反复通过非人类细胞传代而致弱[51]。作为全球根除脊髓灰质炎倡议的一部分，这种疫苗已经消除了野生型脊髓灰质炎病毒在美洲、西太平洋和欧洲的传播[65]。截至1999年，2型脊髓灰质炎病毒已在全球根除。然而，脊髓灰质炎病毒仍然在巴基斯坦、阿富汗和尼日利亚流行[66]。

Sabin's tOPV 疫苗一直是脊髓灰质炎病毒流行国家的首选疫苗，因为它易于递呈，能够在肠道（病毒进入的主要部位）诱导终身免疫[67]，并且事实上它可以从大便中脱落间接免疫社区的其他人[68]。然而，尽管总体上取得了成功，tOPV 在有些感染区效率较低[69,70]。在印度北方邦和比哈尔邦，一些5岁以下的儿童接受了大约15剂tOPV，相比之下印度其他地方的儿童接受了10剂。尽管覆盖率很高，但前一组对疫苗的免疫应答较低，这被归因于卫生条件差和人口密度高造成的腹泻和其他肠道感染[62,71,72]。

疫苗组分也可能导致免疫原性降低。与1型和3型菌株相比，2型菌株能有效地在肠道内定居并引发更强的免疫反应。由于2型病毒有可能比其他毒株更具竞争力，每剂tOPV含有≥$10^{6.0}$个1型病毒感染单位、$10^{5.0}$个2型感染单位和$10^{5.8}$个3型病毒感染单位[65]。在印度北方邦和比哈尔邦以及其他发展中国家，单价1型（mOPV1）和单价3型（mOPV3）疫苗以及二价1+3型（bOPV）疫苗被用来阻断脊髓灰质炎1型和3型病毒的传播[73]。由于这些努力，印度自2011年1月13日起再没有发现野生型脊髓灰质炎病毒[74]。tOPV的缺点之一是疫苗毒株偶尔恢复毒性，在250剂疫苗中有1剂会导致与疫苗相关的麻痹性脊髓灰质炎（VAPP）[75]。因此，在一些成功阻断野生型脊髓灰质炎病毒传播的国家，tOPV已经被脊髓灰质炎灭活病毒疫苗（IPV）取代[76]。

3.5.2 轮状病毒疫苗

轮状病毒是一种双股 RNA 病毒，是引起 5 岁以下儿童严重胃肠炎的主要病因。发展中国家和工业化国家都受该病毒的影响，估计每年造成 52.7 万人死亡，其中大多数发生在发展中国家的儿童中[77]。有效疫苗被认为是减少疾病负担的最佳预防性干预措施[78]，世界卫生组织战略咨询小组建议将其纳入国家免疫计划[79]。

血清分型是基于 VP7 糖蛋白（G 血清型）和 VP4 蛋白酶敏感蛋白（P 血清型）。通常检测到 5 株菌株（P[8]，G1；P[4]，G2；P[8]，G3；P[8]，G4；和 P[8]，G9），构成了主要的疫苗靶点[80]。

RotaTeq® 和 Rotarix® 两种口服疫苗目前已在大多数国家获得许可。先前在美国获得许可的一种口服疫苗 RotaShield® 因与肠套叠有关而退出市场，这是一种在第一次接种后不久出现的意外并发症，其中一部分肠内陷到另一部分肠；该疫苗不再可用[81]。血清 IgA 的诱导与防止儿童轮状病毒感染有关[82]。血清 IgA 水平，特别是循环轮状病毒特异性 IgA-ASC 的频率与活检中测量的肠道 ASC 相关[83]。

3.5.3 Rotarix®

Rotarix®（也称为 RV1）是葛兰素史克生物制品公司开发的一种单价轮状病毒减毒活疫苗。它在 6~14 周时口服，过 4 周后再次口服。该疫苗含有一种 G1P[8] 血清型的 RIX4414 菌株，其来源于亲本疫苗菌株 89-12[53,54]。该疫苗经 43 次传代后致弱。在 11 个拉丁美洲国家和芬兰进行的一项大型跨国双盲 3 期试验表明，Rotarix® 在保护婴儿免受严重轮状病毒胃肠炎方面非常有效（疫苗效力为 85%，与安慰剂相比 $P<0.001$），显著降低了任何原因引起的严重胃肠炎的发生率（发病率下降 40%，$P<0.001$），与肠套叠风险增加无关[57]。墨西哥和巴西进行的一项特许研究表明，在每 51 000~68 000名接种疫苗的婴儿中约有 1 名婴儿其 Rotarix® 与肠套叠的短期风险相关[84]。然而疫苗的好处远远大于风险，因为它每年在这两个国家预防了约 80 000例住院和 1 300 例腹泻死亡。

与欧洲和拉丁美洲的高收入和中等收入国家的卓越疗效相比，在非洲（南非和马拉维）的一项随机、安慰剂对照、多中心试验显示疫苗的疗效为 61.2%[61]。对任何原因引起的严重胃肠炎的疗效仅为 30.2%。非洲婴儿疗效下降的可能原因包括营养不良、与其他肠道病原体共感染、母乳中的抗轮状病毒抗体以及口服脊髓灰质炎疫苗干扰。本章稍后将讨论这些因素。

3.5.4 RotaTeq®

RotaTeq®（也称为 RV5）是默克公司通过重组牛轮状病毒株 WC3 的 10 个基因和 5 个最常见人类血清型（G1、G2、G3、G4 和 PIA）的单个衣壳基因而开发的一种五价减毒轮状病毒疫苗[52]。从 6~12 周开始口服三剂疫苗，随后每隔 4~10 周给予 1 次。在一项对 70 000名婴儿（主要来自美国和芬兰，也来自中美洲和南美、欧洲和亚洲）进行的大型现场试验中，对每剂 12×10^7个感染单位的最终配方进行了评估[58]。该疫苗在预防严重轮状病毒性胃肠炎方面有 98%的疗效，并且在疫苗接种者和安慰剂接种者的肠套叠风险方面没有差异。在美国的一项特许研究中，对 4~34 周大的婴儿给予 786 725次总剂量接种，没有显示由于接种疫苗而增加肠套叠的风险[85]。在最近对非洲和亚洲两个多中心、双盲、安慰剂对照疫苗试验中，报告了对严重轮状病毒胃肠炎的联合疗效为 33.9%，对非常严重胃肠炎联合疗效为 51.2%。这两个试验是在整个随访期间对 4~12 周大的婴儿进行的。这些研究也显示了对非疫苗血清型的交叉保护。尽管这些试验中的疫苗效力很强，但获得的数值低于在发达国家或拉丁美洲发展中国家进行的试验[58,64,87-89]。非洲试验涉及加纳和肯尼亚农村地区和马里城市地区的5 468名婴儿。针对严重轮状病毒肠胃炎的疫苗效力从出生第一年的64.2%下降到出生第二年的19.6%（总有效率为39.3%）。这种效力下降很可能是由于免疫原性减弱。在第3次接种后14天测得的血清IgA几何平均滴度（GMT）比发达国家进行试验的受试者低5~10倍[64]。同样在非洲受试者中，免疫后血清中和抗体对G1的应答降低了4倍，对P1A的应答降低了3倍。亚洲试验涉及2 036名婴儿，分别在孟加拉国的Matlab和越南的城市和城郊的芽庄进行[90]。三剂RotaTeq®疫苗对严重轮状病毒病的疗效为48.3%。然而与非洲的研究不同，疫苗效力在生命的第一年和第二年是相似的（分别为51%和45.5%）。

与其他研究相比，轮状病毒疫苗在非洲和亚洲婴儿中效力较低的可能原因与Rotarix®以及本研究中登记的婴儿年龄较小的原因相同。此外，这些试验中许多婴儿同时接受了口服脊髓灰质炎病毒疫苗和Rotarix®疫苗。先前有研究表明，口服脊髓灰质炎病毒疫苗的联合使用导致口服轮状病毒疫苗的免疫原性降低[63,91,92]。

3.5.5　霍乱疫苗

霍乱仍然是全球主要的公共卫生问题，主要发生在发展中国家。2010年，全世界向世界卫生组织（WHO）报告了317 534例霍乱病例和7 543例死亡病例，据估计全球每年发生300万~500万例病例和10万~13万例死亡病例[93]。建议在高危人群流行地区和暴发期间，接种可用的口服灭活疫苗[93]。霍乱弧菌血清群O1和O139导致世界范围内霍乱流行[94]。血清型O1进一步分为生物型（El Tor和经典型）以及血清型（Inaba型和Ogawa型）。有效的霍乱疫苗需要同时预防生物型和血清型。

霍乱弧菌具有很高的感染性剂量，通常通过食物和水传播，经粪便和口腔污染物获得。该病由霍乱毒素（CT）引起，由5个B（结合）亚单位和1个A（活性）亚单位组成。B亚单位与GM1神经节苷脂受体结合，作为A亚单位进入上皮细胞导管，导致大量的水和电解质分泌。与其他肠道感染不同，以前受到过野生型霍乱弧菌感染可诱导针对临床疾病的血清组特异性保护，这种保护是强大而持久的[95,96]。负责这种保护的精确免疫效应因子还不完全清楚。来自霍乱弧菌感染流行地区的个体产生LPS特异性IgA-ASC，在发病后7天达到高峰[97,98]，并表达肠道归巢整合素受体α4β7[23]。血清杀弧菌抗体与感染易感性呈负相关，但不太可能是保护效应机制[99-101]。另一方面，肠道IgA被认为在黏膜表面阻断病原体发挥了关键作用。在霍乱感染患者十二指肠活检中，检测到了黏膜LPS特异性IgA-ASC，即使在分泌物中没有检测到抗LPS-IgA。当适当刺激这些细胞时，可能恢复抗体产生[102]。霍乱患者发病后1年内发现了针对霍乱弧菌LPS、CTB和TcpA的特异性IgG和IgA B_M细胞[40,103]，并且LPS特异性IgG B_M细胞的频率与预防霍乱患者家庭接触者感染有关[104]。由B_M细胞介导的黏膜记忆应答（主要针对LPS）似乎对长期预防霍乱是必要的[105]。

市面上有两种口服霍乱疫苗，都是由完全灭活的生物体组成。Dukoral™（Crocell公司，瑞典）含有热灭活或福尔马林灭活的霍乱弧菌O1Inaba和Ogawa菌株，这两种菌株均为El Tor和经典生物型，与重组霍乱毒素B亚单位（CTB）混合。尽管它是国际许可的，但由于其高昂的成本，主要用于旅行者[56]。建议服用两剂，间隔10~14天。在一项随机双盲试验中，孟加拉国农村2~15岁儿童和15岁以上妇女的保护效果得到了确定。在62 285名受试者中，3年后疫苗效力为50%（2~5岁的受试者为26%，老年受试者为63%）[60]。经过5年的随访，保护效果仍然是49%[106]。Shanchol™（Shantha赛洛菲巴斯德生物技术，印度）或mORCVAX（VaBiotech，越南）含有几株O1型霍乱弧菌和1株O139型霍乱弧菌，不含CTB，是一种较便宜的替代品。在印度加尔各答霍乱流行区，对101名成人和100名儿童（1~17岁）进行了Shanchol™测试。接种两剂疫苗后，53%的成年人和80%的儿童显示出针对霍乱弧菌O1的血清转化（定义为接种疫苗后血清杀弧菌抗体滴度比基线提高4倍），只有10%的成年人和27%的儿童血清转化为霍乱弧菌 O139[107]。

在霍乱流行地区进行的另一项研究中，孟加拉国达卡米尔普尔地区的2~5岁幼儿和12~23个月大的婴儿接种了两剂疫苗，这两种疫苗在所有年龄组都是安全的，并具有免疫原性[108]。然而，这两项在霍乱流行地区进行的研究，成人中的杀弧菌抗体几何平均倍数上升低于非霍乱流行地区（山罗市，越南）的水平[109]。这很可能是由于在加尔各答和达卡观察到的高基线滴度，表明由于自然感染而存在的免疫可能会对口服减毒活疫苗的免疫应答产生负面影响。幼儿和婴儿的基线水平较低可能是因为他们没有接触 O1 型霍乱弧菌。在印度加尔各答的一项大型双盲试验中，Shanchol® 对1岁及1岁以上受试者中霍乱弧菌 O1 腹泻的疗效为67%（$P<0.0001$）。最近的研究表明，单剂量的 Dukoral® 或 Shanchol® 可能足以引起保护[110,111]。

与野生型霍乱弧菌感染相比，灭活的全细胞霍乱疫苗在成年人、尤其在儿童中诱导了更为有限和短暂的保护[95,112,113]。因此人们寻求能够模拟自然感染的活疫苗菌株来利用自然免疫的免疫效应，理想的情况是在单一免疫后。

CVD 103-HgR 是一种减毒活疫苗，源于霍乱弧菌 O1 经典生物型，Inaba 血清型其基因中含有编码霍乱毒素 A 亚单位的缺失基因，并携带编码抗汞基因［插入溶血素 A（hlyA）基因位点］。在美国[114]和欧洲[115]进行的临床试验表明，单一口服 5×10^8 菌落形成单位（CFU）的 CVD 103-HgR 可分别导致90%以上（血清杀弧菌抗体升高4倍或更高）和88%以上的接种者发生血清转化。在攻毒研究中，该疫苗还能够保护志愿者免受由霍乱弧菌 O1 的生物型（El Tor 和经典）和血清型（Inaba 和 Ogawa）引起的霍乱侵害[59,114,116]。

然而，在几个发展中国家进行的临床试验显示免疫原性降低。在印度尼西亚，5×10^8 CFU 的 CVD 103-HgR 仅在5~9岁儿童中引起血清转化[117]。在2~4岁儿童中也观察到了类似结果[118]。秘鲁[119]和泰国[120]的临床试验结果相似。然而，5×10^9 CFU 的 CVD 103-HgR 能够在印度尼西亚儿童和秘鲁及泰国成人中引发高血清转化率[117,119,120]。CVD 103-HgR 也在雅加达北部一个高度流行地区进行了随机大规模对照现场试验[121]。与美国志愿者的结果相反，由于研究结束时发病率下降，疗效较低[121]。一种可能的解释是，广泛的疫苗接种导致了间接的保护，当疾病消失时，对疗效的估计变得微不足道[122]。

在马里的一项随机、有安慰剂对照、双盲、交叉临床试验中，在 HIV 血清阳性和 HIV 血清阴性受试者中观察到接种 CVD 103-HgR 后较低的弧菌血清转化率[123]。CVD 103-HgR 以前由瑞士血清和疫苗研究所（Berna，瑞士）以商品 Orochol® 和 Mutacol® 制造，并被批准为单剂量霍乱疫苗[124-126]。paxvax 正在恢复生产；目前正在进行临床试验以便在美国上市（NCT01585181；ClinicalTrials. gov）。一线其他活疫苗包括 Peru 15（中国海口 VTI 生物研究所），也是一种基因工程毒素缺陷株，在孟加拉国成人和儿童中发现是安全和有免疫原性的。尽管在第3期的研究中还没有测试以及减毒活霍乱弧菌 638[127]和 CTB-表达菌株 VA 1. 3[128]，两者都是减毒 El Tor 衍生物。

3.5.6 口服伤寒疫苗

伤寒沙门氏菌（*S. Typhi*）引起以持续高热为特征的急性全身感染。它是一种以食物和水传播为主的传染病，在现代工业化国家并不常见，但在缺乏足够卫生条件的不发达地区流行。伤寒在学龄儿童中发病率最高[129,130]。在南亚的城市贫民窟环境中，基于家庭和健康中心的系统性主动监测显示，发热的幼儿和学龄前儿童感染伤寒杆菌的概率很高[130-132]。从工业化国家到发展中国家的旅行者患伤寒的风险也在增加[133]。

自然伤寒沙门氏菌感染可诱导血清抗体对细菌脂多糖、鞭毛、外膜蛋白和其他细菌抗原、肠源性 ASC（抗原特异性抗体分泌细胞）和 sIgA 的应答[134-136]。慢性胆源性伤寒病毒携带者中80%~90%的人血清中，发现了针对荚膜 Vi 抗原的高滴度 IgG，12%~38%未成为慢性携带者的急性伤寒患者中也发现了高滴度 IgG，但来自流行地区的健康个体很少发现这种抗体[137-139]。生活在伤寒流行地区的成年人中发现了针对伤寒抗原的增殖应答，这些人没有临床上明显的伤寒史[140]。培养证实伤寒患者血清中 Th1 型细胞因子（如 IFN-γ、TNF-α、IL-1 和 IL-6）水平升高[141]。

一种口服伤寒减毒活疫苗（菌株 Ty21a）是在 20 世纪 70 年代初通过化学诱变技术开发的，已获准在美国供前往流行地区和世界其他国家的旅行者使用[55]。Ty21a 提供了显著的保护，不会引起不良反应。在对 Ty21a 进行的大规模疗效现场试验涉及智利约 514 000名学童[142-144]、埃及 32 388名[145]、印度尼西亚约 20 543名 3 岁至成人受试者[146]，通过被动监测未发现疫苗引起的不良反应[142-144,147]。有两种制剂获得许可：肠溶胶囊和用缓冲液重新配制的冻干疫苗。然而，近年来只有肠溶胶囊实现了商业化。智利圣地亚哥的一项实地试验表明，每隔 1 天服用三剂肠溶胶囊 Ty21a，在 3 年的随访中可获得 67%的保护[147,148]。基于这些结果，除了美国和加拿大使用四剂量方案，在世界各地都使用三剂量方案。四剂量北美方案是基于在智利进行的另一项大规模随机比较试验的结果，其中四剂量肠溶胶囊 Ty21a 的保护性明显高于两剂量或三剂量肠溶胶囊 Ty21a[149]。

Ty21a 能提供长期的保护[148]。对智利两个实地试验中接受了 Ty21a 的个体进行了总共 7 年的随访，在此期间，Ty21a 肠溶胶囊制剂提供了 62%的疫苗效力[148]，而液体制剂在 5 年的随访中对疫苗效力的点估计为 78%。Ty21a 受者的免疫应答包括针对 O 多糖的血清 IgG[142,150]和在接种后 7～10 天检测到的黏膜启动的 O-特异性 ASC[20]。大多数 ASC 携带肠道归巢整合素 α4β7[20,24]、O-特异性血清 IgG[142]。肠道来源的 ASC 被发现与 Ty21a 不同制剂和免疫方案在实地试验中提供的保护有关。口服 Ty21a 后[152,153]描述了肠道 sIgA 抗体应答[151]和抗体依赖性细胞毒性。此外 Ty21a 刺激强 T 细胞增殖反应、Th1 型细胞因子分泌（例如 IFN-γ、TNF-α）和多功能 $CD8^+$细胞毒性 T 淋巴细胞[154-157]。具有 T_{EM}表型（即 $CCR7^-$、$CD27^-$ $CD45RO^+$ $CD62L^-$）和共表达肠道归巢分子（即 α4β7 和 CCR9）的 T 细胞克隆，源于 Ty21a 疫苗接种者外周单核细胞[158]。

3.6 与口服疫苗接种有关的障碍

人们越来越认识到，与工业化国家相比，口腔疫苗在不发达地区人口中的宿主非常不同。在全球范围内，现代疫苗的最大挑战之一是识别和克服导致贫困人群免疫原性降低和口腔疫苗保护减少的障碍[159-162]。能解释这种现象的一些因素包括小肠细菌过度生长，这表明生活贫困的个体存在环境性肠病、严重肠道蠕虫感染[163]、营养不良、母体抗体、与微生物群的相互作用以及与其他疫苗的相互作用。这些因素在表 3.2 中进行了总结，并在此单独描述，但它们是相互关联的[170]。

3.6.1 小肠细菌过度生长（SBBO）与环境性肠病

许多贫困儿童小肠细菌过度生长，表现为近端肠过度定植和环境性（或热带）肠病，以小肠组织病理变化为特征，包括炎症和绒毛变钝。人们认为，生活在发展中国家的贫困儿童（和成人）经常接触粪便污染物从而导致 SBBO（小肠细菌过度生长）和环境性肠病。有趣的是，环境性肠病可以逆转并最终消失，患者重新转到干净的环境中[171]。

SBBO 可能是 CVD 103-HgR 在发展中国家进行的临床试验中显示疗效降低的原因之一。在智利进行的一项临床研究中，178 名 5 岁至 9 岁儿童中有 10 人检测到 SBBO（通过乳果糖呼吸 H_2进行测量）。这些儿童的杀弧菌几何平均滴度低于其他儿童，尽管这并不显著（160 对 368，$P=0.25$）。逻辑回归显示，H_2峰值（在摄入乳果糖后被细菌酶切割）与杀弧菌血清转化率降低相关（$P=0.04$）。

SBBO 和环境性肠病可能通过产生短链脂肪酸或其他小分子对口服活疫苗产生直接抑制作

用，或者通过激活天然免疫细胞（如黏膜中增多的淋巴细胞）和破坏疫苗的促炎环境产生间接作用。或者，粪便污染物可能会与口服活疫苗竞争肠道中的定植位点，并阻止疫苗进入关键受体或细胞[172]。

在厄瓜多尔的一个蠕虫高发区，用 CVD 103-HgR 进行临床试验显示，肠道共感染影响了活疫苗产生的免疫应答。139 名儿童服用了两剂阿苯达唑（一种广谱抗蠕虫药）或安慰剂，在第二剂后服用了 CVD 103-HgR。与安慰剂组相比，阿苯达唑组的血清转化率（血清杀弧菌反应）更高（29.3%对 15.5%），尽管这一差异不显著（$P=0.06$）。

表 3.2　口服疫苗接种障碍

障碍	因屏障使免疫原性降低或当屏障被影响时提高免疫原性的例子	参考文献
小肠细菌过度生长（SBBO）与环境性肠病	在智利，接种口服霍乱活疫苗 CVD103-H 后，患有 SBBO 的儿童的杀弧菌血清转化率降低	[164]
营养不良	在孟加拉国，补充锌可提高 Dukoral® 引起的杀弧菌血清转化	[165，166]
母体抗体和母乳免疫成分	在孟加拉国，Dukoral® 免疫前暂停母乳喂养 3 小时，导致杀弧菌抗体数量增加	[166]
微生物群	在芬兰，当口服轮状病毒活疫苗 RotaShield® 与益生干酪乳杆菌 GG 联合使用时，IgA 血清转化得到提高	[167]
新生儿口服免疫	与三价口服脊髓灰质炎病毒疫苗相比，给予单价口服脊髓灰质炎病毒疫苗免疫的婴儿血清转化率有所提高	[168]
疫苗的联合应用	口轮状病毒疫苗联合应用后，tOPV 可影响其免疫原性	[169]

3.6.2　营养不良

母婴营养不良在低收入和中等收入国家非常普遍[173]。例如，中南亚 33%的儿童和东非 28%的儿童体重不足。这些儿童中有许多缺乏维生素 A 和锌。锌对许多细胞功能和有效的天然免疫和后天免疫都是必需的[174]。众所周知，维生素 A 缺乏会导致肠道免疫缺陷，影响不同人群的肠道 DC。在小鼠体内，维生素 A 代谢物维甲酸对于将疫苗诱导的 T 细胞运输到胃肠道黏膜和保护疫苗效力至关重要[175]。事实上，维生素 A 缺乏的儿童与补充了正常维生素 A 水平的儿童相比，免疫反应降低了[176]。

一些研究表明，补充锌或维生素 A 可以改善口服疫苗诱导的免疫原性。在孟加拉国进行的一项研究中，2~5 岁的儿童在接种霍乱疫苗 Cholerix/Dukora 前 1 周补充锌、维生素 A 或两者同时补充，2 周后测量杀弧菌滴度[165]。研究表明，补充锌而不是维生素 A 可以提高杀弧菌血清转化率。相反地，当对这些儿童的血清进行霍乱毒素抗体（CT）检测时，接受锌治疗的儿童的 CT 抗体水平明显低于未接受微量营养素补充治疗的儿童[177]。因此，锌增强了霍乱疫苗的杀弧菌作用，但抑制了霍乱疫苗的 CT 抗体应答反应。孟加拉国的第二项研究证实，补充锌可以增加 Dukoral 诱导的杀弧菌应答程度[166]。

在加纳，对 1 085 名婴儿进行的一项研究中确定了补充维生素 A 对口服脊髓灰质炎疫苗免疫应答的影响[178]。婴儿接种白喉/百日咳/破伤风（DPT）疫苗时，维生素 A 或安慰剂分别在 6 周、10 周和 14 周龄服用。测定了 1 型、2 型和 3 型脊髓灰质炎病毒的中和滴度，维生素 A 补充组和安慰剂组之间没有观察到显著差异。在印度新德里进行的一项类似研究中发现，与安慰剂组（未接受维生素 A 补充）相比，在 6 周、10 周和 14 周服用维生素 A 的婴儿对脊髓灰质炎病毒1型（但不是2型或3型）的中和抗体滴度有所增加[179]。这些结果与之前在印度尼西亚进行的一项研究一致，该研究发现维生素A不影响任何脊髓灰质炎病毒的免疫应答[180]。

总的来说，数据表明，补充锌和维生素A可能会改善口服疫苗的免疫应答，但还需要更多的研究来确定对特定疫苗的影响。

3.6.3　母乳抗体和母乳免疫成分

一些口服活疫苗可能由于母乳中抗体水平升高而被抑制。此外，母乳中含有促进肠道免疫系统发育的免疫调节分子和细胞因子，其中一些具有耐受功能，如TGF-β、IL-10和维生素A[181]。众所周知，母乳干扰疫苗接种的例子是轮状病毒疫苗[161]。母乳中含有能中和轮状病毒的IgA抗体，以及能与病毒结合并防止其附着的乳凝集素（一种受体类似物）[182]。为了研究母乳对轮状病毒疫苗免疫原性的影响，人们进行了各种各样的研究。Goveia等[183]研究了来自欧洲、美国和拉丁美洲/加勒比3个地区的11个国家的5 098名婴儿的母乳喂养频率（不母乳喂养或有时母乳喂养或纯母乳喂养）对RotaTeq® 预防轮状病毒胃肠炎能力的影响[88]。母乳喂养不影响疫苗的效力。另一方面，对印度妇女母乳的分析表明，它具有高滴度的IgA和中和活性，理论上可以降低疫苗效价[184]。韩国和越南妇女的母乳IgA和中和滴度略低，美国妇女的滴度最低。孟加拉国进行的一项临床试验表明，与刚接种疫苗的受试者相比，在使用RotaTeq® 免疫前暂时停止母乳喂养3小时，导致了杀弧菌抗体的数量和应答频率增加[166]。这一差异在10~18个月的婴儿中可见，而在6~9个月的婴儿中则没有。

3.6.4　微生物群

目前尚不清楚肠道微生物菌群如何与口服疫苗相互作用并影响免疫。然而众所周知，社会经济、地理和文化环境影响着肠道微生物群特征方面的差异[185]。Turnbaugh等[186]研究美国肥胖和瘦的双胞胎肠道微生物群发现，人类肠道中没有单一丰富细菌种类。相反地，似乎在代谢功能水平上存在核心肠道微生物菌群。这表明，肠道微生物群不仅在不同的种群之间不同，而且在一个种群内也不同（就细菌系统类型方面而言）。芬兰的一项研究表明，微生物菌群可以改变口服疫苗诱导的免疫原性[167]。研究人员将干酪乳杆菌GG与RotaShield® 疫苗一起给2~5个月的婴儿口服，观察到93%的病例发生了轮状病毒IgA血清转化，相比之下，没有接受干酪乳杆菌的安慰剂对照组有74%病例发生了轮状病毒IgA血清转化（$P=0.05$）。目前正在探索通过改变婴儿肠道微生物菌群来改善疫苗应答的可能性，以及使用益生菌作为佐剂和/或运载工具的可能性[187]。

3.6.5　新生儿口服免疫和疫苗联合接种

新生儿和较大婴儿的免疫系统在组成和功能上有所不同。然而人们普遍认为，新生儿免疫系统并不是不发达，而是“缺乏经验”，传统疫苗无法激活与抗原递呈和免疫刺激有关的关键细胞。新生儿已被证明对口服脊髓灰质炎疫苗有反应，出生时一次免疫即可获得保护性免疫。有趣的是，与tOPV疫苗相比，新生儿服用单价疫苗时血清转化率更高[168]。最近的一项研究报告称，同时接种了OPV的新生儿对皮内卡介苗接种的Th1和Th2免疫应答降低[188]，但对潜在的混杂因素提出了警告。

如上所述，口服活疫苗效力会受到共同接种疫苗的影响。已知当这两种疫苗都给婴儿接种时，tOPV疫苗会影响口服轮状病毒疫苗的免疫原性[169]。随着越来越多的口服疫苗获得批准并加入扩大免疫计划（EPI），需要精心设计临床试验来衡量不同疫苗组合的效果。

3.7　新型口服疫苗和肠道病原体免疫方法：伤寒和副伤寒疫苗

具有诱导抗体和细胞免疫能力的活疫苗是预防由伤寒、非伤寒和副伤寒沙门氏菌引起疾病的最有效方法之一。如何产生既安全又具有足够免疫原性的菌株仍然具有挑战性。一种单剂量活减毒量伤寒疫苗，Typhella™（也称为 M01ZH09，最近由英国 Prokarium 有限公司从美国 Emergent BioSolutions™收购）在美国成人和越南儿童中被证明是安全和有免疫原性的[189-192]。这种疫苗在 *aroC*（芳香生物合成途径）和 *ssaV*（SPI-2 Ⅲ型分泌系统）[193]中有突变。需要更大的现场试验来确定其保护效力。一系列减毒伤寒杆菌疫苗，CVD 908-*htrA*（在 *aroC*、*aroD* 和 *hTRA* 中有缺失）和 CVD 909（CVD 908-*htrA* 组成性表达 Vi）已经证明在单剂量接种后是安全和有免疫原性的[194]。

虽然由伤寒引起的肠热病已经大幅降低，但副伤寒热仍然存在。一种由马里兰州巴尔的摩大学疫苗开发中心研发的减毒甲型副伤寒活疫苗 CVD 1902（含有 *guaBA* 和 *clpX* 基因位点缺失），目前正在进行临床 I 期试验测试（NCT01129452；ClinicalTrials. gov）。迄今为止获得的数据表明，该疫苗具有良好的耐受性和免疫原性（Kotloff，个人交流）。

3.7.1 志贺氏菌和产肠毒素大肠杆菌（ETEC）疫苗

由志贺氏菌和产肠毒素大肠杆菌（ETEC）引起的腹泻仍然是一个显著的疾病负担，与全球范围内的大量死亡率有关，尤其与引起许多发展中国家儿童的死亡有关。志贺氏菌属包含 4 种（或 4 个群）：志贺氏痢疾杆菌、福氏痢疾杆菌、鲍氏痢疾杆菌和宋内氏痢疾杆菌，每个群又包括一种或多种血清型[195]。这些侵袭性病原体经常引起大量痢疾，用口服补液通常很难治愈。非常需要针对这些病原体的疫苗，这样的疫苗应该能诱导对志贺氏痢疾杆菌 I（引起严重痢疾流行病）、所有 14 种血清型的福氏痢疾杆菌（引起小儿痢疾流行）和宋内氏痢疾杆菌的广泛保护。在人类研究中已经开发和评估了许多基于死亡或减毒活生物体的志贺氏菌候选疫苗。一种宋内氏痢疾杆菌全细胞灭活疫苗在志愿者中被证明是安全的并且具有免疫原性，他们摄入了 3～5 剂的 10^{10}个死亡的生物体[196]。沃尔特·里德陆军（Walter Reed Army）研究所开发的一系列减毒活疫苗已经在实验性攻毒研究中证明可以保护接种过疫苗的成年志愿者，并在控制性现场试验中保护流行地区的成人和儿童[197,198]。活的减毒福氏痢疾杆菌 2a SC602 菌株在质粒编码基因 *icsA* 和染色体基因 *iuc* 均有突变[199]。在美国进行的 5 项临床试验显示 SC602 耐受性良好，引起黏膜免疫应答[200]。然而，当疫苗在孟加拉国成人和儿童中测试时，尽管它显示出很少的疫苗脱落和反应原性，但也产生非常少的免疫反应[201]。孟加拉国志愿者食物中缺铁被认为是该疫苗免疫原性较低的原因之一，该疫苗在铁摄取方面存在突变。CVD 的研究人员正在研究一种由多种减毒菌株（宋内氏痢疾杆菌、志贺氏痢疾杆菌 1、福氏痢疾杆菌 2a、福氏痢疾杆菌 3a 和福氏痢疾杆菌 6）组成的五价志贺菌疫苗，以提供广谱保护[195]。在成人志愿者第 2 阶段的研究中，已经评估了几种含有 *guaBA* 和 *sen/set* 基因位点突变的候选菌株（例如 CVD 1204、CVD 1208），这种方法还没有在流行人群中试验过。

一种口服活疫苗 ACE527，包含 3 种减弱的 ETEC 毒株，当在成人志愿者中剂量达到 10^{11} CFU 时，在人体内是安全的且具有良好的耐受性[202]。在 2b 期效力研究中用 ETEC 10407 在接种和未接种的志愿者进行攻毒试验，该疫苗降低了腹泻的发生率和严重程度，但没有达到减少中等到严重腹泻的研究终点[203]。

3.7.2 诺如病毒病毒样颗粒

诺如病毒（NoV）是一种引起人类胃肠炎的单股正链 RNA 病毒。NoV 衣壳可以在昆虫或植物细胞中表达，已证明 VP1 衣壳蛋白可以自我组装类似于 NoV 衣壳的病毒样颗粒（VLPs）。使用诺如病毒原型的 VLPs，进行的临床 I 期试验表明，口服[204]和鼻内接种后[205]，诺如病毒病毒样颗粒对人体是安全的，且具有免疫原性，且对 NoV 实验性攻毒有保护作用[206]。这个系统令人感兴趣的特征包括在植物中表达 VLPs 的可能性，这些 VLPs 可以被摄取[207]，以及在植物表面显示疫苗抗原表位进行异源疫苗接种的可行性[208]。

3.7.3 黏膜免疫新佐剂

选择与黏膜传递抗原一起使用的佐剂是非常关键的，因为它可能对所诱导免疫反应的大小、质量和持续时间产生重大影响。Toll 样受体（TLR）激动剂、解毒细菌肠毒素、黏膜黏着剂、皂苷和颗粒被广泛地测试[34]。大肠杆菌热不稳定肠毒素（LT）的一种遗传解毒双突变体，在192G 和 L211A 上有缺失，也称为 dmLT，是目前最有前景的口服候选佐剂之一[209]。小鼠口服 dmLT 能够增强对联合服用破伤风类毒素的应答[209]。正在进行临床试验以评估成人志愿者服用 dmLT 的安全性和免疫原性（NCT01147445；ClinicalTrials. gov)。虽然口服佐剂可能是提高口服抗原免疫原性的有力工具，但它们也可能干扰免疫刺激和耐受之间的微妙平衡，特别是在生命的第一年，肠道免疫系统尚未发育完全[210]。鼻腔内输送解毒的大肠杆菌 LT 的单一缺失突变体与贝尔麻痹（短暂性面神经麻痹）有关，强调需要仔细评估新候选者的安全性[211]。

3.7.4 改善免疫策略

通过不同途径（例如黏膜和肠胃外途径）以相同或不同的制剂提供疫苗抗原，也称为初免-加强免疫策略[212]，能够提供一种方法以克服口服疫苗效力的有限性，并改善免疫原性差的抗原反应。这种方法有助于增强突破黏膜屏障并成为全身性病原体的免疫应答的广度和程度。黏膜初免-胃肠外加强方案在各种动物模型中成功诱导了保护性免疫[213,214]。在人类中口服伤寒杆菌疫苗株 CVD909，然后用伤寒杆菌荚膜抗原胃肠外增强诱导 Vi 血清 IgG 的应答，和对细菌抗原特异的经典（$CD19^{+}IgD^{-}$ $CD27^{+}$）IgA 和 IgG B_M 细胞[43]。OPV（口服脊髓灰质炎疫苗）和胃肠外 IPV（脊髓灰质炎灭活疫苗）（或按相反顺序）的组合已经在几个国家实施，以获得肠道和全身免疫，最大限度地降低疫苗相关麻痹性脊髓灰质炎（VAPP）风险[215]。事实上，只有在接种了 OPV 疫苗后，IPV 才能诱导黏膜 IgA[26]。以色列士兵接受胃肠外 O 型多糖志贺氏菌疫苗的保护才避免被感染，这可能是由于之前自然（口服）接触过生物体而促成的[216]。人们对初免-加强免疫策略的兴趣与日俱增，预计在不久的将来会有更多在人类上的研究。黏膜免疫更有效的替代和潜在的途径包括舌下、气溶胶、眼部和阴道免疫；[217]它们的有效性和实用性将最终决定在人类中的使用。

3.8 结论和未来方向

尽管存在障碍，但迄今为止，根除 2 型脊髓灰质炎和显著减少腹泻疾病负担的现有许可疫苗取得的全球成功表明，口服免疫确实可以预防肠道感染。需要进一步研究来发现新的广谱保护性抗原和有效的递送系统及佐剂，以改善口服疫苗的疗效。还应重点阐明肠道中与免疫启动相关的事件，这些事件导致产生强大的黏膜和系统免疫以及免疫记忆。了解妨碍口服免疫的障碍，对于满足贫穷国家的需求和帮助减轻最需要的疾病负担也是至关重要的。现代技术（例如基因组学、用于抗原发现的蛋白质组学、使用非侵入性样本的高通量免疫检测以及仔细分析保护相关性的数学模型）可以帮助开发候选疫苗和表征免疫结果。多年来，科学家们已经评估了许多通过肠道有效地转运疫苗抗原的方法，如重组活载体、非活颗粒、复制缺陷病毒和抗原表达植物等。需要进行良好的临床试验来确定动物模型中的候选抗原的安全性和有效性。未来的成功可能在于实施新的免疫策略，例如将模拟自然感染的减毒活生物体与保护性亚单位疫苗抗原结合的初免-加强方案。为全球使用创造成功的口服疫苗的主要障碍仍然是需要为工业化国家人口实现足够的安全性，同时保持发展中国家人口的免疫原性和效力。

致谢

作者感谢 Emily DeBoy 为图 3.1 的设计和构造所做的贡献。

参考文献

[1] Suzuki, K., Fagarasan, S.: Diverse regulatory path-ways for IgA synthesis in the gut. Mucosal Immunol. 2, 468-471 (2009).

[2] Yamamoto, M., Pascual, D. W., Kiyono, H.: M cell-targeted mucosal vaccine strategies. Curr. Top. Microbiol. Immunol. 354, 39-52 (2012).

[3] Brandtzaeg, P.: Induction of secretory immunity and memory at mucosal surfaces. Vaccine 2, 5467-5484 (2007).

[4] Soloff, A. C., Barratt-Boyes, S. M.: Enemy at the gates: dendritic cells and immunity to mucosal pathogens. Cell Res. 20, 872-885 (2010).

[5] Sutherland, D. B., Fagarasan, S.: IgA synthesis: a form of functional immune adaptation extending beyond gut. Curr. Opin. Immunol. 24, 261-268 (2012).

[6] Cerutti, A.: Location, location, location: B-cell differentiation in the gut lamina propria. Mucosal Immunol. 1, 8-10 (2008).

[7] Macpherson, A. J., Geuking, M. B., McCoy, K. D.: Homeland security: IgA immunity at the frontiers of the body. Trends Immunol. 33, 160-167 (2012).

[8] Gibbons, D. L., Spencer, J.: Mouse and human intestinal immunity: same ballpark, different players; different rules, same score. Mucosal Immunol. 4, 148-157 (2011).

[9] B emark, M., Boysen, P., Lycke, N. Y.: Induction of gut IgA production through T cell-dependent and T cell-independent pathways. Ann. N. Y. Acad. Sci. 1247, 97-116 (2012).

[10] Slack, E., Balmer, M. L., Fritz, J. H., Hapfelmeier, S.: Functional flexibility of intestinal IgA-broadening the fine line. Front Immunol. 3, 100 (2012).

[11] Mantis, N. J., Rol, N., Corthesy, B.: Secretory IgA's complex roles in immunity and mucosal homeostasis in the gut. Mucosal Immunol. 4, 603-611 (2011).

[12] Pabst, O.: New concepts in the generation and functions of IgA. Nat. Rev. Immunol. 12, 821-832 (2012).

[13] Kett, K., Brandtzaeg, P., Radl, J., Haaijman, J. J.: Different subclass distribution of IgA-producing cells in human lymphoid organs and varioussecretory tissues. J. Immunol. 136, 3631-3635 (1986).

[14] Israel, E. J., et al.: Expression of the neonatal Fc receptor, FcRn, on human intestinal epithelial cells. Immunology 92, 69-74 (1997).

[15] Neutra, M. R., Kozlowski, P. A.: Mucosal vaccines: the promise and the challenge. Nat. Rev. Immunol. 6, 148-158 (2006).

[16] Robbins, J. B., Chu, C., Schneerson, R.: Hypothesis for vaccine development: protective immunity to enteric diseases caused by nontyphoidal salmonellae and shigellae may be conferred by serum IgG antibodies to the O-specific polysaccharide of their lipopolysaccharides. Clin. Infect. Dis. 15, 346-361 (1992).

[17] Yoshida, M., et al.: Human neonatal fc receptor mediates transport of IgG into luminal secretions for delivery of antigens to mucosal dendritic cells. Immunity20, 769-783 (2004).

[18] Boullier, S., et al.: Secretory IgA-mediated neutralization of *Shigellaflexneri* prevents intestinal tissue destruction by down-regulating inflammatory circuits. J. Immunol. 183, 5879-5885 (2009).

[19] Devriendt, B., De Geest, B. G., Goddeeris, B. M., Cox, E.: Crossing the barrier: targeting epithelial receptors for enhanced oral vaccine delivery. J. Control. Release 160, 431-439 (2012).

[20] Kantele, A.: Peripheral blood antibody-secreting cells in the evaluation of the immune response to an oral vaccine. J. Biotechnol. 44, 217-224 (1996).

[21] Sigmundsdottir, H., Butcher, E. C.: Environmental cues, dendritic cells and the programming of tissue-selective lymphocyte trafficking. Nat. Immunol. 9, 981-987 (2008).

[22] Quiding-Jarbrink, M., et al.: Differential expressionof tissue-specific adhesion molecules on human circu-

lating antibody-forming cells after systemic, enteric, and nasal immunizations. A molecular basis for the compartmentalization of effector B cell responses. J. Clin. Invest. 99, 1281-1286 (1997).

[23] Qadri, F., et al.: Enteric infections in an endemic area induce a circulating antibody-secreting cell response with homing potentials to both mucosal and systemic tissues. J. Infect. Dis. 177, 1594-1599 (1998).

[24] Kantele, A., et al.: Homing potentials of circulating lymphocytes in humans depend on the site of activation: oral, but not parenteral, typhoid vaccination induces circulating antibody-secreting cells that all bear homing receptors directing them to the gut. J. Immunol. 158, 574-579 (1997).

[25] Kantele, A., et al.: Differences in immune responses induced by oral and rectal immunizations with*Salmonella typhi* Ty21a: evidence for compartmentalization within the common mucosal immune system in humans. Infect. Immun. 66, 5630-5635 (1998).

[26] Herremans, T. M., Reimerink, J. H., Buisman, A. M., Kimman, T. G., Koopmans, M. P.: Induction of mucosal immunity by inactivated poliovirus vaccine is dependent on previous mucosal contact with live virus. J. Immunol. 162, 5011-5018 (1999).

[27] Sallusto, F., Geginat, J., Lanzavecchia, A.: Central memory and effector memory T cell subsets: function, generation, and maintenance. Annu. Rev. Immunol. 22, 745-763 (2004).

[28] Kunkel, E. J., et al.: Lymphocyte CC chemokine receptor 9 and epithelial thymus-expressed chemokine (TECK) expression distinguish the small intestinal immune compartment: epithelial expression of tissue-specific chemokines as an organizing principle in regional immunity. J. Exp. Med. 192, 761-768 (2000).

[29] Rabinowitz, K., Mayer, L.: Working out mechanisms of controlled/physiologic inflammation in the GI tract. Immunol. Res. 54, 14-24 (2012).

[30] Saurer, L., Mueller, C.: T cell-mediated immunoregulation in the gastrointestinal tract. Allergy 64, 505-519 (2009).

[31] Nagler-Anderson, C., Bhan, A. K., Podolsky, D. K., Terhorst, C.: Control freaks: immune regulatory cells. Nat. Immunol. 5, 119-122 (2004).

[32] Mucida, D., Salek-Ardakani, S.: Regulation of TH17 cells in the mucosal surfaces. J. Allergy Clin. Immunol. 123, 997-1003 (2009).

[33] Ivanov, I. I., et al.: Specific microbiota direct the differentiation of IL-17-producing T-helper cells in the mucosa of the small intestine. Cell Host Microbe 4, 337-349 (2008).

[34] Lycke, N.: Recent progress in mucosal vaccine development: potential and limitations. Nat. Rev. Immunol. 12, 592-605 (2012).

[35] Mucida, D.: T-helping colitis. Gastroenterology 141, 801-805 (2011).

[36] Sarra, M., Pallone, F., MacDonald, T. T., Monteleone, G.: IL-23/IL-17 axis in IBD. Inflamm. Bowel Dis. 16,%1-1813 (2010).

[37] McArthur, M. A., Sztein, M. B.: Heterogeneity of multifunctional IL-17A producing *S.* Typhi-specific $CD8^+$T cells in volunteers following Ty21a typhoid immunization. PLoS One 7, e38408 (2012).

[38] Crotty, S., Aubert, R. D., Glidewell, J., Ahmed, R.: Tracking human antigen-specific memory B cells: a sensitive and generalized ELISPOT system. J. Immunol. Methods 286, 111-122 (2004).

[39] Rojas, O. L., et al.: Evaluation of circulating intestinally committed memory B cells in children vaccinated with attenuated human rotavirus vaccine. Viral Immunol. 20, 300-311 (2007).

[40] Harris, A. M., et al.: Antigen-specific memory B-cell responses to *Vibrio cholerae* O1 infection in Bangladesh. Infect. Immun. 77, 3850-3856 (2009).

[41] Simon, J. K., et al.: Antigen-specific B memory cell responses to lipopolysaccharide (LPS) and invasion plasmid antigen (Ipa) B elicited in volunteers vaccinated with live-attenuated *Shigellaflexneri* 2a vaccine candidates. Vaccine 27, 565-572 (2009).

[42] Wahid, R., Simon, R., Zafar, S. J., Levine, M. M., Sztein, M. B.: Live oral typhoid vaccine Ty21a induces cross-reactive humoral immune responses against *Salmonella enterica* serovarParatyphi A and *S.* Para-

typhi B in humans. Clin. Vaccine Immunol. 19, 825–834 (2012).

[43] Wahid, R., et al.: Oral priming with *Salmonella* Typhi vaccine strain CVD 909 followed by parenteral boost with the *S.* Typhi Vi capsular polysaccharide vaccine induces CD27+IgD–*S.* Typhi–specifi c IgA and IgG B memory cells in humans. Clin. Immunol. 138, 187–200 (2011).

[44] Plotkin, S. A.: Correlates of protection induced by vaccination. Clin. Vaccine Immunol. 17, 1055–1065 (2010).

[45] Chen, R. T., et al.: Measles antibody: reevaluation of protective titers. J. Infect. Dis. 162, 1036–1042 (1990).

[46] Ilan, Y.: Oral tolerance: can we make it work? Hum. Immunol. 70, 768–776 (2009).

[47] Tsuji, N. M., Kosaka, A.: Oral tolerance: intestinal homeostasis and antigen–specific regulatory T cells. Trends Immunol. 29, 532–540 (2008).

[48] Pabst, O., Mowat, A. M.: Oral tolerance to food protein. Mucosal Immunol. 5, 232–239 (2012).

[49] Sun, C. M., et al.: Small intestine lamina propria den–dritic cells promote de novo generation of Foxp3 T reg cells via retinoic acid. J. Exp. Med. 204, 1775–1785 (2007).

[50] Scott, C. L., Aumeunier, A. M., Mowat, A. M.: Intestinal $CD103^{+}$dendritic cells: master regulators of tolerance? Trends Immunol. 32, 412–419 (2011).

[51] Sabin, A. B.: Oral poliovirus vaccine: history of its development and use and current challenge to eliminate poliomyelitis from the world. J. Infect. Dis. 151, 420–436 (1985).

[52] Heaton, P. M., Goveia, M. G., Miller, J. M., Offi t, P., Clark, H. F.: Development of a pentavalent rotavirus vaccine against prevalent serotypes of rotavirus gastroenteritis. J. Infect. Dis. 192 (Suppl 1), S17–S21 (2005).

[53] De, V. B., et al.: A rotavirus vaccine for prophylaxis of infants against rotavirus gastroenteritis. Pediatr. Infect. Dis. J. 23, S179–S182 (2004).

[54] Vesikari, T., et al.: Efficacy of human rotavirus vaccine against rotavirus gastroenteritis during the first 2 years of life in European infants: randomised, doubleblind controlled study. Lancet 370, 1757–1763 (2007).

[55] Germanier, R., Fuer, E.: Isolation and characterization of Gal E mutant Ty 21a of *Salmonella* Typhi: a candidate strain for a live, oral typhoid vaccine. J. Infect. Dis. 131, 553–558 (1975).

[56] Desai, S. N., Clemens, J. D.: An overview of cholera vaccines and their public health implications. Curr. Opin. Pediatr. 24, 85–91 (2012).

[57] Ruiz–Palacios, G. M., et al.: Safety and efficacy of an attenuated vaccine against severe rotavirus gastroenteritis. N. Engl. J. Med. 354, 11–22 (2006).

[58] Vesikari, T., et al.: Safety and efficacy of a pentavalent human–bovine (WC3) reassortant rotavirus vaccine. N. Engl. J. Med. 354, 23–33 (2006).

[59] Levine, M. M., et al.: Safety, immunogenicity, and efficacy of recombinant live oral cholera vaccines, CVD 103 and CVD 103–HgR. Lancet 2, 467–470 (1988).

[60] Clemens, J. D., et al.: Field trial of oral cholera vaccines in Bangladesh: results from three–year follow up. Lancet 335, 270–273 (1990).

[61] Madhi, S. A., et al.: Effect of human rotavirus vaccine on severe diarrhea in African infants. N. Engl. J. Med. 362, 289–298 (2010).

[62] Patriarca, P. A., Wright, P. F., John, T. J.: Factors affecting the immunogenicity of oral poliovirus vaccine in developing countries: review. Rev. Infect. Dis. 13, 926–939 (1991).

[63] Zaman, K., et al.: Successful co–administration of a human rotavirus and oral poliovirus vaccines in Bangladeshi infants in a 2–dose schedule at 12 and 16 weeks of age. Vaccine 27, 1333–1339 (2009).

[64] Armah, G. E., et al.: Efficacy of pentavalent rotavirus vaccine against severe rotavirus gastroenteritis in infants in developing countries in sub–Saharan Africa: a randomised, double–blind, placebo–controlled trial.

Lancet 376, 606-614 (2010).

[65] World Health Organization: Polio vaccines and polio immunization in the pre-eradication era: WHO position paper. Wkly. Epidemiol. Rec. 85, 213-228 (2010).

[66] Kew, O.: Reaching the last one percent: progress and challenges in global polio eradication. Curr. Opin. Virol. 2, 188-198 (2012).

[67] Onorato, I. M., et al.: Mucosal immunity induced by enhance-potency inactivated and oral polio vaccines. J. Infect. Dis. 163, 1-6 (1991).

[68] Laassri, M., et al.: Effect of different vaccination schedules on excretion of oral poliovirus vaccine strains. J. Infect. Dis. 192, 2092-2098 (2005).

[69] Paul, Y.: Why polio has not been eradicated in India despite many remedial interventions? Vaccine 27, 3700-3703 (2009).

[70] John, T. J.: Antibody response of infants in tropics to five doses of oral polio vaccine. Br. Med. J. 1, 812 (1976).

[71] Posey, D. L., Linkins, R. W., Oliveria, M. J., Monteiro, D., Patriarca, P. A.: The effect of diarrhea on oral poliovirus vaccine failure in Brazil. J. Infect. Dis. 175 (Suppl 1), S258-S263 (1997).

[72] Grassly, N. C., et al.: New strategies for the elimination of polio from India. Science 314, 1150-1153 (2006).

[73] Vashishtha, V. M., Kalra, A., John, T. J., Thacker, N., Agarwal, R. K.: Recommendations of 2nd National Consultative Meeting of Indian Academy of Pediatrics (IAP) on polio eradication and improvement of routine immunization. Indian Pediatr. 45, 367-378 (2008).

[74] Kaura, G., Biswas, T.: India reaches milestone of no cases of wild poliovirus for 12 months. BMJ344, e1328 (2012).

[75] Strebel, P. M., et al.: Epidemiology of poliomyelitis in the United States one decade after the last reported case of indigenous wild virus-associated disease. Clin. Infect. Dis. 14, 568-579 (1992).

[76] Prevots, D. R., Burr, R. K., Sutter, R. W., Murphy, T. V.: Poliomyelitis prevention in the United States. Updated recommendations of the Advisory Committee on Immunization Practices (ACIP). MM-WRRecomm. Rep. 49, 1-22 (2000).

[77] Parashar, U. D., et al.: Global mortality associated with rotavirus disease among children in 2004. J. Infect. Dis. 200 (Suppl 1), S9-S15 (2009).

[78] Steele, A. D., et al.: Rotavirus vaccines for infants in developing countries in Africa and Asia: considerations from a world health organization-sponsored consultation. J. Infect. Dis. 200 (Suppl 1), S63-S69 (2009).

[79] Meeting of the immunization Strategic Advisory Group of Experts, April 2009-conclusions and recommendations. Wkly. Epidemiol. Rec .84, 220-236 (2009).

[80] Glass, R. I., et al.: Rotavirus vaccines: current prospects and future challenges. Lancet 368, 323-332 (2006).

[81] Murphy, T. V., Smith, P. J., Gargiullo, P. M., Schwartz, B.: The first rotavirus vaccine and intussusception: epidemiological studies and policy decisions. J. Infect. Dis. 187, 1309-1313 (2003).

[82] Franco, M. A., Angel, J., Greenberg, H. B.: Immunity and correlates of protection for rotavirus vaccines. Vaccine 24, 2718-2731 (2006).

[83] Brown, K. A., Kriss, J. A., Moser, C. A., Wenner, W. J., Offit, P. A.: Circulating rotavirus-specific antibody-secreting cells (ASCs) predict the presence of rotavirus-specific ASCs in the human small intestinal lamina propria. J. Infect. Dis. 182, 1039-1043 (2000).

[84] Patel, M. M., et al.: Intussusception risk and health benefits of rotavirus vaccination in Mexico and Brazil. N. Engl. J. Med. 364, 2283-2292 (2011).

[85] Shui, I. M., et al.: Risk of intussusception following administration of a pentavalent rotavirus vaccine in US

infants. JAMA 307, 598-604 (2012).

[86] Breiman, R. F., et al.: Analyses of health outcomes from the 5 sites participating in the Africa and Asia clinical effi-cacy trials of the oral pentavalent rotavirus vaccine. Vaccine 30 (Suppl 1), A24-A29 (2012).

[87] Block, S. L., et al.: Efficacy, immunogenicity, and safety of a pentavalent human-bovine (WC3) reassortant rotavirus vaccine at the end of shelf life. Pediatrics 119, 11-18 (2007).

[88] Vesikari, T., et al.: Efficacy of a pentavalent rotavirus vaccine in reducing rotavirus-associated health care utilization across three regions (11 countries). Int. J. Infect. Dis. 11 (Suppl 2), S29-S35 (2007).

[89] Ciarlet, M., Schodel, F.: Development of a rotavirus vaccine: clinical safety, immunogenicity, and efficacy of the pentavalent rotavirus vaccine. RotaTeq. Vaccine 27 (Suppl 6), G72-G81 (2009).

[90] Zaman, K., et al.: Efficacy of pentavalent rotavirus vaccine against severe rotavirus gastroenteritis in infants in developing countries in Asia: a randomised, double-blind, placebo-controlled trial. Lancet 376, 615-623 (2010).

[91] Rennels, M. B., Ward, R. L., Mack, M. E., Zito, E. T.: Concurrent oral poliovirus and rhesus-human reassortant rotavirus vaccination: effects on immune responses to both vaccines and on efficacy of rotavirus vaccines. The US Rotavirus Vaccine Efficacy Group. J. Infect. Dis. 173, 306-313 (1996).

[92] Ciarlet, M., et al.: Concomitant use of the oral pentavalent human-bovine reassortant rotavirus vaccine and oral poliovirus vaccine. Pediatr. Infect. Dis. J. 27, 874-880 (2008).

[93] Cholera vaccines: WHO position paper. Wkly. Epidemiol. Rec. 85, 117-128 (2010).

[94] Harris, J. B., LaRocque, R. C., Qadri, F., Ryan, E. T., Calderwood, S. B.: Cholera. Lancet 379, 2466-2476 (2012).

[95] Levine, M. M., et al.: Duration of infection-derived immunity to cholera. J. Infect. Dis. 143, 818-820 (1981).

[96] Levine, M. M., et al.: Immunity of cholera in man: relative role of antibacterial versus antitoxic immunity. Trans. R. Soc. Trop. Med. Hyg. 73, 3-9 (1979).

[97] Qadri, F., et al.: Antigen-specific immunoglobulin A antibodies secreted from circulating B cells are an effective marker for recent local immune responses in patients with cholera: comparison to antibody-secreting cell responses and other immunological markers. Infect. Immun. 71, 4808-4814 (2003).

[98] Qadri, F., et al.: Comparison of immune responses in patients infected with Vibrio cholerae O139 and O1. Infect. Immun. 65, 3571-3576 (1997).

[99] Glass, R. I., et al.: Seroepidemiological studies of El Tor cholera in Bangladesh: association of serum antibody levels with protection. J. Infect. Dis. 151, 236-242 (1985).

[100] Mosley, W. H., McCormack, W. M., Ahmed, A., Chowdhury, A. K., Barui, R. K.: Report of the 1966-67 cholera vaccine field trial in rural East Pakistan. 2. Results of the serological surveys in the study population-the relationship of case rate to antibody titre and an estimate of the inapparent infection rate with *Vibrio cholerae*. Bull. World Health Organ. 40, 187-197 (1969).

[101] Mosley, W. H., Ahmad, S., Benenson, A. S., Ahmed, A.: The relationship of vibriocidal antibody titre to susceptibility to cholera in family contacts of cholera patients. Bull. World Health Organ. 38, 777-785 (1968).

[102] Uddin, T., et al.: Mucosal immunologic responses in cholera patients in Bangladesh. Clin. Vaccine Immunol. 18, 506-512 (2011).

[103] Jayasekera, C. R., et al.: Cholera toxin-specific mem-ory B cell responses are induced in patients with dehydrating diarrhea caused by *Vibrio cholerae* O1. J. Infect. Dis. 198, 1055-1061 (2008).

[104] Patel, S. M., et al.: Memory B cell responses to *Vibrio cholerae* O1 lipopolysaccharide are associated with protection against infection from household contacts of patients with cholera in Bangladesh. Clin. Vaccine Immunol. 19, 842-848 (2012).

[105] Charles, R. C., Ryan, E. T.: Cholera in the 21st century. Curr. Opin. Infect. Dis. 24, 472-477

(2011).

[106] van Loon, F. P., et al.: Field trial of inactivated oral cholera vaccines in Bangladesh: results from 5 years of follow-up. Vaccine 14, 162-166 (1996).

[107] Mahalanabis, D., et al.: A randomized, placebo-controlled trial of the bivalent killed, whole-cell, oral cholera vaccine in adults and children in a cholera endemic area in Kolkata, India. PLoS One 3, e2323 (2008).

[108] Saha, A., et al.: Safety and immunogenicity study of a killed bivalent (O1 and O139) whole-cell oral cholera vaccine Shanchol, in Bangladeshi adults and children as young as 1 year of age. Vaccine 29, 8285-8292 (2011).

[109] Anh, D. D., et al.: Safety and immunogenicity of a reformulated Vietnamese bivalent killed, whole-cell, oral cholera vaccine in adults. Vaccine 25, 1149-1155 (2007).

[110] Kanungo, S., et al.: Immune responses following one and two doses of the reformulated, bivalent, killed, whole-cell, oral cholera vaccine among adults and children in Kolkata, India: a randomized, placebo-controlled trial. Vaccine 27, 6887-6893 (2009).

[111] Alam, M. M., et al.: Antigen-specific memory B-cell responses in Bangladeshi adults after one-or two-dose oral killed cholera vaccination and comparison with responses in patients with naturally acquired cholera. Clin. Vaccine Immunol. 18, 844-850 (2011).

[112] Clemens, J. D., et al.: Biotype as determinant of natural immunisingeffect of cholera. Lancet 337, 883-884 (1991).

[113] Ali, M., Emch, M., Park, J. K., Yunus, M., Clemens, J.: Natural cholera infection - derived immunity in an endemic setting. J. Infect. Dis. 204, 912-918 (2011).

[114] Tacket, C. O., et al.: Randomized, double-blind, placebo-controlled, multicentered trial of the efficacy of a single dose of live oral cholera vaccine CVD 103-HgR in preventing cholera following challenge with *Vibrio cholerae* O1 El tor inaba three months after vaccination. Infect. Immun. 67, 6341-6345 (1999).

[115] Cryz, S. J., Levine, M. M., Kaper, J. B., Furer, E., Althaus, B.: Randomized double - blind placebo controlled trial to evaluate the safety and immunogenicity of the live oral cholera vaccine strain CVD 103-HgR in Swiss adults. Vaccine 8, 577-580 (1990).

[116] Tacket, C. O., et al.: Onset and duration of protective immunity in challenged volunteers after vaccination with live oral cholera vaccine CVD 103-HgR. J. Infect. Dis. 166, 837-841 (1992).

[117] Suharyono, S. C., et al.: Safety and immunogenicity of single-dose live oral cholera vaccine CVD 103HgR in 5-9-year-old Indonesian children. Lancet 340, 689-694 (1992).

[118] Simanjuntak, C. H., et al.: Safety, immunogenicity, and transmissibility of single-dose live oral cholera vaccine strain CVD 103-HgR in 24-to 59-month-old Indonesian children. J. Infect. Dis. 168, 1169-1176 (1993).

[119] Gotuzzo, E., et al.: Safety, immunogenicity, and excretion pattern of single - dose live oral cholera vaccine CVD 103-HgR in Peruvian adults of high and low socioeconomic levels. Infect. Immun. 61, 3994-3997 (1993).

[120] Su - Arehawaratana, P., et al.: Safety and immunoge - nicity of different immunization regimens of CVD 103-HgR live oral cholera vaccine in soldiers and civilians in Thailand. J. Infect. Dis. 165, 1042 - 1048 (1992).

[121] Richie, E. E., et al.: Efficacy trial of single-dose live oral cholera vaccine CVD 103-HgR in North Jakarta, Indonesia, a cholera-endemic area. Vaccine 18, 2399-2410 (2000).

[122] Ali, M., et al.: Herd immunity conferred by killed oral cholera vaccines in Bangladesh: a reanalysis. Lancet 366, 44-49 (2005).

[123] Perry, R. T., et al.: A single dose of live oral cholera vaccine CVD 103-HgR is safe and immunogenic in HIV-infected and HIV-noninfected adults in Mali. Bull. World Health Organ. 76, 63-71 (1998).

[124] Shin, S., Desai, S. N., Sah, B. K., Clemens, J. D.: Oral vaccines against cholera. Clin. Infect. Dis. 52, 1343-1349 (2011).

[125] Levine, M. M.: Enteric infections and the vaccines to counter them: future directions. Vaccine 24, 3865-3873 (2006).

[126] Calain, P., et al.: Can oral cholera vaccination play a role in controlling a cholera outbreak? Vaccine 22, 2444-2451 (2004).

[127] Garcia, L., et al.: The vaccine candidate *Vibrio cholerae* 638 is protective against cholera in healthy volunteers. Infect. Immun. 73, 3018-3024 (2005).

[128] Mahalanabis, D., et al.: Randomized placebo con-trolled human volunteer trial of a live oral cholera vaccine VA1. 3 for safety and immune response. Vaccine 27, 4850-4856 (2009).

[129] Ochiai, R. L., et al.: A study of typhoid fever in five Asian countries: disease burden and implications for controls. Bull. World Health Organ. 86, 260-268 (2008).

[130] Lin, F. Y., et al.: The epidemiology of typhoid fever in the Dong Thap Province. Mekong Delta region ofVietnam. Am. J. Trop. Med. Hyg. 62, 644-648 (2000).

[131] Sinha, A., et al.: Typhoid fever in children aged less than 5 years. Lancet 354, 734-737 (1999).

[132] Brooks, W. A., et al.: Bacteremic typhoid fever in children in an urban slum, Bangladesh. Emerg. Infect. Dis. 11, 326-329 (2005).

[133] Connor, B. A., Schwartz, E.: Typhoid and paraty-phoid fever in travellers. Lancet Infect. Dis. 5, 623-628 (2005).

[134] Charles, R. C., et al.: Characterization of anti-*Salmonella enterica* serotype Typhi antibody responses in bacteremic Bangladeshi patients using Immuno-affinity Proteomic-based Technology (IPT). Clin. Vaccine Immunol. 17, 1188-1195 (2010).

[135] Ortiz, V., Isibasi, A., Garcia Ortigoza, E., Kumate, J.: Immunoblot detection of class - specific humoral immune response to outer membrane proteins isolated from *Salmonella typhi* in humans with typhoid fever. J. Clin. Microbiol. 27, 1640-1645 (1989).

[136] Sheikh, A., et al.: *Salmonella enterica* serovarTyphi - specific immunoglobulin A antibody responses in plasma and antibody in lymphocyte supernatant specimens in Bangladeshi patients with suspected typhoid fever. Clin. Vaccine Immunol. 16, 1587-1594 (2009).

[137] Lanata, C. F., et al.: Vi serology in detection of chronic *Salmonella typhi* carriers in an endemic area. Lancet 2, 441-443 (1983).

[138] Losonsky, G. A., et al.: Development and evaluation of an enzyme-linked immunosorbent assay for serum Vi antibodies for detection of chronic *Salmonella typhi* carriers. J. Clin. Microbiol. 25, 2266-2269 (1987).

[139] Mirza, N. B., Wamola, I. A., Estambale, B. A., Mbithi, E., Poillet, M.: Typhim Vi vaccine against typhoid fever: a clinical trial in Kenya. East Afr. Med. J. 72, 162-164 (1995).

[140] Murphy, J. R., et al.: Characteristics of humoral and cellular immunity to *Salmonella typhi* in residents of typhoid-endemic and typhoid-free regions. J. Infect. Dis. 156, 1005-1009 (1987).

[141] Butler, T., Ho, M., Acharya, G., Tiwari, M., Gallati, H.: Interleukin-6, gamma interferon, and tumor necrosis factor receptors in typhoid fever related to outcome of antimicrobial therapy. Antimicrob. Agents Chemother. 37, 2418-2421 (1993).

[142] Levine, M. M., et al.: Progress in vaccines to prevent typhoid fever. Rev. Infect. Dis. 11, S552-S567 (1989).

[143] Black, R. E., et al.: Efficacy of one or two doses of Ty21a *Salmonella typhi* vaccine in enteric-coated capsules in a controlled field trial. Chilean Typhoid Committee. Vaccine 8, 81-84 (1990).

[144] Levine, M. M., Ferreccio, C., Cryz, S., Ortiz, E.: Comparison of enteric-coated capsules and liquid formulation of Ty21atyphoid vaccine inrandomised controlled field trial. Lancet 336, 891-894 (1990).

[145] Wahdan, M. H., Serie, C., Cerisier, Y., Sallam, S., Germanier, R.: A controlled field trial of live

Salmonella Typhi strain Ty 21a oral vaccine against typhoid: three-year results. J. Infect. Dis. 145, 292-295 (1982).
[146] Simanjuntak, C., et al.: Oral immunisation against typhoid fever in Indonesia with Ty21a vaccine. Lancet 338, 1055-1059 (1991).
[147] Levine, M. M., Ferreccio, C., Black, R. E., Germanier, R.: Large-scale field trial of Ty21a live oral typhoid vaccine in enteric-coated capsule formulation. Lancet 1, 1049-1052 (1987).
[148] Levine, M. M., et al.: Duration of efficacy of Ty21a, attenuated *Salmonella* Typhi live oral vaccine. Vaccine 17 (Suppl 2), S22-S27 (1999).
[149] Ferreccio, C., Levine, M. M., Rodriguez, H., Contreras, R.: Comparative efficacy of two, three, or four doses of TY21a live oral typhoid vaccine in enteric-coated capsules: a field trial in an endemic area. J. Infect. Dis. 159, 766-769 (1989).
[150] Black, R., et al.: Immunogenicity of Ty21a attenuated *Salmonella* Typhi given with sodium bicarbonate or in enteric-coated capsules. Dev. Biol. Stand. 53, 9-14 (1983).
[151] Kantele, A.: Antibody-secreting cells in the evaluation of the immunogenicity of an oral vaccine. Vaccine 8, 321-326 (1990).
[152] D'Amelio, R., et al.: Comparative analysis of immunological responses to oral (Ty21a) and parenteral (TAB) typhoid vaccines. Infect. Immun. 56, 2731-2735 (1988).
[153] Tagliabue, A., et al.: Cellular immunity against *Salmonella* Typhi after live oral vaccine. Clin. Exp. Immunol. 62, 242-247 (1985).
[154] Salerno-Goncalves, R., Fernandez-Vina, M., Lewinsohn, D. M., Sztein, M. B.: Identification of a human HLA-E-restricted $CD8^+$T cell subset in volunteers immunized with *Salmonella enterica* serovarTyphi strain Ty21a typhoid vaccine. J. Immunol. 173, 5852-5862 (2004).
[155] Salerno-Goncalves, R., Pasetti, M. F., Sztein, M. B.: Characterization of $CD8^+$ effector T cell responses in volunteers immunized with *Salmonella enterica* serovarTyphi strain Ty21a typhoid vaccine. J. Immunol. 169, 2196-2203 (2002).
[156] Sztein, M. B.: Cell-mediated immunity and antibody responses elicited by attenuated *Salmonella enterica* SerovarTyphi strains used as live oral vaccines in humans. Clin. Infect. Dis. 45 (Suppl 1), S15-S19 (2007).
[157] Salerno-Goncalves, R., Wahid, R., Sztein, M. B.: Ex Vivo kinetics of early and long-term multifunctional human leukocyte antigen E-specific $CD8^+$cells in volunteers immunized with the Ty21a typhoid vaccine. Clin. Vaccine Immunol. 17, 1305-1314 (2010).
[158] Salerno-Goncalves, R., Wahid, R., Sztein, M. B.: Immunization of volunteers with *Salmonella enterica* serovarTyphi strain Ty21a elicits the oligoclonal expansion of $CD8^+$ T cells with predominant Vbeta repertoires. Infect. Immun. 73, 3521-3530 (2005).
[159] Czerkinsky, C., Holmgren, J.: Enteric vaccines for the developing world: a challenge for mucosal immunology. Mucosal Immunol. 2, 284-287 (2009).
[160] Holmgren, J., Svennerholm, A. M.: Vaccines against mucosal infections. Curr. Opin. Immunol. 24, 343-353 (2012).
[161] Patel, M., et al.: Oral rotavirus vaccines: how well will they work where they are needed most? J. Infect. Dis. 200 (Suppl 1), S39-S48 (2009).
[162] Qadri, F., Bhuiyan, T. R., Sack, D. A., Svennerholm, A. M.: Immune responses and protection in children in developing countries induced by oral vaccines. Vaccine 31, 452-460 (2012).
[163] Cooper, P. J., et al.: Albendazole treatment of children with ascariasis enhances the vibriocidal antibody response to the live attenuated oral cholera vaccine CVD 103-HgR. J. Infect. Dis. 182, 1199-1206 (2000).
[164] Lagos, R., et al.: Effect of small bowel bacterial overgrowth on the immunogenicity of single-dose live oral

cholera vaccine CVD 103-HgR. J. Infect. Dis. 180, 1709-1712 (1999).
[165] Albert, M. J., et al.: Supplementation with zinc, but not vitamin A, improves seroconversion to vibriocidal antibody in children given an oral cholera vaccine. J. Infect. Dis. 187, 909-913 (2003).
[166] Ahmed, T., Svennerholm, A. M., Al, T. A., Sultana, G. N., Qadri, F.: Enhanced immunogenicity of an oral inactivated cholera vaccine in infants in Bangladesh obtained by zinc supplementation and by temporary withholding breast-feeding. Vaccine 27, 1433-1439 (2009).
[167] Isolauri, E., Joensuu, J., Suomalainen, H., Luomala, M., Vesikari, T.: Improved immunogenicity of oral D x RRVreassortant rotavirus vaccine by *Lactobacillus casei* GG. Vaccine 13, 310-312 (1995).
[168] Waggie, Z., et al.: Randomized trial of type 1 and type 3 oral monovalent poliovirus vaccines in newborns in Africa. J. Infect. Dis. 205, 228-236 (2012).
[169] Patel, M., Steele, A. D., Parashar, U. D.: Influence of oral polio vaccines on performance of the monovalent and pentavalent rotavirus vaccines. Vaccine 30 (Suppl 1), A30-A35 (2012).
[170] Kau, A. L., Ahern, P. P., Griffin, N. W., Goodman, A. L., Gordon, J. I.: Human nutrition, the gut microbiome and the immune system. Nature 474, 327-336 (2011).
[171] Gerson, C. D., Kent, T. H., Saha, J. R., Siddiqi, N., Lindenbaum, J.: Recovery of small-intestinal structure and function after residence in the tropics. II. Studies in Indians and Pakistanis living in New York City. Ann. Intern. Med. 75, 41-48 (1971).
[172] Levine, M. M.: Immunogenicity and efficacy of oral vaccines in developing countries: lessons from a live cholera vaccine. BMC Biol. 8, 129 (2010).
[173] Black, R. E., et al.: Maternal and child undernutrition: global and regional exposures and health consequences. Lancet 371, 243-260 (2008).
[174] Overbeck, S., Rink, L., Haase, H.: Modulating the immune response by oral zinc supplementation: a single approach for multiple diseases. Arch. Immunol. Ther. Exp. (Warsz.) 56, 15-30 (2008).
[175] Kaufman, D. R., et al.: Vitamin A deficiency impairs vaccine-elicited gastrointestinal immunity. J. Immunol. 187, 1877-1883 (2011).
[176] Semba, R. D., et al.: Depressed immune response to tetanus in children with vitamin A deficiency. J. Nutr. 122, 101-107 (1992).
[177] Qadri, F., et al.: Suppressive effect of zinc on antibody response to cholera toxin in children given the killed, B subunit-whole cell, oral cholera vaccine. Vaccine 22, 416-421 (2004).
[178] Newton, S., et al.: Vitamin a supplementation does not affect infants' immune responses to polio and tetanus vaccines. J. Nutr. 135, 2669-2673 (2005).
[179] Bahl, R., et al.: Effect of vitamin A administered at Expanded Program on Immunization contacts on antibody response to oral polio vaccine. Eur. J. Clin. Nutr. 56, 321-325 (2002).
[180] Semba, R. D., et al.: Integration of vitamin A supplementation with the expanded program on immunization does not affect seroconversion to oral poliovirus vaccine in infants. J. Nutr. 129, 2203-2205 (1999).
[181] Walker, A.: Breast milk as the gold standard for protective nutrients. J. Pediatr. 156, S3-S7 (2010).
[182] Newburg, D. S., et al.: Role of human-milk lactadherin in protection against symptomatic rotavirus infection. Lancet 351, 1160-1164 (1998).
[183] Goveia, M. G., DiNubile, M. J., Dallas, M. J., Heaton, P. M., Kuter, B. J.: Efficacy of pentavalent human-bovine (WC3) reassortant rotavirus vaccine based on breastfeeding frequency. Pediatr. Infect. Dis. J. 27, 656-658 (2008).
[184] Moon, S. S., et al.: Inhibitory effect of breast milk on infectivity of live oral rotavirus vaccines. Pediatr. Infect. Dis. J. 29, 919-923 (2010).
[185] Yatsunenko, T., et al.: Human gut microbiome viewed across age and geography. Nature 486, 222-227 (2012).

[186] Turnbaugh, P. J., et al.: A core gut microbiome in obese and lean twins. Nature 457, 480–484 (2009).

[187] Bjorksten, B.: Diverse microbial exposure–consequences for vaccine development. Vaccine 30, 4336–4340 (2012).

[188] Sartono, E., et al.: Oral polio vaccine influences the immune response to BCG vaccination. A natural experiment. PLoS One 5, e10328 (2010).

[189] Kirkpatrick, B. D., et al.: The novel oral typhoid vaccine M01ZH09 is well tolerated and highly immunogenic in 2 vaccine presentations. J. Infect. Dis. 192, 360–366 (2005).

[190] Kirkpatrick, B. D., et al.: Evaluation of *Salmonella enterica* serovarTyphi (Ty2 aroC–ssaV–) M01ZH09, with a defined mutation in the *Salmonella* pathogenicity island 2, as a live, oral typhoid vaccine in human volunteers. Vaccine 24, 116–123 (2006).

[191] Lyon, C. E., et al.: In a randomized, double–blinded, placebo–controlled trial, the single oral dose typhoid vaccine, M01ZH09, is safe and immunogenic at doses up to 1. 7 x 10 (10) colony–forming units. Vaccine 28, 3602–3608 (2010).

[192] Tran, T. H., et al.: A randomised trial evaluating the safety and immunogenicity of the novel single oral dose typhoid vaccine M01ZH09 in healthy Vietnamese children. PLoS One 5, e11778 (2010).

[193] Puzzling diversity of rotaviruses. Lancet 335, 573–575 (1990).

[194] Tacket, C. O., Levine, M. M.: CVD 908, CVD 908–htrA, and CVD 909 live oral typhoid vaccines: a logical progression. Clin. Infect. Dis. 45 (Suppl 1), S20–S23 (2007).

[195] Levine, M. M., Kotloff, K. L., Barry, E. M., Pasetti, M. F., Sztein, M. B.: Clinical trials of *Shigella* vaccines: two steps forward and one step back on a long, hard road. Nat. Rev. Microbiol. 5, 540–553 (2007).

[196] McKenzie, R., et al.: Safety and immunogenicity of an oral, inactivated, whole–cell vaccine for *Shigellasonnei*: preclinical studies and a Phase I trial. Vaccine 24, 3735–3745 (2006).

[197] Mel, D. M., Terzin, A. L., Vuksic, L.: Studies on vaccination against bacillary dysentery. 3. Effective oral immunization against *Shigellaflexneri* 2a in a field trial. Bull. World. Health Organ. 32, 647–655 (1965).

[198] Porter, C. K., Thura, N., Ranallo, R. T., Riddle, M. S.: The Shigella human challenge model. Epidemiol. Infect. 141, 223–232 (2013). doi: 10. 1017/S0950268812001677.

[199] Barzu, S., Fontaine, A., Sansonetti, P., Phalipon, A.: Induction of a local anti–IpaC antibody response in mice by use of a *Shigellaflexneri* 2a vaccine candidate: implications for use of IpaC as a protein carrier. Infect. Immun. 64, 1190–1196 (1996).

[200] Coster, T. S., et al.: Vaccination against shigellosis with attenuated *Shigellaflexneri* 2a strain SC602. Infect. Immun. 67, 3437–3443 (1999).

[201] Rahman, K. M., et al.: Safety, dose, immunogenicity, and transmissibility of an oral live attenuated *Shigellaflexneri 2a* vaccine candidate (SC602) among healthy adults and school children in Matlab, Bangladesh. Vaccine 29, 1347–1354 (2011).

[202] Harro, C., et al.: A combination vaccine consisting of three live attenuated enterotoxigenic *Escherichia coli* strains expressing a range of colonization factors and heat–labile toxin subunit B is well tolerated and immunogenic in a placebo–controlled double–blind phase I trial in healthy adults. Clin. Vaccine Immunol. 18, 2118–2127 (2011).

[203] Darsley, M. J., et al.: The oral, live attenuated enterotoxigenic *Escherichia coli* vaccine ACE527 reduces the incidence and severity of diarrhea in a human challenge model of diarrheal disease. Clin. Vaccine Immunol. 19, 1921–1931 (2012).

[204] Tacket, C. O., Sztein, M. B., Losonsky, G. A., Wasserman, S. S., Estes, M. K.: Humoral, mucosal, and cellular immune responses to oral Norwalk virus–like particles in volunteers. Clin. Immunol. 108, 241–247 (2003).

[205] el-Kamary, S. S., et al.: Adjuvanted intranasal Norwalk virus-like particle vaccine elicits antibodies and antibody-secreting cells that express homing receptors for mucosal and peripheral lymphoid tissues. J. Infect. Dis. 202, 1649-1658 (2010).

[206] Atmar, R. L., et al.: Norovirus vaccine against experimental human Norwalk Virus illness. N. Engl. J. Med. 365, 2178-2187 (2011).

[207] Herbst-Kralovetz, M., Mason, H. S., Chen, Q.: Norwalk virus-like particles as vaccines. Expert Rev. Vaccines 9, 299-307 (2010).

[208] Tan, M., Jiang, X.: Norovirus P particle: a subviral nanoparticle for vaccine development against norovirus, rotavirus and influenza virus. Nanomedicine (Lond.) 7, 889-897 (2012).

[209] Norton, E. B., Lawson, L. B., Freytag, L. C., Clements, J. D.: Characterization of a mutant *Escherichia coli* heat-labile toxin, LT (R192G/L211A), as a safe and effective oral adjuvant. Clin. Vaccine Immunol. 18, 546-551 (2011).

[210] Brandtzaeg, P.: Food allergy: separating the science from the mythology. Nat. Rev. Gastroenterol. Hepatol. 7, 380-400 (2010).

[211] Lewis, D. J., et al.: Transient facial nerve paralysis (Bell's palsy) following intranasal delivery of a genetically detoxified mutant of *Escherichia coli* heat labile toxin. PLoS One 4, e6999 (2009).

[212] Woodland, D. L.: Jump-starting the immune system: prime-boosting comes of age. Trends Immunol. 25, 98-104 (2004).

[213] Galen, J. E., et al.: Mucosal immunization with attenuated *Salmonella enterica* serovarTyphi expressing protective antigen of anthrax toxin (PA83) primes monkeys for accelerated serum antibody responses to parenteral PA83 vaccine. J. Infect. Dis. 199, 326-335 (2009).

[214] Azevedo, M. S., et al.: An oral versus intranasal prime/boost regimen using attenuated human rotavirus or VP2 and VP6 virus-like particles with immunostimulating complexes influences protection and antibody-secreting cell responses to rotavirus in a neonatal gnotobiotic pig model. Clin. Vaccine Immunol. 17, 420-428 (2010).

[215] Schoub, B. D., Mphahlele, M. J., Ngcobo, N. J., Hoosen, A. A., Meheus, A.: Introducing new vaccines into the South African national immunisationprogramme-a case study. Vaccine 30 (Suppl 3), C1-C2 (2012).

[216] Cohen, D., et al.: Double-blind vaccine-controlled randomised efficacy trial of an investigational *Shigellasonnei* conjugate vaccine in young adults. Lancet 349, 155-159 (1997).

[217] Pavot, V., Rochereau, N., Genin, C., Verrier, B., Paul, S.: New insights in mucosal vaccine development. Vaccine 30, 142-154 (2012).

(马维民译)

第4章 儿科免疫学和疫苗学

Sofia Ygberg 和 Anna Nilsson①

摘要

在婴儿中诱导保护性免疫有可能降低儿童感染的发病率和死亡率，但在新生儿中很难诱导持久的保护性免疫。本章介绍了新生儿获得性免疫系统的基础。提供了一份关于目前全球范围内儿童免疫方案的最新情况，以及已知的儿童和青年人疫苗的副作用。最后还提供了关于如何改进儿童疫苗接种和未来急需疫苗的数据。

4.1 引言

新生儿出生后的头几个月是感染高风险时期，因此在生命早期通过免疫诱导获得性免疫是有用的。然而长期以来人们都知道，无论是在免疫后还是在感染后，都很难诱导新生儿产生持久的保护性免疫。理论上，这些问题中的一些可以通过对孕妇进行疫苗接种来提高母体保护性抗体对孩子的传递，但这可能会影响婴儿的获得性免疫应答。本章将描述诱导适当疫苗应答的困难，并将其置于新生儿免疫系统的环境中，该系统在许多其他方面能够保护婴儿免受疾病侵害[1,2]。

4.2 天然免疫和获得性免疫的细胞成分

新生儿免疫系统的一个主要缺点是大多数B细胞和T细胞都是幼稚的，与物种无关。尽管大多数细胞和可溶性因子出现在胎儿早期，但数量、相对比例和活性状态与成人不同[3]。目前对人类婴儿的免疫缺乏认识，因为大多数新生儿免疫研究都是从啮齿动物模型中产生的，很难将这些发现转化为人类的研究。因为与人类婴儿相比，新生小鼠发育不良[4]。从脐带血中分离的细胞研究提供了新的见解，但是这些数据可能不能代表新生儿免疫系统中的循环细胞。因此，从脐带血获得的数据应该与从婴儿获得的数据相结合[5]。

4.3 新生儿和儿童抗原呈递细胞

单核细胞和树突状细胞（DCs）作为抗原呈递细胞发挥作用，是天然免疫的关键参与者，但也负责启动获得性免疫反应。外周血中发现两种树突状细胞亚型：髓样DC（mDCs）和浆细胞样DC（pDCs）。mDC是主要的抗原呈递细胞，通过释放细胞因子，如IL-12、IL-6、BAFF和April在B细胞分化中发挥关键作用，这些细胞因子驱动产生抗体B细胞形成。另一方面，pDCs产生干扰素，从而在抗病毒免疫中发挥重要作用[6]。儿童外周血中树突状细胞比例存在与年龄相关的差异，婴儿的树突状细胞数量远远高于较大儿童，这可能反映了在保护性获得性免疫反

① S. Ygberg，MD，博士 · A. Nilsson，MD，博士（✉）（瑞典斯德哥尔摩阿斯特里德 · 林德格伦儿童医院卡罗林斯卡研究所妇女和儿童健康系，E-mail：anna. nilsson. 1@ ki. se）

应开始之前，树突状细胞对早期预防病毒性疾病的重要性[7]。然而，这是以生命早期效率较低的B细胞活化为代价的。

一些脐带血树突状细胞的研究已经证实了一种不成熟的表型，其CD40、协同刺激分子CD80/CD86或MHC Ⅱ类分子的基础表达较低或无基础表达[8-10]。在功能上，其转化为人类新生儿对大多数刺激的次优DC应答[11]。此外，树突状细胞上的Toll样受体（TLR）途径对于诱导获得性免疫应答的重要性是显而易见的，并且已得到很好的证实。尽管在脐带血树突状细胞和成人细胞上有类似的TLRs表达，但脐带血树突状细胞对TLR激动剂的反应能力也显著降低，其特点是促炎性Th1细胞因子TNF-α和IFN产量低[12,13]。然而，最近的数据表明，新生儿TLR介导的损伤是选择性的，因为TLR8激动剂R848能够在脐带血树突状细胞中诱导与成人细胞相当强大的免疫应答。这一发现也已在婴儿细胞中得到证实，可能对新生儿疫苗研究中佐剂的选择具有重要意义[14,15]。

单核细胞和抗原呈递细胞的先天反应在生命的第一年内发育[15,16]。外周血单核细胞和树突状细胞表型分析表明，循环树突状细胞在6个月时具有成人样表型。出生时TLR刺激后的细胞因子产生偏向于Th2反应，产生IL-6、IL-8和IL-10，并且Th1-极化的IL-12p70细胞因子水平较低[17]。然而3个月时IL-6水平与成年人相当。对于IL-10，12个月时的产量也明显较高，IL-8产量也呈现同样趋势[16]。

因此除了TLR8外，幼儿在抗原呈递细胞成熟和这些细胞对细菌和病毒抗原的应答能力方面都受到损害。使用TLR8配体作为疫苗佐剂可能是一个机会窗口。

4.4 新生儿和儿童T细胞

婴儿外周血T淋巴细胞亚群与成人不同。出生时，$CD3^+$T淋巴细胞的绝对数量逐渐增加，从2岁开始下降到与成人相似的水平[18]。辅助$CD4^+$T细胞和细胞毒性$CD8^+$T细胞在出生后头几个月也会增加，在9~15个月后会下降[19]。流式细胞仪分析显示，血液中有几个$CD4^+$记忆性细胞群，出生时只有中央记忆性T细胞群。由于抗原刺激，效应记忆性辅助T细胞在第一年增加到与成人相当的水平，并在儿童时期保持稳定。最近描述的CXCR5+记忆T细胞群，也被定义为滤泡辅助T细胞，出生时不存在，但在出生后第一年随着血清IgA和IgG增加，数量也在增加[19]。滤泡辅助T细胞首先被描述为能够有效地支持次级淋巴器官中转换B细胞分化细胞，随后产生IgA和IgG[20]。调节性T细胞（Treg）绝对数量在出生后第一个月增加，类似于$CD4^+$记忆性T细胞，此后保持稳定。新生儿Treg具有很强的免疫抑制活性，抑制抗原特异性T细胞增殖和IFN-γ产生[21]，它们可能在以后的生活中调节记忆性$CD4^+$T细胞库发育[22]。

新生儿T细胞免疫功能存在固有缺陷。初始$CD4^+$T细胞通过T细胞受体（TCR）-CD3发送信号的一个关键特征是细胞表面CD40配体的上调。新生儿$CD4^+$T细胞在TCR-CD3激活后表达CD40配体能力降低，从而对抗体产生、免疫球蛋白转换和记忆性B细胞生成产生负面影响[23]。在许多环境中对新生儿免疫后辅助T细胞应答进行了调查[1]，一些因素（抗原剂量、佐剂、免疫途径）是否会引起以Th1或Th2为主的应答。除了卡介苗和全细胞百日咳疫苗外，目前大多数儿童疫苗都会引起以Th2为主的应答[24]。

因此，早期T细胞应答偏向Th2应答，新生儿辅助T细胞支持B细胞分化和抗体产生的能力降低。

4.5 新生儿和儿童B细胞

几项研究表明，在生命前5年，外周血B细胞亚群随年龄变化。随着年龄增长，B细胞总数显著减少。最引人注目的是，从婴儿期主要是初始和过渡的B细胞池转变为较大的儿童和成人记忆性B细胞比例增加。与成年人相比，婴儿体内过渡B细胞增多，这可能会在生命早期弥合先天免疫和获得性免疫之间的差距。过渡B细胞在TLR 9刺激下产生IgM，这可能是出生时抵御细菌第一道防线的重要机制[25]。记忆性B细胞池扩展在生命的第一年最为明显，之后，绝对数量随着时间推移是稳定的。总的来说，这些发现表明，随着年龄增长，B细胞总数减少主要与骨髓（BM）中过渡细胞和幼稚细胞的输出量减少有关[26,27]。

用CD27作为人记忆B细胞的替代标志，结合IgD表面表达，对几种记忆B细胞群进行了表征。经典转换记忆B细胞在婴儿期增加，在5~10岁达到高峰[27,28]。经典转换记忆B细胞分化发生在次级淋巴器官生发中心（GC）；免疫组织化学研究表明，在出生时GCs不存在，在12~24个月时逐渐发育到成人大小[29]。有趣的是，大肠杆菌肠道定居促进了婴儿$CD27^+$记忆池的早期发育（0~4个月）。IgM记忆亚群在出生前后逐渐出现在血液循环中，2岁时达到成人水平[31]。几项研究表明，IgM记忆B细胞在感染和免疫后都能对肺炎链球菌起到保护作用[32]。

对婴儿和儿童的终末分化浆细胞池研究较少。在学龄前儿童中，浆细胞室大小与成人相似[33]。然而，在KLH-NP免疫的小鼠模型中已经发现，与成年鼠相比，新生小鼠浆细胞存活率受损[34]。与成年鼠相比，BM基质细胞的支持网络对新生小鼠浆细胞存活因子支持能力较弱。

总之，出生后B细胞室在微生物抗原的刺激下成熟，也需要淋巴器官和骨髓成熟（图4.1）。

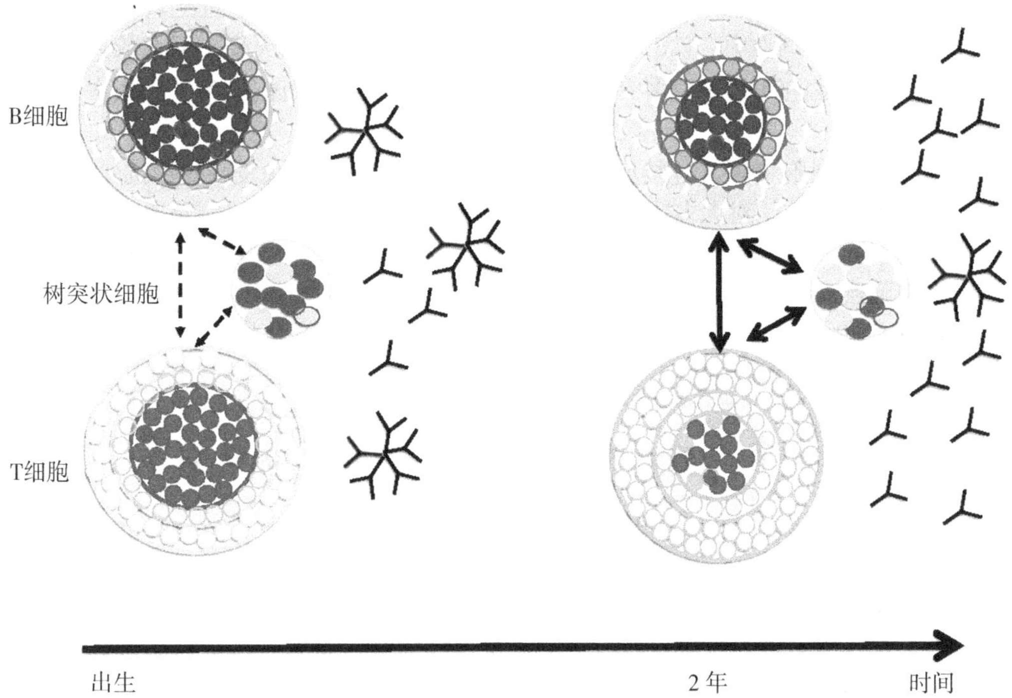

图4.1 参与获得性免疫反应的外周血细胞群体的成熟发生在儿童的头两年

出生时，转化（橙色）和幼稚B细胞（深橙色）最为丰富，T细胞室由幼稚T细胞（深绿色）占据主导地位。环境抗原刺激后，记忆B细胞（浅橙色）、记忆CD4 T细胞（浅绿色）和滤泡T细胞（绿色）池增加。因此，树突状细胞、B细胞和T细胞之间的相互作用增加，抗体反应随着IgG和IgA的产生而成熟。出生时树突状细胞（DCs）主要是浆细胞样树突状细胞（蓝色），但1年后髓样树突状细胞（浅蓝色）增多。

4.6 婴儿抗体应答质量

在胎儿和早期新生儿个体发育过程中，外周B细胞群多样性远低于成人。早期研究表明，在生命早期表达的 B 细胞库偏向于特定的 V_H基因，早期新生儿细胞缺乏成人细胞用于多样化的分子机制[35]。众所周知，与成人相比，新生儿体内抗体（Ab）反应具有较低的亲和力和有限的异质性。

B 细胞活化结果的重要差异是，在抗原特异性活化后，新生儿 B 细胞产生的 Abs 量低于成人[36]。Ab 分泌差异可能是由于树突状细胞或巨噬细胞抗原递呈受损，以及 T 细胞分泌细胞因子不理想。然而也存在内在的 B 细胞差异，即在 B 细胞受体（BCR）交联后[37]，新生儿 B 细胞很少或根本没有增殖，尽管信号转导发生在 Ig 连接上[38]。已经证明，新生儿 B 细胞在 BCR 连接后更容易诱导耐受性和/或凋亡。新生儿 B 细胞也表达较少的 MHC Ⅱ类分子，BCR 触发后协同刺激分子 CD80 / CD86 没有上调[39]。虽然这些损伤可能导致新生儿 B 细胞反应低下，但 CD40 连接和 IL-4 刺激导致 B 细胞活化和增殖，从而允许 B 细胞分化。因此，在 T 辅助机制存在情况下，新生儿 B 细胞应答是足够的，尽管可能需要更多刺激信号来获得与成人 B 细胞相似的结果[35]。

体细胞超突变（SHM）主要发生在脾脏或淋巴结生发中心，由于其引起高亲和力抗体的选择，因此对多样化和改善抗体生成至关重要[40]。这一过程依赖于酶激活诱导脱氨酶（AID），它将点突变插入到 Ig 重链和轻链基因中，从而在库多样化和亲和力成熟中发挥重要作用[41]。关于人类婴儿 SHM 的数据很少，但一项早期研究报告了脐带血中 IgG 和 IgA 重链转录物中的体细胞超突变[42]。在新生儿外周血中，V_H6 基因测序时很少或没有检测到突变，但在较大婴儿（10~60 天）中，在同一基因座中发现更多突变[43]。到 8 个月时，突变范围达到了成人水平，并且有选择库的迹象[44]。

因此，新生儿 B 细胞对 BCR 连接反应较低，更容易发生凋亡或免疫耐受，抗体成熟度有限。由于 T 细胞和 B 细胞无能，对大多数抗原的反应效率将会降低（图 4.2）。对于能够激活未成熟 B 细胞的 1 型 T 细胞独立抗原，由于缺乏 TLR 和 BCR 信号转导，这些抗原部分受到阻碍（图 4.2）。2 型 T 细胞独立抗原是一种重复结构，可以通过与成熟 B 细胞反应多重结合而交联 BCR，但由于成熟 B 细胞少，且 BCR 功能低下，因而在幼儿中非常有限。因此，儿童对由这种抗原（多糖疫苗）组成的疫苗反应不佳。

4.7 新生儿血液中影响获得性免疫应答的可溶性因子

虽然免疫系统的大部分成分出现在胎儿发育过程中，但可溶性成分的浓度可能与成年人有显著差异，特别是婴儿血浆补体蛋白及其活性较低。补体系统是先天免疫的重要组成部分，但它也可能影响获得性免疫应答。它增强了特异性免疫球蛋白的作用，启动抗原呈递细胞并帮助其成熟。最后它增强了 B 淋巴细胞抗原驱动的抗体应答成熟。新生儿经典补体成分水平低于成人，这可能是早期获得性免疫应答缺失原因之一[45]。在生命前 6 个月，有几个补体蛋白向成人水平进化[46]。

近年来新生儿血浆证明含有其他具有免疫调节功能分子，主要影响抗原呈递细胞 TLR 活化结果。内源性嘌呤代谢物腺苷选择性抑制 TLR2 活化单核细胞产生 TNF，同时保持产生 IL-6。因此，腺苷有助于新生儿血浆的 Th2 极化特性[47]。此外，新生儿血浆中仍未确定的因子有能力使 TLR4 介导的细胞因子反应极化，产生低水平的 IL-12p70 和高水平的 IL-10，从而介导生命第一个月的免疫抑制[48]。

母体抗体对获得性免疫应答的影响存在争议。母体抗体影响婴儿免疫应答的潜在机制包括母体抗体特异性屏蔽婴儿 B 细胞表位和母体抗体摄取：APC 递呈的抗原复合物[49]。文献中大量数据支持这些模型，也与母体抗体缺乏在体内干扰婴儿 T 细胞启动的能力这一观察结果相吻合。这个问题将在下面进一步讨论。

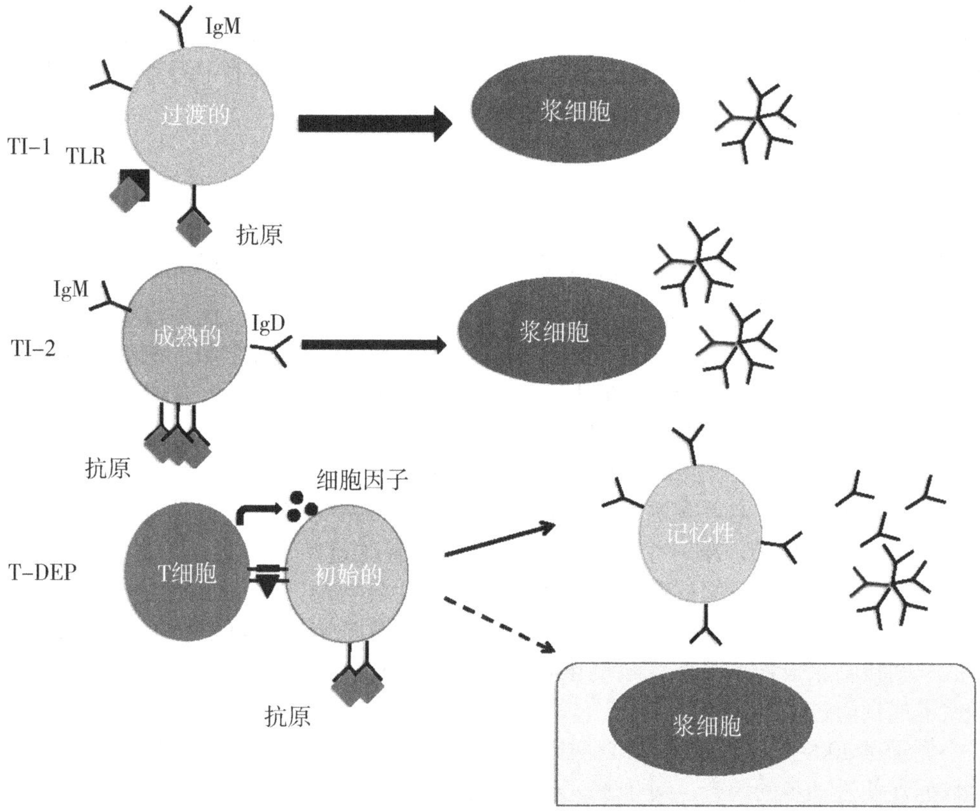

图 4.2　幼儿获得性免疫应答示意图

在出生时，翻译性 B 细胞对某些 T-独立抗原（TI-1）作出应答，主要产生 IgM 作为一线防御的短命浆细胞。IgM+成熟 B 细胞也能对 T 细胞独立抗原（TI-2）作出应答，T 细胞独立抗原通过重复性抗原结构的结合交叉连接数个 BCR。对 T 依赖性抗原的应答在出生时不存在，但在最初 2 年成熟。因此婴儿的交换记忆 B 细胞的产生和浆细胞向骨髓的归巢受到损害。

4.8　目前全球儿童疫苗

全世界 5 岁以下死亡儿童中有一半以上是传染病引起的。这些疾病中有许多是可以用疫苗预防，世卫组织估计 2008 年约有 150 万 5 岁以下儿童死于这些疾病。肺炎链球菌（肺炎球菌）感染和轮状病毒感染是主要原因，其次是新生儿期流感嗜血杆菌 B（HIB）、百日咳博德特氏菌（百日咳）、麻疹病毒和破伤风梭菌感染（新生儿破伤风）。

由于经济问题和该区域的地方性感染状况，世界上使用的疫苗计划各不相同。世卫组织（WHO）的目标是使每个国家的疫苗覆盖率达到 90%（作为国家政策）。在世界上几乎所有地区为预防白喉、破伤风、百日咳和脊髓灰质炎而进行的疫苗接种使这些疾病的发病率在过去 30 年急剧下降。新生儿破伤风仍然很普遍，主要是由于母亲的疫苗覆盖率较低，因此在子宫内转移的抗体较少。MMR 疫苗（麻疹、腮腺炎、风疹）在欧洲、美洲、澳大利亚和一些东地中海国家使用，而普通麻疹疫苗在大多数非洲国家和东南亚国家使用。世界范围内麻疹疫苗接种覆盖率为 85%；然而非洲中部和南部的麻疹疫苗接种覆盖率很低。此外，最近欧洲、北美、澳大利亚和许多非洲和南美国家都引进了肺炎球菌疫苗。预防 HIB 的疫苗在世界范围内也非常普遍，但也只是最近才开始，因此 2010 年世界上只有不到一半的人口受到了保护，尽管这个数字可能会上升。水痘疫苗的引进并不成功，它主要用于美国，也用于其他国家的风险群体。

在未来几年轮状病毒疫苗将被引入非洲、美洲、欧洲和东地中海地区。儿童疫苗接种计划中还可能包括针对当地地方性感染的疫苗，包括脑膜炎奈瑟菌（脑膜炎球菌）、日本脑炎、结核分枝杆菌、肝炎、狂犬病、伤寒沙门氏菌（伤寒）和黄热病。除东地中海地区外，世界各地较大儿童使用人类乳头瘤病毒疫苗减少宫颈癌的情况也在增加。

4.9 疫苗副作用

接种疫苗的副作用是一个有争议的领域。即时反应，如过敏反应[50]和局部反应，注射部位肿胀、发酸和疼痛很容易判断和描述。此外常见的早期全身副作用，包括过敏性和发热可以在人群基础上进行测量。还描述了与疫苗接种密切相关的较不常见副作用，例如长期以来已经确认 MMR 疫苗接种后 7～10 天，600 剂中会有 1 剂发生发热性癫痫[51]。同样 MMR 疫苗接种与 1/50 000剂量免疫血小板减少性紫癜的风险相关[52]。

当涉及罕见和长期的副作用时，举证责任就更具挑战性了。此外这些反应/疾病往往是多因素的。长期副作用调查基本上有两种方法：一种是基于疫苗免疫学数据的纯逻辑假设，另一种是对流行病学数据的怀疑。第一个例子是儿童接种疫苗对过敏和特异反应的作用。众所周知，保留新生儿 Th2 特征较长时间的儿童得过敏症的可能性会增加，而 Th1 基因早期倾斜应答将降低过敏和哮喘的发病率。主要产生 Th1 或 Th2 应答的儿童疫苗因此可能会影响过敏和特异性反应的发展。间接临床报告表明百日咳疫苗存在这种相关性，而对照临床试验表明，百日咳疫苗不具有促进或预防过敏的作用。

有个例子是 2009 年在斯堪的纳维亚接种甲型 H1N1 流感疫苗后儿童嗜睡症发病率增加，流行病学数据促使进一步的调查。对于 11 岁以下人群，发病率从大约 0 例/10 万居民上升到 2010 年的大约 3.4 例/10 万。对于 11～16 岁的孩子来说，2010 年这个数字从 1/100 000上升到了 8.7/100 000左右[53]。这几乎只发生在芬兰和瑞典，归因于遗传背景和可能的其他未知因素。根据遗传数据和发病年龄降低，有人提出疫苗接种会引起一种疾病发病，这种疾病通常会在以后发生[53]。同样百日咳疫苗也被描述为遗传易感个体严重肌阵挛性癫痫（dravet 综合征）的诱因。然而，接种疫苗的时间与该病的典型发病时间一致，已知发热（可能与接种疫苗有关）会引发首发事件。与未接种的遗传易感个体相比，百日咳疫苗接种与疾病的任何改变结果均无关[54]。孤独症谱系疾病与疫苗接种（特别是 MMR）之间的联系尚未得到证实[55,56]。

与此同时，数据表明如果在疫苗接种活动的早期接种疫苗，更有可能产生副作用。对此一种解释是有潜在疾病和/或风险因素的人较早接种疫苗，这些人更容易产生副作用。这在 H1N1 疫苗中得到了证明，早期的患者出现贝尔麻痹、感觉异常和炎症性肠病的频率增加[57]。

最后，不仅抗原本身对副作用很重要，配方也很重要。这一点已经通过比较在同一天分别接种 MMR-水痘（MMRV）疫苗或 MMR 和水痘（MMR+V）疫苗时的发作频率得到证实。与 MMR+V 相比，使用 MMRV 癫痫发作频率上升到 1/2 300剂量；在 MMR+V 中，癫痫发作风险可与单独接种 MMR 差不多[58]。

因此，当谈到导致长期发病率的副作用时，这些副作用与一种疾病的发病有关，这种疾病很可能发生在未接种疫苗的儿童身上，患有潜在疾病的儿童可能有更高副作用风险。对于短期副作用配方似乎起了作用。

4.10 孕期疫苗接种

由于接种疫苗对年幼的儿童效率较低，保护新生儿免受严重感染的一种方法是在怀孕期给母亲接种疫苗，利用母亲子宫将抗体传递给孩子。这一策略已经在一些国家使用，美国卫生当局推荐季节性三价灭活流感疫苗和破伤风/白喉/无细胞百日咳疫苗在妊娠中期和晚期使用。此外，一些其他灭活疫苗被推荐给孕妇风险群体，包括甲型肝炎、乙型肝炎、脑膜炎球菌疫苗和23价肺炎球菌多糖疫苗。孕妇禁用活病毒疫苗。

普遍关注的问题包括母胎免疫的安全性和有效性。使用针对季节性流感和H1N1流感灭活疫苗已经证明是安全的[59]。此外，成人型破伤风、减少的白喉类毒素和无细胞百日咳疫苗也有类似数据[60]。

关于疗效的讨论，有以下几个方面需要考虑。首先，胎儿免疫在预防胎儿疾病是否有效，无论是从社区还是从胎儿时期的母亲。此外，转移的被动抗体是否会对儿童以后的免疫产生负面影响。为了提高效力，我们将讨论两种类型疫苗：含有T-依赖性抗原的疫苗和含有T-非依赖性抗原的疫苗。人们可以假设T-依赖性抗原会更有效，因为它们会产生抗体，在胎盘上更有效地富集。对于流感疫苗等T-依赖性抗原，流行病学数据显示，接种疫苗的母亲所生的孩子有较轻微的流感样症状，经确诊的流感发病率较低[61]。

对于T-非依赖性抗原，疗效尚不明确。有研究表明，母体用肺炎球菌多糖疫苗免疫后，肺炎球菌抗体以足够的水平在子宫内转移[62]。这些是血清型1型和5型的结果，由于血清型1型多糖的两性离子性质，只有5型被认为是T-非依赖性的[63]。在这种情况下，母亲免疫确实会干扰幼儿疫苗接种（产后7~17周），因为疫苗接种不会增加特定抗体的数量（如果婴儿体内已经存在高浓度抗体）。母亲接种疫苗对3岁时的疫苗接种效果没有影响[64]。这些抗体在预防疾病中的作用几乎没有证据，新生儿肺炎球菌定植也不受影响[65]。

反对胎儿免疫的一个论点是可能会干扰儿童后期免疫。这可以通过中和抗体干扰特异T细胞应答的能力和中和抗体干扰免疫时体液应答能力来介导。至于T细胞应答，如上所述，与成人相比，幼儿T细胞应答明显不那么突出。以IFN-γ产生量来衡量，母体中和抗体的存在不影响T细胞应答效力[49,66]。

当谈到体液应答时，情况更加复杂了。很明显地，在9个月时仍保留母体抗体的儿童接种疫苗后会产生较低水平的中和抗体，而在6个月时，无论母体抗体如何，接种疫苗后产生抗体的能力一般都很差。9个月时对儿童免疫的潜在干扰将取决于母体抗体的持久性。针对不同抗原的抗体在体内显示出不同的半衰期，针对麻疹病毒和风疹病毒的抗体比针对腮腺炎病毒的抗体持续时间更长。与自然感染相比，如果母亲接种了腮腺炎疫苗，其抗体持续时间会更短[67]。此外，胎盘转运将不仅仅取决于亚类，因为针对肺炎球菌的IgG1和IgG2抗体转运不及类似亚类如抗破伤风抗体[68]。显然每种疫苗都必须判断潜在的干扰。然而，大多数接种过疫苗但仍从母亲获得免疫的儿童在第二次接种疫苗时反应良好[66]。一些研究发现，在第一次早期接种疫苗失败的情况下，第二次接种疫苗失败率略高[69]。这是否是儿童对疫苗接种反应的内在能力造成的，还是母体抗体的干扰造成的，目前尚不清楚。

在妊娠晚期用灭活疫苗接种似乎是安全的，至少对某些疫苗有效。与儿童免疫接种干扰的数据有冲突，但大多数数据表明干扰很少。每种疫苗都必须给出具体建议，同时还需要进行更多的研究。

4.11 原发性或获得性免疫缺陷儿童的免疫应答

尽管儿童存在严重的免疫缺陷，包括移植和化疗、原发性遗传免疫缺陷疾病或先天性艾滋病病毒感染，但越来越多的儿童在婴儿期和幼儿期能够存活下来。这些儿童易受感染，因此将受益于有效的免疫接种。更好地了解这些患者的受损疫苗反应背后的分子缺陷可能有助于开发更好的疫苗。

常见可变免疫缺陷病（CVID）影响抗体的产生，其特征是血清中IgG和IgA和/或IgM浓度较低，对包膜细菌（流感嗜血杆菌、肺炎链球菌）呼吸道感染的敏感性增加。CVID是一种具有多种遗传缺陷的异质性疾病，涉及B细胞信号转导和/或T-B细胞相互作用的重要分子。在免疫前已经切换了外周血中记忆性B细胞（$CD27^+$）的成年CVID患者中，可以检测到针对几种抗原的保护性抗体应答[70]。同样在一项儿科研究中，发现11/16的儿童对C群脑膜炎球菌多糖疫苗有反应[71]。此外，针对多糖疫苗的疫苗反应与这些患者中IgM记忆性B细胞的存在有关。对CVID患者的类似研究表明，在针对破伤风梭状芽孢杆菌和肺炎链球菌免疫接种后，血浆母细胞形成受到阻碍[72]。

对于大多数抗原，HIV-1感染者的免疫应答在成人和儿童中均受到严重损害[73]。采用高效抗逆转录病毒疗法（HAART）改善了大多数患者的疫苗接种效果[74]。针对这一弱势儿童群体的免疫指南最近已发表[75]。接种疫苗是安全的，几乎没有副作用，目前唯一禁忌的疫苗是针对HIV-1感染儿童的卡介苗。然而关于儿童HIV-1感染的免疫接种，特别是HAART治疗的儿童，其抗体应答与健康个体相比有多持久，仍然存在一些未解决的问题。

由于不同的治疗方式根除了保护性抗体介导的记忆，移植后和化疗后儿童需要重新免疫[76]。对于HIV-1感染儿童，大多数疫苗是安全的，尽管在治疗结束后12~24个月应推迟用活疫苗再接种。目前对于开始再接种的最佳时间或者应该给多少剂量来实现长期保护还没有共识[77,78]。

4.12 开发新儿科疫苗

传统疫苗通常由全灭活的病毒组成，通过肌内注射免疫。对于一些重要的儿童病原体，例如呼吸道合胞病毒（RSV），这种尝试失败了。RSV疫苗的开发将作为婴儿新型疫苗策略的一个例子进行讨论。最初临床研究使用福尔马林灭活RSV接种幼儿，导致感染、住院以及在某些情况下疾病恶化而死亡[79]。这后来被归因于毁灭性的Th2反应，导致肺部病变。RSV引起局部呼吸道疾病，无一般病毒血症，导致显著的住院率、发病率和死亡率。从免疫学角度来看，RSV具有挑战性，因为获得性免疫无法防止再次感染。

这是由于T细胞应答质量差和抗体应答持久性差。一个假设是黏膜免疫会减缓这种快速病毒的速度，并且血清抗体在组织中的含量很低。寻找RSV疫苗接种的替代策略是关键。一种方法是在黏膜内层递送疫苗，例如鼻腔接种（而不是肌肉内）。然后免疫激活将在黏膜相关淋巴组织（MALT）内发生，同时激活微皱上皮细胞（M细胞）以及随后激活潜在的抗原呈递细胞，例如树突状细胞。这将诱导局部IgA和全身IgG。

根据使用路径将会产生不同的局部免疫。例如使用鼻腔接种，上呼吸道和宫颈阴道的免疫将占优势，而口服给药将主要在小肠和乳腺中诱导IgA产生。使用灭活RSV纳米乳剂在小鼠鼻腔接种实验显示，肺中有IgA反应以及保护能力[80]。目前最有希望的结果来自临床II期试验，在试验中经鼻腔接种减毒活RSV菌株[79]。尽管如此，RSV疫苗开发仍受到最初失败的困扰，目前还没有可用的好疫苗。儿童群体中裸DNA和载体表达DNA疫苗临床研究尚待结果。

新疫苗开发还包括引入具有增强免疫刺激能力的佐剂，以补偿抗原内在的低免疫原性[81]。新型佐剂旨在优化B细胞应答，并产生适当T细胞应答，这在儿童疫苗中可能特别重要。其中一种新佐剂是TLR4激动剂单磷酰脂A（MPL），目前已获得许可或正在进行临床III期试验[82]。当MPL与明矾结合时，作用于DCs并促进IFN-γ产生，从而克服与明矾相关的Th2偏向反应。目前已在人乳头瘤病毒（HPV）-16/18疫苗中获得许可，并且证明可诱导持久的B细胞记忆和持久性抗体[83]。对1~4岁的儿童进行恶性疟原虫免疫，一种含有MPL佐剂的高抗疟抗体可长期预防临床疾病[84]。另一种潜在的新佐剂，大肠杆菌（LTK63）热不稳定肠毒素（LT）的无毒突变体证明能够克服滤泡树突状细胞的延迟成熟，从而在小鼠中与多糖共轭疫苗一起接种时诱导生发中心[85]。除了改善B细胞应答外，LTK63与巨噬细胞结合后还能诱导平衡的Th1/Th2细胞因子以及细胞毒性T细胞应答[86]。

因此，为生命早期开发新的高效疫苗可能会利用新的佐剂来规避新生儿获得性免疫的低反应性。最后，本章概述的要点如下。

（1）在生命的前2年，免疫系统逐步发展。这些与年龄相关的步骤在设计新疫苗或保证接种新年龄组疫苗时是很重要的。

（2）儿童接种疫苗的长期副作用风险很低。

（3）需要为患有潜在疾病的儿童制定疫苗接种指南。

（4）孕期接种疫苗，随后通过转移抗体进行被动保护是保护新生儿的一种方法。必须对每种疫苗在未来儿童接种时受到干扰的潜在风险进行判断，但总体而言似乎较低。

（5）未来的儿童疫苗将以新的佐剂、新的免疫途径为基础，并涉及诸如RSV感染和疟疾等重要疾病。

参考文献

[1] Wood, N., Siegrist, C. A.: Neonatal immunization: where do we stand? Curr. Opin. Infect. Dis. 24, 190-195 (2011). doi: 10. 1097/QCO. 0b013e328345d563.

[2] Adkins, B., Leclerc, C., Marshall-Clarke, S.: Neonatal adaptive immunity comes of age. Nat. Rev. Immunol. 4, 553-564 (2004). doi: 10. 1038/nri 1394.

[3] Ygberg, S., Nilsson, A.: The developing immune system-from foetus to toddler. ActaPaediatr. 101, 120-127 (2012). doi: 10. 1111/j. 1651-2227. 2011. 02494. x.

[4] Siegrist, C. A.: Neonatal and early life vaccinology. Vaccine 19, 3331-3346 (2001).

[5] Hodgins, D. C., Shewen, P. E.: Vaccination of neonates: problem and issues. Vaccine 30, 1541-1559 (2012). doi: 10. 1016/j. vaccine. 2011. 12. 047.

[6] Ueno, H., et al.: Dendritic cell subsets in health and disease. Immunol. Rev. 219, 118-142 (2007). doi: 10. 1111/j. 1600-065X. 2007. 00551. x.

[7] Teig, N., Moses, D., Gieseler, S., Schauer, U.: Age-related changes in human blood dendritic cell subpopulations. Scand. J. Immunol. 55, 453-457 (2002).

[8] Sorg, R. V., Kogler, G., Wernet, P.: Identification of cord blood dendritic cells as an immature CD11c-population. Blood 93, 2302-2307 (1999).

[9] Jones, C. A., Holloway, J. A., Warner, J. O.: Fetal immune responsiveness and routes of allergic sensitization. Pediatr. Allergy Immunol. 13 (Suppl 15), 19-22 (2002).

[10] Liu, E., Tu, W., Law, H. K., Lau, Y. L.: Decreased yield, phenotypic expression and function of immature monocyte-derived dendritic cells in cord blood. Br. J. Haematol. 113, 240-246 (2001).

[11] Langrish, C. L., Buddle, J. C., Thrasher, A. J., Goldblatt, D.: Neonatal dendritic cells are intrinsically biased against Th-1 immune responses. Clin. Exp. Immunol. 128, 118-123 (2002).

[12] De Wit, D., et al.: Impaired responses to toll-like receptor 4 and toll-like receptor 3 ligands in human cord blood. J. Autoimmun. 21, 277-281 (2003).

[13] De Wit, D., et al.: Blood plasmacytoid dendritic cell responses to CpG oligodeoxynucleotides are impaired in human newborns. Blood 103, 1030-1032 (2004). doi: 10. 1182/blood-2003-04-1216.

[14] Levy, O., et al.: Selective impairment of TLR-mediated innate immunity in human newborns: neonatal blood plasma reduces monocyte TNF-alpha induction by bacterial lipopeptides, lipopolysaccharide, and imiquimod, but preserves the response to R-848. J. Immunol. 173, 4627-4634 (2004).

[15] Burl, S., et al.: Age-dependent maturation of Toll-like receptor-mediated cytokine responses in Gambian infants. PLoSOne 6, e18185 (2011). doi: 10. 1371/ journal. pone. 0018185.

[16] Nguyen, M., et al.: Acquisition of adult-like TLR4 and TLR9 responses during the first year of life. PLoS One 5, e10407 (2010). doi: 10. 1371/journal. pone. 0010407.

[17] Belderbos, M. E., et al.: Skewed pattern of Toll-like receptor 4-mediated cytokine production in human neonatal blood: low LPS-induced IL-12p70 and high IL-10 persist throughout the first month of life. Clin. Immunol. 133, 228-237 (2009). doi: 10. 1016/j. clim. 2009. 07. 003.

[18] de Vries, E., et al.: Longitudinal survey of lymphocyte subpopulations in the first year of life. Pediatr. Res. 47, 528-537 (2000).

[19] Schatorje, E. J., et al.: Pediatric reference values for the peripheral T-cell compartment. Scand. J. Immunol. (2011). doi: 10. 1111/j. 1365-3083. 2011. 02671. x.

[20] Schaerli, P., et al.: CXC chemokine receptor 5 expressiondefines follicular homing T cells with B cell helper function. J. Exp. Med. 192, 1553-1562 (2000).

[21] Wing, K., et al.: $CD4^{+}$CD25+FOXP3+regulatory T cells from human thymus and cord blood suppress antigen-specific T cell responses. Immunology 115, 516-525 (2005). doi: 10. 1111/j. 1365-2567. 2005. 02186. x.

[22] Rabe, H., et al.: Higher proportions of circulating FOXP3+and CTLA-4+regulatory T cells are associated with lower fractions of memory $CD4^{+}$T cells in infants. J. Leukoc. Biol. 90, 1133-1140 (2011). doi: 10. 1189/jlb. 0511244.

[23] Jullien, P., et al.: Decreased CD154 expression by neonatal $CD4^{+}$T cells is due to limitations in both proximal and distal events of T cell activation. Int. Immunol. 15, 1461-1472 (2003).

[24] Fadel, S., Sarzotti, M.: Cellular immune responses in neonates. Int. Rev. Immunol. 19, 173-193 (2000).

[25] Capolunghi, F., et al.: CpG drives human transitional B cells to terminal differentiation and production of natural antibodies. J. Immunol. 180, 800-808 (2008).

[26] Morbach, H., Eichhorn, E. M., Liese, J. G., Girschick, H. J.: Reference values for B cell subpopulations from infancy to adulthood. Clin. Exp. Immunol. 162, 271-279 (2010). doi: 10. 1111/j. 1365-2249. 2010. 04206. x.

[27] Smet, J., Mascart, F., Schandene, L.: Are the reference values of B cell subpopulations used in adults for classification of common variable immunodeficiencies appropriate for children? Clin. Immunol. 138, 266-273 (2011). doi: 10. 1016/j. clim. 2010. 12. 001.

[28] H uck, K., et al.: Memory B-cells in healthy and antibody-deficient children. Clin. Immunol. 131, 50-59 (2009). doi: 10. 1016/j. clim. 2008. 11. 008.

[29] Kruschinski, C., Zidan, M., Debertin, A. S., von Horsten, S., Pabst, R.: Age-dependent development of the splenic marginal zone in human infants is associated with different causes of death. Hum. Pathol. 35, 113-121 (2004).

[30] Lundell, A. C., et al.: Infant B cell memory differentiation and early gut bacterial colonization. J. Immunol. 188,%1-4322 (2012). doi: 10. 4049/jimmunol. 1103223.

[31] Weller, S., et al.: Human blood IgM "memory" B cells are circulating splenic marginal zone B cells harboring a prediversified immunoglobulin repertoire. Blood 104, 3647-3654 (2004). doi: 10. 1182/blood-2004-01-0346.

[32] Kruetzmann, S., et al.: Human immunoglobulin M memory B cells controlling Streptococcus pneumoniae infections are generated in the spleen. J. Exp. Med. 197, 939-945 (2003). doi: 10. 1084/ jem. 20022020.

[33] Nilsson, A., et al.: Current chemotherapy protocols for childhood acute lymphoblastic leukemia induce loss of humoral immunity to viral vaccination antigens. Pediatrics 109, e91 (2002).

[34] Pihlgren, M., et al.: Reduced ability of neonatal and early-life bone marrow stromal cells to support plasmablast survival. J. Immunol. 176, 165-172 (2006).

[35] Press, J. L.: Neonatal immunity and somatic mutation. Int. Rev. Immunol. 19, 265-287 (2000).

[36] Siegrist, C. A.: The challenges of vaccine responses in early life: selected examples. J. Comp. Pathol. 137 (Suppl 1), S4-S9 (2007). doi: 10. 1016/j. jcpa. 2007. 04. 004.

[37] Tasker, L., Marshall-Clarke, S.: Immature B cells from neonatal mice show a selective inability to up-regulate MHC class II expression in response to antigen receptor ligation. Int. Immunol. 9, 475-484 (1997).

[38] Chang, T. L., Capraro, G., Kleinman, R. E., Abbas, A. K.: Anergy in immature B lymphocytes. Differential responses to receptor-mediated stimulation and T helper cells. J. Immunol. 147, 750-756 (1991).

[39] Marshall-Clarke, S., Reen, D., Tasker, L., Hassan, J.: Neonatal immunity: how well has it grown up? Immunol. Today 21, 35-41 (2000).

[40] McHeyzer-Williams, L. J., McHeyzer-Williams, M. G.: Antigen-specific memory B cell development. Annu. Rev. Immunol. 23, 487-513 (2005). doi: 10. 1146/annurev. immunol. 23. 021704. 115732.

[41] Pan-Hammarstrom, Q., Zhao, Y., Hammarstrom, L.: Class switch recombination: a comparison between mouse and human. Adv. Immunol. 93, 1-61 (2007). doi: 10. 1016/S0065-2776 (06) 93001-6.

[42] Mortari, F., Wang, J. Y., Schroeder Jr., H. W.: Human cord blood antibody repertoire. Mixed population of VH gene segments and CDR3 distribution in the expressed C alpha and C gamma repertoires. J. Immunol. 150, 1348-1357 (1993).

[43] Ridings, J., et al.: Somatic hypermutation of immunoglobulin genes in human neonates. Clin. Exp. Immunol. 108, 366-374 (1997).

[44] Ridings, J., Dinan, L., Williams, R., Roberton, D., Zola, H.: Somatic mutation of immunoglobulin V (H) 6 genes in human infants. Clin. Exp. Immunol. 114, 33-39 (1998).

[45] McGreal, E. P., Hearne, K., Spiller, O. B.: Off to a slow start: under-development of the complement system in term newborns is more substantial following premature birth. Immunobiology217, 176-186 (2012). doi: 10. 1016/j. imbio. 2011. 07. 027.

[46] Davis, C. A., Vallota, E. H., Forristal, J.: Serum complement levels in infancy: age related changes. Pediatr. Res. 13, 1043-1046 (1979).

[47] Levy, O., et al.: The adenosine system selectively inhibits TLR-mediated TNF-alpha production in the human newborn. J. Immunol. 177, 1956-1966 (2006).

[48] Belderbos, M. E., et al.: Neonatal plasma polarizes TLR4-mediated cytokine responses towards low IL-12p70 and high IL-10 production via distinct factors. PLoS One 7, e33419 (2012). doi: 10. 1371/journal. pone. 0033419.

[49] Siegrist, C. A.: Mechanisms by which maternal antibodies influence infant vaccine responses: review of hypotheses and definition of main determinants. Vaccine 21, 3406-3412 (2003).

[50] Gruber, C., Nilsson, L., Bjorksten, B.: Do early child-hood immunizations influence the development of atopy and do they cause allergic reactions? Pediatr. Allergy Immunol. 12, 296-311 (2001).

[51] Klein, N. P., et al.: Measles-mumps-rubella-varicella combination vaccine and the risk of febrile seizures. Pediatrics 126, e1-e8 (2010). doi: 10. 1542/ peds. 2010-0665.

[52] O'Leary, S. T., et al.: The risk of immune thrombocytopenic purpura after vaccination in children and adolescents. Pediatrics 129, 248-255 (2012). doi: 10. 1542/peds. 2011-1111.

[53] Partinen, M., et al.: Increased incidence and clinical picture of childhood narcolepsy following the 2009 H1N1 pandemic vaccination campaign in Finland. PLoSOne 7, e33723 (2012). doi: 10. 1371/journal. pone. 0033723.

[54] McIntosh, A. M., et al.: Effects of vaccination on onset and outcome of Dravet syndrome: a retrospective study. Lancet Neurol. 9, 592-598 (2010). doi: 10. 1016/ S1474-4422 (10) 70107-1.

[55] Uno, Y., Uchiyama, T., Kurosawa, M., Aleksic, B., Ozaki, N.: The combined measles, mumps, and rubella vaccines and the total number of vaccines are not associated with development of autism spectrum disorder: the first case-control study in Asia. Vaccine 30, 4292-4298 (2012). doi: 10. 1016/j. vaccine. 2012. 01. 093.

[56] Demicheli, V., Rivetti, A., Debalini, M. G., Di Pietrantonj, C.: Vaccines for measles, mumps and rubella in children. Cochrane Database Syst. Rev. 2, CD004407 (2012). doi: 1 0. 1002/14651858. CD004407. pub3.

[57] Bardage, C., et al.: Neurological and autoimmune disorders after vaccination against pandemic influenza A (H1N1) with a monovalent adjuvanted vaccine: population based cohort study in Stockholm, Sweden. BMJ343, d5956 (2011). doi: 10. 1136/bmj. d5956.

[58] Klein, N. P., et al.: Measles-containing vaccines and febrile seizures in children age 4 to 6 years. Pediatrics 129, 809-814 (2012). doi: 10. 1542/peds. 2011-3198.

[59] Oppermann, M., et al.: A (H1N1) v2009: a controlled observational prospective cohort study on vaccine safety in pregnancy. Vaccine 30, 4445-4452 (2012). doi: 10. 1016/j. vaccine. 2012. 04. 081.

[60] Fortner, K. B., Kuller, J. A., Rhee, E. J., Edwards, K. M.: Influenza and tetanus, diphtheria, and acellular pertussis vaccinations during pregnancy. Obstet. Gynecol. Surv. 67, 251-257 (2012). doi: 10.

1097/OGX. 0b013e3182524cee.

[61] Eick, A. A., et al.: Maternal influenza vaccination and effect on influenza virus infection in young infants. Arch. Pediatr. Adolesc. Med. 165, 104-111 (2011). doi: 10. 1001/archpediatrics. 2010. 192.

[62] Quiambao, B. P., et al.: Immunogenicity and reactogenicity of 23-valent pneumococcal polysaccharide vaccine among pregnant Filipino women and placental transfer of antibodies. Vaccine25, 4470-4477 (2007). doi: 10. 1016/j. vaccine. 2007. 03. 021.

[63] Groneck, L., et al.: Oligoclonal $CD4^{+}T$ cells promote host memory immune responses to Zwitterionic polysaccharide of Streptococcus pneumoniae. Infect. Immun. 77, 3705 - 3712 (2009). doi: 10. 1128/IAI. 01492-08.

[64] Holmlund, E., Nohynek, H., Quiambao, B., Ollgren, J., Kayhty, H.: Mother-infant vaccination with pneumococcal polysaccharide vaccine: persistence of maternal antibodies and responses of infants to vaccination. Vaccine 29, 4565-4575 (2011). doi: 10. 1016/j. vaccine. 2011. 04. 068.

[65] Lopes, C. R., et al.: Ineffectiveness for infants of immunization of mothers with pneumococcal capsular polysaccharide vaccine during pregnancy. Braz. J. Infect. Dis. 13, 104-106 (2009).

[66] Gans, H., et al.: Measles and mumps vaccination as a model to investigate the developing immune system: passive and active immunity during the first year of life. Vaccine 21, 3398-3405 (2003).

[67] Leuridan, E., Goeyvaerts, N., Hens, N., Hutse, V., Van Damme, P.: Maternal mumps antibodies in a cohort of children up to the age of 1 year. Eur. J. Pediatr. 171, 1167-1173 (2012). doi: 10. 1007/s00431-012-1691-y.

[68] Shahid, N. S., et al.: Serum, breast milk, and infant antibody after maternal immunisation with pneumococcal vaccine. Lancet 346, 1252-1257 (1995).

[69] Redd, S. C., et al.: Comparison of vaccination with measles-mumps-rubella vaccine at 9, 12, and 15 months of age. J. Infect. Dis. 189 (Suppl 1), S116-S122 (2004). doi: 10. 1086/378691.

[70] Goldacker, S., et al.: Active vaccination in patients with common variable immunodeficiency (CVID). Clin. Immunol. 124, 294-303 (2007). doi: 10. 1016/j. clim. 2007. 04. 011.

[71] Rezaei, N., et al.: Serum bactericidal antibody response to serogroup C polysaccharide meningococcal vaccination in children with primary antibody deficiencies. Vaccine 25, 5308-5314 (2007). doi: 10. 1016/j. vaccine. 2007. 05. 021.

[72] Chovancova, Z., Vlkova, M., Litzman, J., Lokaj, J., Thon, V.: Antibody forming cells and plasmablasts in peripheral blood in CVID patients after vaccination. Vaccine 29, 4142-4150 (2011). doi: 10. 1016/j. vaccine. 2011. 03. 087.

[73] Cagigi, A., Nilsson, A., Pensieroso, S., Chiodi, F.: Dysfunctional B-cell responses during HIV-1 infection: implication for influenza vaccination and highly active antiretroviral therapy. Lancet Infect. Dis. 10, 499-503 (2010). doi: 10. 1016/S1473-3099 (10) 70117-1.

[74] Sutcliffe, C. G., Moss, W. J.: Do children infected with HIV receiving HAART need to be revaccinated? Lancet Infect. Dis. 10, 630-642 (2010). doi: 10. 1016/S1473-3099 (10) 70116-X.

[75] Menson, E. N., et al.: Guidance on vaccination of HIV-infected children in Europe. HIV Med. 13, 333-336; e331-e314 (2012). doi: 10. 1111/j. 1468-1293. 2011. 00982. x.

[76] Patel, S. R., Chisholm, J. C., Heath, P. T.: Vaccinations in children treated with standard-dose cancer therapy or hematopoietic stem cell transplantation. Pediatr. Clin. North. Am. 55, 169-186, xi (2008). doi: 10. 1016/j. pcl. 2007. 10. 012.

[77] Brodtman, D. H., Rosenthal, D. W., Redner, A., Lanzkowsky, P., Bonagura, V. R.: Immunodeficiency in children with acute lymphoblastic leukemia after completion of modern aggressive chemotherapeutic regimens. J. Pediatr. 146, 654-661 (2005). doi: 10. 1016/j. jpeds. 2004. 12. 043.

[78] Lehrnbecher, T., et al.: Revaccination of children after completion of standard chemotherapy for acute lymphoblastic leukaemia: a pilot study comparing different schedules. Br. J. Haematol. 152, 754-757 (2011). doi: 10. 1111/j. 1365-2141. 2010. 08522. x.

[79] Graham, B. S.: Biological challenges and technological opportunities for respiratory syncytial virus vaccine development. Immunol. Rev. 239, 149-166 (2011). doi: 10. 1111/j. 1600-065X. 2010. 00972. x.

[80] Lindell, D. M., et al.: A novel inactivated intranasal respiratory syncytial virus vaccine promotes viral clearance without Th2 associated vaccine-enhanced disease. PLoS One 6, e21823 (2011). doi: 10. 1371/ journal. pone. 0021823.

[81] Mastelic, B., et al.: Mode of action of adjuvants: implications for vaccine safety and design. Biologicals 38, 594-601 (2010). doi: 10. 1016/j. biologicals. 2010. 06. 002.

[82] Garcon, N., Segal, L., Tavares, F., Van Mechelen, M.: The safety evaluation of adjuvants during vaccine development: the AS04 experience. Vaccine 29, 4453-4459 (2011). doi: 10. 1016/j. vaccine. 2011. 04. 046.

[83] Giannini, S. L., et al.: Enhanced humoral and memory B cellular immunity using HPV16/18 L1 VLP vaccine formulated with the MPL/aluminium salt combination (AS04) compared to aluminium salt only. Vaccine 24, 5937-5949 (2006). doi: 10. 1016/j. vaccine. 2006. 06. 005.

[84] Sacarlal, J., et al.: Long-term safety and efficacy of the RTS, S/AS02A malaria vaccine in Mozambican children. J. Infect. Dis. 200, 329-336 (2009). doi: 10. 1086/600119.

[85] Bjarnarson, S. P., Adarna, B. C., Benonisson, H., Del Giudice, G., Jonsdottir, I.: The adjuvant LT-K63 can restore delayed maturation of follicular dendritic cells and poor persistence of both protein-and polysaccharide-specific antibody-secreting cells in neonatal mice. J. Immunol. 189, 1265-1273 (2012). doi: 10. 4049/jimmunol. 1200761.

[86] da Hora, V. P., Conceicao, F. R., Dellagostin, O. A., Doolan, D. L.: Non-toxic derivatives of LT as potent adjuvants. Vaccine 29, 1538-1544 (2011). doi: 10. 1016/j. vaccine. 2010. 11. 091.

（马维民，吕律译）

第三部分

传染病疫苗

概　述

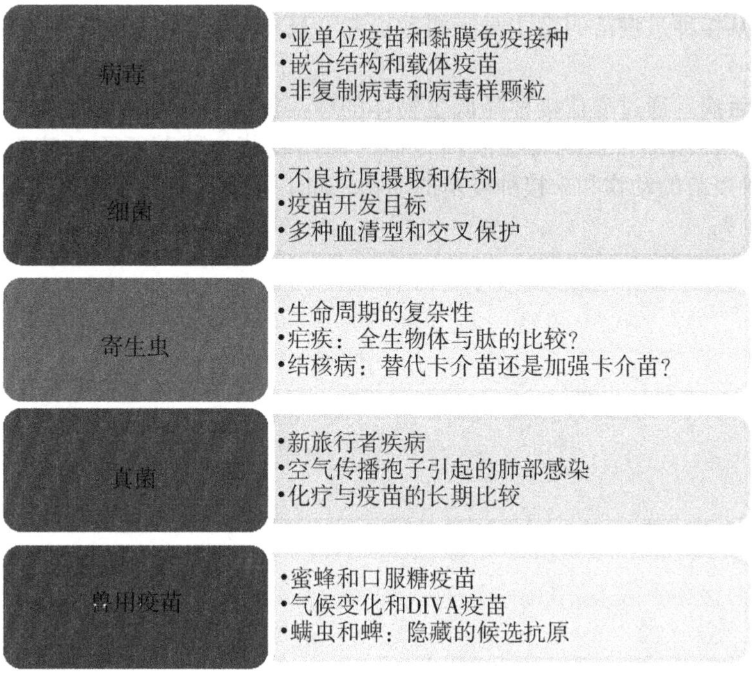

在西方世界，大约 26%的人类新兴疾病是由传染因子引起的：病毒、细菌、寄生虫和真菌。大约 60%的感染源于人畜共患病。在已知的 1 400种病原体中，约有 2/3 是由动物传播的，并且是人畜共患病病原体。但全球人类健康产品中只有 2%是疫苗（其中，20%是兽用疫苗）。90 年前引进的最古老结核病许可疫苗 BCG（另见第 13 章）目前仍在使用中。毫无疑问，目前的疫苗成功地预防了不变病原体。但目前还没有有效的疫苗来预防可变病原体或潜在感染。

呼吸道合胞病毒（RSV）　RSV 发现后的半个多世纪，仍然没有疫苗可用。免疫学和疫苗学的最新进展有助于更好地理解 RSV 是如何逃避免疫应答的，以及它是如何在人类中引起病理反应的。

埃博拉病毒　研究表明，以重组腺病毒为基础的疫苗、基于重组水泡性口炎病毒（VSV）的疫苗、使用病毒样颗粒的疫苗和基于重组人副流感病毒的疫苗具有保护作用。

登革热　减毒活疫苗正处于临床开发进展期。为了避免疾病的免疫增强，需要保护所有 4 种血清型，这使得登革热疫苗的开发成为一项具有挑战性的任务。

诺如病毒　实验研究已证明，植物中产生的诺如病毒病毒样颗粒可诱导有效的细胞和体液免疫应答。由于诺如病毒自然引起胃肠道感染，病毒样颗粒在低 pH 下稳定，对消化酶有抵抗力。

麻疹　麻疹仍然是 5 岁以下儿童死亡的主要原因之一。麻疹是最具传染性的人类疾病之一。为了克服活疫苗的局限性，目前正在研究几种替代疫苗，如载体驱动疫苗。

志贺氏菌—链球菌—沙门氏菌病　一种有希望的方法是使用跨多个血清型的保守蛋白质作为疫苗配方的一部分。在疫苗化合物中加入保守蛋白有可能解决血清型特异性问题。多价亚单位疫苗诱导交叉保护。

疟疾和结核病 最近的工作提高了我们对宿主-疟原虫寄生虫相互作用的理解，并为通过预防性疫苗接种增强抗疟疾免疫新策略提供了关键性见解。卡介苗是世界卫生组织推荐的唯一一种抗结核疫苗，自1921年推出以来，已经注射了30多亿剂。

副球芽孢菌病 抗真菌化学疗法可以持续2年或更长时间，复发率很高。多肽疫苗在严重病例和化疗反应不佳的病例中是一种强有力的药物。

兽用疫苗 为人类健康产品铺平道路?

螨虫和蜱 从接种过疫苗的宿主身上摄取血粉，肠壁会受到损伤，这可能会导致死亡或生殖能力下降。

莱姆病和蓝舌病 通过疫苗接种降低生物体的载荷能力，宿主可能是预防人类莱姆病的有效策略。目前用于控制蓝舌病病毒在欧洲传播的商用疫苗主要基于灭活病毒；这些都不符合DIVA标准。接种疫苗的动物和未接种疫苗的动物之间不可能有血清学上的区别。

第5章 从RSV G蛋白中央保守区构建用于肠外或黏膜呈递的亚单位候选疫苗

Thien N. Nguyen，Christine Libon 和 Stefan Ståhl①

摘要

呼吸道合胞病毒（Respiratory syncytial virus，RSV）是引起全球婴幼儿和老年人严重上下呼吸道疾病的主要病原体。世卫组织称，目前急需 RSV 疫苗。本章描述了基于分子工程的各种亚单位疫苗概念的设计，目的是递呈 RSV 抗原。制备了编码 RSV G 蛋白（G2Na）保守中心区部分的基因片段，用于各种表达和传递方式：①原核表达和单独纯化 G2Na 或融合到不同的载体蛋白中，其中一种即 BBG2Na（明矾），已在老年人中进行临床试验；②G 蛋白衍生抗原在活载体（非致病细菌）；③核酸载体表面展示。

这些亚单位疫苗通过不同途径（胃肠外或黏膜）在啮齿动物和非人灵长类动物中（使用或不使用佐剂）进行试验。我们总结并比较了 RSV 实验动物模型中每种免疫方式所赋予的免疫原性、保护效力和安全性。其中，G2Na 被证明是 RSV 亚单位疫苗最有前景的成分。

呼吸道合胞病毒（RSV）是一种高度传染性病原体，可导致新生儿、幼儿、免疫功能低下者和老年人严重的细支气管炎和肺炎。尽管对 RSV 进行了 60 年的研究，但目前还没有人用疫苗。一项引人注目的福尔马林灭活 RSV（FI-RSV）临床试验“标记”了 RSV 疫苗的发展历史：疫苗接种者中两名儿童因肺部疾病意外恶化而死亡。增强免疫病理学仍未完全阐明[11]。变性 FI-RSV 抗原（Ags）诱导非中性抗体（Abs）、先天性免疫 TLR 缺失启动和 Th2 诱导异常获得性免疫是似乎可信的假设原因之一[16]。高风险婴儿可使用人源化单克隆抗体（hMAb）（Synagis®）进行预防性治疗，但成本效益仍存在争议。

安全有效的人呼吸道合胞病毒（hRSV）疫苗仍然是每年拯救数十万儿童生命的迫切需要。理想情况下，对于儿童来说，RSV 疫苗应与现有的儿科疫苗结合使用。在冬季老年人是另一个易感目标人群，其 RSV 临床症状与流感感染非常相似[19]。一种 RSV 亚单位疫苗有望与目前的流感疫苗相结合。在法国 Pierre Fabre（皮埃尔·法布尔）研究所，我们花费 12 年多时间与瑞典皇家理工学院和加拿大 Armand Frappier 研究所合作研发了 hRSV 疫苗。在本章总结了我们设计亚单位疫苗的不同方法，这些亚单位疫苗以 RSV G 蛋白中心域为基础，以各种形式（非复制疫苗和活疫苗）用于胃肠外或黏膜呈递。其中一种亚单位疫苗候选物，即以明矾为佐剂的 BBG2Na，在老年人中进行了临床试验[34,58,61]。

5.1 疾病

RSV 的特点是能够在人的一生中反复感染上呼吸道（URT）。此外，导致严重下呼吸道（LRT）病理的婴儿，原发性呼吸道合胞病毒感染不一定能预防第二次严重感染[11]。经过 4~7

① T. N. Nguyen，博士（✉）·C. Libon，博士（法国图卢兹皮埃尔·法布尔研究所微生物技术系，E-mail：thien. nguyen@ pierre-fabre. com）·S. Ståhl，博士（瑞典斯德哥尔摩 KTH 皇家理工学院阿尔巴诺瓦大学中心生物技术院分子生物技术室主任）

天的潜伏期后，RSV 第一个症状是流鼻涕，通常伴有发热。1~3 天后，婴儿咳嗽加剧，随着病情发展，出现呼吸急促和呼吸困难。呼吸道合胞病毒是毛细支气管炎最常见的病因之一，呼吸道合胞病毒感染也与儿童哮喘的发展和长达 7 年的反复喘息有关。在成人中，特征是轻度到中度的上呼吸道疾病。然而，严重的肺炎可能会发生，特别是在附发病变或免疫状态下降的老年人中[73]。老年人比成人更容易患上呼吸道合胞病毒感染后的下呼吸道疾病，以啰音和喘息为最常见症状，但也会出现发热、咳嗽和鼻涕。

5.2 病原：RSV 分类和结构

hRSV 是肺病毒属、副黏病毒科的成员。其基因组是一个 15 222个核苷酸长的负链 RNA 分子，编码 11 种病毒蛋白，包括核蛋白（N）、融合蛋白（F）、表面糖蛋白（G）、小疏水蛋白（SH）、基质蛋白（M）和包括 L 蛋白（复制酶）和介导 α 干扰素抗性的毒力因子 NS1 和 NS2。根据与单克隆抗体（mAbs）板的反应模式和 G 蛋白之间的氨基酸（aa）序列差异，将 RSV 菌株分为两种不同的亚型，RSV-A 和 RSV-B。

融合蛋白 F 参与病毒进入宿主细胞的过程，随后参与合胞体形成。G 蛋白以高亲和力结合细胞表面蛋白多糖，但不是感染所必需的。总的来说，F 和 G 在介导受体结合中的作用尚未阐明，尽管两者都被证明与肝素相互作用[11,20]。最近，核仁蛋白被鉴定为 RSV 的细胞受体[67]。G 蛋白被证明是趋化因子的一种结构和功能模拟物，趋化因子是一种介导白细胞迁移和黏附的促炎性 CX3C 趋化因子，这解释了其在发病机制中的作用[69]。F 和 G 是已知的唯一两种诱导 RSV 中和抗体的成分[11,12]。

细胞毒性 T 淋巴细胞（CTL）反应对根除 RSV 感染细胞和下调 Th2 细胞因子都很重要。另一方面，RSV 特异性 $CD8^+T$ 细胞也能增加 RSV 感染小鼠的肺病理[6]。儿童[27]和老年 RSV 感染患者[14]的 CTL 反应差可能是导致更严重疾病的原因之一。在人类中，N、F、M 和 SH 蛋白中发现了许多 CTL 表位。迄今为止，G 蛋白中还没有发现人 CTL 表位。用表达 $H-2^d$ 限制性表位 $M2_{82-90}$ 蛋白的痘苗病毒免疫小鼠，可预防 RSV 感染。然而，这种由 CTL 介导的保护作用在几周内减弱，似乎对长期保护没有贡献[12]。虽然中和抗体是 RSV 疫苗接种的一个主要目标，但一个有趣的策略可能是将长效抗体和 T 细胞免疫结合起来，以防止再感染和/或增强疾病的强有力保护。

5.3 传播

RSV 是一种高度传染性病原体。受感染儿童从发病之初起有 2 周的传染性。通常在喘息消失时，这种疾病被认为是非传染性的，当含有病毒的液滴被感染者打喷嚏或咳嗽到空气中时就会传播。如果有人吸入病毒粒子或病毒粒子接触他们的鼻子、嘴或眼睛，可能会被感染。RSV 通常由感染了 RSV、有轻微上呼吸道感染（如感冒）的学龄儿童引入家庭。呼吸道合胞病毒可以迅速传播给其他家庭成员。

5.4 诊断和经典疗法

毛细支气管炎诊断通常是根据病史和身体检查作出的：临床症状和体征可能包括呼吸困难或呼吸短促、气喘、流鼻涕、呼吸急促和鼻腔扩张。快速的 RSV 诊断可以依靠免疫荧光（IF）或酶联免疫吸附试验（ELISA）。IF 技术快速且易于实施，但结果解释是主观的，标本必须含有足够的鼻咽细胞。ELISA 具有解释更客观、速度快、筛选大量标本的可能性大等优势。

诸如逆转录聚合酶链式反应（RT-PCR）检测的核酸技术现在可用于 RSV 检测。有多种试剂盒，包括其他呼吸道病原体检测。这些检测方法的灵敏度通常超过病毒分离和抗原检测方法的灵敏度[57]。应考虑使用高度灵敏的RT-PCR检测方法，特别是对低病毒载量呼吸样本。

对于患有呼吸窘迫的幼儿，治疗LRT需要相当多的支持性护理：分泌物的机械抽吸、湿氧供应和大多数情况下的呼吸辅助。病毒唑是唯一一种抗病毒化合物，对RSV疾病有疗效。然而其毒性限制了在儿科患者中的应用。预防性给予RSV中和免疫球蛋白（RespiGam®）或hMAbs（Synagis®）形式的被动免疫已被证明可预防患有潜在心肺疾病的新生儿和早产儿感染RSV。然而这些昂贵的治疗方法并非全世界所有儿童都能享受。疫苗接种是减少呼吸道合胞病毒传播和感染的最佳选择。

5.5 疫苗接种策略

自20世纪60年代以来，臭名昭著的FI-RSV疫苗“增强疾病”阻碍了RSV疫苗的开发。因此在研究中RSV疫苗不仅要达到保护作用，而且首先要绝对安全。长期以来，认为Th2型细胞因子是FI-RSV诱导增强型疾病的主要原因的研究范式最近已经失效。不仅Th2细胞因子显著增强，而且Th1型细胞因子和趋化因子也显著增强，这表明免疫反应更加复杂和普遍失调[3,56]。在设计RSV疫苗时，研究了几种方法：①采用鼻内（in）途径接种活疫苗，可以模拟自然感染，而不产生天然RSV株的致病性。②由F和/或G蛋白组成的非复制疫苗，当用适当的佐剂配制时，既能获得良好的体液免疫应答（中和RSV抗体），又能获得平衡的细胞免疫应答。

5.5.1 活疫苗

减毒和温度敏感（*ts*）活毒株的开发构成了产生RSV毒株的方法，该毒株能够在上呼吸道（URT）复制，而不能在下呼吸道（LRT）复制。经鼻腔内给予减毒RSV，可对血清阴性的婴儿同时给予URT和LRT保护。获得有效（免疫原性）和遗传稳定（不可逆的）的*ts*毒株仍有待努力。由于自然感染不能保证婴儿持续的保护性免疫，问题仍然是：减毒活RSV能比野生型RSV做得更好吗？

还有其他方法可以利用异源活载体宿主表达RSV F/G蛋白，如仙台病毒、副流感病毒（hPIV 3）或牛RSV表达人类F/G蛋白[9,11]。作为病毒载体的替代，我们探索了使用共生或食品级革兰氏阳性菌作为G蛋白衍生载体的可能性。这里我们从基于G蛋白中心结构域的Ag亚单位开始，其特征在于胃肠外免疫后诱导高水平IgG抗体应答的能力，或者在黏膜免疫后诱导IgA和IgG抗体应答的能力。

为了简单明了，我们提供了如下数据：①选择G蛋白中央结构域的原理（图5.1）；②以不同形式产生的不同G构建体（图5.2）；③针对胃肠外和黏膜免疫的G蛋白亚单位疫苗候选物（图5.3）；④针对黏膜免疫的革兰氏阳性菌表面展示的G蛋白衍生物（图5.4）。我们以表格形式总结了实验数据。

5.5.2 亚单位疫苗

理论上，天然F和G糖蛋白被认为是诱导抗体介导RSV保护作用中最有效的抗原。在我们手中，除了病毒粒子来源的天然F提取物，重组片段或合成肽并不能复制F蛋白的真实构象结构。

对两种蛋白质的大量肽序列进行广泛筛选后，我们的研究集中在RSV-A $G_{130-230}$蛋白质的中央保守结构域G2Na上（图5.1）。G蛋白胞外结构域分为3个区域：N末端G_{66-130}、中央结构域$G_{130-230}$和C末端$G_{230-298}$。在G2Na片段（黄色）中，糖基化显著减少，RSV A亚型和B亚型之间的序列相对保守。相反地，N和C末端序列（橙色）是高度可变的，并且高度糖基化。G2Na有几个特别的特点。

- 疏水结构域$G_{164-176}$在RSV亚群A和B之间完全相同，可诱导交叉保护免疫反应。

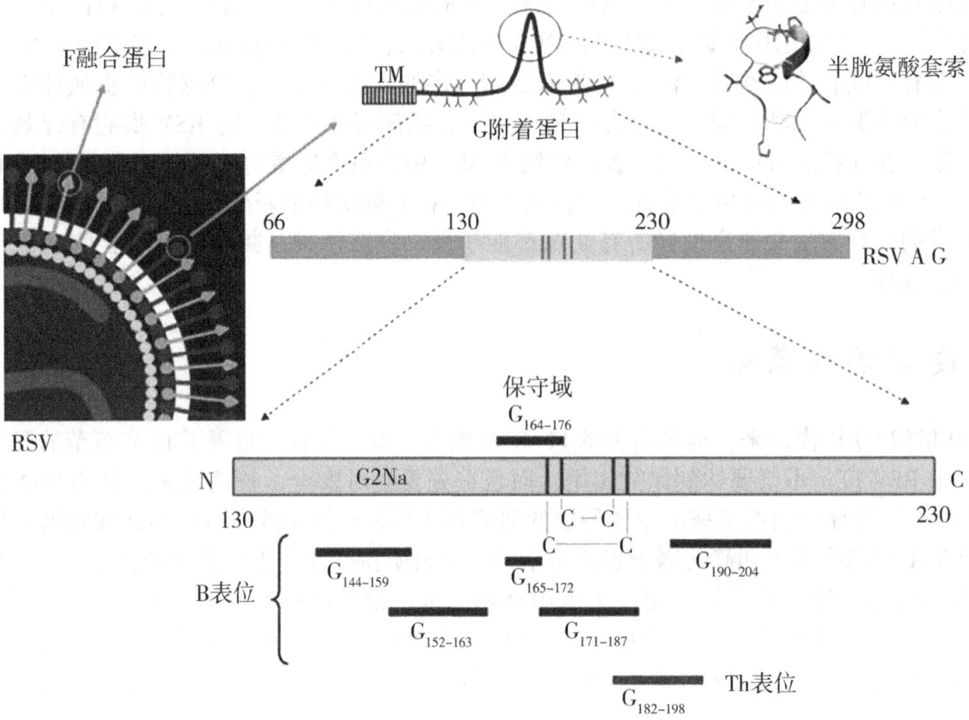

图 5.1 带有表面糖蛋白 F 和 G 的 RSV

在 N 末端带有跨膜（*TM*）区域的附着糖蛋白 G。胞外结构域的 N 端和 C 端都是高度糖基化的（Y），而中心部位是非糖基化的。注意右侧放大的半胱氨酸桥环结构。RSV-AG2Na$_{130-230}$的 B 细胞保护表位和 Th 细胞表位的线性表示。注意 4 个保守的 Cys 残基（垂直蓝色条纹）和它们之间的二硫键。

- 半胱氨酸富集区：

4 个形成构象半胱氨酸环结构的保守半胱氨酸残基（二硫键出现在 Cys_{173}和 Cys_{186}之间以及 Cys_{176}和 Cys_{182}之间）：这对诱导保护性抗体应答至关重要[2,70]。Trudel 等首次鉴定了一种有效的保护肽 G1a$_{174-187}$[70]。

Cys_{182}和 Cys_{186}位于 CX3C 趋化因子基序中[25,69]。含有这一基序的肽产生抑制 RSV G 蛋白结合的抗体，降低肺病毒滴度[76]并中和 RSV 的 A 和 B 亚型[10]。

作为免疫调节剂发挥佐剂作用[16,30,41]。

如果将半胱氨酸替换为丝氨酸，会影响免疫原性、亲和力成熟和保护效力[16,53]。

Th1/Th2 表位（CD4$^+$ Th 表位）。$G_{182-198}$负责小鼠的 URT 保护[26,53,54,68,72]。

- 富含脯氨酸保守片段 $G_{184-198}$具有保守的二级结构。
- 一个富含赖氨酸的肝素结合域对 RSV G 蛋白附着到细胞很重要[20]。
- G2Na 结构域识别并诱导保护性 Abs（保护体）的 5 个表位：$G_{150-157}$、$G_{163-174}$、$G_{178-185}$、$G_{171-187}$和 $G_{190-204}$[52,59]。

这一理论鼓励我们使用 G2Na 作为 RSV 疫苗的主要抗原。为了确保人类具有更好的免疫原性，我们测试了不同的载体蛋白：白蛋白结合域 BB[35]、白喉毒素衍生物 DT[48]和外膜蛋白 P40[24,47]。以融合蛋白或结合疫苗的形式对其进行评估（图 5.2，表 5.1 和表 5.2）。最后选择 BB 作为载体。首先，由于其对人血清白蛋白的高亲和力，它延长了融合蛋白在体内的半衰期[50]；其次，BB 主要通过诱导早期抗体应答来增强 G2Na 的免疫原性[35]。在临床试验前，在动物模型中广泛记录了 BBG2Na 的免疫原性、保护效力和安全性[13,56,58]。

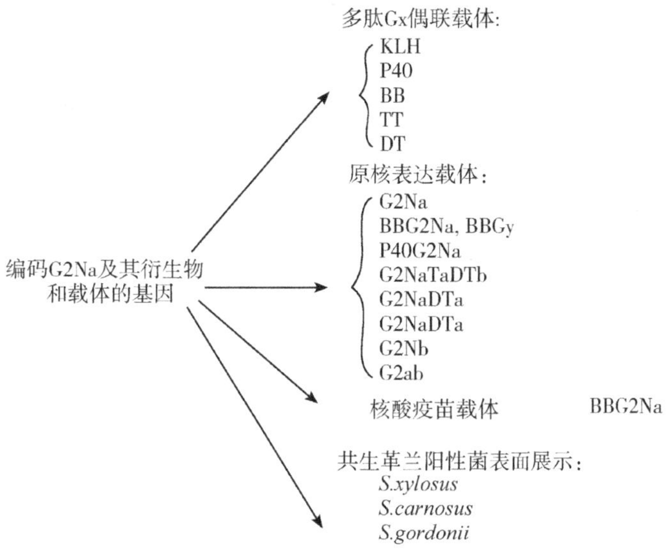

图 5.2 不同的 G2Na 衍生物和 RSV 肽抗原

（1）肽 Gx 与各种载体蛋白化学偶联——G2Na 和 G2Na 与 BB、P40 或 DT 衍生物载体蛋白融合；（2）RSV-B 的 $G2Nb_{130-230}$，RSV-A 和 RSV-B $G_{130-230}$蛋白的融合蛋白 G2ab；（3）核酸疫苗载体中的 BBG2Na；（4）革兰氏阳性菌表面表达的 G 肽或 G2Na。

表 5.1 不同 G2Na 衍生物和肽

抗原	名称	G(aa-aa)	氨基酸序列的性质	文献
RSV A	G2Na	130-230		[48,58]
	G200a	140-200		[53]
	G198a	140-198	天然的	[53]
	G196a	140-196		[53]
	G194a	140-194		[53]
	G192a	140-192		[53]
	G190a	140-190		[53]
	Gcf	131-230		[29,30]
	G2DCa	130-230	C_{173}和C_{186}用S替换	[53]
	G2Sera	130-230	所有的4C用S替换	[53]
	G1a	174-187	C_{186}用S替换	[70]
	G3a	3 X G1a	三聚体G1a肽	[45]
	G4a	172-187		[52,59]
	VG4a	171-187		[52,59]
	G5a	144-159		[52,59]
	G7a	158-190	天然的	[32]
	G8a	158-200		[32]
	G9a	190-204		[6,52,59]
	G11a	164-176		[52,59]
	G2Dela	G130-161 G171-230	嵌合体	[46]
	G20a	G190a G5a	嵌合体	[32]
	G4S	171-187	所有的4C用S替换	[8,65]
RSV B	G2DCb	130-230	C_{173}和C_{186}用S替换	[48]
	G1b	174-187	C_{186}用S替换	[65]
	G2Nb	130-230	天然的	[48]
	G4b	172-187		[2]
	VG4b	171-187		[2,65]
AB	G2ab	G2NaG2DCb	嵌合体	[48]

从左到右列：(i) 抗原分为 A 亚组（黄色条）、B 亚组（绿色条）和 A、B 亚组（蓝色条）。(ii) G2Na 和衍生抗原的名称。(iii) 包括的氨基酸。(iv) 氨基酸序列的性质：4 个天然 Cys_{173}，Cys_{176}，Cys_{182}和 Cys_{186}（红色条带 4 个黄色条纹）或只有两个中央 Cys_{176}和 Cys_{182}（或全部 4 个半胱氨酸被丝氨酸取代）。(v) 参考文献。

表 5.2 载体蛋白

载体蛋白名称	来源	参考文献
BB	链球菌 G 蛋白的白蛋白结合域	[35, 49]
DTa	白喉毒素 A 亚单位	[48]
DTb	白喉毒素 B 亚单位	[48]
DTaDTb	DTa 与 DTb 融合	[48]
KLH	钥孔戚血蓝蛋白	[70]
P40	肺炎克雷伯菌外膜蛋白 A	[24, 47, 59]

从左至右的列：(i) 载体蛋白的不同名称。(ii) 载体来源。(iii) 参考文献。

5.5.3 动物模型

5.5.3.1 BALB/c 小鼠

BALB/c 小鼠是一种白化近交品系，对 RSV 感染是半许可的，通常用于疫苗抗原的初步筛选。我们使用新生 BALB/c 小鼠来记录母源 RSV 抗体存在或不存在时的 BBG2Na 免疫原性[4,64]。相反地，为了模拟老年人的情况，RSV 免疫的小鼠被用来记录血清阳性人群的抗体应答[23,48]。这种方便的模型使我们能够研究 SPF 近交动物中的 RSV 感染（保护和免疫病理），有多种免疫试剂可供使用。

5.5.3.2 棉鼠

棉鼠比较容易感染 RSV。为了比较抗原制剂免疫原性，与 BALB/c 小鼠相比，棉鼠在 RSV 抗体应答方面更具鉴别性，棉鼠也非常有助于记录增强免疫病理[63]。唯一缺点是现有免疫试剂的数量有限。

5.5.3.3 非人灵长类动物

非洲绿猴和猕猴被用于评估我们的候选疫苗 BBG2Na[15,71]。猕猴似乎不太容易受到 hRSV 感染，因此该模型尚不可靠。这些模型的缺点是，在伦理和经济上常常限制了每组动物的数量，这可能导致毫无意义的统计数据。

5.5.3.4 人类灵长类

黑猩猩是最好的模型，极易感染 RSV，呼吸症状大多接近人类，但它们属于受保护物种：不能牺牲黑猩猩[11]。

5.5.4 肠胃外免疫接种

RSV 疫苗最重要的特点是保护下呼吸道（LRT），下呼吸道能够引起最致命的儿童病例。如经许可的 Synagis® 疫苗所示，血清 IgG 抑制 RSV 复制是预防肺部疾病的主要成分。大多数获得许可的人类疫苗是通过肠胃外途径注射的，主要是肌内注射（*im*）。在啮齿类动物模型中，对肌内注射（*im*）、腹腔注射（*ip*）和皮下注射（*sc*）进行了测试[22,59]。

抗原制备 G2 蛋白衍生物（>60aa）单独表达或与载体蛋白融合，即 BB 或 DT 亚单位。这些重组蛋白在大肠杆菌中高产率表达，在包涵体变性和复性后通过亲和色谱法纯化[48]。临床等级的 G2Na 和 BBG2Na 是从 Pierre Fabre 免疫中心中试规模生产获得的。肽段（<20aa）与载体蛋白化学偶联：BB 或 P40 或 KLH[52,59,70]。

表 5.3 是用于肠胃外免疫的肽/蛋白质的不同组合。对于结合肽 Gx、$G1a_{174-187}$，$G5a_{144-159}$、$VG4a_{171-187}$和 $G9a_{190-204}$，保护蛋白与 KLH 或 P40 或 BB 载体蛋白偶联；它们中大多数诱导小鼠对 RSV-A 的中度到高度保护性抗体应答。单独使用 G2Na 或与辅以明矾的 BB 或 DT 衍生物融合；经 *ip*、*im* 或 *sc* 免疫后可诱导高 RSV-A 抗体应答。

诱导的抗体应答始终保护了所有受 RSV-A 攻击的免疫动物的肺部。产生了从 C 端删除两个氨基酸残基的 G2Na（Gy）重组片段，使我们能够将 $G194a_{140-194}$ 定义为最小的保护性氨基酸序列[53]。另一种嵌合蛋白 G20a（$G_{140-190}$ 与 $G5a_{144-159}$ 融合）在小鼠中显示出保护作用[32]，但在棉鼠中则显示没有保护作用，因为棉鼠需要 G20a 融合到 DT 载体以诱导保护性免疫应答[44]。半胱氨酸残基对免疫活性至关重要。当 4 种半胱氨酸全部被丝氨酸（BBG2 Sera）取代时，免疫原性显著下降[53]。当 Cys_{173} 和 Cys_{186} 被取代时，免疫原性降低，但不足以使肺保护受损。

相比之下，上呼吸道（URT）保护受到影响，很可能是因为含有 Cys_{186} 的 Th 表位 $G_{184-198}$ 的改变。BBG2Na 在新生 BALB/c 小鼠中显示出免疫原性，即使存在母体 RSV 抗体，也没有任何增强的 FI-RSV 样疾病症状[4,54,56,64]。BBG2Na 和 G2Na 在 RSV 引发的小鼠中诱导持续（≥24 周）和高水平 RSV-A 抗体应答[23,48]。应该注意的是，诱导的抗体也与最近的临床分离株 BT2a 结合[48]。类似地，Jang 等人研究显示三聚 G2Na 抗原诱导抗体识别其他最近的临床分离株 hRSV-A 和 hRSV-B[28,74]。BBG2Na 也以核酸疫苗的形式进行评估。

当在小鼠体内肌内注射（*im*）时，载体诱导了用 RSV 攻毒的部分免疫保护反应。为了提高效率，需要根据 G2Na 的体内表达来优化核酸载体[1]。

表 5.4：异源保护功效。正如预期，$G1a_{174-187}$ 和 $VG4a_{171-187}$ A 亚组肽没有诱导抗 RSV-B 的交叉反应抗体。相反地，B 组 $G1b_{174-187}$ 肽诱导中度 RSV-B IgG 水平，保护动物肺免受 RSV-B 攻击[65]。令人惊讶的是，G2Nb（RSV-B）相当于 G2Na，它诱导了较弱的应答，并没有增强 RSV-B 抗体应答，既不单独存在，也不与 G2Na 混合或融合[48]。这可能是由于 G2Nb 结构的构象问题。与之形成鲜明对比的是，G2Na 和 BBG2Na 均能诱导抗 RSV-B 的中度交叉反应抗体：在 RSV-B 攻击后，免疫动物肺中的病毒滴度大大降低[48,58,62]。其他研究小组也证实了这一点，他们的研究表明，G2Na 中心保守半胱氨酸富集区在诱导对 RSV-B 株的交叉反应性抗体反应中很重要[10,28,43,76]。选择 BBG2Na 作为人类疫苗候选物有以下几个原因：BB 主要用作纯化目的亲和标记物[48]，显示其作为载体蛋白[35]，并延长了体内融合分子的半衰期[50]。

BBG2Na 在幼稚新生小鼠模型中具有免疫原性，即使在母亲抗体存在的情况下也是如此[4,64]。对孕妇接种疫苗的策略进行了研究，用 F 亚单位蛋白进行临床试验表明是可行的[42]。与 FI-RSV 疫苗接种的动物相比，在攻毒后 BBG2Na 没有引起增强型免疫病理[51]。虽然 BBG2Na /明矾引发 Th2 型应答，RSV 攻毒引起了 Th1/Th2 混合应答[13,56,60]。此外当用 DDA、CpG、MPL、弗氏佐剂或 TiterMax 作为 BBG2a 的佐剂时，与由 BBG2Na /明矾诱导的唯一 IgG1 抗体反应相比，混合 IgG1/IgG2a 抗体反应提高，表明 Th 应答可由几种人类佐剂调节。无论使用何种佐剂，保护效果都没有改变[4,31,44]。

表 5.3　肠外免疫

	抗原	佐剂	免疫途径	动物模型	RSV A ELISA 滴度	下呼吸道防止 RSVA 感染	参考文献
肽 Gx	KLH-G1a，KLH-G5a	弗氏佐剂	*ip*	小鼠	+++	+++	[70]
	P40-G1a，P40-VG4a，P40-G5a，P40-G9a	明矾	*ip*，*im*	小鼠	++	+++	[52]
	BB-G1a. BB-G5a，	明矾	*ip*	小鼠		+++	[59]
	G20a	明矾	*im*	小鼠	+++	+++	[32]
	G20aDTa	明矾	*im*	大鼠	+++	+++	[44]
	G125-203	明矾	*ip*	小鼠	+++	+++	[65]

（续表）

	抗原	佐剂	免疫途径	动物模型	RSV A ELISA 滴度	下呼吸道防止 RSVA 感染	参考文献
G2Na	G2Na	明矾	*ip*，*im*	小鼠和大鼠 +++	+++	+++	[48]
	G2Na	PLGA	*im*	小鼠和 RSV⁺ +++小鼠	+++	+++	[48]
	G2Na	PLGA 和/或明矾	*im*	小鼠和大鼠 +++	+++	+++	[44，48]
	G2Na+F	明矾	*im*	小鼠	+++	+++	[44]
	G2Na+G2Nb	明矾	*im*	小鼠	+++	+++	[48]
	G2ab	明矾	*im*	小鼠	+++	+++	[48]
	G2NaDTa，G2NaDTb，G2NaDTaDTb	明矾	*im*	小鼠	+++	+++	[48]
BBGy	BBG200a，BBG198a，BBG196a 和 BBG194a	明矾	*ip*	小鼠	+++	+++	[53]
	BBG192a and BBG190a	明矾	*ip*	小鼠	++	++	[53]
BBG2Na	BBG2Sera	明矾	*ip*	小鼠	+	++	[53]
	BBG2DCa	明矾	*ip*	小鼠	++	+++	[53]
	BBG2Na	明矾	*ip*，*im*，*sc*	小鼠和 RSV⁺小鼠	+++	+++	[23，58]
	BBG2Na	明矾或弗氏佐剂	*im*	非洲绿猴	+++	+++	[71]
	BBG2Na	明矾	*im*	猕猴	+	-	[15]
	BBG2Na	DDA	*im*	小鼠和大鼠 +++	+++	+++	[31]
	BBG2Na	DDA+明矾	*im*	小鼠和大鼠	+++	+++	[31]
	BBG2Na	CpG+明矾	*im*	小鼠	+++	+++	[44]
	BBG2Na	MPL+明矾	*im*	小鼠	+++	+++	[44]
	BBG2Na	Titer Max	*ip*，*im*	新生小鼠	+++	+++	[4，64]
	BBG2Na（核酸疫苗）	无佐剂	*im*	小鼠	-	+	[1]

列从左至右：(i) 抗原（G1a、G4a、G5a 和 G9a 肽与 KLH、P40 或 BB 载体蛋白偶联）。G2Na 单独或与白喉衍生物 DTa，DTb 或 DTaDTb 融合；Gy 片段融合于 BB 和 BBG2Sera，BBG2DCa，BBG2Na。(ii) 佐剂：明矾铝胶或佐剂-磷，*PLGA* 聚合物（D，L-丙交酯-共-乙交酯），*MPL* 单磷酰脂质 A，*DDA* 二甲基二十八烷基溴化物，*CpG* 胞嘧啶-磷酸盐-鸟嘌呤寡聚脱氧核苷酸。(iii) 免疫途径：*ip* 腹腔注射、*im* 肌内注射、*sc* 皮下注射。(iv) 动物模型：小鼠（BALB/c 幼稚动物，如果不说明，RSV⁺小鼠：RSV 初免的小鼠），大鼠（棉鼠）、猴子（非洲绿猴）。(v) RSV-A IgG ELISA 滴度：+++高、++中、+低，-检测不到。(vi) LRT 防止 RSV-A 感染：+++完整，++良好，+部分，-无保护。(vii) 参考资料。

表 5.4 肠外免疫—交叉反应性 RSV A 和 RSV B

抗原	佐剂	免疫途径	动物模型	LRT RSV A	保护 RSV B ELISA	vs RSV A ELISA	LRT 保护 vs RSV B	参考文献
KLH-G1a	弗氏佐剂	*ip*	小鼠	+++	+++	–	–	[65, 70]
KLH-VG4a	弗氏佐剂	*ip*	小鼠	+++	+++	–	–	[65]
KLH-G1b	弗氏佐剂	*ip*	小鼠	–	–	++	++	[65]
G2Na	明矾	*im*	小鼠和大鼠	+++	+++	++	++	[48]
G2Nb	明矾	*im*	小鼠	+	NT	+	NT	[48]
G2aba	明矾	*im*	小鼠	+++	NT	++	NT	[48]
G2Na+G2Nb	明矾	*ip*, *im*	小鼠	+++	NT	++	NT	[48]
G2Na+F	明矾	*im*	小鼠和大鼠	+++	+++	+++	NT	[44]
BBG2Na	明矾	*ip*, *im*, *sc*	小鼠和大鼠	+++	+++	++	++	[58, 62]
BBG2Na	明矾或 FA	*im*	非洲绿猴	+++	+++	++	NT	[71]
G2 (aa148–198)	TiterMax	*im*	小鼠	+++	+++	++	NT	[10, 76]

从左至右：(i) 抗原名称。(ii) 佐剂。(iii) 免疫途径：*ip* 腹腔注射、*im* 肌内注射、*sc* 皮下注射。(iv) 动物模型：BALB/c 小鼠或棉鼠。(v) RSV-A IgG ELISA 滴度：+++高、++中、+低。(vi) LRT 保护（肺）免受 RSV 感染：+++完整，++良好，+部分，–无保护。(vii) 参考文献。(viii) *NT* 未测。

在非洲绿猴中，通过 *im* 注射，当添加不完全弗氏佐剂（IFA）时，BBG2Na 具有高度免疫原性，而用明矾诱导了中等高的 RSV-A 抗体滴度。诱导的抗体也与 RSV-B 交叉反应。在随后的 RSV-A 攻毒研究中，发现用明矾或 IFA 作为佐剂的 BBG2Na 疫苗显示出保护功效[71]。在猕猴中 BBG2Na 显示出弱免疫原性[15]。模型中 RSV 敏感性和有限数量的动物使得对数据的解释不可靠。为了模拟老年人的情况，我们发现 BBG2Na 在 RSV 感染的小鼠中具有免疫原性，没有观察到抗体应答的抑制现象[23,48]。就安全性和免疫原性而言，BBG2Na 成功通过了临床Ⅰ期和Ⅱ期试验[34,55,61]。不幸的是，由于极少数疫苗接种者发生罕见的不良反应，临床Ⅲ期试验被终止。

事件的时间顺序、延迟发病和症状提示与疫苗相关的Ⅲ型超敏反应有关。当在一个Ⅲ型超敏家兔模型中进行试验时，我们发现 BBG2Na 诱导了 Arthus 反应（血管炎型变态反应），是 BB 组分（而不是 G2Na）导致此效果[36]。这为进一步研究 G2Na 的免疫原性和保护作用提供了动力。

尽管在我们之前的研究中，我们发现 BBG2Na 诱导的 G2Na 抗体应答比 G2Na 单独诱导的早且高[35]，但我们最近发现，无论有无载体蛋白（DT 衍生物或 BB），G2Na 诱导的小鼠和棉鼠的 RSV-A 和 RSV-B IgG 应答水平相似[48]。在棉鼠中 G2Na/明矾诱导对 RSV-A 和 RSV-B 的高持续抗体应答达 148 天。这些抗体保护动物肺部免受 RSV-A 感染，相当于用 Synagis® 对照治疗的动物。重要的是，我们提供了强有力的证据，证明在存在抗 RSV 抗体的情况下，用不同的 G2Na 制剂进行免疫并没有随着时间的推移损害抗体的亲和力成熟[48]。为了获得更好的交叉反应性和长期的 CTL 反应，理想情况下，G2Na 应该与 F 和/或 M 蛋白联系在一起使用。事实上 G 半胱氨酸富集结构域在与 RSV 基质 M 蛋白 Ag 共用[5]或与 F 和 M2 蛋白融合时均能增强 CTL 反应：

$G_{125-225}$ F/M_{81-95}[75]。这一策略的缺点是从 RSV 病毒中纯化 F 蛋白的产量仍然很低，目前还没有在工业规模上成功地生产出重组 F 蛋白。我们目前正在研究是否有可能将 RSV-B G 肽与 G2Na 结合（图 5.3）。

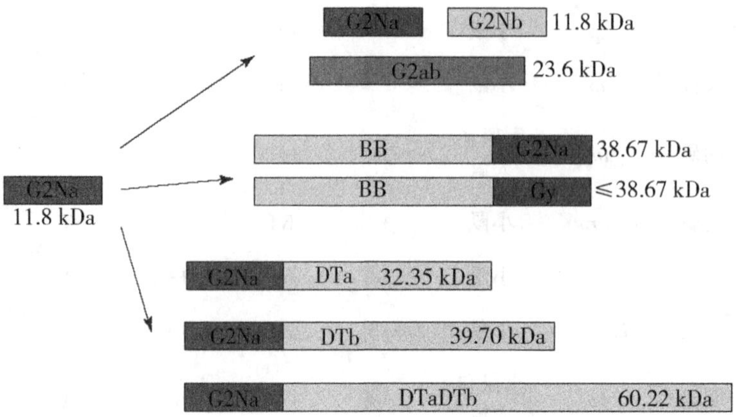

图 5.3　重组蛋白的不同结构

G2Na、G2Nb、G2ab、BBG2Na、BBGy（Gy：G2Sera，G2DCa，$G_{140-190}$，$G_{140-192}$，$G_{140-194}$，$G_{140-196}$，$G_{140-198}$，$G_{140-200}$）-G2Na 与 DT 的衍生物融合。

5.5.5　黏膜免疫

除了肠胃外途径，我们还探索了替代途径即黏膜免疫（表 5.5）。在动物模型中，黏膜免疫接种并不容易增强免疫病理。通过诱导黏膜 IgA 作为对抗病毒的一线免疫防御，黏膜免疫接种可能进一步提高疫苗效力。虽然 RSV 特异性 IgA 抗体在人类保护中的作用仍不清楚，但在小鼠发生原发性 RSV 感染后，其上呼吸道会迅速分泌 RSV 特异性 IgA 抗体，从而起到保护作用。G2Na 用霍乱毒素（CT）作为佐剂经皮免疫（*tsc*）后，诱导了低 RSV-A IgG 滴度，给予部分保护[21]。然而鼻内接种是实现保护的常用方法。这种途径的主要优点是易于接种（无针头），因此更容易被父母接受。但是需要使用黏膜佐剂，如霍乱毒素 B 亚单位（CTB）或二甲基十八烷基溴化铵（DDA）[33]。CTB 具有重要意义：作为载体分子的免疫增强能力被认为与其与黏膜上皮细胞中单唾液酸神经节苷脂 GM1 的结合能力有关[7,37]。当在小鼠中接种时，用 CTB 或 DDA 辅助的 BG2Na 诱导局部 IgA 和全身 IgG，保护肺免受 RSV-A 侵袭。另一组显示，$G2_{131-230}$通过舌下注射（*sl*）或接种（无论有无 CT 佐剂），均可诱导小鼠中 IgA 和 IgG 的保护性作用[30]。这些数据证实并加强了在黏膜免疫中，G2Na 是潜在的 RSV 亚单位候选物。

活疫苗构成了另一种替代疫苗接种原则，这些疫苗可以通过口服或口服途径接种。目前正在开发减毒 RSV 活疫苗[9,11]。我们选择了一种非常不同的方法即使用食品级重组葡萄球菌，即木糖葡萄球菌和肉葡萄球菌作为疫苗递送载体（图 5.4 和表 5.5）。这两种非致病性菌株均证明高剂量黏膜或皮下注射是安全的[66]。在小鼠中，口服表达 G3 的木糖酵母菌（一种修剪过的 $G1a_{174-187}$表位）能够触发血清 G3 特异性 IgG，这种 IgG 在免疫后第 143 天都能检测到[45]。利用 G2Na 保守域 $G_{164-176}$的疏水性工程，分泌和表面展示木糖表面的 G2Sub（苯丙氨酸替代丝氨酸）或 G2Del（疏水性拉伸缺失）[46]。然而，皮下注射（*sc*）重组细菌免疫诱导的 RSV 抗体水平很低。

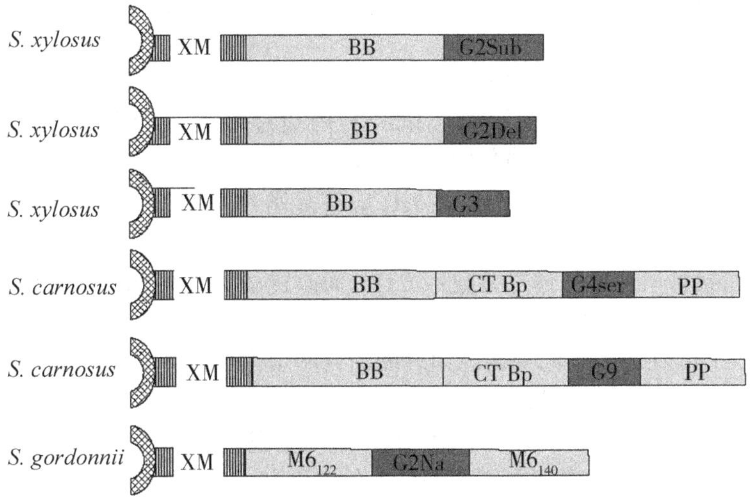

图 5.4 革兰氏阳性菌不同 G 片段抗原结构的表面展示

表 5.5 黏膜免疫

抗原名称	抗原	佐剂	给药途径	动物模型	RSV A IgG	ELISA IgA	LRT 保护 VS. RSV A	参考文献
蛋白	G2Na	CT	*tsc*	小鼠	+	*ND*	++	[21]
	G2cf (131-230)	±CT	*sl and in*	小鼠	++	++	+++	[30]
	BBG2Na	CTB	*in*	小鼠和大鼠	+++	++	+++	[33]
	BBG2Na	DDA	*in*	小鼠和大鼠	+++	++	+++	[33]
S. xylosus 木葡萄球菌	XM | BB | G2Del		*in*	小鼠	-	*ND*	NT	[46]
	XM | BB | G2Sub		*in*	小鼠	-	*ND*	NT	[46]
	XM | BB | G3		口服	小鼠	+	*ND*	NT	[45]
S. carnosus 肉葡萄球菌	XM | BB | CTBp | G4ser | PP		*in*	小鼠	+	*ND*	+	[8]
	XM | BB | CTBp | G9 | PP		*in*	小鼠	+	*ND*	++	[8]
S. gordonii 戈登氏葡萄球菌	S. gordonii XM | M6122 | G2Na | M6141		*in*	小鼠	+/-	+	++	[18]
BAC 杆状病毒	杆状病毒 BAC-G2Na		*in*	小鼠	++	++	+++	[29]
AD 腺病毒	重组腺病毒 rAD-3xG2Na		*in*	小鼠	+++	*ND*	+++	[28]

从左至右的列：(i) G 蛋白衍生物黏膜佐剂在革兰氏阳性细菌表面显示的抗原名称：*S. xylosus* 木糖葡萄球菌；*S. carnosus* 肉葡萄球菌和 *S. gordonii* 戈登氏葡萄球菌；表达 G2Na 的杆状病毒载体和腺病毒载体。(ii) 蛋白抗原佐剂：*CT* 霍乱毒素、*CTBp* 霍乱毒素亚单位 B、*DDA* 二甲基十八烷基溴化铵。(iii) 给药途径：*in* 鼻内给药、*sl* 舌下给药。(iv) 动物模型：BALB/c 小鼠和/或棉鼠。(v) RSV-AIgGELISA 滴定度或 RSV-A IgA ELISA 滴定度：*ND* 不确定，+++高，++中，+低，+/-非常低，-探测不到。(vi) LRT 对 RSV-A 的防护：+++完全保护，++良好，+部分，NT 未测试。(vii) 参考文献。

尽管有报道称用肉葡萄球菌系统作为递送口服载体，表面暴露的抗原会引起全身抗体应答[38,40,66]，但在通过黏膜途径免疫接种后，根据暴露于表面的抗原决定簇产生的抗体反应来改善肉葡萄球菌系统。

为了增加抗体应答，在细菌表面共表达一种细菌黏附因子，一种来自乳链球菌的纤连蛋白结合域将疫苗靶向黏膜[7,37-39]。这导致小鼠经鼻内免疫后对共显抗原的血清 IgG 应答增加 1.5 $\log_{10}$[39]。另一种方法中，在肉葡萄球菌表面共表达的 CTB_{50-75}（CTBp）被证明具有免疫增强作用[17]。鼻内给药后，其血清抗原特异性 IgG 应答显著增加（接近 2$\log_{10}$）[7]。进一步研究表明，需要活的葡萄球菌来获得这种效果。

在随后的疫苗接种研究中，通过共显示 CTBp 而改进的肉葡萄球菌系统被用于递送 G 肽[8]。3 种肽即 G 144-159（G5）、G 190-203（G9）和 G 171-188（G4S）在肉葡萄球菌系统中表达，通过重组方法使 CTBp 在 3 种不同葡萄球菌表面的共展示得到改善。用活的重组葡萄球菌对小鼠进行鼻内免疫，可引起显著的抗肽和抗 RSV-A IgG2a/IgG1 反应。在 G9 和 G4S 免疫组中，经 RSV-A 攻毒后，大约一半的小鼠表现出肺保护作用。据我们所知，这是第一次使用食品级细菌作为疫苗递送载体来激发对病毒病原体的保护性免疫研究。

后来在类似方法中，共生细菌格氏链球菌被成功用于在细胞表面表达整个 G2Na。用活细菌鼻内免疫诱导 IgA 对抗 RSV-A 和 G2Na。RSV-A 攻毒后，免疫动物的肺部病毒滴度明显低于对照动物[18]。

另一组研究最近证明，在鼻内用重组复制缺陷型腺病毒疫苗（血清型 5）表达三聚体 G2Na 在小鼠中诱导保护性抗体[28,74]。在另一项研究中，用表面展示 $G_{131-230}$的重组杆状病毒对小鼠鼻内接种，可诱导保护性 IgG 和 IgA 抗体应答[29]。这些数据证实并加强了 G2Na 作为通过活载体进行黏膜输送合适 RSV 抗原候选物的潜力。

5.6 优点和缺点

有效的 hRSV 疫苗似乎很难开发。RSV 发现后的半个多世纪，仍然没有疫苗可用。免疫学和疫苗学的最新进展有助于更好地理解 RSV 是如何逃避免疫应答的，以及它是如何在人类中引起病理的。动物模型可以用来评估各种抗原的安全性和有效性。由于开发了替代人类佐剂以替代明矾，非复制疫苗有了新的希望。挑战是双重的：首先，抗原在人类中应该是非常安全和有免疫原性的；其次，RSV 疫苗的生产成本应该在经济上是可行的，因为这是世卫组织优先考虑的，应该在全世界范围内分发和获取。

总之，对于非肠道疫苗，我们的研究已经清楚地表明，G2Na 具有许多适合老年人预防 RSV 疫苗抗原的特性。在动物身上它诱导出高度保护性的 RSV-A 抗体，这种抗体也与 RSV-B 交叉反应。这种疫苗对于限制医护人员（如流感疫苗）和患者之间的 RSV 传播非常有用。对于血清阴性的儿童，应该在动物模型中从免疫病理学角度对其进行仔细监测。G2Na 可以与 RSV-B G 肽组分或 F 蛋白衍生物一起作用，和/或与特异性免疫应答驱动佐剂（例如 TLR 激动剂）一起配制，以实现安全性和保护效果。这种 G2Na 抗原可以以优异的产量重组产生。工业流程（上游和下游）已经被优化，以提供合格的注射用人疫苗。

未来的挑战是提供一种安全、高效、廉价且易于接种的疫苗，该疫苗可以经黏膜免疫（鼻内、皮下或口服）。只有活的载体才能达到这些标准。我们已经证明，食品级革兰氏阳性细菌可以引起保护性局部和全身免疫应答。其他组的研究也证实了我们的结果，其他细菌和病毒载体表面显示相同的 G2Na 抗原。然而仍有优化的空间：提高人类黏膜 RSV 疫苗的免疫原性、稳定性和安全性。

致谢

C Andersson, M Azin, T Baussant, A Beck, H Binz, JF Boe, JY Bonnefoy, N Bouveret-Le-Cam, JF Cano, B Chol, JC Corbiere, N Corvaia, N Dagouassat, F Farinole, L Goetsch, M Hansson, JF Haeuw, K Helffer, T Huss, C Issac, C Klinguer, PH Lambert, M Lokteff, S Liljeqvist, M Maitre, C Nadon, Q Nguyen, PA Nygren, H Plotnicky-Gilquin, A Perez, U Power, A Robert, P Samuelsson, CA Siegrist, C Simard, M Trudel, M Uhlén, D Velin and L Zanna.

参考文献

[1] Andersson, C., Liljestrom, P., Stahl, S., Power, U. F.: Protection against RSV elicited in mice by plasmid DNA immunisation encoding a secreted RSV G protein-derived antigen. FEMS Immunol. Med. Microbiol. 29, 247-253 (2000).

[2] Beck, A., et al.: Synthesis and characterization of RSV protein G related peptides containing two disulfide bridges. J. Pept. Res. 55, 24-35 (2000).

[3] Blanco, J. C., Boukhvalova, M. S., Shirey, K. A., Prince, G. A., Vogel, S. N.: New insights for development of a safe and protective RSV vaccine. Hum. Vaccin. 6, 482-492 (2010).

[4] Brandt, C., et al.: Protective immunity against RSV in early life after murine maternal or neonatal vaccination with the recombinant G fusion protein BBG2Na. J. Infect. Dis. 176, 884-891 (1997).

[5] Bukreyev, A., et al.: The cysteine-rich region and secreted form of the attachment G glycoprotein of RSV enhance the cytotoxic T-lymphocyte response despite lacking major histocompatibility complex class I-restricted epitopes. J. Virol. 80, 5854-5861 (2006).

[6] Cannon, M. J., Openshaw, P. J., Askonas, B. A.: Cytotoxic T cells clear virus but augment lung pathology in mice infected with RSV. J. Exp. Med. 168, 1163-1168 (1988).

[7] Cano, F., et al.: A surface-displayed cholera toxin B peptide improves antibody responses using food-grade staphylococci for mucosal subunit vaccine delivery. FEMSImmunol. Med. Microbiol. 25, 289-298 (1999).

[8] Cano, F., et al.: Partial protection to RSV elicited in mice by intranasal immunization using live staphylococci with surface-displayed RSV-peptides. Vaccine18, 2743-2752 (2000).

[9] Chang, J.: Current progress on development of RSV vaccine. BMB Rep. 44, 232-237 (2011).

[10] Choi, Y., et al.: Antibodies to the central conserved region of respiratory syncytial virus (RSV) G protein block RSV G protein CX3C-CX3CR1 binding and cross-neutralize RSV A and B strains. Viral Immunol. 25, 193-203 (2012).

[11] Collins, P. L., Crowe, J. E., Jr.: Respiratory syncytial virus and metapneumovirus. In D. M. Knipe and P. M. Howley (eds.), Fields Virology. Lippincott Williams &Wilkins, Philadelphia, Vol. 2. 5th ed. 1601-1646 (2007).

[12] Connors, M., et al.: Resistance to RSV challenge induced by infection with a vaccinia virus recombinant expressing the RSV M2 protein (Vac-M2) is mediated by $CD8^+$T cells, while that induced by Vac-F or Vac-G recombinants is mediated by antibodies. J. Virol. 66, 1277-1281 (1992).

[13] Corvaia, N., et al.: Challenge of BALB/c mice with RSV does not enhance the Th2 pathway induced after immunization with a recombinant G fusion protein, BBG2Na, in aluminum hydroxide. J. Infect. Dis. 176, 560-569 (1997).

[14] de Bree, G. J., et al.: Respiratory syncytial virus-specific $CD8^+$memory T cell responses in elderly persons.

J. Infect. Dis. 191, 1710–1718 (2005).

[15] de Waal, L., et al.: Evaluation of BBG2Na in infant macaques: specific immune responses after vaccination and RSV challenge. Vaccine22, 915–922 (2004).

[16] Delgado, M. F., et al.: Lack of antibody affinity maturation due to poor Toll-like receptor stimulation leads to enhanced RSV disease. Nat. Med. 15, 34–41 (2009).

[17] Delmas, A., Partidos, C. D.: The binding of chimeric peptides to GM1 ganglioside enables induction of antibody responses after intranasal immunization. Vaccine14, 1077–1082 (1996).

[18] Falcone, V., et al.: Systemic and mucosal immunity to RSV induced by recombinant Streptococcus gordonii surface-displaying a domain of viral glycoprotein G. FEMS Immunol. Med. Microbiol. 48, 116–122 (2006).

[19] Falsey, A. R., et al.: RSV and influenza A infections in the hospitalized elderly. J. Infect. Dis. 172, 389–394 (1995).

[20] Feldman, S. A., Hendry, R. M., Beeler, J. A.: Identification of a linear heparin binding domain for human RSV attachment glycoprotein G. J. Virol. 73, 6610–6617 (1999).

[21] Godefroy, S., et al.: Immunization onto shaved skin with a bacterial enterotoxin adjuvant protects mice against RSV. Vaccine21, 1665–1671 (2003).

[22] Goetsch, L., et al.: Influence of administration dose and route on the immunogenicity and protective efficacy of BBG2Na, a recombinant RSV subunit vaccine candidate. Vaccine18, 2735–2742 (2000).

[23] Goetsch, L., et al.: BBG2Na an RSV subunit vaccine candidate intramuscularly injected to human confers protection against viral challenge after nasal immunization in mice. Vaccine19, 4036–4042 (2001).

[24] Haeuw, J. F., et al.: The recombinant Klebsiella pneumoniae outer membrane protein OmpA has carrier properties for conjugated antigenic peptides. Eur. J. Biochem. 255, 446–454 (1998).

[25] Harcourt, J., et al.: RSV G protein and G protein CX3C motif adversely affect CX3CR1+T cell responses. J. Immunol. 176, 1600–1608 (2006).

[26] Huang, Y., Anderson, R.: A single amino acid substitution in a recombinant G protein vaccine drastically curtails protective immunity against RSV. Vaccine21, 2500–2505 (2003).

[27] Isaacs, D., Bangham, C. R., McMichael, A. J.: Cell-mediated cytotoxic response to RSV in infants with bronchiolitis. Lancet 2, 769–771 (1987).

[28] Jang, J. E., et al.: Evaluation of protective efficacy of RSV vaccine against A and B subgroup human isolates in Korea. PLoS One 6, e23797 (2011).

[29] Kim, S., Chang, J.: Baculovirus-based vaccine displaying RSV glycoprotein induces protective immunity against RSV infection without vaccine-enhanced disease. Immune. Netw. 12, 8–17 (2012).

[30] Kim, S., et al.: Dual role of RSV glycoprotein fragment as a mucosal immunogen and chemotactic adjuvant. PLoS One 7, e32226 (2012).

[31] Klinguer-Hamour, C., et al.: DDA adjuvant induces a mixed Th1/Th2 immune response when associated with BBG2Na, a RSV potential vaccine. Vaccine20, 2743–2751 (2002).

[32] Klinguer-Hamour, C., et al.: Synthesis, refolding and protective immune responses of a potential antigen for human RSV vaccine. J. Pept. Res. 62, 27–36 (2003).

[33] Klinguer, C., et al.: Lipophilic quaternary ammonium salt acts as a mucosal adjuvant when co-administered by the nasal route with vaccine antigens. Vaccine19, 4236–4244 (2001).

[34] Le Cam, N. B., et al.: Safety and immunogenicity of BBG2Na a RSV subunit vaccine based on G derived recombinant protein in the elderly: a phase II study. In: 40th ICAAC meeting. Toronto. 21–26 Sept 1999.

[35] Libon, C., et al.: The serum albumin-binding region of Streptococcal protein G (BB) potentiates the im-

munogenicity of the G130-230 RSV-A protein. Vaccine17, 406-414 (1998).

[36] Libon, C., et al.: Identification of the component responsible for the type III hypersensitivity reaction (HS III) induced by BBG2Na, a subunit RSV vaccine. In: Poster-Sixth International RSV meeting, Marco Island, FL, 25-28 Oct 2007.

[37] Liljeqvist, S., et al.: Surface display of the cholera toxin B subunit on Staphylococcus xylosus and Staphylococcus carnosus. Appl. Environ. Microbiol. 63, 2481-2488 (1997).

[38] Liljeqvist, S., et al.: Fusions to the cholera toxin B subunit: influence on pentamerization and GM1 binding. J. Immunol. Methods210, 125-135 (1997).

[39] Liljeqvist, S., et al.: Surface display of functional fibronectin-binding domains on Staphylococcus carnosus. FEBS Lett. 446, 299-304 (1999).

[40] Liljeqvist, S., Stahl, S.: Production of recombinant subunit vaccines: protein immunogens, live delivery systems and nucleic acid vaccines. J. Biotechnol. 73, 1-33 (1999).

[41] Melendi, G. A., et al.: Conserved cysteine residues within the attachment G glycoprotein of RSV play a critical role in the enhancement of cytotoxic T-lymphocyte responses. Virus Genes42, 46-54 (2011).

[42] Munoz, F. M., Piedra, P. A., Glezen, W. P.: Safety and immunogenicity of RSV purified fusion protein-2 vaccine in pregnant women. Vaccine21, 3465-3467 (2003).

[43] Murata, Y., Catherman, S. C.: Antibody response to the central unglycosylated region of the RSV attachment protein in mice. Vaccine30, 5382-5388 (2012).

[44] Nguyen, T. N., Libon, C.: G2Na & BBG2Na. Unpublished.

[45] Nguyen, T. N., et al.: Cell-surface display of heterologous epitopes on Staphylococcus xylosus as a potential delivery system for oral vaccination. Gene128, 89-94 (1993).

[46] Nguyen, T. N., et al.: Hydrophobicity engineering to facilitate surface display of heterologous gene products in Staphylococcus xylosus. J. Biotechnol. 42, 207-219 (1995).

[47] Nguyen, T. N., et al.: Chromosomal sequencing using a PCR-based biotin-capture method allowed isolation of the complete gene for the outer membrane protein A of Klebsiella pneumoniae. Gene210, 93-101 (1998).

[48] Nguyen, T. N., et al.: The RSV G protein conserved domain induces a persistent and protective antibody response in rodents. PLoS One 7, e34331 (2012).

[49] Nygren, P. A., Eliasson, M., Abrahmsen, L., Uhlen, M., Palmcrantz, E.: Analysis and use of the serum albumin binding domains of streptococcal protein G. J. Mol. Recognit. 1, 69-74 (1988).

[50] Nygren, P. A., Uhlen, M., Flodby, P., Wigzell, H.: In vivo stabilization of human recombinant CD4 derivative by fusion to a serum-albumin-binding receptor. Vaccines91 Cold Spring Habor Laboratory Press, 363-368 (1991).

[51] Plotnicky-Gilquin, H., et al.: Absence of lung immunopathology following RSV challenge in mice immunized with a recombinant RSV G protein fragment. Virology258, 128-140 (1999).

[52] Plotnicky-Gilquin, H., et al.: Identification of multiple protective epitopes (protectopes) in the central conserved domain of a prototype human RSV G protein. J. Virol. 73, 5637-5645 (1999).

[53] Plotnicky-Gilquin, H., et al.: CD4 (+) T-cell-mediated antiviral protection of the upper respiratory tract in BALB/c mice following parenteral immunization with a recombinant RSV G protein fragment. J. Virol. 74, 3455-3463 (2000).

[54] Plotnicky-Gilquin, H., et al.: Gamma interferon-dependent protection of the mouse upper respiratory tract following parenteral immunization with a RSV G protein fragment. J. Virol. 76, 10203-10210 (2002).

[55] Plotnicky - Gilquin, H., et al.: Passive transfer of serum antibodies induced by BBG2Na, a subunit vaccine, in the elderly protects SCID mouse lungs against RSV challenge. Virology303, 130-137 (2002).

[56] Plotnicky-Gilquin, H., et al.: Enhanced pulmonary immunopathology following neonatal priming with formalin-inactivated RSV but not with the BBG2NA vaccine candidate. Vaccine21, 2651-2660 (2003).

[57] Popow-Kraupp, T., Aberle, J. H.: Diagnosis of RSV infection. Open Microbiol. J. 5, 128-134 (2011).

[58] Power, U. F., et al.: Induction of protective immunity in rodents by vaccination with a prokaryotically expressed recombinant fusion protein containing a RSV G protein fragment. Virology230, 155-166 (1997).

[59] Power, U. F., et al.: Identification and characterisation of multiple linear B cell protectopes in the RSV G protein. Vaccine19, 2345-2351 (2001).

[60] Power, U. F., et al.: Differential histopathology and chemokine gene expression in lung tissues following RSV challenge of formalin - inactivated RSV - or BBG2Na - immunized mice. J. Virol. 75, 12421 - 12430 (2001).

[61] Power, U. F., et al.: Safety and immunogenicity of a novel recombinant subunit RSV vaccine (BBG2Na) in healthy young adults. J. Infect. Dis. 184, 1456-1460 (2001).

[62] Power, U. F., Plotnicky, H., Blaecke, A., Nguyen, T. N.: The immunogenicity, protective efficacy and safety of BBG2Na, a subunit RSV vaccine candidate, against RSV-B. Vaccine22, 168-176 (2003).

[63] Prince, G. A., Curtis, S. J., Yim, K. C., Porter, D. D.: Vaccine-enhanced RSV disease in cotton rats following immunization with Lot 100 or a newly prepared reference vaccine. J. Gen. Virol. 82, 2881 - 2888 (2001).

[64] Siegrist, C. A., et al.: Protective efficacy against RSV following murine neonatal immunization with BBG2Na vaccine: influence of adjuvants and maternal antibodies. J. Infect. Dis. 179, 1326-1333 (1999).

[65] Simard, C., et al.: Subgroup specific protection of mice from RSV infection with peptides encompassing the amino acid region 174-187 from the G glycoprotein: the role of cysteinyl residues in protection. Vaccine15, 423-432 (1997).

[66] Ståhl, S. et al. In: Pozzi, G., Wells, J. M. (eds.) Gram-Positive Bacteria: Chapter 4: Development of nonpathogenic Staphylococci as vaccine delivery vehicles. Vaccine vehicles for Mucosal Immunization, pp. 62-81. Landes Bioscience, Georgetown (1997).

[67] Tayyari, F., et al.: Identification of nucleolin as a cellular receptor for human RSV. Nat. Med. 17, 1132-1135 (2011).

[68] Tebbey, P. W., Hagen, M., Hancock, G. E.: Atypical pulmonary eosinophilia is mediated by a specific amino acid sequence of the attachment (G) protein of RSV. J. Exp. Med. 188, 1967-1972 (1998).

[69] Tripp, R. A., et al.: CX3C chemokine mimicry by RSV G glycoprotein. Nat. Immunol. 2, 732 - 738 (2001).

[70] Trudel, M., Nadon, F., Seguin, C., Binz, H.: Protection of BALB/c mice from RSV infection by immunization with a synthetic peptide derived from the G glycoprotein. Virology185, 749-757 (1991).

[71] Trudel, M., Power, U. F., Stahl, S., Bonnefoy, J. Y., Nguyen, N. T.: Immunogenicity and protective efficacy of a RSV subunit vaccine candidate BBG2Na in African green monkeys. In: 38th ICAAC meeting, San Diego. 24-27 Sept 1998.

[72] Varga, S. M., Wissinger, E. L., Braciale, T. J.: The attachment (G) glycoprotein of respiratory syncytial virus contains a single immunodominant epitope that elicits both Th1 and Th2 $CD4^+$T cell responses. J. Immunol. 165, 6487-6495 (2000).

[73] Walsh, E. E.: RSV infection in adults. Semin. Respir. Crit. Care Med. 32, 423-432 (2011).

[74] Yu, J. R., Kim, S., Lee, J. B., Chang, J.: Single intranasal immunization with recombinant adenovirus-based vaccine induces protective immunity against RSV infection. J. Virol. 82, 2350–2357 (2008).

[75] Zeng, R., Zhang, Z., Mei, X., Gong, W., Wei, L.: Protective effect of a RSV subunit vaccine candidate G1F/M2 was enhanced by a HSP70-Like protein in mice. Biochem. Biophys. Res. Commun. 377, 495–499 (2008).

[76] Zhang, W., et al.: Vaccination to induce antibodies blocking the CX3C-CX3CR1 interaction of RSV G protein reduces pulmonary inflammation and virus replication in mice. J. Virol. 84, 1148–1157 (2010).

（何继军译）

第6章　埃博拉病毒疫苗

Thomas Hoenen①

摘要

埃博拉病毒在人类和非人灵长类动物中引起严重出血热，病死率高达90%。目前还没有许可的治疗方法或疫苗可用于埃博拉病毒。然而在过去10年中，实验性疫苗的开发取得了重大进展，这些疫苗在非人灵长类动物身上显示出巨大的应用前景，而灵长类动物是埃博拉病毒出血热最严格的疾病模型。本章将讨论目前正在开发的实验性疫苗以及在不同情况下有效接种疫苗的要求。特别是基于腺病毒和水疱性口炎病毒的重组疫苗将详细讨论，因为重组疫苗在保护非人灵长类动物免受埃博拉病毒出血热感染方面非常有效。

6.1　疾病

埃博拉病毒引起埃博拉出血热（EHF），这是1976/1977年首次在前扎伊尔（现刚果民主共和国）和前苏丹（现南苏丹）两次同时暴发时发现的疾病[1]。这种疾病的特点是病例的死亡率非常高，根据病毒种类，这一比率可高达90%。初始症状通常发生在潜伏期后，潜伏期一般为3~13天(最长21天)[2,3]，相对来说没有特异性，包括发热、头痛、疲劳、肌肉痛和关节痛、喉咙痛和胸痛（在[2-4]中回顾）等症状。疾病发病后2~7天，会出现胃肠道症状，包括腹泻、恶心、呕吐和厌食症。大约1/3的病例出现斑丘疹[4]，尽管这在深色皮肤患者中可能很难察觉。出血症状最常见于胃肠道，表现为黑便、便血和吐血，尽管也可能发生其他出血症状，如静脉穿刺部位出血、牙龈出血和结膜出血[2-4]。就实验室参数而言，描述了血小板减少、淋巴细胞减少、血清丙氨酸和天冬氨酸氨基转移酶水平升高、凝血时间延长以及D-二聚体水平升高[3,4]。

在致死病例中，死亡通常发生在发病的第二周，由类似于败血性休克综合征引起[5]。在非致死病例中，恢复时间较长（长达3个月），后遗症可能包括关节痛、肌痛、头痛、乏力和贫血[3,4]。此外，在一些幸存者中还描述了脱发、耳鸣、暂时性听力损失、葡萄膜炎、结膜炎和睾丸炎。

EHF发病机制是复杂的（在［5-7］中回顾）。巨噬细胞和树突状细胞被认为是早期的靶细胞，尽管包括内皮细胞在内的许多其他类型细胞被埃博拉病毒感染，很可能是在疾病过程后期[5]。埃博拉病毒已经被证明极大地削弱了免疫反应，特别是两种病毒蛋白VP35和VP24削弱天然免疫应答。VP35干扰干扰素应答，并拮抗蛋白激酶R（PKR）活性[8,9]，而VP24阻断信号转导和转录激活剂（STAT）介导和p38介导的干扰素信号通路[10,11]。此外，感染埃博拉病毒的树突状细胞在支持T细胞增殖的能力受到损害，不产生促炎细胞因子[12,13]。

此外，就获得性免疫而言，EHF幸存者和非幸存者之间似乎存在巨大差异，幸存者在症状发作后2天就显示IgM抗体，在症状发作后5~8天出现IgG抗体，而只有30%的非幸存者显示IgM抗体，没有人出现IgG应答[14,15]。此外，埃博拉病毒已经被证明会导致患者淋巴细胞大量流失，即使这些细胞没有被直接感染，而可能是通过旁观者凋亡[16]。

---

① T. Hoenen，博士（美国蒙大拿州汉密尔顿落基山实验室，国立卫生研究院过敏和传染病研究所室内研究部病毒学实验室，E-mail：thomas. hoenen@ nih. gov）

EHF 发病机制的另一个标志是产生高水平的促炎细胞因子和趋化因子，有时被称为“细胞因子风暴”[16]。此外，组织因子在埃博拉病毒感染期间大量产生[17]，并参与弥漫性血管内凝血（即血液凝固系统的系统激活，导致纤维蛋白沉积和微血管血栓）的发展，这是 EHF 的一个显著特征[18]。

6.2 病原

埃博拉病毒和相关的马尔堡病毒属于丝状病毒科。目前有 5 种公认的埃博拉病毒：扎伊尔埃博拉病毒（ZEBOV）、苏丹埃博拉病毒（SEBOV）、本迪布焦埃博拉病毒（BEBOV）、科特迪瓦埃博拉病毒（CIEBOV）和雷斯顿埃博拉病毒（REBOV）[19]（表 6.1）。

由于丝状病毒包含 1 个不分节段的单股负链 RNA 基因组，因此按单股负链病毒目顺序分组。这个基因组长约 19kb，包含 7 个基因，编码至少 9 种蛋白质（图 6.1）[1]。这些蛋白质中有 7 种是结构性蛋白，存在于病毒颗粒中，具有典型的线状外观。颗粒平均直径为 98 nm（不测量糖蛋白突起），其中大多数颗粒长度约为 980 nm[20]。

螺旋核衣壳位于病毒颗粒中心，由被核蛋白 NP 包裹的 RNA 基因组组成，与病毒聚合酶 L、聚合酶辅因子 VP35、转录激活因子 VP30 和辅助蛋白 VP24 结合，后者参与核衣壳形成和初级转录[20-22]。核衣壳周围是含有基质蛋白 VP40 的基质空间，VP40 负责新病毒粒子的形态发生和出芽[23]。病毒颗粒被宿主细胞衍生的脂质双层包裹，表面蛋白 $GP_{1,2}$ 嵌入其中，介导进入宿主细胞膜，并与宿主细胞膜融合[24]。

由于基因组尺寸紧凑，许多埃博拉病毒蛋白具有多种功能。例如，聚合酶辅助因子 VP35 和辅助蛋白 VP24 也充当干扰素拮抗剂[8,11]，VP24 和基质蛋白 VP40 都参与病毒基因组复制和转录调节[25]。作为使编码能力最大化的第二种策略，GP 基因的病毒转录主要导致糖蛋白的非结构可溶性变体 sGP 产生。然而在转录过程中，非模板化的腺苷残基可以通过病毒聚合酶在 GP 基因多聚 A 编辑位点掺入新生 mRNAs 中[26]。这导致了开放阅读框架的转变，并产生了结构性膜嵌入 $GP_{1,2}$ 蛋白，或者第二种非结构性 GP 变体，称为 ssGP[27]。非结构蛋白（sGP 和 ssGP）功能目前尚不清楚。

表 6.1 埃博拉病毒种类概况

概况	扎伊尔埃博拉病毒	苏丹埃博拉病毒	本迪布焦埃博拉病毒	科特迪瓦埃博拉病毒	雷斯顿埃博拉病毒
缩略词	ZEBOV	SEBOV	BEBOV	CIEBOV	REBOV
病死率[a]	44%~90%（平均 79%）	41%~71%（平均 54%）	25%~49%（平均 32%）	0%	0%
发病数	1 388	785	214	1	0
流行国家	加蓬，刚果共和国，刚果民主共和国	南苏丹，乌干达	乌干达，刚果民主共和国	象牙海岸	菲律宾
疾病特征	严重出血热	严重出血热	严重出血热	严重出血热	人类发病
试验性疫苗平台[b]	rAdV，rVSV，rhPIV，VLPs	rVSV，rAdV	rAdV[c]，rVSV[c,d]	rVSV[c]	无

列出了 5 种公认的埃博拉病毒的特征。

缩略词：*rAdV* 重组腺病毒，*rVSV* 重组水疱性口炎病毒，*rhPIV* 重组人副流感病毒，*VLPs* 病毒样颗粒。

[a]病死率的范围是基于 1 个以上病人的暴发；平均病死率考虑了所有暴发。

[b]所列的只是疫苗平台，已经证明对非人灵长类动物具有保护作用。

[c]通过针对不同埃博拉病毒物种的疫苗进行交叉保护。

[d]到目前为止，只证明了部分保护。

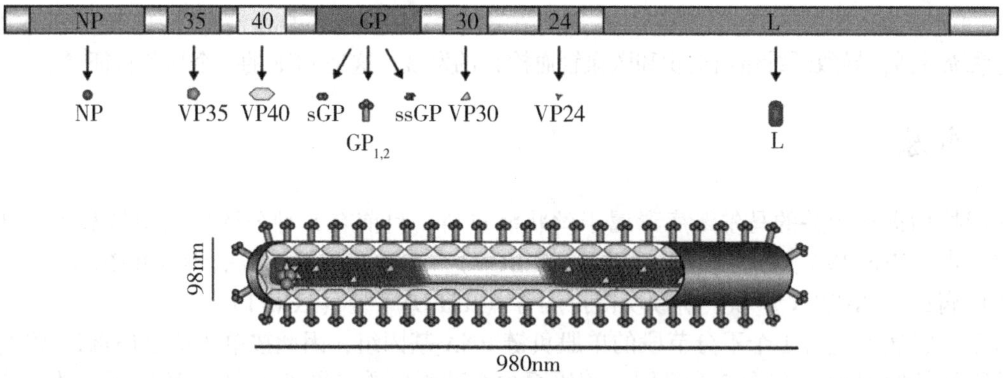

图 6.1　埃博拉病毒的结构

图中显示了埃博拉病毒基因组结构（上图），以及埃博拉病毒粒子示意图（下图）。核糖核蛋白复合蛋白（NP、VP35、VP30、L 和 VP24）以红色显示，基质蛋白 VP40 以黄色显示，糖蛋白（结构糖蛋白 $GP_{1,2}$和非结构糖蛋白 sGP 和 ssGP）是蓝色的。

埃博拉病毒的生命周期遵循与其他不分节段负链 RNA 病毒相同的一般原则。埃博拉病毒可以通过许多细胞表面分子附着在宿主细胞，包括各种 C 型凝集素、TIM-1 和 Tyro3 受体酪氨酸激酶家族成员[28]。病毒粒子通过大胞吐作用被内化[24,29]，将其递送到核内体，在核内体中黏蛋白样结构域以及聚糖帽被宿主细胞蛋白酶如组织蛋白酶从 $GP_{1,2}$上切割下来[30]。这样就暴露了受体结合结构域，并使 $GP_{1,2}$与其受体 Niemann-Pick C1（NPC1）相互作用[31,32]，最终导致病毒膜与内体膜融合，从而将核衣壳递送到细胞质中，完成剩余复制周期。最初 mRNAs 是从病毒基因组中转录出来的，这一过程被称为初级转录，只涉及病毒颗粒中带入细胞的成分。随后利用宿主细胞中新合成的病毒成分进行二次转录，并通过抗基因组中间体复制病毒基因组。

病毒基因组复制发生在被称为包涵体的病毒工厂，也可能涉及病毒生命周期的其他部分[33]。新形成的核衣壳被运输到细胞表面，在那里 VP40 促进了一个发芽过程，这个过程涉及 VP40 的寡聚[34,35]和细胞 ESCT 机械篡位[36-38]。

6.3　传播

到目前为止，所有人类 EHF 病例都起源于中非（图 6.2）。然而在美国和欧洲一些与马尔堡病毒密切相关的输入性感染突显了在世界范围内输入性 EHF 病例的潜力[39,40]。此外，从加蓬输入南非的 ZEBOV（扎伊尔埃博拉病毒），导致约翰内斯堡一名医生致命感染，证明了风险性[41]。REBOV（雷斯顿埃博拉病毒）是非洲以外唯一已知的埃博拉病毒物种，起源于菲律宾，具有感染人类的能力，但迄今为止，在感染 REBOV 后未发现人类发病症状。有趣的是，一些报告指出，根据血清学证据，埃博拉病毒或相关丝状病毒的亚临床感染可能在已确认流行地区以外的地区发生[42,43]。此外一些报告还表明，亚临床或轻度埃博拉病毒感染也在地方病流行地区发生，其发生频率远高于暴发时确认的频率，这也是基于血清学的证据[44-46]。

EHF 病例与暴发相关联，在过去 10 年中，暴发频率约为每年 1 次（图 6.2，略）。对于一些暴发，人类和非人灵长类动物（NHPs）之间接触被认为是一个可能原因[47]。然而蝙蝠（通常认为是埃博拉病毒天然宿主[48]）直接将病毒传染给人类也被认为是可能的传染源[49]。人与人之间传播主要是通过与病人或死亡患者的密切接触，尤其是通过体液接触[50,51]。屏障护理技术被认为足以最小化人与人之间传播风险[51]。

6.4 诊断和经典疗法

由于EHF早期的非特异性症状，因此诊断很困难，特别是疫情被确认之前。引起类似临床症状的最常见疾病是疟疾和伤寒，尽管还必须考虑一些其他传染病[52]。为了确定诊断，需要检测病毒材料或检测对埃博拉病毒的免疫反应。病毒分离是检测埃博拉病毒的敏感方法；然而需要四级生物安全设施，速度相对较慢。相比之下，RT-PCR是高度灵敏和快速的方法，根据所用的特定检测方法，在症状发作后0~3天，可以在患者血液中检测到病毒核酸[53,54]。

由于离液剂用作RNA分离的第一步，样品被灭活，这使得RT-PCR成为一种安全方法，无需实验室的控制。此外，RT-PCR非常适合现场诊断，因为该方法可由疫情应对团队在现场实施，对可用基础设施要求最低[52]。在许多疫情中，RT-PCR被用作诊断依据[53-56]。然而RT-PCR有可能产生假阳性或假阴性结果，因此应始终进行确认性RT-PCR分析，如抗原捕获ELISA，或者如果没有，最好使用独立的靶基因和/或独立样本进行确认性RT-PCR分析[52,53]。

已经建立了抗原捕获ELISAs，并已在一些疫情暴发中用作诊断方法[53-57]。虽然ELISAs通常比RT-PCR更不敏感，尤其是在疾病发作早期[53,54]，但ELISAs方法对于补充基于RT-PCR的检测是非常有价值的。最后，旨在检测IgM的血清学也用于诊断目的，有些患者在感染后2天就可以检测到IgM抗体，尽管其他患者没有出现显著的IgM反应，或者只是在疾病晚期才出现，因此这种方法只能与其他诊断方法结合使用。

在确认疫情暴发后，最重要的问题是通过建立隔离程序来最大限度地减少传播，确保在治疗患者时使用适当的隔离护理技术，并启动追踪接触者，尽早识别疑似EHF病例。不幸的是，到目前为止还没有针对EHF的具体治疗方法，因此治疗仅限于支持性护理，包括补液和提供止痛药[2,58]。然而在代表EHF金标准动物模型的非人灵长类动物中，一些实验治疗策略非常成功。

目前最有效的实验治疗方法是抗体[59,60]和siRNAs[61]，这两种抗体都能100%保护的受感染动物免受致命攻击。其他旨在调节宿主应答的策略也显示出希望，尽管它们不能完全保护NHPs。其中包括重组线虫抗凝蛋白C2（rNAPc2），一种有效的组织因子启动凝血抑制剂[18]和活化蛋白C[62]，分别在非人灵长类动物中显示33%和18%的保护作用。

不幸的是，尽管动物模型结果很有希望，但将特定EHF疗法推进到经典（前期）临床试验仍然面临许多挑战，包括缺乏商业利益，受影响地区可用基础设施不足，暴发的零星性质和病例数量相对较少，因此需要替代方法将这些疗法引入[63]。此外，评估已经批准用于其他医疗条件的药物是否适合治疗EHF，可能是一种非常谨慎的方法，因为这将大大加快批准过程，并降低药物开发成本。

6.5 疫苗

自从35年前发现埃博拉病毒以来，已经开发和评估了许多不同的疫苗。为了评估这些疫苗，目前有许多动物模型，包括小鼠、仓鼠、豚鼠和非人灵长类动物。虽然所有啮齿动物模型都需要使用啮齿动物适应的病毒株，但非人灵长类动物对未适应的病毒很敏感。非人灵长类动物也代表了最接近人类的EHF模型，显示了人类感染中的典型病理和临床特征，尽管疾病进展似乎比人类更快，结果实际上在非人灵长类动物上比人类更糟，因此它们可以被认为是疫苗和药物评估极其严格的模型[64,65]。

埃博拉病毒疫苗可大致分为复制疫苗和非复制疫苗，以及灭活疫苗、亚单位疫苗和载体疫苗（图6.2）。使用灭活埃博拉病毒、表达埃博拉病毒蛋白的复制子或重组表达埃博拉病毒蛋白的早期疫苗接种方法，虽然在小鼠和豚鼠中很有希望，但在非人灵长类动物中并不成功[66]。相比之下，最近的一些疫苗证明对非人灵长类动物具有100%的保护作用，应该被认为是人类使用的候选疫苗。这些疫苗包括基于重组腺病毒的疫苗、基于重组水疱性口炎病毒（VSV）的疫苗、使用病毒样颗粒的疫苗和基于重组人副流感病毒的疫苗（图6.2）。不幸的是，目前这些疫苗都没有被批准用于人类，只有腺病毒疫苗已经进入临床试验。然而，在实验室暴露于ZEBOV后，VSV疫苗已经在一个人身上紧急使用。因此本章将重点介绍这两种疫苗，这两种疫苗代表了研究最广泛的埃博拉病毒疫苗，在人类中至少有一些经验。

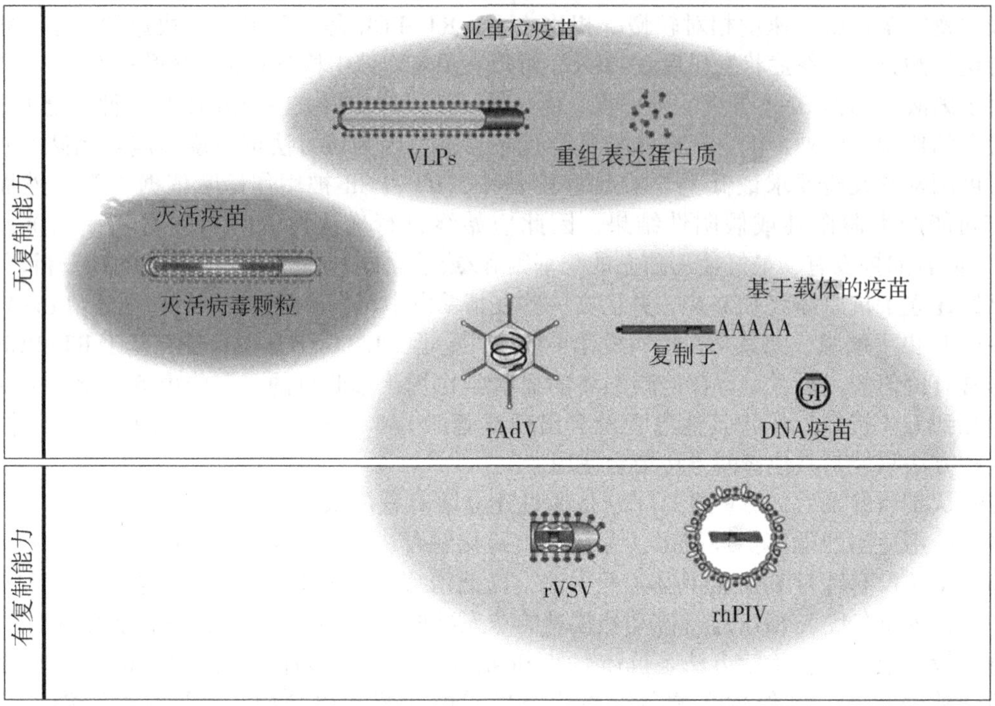

图6.2 EHF疫苗概述

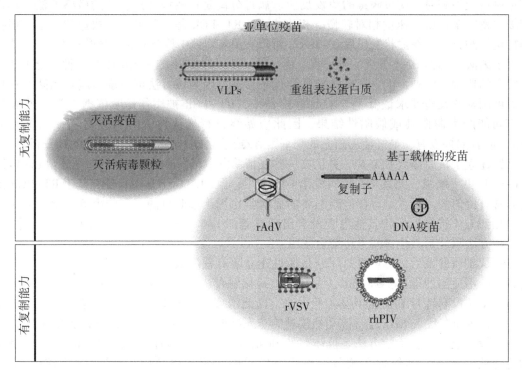

EHF疫苗可大致分为可复制疫苗和非复制疫苗。此外，它们可分为灭活疫苗、亚单位疫苗（包括重组表达蛋白以及病毒样颗粒（*VLPs*））和载体疫苗。基于载体的疫苗包括DNA疫苗和复制子，以及重组病毒，如重组腺病毒（*rAdVs*），重组水疱性口炎病毒（*rVSVs*）和重组人副流感病毒（*rhPIVs*）。天然埃博拉病毒组分以颜色显示，而灭活的埃博拉病毒组分和载体组分以灰色显示。

6.5.1 重组腺病毒疫苗

第一次完全保护非人灵长类动物免受致命埃博拉病毒攻击的疫苗是重组腺病毒疫苗（rAdV），与DNA疫苗结合使用[67]（图6.3）。之所以选择rAdV疫苗接种平台，是因为重组病毒（尽管不是腺病毒）已被证明能够将DNA疫苗产生的免疫提高到比单独使用DNA疫苗更高的水平。与DNA疫苗组分类似，rAdV疫苗组分作为基因传递载体，包含一个编码埃博拉病毒抗原的基因，该抗原在接种者细胞内表达，但在腺病毒表面不存在。最初DNA/rAdV疫苗接种策略共包括4次免疫（3次用DNA组分，1次用腺病毒组分），第一次免疫和攻毒之间需要6个月以上。然而在一项后续研究中显示，在没有任何DNA成分的情况下，仅用rAdV进行一次疫苗接种，在疫苗接种后4周内就能对非人灵长类动物提供100%的保护[68]。

埃博拉病毒疫苗中使用的rAdVs在E1区域带有缺失，有时还在E3和E4区域带有缺失，因此不能复制。出于这个原因，必须使用相对高剂量的疫苗（在非人灵长类动物中需要1×10^{10}个病毒颗粒才能实现100%的保护），并且接种剂量对疫苗效力至关重要[69]。在小鼠中，可以通过使用优化的表达盒来降低所需的接种剂量，但这在非人灵长类动物中尚未得到评估[70]。

就血清型而言，人类腺病毒5型（HAdV-5）通常用于重组埃博拉病毒疫苗。然而由于这种

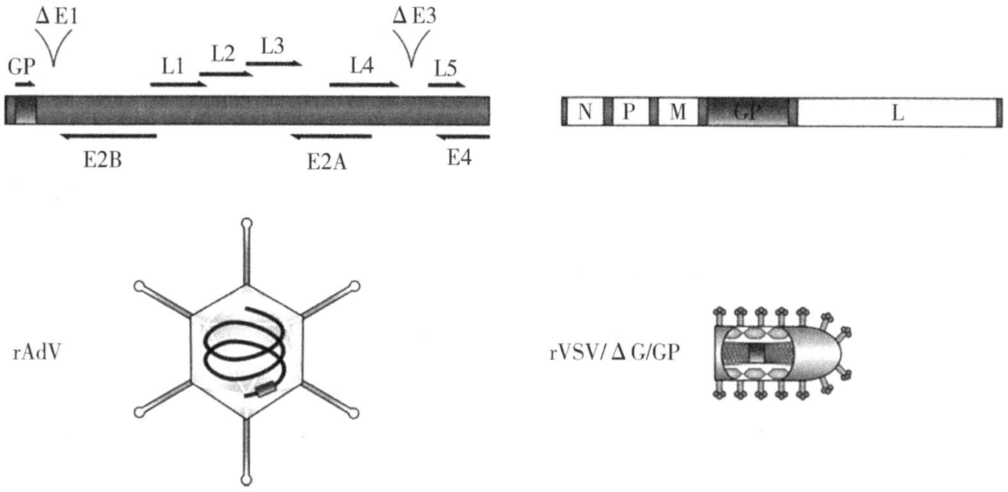

图 6.3　重组腺病毒和基于水泡性口炎病毒的 EHF 疫苗结构

显示了重组腺病毒（rAdV）和水疱性口炎病毒（rVSV/ΔG/GP）疫苗的基因组结构（上部）和病毒颗粒（下部）示意图。埃博拉病毒成分用颜色描述，载体成分用灰色描述。指出了 rAdV 基因组中的缺失（ΔE1 and ΔE3）。

血清型的血清流行率很高，60%~90%的成人是血清阳性的，这取决于他们的来源国[71]，先前存在的载体免疫力是一个潜在问题。事实上研究表明，在啮齿动物[72-74]和非人灵长类动物中[75]预先存在的免疫力可显著损害 HADV-5 疫苗平台的效力，并且在人类中预先存在免疫力的个体抗体应答低于 HADV-5-幼稚个体的抗体应答[76]。

已经开发了几种方法来解决这个问题。基于不同人类血清型（即 HAdV-26 和 HAdV-35）的 rAdVs 在非人灵长类动物中被证明是完全保护的，尽管需要一种主要的加强免疫方法来获得保护，并克服先前存在的免疫力问题[75]。此外，基于类人猿血清型 SAdV-21 和 SAdV-24 的疫苗在小鼠和豚鼠身上显示出保护作用，这些疫苗在人类身上不存在或只有很少预先存在的免疫力。同样地，将接种途径改为黏膜途径增加了对已有腺病毒载体免疫的小鼠的保护[72,78]，增加了对 HADV5 免疫豚鼠的免疫应答[74]。最后增加疫苗接种数量已被证明可以克服非人灵长类动物中先前存在的免疫力[79]。

最初研究使用了表达两种抗原的不同 rAdVs 混合物，即 $GP_{1,2}$ 和 NP[68]，但随后研究表明，在疫苗接种中省略 NP 而不丧失保护效力[69]。此外，rAdV 载体已被开发出来，允许表达来自单个 rAdVs 的多种抗原已被用于使用单一疫苗接种非人灵长类动物，以对抗多种埃博拉病毒[69,80]。另一种保护多种埃博拉病毒感染的方法是使用混合疫苗，也被证明是成功的[69,81]。事实上，使用来自两种埃博拉病毒（ZEBOV 和 SEBOV）rAdVs（含有 GPs）的混合方法，可以保护非人灵长类动物免受第三种病毒（BEBOV）的攻击，证明这种方法可以诱导对埃博拉病毒的广泛免疫[81]。

一些关于 GP 潜在毒性理论问题已经提出，因为这种蛋白的高水平表达会在体外引起细胞毒性[82]。为解决这些问题，开发了 GP（E71D）点突变体，该突变体不再具有细胞毒性，但仍具有免疫原性，可以在非人灵长类动物中诱导保护性免疫，尽管至少与 NP 结合后，诱导的保护性较弱[69]。

rAdV 疫苗诱导强烈的埃博拉病毒特异性 $CD8^+$ T 细胞反应[68,83]。然而这些疫苗接种后也观察到中和抗体反应[68,84]，抗体滴度与保护作用相关[69,85]。为了研究细胞应答和体液应答对保护机制的贡献，进行了抗体转移和 T 细胞耗竭研究。有趣的是，仅从接种 rAdV 的动物身上转移抗体并不能保护接受者即非人灵长类动物。尽管接受者的抗体滴度超过了接种动物的抗体滴度，而且抗体显示出中和活性[84]。

同样 B 细胞耗竭并没有降低疫苗在非人灵长类动物中的效力，进一步说明了抗体在这种疫苗接种方法中的主要作用。相比之下，CD8$^+$T 细胞的耗竭显著损害了 rAdV 疫苗的保护效力，与接种疫苗对照组 100%存活率相比，接种 CD8$^+$T 细胞的动物存活率只有 20%。然而 CD8$^+$T 细胞耗竭动物的死亡时间延长，表明 CD8$^+$T 细胞以外的免疫反应仍然在保护中发挥着一定作用。有趣的是，最近另外两项研究表明，在某些情况下抗体足以保护非人灵长类动物免受其他致命埃博拉病毒的攻击[59,60]，说明不同疫苗和治疗方法对非人灵长类动物的保护机制可能不同，必须为每个疫苗平台分别建立相关的保护机制。

埃博拉病毒暴发是零星和不可预测的，使得经典的临床疗效试验无法进行。在这种情况下根据动物规则，有可能获得 FDA（美国食品和药物管理局）对新型疫苗的批准，这要求在动物模型中显示疗效，在人类中显示保护作用的安全性和免疫原性[85]。

为了获得安全性和免疫原性数据，对 rAdV 埃博拉病毒疫苗进行了 I 期临床试验，该疫苗在 22 名个体中显示了总体安全性[76]。这项试验旨在评估 rAdV 疫苗作为 DNA/rAdV 联合疫苗的一个组成部分，也评估了 rAdV 免疫原性。用表达来自 ZBOV 和 SEBOV 的 $GP_{1,2}$（E71D 突变）的 10^{10}rAdVs 颗粒单次接种疫苗后，所有疫苗都产生了抗 SEBOV GP 的抗体，超过一半的疫苗产生了抗 ZEBOV GP 的抗体。不幸的是，CD8$^+$T 细胞应答被认为是基于 rAdV 埃博拉病毒疫苗保护的主要机制，但应答率较低，只有 9%的疫苗显示 CD8$^+$T 细胞对 SEBOV GP 应答，21%的疫苗显示对 ZBOV GP 应答[76]，这表明该疫苗平台需要进一步研究，旨在提高免疫原性并可能涉及野生型 GP。

6.5.2 重组水疱性口炎病毒疫苗

非人灵长类动物的第二个高效疫苗平台是基于重组水疱性口炎病毒（rVSV）的疫苗。水泡性口炎病毒属于弹状病毒科，在家畜中引起自我限制性疾病，而人类感染很少见，没有症状或伴有轻微流感样症状[86]。用于埃博拉病毒疫苗接种的 rVSV 糖蛋白基因被编码 $GP_{1,2}$（rVSV/ΔG/GP）的埃博拉病毒 GP 基因取代（图 6.3）。在非人灵长类动物中，rVSV/ΔG/GP 在攻毒前 4 周给予单次疫苗接种后获得了 100%的保护，攻击后没有检测到埃博拉病毒血症[87]。

与用于埃博拉病毒疫苗接种的 rAdV 相比，rVSV 仍然具有复制能力，因此，成功接种非人灵长类动物只需要相对较低的剂量（10^7 PFU rVSV 比 10^{10}个 rAdV 粒子）[87]，至少在小鼠中仅用两个 PFU 就能提供完全保护[88]。然而 VSV 的复制能力也引起了对疫苗安全性的关注。事实上接种过疫苗的动物有时会出现短暂病毒血症；但是在 100 多个接种过疫苗的非人灵长类动物中，没有观察到不良反应[66]。在 VSV 感染的动物模型中，致病性尤其是神经毒力与 G 基因有关[89,90]，该基因在 rVSV/ΔG/ GP 中不存在。

因此，这种病毒在非人灵长类动物中不表现出神经毒性，即使是在鼻内接种[91]。为了进一步确认 rVSV 疫苗平台的安全性，感染猿猴人类免疫缺陷病毒（SHIV）的非人灵长类动物（NHPs）接种了 rVSV/ΔG/ GP，尽管免疫功能低下，但在疫苗接种点没有出现任何临床疾病、发烧或局部反应迹象[92]。这项研究具有特别的相关性，不仅因为证明了 rVSV 的安全性，而且因为在埃博拉病毒暴发地区，艾滋病病毒流行率很高。除了在猿猴人类免疫缺陷病毒感染的非人灵长类动物中是安全的，rVSV 疫苗还能够保护 6 个猿猴人类免疫缺陷病毒感染非人灵长类动物中的 4 个免受致命的埃博拉病毒攻击[92]，尽管免疫功能低下。最后在非人灵长类动物中没有观察到明显的疫苗脱落。

许多埃博拉病毒物种的 rVSV 疫苗已在非人灵长类动物中开发和测试[93,94]，与 rAdV 平台类似，混合的 rVSV 疫苗已被证明在一次疫苗接种后可抵御多种物种。特别是 3 种编码来自 ZEBOV、SEBOV 和相关马尔堡病毒糖蛋白的 rVSV/ΔG/GP 混合物不仅对这些病毒物种具有保护作用，而且对 CIEBOV 的攻击也具有保护作用[95]。

相比之下，在几个物种之间开发单一（非混合）rVSV/ΔG/GP 交叉保护更具挑战性。使用 ZEBOV 编码的 rVSV/ΔG/GP 疫苗接种并不能保护非人灵长类动物免受 SEBOV 致命攻击[87]，尽管 rVSV/ΔG/GP 疫苗对本迪布焦埃博拉病毒的攻击具有部分保护作用[93]。目前尝试生产基于 rVSV 交叉保护的埃博拉病毒疫苗，主要集中在第二代 rVSV 上，后者编码多种抗原，例如 $GP_{1,2}$ 和 VP40[96]。这种方法在豚鼠模型中显示出了前景；但是还没有在非人灵长类动物中进行评估，目前需要进行几次疫苗接种，以提供对致命攻毒的完全保护。

与 rAdV 疫苗相比，rVSV 疫苗的保护机制和相互关系研究较少。然而研究表明，rVSV/ΔG/GP 在接种的非人灵长类动物中诱导强烈的抗体反应，尽管这些抗体大多是非中和性的[97]。此外，免疫接种后产生 IFN-γ 和 IL-2 产生的效应细胞（T 细胞和/或 NK 细胞）。

在小鼠中已经显示，$CD8^+$T 细胞耗竭并不抑制用 rVSV/ΔG/GP 免疫的动物对致命埃博拉病毒攻击的保护，而从免疫小鼠到幼稚小鼠的血清转移则保护了 80%的这些动物免受其他致命攻击。这表明至少在小鼠中 rVSV/ΔG/GP 疫苗的保护机制是体液免疫应答，与细胞免疫应答无关。然而迄今为止，在非人灵长类动物中还缺乏类似研究，需要对这种疫苗平台的保护机制和相互关系给出更明确的答案。

rVSV 疫苗一个非常显著的特点是可以用于暴露后保护。如果非人灵长类动物在攻击后 30 分钟接种了表达同源 GP 的 rVSV 疫苗，至少 50%的动物免受其他致命 ZEBOV 攻击，100%免受其他致命 SEBOV 攻击。

对于相关的马尔堡病毒，疫苗接种甚至可以在暴露攻击后 48 小时进行，33%的受攻击动物存活，而如果暴露后疫苗接种分别在攻击后 24 小时或 30 分钟进行，则 83%或 100%的动物存活于马尔堡病毒攻击[99]。与预防性疫苗接种相似，暴露后保护机制还不清楚。然而据观察，幸存者和非幸存者之间的埃博拉病毒血症以及体液免疫应答有显著差异[94,98]。幸存者仅表现出低水平病毒血症，而非幸存者则发展成病毒血症，超过 10^4 PFU/mL（最高滴度达到 10^8 PFU/mL）。有趣的是这反映了人类的情况，在人类中病毒载量已经被证明是 EHF 的预测结果[53,100]。就体液应答而言，幸存者在攻击后约 6 天显示 IgM 滴度，在攻击后 10 天显示 IgG 滴度，而非幸存者没有产生可检测的抗体反应[94,98]。

与 rAdV 疫苗相比，rVSV 疫苗还没有进入临床研究阶段。然而基于其在非人灵长类动物中能提供暴露后保护的能力，在 BSL-4 级实验室发生埃博拉病毒污染针头针刺事故后，该疫苗已被用作一名实验室工作人员暴露后实验疫苗[101]。事故发生后约 48 小时，患者接受了 5×10^7 PFU 的疫苗接种，接种后 1 天出现轻度发烧和肌痛。这些症状在不到一天内就有所缓解，与此同时 RT-PCR 可以检测到 rVSV 疫苗低水平病毒血症。除此之外没有观察到不良反应，最重要的是患者没有出现任何可检测到的埃博拉病毒血症。虽然无法确定他们是否受到感染，是否完全受到疫苗保护或者接触疫苗是否实际上没有导致 EBOV 感染，但这一事件首次表明 rVSV 疫苗对人类是安全的。

6.6 优点和缺点

埃博拉出血热（EHF）呈零星暴发，而且经常在非洲的偏远地区发生。与其他疾病相比，总体病例数相对较低，尽管仅根据病例数对公共卫生系统的影响往往高于预期，这是由于受影响的卫生设施关闭，以及健康保健提供者经常受到 EHF 严重影响的事实，特别是在疫情被确认之前。此外，埃博拉病毒被认为是潜在的生物恐怖媒介，因而被公众视为重大威胁。基于这些特性有许多可能的暴露场景，每种场景对理想疫苗有独特要求[66]。

对流行地区全部人口进行预防性疫苗接种，虽然理论上是最严格的解决办法，但由于受影响地区的规模以及后勤方面挑战，这种办法不切实际。然而对高危人群如医疗保健提供者、疫情应对人员或从事埃博拉病毒工作的实验室工作人员进行预防性疫苗接种应该更加可行。在这种情况下，对尽可能多的埃博拉病毒种（以及相关的马尔堡病毒）进行长期和广泛免疫至关重要。然而快速保护只是次要的，因为疫苗接种很可能早在潜在接触之前就已经发生了。此外，先前存在的对疫苗载体的免疫不是问题，因为目前数据表明，即使存在载体免疫，一种主要的免疫加强方法也能够提供保护[79]。

相比之下，在疫情暴发（或生物恐怖袭击）期间为了通过限制人与人之间传播来控制疫情暴发，需要对受影响地区的人进行疫苗接种，最好是在接种一次疫苗之后。在这种情况下，预先存在的载体免疫将构成一个重大问题。相比之下，广泛的保护只是次要的，因为它最有可能确定导致暴发的确切病毒种类。唯一例外是如果暴发是由一种新的埃博拉病毒引起，则没有针对这种病毒的特异性疫苗可用。因此实现快速保护的唯一途径是现有疫苗的交叉保护特性。

这里讨论的两种疫苗平台都具有使其适应这些情况的特性（表 6.2）。基于 rAdV 的疫苗提供了广泛的保护，预先存在的载体免疫可以用一种主要免疫加强方法来克服。该疫苗在临床试验方面进展最快，是在 GMP 条件下生产的。其他病毒性疾病 rAdV 疫苗经验表明，大量生产该疫苗是可能的。因此是预防性疫苗一个非常可行的候选方案，尽管第一次临床试验结果表明为了获得最大免疫原性，可能需要进一步优化这种疫苗。

表 6.2　重组腺病毒和重组水疱性口炎病毒疫苗的特征

特征	重组腺病毒	重组水疱性口炎病毒
复制能力	无复制能力	可以复制
所需接种剂量	10^{10}个粒子	$\leqslant 10^7$ PFU
NHPs /豚鼠/小鼠接种疫苗和攻毒之间		
所需的时间	4 周/ 4 周/ 4 周	4 周/ 3 周/ 1 天
被预先存在的免疫力损害	是的，但是多次接种疫苗会克服预先存在的免疫力	没有
对多种物种的保护	可以，单一疫苗载体	可以，混合的疫苗混合物
针对其他/新埃博拉病毒种类的交叉保护	可以，取决于疫苗配方	可以，取决于疫苗配方
暴露后保护	不行	可以
保护机制	主要是 T 细胞介导的	未知
在 GMP 条件下生产	是	没有
开发期	I 期临床试验	尚未进入临床试验

缩略词：PFUs 斑块形成单位，NHPs 非人类灵长类，GMP 良好制造规范。

基于 rVSV 的疫苗提供了非常快速的保护，甚至可以在暴露后使用，这使得该疫苗非常适合控制疫情暴发或者在实验室暴露的情况下使用。预先存在的免疫力是无关紧要的，所有相关埃博拉病毒种类的基于 rVSV 疫苗都是可用的。不幸的是，这种疫苗还没有进入临床试验阶段，这可能部分是由于安全风险，尽管在 100 多个非人灵长类动物（包括免疫功能低下的动物）和一个人中没有不良反应。此外，这种疫苗的保护机制和相互关系尚未阐明，在将这种疫苗投入临床试验之前研究清楚 rVSV 疫苗的保护机制和相互关系是必要的。

总的来说，目前已经有几个非常有希望的预防和治疗 EHF 候选疫苗，虽然进一步优化和研究这些疫苗是必要的，但现在要克服的最重要障碍是将这些疫苗转入临床试验，然后进一步进入该领域，这些障碍有政治、监管和经济方面的因素。

致谢

这项研究得到了NIH、NIAID内部研究项目的支持。

参考文献

[1] Sanchez, A., Geisbert, T. W., Feldmann, H.: Filoviridae: Marburg and Ebola Viruses. In: Fields, B. N., Knipe, D. M., Howley, P. M. (eds.) Fields Virology, pp. 1410 - 1448. Lippincott Williams & Wilkins, Philadelphia (2007).

[2] Jeffs, B.: A clinical guide to viral haemorrhagic fevers: Ebola, Marburg and Lassa. Trop. Doct. 36, 1-4 (2006). doi: 10. 1258/004947506775598914.

[3] Kortepeter, M. G., Bausch, D. G., Bray, M.: Basic clinical and laboratory features of filoviral hemorrhagic fever. J. Infect. Dis. 204 (Suppl 3), S810-S816 (2011). doi: 10. 1093/infdis/jir299.

[4] Hoenen, T., Gottschalk, R., Becker, S.: In: Hofmann F. (ed.) Manual of infectious diseases, VIII6. 11. 11-15. Ecomed Medizin, Landsberg/Lech, Germany (2007).

[5] Hoenen, T., Groseth, A., Falzarano, D., Feldmann, H.: Ebola virus: unravelling pathogenesis to combat a deadly disease. Trends Mol. Med. 12, 206-215 (2006). doi: 10. 1016/j. molmed. 2006. 03. 006.

[6] Aleksandrowicz, P., et al.: Viral haemorrhagic fever and vascular alterations. Hamostaseologie28, 77 - 84 (2008).

[7] Hensley, L. E., Jones, S. M., Feldmann, H., Jahrling, P. B., Geisbert, T. W.: Ebola and Marburg viruses: pathogenesis and development of countermeasures. Curr. Mol. Med. 5, 761-772 (2005).

[8] Basler, C. F., et al.: The Ebola virus VP35 protein functions as a type I IFN antagonist. Proc. Natl. Acad. Sci. U. S. A. 97, 12289-12294 (2000). doi: 10. 1073/pnas. 220398297.

[9] Schumann, M., Gantke, T., Muhlberger, E.: Ebola virus VP35 antagonizes PKR activity through its C-terminal interferon inhibitory domain. J. Virol. 83, 8993-8997 (2009). doi: 10. 1128/JVI. 00523-09.

[10] Halfmann, P., Neumann, G., Kawaoka, Y.: The Ebolavirus VP24 protein blocks phosphorylation of p38 mitogen-activated protein kinase. J. Infect. Dis. 204 (Suppl 3), S953-S956 (2011). doi: 10. 1093/ infdis/jir325.

[11] Reid, S. P., et al.: Ebola virus VP24 binds karyopherin alpha1 and blocks STAT1 nuclear accumulation. J. Virol. 80, 5156-5167 (2006). doi: 10. 1128/ JVI. 02349-05.

[12] Bosio, C. M., et al.: Ebola and Marburg viruses replicate in monocyte-derived dendritic cells without inducing the production of cytokines and full maturation. J. Infect. Dis. 188, 1630 - 1638 (2003). doi: 10. 1086/379199.

[13] Mahanty, S., et al.: Cutting edge: impairment of dendritic cells and adaptive immunity by Ebola and Lassa viruses. J. Immunol. 170, 2797-2801 (2003).

[14] Baize, S., et al.: Defective humoral responses and extensive intravascular apoptosis are associated with fatal outcome in Ebola virus-infected patients. Nat. Med. 5, 423-426 (1999). doi: 10. 1038/7422.

[15] Ksiazek, T. G., et al.: Clinical virology of Ebola hemorrhagic fever (EHF): virus, virus antigen, and IgG and IgM antibody findings among EHF patients in Kikwit, Democratic Republic of the Congo, 1995. J. Infect. Dis. 179 (Suppl 1), S177-S187 (1999). doi: 10. 1086/514321.

[16] Wauquier, N., Becquart, P., Padilla, C., Baize, S., Leroy, E. M.: Human fatal zaireebola virus infection is associated with an aberrant innate immunity and with massive lymphocyte apoptosis. PLoS Negl. Trop. Dis. 4, (2010). doi: 10. 1371/journal. pntd. 0000837.

[17] Geisbert, T. W., et al.: Mechanisms underlying coagulation abnormalities in ebola hemorrhagic fever: overexpression of tissue factor in primate monocytes/macrophages is a key event. J. Infect. Dis. 188, 1618-1629 (2003). doi: 10. 1086/379724.

[18] Geisbert, T. W., et al.: Treatment of Ebola virus infection with a recombinant inhibitor of factor VIIa/tissue

factor: a study in rhesus monkeys. Lancet362, 1953-1958 (2003). doi: 10. 1016/S0140-6736 (03) 15012-X.

[19] Kuhn, J. H. et al.: Family Filoviridae. In: King, A. M. Q., Adams, M. J., Carstens, E. B., Lefkowitz, E. J. (eds.) Virus Taxonomy-Ninth Report of the International Committee on Taxonomy of Viruses, pp. 665-671. Elsevier, London, United Kingdom (2012).

[20] Beniac, D. R., et al.: The organisation of ebola virus reveals a capacity for extensive, modular polyploidy. PLoSOne7, e29608 (2012). doi: 10. 1371/journal. pone. 0029608.

[21] Hoenen, T., et al.: Infection of naive target cells with virus-like particles: implications for the function of ebola virus VP24. J. Virol. 80, 7260-7264 (2006). doi: 10. 1128/JVI. 00051-06.

[22] Muhlberger, E., Weik, M., Volchkov, V. E., Klenk, H. D., Becker, S.: Comparison of the transcription and replication strategies of marburg virus and Ebola virus by using artificial replication systems. J. Virol. 73, 2333-2342 (1999).

[23] Jasenosky, L. D., Neumann, G., Lukashevich, I., Kawaoka, Y.: Ebola virus VP40-induced particle formation and association with the lipid bilayer. J. Virol. 75, 5205-5214 (2001). doi: 10. 1128/JVI. 75. 11. 5205-5214. 2001.

[24] Nanbo, A., et al.: Ebolavirus is internalized into host cells via macropinocytosis in a viral glycoprotein-dependent manner. PLoSPathog. 6, e1001121 (2010). doi: 10. 1371/journal. ppat. 1001121.

[25] Hoenen, T., Jung, S., Herwig, A., Groseth, A., Becker, S.: Both matrix proteins of Ebola virus contribute to the regulation of viral genome replication and transcription. Virology403, 56-66 (2010). doi: 10. 1016/j. virol. 2010. 04. 002.

[26] Volchkov, V. E., et al.: GP mRNA of Ebola virus is edited by the Ebola virus polymerase and by T7 and vaccinia virus polymerases. Virology214, 421-430 (1995).

[27] Mehedi, M., et al.: A new Ebola virus nonstructural glycoprotein expressed through RNA editing. J. Virol. 85, 5406-5414 (2011). doi: 10. 1128/JVI. 02190-10.

[28] Takada, A.: Filovirus tropism: cellular molecules for viral entry. Front. Microbiol. 3, 34 (2012). doi: 10. 3389/fmicb. 2012. 00034.

[29] Saeed, M. F., Kolokoltsov, A. A., Albrecht, T., Davey, R. A.: Cellular entry of ebola virus involves uptake by a macropinocytosis-like mechanism and subsequent trafficking through early and late endosomes. PLoSPathog. 6, e1001110 (2010). doi: 10. 1371/journal. ppat. 1001110.

[30] Chandran, K., Sullivan, N. J., Felbor, U., Whelan, S. P., Cunningham, J. M.: Endosomal proteolysis of the Ebola virus glycoprotein is necessary for infection. Science308, 1643-1645 (2005). doi: 10. 1126/ science. 1110656.

[31] Carette, J. E., et al.: Ebola virus entry requires the cholesterol transporter Niemann-Pick C1. Nature477, 340-343 (2011). doi: 10. 1038/nature10348.

[32] Cote, M., et al.: Small molecule inhibitors reveal Niemann-Pick C1 is essential for Ebola virus infection. Nature477, 344-348 (2011). doi: 10. 1038/ nature10380.

[33] Hoenen, T., et al.: Inclusion bodies are a site of Ebola virus replication. J. Virol. 86, 11779-11788 (2012).

[34] Adu-Gyamfi, E., Digman, M. A., Gratton, E., Stahelin, R. V.: Investigation of ebola VP40 assembly and oligomerization in live cells using number and brightness analysis. Biophys. J. 102, 2517-2525 (2012). doi: 10. 1016/j. bpj. 2012. 04. 022.

[35] Hoenen, T., et al.: Oligomerization of Ebola virus VP40 is essential for particle morphogenesis and regulation of viral transcription. J. Virol. 84, 7053-7063 (2010). doi: 10. 1128/JVI. 00737-10.

[36] Harty, R. N., Brown, M. E., Wang, G., Huibregtse, J., Hayes, F. P.: A PPxY motif within the VP40 protein of Ebola virus interacts physically and functionally with a ubiquitin ligase: implications for filovirus budding. Proc. Natl. Acad. Sci. U. S. A. 97, 13871-13876 (2000). doi: 10. 1073/pnas. 250277297.

[37] Licata, J. M., et al.: Overlapping motifs (PTAP and PPEY) within the Ebola virus VP40 protein function independently as late budding domains: involvement of host proteins TSG101 and VPS-4. J. Virol. 77, 1812-1819 (2003).

[38] Timmins, J. et al.: Ebola virus matrix protein VP40 interaction with human cellular factors Tsg101 and Nedd4. J Mol Biol. 326, 493-502 (2003). doi: S0022283602014067 [pii].

[39] Timen, A., et al.: Response to imported case of Marburg hemorrhagic fever, the Netherland. Emerg. Infect. Dis. 15, 1171-1175 (2009). doi: 10. 3201/ eid1508. 090015.

[40] Centers for Disease Control and Prevention: Imported case of Marburg hemorrhagic fever-Colorado, 2008. MMWRMorb. Mortal. Wkly Rep. 58, 1377-1381 (2009).

[41] Richards, G. A., et al.: Unexpected Ebola virus in a tertiary setting: clinical and epidemiologic aspects. Crit. Care Med. 28, 240-244 (2000).

[42] Becker, S., Feldmann, H., Will, C., Slenczka, W.: Evidence for occurrence of filovirus antibodies in humans and imported monkeys: do subclinical filovirus infections occur worldwide? Med. Microbiol. Immunol. 181, 43-55 (1992).

[43] Nidom, C. A., et al.: Serological evidence of ebola virus infection in Indonesian orangutans. PLoS One7, e40740 (2012). doi: 10. 1371/journal. pone. 0040740.

[44] Becquart, P., et al.: High prevalence of both humoral and cellular immunity to Zaire ebolavirus among rural populations in Gabon. PLoSOne5, e9126 (2010). doi: 10. 1371/journal. pone. 0009126.

[45] Busico, K. M., et al.: Prevalence of IgG antibodies to Ebola virus in individuals during an Ebola outbreak, Democratic Republic of the Congo, 1995. J. Infect. Dis. 179 (Suppl1), S102-S107 (1999). doi: 10. 1086/514309.

[46] Gonzalez, J. P., Nakoune, E., Slenczka, W., Vidal, P., Morvan, J. M.: Ebola and Marburg virus antibody prevalence in selected populations of the Central African Republic. Microbes Infect. 2, 39 - 44 (2000).

[47] Leroy, E. M., et al.: Multiple Ebola virus transmission events and rapid decline of central African wildlife. Science303, 387-390 (2004). doi: 10. 1126/science. 1092528.

[48] Leroy, E. M., et al.: Fruit bats as reservoirs of Ebola virus. Nature 438, 575 - 576 (2005). doi: 10. 1038/438575a.

[49] Leroy, E. M., et al.: Human Ebola outbreak resulting from direct exposure to fruit bats in Luebo, Democratic Republic of Congo. Vector Borne Zoonotic Dis. 9, 723-728 (2007). doi: 10. 1089/ vbz. 2008. 0167 (2009).

[50] Francesconi, P., et al.: Ebola hemorrhagic fever transmission and risk factors of contacts, Emerg. Infect. Dis. 9, 1430-1437 (2003). doi: 10. 3201/ eid0911. 030339.

[51] Bausch, D. G., et al.: Assessment of the risk of Ebola virus transmission from bodily fluids and fomites. J. Infect. Dis. 196 (Suppl 2), S142-S147 (2007). doi: 10. 1086/520545.

[52] Grolla, A., Lucht, A., Dick, D., Strong, J. E., Feldmann, H.: Laboratory diagnosis of Ebola and Marburg hemorrhagic fever. Bull. Soc. Pathol. Exot. 98, 205-209 (2005).

[53] Towner, J. S., et al.: Rapid diagnosis of Ebola hemorrhagic fever by reverse transcription-PCR in an outbreak setting and assessment of patient viral load as a predictor of outcome. J. Virol. 78, 4330 - 4341 (2004).

[54] Onyango, C. O., et al.: Laboratory diagnosis of Ebola hemorrhagic fever during an outbreak in Yambio, Sudan, 2004. J. Infect. Dis. 196 (Suppl 2), S193-S198 (2007). doi: 10. 1086/520609.

[55] Nkoghe, D., Kone, M. L., Yada, A., Leroy, E.: A limited outbreak of Ebola haemorrhagic fever in Etoumbi, Republic of Congo. Trans. R. Soc. Trop. Med. Hyg. 105, 466-472 (2005). doi: 10. 1016/ j. trstmh. 2011. 04. 011 (2011).

[56] Borchert, M., et al.: Ebola hemorrhagic fever outbreak in Masindi District, Uganda: outbreak description and lessons learned. BMC Infect. Dis. 11, 357 (2011). doi: 10. 1186/1471-2334-11-357.

[57] Saijo, M., et al.: Laboratory diagnostic systems for Ebola and Marburg hemorrhagic fevers developed with recombinant proteins. Clin. Vaccine Immunol. 13, 444 - 451 (2006). doi: 10. 1128/CVI. 13. 4. 444 - 451. 2006.

[58] Nkoghe, D., et al.: Practical guidelines for the management of Ebola infected patients in the field. Med. Trop. (Mars) 64, 199-204 (2004).

[59] Qiu, X., et al.: Successful treatment of ebola virus-infected cynomolgus macaques with monoclonal antibodies. Sci. Transl. Med. 4, 138ra181 (2012). doi: 10. 1126/scitranslmed. 3003876.

[60] Dye, J. M., et al.: Postexposure antibody prophylaxis protects nonhuman primates from filovirus disease. Proc. Natl. Acad. Sci. U. S. A. 109, 5034-5039 (2012). doi: 10. 1073/pnas. 1200409109.

[61] Geisbert, T. W., et al.: Postexposure protection of non-human primates against a lethal Ebola virus challenge with RNA interference: a proof-of-concept study. Lancet375, 1896-1905 (2010). doi: 10. 1016/S0140-6736 (10) 60357-1.

[62] Hensley, L. E., et al.: Recombinant human activated protein C for the postexposure treatment of Ebola hemorrhagic fever. J. Infect. Dis. 196 (Suppl 2), S390-S399 (2007). doi: 10. 1086/520598.

[63] Bausch, D. G., Sprecher, A. G., Jeffs, B., Boumandouki, P.: Treatment of Marburg and Ebola hemorrhagic fevers: a strategy for testing new drugs and vaccines under outbreak conditions. Antiviral Res. 78, 150-161 (2008). doi: 10. 1016/j. antiviral. 2008. 01. 152.

[64] Baskerville, A., Bowen, E. T., Platt, G. S., McArdell, L. B., Simpson, D. I.: The pathology of experimental Ebola virus infection in monkeys. J. Pathol. 125, 131 - 138 (1978). doi: 10. 1002/path. 1711250303.

[65] Bowen, E. T., Platt, G. S., Simpson, D. I., McArdell, L. B., Raymond, R. T.: Ebola haemorrhagic fever: experimental infection of monkeys. Trans. R. Soc. Trop. Med. Hyg. 72, 188-191 (1978).

[66] Hoenen, T., Groseth, A., Feldmann, H.: Current ebola vaccines. Expert Opin. Biol. Ther. 12, 859-872 (2012). doi: 10. 1517/14712598. 2012. 685152.

[67] Sullivan, N. J., Sanchez, A., Rollin, P. E., Yang, Z. Y., Nabel, G. J.: Development of a preventive vaccine for Ebola virus infection in primates. Nature408, 605-609 (2000). doi: 10. 1038/35046108.

[68] Sullivan, N. J., et al.: Accelerated vaccination for Ebola virus hemorrhagic fever in non-human primates. Nature424, 681-684 (2003). doi: 10. 1038/ nature01876.

[69] Sullivan, N. J., et al.: Immune protection of nonhuman primates against Ebola virus with single low-dose adenovirus vectors encoding modified GPs. PLoS Med. 3, e177 (2006). doi: 10. 1371/journal. pmed. 0030177.

[70] Richardson, J. S., et al.: Enhanced protection against Ebola virus mediated by an improved adenovirus-based vaccine. PLoS One 4, e5308 (2009). doi: 10. 1371/journal. pone. 0005308.

[71] Mast, T. C., et al.: International epidemiology of human preexisting adenovirus (Ad) type-5, type-6, type-26 and type-36 neutralizing antibodies: correlates of high Ad5 titers and implications for potential HIV vaccine trials. Vaccine28, 950-957 (2010). doi: 10. 1016/j. vaccine. 2009. 10. 145.

[72] Croyle, M. A., et al.: Nasal delivery of an adenovirus-based vaccine bypasses preexisting immunity to the vaccine carrier and improves the immune response in mice. PLoS One 3, e3548 (2008). doi: 10. 1371/journal. pone. 0003548.

[73] Kobinger, G. P., et al.: Chimpanzee adenovirus vaccine protects against Zaire Ebola virus. Virology346, 394-401 (2006). doi: 10. 1016/j. virol. 2005. 10. 042.

[74] Richardson, J. S., et al.: Impact of systemic or mucosal immunity to adenovirus on Ad-based Ebola virus vaccine efficacy in guinea pigs. J. Infect. Dis. 204 (Suppl 3), S1032-S1042 (2011). doi: 10. 1093/ infdis/jir332.

[75] Geisbert, T. W., et al.: Recombinant adenovirus serotype 26 (Ad26) and Ad35 vaccine vectors bypass immunity to Ad5 and protect nonhuman primates against ebolavirus challenge. J. Virol. 85, 4222 - 4233 (2011). doi: 10. 1128/JVI. 02407-10.

[76] Ledgerwood, J. E., et al.: A replication defective recombinant Ad5 vaccine expressing Ebola virus GP is safe and immunogenic in healthy adults. Vaccine29, 304 - 313 (2010). doi: 10. 1016/j. vaccine. 2010. 10. 037.

[77] Roy, S., et al.: Generation of an adenoviral vaccine vector based on simian adenovirus 21. J. Gen. Virol. 87, 2477-2485 (2006). doi: 10. 1099/vir. 0. 81989-0.

[78] Choi, J. H., et al.: A single sublingual dose of an adenovirus-based vaccine protects against lethal Ebola challenge in mice and guinea pigs. Mol. Pharm. 9, 156-167 (2012). doi: 10. 1021/mp200392g.

[79] Pratt, W. D., et al.: Protection of nonhuman primates against two species of Ebola virus infection with a single complex adenovirus vector. Clin. VaccineImmunol. 17, 572-581 (2010). doi: 10. 1128/ CVI. 00467-09.

[80] Swenson, D. L., et al.: Vaccine to confer to nonhuman primates complete protection against multi-strain Ebola and Marburg virus infections. Clin. Vaccine Immunol. 15, 460-467 (2008). doi: 10. 1128/CVI. 00431-07.

[81] Hensley, L. E., et al.: Demonstration of cross-protective vaccine immunity against an emerging pathogenic Ebolavirus Species. PLoSPathog. 6, e1000904 (2010). doi: 10. 1371/journal. ppat. 1000904.

[82] Yang, Z. Y., et al.: Identification of the Ebola virus glycoprotein as the main viral determinant of vascular cell cytotoxicity and injury. Nat. Med. 6, 886-889 (2000). doi: 10. 1038/78645.

[83] Geisbert, T. W., et al.: Vector choice determines immunogenicity and potency of genetic vaccines against Angola Marburg virus in nonhuman primates. J. Virol. 84, 10386-10394 (2010). doi: 10. 1128/JVI. 00594-10.

[84] Sullivan, N. J., et al.: $CD8^+$ cellular immunity mediates rAd5 vaccine protection against Ebola virus infection of nonhuman primates. Nat. Med. 17, 1128-1131 (2011). doi: 10. 1038/nm. 2447.

[85] Sullivan, N. J., Martin, J. E., Graham, B. S., Nabel, G. J.: Correlates of protective immunity for Ebola vaccines: implications for regulatory approval by the animal rule. Nat. Rev. Microbiol. 7, 393-400 (2009). doi: 10. 1038/nrmicro2129.

[86] Geisbert, T. W., Feldmann, H.: Recombinant vesicular stomatitis virus-based vaccines against Ebola and Marburg virus infections. J. Infect. Dis. 204 (Suppl 3), S1075-S1081 (2011). doi: 1 0. 1093/ infdis/jir349.

[87] Jones, S. M., et al.: Live attenuated recombinant vaccine protects nonhuman primates against Ebola and Marburg viruses. Nat. Med. 11, 786-790 (2005). doi: 10. 1038/nm1258.

[88] Jones, S. M., et al.: Assessment of a vesicular stomatitis virus-based vaccine by use of the mouse model of Ebola virus hemorrhagic fever. J. Infect. Dis. 196 (Suppl 2), S404-S412 (2007). doi: 10. 1086/520591.

[89] Robain, O., Chany-Fournier, F., Cerutti, I., Mazlo, M., Chany, C.: Role of VSV G antigen in the development of experimental spongiform encephalopathy in mice. ActaNeuropathol. 70, 220-226 (1986).

[90] Martinez, I., Rodriguez, L. L., Jimenez, C., Pauszek, S. J., Wertz, G. W.: Vesicular stomatitis virus glycoprotein is a determinant of pathogenesis in swine, a natural host. J. Virol. 77, 8039-8047 (2003).

[91] Mire, C. E., et al.: Recombinant vesicular stomatitis virus vaccine vectors expressing fi lovirus glycoproteins lack neurovirulence in nonhuman primates. PLoSNegl. Trop. Dis. 6, e1567 (2012). doi: 10. 1371/ journal. pntd. 0001567.

[92] Geisbert, T. W., et al.: Vesicular stomatitis virus-based ebola vaccine is well-tolerated and protects immunocompromised nonhuman primates. PLoSPathog. 4, e1000225 (2008). doi: 10. 1371/journal. ppat. 1000225.

[93] Falzarano, D., et al.: Single immunization with a monovalent vesicular stomatitis virus-based vaccine protects nonhuman primates against heterologous challenge with Bundibugyo ebolavirus. J. Infect. Dis. 204 (Suppl 3), S1082-S1089 (2011). doi: 10. 1093/infdis/jir350.

[94] Geisbert, T. W., et al.: Recombinant vesicular stomatitis virus vector mediates postexposure protection against Sudan Ebola hemorrhagic fever in nonhuman primates. J. Virol. 82, 5664-5668 (2008). doi: 10. 1128/JVI. 00456-08.

[95] Geisbert, T. W., et al.: Single-injection vaccine protects nonhuman primates against infection with marburg virus and three species of ebola virus. J. Virol. 83, 7296-7304 (2009). doi: 10. 1128/JVI. 00561-09.

[96] Marzi, A., et al.: Vesicular stomatitis virus-based Ebola vaccines with improved cross-protective efficacy. J. Infect. Dis. 204 (Suppl 3), S1066-S1074 (2011). doi: 10. 1093/infdis/jir348.

[97] Qiu, X., et al.: Mucosal immunization of cynomolgus macaques with the VSVDeltaG/ZEBOVGP vaccine stimulates strong ebola GP-specific immune responses. PLoS One 4, e5547 (2009). doi: 10. 1371/ journal. pone. 0005547.

[98] Feldmann, H., et al.: Effective post-exposure treatment of Ebola infection. PLoS Pathog. 3, e2 (2007).

doi：10. 1371/journal. ppat. 0030002.

[99] Geisbert, T. W., et al.: Postexposure treatment of Marburg virus infection. Emerg. Infect. Dis. 16, 1119-1122 (2010). doi: 10. 3201/eid1607. 100159.

[100] Sanchez, A., et al.: Analysis of human peripheral blood samples from fatal and nonfatal cases of Ebola (Sudan) hemorrhagic fever: cellular responses, virus load, and nitric oxide levels. J. Virol. 78, 10370-10377 (2004). doi: 10. 1128/JVI. 78. 19. 10370-10377. 2004.

[101] Gunther, S., et al.: Management of accidental exposure to Ebola virus in the biosafety level 4 laboratory, Hamburg, Germany. J. Infect. Dis. 204 (Suppl 3), S785-S790 (2011). doi: 10. 1093/infdis/jir298.

(何继军，郭建宏译)

第7章　实验性登革热疫苗

Sathyamangalam Swaminathan 和 Navin Khanna①

摘要

全球近一半人口面临感染登革热风险，登革热是一种由伊蚊传播给人类的病毒性疾病。黄病毒科4种密切相关的登革热病毒中，每一种都能引起广泛的疾病，从轻微的登革热到可能致命的登革热出血热和休克。登革热临床明确诊断很困难，需要实验室确诊。尽管近年来已开发出广泛的诊断工具，包括抗原检测试验，但缺乏特征明确的血清盘，妨碍了其验证。一旦确诊，严重登革热患者治疗的唯一选项是症状性和支持性医疗。病媒控制措施失败以及疫苗和药物缺乏已使登革热成为世界范围内一个重大公共健康问题。

接种登革热疫苗是阻止登革热传播的有效途径。近年来，意识的提高有助于加强多个参与方正在进行的疫苗研发工作。因此，许多候选疫苗都处于不同的发展阶段。其中减毒活疫苗正处于临床开发阶段，引发了人们对登革热疫苗即将问世的预期。登革热疫苗需要保护所有4种血清型，以便避免疾病免疫增强。由于缺乏临床前动物模型来可靠的预测疫苗在人体中的疗效，使得登革热疫苗开发成为一项具有挑战性的任务。

最近的功效试验数据强调，需要更好地理解免疫系统在发病机制和保护中的作用，以便能够设计潜在有效的登革热疫苗。

登革热通过蚊子传播给人类，是21世纪初最重要的虫媒病毒性疾病。近几十年来登革热流行率和发病率急剧上升[3]，似乎是多种因素综合作用的结果，包括大规模计划外城市化、人口过剩、全球旅行增加，重要的是未能根除蚊子[4]。目前登革热流行于亚太地区、美洲、中东和非洲的100多个国家。世界卫生组织（WHO）估计，全球超过25亿人生活在登革热和非登革热地区，每年约有5 000万人感染，其中50万人病情严重，导致超过2万人死亡[5]。最近研究表明，全球登革热负担可能要高得多[1]。例如中国大约1/5地区属于热带纬度，可能有利于登革热传播。尽管中国在20世纪80年代和90年代向世卫组织报告了登革热疫情，但自那以后就停止了报告。据报道，美国已经暴发了登革热，来自返回的旅行者输入。在欧洲发现了一种登革热蚊子媒介，预示着未来登革热也会传播到这一大陆[3]。

尽管登革热在历史上没有被认为是一个重大的公共卫生问题，但人们认识到登革热可以传播到热带世界边界以外，这激发了人们对了解登革热和开发预防登革热的疫苗的兴趣[2]。本章将力求提供有关登革热的简明背景信息，然后概述当前疫苗开发工作和必须解决的问题和挑战。

7.1　登革热病

登革热感染导致一系列症状。传统的世卫组织病例定义基于4个标准，即发热、出血、血小板减少和血浆渗漏，认识以下不同的临床条件：典型登革热（DF）、登革热出血热（DHF）和登革热休克综合征（DSS）[6]。与不连续的临床实体不同，这3种疾病可能代表持续

① S. Swaminathan，博士（✉）·N. Khanna，博士（皮拉尼理工学院生物科学系，海得拉巴校区，Jawahar Nagar，shaerpetmandal，海得拉巴 500078，安得拉邦，印度，E-mail：swaminathan@ hyderabad. bits-pilani. ac. in）

的登革热病谱中逐渐严重的阶段（图 7.1，左图）。

典型登革热（DF）表现为突然发高热，通常持续 5~7 天。这种与病毒血症同时出现的发热期与许多症状有关。通常在大多数情况下会完全恢复。一小部分典型登革热患者可能会迅速发展为登革热出血热（DHF）。当发烧下降时，可能有出汗、烦躁不安和循环紊乱的迹象。血浆渗漏是区分典型登革热和登革热出血热的标志。当这种渗漏很严重时，会导致危及生命的循环衰竭（低血容量性休克）。使典型登革热患者易患严重疾病的因素尚不清楚。临床上退热被认为是一个关键阶段。出血、严重血小板减少、血浓度≥20%等症状，伴有严重持续腹痛、烦躁不安、持续呕吐、突然体温过低伴随大量出汗、虚弱、昏厥，预示着休克临近。与典型登革热相比，登革热出血热/登革热休克综合征中，病毒血症比典型登革热高 2 个数量级[7,8]。世卫组织最近的一项分类旨在根据单个参数识别严重登革热，已用以下 3 个实体取代了典型登革热、登革热出血热和登革热休克综合征：无预警信号登革热、有预警信号登革热和严重登革热[9]。虽然修订后的分类法能够更好地捕获病例[10]，但人们认为在疫情暴发期间，会使健康保健设施超负荷[10,11]。

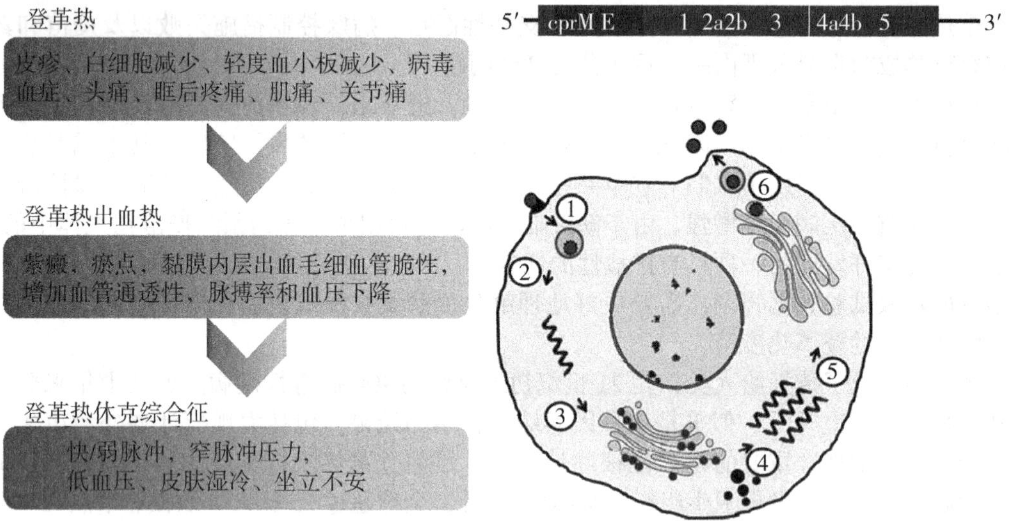

图 7.1　登革热病、病毒及其生命周期

左图显示了登革热疾病进展过程中症状的示意图。右面板顶部显示的是 DENV 基因组。基因组两端的短黑线表示 5’和 3’NTRs。由单个 ORF 编码的 10 种病毒蛋白由红色（衣壳蛋白和 NS 蛋白）和蓝色框（*prM* 和 *E* 蛋白）显示。右侧面板的底部描述了 DENV 的生命周期。生命周期步骤由圈出的数字表示：①受体识别和进入；②病毒/宿主膜融合和脱壳；③基因组 RNA 在粗面内质网翻译；④利用 NS 蛋白复制基因组 RNA 进一步翻译；⑤病毒体组装和成熟；⑥最终成熟并离开细胞。

7.2　登革热病毒

登革热由 4 种密切相关但抗原性不同的登革热病毒（登革热病毒-1、-2、-3 和-4）中的任何一种引起。登革热病毒（DENV）属于黄病毒科黄病毒属。成熟的 DENV 粒子尺寸约为 50 nm，由核衣壳核心组成，核衣壳核心被包被有病毒结构蛋白的宿主衍生脂质双层包围[12]。登革热病毒与包括黄热病（YF）、日本脑炎（JE）和蜱传脑炎（TBE）病毒在内的黄病毒属的其他成员具有相似遗传组织[13]。黄病毒基因组由约 11 000个碱基组成，5’端有帽子结构、缺少聚 A 尾、单股正链（+）RNA。该 RNA 有一个开放阅读框（ORF），两侧分别有 5’和 3’非翻译区（NTRs），长度约为100和450个核苷酸（nt）（图7.1，右上角）。NTRs包含独特的序列和结构原件，介导基因组RNA循环和RNA/蛋白质相互作用，这对病毒生命周期至关重要[2,14-16]。ORF编码10种病毒蛋白，其中3种是结构蛋白［衣壳，C，膜，M（合成为较大的前体prM）和包膜，E］以及其余的非结构蛋白（NS）NS1，2a，2b，3，4a，4b和5[13]。

锚定在脂质双层上的E蛋白代表病毒的主要结构抗原。90个E蛋白的同源二聚体形成一个“人”字形阵列，覆盖整个病毒粒子表面[12]。它与一个尚未确定的宿主细胞表面受体结合，介导膜融合并作为宿主中和抗体应答的主要靶标[13,17]。每个E蛋白单体由3个不同的结构域组成[18]，其中从病毒体表面突出的羧基末端包膜结构域Ⅲ（EDⅢ）[12]和宿主受体结合有关[19]，包含多个血清型特异性构象中和表位[17,20]。受体结合和内化后，E蛋白同源二聚体响应核内体的酸性环境重新排列成同源三聚体。这对于膜融合和病毒基因组RNA释放到细胞质中是必要的[21]。

病毒RNA的单个ORF指导合成一个大于3 000个氨基酸（aa）的多聚蛋白前体。病毒和宿主蛋白酶联合作用将这种前体蛋白加工成上述成熟病毒蛋白[13]。病毒RNA复制也发生在细胞质中，通过互补负链RNA中间产物进行，主要由NS3和NS5介导，它们共同提供病毒RNA复制酶功能。病毒RNA和蛋白质合成后，病毒形态形成，开始产生覆盖有prM-E异二聚体的不成熟病毒粒子[22]。prM与E蛋白的关联被认为有助于通过酸化分类室防止病毒成熟过程中后者的膜过早融合。在通过胞吐作用释放前，宿主弗林蛋白酶介导的prM分裂触发最终成熟步骤，同时释放成熟病毒粒子[21]（图7.1）。

7.3　传播

登革热病毒由伊蚊、埃及伊蚊、白纹伊蚊和波利尼西亚伊蚊传播给人类[23]。世界上许多城市人口稠密，蚊子传播媒介盛行以及多种登革热病毒血清型的共同传播，为维持登革热病毒在人类和蚊子之间有效传播提供了理想的环境[4]（图7.2）。埃及伊蚊是登革热的主要传播媒介，因为它高度适应城市环境。登革热病毒是由一只感染埃及伊蚊的雌性蚊子叮咬传染给人类宿主的。一旦进入人类宿主，病毒就会开始其生命周期的内在阶段。当病毒出现在血液中时，经过4~7天的潜伏期，疾病症状变得明显。持续约5天的病毒血症与持续5~7天的发热期一致。人类因此成为一个放大宿主。

病毒血症阶段的病毒载量似乎与疾病严重程度相关[7,8]。高病毒血症可促进有效传播。当未受感染的蚊子以病毒性宿主为食时，它会获得病毒，现在病毒在蚊子宿主中经历其生命周期的外在阶段。在这一阶段结束时（8~12天），蚊子在其余生中会变得具有传染性。当这种蚊子叮咬另一个人类宿主重新觅食时，这种循环会继续。一只感染性蚊子在一次血餐中叮咬几次，可以在短时间内将病毒传播给几个人[4]。

7.4　诊断和治疗

由于临床诊断可能不明确，实验室诊断对确认登革热感染至关重要。根据登革热病毒，其基因组RNA、其抗原或其引发的抗体[2,4,24]的鉴定可确认感染（图7.2）。目前没有针对登革热的抗病毒治疗方法。症状性和支持性医疗是治疗的唯一方法[6]。

7.4.1　实验室诊断

通过蚊子接种或细胞培养进行登革热病毒分离，如果在发热期采集患者血清，则可进行登革热病毒鉴定。病毒RNA检测和血清型鉴定可以使用耦合逆转录聚合酶链反应（RT-PCR）或恒温RNA特异性扩增[28]。多种形式的抗原检测试验，旨在检测早期出现在血液中的病毒NS1蛋

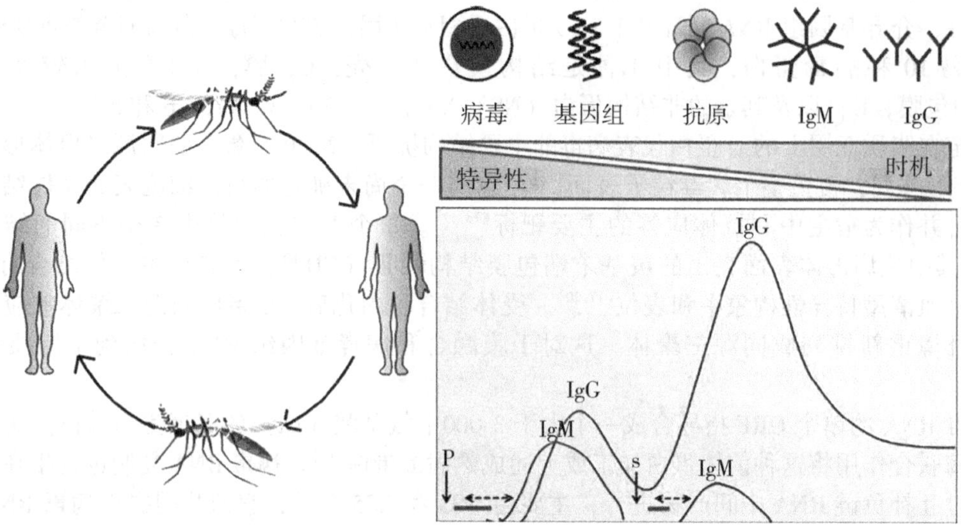

图 7.2 登革热传播、诊断和抗体反应的各个方面

左图描绘了埃及伊蚊传播登革热的城市流行周期。当人类作为扩增宿主时，人类和蚊子之间的 DENV 传播可以有效地进行，而不需要森林扩增宿主。右图顶部是使用直接（病毒、其基因组及其编码的抗原）和间接（由登革病毒引发的抗体）标志对登革病毒感染进行实验室诊断的相对优缺点示意图。只能在有限的时间窗口内检测到直接标志。IgM 和 IgG 是更实用的诊断确认标志（特异性丧失是由黄病毒交叉反应引起的）。右图底部是抗体反应定性的图示（垂直轴和水平轴分别代表登革病毒感染后的抗体水平和时间）。黄病毒幼稚个体的第一次感染（P）引发了一次抗体应答，其特征是在患病 3~5 天诱导 IgM 抗体，在大约 2 周达到高峰，并在接下来的几个月下降。IgG 抗体在此后不久出现，并持续终生。在重复感染（S）期间，开始次级抗体应答，其中高滴度 IgG 抗体出现在 IgM 抗体应答之前或与相对较弱的 IgM 抗体应答一起出现。继发感染期间，与发热相一致的病毒血症窗口（由水平轴正上方的实心红色条表示）缩短。NS1 抗原检测窗口（由双箭头虚线表示）与病毒血症重叠，在继发感染中也被缩短。

白[29,30]，可作为严重疾病的预测标志物[31]，最近开发了 NS1 特异性 IgM 和 IgG 检测方法[24]。近年来，不同形式的商用 NS1 抗原检测试验已经开始使用[32,33]。

历史上，血凝抑制（HI）试验是世卫组织确认近期登革病毒感染的标准试验，它测量成对的急性和恢复期血清中抗体抑制病毒糖蛋白介导的红细胞凝集能力。成对血清 HI 滴度≥4 倍差异可以诊断为近期感染[6]。噬斑减少中和试验（PRNT）[34]和其他 ELISA[35]以及基于流式细胞术的病毒中和试验[36]主要用于疫苗研究，这些试验可用于估计中和抗体滴度和鉴定登革热病毒血清型。

登革热常规血清学检测使用基于 ELISA 的抗体检测，因为这些检测相对便宜且易于实施[4,24]。由于 IgM 抗体滴度往往相对较低（图 7.2，右下方），它们的识别基于使用结合到固相的抗人 IgM 特异性抗体的捕获，然后使用 4 种血清型登革热病毒抗原混合物进行检测。一个主要限制是，登革病毒特异性 IgG 和与其他黄病毒交叉反应引起的假阳性范围[4,24]。另一方面 IgG 抗体使用相同的登革热病毒抗原用间接 ELISA 检测。配对血清的 IgG ELISA 可用于确认登革热感染。由于黄病毒血清复合体[17]中广泛的交叉反应性，IgG ELISA 缺乏登革热病毒特异性。消除这一问题的新策略是基于合成多表位蛋白（MEP）抗原，该抗原结合了精心选择的登革热病毒特异性表位。MEP 抗原在对其诊断效用的初步研究中显示出有希望的结果[37-40]。IgM 与 IgG 抗体比值（两者都用捕获方式检测）可用于区分原发性和继发性登革病毒感染[41]。市面上有几种具有不同敏感性和特异性商业登革热病毒抗体检测试剂盒。由于缺乏来自不同流行地区的原发性和继发性登革病毒感染，以及非登革病毒黄病毒感染和其他发热性感染的明确血清样品，这些试剂盒无法得到有意义的验证。

7.4.2 临床管理

对于不复杂的典型登革热病例治疗包括卧床休息、口服补液和作为解热镇痛药的扑热息痛。阿司匹林不推荐使用，因为会导致血小板功能障碍。从发热第三天开始，在退热前后，临床和实验室监测都是必不可少的。表明病情发展为严重疾病的临床症状包括四肢冰冷、脉搏低、尿量低、黏膜出血和腹痛。血细胞比容上升（≥20%）和血小板计数下降（$10^6/\mu L$）强烈表明发展到了登革热出血热阶段。如果发现这些迹象，必须立即住院治疗。

登革热出血热患者临床护理的关键组成部分是静脉输液治疗，以维持血浆渗漏期间的有效循环，并对血细胞比容、血小板计数、脉搏率和血压、温度、尿量、输液量和其他休克症状进行仔细的临床监测。由于血浆渗漏率可能不同，因此有必要相应地调整液体疗法。许多患者在液体治疗后12~48小时恢复很快。长期液体治疗是危险的，因为当丢失的血浆被重新吸收时，会引起肺水肿。在登革热休克综合征阶段，拯救患者的关键是尽早认识到患者处于休克状态。治疗包括立即和积极的液体胶体疗法，以恢复有效的循环量，细致和全面的监测以及对任何并发症的管理。尽管进行了充分的液体治疗，但患者的临床状况不佳，加之血细胞比容下降，表明存在内出血，必须进行全血输血治疗。及时、最佳的临床护理可显著降低病死率[6]。

7.5 疫苗开发

这4种登革热病毒血清型在世界各地几个国家都是同时流行的，而且每一种都能引起登革热。对任何一种登革热病毒的免疫都是持久和血清型特异性的，只有对其余3种登革热病毒血清型的短暂交叉保护[42]。重要的是，以前对不同登革热病毒血清型有免疫力的人群，登革热病毒感染往往很严重，这些人要么通过母亲转移被动获得，要么通过原发感染主动获得[43-45]。在严重登革热病毒感染中发现的高病毒血症与引发强烈的细胞因子风暴有关[7,8]，据信是内皮损伤和毛细血管渗漏的基础[2,17,46]。假设交叉反应性抗体能与异型登革热病毒结合，促进其对含Fc受体细胞（如单核细胞和巨噬细胞）的吸收，这些细胞被认为是登革热病毒复制的体内位点。这种称为抗体依赖性增强（ADE）的现象可能是导致严重疾病病毒复制增加的原因[17,45]。

最近的体外和体内研究揭示了prM蛋白的高交叉反应抗体在调节增强和引起严重疾病中的作用[47,48]。原发性抗原痕迹导致低亲和力的扩张，即先前存在的记忆性T细胞不能有效清除异型病毒，可能是导致继发性登革热病毒感染病毒载量增加的另一个因素[49]。因此，一种提供不完全免疫的登革热疫苗可能使疫苗接种者对严重的登革热病敏感，因为这4种登革热病毒血清型在全球范围传播。所以有人认为登革热疫苗必须是4价的，能够同时提供持久的保护，防止4种血清型同时感染。

7.5.1 实验性登革热疫苗

从临床前到临床发展阶段，采用许多不同的策略寻求一系列候选试验疫苗[2,24,50-54]。但除了少数例外，大多数疫苗都是基于制造单价疫苗，针对每一种登革热病毒血清型，然后将其混合成一种4价的登革热疫苗配方。历史上，根据经验致弱一直是登革热疫苗努力的主要焦点。有两个试验小组在进行疫苗开发，一组在泰国的马希德大学（Mahidol University）[55]，另一组在美国的沃尔特里德陆军研究所（WRAR）[56]，通过在原代狗肾细胞（PDK）中连续传代，分别为4种血清型独立开发了单价登革热病毒疫苗（表7.1）。这两种疫苗都已不研究，已经让位于用感染性克隆技术即反向遗传技术实现弱化的方法[57]。重组技术也产生了疫苗候选抗原，这些抗原是基于登革热病毒的结构抗原，即基因载体编码或利用异源宿主表达系统产生的重组抗原。

表 7.1 复制病毒疫苗[a]中使用的登革热病毒株

DENV[b]	Mahidol[c]	WRAIR/GSK[d]	CYD[e]	DENVax[f]	TetraVax[g]
1	16007（PDK-13）	45AZ5（PDK-20/27）	PUO-359	16007	West Pac
2	16681（PDK-53）	S16803（PDK-50）	PUO-218	16681	New Guinea C
3	16562（PGMK-30）	CH53489（PDK-20）	PaH881/88	16562	Sleman/78
4	1036（PDK-48）	341750（PDK-20/6）	1228	1036	814669

[a]这些毒株表明 DENVs 用于经验减毒的 Mahidol 和 WRAIR/GSK 疫苗，用于 CYD、DENVax 和 TetraVax 疫苗；这些毒株表明了供体病毒，*prM* 和 *E* 基因被用于疫苗病毒构建。

[b]本栏表示 DENV 血清型。

[c]减毒毒株的传代次数在括号中表示；DENV-1、-2 和-4 在原代犬肾细胞（PDK）中传代，而 DENV-3 在绿猴肾细胞（PGMK）中传代，因为它不能在 PDK 细胞中生长（数据来自参考文献［50］）。

[d]减毒毒株的传代次数用括号表示；DENV-1/PDK-20 被 DENV-1/PDK-27 取代，DENV-4/PDK-20 被 DENV-4/PDK-6 取代（17 号）。

[e]赛诺菲嵌合黄热病 17D 载体登革热疫苗。

[f]基于 Mahidol DENV-2/PDK-53 载体骨架的 CDC/Inviragen 型间嵌合疫苗；特异于 DENV-1，-DENV-3 和 DENV-4 的型间嵌合结构携带相应 Mahidol 疫苗株的 *prM* 和 *E* 基因。

[g]基于 Δ30 突变载体骨架的 NIH 疫苗。

7.5.1.1 复制型病毒疫苗

利用反向遗传学，美国 Acambis 公司的科学家已经用相应的登革热病毒基因替换了经验致弱黄热病疫苗病毒（YF17D）载体的 *prM* 和 *E* 基因，开发了 4 种嵌合构建体，每种构建体携带一种登革热病毒血清型 *prM* 和 *E* 基因[58-60]。这些嵌合型黄热病——登革热（CYD）疫苗是登革热疫苗开发流程中的领先者。反向遗传学的使用揭示了 3’NTR 改变会导致衰减。

例如这种方法导致了登革热病毒-4 的 30 个核苷酸缺失突变体的鉴定[61]。这种缺失（Δ30 突变）去除了 3’NTR 中的核苷酸 10 478～10 507，在猴子[61]和人类[62]中提供了可接受的致弱和免疫原性之间的平衡。美国国立卫生研究院（NIH）研究人员随后的研究显示，尽管 Δ30 突变导致登革热病毒-1 减弱[63]，但由于未知原因，登革热病毒-2[64]和登革热病毒-3 并未减弱[65]。使用早先为 CYD 疫苗创造而设计的抗原嵌合策略，登革热病毒-4 Δ30 突变体 prM 和 E 基因已经被登革热病毒-2[66]和登革热病毒-3[65]的相应基因取代。后两种变体是基因型间嵌合结构，其血清型对应于同源 *prM+E* 基因供体的血清型。美国疾病控制和预防中心（CDC）科学家已经开发出了基于登革热病毒 2 PDK-53 Mahidol 疫苗病毒作为骨架，携带登革热病毒-1、登革热病毒-3 和登革热病毒-4 的 *prM+E* 基因异型嵌合体[67]。由于亲本登革热病毒-2 PDK-53 病毒在其 *prM* 和 *E* 基因中没有携带任何减毒突变[68]，它可以直接作为登革热病毒-2 疫苗候选抗原。

最近报道了一种开发宿主范围登革热病毒突变体的新方法[69]。这些在 E 蛋白跨膜域中缺失的突变体在昆虫细胞中表现出优先生长，在哺乳动物细胞中受损的传染性，并保留诱导病毒特异性中和抗体的能力。

7.5.1.2 载体疫苗

这类实验性登革热疫苗是基于使用质粒[70,71]或病毒[72-75]载体编码登革热抗原基因。尽管在过去已经描述了编码登革热病毒抗原的减毒痘病毒载体[72]，但最近西尼罗河（WN）病毒[75]和人类腺病毒 5 型（hAd5）[74]载体已被用于表达登革热病毒抗原。尽管基于质粒的疫苗消除了上述复制病毒疫苗的安全风险，但并不能诱导强大的免疫应答。通过将质粒载体替换为具有内置安全特性的病毒载体来解决这一缺点。例如上述的WN载体通过删除其衣壳基因在接种的宿主中进行一个感染周期，而hAd5载体则通过删除其对复制至关重要的早期区域而导致复制缺陷。

许多质粒和病毒载体疫苗候选物编码*prM*和*E*基因具有在体内产生病毒样颗粒（VLPs）的能力，并且基于4种单价疫苗的物理混合物来开发4价疫苗[71,75]，类似于复制型病毒疫苗所采用的方法。然而有些是基于独特的方法，以引起对所有登革热病毒血清型的免疫。例如一种基于质粒的策略利用单个无序的*E*基因，该基因包含代表所有4种登革热病毒血清型的表位[70]，而另一种策略利用两种复制缺陷hAd5载体，每种载体编码两种登革热病毒血清型的*prM+E*基因[74]。文献中也报道了编码基于EDⅢ抗原的质粒载体[76,77]和病毒载体疫苗[78-80]。其中一些基于单一4价结构结合了所有4种登革热病毒血清型的EDⅢs（包膜结构域Ⅲ）[77,80,81]。

7.5.1.3　非复制疫苗

除上述质粒疫苗外，开发非复制登革热疫苗的传统方法是基于使用灭活病毒制剂。纯化灭活病毒（PIV）作为一种疫苗，具有提供免疫而不产生感染风险的优点。然而必须在良好的佐剂中制备以获得强大的免疫。过去有研究表明，低浓度福尔马林可以使登革热病毒-2失活，而不会影响其免疫原性[82]。直到最近，对低生长滴度和灭活对疫苗抗原性潜在副作用的关注，使纯化灭活病毒退居幕后[83]。近年来，人们对发展登革热病毒纯化灭活病毒重新产生了兴趣[84,85]。沃尔特里德陆军研究所（WRAIR）与葛兰素史克公司（GSK）合作，加速了4种登革热病毒血清型纯化灭活病毒疫苗的开发（表7.2）。

重组亚单位疫苗提供了另一种像纯化灭活病毒（PIVs）的非感染性替代品。与纯化灭活病毒相似，重组亚单位疫苗也需要佐剂来增强免疫原性，但消除了纯化灭活病毒的产量限制，因为可以在异源宿主系统中高水平表达。由于前面列举的各种原因，E蛋白和EDⅢ一直是重组疫苗工作的重点。重组登革热病毒E蛋白已在酵母[88-90]和昆虫[91,92]表达系统中表达，纯化到不同程度的同质性，并作为小鼠和猴子的免疫原进行分析。

所有这些研究都集中在被称为登革热病毒胞外域分子的80%的E氨基末端。其中使用果蝇表达系统[92]产生的4种血清型登革热病毒E胞外域抗原得到了工业开发人员的支持（表7.2）。利用大肠杆菌和酵母菌表达宿主，研究了不同群体基于登革热病毒EDⅢ的重组抗原。在不同载体（如麦芽糖结合蛋白[93]和脑膜炎奈瑟菌p64k蛋白[94-96]）上，独立表达或融合表达的重组EDⅢ抗原，已证明了在诱发抗登革热病毒免疫应答方面的潜力。一些研究小组已经将4种登革热病毒血清型EDⅢ整合到一个单一结构中，以设计4价免疫原[97,98]。最近有报道称，在自然登革热病毒感染中，针对EDⅢ的抗体比例非常小，这削弱了开发EDⅢ疫苗的热情[99]。

基于病毒样颗粒（VLP）平台的登革热病毒亚单位重组疫苗是新出现的一种研究方法。许多病毒的重组结构抗原具有组装成VLPs的固有倾向[100]。这些显示表位有序重复排列的VLPs是强大的免疫原[101,102]。在异源宿主中共表达的登革热病毒prM和E蛋白共同组装成VLPs。文献表明，哺乳动物[103]、昆虫[104]和酵母[105,106]细胞可以作为宿主产生登革热病毒VLPs。

从疫苗用VLPs的角度看，酵母系统可能更合适，因为酵母系统有可能获得更高的产量。最近的研究表明，在缺乏prM的情况下，酵母表达的登革热病毒E胞外域形成了VLPs[53]。另一种方法是在由乙型肝炎病毒表面或核心抗原形成的VLPs上[53,107]或QβVLPs上展示登革热病毒EDⅢ[53]。

7.5.2　登革热疫苗发展现状

对于大多数登革热疫苗开发方法来说，标准策略是分别开发4种单价候选疫苗，首先单独评估然后再组合评估。结果证明，登革热疫苗的开发过程是一个漫长的过程。一些候选疫苗已达到临床试验阶段[108]，表7.2总结了作为4价制剂进行试验的疫苗。所有这些疫苗都需要至少

2 剂才能充分转化为所有 4 种登革热病毒血清型。

在开发中的 3 种复制病毒疫苗中，赛诺菲的 CYD 疫苗是领先的，在多个登革热流行地区进行了多达 6 次不同的Ⅲ期试验，涉及数千名试验志愿者。几项Ⅱ期研究已经完成，来自秘鲁儿童试验的结果得出结论，4 价 CYD 疫苗的三剂量方案对之前接种过黄热病疫苗的儿童具有良好的安全性，引发了平衡的 4 价免疫反应[87]。一项概念验证 CYD 疫苗效力试验中期数据显示总体效力约为 30%[86]，该试验涉及登革热流行国家泰国的约 4 000名儿童。CYD 疫苗未能防止登革病毒-2 感染，引发登革热疫苗研究界的激烈争论。

基于登革热病毒-2（DENV-2）PDK-53 骨架的减毒疫苗 DENVax，在 AG129 小鼠攻击模型中，显示出对 DENV-1 和 DENV-2 的保护效力[109]。携带干扰素-α/β 和 γ-双受体敲除突变的 AG129 小鼠在受到某些登革病毒强毒株的攻击时，表现出登革样症状[110]。DENVax 在一个地方进行了Ⅰ期测试，并在拉丁美洲和东南亚进行了 Ⅱ期测试。针对每种登革病毒血清型的多种单价 Δ30 疫苗病毒，已经在临床试验中得到广泛评估，以设计不同的 4 价制剂[111]，称为 4 价病毒，目前正处于Ⅰ期试验中。

值得注意的是，葛兰素史克公司（GSK）与沃尔特里德陆军研究所（WRAIR）积极合作开发的基于经验减毒复制病毒疫苗（已经到了Ⅱ期）已经停止了进一步的开发。WRAIR / GSK 疫苗Ⅰ/Ⅱ阶段试验数据显示，4 价铅制剂（17 号）在 0 个月和 6 个月接种两次剂量后，在婴儿[112]和成人[113]中分别引发了 54%和 63%的 4 价血清转化。处于第一阶段的非复制疫苗包括 WRAIR 的灭活病毒（TDENV-PIV）和质粒 DNA（TVDV）疫苗以及默克公司（Merck's）的重组蛋白疫苗（V180）。所有其他候选疫苗都处于临床前开发阶段，基于 hAd5 的 4 价疫苗正在非人灵长类动物上进行临床前疗效试验。

表 7.2 临床试验中的 4 价登革热疫苗摘要

疫苗[a]	剂量[b]	试验[c]	阶段	试验地点	当前状态（截止日期）
CYD（Sanofi）	3（0，6，12）	6（>33 000）	Ⅲ	澳大利亚，拉丁美洲和东南亚	正在进行中（2016 年 8 月）
		1（~4 000）	Ⅱb	泰国	公布中期结果（2014 年 3 月）[d]
		11（~3 600）	Ⅱ	美国，澳大利亚，拉丁美洲，东南亚，印度	5 个已完成，[e]6 个正在进行中（2015 年 6 月）
TetraVax-DV（NIH）	2（0，6）	1（300）	Ⅱ	巴西	尚未开始（2018 年 5 月）
		3（374）	Ⅰ	美国	正在进行中（2015 年 5 月）
DENVax（CDC-Inviragen）	2（0，3）	1（344）	Ⅱ	拉丁美洲，东南亚	正在进行中（2016 年 10 月）
		4（416）	Ⅰ	哥伦比亚和美国	1 项已完成；3 个正在进行中（2013 年 9 月）
LAV（WRAIR/GSK）	2（0，6）	7（~1 100）	Ⅰ/Ⅱ	美国，泰国，波多黎各	全部完成（2010 年 4 月）[6]搁置
TDENV-PIV（WRAIR/GSK）	2（0，1）	2（200）	Ⅰ	美国和波多黎各	正在进行中（2015 年 12 月）
TVDV（WRAIR）	3（0，1，3）	1（40）	Ⅰ	美国	正在进行中（2012 年 12 月）

（续表）

疫苗[a]	剂量[b]	试验[c]	阶段	试验地点	当前状态（截止日期）
V180（Merck）	3（0，1，2）	1（120）	I	澳大利亚	正在进行中（2014年12月）

[a] 开发人员提及的疫苗名称（括号内显示）：YF17D 骨架中的 *CYD* 嵌合性登革热疫苗病毒、携带 DENVΔ30 缺失的 *TetraVax-DV* 登革热疫苗病毒、DENV-2 PDK-53 背景中的 *DENVax* 登革热疫苗病毒、经连续传代经验衰减的 *LAV* 登革热疫苗病毒、*TDENV-PIV* 纯化灭活病毒疫苗，用 Iscomatrix 或明矾配制的 *TVDV* 质粒 DNA 疫苗，*V180* 重组 E 胞外域（昆虫细胞表达）亚单位疫苗。

[b] 疫苗剂量数（括号内数字表示给药月份）。

[c] 每个候选疫苗的试验次数（括号中显示了参与试验的受试者总数）。

[d] 参考文献中发布的临时数据[86]。

[e] 参考文献中公布的一项试验数据[87]。

7.6 挑战

登革热疫苗开发面临几个重大障碍。与针对单一黄病毒的黄热病毒（YF）、日本脑炎（JE）和蜱传脑炎（TBE）疫苗不同[17,83]，登革热疫苗需要4价针对所有四种DENV血清型。抗体参与保护和增强。开发一种安全有效的疫苗，能够同时预防所有四种登革热病毒，而不使疫苗接种者对严重疾病敏感，这是一个重大挑战。当混合到4价制剂中时，单价复制疫苗病毒之间的干扰是一个意想不到的障碍。这种干扰首次出现在10年前的 Mahidol 疫苗中[114]，也困扰着 WRAIR/GSK 疫苗[115,116]。

在多次尝试替换某些单价疫苗病毒和几个重新调整的4价配方之后[113,115-120]，这两种疫苗都被搁置。病毒干扰的挑战不仅是经验减毒病毒疫苗所特有的，这一点可以从通过反向遗传学方法制造的复制病毒疫苗的易感性中明显看出。CYD 疫苗[121]、TetraVax-TV 疫苗[111]和 DENVax 疫苗[109]都显示了潜在的干扰。理解和解决干扰对这些复制病毒疫苗的成功至关重要。

另一个挑战是确定保护性免疫的相关性，以便可靠地评估实验疫苗效力。中和抗体被认为是疫苗保护效果的指标，但尚未确定。一些证据表明中和抗体在预防登革热中的重要性。例如在登革热流行地区，婴儿在其生命的最初几个月内，抵抗登革热病毒感染与母亲在其循环中存在登革热病毒中和抗体有关[122]。关于细胞介导的免疫应答对保护的作用，人们知之甚少。最后，缺乏预测性动物模型是疫苗候选物临床前评估的重要障碍之一。用于实验疫苗临床前评估的小鼠和猴子模型均未表现出登革热，因此在预测疫苗对人类的疗效方面价值有限。最近，CYD 疫苗Ⅱb期试验数据强调了这一点[86]。

7.7 结论

登革热在全世界100多个国家流行，主要是热带/亚热带国家。这种疾病有可能传播到热带世界以外，这促使人们为预防这种疾病而大力开发疫苗。不断增长的疫苗渠道和许多病毒候选活疫苗临床试验的进展，促使人们期望在不久的将来能获得登革热疫苗。然而疫苗行动面临着巨大的挑战，这些挑战源于导致这种疾病的登革热病毒的独特生物学和流行病学。近期一个主要减毒候选活疫苗疗效试验结果显示，疫苗免疫原性的预测未必能预测疫苗的保护效果。这激发了人们对几种非复制疫苗方法的兴趣。然而越来越明显的是，目前对免疫增强、病毒干扰和防止感染的机制的理解上存在巨大知识空白。由于人类是检验登革热候选疫苗的唯一模型，因此有必要更好地了解这些现象，以便能够合理设计疫苗，提供更大的保护机会。

致谢

作者感谢印度政府生物技术部支持他们在登革热疫苗方面的工作。两个图形中的一部分是使用 Servier Medical Art 制作的。

参考文献

[1] Gubler, D. J.: The economic burden of dengue. Am. J. Trop. Med. Hyg. 86, 743-744 (2012).

[2] Swaminathan, S., Khanna, N.: Dengue: recent advances in biology and current status of translational research. Curr. Mol. Med. 9, 152-173 (2009).

[3] Guzman, A., Istúriz, R. E.: Update on the global spread of dengue. Int. J. Antimicrob. Agents36S, S40-S42 (2010).

[4] Gubler, D. J.: Dengue and dengue hemorrhagic fever. Clin. Microbiol. Rev. 11, 480-496 (1998).

[5] WHO.: Dengue and dengue haemorrhagicfever. Factsheet N°1 17 [online] WHO, GenevaSwitzerland. www. who. int/mediacentre/factsheets/ fs117/en/ (2012). Accessed 8 Dec 2012.

[6] WHO: Dengue haemorrhagic fever: diagnosis, treatment, prevention and control. WHO, Geneva, Switzerland (1997).

[7] Vaughn, D. W., Greene, S., Kalayanarooj, S., Innis, B. L., Nimmannitya, S., Suntayakorn, S., Endy, T. P., Raengsakulrach, B., Rothman, A. L., Ennis, F. A., Nisalak, A.: Dengue viremia titer, antibody response pattern, and virus serotype correlate with disease severity. J. Infect. Dis. 181, 2-9 (2000).

[8] Libraty, D. H., Endy, T. P., Houng, H. S. H., Green, S., Kalayanarooj, S., Suntayakorn, S., Chansiriwongs, W., Vaughn, D. W., Nisalak, A., Ennis, F. A., Rothman, A. L.: Differing influences of virus burden and immune activation on disease severity in secondary dengue-3 virus infections. J. Infect. Dis. 185, 1213-1221 (2002).

[9] WHO: Dengue guidelines for diagnosis, treatment, prevention and control, 3rd edn. WHO, Geneva, Switzerland (2009).

[10] Narvaez, F., Gutierrez, G., Pérez, M. A., Elizondo, D., Nunez, A., Balmaseda, A., Harris, E.: Evaluation of the traditional and revised WHO classifications of dengue disease severity. PLoSNegl. Trop. Dis. 5, e1397 (2011).

[11] Kalayanarooj, S.: Dengue classification: current WHO vs. the newly suggested classification for better clinical application? J. Med. Assoc. Thai. 94 (Suppl. 3), S74-S84 (2011).

[12] Kuhn, R. J., Zhang, W., Rossman, M. G., Pletnev, S. V., Corver, J., Lenches, E., Jones, C. T., Mukhopadhyay, S., Chipman, P. R., Strauss, E. G., Baker, T. S., Strauss, J. H.: Structure of dengue virus: implications for flavivirus organization, maturation, and fusion. Cell 108, 717-725 (2002).

[13] Lindenbach, B. D., Thiel, H. J., Rice, C. M.: *Flaviviridae*: the viruses and their replication. In: Knipe, D. M., Howley, P. M. (eds.) Fields of virology, 5th edn, pp. 1101-1152. Wolters Kluwer and Lippincott Williams & Wilkins, Philadelphia (2007).

[14] Filomatori, C. V., Lodeiro, M. F., Alvarez, D. E., Samsa, M. M., Pietrasanta, L., Gamarnik, A. V.: A 5′ RNA element promotes dengue virus RNA synthesis on a circular genome. Genes Dev. 20, 2238-2249 (2006).

[15] Zeng, L., Falgout, B., Markoff, L.: Identification of specific nucleotide sequences within the conserved 3′-SL in the dengue type 2 virus genome required for replication. J. Virol. 72, 7510-7522 (1998).

[16] Alvarez, D. E., Lodeiro, M. F., Ludueňa, S. J., Pietrasanta, L. I., Gamarnik, A. V.: Long-range RNA-RNA interactions circularize the dengue virus genome. J. Virol. 79, 6631-6643 (2005).

[17] Gubler, D. J., Kuno, G., Markoff, L.: Flaviviruses. In: Knipe, D. M., Howley, P. M. (eds.) Fields of virology, 5th edn, pp. 1153-1252. Wolters Kluwer and Lippincott Williams & Wilkins, Philadelphia

(2007).
[18] Modis, Y., Ogata, S., Clements, D., Harrison, S. C.: A ligand-binding pocket in the dengue virus envelope glycoprotein. Proc. Natl. Acad. Sci. USA 100, 6986-6991 (2003).
[19] Crill, W. D., Roehrig, R. T.: Monoclonal antibodies that bind to domain III of dengue virus E glycoprotein are the most efficient blockers of virus adsorption to Vero cells. J. Virol. 75, 7769-7773 (2001).
[20] Sukupolvi-Petty, S., Austin, S. K., Purtha, W. E., Oliphant, T., Nybakken, G. E., Schlesinger, J. J., Roehrig, J. T., Gromowski, G. D., Barrett, A. D., Fremont, D. H., Diamond, M. S.: Type-and subcomplex-specific neutralizing antibodies against domain III of dengue virus type 2 envelope protein recognize adjacent epitopes. J. Virol. 81, 12816-12826 (2007).
[21] Mukhopadhyay, S., Kuhn, R. J., Rossmann, M. G.: A structural perspective of the *flavivirus* life cycle. Nat. Rev. Microbiol. 3, 13-22 (2005).
[22] Zhang, Y., Corver, J., Chipman, P. R., Zhang, W., Pletnev, S. V., Sedlak, D., Baker, T. S., Strauss, J. H., Kuhn, R. J., Rossmann, M. G.: Structure of immature flavivirus particle. EMBO J. 22, 2604-2613 (2003).
[23] Rodhain, F., Rosen, L.: Mosquito vectors and dengue virus-vector relationships. In: Gubler, D. J., Kuno, G. (eds.) Dengue and dengue hemorrhagic fever, pp. 45-60. CAB International, New York (1997).
[24] Guzman, M. G., Halstead, S. B., Artsob, H., Buchy, P., Farrar, J., Gubler, D. J., Hunsperger, E., Kroeger, A., Margolis, H. S., Martínez, E., Nathan, M. B., Pelegrino, J. L., Simmons, C., Yoksan, S., Peeling, R. W.: Dengue: a continuing global threat. Nat. Rev. Microbiol. 8, S7-S16 (2010).
[25] Lanciotti, R. S., Calisher, C. H., Gubler, D. J., Chang, G. J., Vorndam, A. V.: Rapid detection and typing of dengue viruses from clinical samples by using reverse transcriptase-polymerase chain reaction. J. Clin. Microbiol. 30, 545-551 (1992).
[26] Harris, E., Roberts, T. G., Smith, L., Selle, J., Kramer, L. D., Valle, S., Sandoval, E., Balmaseda, A.: Typing of dengue viruses in clinical specimens and mosquitoes by single-tube multiplex reverse transcriptase PCR. J. Clin. Microbiol. 36, 2634-2639 (1998).
[27] Johnson, B. W., Russell, B. J., Lanciotti, R. S.: Serotype-specific detection of dengue viruses in a fourplex real-time reverse transcriptase PCR assay. J. Clin. Microbiol. 43, 4977-4983 (2005).
[28] Wu, S. J., Lee, E. M., Putvatana, R., Shurtliff, R. N., Porter, K. R., Suharyono, W., Watts, D. M., King, C. C., Murphy, G. S., Hayes, C. G., Romano, J. W.: Detection of dengue viral RNA using a nucleic acid sequence-based amplification assay. J. Clin. Microbiol. 39, 2794-2798 (2001).
[29] Alcon, S., Talarmin, A., Debruyne, M., Falconar, A., Deubel, V., Flamand, M.: Enzyme-linked immunosorbent assay specific to dengue virus type 1 nonstructural protein NS1 reveals circulation of the antigen in the blood during the acute phase of disease in patients experiencing primary or secondary infections. J. Clin. Microbiol. 40, 376-381 (2002).
[30] Young, P. R., Hilditch, P. A., Bletchly, C., Halloran, W.: An antigen capture enzyme-linked immunosorbent assay reveals high levels of the dengue virus protein NS1 in the sera of infected patients. J. Clin. Microbiol. 38, 1053-1057 (2000).
[31] Libraty, D. H., Young, P. R., Pickering, D., Endy, T. P., Kalayanarooj, S., Green, S., Vaughn, D. W., Nisalak, A., Ennis, F. A., Rothman, A. L.: High circulating levels of the dengue virus nonstructural protein NS1 early in dengue illness correlate with the development of dengue hemorrhagic fever. J. Infect. Dis. 186, 1165-1168 (2002).
[32] Dussart, P., Labeau, B., Lagathu, G., Louis, P., Nunes, M. R. T., Rodrigues, S. G., Storck-Herrmann, C., Cesaire, R., Morvan, J., Flamand, M., Baril, L.: Evaluation of an enzyme immunoassay for detection of dengue virus NS1 antigen in human serum. Clin. Vaccine Immunol. 13, 1185-1189 (2006).
[33] Dussart, P., Petit, L., Labeau, B., Bremand, L., Leduc, A., Moua, D., Matheus, S., Baril, L.: Evaluation of two new commercial tests for the diagnosis of acute dengue virus infection using NS1 antigen detection in human serum. PLoSNegl. Trop. Dis. 2, e280 (2008).
[34] Roehrig, J. T., Hombach, J., Barrett, A. D. T.: Guidelines for plaque-reduction neutralization testing of

human antibodies to dengue viruses. Viral Immunol. 21, 123-132 (2008).

[35] Vorndam, V., Beltran, M.: Enzyme-linked immuno-sorbent assay-format microneutralization test for dengue viruses. Am. J. Trop. Med. Hyg. 66, 208-212 (2002).

[36] Lambeth, C. R., White, L. J., Johnston, R. E., de Silva, A. M.: Flow cytometry-based assay for titrating d engue virus. J. Clin. Microbiol. 43, 3267-3272 (2005).

[37] AnandaRao, R., Swaminathan, S., Fernando, S., Jana, A. M., Khanna, N.: A custom-designed recombinant multiepitope protein as a dengue diagnostic reagent. Protein Expr. Purif. 41, 136-147 (2005).

[38] AnandaRao, R., Swaminathan, S., Fernando, S., Jana, A. M., Khanna, N.: Recombinant multiepitope protein for early detection of dengue infections. Clin. Vaccine Immunol. 13, 59-67 (2006).

[39] Hapugoda, M. D., Batra, G., Abeyewickreme, W., Swaminathan, S., Khanna, N.: Single antigen detects both immunoglobulin M (IgM) and IgG antibodies elicited by all four dengue virus serotypes. Clin. Vaccine Immunol. 14, 1505-1514 (2007).

[40] Batra, G., Nemani, S. K., Tyagi, P., Swaminathan, S., Khanna, N.: Evaluation of envelope domain III-based single chimeric tetravalent antigen and monovalent antigen mixtures for detection of anti-dengue antibodies in human sera. BMC Infect. Dis. 11, 64 (2011).

[41] Shu, P. Y., Chen, L. K., Chang, S. F., Yueh, Y. Y., Chow, L., Chien, L. J., Chin, C., Lin, T. H., Huang, J. H.: Comparison of capture immunoglobulin M (IgM) and IgG enzyme - linked immunosorbent assay (ELISA) and nonstructural protein NS1 serotype-specific IgG ELISA for differentiation of primary and secondary dengue virus infections. Clin. Diagn. Lab. Immunol. 10, 622-630 (2003).

[42] Innis, B. L., Kuno, G.: Antibody responses to dengue virus infection. In: Gubler, D. J. (ed.) Dengue and dengue hemorrhagic fever, pp. 221-243. CAB International, Wallingford (1997).

[43] Burke, D. S., Nisalak, A., Johnson, D. E., Scott, R. M.: A prospective study of dengue infections in Bangkok. Am. J. Trop. Med. Hyg. 38, 172-180 (1988).

[44] Guzmán, M. G., Kouri, G. P., Bravo, J., Soler, M., Vazquez, S., Morier, L.: Dengue hemorrhagic fever in Cuba, 1981: a retrospective seroepidemiologic study. Am. J. Trop. Med. Hyg. 42, 179 - 184 (1990).

[45] Halstead, S. B.: Neutralization and antibody dependent enhancement of dengue viruses. Adv. Virus Res. 60, 421-467 (2003).

[46] Rothman, A.,. L.: Dengue: defining protective versus pathologic immunity. J. Clin. Invest. 113, 946-951 (2004).

[47] Dejnirattisai, W., Jumnainsong, A., Onsirisakul, N., Fitton, P., Vasanawathana, S., Limpitikul, W., Puttikhunt, C., Edwards, C., Duangchinda, T., Supasa, S., Chawansuntati, K., Malasit, P., Mongkolsapaya, J., Screaton, G.: Cross-reacting antibodies enhance dengue virus infection in humans. Science 328, 745-748 (2010).

[48] Rodenhuis-Zybert, I. A., van der Schaar, H. M., da Silva Voorham, J. M., van der Ende-Metselaar, H., Lei, H. Y., Wilschut, J., Smit, J. M.: Immature dengue virus: a veiled pathogen? PLoS Pathog. 6, e1000718 (2010).

[49] Mongkolsapaya, J., Dejnirattisai, W., Xu, X. N., Vasanawathana, S., Tangthawornchaikul, N., Chairunsri, A., Sawasdivorn, S., Duangchinda, T., Dong, T., Rowland-Jones, S., Yenchitsomanus, P. T., McMichael, A., Malasit, P., Screaton, G.: Original antigenic sin and apoptosis in the pathogenesis of dengue hemorrhagic fever. Nat. Med. 9, 921-927 (2003).

[50] Swaminathan, S., Batra, G., Khanna, N.: Dengue vaccines: state of the art. Expert Opin. Ther. Pat. 20, 819-835 (2010).

[51] Durbin, A. P., Whitehead, S. S.: Dengue vaccine candidates in development. Curr. Top. Microbiol. Immunol. 338, 129-143 (2010).

[52] Coller, B. A. G., Clements, D. E.: Dengue vaccines: progress and challenges. Curr. Opin. Immunol. 23, 391-398 (2011).

[53] Schmitz, J., Roehrig, J., Barrett, A., Hombach, J.: Next generation dengue vaccines: a review of candidates in preclinical development. Vaccine 29, 7276-7284 (2011).

[54] Thomas, S. J., Endy, T. P.: Vaccines for the prevention of dengue: development update. Hum. Vaccin. 7, 674-684 (2011).

[55] Bhamarapravati, N., Sutee, Y.: Live attenuated tetra - valent dengue vaccine. Vaccine 18, 44 - 47 (2000).

[56] Eckels, K. H., Dubois, D. R., Putnak, R., Vaughn, D. W., Innis, B. L., Henchal, E. A., Hoke Jr., C. H.: Modification of dengue virus strains by passage in primary dog kidney cells: preparation of candidate vaccines and immunization of monkeys. Am. J. Trop. Med. Hyg. 69 (suppl 6), 12-16 (2003).

[57] Lai, C. J., Monath, T. P.: Chimeric flaviviruses: novel vaccines against dengue fever, tick-borne encephalitis, and Japanese encephalitis. Adv. Virus Res. 61, 469-509 (2003).

[58] Guirakhoo, F., Weltzin, R., Chambers, T. J., Zhang, Z. X., Soike, K., Ratterree, M., Arroyo, J., Georgakopoulos, K., Catalan, J., Monath, T. P.: Recombinant chimeric yellow fever-dengue type 2 virus is immunogenic and protective in nonhuman primates. J. Virol. 74, 5477-5485 (2000).

[59] Guirakhoo, F., Arroyo, J., Pugachev, K. V., Miller, C., Zhang, Z. -X., Weltzin, R., Georgakopoulos, K., Catalan, J., Ocran, S., Soike, K., Ratterree, M., Monath, T. P.: Construction, safety, and immunogenicity in nonhuman primates of a chimeric yellow fever-dengue virus tetravalent vaccine. J. Virol. 75, 7290-7304 (2001).

[60] Guirakhoo, F., Pugachev, K., Arroyo, J., Miller, C., Zhang, Z. -X., Weltzin, R., Georgakopoulos, K., Catalan, J., Ocran, S., Draper, K., Monath, T. P.: Viraemia and immunogenicity in nonhuman primates of a tetravalent yellow fever-dengue chimeric vaccine: genetic reconstructions, dose adjustment, and antibody responses against wild-type dengue virus isolates. Virology 298, 146-159 (2002).

[61] Men, R., Bray, M., Clark, D., Chanock, R., Lai, C. J.: Dengue type 4 virus mutants containing deletions in the 3′ noncoding region of the RNA genome: analysis of growth restriction in cell culture and altered viremia pattern and immunogenicity in rhesus monkeys. J. Virol. 70, 3930-3937 (1996).

[62] Durbin, A. P., Karron, R. A., Sun, W., Vaughn, D. W., Reynolds, M. J., Perreault, J. R., Thumar, B., Men, R., Lai, C. -J., Elkins, W. R., Chanock, R. M., Murphy, B. R., Whitehead, S. S.: Attenuation and immunogenicity in humans of a live dengue virus type-4 vaccine candidate with a 30-nucleotide deletion in its 3′-untranslated region. Am. J. Trop. Med. Hyg. 65, 405-413 (2001).

[63] Whitehead, S. S., Falgout, B., Hanley, K. A., Blaney Jr., J. E., Markoff, L., Murphy, B. R.: A live, attenuated dengue virus type 1 vaccine candidate with a 30-nucleotide deletion in the 3′ untranslated region is highly attenuated and immunogenic in monkeys. J. Virol. 77, 1653-1657 (2003).

[64] Blaney Jr., J. E., Hanson, C. T., Hanley, K. A., Murphy, B. R., Whitehead, S. S.: Vaccine candidates derived from a novel infectious cDNA clone of an American genotype dengue virus type 2. BMC Infect. Dis. 4, 39 (2004).

[65] Blaney Jr., J. E., Hanson, C. T., Firestone, C. Y., Hanley, K. A., Murphy, B. R., Whitehead, S. S.: Genetically modified, live attenuated dengue virus type 3 vaccine candidates. Am. J. Trop. Med. Hyg. 71, 811-821 (2004).

[66] Whitehead, S. S., Hanley, K. A., Blaney Jr., J. E., Gilomre, L. E., Elkins, W. R., Murphy, B. R.: Substitution of the structural genes of dengue virus type 4 with those of type 2 results in chimeric vaccine candidates which are attenuated for mosquitoes, mice, and rhesus monkeys. Vaccine 21, 4307-4316 (2003).

[67] Huang, C. Y. -H., Butrapet, S., Pierro, D. J., Chang, G. -J.,. J., Hunt, A. R., Bhamarapravati, N., Gubler, D. J., Kinney, R. M.: Chimeric dengue type 2 (vaccine strain PDK-53) /dengue type 1 virus as a potential candidate dengue type 1 virus vaccine. J. Virol. 74, 3020-3028 (2000).

[68] Butrapet, S., Huang, C. Y. -H., Pierro, D. J., Bhamarapravati, N., Gubler, D. J., Kinney, R. M.: Attenuation markers of a candidate dengue type 2 vaccine virus, strain 16681 (PDK-53), are defined by mutations in the 5′ noncoding region and non-structural proteins 1 and 3. J. Virol. 74, 3011-3019 (2000).

[69] Smith, K. M., Nanda, K., Spears, C. J., Ribeiro, M., Vancini, R., Piper, A., Thomas, G. S., Thomas, M. E., Brown, D. T., Hernandez, R.: Structural mutants of dengue virus 2 transmembrane domains exhibit host-range phenotype. Virol. J. 8, 289 (2011).

[70] Apt, D., Raviprakash, K., Brinkman, A., Semyonov, A., Yang, S., Skinner, C., Diehl, L.,

Lyons, R., Porter, K., Punnonen, J.: Tetravalent neutralizing antibody response against four dengue serotypes by a single chimeric dengue envelope antigen. Vaccine 24, 335-344 (2006).

[71] Konishi, E., Kosugi, S., Imoto, J. I.: Dengue tetravalent DNA vaccine inducing neutralizing antibody and anamnestic responses to four serotypes in mice. Vaccine 24, 2200-2207 (2006).

[72] Men, R., Wyatt, L., Tokimatsu, I., Arakaki, S., Shameem, G., Elkins, R., Chanock, R., Moss, B., Lai, C. J.: Immunization of rhesus monkeys with a recombinant of modified vaccinia virus Ankara expressing a truncated envelope glycoprotein of dengue type 2 virus induced resistance to dengue type 2 virus challenge. Vaccine 18, 3113-3122 (2000).

[73] Jaiswal, S., Khanna, N., Swaminathan, S.: Replication-defective adenoviral vaccine vector for the induction of immune responses to dengue virus type 2. J. Virol. 77, 12907-12913 (2003).

[74] Raviprakash, K., Wang, D., Ewing, D., Holman, D. H., Block, K., Woraratanadharm, J., Chen, L., Hayes, C., Dong, J. Y., Porter, K.: A tetravalent dengue vaccine based on a complex adenovirus vector provides significant protection in rhesus monkeys against all four serotypes of dengue virus. J. Virol. 82, 6927-6934 (2008).

[75] Suzuki, R., Winkelmann, E. R., Mason, P. W.: Construction and characterization of a single-cycle chimeric flavivirus vaccine candidate that protects mice against lethal challenge with dengue virus type 2. J. Virol. 83, 1870-1880 (2009).

[76] Mota, J., Acosta, M., Argotte, R., Figueroa, R., Méndez, A., Ramos, C.: Induction of protective antibodies against dengue virus by tetravalent DNA immunization of mice with domain III of the envelope protein. Vaccine 23, 3469-3476 (2005).

[77] Ramanathan, M. P., Kuo, Y. C., Selling, B. H., Li, Q., Sardesai, N. Y., Kim, J. J., Weiner, D. B.: Development of a novel DNA SynCon tetravalent dengue vaccine that elicits immune responses against four serotypes. Vaccine 27, 6444-6453 (2009).

[78] Khanam, S., Khanna, N., Swaminathan, S.: Induction of antibodies and T cell responses by dengue virus type 2 envelope domain III encoded by plasmid and adenoviral vectors. Vaccine 24, 6513-6525 (2006).

[79] Khanam, S., Rajendra, P., Khanna, N., Swaminathan, S.: An adenovirus prime/plasmid boost strategy for induction of equipotent immune responses to two dengue virus serotypes. BMC Biotechnol. 7, 10 (2007).

[80] Khanam, S., Pilankatta, R., Khanna, N., Swaminathan, S.: An adenovirus type 5 (AdV5) vector encoding an envelope domain III-based tetravalent antigen elicits immune responses against all four dengue viruses in the presence of prior AdV5 immunity. Vaccine 27, 6011-6021 (2009).

[81] Brandler, S., Ruffi e, C., Najburg, V., Frenkiel, M. P., Bedouelle, H., Desprès, P., Tangy, F.: Pediatric measles vaccine expressing a dengue tetravalent antigen elicits neutralizing antibodies against all four dengue viruses. Vaccine 28, 6730-6739 (2010).

[82] Putnak, R., Barvir, D. A., Burrous, J. M., Dubois, D. R., D'Andrea, V. M., Hoke, C. H., Sadoff, J. C., Eckels, K. H.: Development of a purified, inactivated, dengue-2 virus vaccine prototype in Vero cells: immunogenicity and protection in mice and rhesus monkeys. J. Infect. Dis. 174, 1176-1184 (1996).

[83] Eckels, K. H., Putnak, R.: Formalin-inactivated whole virus and recombinant subunit flavivirus vaccines. Adv. Virus Res. 61, 395-418 (2003).

[84] Putnak, J. R., Coller, B. A., Voss, G., Vaughn, D. W., Clements, D., Peters, I., Bignami, G., Houng, H. S., Chen, R. C. M., Barvir, D. A., Seriwatana, J., Cayphas, S., Garcon, N., Gheysen, D., Kanesa-thasan, N., McDonnel, M., Humphreys, T., Eckels, K. H., Innis, J. P., Prieels, B. L.: An evaluation of dengue type-2 inactivated, recombinant subunit, and live-attenuated vaccine candidates in the rhesus macaque model. Vaccine 23, 4442-4452 (2005).

[85] Maves, R. C., Ore, R. M., Porter, K. R., Kochel, T. J.: Immunogenicity and protective efficacy of a psoralen-inactivated dengue-1 virus vaccine candidate in *Aotusnancymaae* monkeys. Vaccine 29, 2691-2696 (2011).

[86] Sabchareon, A., Wallace, D., Sirivichayakul, C., Limkittikul, K., Chanthavanich, P., Suvannadabba, S., Jiwariyavej, V., Dulyachai, W., Pengsaa, K., Wartel, T. A., Moureau, A., Saville, M., Bouckenooghe, A., Viviani, S., Tornieporth, N. G., Lang, J.: Protective efficacy of the recombinant, live-at-

tenuated, CYD tetravalent dengue vaccine in Thai schoolchildren: a randomised, controlled phase 2b trial. Lancet 380, 1559–1567 (2012).

[87] Lanata, C. F., Andrade, T., Gil, A. I., Terrones, C., Valladolid, O., Zambrano, B., Saville, M., Crevat, D.: Immunogenicity and safety of tetravalent dengue vaccine in 2 – 11 year – olds previously vaccinated against yellow fever: randomized, controlled, phase II study in Piura Peru. Vaccine 30, 5935–5941 (2012).

[88] Muné, M., Rodríguez, R., Ramírez, R., Soto, Y., Sierra, B., Roche, R. R., Marquez, G., Garcia, J., Guillén, G., Guzmán, M. G.: Carboxy–terminally truncated dengue 4 virus envelope glycoprotein expressed in *Pichia pastoris* induced neutralizing antibodies and resistance to dengue 4 virus challenge in mice. Arch. Virol. 148, 2267–2273 (2003).

[89] Guzmán, M. G., Rodríguez, R., Rodríguez, R., Hermida, L., Alvarez, M., Lazo, L., Mune, M., Rosario, D., Valdés, K., Vazquez, S., Martinez, R., Serranao, T., Paez, J., Espinosa, R., Pumariega, T., Guillén, G.: Induction of neutralizing antibodies and partial protection from viral challenge in *Macacafascicularis* immunized with recombinant dengue 4 virus envelope glycoprotein expressed in *Pichia pastoris*. Am. J. Trop. Med. Hyg. 69, 129–134 (2003).

[90] Valdés, I., Hermida, L., Zulueta, A., Martín, J., Silva, R., Álvarez, M., Guzmán, M. G., Guillén, G.: Expression in *Pichia pastoris* and immunological evaluation of a truncated dengue envelope protein. Mol. Biotechnol. 35, 23–30 (2007).

[91] Kelly, E. P., Greene, J. J., King, A. D., Innis, B. L.: Purified dengue 2 virus envelope glycoprotein aggregates produced by baculovirus are immunogenic in mice. Vaccine 18, 2549–2559 (2000).

[92] Clements, D. E., Coller, B. G., Lieberman, M. M., Ogata, S., Wang, G., Harada, K. E., Putnak, J. R., Ivy, J. M., McDonnell, M., Bignami, G. S., Peters, I. D., Leung, J., Weeks–Levy, C., Nakano, E. T., Humphreys, T.: Development of a recombinant tetravalent dengue virus vaccine: immunogenicity and efficacy studies in mice and monkeys. Vaccine 28, 2705–2715 (2010).

[93] Simmons, M., Murphy, G. S., Hayes, C. G.: Short report: antibody responses of mice immunized with a tetravalent dengue recombinant protein subunit vaccine. Am. J. Trop. Med. Hyg. 65, 159–161 (2001).

[94] Hermida, L., Rodríguez, R., Lazo, L., Bernardo, L., Silva, R., Zulueta, A., López, C., Martín, J., Valdes, I., del Rosario, D., Guillén, G., Guzmán, M. G.: A fragment of the envelope protein from dengue–1 virus, fused in two different sites of the meningococcal P64k protein carrier, induces a functional immune response in mice. Biotechnol. Appl. Biochem. 39, 107–114 (2004).

[95] Hermida, L., Rodríguez, R., Lazo, L., Silva, R., Zulueta, A., Chinea, G., López, C., Guzmán, M. G., Guillén, G.: A dengue–2 envelope fragment inserted within the structure of the P64k meningococcal protein carrier enables a functional immune response against the virus in mice. J. Virol. Methods 115, 41–49 (2004).

[96] Hermida, L., Bernardo, L., Martín, J., Alvarez, M., Prado, I., López, C., de la Sierra, B. C., Martínez, R., Rodríguez, R., Zulueta, A., Pérez, A. B., Lazo, L., Rosario, D., Guillén, G., Guzmán, M.: A recombinant fusion protein containing the domain III of the dengue–2 envelope protein is immunogenic and protective in nonhuman primates. Vaccine 24, 3165–3171 (2006).

[97] Chen, S., Yu, M., Jiang, T., Deng, Y., Qin, C., Qin, E.: Induction of tetravalent protective immunity against four dengue serotypes by the tandem domain III of the envelope protein. DNA Cell Biol. 26, 361–(2007).

[98] Etemad, B., Batra, G., Raut, R., Dahiya, S., Khanam, S., Swaminathan, S., Khanna, N.: An envelope domain III–based chimeric antigen produced in *Pichia pastoris* elicits neutralizing antibodies against all four dengue virus serotypes. Am. J. Trop. Med. Hyg. 79, 353–363 (2008).

[99] Wahala, W. M. P. B., Kraus, A. A., Haymore, L. B., Accavitti–Loper, M. A., de Silva, A. M.: Dengue virus neutralization by human immune sera: role of envelope protein domain III–reactive antibody. Virology 392, 103–113 (2009).

[100] Chackerian, B.: Virus–like particles: flexible platforms for vaccine development. Expert Rev. Vaccines 6, 381–390 (2007).

[101] Bachmann, M. F., Jennings, G. T.: Vaccine delivery: a matter of size, geometry, kinetics and molecular patterns. Nat. Rev. Immunol. 10, 787-796 (2010).

[102] Yildiz, I., Shukla, S., Steinmetz, N. F.: Applications of viral nanoparticles in medicine. Curr. Opin. Biotechnol. 22, 901-908 (2011).

[103] Wang, P. G., Kudelko, M., Lo, J., Siu, L. Y. L., Kwok, K. T. H., Sachse, M., Nicholls, J. M., Bruzzone, R., Altmeyer, R. M., Nal, B.: Efficient assembly and secretion of recombinant subviral particles of the four dengue serotypes using native prM and E proteins. PLoS One 4, e8325 (2009).

[104] Kuwahara, M., Konishi, E.: Evaluation of extracellularsubviral particles of dengue virus type 2 and Japanese encephalitis virus produced by *Spodopterafrugiperda* cells for use as vaccine and diagnostic antigens. Clin. Vaccine Immunol. 17, 1560-1566 (2010).

[105] Liu, W., Jiang, H., Zhou, J., Yang, X., Tang, Y., Fang, D., Jiang, L.: Recombinant dengue virus-like particles from *Pichia pastoris*: efficient production and immunological properties. Virus Genes 40, 53-59 (2010).

[106] Tang, Y., Jiang, L., Zhou, J., Yin, Y., Yang, X., Liu, W., Fang, D.: Induction of virus-neutralizing antibodies and T cell responses by dengue virus type 1 virus-like particles prepared from *Pichia pastoris*. Chin. Med. J. 125, 1986-1992 (2012).

[107] Arora, U., Tyagi, P., Swaminathan, S., Khanna, N.: Chimeric hepatitis B core antigen virus-like particles displaying the envelope domain III of dengue virus type 2. J. Nanobiotechnol. 10, 30 (2012).

[108] Clinical trials database of the US NIH [online] http: // clinicaltrials. gov/ct2/results? term=dengue. Accessed 10Dec 2012.

[109] Brewoo, J. N., Kinney, R. M., Powell, T. D., Arguello, J. J., Silengo, S. J., Partidos, C. D., Huang, C. Y. H., Stinchcomb, D. T., Osorio, J. E.: Immunogenicity and efficacy of chimeric dengue vaccine (DENVax) formulations in interferon-deficient AG129 mice. Vaccine 30, 1513-1520 (2012).

[110] Shresta, S., Sharar, K. L., Prigozhin, D. M., Beatty, P. R., Harris, E.: Murine model for dengue virus-induced lethal disease with increased vascular permeability. J. Virol. 80, 10208-10217 (2006).

[111] Durbin, A. P., Kirkpatrick, B. D., Pierce, K. K., Schmidt, A. C., Whitehead, S. S.: Development and clinical evaluation of multiple investigational monovalent DENV vaccines to identify components for inclusion in a live attenuated tetravalent DENV vaccine. Vaccine 29, 7242-7250 (2011).

[112] Watanaveeradej, V., Simasathien, S., Nisalak, A., Endy, T. P., Jarman, R. G., Innis, B. L., Thomas, S. J., Gibbons, R. V., Hengprasert, S., Samakoses, R., Kerdpanich, A., Vaughn, D. W., Putnak, J. R., Eckels, K. H., Barrera, R. D. L., Mammen Jr., M. P.: Safety and immunogenicity of a tetravalent live-attenuated dengue vaccine in flavivirus-naive infants. Am. J. Trop. Med. Hyg. 85, 341-351 (2011).

[113] Sun, W., Cunningham, D., Wasserman, S. S., Perry, J., Putnak, J. R., Eckels, K. H., Vaughn, D. W., Thomas, S. J., Kanesa-Thasan, N., Innis, B. L., Edelman, R.: Phase 2 clinical trials of three formulations of tetravalent live attenuated dengue vaccine in flavivirus-naïve adults. Hum. Vaccin. 5, 33-40 (2009).

[114] Kanesa-thasan, N., Sun, W., Kim-Ahn, G., Van Albert, S., Putnak, J. R., King, A., Raengsakulsrach, B., Christ-Schmidt, H., Gilson, K., Zahradnik, J. M., Vaughn, D. W., Innis, B. L., Saluzzo, J. -F., Hoke Jr., C. H.: Safety and immunogenicity of attenuated dengue virus vaccines (Aventis Pasteur) in human volunteers. Vaccine 19, 3179-3188 (2001).

[115] Sun, W., Edelman, R., Kanesa-Thasan, N., Eckels, K. H., Putnak, J. R., King, A. D., Houng, H. S., Tang, D., Scherer, J. M., Hoke Jr., C. H., Innis, B. L.: Vaccination of human volunteers with monovalent and tetravalent live-attenuated dengue vaccine candidates. Am. J. Trop. Med. Hyg. 69 (Suppl 6), 24-31 (2003).

[116] Edelman, R., Wasserman, S. S., Bodison, S. A., Putnak, R. J., Eckels, K. H., Tang, D., Kanesa-Thasan, N., Vaughn, D. W., Innis, B. L., Sun, W.: Phase I trial of 16 formulations of a tetravalent live-attenuated dengue vaccine. Am. J. Trop. Med. Hyg. 69 (Suppl 6), 48-60 (2003).

[117] Kitchener, S., Nissen, M., Nasveld, P., Forrat, R., Yoksan, S., Lang, J., Saluzzo, J. F.: Immu-

nogenicity and safety of two live-attenuated tetravalent dengue vaccine formulations in healthy Australian adults. Vaccine 24, 1238-1241 (2006).

[118] Sanchez, V., Gimenez, S., Tomlinson, B., Chan, P. K. S., Thomas, G. N., Forrat, R., Chambonneau, L., Deauvieau, F., Lang, J., Guy, B.: Innate and adaptive cellular immunity in flavivirus-naïve human recipients of a live-attenuated dengue serotype 3 vaccine produced in Vero cells (VDV3). Vaccine 24, 4914-4926 (2006).

[119] Simasathien, S., Thomas, S. J., Watanaveeradej, V., Nisalak, A., Barberousse, C., Innis, B. L., Sun, W., Putnak, J. R., Eckels, K. H., Hutagalung, Y., Gibbons, R. V., Zhang, C., Barrera, R. D. L., Jarman, R. G., Chawachalasai, W., Mammen Jr., M. P.: Safety and immunogenicity of tetravalent live-attenuated dengue vaccine in flavivirus naive children. Am. J. Trop. Med. Hyg. 78, 426-433 (2008).

[120] Anderson, K. B., Gibbons, R. V., Edelman, R., Eckels, K. H., Putnak, R. J., Innis, B. L., Sun, W.: Interference and facilitation between dengue serotypes in a tetravalent live dengue virus vaccine candidate. J. Infect. Dis. 204, 442-450 (2011).

[121] Guy, B., Barban, V., Mantel, N., Aguirre, M., Gulia, S., Pontvianne, J., Jourdier, T. M., Ramirez, L., Gregoire, V., Charnay, C., Burdin, N., Dumas, R., Lang, J.: Evaluation of interferences between dengue vaccine serotypes in a monkey model. Am. J. Trop. Med. Hyg. 80, 302-311 (2009).

[122] Pengsaa, K., Luxemburger, C., Sabchareon, A., Limkittikul, K., Yoksan, S., Chambonneau, L., Chaovarind, U., Sirivichayakul, C., Lapphra, K., Chanthavanich, P., Lang, J.: Dengue virus infections in the first 2 years of life and the kinetics of transplacentally transferred dengue neutralizing antibodies in Thai children. J. Infect. Dis. 194, 1570-1576 (2006).

（何继军译）

第8章　预防胃肠炎的诺如病毒病毒样颗粒疫苗

Qiang Chen①

摘要

胃肠炎（Gastroenteritis，GE）及其相关腹泻疾病仍然是引起世界上人类死亡最主要的原因之一。诺如病毒（NoVs）是一组遗传性多样的RNA病毒，导致人类绝大多数非细菌性胃肠炎。然而，目前仍然没有获得批准的人类疫苗来预防诺如病毒胃肠炎。缺乏组织培养系统和小动物模型进一步阻碍了诺如病毒疫苗的开发。

然而，可以在昆虫、哺乳动物和植物细胞中通过衣壳蛋白的表达产生病毒样颗粒（VLPs），以模拟真实病毒的抗原结构。病毒结构蛋白表面的颗粒性质和高密度表现使得VLPs成为具有卓越安全性、免疫原性和可制造性的首要疫苗平台。因此本章将重点放在基于衣壳蛋白VLPs的有效诺如病毒疫苗开发上。广泛讨论了诺如病毒VLPs的表达和结构，特别是诺如病毒原型诺瓦克病毒VLPs。描述了诺如病毒VLPs通过口服和鼻内接种激活小鼠体内有效的系统和黏膜抗诺如病毒免疫能力。

从成本效益、生产速度和可扩展性的角度讨论了植物表达系统作为基于VLP的诺如病毒疫苗的新生产平台优势。详细介绍了在美国食品和药物管理局（FDA）现行的良好生产规范（cGMP）监管下，首次在植物表达系统中成功展示诺如病毒VLP生产的最新成果。

此外，还展示了人类临床试验的结果，证明昆虫和植物来源诺如病毒VLPs的安全性和有效性。由于不同的诺如病毒毒株中衣壳蛋白的多样性及其快速的抗原漂移，我们推测疫苗开发应该集中在从流行毒株衣壳蛋白衍生的多价VLP疫苗上。随着FDA最近批准第一批植物制造的生物制品，我们也推测植物生产系统将在制造这种基于VLP的多价诺如病毒疫苗中发挥重要作用。

8.1　胃肠炎

胃肠炎（GE）是一个世界性的健康问题，影响着所有年龄段的人。顾名思义，胃肠炎的特征就是胃肠道发炎，通常伴有腹泻、恶心、呕吐、腹部绞痛和疼痛症状。胃肠炎及其相关腹泻疾病仍然是世界上引起人类死亡的主要原因之一，特别是在发展中国家和幼儿中，估计每年死亡人数为400万~600万[1,2]。胃肠炎也是发达国家最常见的疾病。例如有人提出，在美国可能每个人每年至少感染一次胃肠炎病毒，610 000人将住院治疗，至少4 000人死于胃肠炎引起的疾病[1]。胃肠炎可由多种病原引起，包括病毒、细菌和寄生虫以及摄入非传染性毒素或药物，其中病毒是最常见的致病因子[1]。诺如病毒（NoV）和轮状病毒是引起胃肠炎的最常见病毒，而腺病毒、星状病毒、冠状病毒和副肠孤病毒也是已知引起胃肠炎的病毒[1-3]。

---

① Q. Chen，博士（亚利桑那州立大学生物设计研究所传染病和疫苗学中心，美国亚利桑那州坦佩市，亚利桑那州立大学技术与创新学院，美国亚利桑那州梅萨市，E-mail：qiang. chen. 4@ asu. edu）

8.2 诺如病毒

诺如病毒是一组遗传性多样的RNA病毒，属于杯状病毒科、诺如病毒属[4,5]。1972年首次发现诺如病毒，以原型病毒诺瓦克病毒（NV）为特征[6,7]。对诺瓦克病毒的研究表明，诺如病毒是一种非包膜病毒，其RNA基因组被一个直径约为38 nm的圆形外壳蛋白所包围[8]。然而直到最近，随着分子生物学的发展，诺如病毒分类已经被证明是困难和有争议的。由于在体外没有培养系统来培养这些病毒，因此基于中和作用的准确血清分型是不可能的[9]。相反诺如病毒分类必须依赖于志愿者的交叉攻毒研究和电子显微镜下免疫交叉反应性分析[10-13]，由于抗体的交叉反应性[14]，这种分析缺乏准确性和再现性。近年来，包括反转录聚合酶链式反应（RT-PCR）在内的复杂分子方法的发展，使得基于主要衣壳蛋白氨基酸序列的诺如病毒分类更加可靠[15]。在这一分类系统中，诺如病毒分为5个基因群和29个簇，其中8个簇在基因群Ⅰ（GⅠ），17个簇在GⅡ，2个簇在GⅢ，GⅣ和GⅤ各有1个簇（表8.1）[15]。在这5种基因型中，GⅠ和GⅣ毒株只感染人类，GⅡ在人类和猪身上都存在，而GⅢ和GⅤ毒株分别感染牛和鼠[15,16]。目前GⅡ组中第4簇（GⅡ.4）毒株是人类群体中最普遍的诺如病毒[17,18]。

以诺瓦克病毒为特征的诺如病毒基因组包含1个7.5~7.7 kb的单股正链RNA，有3个开放阅读框（ORFs）和1个位于3’末端的聚A尾[19]。ORF1编码一种多聚蛋白，该多聚蛋白被病毒蛋白酶3CLpro加工成RNA依赖的RNA聚合酶和大约5种其他非结构蛋白，包括p48、核苷三磷酸酶、p22、VPg和3CLpro[20,21]。这两种结构蛋白，主要衣壳蛋白（VP1）和次要蛋白（VP2）分别由ORF2和ORF3编码[19,22]。对诺如病毒结构的分析显示，每一个病毒衣壳由90个VP1二聚体组成，$T=3$二十面体对称[8,23]。VP1折叠成两个结构域：负责引发衣壳组装和二十面体接触的壳（S）结构域和包含P1和P2两个子结构域的突出结构域（P），这两个子结构域通过在VP1二聚体之间提供分子间接触来增强衣壳的稳定性[24]。

表8.1 诺如病毒分类

基因组	动物宿主	基因型簇	代表性病毒株
GⅠ	人	GⅠ.1	NV-USA93、Wtchest-USA、KY89-JPN
		GⅠ.2	SOV-GBR93、C59-USA、FB258-JPN
		GⅠ.3	DSV-USA93、LR316USA、VA98115-USA
		GⅠ.4	Chiba-JPN00、Valetta-MLT、NO266-USA
		GⅠ.5	Musgrov- GBR00、AB318-USA、SzUG1-JPN
		GⅠ.6	Hesse-DEU98、CS841-USA、WUG1-JPN
		GⅠ.7	Wnchest- GBR00
		GⅠ.8	Boxer-USA02
GⅡ	人和猪	GⅡ.1	Hawaii-USA94、Miami81-USA、DG391-DEU
		GⅡ.2	Msham-GBR95、CF434-USA、SMV1-USA
		GⅡ.3	Toronto- CAN93、MD102-USA、NLV2004-SWE
		GⅡ.4	Bristol-GBR93、Minerva、NT104-JPN
		GⅡ.5	Hilingd-GBR00、MOH99-HUN、NO306-USA
		GⅡ.6	Seacrof-GBR00、SU3-JPN、Miami292-USA
		GⅡ.7	Leeds-GBR00、GN273-USA
		GⅡ.8	Amstdam- NLD99、SU25-JPN
		GⅡ.9	VABeach- USA01、Idafall-USA

（续表）

基因组	动物宿主	基因型簇	代表性病毒株
		GⅡ.10	Erfurt-DEU01
		GⅡ.11	SW918-JPN01、SWVA34-USA、SW43-JPN
		GⅡ.12	Wortley- GBR00、U1GII-JPN、Pirna110-DEU
		GⅡ.13	Faytvil-USA02、KSW47-JPN
		GⅡ.14	M7-USA03
		GⅡ.15	J23-USA02、Mex7076-USA
		GⅡ.16	Tiffin-USA03、Fayett-USA、Tonto-USA
		GⅡ.17	CSE1-USA03
GⅢ	牛	GⅢ.1	BoJena-DEU98
		GⅢ.2	BoCH126- NLD00、BoNA2-GBR、BoCV95-USA
GⅣ	人	GⅣ.1	Alphatn- NLD99、FLD560-USA、SCD624-USA
GⅣ	鼠科动物	GⅤ.1	Murine1- USA03

表是基于参考文献［13］中衣壳蛋白 VP1 序列的系统发育分析结果生成的。

除了提供诺如病毒衣壳的唯一结构外壳成分，VP1 还必须执行其他功能，包括免疫原性和传染性。有人提出，P 结构域可能包含细胞受体结合位点和病毒表型或血清型决定簇[23,25,26]。P2 子域在不同的诺如病毒株中具有最大的可变序列，突出到衣壳表面，这表明它负责免疫识别和细胞相互作用[27,28]。诺如病毒的研究还表明，VP2 蛋白提高了 VP1 的表达水平，稳定了病毒衣壳中的 VP1[24]。

8.3 诺如病毒的传播和感染周期

诺如病毒是所有年龄段人群中引起胃肠炎的主要原因之一，在成人中超过 95%的病毒性胃肠炎是由诺如病毒引起的[29]。随着新诊断方法的发展，人们现在认识到病毒的影响被大大低估了，因为引起的暴发和感染远多于以前的认识[30,31]。诺如病毒只需要极低的感染剂量，因为诺瓦克病毒攻毒研究表明，在易感人群中，感染单个病毒粒子的概率约为 50%[32]。病毒可以通过几种途径传播，包括粪便污染的食物或水、人与人直接接触、通过液滴间接接触或感染者的污染物[33]。

诺如病毒的高度环境稳定性进一步促进了其在人类中的传播。例如诺如病毒在零下至 60℃的温度范围内稳定存在，其 pHs 甚至可以小于 2.7，并且耐氯（0.5~1.0 mg/L）、乙醇、季铵化合物或基于洗涤剂的清洁剂处理[34,35]。因此，诺如病毒具有很强的传染性，传播迅速，暴发通常发生在人们共享共同食物和水源的各种社会场所和环境中，或发生在人群密集的地方，如游轮、学校、军事单位、疗养院、日托中心、医院、餐馆和餐饮活动的地方[29,36-39]。

病毒脱落可能在症状出现之前开始，在症状消失后继续。这种延长的诺如病毒脱落期（数周）进一步提高了二次攻击率，通常导致大规模暴发[38,40-42]。因此，这些疫情往往导致数千人感染以及设施和企业关闭[43,44]。例如诺如病毒被确定为医院病房关闭最常见的病原之一[45]。诺如病毒胃肠炎疫情在全世界和全年都有分布，但正如其旧名称“冬季呕吐病”所暗示的那样，在一年中冬季暴发的次数更多[46]。有人认为，这种季节性是由于气候条件促进了人与人之间的密切接触，有利于病毒存活[47]。进一步研究表明，不同诺如病毒毒株具有不同的季节性暴发周期，因为 GⅡ 毒株主要出现在冬季，而 GⅠ 病毒在一年中传播更为均匀[48-50]。

由于传播途径不同，幼儿、养老院老人、学生、军事人员、旅行者和免疫功能低下的人特别容易受到诺如病毒感染。最近的研究表明，诺如病毒感染的易感性可能有遗传基础。例如发现人类群体中的易感性与个体的ABH组织血型、黏膜细胞表达的糖类、分泌状态以及组织血型抗原（HBGA）受体的毒株结合偏好有关[26,51-55]。此外已经发现，不同病毒基因型的P2s对某些ABO HBGA具有特异性亲和力，GI基因群诺如病毒优先结合群抗原A和O，GII基因群诺如病毒优先结合群抗原A和B[56,57]。因此暗示HBGA可能作为诺如病毒的假定受体[54,55,58-60]。这些发现可能有助于解释在某些个体中“无症状感染”的现象，这些个体出现了诺如病毒特异性抗体反应和脱落病毒，但在诺瓦克病毒感染后没有表现出任何典型症状。除了急性胃肠炎外，免疫功能低下的患者也有慢性诺如病毒感染的记录[61,62]。

由于缺乏体外细胞培养系统和小动物感染模型，人们对诺如病毒的生命周期还没有完全了解。然而Asanaka及其同事的一项研究表明，表达的诺如病毒基因组RNA具有复制、转录到诺如病毒亚基因组RNA并随后在哺乳动物细胞中转化为VP1的能力[63]。此外还发现，VP1表达可以在这个特定的细胞中产生充满诺如病毒基因组RNA的病毒粒子[63]。这些结果清楚地证明了哺乳动物细胞支持诺如病毒基因组RNA复制和病毒RNA组装成病毒粒子的能力。因此在哺乳动物细胞培养中，诺如病毒复制失败不是由于缺乏宿主因子来支持诺如病毒RNA细胞内表达。相反问题可能在于病毒与细胞受体结合的步骤，病毒在受体结合后进入细胞，或解开病毒粒子将基因组RNA释放到细胞质中。

在摄入病毒粒子后，呕吐和腹泻的症状通常出现在12~48小时的潜伏期之后。这些症状在疾病过程中（12~72小时）持续存在，通常伴有恶心、腹部绞痛，偶尔伴有不适、发冷、肌肉疼痛、头痛和低热[34]。诺如病毒引起腹泻和呕吐的分子和细胞机制尚未完全揭示。此外，人类诺如病毒靶细胞还没有确定。然而据观察，诺如病毒在感染后开始在小肠内繁殖，在空肠内引起组织病理损伤，在有症状和无症状的个体中，空肠绒毛可逆性变宽和变钝[64]。此外还观察到，上皮细胞中有独特的$CD8^+$淋巴细胞群的细胞质空泡化和浸润[65,66]。推测这些变化可能与绒毛变钝有关。尽管仍不可能在攻毒研究活检中看到诺如病毒粒子，但体外结合实验表明病毒结合主要发生在绒毛水平[67]。

虽然诺如病毒引起的急性胃肠炎通常是轻微的和自我限制的，可以用水合液体治疗，但是坏死性小肠结肠炎和脱水致死的情况确实发生了。据估计，在发展中国家每年造成20万名儿童死亡，在美国每年造成300人死亡[68-70]。此外，诺如病毒疫情的发生频率和规模已经成为医疗保健系统的主要负担，在发达国家造成了巨大的经济损失[45]。自从2002年新的诺如病毒变种GII.4出现以来，暴发的次数显著增加，几乎每两年全世界就发生一次大规模流行[71,72]。因此诺如病毒被认为是新出现的病原。此外，由于最近认识到诺如病毒在引起胃肠炎方面的突出作用以及可能引起迅速和大规模的水、食物和可能的空气传播疾病，其中只有10~100个病毒粒子，因此诺如病毒也被归类为B类生物防御病原[73]。目前没有有效的药物或疫苗来治疗或预防诺如病毒感染。因此迫切需要开发针对这些病毒的疫苗和疗法。

8.4 诺如病毒胃肠炎目前的诊断和治疗方法

已为诺如病毒开发了各种诊断方法[74]。虽然诺如病毒的血清抗体可以很容易地检测到，但由于抗体的交叉反应性，该方法的临床相关性很小。由于诺如病毒没有可用的培养系统，粪便样本中的病毒检测已成为首选的诊断方法。传统上诺如病毒感染是通过免疫透射电镜（TEM）诊断[11]。透射电镜提供了直接观察粪便样本中任何潜在可靠病毒粒子的优势。但是它的缺点是需要复杂且昂贵的设备和高度专业的技术人员来操作。

此外，样本中完整病毒粒子的浓度必须至少为10^5~10^6颗粒/mL才能被检测到。透射电镜能对单个样品提供快速的诊断结果，通常在送样3小时内完成[75]。然而由于操作人员必须花费大量时间检查每个样本，因此免疫透射电镜不是快速处理大量样本的高通量分析方法。免疫透射电镜依赖于对具有典型诺如病毒形态的病毒粒子的鉴定，因此不允许在诺如病毒属中进行可靠的形态鉴定，这降低了其特异性，可能阻碍诊断。这些挑战可能导致了对诺如病毒流行病的历史性低估。

后来开发了几种检测诺如病毒抗原的酶联免疫吸附试验（ELISAs）方法，用于诺如病毒诊断[76,77]。基于抗原的 ELISAs 是高通量的，相对容易操作，可以在 1 个工作日内快速提供诊断结果[78]，因此在需要快速经济筛选大量样本的情况下，ELISAs 通常是首选方法。研究表明，检测诺如病毒抗原的 ELISAs 具有高特异性（94%~96%），但敏感性差（40%~60%），这很可能是由于诺如病毒毒株的抗原多样性所致[4,79]。因此，它们对于特定诺如病毒的特定检测是有用的，但是对于需要检测广谱诺如病毒的应用来说不够敏感。然而针对多种抗原的集合抗体或针对更广泛重组病毒抗原的多价抗体的使用提高了 ELISAs 的敏感性[80]。

随着 1990 年诺瓦克病毒基因组的克隆和分子生物学方法的发展，粪便样本 RT-PCR（检测病毒 RNA）越来越成为诺如病毒诊断的主要检测方法[81-83]。类似于基于 ELISA 方法的检测，RT-PCR 快速且可靠，因为可以同时处理大量样品并且可以在 1 个工作日内获得结果。然而这需要从粪便样本中提取 RNA，需要昂贵的设备和熟练的工人来操作。因此，RT-PCR 比 ELISAs 更为劳动密集，也不太经济。使用为基因组高度保守区域或毒株特异性区域设计的引物，且 RT-PCR 是通用的，既能提供与 ELISAs 相当的高特异性，又能提供高得多的敏感性。RT-PCR 检测的高灵敏度允许检测病毒浓度太低而无法通过其他方法检测的样品中的诺如病毒，包括病毒载量低的临床样本和环境样本，如食物和水[83]。

RT-PCR 的另一个优点是，当与随后的核苷酸测序结合时，允许基因分型来追踪暴发源头[84]。另一方面检测的高灵敏度可能会因非特异性扩增或潜在的样品交叉污染而产生假阳性结果。这一挑战已经通过设计更特异的引物来解决，以避免非特异性 RNA 扩增，通过实施更严格的程序来防止样品的交叉污染。由于该属病毒之间异常的序列多样性，没有单一的通用引物可检测到诺如病毒的所有菌株。然而已经开发了几个引物可以检测 GI 和 GII 中超过 90% 的所有菌株[85]。基于 RT-PCR 的诊断方法的稳健性正在通过采用更快和更灵敏的检测方法进一步优化，例如实时定量 RT-PCR，其目的是在疫情暴发期间快速检测大量粪便样品中甚至更低拷贝数的诺如病毒[86]。这些研究表明，基于 TaqMan 的实时荧光定量 PCR 在偶发病例和暴发背景下，都能提供对诺如病毒的更灵敏的实时诊断，其结果易于解读且在工作日内即可获得结果[87]。

总的来说，TEM、ELISA 和基于 RT-PCR 的方法都有其优势和挑战。TEM 允许检测所有可靠的病毒制剂，但价格昂贵，通量相对低。相比之下，ELISA 具有高通量、经济和易于使用的优点，但需要提高灵敏度。RT-PCR 结合了敏感性和特异性的优点，能够进行分子流行病学研究。这 3 种方法检测病毒的不同成分，因此是相互补充的。

对于特定的诊断应用，3 个因素决定了 TEM、ELISA 和 RT-PCR 之间的选择。包括：①诊断类型，如流行病调查与临床诊断；②每种试验分析的敏感性和特异性；③分析的速度、稳健性和易用性。还必须记住粪便样品的异质性和复杂性以及储存条件常常会导致病毒粒子和/或 RNA 完整性的变化，从而影响这些方法在酶和免疫分析中的反应性。例如，如果基于 ELISA 的方法被用作临床样本的唯一诊断方法，那么 ELISA 的低灵敏度将导致大量的假阴性样本。相反这些方法仅用作临床诊断的“筛选”试验，以利用它们在靠近患者的位置进行的速度和有效性。然而，ELISAs 的高度特异性使其成为一种疫情调查中合适的分析方法，因为多种样本的可用性提高了其确认疫情原因的可靠性[79,80]。

虽然如此，RT-PCR 应与 ELISAs 一起用于临床诊断和疫情调查，以确认阳性结果和评估阴性样品。相比之下，如果诊断目标是在流行病学研究中追踪感染的起源，RT-PCR 和 DNA 测序应该是选择的方法，因为 RT-PCR 的高敏感性和特异性更有优势[88,89]。TEM 已被公认为诺如病毒诊断的“金标准”。然而随着更多的比较结果可用，有人建议 RT-PCR 或上述 3 种检测中的两种阳性结果应被视为“金标准”[78,79,90]。无论如何，建议 TEM、ELISA 和 RT-PCR 中的 3 种方法中至少两种结合起来，以获得可靠的临床诊断结果，而这 3 种方法都适用于诺如病毒疫情流行病学调查，具体取决于样本数量[79]。

目前，尚无针对诺如病毒胃肠炎预防或治疗的有效疫苗或治疗方法[34]。一般临床管理包括口服含有必需电解质和糖的补液溶液来治疗腹泻。对于有明显脱水症状的患者或不能忍受口服补液的患者，可能需要进行肠外液体和电解质置换[4]。目前的工作重点是开发有效的疫苗和可能的附着抑制剂，通过糖模拟物进行抗黏附治疗[91]。

8.5 诺如病毒疫苗开发

8.5.1 诺如病毒感染免疫学

由于缺乏小动物模型，诺如病毒的免疫学知识主要来自人类攻毒研究和自然暴发。这些研究表明，受感染的志愿者在诺如病毒攻击后确实产生了免疫力[13,92,93]。然而对一种毒株的免疫并不能完全保护其免受异源毒株感染，有症状个体在 2～4 年后暴露于同一个诺如病毒毒株时往往会再次感染，这表明免疫似乎是毒株或基因组特异性的并且寿命很短[13,92-94]。攻毒性研究并未得出关于长期免疫的结论性结果，这可能与志愿者对各种循环的诺如病毒毒株的预暴露相混淆[71]。对成人反复感染观察表明，对这些病毒缺乏长期免疫力[38,84]。然而其他研究表明，近 50%的遗传易感受试者没有受到诺如病毒感染，这支持了长期免疫的可能性[95]。此外，群落试验研究表明，症状持续时间通常随着年龄的增长而缩短，这表明至少对诺如病毒的发病起到了部分保护作用[38,93]。

虽然目前还没有可用于预防人类诺如病毒的疫苗，但这些研究支持开发能够诱导保护性免疫和减轻病毒疾病负担的疫苗可行性。

8.5.2 有效预防诺如病毒的病毒样颗粒疫苗

组织培养系统的缺乏也阻碍了诺如病毒疫苗开发。幸运的是，表达的 VP1 自发组装成病毒样颗粒（VLPs），其形态和抗原与天然病毒相似，这促进了疫苗开发[73]。VLPs 结合全病毒和亚单位抗原的最佳特性用于疫苗开发。由于没有病毒核酸基因组，VLPs 比灭活或减毒病毒更安全。至少在理论上，VLPs 免疫原性甚至可以通过排除其组成中的免疫抑制病毒蛋白而增强。此外对于 VLP 生产来说，不需要一个灭活过程。因此不会发生意想不到的表位修饰，进一步确保 VLPs 的免疫原性。重要的是，VLPs 可以在不使用佐剂的情况下诱导有效的细胞和体液免疫应答，并且比其他亚单位抗原更有效，因为 VLPs 的结构模仿感染性病毒。VLPs 可以通过重组技术在异源表达系统中产生，而无须支持病毒复制[31]。这对诺如病毒尤为重要，因为还没有开发出这样的培养系统来支持病毒生长[96]。

粒子性质和表面上表位的密集重复排列使得 VLP 比其他亚单位疫苗更具免疫原性。VLPs 可以通过与抗原呈递细胞（APCs），特别是树突状细胞（DCs）的相互作用，有效诱导 T 细胞介导的免疫应答。具体来说，VLPs 可以模拟自然病毒感染过程，被树突状细胞特异性识别和吸收，随后被处理并递呈给细胞毒性 T 细胞，以触发它们的激活和增殖[97]。研究表明，病毒和相应 VLPs 粒子大小对于树突状细胞和巨噬细胞摄取来说是理想的，可以启动抗原处理[98,99]。因此，VLPs 的粒子特性有利于靶向相关 APCs，以诱导最佳的 T 细胞介导的免疫应答。

VLPs 也可以有效地递呈给 B 细胞，诱导强烈的抗体应答。像活病毒一样，VLPs 的准晶表面及其重复表位阵列是脊椎动物 B 细胞进化来特异性识别的主要目标[100]。这种识别触发 B 细胞表面膜相关免疫球蛋白（Ig）的交联[101-103]，并导致它们的增殖和迁移、T 辅助细胞活化、抗体产生和分泌以及记忆 B 细胞产生[101]。因此，VLPs 可以比其他亚单位抗原低得多的浓度直接激活 B 细胞，并在没有佐剂的情况下诱导高滴度和持久的 B 细胞应答。

由于诺如病毒是一种肠道致病性病毒，一种有效的疫苗也应该通过口服等方式诱导诺如病毒特异性肠道黏膜免疫。一般来说，由于胃酸和消化酶可能使抗原变性和降解、向肠道相关淋巴组织（GALT）转运抗原，以进行抗原加工和递呈，可能刺激系统免疫耐受，口服亚单位蛋白疫苗可能无效[102]。然而 VLPs 紧凑且高度有序的结构使其比其他蛋白疫苗更能抵抗消化道中的降解酶。VLPs 与真实病毒粒子的相似性也可能产生一种“危险信号”，这种信号会克服肠道抗原是良性的感觉，从而阻止免疫耐受的发展[55]。这两个特点对诺如病毒 VLPs 尤其适用，因为它们的同源病毒是天然的胃肠道病原体。此外，VLPs 也被自然识别，并有效地运入 GALT[4]。因此，口服疫苗呈递的挑战可能被 VLPs 的独特结构所克服，这种结构允许 VLPs 引起有效的肠道免疫反应。

VLPs 的这些固有优势使其成为最成功的重组疫苗平台之一。例如，5 种基于 VLP 的乙型肝炎病毒（HBV）和人乳头瘤病（HPV）疫苗已获得商业许可，所有这些疫苗都证明了良好的安全性和对人体感染的长期保护[31]。这些成功以及在口服后激发肠道黏膜免疫反应的潜力鼓励了诺如病毒 VLP 候选疫苗临床前和临床开发测试。

8.5.3 诺如病毒 VLPs 特性

诺如病毒 VLPs 首先在昆虫细胞中使用杆状病毒载体产生[23,104]，然后在植物中使用烟草花叶病毒载体[101]和双生病毒载体[102]，以及在哺乳动物细胞中使用委内瑞拉马脑炎复制子系统[105]。这些研究表明，主要衣壳蛋白 VP1 单独表达就可以驱动形态和抗原性类似于天然病毒粒子的 VLPs 自我组装（图 8.1）。所有 3 个表达系统生成的 VLPs 彼此相似。

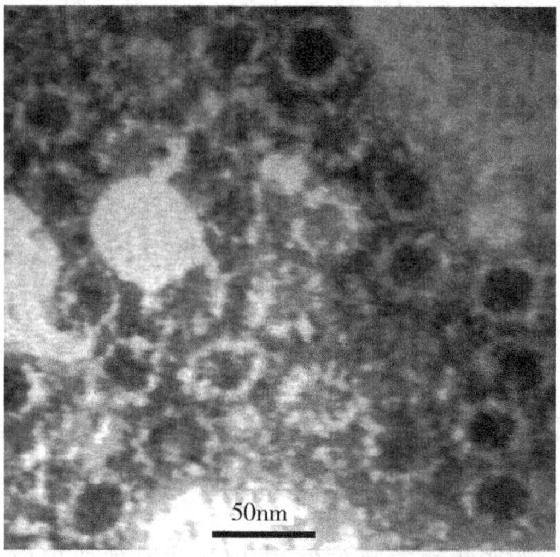

图 8.1　在植物中生产的诺瓦克病毒衣壳蛋白 VLPs

用双生病毒复制载体在莴苣叶片中表达诺瓦克病毒衣壳蛋白。从植物中纯化出诺瓦克病毒衣壳蛋白 VLPs，用 0.5%醋酸铀酰进行负染，并进行电镜检查。Bar = 50 nm。显示了 3 个试验中的代表性试验之一。

诺瓦克病毒衣壳蛋白（NVCP）的 VLP 是诺如病毒 VLPs 结构的例证。通过低温电子显微镜和 X 射线晶体学对昆虫细胞杆状病毒衍生的诺瓦克病毒衣壳蛋白 VLPs 研究表明，诺瓦克病毒衣壳是由 180 个拷贝的衣壳蛋白 VP1（58 kDa）组成的 38 nm 二十面体排列，以 $T=3$ 对称排列成 90 个二聚体[104]。虽然所有二聚体都由两种相同的诺瓦克病毒衣壳蛋白单体形成，但是需要两种不同的二聚体构型来正确形成完整的组装衣壳[106,107]。

如同天然诺瓦克病毒粒子一样，诺瓦克病毒衣壳蛋白也折叠成 VLPs 中的两个不同结构域，

S结构域形成外壳的内核，P结构域从衣壳伸出[108]。类似地，P2子域也是诺瓦克病毒VLPs中最大的表面暴露区域，可能包含组织血型抗原（HBGA）和中和抗体结合位点以及毒株特异性决定簇[27,56,108-111]。已经证明了诺如病毒和包括GII.4病毒在内的其他诺如病毒VLPs之间的相似性[112]。

8.6 诺如病毒VLPs临床前开发

8.6.1 昆虫细胞杆状病毒载体产生的NVCP VLPs

如前所述，对于诺如病毒疫苗来说引发肠黏膜免疫是可取的。口服接种是诺瓦克病毒衣壳蛋白VLPs诱导免疫的可行策略，因为肠道感染的进化选择使其在口服-胃肠道环境中保持稳定，有效地转运到肠道相关淋巴组织（GALT）进行抗原处理和递呈[13]。因此，给予小鼠5~500 μg范围内的各种口服诺瓦克病毒衣壳蛋白VLPs以检查引发系统和黏膜抗体反应的能力。

研究表明，在大部分（8/11）用VLP喂养的iCD1杂交小鼠中，4次口服剂量少至5 μg无佐剂诺瓦克病毒衣壳蛋白VLPs诱导了诺瓦克病毒特异性抗血清IgG应答[106]。两次口服剂量后观察到系统IgG应答，4次剂量的200 μg VLPs诱导了最高滴度。此外200 μg剂量组中的小鼠产生了诺瓦克病毒特异性肠IgA，水平高达总IgA的0.1%。加入黏膜佐剂霍乱毒素（CT）并没有显著改变血清IgG或肠IgA阳性应答者的数量，但是显著增加了血清IgG应答的幅度，尤其是对于更高剂量的VLPs[106]。因此，诺瓦克病毒衣壳蛋白VLP显然是一种有效的口服免疫原，可以诱导系统和肠黏膜抗体应答。

诺瓦克病毒衣壳蛋白VLPs口服接种成功鼓励了通过替代黏膜途径接种的探索。Guerrero及其同事的一项研究表明，鼻内（in）接种比口服接种能更有效地激发低剂量VLPs对诺瓦克病毒衣壳蛋白特异性血清IgG和肠道IgA的应答[114]。例如在黏膜佐剂（突变大肠杆菌热敏毒素LTR192G）存在的情况下，两次10 μg剂量的昆虫细胞产生的诺瓦克病毒衣壳蛋白VLPs递呈，引发了相当于两次200 μg口服递送的有佐剂VLPs抗诺瓦克病毒衣壳蛋白滴度[114]。

除了粪便样品中的肠道IgA外，阴道冲洗液中也检测到强烈的抗诺瓦克病毒衣壳蛋白IgA应答。此外，这些黏膜IgA应答持续时间很长，可以在鼻内（in）免疫后1年中都检测到[114]。这些数据不仅表明诺瓦克病毒衣壳蛋白VLP是一种有效的黏膜抗原，可刺激全身和局部黏膜抗体反应，还表明其在远端黏膜引发抗体应答的能力。这些发现清楚地证明了昆虫细胞产生的诺瓦克病毒衣壳蛋白VLPs引发系统和黏膜B细胞应答以潜在地中和诺瓦克病毒并抑制其感染方面的能力，也表明了其作为异源表位载体在抗击STI（性传播感染）方面的应用。例如用STI病原表位修饰的嵌合诺瓦克病毒衣壳蛋白VLPs，可诱导生殖黏膜中产生中和IgA。

8.6.2 哺乳动物细胞VEE复制子产生诺如病毒VLPs

VEE是一种甲病毒，已经被开发为复制子疫苗载体。为了表达特定疫苗，将感兴趣的抗原编码序列克隆到VEE复制子cDNA中26S启动子下游的VEE结构基因中，以驱动其高表达水平[105]。cDNA复制子构建体与携带VEE结构基因的另一构建体共转染到哺乳动物细胞中，将允许重组病毒RNA包装在VEE病毒衣壳中，形成病毒复制子颗粒（VRPs）[57,105,115]。这些VEE VRPs可以感染哺乳动物细胞，并积累大量疫苗蛋白[116]。由于VEE结构基因是以反式方式提供用于VRP的形成，而不是重组基因组的一部分，所以VEE VRPs对细胞的感染是一次性事件[116]。

为了生产诺瓦克病毒衣壳蛋白VLPs，诺瓦克病毒VP1基因在VEE复制子cDNA中克隆，并在哺乳动物细胞中表达[105,115,117,118]。VEE-VRP/哺乳动物细胞源性VLPs可以被纯化并作为免疫原通过亲本或黏膜途径递送。此外，VRPs本身可用作疫苗在体内产生VLPs。因此，该系统可通过两种方式诱导免疫，并具有潜在的优势，因为VRPs可以直接用于感染哺乳动物宿主靶细胞，其中大量诺瓦克病毒衣壳蛋白VLPs被组装用于向B细胞和T细胞递呈抗原[117]。例如通过小鼠脚垫直接皮下接种两剂重组NVCP-VRPs（诺瓦克病毒衣壳蛋白-病毒复制子颗粒）（每剂10个感染单位[7]），引发了针对诺瓦克病毒VLP强烈的全身IgG和肠道IgA应答，以及针对另一个GI诺如病毒的VLPs异型应答。此外，接种两剂NVCP-VRPs的小鼠其血清和肠内免疫应答明显强于口服两剂75 μg或200 μg VEE衍生的NVCP VLPs小鼠[117]。

在VEE-VRP系统中也生产了多价诺如病毒候选疫苗并且在小鼠中进行了测试[115,119]。这些研究表明，在小鼠体内接种表达毒株特异性VLPs3个或4个病毒复制子颗粒混合物，不仅能诱导对所有接种毒株的强血清抗体应答，而且还能诱导对新毒株的受体阻断异型抗体应答[115]。此外，一个用诺如病毒VLPs共同接种的8人小组，用涵盖超过95%的所有诺如病毒感染，研究表明，疫苗混合物包括来自两个基因群（GI和GII）的VLPs并没有降低任何一个基因群的特异性反应，而是诱导了针对基因群间毒株的受体阻断抗体[119]。

与昆虫细胞杆状病毒系统相比，VEE-VRP系统由于需要共感染和BSL-3设施来生产病毒复制子颗粒，因此比较烦琐和昂贵。然而胃肠外接种VRP疫苗已被证明是针对树突状细胞并刺激有效的体液免疫和细胞免疫，包括保护黏膜表面免受感染[113,120,121]。此外，已证明VRP具有固有的佐剂活性，这很可能是由于哺乳动物细胞中单轮病毒RNA复制刺激免疫细胞所致[119,122]。总的来说，当VRP被直接用作诱导对诺如病毒潜在保护性免疫的疫苗时，这些优点可能使VEE-VRP成为优于杆状病毒系统的一种系统。

8.7 植物源诺如病毒VLPs的生产和免疫原性

8.7.1 植物作为诺如病毒VLPs生产平台

尽管诺如病毒VLPs或VRPs作为诺如病毒候选疫苗显示出有希望的结果，但是基于昆虫和哺乳动物细胞培养物的生产系统有几个限制，可能会阻碍这些疫苗的商业开发[54]。例如在杆状病毒/昆虫细胞系统中，杆状病毒颗粒与诺如病毒VLPs的共同生产可能会在VLP纯化、免疫原性和监管批准方面产生问题。由于残留杆状病毒可能会改变诺如病毒VLP制剂的整体免疫原性并引起安全问题，因此必须通过烦琐而昂贵的纯化过程或化学处理去除或灭活残留杆状病毒及其传染性。这些程序不仅增加了总生产成本，还可能损害最终VLPs质量[31]。

VEE/哺乳动物细胞系统的生产成本明显高于昆虫细胞培养物。此外建造BSL-3制造设施也需要大量资本投资[123-126]。昆虫和哺乳动物细胞的生产系统也面临着可扩展性方面的挑战，因为必须建造新的发酵罐和设施来进行大规模生产。这些挑战可能会阻碍诺如病毒VLPs健康效益潜力的充分实现，尤其是在发展中国家。因此已经探索植物作为诺如病毒VLPs安全、经济、可扩展的生产平台。

植物被认为是一种替代性的蛋白质疫苗生产系统，因为植物可以以低成本生产高水平疫苗蛋白，而生物量生产不需要在细胞培养设施或设施重复方面进行昂贵投资来扩大生产[124,127]。因此，植物生物量产生和扩大的灵活性和资本效率远优于目前的发酵技术[128,129]。此外，植物还具有真核加工机制，能够对蛋白质进行适当地翻译后修饰和组装，将外来病原体引入人体的风险很低[124,127]。尽管有这些潜在的优势，但早期使用稳定转基因植物生产的蛋白质疫苗导致靶蛋白积累缓慢且水平很低[54,126]。产生转基因植物的长时间框架（几个月到1年）、缺乏强启动子以及转基因随机插入的位置效应是导致这些问题的原因，这些问题降低了植物作为生产系统的成本节约效益[102]。

VLP生产速度和产量的挑战已经被基于植物病毒的瞬时植物表达系统发展所克服[130,131]。例如我们研究小组和其他研究人员报告称，植物源性VLPs的克隆和高水平瞬时表达可以用MagnI-

CON 系统载体渗透在 1~2 周内快速实现，MagnICON 系统是基于 TMV（烟草花叶病毒）RNA 复制子系统或源自 BeYDV（豆黄矮病毒）双生病毒 DNA 复制子系统[126,132-134]。VLP 表达速度和产量的提高也提供了植物表达系统，这是生产针对诺如病毒和其他病毒 VLP 疫苗的一个额外优势，这些病毒具有快速抗原漂移，在世界范围内有多个基因群和具有不可预测的流行毒株。

一种有效的诺如病毒疫苗需要在毒种鉴定后的最短时间内生产出来，以阻止新毒种的传播，最好是通过低成本平台，包括发展中国家在内的地方生产出负担得起的疫苗。基于瞬时植物表达的平台可能会解决这样的成本和时间问题，提供关键的多功能性，允许快速生产毒株特异性疫苗，以及时控制潜在的诺如疫情暴发。

8.7.2　植物作为诺如病毒 VLP 疫苗递送工具

由昆虫或哺乳动物细胞培养物生产的 VLP 疫苗是纯化产品，需要昂贵的下游加工、冷藏和运输温度[102]。通过口服递送昆虫细胞产生的诺如病毒 VLPs 成功诱导系统和黏膜抗体应答，这表明口服免疫也可以通过摄入含有诺如病毒 VLPs 的可食用植物部分来实现[124]。这是一种有吸引力的方法，因为可以减少对昂贵的纯化步骤需求，且可以规避后勤挑战，允许在冷藏和其他医疗用品有限的地区实施免疫计划。

这种方法最有可能在诺如病毒 VLPs 中获得成功，因为诺如病毒会自然感染胃肠道系统，对胃肠道口服的变性和消化有抵抗力[31,107]。此外，VLPs 与天然病毒的相似性允许被肠上皮的“M”细胞有效地取样，并被运送到肠道相关淋巴组织（GALT）中，用于抗原处理和递呈[102]。因此，食用植物中产生 VLPs 代表了一种通过口服接种，建立肠黏膜免疫的新颖且经济有效的方法[135-137]。由于疫苗需要有明确的剂量单位，这一策略在发达国家可能面临商业化的监管障碍[102,123,124]。然而这一策略最终可能为口服途径商业疫苗递送提供一个可行选择，特别是因为新一代表达载体正在实现每单位植物组织更一致的 VLP 积累。

总的来说，目前的植物表达系统提供的优势远远超过了传统的真核蛋白质修饰和组装，成本低、可扩展性强、安全性更高。例如植物表达系统允许 VLP 以前所未有的速度生产，以控制潜在的流行病或瘟疫。

8.7.3　稳定的转基因植物源性诺如病毒 VLPs

我们实验室研究了植物中 VLPs 的表达和组装，成功生产了几种无包膜和包膜 VLPs，包括基于诺瓦克病毒衣壳蛋白、乙型肝炎病毒核心抗原（HBcAg）和西尼罗病毒包膜蛋白 VLPs[31]。基于诺瓦克病毒衣壳蛋白的 VLP 是植物中研究最多的 VLPs 之一，我们团队和合作者已经成功地在许多植物物种中进行了表达，包括烟草、马铃薯、本氏烟、番茄和莴苣等[103,129,132,133]。

至于其他疫苗蛋白，诺瓦克病毒衣壳蛋白首先在转基因烟草和马铃薯植株中表达[103]。通常需要几个月的时间来生产和选择表达诺瓦克病毒衣壳蛋白的转基因烟草和马铃薯植株。诺瓦克病毒衣壳蛋白在这些转基因植物中表达约为 10 μg/g 新鲜组织重量，与昆虫或哺乳动物细胞表达系统相比，这是相当低的[101]。然而在转基因烟草叶片和马铃薯块茎中观察到类似于昆虫细胞源性 VLP 或天然诺瓦克病毒颗粒的病毒大小二十面体 VLP 组装[101]。在另一个实验中，对诺瓦克病毒 VP1 基因进行了密码子优化，在转基因番茄植株中表达[138]。比较研究表明，VLP 产量和组装取决于密码子使用、寄主植物种类以及诺瓦克病毒衣壳蛋白积累的目标组织[101]。例如诺瓦克病毒衣壳蛋白表达水平较低，只有 25%~50% 的诺瓦克病毒衣壳蛋白在马铃薯块茎中被组装成 VLP，而在番茄果实中至少实现了 10 倍的高表达和更有效的 VLP 组装[101]。

为了获得对诺瓦克病毒的肠道黏膜免疫，在第 1 天、第 2 天、第 11 天和第 28 天给小鼠喂食 4g 含有 40~80 μg 未煮熟的诺瓦克病毒衣壳蛋白马铃薯块茎。这种免疫机制诱导小鼠特异性血清 IgG 和肠道 IgA 应答[101]。当 4 剂 50 μg 纯化的诺瓦克病毒衣壳蛋白 VLPs 口服给同一免疫方案的小鼠时，血清抗诺瓦克病毒衣壳蛋白反应比用表达诺瓦克病毒衣壳蛋白的块茎喂养的小鼠高 4

倍[101]。当与佐剂（CT）合用时，体液免疫应答可进一步增强到16倍以上[101]。可能是由于马铃薯块茎中的诺瓦克病毒衣壳蛋白只有约50%被组装成VLPs，所以纯化样品中组装的VLPs富集有助于提高体液免疫能力[31]。也有可能诺瓦克病毒衣壳蛋白不能有效地从马铃薯组织中释放出来，被运送到肠道相关淋巴组织（GALT）进行抗原加工和展示。

番茄果实比马铃薯块茎更可口，是开发口服诺如病毒疫苗更可行的植物材料，食品工业已经很好地建立了番茄果实的生产和加工[31]。口服4剂0.4g冻干番茄（含40 μg VLP），刺激80%以上小鼠的强血清抗诺瓦克病毒衣壳蛋白IgG和肠黏膜IgA应答[139]。此外，当使用更高剂量（每剂0.8g）的转基因番茄时，100%小鼠出现强烈的全身和黏膜抗体应答[138]。

同一项研究还表明，诺瓦克病毒衣壳蛋白在冷冻干燥的番茄中的免疫原性比在冷冻干燥的马铃薯块茎中的免疫原性强[138]。番茄果实中氧化环境较少或独特的组织结构可能使VLPs具有更好的稳定性或更有效的释放，进而具有更好的免疫原性。人们还注意到，与诺瓦克病毒衣壳蛋白冻干番茄相比，摄入诺瓦克病毒衣壳蛋白风干番茄会引起更强的血清IgG和肠道IgA应答。冷冻干燥过程有可能改变番茄果实中VLPs的组装状态[138]。或者空气干燥可以通过保持VLPs的结构和组织结构而导致更好的VLP稳定性，从而VLPs更好地免受口腔-胃肠道中消化酶的影响，或者更有效地从组织中释放出来供GALT吸收。总的来说，这些数据支持番茄植物诺如病毒口服疫苗的开发。

8.7.4 植物病毒载体瞬时表达诺如病毒VLPs

到目前为止，获得表达VLP的植物株系的速度慢和VLP产量低是植物诺如病毒VLP生产面临的主要挑战。克服这些挑战的策略之一是使用基于植物病毒载体的瞬时表达系统。在这些系统中转基因不整合到植物基因组中，而是在转录的同时短暂地存在于植物细胞核中，随后转录物被转运到细胞质中，翻译成转基因蛋白质[140,141]。

这些系统专注于生产速度和产量，并随着病毒的速度和表达扩增而获得核基因表达的灵活性。例如基于复制能力强的烟草花叶病毒（TMV）和马铃薯病毒X（PVX）的MagnICON瞬时表达系统，允许在载体递送后7~10天内，生产高水平的重组蛋白[141,142]。事实上，我们的结果表明，MagnICON系统允许我们在本氏烟植物渗透的12天内，以0.8mg/g FLW（鲜叶重量）的水平生产完全组装的诺瓦克病毒衣壳蛋白VLPs，比转基因烟草和番茄产量高出至少80倍[133]。

当通过口服途径递送时，来自瞬时表达植物材料的部分纯化VLPs（每剂100 μg）在没有任何佐剂的情况下，引发有效和平衡的全身性IgG1/IgG2a应答[133]。同样的免疫制度也在100%免疫小鼠中检测到显著的诺瓦克病毒衣壳蛋白特异性阴道和粪便黏膜IgA应答[133]。此外，通过在口服免疫中加入佐剂，诺瓦克病毒衣壳蛋白特异性免疫显著增强[103]。因此基于“被破坏的”病毒载体的瞬时表达系统，使我们能够克服与转基因植物系统相关的挑战，并为诺如病毒的生产提供了一个强大植物系统。

为了进一步优化诺如病毒VLP疫苗商业化生产的瞬时表达系统，我们开发了另一个基于双生病毒BeYDV DNA复制子载体和市售莴苣的强健表达系统[129,143]。这种瞬时表达系统有两个主要优势：BeYDV DNA复制子载体非竞争性和商业生产莴苣的使用。MagnICON系统可以生产最多含有两种不同异质亚单位的蛋白质，与MagnICON系统不同，双生病毒载体可以生产含有至少5种异质亚单位的VLPs[143]。

莴苣可以在成熟的商业温室中快速、方便地大量种植。不同于烟草和烟草属的相关物种（如本氏烟），莴苣是一种可口的植物，可以生吃，用于口服递送诺如病毒VLPs。当诺瓦克病毒衣壳蛋白在含有BeYDV复制子载体的本氏烟中表达时，组装的VLPs在叶中累积到0.4 mg/g FLW（鲜叶重量）[132]。我们的数据还表明，BeYDV复制子系统允许莴苣中高水平的诺瓦克病毒衣壳蛋白VLP表达和组装，就像本氏烟MagnICON系统驱动的那样[129]。

事实上，VLP在载体导入后的第4天积累到最高水平，表明这种表达系统可以在莴苣中产生类似水平的VLPs，但比烟草MagnICON系统要短得多。重要的是，这项研究是第一次使用商业生产的莴苣快速、高水平生产 VLPs 的展示之一[129]。这对于最终商业化生产基于诺如病毒 VLP 疫苗来说是一个重要里程碑，因为这将允许我们的生产系统从现有商业来源获得无限量的低成本材料。结合莴苣生产的几乎无限性，BeYDV 复制子载体的高效率和可扩展性为我们提供了一个诺如病毒 VLP 新生产平台，该平台成本低、强健、安全，适合大规模生产。

8.8 在监管条件下生产植物源性诺如病毒 VLP 疫苗

尽管在植物中成功地表达了诺瓦克病毒衣壳蛋白 VLPs 和多种其他亚单位疫苗，并且最近通过瞬时表达系统提高了疫苗生产水平，但还没有植物源性疫苗被批准用于人类。商业上缺乏成功在于几个有待克服的技术和监管障碍。这些挑战包括：缺乏可扩展的下游加工程序、生产过程监管合规性的不确定性，以及迄今为止缺乏在身份、纯度、效力和安全性方面满足监管机构要求标准的植物源性疫苗的证明[123,24,127]。

虽然通过摄取植物组织进行免疫仍然是口服递送植物生产 VLPs 的可行方法，但产品监管方面的考虑使得有必要开发下游加工技术，以生产具有规定单位剂量的纯化 VLPs 疫苗[123]。因此，要使植物作为疫苗生产替代品的承诺成为现实，这项技术必须能够生产足够数量相关规模的 VLPs，且产品质量符合 FDA 等监管机构的所有要求标准。

8.9 根据 FDA 法规建立诺如病毒 VLPs 植物生产设施

为了克服这些挑战，必须首先解决 cGMP（现行良好生产规范）下诺如病毒 VLPs 生产设施、材料和程序的合规性。我们团队主持了这一领域的工作，在学术界建立了第一个符合 cGMP 标准的生物处理设施，用于在植物中生产诺如病毒 VLPs[126]。

如图 8.2 所示，我校专门修建的设施允许在 cGMP 质量管理系统（QMS）下对植物源性诺如病毒 VLPs 疫苗进行生物量生成和生物处理。例如中央生物处理套件中专门设计了用于缓冲液制备、进入的植物样品、色谱法和无菌独立房间（图 8.2a）。

如图中箭头所示该套间在压差下运行，无菌室为100级环境，设计为具有单独的单向流动的工人和生物材料流程，以使产品满足美国和欧盟（EU）监管机构对人类药物制造的要求。作为质量管理体系的一部分，质量控制（QC）/质量保证（QA）实验室是毗邻的，该生物处理套间用于产品测试和质量管理（图 8.2a）。在 cGMP 下还建造了一个 3 600ft^2 的生物安全 2 级（BSL-2）温室设施，用于生产植物生物量（图 8.2b）。

BSL-2 温室配备了先进的控制植物或病原体技术，这对于维持植物源性 VLPs 完整性和减少外部环境意外基因污染风险是必要的。

已建立的生物加工设施和质量保证/质量控制（QA/QC）实验室也已获得生产人类疫苗的许可，这为遵守 VLP 生产程序和最终 VLP 疫苗产品的 cGMP 奠定了坚实基础。

8.9.1 根癌农杆菌、本氏烟主库和工作库

作为上游过程 cGMP 合规性的一部分，生物材料包括相关的农杆菌菌株和本氏烟植物系，都必须通过质量管理体系认证。这一过程始于建立包含诺瓦克病毒衣壳蛋白表达载体的根癌农杆菌菌株主库和工作库以及野生型本氏烟的种子库，用于生长生物量和瞬时表达。为了生产符合 cGMP 要求的诺瓦克病毒衣壳蛋白 VLPs 植物材料，在 GV3101 细菌株中建立了 3 个根癌农杆菌主库[126]，每个主库包含用于在 TMV 3’模块、TMV 5’模块中生产 VLP 表达盒和 MagnICON 系统整合酶[141]，充分验证了根癌农杆菌宿主菌株及其相关功能基因和载体特性。这些库的基因在 4 年内保持稳定[126]。同时还建立了野生型本氏烟主种子库。我们的试验表明，这种种子库的发芽率为99%，即使在4℃±2℃条件下保存4 年后，也能产生具有典型本氏烟形态的均匀幼苗（图

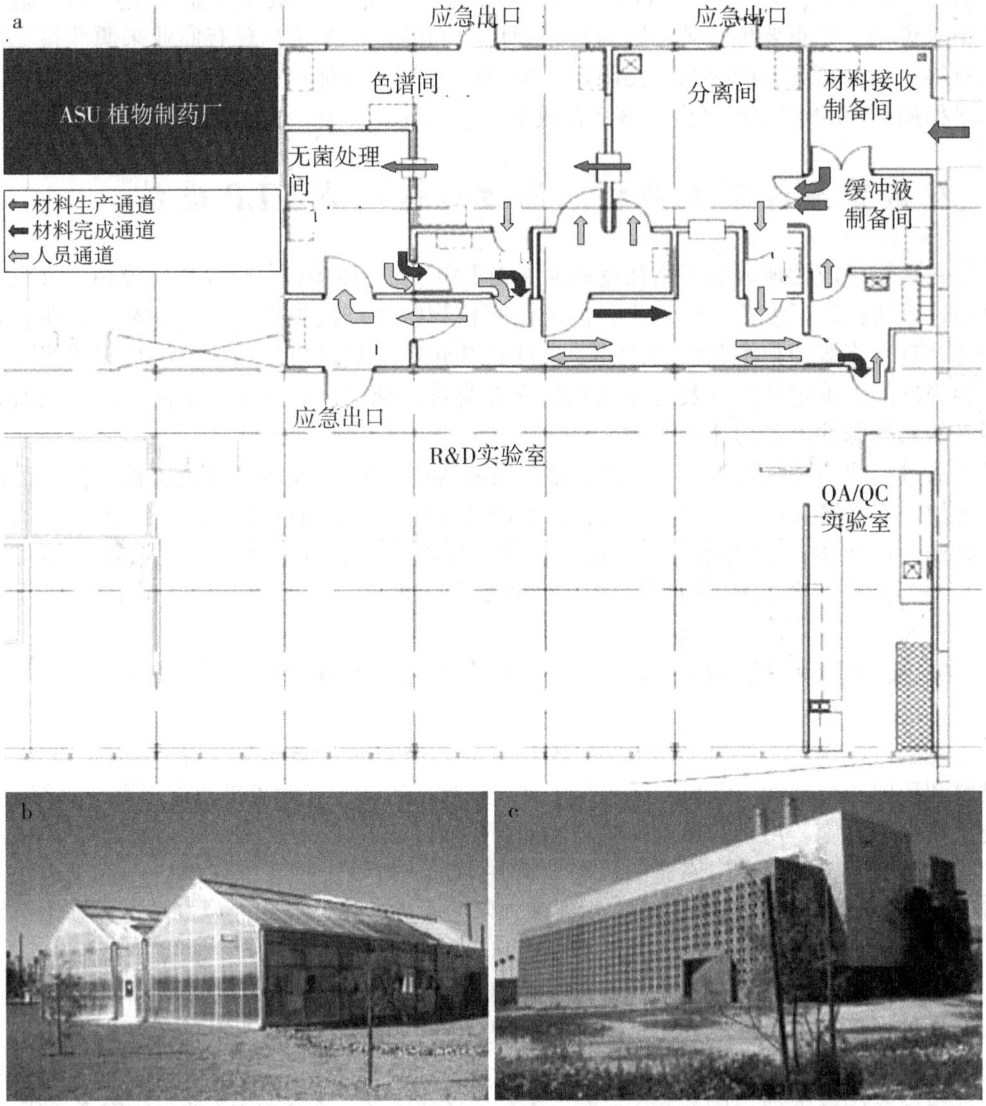

图 8.2　NVCP VLP 疫苗的 cGMP 生产设施

(a) 中央生物处理套间的设计。箭头表示过程中材料（绿色）、纯化的最终产品（红色）和人（黄色）的单向流动。(b) 用于植物生物量产生和 NVCP 表达的 BSL-2 温室设施。(c) 植物生物制药中心，包括中央生物加工套间、QA/QC 实验室和工艺开发实验室（经斯普林格科学+商业媒体的善意许可下改编：植物细胞报道，Lai 和 Chen[126]，图 1)。

8.3)[126]。这些结果表明，建立符合诸如病毒 VLPs cGMP 生产特性和稳定性鉴定标准的根癌农杆菌菌株和本氏烟种子主库和工作库是可行的。

8.9.2　中试规模生物量生产和渗透

在上游过程这一部分，目标是优化允许每平方米温室空间最大限度生产叶子生物量和 VLPs 的条件。我们小组和其他人研究表明，温度、光源和强度、植物年龄和叶片渗透后的培养时间都对生物量产生和 VLP 积累有显著影响[126]。例如，在自然光下生长的本氏烟比在人造光下生长的植物产生更多的生物量，但积累的诺瓦克病毒衣壳蛋白 VLPs 要少得多（图 8.4a)。此外，在自然光下生长的叶子在提取 VLP 的过程中也会产生更多的固体碎片，给 VLP 纯化带来问题[126]。数据还表明在 25℃ 下，16 小时光照/ 8 小时黑暗循环是在人工照明下产生生物量的最佳条

图 8.3　本研究生产条件下的生物量生成

野生型本氏烟在种植后 1 周（a）、3 周（b）、4 周（c）和 5 周（d）（经斯普林格科学+商业媒体的善意许可下改编：植物细胞报道，Lai 和 Chen[126]，图 2）

件[126]。当在这些条件下，生长本氏烟植物并在不同年龄对其进行诺瓦克病毒衣壳蛋白 VLP 表达取样时，发现 5 周的植物为 VLP 生产提供了最佳材料，因为已经积累了足够的生物量并产生了最高水平 VLP（图 8.4b）。较年轻植物产生的生物量不足，较老的植物太高而无法渗透，开始产生更多的次生代谢物使 VLP 纯化变得复杂。因此 5 周的本氏烟最大限度地平衡了生物量产生、过滤操作和 VLP 积累的综合需求。我们的数据表明在这些优化的条件下，在 5 周时间内，在 5.6m^2 的温室空间内，可以产生足够量的生物量，用于第一阶段临床试验的纯化诺瓦克病毒衣壳蛋白 VLPs[126]。通过垂直堆叠多层生长托盘，这类产品的水平空间需求可以显著降低。

还开发了通过真空渗透将农杆菌有效导入叶片的方法[126]。如绿色荧光蛋白（GFP）和诺瓦克病毒衣壳蛋白（NVCP）VLPs 的表达所示，基于这些方法，在放大条件下渗透的植物允许携带载体的农杆菌，在整个叶面积中有效渗透（图 8.4c），并在 7 dpi 的叶中积累最高水平的诺瓦克病毒衣壳蛋白 VLPs（图 8.4a，图 8.4b）。此外在这些放大条件下，诺瓦克病毒衣壳蛋白 VLPs 产量和时间表达模式类似于我们之前的小规模实验[126,133]。这些结果清楚地表明，植物中诺瓦克病毒 VLP 生产的上游工艺是可扩展的，可以进一步扩展用于商业生产。

8.9.3　诺如病毒 VLPs 下游处理

尽管食用未加工或部分加工的含诺如病毒 VLP 植物组织仍然是疫苗递送的可行选择，但由于监管要求，单位剂量明确纯化 VLPs 最有可能成为第一代成功的商业化候选疫苗[123,124]。因此，为了使植物生产的诺如病毒 VLP 疫苗成为商业现实，必须有一个强大且可扩展的下游流程，能够有效地回收和纯化 VLP。优化的下游工艺应该能够保持 VLPs 完整性，并以高回收率和最少的加工步骤生产高度纯化的 VLPs。

还应具有可扩展性、经济性，并符合 FDA 的 cGMP 法规。诺如病毒 VLP 纯化最常见的方法是密度梯度超离心法[106,108,133,144]。虽然这些方法对于描述 VLP 尺寸和组装特征很有用[107,133,144]，但不适用于商业 VLP 制造，因为很难扩大规模和耗费时间[143,145]。由于高渗梯度剂和严重的离心力通常会剪切 VLP，因此密度梯度超离心法在保持 VLPs 完整性方面也可能存在问题[146]。为了满足诺如病毒 VLP 加工更具可扩展性和强健的方法需求，一种新趋势是开始探索更复杂的方

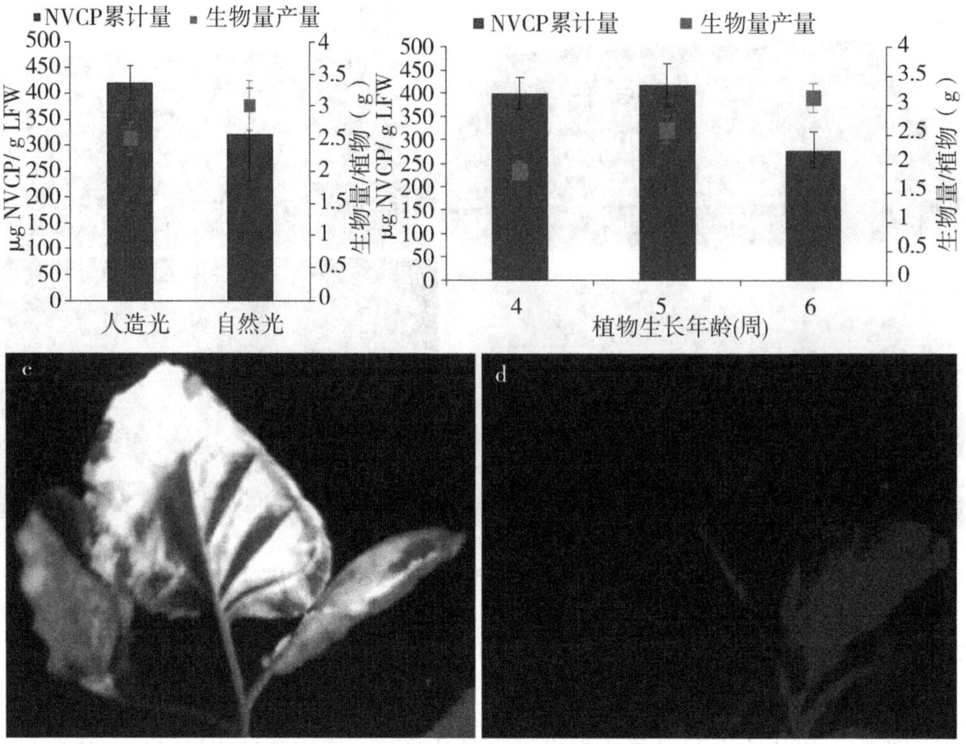

图 8.4 生物量和 NVCP VLP 积累的优化

(a) 在自然光和人造光下的生物量和 NVCP 产量。本氏烟植物在自然光或人造光下生长 5 周。通过称重和 ELISA 分别测量叶片生物量（绿色正方形）和 NVCP 表达水平（红色柱）。(b) 生物量产量和 NVCP 表达的时间模式。植物在人造光下生长 5 周，检查生物量产量（绿色正方形）和 NVCP 累积量（红色柱）。对于（a）和（b），给出了来自 3 个独立渗透实验的样品（$N>10$）的平均标准偏差（SD）。(c) 和（d）农业渗透叶中 GFP 表达的可视化。用携带 GFP 表达的 3 种 MagnICON 载体的 GV3101 培养物（c）或用渗透缓冲液作为阴性对照渗透本氏烟植物（d）。在紫外光下检查叶子并拍摄 7 dpi。展示了至少 3 个独立实验之一（经斯普林格科学+商业媒体的善意许可下改编：植物细胞报道，Lai 和 Chen[126]，图 3）

法，如过滤和色谱法[31]。例如我们团队最近开发了一个强大且可扩展的下游处理方案，用于从植物中回收诺瓦克病毒衣壳蛋白 VLPs（图 8.5）。

在这 3 个步骤中，植物提取物通过低 pH 沉淀、超滤和渗滤（UF/DF），以及切向流过滤（TFF）膜和离子交换（IEX）色谱法进行处理[126]。我们发现，低 pH 和 UF/DF 可以消除大多数植物宿主蛋白，IEX 色谱法可以将诺瓦克病毒衣壳蛋白 VLPs 纯化到 95%以上的纯度（图 8.6，第 5 道）[126]。不同规模的植物生物量生产运行表明，该下游工艺具有很高的可扩展性，能够持续生产高纯度、高回收率的诺瓦克病毒衣壳蛋白 VLPs[126]。与费时费力的梯度离心法相比，新方法更稳健，可扩展性更强，可以在 12 小时内完成，而不是几天[126,143]。重要的是，这一新工艺完全符合 cGMP 标准，并已为人类临床试验生产 cGMP 级 VLPs[31]。

微阵列等新技术也被用于进一步优化诺如病毒植物的 VLP 纯化[143]。例如我们实验室及其合作者已经从微阵列中鉴定出与诺瓦克病毒衣壳蛋白 VLP 具有特异亲和力的肽配体。当与层析珠结合时，这些亲和配体允许从本氏烟植物提取物中回收高度纯化的诺瓦克病毒衣壳蛋白 VLPs[31]。这种方法优于传统亲和色谱法，因为我们的亲和配体是完全合成的，对配体变性或降解不敏感。

基于上述优点，加之快速的配体发现过程和低成本的肽生产，将使其应用于大规模诺如病

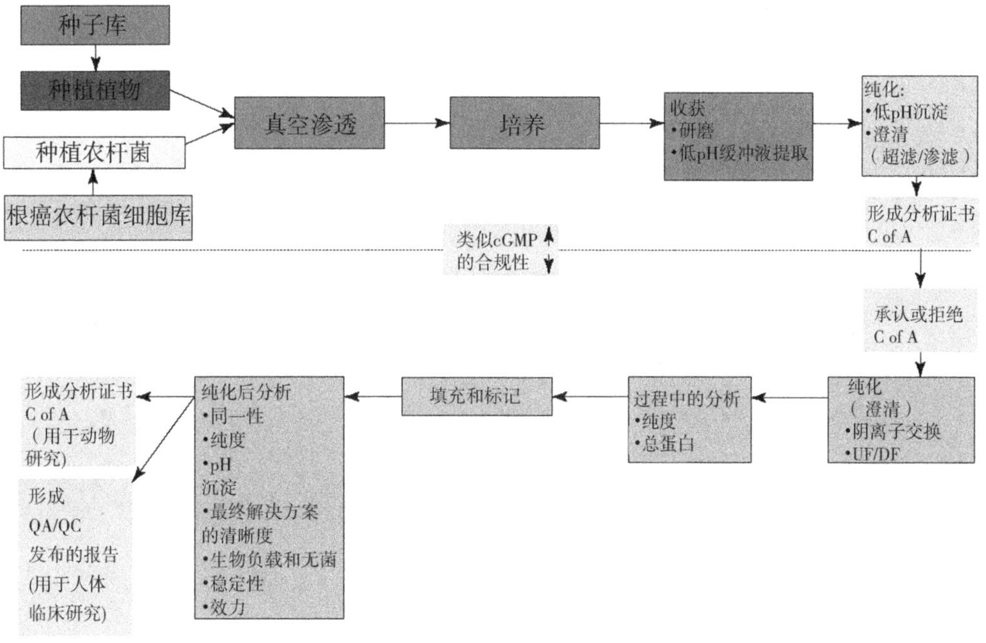

图 8.5　根据 cGMP 法规生产 NVCP VLP 疫苗的工艺流程

C of A 分析证书；*UF/DF* 超滤和渗滤（经斯普林格科学+商业媒体的善意许可下改编：植物细胞报道，Lai 和 Chen[126]，图 4）。

毒 VLP 生产。

8.9.4　植物源性诺如病毒 VLPs 质量控制

植物源疫苗商业化的其余挑战之一是缺乏在身份、纯度、效力和安全性方面符合监管机构要求的标准范例[123,124,127]。作为克服这一挑战努力的一部分，我们实验室已经确定并开发了分析测定法以监控过程中的样品，并确保最终 VLP 产品在身份、纯度、浓度、三级结构、功能性，以及对 cGMP 合规性至关重要的宿主污染分子浓度方面，符合人类药物释放的预设规范（表 8.2）。我们的分析表明，最终 VLP 产物的身份和组装得到确认，其纯度、浓度、外观、残留宿主 DNA 浓度和稳定性均符合昆虫细胞衍生的参考标准[126]。最终 VLP 产物在小鼠体内引发了强烈的诺瓦克病毒衣壳蛋白特异性全身、局部和远端黏膜免疫应答，证明了其效力[126]。这些结果提供了第一个来自符合预定发布规范的放大过程的植物源诺如病毒 VLPs 例子。

总之，我们实验室的生产研究已经成功开发了植物生物质产生、渗透和强劲的诺如病毒 VLP 积累上游工艺，以及从植物组织中高效回收 VLPs 的新型下游工艺[126]。此外这些工艺已经在 cGMP 法规下成功运行并生产出高质量的 VLPs，这些 VLPs 在身份、纯度、效力和安全性方面符合所有预设的发布规范[126]。因此，这些研究在学术背景下提供了第一个根据 cGMP 法规大规模生产植物源疫苗的先例，也是植物生产诺如病毒 VLP 疫苗成为商业现实的重要一步。

我们团队和合作者正在研究评估 cGMP 纯化的 VLPs 与各种佐剂共同递送时的全身免疫和黏膜免疫[31]。初步数据表明，与单独使用 VLP 的相比，使用几种佐剂的黏膜免疫引起更强的血清 IgG 和黏膜 IgA 应答[31]。我们预计来自 cGMP 工厂生产运行的纯化诺瓦克病毒衣壳蛋白 VLPs 和我们当前研究中定义的佐剂，将在不久的将来用于新的临床 I 期试验。

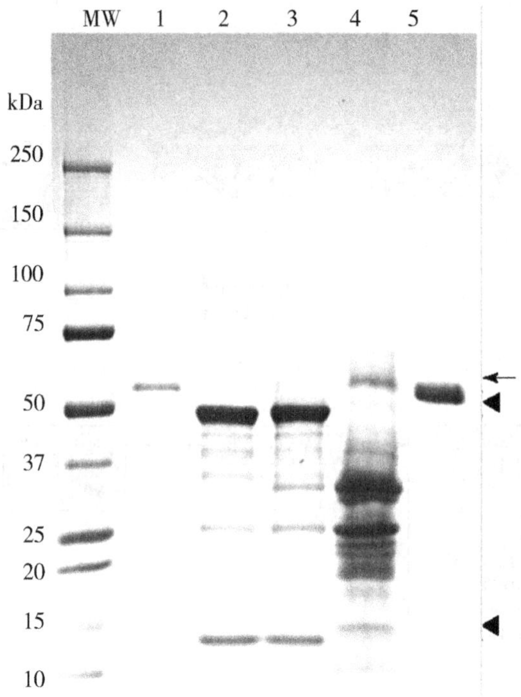

图 8.6 从本氏烟植物中纯化 NVCP

在还原条件下，在4%~20% SDS-PAGE 凝胶上纯化和分析叶蛋白提取物。泳道1昆虫细胞产生的 NVCP 作为参考标准；泳道2来自未过滤植物的澄清叶提取物；泳道3来自 NVCP 产生植物的澄清叶提取物；泳道4低 pH 沉淀上清液；泳道5来自 DEAE 阴离子交换层析的纯化植物源性 NVCP。◀：Rubisco 大小亚单位；← NVCP（经斯普林格科学+商业媒体的善意许可下改编：植物细胞报道，Lai 和 Chen[126]，图 5）。

表 8.2 NVCP VLP 进程样品和最终纯化产品的分析测定

产品属性	分析方法
产品标识	蛋白质印迹法
产品标识	具有自动化 Edman 测序的 N-末端序列
纯度和分子	SDS-PAGE /图谱考马斯和银染
蛋白质纯度和浓度	反相高压液相色谱（RP-HPLC）
蛋白质浓度	波长吸收 280 nm（A280）
蛋白质浓度	氨基酸分析（AAA）
Mass	ESI-TOF 质谱法（ESI-TOF MS）
四级结构	SEC HPLC 多角度激光散射（MALLS）
四级结构	蔗糖梯度离心，TEM
PI 测定	等电聚焦（IEF）-PAGE
聚糖分析	MALDI-TOF MS
内毒素	鲎变形细胞溶解物

（续表）

产品属性	分析方法
生物负荷	浑浊度和菌落形成
残留宿主 DNA	PicoGreen 荧光染料法和 PCR
小分子，尼古丁	LC-MS/MS
残留宿主细胞蛋白	多克隆抗体转野生型本氏烟植物蛋白

经斯普林格科学+商业媒体的善意许可下改编：植物细胞报道，Lai 和 Chen[126]，表 2。SDS-PAGE 十二烷基硫酸钠聚丙烯酰胺凝胶电泳，RP-HPLC 反相高压液相色谱，ESI-TOF 电喷雾电离飞行时间，TEM 透射电子显微镜，MALDI 基质辅助激光解吸电离，PCR 聚合酶链反应，LC-MS/MS 液相色谱-串联质谱。

8.9.5　用诺如病毒 VLPs 进行的人体临床试验

临床前研究成功地证明了强有力的全身和黏膜免疫原性，这导致了几项临床试验，以检验基于诺如病毒 VLP 疫苗在人体中的安全性和免疫原性（表 8.3）。

昆虫细胞源诺瓦克病毒衣壳蛋白 VLPs 最初在两个 I 期试验中进行了测试[144,147]。在第一个试验中，给予 20 名抗体阳性的成年志愿者两次口服剂量（第 1 天和第 21 天），即在水中配制无佐剂的 100 μg 或 250 μg VLPs。据观察血清 IgG 反应是剂量依赖性的，100%接种疫苗的受试者接受 250 μg VLPs 后，其诺如病毒特异性滴度至少增加了 4 倍[147]。18 名受试者中，有 15 名在接种第一次 VLP 剂量后对血清 IgG 滴度有应答，第二次剂量后没有增加[147]。重要的是，接种疫苗的志愿者没有观察到副作用[147]。

在第二次试验中，36 名年龄在 18～40 岁的成人血清阳性健康个体中，使用相同的免疫方案，在没有佐剂的情况下，用两次剂量递增的抗原（250 μg、500 μg 和 2 000 μg）进一步测试诺瓦克病毒衣壳蛋白 VLPs 的安全性和免疫原性[144]。在所有接种志愿者中，诺如病毒特异性 IgA 抗体分泌细胞（ASC）的数量显著增加，30%～40%的志愿者产生唾液、粪便或生殖器液 IgA 抗体。在接受 250 μg VLPs 志愿者中，90%的人检测到诺如病毒特异性 IgG 滴度增加[144]。然而 VLP 剂量的进一步增加（500 μg 和 2 000 μg），并没有增加血清转化率或 IgG 滴度[144]。

尽管前两次试验结果令人鼓舞，但口服免疫诱导的最大血清 IgG 滴度低于实验性诺如病毒感染后水平。因此进行了另外两项 I 期研究，研究是否可以通过使用黏膜佐剂进一步增强诺如病毒衣壳蛋白 VLPs 免疫原性[148]。昆虫细胞产生的诺如病毒衣壳蛋白 VLPs 配制成干粉，含有 TLR4 激动剂佐剂、单磷酰脂质 A（MPL）和黏着性壳聚糖，通过鼻内途径递送给健康受试者[148]。

表 8.3　NVCP VLP 疫苗 I 期人类临床试验

生产系统	志愿者人数	配方	剂量数和剂量范围	递送路线	免疫原性
昆虫细胞/杆状病毒	20	水，无佐剂	第 1 天和第 21 天两剂 100 ～ 250 μg	口服	血清 IgG 和 IgA 反应呈剂量依赖性，250 μg 组 100%志愿者的诺如病毒特异性滴度至少增加了 4 倍。在第一次 VLP 剂量后，83%的血清 IgG 滴度有反应，第二次剂量后没有增加[147]

（续表）

生产系统	志愿者人数	配方	剂量数和剂量范围	递送路线	免疫原性
昆虫细胞/杆状病毒	36	水，无佐剂	第 1 天和第 21 天 两次剂量分别为 250 μg、500 μg 和 2 000 μg	口服	在 250 μg 组中，90%的参与者检测到诺如病毒特异性 IgG 滴度增加。500 μg 和 2 000 μg 组的血清转化率或 IgG 滴度没有进一步增加。观察到 100%接种疫苗的志愿者中诺如病毒特异性 IgA ASC 数量增加，30%～40%的志愿者出现唾液、粪便或生殖器液 IgA 抗体[144]
昆虫细胞/杆状病毒	28	壳聚糖和 MPL 的干粉，MPL 作为佐剂	第 0 天和第 21 天 两次剂量分别为 5 μg、15 μg 和 50 μg	鼻内	血清 IgG 和 IgA 应答呈剂量依赖性，50 μg 组的滴度分别增加 4.7 倍和 4.5 倍。53%的受试者出现了诺如病毒特异性 IgA-ASCs 升高[148]
昆虫细胞/杆状病毒	61	壳聚糖和 MPL 的干粉，MPL 作为佐剂	第 0 天和第 21 天 两剂 50 μg 和 100 μg	鼻内	100 μg 组 63%和 79%的接种受试者的 IgG 和 IgA 滴度分别增加了 4.8 倍和 9.1 倍。100 μg 组的 IgG 和 IgA 滴度高于 50 μg 组，但无统计学差异。在 50 μg 和 100 μg 组中，所有接种过疫苗的个体都出现了 IgA-ASCs。在这些 ASCs 上检测到黏膜和周围淋巴组织的归巢分子[148]
转基因马铃薯	20	生马铃薯块[a]	150 g 马铃薯块茎块[a]含有 215～751 μg 抗原，第 0 天和第 21 天或在第 0 天、第 7 天和第 21 天 分 2～3 剂服用	口服	20%和 30%的接种志愿者产生了 NV 特异性血清 IgG（增加 12 倍）和粪便 IgA（增加 17 倍）。95%的受试者产生了诺如病毒 IgA ASCs[149]

IN intranasal，鼻内；*ASCs* antibody-secreting cells，抗体分泌细胞。

[a]每 150 g 生马铃薯块茎中含有 215～751 μg 可变量的诺瓦克病毒衣壳蛋白抗原。

研究 1 是一项 5 μg、15 μg 和 50 μg 诺瓦克病毒衣壳蛋白 VLPs 剂量逐步递增试验；研究 2 是两种最高剂量（50 μg 和 100 μg）VLPs 剂量比较研究。这些研究表明用单磷酰脂质 A（MPL）和壳聚糖以干粉配方递送诺瓦克病毒衣壳蛋白 VLPs 的耐受性良好，没有发生与疫苗相关的严重不良事件[148]。在研究 1 中，观察到 50 μg 组的剂量依赖性血清 IgG 和 IgA 反应，其滴度分别增加了 4.7 倍和 4.5 倍[148]；在研究 2 中，100 μg 组的 IgG 和 IgA 效价比 50 μg 组高（分别增加了 4.8 倍和 9.1 倍），但统计学上差异不显著[148]。

50 μg 和 100 μg 组中，所有疫苗接种者都产生 IgA 抗体分泌细胞（ASC）。此外在这些 ASCs 中检测到靶向黏膜和外周淋巴组织归巢分子表达。与先前描述的口服无佐剂 VLPs 相比，单磷酰脂质 A（MPL）作为佐剂的 VLPs 通过鼻内递送诱导了更高数量的 ASCs[148]。虽然对诺如病毒疾病的保护性免疫的相关性还不清楚，但是这些黏膜引发的 ASCs 与血清 IgG 和 IgA 抗体结合可能有助于保护。

一项临床Ⅰ/Ⅱ期试验已经开始，以评估由鼻内途径接种诺瓦克病毒衣壳蛋白 VLPs 的安全性和免疫原性，随后是一项活毒攻毒试验，以确定这种方法在预防或限制人类诺如病毒感染方面的有效性[103]。

为了证明植物源性诺瓦克病毒衣壳蛋白 VLPs 的安全性和有效性，用马铃薯生产的 NVCP VLPs 进行了临床 I 期试验。在该试验中 20 名人类受试者在第 0 天和第 7 天，或在第 0 天、第 7 天和第 21 天口服了 2～3 剂 150g 未煮熟诺瓦克病毒衣壳蛋白转基因马铃薯块茎（每剂 215～751 μg VLPs）。

除 1 名志愿者外，所有志愿者的诺如病毒特异性 IgA 抗体分泌细胞（ASC）增加，而大多数志愿者在第一次接种后都有反应。4 名志愿者产生了诺瓦克病毒衣壳蛋白特异性血清 IgG 应答，6 名志愿者发生了特异性肠道 IgA 应答，平均滴度分别提高了 12 倍和 17 倍[149]。食用重组或对照块茎的志愿者恶心、呕吐、轻度痉挛、发热或腹泻的发生率相似，表明摄入产生 VLP 的马铃薯块茎似乎是安全的[149]。

总之，这些结果表明了食用含 VLP 的植物部分作为人类诺如病毒口服疫苗的免疫原性和安全性。然而，整体抗体反应不如通过在昆虫细胞中产生、经口递送的纯化诺瓦克病毒衣壳蛋白 VLPs（每剂 250 μg）获得的抗体反应强[144]。可能是由于在马铃薯块茎中，诺瓦克病毒衣壳蛋白含量不一致和 VLP 装配不良引起的可变有效 VLP 剂量，导致这种弱抗原性。植物源性 VLP 效力也可能因其在肠腔中从马铃薯组织中不良释放而进一步降低。鉴于这一点以及最近使用昆虫细胞源性 VLP 进行的人类临床试验结果，在不久的将来，我们正在计划一项新的人类临床试验，使用来自我们 cGMP 运行的纯化诺瓦克病毒衣壳蛋白 VLPs 和我们当前研究中确定的佐剂（见上文质量控制部分）[126]。

总的来说，这些临床研究已经证明，在昆虫细胞和植物中生产的基于诺如病毒 VLP 候选疫苗，在人类中是安全和有免疫原性的。由于最近建立了人类志愿者诺如病毒感染模型[40,150]，因此，应在该人体攻毒模型中，进一步研究通过鼻内和口服接种含有佐剂的诺瓦克病毒衣壳蛋白 VLP 疫苗疗效。

8.10　诺如病毒 VLP 疫苗的优势和挑战

对于无法培养的诺如病毒，基于衣壳蛋白的 VLP 具有最佳的免疫原性、安全性、稳定性和可制造性，可作为预防诺如病毒胃肠炎的候选疫苗。与其他基于 VLP 的疫苗一样，由于其与感染性病毒的相似性，诺如病毒 VLP 已被证明能够在无佐剂的情况下诱导有效的细胞免疫应答和体液免疫应答。由于诺如病毒自然引起胃肠道感染，VLP 在低 pH 下稳定，对消化酶有抵抗力，因此可以口服，并在小鼠和人类中引起有效的黏膜抗体应答和全身应答。例如口服诺瓦克病毒衣壳蛋白 VLPs，可诱导外周血单核细胞产生大量的抗诺如病毒血清和肠道抗体以及 IF-γ[144]。

类似地通过鼻内递送，小鼠和人类产生了强有力的全身免疫反应和黏膜免疫反应。通过黏膜递送，诱导诺如病毒特异性肠黏膜免疫的能力，提供了 VLPs 作为这些肠道致病性病毒疫苗的显著优势。此外在几项人体临床试验中，已经证明诺如病毒 VLPs 的安全性。诺如病毒 VLPs 稳定性也有利于作为商业疫苗，因为可以冷冻干燥或在 4℃ 的简单缓冲液中保存多年而不会降解[73]。值得注意的是，我们小组和其他人研究表明，诺如病毒 VLPs 可以通过重组技术，在植物中低成本大规模生产[31]。

我们小组开发了一种新的下游工艺，成功生产了符合所有监管部门发布的关于诺瓦克病毒衣壳蛋白 VLPs 规范的疫苗，这些规范包括同一性、纯度、效力和安全性，为诺如病毒 VLP 疫苗成为商业现实提供了一个额外选项[126]。这些成功证明了 VLPs 作为诺如病毒疫苗的潜力和可行性，表明可以作为开发具有非复制抗原黏膜免疫有效策略的良好模型。

开发有效的诺如病毒疫苗仍然存在挑战。例如基于 VLP 的疫苗能否保护人类免受诺如活病毒攻击的问题仍然没有答案。此外，由于缺乏对保护免疫相关性的全面理解，很难预测新候选疫苗能够提供的保护水平。由于缺乏体外培养诺如病毒的培养系统和诺如病毒发病机制的小动物模型，这一问题进一步加剧。此外，病毒不同基因型和基因群之间缺乏完全的交叉保护，以及新变异株的快速进化进一步阻碍了针对多种诺如病毒毒株有效疫苗的开发。

总的来说，有效的全身免疫原性和黏膜免疫原性、植物中稳健和低成本 cGMP 可制造性以及人类临床试验中的安全性，都支持基于 VLP 的疫苗进一步开发，以预防与诺如病毒相关的胃肠炎。由于不同诺如病毒毒株中衣壳蛋白的多样性及其快速的抗原漂移，疫苗的开发应该集中在多价 VLP 疫苗上，这些疫苗来源于优势流行毒株的衣壳蛋白。此外流行病学监测必须与诺如病毒疫苗生产相结合，以确定这一移动目标，并在配方中包括优势流行循环菌株以获得最佳保护。正在进行和计划中的人类临床试验新数据特别是攻毒试验，应该会在未来几年中揭示 VLPs 在预防或限制诺如病毒感染方面的功效。它们也可能为免疫相关的保护提供线索。这一知识将有助于开发有效的诺如病毒疫苗。

经过 25 年的积极研究和开发，对以植物为生产平台的持续批评是美国缺乏疫苗的原因[125]。令人兴奋的是，最近 FDA 批准了一种植物生产的葡萄糖脑苷酶（商品名：ELEYSO™）治疗高雪病，从而克服了最后一个障碍，预示着植物制药学领域的新纪元[31]。我们推测基于植物的生产系统将为诺如病毒 VLP 生产提供卓越的可扩展性、安全性、时间和经济效益。

致谢

感谢 Huafang Lai，Junyun He，Gary Morris，Bingying Jiang，Linh Nguyen，Julie Featherston，Kadie Gavan，Darin Stone，Yihuang Lu 以及我们实验室的其他成员对植物材料生产、叶子渗透、蛋白质分析和显微镜检查做出的贡献。感谢 Joseph Caspermeyer 对手稿的批判性阅读。这项研究得到了美国国立过敏与传染病研究所（NIAID）给 QC 的 1U01AI075549 号拨款的部分支持。

参考文献

[1] Clark，B.，McKendrick，M.：A review of viral gastroenteritis. Curr. Opin. Infect. Dis. 17，461－469（2004）.

[2] Kapikian，A. Z.：Overview of gastroenteritis. Arch. Virol. Suppl. 12，7-19（1996）.

[3] Greenberg，H. B.，et al.：Role of Norwalk virus in outbreaks of nonbacterial gastroenteritis. J. Infect. Dis. 139，564-568（1979）.

[4] Patel，M. M.，Hall，A. J.，Vinje，J.，Parashar，U. D.：Noroviruses：a comprehensive review. J. Clin. Virol. 44，1-8（2009）.

[5] Green，K. Y.，Chanock，R. M.，Kapikian，A. Z.：Caliciviridae：The Noroviruses. In Howley，P. M.，Knipe，D. M.（eds.）Fields virology，Ch. 28，pp. 949-979. Lippincott，Williams & Wilkins，（2007）.

[6] Dolin，R.，et al.：Biological properties of Norwalk agent of acute infectious nonbacterial gastroenteritis. Proc. Soc. Exp. Biol. Med. 140，578-583（1972）.

[7] Kapikian，A. Z.，et al.：Visualization by immune electron microscopy of a 27-nm particle associated with acute infectious nonbacterial gastroenteritis. J. Virol. 10，1075-1081（1972）.

[8] Chen，R.，Neill，J. D.，Estes，M. K.，Prasad，B. V.：X-ray structure of a native calicivirus：structural insights into antigenic diversity and host specificity. Proc. Natl. Acad. Sci. U. S. A. 103，8048-8053（2006）.

[9] Duizer，E.，et al.：Laboratory efforts to cultivate noroviruses. J. Gen. Virol. 85，79-87（2004）.

[10] Green，S. M.，Lambden，P. R.，Caul，E. O.，Ashley，C. R.，Clarke，I. N.：Capsid diversity in small round-structured viruses：molecular characterization of an antigenically distinct human enteric calicivirus. Virus Res. 37，271-283（1995）.

[11] Lewis，D.，Ando，T.，Humphrey，C. D.，Monroe，S. S.，Glass，R. I.：Use of solid-phase immune electron microscopy for classification of Norwalk-like viruses into six antigenic groups from 10 outbreaks of gastroenteritis in the United States. J. Clin. Microbiol. 33，501-504（1995）.

[12] Okada, S., et al.: Antigenic characterization of small, round-structured viruses by immune electron microscopy. J. Clin. Microbiol. 28, 1244-1248 (1990).

[13] Wyatt, R. G., et al.: Comparison of three agents of acute infectious nonbacterial gastroenteritis by cross-challenge in volunteers. J. Infect. Dis. 129, 709-714 (1974).

[14] Ando, T., Noel, J. S., Fankhauser, R. L.: Genetic classification of Norwalk-like viruses. J. Infect. Dis. 181 (Suppl 2), S336-S348 (2000).

[15] Zheng, D. P., et al.: Norovirus classification and proposed strain nomenclature. Virology 346, 312-323 (2006).

[16] Scipioni, A., Mauroy, A., Vinje, J., Thiry, E.: Animal noroviruses. Vet. J. 178, 32-45 (2008).

[17] Zheng, D. P., Widdowson, M. A., Glass, R. I., Vinje, J.: Molecular epidemiology of genogroup II-genotype 4 noroviruses in the United States between 1994 and 2006. J. Clin. Microbiol. 48, 168-177 (2010).

[18] Lindesmith, L. C., et al.: Mechanisms of GII. 4 norovirus persistence in human populations. PLoS Med. 5, e31 (2008).

[19] Jiang, X., Wang, M., Wang, K., Estes, M. K.: Sequence and genomic organization of Norwalk virus. Virology 195, 51-61 (1993).

[20] Hardy, M. E.: Norovirus protein structure and function. FEMS Microbiol. Lett. 253, 1-8 (2005).

[21] Chaudhry, Y., et al.: Caliciviruses differ in their functional requirements for eIF4F components. J. Biol. Chem. 281, 25315-25325 (2006).

[22] Jiang, X., Graham, D. Y., Wang, K. N., Estes, M. K.: Norwalk virus genome cloning and characterization. Science 250, 1580-1583 (1990).

[23] Prasad, B. V., et al.: X-ray crystallographic structure of the Norwalk virus capsid. Science 286, 287-290 (1999).

[24] Bertolotti-Ciarlet, A., Crawford, S. E., Hutson, A. M., Estes, M. K.: The 3′ end of Norwalk virus mRNA contains determinants that regulate the expression and stability of the viral capsid protein VP1: a novel function for the VP2 protein. J. Virol. 77, 11603-11615 (2003).

[25] Chakravarty, S., Hutson, A. M., Estes, M. K., Prasad, B. V.: Evolutionary trace residues in noroviruses: importance in receptor binding, antigenicity, virion assembly, and strain diversity. J. Virol. 79, 554-568 (2005).

[26] Tan, M., Hegde, R. S., Jiang, X.: The P domain of norovirus capsid protein forms dimer and binds to histo-blood group antigen receptors. J. Virol. 78, 6233-6242 (2004).

[27] Lochridge, V. P., Jutila, K. L., Graff, J. W., Hardy, M. E.: Epitopes in the P2 domain of norovirus VP1 recognized by monoclonal antibodies that block cell interactions. J. Gen. Virol. 86, 2799-2806 (2005).

[28] Nilsson, M., et al.: Evolution of human calicivirus RNA in vivo: accumulation of mutations in the protruding P2 domain of the capsid leads to structural changes and possibly a new phenotype. J. Virol. 77, 13117-13124 (2003).

[29] Glass, R. I., Parashar, U. D., Estes, M. K.: Norovirus gastroenteritis. N. Engl. J. Med. 361, 1776-1785 (2009).

[30] Beersma, M. F., et al.: Unrecognized norovirus infections in health care institutions and their clinical impact. J. Clin. Microbiol. 50, 3040-3045 (2012).

[31] Chen, Q., Lai, H.: Plant-derived virus-like particles as vaccines. Hum. Vaccin. Immunother. 9 (2013).

[32] Teunis, P. F., et al.: Norwalk virus: Virus-like particle (VLP) Vaccines how infectious is it? J. Med. Virol. Marcel Dekker80, 1468-1476 (2008).

[33] Estes, M. K.: Levine, M. M. (ed.) New generation vaccines, pp. 283-294. Academic Press (2004).

[34] Dolin, R.: Noroviruses-challenges to control. N. Engl. J. Med. 357, 1072-1073 (2007).

[35] Bresee, J. S., Widdowson, M. A., Monroe, S. S., Glass, R. I.: Foodborne viral gastroenteritis: challenges and opportunities. Clin. Infect. Dis. 35, 748-753 (2002).

[36] Fankhauser, R. L., et al.: Epidemiologic and molecular trends of "Norwalk-like viruses" associated with outbreaks of gastroenteritis in the United States. J. Infect. Dis. 186, 1-7 (2002).

[37] Widdowson, M. A., et al.: Outbreaks of acute gastroenteritis on cruise ships and on land: identification of a predominant circulating strain of norovirus-United States, 2002. J. Infect. Dis. 190, 27-36 (2004).

[38] Rockx, B., et al.: Natural history of human calicivirus infection: a prospective cohort study. Clin. Infect. Dis. 35, 246-253 (2002).

[39] Anderson, A. D., et al.: Multistate outbreak of Norwalk-like virus gastroenteritis associated with a common caterer. Am. J. Epidemiol. 154, 1013-1019 (2001).

[40] Atmar, R. L., et al.: Norwalk virus shedding after experimental human infection. Emerg. Infect. Dis. 14, 1553-1557 (2008). doi: 10. 3201/eid1410. 080117.

[41] Siebenga, J. J., et al.: High prevalence of prolonged norovirus shedding and illness among hospitalized patients: a model for in vivo molecular evolution. J. Infect. Dis. 198, 994-1001 (2008).

[42] Patterson, T., Hutchings, P., Palmer, S.: Outbreak of SRSV gastroenteritis at an international conference traced to food handled by a post-symptomatic caterer. Epidemiol. Infect. 111, 157-162 (1993).

[43] Khanna, N., Goldenberger, D., Graber, P., Battegay, M., Widmer, A. F.: Gastroenteritis outbreak with norovirus in a Swiss university hospital with a newly identified virus strain. J. Hosp. Infect. 55, 131-136 (2003).

[44] Nygard, K., et al.: Emerging genotype (GGIIb) of norovirus in drinking water, Sweden. Emerg. Infect. Dis. 9, 1548-1552 (2003).

[45] Hansen, S., et al.: Closure of medical departments during nosocomial outbreaks: data from a systematic analysis of the literature. J. Hosp. Infect. 65, 348-353 (2007).

[46] Mounts, A. W., et al.: Cold weather seasonality of gastroenteritis associated with Norwalk-like viruses. J. Infect. Dis. 181 (Suppl 2), S284-S287 (2000).

[47] Lopman, B., et al.: Increase in viral gastroenteritis outbreaks in Europe and epidemic spread of new norovirus variant. Lancet 363, 682-688 (2004).

[48] Marshall, J. A., et al.: Incidence and characteristics of endemic Norwalk-like virus-associated gastroenteritis. J. Med. Virol. 69, 568-578 (2003).

[49] Nordgren, J., Bucardo, F., Dienus, O., Svensson, L., Lindgren, P. E.: Novel light-upon-extension real-time PCR assays for detection and quantification of genogroup I and II noroviruses in clinical specimens. J. Clin. Microbiol. 46, 164-170 (2008).

[50] Gallimore, C. I., Iturriza-Gomara, M., Xerry, J., Adigwe, J., Gray, J. J.: Inter-seasonal diversity of norovirus genotypes: emergence and selection of virus variants. Arch. Virol. 152, 1295-1303 (2007).

[51] Harrington, P. R., Vinje, J., Moe, C. L., Baric, R. S.: Norovirus capture with histo-blood group antigens reveals novel virus-ligand interactions. J. Virol. 78, 3035-3045 (2004).

[52] Hutson, A. M., Atmar, R. L., Estes, M. K.: Norovirus disease: changing epidemiology and host susceptibility factors. Trends Microbiol. 12, 279-287 (2004).

[53] Meyer, E., Ebner, W., Scholz, R., Dettenkofer, M., Daschner, F. D.: Nosocomial outbreak of norovirus gastroenteritis and investigation of ABO histo-blood group type in infected staff and patients. J. Hosp. Infect. 56, 64-66 (2004).

[54] Roldao, A., Mellado, M. C., Castilho, L. R., Carrondo, M. J., Alves, P. M.: Virus-like particles in vaccine development. Expert Rev. Vaccines 9, 1149-1176 (2010).

[55] Chackerian, B.: Virus-like particles: flexible platforms for vaccine development. Expert Rev. Vaccines 6, 381-390 (2007).

[56] Tan, M., et al.: Mutations within the P2 domain of norovirus capsid affect binding to human histo-blood group antigens: evidence for a binding pocket. J. Virol. 77, 12562-12571 (2003).

[57] Harrington, P. R., Lindesmith, L., Yount, B., Moe, C. L., Baric, R. S.: Binding of Norwalk virus-like particles to ABHhisto-blood group antigens is blocked by antisera from infected human volunteers or experimentally vaccinated mice. J. Virol. 76, 12335-12343 (2002).

[58] Rockx, B. H., Vennema, H., Hoebe, C. J., Duizer, E., Koopmans, M. P.: Association of histo-blood group antigens and susceptibility to norovirus infections. J. Infect. Dis. 191, 749-754 (2005).

[59] Kovacsovics-Bankowski, M., Clark, K., Benacerraf, B., Rock, K. L.: Efficient major histocompatibility

complex class I presentation of exogenous antigen upon phagocytosis by macrophages. Proc. Natl. Acad. Sci. U. S. A. 90, 4942-4946 (1993).

[60] Bachmann, M. F., et al.: Dendritic cells process exogenous viral proteins and virus-like particles for class I presentation to $CD8^+$cytotoxic T lymphocytes. Eur. J. Immunol. 26, 2595-2600 (1996).

[61] Westhoff, T. H., et al.: Chronic norovirus infection in renal transplant recipients. Nephrol. Dial. Transplant. 24, 1051-1053 (2009).

[62] Gallimore, C. I., Cubitt, D., du Plessis, N., Gray, J. J.: Asymptomatic and symptomatic excretion of noroviruses during a hospital outbreak of gastroenteritis. J. Clin. Microbiol. 42, 2271-2274 (2004).

[63] Asanaka, M., et al.: Replication and packaging of Norwalk virus RNA in cultured mammalian cells. Proc. Natl. Acad. Sci. U. S. A. 102, 10327-10332 (2005).

[64] Schreiber, D. S., Blacklow, N. R., Trier, J. S.: The small intestinal lesion induced by Hawaii agent acute infectious nonbacterial gastroenteritis. J. Infect. Dis. 129, 705-708 (1974). doi: 10. 1093/ infdis/ 129. 6. 705.

[65] Schreiber, D. S., Blacklow, N. R., Trier, J. S.: The mucosal lesion of the proximal small intestine in acute infectious nonbacterial gastroenteritis. N. Engl. J. Med. 288, 1318-1323 (1973). doi: 10. 1056/NEJM197306212882503.

[66] Troeger, H., et al.: Structural and functional changes of the duodenum in human norovirus infection. Gut 58, 1070-1077 (2009).

[67] Marionneau, S., et al.: Norwalk virus binds to histoblood group antigens present on gastroduodenal epithelial cells of secretor individuals. Gastroenterology 122, 1967-1977 (2002).

[68] Harris, J. P., Edmunds, W. J., Pebody, R., Brown, D. W., Lopman, B. A.: Deaths from norovirus among the elderly, England and Wales. Emerg. Infect. Dis. 14, 1546-1552 (2008).

[69] Turcios-Ruiz, R. M., et al.: Outbreak of necrotizing enterocolitis caused by norovirus in a neonatal intensive care unit. J. Pediatr. 153, 339-344 (2008).

[70] Patel, M. M., et al.: Systematic literature review of role of noroviruses in sporadic gastroenteritis. Emerg. Infect. Dis. 14, 1224-1231 (2008).

[71] Donaldson, E. F., Lindesmith, L. C., Lobue, A. D., Baric, R. S.: Norovirus pathogenesis: mechanisms of persistence and immune evasion in human populations. Immunol. Rev. 225, 190-211 (2008).

[72] Koopmans, M.: Progress in understanding norovirus epidemiology. Curr. Opin. Infect. Dis. 21, 544-552 (2008).

[73] Jansen, K. U., Conner, M. E., Estes, M. K.: Virus-Like Particles as Vaccines and Vaccine Delivery Systems. In Levine, M. M. (ed.) New generation vaccines, Ch. 29, pp. 298-305. Informa Healthcare USA, Inc. (2009).

[74] Gallimore, C. I., et al.: Methods for the detection and characterisation of noroviruses associated with outbreaks of gastroenteritis: outbreaks occurring in the north-west of England during two norovirus seasons. J. Med. Virol. 73, 280-288 (2004).

[75] Biel, S. S., Gelderblom, H. R.: Diagnostic electron microscopy is still a timely and rewarding method. J. Clin. Virol. 13, 105-119 (1999).

[76] Richards, A. F., et al.: Evaluation of a commercial ELISA for detecting Norwalk-like virus antigen in faeces. J. Clin. Virol. 26, 109-115 (2003).

[77] Atmar, R. L., Estes, M. K.: The epidemiologic and clinical importance of norovirus infection. Gastroenterol. Clin. North Am. 35, 275-290 (2006), viii.

[78] Rabenau, H. F., et al.: Laboratory diagnosis of norovirus: which method is the best? Inter-virology 46, 232-238 (2003).

[79] Gray, J. J., et al.: European multicenter evaluation of commercial enzyme immunoassays for detecting norovirus antigen in fecal samples. Clin. Vaccine Immunol. 14, 1349-1355 (2007).

[80] de Bruin, E., Duizer, E., Vennema, H., Koopmans, M. P.: Diagnosis of Norovirus outbreaks by commercial ELISA or RT-PCR. J. Virol. Methods 137, 259-264 (2006).

[81] Moore, C., et al.: Evaluation of a broadly reactive nucleic acid sequence based amplification assay for the detection of noroviruses in faecal material. J. Clin. Virol. 29, 290-296 (2004).

[82] Amar, C. F., et al.: Detection by PCR of eight groups of enteric pathogens in 4, 627 faecal samples: re-examination of the English case-control Infectious Intestinal Disease Study (1993-1996). Eur. J. Clin. Microbiol. Infect. Dis. 26, 311-323 (2007).

[83] Vinje, J., et al.: International collaborative study to compare reverse transcriptase PCR assays for detection and genotyping of noroviruses. J. Clin. Microbiol. 41, 1423-1433 (2003).

[84] Fankhauser, R. L., et al.: Epidemiologic and molecular trends of "Norwalk-like viruses" associated with outbreaks of gastroenteritis in the United States. J. Infect. Dis. 186, 1-7 (2002). doi: 10. 1086/341085.

[85] Medici, M. C., et al.: Broadly reactive nested reverse transcription-PCR using an internal RNA standard control for detection of noroviruses in stool samples. J. Clin. Microbiol. 43, 3772-3778 (2005). doi: 10. 1128/jcm. 43. 8. 3772-3778. 2005.

[86] Trujillo, A. A., et al.: Use of TaqMan real-time reverse transcription-PCR for rapid detection, quantification, and typing of norovirus. J. Clin. Microbiol. 44, 1405-1412 (2006).

[87] Gunson, R. N., Carman, W. F.: Comparison of two real-time PCR methods for diagnosis of norovirus infection in outbreak and community settings. J. Clin. Microbiol. 43, 2030-2031 (2005).

[88] Dowell, S. F., et al.: A multistate outbreak of oyster-associated gastroenteritis: implications for interstate tracing of contaminated shellfish. J. Infect. Dis. 171, 1497-1503 (1995). doi: 1 0. 1093/ infdis/171. 6. 1497.

[89] Parashar, U. D., et al.: An outbreak of viral gastroenteritis associated with consumption of sandwiches: implications for the control of transmission by food handlers. Epidemiol. Infect. 121, 615-621 (1998).

[90] Fisman, D. N., Greer, A. L., Brouhanski, G., Drews, S. J.: Of gastro and the gold standard: evaluation and policy implications of norovirus test performance for outbreak detection. J. Transl. Med. 7, 23 (2009).

[91] Hansman, G. S., et al.: Structural basis for norovirus inhibition and fucose mimicry by citrate. J. Virol. 86, 284-292 (2012). doi: 10. 1128/jvi. 05909-11.

[92] Parrino, T. A., Schreiber, D. S., Trier, J. S., Kapikian, A. Z., Blacklow, N. R.: Clinical immunity in acute gastroenteritis caused by Norwalk agent. N. Engl. J. Med. 297, 86-89 (1977).

[93] Johnson, P. C., Mathewson, J. J., DuPont, H. L., Greenberg, H. B.: Multiple-challenge study of host susceptibility to Norwalk gastroenteritis in US adults. J. Infect. Dis. 161, 18-21 (1990).

[94] Matsui, S. M., Greenberg, H. B.: Immunity to calicivirus infection. J. Infect. Dis. 181 (Suppl 2), S331-S335 (2000).

[95] Lindesmith, L., et al.: Human susceptibility and resistance to Norwalk virus infection. Nat. Med. 9, 548-553 (2003).

[96] Fifis, T., et al.: Size-dependent immunogenicity: therapeutic and protective properties of nano-vaccines against tumors. J. Immunol. 173, 3148-3154 (2004).

[97] Bachmann, M. F., et al.: The influence of antigen organization on B cell responsiveness. Science 262, 1448-1451 (1993).

[98] Bachmann, M. F., Zinkernagel, R. M.: The influence of virus structure on antibody responses and virus serotype formation. Immunol. Today 17, 553-558 (1996).

[99] Fehr, T., Skrastina, D., Pumpens, P., Zinkernagel, R. M.: T cell - independent type I antibody response against B cell epitopes expressed repetitively on recombinant virus particles. Proc. Natl. Acad. Sci. U. S. A. 95, 9477-9481 (1998).

[100] Bachmann, M. F., Zinkernagel, R. M.: Neutralizing antiviral B cell responses. Annu. Rev. Immunol. 15, 235-270 (1997).

[101] Santi, L., Huang, Z., Mason, H.: Virus-like particles production in green plants. Methods 40, 66-76 (2006).

[102] Chen, Q., et al.: Subunit Vaccines Producing Using Plant Biotechnology. In Levine M. M. (ed.) New generation vaccines, Ch. 30, pp. 306-315. Informa Healthcare USA, Inc. (2009).

[103] Herbst-Kralovetz, M., Mason, H. S., Chen, Q.: Norwalk virus-like particles as vaccines. Expert Rev. Vaccines 9, 299-307 (2010). doi: 10. 1586/erv. 09. 163.

[104] Prasad, B. V., Rothnagel, R., Jiang, X., Estes, M. K.: Three-dimensional structure of baculovirus-expressed Norwalk virus capsids. J. Virol. 68, 5117-5125 (1994).

[105] Baric, R. S., et al.: Expression and self-assembly of Norwalk virus capsid protein from Venezuelan equine encephalitis virus replicons. J. Virol. 76, 3023-3030 (2002).

[106] Jiang, X., Wang, M., Graham, D. Y., Estes, M. K.: Expression, self-assembly, and antigenicity of the Norwalk virus capsid protein. J. Virol. 66, 6527-6532 (1992).

[107] Ausar, S. F., Foubert, T. R., Hudson, M. H., Vedvick, T. S., Middaugh, C. R.: Conformational stability and disassembly of Norwalk virus like particles: effect of pH and temperature. J. Biol. Chem. 281, 19478-19488 (2006). doi: 10. 1074/jbc. M603313200.

[108] Prasad, B. V. V., Hardy, D., Estes, M.: Structural studies of recombinant Norwalk capsids. J. Infect. Dis. 181, S317-S321 (2000). doi: 10. 1086/315576.

[109] Tan, M., Jiang, X.: The p domain of norovirus capsid protein forms a subviral particle that binds to histo-blood group antigen receptors. J. Virol. 79, 14017-14030 (2005).

[110] Tan, M., Zhong, W., Song, D., Thornton, S., Jiang, X.: E. coli-expressed recombinant norovirus capsid proteins maintain authentic antigenicity and receptor binding capability. J. Med. Virol. 74, 641-649 (2004).

[111] Tan, M., Meller, J., Jiang, X.: C-terminal arginine cluster is essential for receptor binding of norovirus capsid protein. J. Virol. 80, 7322-7331 (2006).

[112] Cao, S., et al.: Structural basis for the recognition of blood group trisaccharides by norovirus. J. Virol. 81, 5949-5957 (2007).

[113] Pushko, P., et al.: Replicon-helper systems from attenuated Venezuelan equine encephalitis virus: expression of heterologous genes in vitro and immunization against heterologous pathogens in vivo. Virology 239, 389-401 (1997).

[114] Guerrero, R. A., et al.: Recombinant Norwalk virus-like particles administered intranasally to mice induce systemic and mucosal (fecal and vaginal) immune responses. J. Virol. 75, 9713-9722 (2001).

[115] LoBue, A. D., et al.: Multivalent norovirus vaccines induce strong mucosal and systemic blocking antibodies against multiple strains. Vaccine 24, 5220-5234 (2006).

[116] Pushko, P., et al.: Recombinant RNA replicons derived from attenuated Venezuelan equine encephalitis virus protect guinea pigs and mice from Ebola hemorrhagic fever virus. Vaccine 19, 142-153 (2000).

[117] Harrington, P. R., et al.: Systemic, mucosal, and heterotypic immune induction in mice inoculated with Venezuelan equine encephalitis replicons expressing Norwalk virus-like particles. J. Virol. 76, 730-742 (2002).

[118] LoBue, A. D., Lindesmith, L. C., Baric, R. S.: Identification of cross-reactive norovirus $CD4^+$ T cell epitopes. J. Virol. 84, 8530-8538 (2010).

[119] LoBue, A. D., Thompson, J. M., Lindesmith, L., Johnston, R. E., Baric, R. S.: Alphavirus-adjuvanted norovirus-like particle vaccines: heterologous, humoral, and mucosal immune responses protect against murine norovirus challenge. J. Virol. 83, 3212-3227 (2009).

[120] Davis, N. L., Brown, K. W., Johnston, R. E.: A viral vaccine vector that expresses foreign genes in lymph nodes and protects against mucosal challenge. J. Virol. 70, 3781-3787 (1996).

[121] MacDonald, G. H., Johnston, R. E.: Role of dendritic cell targeting in Venezuelan equine encephalitis virus pathogenesis. J. Virol. 74, 914-922 (2000).

[122] Thompson, J. M., et al.: Mucosal and systemic adjuvant activity of alphavirus replicon particles. Proc. Natl. Acad. Sci. U. S. A. 103, 3722-3727 (2006).

[123] Chen, Q.: Expression and purification of pharmaceutical proteins in plants. Biol. Eng. 1, 291-321 (2008).

[124] Chen, Q.: Mou, B., Scorza, R. (eds.) Transgenic horticultural crops: Expression and Manufacture of Pharmaceutical Proteins in Genetically Engineered Horticultural Plants challenges and opportunities-essays

by experts, Ch. 4, pp. 83-124. Taylor & Francis (2011).

[125] Chen, Q.: Turning a new leaf. European Biopharm. Rev. 2, 64-68 (2011).

[126] Lai, H., Chen, Q.: Bioprocessing of plant-derived virus-like particles of Norwalk virus capsid protein under current Good Manufacture Practice regulations. Plant Cell Rep. 31, 573-584 (2012).

[127] Faye, L., Gomord, V.: Success stories in molecular farming-a brief overview. Plant Biotechnol. J. 8, 525-528 (2010).

[128] Lai, H., et al.: Monoclonal antibody produced in plants efficiently treats West Nile virus infection in mice. Proc. Natl. Acad. Sci. U. S. A. 107, 2419-2424 (2010). doi: 10. 1073/pnas. 0914503107.

[129] Lai, H., He, J., Engle, M., Diamond, M. S., Chen, Q.: Robust production of virus-like particles and monoclonal antibodies with geminiviral replicon vectors in lettuce. Plant Biotechnol. J. 10, 95-104 (2012).

[130] Komarova, T. V., et al.: Transient expression systems for plant-derived biopharmaceuticals. Expert Rev. Vaccines 9, 859-876 (2010). doi: 10. 1586/erv. 10. 85.

[131] Lico, C., Chen, Q., Santi, L.: Viral vectors for production of recombinant proteins in plants. J. Cell. Physiol. 216, 366-377 (2008).

[132] Huang, Z., Chen, Q., Hjelm, B., Arntzen, C., Mason, H.: A DNA replicon system for rapid high-level production of virus-like particles in plants. Biotechnol. Bioeng. 103, 706-714 (2009).

[133] Santi, L., et al.: An efficient plant viral expression system generating orally immunogenic Norwalk virus-like particles. Vaccine 26, 1846-1854 (2008).

[134] He, J., Lai, H., Brock, C., Chen, Q.: A novel system for rapid and cost-effective production of detection and diagnostic reagents of west Nile virus in plants. J. Biomed. Biotechnol. 2012, 1-10 (2012). doi: 10. 1155/2012/106783.

[135] Tacket, C. O., Pasetti, M. F., Edelman, R., Howard, J. A., Streatfield, S.: Immunogenicity of recombinant LT-B delivered orally to humans in transgenic corn. Vaccine 22, 4385-4389 (2004).

[136] Kapusta, J., et al.: A plant-derived edible vaccine against hepatitis B virus. FASEB J. 13, 1796-1799 (1999).

[137] Yusibov, V., et al.: Expression in plants and immunogenicity of plant virus-based experimental rabies vaccine. Vaccine 20, 3155-3164 (2002).

[138] Zhang, X., Buehner, N. A., Hutson, A. M., Estes, M. K., Mason, H. S.: Tomato is a highly effective vehicle for expression and oral immunization with Norwalk virus capsid protein. Plant Biotechnol. J. 4, 419-432 (2006).

[139] Huang, Z., et al.: Virus-like particle expression and assembly in plants: hepatitis B and Norwalk viruses. Vaccine 23, 1851-1858 (2005).

[140] Huang, Z., et al.: High-level rapid production of full-size monoclonal antibodies in plants by a single-vector DNA replicon system. Biotechnol. Bioeng. 106, 9-17 (2010).

[141] Giritch, A., et al.: Rapid high-yield expression of full-size IgG antibodies in plants coinfected with non-competing viral vectors. Proc. Natl. Acad. Sci. U. S. A. 103, 14701-14706 (2006).

[142] Marillonnet, S., et al.: In planta engineering of viral RNA replicons: efficient assembly by recombination of DNA modules delivered by Agrobacterium. Proc. Natl. Acad. Sci. U. S. A. 101, 6852-6857 (2004).

[143] Chen, Q., He, J., Phoolcharoen, W., Mason, H. S.: Geminiviral vectors based on bean yellow dwarf virus for production of vaccine antigens and monoclonal antibodies in plants. Hum. Vaccin. 7, 331-338 (2011).

[144] Tacket, C. O., Sztein, M. B., Losonsky, G. A., Wasserman, S. S., Estes, M. K.: Humoral, mucosal, and cellular immune responses to oral Norwalk virus-like particles in volunteers. Clin. Immunol. 108, 241-247 (2003).

[145] Rolland, D., et al.: Purification of recombinant HBc antigen expressed in Escherichia coli and Pichia pastoris: comparison of size-exclusion chromatography and ultracentrifugation. J. Chromatogr. B Biomed. Sci. Appl. 753, 51-65 (2001).

[146] Rodrigues, T., et al.: Screening anion-exchange chromatographic matrices for isolation of oncoretroviral

vectors. J. Chromatogr. B Analyt. Technol. Biomed. Life Sci. 837, 59-68 (2006).

[147] Ball, J. M., et al.: Recombinant Norwalk virus-like particles as an oral vaccine. Arch. Virol. Suppl. 12, 243-249 (1996).

[148] El-Kamary, S. S., et al.: Adjuvanted intranasal Norwalk virus-like particle vaccine elicits antibodies and antibody-secreting cells that express homing receptors for mucosal and peripheral lymphoid tissues. J. Infect. Dis. 202, 1649-1658 (2010). doi: 10. 1086/657087.

[149] Tacket, C. O., et al.: Human immune responses to a novel Norwalk virus vaccine delivered in transgenic potatoes. J. Infect. Dis. 182, 302-305 (2000).

[150] Frenck, R., et al.: Predicting susceptibility to norovirus GII. 4 by use of a challenge model involving humans. J. Infect. Dis. 206 (9), 1386-1393 (2012). doi: 10. 1093/infdis/jis514.

（马维民译）

第 9 章　研制一种新麻疹疫苗

Alexander N. Zakhartchouk 和 George K. Mutwiri①

摘要

麻疹是由麻疹病毒（measles virus，MV）引起的一种高度传染性疾病。自减毒活疫苗在 1963 年获得许可，麻疹发病率就下降了 98%以上。然而麻疹仍然是发展中国家儿童死亡的主要原因。部分原因是由于许可疫苗的几个局限性，包括热不稳定性和对幼儿的低保护效力。因此，我们正在开发一种改进的新麻疹疫苗，该疫苗基于重组麻疹病毒血细胞凝集素（H）蛋白的一部分（球状头部结构域）。H 蛋白的主要功能是与宿主细胞受体结合，大多数麻疹病毒中和抗体直接针对麻疹病毒 H 蛋白。重组蛋白在稳定转染的人细胞中产生，并从细胞培养基中纯化。我们选择了稳定转染的人类细胞作为疫苗抗原生产平台，因为不存在与病毒和病毒载体使用相关的安全问题。此外，在人类细胞中表达的蛋白质翻译后修饰与病毒蛋白质的天然修饰最为相似。在一项可行性研究中，我们证明了用佐剂配制的纯化重组蛋白对小鼠进行两次皮下免疫，导致血清产生高的麻疹病毒特异性中和抗体滴度。在疫苗配方中加入聚磷腈佐剂增加了 Th1 型免疫应答，这对于避免称为非典型麻疹的麻疹病毒疫苗并发症非常重要。此外，我们的数据表明，纯化重组蛋白是热稳定的。需要更多研究来检验我们的新疫苗在猴模型中以及麻疹病毒特异性（母体）抗体存在下的免疫原性和保护效力。

9.1　疾病

尽管有减毒活疫苗，全球麻疹疫苗接种覆盖率超过 80%，但麻疹仍然是 5 岁以下儿童死亡的主要原因之一。麻疹是最具传染性的人类疾病之一。该疾病的早期临床症状是轻度发热、咳嗽、鼻炎、结膜炎和畏光症。这些症状先于 Koplik 斑点，即脸颊内的小白点。在皮疹发生前 24 小时内斑点会迅速出现和消失。典型的斑丘疹通常从颈部和脸颊开始，并在 3 天内遍布全身。随着皮疹出现，体温通常会上升到 39~40℃，在不复杂的情况下会迅速下降。皮疹持续 3~4 天，并且皮疹消退按照与外观相同的顺序向下进行[1]。皮疹发生后不久就开始临床恢复。

麻疹病毒引起淋巴细胞减少和 T 淋巴细胞反应的暂时抑制，持续数周，使感染者易受其他感染。因此，继发性细菌和病毒感染是麻疹相关发病率和死亡率的主要原因[2]。

并发症发生率高达 40%。麻疹的主要并发症是中耳炎、肺炎和脑炎[3]。肺部感染导致间质性病变，病理学上认为是巨细胞性肺炎。麻疹后脑脊髓炎 1 000例中有 1 例复杂化[4]，主要发生在年龄较大的儿童和成人。最严重但罕见的并发症是麻疹包涵体脑炎（MIBE）和亚急性硬化性全脑炎（SSPE），是由中枢神经系统持续性麻疹病毒感染引起的。

---

① A. N. Zakhartchouk，DVM，博士（✉）（加拿大萨斯喀彻温大学疫苗和传染病组织-国际疫苗中心（VidoIntervac），120 兽医路，萨斯喀通，SK S7N 5E3 E-mail：alex. zak@ usask. ca ）.

G. K. Mutwiri，DVM，PhD（加拿大萨斯喀彻温大学疫苗和传染病组织-国际疫苗中心（VidoIntervac），120 兽医路，萨斯喀通，SK S7N 5E3，加拿大 SK 萨斯喀通萨斯喀彻温大学公共卫生学院）

9.2 病原

麻疹病毒是引起麻疹的病原，是副粘病毒科麻疹病毒属成员。这是一种球形、不分段、单股负链 RNA 病毒。麻疹病毒 RNA 基因组包含约 16kb 的核苷酸，被包裹在含脂质的包膜中。基因组编码 8 种蛋白质，其中两种（V 和 C）是非结构性蛋白质，或者从 RNA 中翻译出来，编码磷蛋白（P）。在 6 种结构蛋白中，P、大蛋白（L）和核蛋白（N）构成包围病毒 RNA 的核衣壳。血凝素蛋白（H）、融合蛋白（F）和基质蛋白（M）与来自宿主细胞膜脂质，一起形成病毒包膜。

H 和 F 蛋白负责病毒包膜与宿主细胞膜黏附和融合。受体结合残基被定位到 H 头部区域[5-7]，表明 H 蛋白的主要功能是结合到宿主细胞受体。到目前为止已经鉴定出其中两种（CD46 和 SLAM，也称为 CD150），另外一种是假定的上皮细胞受体（EpR）。最初被鉴定为疫苗菌株受体的 CD46 蛋白似乎与野生型麻疹病毒感染无关。

9.3 传播

感染者的呼吸液滴起到麻疹病毒传播媒介的作用。麻疹病毒感染呼吸道中的淋巴细胞、树突状细胞和巨噬细胞，进而复制，然后扩散到区域淋巴结。在病毒血症期间，病毒传播到各种器官包括淋巴结、皮肤、肾脏、胃肠道和肝脏。最后麻疹病毒感染极化的气道上皮细胞并离开宿主。目前麻疹病毒感染模型，假设野生型麻疹病毒系统性传播仅依赖于表达 CD150 的淋巴细胞感染，而没有呼吸道上皮中第一次病毒扩增。只有当病毒离开宿主时才会穿过呼吸上皮[6,10]。

9.4 诊断、治疗和目前的疫苗

麻疹的临床诊断基于临床症状，包括发热、咳嗽、感冒、结膜炎和 Koplik 斑，随后出现持续至少 3 天的全身性斑疹。然而临床诊断需要通过实验室诊断确认。实验室检测的样本包括血液、鼻咽拭子和尿液。

实验室诊断麻疹的“金标准”是在皮疹发病后 7 天内出现特异性血清免疫球蛋白 M（IgM）抗体。如果有临床症状的人急性（初始）血清学结果显示 IgM 滴度较低或为阴性，则在 7~10 天后抽取第二份血样分析麻疹特异性 IgM 和 IgG 滴度。急性和康复血清免疫球蛋白 G（IgG）抗体滴度显著增加表明已感染。

其他可以在实验室进行的分析包括逆转录酶（RT）PCR 和病毒分离。稳定表达人 SLAM（Vero/ hSLAM）的猴肾 Vero 细胞现在通常用于麻疹病毒分离[11]。

急性麻疹感染无法治愈。幼儿可以用人类免疫球蛋白制剂治疗。发烧可以通过降温剂控制，继发性细菌感染，如肺炎或耳部感染，可以用抗生素治疗。维生素 A 可以减轻麻疹的严重程度。

预防麻疹最好的方法是接种疫苗。全世界有几种减毒活麻疹疫苗，可以作为单一病毒疫苗，也可以与其他疫苗病毒（通常是风疹和腮腺炎，MMR）结合使用。这些疫苗大多来自 1954 年分离的埃德蒙顿（Edmonston）病毒株。

麻疹疫苗诱导体液免疫应答和细胞免疫应答。循环抗体首先在接种疫苗后 12~15 天出现，在 21~28 天达到高峰，然后持续数年。疫苗接种还诱导麻疹特异性细胞免疫应答[12]。麻疹疫苗接种后产生保护性抗体滴度的儿童比例取决于母体抗体的存在：大约 85%的儿童在 9 个月大时产生保护性抗体滴度，90%~95%儿童在 12 个月大时接种疫苗后产生保护性抗体应答。根据这些数据，CDC 建议用 MMR 接种两剂疫苗，第一剂在 12~15 个月，第二剂在 4~6 岁。对于发展中国家，世卫组织建议 9 个月的儿童开始接种疫苗。

9.5 新疫苗

9.5.1 新疫苗原理

尽管麻疹疫苗接种对公众健康有益，但获得许可的疫苗有许多局限性。第一，麻疹减毒活疫苗容易被光和热灭活；因此必须保持冷链。第二，麻疹疫苗必须皮下注射或肌内注射，这就需要训练有素的医护人员、针头、注射器以及危险废物的妥善处置。第三，母亲抗体降低了婴儿早期活疫苗的保护效力。第四，麻疹减毒活疫苗有可能对免疫功能严重受损的个体造成严重不良后果，如肺部或脑部疾病。第五，必须注射两剂疫苗，以达到足够的人群免疫水平，从而阻断麻疹病毒传播，尽管最近的数据表明即使注射两剂也不够[13]。最后在全球消灭野生型病毒后，麻疹减毒活疫苗不能在普通人群中使用。理想的麻疹疫苗的基本特征有：只需要单剂量给药、对免疫受损的个体是安全的、保护幼儿和新生儿、接种疫苗后不久就提供长期的保护、不用针和注射器注射、不需要冷链、与其他疫苗结合使用、低收入国家负担得起等。

因此，我们[14,15]和其他人（表 9.1）正在开发一种新的抗麻疹疫苗。四大类麻疹病毒疫苗已经或正在开发中[16]：①含有病毒糖蛋白的免疫刺激复合物；②含 Protollin 的辅助分裂麻疹病毒疫苗；③DNA 疫苗和④病毒载体疫苗。我们选择稳定转染的人类细胞作为疫苗抗原生产平台，因为没有与病毒和病毒载体使用相关的安全问题。此外，在人类细胞中表达的蛋白质翻译后修饰与病毒蛋白质的天然修饰最为相似。

表 9.1 对灵长类动物有保护作用的重组麻疹疫苗

疫苗	接种方式	参考文献	备注
免疫刺激复合物	im	van Binnendijk 等[17,18]	在麻疹病毒特异性抗体存在的情况下具有免疫原性
痘苗病毒	id，im	van Binnendijk 等[17]	在麻疹病毒特异性抗体存在的情况下不产生免疫原性
改良痘苗病毒	im，in	Stittelaar 等[19,20]	在麻疹病毒特异性抗体存在的情况下具有免疫原性，对免疫抑制动物是安全的
甲病毒复制子颗粒	im，id	Pan 等[21,22]	保护幼年猕猴
DNA	基因枪，id	Polack 等[23-25]	对新生儿保护有限，且存在麻疹病毒特异性抗体
甲病毒复制子 DNA+辅助蛋白	id，in	Premenko-Lanier 等[26]	保护幼年猕猴

im 肌内注射，id 皮内注射，in 鼻内接种

关于实验性麻疹疫苗研究的更全面的综述可以在其他地方找到[16,27]。

9.5.2 抗原

我们在研究中使用的抗原是麻疹病毒 H 蛋白的球状头部结构域（残基 156~617）。麻疹病毒 H 蛋白是一种 617 个氨基酸（78 kDa）的Ⅱ型跨膜糖蛋白（图 9.1a），由 N 端细胞质尾、跨膜结构域和与大 C 端球状头部相连的细胞外膜近端柄区组成[28]。麻疹病毒 H 蛋白在病毒的趋性、受体结合和血凝活性中起着至关重要的作用。由于大多数麻疹病毒中和抗体直接针对麻疹病毒 H 蛋白，在诱导针对麻疹的保护性免疫中也起着至关重要的作用[29]。

全长麻疹病毒 H 形成调光器[30]，球状头部区域也结晶为二聚体[31]。球状头部区域的晶体

结构分析显示，它覆盖着 N-连接糖[31,32]。因此需要在哺乳动物细胞中生产重组蛋白，因为其抗原性可能取决于其糖基化[33]。

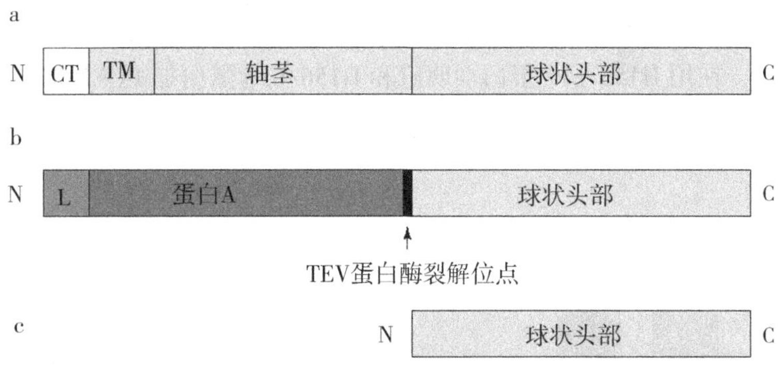

图 9.1　蛋白质示意图

（a）MV H 蛋白（78-kDa）。CT，细胞质尾；TM，跨膜域；N 和 C、N-和 C-端。（b）由稳定转染的 HEK 293 T 细胞产生的重组 ProtA-MV-H156/617 蛋白（95 kDa）。L，前导序列；TEV，烟草蚀刻病毒。（c）在可行性研究中用作抗原的重组 H 156/617 蛋白（60 kDa）。

9.5.3　抗原产生

编码麻疹病毒 H 蛋白球状头部结构域（残基 156～617）密码子优化基因被克隆到表达载体 pPA-TEV 中[34]，与转移素[35]先导、蛋白 A 纯化标签和烟草蚀刻病毒（TEV）蛋白酶切割位点一起形成 pProtA-MV H156/617。接下来，用磷酸钙法转染人胚胎肾（HEK）293 T 细胞（ATCC CRL-1573），扩大对嘌呤霉素（5 μg/mL）的大量培养物，并用抗 ProtA 抗体通过 Western 印迹分析方法检测培养基中分泌的蛋白质水平。

重组 ProtA-MV-H156/617 蛋白（图 9.1b）在含有 3% FBS（Invitrogen）、100 U/mL 青霉素、0.1 mg/mL 链霉素、1 mg/L 嘌呤霉素和 1 mg/L 抑肽酶的 DMEM/F12（50 ：50，Invitrogen）培养的 HEK 293 T 细胞中产生。将培养基过滤并在 TIFF 预规模浓缩器（Millipore）上浓缩 10 倍。浓缩培养基与 IgG 琼脂糖珠在 4℃ 孵育过夜。然后用缓冲液（50 mmol/L Tris pH 7.5，150 mmol/L NaCl）洗涤珠，随后用 TEV 蛋白酶在柱上切割过夜以释放 H 蛋白片段（H 156/617，图 9.1c）。将 H 蛋白片段交换到 50 mmol/L Tris pH7.5，50 mmol/L NaCl 中，并在 HiTrap Q（GE 医疗保健）柱上纯化。将得到的蛋白质浓缩（至 1 mg/mL）并用 10 K 浓缩器（Millipore）交换成无菌 PBS 用于免疫研究。每 2L 收获的细胞培养基中纯化的蛋白质产量约为 1.8 mg。

9.5.4　动物模型及可行性研究

非人灵长类动物已经被用作麻疹病毒疫苗开发和评估的选择模型，并被用于解决麻疹病毒发病机制中的许多问题。食蟹猴[36]和恒河猴[37]感染麻疹病毒产生了一种疾病，与受感染的人类有许多相似之处。实验感染食蟹猴会发生病毒血症、对病毒的细胞免疫应答以及免疫抑制的证据，但通常缺乏麻疹的临床症状。实验性通过呼吸道接种感染麻疹病毒的猕猴会出现斑丘疹、PBMC 感染、淋巴细胞减少和免疫抑制。

棉鼠是麻疹病毒的另一种模型，因为棉鼠可以通过鼻内途径感染麻疹病毒，随后显示出免疫抑制证据[38-40]。发现肺部感染后第 4 天出现肺滴度峰值，肺部切片间隙可见散在的炎症区域。白细胞是感染的主要靶细胞，肺外病毒传播很常见。尽管棉鼠不能取代非人灵长类动物作为最佳模型，但棉鼠为麻疹病毒候选疫苗的初步评估提供了更实用的模型。

在可行性研究中，我们没有检查疫苗的保护效果；因此，我们用小鼠模型来研究疫苗的免疫原性。在本实验中，C57Bl/6 小鼠被随机分配到 4 组，每组 5 只，间隔 4 周皮下接种两次。用磷酸盐缓冲盐水（PBS）、单独抗原（每剂含有 H156/617 蛋白 3 μg）、用明矾配制的相同剂量的蛋白或用聚［二（羧甲酰基乙基苯氧基钠）磷腈］（PCEP）配制的相同剂量的蛋白免疫各组小鼠。如图 9.2a 所示，在用 H156/617 蛋白加明矾和 H156/617 蛋白加 PCEP 接种两次后，产生了高水平的病毒中和抗体滴度。单独 H156/617 蛋白在该剂量下不诱导任何中和抗体滴度。

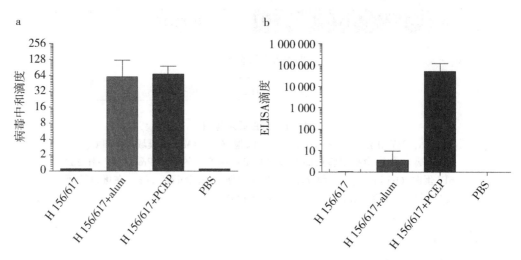

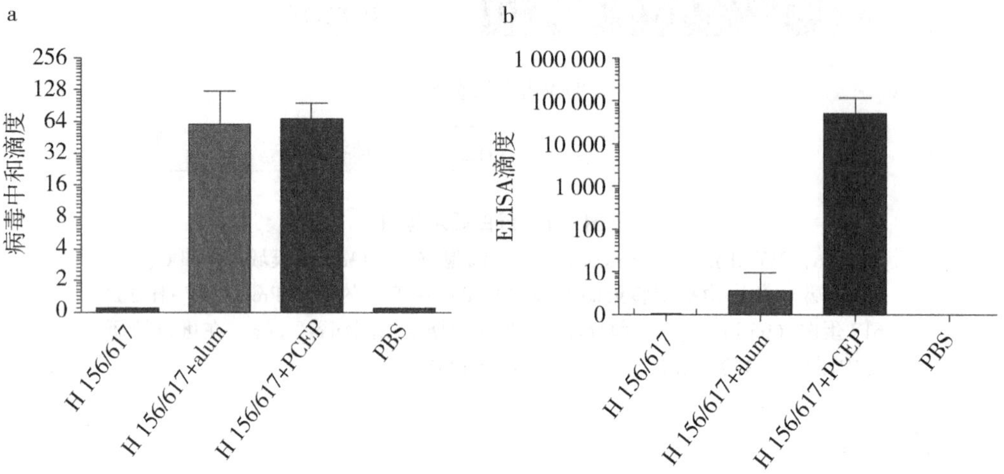

图 9.2　（a）两次免疫后小鼠血清中 MV 中和滴度；（b）对照和接种小鼠血清中 MV H 特异性 IgG2a 抗体的滴度，误差条代表每组 5 只老鼠平均值的标准差

儿童通过菌斑减少中和（PRNT）试验中测得 120 或更高中和滴度显示具有保护作用[41]。因为 PRNT 滴度比中和滴度高 10 倍，所以 PRNT 滴度 120 对应于中和滴度 12。在我们的实验中，接种过疫苗的小鼠平均中和效价约为 60，这是为儿童确定的保护量的 5 倍。

Th2 型占主导地位的免疫应答可能是非典型麻疹的基础，这是一种通过福尔马林灭活的麻疹病毒疫苗接种引发的严重疾病[42]。由于明矾有利于 Th2 型偏向[43]，我们测试了疫苗配方中的 PCEP 佐剂，已知该佐剂促进 Th1/Th2 型混合应答[44]。为了确定产生的免疫应答类型，用 ELISA 检测小鼠血清中麻疹病毒 H 特异性 IgG2a 滴度。如图 9.2b 所示，与明矾作为佐剂的蛋白相比，用 PCEP 配制的蛋白进行疫苗接种可显著提高（$P < 0.05$）IgG2a 滴度；因此正如所料，PCEP 可增强 Th1 型免疫应答。

9.6　优点和缺点

使用人类细胞系作为抗原生产系统是我们疫苗开发策略的优势。人类细胞提供了适当的病毒蛋白翻译后修饰潜力，例如糖基化，这已被证明对麻疹病毒 H 蛋白的加工和抗原性很重要[33]。该系统其他优点是安全性，因为不含病毒和病毒载体，而且具有可扩展性和生产能力。

在可行性研究中，我们使用 HEK 293 T 细胞建立一个稳定表达麻疹病毒 H 蛋白分泌片段的细胞系。在将来必须使用另一种细胞系即 PER. C6®。PER. C6® 细胞的一大优点是比其他细胞系生长密度更高，这意味着可以从更小的生物反应器中收获更多生物产品。2008 年有报告显示使用这些细胞（www. crucell. com）生产了创纪录的生物制品产量（27 g/L）。此外来自许多临床前和临床研究显示，PER. C6® 细胞系具有良好的安全性记录。这些数据连同 PER. C6© 细胞系起源和发展的完整描述，都包含在提交给 FDA 的细胞基质生物制品主文件中。

麻疹活疫苗局限性之一是依赖于冷链维护。重建后，如果在 20℃ 下储存 1 小时它们的效力会降低一半[45]。我们的数据[15]表明，重组H 156/617蛋白在37℃下储存至少2周是稳定的，因此我们的新疫苗具有极好的热稳定性，可以避免目前麻疹疫苗中的冷链供应。

麻疹疫苗的发展史被1963年引进福尔马林灭活的麻疹病毒疫苗的不幸事件所掩盖。这种疫苗提供短暂的保护，暴露于麻疹病毒的儿童通常会出现非典型麻疹，这是一种以出血性皮疹和肺炎为特征的严重疾病。如今越来越多的证据表明，非典型麻疹是由非保护性Th2细胞启动引起的[46]。因此，在麻疹疫苗接种中必须避免诱导Th2偏斜应答。克服这一问题的方法之一是在疫苗配方中使用Th1型佐剂。我们测试了这些佐剂其中之一即PCEP，证明它增强了作为Th1型免疫反应指标的IgG2a抗体的产生。然而需要用恒河猴疾病模型来评估非典型麻疹的可能风险。

胎盘和初乳转移的母体抗体提供了出生时预防传染病的主要保护。这些母体抗体在新生儿出生的头几个月里逐渐减弱，而新生儿自身免疫系统正逐步发育。然而母体抗体阻碍了麻疹活疫苗的成功免疫，婴儿在母体抗体减弱后直到9个月或15个月时才接种疫苗，因此，这期间婴儿没有任何保护。因此，需要开发一种在婴儿体内存在母源性麻疹病毒特异性抗体的情况下，诱导保护性免疫的疫苗。在麻疹病毒特异性（母体）抗体存在的情况下，需要更多实验来评估新疫苗的免疫原性和保护效力。这些实验可以使用猴模型[17]或棉鼠模型进行[47]。

另一个需要日益关注的问题是疫苗注射装置的安全性和危险废物的安全处置。目前麻疹疫苗是通过针头注射接种的。在可行性研究中，我们也用针头和注射器给小鼠皮下注射疫苗。未来，我们将评估疫苗接种的黏膜途径。黏膜免疫将克服和使用针头的必要性，并且在母体抗体存在的情况下也可能是有效的。

致谢

感谢VIDO-InterVac的工作人员，他们参与了麻疹亚单位疫苗项目：LiubovM. Lobanova、Nelson F. Eng和Barry Carroll。感谢多伦多大学的Malathy Satkunarajah和James M. Rini生产和纯化抗原。这项工作是由加拿大健康研究所和萨斯喀彻温健康研究基金会（Sask-CIHR RPP经营赠款给A. Z.）资助的。发表于《VIDO》期刊第653期。

参考文献

[1] Katz, M.: Clinical spectrum of measles. Curr. Top. Microbiol. Immunol. 191, 1-12 (1995).

[2] Beckford, A. P., Kaschula, R. O., Stephen, C.: Factors associated with fatal cases of measles. A retrospective autopsy study. S. Afr. Med. J. 68, 858-863 (1985).

[3] Ibrahim, S.A., et al.: Measles in suburban Khartoum: an epidemiological and clinical study. Trop. Med. Int. Health 7, 442-449 (2002).

[4] Miller, D. L.: Frequency of complications of measles, 1963. Report on a National Inquiry by the Public Health Laboratory Service in Collaboration with the Society of Medical Officers of Health. Br. Med. J. 2, 75-78 (1964).

[5] Leonard, V. H., et al.: Measles virus blind to its epithelial cell receptor remains virulent in rhesus monkeys but cannot cross the airway epithelium and is not shed. J. Clin. Invest. 118, 2448-2458 (2008).

[6] Navaratnarajah, C. K., et al.: Dynamic interaction of the measles virus hemagglutinin with its receptor signaling lymphocytic activation molecule (SLAM, CD150). J. Biol. Chem. 283, 11763-11771 (2008).

[7] Tahara, M., et al.: Measles virus infects both polarized epithelial and immune cells by using distinctive receptor-binding sites on its hemagglutinin. J. Virol. 82, 4630-4637 (2008).

[8] Dorig, R. E., Marcil, A., Richardson, C. D.: CD46, a primate - specific receptor for measles

virus. Trends Microbiol. 2, 312–318 (1994).

[9] Yanagi, Y., Takeda, M., Ohno, S.: Measles virus: cellular receptors, tropism and pathogenesis. J. Gen. Virol.87, 2767–2779 (2006).

[10] de Swart, R. L., et al.: Predominant infection of CD150+lymphocytes and dendritic cells during measles virus infection of macaques. PLoS Pathog. 3, e178 (2007).

[11] Ono, N., et al.: Measles viruses on throat swabs from measles patients use signaling lymphocytic activation molecule (CDw150) but not CD46 as a cellular receptor. J. Virol. 75, 4399–4401 (2001).

[12] Ovsyannikova, I. G., Dhiman, N., Jacobson, R. M., Vierkant, R. A., Poland, G. A.: Frequency of measles virus-specific $CD4^+$ and $CD8^+$ T cells in subjects seronegative or highly seropositive for measles vaccine. Clin. Diagn. Lab. Immunol. 10, 411–416 (2003).

[13] Haralambieva, I. H., et al.: A large observational study to concurrently assess persistence of measles specific B-cell and T-cell immunity in individuals following two doses of MMR vaccine. Vaccine 29, 4485–4491 (2011).

[14] Lobanova, L. M., Baig, T. T., Tikoo, S. K., Zakhartchouk, A. N.: Mucosal adenovirus-vectored vaccine for measles. Vaccine 28, 7613–7619 (2010).

[15] Lobanova, L. M., et al.: The recombinant globular head domain of the measles virus hemagglutinin protein as a subunit vaccine against measles. Vaccine 30, 3061–3067 (2012).

[16] de Vries, R. D., Stittelaar, K. J., Osterhaus, A. D., de Swart, R. L.: Measles vaccination: new strategies and formulations. Expert Rev. Vaccines 7, 1215–1223 (2008).

[17] van Binnendijk, R. S., Poelen, M. C., van Amerongen, G., de Vries, P., Osterhaus, A. D.: Protective immunity in macaques vaccinated with live attenuated, recombinant, and subunit measles vaccines in the presence of passively acquired antibodies. J. Infect. Dis. 175, 524–532 (1997).

[18] Stittelaar, K. J., et al.: Longevity of neutralizing antibody levels in macaques vaccinated with Quil A-adjuvanted measles vaccine candidates. Vaccine 21, 155–157 (2002).

[19] Stittelaar, K. J., et al.: Protective immunity in macaques vaccinated with a modified vaccinia virus Ankara-based measles virus vaccine in the presence of passively acquired antibodies. J. Virol. 74, 4236–4243 (2000).

[20] Stittelaar, K. J., et al.: Safety of modified vaccinia virus Ankara (MVA) in immune-suppressed macaques. Vaccine 19, 3700–3709 (2001).

[21] Pan, C. H., et al.: A chimeric alphavirus replicon particle vaccine expressing the hemagglutinin and fusion proteins protects juvenile and infant rhesus macaques from measles. J. Virol. 84, 3798–3807 (2010).

[22] Pan, C. H., et al.: Modulation of disease, T cell responses, and measles virus clearance in monkeys vaccinated with H-encoding alphavirus replicon particles. Proc. Natl. Acad. Sci. U. S. A. 102, 11581–11588 (2005).

[23] Polack, F. P., et al.: Successful DNA immunization against measles: neutralizing antibody against either the hemagglutinin or fusion glycoprotein protects rhesus macaques without evidence of atypical measles. Nat. Med. 6, 776–781 (2000).

[24] Stittelaar, K. J., et al.: Priming of measles virus-specific humoral- and cellular-immune responses in macaques by DNA vaccination. Vaccine 20, 2022–2026 (2002).

[25] Premenko-Lanier, M., Rota, P. A., Rhodes, G. H., Bellini, W. J., McChesney, M. B.: Protection against challenge with measles virus (MV) in infant macaques by an MV DNA vaccine administered in the presence of neutralizing antibody. J. Infect. Dis. 189, 2064–2071 (2004).

[26] Pasetti, M. F. , et al. : Heterologous prime-boost strategy to immunize very young infants against measles: pre-clinical studies in rhesus macaques. Clin. Pharmacol. Ther. 82, 672-685 (2007).

[27] Putz, M. M. , Bouche, F. B. , de Swart, R. L. , Muller, C. P. : Experimental vaccines against measles in a world of changing epidemiology. Int. J. Parasitol. 33, 525-545 (2003).

[28] Alkhatib, G. , Briedis, D. J. : The predicted primary structure of the measles virus hemagglutinin. Virology 150, 479-490 (1986).

[29] de Swart, R. L. , Yuksel, S. , Osterhaus, A. D. : Relative contributions of measles virus hemagglutinin- and fusion protein-specific serum antibodies to virus neutralization. J. Virol. 79, 11547-11551 (2005).

[30] Plemper, R. K. , Hammond, A. L. , Cattaneo, R. : Characterization of a region of the measles virus hemagglutinin sufficient for its dimerization. J. Virol. 74, 6485-6493 (2000).

[31] Hashiguchi, T. , et al. : Crystal structure of measles virus hemagglutinin provides insight into effective vaccines. Proc. Natl. Acad. Sci. U. S. A. 104, 19535-19540 (2007).

[32] Colf, L.A., Juo, Z.S., Garcia, K.C.: Structure of the measles virus hemagglutinin. Nat. Struct. Mol. Biol. 14, 1227-1228 (2007).

[33] Hu, A. , Cattaneo, R. , Schwartz, S. , Norrby, E. : Role of N-linked oligosaccharide chains in the processing and antigenicity of measles virus haemagglutinin protein. J. Gen. Virol. 75 (Pt 5), 1043 - 1052 (1994).

[34] Pak, J. E. , Rini, J. M. : X-ray crystal structure determination of mammalian glycosyltransferases. Methods Enzymol. 416, 30-48 (2006).

[35] Sanchez-Lopez, R. , Nicholson, R. , Gesnel, M. C. , Matrisian, L. M. , Breathnach, R. : Structure-function relationships in the collagenase family member transin. J. Biol. Chem. 263, 11892-11899 (1988).

[36] van Binnendijk, R. S. , van der Heijden, R. W. , van Amerongen, G. , UytdeHaag, F. G. , Osterhaus, A. D. : Viral replication and development of specific immunity in macaques after infection with different measles virus strains. J. Infect. Dis. 170, 443-448 (1994).

[37] Zhu, Y. D. , et al. : Experimental measles. II. Infection and immunity in the rhesus macaque. Virology 233, 85-92 (1997).

[38] Wyde, P.R., Ambrose, M.W., Voss, T.G., Meyer, H.L., Gilbert, B.E.: Measles virus replication in lungs of hispid cotton rats after intranasal inoculation. Proc. Soc. Exp. Biol. Med. 201, 80-87 (1992).

[39] Wyde, P. R. , Moore-Poveda, D. K. , Daley, N. J. , Oshitani, H. : Replication of clinical measles virus strains in hispid cotton rats. Proc. Soc. Exp. Biol. Med. 221, 53-62 (1999).

[40] Niewiesk, S., et al.: Measles virus-induced immune suppression in the cotton rat (Sigmodonhispidus) model depends on viral glycoproteins. J. Virol. 71, 7214-7219 (1997).

[41] Chen, R. T. , et al. : Measles antibody: reevaluation of protective titers. J. Infect. Dis. 162, 1036 - 1042 (1990).

[42] Polack, F. P. , Hoffman, S. J. , Moss, W. J. , Griffin, D. E. : Altered synthesis of interleukin-12 and type 1 and type 2 cytokines in rhesus macaques during measles and atypical measles. J. Infect. Dis. 185, 13-19 (2002).

[43] Lambrecht, B.N., Kool, M., Willart, M.A., Hammad, H.: Mechanism of action of clinically approved adjuvants. Curr. Opin. Immunol. 21, 23-29 (2009).

[44] Mutwiri, G. , et al. : Poly [di (sodium carboxylatoethyl-phenoxy) phosphazene] (PCEP) is a potent enhancer of mixed Th1/Th2 immune responses in mice immunized with influenza virus antigens. Vaccine 25, 1204-1213 (2007).

[45] Moss, W. J., Griffin, D. E.: Global measles elimination. Nat. Rev. Microbiol. 4, 900-908 (2006).

[46] Polack, F. P.: Atypical measles and enhanced respiratory syncytial virus disease (ERD) made simple. Pediatr. Res. 62, 111-115 (2007).

[47] Schlereth, B., et al.: Successful mucosal immunization of cotton rats in the presence of measles virus-specific antibodies depends on degree of attenuation of vaccine vector and virus dose. J. Gen. Virol. 84, 2145-2151 (2003).

（何继军，靳野译）

第 10 章 预防志贺氏菌病亚单位疫苗的开发：最新进展

Francisco J. Martinez-Becerra，Olivia Arizmendi，
Jamie C. Greenwood II 和 Wendy L. Picking①

摘要

志贺氏菌病是一种胃肠道疾病，发病率和死亡率都很高；发展中国家的婴儿是最危险的群体。该病是由志贺氏菌引起的，志贺氏菌属细菌可以分为 4 个群 50 种不同的血清型。虽然已经提出了几种配方，但目前还没有一种广泛的保护性疫苗上市。志贺氏菌亚单位疫苗的制备和鉴定为疫苗开发提供了一个很好的机会。

志贺氏菌病仍然是引起人类发病和死亡的一个重要原因，每年大约发生 9 000万次，每年大约有 10 万人死亡。大约 60%的死亡发生在生活在第三世界国家 5 岁以下婴儿中。对志贺氏菌疫苗的研究已经持续了多年，虽然已经取得了一些进展，但是预防志贺氏菌病的广谱保护性疫苗仍然没有研发成功。已经描述 4 个不同的群（福氏痢疾杆菌、宋内氏痢疾杆菌、鲍氏痢疾杆菌和志贺氏痢疾杆菌）以及 50 多种 O 抗原血清型变种。

大量血清型的存在是一个重要问题，这引起了对血清型限制性疫苗（如减毒活疫苗或基于脂多糖的制剂）有效性的质疑。使用跨多个血清型的保守蛋白质作为疫苗配方的一部分是很有前途的方法。在疫苗化合物中加入保守蛋白可能解决血清型特异性问题，从而产生一种非常理想的泛志贺氏菌疫苗。此外，重组蛋白通常具有更高的安全性。本章综述了一种针对志贺氏杆菌重组蛋白疫苗的优点，分析迄今为止所测试的候选蛋白，并讨论在这一扩展领域中所取得的进展。

10.1 志贺氏菌病

儿童腹泻是 5 岁以下儿童死亡的主要原因之一，每年造成 75.9 万人死亡[1]。这大约占全世界每年发生的婴儿死亡人数的 11%。在腹泻病病原菌中，志贺氏菌属因其在儿童中的高发病率而非常重要。志贺氏菌病的特点是血性腹泻、便秘、呕吐、肠痉挛和发热；然而症状和严重程度因人而异。例如，2000—2004 年在亚洲进行的一项多中心研究发现，60%~70%的志贺氏菌感染阳性患者出现水样大便，而只有约 40%的患者出现发热[2]。除了死亡，这些事件还导致了严重的营养缺陷，感染儿童往往发育迟缓和认知发育迟缓，这又增加了该疾病的严重程度[3,4]。此外，一些感染导致慢性关节炎，可能是由于 MHC Ⅰ类分子介导的自身反应多肽表现[5]。

1999 年进行的统计将志贺氏菌介导的发病数定为 1.65 亿例，死亡数为 110 万例[6]。其中发病数的 69%和死亡数的 61%发生在 5 岁以下的儿童身上，这突出表明这一年龄段的儿童最易患志贺氏病。最近全球腹泻负担有所下降[7,8]。这些最新估计将发病数定为 9 000万例和 108 000例死亡[9]。然而这些统计数字可能不足以反映志贺氏菌感染的实际数量，因为一些新的分析排除

① F. J. Martinez-Becerra，博士 · O. Arizmendi · J. C. Greenwood Ⅱ · W. L. Picking，博士（✉）（俄克拉何马州立大学微生物学和分子遗传学系，斯蒂尔沃特市，OK，美国，E-mail：wendy. picking@ okstate. edu）

了志贺氏菌暴发（这对流行病学有重大影响），没有估计未在医院治疗的感染数量。预计这些遗漏将对发展中国家志贺氏菌病的负担产生重大影响。不幸的是，这些减少的数字掩盖了保护性疫苗的迫切需要。

2010 年对美国 10 个州的监测显示，有 1 780例志贺氏菌感染病例[10]，其中大多数出现在 5 岁以下儿童身上，发病率为 16.4 例/10 万人。这种发病率使志贺氏菌感染成为仅次于沙门氏菌和弯曲杆菌的第三大实验室确认的病原体。志贺氏菌病在某些高风险人群中尤其重要，如难民。已经在这些人群中发生了大规模暴发。最近对位于中非难民营中志贺氏菌暴发的分析描述了与志贺氏菌感染相关的发病率（从 6.3%到 39.1%）和死亡率（高达 9%）[11]。另一个易感群体是部署的军事人员，志贺氏菌是几个国家军队中最孤立的病原体。在国外的军事人员中，腹泻症状也非常普遍，据报道在一些国家发病率高达 10%[12,13]。

10.2 志贺氏菌是志贺氏菌病的病原

志贺氏菌首先在日本被志贺氏定义为细菌性痢疾的病原体，是一种革兰氏阴性杆菌，无包膜，不能运动。诊断通常基于症状[14]，因为血样、黏液样大便提示志贺氏菌感染。然而由于其他微生物引起的一些腹泻感染具有这些症状（肠侵袭性大肠杆菌和弯曲杆菌等），因此仅对症状进行分析不足以作出准确诊断。临床诊断必须辅以从培养物中分离微生物。

由于志贺氏菌在人类宿主之外无法长时间存活，因此培养存在一定困难。然而，培养物可以在选择性培养基如 MacConkey 和 Hektoen 琼脂中有氧生长。最近几个研究小组提议使用分子技术来确认志贺氏菌感染[15-17]。例如，PCR 扩增 ipaH 基因使痢疾患者中志贺氏菌感染的诊断数量增加了 45%。在中国农村进行的一项分析表明，58%的培养阴性样品通过 PCR 扩增 ipaH 而呈阳性，而 97%的志贺氏菌培养阳性样品通过 PCR 扩增均呈阳性。与培养方法相比，这种基于 PCR 的方法在使用抗生素后不会失去敏感性，在不影响患者治疗的情况下，增加了正确诊断的时间范围。

因此，将分子诊断方法应用于志贺氏菌诊断可以更准确地描述世界范围内志贺氏菌介导感染的疾病严重程度。

10.3 志贺氏菌侵袭与发病机制

志贺氏菌通过食用受污染的食物和水经粪-口途径传播。摄入后耐酸志贺氏菌通过胃和小肠进入大肠[18]（图 10.1a）。在这里它们被 M 细胞吸收，转移到结肠上皮的基底外侧表面，并递呈给常驻巨噬细胞（图 10.1b），其中三型分泌系统（T3SS）的 IpaB 通过半胱天冬酶 1 激活诱导凋亡，从而避免被巨噬细胞杀死（图 10.1c）[19]。然后志贺氏菌利用其 T3SS 侵入上皮细胞，在宿主细胞膜中形成易位孔，启动效应器协调流进入宿主细胞胞质，诱导肌动蛋白重排，最终导致细菌吸收（图 10.1d）。一旦进入，志贺氏菌迅速逃离液泡，开始复制，并通过基于肌动蛋白的运动在细胞质中移动。然后志贺氏菌以 T3SS 依赖性的方式向邻近的未受感染的细胞伸出，所产生的液泡迅速溶解，完成细胞间扩散过程。

与 T3SS 相关的基因编码在一个 220kB 的质粒上，在志贺氏菌中高度保守。T3SS 的核心是三型分泌装置（T3SA），由一个类似于鞭毛系统的基体和一个细胞外针状物组成[20]。入侵质粒抗原 D（IpaD）是一种 37 kDa 的蛋白质，在针尖形成一个五聚环[21]。它控制效应蛋白的分泌并且是环境传感器，用于将 IpaB 调动到 T3SA 尖端复合体[22]。IpaB 是一个 64 kDa 的转运蛋白，它在 IpaD 环上形成一个环，负责与宿主细胞接触。这种接触需要将 IpaC 移动到针尖[23]，并形成从细菌细胞质到宿主细胞细胞质的完整单向导管。炎症和侵袭过程开始只发生在宿主细胞基底外侧，这突出了之前在肠道志贺氏菌定植中巨噬细胞破坏的重要性。

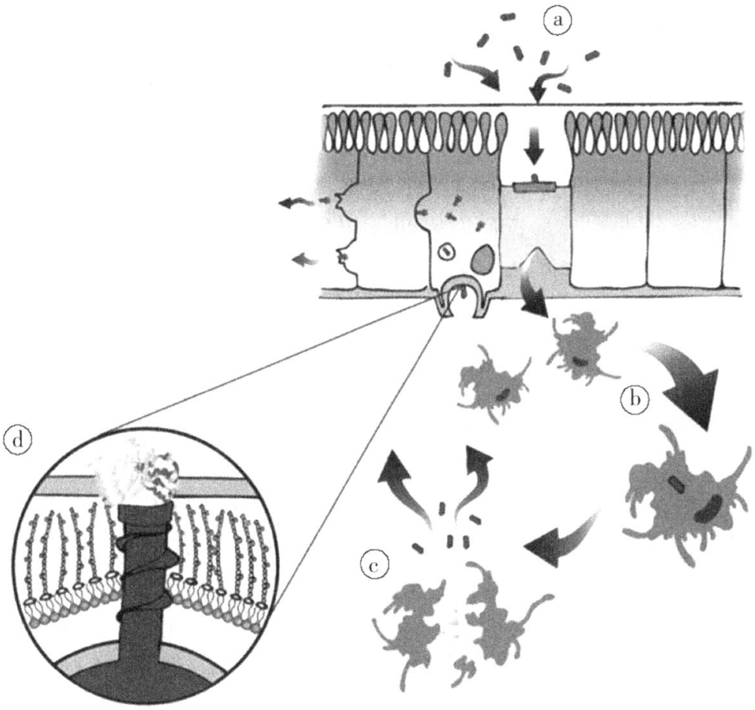

图 10.1　志贺氏菌侵袭上皮细胞的模型

志贺氏菌到达肠腔，被 M 细胞吸收并释放到基底侧（a）。随后志贺氏菌被 M 细胞下的巨噬细胞吞噬（b）。通过诱导凋亡逃脱后（c）。志贺氏菌利用其三型分泌系统（T3SS）侵入上皮细胞（d）。

10.4　血清型变异性和抗生素耐药性：突出疫苗的必要性

志贺氏菌包括 4 种不同的种类或血清群：福氏痢疾杆菌、宋内氏痢疾杆菌、志贺氏痢疾杆菌和鲍氏痢疾杆菌。根据 O 抗原变异性进一步将这 4 个群体划分为超过 50 种不同的血清型（福氏痢疾杆菌 14 种，宋内氏痢疾杆菌 1 种，鲍氏痢疾杆菌 20 种，志贺氏痢疾杆菌 15 种）[24]。流行病学研究显示，每一种类和血清型的疾病负担有很大差异，部分是基于特定的地理位置[2,6]。案例研究综合分析显示在 2007—2009 年期间，在亚洲和非洲，福氏痢疾杆菌是引起大多数感染病例（83.5%）的病原，而宋内氏痢疾杆菌占所有病例的 14.9%，位居第二。

在美国和欧洲，两个主要血清学群的流行频率是相反的：宋内氏痢疾杆菌是主要的血清学群，占 62.4%，福氏痢疾杆菌次之，占 31.7%。志贺氏痢疾杆菌和鲍氏痢疾杆菌的比例通常较低 4%[25]。另一个影响流行血清型的变量是受影响个体的年龄组。在以色列一项为期 2 年的研究发现，1~4 岁儿童的志贺氏菌感染主要是由福氏痢疾杆菌血清型 2a 引起的，而福氏痢疾杆菌 6 型主要引起 5~14 岁儿童发病。

血清型变异是由脂多糖 O 抗原部分的修饰决定的。O 抗原在寡糖单位重复数量、糖类类型和分布，以及分子内和分子间的连接方面有所不同[26]。在福氏痢疾杆菌中，这些基因编码在细菌染色体中。相比之下，宋内氏链球菌没有血清型变异，表达质粒编码的 O 抗原修饰酶。O 抗原是志贺氏菌的主要免疫原性成分之一，也是一种毒力因子，部分原因是掩盖了三型分泌装置暴露[27]。

此外，在噬菌体感染和重组后，通过获得编码葡萄糖基转移酶和乙酰转移酶的基因，可以对这些分子进行修饰[28]。这些修饰在免疫逃避中起着关键作用[27]。因此 O 抗原变异性是疫苗开发需要考虑的一个重要因素，因为它是血清型特异性保护的基础——暴露于特定血清型，可防止同一血清型后续感染，但不能防止其他血清型感染，这突出了感染期间脂多糖的免疫显性作用。这种优势是如此之强以至接触志贺氏菌会产生大多数血清型受限的黏膜反应，掩盖了异源识别[29]。

志贺氏菌流行病学中促使疫苗研发的另一个现象是抗生素抗性菌株的频率不断增加。这种病原的抗生素耐药性持续上升[30]。最近研究表明，超过 30% 的被分析的志贺氏菌对环丙沙星耐药，环丙沙星是目前治疗志贺氏菌病的一线药物[25]。耐药性菌株的不断增加加大了研发新型志贺氏菌疫苗候选株的紧迫性，这种疫苗候选株可以抵御多种血清型，在婴儿群体中是安全的——婴儿群体是我们社会中风险最大的群体。

10.5 候选疫苗试验：动物模型

志贺氏菌病严格来说是一种人类疾病。虽然这种限制的基础尚不清楚，但使研究志贺氏菌发病机制的能力变得复杂。然而，已经建立了一些动物模型来研究志贺氏菌发病机制、对志贺氏菌抗原产生的免疫应答以及候选疫苗对志贺氏菌病的保护效果。

小鼠致死性肺攻毒试验 最近许多疫苗研发工作已经在小鼠致死性肺模型中进行了测试。在这个模型中，大剂量的烈性志贺氏菌鼻内接种小鼠。志贺氏菌随后侵入肺部上皮细胞，几天后，小鼠死于感染[31,32]。尽管这是一个不自然的模型，但它利用了对小鼠免疫系统日益增长的知识以及大量可用于分析免疫反应的试剂和方案。相反地，一个显著的缺点是感染部位大相径庭，这就提出了一个问题，即我们是否有能力将在该模型中测试的疫苗所观察到的保护作用外推到人类身上。

小鼠结肠感染 最近开发的另一种小鼠模型涉及使用链霉素清除肠道共生菌。经过这样处理，志贺氏菌能够在结肠内定植，从粪便样本中分离出活志贺氏菌长达 30 天[33]。由 CXCL8 介导的中性粒细胞浸润是人类存在的一种重要免疫应答，在这些动物中是缺失的；然而与志贺氏菌和 CXCL8 联合使用模拟了细菌性痢疾中描述的病理学特征[34]。

Sereny 试验 Sereny 试验模型[35]长期用于测试志贺氏菌的入侵能力。在这个模型中，在豚鼠眼睛中接种志贺氏菌，会导致角膜结膜炎。这个模型可以检测志贺氏菌的入侵和候选疫苗的保护效果。然而与肺模型一样，感染部位与人体肠道无关。

家兔盲肠结扎模型 第一种建立志贺氏菌病家兔模型的方法是长期禁食，用抗生素和有毒药物治疗，然后口服毒性细菌。然而这种感染通常局限于小肠，症状并不总是像人类志贺氏菌病。随后建立了一个家兔结肠感染模型，包括在盲肠区域结扎（盲肠旁路），然后将毒志贺氏菌直接接种到结肠[36]。在这个模型中，腹泻被用作感染和疾病的指标。尽管该模型可用于描述志贺氏菌与感染自然部位肠黏膜的相互作用，但在实验动物中，手术引入志贺氏菌难度较大。

非人灵长类动物（NHP）模型 NHP 模型已经被用来定义疫苗引发免疫反应和保护的能力（恒河猴和食蟹猴）[37,38]。这个模型的主要优点是志贺氏菌能够在大肠中定居，产生这些细菌在人类感染中产生的症状，还可以比较高度相似的免疫系统。然而使用这种模式需要训练有素的人员和专门的动物设施。该模型的另一个缺点是 NHPs 实验成本非常高。

10.6 寻找疫苗：亚单位疫苗

在寻找抗志贺氏菌候选活疫苗的过程中，使用了两种主要方法：①产生弱毒菌株；②亚单位疫苗。一些候选疫苗是基于弱毒菌株，设计突变以减少毒力。这种方法的原理是用活的但没有致病性的细菌，以模拟由强毒对应物自然引起的感染。一个或多个与志贺氏菌毒力有关的基因突变如*virG*和*ShET*1/2，或代谢途径（*aroA*，*guaBA*）已经产生[39,40]，在动物模型中显示出良好的保护作用。然而当在人类身上进行试验时，许多弱毒菌株反应性太强，或免疫原性较差。因此与口服活疫苗相比，亚单位疫苗更安全，且通常具有更好的长期稳定性，因此是比弱毒菌株更好的选择[24,41]。

10.7 寻找疫苗开发目标

几项研究集中在寻找亚单位疫苗靶点上。例如，对福氏痢疾杆菌的免疫蛋白质组分析显示，可溶性和膜结合蛋白都是由血清识别的，来自受感染个体[42]或用福尔马林灭活细菌皮下免疫的小鼠[43]。不幸的是，在进行这些研究时，没有直接检测出志贺氏菌测序蛋白。观察到的攻击倾向于识别膜蛋白和在其他微生物中发现的鞭毛蛋白。因此，这些新的免疫原性蛋白是疫苗开发的一种有趣的替代品，对其进一步分析和鉴定可以阐明它们在志贺氏菌中的作用。

另一个有趣的疫苗候选物是OspE效应分子家族。用胆汁盐（模拟宿主肠道环境信号）治疗后，志贺氏菌与结肠上皮细胞[44]结合能力增强。微阵列实验表明，暴露于胆汁盐后，OspE1和OspE2表达上调。此外在OspE缺失突变体中，胆汁诱导的黏附被消除。这些分子介导了对靶细胞的第一次粘附，这一事实使它们成为疫苗的一个值得关注的靶标物。一种有免疫调节能力的蛋白质是主要的外膜蛋白（MOMP）[45]，是一种来自志贺氏痢疾杆菌的具有孔蛋白活性的蛋白质。主要的外膜蛋白（MOMP）能够通过Toll样受体（TLR）激活T细胞[46]，导致T细胞增殖并释放促炎细胞因子和趋化因子，这可能是针对志贺氏菌的免疫反应的重要影响因素。这些分子和其他与发病机制有关的分子的进一步特性描述拓宽了疫苗开发的可用目标。其他候选疫苗已经在动物模型中进行了测试，对其安全性和保护效力进行了评估，临床试验也对其中一些进行了描述[47,48]。

O抗原/蛋白体 如上所述，O抗原代表志贺氏菌LPS（脂多糖）的可变部分（图10.2a）。在动物模型中单独给予LPS或O抗原不足以引发免疫应答，从而使免疫原无效。为了解决这一限制，这些分子已经与不同的蛋白质一起用作载体。已经开发和描述了LPS / O抗原混合物的几种变体，其中一种蛋白质组合方法是使用福氏痢疾杆菌和宋内氏痢疾杆菌LPS与脑膜炎奈瑟菌外膜蛋白蛋白体复合物[49-51]（图10.2b）。LPS通过热酚提取从福氏痢疾杆菌或宋内氏痢疾杆菌中提取，并与从脑膜炎奈瑟氏球菌中提取的去污剂外膜蛋白混合，然后通过凝胶过滤色谱法从混合物中存在的游离LPS中分离复合物。这种疫苗背后的概念是脑膜炎奈瑟菌蛋白体中存在的蛋白质能够作为T细胞刺激的载体，从而允许识别LPS。

当这种复合物口服和鼻内给药于小鼠和豚鼠[49]时，肠和肺中产生血清IgG和黏膜IgA，并且在Sereny试验以及小鼠致死性肺模式[51]中显示出保护作用。尽管这种疫苗的优点是使用脑膜炎奈瑟菌蛋白体作为载体/佐剂系统，允许识别糖类抗原，但应答受到血清型限制。因此当在小鼠中施用蛋白体/脂多糖混合物时，没有观察到交叉保护，并且异源生物被用于攻击。

蛋白体/脂多糖候选疫苗已经在成人志愿者中经鼻内给药进行了I期临床试验。疫苗耐受良好，有一些与剂量无关的症状，如头痛和肌肉疼痛，以及轻微的与剂量有关的反应，如鼻腔分泌物。两个最高剂量（1 mg和1.5 mg）对志贺氏菌脂多糖的血清IgG抗体反应具有统计学意义，但是当分析黏膜IgA时，剂量的影响并不明确[52]。在临床试验期间，所测量的免疫反应总体上被认为是轻微的。最后含脂多糖疫苗的安全性经常在现场受到质疑。

rEPA-LPS 另一种脂多糖（LPS）载体包括重组绿脓杆菌外蛋白A（rEPA），一种具有佐剂性质的ADP核糖转移酶[53]。对于这种疫苗，O抗原是通过细菌脂多糖的酸水解产生的，然后进行DNAse、RNase和蛋白酶处理，最后进行尺寸排阻色谱分离。产生了福氏痢疾杆菌、宋内氏痢疾杆菌和志贺氏痢疾杆菌的O抗原的反应形式，与重组绿脓杆菌外蛋白A（rEPA）偶联，并通过色谱法纯化（图10.2c）。通过皮下注射将O抗原偶联物递送给小鼠，测量针对O抗原的血

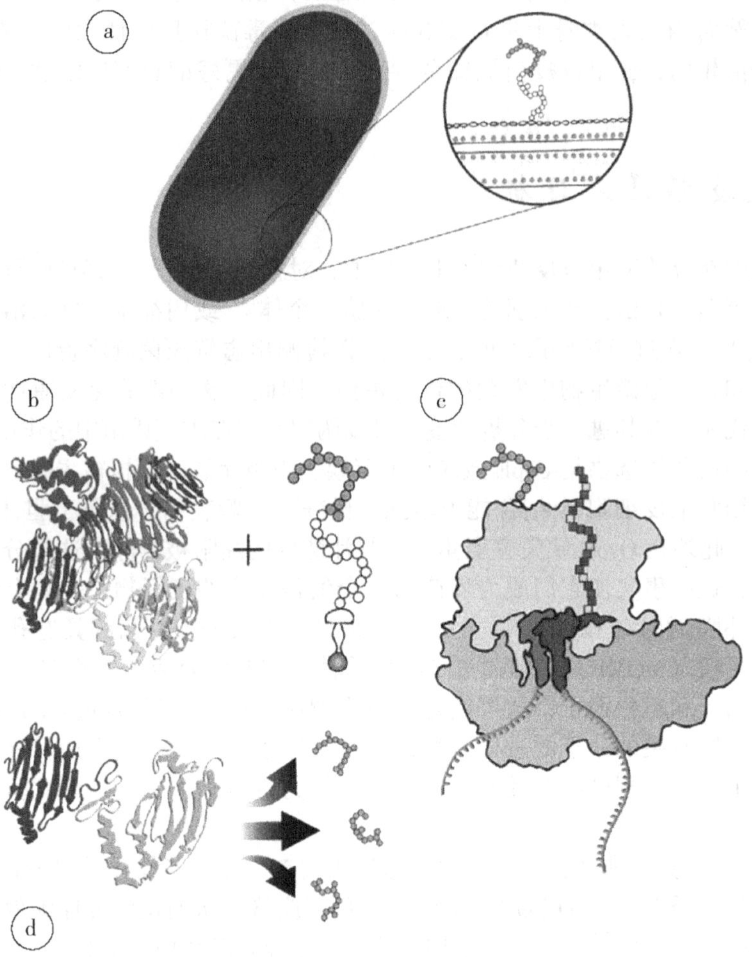

图 10.2　LPS/O 抗原疫苗的描述

将从福氏痢疾杆菌或宋内氏痢疾杆菌提取的 LPS（脂多糖）（a）与脑膜炎奈瑟菌的蛋白质制剂混合，用作疫苗化合物（b）。从 LPS 纯化的 O 抗原与绿脓杆菌的外蛋白 A 结合传递（c）。最后将来自不同志贺氏菌血清群的 O 抗原与来自志贺氏菌的核糖体结合在一起，如（d）所示。

清 IgG。用与明矾一起递送的 O 抗原/重组绿脓杆菌外蛋白 A（rEPA）结合物免疫的小鼠其抗体水平更高[53]。

当给成年志愿者服用这些结合物时，血清中的 O 抗原 IgG 水平升高[53]。该疫苗的安全性和免疫原性在另一项研究中进行了评估，再次显示出强烈的血清 IgG 和 IgA 反应[54]。一项针对年轻人的疗效试验表明，该疫苗对宋内氏痢疾杆菌感染具有 74%的保护作用，血清抗体水平与保护作用相关[55]。这种疫苗在 4~7 岁的儿童进行了试验，只发现了轻微的、短暂的副作用。接种后 6 周（或 1~4 岁儿童免疫后 2 年）O 抗原特异性抗体水平仍保持不变[56]。虽然证明这种疫苗在成人志愿者中是安全和有保护作用的，但与该疫苗能够激发儿童的免疫应答并只产生轻微的不良反应有很大的相关性。然而血清型限制仍然是这种疫苗的一个可能缺点，进一步开发和产生结合混合物可以解决这个问题。

LPS-核糖体　志贺氏菌的核糖体也与 O 抗原结合使用（图 10.2d）。这些颗粒佐剂活性的确切机制尚不清楚；然而有人提出它们充当较弱抗原的载体[58]。核糖体是从无毒福氏痢疾杆菌的细胞质提取物中分离出来的。这些制剂在豚鼠、猴子和小鼠中显示出保护作用[59,60]。鼻内接种这种疫苗后，产生了强烈的血清 IgG 应答[61]。黏膜 IgA 分泌也在多个黏膜区室中出现。发现诱导 IgA 应答的黏膜接种是核糖体/O 抗原混合物产生保护所必需的，胃肠外接种没有保护作用。

从头合成　用于提高 O 抗原疫苗安全性的一种替代方法是从头合成不同链长和组分的 O 抗原。这些分子由多步化学合成产生，并与马来酰亚胺活化的破伤风类毒素结合，生成免疫原性结合物[62-64]。这种方法允许选择该分子的最小抗原部分，以最大限度地提高免疫识别，同时避免使用全长分子的任何副作用。

通过肌内途径用结合物免疫小鼠产生对合成 O 抗原所基于的 LPS 特异抗体滴度。化学合成 O 抗原类似物的优点之一是避免志贺氏菌的生长和加工，避免了完全有毒的 LPS 和其他污染物的污染。缺点是缺乏异源保护；然而这种疫苗中使用的 O 抗原部分可以结合形成多血清型疫苗。此外合成这些类似物的成本可能是其使用的阻碍。

志贺氏菌外膜囊泡　外膜囊泡（OMVs）是由脂多糖、蛋白质和核酸组成的颗粒。在提出的疫苗配方中，通过离心和随后的过滤，从鲍氏痢疾杆菌液体培养物中纯化这些颗粒（图 10.3a）。目前还不清楚该制剂中所含蛋白质的确切身份和数量，尽管存在与 IpaB、IpaC 和 IpaD 质量相同的蛋白质，表明其成分中包括这些蛋白质。当这些 OMV 口服给小鼠时，会产生抗 OMV 溶解物的抗体。

免疫被动地从接种过疫苗的雌性小鼠转移到后代小鼠，然后这些小鼠就能够在志贺氏菌的口腔感染中存活下来。有趣的是，在这些免疫动物身上实现了异源保护[65]。这可能是由于囊泡中存在跨血清群保守蛋白质。这些囊泡已经通过鼻内、口服和眼部给药途径进行了测试，并且在小鼠致死性肺模型中发现单独使用或与基于聚合物的纳米颗粒结合使用都是有效的[66]。这种疫苗具有异源保护的优点（如对每个志贺氏菌血清群菌株的攻击所示），并且没有佐剂依赖性。此外免疫可以被动地传递给后代，这表明保护机制涉及抗体，提高了这种疫苗可用于婴儿的可能性，婴儿是志贺氏菌疫苗的主要目标群体。在制备过程中使用活的、完全有毒的志贺氏菌、存在 LPS 和逐批一致性可能是该制剂的缺点。

invaplex　另一种将 T3SS 蛋白质和 LPS 作为配方组成部分的候选疫苗是 invaplex 疫苗[67]。这些复合物通过水萃取和离子交换色谱法获得（图 10.3b）。这些提取物的精确成分尚未完全确定，但包括 LPS、IpaB 和 IpaC[68]。这些复合物能够诱导小鼠和豚鼠对 Ipa 蛋白以及 LPS 的 IgG 和 IgA 应答。此外，它们对小鼠和豚鼠攻击模型中用于提取产生的志贺氏菌种/血清型具有保护作用[69]。

在成人志愿者中使用 invaplex 疫苗进行了两个第一阶段的研究[47,48]，结果显示鼻内给药剂量高达 690 μg，没有重大副作用。这些研究中使用的最高剂量在 58%的志愿者中产生了对脂多糖的抗体分泌细胞（ASC）反应。这种方法的优点之一是除了 invaplex 本身，不需要额外的佐剂。这种疫苗的缺点在于生产过程具有挑战性，包括毒性志贺氏菌的培养以及中间步骤和最终配方中存在细菌 LPS 产物。另一个可能的警告是通过批量生产，这些复合物中的蛋白质组分是一致的。

最后这种疫苗的设计不是为了抵御多种血清型。然而解决这一可能缺点的一种方法是生成包含由 1 种以上特定血清型生成的 invaplex 复合物制剂，这增加了本已困难的制造过程。这将有助于针对特定地区流行的血清型产生疫苗制剂。

重组 T3SS 蛋白　针对保守志贺氏菌毒力蛋白的候选疫苗包括一些 T3SS Ipa 蛋白（图 10.4a）。重组 IpaB 和 IpaD 可以在大肠杆菌中高水平表达。然后 IpaD 很容易从大肠杆菌胞液中纯化出来，而 IpaB 必须与其同源伴侣 IpgC 以复合物的形式纯化出来。需要伴侣来保持疏水性 IpaB 处于可溶性状态，并为 IpaB 提供抗蛋白水解降解的稳定性。然后在低浓度去垢剂中，IpaB 可以从 IpgC 中分离后进一步纯化。分析表明按照这个方案，IpaB 的纯度超过 90%。在其最终配方中这种基于 Ipa 的疫苗还含有来自大肠杆菌的热不稳定肠毒素的双突变体（dmLT 作为佐剂）[70]。这种疫苗的保护机制尚未确定，然而它在小鼠致死性肺模型中进行了测试[71]，在该模型中，表现出超过 90%的同源保护（针对福氏痢疾杆菌）和超过 60%的异源保护（在攻毒实验

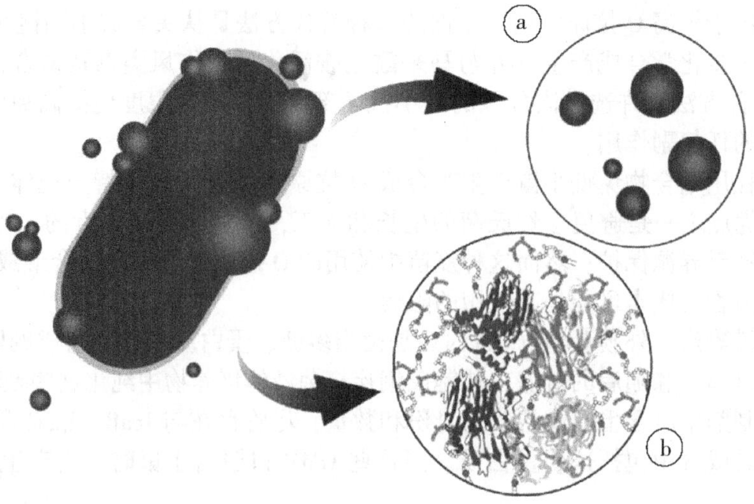

图 10.3　基于囊泡疫苗的描述

（a）疫苗制剂由纯化自鲍氏痢疾杆菌的外膜囊泡组成，这些囊泡是由所有志贺氏菌自然产生的。（b）另一种称为 Invaplex 的结合物由福氏痢疾杆菌或宋内氏痢疾杆菌的水提取物组成。主要由脂多糖和蛋白质组成，尤其是 IpaB 和 IpaC。

中使用宋内氏痢疾杆菌）。鼻内给药后产生 IgG 和黏膜 IgA 以及抗原特异性 IFN-γ 分泌细胞。

抗 IpaB 和 IpaD 的抗体已经被用于在体外直接阻断福氏痢疾杆菌入侵，因此这种疫苗可以利用这种机制；然而不能排除其他机制，如细胞因子分泌细胞产生或吞噬细胞激活。这种疫苗的主要优势在于针对所有致病性志贺氏菌的主要毒力机制，从而可以防护志贺氏菌，而且不受血清型限制。这种疫苗的另一个优点是使用两种不同的蛋白质作为靶标，降低了其中一种蛋白质发生突变的可能性，这种突变会使细菌避免识别这种疫苗引发的免疫反应。缺点包括在蛋白质生产过程中使用去垢剂混合物，以及重组蛋白质表达和纯化的可能成本。

OmpA　使用离子交换色谱法从福氏痢疾杆菌 2a 中纯化出一种 34 kDa 的外膜蛋白（OMP）（图 10.4b）。巨噬细胞与这种 34 kDa 蛋白的孵育诱导了一氧化氮的产生，增加了 IL-12 和 TNF-α 的产生。这种蛋白质在兔体内经肠外给药 5 次，在兔盲肠结扎模型中提供了由福氏痢疾杆菌攻击的保护[72]。随后使用亲和层析纯化的重组蛋白进行的工作将这种 34 kDa OMP 鉴定为 OmpA，这是存在于许多革兰氏阴性菌中的免疫调节蛋白家族的一部分。这种蛋白在小鼠致死性肺模型中显示出很高的保护作用[73]，在该模型中能诱导血清 IgG 应答和黏膜 IgA 应答。

此外，该蛋白能够直接激活 TLR2 并诱导 Th1 应答[74]。使用这种蛋白质作为疫苗的优势包括在疫苗制备步骤中不存在 LPS，OmpA 固有的免疫活化活性允许其在不添加佐剂的情况下使用。确定这种疫苗是否能提供异源保护是很有意义的，在不同志贺氏菌血清型之间和表达密切相关 OmpA 同源物的其他致病菌之间都是异源的。

10.8　结语

志贺氏菌病仍然是一个重要的全球公共健康问题，特别是在发展中国家，5 岁以下的儿童特别容易受到多次志贺氏菌病的感染。志贺氏菌血清型的高度变异性和抗生素耐药性增加强调了加大疫苗研发力度的必要性。在理想的情况下，这种疫苗应该能预防多种血清型，并且由于婴儿志贺氏菌病的高发病率，针对这一群体的疫苗应该有可靠的安全性。

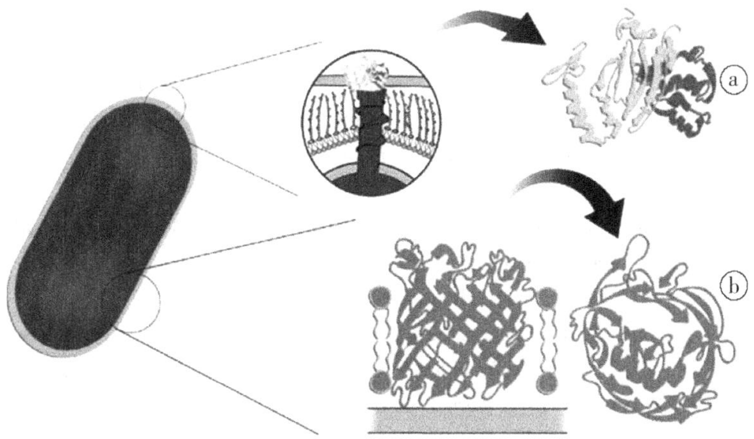

图 10.4　志贺氏菌重组蛋白疫苗的描述

IpaB 和 IpaD 是志贺氏菌 T3SS 的关键成分，在大肠杆菌中表达并与佐剂 dmLT 一起免疫（a）。外膜蛋白 OmpA 在大肠杆菌中表达，且在没有额外佐剂的情况下递送（b）。

亚单位疫苗的使用代表了开发这种疫苗的重要努力，与弱毒志贺氏菌菌株相比，这种疫苗具有广泛的血清型保护和安全性的提高，这对于候选重组蛋白疫苗尤其有效。目前几种重组蛋白疫苗和含 O 抗原的制剂已经在动物模型中进行了测试，其中一些正在临床试验中，显示出良好的免疫原性和保护力[24]。

不幸的是，对于许多候选疫苗来说，免疫保护机制的全部范围以及如何能够阻断不同动物模型中的疾病仍不清楚。志贺氏菌致病研究过程中产生的知识将进一步促进疫苗学工作，确定疫苗开发的新目标。此外，对这些化合物的保护机制的全面分析将使我们能够进一步理解这种病原体与其宿主的相互作用，进而使我们能够将这种理解应用于其他复杂的病原体。

致谢

感谢 Dr. Nicholas Dickenson 和 Dr. William Picking 对这一章的审阅。

参考文献

[1] WHO：Children：Reducing mortality. Fact sheet N°178. < http：//www. who. int/mediacentre/factsheets/fs178/en/index. html>（2012）.

[2] von Seidlein，L.，et al.：A multicentre study of Shigella diarrhoea in six Asian countries：disease burden，clinical manifestations，and microbiology. PLoS Med. 3，e353（2006）. doi：10. 1371/journal. pmed. 0030353.

[3] Kolling，G.，Wu，M.，Guerrant，R. L.：Enteric pathogens through life stages. Front. Cell. Infect. Microbiol. 2，114（2012）. doi：10. 3389/fcimb. 2012. 00114.

[4] Black，R. E.，Brown，K. H.，Becker，S.：Effects of diarrhea associated with specific enteropathogens on the growth of children in rural Bangladesh. Pediatrics73，799-805（1984）.

[5] Boisgerault，F.，et al.：Alteration of HLA-B27 peptide presentation after infection of transfected murine L cells by shigellaflexneri. Infect. Immun. 66，4484-4490（1998）.

[6] Kotloff，K. L.，et al.：Global burden of Shigella infections：implications for vaccine development and implementation of control strategies. Bull. World Health Organ. 77，651-666（1999）.

[7] Bardhan, P., Faruque, A. S., Naheed, A., Sack, D. A.: Decrease in shigellosis-related deaths without Shigella spp. -specific interventions, Asia. Emerg. Infect. Dis. 16, 1718-1723 (2010). doi: 10. 3201/eid1611. 090934.

[8] Kosek, M., Bern, C., Guerrant, R. L.: The global burden of diarrhoeal disease, as estimated from studies published between 1992 and 2000. Bull. World Health Organ. 81, 197-204 (2003).

[9] WHO: Shigellosis. < http: //www. who. int/vaccine _ research/diseases/diarrhoeal/en/index6. html > (2009).

[10] Henao, O. L., et al.: Centers for Disease Control and Prevention: Vital signs: incidence and trends of infection with pathogens transmitted commonly through food-foodborne diseases active surveillance network, 10 U. S. sites1996-2010. MMWRMorb. Mortal. Wkly Rep. 60, 749-755 (2011).

[11] Kerneis, S., Guerin, P. J., von Seidlein, L., Legros, D., Grais, R. F.: A look back at an ongoing problem: Shigella dysenteriae type 1 epidemics in refugee settings in Central Africa (1993-1995). PLoSOne 4, e4494 (2009). doi: 10. 1371/journal. pone. 0004494.

[12] Huerta, M., et al.: Trends in Shigella outbreaks in the Israeli military over 15 years. Eur. J. Clin. Microbiol. Infect. Dis. 24, 71-73 (2005). doi: 10. 1007/s10096-004-1252-z.

[13] Mayet, A., et al.: Epidemiology of food-borne disease outbreaks in the French armed forces: a review of investigations conducted from 1999 to 2009. J. Infect. 63, 370-374 (2011). doi: 10. 1016/j. jinf. 2011. 08. 003.

[14] Niyogi, S. K.: Shigellosis. J. Microbiol. 43, 133-143 (2005).

[15] Dutta, S., et al.: Sensitivity and performance characteristics of a direct PCR with stool samples in comparison to conventional techniques for diagnosis of Shigella and enteroinvasive Escherichia coli infection in children with acute diarrhoea in Calcutta, India. J. Med. Microbiol. 50, 667-674 (2001).

[16] Wang, S. M., et al.: Surveillance of shigellosis by real-time PCR suggests underestimation of shigellosis prevalence by culture-based methods in a population of rural China. J. Infect. 61, 471-475 (2010). doi: 10. 1016/j. jinf. 2010. 10. 004.

[17] Gaudio, P. A., Sethabutr, O., Echeverria, P., Hoge, C. W.: Utility of a polymerase chain reaction diagnostic system in a study of the epidemiology of shigellosis among dysentery patients, family contacts, and well controls living in a shigellosis-endemic area. J. Infect. Dis. 176, 1013-1018 (1997).

[18] Wassef, J. S., Keren, D. F., Mailloux, J. L.: Role of M cells in initial antigen uptake and in ulcer formation in the rabbit intestinal loop model of shigellosis. Infect. Immun. 57, 858-863 (1989).

[19] Zychlinsky, A., et al.: IpaB mediates macrophage apoptosis induced by Shigella flexneri. Mol. Microbiol. 11, 619-627 (1994).

[20] Schroeder, G. N., Hilbi, H.: Molecular pathogenesis of Shigella spp.: controlling host cell signaling, invasion, and death by type III secretion. Clin. Microbiol. Rev. 21, 134-156 (2008). doi: 10. 1128/CMR. 00032-07.

[21] Epler, C. R., Dickenson, N. E., Bullitt, E., Picking, W. L.: Ultrastructural analysis of IpaD at the tip of the nascent MxiH type III secretion apparatus of Shigellaflexneri. J. Mol. Biol. 420, 29-39 (2012). doi: 1 0. 1016/j. jmb. 2012. 03. 025.

[22] Stensrud, K. F., et al.: Deoxycholate interacts with IpaD of Shigellaflexneri in inducing the recruitment of IpaB to the type III secretion apparatus needle tip. J. Biol. Chem. 283, 18646-18654 (2008). doi: 10. 1074/ jbc. M802799200.

[23] Epler, C. R., Dickenson, N. E., Olive, A. J., Picking, W. L., Picking, W. D.: Liposomes recruit IpaC to the Shigella flexneri type III secretion apparatus needle as a final step in secretion induction. Infect. Immun. 77, 2754-2761 (2009). doi: 10. 1128/IAI. 00190-09.

[24] Levine, M. M., Kotloff, K. L., Barry, E. M., Pasetti, M. F., Sztein, M. B.: Clinical trials of Shigella vaccines: two steps forward and one step back on a long, hard road. Nat. Rev. Microbiol. 5, 540-553 (2007). doi: 10. 1038/nrmicro1662.

[25] Gu, B., et al.: Comparison of the prevalence and changing resistance to nalidixic acid and ciprofloxacin of Shigella between Europe-America and Asia-Africa from 1998 to 2009. Int. J. Antimicrob. Agents40, 9-17

(2012). doi: 10. 1016/j. ijantimicag. 2012. 02. 005.
[26] Liu, B., et al.: Structure and genetics of Shigella O antigens. FEMS Microbiol. Rev. 32, 627-653 (2008). doi: 10. 1111/j. 1574-6976. 2008. 00114. x.
[27] West, N. P., et al.: Optimization of virulence functions through glucosylation of Shigella LPS. Science307,%1-131705). doi: 10. 1126/science. 1108472.
[28] Allison, G. E., Verma, N. K.: Serotype-converting bacteriophages and O-antigen modification in Shigella flexneri. Trends Microbiol. 8, 17-23 (2000).
[29] Rasolofo-Razanamparany, V., Cassel-Beraud, A. M., Roux, J., Sansonetti, P. J., Phalipon, A.: Predominance of serotype-specific mucosal antibody response in Shigella flexneri-infected humans living in an area of endemicity. Infect. Immun. 69, 5230-5234 (2001).
[30] Ashkenazi, S., Levy, I., Kazaronovski, V., Samra, Z.: Growing antimicrobial resistance of Shigella isolates. J. Antimicrob. Chemother. 51, 427-429 (2003).
[31] Mallett, C. P., VanDeVerg, L., Collins, H. H., Hale, T. L.: Evaluation of Shigella vaccine safety and efficacy in an intranasally challenged mouse model. Vaccine11, 190-196 (1993).
[32] van de Verg, L. L., et al.: Antibody and cytokine responses in a mouse pulmonary model of Shigella flexneri serotype 2a infection. Infect. Immun. 63, 1947-1954 (1995).
[33] Martino, M. C., et al.: Mucosal lymphoid infiltrate dominates colonic pathological changes in murine experimental shigellosis. J. Infect. Dis. 192, 136-148 (2005). doi: 10. 1086/430740.
[34] Singer, M., Sansonetti, P. J.: IL-8 is a key chemokine regulating neutrophil recruitment in a new mouse model of Shigella-induced colitis. J. Immunol. 173, 4197-4206 (2004).
[35] Sereny, B.: Not available. ActaMicrobiol. Acad. Sci. Hung. 4, 367-376 (1957).
[36] Rabbani, G. H., et al.: Development of an improved animal model of shigellosis in the adult rabbit by colonic infection with Shigella flexneri 2a. Infect. Immun. 63, 4350-4357 (1995).
[37] Formal, S. B., et al.: Effect of prior infection with virulent Shigella flexneri 2a on the resistance of monkeys to subsequent infection with Shigella sonnei. J. Infect. Dis. 164, 533-537 (1991).
[38] Shipley, S. T., et al.: A challenge model for Shigelladysenteriae 1 in cynomolgus monkeys (Macacafascicularis). Comp. Med. 60, 54-61 (2010).
[39] Kotloff, K. L., et al.: Shigellaflexneri 2a strain CVD 1207, with specific deletions in virG, sen, set, and guaBA, is highly attenuated in humans. Infect. Immun. 68, 1034-1039 (2000).
[40] Kotloff, K. L., et al.: Safety, immunogenicity, and transmissibility in humans of CVD 1203, a live oral Shigella flexneri 2a vaccine candidate attenuated by deletions inaroA and virG. Infect. Immun. 64, 4542-4548 (1996).
[41] Phalipon, A., Mulard, L. A., Sansonetti, P. J.: Vaccination against shigellosis: is it the path that is difficult or is it the difficult that is the path? Microbes Infect. 10, 1057-1062 (2008). doi: 10. 1016/j. micinf. 2008. 07. 016.
[42] Jennison, A. V., Raqib, R., Verma, N. K.: Immunoproteome analysis of soluble and membrane proteins of Shigellaflexneri2457T. World J. Gastroenterol. 12, 6683-6688 (2006).
[43] Peng, X., Ye, X., Wang, S.: Identification of novel immunogenic proteins of Shigella flexneri 2a by proteomic methodologies. Vaccine 22, 2750-2756 (2004). doi: 10. 1016/j. vaccine. 2004. 01. 038.
[44] Faherty, C. S., Redman, J. C., Rasko, D. A., Barry, E. M., Nataro, J. P.: Shigella flexneri effectors OspE1 and OspE2 mediate induced adherence to the colonic epithelium following bile salts exposure. Mol. Microbiol. 85, 107-121 (2012). doi: 1 0. 1111/j. 1365-2958. 2012. 08092. x.
[45] Roy, S., Das, A. B., Ghosh, A. N., Biswas, T.: Purification, pore-forming ability, and antigenic relatedness of the major outer membrane protein of Shigella dysenteriae type 1. Infect. Immun. 62, 4333-4338 (1994).
[46] Biswas, A., Banerjee, P., Biswas, T.: Porin of Shigelladysenteriae directly promotes toll-like receptor 2-mediated $CD4^+$ T cell survival and effector function. Mol. Immunol. 46, 3076-3085 (2009). doi: 10. 1016/j. molimm. 2009. 06. 006.
[47] Tribble, D., et al.: Safety and immunogenicity of a Shigella flexneri 2a Invaplex 50 intranasal vaccine in a-

dult volunteers. Vaccine 28, 6076-6085 (2010). doi: 10. 1016/j. vaccine. 2010. 06. 086.

[48] Riddle, M. S., et al.: Safety and immunogenicity of an intranasal Shigella flexneri 2a Invaplex 50 vaccine. Vaccine 29, 7009-7019 (2011). doi: 0. 1016/j1. vaccine. 2011. 07. 033.

[49] Orr, N., Robin, G., Cohen, D., Arnon, R., Lowell, G. H.: Immunogenicity and efficacy of oral or intranasal Shigella flexneri 2a and Shigella sonneiproteosome-lipopolysaccharide vaccines in animal models. Infect. Immun. 61, 2390-2395 (1993).

[50] Orr, N., et al.: Enhancement of anti-Shigella lipopoly-saccharide (LPS) response by addition of the cholera toxin B subunit to oral and intranasal proteosome-Shigella flexneri 2a LPS vaccines. Infect. Immun. 62, 5198-5200 (1994).

[51] Mallett, C. P., et al.: Intransal or intragastric immunization with proteosome-Shigella lipopolysaccharide vaccines protects against lethal pneumonia in a murine model of Shigella infection. Infect. Immun. 63, 2382-2386 (1995).

[52] Fries, L. F., et al.: Safety and immunogenicity of a proteosome-Shigellaflexneri 2a lipopolysaccharide vaccine administered intranasally to healthy adults. Infect. Immun. 69, 4545-4553 (2001). doi: 10. 1128/IAI. 69. 7. 4545-4553. 2001.

[53] Taylor, D. N., et al.: Synthesis, characterization, and clinical evaluation of conjugate vaccines composed of the O-specific polysaccharides of Shigella dysenteriae type 1, Shigella flexneri type 2a, and Shigella sonnei (Plesiomonas shigelloides) bound to bacterial toxoids. Infect. Immun. 61, 3678-3687 (1993).

[54] Cohen, D., et al.: Safety and immunogenicity of investigational Shigella conjugate vaccines in Israeli volunteers. Infect. Immun. 64, 4074-4077 (1996).

[55] Cohen, D., et al.: Double-blind vaccine-controlled randomised efficacy trial of an investigational Shigella sonnei conjugate vaccine in young adults. Lancet 349, 155-159 (1997). doi: 10. 1016/S0140-6736 (96) 06255-1.

[56] Ashkenazi, S., et al.: Safety and immunogenicity of Shigella sonnei and Shigella flexneri 2a O-specific polysaccharide conjugates in children. J. Infect. Dis. 179, 1565-1568 (1999). doi: 10. 1086/314759.

[57] Passwell, J. H., et al.: Safety and immunogenicity of Shigella sonnei-CRM9 and Shigella flexneri type 2a-rEPAsucc conjugate vaccines in one-to four-year-old children. Pediatr. Infect. Dis. J. 22, 701-706 (2003). doi: 10. 1097/01. inf. 0000078156. 03697. a5.

[58] Gregory, R. L.: Microbial ribosomal vaccines. Rev. Infect. Dis. 8, 208-217 (1986).

[59] Levenson, V. J., Mallett, C. P., Hale, T. L.: Protection against local Shigella sonnei infection in mice by parenteral immunization with a nucleoprotein subcellular vaccine. Infect. Immun. 63, 2762-2765 (1995).

[60] Levenson, V. I., et al.: Parenteral immunization with Shigella ribosomal vaccine elicits local IgA response and primes for mucosal memory. Int. Arch. Allergy Appl. Immunol. 87, 25-31 (1988).

[61] Shim, D. H., et al.: Immunogenicity and protective efficacy offered by a ribosomal-based vaccine from Shigella flexneri 2a. Vaccine 25, 4828-4836 (2007). doi: 10. 1016/j. vaccine. 2007. 03. 050.

[62] Theillet, F. X., et al.: Effects of backbone substitutions on the conformational behavior of Shigella flexneri O-antigens: implications for vaccine strategy. Glycobiology21, 109-121 (2011). doi: 1 0. 1093/glycob/cwq136.

[63] Phalipon, A., et al.: A synthetic carbohydrate-protein conjugate vaccine candidate against Shigella flexneri 2a infection. J. Immunol. 182, 2241-2247 (2009). doi: 10. 4049/jimmunol. 0803141.

[64] Phalipon, A., et al.: Characterization of functional oligosaccharide mimics of the Shigella flexneri serotype 2a O-antigen: implications for the development of a chemically defined glycoconjugate vaccine. J. Immunol. 176, 1686-1694 (2006).

[65] Mitra, S., et al.: Outer membrane vesicles of Shigellaboydii type 4 induce passive immunity in neonatal mice. FEMS Immunol. Med. Microbiol. (2012). doi: 10. 1111/j. 1574-695X. 2012. 01004. x.

[66] Camacho, A. I., et al.: Mucosal immunization with Shigella flexneri outer membrane vesicles induced protection in mice. Vaccine 29, 8222-8229 (2011). doi: 10. 1016/j. vaccine. 2011. 08. 121.

[67] Turbyfill, K. R., Hartman, A. B., Oaks, E. V.: Isolation and characterization of a Shigella flexneri invasin complex subunit vaccine. Infect. Immun. 68, 6624-6632 (2000).

[68] Turbyfill, K. R., Kaminski, R. W., Oaks, E. V.: Immunogenicity and efficacy of highly purified invasin complex vaccine from Shigella flexneri 2a. Vaccine 26, 1353-1364 (2008). doi: 10. 1016/j. vaccine. 2007. 12. 040.

[69] Oaks, E. V., Turbyfill, K. R.: Development and evaluation of a Shigella flexneri 2a and S. sonnei bivalent invasin complex (Invaplex) vaccine. Vaccine 24, 2290-2301 (2006). doi: 10. 1016/j. vaccine. 2005. 11. 040.

[70] Norton, E. B., Lawson, L. B., Freytag, L. C., Clements, J. D.: Characterization of a mutant Escherichia coli heat-labile toxin, LT (R192G/L211A), as a safe and effective oral adjuvant. Clin. Vaccine Immunol. 18, 546-551 (2011). doi: 10. 1128/CVI. 00538-10.

[71] Martinez-Becerra, F. J., et al.: Broadly protective Shigella vaccine based on type III secretion apparatus proteins. Infect. Immun. 80, 1222-1231 (2012). doi: 10. 1128/IAI. 06174-11.

[72] Pore, D., et al.: Purification and characterization of an immunogenic outer membrane protein of Shigella flexneri 2a. Vaccine 27, 5855-5864 (2009). doi: 10. 1016/j. vaccine. 2009. 07. 054.

[73] Pore, D., Mahata, N., Pal, A., Chakrabarti, M. K.: Outer membrane protein A (OmpA) of Shigella flexneri 2a, induces protective immune response in a mouse model. PLoS One 6, e22663 (2011). doi: 10. 1371/journal. pone. 0022663.

[74] Pore, D., Mahata, N., Chakrabarti, M. K.: Outer mem-brane protein A (OmpA) of Shigella flexneri 2a links innate and adaptive immunity in a TLR2-dependent manner and involvement of IL-12 and nitric oxide. J. Biol. Chem. 287, 12589-12601 (2012). doi: 10. 1074/ jbc. M111. 335554.

（马维民，杨亚民译）

第 11 章　A 群链球菌亚单位疫苗的研制

Colleen Olive①

摘要

目前尚无预防化脓性链球菌（也称为 A 群链球菌，GAS）感染的疫苗，GAS 可导致危及生命的风湿性热（RF）和风湿性心脏病（RHD）。RF 和 RHD 在世界范围内发生，许多发展中国家的人口和发达国家的土著人口都有特别高的此类疾病发病率。大多数 GAS 疫苗的研发工作都是把细菌表面抗原 M 蛋白作为靶标。然而由于 M 蛋白含有宿主组织交叉反应性 B 细胞和 T 细胞表位，因此整个 M 蛋白抗原疫苗方法并不是可行的选择，因为 M 蛋白可能诱导自身免疫。本章将讨论基于 M 蛋白的各种亚单位疫苗方法，正在研究开发一种高效的 GAS 疫苗，包括用于临床前研究和临床开发中候选疫苗接种的不同递送技术、制剂和佐剂。

11.1　疾病

A 群链球菌（化脓性链球菌）（GAS）是一种重要的人类黏膜病原体，导致各种不同的临床表现和严重程度的疾病[1,2]（表 11.1）。咽炎（链球菌性喉炎）是 GAS 感染常见的次要并发症，但不治疗会导致危及生命的疾病，包括自身免疫后遗症风湿热（RF）和风湿性心脏病（RHD）。在每年诊断的 6.16 亿新的 GAS 咽炎病例中，估计高达 3%的病例会引发 RF，并有可能发展到 RHD[3,4]。RF 主要影响关节、大脑和心脏，导致多发性关节炎、舞蹈病和心脏炎症状[5]。RHD 导致心脏组织和瓣膜永久性损伤。

表 11.1　GAS 感染的主要临床表现

非侵袭性疾病	侵袭性疾病	感染后疾病
咽炎	蜂窝组织炎	风湿热
扁桃体炎	肺炎	风湿性心脏病
脓疱病	坏死性筋膜炎	链球菌感染后肾小球肾炎
脓皮病	中毒性休克综合征	
猩红热	丹毒	
中耳炎	脑膜炎	
	败血病	

GAS 感染每年导致超过 500 000人死亡，主要发生在发展中国家和发达国家的土著人口中，在这些国家，恶劣的社会经济条件和过度拥挤是 GAS 疾病高发的原因[3]。在发展中国家，RF 是儿童心脏病的主要原因[6]。目前还没有可用的疫苗来预防 GAS 感染，从而预防 GAS 疾病。

① C. Olive，Dr（澳大利亚昆士兰州赫斯顿昆士兰州医学研究所免疫学系，E－mail：colleen. olive@qimr. edu. au）

开发一种安全有效的 GAS 疫苗，诱导黏膜免疫以根除上呼吸道 GAS，是减轻全球 GAS 疾病负担的最佳策略，也是对公共健康的重大贡献。黏膜表面的免疫需要诱导分泌性 IgA 抗体[7]。

除了抗体，最近报道 Th17 应答对于宿主抵御黏膜 GAS 感染至关重要[8]。因此，成功的黏膜 GAS 疫苗需要刺激适当的体液免疫和细胞免疫，以防止 GAS 感染（图 11.1）。缺乏增强免疫应答至关重要的人类兼容的黏膜疫苗佐剂，这一点尤为困难。因此，研究人员主要关注肠胃外 GAS 疫苗的递送方法，这种方法有合适的佐剂，通过诱导调理 IgG 抗体来预防全身感染。

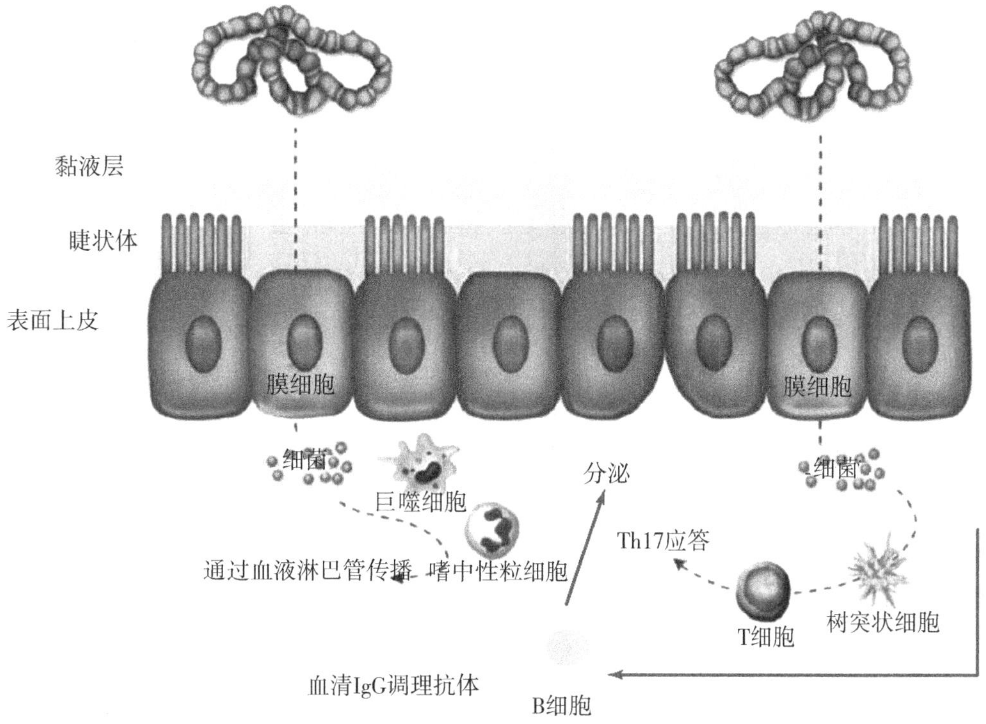

图 11.1　GAS 疫苗接种方法

GAS 破坏鼻腔相关淋巴组织黏膜上皮物理屏障，其功能与人类扁桃体相似，据称通过与膜（M）细胞的结合被转运到潜在的淋巴组织。先天免疫系统细胞感应 GAS，产生细胞因子和趋化因子来抑制黏膜感染。GAS 抗原被递送到抗原呈递细胞，如 DCs 细胞和 B 细胞。IgA 作用的 B 细胞被激活并引发抗原特异性 IgA 应答。DCs 细胞对 GAS 免疫的发展起着重要作用，向 T 细胞递呈抗原以诱导 Th17 应答，这种应答与 IgA 结合在一起用于黏膜防御咽部 GAS 定植。黏膜接种旨在模拟这些反应，并在感染时有效清除黏膜表面的 GAS，以防止 GAS 定植和运输。逃离宿主防御机制的 GAS 会扩散到淋巴管和血液中，导致全身感染。黏膜接种也能诱导全身免疫应答，其特征是诱导调理 IgG 抗体，通过调理吞噬作用破坏病原体。胃肠外 GAS 疫苗接种可诱导血清 IgG，但不能诱导黏膜免疫。

11.2　疫苗递送

有效的 GAS 疫苗需要有广泛的抗原覆盖面，因为在人群中存在许多不同的 GAS 菌株（>150 种不同的 M 型），并且不应诱导可能与自身组织蛋白交叉反应的免疫反应。GAS M 蛋白是主要的保护性抗原，是疫苗开发的理想目标；然而 M 蛋白包含心脏组织交叉反应表位，特别是在保守区域[9-11]。有证据表明，交叉反应性 T 细胞尤其在 RHD 的发病机制中起着关键作用[12]。

M 蛋白是一种 α 螺旋盘绕表面蛋白，由一个高度可变氨基末端区域和一个高度保守的（>

98%序列同一性）羧基末端C重复区域组成[13]（图11.2）。功能上，M蛋白在通过补体介导的吞噬作用防止细菌清除方面很重要，这限制了宿主防御机制[14]。先前的研究表明，针对M型特异性氨基末端区域血清型表位的调理抗体可以诱发对GAS的保护性免疫[15]。

C-重复区表位诱导的抗体通常没有调理性质，可以通过阻断细菌细胞黏附和黏膜表面的定植，而对非血清型GAS起到保护作用[16]。研究人员一直专注于使用含有保护性表位的M蛋白片段或肽作为潜在的GAS亚单位候选疫苗，同时确定合适的递送策略。

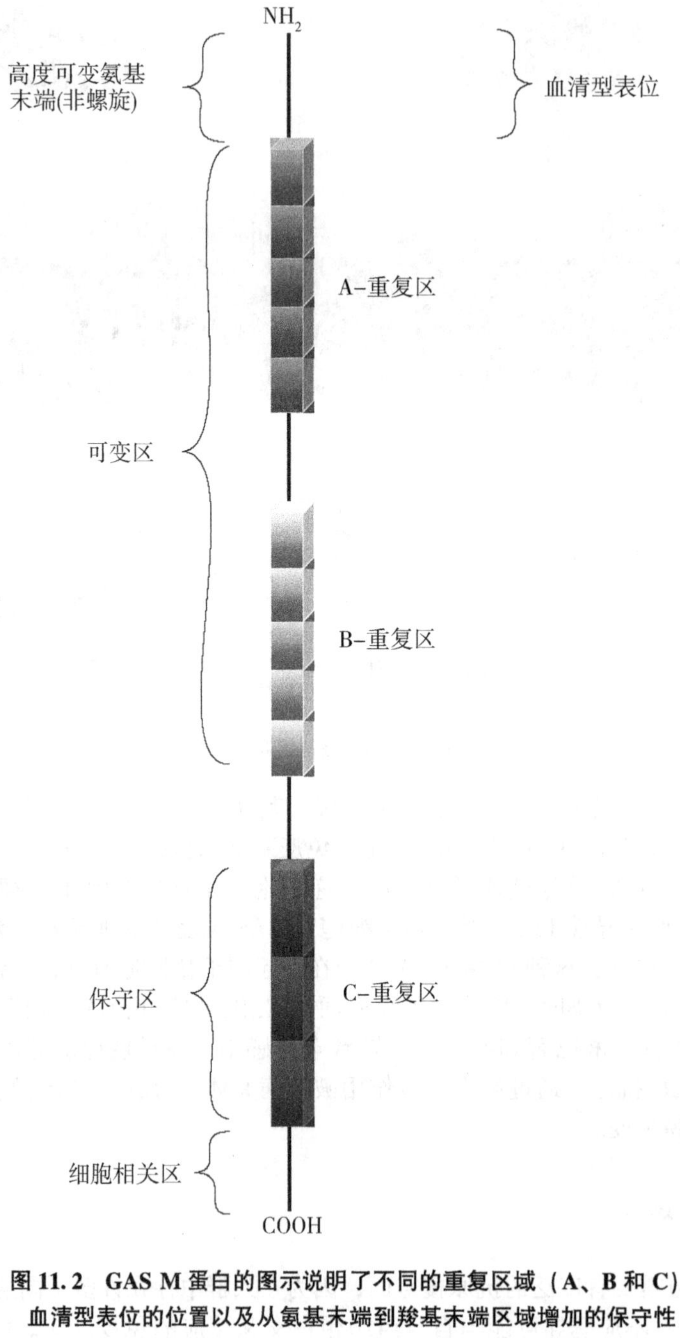

图11.2　GAS M蛋白的图示说明了不同的重复区域（A、B和C），血清型表位的位置以及从氨基末端到羧基末端区域增加的保守性

11.3 佐剂配方/化学组成

亚单位疫苗虽然安全性更高但本质上免疫原性差。免疫增强剂对于增强免疫应答的激活至关重要。目前许可用于人类疫苗的疫苗佐剂仅限于明矾、明矾-单磷酰脂质A（MPL）组合佐剂系统（AS）04、水包油乳剂MF59和AS03以及重组流感病毒体[17]。佐剂功能是增加抗原呈递细胞（APCs）对抗原的摄取，刺激先天免疫系统细胞上的模式识别受体（PRRs），从而引发先天性免疫应答。树突状细胞（DCs）将先天性免疫系统的适当信号转化为获得性免疫系统，以刺激B细胞和T细胞应答，因此在疫苗效力中发挥着关键作用。

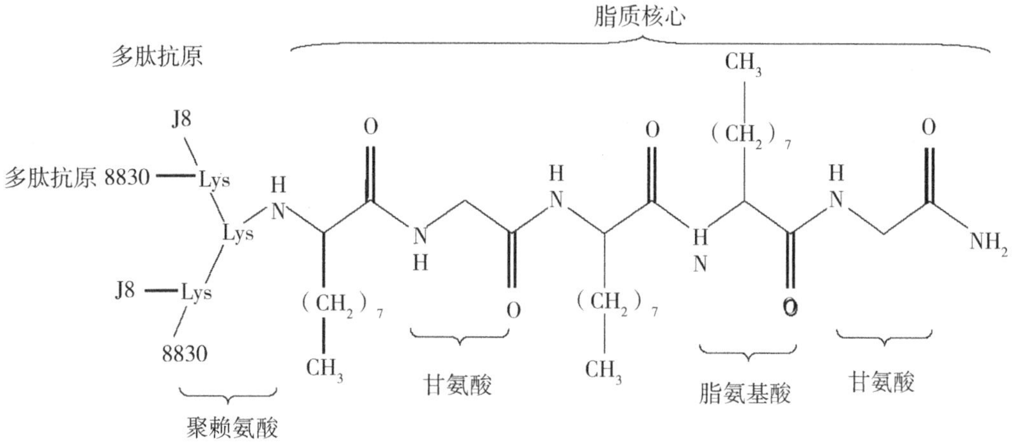

图11.3 脂质核心肽（LCP）系统

11.4 微技术/纳米技术

疫苗/佐剂递送系统提供了黏膜疫苗递送的潜力。例如脂质核心肽（LCP）系统是一种新型的、合成的、自佐剂疫苗递送系统，它将疫苗的佐剂（PRR激动剂）、载体和抗原性肽整合到单个分子实体中[18]（图11.3）。这个系统以前已经被证明能有效地递送GAS疫苗并诱导免疫[19]。有证据表明，LCP的佐剂活性涉及DC激活的诱导[20,21]。

11.5 工作机制

明矾广泛用于人类疫苗接种，但其作用机制尚不清楚。有报道表明，明矾佐剂作用是通过NLRP3炎症体的激活[22-24]、膜脂质结构的改变[25]以及细胞死亡后释放的宿主细胞DNA来介导的[26]。MF59佐剂性似乎独立于NLRP3炎症体[27]，AS03和病毒体佐剂的作用机制不太清楚。与促进Th2反应的明矾不同[28]，MPL调节免疫反应的质量有助于Th1/Th2反应达到平衡[29,30]。MPL是明尼苏达州沙门氏菌LPS的解毒衍生物，自明矾获得批准以来，MPL是第一个在人群中广泛使用的新一代明确的疫苗佐剂[17]。

MPL能增强获得性免疫而不会引起过度炎症。MPL实现有效且安全的佐剂性的机制似乎是偏置细胞信号通路的结果[31,32]。

LCP系统包含佐剂成分（由脂氨基酸制成的脂质核）和其上附着肽表位的聚赖氨酸载体。这个例子显示了一种GAS疫苗候选物，含有J8和一个名为8830的氨基末端血清型表位。该佐剂具有 3 个由甘氨酸氨基酸间隔物分离的 2-氨基十二烷酸（$n=9$）脂氨基酸。

最近也有报道称，MPL 未能激活半胱天冬酶-1，导致促炎细胞因子 IL-1β 产生缺陷和 NLRP3 炎症体激活受损[33]。

11.6 动物模型和可行性研究

在小鼠被动转移实验中已经证明，M 蛋白特异性分泌 IgA（而非 IgG）保护小鼠免受黏膜 GAS 感染，鼻咽 GAS 定植的减少表明了这一点[34]。这些发现表明，局部 IgA 免疫应答在预防 GAS 在黏膜表面定植的重要性。因此通过黏膜途径（如鼻内递送）是预防 GAS 感染的合理方法。黏膜 GAS 疫苗也可以通过诱导调理抗体引发全身免疫。

11.7 临床前开发

目前正在研究 3 种基于 M 蛋白的 GAS 亚单位疫苗的开发方法。多价方法使用代表不同 M 型的氨基末端蛋白片段组合，被设计成人群中流行的 GAS 菌株。使用这种方法，一种含有来自北美流行的 26 种不同 GAS 血清型 M 蛋白肽的重组多价 GAS 疫苗证明在动物中引发了调理抗体[35]。从流行病学数据来看，26 价疫苗将覆盖大多数咽炎和侵袭性 GAS 疾病，包括 RF、侵袭性筋膜炎和中毒性休克综合征。最近一种新的 30 价 GAS 疫苗被证明在兔子体内具有免疫原性，引发了针对“非疫苗”血清型的调理抗体，有可能产生覆盖面更广的疫苗[36]。这种疫苗针对特定人群，因此可能不是普遍有效的，也可能需要定期重新设计以反映 GAS 感染流行病学变化。

使用 M 蛋白保守 C-重复区域的肽表位 GAS 疫苗是第二种方法，在理论上，有可能扩大 M 型疫苗的覆盖面。用与载体蛋白白喉类毒素（dT）偶联的 C 区肽 GAS 疫苗候选物 J8 免疫小鼠，并与合适的佐剂联合给药，可防止全身和黏膜 GAS 感染[37-39]（图 11.4）。当 J8 与脂肽相关联时，J8 也引发了针对 GAS 的保护性免疫[40,41]。其他研究表明，用与实验性黏膜佐剂霍乱毒素 B 亚单位（CTB）结合的 C 区肽对小鼠进行鼻内免疫，在黏膜水平诱发了对 GAS 的保护性免疫[16,42,43]。CTB 可能进入中枢神经系统的嗅觉区域，在鼻内递送后引起神经元损伤[44]，因此不适合人类使用。载体递送方法包括在痘苗病毒[45]、共生细菌乳酸乳球菌[46,47]或 *gordonii* 链球菌上表达 C-重复区域[48]。

第三种组合疫苗方法使用血清型和保守的 M 蛋白肽表位。最初，通过自由基诱导丙烯酰肽聚合，合成了一种异质共聚物 GAS 疫苗构建体。将 7 个血清型表位和一个高度保守的 C 区肽表位（称为 J14）结合起来[49]。异质共聚物中的 M 型代表了澳大利亚北部地区流行的 GAS 感染——该地区高度流行 GAS。用异质共聚物免疫小鼠显示出优异的免疫原性，对同源和异源 GAS 菌株均能保护，表明有潜力提供广泛的覆盖范围[49,50]。然而批次之间的差异导致免疫反应的改变（C. Olive，未公布的数据），这限制了对人类的适用性。疫苗还需要添加有效的佐剂，由于缺乏安全有效的黏膜佐剂，进一步限制了其作为黏膜疫苗的用途。后来多表位 GAS 候选疫苗基于 LCP 系统合成，LCP 系统在注射给小鼠后诱导高调理性抗体[51,52]，在鼻内免疫后防止黏膜 GAS 感染[53]（图 11.4）。

11.8 安全性和有效性

当在 GAS 疫苗中使用大面积 M 蛋白时，主要关注的是由于与宿主蛋白的免疫交叉反应诱导自身免疫反应的可能性。因此，识别保护性抗原决定簇，并将生物相关表位与宿主组织交叉反应和潜在有害的表位分开是非常重要的。表位定位研究被用于鉴定保守的 GAS 疫苗候选 J8，该候选 J8 包含构象保护性 B 细胞表位，设计用于缺少人心脏交叉反应性 T 细胞表位[54,55]。

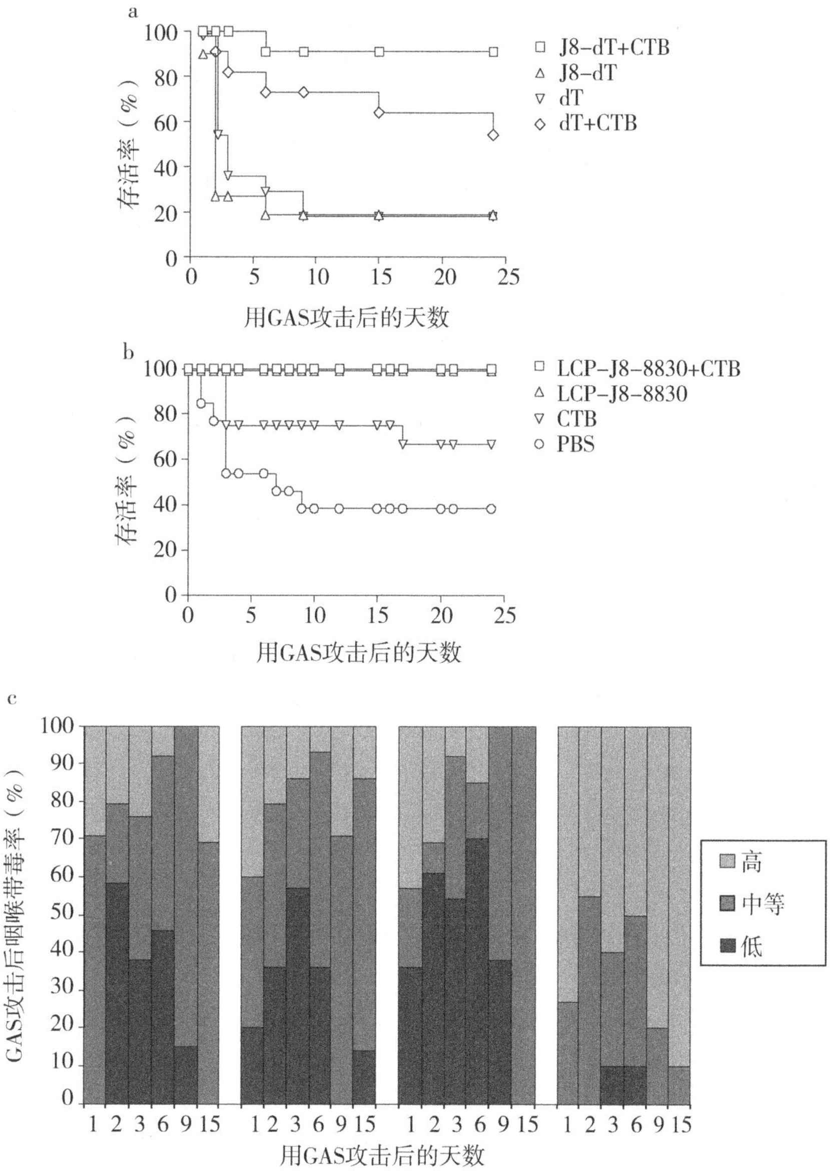

图 11.4 GAS 候选疫苗临床前评估

用 J8-dT 候选疫苗对小鼠进行鼻内免疫，与对照组相比，在用 GAS 鼻内攻击后，小鼠的存活率显著提高，但是黏膜佐剂 CTB 对于保护是必不可少的（a）。一种基于 LCP 的多表位 GAS 疫苗候选物引发了针对黏膜 GAS 感染的保护性免疫，即使没有 CTB（b）。LCP-GAS 疫苗候选物和 J8-dT 在减少咽拭子中检测到的 GAS 负担方面也很有效，但与 J8-dT 不同，LCP-GAS 疫苗候选物的功效不需要 CTB（c）（转载自 Olive 等[39]，得到 Elsevier 的许可。（b）改编自 Olive 等[53]，得到牛津大学出版社的许可。（c）转载自 Olive 等[53]，得到牛津大学出版社的许可。Olive）

11.9 临床开发

早期的人类临床试验使用矿物油乳剂中整个 M 蛋白，对患有各种胸部并发症的成年患者进行治疗[56]。虽然已诱导抗体应答，支持了 M 蛋白疫苗策略的可行性，但这些配方具有高度的反应性，许多个体在数天内经历了相当大的疼痛、酸痛或僵硬。用纯化的 1 型 M 蛋白诱导的血清调理抗体对健康成年志愿者进行免疫，用 GAS 黏膜攻击后预防临床疾病，但对上呼吸道的 GAS 定植没有影响[57]。相反地，在 21 例健康成人的鼻咽中，在用 GAS 进行黏膜攻击之前，通过雾化喷雾纯化 1 型 M 蛋白，防止了 GAS 定植和临床疾病[58]。

同样地，将纯化的 M 蛋白（3 型或 12 型）经鼻内递送（明矾不经胃肠外给药）给健康个体，导致 GAS 咽喉定植减少[59]。在这些研究中没有发生 GAS 感染的后遗症（Massell 等）[60]。报道称 21 名接种了部分纯化的 3 型 M 蛋白质的儿童，有两例出现 RF，1 例可能出现 RF。与未接种的个体相比，这种高频率的 RF 增加了用 M 蛋白接种可能导致发生 RF 的可能性[60]。因此认为完全 M 蛋白的使用不是 GAS 疫苗的可行选择，已经被只含有保护性表位的亚单位疫苗的开发所取代。

Dale 的多价亚单位 GAS 疫苗由表面 M 蛋白的末端片段制成，旨在产生针对不同 GAS 菌株的保护性抗体应答。临床试验显示，用 26 价 StreptAvax™（ID Biomedical Corporation）接种的受试者对疫苗产生了抗体反应，这些抗体没有一种与人体组织交叉反应[61,62]。接种疫苗的受试者血清显示疫苗中的 26 种血清型都有显著的杀灭作用。StreptAvax™表现出广泛的免疫原性，志愿者对其耐受性良好。

临床试验计划使用明矾配制的 J8-dT（个人交流，良好的 MF）。

11.10 优点和缺点

黏膜接种疫苗对于诱导黏膜免疫对抗黏膜病原体是非常理想的，但却是一项相当困难的任务。由于抗原摄取效率低下而导致免疫原性差，以及鉴定人类可耐受的有效黏膜佐剂，是该方法可行所需克服的关键参数。

致谢

感谢澳大利亚国家卫生和医学研究委员会、澳大利亚国家心脏基金会和查尔斯王子医院基金会提供财政支持。作者宣称她没有利益冲突。

参考文献

[1] Cunningham, M. W.: Pathogenesis of group A streptococcal infections. Clin. Microbiol. Rev. 13, 470-511 (2000).

[2] Bisno, A. L.: Group A streptococcal infections and acute rheumatic fever. N. Eng. J. Med. 325, 783-793 (1994).

[3] Carapetis, J. R., Steer, A. C., Mulholland, E. K., Weber, M.: The global burden of group A streptococcal diseases. Lancet Infect. Dis. 5, 685-694 (2005).

[4] Martin, D. R., Sriprakash, K. S.: Epidemiology of group A streptococcal disease in Australia and New Zealand. Recent. Adv. Microbiol. 4, 1-40 (1996).

[5] Stollerman, G. H.: Rheumatic fever. Lancet349, 935-942 (1997).

[6] Vijakumar, M., Narula, J., Reddy, K. S., Kaplan, E. L.: Incidence of rheumatic fever and prevalence of rheumatic heart disease in India. Int. J. Cardiol. 43, 221-228 (1994).

[7] Macpherson, A. J., Geuking, M. B., McCoy, K. D.: Homeland security: IgA immunity at the frontiers of the body. Trends Immunol. 33, 160-167 (2012).

[8] Wang, B., Dileepan, T., Briscoe, S., Hyland, K. A., Kang, J., Khoruts, A., Cleary, P. P.: Induction of TGF-β1 and TGF-β1-dependent predominant Th17 differentiation by group A streptococcal infection. Proc. Natl. Acad. Sci. U. S. A. 107, 5937-5942 (2010).

[9] Cunningham, M. W., Antone, S. M., Smart, M., Liu, R., Kosanke, S.: Molecular analysis of human

cardiac myosin-cross reactive B-and T-cell epitopes of the group A streptococcal M5 protein. Infect. Immun. 65, 3913-3923 (1997).

[10] Dale, J. B., Beachey, E. H.: Multiple cross reactive epitopes of streptococcal M proteins. J. Exp. Med. 161, 113-122 (1985).

[11] Pruksakorn, S., et al.: Identification of T cell auto-epitopes that cross-react with the C-terminal segment of the M protein of group A streptococci. Int. Immunol. 6, 1235-1244 (1994).

[12] Guilherme, L., et al.: Human heart-infiltrating T-cell clones from rheumatic heart disease patients recognize both streptococcal and cardiac proteins. Circulation92, 415-420 (1995).

[13] Fischetti, V. A.: Streptococcal M protein: molecular design and biological behavior. Clin. Microbiol. Rev. 2, 285-314 (1989).

[14] Horstmann, R. D., Sieversten, H. J., Knobloch, J., Fischetti, V. A.: Antiphagocytic activity of streptococcal M protein: selective binding of complement control protein factor H. Proc. Natl. Acad. Sci. U. S. A. 85, 1657-1661 (1988).

[15] Dale, J. B., Beachey, E. H.: Localization of protective epitopes of the amino terminus of type 5 streptococcal M protein. J. Exp. Med. 163, 1191-1202 (1986).

[16] Bessen, D., Fischetti, V. A.: Synthetic peptide vaccine against mucosal colonization by group A streptococci. I. Protection against a heterologous M serotype with shared C repeat region epitopes. J. Immunol. 145, 1251-1256 (1990).

[17] Montomoli, E., Piccirella, S., Khadang, B., Mennitto, E., Camerini, R., De Rosa, A.: Current adjuvants and new perspectives in vaccine formulation. Expert Rev. Vaccines10, 1053-1061 (2011).

[18] Toth, I., Danton, M., Flinn, N., Gibbons, W. A.: A combined adjuvant and carrier system for enhancing synthetic peptides immunogenicity utilising lipidic amino acids. Tetrahedron Lett. 34, 3925-3928 (1993).

[19] Olive, C., Moyle, P. M., Toth, I.: Towards the development of a broadly protective group a streptococcal vaccine based on the Lipid-Core Peptide system. Curr. Med. Chem. 14, 2976-2988 (2007).

[20] Yong, M., Mitchell, D., Caudron, A., Toth, I., Olive, C.: Expression of maturation markers on murine dendritic cells in response to group A streptococcal lipopeptide vaccines. Vaccine 27, 3313-3318 (2009).

[21] Phillipps, K. S., Wykes, M. N., Liu, X. Q., Brown, M., Blanchfield, J., Toth, I.: A novel synthetic adjuvant enhances dendritic cell function. Immunology128 (1 Suppl), e582-e588 (2009).

[22] Eisenbarth, S. C., Colegio, O. R., O'Connor, W., Sutterwala, F. S., Flavell, R. A.: Crucial role for the Nalp3 inflammasome in the immunostimulatory properties of aluminium adjuvants. Nature453, 1122-1126 (2008).

[23] Kool, M., et al.: Cutting edge: alum adjuvant stimulates inflammatory dendritic cells through activation of the NALP3 inflammasome. J. Immunol. 181, 3755-3759 (2008).

[24] Li, H., Willingham, S. B., Ting, J. P., Re, F.: Cutting edge: inflammasome activation by alum and alum's adjuvant effect are mediated by NLRP3. J. Immunol. 181, 17-21 (2008).

[25] Flach, T. L., et al.: Alum interaction with dendritic cell membrane lipids is essential for its adjuvanticity. Nat. Med. 17, 479-487 (2011).

[26] Marichal, T., et al.: DNA released from dying host cells mediates aluminum adjuvant activity. Nat. Med. 17, 996-1002 (2011).

[27] Seubert, A., et al.: Adjuvanticity of the oil-in-water emulsion MF59 is independent of Nlrp3 inflammasome but requires the adaptor protein MyD88. Proc. Natl. Acad. Sci. U. S. A. 108, 11169-11174 (2011).

[28] Brewer, J. M.: (How) do aluminium adjuvants work? Immunol. Lett. 102 (10-15) (2006).

[29] Garçon, N., Van Mechelen, M.: Recent clinical experience with vaccines using MPL-and QS-21-containing adjuvant systems. Expert Rev. Vaccines10, 471-486 (2011).

[30] Casella, C. R., Mitchell, T. C.: Putting endotoxin to work for us: monophosphoryl lipid A as a safe and effective vaccine adjuvant. Cell. Mol. Life Sci. 65, 3231-3240 (2008).

[31] Mato-Haro, V., Cekic, C., Martin, M., Chilton, P. M., Casella, C. R., Mitchell, T. C.: The vaccine adjuvant monophosphoryl lipid A as a TRIF-biased agonist of TLR4. Science316, 1628-1632 (2007).

[32] Cekic, C., Casella, C. R., Eaves, C. A., Matsuzawa, A., Ichijo, H., Mitchell, T. C.: Selective activation of the p38 MAPK pathway by synthetic monophosphoryl lipid A. J. Biol. Chem. 284, 31982-31991 (2009).

[33] Embry, C. A., Franchi, L., Nuñez, G., Mitchell, T. C.: Mechanism of impaired NLRP3 inflammasome priming by monophosphoryl lipid A. Sci. Signal. 4, ra28 (2011).

[34] Bessen, D., Fischetti, V. A.: Passive acquired mucosal immunity to group A streptococci by secretory immunoglobulin A. J. Exp. Med. 167, 1945-1950 (1988).

[35] Hu, M. C., Walls, M. A., Stroop, S. D., Reddish, M. A., Beaull, B., Dale, J. B.: Immunogenicity of a 26-valent group A streptococcal vaccine. Infect. Immun. 70, 2171-2177 (2002).

[36] Dale, J. B., Penfound, T. A., Chiang, E. Y., Walton, W. J.: New 30-valent M protein-based vaccine evokes cross-opsonic antibodies against non-vaccine serotypes of group A streptococci. Vaccine29, 8175-8178 (2011).

[37] Batzloff, M. R., et al.: Towards the development of an anti-disease, transmission blocking vaccine for *Streptococcus pyogenes*. J. Infect. Dis. 192, 1450-1455 (2005).

[38] Batzloff, M. R., Hayman, W. A., Davies, M. R., Zeng, M., Pruksakorn, S., Brandt, E. R., Good, M. F.: Protection against group A streptococcus by immunization with J8-diphtheria toxoid: Contribution of J8-and diphtheria toxoid-specific antibodies to protection. J. Infect. Dis. 187, 1598-1608 (2003).

[39] Olive, C., Clair, T., Yarwood, P., Good, M. F.: Protection of mice from group A streptococcal infection by intranasal immunization with a peptide that contains a conserved M protein B-cell epitope and lacks a T-cell autoepitope. Vaccine20, 2816-2825 (2002).

[40] Batzloff, M. R., Hartas, J., Zeng, W., Jackson, D. C., Good, M. F.: Intranasal vaccination with a lipopeptide containing a minimal, conformationally constrained conserved peptide, a universal T-cell epitope and a self-adjuvanting lipid protects mice from*Streptococcus pyogenes* and reduces throat carriage. J. Infect. Dis. 194, 325-330 (2006).

[41] Olive, C., et al.: Immunisation of mice with a lipid core peptide construct containing a conserved region determinant of group A streptococcal M protein elicits heterologous opsonic antibodies in the absence of adjuvant. Infect. Immun. 70, 2734-2738 (2002).

[42] Bessen, D., Fischetti, V. A.: Influence of intranasal immunization with synthetic peptides corresponding to conserved epitopes of M protein on mucosal colonization by group A streptococci. Infect. Immun. 56, 2666-2672 (1988).

[43] Bronze, M. S., Courtney, H. S., Dale, J. B.: Epitopes of group A streptococcal M protein that evoke cross-protective local immune responses. J. Immunol. 148, 888-893 (1992).

[44] van Ginkel, F. W., Jackson, R. J., Yuki, Y., McGhee, J. R.: Cutting edge: the mucosal adjuvant cholera toxin redirects vaccine proteins into olfactory tissues. J. Immunol. 165, 4778-4782 (2000).

[45] Fischetti, V. A., Hodges, W. M., Hruby, D. E.: Protection against streptococcal pharyngeal colonization with a vaccinia: M protein recombinant. Science244, 1487-1490 (1989).

[46] Mannam, P., Jones, K. F., Geller, B. L.: Mucosal vaccine made from live, recombinant *Lactococcuslactis* protects mice against pharyngeal infection with *Streptococcus pyogenes*. Infect. Immun. 72, 3444-3450 (2004).

[47] Bolken, T. C., et al.: Analysis of factors affecting sur-face expression and immunogenicity of recombinant proteins expressed by gram-positive commensal vectors. Infect. Immun. 70, 2487-2491 (2002).

[48] Kotloff, K. L., Wasserman, S. S., Jones, K. F., Livio, S., Hruby, D. E., Franke, C. A., Fischetti, V. A.: Clinical and microbiological responses of volunteers to combined intranasal and oral inoculation with a *Streptococcusgordonii* carrier strain intended for future use as a group A streptococcus vaccine. Infect. Immun. 73, 2360-2366 (2005).

[49] Brandt, E. R., et al.: New multi-determinant strategy for a group A streptococcal vaccine designed for the Australian Aboriginal population. Nat. Med. 6, 455-459 (2000).

[50] Dunn, L. A., McMillan, D. J., Batzloff, M., Zeng, W., Jackson, D. C., Upcroft, J. A., Upcroft, P., Olive, C.: Parenteral and mucosal delivery of a novel multi-epitope M protein-based group A streptococcal vaccine construct: investigation of immunogenicity in mice. Vaccine20, 2635-2640 (2002).

[51] Olive, C., Ho, M. -F., Dyer, J., Lincoln, D., Barozzi, N., Toth, I., Good, M. F.: Immunization with a tetraepitopic lipid core peptide vaccine construct induces broadly protective immune responses against group A streptococcus. J. Infect. Dis. 193 (1666-1676) (2006).

[52] Olive, C., Batzloff, M. R., Horváth, A., Clair, T., Yarwood, P., Toth, I., Good, M. F.: The potential of lipid core peptide technology as a novel self-adjuvanting vaccine delivery system for multiple different synthetic peptide immunogens. Infect. Immun. 71, 2373-2383 (2003).

[53] Olive, C., Kuo Sun, H., Ho, M. -F., Dyer, J., Horvath, A., Toth, I., Good, M. F.: Intranasal administration is an effective mucosal vaccine delivery route for self-adjuvanting lipid core peptides targeting the group A streptococcal M protein. J. Infect. Dis. 194 (316-324) (2006).

[54] Relf, W. A., et al.: Mapping a conserved conformational epitope from the M protein of group A streptococci. Pept. Res. 9, 12-20 (1996).

[55] Hayman, W. A., Brandt, E. R., Relf, W. A., Cooper, J., Saul, A., Good, M. F.: Mapping the minimal murine T cell and B cell epitopes within a peptide vaccine candidate from the conserved region of the M protein of group A streptococcus. Int. Immunol. 9, 1723-1733 (1997).

[56] Wolfe, C. K., Hayashi, J. A., Walsh, G., Barkulis, S. S.: Type-specific antibody response in man to infections of cells walls and M protein from group A, type 14 streptococci. J. Lab. Clin. Med. 61, 459-468 (1963).

[57] Fox, E. N., Waldman, R. H., Wittner, M. K., Mauceri, A. A., Dorfman, A.: Protective study with a group A streptococcal M protein vaccine. Infectivity challenge of human volunteers. J. Clin. Invest. 52, 1885-1892 (1973).

[58] Polly, S. M., Waldman, R. H., High, P., Wittner, M. K., Dorfman, A., Fox, E. N.: Protective studies with a group A streptococcal M protein vaccine. II. Challenge of volunteers after local immunisation in the upper respiratory tract. J. Infect. Dis. 131, 217-224 (1975).

[59] D'Alessandri, R., Plotkin, G., Kluge, R. M., Wittner, M. K., Fox, E. N., Dorfman, A., Waldman, R. H.: Protective studies with group A streptococcal M protein vaccine. III. Challenge of volunteers after systemic or intranasal immunization with type 3 or type 12 group A streptococcus. J. Infect. Dis. 138, 712-718 (1978).

[60] Massell, B. F., Honikman, L. H., Amezcua, J.: Rheumatic fever following streptococcal vaccination. Report of three cases. JAMA207, 1115-1119 (1969).

[61] Kotloff, K. L., et al.: Safety and immunogenicity of a recombinant multivalent group A streptococcal vaccine in healthy adults. Phase 1 Trial. JAMA292, 709-715 (2004).

[62] McNeil, S. A., et al.: Safety and immunogenicity of 26-valent group A streptococcus vaccine in healthy adult volunteers. Clin. Infect. Dis. 41, 1114-1122 (2005).

（张杰译）

第 12 章　疟疾寄生虫疫苗接种：模式、风险和进展

Noah S. Butler①

摘要

人类感染疟原虫引起的疟疾仍然是一个全球性的卫生紧急事件，每年有超过 2 亿的新病例以及数十万人死亡。尽管抗严重疟疾的天然获得性耐药性会随着年龄的增长和反复接触疟原虫而产生，但目前控制疟原虫感染仍然严重依赖于使用抗疟疾化学疗法。然而持续选择耐药性寄生虫会继续阻碍有效控制这一主要人类病原体引起的疾病的发病率和患病率。

因此，开发有效的疟原虫疫苗仍然是改善全球公共卫生的主要目标。尽管几十年来作出了巨大努力，但目前还没有获得许可的疟疾疫苗。有几个因素导致了研发抗疟疾疫苗的困难，最显著的是疟原虫复杂的多宿主、多阶段发育生命周期。此外，我们对宿主控制或清除疟原虫所必需的免疫需求了解有限，目前仍不清楚。重要的是，最近的工作提高了我们对宿主-疟原虫相互作用的理解，为通过预防性疫苗接种增强抗疟疾免疫的新策略提供了重要的见解。

基于这些新概念和范例，亚单位、载体和全寄生虫疫苗接种领域已经取得了令人兴奋的进展，可以预防肝期和红内期疟原虫感染。本章综述了疟原虫感染的生物学和发病机制，强调最新的进展，并举例说明疟疾寄生虫疫苗接种的剩余障碍。

12.1　引言

疟疾是在感染疟原虫后发生的。这种疾病仍然是全球性的健康危机，每年有超过 65 万人死亡，超过 2.15 亿的新临床病例[1,2]。迄今为止，超过 80%与疟疾相关的死亡发生在撒哈拉以南的非洲地区[2]。用有效的工具对抗感染，包括治疗疟疾的药物和控制疟原虫传播的蚊子媒介杀虫剂。然而继续选择耐药性寄生虫[3]和耐化学性蚊子[4]，这些药物的广泛使用进一步使控制疟疾的努力复杂化[5]。基于这些原因，国际社会正在接受通过生产安全有效的疫苗来实现根除疟疾的长期目标。

事实证明，疫苗对多种致病微生物有效，包括导致天花、风疹和百日咳的病原体[6]。事实上，获得许可的抗疟疾疫苗将是对抗这种毁灭性疾病的强有力武器，大大缩小使用治疗性和预防性药物、杀虫剂和蚊帐的差距。鉴于世界上近 50%的人口生活在疟疾传播流行地区[2]，人们很容易理解，即使是一种稍微有效的疫苗，每年也能保护成千上万的人免受疟疾侵害。

显然，研发抗疟疾疫苗的动机很强。然而尽管经过 50 多年的努力，世界上仍然没有一种获得完全批准的疫苗能够提供对疟原虫感染的杀菌免疫。如果没有一种疫苗能够完全防止寄生虫感染或传播，就不可能实现消灭疟疾的目标。目前缺乏有效的疟疾疫苗是由于许多可能因素造成的，包括资金不足、寄生虫生命周期的复杂性[7,8]以及田间分离株之间广泛的抗原变异[9]。虽然这些都是重大障碍，但开发抗疟疾疫苗的最大障碍可能是我们对寄生虫和脊椎动物宿主免疫

① N. S. Butler，博士（俄克拉何马大学健康科学中心微生物学与免疫学系，俄克拉荷马市，俄克拉荷马州，美国，E-mail：noah-butler@ouhsc.edu）

系统之间复杂关系的不完全理解。然而随着疟原虫分子生物学和反向遗传学工具的出现，再加上抗疟疾疫苗新方法的发展，我们正朝着有效的、得到许可的抗疟疾疫苗迈进。在这一章中我们将重点关注寄生虫的生物学/生命周期、疟疾发病机制和疾病之间的关系，以及对这些过程相互联系的理解如何既识别了陷阱又强化了抗疟疾疫苗接种策略。

12.2　疟原虫的生命周期和传播

疟原虫属顶复亚门、疟原虫属。包括隐芽孢虫和弓形虫在内的所有顶复门寄生虫的定义特征是“顶端复合体”的存在，包括介导宿主细胞入侵的生物机制[10]。疟原虫可以存在于两个宿主中：脊椎动物是中间宿主，其繁殖是无性的；按蚊是其终末宿主，在按蚊内是有性繁殖的[11]。雌性按蚊也负责宿主间的传播，因此起到疟疾传播媒介的作用。当按蚊在吸血期间传播时，疟原虫芽孢首先沉积在脊椎动物宿主真皮中[12]（图 12.1）。

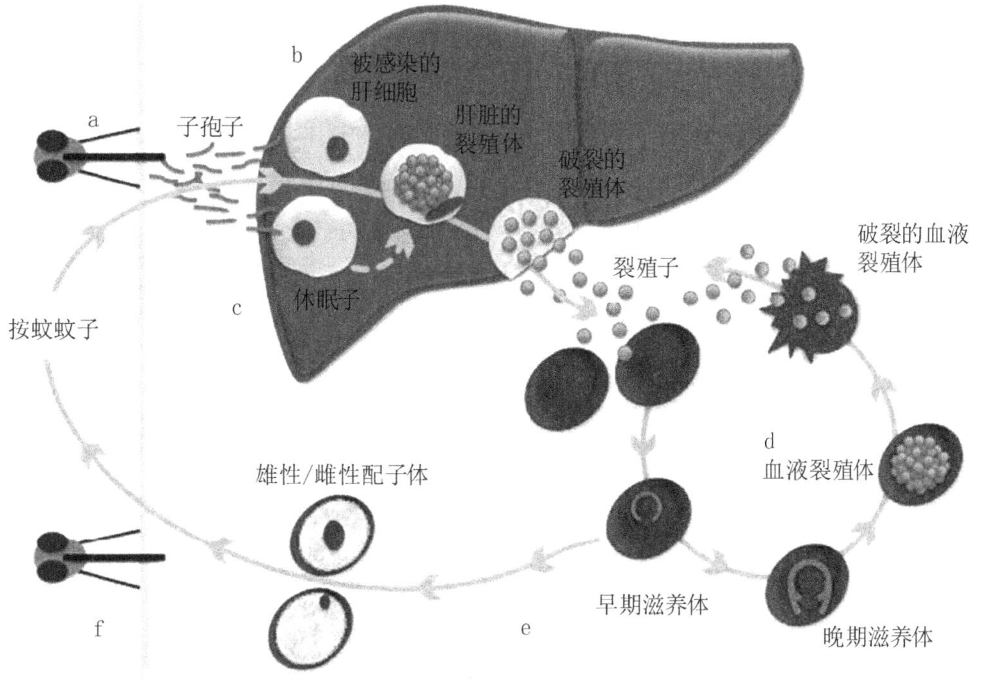

图 12.1　疟原虫的生命周期

（a）在吸血期间，受疟原虫感染的按蚊将子芽孢接种到脊椎动物宿主的真皮中。（b）子芽孢从真皮迁移到血流中，然后被动转移到肝脏，侵入肝细胞。寄生虫经历了数量巨大扩张的阶段，形成多核裂殖体并分化成裂殖子。（c）休眠体是静止的肝期形式，仅在间日疟原虫或卵圆形疟原虫感染期间发育。肝期的疟原虫感染不会引起临床症状，但是随着重新激活并释放到循环系统中，延迟发作或复发的疾病会在最初感染后数月内发生。（d）裂殖体破裂，释放裂殖子进入循环，侵入红细胞，导致脊椎动物宿主的临床疾病。（e）一些裂殖子分化成雄性或雌性配子体。（f）如果每种配子体在喂养过程中被一只新的按蚊摄入，这些细胞将融合形成卵囊，随后成熟并在中肠进一步分化为子芽孢。中肠子芽孢对脊椎动物没有感染性。然而一旦它们离开蚊子的中肠，会随着迁移到唾液腺而成熟为传染性的形式。这种按蚊现在有能力在新的脊椎动物宿主中完成传播周期。

在蚊子叮咬的过程中，只接种了几十到几百个子芽孢[13]，因此初次传播后寄生虫的负担相对较低。值得注意的是，这一数字瓶颈是通过接种疟原虫子芽孢或子芽孢表达抗原来努力阻止或限制感染的主要原因（下面讨论）。在没有预先存在体液免疫的情况下，子芽孢通过循环畅通无阻地进入肝脏。到达肝脏后寄生虫会沿着（开放的内皮网络）有窗孔的肝窦“滑动”[14,15]。子芽孢滑动活性是由两种主要抗疟疾疫苗候选抗原之间的特异性相互作用介导的，这两种抗原是环子芽孢蛋白（CSP）和血小板反应蛋白相关黏附蛋白（TRAP）以及硫酸乙酰肝素蛋白多糖，它们通过内皮细胞网络中的窗孔从亚正弦细胞外基质中突出[16,17]。

值得注意的是，CSP是一种主要的、高度表达的表面蛋白，在芽孢穿过宿主组织时从芽孢中分离出来。因此几十年来CSP一直是抗肝期疫苗的主要靶蛋白（表12.1、表12.2和表12.3）。当一个滑动的子芽孢遇到肝脏内的巨噬细胞，即Kupffer细胞（KC），寄生虫就会停止活动，然后通过涉及额外的疫苗抗原靶点、动合子和子芽孢的细胞穿过蛋白（CelTOS）机制[18,19]，继续活跃地穿过KC[14]。KC穿过的第一步最近被证明可以增强肝细胞的芽孢感染性，肝细胞是疟原虫芽孢的最终靶细胞。有趣的是，在最终停留在肝脏中之前，子孢子穿过一些肝细胞。单个子芽孢对肝细胞的每一次有效感染都会产生数万个裂殖子，这是一种可感染红细胞的寄生虫分化形式[20]（RBC）。因此，疟原虫感染肝期代表着寄生虫数量的巨大扩张期，对感染人类的疟原虫物种来说，这一时期持续6~16天[21]，完全没有症状[22]。

表12.1　目前正在临床评估的主要疟原虫候选疫苗抗原

肝期抗原	基因功能	免疫机制	参考文献
CSP	子芽孢运动性、感染性	CD4 T细胞和抗体	[108，109]
TRAP	子芽孢运动性	CD4 T细胞和抗体	[110]
CelTOS	子芽孢细胞穿越，感染性	CD8 T细胞和抗体	[19]
红内期抗原AMA-1	红细胞结合	抗体	[25，26，108，109，111]
MSP-1	红细胞结合/感染性	抗体	[27，28，112，113]
MSP-3	红细胞结合/感染性	抗体	[29，30，114]
EBA175	红细胞感染性	抗体	[31，115]
SERA5（SE36）	红细胞流出	抗体	[116]
GLURP/MSP-3	红细胞感染性（GMZ2）	抗体	[117，118]

表12.2　严重疟原虫感染引起的疾病表现

疾病名称	临床特征	机制
脑疟疾	昏迷、认知功能受损、无意识、长期神经后遗症	脑寄生虫隔离，中枢神经系统微血管阻塞，内皮细胞活化，细胞黏附、炎症、中性粒细胞和T细胞募集和激活
胎盘疟疾	初生体重低、早产、胎儿丢失	PfEMP-1（VAR2CSA）介导的胎盘内皮细胞黏附、炎症、中性粒细胞和T细胞募集和激活
重度贫血	昏睡	抑制造血，破坏红细胞，血红蛋白损失，清除未受感染的红细胞
酸中毒	呼吸窘迫，低血容量	广泛的寄生虫隔离、高寄生虫负担和相关的寄生虫糖酵解、糖原异生减少、血管通透性增加、贫血

表12.3　目前临床评价中的分子抗疟疾疫苗

亚单位疫苗	目标生命周期阶段	抗原	佐剂	免疫途径
RTS，S/AS01	肝期	CSP，重组蛋白（病毒样颗粒）	AS01[a]	肌内注射

（续表）

亚单位疫苗	目标生命周期阶段	抗原	佐剂	免疫途径
PfCelTOS FMP012	肝期	CelTOS，重组蛋白	GLA-SE[b]	肌内注射
EBA175 RII	红内期	EBA175，重组蛋白	Aluminum phosphate	肌内注射
FMP2.1/ASO2A	红内期	AMA-1，重组蛋白	ASO2A[c]	肌内注射
GMZ2	红内期	GLURP + MSP - 3，重组蛋白	Alum，DDA-TDB[d]	肌内注射
MSP-3$_{181-276}$	红内期	MSP-3，合成肽	Alum，Montanide ISA 720[e]	皮下注射
SE36	红内期	SERA5，重组蛋白	Aluminum hydroxide gel	皮下注射
JAIVAC	红内期	MSP - 1 + EBA175，重组蛋白	Montanide ISA 720	肌内注射
AMA1-C1/Alhydrogel ® + CPG 7909	红内期	AMA1 - C1，重组蛋白	Alhydrogel ® [f] CPG 7909[g]	肌内注射
BSAM-2/Alhydrogel ® +CPG 7909	红内期	MSP - 1 + AMA - 1，重组蛋白	Alhydrogel ® CPG 7909	肌内注射
CSP，AMA - 1 virosomes	肝期和红内期	CSP+AMA-1，病毒体平中的模拟表位	N/A	肌内注射
DNA 疫苗	目标生命周期阶段	抗原	佐剂	免疫途径
多表位 DNA EP1300	肝期	CSP，TRAP，LSA-1，EXP-1 带有接头的多个表位	N/A	肌内注射（电穿孔）
NMRC - M3V - D/Ad-PfCA（加强免疫）	肝期和红内期	CSP+AMA-1，DNA 初免+腺载体加强（Ad5 血清型）	N/A	肌内注射（喷射注射）/肌内注射
载体疫苗	**目标生命周期阶段**	**抗原**	**佐剂**	**免疫途径**
Adenovirus（Ad35）CS 和 RTS，S（加强免疫）	肝期	CSP，重组蛋白和腺病毒载体	N/A	肌内注射/肌内注射
ChAd63/MVA ME-TRAP	肝期	TRAP，CSP，LSA-1，LSA - 3，STARP，EXP-1 黑猩猩腺病毒启动和改良的安卡拉痘苗病毒加强	N/A	肌内注射/肌内注射
Ad35 vectored CS	肝期	CSP，复制缺陷型腺病毒血清型 35	N/A	肌内注射
Ad35 CS prime Ad26 CS boost	肝期	CSP，复制缺陷型腺病毒血清型 35 和 26 的免疫加强	N/A	肌内注射

（续表）

载体疫苗	目标生命周期阶段	抗原	佐剂	免疫途径
ChAd63/MVA MSP-1	红内期	MSP-1，黑猩猩腺病毒启动和改良的安卡拉痘苗病毒加强	N/A	肌内注射
减毒子芽孢疫苗	**目标生命周期阶段**	**抗原**	**佐剂**	**免疫途径**
PfSPZ	肝期	全子芽孢，辐射减毒	N/A	皮下或皮内注射（最近用静脉注射尝试）
Pf GAP p52-/p32-	肝期	全子芽孢，遗传减毒	N/A	蚊虫叮咬

[a]脂质体和免疫刺激剂 MPL 和 QS21 的混合物：MPL（3-O-去酰基-4′-单磷酰基脂质 A）、QS21（从皂树树皮中纯化的糖苷）

[b]合成 Toll 样受体（TLR）4 激动剂制剂、吡喃葡萄糖基脂佐剂稳定乳剂（GLA-SE）

[c]油包水乳液中的单磷酰脂质 A 和 QS21

[d]角鲨烯基油包水佐剂制剂

[e]二甲基二十八烷基铵（DDA）和海藻糖二苯甲酸酯（TDB）

[f]氢氧化铝凝胶® 是一种灭菌氢氧化铝湿凝胶悬浮液

[g]CPG 7909（VaxImmune™），一种 CPG 寡脱氧核苷酸（ODN）

受感染的肝细胞最终破裂或起出包裹着数百个裂殖子的多个宿主细胞膜泡[23,24]。这些所谓的裂殖子进入循环，似乎可以防止新释放的裂殖子接触宿主免疫清除途径[24]。肝脏中的裂殖小体和裂殖子的释放标志着疟原虫感染和疟疾疾病临床阶段的开始。每个裂殖子可以附着并感染一个红细胞。这些相互作用再次被宿主和寄生虫之间的特定受体配体相互作用所介导，并在很大程度上决定特定种类的寄生虫能否在特定种类的脊椎动物宿主中复制并引起疾病。毫不奇怪，许多裂殖子表达的红细胞结合蛋白也是目前临床试验中正在评估的主要疫苗候选抗原，包括顶膜蛋白-1（AMA-1）[25,26]、裂殖子表面蛋白-1（MSP-1）[27,28]、MSP-3[29,30]和红细胞结合抗原175（EBA175）[31]（表 12.1、表 12.2 和表 12.3）。红细胞侵入后，裂殖子发育成环状，再发育成滋养体，最后分裂成子裂殖子。这种成熟和分裂通常需要 48 小时，之后红细胞会破裂，释放出裂殖子，无性繁殖循环会重复。通过不太为人知的过程，一些环状滋养体发展成有性阶段的寄生虫，雄性和雌性配子体。值得注意的是，配子体表达了一系列独特的细胞表面抗原，如相对于无性裂殖子的 pfs230 和 pfs48/45[32]或 pfs25[33]。因此几个性阶段抗原可能作为候选疫苗抗原，旨在通过防止配子体存活或在蚊子宿主中融合来限制疟疾传播[34]。最后如果雄性和雌性配子体都被一只蚊子在吸血过程吸收，性阶段就可以开始了。性阶段包括雄性和雌性配子体的融合，在蚊子的肠壁形成卵母细胞。卵母细胞最终成熟为卵囊，随后发育成子芽孢，最终从肠道迁移到蚊子的唾液腺，有效地完成了疟原虫的生命周期。

12.3 临床疾病和发病机制

有 5 种疟原虫感染人类并引起临床疾病。尽管间日疟原虫、卵圆疟原虫和诺氏疟原虫致病性较低，临床感染较少，对公共健康的负担较小，但间日疟原虫和恶性疟原虫会导致显著的发病率。间日疟原虫经常与伴有或不伴有贫血的急性热病相关，但这种感染通常不会致命。尽管如此，与恶性疟原虫疟疾相比，地理上有更多的人有患间日疟原虫疟疾的风险[35]。相比之下，恶性疟原虫是最严重临床疾病（严重疟疾）的罪魁祸首，也是疟原虫相关死亡率的主要原因[2]。

如上所述，疟原虫表现出复杂的多宿主、多阶段生命周期，疟疾与寄生虫的生物学和生命周期有关。子芽孢接种及其随后对肝脏的感染完全没有症状。然而红内期裂殖子通过其周期性感染波和红细胞破坏驱动了临床疾病的整个范围。此外，已知引起人类发病率的疟原虫种类以同步方式复制，大约每48小时从红细胞中分裂出来。红细胞同步破裂通常表现为回归热、不适、寒战和出汗。疟疾寄生虫对宿主红细胞不同成熟阶段的趋性与临床严重程度之间也有很好的关系。

例如，间日疟原虫优先感染Duffy血抗原阳性[36]网织红细胞（占所有红细胞的<1%未成熟红细胞），并导致较轻的疾病，而恶性疟原虫同时感染网织红细胞和正常细胞（成熟红细胞），导致更严重的疾病[37]。

疾病模式和表现也在很大程度上取决于暴露史和宿主免疫状态。事实上，随着恶性疟原虫流行区成人的大部分接触，寄生虫有效地感染导致明显的红内期感染，但感染仍保持临床沉默[38]。后一项观察强调，暴露诱导的自然获得性免疫（NAI）可以随着时间而发展。然而这种免疫形式被广泛认为是为了预防严重疾病，而不是感染本身[38-40]。与此相反，恶性疟原虫个体发展为严重疾病的风险最大，包括单器官、多器官或系统性表现（表12.2）。最严重的疾病表现包括代谢性酸中毒、脑疟疾和严重疟疾性贫血。

疟疾发病机制不统一，差异很大。代谢性酸中毒在很大程度上与寄生虫从葡萄糖厌氧糖酵解中获取能量的能力有关[41]。由于寄生虫密度通常超过50 000或更多寄生虫/μL血液，因此低血糖和代谢性酸中毒是严重疟疾的常见表现就不足为奇了[42,43]。寄生虫密度高也是导致恶性疟原虫感染者出现严重贫血的主要因素。疟疾感染期间严重贫血通常与受感染红细胞中血红蛋白的降解有关，红细胞是寄生虫的主要能源。然而贫血也源于受感染和未受感染红细胞的清除：未受感染红细胞可以获得寄生虫糖基磷脂酰肌醇（GPI），有效地靶向它们以在脾脏中清除[44]。红细胞溶解也会释放寄生虫衍生的GPI，GPI会引起系统生成促炎性细胞因子，如TNF和IL-1等[45-47]。已知这些促炎细胞因子可抑制造血，从而进一步加重了贫血的严重程度。

除了寄生虫对宿主红细胞的代谢和物理修饰外，大量研究支持受感染红细胞（iRBC）的细胞黏附和隔离是疾病的主要机制[48,49]。寄生虫感染的红细胞在特定靶器官中隔离会导致血流受阻、炎症级联反应和内皮细胞破坏。中枢神经系统内的炎症和缺血性闭塞是这种疾病最致命形式——脑疟疾的原因。感染和未感染的红细胞和血管内皮细胞之间的细胞黏附是由一类寄生虫表达的蛋白质介导的，包括恶性疟原虫红细胞膜蛋白-1（PfEMP-1/ VAR2CSA）[50]。PfEMP-1蛋白结合宿主蛋白，如血管内皮细胞上的ICAM-1、内皮细胞和血小板上的CD36以及主要在胎盘中表达的CSA。事实上，孕妇和胎儿极易患疟疾，疟疾导致贫血、初生体重过低、早产和大量婴儿死亡[51]。通过疫苗接种以PfEMP-1蛋白作为目标似乎是预防严重妊娠相关疟疾的合理策略，但是这一蛋白家族的广泛多态性使努力受挫[52]。总的来说，疟原虫寄生虫会诱导红细胞膜发生一系列深刻变化，使红细胞膜不易变形，并具有高度黏附性[53,54]，这就加剧了与微血管堵塞和脑疟疾发展相关的问题。

12.4　诊断标准和经典疗法

疟疾的诊断可以通过4种不同的模式进行：临床诊断、显微镜诊断、抗原检测和分子诊断。临床诊断标准包括患者出现发热、寒战、出汗、头痛、肌肉疼痛以及可能的恶心和呕吐等症状。然而值得注意的是，疟疾临床症状与疟疾流行地区其他儿童期感染的症状非常相似。基于这些原因，重要的是这些临床标准得到额外的诊断试验支持或确认。

姬姆萨染色的薄且稠的外周血涂片显微镜检查仍然是诊断疟疾的金标准。然而这种类型的诊断需要高质量的试剂和训练有素的人员，他们既能准确识别疟原虫感染的红细胞，又能识别细胞内疟原虫的特定种类。最近基于寄生虫抗原的检测已经成功地用于快速准确地诊断疟原虫感染。这些检测依赖于通过固定在小卡片上的抗体检测或“捕获”患者血液中的疟原虫抗原[55,56]。虽然基于抗原的诊断只需要15分钟，但这种检测并不便宜，因此在疟疾流行地区不太普遍使用。此外，检测的灵敏度（即检测少量寄生虫的能力）仍然是一个限制，必须在显微镜下确认阳性结果。与抗原检测不同，分子检测代表了对疟疾寄生虫高度敏感和特异的检测。这些分析依赖于使用PCR从患者血液中检测和扩增寄生虫基因组DNA[57]。分子检测非常敏感，也可以用来区分各种疟原虫。但是由于对专用试剂、设备和培训的需求，广泛使用受到限制。

关于治疗，病例管理取决于疾病是否表现为简单或严重的疟疾。患有严重恶性疟原虫疟疾的患者通常不能口服药物，因此在医院环境中接受持续静脉注射治疗。相比之下，患有简单疟疾的患者通常在门诊接受治疗。存在多种具有抗寄生虫生命周期关键阶段活性的药剂。然而，包括氯喹、甲氟喹和乙胺嘧啶在内的许多药物在世界许多地方不再有效，因为过度应用导致了耐药性寄生虫菌株的选择。尽管如此，一些药物仍然有效并且在今天被广泛使用。最有效的疗法是包含青蒿素的联合疗法。例如，包括青蒿素（或青蒿琥酯）与阿莫地喹、阿托伐醌、多西环素或奎宁联合治疗。为了扩大青蒿素及其衍生物的用途，不建议使用青蒿素单一疗法[5,58]。

12.5 疫苗原理

一些证据表明，为人类提供预防性抗疟疾疫苗是可行的。首先，近40年前观察到在接种减毒的全疟原虫子芽孢疫苗后，人类志愿者可以获得针对子芽孢攻击的杀菌保护[59]。其次，如上所述，自然获得性免疫（NAI）在疟疾流行国家的人们生命头20年中已经逐步建立起来[60,61]。

因此，与包括人类免疫缺陷病毒（HIV）在内的许多病原体不同，疟原虫感染可引起“临床免疫”或对严重疾病的保护。最后，免疫球蛋白从半免疫的成年人被动转移到发热和高寄生虫血症儿童身上，从而导致寄生虫负担的减轻[62]。总的来说，这些观察结果表明，诱导对疟原虫产生保护性免疫是可能的，这为最终开发出一种获得许可的疫苗带来了巨大希望。虽然过去几十年里已经取得了重大进展，但正如下文所讨论的，目前只有一种候选疫苗在临床试验中显示出部分疗效，即RTS，S。

然而对疟原虫病理学的新认识和对自然获得性免疫（NAI）的更好理解将继续开发抗疟疾疫苗的新方法。目前抗疟疾疫苗的工作重点是开发三种主要类型的疫苗：肝期、红内期和阻断传播的疫苗。阻断传播的疫苗在其他地方已详细评估[63]。此外由于阻断传播的疫苗不能直接保护接种者，我们的讨论将只集中在肝期和红内期的抗疟疾疫苗。值得注意的是，世界卫生组织保留了一份目前正在进行临床前或临床试验的有效、全面和最新抗疟疾疫苗清单[52]。

12.6 红细胞前期（肝期）疫苗

疟原虫在肝脏中完全发育和分化之前，阻止疟原虫感染将完全预防临床疾病。针对这一生命周期阶段的疫苗被设计成引发有效的抗体应答，阻止子芽孢有效感染肝细胞，或者引发有效的T细胞免疫，从而识别和消除疟原虫感染的肝细胞。这些方法一个潜在缺点包括在子芽孢作为细胞外寄生虫存在非常短暂的时期内，有效靶向子芽孢需要非常高的抗体滴度。

此外，即使是一个有生产力的肝细胞也可能导致红内期疟疾。因此，除非所有受感染的肝细胞从宿主体内有效清除，否则临床疾病将会发展。与此相一致，最近的数据表明为了消除疟原虫感染的肝细胞和防止红内期感染，需要大量寄生虫特异性T细胞[64]。尽管有这些潜在限制，人们普遍认为，即使疫苗只减少子芽孢或受感染肝细胞数量，也应该能减少红内期寄生虫感染发生率或程度[65]。

多年来，几种分子肝期疫苗平台已经开发出来，包括使用肽和重组蛋白为基础的疫苗、DNA疫苗、病毒载体疫苗，以及持续开发基于全子芽孢疫苗[52]（表12.3）。在这里我们描述了几种候选方法，强调那些在临床试验中显示出重大前景或对改进当前抗疟疾疫苗方法有重要见解的方法。

基于重组蛋白的RTS,S。24年前开发了一种新的疫苗平台，专门针对疟原虫感染的红细胞前期。这种疫苗被称为RTS,S，是恶性疟原虫环子芽孢蛋白（CSP）亚单位的配方。该疫苗含有CSP中央重复序列（R）成分，具有已知的B细胞表位，与CSP的C末端（T）区域相连，已知该区域含有T细胞表位。这些CSP多肽与乙型肝炎表面抗原融合，产生RTS。该构建体在酵母中与游离乙肝表面抗原（S）共表达，产生病毒样抗原颗粒RTS,S[66]。

在II期初步试验中，已经确定RTS,S在30%~50%疟疾幼稚健康个体中诱导了保护[66,67]，但仅在使用有效的佐剂特别是与专有佐剂AS01一起给药时[68]。这些效力水平是迈向获得许可的抗疟疾疫苗发展道路上的重要一步，RTS,S总体效力还没有被另一种基于重组蛋白的疫苗制剂超越。从那时起，RTS,S已经在非洲7个国家进入了大规模III期试验。最近公布的结果显示，该疫苗为5~17个月儿童提供了预防临床疾病（56%有效）和严重疟疾（47%有效）的保护[69]。

RTS,S疫苗接种引发的保护机制尚未完全确定，但是很明显，当在AS01佐剂存在下给药时，RTS,S疫苗接种会引发非常高的抗CSP抗体滴度，这与保护相关[70]，可能减少感染子芽孢的数量[65]。重要的是，已知抗体直接干扰肝细胞的子芽孢感染[71,72]，并且典型地参与病原体的调理作用，随后增强了吞噬细胞，如巨噬细胞和树突细胞（DC），对病原体的吸收作用。因此，可以想象，在RTS,S免疫个体中，调理作用途径也能加速子芽孢的宿主清除。除了抗体，靶向CSP末端区域的CD4 T细胞也被认为有助于保护。然而，在接种疫苗的个体中测量到的T细胞应答是适度的，CD4 T细胞（或CD8 T细胞）参与的证据只有极低的相关性[73]。

尽管RTS,S已经成为当前实验性抗疟疾亚单位疫苗的效力基准，但仍存在几个问题。首先，迄今为止大规模临床试验仅限于针对非洲儿童研究，恶性疟原虫疟疾是一个全球公共健康问题，特别是对南美和亚洲来说。因此RTS,S在其他人群中是否同样有效（或更有效）仍有待确定。

其次，保护的耐久性还不清楚；到目前为止，III期试验都有相对较短的随访期（14个月）[69]，因此确定RTS,S的表观保护能力持续多长时间至关重要。最后，了解RTS,S如何或是否影响接种人群之间的传染率是很重要的。事实上，即使RTS,S的适度保护效果也可能对疟原虫传播率产生显著影响。另一方面RTS,S也有可能干扰自然暴露诱发的自然获得性免疫（NAI）的发展，从而可能使年龄较大儿童在以后的生活中面临更大的严重疾病风险。尽管有这些问题和潜在的问题，但完全批准RTS,S有助于降低全球疟疾发病率和流行率的希望依然很高。

质粒和病毒载体疫苗 诱导肝期免疫的第二个主要方法是使用DNA初免+病毒载体加强或病毒载体加强免疫方法（表12.3）。基于CSP的实验性质粒DNA疫苗最初是在20世纪90年代末开发的[74]，但其相对较低的刺激能力需要用编码CSP[75]的非复制病毒载体，包括痘病毒载体（如改良的疫苗病毒安卡拉或鸡痘）和腺病毒载体进行二次强化免疫。

与增强的免疫原性相一致，病毒载体通常被认为能够引发针对从病毒骨架表达编码转基因的有效T细胞免疫。事实上，在被感染肝细胞内表达的疟原虫蛋白载体递送，仍然是产生大量寄生虫特异性细胞毒性T细胞的中心策略，这可以消除那些被感染的细胞[76]。除了CSP，其他子芽孢/肝期抗原已经通过DNA和/或病毒载体途径进行了临床评估，包括TRAP、肝期抗原-1（LSA-1）、LSA-3和输出蛋白-1（EXP-1）[52]。当对在没有感染过疟疾的人类志愿者中进行保护评估时，病毒载体疗法效果最初是有希望的[65,75,77,78]。不幸的是，当几个过去的配方的临床试验转移到冈比亚成年人和肯尼亚儿童保护检查时，结果并不令人鼓舞[79,80]，这些试验已经被叫停[52]。

尽管最初的结果令人失望，但扩大病毒载体方法发展的工作仍在继续。例如目前正在试验的是异源病毒载体的免疫增强方法，该策略结合了来自肝期（CSP、TRAP）和红内期（AMA-1、MSP-1）抗原[52]，目的是引发跨阶段保护（表12.3）。此外，提高载体免疫原性的努力仍在继续。事实上这种方法的一个主要障碍是接种疫苗的个体迅速发展或具有预先存在的抗载体免疫。

因此，任何“增强”免疫反应强度的努力都会被体液免疫应答和细胞免疫应答所抵消，体液免疫应答和细胞免疫应答能有效中和载体，并在二次接触时加速其与宿主的清除。这种载体的快速清除随后抑制了对编码免疫原的应答。为了增强免疫原性，正在研究其他类型和组合的病毒载体，包括各种痘病毒[81]。为了对抗先前存在的免疫，研究人员继续评估黑猩猩腺病毒和不太常见的人类腺病毒血清型（如 Ad26 和 Ad35）的刺激能力[82-84]。

总体而言，尽管早期临床结果不佳并有明显的局限性，但病毒载体初免-加强方法很可能被证明是有用的疫苗平台，可以与其他方式（例如 RTS,S）组合使用，以增强对疟原虫感染的保护。事实上，这些平台可以相对容易地扩大规模并部署到疟疾流行地区，这突出表明我们仍然需要在更大的临床试验中进一步开发和评估这些平台。

减毒活寄生虫疫苗　虽然不是正式的分子疫苗平台，但有必要对全寄生虫（子芽孢）疫苗接种方法进行简要讨论（表 12.3），因为这些平台是唯一能够在人体内提供高水平灭菌免疫的实验性疫苗[85,86]。40 多年来，已知接种大量 RAS（辐射衰减子芽孢）可诱导对肝期特异性免疫灭菌。RAS 寄生虫通过辐射诱导随机 DNA 损伤而减弱，在复制开始时经历肝细胞内发育停滞。在临床前模型中，RAS 诱导的保护主要由 CD8 T 细胞介导，其活性针对疟原虫感染的肝细胞[87-90]。

与转化 RAS 疫苗相关的一个主要问题是，需要用一定剂量的辐射来减弱疟原虫子芽孢来诱导 DNA 损伤，以防止寄生虫复制和分化（即进展到红内期感染），保持子芽孢的传染性和免疫原性[91,92]。尽管存在安全性和部署方面的潜在问题，RAS 最近已经发展到临床评估阶段[52,93,94]，但迄今为止在人体内几乎没有显示出保护效果[95]。这些初步试验的中心结论是，效力有限是由于皮下或皮内注射 RAS 而导致免疫原性差，而通过静脉注射免疫的非人灵长类动物免疫原性高[95]。

与 RAS 相比，通过靶向基因缺失产生的 GAP（遗传性减毒寄生虫）是一种有吸引力的替代物，因为寄生虫可以被“设计”在肝脏发育过程中特定点停止[96]，从而显示出均匀的生物活性，这可能有利于疫苗接种。在广泛寄生虫复制之前[97]或在晚期肝脏（裂殖体）阶段之前[98]，在肝脏发育早期已经产生了 GAP 来阻止寄生虫发育。在临床前模型中，与早期阻止 GAP 相比，晚期阻止 GAP 引发更高水平的保护，这与针对多种寄生虫抗原的更大 CD8 T 细胞反应相关[99]。恶性疟原虫 GAP 作为一种潜在的抗疟疾疫苗的临床评估已经开始[52]，当通过皮内、皮下或蚊虫叮咬接种时，与 RAS 相比，GAP 是否能引发类似（或更高）水平的保护将是非常有意义的。

与接种数量相对较大的减毒 RAS 和 GAP 子芽孢形成对照的是感染治疗免疫（ITI）。在这个实验平台上，少量恶性疟原虫子芽孢被用作免疫原通过蚊子叮咬传播。当给从未感染过疟疾的志愿者同时服用氯喹（以防止红内期寄生虫的复制）时，有毒子芽孢会引发高度保护，防止随后蚊子叮咬传播的子芽孢攻击[100]。在 ITI 情况下，子芽孢会有效地感染肝脏，寄生虫会经历完整的肝脏发育阶段，最终释放出裂殖子。然而在氯喹存在的情况下，氯喹只对红内期的寄生虫有效，它防止了临床疾病的发生，裂殖子感染的红细胞被迅速清除。临床前模型中的保护与针对肝期寄生虫的 T 细胞反应相关，然而在最近的临床试验中，红内期免疫是否对人类 ITI 疫苗的保护发挥了实质性作用，仍是一个公开且至关重要的问题[100]。

第二个主要问题涉及接种如此少的未减毒子芽孢后的极高效力。一个假设是，允许寄生虫完成肝期发育可以有效地增加寄生虫抗原的生物量，从而提高有效的消除性免疫。此外，类似于晚期肝期抑制子芽孢 GAP 疫苗[99]，当寄生虫成熟为血液阶段裂殖子时，寄生虫表达的抗原多样性增加可能有助于保护。虽然在人体试验中没有提供抗原靶点多样化证据，但是这些研究提供了明确证据，证明延长肝期分化和抗原负荷增加与子芽孢疫苗接种后增强保护性免疫相关。

显然向疟疾流行人群转化和部署活子芽孢疫苗存在重大障碍。然而这些实验疫苗可能揭示了疟原虫和宿主免疫系统相互作用的重要信息。了解全子芽孢疫苗有效免疫原性的精确细胞和分子基础可能会揭示一些线索，有助于开发易于部署的亚单位或病毒载体疫苗。例如，活的子芽孢可能在增强其免疫原性的独特环境中将抗原暴露于宿主免疫系统；抗原递呈或T细胞或B细胞启动，可能以完全不同的方式发生，当相同的抗原通过重组蛋白或载体制剂给药时，这些方式是不被模仿的。

或者，当前的亚单位和载体方法可能不针对免疫原性或保护性最强的抗原或保护性免疫所必需抗原的正确组合（下文讨论）。解决这些问题应该提供亚单位疟疾疫苗接种期间抗原递呈的关键信息，而全子芽孢实验疫苗应该继续证明是剖析疟疾免疫关键机制的有用工具。

12.7　红细胞（红内期）疫苗

针对疟原虫感染红内期的疫苗有望减少寄生虫负担，从而降低疾病的严重程度。因此，除了预防未感染过疟疾个体的严重感染外，红内期疫苗还可以起到治疗作用，降低疟疾接触者的发病率和死亡率。与肝期相似，用于诱导预防红内期感染的主要疫苗平台包括肽、重组蛋白和载体技术[52]（表12.3）。尽管目前临床试验中，红内期疫苗数量相对较多，但与肝期实验疫苗相比[52]，开发红内期疫苗的进展要慢得多，而且成功率有限[101]。由于没有任何候选物能够证明其对红内期能够有效感染，因此我们的讨论将集中在继续努力开发和许可红内期抗疟疾疫苗的潜在原因上。

在开发红内期疫苗方面，可能有几个原因导致相对较少的成功。第一，红内期疫苗重点是引发对裂殖子或受感染红细胞表面表达的数量极其有限的蛋白质免疫。临床疫苗试验中目前候选抗原仅包含6种蛋白质：MSP-1、MSP-3、AMA-1、EBA175、SERALP 5（丝氨酸重复抗原-5）和富含谷氨酰胺的蛋白质（GLURP）（表12.1、表12.2和表12.3）。值得注意的是，与肝期表达的蛋白质相比，恶性疟原虫田间分离株中红内期蛋白质具有高度多态性[102]。

事实上在自然暴露和感染期间，抗体针对这些蛋白质，似乎对寄生虫施加了高度的选择性压力[103]。第二，在亚单位疫苗中表达和配制的重组蛋白不能采用模仿该蛋白在裂殖子或红细胞表面表达时免疫系统所见的必要构象。因此，用可溶性重组或载体疟疾抗原进行疫苗接种通常不能获得保护性免疫，因为相关的B细胞表位是构象依赖性的。第三，目前的策略通常是在有强佐剂的情况下使用重组蛋白（表12.3）。不幸的是，大多数疫苗佐剂的作用机制还不太清楚。此外，目前可用的佐剂通常是疟疾暴露个体抗体反应的不良诱导剂[104]。与此一致的是，对某些重组疟原虫蛋白的疫苗诱导抗体水平通常不高于生活在疟疾流行区个体检测到的反应[104]。

显然，在开发出仅针对裂殖子或受感染红细胞的有效疫苗之前，还有一些重大问题需要解决。尽管有这些限制，现在人们越来越认识到，针对肝期和红内期寄生虫的多阶段疫苗可以协同减少寄生虫负担，从而降低严重疾病的发病率和流行率。然而在这方面，目前在29项抗疟疾疫苗临床试验中，只有3项侧重于确定CSP和AMA-1是否可以同时用于改善保护[52]（表12.3）。尽管多阶段疫苗提供了显著影响疾病负担的潜在机会，但需要更多的工作来确定候选抗原和疫苗递送平台的最佳组合，以获得可观的保护。

12.8　结论和展望

疟疾负担是巨大的，这种疾病仍然是一种全球威胁。每年疟原虫感染导致数十万人死亡，大多数临床病例涉及儿童。疟疾确实是现代世界最紧迫的公共健康问题之一。虽然疟疾发病率和流行率控制最近取得了一些成功，但是随着蚊帐项目的继续扩大和新药的发现，这种疾病的彻底根除将取决于有效疫苗的开发。对脊椎动物宿主体内寄生虫生物学更全面的了解有助于集中新策略来对抗疟原虫感染和传播。然而，许多关键的障碍仍然阻碍着开发下一代有效抗疟疾疫苗。

也许最大的障碍在于我们不完全理解寄生虫和宿主免疫系统之间的相互作用。尽管很明显，重复暴露后自然获得性免疫（NAI）会随着时间的推移而发展，但对于这种免疫的发展知之甚少[105]。未来旨在剖析这些保护途径的免疫学研究应反过来帮助指导抗疟疾疫苗开发。事实上，通过疫苗接种来模仿 NAI 是否应该成为目标或者 NAI 是否在很大程度上功能失调（因此不消毒）仍然是主要问题。

疫苗开发的另一个问题与寄生虫的复杂性有关。寄生虫不同生命周期阶段和红内期感染期间高度多态蛋白表达提出了一个重要的问题[103]，即目前是否通过肽、亚单位和载体策略把“最佳”抗原作为靶标。与此相一致，人们充分认识到用减毒的全寄生虫疫苗接种后可诱导最有效的免疫。在这种情况下免疫系统提供了疟原虫肝期和/或红内期抗原的全部补体。重要的是，要注意疟原虫物种编码 >5 200个已知或预测的基因[106]。值得注意的是，只有约 35 种蛋白质（占预测蛋白质的 0.8%）被评估为恶性疟原虫潜在的疫苗候选物[52]，或者被确定为宿主免疫系统的目标[107]。基于迄今评估的候选抗原数量有限，宿主免疫系统可能会识别许多额外的疟原虫蛋白，包括临床相关疟原虫物种（恶性疟原虫和间日疟原虫）表达的更保守蛋白。

最后仍然迫切需要确定免疫介导保护的有意义的关联，开发新的体内模型并扩大目前临床抗疟疾疫苗测试中心的数量。开发关键的功能性检测和量化抗体中和子芽孢和阻断肝细胞入侵的能力将是一个重大进步。此外，开发和采用标准化检测方法，来评估接种过疫苗个体的血清抗体阻止裂殖子侵入红细胞的能力，将有助于全世界的研究人员更准确地解释和比较他们的数据集。同样重要的是需要更好的体内模型。由于恶性疟原虫和间日疟原虫不会感染小型实验动物，因此，产生表达恶性疟原虫或间日疟原虫抗原的啮齿动物或非人灵长类寄生虫的新重组版本将是非常有价值的工具，有助于集中研究候选抗原的免疫原性和保护能力。

新的体内模型也将显著增加评估发病机制和免疫力的系统数量（表 12.4），其中一些受到重大生物限制和缺点的阻碍。最后需要额外的临床测试设施，类似于最近在西雅图生物医学研究所和疟疾临床试验中心开始的测试，以扩大抗疟疾疫苗研究。当然所有这些需要的进步都是有代价的。在过去 10 年里对疟疾研究和控制项目的投资显著增加[2]，但是如果没有进一步的投资，疟疾防治的重要成果将无法继续。因此，如果国际社会真的致力于开发下一代抗疟疾疫苗，仍需克服一些经济、生物和免疫障碍。

表 12.4　用于研究疟原虫发病机制或评估抗疟疾疫苗效力的动物模型

啮齿动物模型	优点：使用近交或基因敲除小鼠可以进行疾病发病机制和/或免疫的分子和细胞剖析	弱点：实验鼠不是自然宿主（没有共同进化）
疟原虫属[a]	临床特征模型的常见用途	
P. berghei ANKA	感染网织红细胞（未成熟红细胞）和正常细胞（成熟红细胞）。引起与感染引起的中枢神经系统炎症、中性粒细胞和 T 细胞募集和激活相关的快速致死病理	C57BL/6 小鼠脑疟疾、严重疟疾、贫血、药物敏感性和先天免疫研究模型
P. yoeliiyoelii 17XNL	感染网织红细胞并导致持续数周的高负荷寄生虫血症	疫苗疗效、天然和获得性免疫、发病机制和严重贫血模型
P. chabaudi chabaudi AS	感染正常细胞并表现出长期持续感染（数月），以复发为特征	疫苗效力、适应性免疫、寄生虫药物敏感性和发病机制的模型，以及具有可变复发率的红内期恶性疟原虫低级寄生虫负担的模型

（续表）

啮齿动物模型	优点：使用近交或基因敲除小鼠可以进行疾病发病机制和/或免疫的分子和细胞剖析	弱点：实验鼠不是自然宿主（没有共同进化）
P. vinckeivinckei	感染网状细胞和正常细胞。导致与高寄生虫负担和贫血相关的急性致命感染	严重疟疾发病机制和寄生虫药物敏感性模型
P. falciparum in "humanized" mice	恶性疟原虫和人红细胞向免疫缺陷小鼠的共转移。缺乏获得性和天然免疫的小鼠将支持人类红细胞	能够在易处理的系统中研究恶性疟原虫的发病机制或药物敏感性
非人灵长类动物模型	优点：被认为是人类恶性疟原虫感染、寄生虫和宿主物种共同进化的更接近的模拟	弱点：需要专门的设施和专业知识来进行这项研究。成本和伦理考虑也影响适用性
疟原虫属[b]	临床特征	模型的常见用途
P. coatneyi	自然宿主灵长类动物长期慢性感染，但恒河猴会出现严重的疾病	寄生虫感染红细胞的脑隔离
P. knowlesi	对恒河猴具有快速致命性。表现出较不常见的24小时红细胞周期	广泛应用于贫血模型的抗原变异模型及红细胞浸润机制
P. cynomolgi	模拟疟原虫休眠体的间日疟原虫形成，并建立慢性感染	休眠体形成、肝期疫苗和寄生虫药物效力的研究模型
P. fragile	长期慢性感染，以自然宿主猴子的复发为特征，多数猴子有高寄生虫血症和严重感染	红内期疫苗效力模型
P. falciparum in chimpanzees and limited species of New World monkey	寄生虫病急性感染，可迅速消退（10~14天）	肝期和红内期疫苗效力模型

[a]每种啮齿类疟疾都存在大量实验室菌株和分离株，并且呈现出不同的发病模式。例如 *yoelii* 疟原虫的一些实验室菌株（例如，*yoelii* 疟原虫菌株 YM 或菌株 17XL）在小鼠中引起快速致死感染，而其他菌株（例如 *yoelii* 疟原虫 17NL）通常在小鼠中分解。为了简单起见，只记录了每种疟原虫中研究最广泛菌株的一般特征

[b]这些疟原虫引起感染的病理和发病机制在很大程度上取决于非人灵长类宿主的种类

致 谢

感谢 Nathan Schmidt 博士对手稿的批判性阅读，感谢 Jessica Wood 博士对手稿和图 12.1 中插图的批判性阅读。

参考文献

[1] Murray, C. J., et al.: Global malaria mortality between 1980 and 2010: a systematic analysis. Lancet379, 413-431 (2012). doi: 1 0. 1016/S0140-6 736 (12) 60034-8. S0140-6736 (12) 60034-8 [pii].

[2] World Health Organization: World Malaria Report 2010. http: //www. who. int/malaria/world_malaria_report_2010/en/ (2010).

[3] Dondorp, A. M., et al.: Artemisinin resistance: current status and scenarios for containment. Nat. Rev. Microbiol. 8, 272-280 (2010). doi: 10. 1038/nrmicro2331.

[4] N'Guessan, R., Corbel, V., Akogbeto, M., Rowland, M.: Reduced efficacy of insecticide-treated nets and indoor residual spraying for malaria control in pyrethroid resistance area, Benin. Emerg. Infect. Dis. 13, 199-206 (2007). doi: 10. 3201/eid1302. 060631.

[5] Mackinnon, M. J., Marsh, K.: The selection landscape of malaria parasites. Science328, 866 - 871 (2010). doi: 10. 1126/science. 1185410.

[6] Plotkin, S. A.: Correlates of protection induced by vaccination. Clin. Vaccine Immunol. 17, 1055 - 1065 (2010). doi: 10. 1128/CVI. 00131-10.

[7] Florens, L., et al.: A proteomic view of the Plasmodium falciparum life cycle. Nature419, 520 - 526 (2002). doi: 10. 1038/nature01107.

[8] Gardner, M. J., et al.: Genome sequence of the human malaria parasite Plasmodium falciparum. Nature419, 498-511 (2002). doi: 10. 1038/nature01097.

[9] Scherf, A., Lopez-Rubio, J. J., Riviere, L.: Antigenic variation in Plasmodium falciparum. Annu. Rev. Microbiol. 62, 445-470 (2008). doi: 10. 1146/annurev. micro. 61. 080706. 093134.

[10] Tyler, J. S., Treeck, M., Boothroyd, J. C.: Focus on the ringleader: the role of AMA1 in apicomplexan invasion and replication. Trends Parasitol. 27, 410-420 (2011). doi: 10. 1016/j. pt. 2011. 04. 002.

[11] Aly, A. S., Vaughan, A. M., Kappe, S. H.: Malaria parasite development in the mosquito and infection of the mammalian host. Annu. Rev. Microbiol. 63, 195 - 221 (2009). doi: 10. 1146/annurev. micro. 091208. 073403.

[12] Sidjanski, S., Vanderberg, J. P.: Delayed migration of Plasmodium sporozoites from the mosquito bite site to the blood. Am. J. Trop. Med. Hyg. 57, 426-429 (1997).

[13] Kebaier, C., Voza, T., Vanderberg, J.: Kinetics of mosquito-injected Plasmodium sporozoites in mice: fewer sporozoites are injected into sporozoite-immunized mice. PLoSPathog. 5, e1000399 (2009). doi: 10. 1371/ journal. ppat. 1000399.

[14] Frevert, U., et al.: Intravital observation of Plasmodium berghei sporozoite infection of the liver. PLoS Biol. 3, e192 (2005). doi: 10. 1371/journal. pbio. 0030192.

[15] Baum, J., et al.: A conserved molecular motor drives cell invasion and gliding motility across malaria life cycle stages and other apicomplexan parasites. J. Biol. Chem. 281, 5197-5208 (2006). doi: 10. 1074/ jbc. M509807200.

[16] Pradel, G., Garapaty, S., Frevert, U.: Proteoglycans mediate malaria sporozoite targeting to the liver. Mol. Microbiol. 45, 637-651 (2002).

[17] Robson, K. J., et al.: Thrombospondin-related adhesive protein (TRAP) of Plasmodium falciparum: expression during sporozoite ontogeny and binding to human hepatocytes. EMBO J. 14, 3883-3894 (1995).

[18] Kariu, T., Ishino, T., Yano, K., Chinzei, Y., Yuda, M.: CelTOS, a novel malarial protein that mediates transmission to mosquito and vertebrate hosts. Mol. Microbiol. 59, 1369 - 1379 (2006). doi: 10. 1111/j. 1365-2958. 2005. 05024. x.

[19] Bergmann-Leitner, E. S., et al.: Immunization with pre-erythrocytic antigen CelTOS from Plasmodium falciparum elicits cross-species protection against heterologous challenge with Plasmodium berghei. PLoSOne5, e12294 (2010). doi: 10. 1371/journal. pone. 0012294.

[20] Cowman, A. F., Crabb, B. S.: Invasion of red blood cells by malaria parasites. Cell124, 755 - 766 (2006). doi: 10. 1016/j. cell. 2006. 02. 006.

[21] Lindner, S. E., Miller, J. L., Kappe, S. H.: Malaria parasite pre-erythrocytic infection: preparation meets opportunity. Cell. Microbiol. 14, 316-324 (2012). doi: 10. 1111/j. 1462-5822. 2011. 01734. x.

[22] Schofield, L., Grau, G. E.: Immunological cbrsprocesses in malaria pathogenesis. Nat. Rev. Immunol. 5,

722-735 (2005). doi: 10. 1038/nri1686.

[23] Baer, K., Klotz, C., Kappe, S. H., Schnieder, T., Frevert, U.: Release of hepatic Plasmodium yoeliimerozoites into the pulmonary microvasculature. PLoSPathog. 3, e171 (2007). doi: 10. 1371/journal. ppat. 0030171.

[24] Sturm, A., et al.: Manipulation of host hepatocytes by the malaria parasite for delivery into liver sinusoids. Science313, 128-1290 (2006). doi: 10. 1126/ science. 1129720.

[25] Lyke, K. E., et al.: Cell-mediated immunity elicited by the blood stage malaria vaccine apical membrane antigen 1 in Malian adults: results of a Phase I randomized trial. Vaccine27, 2171-2176 (2009). doi: 10. 1016/j. vaccine. 2009. 01. 097.

[26] Thera, M. A., et al.: A field trial to assess a blood-stage malaria vaccine. N. Engl. J. Med. 365, 1004-1013 (2011). doi: 10. 1056/NEJMoa1008115.

[27] Draper, S. J., et al.: Recombinant viral vaccines expressing merozoite surface protein-1 induce antibody- and T cell-mediated multistage protection against malaria. Cell Host Microbe5, 95-105 (2009). doi: 10. 1016/j. chom. 2008. 12. 004.

[28] Duncan, C. J., et al.: Impact on malaria parasite multiplication rates in infected volunteers of the protein-in-adjuvant vaccine AMA1-C1/Alhydrogel+CPG 7909. PLoSOne6, e22271 (2011). doi: 10. 1371/journal. pone. 0022271.

[29] Lusingu, J. P., et al.: Satisfactory safety and immunogenicity of MSP3 malaria vaccine candidate in Tanzanian children aged 12-24 months. Malar. J. 8, 163 (2009). doi: 10. 1186/1475-2875-8-163.

[30] Sirima, S. B., et al.: Safety and immunogenicity of the malaria vaccine candidate MSP3 long synthetic peptide in 12-24 months-old Burkinabe children. PLoS One 4, e7549 (2009). doi: 10. 1371/journal. pone. 0007549.

[31] El Sahly, H. M., et al.: Safety and immunogenicity of a recombinant non-glycosylated erythrocyte binding antigen 175 region II malaria vaccine in healthy adults living in an area where malaria is not endemic. Clin. Vaccine Immunol. 17, 1552-1559 (2010). doi: 10. 1128/CVI. 00082-10.

[32] Chowdhury, D. R., Angov, E., Kariuki, T., Kumar, N.: A potent malaria transmission blocking vaccine based on codon harmonized full length Pfs48/45 expressed in Escherichia coli. PLoS One 4, e6352 (2009). doi: 10. 1371/journal. pone. 0006352.

[33] Kaslow, D. C., Quakyi, I. A., Keister, D. B.: Minimal variation in a vaccine candidate from the sexual stage of Plasmodium falciparum. Mol. Biochem. Parasitol. 32, 101-103 (1989).

[34] Carter, R., Mendis, K. N., Miller, L. H., Molineaux, L., Saul, A.: Malaria transmission-blocking vaccines-how can their development be supported? Nat. Med. 6, 241-244 (2000). doi: 10. 1038/73062.

[35] Carlton, J. M., Sina, B. J., Adams, J. H.: Why is Plasmodium vivax a neglected tropical disease? PLoS Negl. Trop. Dis. 5, e1160 (2011). doi: 10. 1371/journal. pntd. 0001160.

[36] Miller, L. H., Mason, S. J., Clyde, D. F., McGinniss, M. H.: The resistance factor to Plasmodium vivax in blacks. The Duffy-blood-group genotype, FyFy. N. Engl. J. Med. 295, 302-304 (1976). doi: 10. 1056/NEJM197608052950602.

[37] Iyer, J., Gruner, A. C., Renia, L., Snounou, G., Preiser, P. R.: Invasion of host cells by malaria parasites: a tale of two protein families. Mol. Microbiol. 65, 231-249 (2007). doi: 10. 1111/j. 1365-2958. 2007. 05791. x.

[38] Marsh, K., Kinyanjui, S.: Immune effector mechanisms in malaria. Parasite Immunol. 28, 51-60 (2006). doi: 10. 1111/j. 1365-3024. 2006. 00808. x.

[39] Langhorne, J., Ndungu, F. M., Sponaas, A. M., Marsh, K.: Immunity to malaria: more questions than

answers. Nat. Immunol. 9, 725-732 (2008). doi: 10. 1038/ni. f. 205.

[40] Beeson, J. G., Osier, F. H., Engwerda, C. R.: Recent insights into humoral and cellular immune responses against malaria. Trends Parasitol. 24, 578-584 (2008). doi: 10. 1016/j. pt. 2008. 08. 008. S1471-4922 (08) 00226-2 [pii].

[41] Daily, J. P., et al.: Distinct physiological states of Plasmodium falciparum in malaria-infected patients. Nature450, 1091-1095 (2007). doi: 10. 1038/ nature06311.

[42] Lackritz, E. M., et al.: Effect of blood transfusion on survival among children in a Kenyan hospital. Lancet340, 524-528 (1992).

[43] Marsh, K., et al.: Indicators of life-threatening malaria in African children. N. Engl. J. Med. 332, 1399-1404 (1995). doi: 10. 1056/NEJM199505253322102.

[44] Brattig, N. W., et al.: Plasmodium falciparum glycosylphosphatidylinositol toxin interacts with the membrane of non-parasitized red blood cells: a putative mechanism contributing to malaria anemia. Microbes Infect. 10, 885-891 (2008). doi: 10. 1016/j. micinf. 2008. 05. 002.

[45] Naik, R. S., et al.: Glycosylphosphatidylinositol anchors of Plasmodium falciparum: molecular characterization and naturally elicited antibody response that may provide immunity to malaria pathogenesis. J. Exp. Med. 192, 1563-1576 (2000).

[46] Schofield, L., Hackett, F.: Signal transduction in host cells by a glycosylphosphatidylinositol toxin of malaria parasites. J. Exp. Med. 177, 145-153 (1993).

[47] Tachado, S. D., et al.: Signal transduction in macro-phages by glycosylphosphatidylinositols of Plasmodium, Trypanosoma, and Leishmania: activation of protein tyrosine kinases and protein kinase C by inositolglycan and diacylglycerol moieties. Proc. Natl. Acad. Sci. U. S. A. 94, 4022-4027 (1997).

[48] Beeson, J. G., Brown, G. V.: Pathogenesis of Plasmodium falciparum malaria: the roles of parasite adhesion and antigenic variation. Cell. Mol. Life Sci. 59, 258-271 (2002).

[49] Schofield, L., Mueller, I.: Clinical immunity to malaria. Curr. Mol. Med. 6, 205-221 (2006).

[50] Fairhurst, R. M., Bess, C. D., Krause, M. A.: Abnormal PfEMP1/knob display on Plasmodium falciparum-infected erythrocytes containing hemoglobin variants: fresh insights into malaria pathogenesis and protection. Microbes Infect. 14, 851-862 (2012). doi: 10. 1016/j. micinf. 2012. 05. 006.

[51] Umbers, A. J., Aitken, E. H., Rogerson, S. J.: Malaria in pregnancy: small babies, big problem. Trends Parasitol. 27, 168-175 (2011). doi: 10. 1016/j. pt. 2011. 01. 007.

[52] World Health Organization: Malaria Vaccine Rainbow Table. < http: //www. who. int/vaccine_research/links/Rainbow/en/index. html> (2012).

[53] Fedosov, D. A., Caswell, B., Suresh, S., Karniadakis, G. E.: Quantifying the biophysical characteristics of Plasmodium-falciparum-parasitized red blood cells in microcirculation. Proc. Natl. Acad. Sci. U. S. A. 108, 35-39 (2011). doi: 10. 1073/pnas. 1009492108.

[54] Fedosov, D. A., Lei, H., Caswell, B., Suresh, S., Karniadakis, G. E.: Multiscale modeling of red blood cell mechanics and blood flow in malaria. PLoSComput. Biol. 7, e1002270 (2011). doi: 10. 1371/journal. pcbi. 1002270.

[55] Howden, B. P., Vaddadi, G., Manitta, J., Grayson, M. L.: Chronic falciparum malaria causing massive splenomegaly 9 years after leaving an endemic area. Med. J. Aust. 182, 186-188 (2005).

[56] Hendriksen, I. C., et al.: Diagnosing severe falciparum malaria in parasitaemic African children: a prospective evaluation of plasma PfHRP2 measurement. PLoS Med. 9, e1001297 (2012). doi: 10. 1371/journal. pmed. 1001297.

[57] Snounou, G., et al.: High sensitivity of detection of human malaria parasites by the use of nested polymerase

chain reaction. Mol. Biochem. Parasitol. 61, 315-320 (1993).

[58] Fairhurst, R. M., et al.: Artemisinin-resistant malaria: research challenges, opportunities, and public health implications. Am. J. Trop. Med. Hyg. 87, 231-241 (2012). doi: 10. 4269/ajtmh. 2012. 12-0025.

[59] Clyde, D. F., Most, H., McCarthy, V. C., Vanderberg, J. P.: Immunization of man against sporozoite-induced falciparum malaria. Am. J. Med. Sci. 266, 169-177 (1973).

[60] Baird, J. K., et al.: Age-dependent acquired protection against Plasmodium falciparum in people having 2 years exposure to hyperendemic malaria. Am. J. Trop. Med. Hyg. 45, 65-76 (1991).

[61] Gupta, S., Snow, R. W., Donnelly, C. A., Marsh, K., Newbold, C.: Immunity to non-cerebral severe malaria is acquired after one or two infections. Nat. Med. 5, 340-343 (1999). doi: 10. 1038/6560.

[62] Cohen, S., Mc, G. I., Carrington, S.: Gamma-globulin and acquired immunity to human malaria. Nature192, 733-737 (1961).

[63] Bousema, T., Drakeley, C.: Epidemiology and infectivity of Plasmodium falciparum and Plasmodium vivax gametocytes in relation to malaria control and elimination. Clin. Microbiol. Rev. 24, 377-410 (2011). doi: 10. 1128/CMR. 00051-10.

[64] Schmidt, N. W., Butler, N. S., Badovinac, V. P., Harty, J. T.: Extreme CD8 T cell requirements for anti-malarial liver-stage immunity following immunization with radiation attenuated sporozoites. PLoS Pathog. 6, e1000998 (2010). doi: 10. 1371/journal. ppat. 1000998.

[65] Bejon, P., et al.: Calculation of liver-to-blood inocula, parasite growth rates, and preerythrocytic vaccine efficacy, from serial quantitative polymerase chain reaction studies of volunteers challenged with malaria sporozoites. J. Infect. Dis. 191, 619-626 (2005). doi: 10. 1086/427243.

[66] Gordon, D. M., et al.: Safety, immunogenicity, and efficacy of a recombinantly produced Plasmodium falciparum circumsporozoite protein-hepatitis B surface antigen subunit vaccine. J. Infect. Dis. 171, 1576-1585 (1995).

[67] Stoute, J. A., et al.: A preliminary evaluation of a recombinant circumsporozoite protein vaccine against Plasmodium falciparum malaria. RTS,S Malaria Vaccine Evaluation Group. N. Engl. J. Med. 336, 86-91 (1997). doi: 10. 1056/NEJM199701093360202.

[68] Garcon, N., Chomez, P., Van Mechelen, M.: GlaxoSmithKline Adjuvant Systems in vaccines: concepts, achievements and perspectives. Expert Rev. Vaccines6, 723-739 (2007). doi: 10. 1586/14760584. 6. 5. 723.

[69] Agnandji, S. T., et al.: First results of phase 3 trial of RTS, S/AS01 malaria vaccine in African children. N. Engl. J. Med. 365, 1863-1875 (2011). doi: 10. 1056/NEJMoa1102287.

[70] Kester, K. E., et al.: Randomized, double-blind, phase 2a trial of falciparum malaria vaccines RTS, S/AS01B and RTS, S/AS02A in malaria-naive adults: safety, efficacy, and immunologic associates of protection. J. Infect. Dis. 200, 337-346 (2009). doi: 10. 1086/600120.

[71] Gysin, J., Barnwell, J., Schlesinger, D. H., Nussenzweig, V., Nussenzweig, R. S.: Neutralization of the infectivity of sporozoites of Plasmodium knowlesi by antibodies to a synthetic peptide. J. Exp. Med. 160, 935-940 (1984).

[72] Plassmeyer, M. L., et al.: Structure of the Plasmodium falciparum circumsporozoite protein, a leading malaria vaccine candidate. J. Biol. Chem. 284, 26951-26963 (2009). doi: 10. 1074/jbc. M109. 013706.

[73] Schwenk, R., et al.: Immunization with the RTS, S/AS malaria vaccine induces IFN-gamma (+) CD4 T cells that recognize only discrete regions of the circumsporozoite protein and these specificities are maintained following booster immunizations and challenge. Vaccine29, 8847-8854 (2011). doi: 10. 1016/j. vaccine. 2011. 09. 098.

[74] Wang, R., et al.: Induction of antigen-specific cytotoxic T lymphocytes in humans by a malaria DNA vaccine. Science282, 476-480 (1998).

[75] McConkey, S. J., et al.: Enhanced T-cell immunogenicity of plasmid DNA vaccines boosted by recombinant modified vaccinia virus Ankara in humans. Nat. Med. 9, 729-735 (2003). doi: 10. 1038/nm881.

[76] Hill, A. V., et al.: Prime-boost vectored malaria vaccines: progress and prospects. Hum. Vaccin. 6, 78-83 (2010).

[77] Dunachie, S. J., et al.: A DNA prime-modified vaccinia virus Ankara boost vaccine encoding thrombospondin-related adhesion protein but not circumsporozoite protein partially protects healthy malaria-naive adults against Plasmodium falciparum sporozoite challenge. Infect. Immun. 74, 5933-5942 (2006). doi: 10. 1128/IAI. 00590-06.

[78] Ockenhouse, C. F., et al.: Phase I/IIa safety, immunogenicity, and efficacy trial of NYVAC-Pf7, a pox-vectored, multiantigen, multistage vaccine candidate for Plasmodium falciparum malaria. J. Infect. Dis. 177, 1664-1673 (1998).

[79] Moorthy, V. S., et al.: A randomised, double-blind, controlled vaccine efficacy trial of DNA/MVA ME-TRAP against malaria infection in Gambian adults. PLoS Med. 1, e33 (2004). doi: 10. 1371/journal. pmed. 0010033.

[80] Bejon, P., et al.: Extended follow-up following a phase 2b randomized trial of the candidate malaria vaccines FP9 ME-TRAP and MVA ME-TRAP among children in Kenya. PLoSOne2, e707 (2007). doi: 10. 1371/journal. pone. 0000707.

[81] Webster, D. P., et al.: Enhanced T cell-mediated protection against malaria in human challenges by using the recombinant poxviruses FP9 and modified vaccinia virus Ankara. Proc. Natl. Acad. Sci. U. S. A. 102, 4836-4841 (2005). doi: 10. 1073/pnas. 0406381102.

[82] Ophorst, O. J., et al.: An adenoviral type 5 vector carrying a type 35 fiber as a vaccine vehicle: DC targeting, cross neutralization, and immunogenicity. Vaccine22, 3035-3044 (2004). doi: 10. 1016/j. vaccine. 2004. 02. 011.

[83] Lemckert, A. A., et al.: Immunogenicity of heterologous prime-boost regimens involving recombinant adenovirus serotype 11 (Ad11) and Ad35 vaccine vectors in the presence of anti-ad5 immunity. J. Virol. 79, 9694-9701 (2005). doi: 10. 1128/ JVI. 79. 15. 9694-9701. 2005.

[84] Radosevic, K., et al.: The Th1 immune response to Plasmodium falciparum circumsporozoite protein is boosted by adenovirus vectors 35 and 26 with a homologous insert. Clin. Vaccine Immunol. 17, 1687-1694 (2010). doi: 10. 1128/CVI. 00311-10.

[85] Clyde, D. F.: Immunization of man against falciparum and vivax malaria by use of attenuated sporozoites. Am. J. Trop. Med. Hyg. 24, 397-401 (1975).

[86] Hoffman, S. L., et al.: Protection of humans against malaria by immunization with radiation-attenuated Plasmodium falciparum sporozoites. J. Infect. Dis. 185, 1155-1164 (2002). doi: 10. 1086/339409. JID010922 [pii].

[87] Krzych, U., Schwenk, J.: The dissection of CD8 T cells during liver-stage infection. Curr. Top. Microbiol. Immunol. 297, 1-24 (2005).

[88] Tsuji, M.: A retrospective evaluation of the role of T cells in the development of malaria vaccine. Exp. Parasitol. 126, 421-425 (2010). doi: 10. 1016/j. exppara. 2009. 11. 009.

[89] Overstreet, M. G., Cockburn, I. A., Chen, Y. C., Zavala, F.: Protective CD8 T cells against Plasmodium liver stages: immunobiology of an 'unnatural' immune response. Immunol. Rev. 225, 272-283 (2008). doi: 10. 1111/j. 1600-065X. 2008. 00671. x.

[90] Hafalla, J. C., Cockburn, I. A., Zavala, F.: Protective and pathogenic roles of CD8$^+$ T cells during malaria infection. Parasite Immunol. 28, 15-24 (2006). doi: 10. 1111/j. 1365-3024. 2006. 00777. x. PIM777 [pii].

[91] Mellouk, S., Lunel, F., Sedegah, M., Beaudoin, R. L., Druilhe, P.: Protection against malaria induced by irradiated sporozoites. Lancet335, 721 (1990). 0140-6736 (90) 90832-P [pii].

[92] Chattopadhyay, R., et al.: The effects of radiation on the safety and protective efficacy of an attenuated Plasmodium yoelii sporozoite malaria vaccine. Vaccine 27, 3675-3680 (2006). doi: 10. 1016/j. vaccine. 2008. 11. 073. S0264-410X (08) 01596-X [pii].

[93] Schmidt, N. W., Butler, N. S., Harty, J. T.: CD8 T cell immunity to Plasmodium permits generation of protective antibodies after repeated sporozoite challenge. Vaccine 27, 6103-6106 (2009). doi: 10. 1016/j. vaccine. 2009. 08. 025. S0264-410X (09) 01203-1 [pii].

[94] Crompton, P. D., Pierce, S. K., Miller, L. H.: Advances and challenges in malaria vaccine development. J. Clin. Invest. 120, 4168-4178 (2010). doi: 10. 1172/JCI44423. 44423 [pii].

[95] Epstein, J. E., et al.: Live attenuated malaria vaccine designed to protect through hepatic CD8 (+) T cell immunity. Science334, 475-480 (2011). doi: 10. 1126/ science. 1211548.

[96] Kappe, S. H., Vaughan, A. M., Boddey, J. A., Cowman, A. F.: That was then but this is now: malaria research in the time of an eradication agenda. Science328, 862-866 (2010). doi: 10. 1126/ science. 1184785. 328/5980/862 [pii].

[97] Aly, A. S., et al.: Targeted deletion of SAP1 abolishes the expression of infectivity factors necessary for successful malaria parasite liver infection. Mol. Microbiol. 69, 152-163 (2008). doi: 10. 1111/j. 1365-2958. 2008. 06271. x. MMI6271 [pii].

[98] Vaughan, A. M., et al.: Type II fatty acid synthesis is essential only for malaria parasite late liver stage development. Cell. Microbiol. 11, 506-520 (2009). doi: 10. 1111/j. 1462-5822. 2008. 01270. x.

[99] Butler, N. S., et al.: Superior antimalarial immunity after vaccination with late liver stage-arresting genetically attenuated parasites. Cell Host Microbe9, 451-462 (2011). doi: 10. 1016/j. chom. 2011. 05. 008. S1931-3128 (11) 00172-7 [pii].

[100] Roestenberg, M., et al.: Protection against a malaria challenge by sporozoite inoculation. N. Engl. J. Med. 361, 468-477 (2009). doi: 10. 1056/NEJMoa0805832.

[101] Goodman, A. L., Draper, S. J.: Blood-stage malaria vaccines-recent progress and future challenges. Ann. Trop. Med. Parasitol. 104, 189-211 (2010). doi: 10. 1179/136485910X12647085215534.

[102] Genton, B., Reed, Z. H.: Asexual blood-stage malaria vaccine development: facing the challenges. Curr. Opin. Infect. Dis. 20, 467-475 (2007). doi: 10. 1097/QCO. 0b013e3282dd7a29.

[103] Greenwood, B. M., Targett, G. A.: Malaria vaccines and the new malaria agenda. Clin. Microbiol. Infect. 17, 1600-1607 (2011). doi: 10. 1111/j. 1469-0691. 2011. 03612. x.

[104] Hill, A. V.: Vaccines against malaria. Philos. Trans. R. Soc. Lond. B Biol. Sci. 366, 2806-2814 (2011). doi: 10. 1098/rstb. 2011. 0091.

[105] Doolan, D. L., Dobano, C., Baird, J. K.: Acquired immunity to malaria. Clin. Microbiol. Rev. 22, 13-36 (2009). doi: 10. 1128/CMR. 00025-08. Table of Contents.

[106] Volkman, S. K., Neafsey, D. E., Schaffner, S. F., Park, D. J., Wirth, D. F.: Harnessing genomics and genome biology to understand malaria biology. Nat. Rev. Genet. 13, 315-328 (2012). doi: 10. 1038/ nrg3187. nrg3187 [pii].

[107] Doolan, D. L.: Plasmodium immunomics. Int. J. Parasitol. 41, 3-20 (2011). doi: 10. 1016/j. ijpara. 2010. 08. 002, S0020-7519 (10) 00289-4 [pii].

[108] Sedegah, M., et al.: Adenovirus 5-vectored P. falciparum vaccine expressing CSP and AMA1. Part A: safety and immunogenicity in seronegative adults. PLoS One 6, e24586 (2011).

[109] Tamminga, C., et al.: Adenovirus-5-vectored P. falciparum vaccine expressing CSP and AMA1. Part B: safety, immunogenicity and protective efficacy of the CSP component. PLoSOne 6, e25868 (2011).

[110] Bakshi, S., Imoukhuede, E. B.: Malaria Vectored Vaccines Consortium (MVVC). Hum. Vaccin. 6, 433-434 (2010).

[111] Spring, M. D., et al.: Phase 1/2a study of the malaria vaccine candidate apical membrane antigen-1 (AMA-1) administered in adjuvant system AS01B or AS02A. PLoSOne 4, e5254 (2009). doi: 10. 1371/ journal. pone. 0005254.

[112] Lyon, J. A., et al.: Protection induced by Plasmodium falciparum MSP1 (42) is strain-specific, antigen and adjuvant dependent, and correlates with antibody responses. PLoS One 3, e2830 (2008). doi: 10. 1371/ journal. pone. 0002830.

[113] Thera, M. A., et al.: Safety and immunogenicity of an AMA-1 malaria vaccine in Malian adults: results of a phase 1 randomized controlled trial. PLoS One 3, e1465 (2008). doi: 10. 1371/journal. pone. 0001465.

[114] Nebie, I., et al.: Humoral and cell-mediated immunity to MSP3 peptides in adults immunized with MSP3 in malaria endemic area, Burkina Faso. Parasite Immunol. 31, 474-480 (2009). doi: 10. 1111/j. 1365-3024. 2009. 01130. x.

[115] Peek, L. J., Brandau, D. T., Jones, L. S., Joshi, S. B., Middaugh, C. R.: A systematic approach to stabilizing EBA-175 RII-NG for use as a malaria vaccine. Vaccine 24, 5839-5851 (2006). doi: 10. 1016/j. vaccine. 2006. 04. 067.

[116] Horii, T., et al.: Evidences of protection against blood-stage infection of Plasmodium falciparum by the novel protein vaccine SE36. Parasitol. Int. 59, 380-386 (2010). doi: 10. 1016/j. parint. 2010. 05. 002.

[117] Belard, S., et al.: A randomized controlled phase Ib trial of the malaria vaccine candidate GMZ2 in African children. PLoS One 6, e22525 (2011). doi: 10. 1371/journal. pone. 0022525.

[118] Mordmuller, B., et al.: Safety and immunogenicity of the malaria vaccine candidate GMZ2 in malariaexposed, adult individuals from Lambarene, Gabon. Vaccine 28, 6698-6703 (2010). doi: 10. 1016/j. vaccine. 2010. 07. 085.

（马维民译）

第 13 章　结核病疫苗：现状与进展

Rodrigo Ferracine Rodrigues，Rogério Silva Rosada，
Fabiani Gai Frantz，Frederico Gonzalez Colombo Arnoldi，
Lucimara Gaziola de la Torre 和 Celio Lopes Silva①

摘要

结核病是人类最古老的疾病之一，每年仍有 200 万人死于结核病。据信世界上 1/3 的人口感染了结核杆菌，这是一个巨大的令人担忧的病原体库。不同的因素导致了这种令人不安的情况：唯一可用的疫苗卡介苗（BCG）（译者）效率低下；治疗时间太长，造成相当大的副作用，产生大量不符合率，有利于产生耐药菌株；诊断不总是准确的。尽管宿主—病原体相互作用和抗结核病免疫反应得到了深入研究，但仍有许多未解决的问题。因此新诊断方法、疫苗和疗法发展受到了极大影响。

在本章中，首先将向读者提供所有结核病方面的概述，以支持对新疫苗开发的详细讨论。通过将疫苗按主题与用于开发每一种疫苗的不同技术分开，解释其设计背后的目标和合理性，并展示直到本书出版为止最相关的结果，突出临床测试阶段的结果。本章将讨论该领域的前景，讨论结核病研究中直接影响新疫苗开发的挑战点。

13.1　疾病的发展和免疫应答

结核病（TB）是严重的全球性公共卫生问题之一，涉及社会和经济方面。但这远不是一个现代问题，因为有迹象表明感染可以追溯到新石器时代[1]。结核病（TB）是 17 世纪和 18 世纪欧洲最严重的传染病之一，被称为“白色瘟疫”。此时，1/4 的成人死亡归因于结核病，据信几乎所有欧洲公民都受到了感染[2]。

今天据估计世界人口中约有 1/3 感染了结核杆菌。每年大约有 5 400万人感染，940 万人发展成这种疾病，170 万人死于这种可治愈的疾病[3]。结核杆菌杀死的人比任何其他传染病病原体都多。遏制 TB 伙伴关系和千年发展目标——一个建立公私伙伴关系的国际组织网络——汇集了减少结核病流行率和发病率的重要国际政策和支持。因此，TB 主要是被忽视的人的一种疾病，他们生活在恶劣的住房和卫生条件下，酗酒和吸烟也增加了感染 TB 的倾向。最重要的是，导致结核病例和易患结核病的因素是与艾滋病病毒（HIV）的共同感染[4]。

在世界范围内，Robert Koch（柯赫）于 1882 年发现的结核分枝杆菌是人类 TB 的主要病原体。在非洲，主要是在西部，非洲分枝杆菌在流行病学上也很重要[2]。此后结核分枝杆菌被称为 Mtb（结合分子杆菌）。Mtb 属于分枝杆菌科，放线菌群，分枝杆菌属，是一种兼性细胞内致病菌，需要氧气才能生长，但据信可以在厌氧环境中以低代谢率生存。Mtb 在原核生物中有独特

---

① R. F. Rodrigues，博士 · R. S. Rosada，博士（✉）F. G. C. Arnoldi，博士 · C. L. Silva，博士（巴西圣保罗大学 Ribeirão Preto 医学院结核病研究中心生物化学与免疫学系 E-mail：rosada@ usp. br ）

F. G. Frantz，博士（DACTB，巴西圣保罗大学 Ribeirão Preto 药物科学学院）

L. G. de la　Torre，博士（巴西坎皮纳斯（UNICAMP）坎皮纳斯大学化学工程学院）

的高脂细胞壁：肽聚糖，在质膜周围，与阿拉伯半乳聚糖多糖相连，后者与分枝杆菌酸（长链脂肪酸）相连。游离的两亲性脂质（如 TDM、海藻糖二霉菌酸酯或索状因子）以及糖类（如葡聚糖）完成了这一惊人结构，赋予了宿主毒性、抗药性以及感染宿主和破坏其免疫反应的工具并使杆菌对革兰氏染色有抵抗力（使用抗酸检测技术代替）[5]。

当患有活动性结核病的人在讲话、咳嗽或打喷嚏时排出含有活菌的小液滴，被其他人吸入时，就会发生 Mtb 感染。结核病主要影响下呼吸道，肺部是细菌滋生的主要场所，尽管也有非肺结核的形式。分枝杆菌在人类宿主中的繁殖速度缓慢，症状可能需要数月才能显现，包括胸痛、持续 3 周以上的剧烈咳嗽或咯血（咳血）。患者可能会出现进一步发烧、发冷、盗汗、食欲不振、体重减轻、面色苍白和疲劳[6]。

据估计，与杆菌接触后只有 5%~10%的个体发展成活动性肺结核。在另外 90%~95%的病例中，个体发展成无症状的潜伏感染并且不传播芽孢杆菌。这是因为免疫反应不能消除芽孢杆菌，但可以控制其传播，限制其生物活性和进入其他细胞和器官。潜伏期可持续个体的整个生命期，构成病原体的重要储库。这一特征导致结核病的高发病率和流行率，因为在任何时候导致免疫抑制的因素都会导致疾病复发。此外，针对 Mtb 的免疫反应以及由此产生的活动性或潜伏性结核病可能与遗传多态性有关[6]。

一旦进入肺部，感染的建立依赖于病原体与专业宿主吞噬细胞（主要是肺泡巨噬细胞）的相互作用，后者将捕获 Mtb。在吞噬体中 Mtb 应该被破坏，但是相反 Mtb 激活允许其在吞噬体中存活和复制的机制，例如改变内体 pH、破坏吞噬体和溶酶体融合、抑制活性形式的氧自由基和氮生成并减少炎性细胞因子的产生。然而有许多先天免疫的受体和细胞，一旦被激活就会协同产生免疫应答，能够包含病原体诱导形成典型的炎症反应，其特征是由免疫系统细胞组成的病理组织球状结构，这些结构被称为肉芽肿[7]。

肉芽肿是肺结核诊断中使用的特征之一，是由被感染的巨噬细胞和树突细胞迁移到邻近的淋巴结引起的，在那里分枝杆菌抗原递呈给淋巴细胞，开始发展适应性免疫应答。此后活化的淋巴细胞迁移到感染部位。因此，肉芽肿主要由 B 淋巴细胞和位于感染巨噬细胞和树突细胞周围的淋巴细胞 $TCD4^+$和 $TCD8^+$ T，以及其他细胞组成。随后，肉芽肿被成纤维细胞产生的胶原纤维包围，可能会在低氧和低营养的环境中长时间限制杆菌的传播，在这种环境中肉芽肿会降低新陈代谢，这是感染的潜伏期[8]。尽管存在差异，主要是由于采用的研究设计，但肉芽肿今天被视为细胞和病原体之间动态相互作用的地方，所有这些都在不断增殖、死亡和更新，都可以保护宿主，有助于病原体对其定植。

事实上如前所述，这种抑制系统可能会被削弱宿主免疫系统的因素破坏而导致疾病的复发，通常伴随坏死、干酪化、肉芽肿液化和细菌释放。因此杆菌可以通过循环系统扩散到局部肺或其他器官，或者通过呼吸和咳嗽释放到环境中导致其他个体受到污染[9]。

因此，结核病免疫应答成败取决于一个复杂的激活和抑制机制网络，其组成部分仍在建立或更详细地了解之中。换句话说，这是一个复杂的生物现象，不能完全理解。许多因素被定义为控制 Mtb 的“必要”。然而关于哪些宿主因素“足以”在对抗病原体时产生有效免疫应答，或者哪些因素导致疾病发展的知识仍然不完整，一直是深入研究的主题。有人承认但没有证明：有些人可以自然消除 Mtb。

导致这种情况的理想免疫应答是进化先天免疫和获得性免疫成分，特别是 T 辅助（Th）1 型。简而言之，一旦吞噬细胞捕获 Mtb，它就开始通过 Toll 样受体（TLR）和/或核苷酸结合和寡聚化结构域（NOD）发出信号，促进因子核 κB（NF-κB）向细胞核易位，导致白细胞介素（IL）-12 产生。反过来，IL-12 的主要作用之一是刺激 T $CD4^+$细胞的生长和活化，有利于产生 Th1 应答。Th1 应答的主要事件是 T-盒转录因子（T-bet）向细胞核易位，这将控制标志性细胞因子干扰素（IFN-γ）的产生。在多种效应中，IFN-γ 增加巨噬细胞的抗原递呈和溶酶体活性，激活诱导型一氧化氮合酶（iNOS）有助于杀死 Mtb。此外，IFN-γ 还激活 T $CD8^+$细胞，T $CD8^+$

细胞可以产生杀死受感染细胞的分子，如穿孔蛋白和颗粒溶蛋白，对于克服结核病感染至关重要[7]。

13.2 诊断和经典疗法

结核病诊断通常基于患者的临床和流行病学史。胸部X射线和结核菌素皮肤试验（TST）是一种常规检查，敏感性好但特异性差，可以确认疾病病因。自从结核分枝杆菌在1882年被定义为结核病的病原体以来，已经描述了几种检测结核分枝杆菌的技术。

一个多世纪后，结核病病因学诊断标准方法仍然基于柯赫阐述的相同基本原则。因此，采用齐尼（Ziehl-Neelsen）抗酸染色法涂片显微镜检查仍然是全世界最常用的结核病诊断方法。最近的一项系统综述表明，尽管涂片显微镜灵敏度可能达到70%，但在不同环境中差异很大，而且在结核病/艾滋病双重感染高发地区，灵敏度要低得多（约35%）[11]。除了灵敏度低之外，涂片显微镜还不能提供结核分枝杆菌种类和芽孢杆菌对抗结核药物敏感性的信息。

分枝杆菌特异性培养基中微生物培养仍然是诊断和评估Mtb对治疗药物耐药性的金标准。尽管与涂片显微镜相比，培养的敏感性更高，80%~90%，但通常仅在治疗大量和更复杂病例的参考服务中进行。培养的主要限制仍然是诊断确认所需的时间，这可能需要长达8~12周的时间[12]。

然而结核病诊断技术在过去30年里取得了显著的进步，这些基于研究资源的转化潜力是巨大的。为评估结核病免疫反应性而开发的最终检测是IFN-γ释放试验（IGRA）。这恰好是一个简单的基于免疫的检测，IFN-γ释放试验评估由Mtb抗原刺激培养物中全血细胞产生的IFN-γ。IGRAs既有优点，也有缺点；IGRAs是高度特异性的，因为之前接种卡介苗不会导致假阳性结果。然而，IGRAs无助于区分潜伏性结核病感染和结核病。当我们主要谈论发展中国家时，IGRAs是非常昂贵的[13]。

血清学试验的发展、特异性Mtb蛋白检测和炎性细胞因子测量提供了进一步的诊断选择。这些方法旨在区分活动性结核病和潜伏性结核病感染。然而很少有生物标志物能够识别和/或区分结核分枝杆菌感染、卡介苗接种和活动性结核疾病，尽管它们可能与需要治疗病例的管理高度相关。核酸扩增技术在结核病诊断中的应用引起了人们的极大兴趣，特别是因为在临床样品中检测和鉴定结核分枝杆菌所需的时间可能会缩短[10]。

不幸的是，开发和实施新的结核病微生物检测的时间没有赶上医学技术发展，也没有赶上与艾滋病大流行相关疾病的灾难性传播，低收入人口被边缘化到大城市中心郊区，以及出现多药耐菌性菌株[10]。新诊断方法的发展还取决于疾病每个阶段生物标志物的可靠测定，如下文所述。

治疗

抗结核化疗有几种方案，包括旨在快速消除活性杆菌的初始强化阶段，随后是旨在消除潜在杆菌的持续阶段[14]。因此在大多数国家第一阶段持续两个月，此时给予利福平、异烟肼、吡嗪酰胺和乙胺丁醇片的组合；第二阶段持续4个月，此时仅给予利福平和异烟肼[14,15]。如果适当实施该方案，在90%以上的病例中效果是有效的，但是其持续时间长、相关的副作用以及在完全消除杆菌之前症状的改善导致许多患者在完成治疗前退出。在这种情况下增加了复发风险，并有利于出现对一线药物特别是利福平和异烟肼耐药的杆菌，导致多耐药性结核病，或在极端情况下，对所有一线和某些二线药物耐药，这种状态被称为广泛耐药结核病（XDR-B）[15-17]。

尽管希望有能使治疗更简单和/或更短的新药，但自20世纪70年代美国食品和药物管理局批准利福平以来，还没有新药上市[14]。世卫组织最新报告[3]列出了目前正在进行临床研究的10种新药，其中7种处于临床Ⅱ期，3种处于临床Ⅲ期。

也许这些新药将允许开发尽可能短且副作用尽可能少的有效方案，从而对患者的治疗依从性产生积极影响。一些持续三四个月的方案同时使用新药物和旧药物，例如利福喷丁联合异烟肼和喹诺酮类药物（如加替沙星和莫西沙星）。尽管这些制度比目前实施的制度短50%，但它们仍然远远不能解决问题。最近我们团队在动物模型中研究了化疗和免疫疗法相结合的新治疗方案[18]。结果表明，这种方法有可能诱导更快的治愈，完全消除杆菌，降低内源性再激活率。同样另一个挑战是开发指示抗结核治疗效果的标志物，优选在开始后尽可能短的时间内，这些标志物目前还没有。

13.3 疫苗

考虑到受感染人群所代表的巨大病原体库，疫苗开发对于预防新病例和降低结核病高流行率至关重要。目前BCG（卡介苗）是世卫组织推荐的唯一疫苗[19]，自1921年推出以来已注射了30多亿剂。卡介苗是由减毒活牛结核杆菌（牛结核分枝杆菌）菌株制备的，该菌株通过在人工培养基中特殊传代培养而在人体中失去毒性。然而，尽管卡介苗保护新生儿和儿童免受严重肺结核的侵袭，但对青少年和成人肺结核的疗效远非最佳，保护率根据地理区域为0%~80%。

由于结核病是一种全球性的疾病，而且迫切需要一种新疫苗，因此在过去20年里，一项涉及所有战略部门的全球努力已经开始启动。现代免疫学工具侧重于特定的细胞激活、分子表达、被抗原呈递细胞（APCs）识别和递呈的最佳肽，以及许多与系统生物学背景相关的其他工具正在帮助实现这一目标。

13.3.1 但是，我们应该从新的结核疫苗中期待什么特性呢?

情况是复杂的，但是一些必需的元素是新生儿和成人的效率、安全性和通用性。这意味着要比卡介苗更好，激发持久的记忆细胞；给免疫缺陷者带来风险，如艾滋病病毒携带者；在不同条件和情况下都是有用的，如单纯的个体、接种卡介苗的个体和暴露于Mtb的个体。也许这个灵丹妙药的想法并不适合挑战，我们需要的不仅仅是一个简单的解决方案。

考虑到这一点，正在开发新的疫苗和制定疫苗接种策略，包括使用减毒活菌、重组微生物和亚单位疫苗；基于在两种不同疫苗载体下连续施用某种细菌抗原的加强免疫策略；以及DNA疫苗。到目前为止，在2012年有12种以上的疫苗正在进行临床试验，其中大多数集中在结核病的预防上。更多的候选疫苗正在进入临床试验，包括针对潜在感染者的疫苗。这些努力可以概括为两个主要的理论依据：取代BCG或改善BCG[20]。表13.1简要介绍了可能用于防治结核病疫苗的主要特点。

13.3.2 BCG改进/更换

重组卡介苗（rBCG）开发集中在去除不需要的基因、候选基因的过度表达或引入有利于对卡介苗的不同免疫反应的外源基因，从而提供免疫记忆和对结核病的有效保护。

VPM1002是一种基于rBCG的抗结核活疫苗，目前正在进行Ⅱ期临床试验。该疫苗在基因水平上有两个重要的修饰，共同提高了疫苗的免疫原性。一种方法是插入编码蛋白李斯特菌溶胞素（Hly）的单核细胞增生李斯特菌基因。这种蛋白质参与细菌内吞摄入后吞噬体形成小孔，允许细菌进入胞质溶胶。这种方法有助于克服卡介苗只进行MHC Ⅱ类抗原递呈的特性。向胞质溶胶的逃逸允许MHC Ⅰ类的递呈和随后的$CD8^+$活化。从免疫学角度来说，这是对抗Mtb免疫应答的理想模式。另一种基因修饰是脲酶C（ureC）卡介苗基因的失活。脲酶C催化尿素水解成二氧化碳和氨，从而创造碱性环境。当细菌失去ureC时，内部环境变成酸性，有利于李斯特菌溶胞素活性[21]。

表 13.1　临床试验中主要的抗结核疫苗

疫苗名称	免疫策略	设计	基本原理	功能作用	进展期	参考文献
VPM1002	改善 BCG	重组 BCG	插入编码李斯特菌溶血素蛋白的单核细胞增生李斯特菌基因。脲酶 C BCG 基因的失活	使进入细胞质，促进 MHC Ⅰ抗原递呈，改善 $CD8^+T$ 应答	Ⅱ期临床试验	21
rBCG30	改善 BCG	重组 BCG	分泌 Ag85，是一种分枝酰转移酶	Mtb 培养物中最丰富的分泌蛋白诱导 $CD4^+$ 和 $CD8^+$ T 效应细胞	Ⅰ期临床试验完成，进入下一期试验	22
MTBVAC01	取代 BCG	Mtb 减毒活株去除	去除两个独立毒力基因，fadD26 和 phoP	破坏 Mtb 阻断吞噬体-溶酶体融合的能力	Ⅰ期临床试验	23
MVA85A/AERAS-485	增强 BCG 免疫应答	重组修饰痘苗病毒 Ankara（MVA）	表达 Ag85A 基因	诱导多功能、持久的 $CD4^+$ T 细胞	Ⅱb 期临床试验完成	24-26
Crucell Ad35/AERAS-402	增强 BCG 免疫应答	复制缺陷型腺病毒 35	表达高免疫原性抗原 Ag85A、Ag85B 以及来自 Mtb 的 TB10.4	诱发 $CD4^+$和 $CD8^+T$ 细胞激活和 IFN-γ 产生	Ⅱb 期临床试验	27
AdAg85A	增强 BCG 免疫应答	复制缺陷型腺病毒	腺病毒疫苗-表达 Mtb Ag85A（AdAg85A）抗原	增加伽玛干扰素阳性 $CD4^+$和 $CD8^+$ T 细胞的数量	Ⅰ期临床试验	28-29
Hybrid 1+IC31vaccine	增强 BCG 免疫应答	重组蛋白和佐剂	结合抗原 Ag85B 和 ESAT-6，含 IC31（或 CFA01）佐剂	诱导 IFN-γ$CD4^+T$ 细胞产生；持久免疫力	Ⅰ期临床试验/计划Ⅱ期临床试验	30-32
SSI/SP H4 - IC31（or Hyvac 4）	增强 BCG 免疫应答	重组蛋白和佐剂	融合抗原 Ag85B 和 TB10.4，含 IC31 佐剂	诱导强劲和持续很久的 $CD4^+T$ 细胞应答	Ⅱ期临床试验	33
GSK M72	增强 BCG 免疫应答	重组蛋白和佐剂	融合抗原 Rv1196 和 Rv0125，含 AS01 佐剂	诱导强大的 $CD4^+T$ 细胞反应，高和持续的抗体水平	Ⅱb 期临床试验	34

（续表）

疫苗名称	免疫策略	设计	基本原理	功能作用	进展期	参考文献
SSIH56 -IC31vaccine	增强 BCG 免疫应答	重组蛋白和佐剂	Ag85B 和 ESAT6 融合抗原加入 Rv2660c 抗原，含 IC31 佐剂	来自疾病不同阶段（活跃期和潜伏期）的抗原可以作为暴露前和暴露后的疫苗使用	I 期临床试验	35–37
ID93+GLA-SE	取代 BCG	重组融合多蛋白和佐剂	Mtb 抗原与毒力或潜伏期有关（Rv2608、Rv3619、Rv3620、Rv1813），含 GLA-SE 佐剂	诱导抗原特异性 Th1 型免疫反应，靶向活性和潜伏性肺结核，可作为免疫治疗	I 期临床试验	38–40
RUTI	治疗性疫苗	亚单位和佐剂	解毒，碎片化结核杆菌细胞，通过脂质体传递	诱导多抗原细胞和体液免疫应答	II 期临床试验	41
Mycobacterium vaccae vaccine	增强 BCG 免疫应答；或许可以进行免疫治疗	灭活株	热灭活 *M. vaccae*	押注共享抗原引起持久的细胞免疫反应，提高 IFN-γ 响应和淋巴细胞增殖	III 期临床试验完成	42–44

卡介苗（BCG）是目前世界上唯一批准使用的抗结核疫苗，基于活的减毒牛结核菌株。
只能提供部分保护（对婴儿的粟疹和脑膜炎结核有效，但对成人的肺结核无效）。基本上，候选疫苗的目的是取代或改善卡介苗。

另一种重组卡介苗策略名为 rBCG30，于 2004 年进入 1 期临床试验。该疫苗通过了安全性研究，现在处于“暂停”状态，还没有进入下一阶段。这种 rBCG 分泌一种 30 kDa 蛋白，是 Mtb 在培养中检测到最丰富的分泌蛋白。这种蛋白质也称为 Ag85B，是一种分枝酰基转移酶，为接种疫苗的豚鼠提供抗结核保护，并在人类中诱导旺盛的免疫应答[22]。

MTBVAC01 设计旨在取代卡介苗。它不是重组卡介苗，而是大胆地基于活的 Mtb 基因减毒株，去除两个独立的毒力基因 fadD26 和 phoP。phoP 基因在导致高死亡率艾滋病毒感染患者结核病暴发的流行株中大量表达。FadD26 与细胞壁毒力脂质结核菌醇二分支菌酸的合成有关。与卡介苗接种相比，临床前结果显示与产生多功能细胞因子分泌的 T $CD4^+$ 细胞和长期维持 T $CD4^+$ 中央记忆细胞有关的抗结核保护作用是一致的。

这些有趣结果背后的基本原理似乎是由于与亲本结核分枝杆菌菌株相比，MTBVAC01 阻断吞噬体-溶酶体融合的能力受损。这种候选疫苗最近进入 I 期临床试验，是第一种开始临床评估的此类疫苗[23]。

13.3.3　病毒载体

病毒载体携带感兴趣的免疫原基因用于提高免疫应答，非常有效。另一方面它们往往具有高度的免疫原性，导致对载体的不良加重免疫反应。为了避免这个问题，一些研究人员正试图研究人类实际上尚未接触的病毒载体。实际上下面我们将讨论一些使用病毒载体的疫苗，这些病毒载体可以满足这一期望。

牛津-结核病联盟构建了一种名为 MVA85A /AERAS-85 的疫苗，该疫苗在冈比亚和南非通过了 II 期临床试验。这种疫苗使用重组改良的安卡拉痘苗病毒（MVA）作为递送系统，诱导 Ag85A 基因表达。其给药诱导了儿童体内高水平抗原特异性 $CD4^+$T 细胞的增殖，这种细胞是多功能的，在接种疫苗后持续了 14 个月[24,25]。然而最近与南非伍斯特开普敦大学的南非结核病疫苗倡议（SATVI）合作，完成了一项评估婴儿 MVA85A 疗效、安全性和免疫原性的 IIb 期临床试验。结果表明，这种疫苗耐受性良好，但诱导了适度的细胞免疫应答[26]。

使用腺病毒作为输送系统，Crucell Ad35 /AERAS-402 在南非、肯尼亚和美国处于 IIb 期临床试验阶段，在莫桑比克、肯尼亚和南非有将近 4 000名婴儿处于 IIb 期临床试验阶段。这种疫苗由来自具有高度免疫原性的 Mtb 抗原 Ag85A、Ag85B 和 TB10.4 组成，能诱导 $CD4^+$ 和 $CD8^+$T 细胞活化以及 IFN-γ 的产生。此外，腺病毒 35 的使用与 $CD8^+$T 细胞增加有关，这被认为是结核病疫苗的基础[27]。使用腺病毒 5 并仅输送 Ag85A，加拿大正在进行一项 I 期临床试验，疫苗名为 AdAg85A，由安大略省汉密尔顿市麦克马斯特大学的周兴教授研发，后来由一家中国公司 CanSino 收购[28,29]。

13.3.4　重组蛋白

重组蛋白的主要原理，包括结合至少两种免疫显性抗原的重组融合蛋白，是为了引发针对病原体的强烈免疫应答，而没有可能限制其应用的活的或灭活的全生物体疫苗的风险，尤其是在免疫功能低下的宿主中（如 HIV 感染者）。作为所有基于蛋白质的疫苗，必须与有效佐剂一起给药以引发足够和充分的免疫应答，例如将 T $CD4^+$细胞极化为 Th1 型，触发 T $CD8^+$细胞毒性细胞，激活树突细胞和其他抗原呈递细胞。下面列出了处于最高级研究阶段的这类疫苗例子。

Staten 血清研究所于 2012 年开始对 Hybrid1+IC31 疫苗的 HIV 阳性患者进行Ⅱ期临床试验，该疫苗结合了抗原 Ag85B（存在于 Mtb 和 BCG 中）和 ESAT-6（存在于 Mtb 中，但不存在于 BCG 中）。该制剂由 Intercell 公司生产的 IC31（含有寡脱氧核苷酸和阳离子抗菌肽的脂质体制剂）作为佐剂。两剂该疫苗的免疫原性和安全性结果预计 2013 年公布。还打算对健康青少年进行 II 期临床试验。临床前和 I 期临床试验表明这种疫苗是安全的；在不同人群中诱导持久免疫，包括 IFN-γ T $CD4^+$产生细胞；能够增强 BCG 诱导的免疫应答。该疫苗也正在用 CFA01 佐剂进行测试，CFA01 佐剂也是脂质体制剂，但含有二甲基二十八烷基铵（DDA）和海藻糖 6，6’-二苯甲酸酯（TDB），这是一种分枝杆菌索状因子海藻糖 6，6’-二霉菌酸酯（TDM）的合成类似物[30-32]。

国际结核疫苗协会组织（Aeras）完成了一种类似疫苗 SSI / SP H4-IC31（或 Hyvac 4）的 I 期临床试验，该疫苗由融合抗原 Ag85B 和 TB10.4 组成，由 IC31 作为佐剂。以往的临床前结果显示这种疫苗诱导了具有效应记忆和中枢记忆表型的 $CD4^+T$ 细胞组成的应答增加和延长。有利的是，Hyvac 4 没有损害基于 ESAT-6 的诊断，这解释了它被 TB10.4 取代的原因[33]。

葛兰素史克（GSK）设计了 M72 融合蛋白，由 Mtb 抗原 Rv1196 和 Rv0125 组成，这两种抗原都由 Mtb 和 BCG 表达。这种疫苗已经用不同的合成佐剂进行了测试，AS01（单磷酰脂质 A 与皂甙 QS21 混合）显示出更好的性能。I/IIa 期临床试验表明，M72 具有临床上可接受的安全性和反应原性特征，并且具有高度免疫原性，诱导 $CD4^+T$ 细胞的强烈应答和持续长达 3 年的高抗体水平。GSK M72（或 M72 / AS01）疫苗因此被选择用于进一步的临床开发即临床 IIb 期，如 Aeras 和（GSK）刚刚宣布的[34]。

SSI H56-IC31 疫苗基本上是由混合 I（Ag85B 和 ESAT6 融合抗原）加上 Rv2660c 抗原组成的。顾名思义，IC31 是佐剂。重要区别是，Rv2660c 是一种强抗原，在疾病潜伏期上调，Mtb 面临饥饿状况，显著改变其基因表达谱，进而影响免疫系统“可见”抗原的可用性和多样性。与其他暴露前疫苗相比，这种策略的优势在于作为暴露前和暴露后（也称为多阶段）疫苗的灵活性，如果我们记住世界人口的 1/3 有潜在感染的可能，这是非常有趣的。与单独使用卡介苗相比，使用 H56-IC31 促进卡介苗接种的临床前研究有效控制了 Mtb 感染，提高了动物的存活率。免疫增强的动物显示肺病理减少，肺外扩散和保护与针对 ESAT-6 和 Rv2660c 的强烈回忆应答相关。重要的是，BCG/ H56-v 疫苗接种的猴子在抗肿瘤坏死因子（TNF）抗体治疗后没有重新激活潜在感染。顺便提一下，当患者重新激活一种以前从未检测到的潜在结核病时，抗 TNF-α 治疗是类风湿性关节炎免疫疗法的常见副作用。出于这个原因，这种疫苗是一种很好的候选疫苗，已进入 I 期临床试验[35-37]。

最近国际结核疫苗协会（Aeras）和传染病研究所（IDRI）宣布开始 ID93+GLA-SE 疫苗的 I 期临床试验。ID93 是一种重组融合多蛋白，由与毒性或潜伏期相关的 Mtb 抗原组成（Rv2608、Rv3619、Rv3620 和 Rv1813），GLA-SE 指的是佐剂吡喃葡萄糖基脂质 A 稳定乳液。这种疫苗既针对活动性结核病，也针对潜伏性结核病，这也使它成为一种免疫治疗工具。以食蟹猴为模型，将这种疫苗与常规结核病化学疗法相结合的临床前研究显示，多能抗原特异性 Th1 型免疫应答的激活与细菌负荷的降低相关。作者还确认化疗的持续时间缩短了，一旦治疗结核病所需的持续周期缩短，有助于患者愿意接受治疗并坚持到结束，这是一个重要的特征[38-40]。

13.3.5 治疗性疫苗

研究了使用治疗性疫苗作为治疗结核病替代策略的可能性，主要是如果可能与传统的结核病化疗有关。如上所述，可以允许采用更短的化疗时间，通过恢复宿主免疫反应的潜力来防止可能的持久杆菌再活化。

RUTI 是一种治疗性疫苗，由解毒、破碎的结核分枝杆菌细胞制成，以脂质体形式递送。在临床前研究中，短期化疗后服用 RUTI。结果表明，T $CD4^+$和 T $CD8^+$细胞对分泌性抗原和结构性抗原的应答增强，诱导多抗原体液应答，这种策略能够满足感染的再激活。基于这些结果，临床 I 期试验已经完成，具有良好的安全性和免疫原性，临床 II 期试验也几乎完成了[41]。

母牛分枝杆菌灭活疫苗。使用母牛分枝杆菌作为疫苗的理论基础是基于数据显示，天然或疫苗诱导的非结核分枝杆菌感染可以防止结核病的发展。这很可能是通过对分枝杆菌共享抗原的反应发生的。这种疫苗完成了 III 期临床试验，首次显示了基于疫苗的有效预防艾滋病毒感染成人的机会性感染。该疫苗增强了 IFN-γ 应答和淋巴细胞对全疫苗超声波处理的增殖，以及对脂肪阿拉伯甘露聚糖的抗体应答。在卡介苗引发的、艾滋病病毒阳性的和艾滋病病毒阴性的受试者中，这种疫苗诱导持久的细胞免疫应答。

这些结果支持利用卡介苗和全细胞灭活分枝杆菌疫苗制定一项初免-加强免疫策略，以预防影响发展中国家艾滋病病毒感染者最重要的机会性感染。考虑到母牛分枝杆菌疫苗的免疫原性，一些研究怀疑这种疫苗作为一种免疫治疗工具是否也可以治疗结核病。在研究、试验和综合分析中，关于母牛分枝杆菌是否是结核病的良好免疫疗法存在争议。可能是不同的途径、分析、方法或包含标准导致了明显的冲突结果。值得注意的是，这些参考的研究包括不同的情况，如注射或口服制剂，是否与化疗相关，以及结核病的类型（药物敏感或耐药，是否已经治疗）。但是总的来说，这些研究给人留下了一种自信的印象，即热致死的母牛分枝杆菌疫苗，口服和每日与化疗一起给药，可能有助于结核病的治疗[42-44]。

13.3.6 未来展望

在这里，我们考虑了一些可能加速结核病控制的疫苗开发技术和方法，包括我们的研究团队在这一领域做出的贡献。

最初开发的一项净化血浆和血小板血液制品的技术，称为“拦截血液系统”，在疫苗领域产生了另一种有趣的方法，即“杀死但代谢活性（KBMA）疫苗”[45]。

KBMA疫苗是通过用补骨脂素交联剂光化学灭活杀死的全致病性或减毒生物体，影响到DNA复制的绝对阻断和可能的疫苗效果。在这一过程中，DNA改变随机发生且是有限的，这使得在一个含有百万单位微生物的疫苗剂量中，几乎所有抗原库的表达都得以实现。在这一策略中最重要的是，在这些条件下，微生物保持足够的代谢活性，以诱导免疫应答，因为它们保持与野生生物自然生命周期相关的特性，对宿主来说被视为活体。

综上所述，该工具的主要应用是使用KBMA作为编码选定抗原的重组载体，并根据病原体的减毒形式设计KBMA疫苗。在结核病的情况下，两种方法都可以使用。在前一种情况下，使用杆菌再活化时表达的Mtb抗原作为阻断或预防潜在结核病发展为活动性结核病的方式是可能的，也是有趣的。后一种策略特别有趣，因为它允许展示病原体的全部抗原谱，这是非常理想的，因为保护的相关因素是未知的或不清楚的。此外，结核病是艾滋病病毒感染者的主要机会性疾病，KBMA消除了其他方法可能带来的一些与安全相关的问题。临床前研究表明，KBMA疫苗能诱导与疗效相关的功能性免疫应答，但是到目前为止还没有进行临床试验[45]。

13.3.7 DNA疫苗

基本上，DNA疫苗包括细菌质粒和编码抗原的基因，细菌质粒利用能够在哺乳动物细胞中起作用的启动子。一旦疫苗被细胞捕获，基因就会被翻译成蛋白质，这种蛋白质可以通过MHC Ⅰ类途径提供给CTL细胞，这是设计这种策略时寻求的主要特征。但DNA疫苗显示出刺激先天免疫和触发获得性免疫的其他分支的能力：抗体和辅助T细胞。此外，还具有预防和治疗性质。所产生的反应偏差，例如Th1或Th2，可以通过选择用于疫苗递送的方法来诱导，例如肌内注射或基因枪分别诱导[46]。

在异源性免疫增强方案中，或通过分子成瘾（例如细胞因子）或使用聚合物微球和脂质体等制剂将DNA疫苗输送到细胞中，可以提高其与其他疫苗一起使用DNA疫苗的性能。实际上人们致力于在通过改善DNA表达、传递和免疫原性上，在临床前模型中显示出同样的效果[47]。

DNA疫苗的效力和安全性可在其与强致病性病原体或使用有害病原体抗原时得到维持。此外，使用质粒可以避免减毒病毒或病毒载体的安全问题。因此，考虑到其相对简单和快速的构建/生产以及在室温下的稳定性，这一战略作为全球疫苗规划的替代方案是有吸引力的，正如结核病所需要的那样[48]。

尽管如此，仍有理由认为BCG赋予的可变保护至少部分是由于其对MHC Ⅰ类限制性反应的弱诱导。因此，寻求提高卡介苗或其突变体的免疫应答以预防结核病感染或激活的策略可能无法完全处理所需的所有方面，使用质粒 DNA 疫苗可以提供有价值的合作伙伴关系[49]。

13.4 优点和缺点

尽管世卫组织最新报告[3]比以前的报告更为乐观，但对于到 2015 年将结核病流行率和死亡率比 1990 年降低 50%，以及到 2050 年将结核病作为一个全球健康问题消除的目标仍存在严重疑问（停止结核病伙伴关系项目）。在这方面，还必须克服一些重大的挑战。

如我们强调的那样，结核病与恶劣的生活条件和贫困之间的关系已经确立。一些国家仅仅通过提高受影响人口的生活水平就能够控制结核病。事实上，贫困地区、高风险群体、共感染流行率、低质量控制措施和不稳定的医疗保健服务等因素造成了当前令人担忧的局面。然而即使社会和运营问题得到解决，在 21 世纪控制结核病仍必须解决几个关键和具有挑战性的问题。

因此，迫切需要研究和开发新的知识、策略和产品来诊断、治疗和预防结核病。此外，应该考虑结核病研究通常是用动物模型进行的，其结果不能总是转移到人类临床实践中。此外，任何治疗、预防或诊断结核病的新工具的开发，都必须得到可靠的生物标记的支持，而这方面的信息是缺乏的。

13.4.1 共感染

结核病疫苗开发中的一个弱点与另一种病原体的共同感染有关。自 20 世纪 80 年代以来，随着艾滋病病毒感染的出现，这种情况变得更加明显。据观察，携带结核杆菌和艾滋病病毒的个体增强了这两种疾病，免疫功能迅速减弱，如果不加以治疗，则会过早死亡（图 13.2）。在这种情况下，预计全球将有 1 400万人会共感染[71]。结核病占艾滋病死亡人数的 26%[72]，其中 99% 发生在发展中国家[73]。

知识点 13.1

DNA-hsp65

文献中描述的第一种结核 DNA 疫苗是 1994 年由我们的小组创建的[50]。这种疫苗名为 DNA-hsp65，包括编码麻风杆菌抗原 hsp65 细胞内表达的基因序列。在这项原始研究的基础上，几名研究人员开发了其他 DNA 疫苗，这些疫苗含有表达其他 Mtb 免疫显性抗原的基因序列，如 Ag85[51]、ESAT-6、MPT-64[52]和 MPT-83[53]，其中许多疫苗能诱导类似 BCG 的保护水平。

我们小组进行的一项最杰出的研究表明，基因疫苗 DNA-hsp65 不仅具有预防 Mtb 感染的作用，还具有治疗作用，可以作为一种免疫调节剂来对抗已确立的疾病。这些初步结果表明，DNA-hsp65 可以治疗具有全身效应的慢性患者、潜伏性结核病和多耐药性结核病，这些结果发表在 1999 年的《自然》杂志[54]上。这种疫苗进一步阻碍免疫抑制动物疾病的复发[18,54]，并且它加入化疗中显著缩短了抗结核治疗时间[18]。在用 DNA-hsp65 进行的几项研究中，主要的研究集中在小鼠、豚鼠[55,56]和非人灵长类动物（Silva 等，未公布的数据）的实验性 Mtb 感染模型的免疫原性和保护上；PLGA 微球制剂[57]、脂质体[58]、基因枪[59]和免疫增强系统[60]的优化；技术发展和扩大生产；动物临床前试验[61,62]；人类 I 期临床试验[63,64]。

用于 DNA-hsp65 递送的阳离子脂质体

考虑到 DNA 疫苗，特别是 DNA-hsp65 的发展带来的有希望的结果，最大的挑战是单剂量给药中 DNA 疫苗的改进。当裸质粒 DNA-hsp65 通过肌内途径[58]以 400 μg 的总量给药 4 次时，在小鼠中实现了抗结核的疫苗接种效率。给药剂量的数量是裸 DNA 必须克服从细胞外基质到细胞核的不同障碍的结果，这需要使用载体来正确递送 DNA[65]。一个安全的策略是使用非病毒载体，这些载体基本上是在DNA与阳离子分子（如壳聚糖和阳离子脂质等）络合后形成的[66]。我们的研究小组对阳离子脂质体进行了广泛的研究，这些胶体系统由两亲性脂质组成，两亲性脂质在双层中自我组装，形成具有内部水腔的囊泡。阳离子特性是通过使用合成脂质获得的，具有毒性。在我们的案例中使用了1，2-二油酰基-3-三甲基铵-丙烷（DOTAP）以及辅助和中性脂质1，2-二油酰-sn-甘油3-磷酸乙醇胺（DOPE）。为了克服阳离子细胞毒性，用鸡蛋L-α-磷脂酰胆碱制备脂质体[58,67]。在不同的参数中DNA掺入脂质体的方式影响免疫反应，反映在不同的核酸释放中[67]。为了评估这一假设我们制备了200~600 nm范围内的脱水水合脂质体（DRV），DNA掺入纳米结构中（常规DRV包封），或者促进DNA和“空”DRV之间的静电络合。络合方案提高了小鼠的预防效果，将给药剂量减少了16倍并使用了非侵入性鼻内途径。

除此之外，众所周知阳离子脂质体能够有效地将核酸传递到胞质溶胶中[66,68]，核膜是一个额外的屏障。为了避免这个问题，我们设计了一种阳离子合成肽，作为核定位信号（NLS）帮助DNA穿过核膜[69]。NLS基本上是一个短肽序列，与特定的蛋白质结合并提供核转运[70]。设计的NLS在静电作用下与DNA-hsp65结合形成一个二元复合物，然后与“空的”DRV脂质体复合（图13.1）。与裸DNA相比这种肽DNA阳离子脂质体复合物对结核的治疗效果相似，但DNA减少了4倍，证明了非病毒载体的保护作用。

阳离子脂质体的使用和添加NLS进行DNA-hsp65递送反映了结核病预防接种和治疗的两种方法，这表明了结合不同策略以提高其他疫苗效力的关键作用。

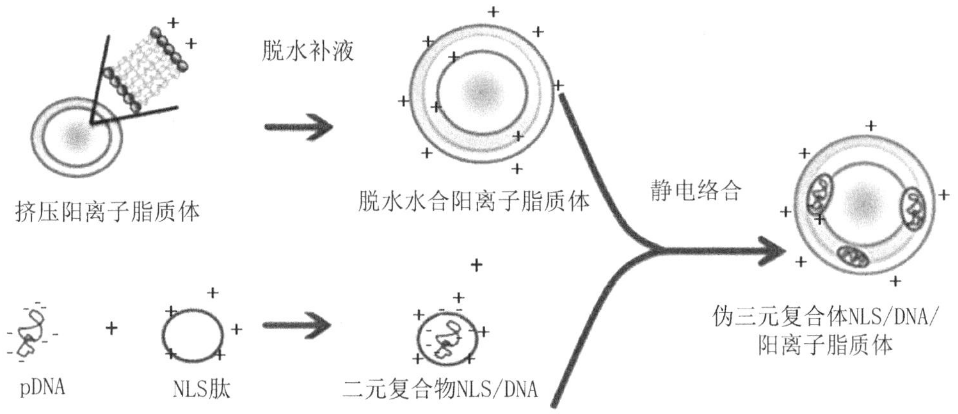

图13.1　肽-DNA-阳离子脂质体复合物的示意图

（转载自Rosada[69]，8页，版权（2012年），获得Elsevier许可）

由艾滋病病毒感染引起的免疫应答系统性失败，可以很容易使某些候选疫苗的保护作用失效。艾滋病病毒病毒杀死效应T $CD4^+$细胞，这是对抗Mtb的一个标志性特征。记忆T细胞也会受到艾滋病病毒细胞溶解效应的影响，使大多数疫苗失去了信誉[74]。避免这种情况的一种较好的替代方法是开发一种TB/HIV联合疫苗，例如携带Mtb和艾滋病病毒抗原组合的重组BCG[75]。

在艾滋病病毒共同感染的情况下，开发新的结核病疫苗时需要考虑的另一个弱点是潜伏因素。

免疫病理学和治疗的意义。黄色方框代表宿主，绿色方框代表结核病，红色方框代表艾滋病毒。箭头表示每种疾病如何影响对另一种病毒的免疫反应[74,76]。

实际上，临床试验或动物模型中最有效的候选疫苗可以预防活动性结核，但是它们不能成功地使宿主免受Mtb感染。因此，如果接种疫苗的人同时感染了艾滋病病毒，免疫应答将注定失败，潜伏库可能会重新激活。

此外，不仅病毒是造成这种情况的原因，寄生虫也是，寄生虫在贫穷国家高度流行，如蠕

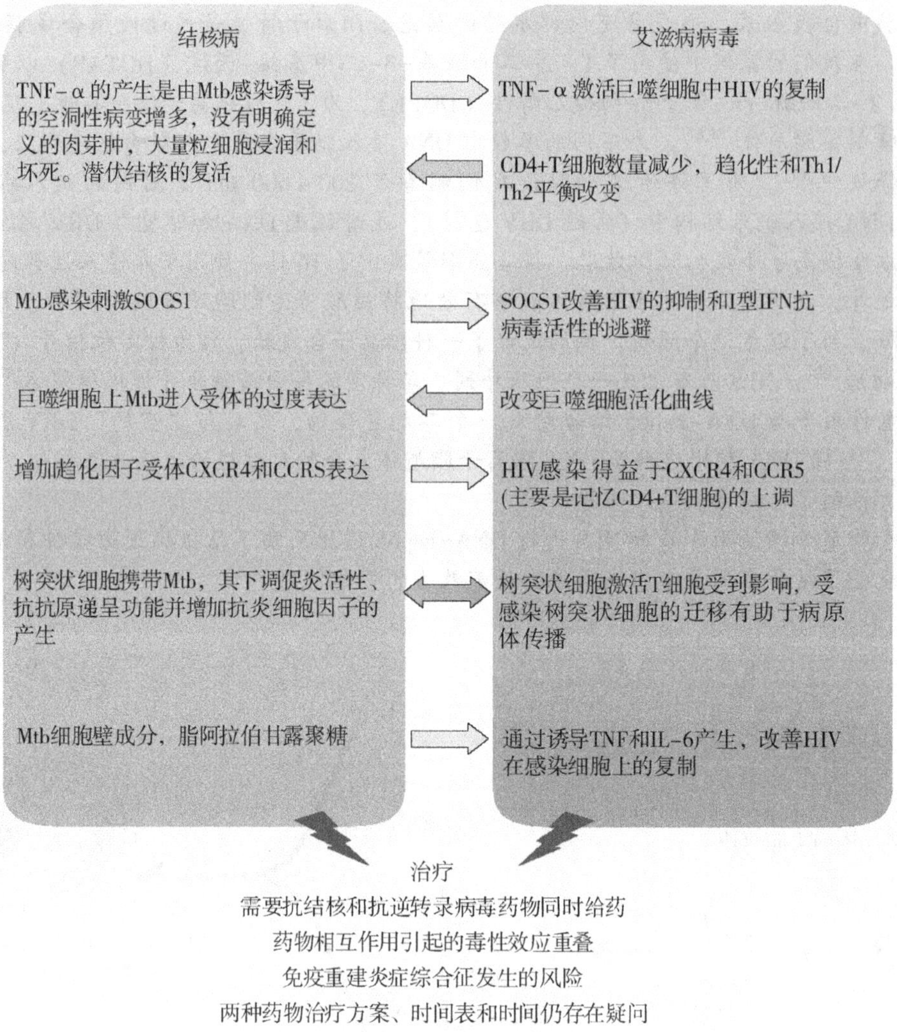

图 13.2 TB/HIV 在共感染宿主中的相互作用

虫。在世界上最贫穷的地区，至少 1/3 的人口感染了一种或多种蠕虫[77,78]。蠕虫感染通常被称为 Th2 反应诱导物，因此，许多研究人员已经证明蠕虫寄生虫在临床前试验中损害了疫苗诱导的 Th1 免疫应答[79-83]。相反，我们先前发现，用犬弓蛔虫感染 BALB/C 小鼠并不能改变随后感染结核分枝杆菌的敏感性[84]。

相比之下，曼氏血吸虫感染会大大增强对结核病的敏感性[82]，并削弱卡介苗接种的保护作用[85]。这些数据提供了结核/蠕虫共感染宿主的特性证据。此外，在埃塞俄比亚的一项研究中观察到卡介苗免疫原性降低，健康志愿者或蠕虫感染志愿者接受抗蠕虫疗法或安慰剂，随后接种卡介苗。因此，在用 BCGa 免疫的蠕虫感染组中，当用 PPD（纯化蛋白衍生物）体外刺激患者细胞时，观察到 IFN-γ 分泌减少，TGF-β 产生增加[86]。这些结果显示了 BCG 预防肺结核明显失败的潜在解释。

然而，并非所有疫苗都受到蠕虫寄生的影响，正如在 DNA-hsp65 中所观察到的。在结核病和血吸虫病、或结核病和弓蛔虫病实验模型中，显示 DNA-hsp65 治疗效果在蠕虫感染引起的不相关 Th2 免疫应答存在的情况下不受影响[87]。因此，必须考虑蠕虫对结核病期间免疫应答的压力，影响针对这种疾病的新疗法和/或疫苗接种方案开发。

13.4.2 动物模型

在结核病研究中使用动物始于 Robert Koch 将老鼠作为研究模型[88]。从那时起在利用动物模型了解结核分枝杆菌感染的动力学方面取得了相当大的进展。针对 Mtb 的细胞免疫应答和产生细胞因子在小鼠中非常强烈，它们与在感染结核病的人身上观察到的一些参数相关对应，如 $CD4^{+}T$ 细胞的参与[89,90]、IL-12[91,92]和 TNF-α[93,94]。此外，小鼠模型具有很好的免疫遗传学特征，并且有广泛的试剂可用于评估这些特征。

尽管在小鼠上进行结核病研究有很多好处，但应该考虑与人类的一些重要差异。从免疫组织病理学角度来看，小鼠感染 Mtb 并不会导致肺部典型肉芽肿形成，而是在炎症过程中心诱导大量淋巴细胞，而这些淋巴细胞正是细菌所在的位置。相反地，在人类中感染引起的炎症浸润表现出典型的肉芽肿结构，感染的巨噬细胞周围有一圈淋巴细胞，占据了结构的中心[95,96]。此外干酪样坏死和空化现象发生在人类而不是小鼠[97]。

为了克服这些问题，几个团队采用了其他实验动物模型，如豚鼠、兔子和猴子，并将结果与小鼠的结果进行了比较，特别是在功效筛选和/或新开发的疫苗和药物方面[98]。然而由于不同的基因组成，在其他实验模型和人类中，小鼠 Mtb 感染中所涉及的几种宿主分子是缺失的[99,100]。此外，肺环境中的免疫反应，即小鼠、猴子和其他模型中 Mtb 感染的部位，不同于外周血中的反应，使得很难将实验模型中从肺获得的数据与人类外周血中的发现相关联[101-103]。相反地，由于伦理原因和程序上的困难，在支气管肺泡灌洗细胞或肺活检样本中，评估肺结核患者肺部环境中的免疫反应并不普遍可用，肺部特征与从患者外周血中获得的数据相关性是有争议的[104,105]。

因此，尽管在结核病研究中使用动物提高了我们对 Mtb 感染动态的理解，但由此获得的数据不能系统地转移到人类临床实践中。没有一项研究在全球范围内显示，在几个被调查的动物物种中实际存在哪些感染或疾病标记。因此我们可以得出结论：评估结核病临床形式和治疗干预效果以及在实验和临床数据之间建立相关性的技术，必须在生物液体样本中进行，例如易于采集的血液和临床实践中最常用的来源。

从这个意义上来说，疾病不同阶段的可靠生物标志对于正确诊断结核病的几种临床形式以及分析治疗和疫苗效力至关重要。然而这种生物标志还没有被鉴定出来。

13.4.3 生物标志

使用标志诊断疾病的第一个参考文献是在公元前400年左右，当时希波克拉底认为，包括血液在内的有机液体的改变与疾病有关。从那时起，疾病的特征就是它们在我们身体的组成部分引起的变化。到了20世纪末21世纪初，为了提高诊断的准确性，开始进行的研究试图确定一个人的正常分子特性及其对疾病的反应改变。目前生物标志被定义为一种生物学特性，它被客观地测量和评估为正常生物或致病过程的指标，或对治疗干预的药理学反应，支持关于个体健康的当前或未来状态，以及诱导的病理学类型或程度的结论。

Mtb 感染进展的复杂性和针对它的免疫反应引起了在一种生理成分中寻找诊断工具的尝试。目前疑似结核感染患者的肺部 X 线片和痰培养物与结核菌素皮肤试验（对 Mtb 抗原的延迟过敏反应）相结合以确定诊断。最近人们尝试利用外周血液中的细胞因子水平、不同细胞群体的频率和细胞表面分子的表达来区分结核病的形式。尽管有机体组成成分的可能组合不计其数，但还没有建立具有足够敏感性和特异性的特定元素集来区分不同的结核病形式，并预测目前和可能的治疗和预防疫苗的成功率[12]。

在结核病全球紧急情况的现代背景下，生物标志可能有不同的应用，包括从诊断、通过预防和治疗到结果测量。在每一种情况下，使用生物标志都有重要的细分。显而易见，它们将是诊断不同类型的结核病（如抗生素耐药性）、评估疾病恢复或复发风险以及开发新药和新的预防和治疗疫苗的重要催化剂，证明它们的有效性并为临床试验提供终点。与此同时，寻找生物标志的研究将提供对疾病本身的深入了解[106]。

这些生物标志还没有被确认，但一些国际研究小组正在努力寻找[15,107]。大量的研究已经测量了产生 IFN-γ 的 T 细胞的频率，但是已经证明它不能很好地预测疫苗的保护效果。同样地，我们也考虑了分泌多种细胞因子的抗原特异性 T 细胞的多功能亚群，但没有得出任何最终结论。此外实验和临床研究也表明，抗体在宿主防御分支杆菌感染方面的重要性仍存在争议[108]。如果检测特定抗体将有助于保护相关性，它仍然有待阐明。由于结核病免疫应答是一个非常复杂的过程，因此应进一步研究其他成分作为保护的潜在关联[109]。

虽然在测试新疫苗之前，保护相关因素应该是可以获得的，但是这些生物标志很有可能来自已经进行的随机对照临床试验[17,106]。这种方法的原理是比较疫苗组和安慰剂组中受保护的个体。这里的主要问题是在传染病流行区建立理想群体以及这一策略的耗时特点。随着越来越多的疫苗进入临床试验，人们乐观地认为应该尽快发现这些相关性。比较不同研究的发现对于定义一组更通用的标记非常有用。

但是，如何寻找生物标志，或者什么是获得可靠结果的最佳工具？如今生物标志已经在分子水平上得到了展望。"组学"时代，即基因组学、蛋白质组学、转录组学和代谢组学，无疑开辟了新的视角，来自两种或更多种方法的数据关联可能会带来更强有力的答案，我们称之为生物标志。然而充分分析使用这些技术调查产生的大量信息同样具有挑战性，生物信息学的进步至关重要，包括系统生物学的丰硕领域[110]。为了支持这种方法，一些存储库数据集有助于加速研究。

结核病数据库（TBDB-http：//www. tbdb. org/）是一个集成数据库，提供对结核病基因组数据和相关资源的访问，以发现和开发结核病药物、疫苗和生物标志。此外，TBDB 提供了对软件的访问，允许比较不同的数据集，构成了一个非常有吸引力的结核病研究平台[111]。TDR 目标数据库（http：//tdrtargets. org）探索病原体特异性基因组信息的可用性并将其与功能数据集相结合，是一个在线、多功能的资源，有助于快速识别不同疾病药物开发的分子目标并确定其优先级，包括结核病[112]。

这些工具的多功能性和查询复杂数据集的能力是至关重要的，因为除了一些在地理上相对成功的研究之外，重要的是要记住，这可能不是"一刀切"的解决方案，我们必须意识到这一点并偶尔调整区域结果。

参考文献

[1] Hershkovitz, I., et al.: Detection and molecular characterization of 9, 000-year-old Mycobacterium tuberculosis from a Neolithic settlement in the Eastern Mediterranean. PLoSOne3, e3426 (2008).

[2] Gagneux, S.: Host-pathogen coevolution in human tuberculosis. Philos. Trans. R. Soc. Lond. B Biol. Sci. 367, 850-859 (2012).

[3] WHO: Global tuberculosis control, p. 246. World Health Organization, Geneva (2011).

[4] Barreira, D., Grangeiro, A.: Evaluation of tuberculosis control strategies in Brazil. Foreword. Rev. Saude-Publica41 (Suppl 1), 4-8 (2007).

[5] Kaur, D., Guerin, M. E., Skovierová, H., Brennan, P. J., Jackson, M.: Chapter 2: Biogenesis of the cell wall and other glycoconjugates of Mycobacterium tuberculosis. Adv. Appl. Microbiol. 69, 23-78 (2009).

[6] Smith, I.: Mycobacterium tuberculosis pathogenesis and molecular determinants of virulence. Clin. Microbiol. Rev. 16, 463-496 (2003).

[7] Kaufmann, S. H.: How can immunology contribute to the control of tuberculosis? Nat. Rev. Immunol. 1, 20-30 (2001).

[8] Dheda, K., et al.: Lung remodeling in pulmonary tuberculosis. J. Infect. Dis. 192, 1201-1209 (2005).

[9] Welsh, K. J., Risin, S. A., Actor, J. K., Hunter, R. L.: Immunopathology of post-primary tuberculosis: increased T-regulatory cells and DEC-205-positive foamy macrophages in cavitary lesions. Clin. Dev. Immunol. 2011, 307631 (2011).

[10] Dorman, S. E.: New diagnostic tests for tuberculosis: bench, bedside, and beyond. Clin. Infect. Dis. 50 (Suppl 3), S173-S177 (2010).

[11] Steingart, K. R., et al.: Fluorescence versus conventional sputum smear microscopy for tuberculosis: a systematic review. Lancet Infect. Dis. 6, 570-581 (2006).

[12] Perrin, F. M., Lipman, M. C., McHugh, T. D., Gillespie, S. H.: Biomarkers of treatment response in clinical trials of novel anti-tuberculosis agents. Lancet Infect. Dis. 7, 481-490 (2007).

[13] Centers for Disease Control and Prevention. Updated Guidelines for Using Interferon Gamma Release Assays to Detect Mycobacterium tuberculosis Infection, United States. (PDF) MMWR 2010; 59 (No. RR-5). http: // www. cdc. gov/mmwr/preview/mmwrhtml/rr5905a1. htm? s_cid=rr5905a1_e.

[14] Ginsberg, A. M., Spigelman, M.: Challenges in tuberculosis drug research and development. Nat. Med. 13, 290-294 (2007).

[15] Parida, S. K., Kaufmann, S. H.: The quest for biomarkers in tuberculosis. Drug Discov. Today15, 148-157 (2010).

[16] Jain, A., Mondal, R.: Extensively drug-resistant tuberculosis: current challenges and threats. FEMSImmunol. Med. Microbiol. 53, 145-150 (2008).

[17] Aagaard, C., Dietrich, J., Doherty, M., Andersen, P.: TB vaccines: current status and future perspectives. Immunol. Cell Biol. 87, 279-286 (2009).

[18] Silva, C. L., et al.: Immunotherapy with plasmid DNA encoding mycobacterial hsp65 in association with chemotherapy is a more rapid and efficient form of treatment for tuberculosis in mice. Gene Ther. 12, 281-287 (2005).

[19] Skeiky, Y. A., Sadoff, J. C.: Advances in tuberculosis vaccine strategies. Nat. Rev. Microbiol. 4, 469-476 (2006).

[20] Ottenhoff, T. H., Kaufmann, S. H.: Vaccines against tuberculosis: where are we and where do we need to go? PLoS Pathog. 8, e1002607 (2012).

[21] Grode, L., et al.: Increased vaccine efficacy against tuberculosis of recombinant Mycobacterium bovis bacilleCalmette-Guerin mutants that secrete listeriolysin. J. Clin. Invest. 115, 2472-2479 (2005).

[22] Tullius, M. V., Harth, G., Maslesa-Galic, S., Dillon, B. J., Horwitz, M. A.: A Replication-Limited Recombinant Mycobacterium bovis BCG vaccine against tuberculosis designed for human immunodeficiency virus-positive persons is safer and more efficacious than BCG. Infect. Immun. 76, 5200-5214 (2008).

[23] Green light for clinical trial of new tuberculosis vaccine candidate. < http: //ec. europa. eu/research/ index. cfm? lg=en&na=na-161012&pg=newsalert&y ear=2012> (2012).

[24] McShane, H., et al.: Recombinant modified vaccinia virus Ankara expressing antigen 85A boosts BCG-primed and naturally acquired anti-mycobacterial immunity in humans. Nat. Med. 10, 1240-1244 (2004).

[25] Scriba, T. J., et al.: Modified vaccinia Ankara-expressing Ag85A, a novel tuberculosis vaccine, is safe in adolescents and children, and induces poly-functional $CD4^+$ T cells. Eur. J. Immunol. 40, 279-290 (2010).

[26] Tameris, M. D., et al.: Safety and efficacy of MVA85A, a new tuberculosis vaccine, in infants previously vaccinated with BCG: a randomised, placebo-controlled phase 2b trial. Lancet 381 (9871), 1021-1028 (2013).

[27] Hoft, D. F., et al.: A recombinant adenovirus expressing immuno-dominant TB antigens can significantly enhance BCG-induced human immunity. Vaccine 30, 2098-2108 (2012).

[28] Wang, J., et al.: Single mucosal, but not parenteral, immunization with recombinant adenoviral-based vaccine provides potent protection from pulmonary tuberculosis. J. Immunol. 173, 6357-6365 (2004).

[29] Santosuosso, M., McCormick, S., Zhang, X., Zganiacz, A., Xing, Z.: Intranasal boosting with an adenovirus - vectored vaccine markedly enhances protection by parenteral Mycobacterium bovis BCG immunization against pulmonary tuberculosis. Infect. Immun. 74, 4634-4643 (2006).

[30] Ottenhoff, T. H., et al.: First in humans: a new molecularly defined vaccine shows excellent safety and strong induction of long-lived Mycobacterium tuberculosis-specific Th1-cell like responses. Hum. Vaccin. 6, 1007-1015 (2010).

[31] van Dissel, J. T., et al.: Ag85B-ESAT-6 adjuvanted with IC31 (R) promotes strong and long-lived Mycobacterium tuberculosis specific T cell responses in volunteers with previous BCG vaccination or tuberculosis infection. Vaccine 29, 2100-2109 (2011).

[32] van Dissel, J. T., et al.: Ag85B-ESAT-6 adjuvanted with IC31 promotes strong and long-lived Mycobacterium tuberculosis specific T cell responses in naive human volunteers. Vaccine 28, 3571-3581 (2010).

[33] Billeskov, R., Elvang, T. T., Andersen, P. L., Dietrich, J.: The HyVac4 subunit vaccine efficiently boosts BCG-primed anti-mycobacterial protective immunity. PLoS One 7, e39909 (2012).

[34] Von Eschen, K., et al.: The candidate tuberculosis vaccine Mtb72F/AS02A: tolerability and immunogenicity in humans. Hum. Vaccin. 5, 475-482 (2009).

[35] Aagaard, C., et al.: A multistage tuberculosis vaccine that confers efficient protection before and after exposure. Nat. Med. 17, 189-194 (2011).

[36] Govender, L., et al.: Higher human CD4 T cell response to novel Mycobacterium tuberculosis latency associated antigens Rv2660 and Rv2659 in latent infection compared with tuberculosis disease. Vaccine 29, 51-57 (2010).

[37] Lin, P. L., et al.: The multistage vaccine H56 boosts the effects of BCG to protect cynomolgus macaques against active tuberculosis and reactivation of latent Mycobacterium tuberculosis infection. J. Clin. Invest. 122, 303-314 (2012).

[38] Bertholet, S., et al.: A defined tuberculosis vaccine candidate boosts BCG and protects against multidrug-resistant mycobacterium tuberculosis. Sci. Transl. Med. 2, 53ra74 (2010).

[39] Baldwin, S. L., et al.: The importance of adjuvant formulation in the development of a tuberculosis vaccine. J. Immunol. 188, 2189-2197 (2012).

[40] Coler, R. N., et al.: Therapeutic immunization against Mycobacterium tuberculosis is an effective adjunct to antibiotic treatment. J. Infect. Dis. 207 (8), 1242-1252 (2013).

[41] Cardona, P. J.: RUTI: a new chance to shorten the treatment of latent tuberculosis infection. Tuberculosis (Edinb.) 86 (273-289) (2006).

[42] Waddell, R. D., et al.: Safety and immunogenicity of a five-dose series of inactivated Mycobacterium vaccae vaccination for the prevention of HIV-associated tuberculosis. Clin. Infect. Dis. 30 (Suppl 3), S309-S315 (2000).

[43] Mayo, R. E., Stanford, J. L.: Double-blind placebo-controlled trial of Mycobacterium vaccae immunotherapy for tuberculosis in KwaZulu, South Africa, 1991—1997. Trans. R. Soc. Trop. Med. Hyg. 94, 563-568 (2000).

[44] de Bruyn, G., Garner, P.: Mycobacterium vaccae immunotherapy for treating tuberculosis. Cochrane Database Syst Rev. (1), CD001166 (2003).

[45] Dubensky Jr., T. W., Skoble, J., Lauer, P., Brockstedt, D. G.: Killed but metabolically active vaccines. Curr. Opin. Biotechnol. 23 (6), 917-923 (2012).

[46] Liu, M.: DNA vaccines: a review. J. Intern. Med. 253, 402-410 (2003).

[47] Liu, M. A.: DNA vaccines: an historical perspective and view to the future. Immunol. Rev. 239, 62-84

(2011).
[48] Ingolotti, M., Kawalekar, O., Shedlock, D. J., Muthumani, K., Weiner, D. B.: DNA vaccines for targeting bacterial infections. Expert Rev. Vaccines 9, 747-763 (2010).
[49] Romano, M., Huygen, K.: An update on vaccines for tuberculosis-there is more to it than just waning of BCG efficacy with time. Expert Opin. Biol. Ther. 12 (12), 1601-1610 (2012).
[50] Lowrie, D. B., Tascon, R. E., Colston, M. J., Silva, C. L.: Towards a DNA vaccine against tuberculosis. Vaccine 12, 1537-1540 (1994).
[51] Ulmer, J. B., et al.: Expression and immunogenicity of Mycobacterium tuberculosis antigen 85 by DNA vaccination. Vaccine 15, 792-794 (1997).
[52] Kamath, A. T., Feng, C. G., Macdonald, M., Briscoe, H., Britton, W. J.: Differential protective efficacy of DNA vaccines expressing secreted proteins of Mycobacterium tuberculosis. Infect. Immun. 67, 1702-1707 (1999).
[53] Zhu, X., et al.: Functions and specificity of T cells following nucleic acid vaccination of mice against Mycobacterium tuberculosis infection. J. Immunol. 158, 5921-5926 (1997).
[54] Lowrie, D. B., et al.: Therapy of tuberculosis in mice by DNA vaccination. Nature 400, 269-271 (1999).
[55] dos Santos, S. A., et al.: A subunit vaccine based on biodegradable microspheres carrying rHsp65 protein and KLK protects BALB/c mice against tuberculosis infection. Hum. Vaccin. 6, 1047-1053 (2010).
[56] de Paula, L., et al.: Comparison of different delivery systems of DNA vaccination for the induction of protection against tuberculosis in mice and guinea pigs. Genet. Vaccines Ther. 5, 2 (2007).
[57] Lima, K. M., et al.: Single dose of a vaccine based on DNA encoding mycobacterial hsp65 protein plus TDM-loaded PLGA microspheres protects mice against a virulent strain of Mycobacterium tuberculosis. Gene Ther. 10, 678-685 (2003).
[58] Rosada, R. S., et al.: Protection against tuberculosis by a single intranasal administration of DNA-hsp65 vaccine complexed with cationic liposomes. BMC Immunol. 9, 38 (2008).
[59] Lima, K. M., et al.: Efficacy of DNA-hsp65 vaccination for tuberculosis varies with method of DNA introduction in vivo. Vaccine 22, 49-56 (2003).
[60] Ruberti, M., et al.: Prime - boost vaccination based on DNA and protein - loaded microspheres for tuberculosis prevention. J. Drug Target. 12, 195-203 (2004).
[61] Morais Fonseca, D., et al.: Experimental tuberculosis: designing a better model to test vaccines against tuberculosis. Tuberculosis (Edinb.) 90 (135-142) (2010).
[62] Zarate-Blades, C. R., et al.: Comprehensive gene expression profiling in lungs of mice infected with Mycobacterium tuberculosis following DNAhsp65 immunotherapy. J. Gene Med. 11, 66-78 (2009).
[63] Michaluart, P., et al.: Phase I trial of DNA-hsp65 immunotherapy for advanced squamous cell carcinoma of the head and neck. Cancer Gene Ther. 15, 676-684 (2008).
[64] Victora, G. D., et al.: Immune response to vaccination with DNA-hsp65 in a phase I clinical trial with head and neck cancer patients. Cancer Gene Ther. 16 (7), 598-608 (2009).
[65] Rolland, A.: Gene medicines: the end of the beginning? Adv. Drug Deliv. Rev. 57, 669-673 (2005).
[66] Mintzer, M. A., Simanek, E. E.: Nonviral vectors for gene delivery. Chem. Rev. 109, 259-302 (2009).
[67] de la Torre, L. G., et al.: The synergy between structural stability and DNA-binding controls the antibody production in EPC/DOTAP/DOPE liposomes and DOTAP/DOPE lipoplexes. Colloids Surf. B Biointerfaces73, 175-184 (2009).
[68] Xu, Y., Szoka Jr., F. C.: Mechanism of DNA release from cationic liposome/DNA complexes used in cell transfection. Biochemistry 35, 5616-5623 (1996).
[69] Rosada, R. S., Silva, C. L., Santana, M. H., Nakaie, C. R., de la Torre, L. G.: Effectiveness, against tuberculosis, of pseudo-ternary complexes: peptide-DNA-cationic liposome. J. Colloid Interface Sci.

373, 102-109 (2012).

[70] Byrnes, C. K., et al.: Novel nuclear shuttle peptide to increase transfection efficiency in esophageal mucosal cells. J. Gastrointest. Surg. 6, 37-42 (2002).

[71] Getahun, H., Gunneberg, C., Granich, R., Nunn, P.: HIV infection-associated tuberculosis: the epidemiology and the response. Clin. Infect. Dis. 50 (Suppl 3), S201-S207 (2010).

[72] Corbett, E. L., et al.: The growing burden of tuberculosis: global trends and interactions with the HIV epidemic. Arch. Intern. Med. 163, 1009-1021 (2003).

[73] Collins, K. R., Quinones-Mateu, M. E., Toossi, Z., Arts, E. J.: Impact of tuberculosis on HIV-1 replication, diversity, and disease progression. AIDS Rev. 4, 165-176 (2002).

[74] Pawlowski, A., Jansson, M., Skold, M., Rottenberg, M. E., Kallenius, G.: Tuberculosis and HIV coinfection. PLoSPathog. 8, e1002464 (2012).

[75] Kaufmann, S. H., McMichael, A. J.: Annulling a dangerous liaison: vaccination strategies against AIDS and tuberculosis. Nat. Med. 11, S33-S44 (2005).

[76] Curran, A., Falco, V., Pahissa, A., Ribera, E.: Management of tuberculosis in HIV-infected patients. AIDS Rev. 14, 231-246 (2012).

[77] de Silva, N. R., et al.: Soil-transmitted helminth infections: updating the global picture. Trends Parasitol. 19, 547-551 (2003).

[78] Hotez, P. J., et al.: Helminth infections: the great neglected tropical diseases. J. Clin. Invest. 118, 1311-1321 (2008).

[79] Actor, J. K., et al.: Helminth infection results in decreased virus-specific $CD8^+$ cytotoxic T-cell and Th1 cytokine responses as well as delayed virus clearance. Proc. Natl. Acad. Sci. U. S. A. 90, 948-952 (1993).

[80] Araujo, M. I., et al.: Interleukin-12 promotes pathologic liver changes and death in mice coinfected with Schistosoma mansoni and Toxoplasma gondii. Infect. Immun. 69, 1454-1462 (2001).

[81] Chen, C. C., Louie, S., McCormick, B., Walker, W. A., Shi, H. N.: Concurrent infection with an intestinal helminth parasite impairs host resistance to enteric Citrobacter rodentium and enhances Citrobacter-induced colitis in mice. Infect. Immun. 73, 5468-5481 (2005).

[82] Elias, D., Akuffo, H., Thors, C., Pawlowski, A., Britton, S.: Low dose chronic Schistosoma mansoni infection increases susceptibility to Mycobacterium bovis BCG infection in mice. Clin. Exp. Immunol. 139, 398-404 (2005).

[83] Mansfield, L. S., et al.: Enhancement of disease and pathology by synergy of Trichurissuis and Campylobacter jejuni in the colon of immunologically naive Swine. Am. J. Trop. Med. Hyg. 68, 70-80 (2003).

[84] Frantz, F. G., et al.: The immune response to toxocariasis does not modify susceptibility to Mycobacterium tuberculosis infection in BALB/c mice. Am. J. Trop. Med. Hyg. 77, 691-698 (2007).

[85] Elias, D., et al.: Schistosoma mansoni infection reduces the protective efficacy of BCG vaccination against virulent Mycobacterium tuberculosis. Vaccine 23, 1326-1334 (2005).

[86] Elias, D., Britton, S., Aseffa, A., Engers, H., Akuffo, H.: Poor immunogenicity of BCG in helminth infected population is associated with increased in vitro TGF-beta production. Vaccine 26, 3897-3902 (2008).

[87] Frantz, F. G., et al.: Helminth coinfection does not affect therapeutic effect of a DNA vaccine in mice harboring tuberculosis. PLoSNegl. Trop. Dis. 4, e700 (2010).

[88] Collins, F. M.: The immunology of tuberculosis. Am. Rev. Respir. Dis. 125, 42-49 (1982).

[89] Caruso, A. M., et al.: Mice deficient in CD4 T cells have only transiently diminished levels of IFN-gamma, yet succumb to tuberculosis. J. Immunol. 162, 5407-5416 (1999).

[90] Scanga, C. A., et al.: Depletion of CD4 (+) T cells causes reactivation of murine persistent tuberculosis despite continued expression of interferon gamma and nitric oxide synthase 2. J. Exp. Med. 192, 347-358

(2000).

[91] Cooper, A. M., Callahan, J. E., Keen, M., Belisle, J. T., Orme, I. M.: Expression of memory immunity in the lung following re-exposure to Mycobacterium tuberculosis. Tuber. Lung Dis. 78, 67-73 (1997).

[92] de Jong, R., et al.: Severe mycobacterial and Salmonella infections in interleukin-12 receptor-deficient patients. Science 280, 1435-1438 (1998).

[93] Flynn, J. L., et al.: Tumor necrosis factor-alpha is required in the protective immune response against Mycobacterium tuberculosis in mice. Immunity 2, 561-572 (1995).

[94] Keane, J., et al.: Tuberculosis associated with infliximab, a tumor necrosis factor alpha-neutralizing agent. N. Engl. J. Med. 345, 1098-1104 (2001).

[95] Ulrichs, T., Kaufmann, S. H.: New insights into the function of granulomas in human tuberculosis. J. Pathol. 208, 261-269 (2006).

[96] Turner, O. C., Basaraba, R. J., Orme, I. M.: Immuno-pathogenesis of pulmonary granulomas in the guinea pig after infection with Mycobacterium tuberculosis. Infect. Immun. 71, 864-871 (2003).

[97] Dannenberg Jr., A. M.: Perspectives on clinical and preclinical testing of new tuberculosis vaccines. Clin. Microbiol. Rev. 23, 781-794 (2010).

[98] Gupta, U. D., Katoch, V. M.: Animal models of tuberculosis for vaccine development. Indian J. Med. Res. 129, 11-18 (2009).

[99] Behar, S. M., Dascher, C. C., Grusby, M. J., Wang, C. R., Brenner, M. B.: Susceptibility of mice deficient in CD1D or TAP1 to infection with Mycobacterium tuberculosis. J. Exp. Med. 189, 1973-1980 (1999).

[100] Carding, S. R., Egan, P. J.: The importance of gamma delta T cells in the resolution of pathogen-induced inflammatory immune responses. Immunol. Rev. 173, 98-108 (2000).

[101] Arriaga, A. K., Orozco, E. H., Aguilar, L. D., Rook, G. A., Hernández Pando, R.: Immunological and pathological comparative analysis between experimental latent tuberculous infection and progressive pulmonary tuberculosis. Clin. Exp. Immunol. 128, 229-237 (2002).

[102] Basaraba, R. J.: Experimental tuberculosis: the role of comparative pathology in the discovery of improved tuberculosis treatment strategies. Tuberculosis (Edinb.) 88 (Suppl 1), S35-S47 (2008).

[103] Helke, K. L., Mankowski, J. L., Manabe, Y. C.: Animal models of cavitation in pulmonary tuberculosis. Tuberculosis (Edinb.) 86 (337-348) (2006).

[104] Sable, S. B., Goyal, D., Verma, I., Behera, D., Khuller, G. K.: Lung and blood mononuclear cell responses of tuberculosis patients to mycobacterial proteins. Eur. Respir. J. 29, 337-346 (2007).

[105] Ainslie, G. M., Solomon, J. A., Bateman, E. D.: Lymphocyte and lymphocyte subset numbers in blood and in bronchoalveolar lavage and pleural fluid in various forms of human pulmonary tuberculosis at presentation and during recovery. Thorax 47, 513-518 (1992).

[106] Wallis, R. S., et al.: Biomarkers and diagnostics for tuberculosis: progress, needs, and translation into practice. Lancet 375, 1920-1937 (2010).

[107] Jacobsen, M., Mattow, J., Repsilber, D., Kaufmann, S. H.: Novel strategies to identify biomarkers in tuberculosis. Biol. Chem. 389, 487-495 (2008).

[108] Glatman-Freedman, A.: The role of antibody-mediated immunity in defense against Mycobacterium tuberculosis: advances toward a novel vaccine strategy. Tuberculosis (Edinb.) 86 (191-197) (2006).

[109] Abebe, F.: Is interferon-gamma the right marker for bacille Calmette-Guérin-induced immune protection? The missing link in our understanding of tuberculosis immunology. Clin. Exp. Immunol. 169, 213-219 (2012).

[110] Rappuoli, R., Aderem, A.: A 2020 vision for vaccines against HIV, tuberculosis and malaria. Nature

473, 463-469 (2011).

[111] Reddy, T. B., et al.: TB database: an integrated platform for tuberculosis research. Nucleic Acids Res. 37, D499-D508 (2009).

[112] Magariños, M. P., et al.: TDR Targets: a chemo-genomics resource for neglected diseases. Nucleic Acids Res. 40, D1118-D1127 (2012).

(马维民译)

第14章 副球芽孢菌病：分子疫苗的研究进展

Luiz R. Travassos，Glauce M. G. Rittner
和 Carlos P. Taborda[①]

摘要

P10肽是巴西副球芽孢菌大多数分离株gp43诊断抗原中的保守序列。P10由来自小鼠单倍型的3种不同MHC类分子以及90%的高加索HLA-DR抗原混合表达的。它引发IFN-γ依赖性保护性抗真菌应答，是候选的肽疫苗。P10免疫的小鼠用剧毒酵母气管内攻毒，通过肺部菌落形成单位测量，显示提供了显著的保护，获得了主要的Th-1免疫应答。不同的佐剂增强了肽活性，而P10诱导树突状细胞在清除真菌感染方面非常有效。用表达P10和IL-12的质粒进行基因治疗，诱导保护性Th-1免疫应答而不会引起过度炎症或纤维化。在一项长期实验中，用pP10和pIL-12接种疫苗实际上消除了所有真菌成分，并保留了正常的肺实质。

真菌性疾病可影响流行地区的大量个体，对免疫功能低下患者具有相当严重甚至致命的影响，迄今为止还没有一种批准的疫苗用于免疫预防或治疗和化疗。然而随着保护性抗原、毒力因子、突变株、完整基因组学的发展以及免疫应答知识的增加，在许多病原真菌研究领域取得了明显进展。最有希望的是那些能引发杀真菌抗体的疫苗。其他疫苗的目的是保护细胞免疫反应，即使在免疫缺陷和中性粒细胞减少的患者中也能起到保护作用。目前我们综述了肽疫苗或等效基因治疗系统性副球芽孢病的研究进展。

14.1 疾病

副球芽孢菌病（PCM），以前称为南美芽生菌病，是一种由热双相巴西副球芽孢菌引起的地方性真菌感染。阿道夫·卢茨于1908年在圣保罗首次描述了这种疾病，它是拉丁美洲普遍存在的系统性真菌感染（从墨西哥到阿根廷）。大多数病例报告在巴西（约80%），其次是哥伦比亚、委内瑞拉和阿根廷。到目前为止，在智利、尼加拉瓜和安的列斯群岛还没有单独的病例报告。

在美国、一些欧洲国家和亚洲报告了非本地病例，所有这些病例均由以前访问过拉丁美洲流行区的个人表现出来的。因此副球芽孢菌病也可被视为在流行地区长期生活的旅行者的一种疾病[1,2]。

向巴西和其他拉丁美洲国家的公共卫生当局报告真菌病不是强制性的。因此，关于流行地区真菌感染的实际发生率数据很少。1996—2006年，巴西3 583例确诊的系统性真菌死亡中，

---

① L. R. Travassos，医学博士，博士（✉）（巴西圣保罗联邦大学微生物、免疫和寄生虫学系细胞生物学室。圣保罗，SP，巴西，E-mail：travassos@ unifesp. br）

G. M. G. Rittner，PhD（巴西圣保罗大学生物医学科学研究所微生物学系，圣保罗，SP，巴西）

C. P. Taborda，PhD（圣保罗大学圣保罗热带医学研究所医学真菌学实验室，圣保罗大学生物医学科学研究所微生物学系，圣保罗，SP，巴西）

1 853例（51.2%）是由 PCM 引起的[3]。

感染是通过吸入空气传播的繁殖体（分生芽孢）而获得的，这种繁殖体以真菌菌丝体形式释放出来，这种菌丝体在流行地区土壤中生长。初始接触后，可以发生不同的进展状态，称为：①PCM 感染；②PCM 病（急性/亚急性或慢性形式）；③单灶性感染；④多灶性感染；⑤残余临床形式或后遗症[4]。

原发性肺部感染通常无症状或症状过少，个体一生都可能受到感染而不会发展成进行性 PCM。少数受感染患者在吸入真菌分生芽孢数周至数月后可能会发展成临床疾病。典型的急性或亚急性临床表现可能代表 3%~5%的 PCM 病例，主要影响儿童和青少年（不分性别）。主要临床症状包括淋巴结肿大、肝脾肿大、消化和/或骨关节受累，以及皮肤损伤[4,5]。

大多数有症状的患者在感染后几年由于静止病灶（慢性形式）重新激活而发展成这种疾病。据报道，活动性疾病主要发生在成年男性（15∶1）和流行地区的农村工人中，年龄在 30~60 岁。当通过阳性皮肤测试诊断时，感染在性别之间平均分布。β-雌二醇似乎可以保护妇女不患上这种疾病。90%的患者有肺部临床表现。然而该病可能涉及其他器官，如口咽黏膜、皮肤、淋巴结、肾上腺和中枢神经系统[4,6]。尽管 PCM 是一种受细胞介导免疫控制的感染，但这种真菌病在艾滋病患者、癌症患者和接受器官移植的患者中鲜有报道[7,8]。

14.2 病原

PCM 病原体是一种真核双相真菌，仅知其无性状态（无性型），称为巴西副球芽孢菌[9]。新鲜标本中检查的特征是直径为 5~25 μm 的圆形细胞，有清晰的折射壁和多芽。在菌丝阶段用扫描电子显微镜观察时，显示菌丝表面光滑，管状有突出的中隔环状物。在培养物中发现厚垣芽孢、关节芽孢和分生芽孢。产孢过程是一个不同步的过程，似乎取决于生长条件，如培养基成分、培养期、氧气和湿度[10]。巴西副球芽孢菌的菌丝阶段没有特定结构，但是多芽酵母菌可以诊断出这种真菌。菌丝阶段菌落（环境温度-22℃左右）很小，白色，不规则，一些分离物的气生菌丝可能会覆盖褐色色素。在酵母菌阶段（37℃）菌落呈米黄色和脑状（图 14.1）。

基于分子系统发育研究，巴西副球芽孢菌与其他双相性真菌（皮炎芽生菌和荚膜组织胞浆菌）按以下分类法分类：真菌界、子囊菌门或科、多形菌纲、爪甲团囊菌目、阿耶洛菌科、副球虫属和巴西种。

利用基于分子系统分类学的方法，将来自 65 个分离物的 8 个基因座核苷酸序列进行分类，确定了 3 个系统发育物种：PS2（巴西和委内瑞拉）、PS3（哥伦比亚）和 S1（巴西、阿根廷、秘鲁、巴拉圭和委内瑞拉）[11]。对 21 个巴西副球芽孢菌分离株（其中 14 个先前被鉴定为 S1 或 PS3）中的 21 个多态位点进行系统发育分析，发现一个非典型分离株与所有其他分离株高度不同，明显远离所描述的 3 个系统发育物种[12]。Teixeira 等[13]表明，这种被称为 Pb01 的分离物以及其他被称为“Pb01 样”的分离物可能属于一种新系统发育物种，命名为“卢氏副球芽孢菌属”。

14.3 传播

副球芽孢菌病是通过吸入可能存在于流行地区土壤中的巴西副球芽孢菌菌丝体阶段释放的芽孢而感染的。然而从这些地区成功地分离出真菌是一件罕见的事情，而且很难确定精确的真菌生态位（Barrozo 等对此进行了评论[14]）。

虽然很少从腐生生物中分离出巴西副球芽孢菌，但经常从人体临床样本和犰狳、狗、双趾树懒和其他野生动物的内部组织中获得。最近 Arantes 等[15]利用巢式 PCR 方法扩增 rRNA 编码的相关序列，包括不同培养技术的 ITS1-5.8S-ITS2 区。虽然传统培养方法不能分离出病原菌，但

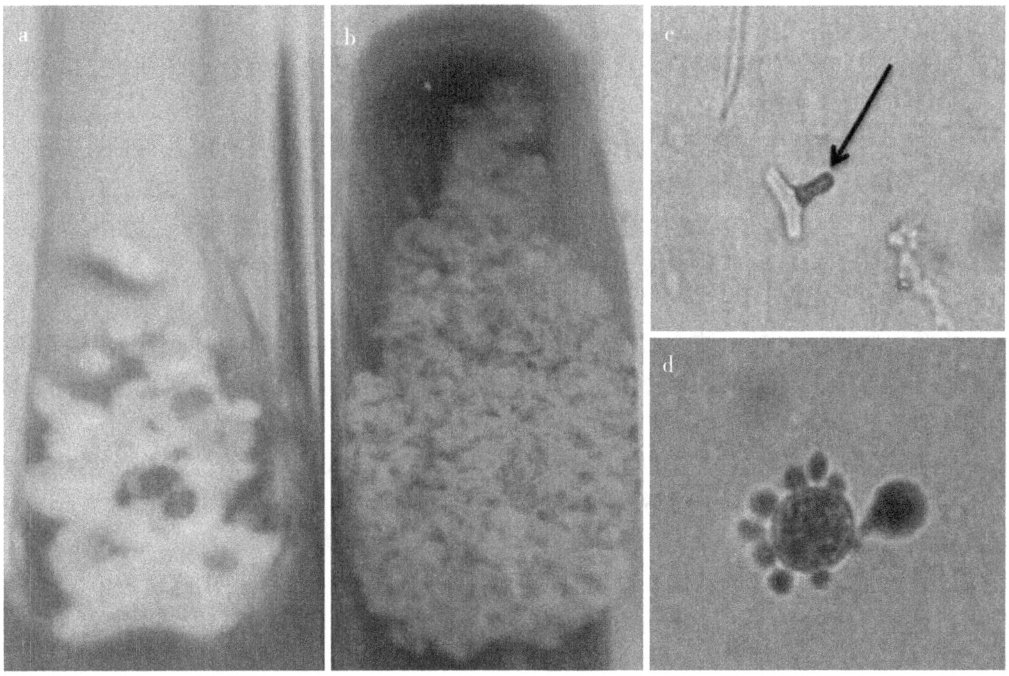

图14.1 巴西副球芽孢菌在室温下的培养显示菌丝形式（a）；在37℃下培养，显示酵母形式（b）；有新分生芽孢的菌丝（黑色箭头）（c）；典型的多芽酵母形式。乳酚棉蓝染色（d）

从犰狳洞穴中获得的气溶胶及其序列扩增结果显示其具有很高的同源性，提示有副球芽孢菌复合物。

目前尚无与环境条件有关的PCM暴发的报告，但1985年在巴西圣保罗州发现了一组急性/亚急性PCM病例[14]，暂时与气候有关。该组包括10例病例，尽管在特定时期该区域预期有2.19例（$P< 0.005$）。作者观察到1982/1983年由于该地区强烈的厄尔尼诺效应，土壤蓄水量异常高，1984年，绝对空气湿度远高于正常值，这些条件可能分别有利于土壤中真菌生长和分生芽孢释放，这可能是患者接触真菌的年份。

14.4 诊断和常规治疗

明确诊断PCM可以通过直接检查痰液、活检标本或化脓淋巴结的结痂/脓来进行，化脓淋巴结含有典型巴西副球芽孢菌酵母菌形式并结合培养的任何临床标本真菌。巴西副球芽孢菌寄生形式的真菌成分形态特征显示，大的球形细胞带有窄颈多芽酵母或只有两个芽的母细胞（Travassos等综述）[16]。湿制剂常规方法使用KOH（氢氧化钾）处理或荧光增白剂染色。组织病理学制剂通常用Grocott-Gomori染色或PAS染色，以更好地鉴定真菌成分。

在含有氯霉素和环己酰亚胺的Sabouraud琼脂或酵母提取物琼脂上进行临床样品中的巴西副球芽孢菌培养。病原体回收率可能受到潜在污染的有机液体（例如，痰和皮肤损伤[16]）上细菌过度生长的限制。

血清学试验对PCM诊断很重要，包括双免疫扩散试验、免疫酶试验和反免疫电泳。在临床实践中，特异性抗体检测被用于筛查疑似感染巴西副球芽孢菌的患者以及监测对治疗的临床反应[16]。然而感染了“Pb01样”分离物（*P. lutzii*）的患者可能会出现假阴性血清学结果，因此可能需要大量抗原来诊断[17]。

一些研究人员试图在PCM患者临床标本中检测特异性巴西假单胞菌抗原。使用免疫酶法检测巴西副球芽孢菌 gp43 和 gp70 特异性抗原。监测巴西副球芽孢菌抗原在 PCM 诊断中有潜在的应用价值，并且可以作为评估特定抗真菌治疗临床反应的工具[16]。

巴西副球芽孢菌粗抗原制剂和纯化抗原（gp43）皮肤测试用于流行病学研究，但没有诊断价值。

DNA 扩增 PCR 方法也被用于鉴别 PCM 患者。不同作者描述了具有潜在诊断用途的巴西副球芽孢菌 DNA 序列，包括 5.8S rRNA 基因及其 ITS 区域[18]和 gp43 基因[19]。San-Blas 等[20]报道了在巴西副球芽孢菌 0.72kb DNA 片段上设计的特异性引物，该引物可用于鉴别 PCM 患者痰和脑脊液中的病原体。由于成本高，常规实验室不使用分子方法诊断 PCM。

评估患者还需要其他实验室检查和图像检测，如 X 射线、腹部超声、血细胞计数、红细胞沉降率、生化肝脏测试（转氨酶、碱性磷酸酶）、蛋白质电泳和肾脏代谢（肌酐、钠和钾）。那些涉及中枢神经系统、胃肠系统、肾上腺功能障碍、呼吸衰竭以及骨骼或肌肉损伤的患者需要更复杂的检查[4]。

PCM 治疗需要抗真菌化疗，即使在治疗后，也不能保证真菌完全清除。初始治疗持续 2~6 个月，包括磺胺类、两性霉素 B 或唑类药物。延长治疗时间通常是必要的，长达 2 年或更长时间，复发频率很高[4,21]。

目前口服伊曲康唑是首选药物。对于重度患者，静脉注射两性霉素 B 或磺胺甲恶唑/甲氧苄啶联合用药是更合适的选择。在疾病最初治疗后，继续维持用药（包括磺胺二甲氧嘧啶或磺胺多辛）2 年[4,21,22]。

14.5 疫苗

14.5.1 疫苗的合理性

尽管 PCM 通常由化疗控制，但治疗效果受到宿主免疫反应状态的限制。免疫功能低下的个体可能发展为致命病例的传播性疾病。因此，与其他系统性真菌疾病一样，真菌化疗需要一个成功的结果并辅以有效免疫应答以实现对感染的长期控制。长期治疗、复发、纤维化后遗症和疾病的严重程度也反映了患者的免疫防御机制和炎症反应。

疫苗可以刺激免疫系统而不会产生不良过度炎症反应，这种疫苗可以有效地补充化疗，缩短治疗周期，减少复发。在严重 PCM 病例和化疗反应不佳的病例中，可能是一种强有力的药物。在 PCM 发病率高的流行地区，可以考虑预防性使用这种疫苗。

14.5.2 抗原特征

1986 年发现，糖蛋白 gp43[23]是巴西副球芽孢菌的主要诊断抗原。PCM 患者具有针对 gp43 的高频率抗体，抗 gp43 IgG、IgA 和 IgM 的滴度降低与临床改善有关[24]。pIs 在 5.8 到 8.5 范围内识别出 gp43 不同亚型，但是它们对患者血清反应不一样[25]。引起高滴度抗体的 B 细胞表位本质上是肽性的[26]，针对 gp43 的单克隆抗体（mAb 3E）与肽 NHVRIPIGYWAV 反应，该序列与烟曲霉菌和米曲霉 A 的 β-1，3-葡聚糖酶内肽共享[27]。事实上，gp43 与来自几种不同真菌物种的外-β-1，3-D-葡聚糖酶，具有 54%~60%的同源性和 50%的同一性，但其本身并不表达任何酶活性。gp43 存在于细胞内的液泡中，但随后在细胞壁中积累并分泌到上清液中，在上清液中 gp43 代表了近 80%的外源抗原蛋白。然后可以使用单克隆抗体 mAb 17c 通过亲和层析法分离。

gp43 基因已经克隆、测序并在大肠杆菌中表达[28]；重组蛋白含有 416 个氨基酸并带有一个 35 个残基前导序列。成熟糖蛋白具有单一 N-糖基化位点，具有高甘露糖 $Hex_{13}GlcNAc_2$ 寡糖[29]。介导迟发型超敏反应和致敏 T $CD4^+$淋巴细胞的 T 细胞表位已被定位到 P10（$gp43^{181-195}$），这是一种序列为 QTLIAIHTLAIRYAN 的 15 聚肽[30]，邻近 N-糖基化位点。用巴西副球芽孢菌

（Pb18）毒株酵母形式在气管内感染 Balb/c 小鼠，显示纯化的 gp43 和具有酰胺化 C 末端的合成 P10 肽在 Balb/c 小鼠中均显示出保护作用（图 14.2）。尽管通过检查来自许多分离株的前体基因显示出 gp43 多态性，但编码 P10 序列的核苷酸没有突变[31]。

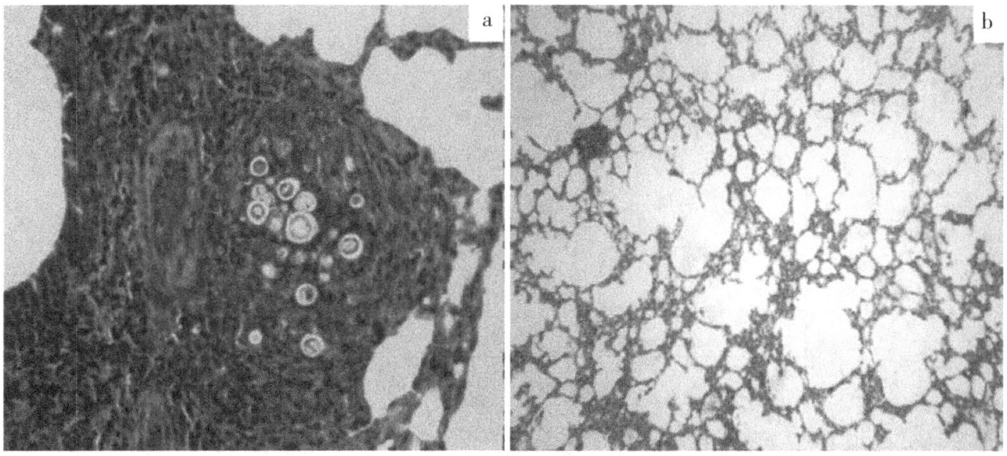

图 14.2　感染 30 天后，接种 p10 的 Balb/c 小鼠肺组织切片病理学观察

（a）感染的非免疫对照小鼠，400 倍；（b）感染的免疫小鼠，100 倍。马森三色染色。蓝色染色显示 I 型胶原纤维。

gp43 可以被树突细胞或 B 细胞加工用于肽递呈，从而引发 Th-1 或 Th-2 免疫应答和相应的细胞因子[32,33]。获得了更有利的抗真菌应答，主要的 Th-1 反应富含 IFN-γ 和 IL-12。人们还普遍认识到，抑制 Th-2 免疫和刺激 Th-1 细胞因子提高了真菌化学疗法疗效[34]。因此，我们将精力集中在肽 P10 上，该肽引发 Th-1 免疫应答，并且没有诱导抗 gp43 抗体。在最激进的 PCM 形式中，患者表现出高抗体滴度和严重的细胞免疫抑制。尽管 gp43 的一些单克隆抗体是保护性的，但有些则不是，占优势的 Th-2 免疫显然不能抵御感染。

14.5.3　验证肽疫苗

如上所述，P10 序列在巴西福球芽孢菌大多数分离株中是保守的。此外，来自 3 种小鼠单倍型的 MHC 类分子呈现合成肽，这是从其保护作用推断的[30]。然后对 HLA-DR 分子进行了直接研究。P10 和肽 $gp43^{180-194}$（无 C-末端天冬酰胺而有 N-末端赖氨酸）与 9 种流行 HLA-DR 分子结合[35]。使用覆盖 25 种高加索 HLD-DR 类型的 TEPITOPE 算法，预测 P10 和 gp43 中相邻肽会结合 90%这类分子。因此，由人类 MHC Ⅱ类分子递呈的 P10 混杂特性是该肽作为疫苗的一个基本特征。

P10 治疗性疫苗应与化疗同时使用；因此抗真菌药物和 P10 免疫接种组合在不同的方案中进行了测试。Balb/c 小鼠气管内感染 3×10^5酵母形式的巴西副球芽孢菌 Pb18，每 24 小时腹腔内注射伊曲康唑、氟康唑、酮康唑或甲氧苄啶-磺胺甲恶唑或每 48 小时注射两性霉素 B。药物治疗在感染后 48 小时或 30 天开始，并持续 30 天。一组动物每周用 P10（20 μg）免疫一次，持续 4 周，第一组用完全弗氏佐剂乳化，其余 3 组用不完全弗氏佐剂乳化。P10 皮下注射和腹腔注射（最后两次）接种。尽管在药物治疗组或 P10 免疫组中通过集落形成单位（CFU）从肺匀浆中测量到了显著的保护作用，但是化疗和 P10 免疫的结合显示出累加效应[36]。使用第二种方案在感染 60～120 天后，联合治疗的肺 CFU 比未治疗小鼠的少 60%～80%，肺细胞因子显示出主要的 Th-1 免疫应答（图 14.3）。

P10 的免疫保护作用依赖于 IFN-γ 产生，P10 含有 T $CD4^+$表位，由 MHC Ⅱ类抗原杂乱地递呈，这一点通过使用纯合子小鼠进行基因无效突变而清楚地显示出来，这些基因编码 IFN-γ、

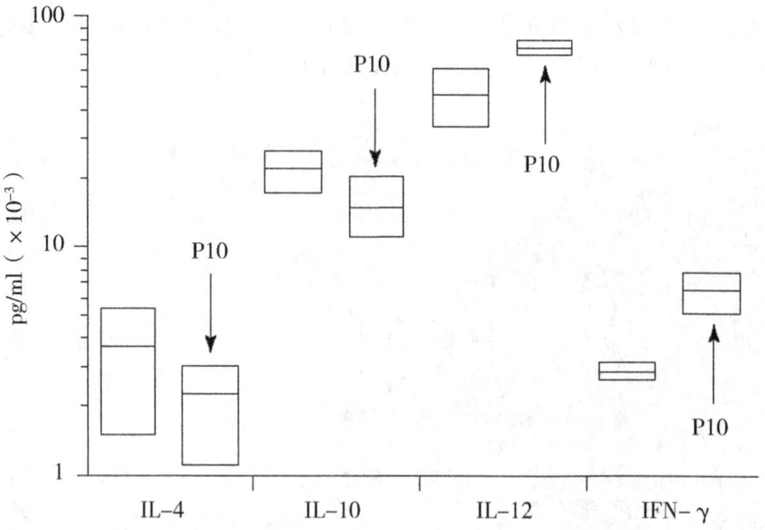

图 14.3 经气管内感染的小鼠接受化疗并使用相同药物和 P10 疫苗后细胞因子产生范围。内部线是考虑到所有测试药物的平均细胞因子浓度

IFN-γ-R 或 IRF-1 转录因子，但不编码 IFN-α-R 或 IFN-β-R。所有消除 IFN-γ 存在或功能的基因敲除对巴西副球芽孢菌气管内感染非常敏感，并伴有早期死亡。用 P10 疫苗接种这些动物未能诱导免疫保护。缺乏 IFN-γ 的受感染小鼠不形成肉芽肿，中性粒细胞大量浸润，同时伴有巴西副球芽孢菌酵母形式[37]。其他小组也进行了 IFN-γ 缺陷小鼠对这种真菌敏感性增加的研究[38,39]。

面对 p10 疫苗有效性的免疫学要求，这种方法对缺乏免疫力的动物有效吗？这一问题是通过在饮用水中使用经过地塞米松 21-磷酸盐预处理的 BALB/C 小鼠来解决的。在 30 天内这些动物对迟发型超敏反应试验呈阴性，如之前在感染动物中测定的，用 0.15 mg/kg 地塞米松治疗，并用巴西副球芽孢菌粗抗原致敏。感染 70 天后，所有免疫无能性感染小鼠死亡。DTH 阴性小鼠与有严重疾病和预后不良的 PCM 患者相似，在气管内感染了一株毒力较强的巴西副球芽孢菌分离株。

感染 15 天后，小鼠给予甲氧苄啶磺胺甲恶唑或伊曲康唑，同时或不同时进行 p10 免疫。在这种情况下，p10 疫苗与化疗联合使用可显著降低肺部 CFU，大体上保留肺泡结构，防止真菌传播到肝脏和脾脏。p10 疫苗成功诱导 Th-1 应答，IL-2、IL-12 和 IFN-γ 水平较高，IL-4 和 IL-10 产量比贫血、感染但未经治疗的动物更低[40]。尽管效果显著，但由于实验设计时间短，部分真菌消除率并不高[41]。

p10 疫苗接种必须在不使用 CFA 作为佐剂的情况下进行测试，这是早期实验中的常规程序。必须测试允许人类免疫的不同配方。除了下文讨论的其他佐剂外，还研究了具有支链赖氨酸核心的多抗原肽（MAP）结构（图 14.4）。由于合成限制，复合链中只允许有 13 个 p10 氨基酸。这被命名为 M10，以与 P10 进行比较。

对照 MAPs 包含 4 条 5-10aa 的 P10 衍生肽链。只需要 1 μg M10 就能使小鼠对 P10 刺激的淋巴结细胞增殖敏感。无佐剂 M10 疫苗对经气管内感染的小鼠提供了显著保护，肺部、脾脏和肝脏 CFU 较少，肺部组织病理学切片中酵母也少得多或没有酵母[42]。

14.5.4 动物模型

大多数用 gp43 或 P10 肽进行的免疫保护实验都是用对巴西副球芽孢菌感染敏感的-BALB/C 小鼠。虽然不同的实验室采用腹腔内注射和静脉内注射感染途径，但我们选择气管内感染途径

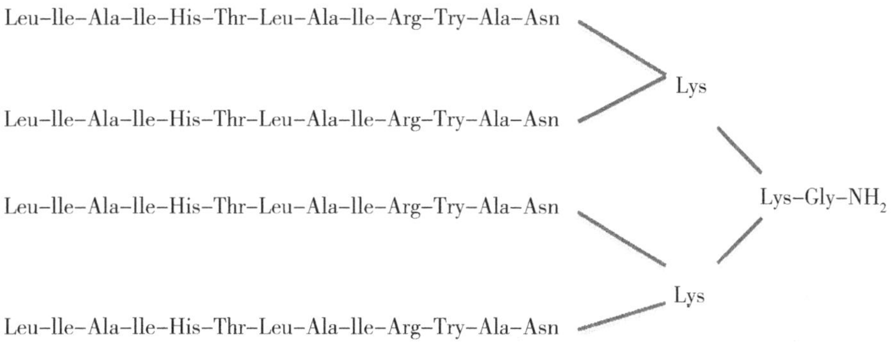

图 14.4　缩短的 P10 四聚体连接到支链赖氨酸核上的多抗原肽 MAP（M10）

作为标准感染方法，并对动物作了适当地麻醉。P10 疫苗经皮下接种。在少数情况下，当鞭毛蛋白作为佐剂时，则采用鼻内途径接种。如在文献[30]中所示，其他小鼠单倍体也受到保护，但 BALB/C 小鼠的重复性最高，Th-1 介导的免疫应答一致。在基因治疗的情况下（见下文），一个更严格的模型被用于高度易受感染的 B10. A 小鼠。与抗 A/Sn 的动物相比，B10. A 小鼠的巨噬细胞产生高浓度和持续的 NO 水平[43]。这些动物也合成了高水平 IL-10，主要通过 T-CD8$^+$细胞控制真菌传播[44]。然而 B10. A 小鼠对用编码 P10 和 IL-12 的质粒进行的长期疫苗接种实验反应良好[45]。其他基因治疗实验在 BALB/C 小鼠中进行，肌内或皮内接种[45,46]。

利用编码 P10 的治疗性 DNA 作为一种有希望的候选疫苗，采用 3 种不同的方案进行免疫预防治疗[45]：（a）对于免疫预防治疗，BALB/C 小鼠每周注射 4 次疫苗，在第 7 周感染，不同组在 11 周或 15 周（感染后 30 或 60 天）处死；（b）对于短期治疗，BALB/C 和 B10. A 小鼠在第 4 周、5 周、6 周和第 7 周经气管内感染并接种疫苗，在第 8 周处死；（c）对于长期治疗，B10. A 小鼠在第 4 周、5 周、6 周和第 7 周经气管内感染并接种疫苗，然后在第 11 周、15 周和第 19 周再次接种疫苗，并在感染后的第 23 周处死。在最后一个长期治疗方案中真菌几乎被消灭了[45]。

14.5.5　基因疗法

在 CMV 启动子控制下，用携带 gp43 基因的哺乳动物表达载体（VR-gp43）对 BALB/C 小鼠进行免疫，诱导 T 细胞介导的免疫应答和抗体[46]。质粒 DNA 能够在 IFN-γ 介导下产生有效、持久的细胞免疫应答。最后一次 DNA 接种 6 个月后体液和细胞特异性反应有效。气管内免疫接种可防止感染毒性巴西副球芽孢菌，肺 CFU 显著降低，免疫小鼠脾脏和肝脏减少或不播散巴西副球芽孢菌[46]。

强毒酵母攻击之前或之后，使用 P10 和 IL-12 插入片段进行质粒接种，测试了质粒在小鼠肺 PCM 病模型中的有效性[45]。P10 核苷酸序列从源自 gp43 的正链 PCR 引物扩增获得[28]。在正链和反链引物中分别添加 HINDIII 位点和 EcoRI 位点，然后克隆到质粒 pcDNA3。用 cDNA 和等量的每种引物进行 PCR 扩增。PCR 产物纯化、用合适的限制性酶消化并克隆到 pcDNA3 中，得到的质粒被称为 pP10。质粒 pORF-mIL-12 是商业获得的，在确认插入片段后使用[45]。在转化的大肠杆菌 XL1 蓝和 DH5α 细胞中制备用于免疫的质粒 DNA 并纯化，通过肌内途径接种 100 μL 质粒用于免疫（含有 100 μg 质粒，即 50 μg pP10+50 μg 空质粒或 50 μg pP10+pIL-12）。

在免疫预防治疗中，用组织 CFUs 测量，pP10 和 pIL-12 的组合完全消除了肺中的真菌[45]。仅用质粒在没有插入物的情况下观察到 CFUs 显著减少，这表明非甲基化的 CpG 基序可能激活树突状细胞的 Toll 样受体 9。保护作用具有协同作用，但明显涉及 pP10 的主要活性。在 Balb/c 和 B10. A 小鼠均被感染并在 1 个月后接种的治疗方案中，空质粒无保护作用，pP10 和 pIL-12 均具有活性，两种质粒结合可达到最大保护作用（表 14.1）。

表 14.1 B10.A 易感小鼠感染巴西副球芽孢菌并使用质粒-P10 治疗疫苗治疗的肺 CFUs

基因疗法	短期治疗方案 （30 天）	长期治疗方案 （6 个月）
未处理的对照	14 630±2 840	22 830±2 232
pCDNA	315 867±204	18 214±2 801
pCDNA3+pP	102 058±381 **	216±25 **
pCDNA3+pIL-12	3 333±111 **	1 261±71 **
pP10+pIL-1	2 372±122 **	18±17 **

治疗方案如前所述[45]。pCDNA3，空载体；pP10 或 pIL-12，插入 P10 或 IL-12 的载体。** $P< 0.001$，与空载体相比有显著差异。

用含有或不含 pIL-12 的 pP10 接种小鼠，肺部显示炎症减少，酵母细胞减少或检测不到。最后在长期治疗方案中，基因免疫在感染后 30 天开始，感染后 6 个月处死小鼠。疫苗接种时间表见本节 14.5.4。尽管空质粒在这些条件下没有明显的保护作用，但是单独用 pP10 接种会使肺部 CFUs 减少 100 倍以上，pP10 和 pIL-12 结合实际上消除了所有真菌成分。与用空质粒接种的小鼠肺中观察到含有巨细胞和大量酵母细胞上皮样肉芽肿相比，pP10 疫苗接种导致较少的肉芽肿和具有大面积正常肺结构的酵母细胞。pP10+pIL-12 组合几乎恢复了肺实质，几乎检测不到酵母形式[45]。

在研制 PCM 疫苗方面，这些最令人鼓舞的结果并不是这种真菌疾病基因治疗的唯一尝试。在受感染 Balb/c 小鼠中还研究了编码 rB27[47] 和麻风分枝杆菌 DNAhsp65 质粒[48] 的 cDNA。

14.5.6 可行性和临床前开发

P10 肽和编码 P10 或 IL-12 的质粒都是很容易获得的疫苗，可以制备高纯度和高重现性的疫苗。人类使用 GMP 扩大生产和提供的成本通常很高，并取决于直接投资。考虑到患者数量和分布以及流行地区位置，这在经济上可能不具吸引力。然而成功的抗真菌肽疫苗或等效基因疗法试点生产可能引起流行地区政府的兴趣，并刺激其他更普遍分布的系统性真菌病的类似程序。

到目前为止，包括灵长类动物毒性在内，已注册的肽疫苗临床前研究尚未开展。多肽疫苗有效性已多次在小鼠气管内感染模型中得到验证，主要集中在多肽佐剂和传递问题上。为替代用于 P10 免疫的完全和不完全弗氏佐剂，对不同佐剂进行了人体免疫试验。在以前感染巴西副球芽孢菌 Pb18 菌株的 Balb/c 模型中，感染 52 天后，P10 与不同佐剂联合使用进行免疫[49]。

P10 肽和阳离子脂质二十八烷基二甲基溴化铵（DODAB）结合，可以使肺部活酵母细胞数量非常小以及肉芽肿被清除。IFN-γ 和 TNF-α 产量高，但 IL-4 和 IL-10 产量低。鞭毛蛋白（来自肠道沙门氏菌的 FliC[50]）也有效，但效果低于 DODAB。氢氧化铝效果最差，因为与 P10 联合没有清除含有大量酵母细胞和胶原纤维的大肉芽肿，类似于未处理的肺，并且 IL-4 产量高。

由于持续的抗原刺激和积极的免疫反应，巴西副球芽孢菌感染以肉芽肿性炎症为特征，肉芽肿性炎症增加了富含 Ⅰ 型和 Ⅲ 型胶原肺结缔组织导致功能改变和纤维化，这是众所周知的肺纤维化后遗症。值得注意的是，DODAB 和 P10 联合使用以及用 pP10 进行基因治疗导致感染的小鼠肺纤维化显著降低。

14.6 优点和缺点

由短尺寸天然酰胺化肽组成的制剂主要问题是药代动力学较差，主要是由于血浆蛋白水解和肾滤过。尽管如此，我们已经证明在皮下或腹腔注射的小鼠体内，除了 C 端酰胺化外，未衍生化的肽具有显著的抗真菌和抗肿瘤作用[51]。据推测这些生物活性肽在进入血液循环之前会被树突状细胞迅速吸收，从而刺激一系列导致免疫保护的事件。佐剂使用肯定会通过树突状细胞活化增强肽活性。已研究了致敏树突状细胞在保护 PCM 方面的效率[52]。气管内感染小鼠在感染后第 31 天和第 38 天接受静脉或皮下注射 P10 致敏树突状细胞治疗。第二次注射树突状细胞后 7 天处死小鼠，真菌清除效果显著。致敏树突状细胞在预防方案中也非常有效。

此外从 P10 药代动力学角度来看，另一种方法可能是将肽包装在纳米颗粒中。这实际上是使用聚乳酸-乙醇酸（PLGA）纳米颗粒进行[53]。在 PLGA 中包埋的 P10 比在弗氏佐剂中乳化的游离 P10 更有效，这是结合化疗进行试验的结果。将 P10 加入 PLGA 可减少所需肽量(1 μg/50 μL)，从而显著降低受感染动物的真菌负荷，并避免疾病复发。聚乳酸-乙醇酸与树突状细胞相互作用可能改善了 P10 的免疫调节作用。

以表达质粒形式的基因治疗是将肽传递到适当抗原呈递细胞的另一种方式，同时也在不使用另一种佐剂情况下为 DC 活化提供 CpG 基序。表达主要 P10 抗原和 IL-12 的质粒组合，获得了保护性 Th-1 免疫应答的正确刺激，而不会引起作为治疗后遗症的高度炎症和纤维化。虽然目前实验处于临床前阶段，但由于达到完全保护，没有同时化疗，也没有明显毒性，因此是有希望的。

致谢

一系列巴西副球芽孢菌肽疫苗和基因治疗研究得到了巴西 the Fundação de Amparo à Pesquisa do Estado de São Paulo 的赞助。

参考文献

[1] Benard, G., Franco, M.: Paracoccidioidomycosis. In: Merz, W. G., Hay, R. J. (eds.) Medical Mycology-Topley & Wilson's Microbiology and Microbial Infections, vol. 28, 10th edn, pp. 541-559. ASMPress, Washington, DC (2005).

[2] Wanke, B., Londero, A. T.: Epidemiology and paracoccidioidomycosis infection. In: Franco, M., Lacaz, C. S., Restrepo-Moreno, A., Del Negro, G. (eds.) Paracoccidioidomycosis, pp. 109-120. CRC Press, Boca Raton (1994).

[3] Prado, M., Silva, M. B., Laurenti, R., Travassos, L. R., Taborda, C. P.: Mortality due to systemic mycoses as a primary cause of death or in association with AIDS in Brazil: a review from 1996 to 2006. Mem. Inst. Oswaldo Cruz104, 513-521 (2009).

[4] Shikanai-Yasuda, M. A., Telles-Filhos, F. Q., Mendes, R. P., Colombo, A. L., Moretti, M. L., et al.: Guidelines in paracoccidioidomycosis. Rev. Soc. Bras. Med. Trop. 39, 297-310 (2006).

[5] Benard, G., Kavakama, J., Mendes-Giannini, M. J. S., Kono, A., Duarte, A. J., et al.: Contribution to the natural history of paracoccidiodomycosis: identification of primary pulmonary infection in the severe acute form of the disease-a case report. Clin. Infect. Dis. 40, 1-4 (2005).

[6] Brummer, L., Castaneda, E., Restrepo, A.: Paracoccidiodomycosis: an update. Clin. Microbiol. Rev. 6, 89-117 (1993).

[7] Goldani, L. Z., Sugar, A. M.: Paracoccidiodomycosis and AIDS: an overview. Clin Infect. Dis. 21, 1275-1281 (1995).

[8] Sarti, E. C., Oliveira, S. M., Santos, L. F., Camargo, Z. P., Paniago, A. M.: Paracoccidioidal infection in HIV patients at an endemic area of paracoccidioidomycosis in brazil. Mycopathologia173, 145-149 (2012).

[9] Lacaz, C. S.: Paracoccidioides brasiliensis: morphology; evolutionary cycle; maintenance during saprophytic life; biology, virulence, taxonomy. In: Franco, M., Lacaz, C. S., Restrepo-Moreno, A., Del Negro, G. (eds.) Paracoccidioidomycosis, pp. 13-25. CRC Press, Boca Raton (1994).

[10] Queiroz-Telles, F.: Paracoccidioidesbrasiliensis ultrastructural findings. In: Franco, M., Lacaz, C. S., RestrepoMoreno, A., Del Negro, G. (eds.) Paracoccidioidomycosis, pp. 27-47. CRC Press, Boca Raton (1994).

[11] Matute, D. R., McEwen, J. G., Puccia, R., Montes, B. A., San-Blas, G., et al.: Cryptic speciation and recombination in the fungus *Paracoccidioides brasiliensis* as revealed by gene genealogies. Mol. Biol. Evol. 23, 65-73 (2006).

[12] Carrero, L. L., Niño-Veja, G., Teixeira, M. M., Carvalho, M. J., Soares, C. M., et al.: New *Paracoccidioides brasiliensis* isolate reveals unexpected genomic variability in this human pathogen. Fungal Genet. Biol. 45, 605-612 (2008).

[13] Teixeira, M. M., Theodoro, R. C., de Carvalho, M. J., Fernandes, L., Paes, H. C., et al.: Phylogenetic analysis reveals a high level of speciation in the *Paracoccidioides* genus. Mol. Phylogenet. Evol. 52, 273-283 (2009).

[14] Barrozo, L. V., Benard, G., Silva, M. E., Bagagli, E., Marques, S. A., et al.: First description of a cluster of acute/subacute paracoccidioidomycosis cases and its association with a climatic anomaly. PLoS Negl. Trop. Dis. 4, e643 (2010).

[15] Arantes, T. D., Theodoro, R. C., Macoris, S. A. G., Bagagli, E.: Detection of *Paracoccidioides* spp. in environmental aerosol samples. Med. Mycol. 51 (1), 83-92 (2013).

[16] Travassos, L. R., Taborda, C. P., Colombo, A. L.: Treatment options for paracoccidioidomycosis and new strategies investigated. Expert Rev. Anti Infect. Ther. 6, 251-262 (2008).

[17] Batista Jr., J., Camargo, Z. P., Fernandes, G. F., Vicentini, A. P., Fontes, C. J., et al.: Is the geographical origin of a *Paracoccidioides brasiliensi* s isolate important for antigen production for regional diagnosis of paracoccidioidomycosis? Mycoses 53, 176-180 (2009).

[18] Montoyama, A. B., Venancio, E. J., Brandao, G. O., Petrofeza-Silva, S., Pereira, I. S., et al.: Molecular identification of *Paracoccidioides brasiliensis* by PCR amplification of ribosomal DNA. J. Clin. Microbiol. 38, 3106-3109 (2000).

[19] Gomes, G. M., Cisalpino, P. S., Taborda, C. P., Camargo, Z. P.: PCR for diagnosis of paracoccidioidomycosis. J. Clin. Microbiol. 33, 1652-1654 (2000).

[20] San Blas, G., Nino-Veja, G., Barreto, L., Hebeler-Barbosa, F., Bagagli, E., et al.: Primers for clinical detection of *Paracoccidioides brasiliensis*. J. Clin. Microbiol. 43, 4255-4257 (2005).

[21] Shikanai-Yasuda, M. A.: Pharmacological management of paracoccidioidomycosis. Expert Opin. Pharmacother. 6, 385-397 (2005).

[22] Travassos, L. R., Taborda, C. P.: Paracoccidioidomycosis: advances in treatment incorporating modulators of the immune response. J. Invasive. Fungal. Infect. 5, 1-6 (2011).

[23] Puccia, R., Schenkman, S., Gorin, P. A., Travassos, L. R.: Exocellular components of *Paracoccidioides brasiliensis*: identification of a specific antigen. Infect. Immun. 53, 199-206 (1986).

[24] Giannini, M. J., Bueno, J. P., Shikanai-Yasuda, M. A., Stolf, A. M., Masuda, A., et al.: Antibody response to the 43 kDa glycoprotein of *Paracoccidioides brasiliensis* as a marker for the evaluation of patients under treatment. Am. J. Trop. Med. Hyg. 43, 200-206 (1990).

[25] Souza, M. C., Gesztesi, J. L., Souza, A. R., Moraes, J. Z., Lopes, J. D., Camargo, Z. P.: Differences in reactivity of paracoccidioidomycosis sera with gp43 isoforms. J. Med. Vet. Mycol. 35, 13-18 (1997).

[26] Puccia, R., Travassos, L. R.: The 43-kDa glycoprotein from the human pathogen *Paracoccidioides brasiliensis* and its deglycosylated form: excretion and susceptibility to proteolysis. Arch. Biochem. Biophys. 28, 298-302 (1991).

[27] Travassos, L. R., Goldman, G., Taborda, C. P., Puccia, R.: Insights in *Paracoccidioides brasiliensis* pathogenicity. In: Kavanagh, K. (ed.) New Insights in Medical Mycology, pp. 241-265. Springer, Dordrecht (2007).

[28] Cisalpino, P. S., Puccia, R., Yamauchi, L. M., Cano, M. I., Silveira, J. F., et al.: Cloning, characterization, and epitope expression of the major diagnostic antigen of *Paracoccidioides brasiliensis*. J. Biol.

Chem. 271, 4553-4560 (1996).

[29] Almeida, I. C., Neville, D. C., Mehlert, A., Treumann, A., Ferguson, M. A., et al.: Structure of the N-linked oligosaccharide of the main diagnostic antigen of the pathogenic fungus *Paracoccidioides brasiliensis*. Glycobiology6, 507-515 (1996).

[30] Taborda, C. P., Juliano, M. A., Puccia, R., Franco, M., Travassos, L. R.: Mapping of the T-cell epitope in the major 43-kilodalton glycoprotein of *Paracoccidioides brasiliensis* which induces a Th-1 response protective against fungal infection in BALB/c mice. Infect. Immun. 66, 786-793 (1998).

[31] Morais, F. V., Barros, T. F., Fukada, M. K., Cisalpino, P. S., Puccia, R.: Polymorphism in the gene coding for the immunodominant antigen gp43 from the pathogenic fungus *Paracoccidioides brasiliensis*. J. Clin. Microbiol. 38, 3960-3966 (2000).

[32] Almeida, S. R., Moraes, J. Z., Camargo, Z. P., Gesztesi, J. L., Mariano, M., Lopes, J. D.: Pattern of immune response to go43 from *Paracoccidioides brasiliensis* in susceptible and resistant mice is influenced by antigen-presenting cells. Cell. Immunol. 190, 68-76 (1998).

[33] Ferreira, K. S., Lopes, J. D., Almeida, S. R.: Regulation of T helper cell differentiation in vivo by GP43 from *Paracoccidioides brasiliensis* provided by different antigen-presenting cells. Scand. J. Immunol. 58, 290-297 (2003).

[34] Romani, L.: Immunity to fungal infections. Nat. Rev. Immunol. 11, 275-288 (2011).

[35] Iwai, L. K., Yoshida, M., Sidney, J., Shikanai-Yasuda, M. A., Goldberg, A. C., et al.: In silico prediction of peptides binding to multiple HLA-DR molecules accurately identifies immunodominant epitopes from gp43 of *Paracoccidioides brasiliensis* frequently recognized in primary peripheral blood mononuclear cell responses from sensitized individuals. Mol. Med. 9, 209-219 (2003).

[36] Marques, A. F., Silva, M. B., Juliano, M. A., Travassos, L. R., Taborda, C. P.: Peptide immunization as an adjuvant to chemotherapy in mice challenged intratracheally with virulent yeast cells of *Paracoccidioides brasiliensis*. Antimicrob. Agents Chemother. 50, 2814-2819 (2006).

[37] Travassos, L. R., Casadevall, A., Taborda, C. P.: Immunomodulation and immunoprotection in fungal infections: humoral and cellular immune responses. In: San - Blas, G., Calderone, R. A. (eds.) Pathogenic fungi. Host interactions and emerging strategies for control, pp. 241 - 283. Caister Academic Press, Wymondham (2004).

[38] Deepe Jr., G. S., Romani, L., Calich, V. L., Huffnagle, G., Arruda, C., et al.: Knockout mice as experimental models of virulence. Med. Mycol. 38 (Suppl 1), 87-98 (2000).

[39] Souto, J. T., Figueiredo, F., Furlanetto, A., Pfeffer, K., Rossi, M. A., Silva, J. S.: Interferon-gamma and tumor necrosis factor-alpha determine resistance to *Paracoccidioides brasiliensis* infection in mice. Am. J. Pathol. 156, 1811-1820 (2000).

[40] Marques, A. F., Silva, M. B., Juliano, M. A., Munhõz, J. E., Travassos, L. R., et al.: Additive effect of P10 immunization and chemotherapy in anergic mice challenged intratracheally with virulent yeasts of *Paracoccidioides brasiliensis*. Microbes Infect. 10, 1251-1258 (2008).

[41] Travassos, L. R., Rodrigues, E. G., Iwai, L. K., Taborda, C. P.: Attempts at a peptide vaccine against paracoccidioidomycosis, adjuvant to chemotherapy. Mycopathologia165, 341-352 (2008).

[42] Taborda, C. P., Nakaie, C. R., Cilli, E. M., Rodrigues, E. G., Silva, L. S., et al.: Synthesis and immunological activity of a branched peptide carrying the T-cell epitope of gp43, the major exocellular antigen of *Paracoccidioides brasiliensis*. Scand. J. Immunol. 59, 58-65 (2004).

[43] Nascimento, F. R., Calich, V. L., Rodríguez, D., Russo, M.: Dual role for nitric oxide in paracoccidioidomycosis: essential for resistance, but overproduction associated with susceptibility. J. Immunol. 168, 4593-4600 (2002).

[44] Cano, L. E., Singer-Vermes, L. M., Costa, T. A., Mengel, J. O., Xidieh, C. F., et al.: Depletion of $CD8^+$T cells in vivo impairs host defense of mice resistant and susceptible to pulmonary paracoccidioidomycosis. Infect. Immun. 68, 352-359 (2000).

[45] Rittner, G. M. G., Munoz, J. E., Marques, A. F., Nosanchuk, J. D., Taborda, C. P., Travassos, L. R.: Therapeutic DNA vaccine encoding peptide P10 against experimental paracoccidioidomycosis. PLoSNegl.

Trop. Dis. 6, e1519 (2012).
[46] Pinto, A. R., Puccia, R., Diniz, S. N., Franco, M. F., Travassos, L. R.: DNA-based vaccination against murine paracoccidioidomycosis using the gp43 gene from *Paracoccidioides brasiliensis*. Vaccine 18, 3050-3058 (2000).
[47] Reis, B. S., Fernandes, V. C., Martins, E. M., Serakides, R., Goes, A. M.: Protective immunity induced by rPb27 of *Paracoccidioides brasiliensis*. Vaccine 26, 5461-5469 (2008).
[48] Ribeiro, A. M., Bocca, A. L., Amaral, A. C., Faccioli, L. H., Galetti, F. C., et al.: DNAhsp65 vaccination induces protection in mice against *Paracoccidioides brasiliensis* infection. Vaccine 27, 606-613 (2009).
[49] Mayorga, O., Muñoz, J. E., Lincopan, N., Teixeira, A. F., Ferreira, L. C. S., et al.: The role of adjuvants in therapeutic protection against paracoccidioidomycosis after immunization with the P10 peptide. Front. Microbiol. 3, 154 (2012).
[50] Braga, C., Rittner, G., Muñoz, J., Teixeira, A., Massis, L., et al.: *Paracoccidioides brasiliensis* vaccine formulations based on the gp43-derived P10 sequence and the *Salmonella enterica* FliC flagellin. Infect. Immun. 77, 1700-1707 (2009).
[51] Arruda, D. C., Santos, L. C., Melo, F. M., Pereira, F. V., Figueiredo, C. R., et al.: β-Actin-binding complementarity-determining region 2 of variable heavy chain from monoclonal antibody C7 induces apoptosis in several human tumor cells and is protective against metastatic melanoma. J. Biol. Chem. 287, 14912-14922 (2012).
[52] Magalhães, A., Ferreira, K. S., Almeida, S. R., Nosanchuk, J. D., Travassos, L. R., Taborda, C. P.: Prophylactic and therapeutic vaccination using dendritic cells primed with peptide 10 derived from the 43-kilodalton glycoprotein of *Paracoccidioides brasiliensis*. Clin. Vaccine Immunol. 19, 23-29 (2012).
[53] Amaral, A. C., Marques, A. F., Munoz, J. E., Bocca, A. L., Simioni, A. R., et al.: Poly (lactic acid-glycolic acid) nanoparticles markedly improve immunological protection provided by peptide P10 against murine paracoccidioidomycosis. Br. J. Pharmacol. 159, 1126-1132 (2010).

（马维民，吕律译）

第 15 章　蜜蜂抗狄斯瓦螨的口服疫苗

Sebastian Giese 和 Matthias Giese[①]

摘要

美国首次报道了一种神秘的蜂群体性瘫痪失调症（CCD），导致那里的蜂群数量减少了50%~90%。作为传粉者蜜蜂的全球的巨大损失，对农业授粉产生了巨大的影响。大约 130 种农作物、坚果、水果、蔬菜都是由意大利蜜蜂授粉的，在美国，总市值超过 150 亿美元，2005 年在欧盟总市值超过 140 亿欧元。在所有蜂群 CCD 病例中，均可检测到吸血瓦螨超载，瓦螨目前被认为是养蜂的主要威胁。蜜蜂具有体液免疫和细胞免疫系统，如多种抗菌肽和免疫反应酶。瓦螨通过大量抑制这些免疫基因来克服蜜蜂的免疫反应。我们开发了一种口服 DNA 疫苗，它能够重建蜜蜂免疫系统的全部功能。这种疫苗绝对是生物安全的，耐受性好，没有副作用，不会对环境造成任何负担。

你读了这篇文章，也许你会认为我们在讲笑话。这些家伙没有更好的事可做吗？那么多人类健康对疫苗研发的严重需求没有得到满足，现在又有了蜜蜂疫苗？针对一种简单昆虫的深入研究和产品开发？如果口服接种成功，这种疫苗将如何发挥作用？顺便问一下，这真的是疫苗吗？我们希望我们让你感到惊讶，就像我们第一次获得辉煌的结果时感到惊讶一样。花点时间看本章，尽情享受吧。

15.1　引言

环境压力源和疾病对欧洲蜜蜂（意大利蜜蜂）的生活有很大的影响和威胁。欧洲蜜蜂在全球范围内进行专业的蜂蜜生产和授粉管理。这种蜜蜂 400 年前随第一批欧洲移民进入美国，印第安人称为“白人的苍蝇”。

最早在美国报道，一种神秘的所谓蜂群体性瘫痪失调症（CCD）摧毁了那里 50%~90%的蜂群，第一次是在 1995—1996 年冬季观察到的，然后是 2000—2001 年，到目前为止还没有中断。欧洲也出现了类似的情况。有 20 000~60 000只蜜蜂生活在一个群体中。

冬季蜜蜂可接受的正常损失为 5%~10%。蜂群体性瘫痪失调症（CCD）的特点是所有成年工蜂突然在蜂巢中消失，而未成熟的蜜蜂则会给下一个群体带来严重的后果。尽管“蜂群体性瘫痪失调症”这一名称是新的，但蜜蜂蜂群大量丧失的现象却由来已久。

第一次描述起源于 1950 年。19 世纪初，英国的蜂群损失被称为“怀特病岛”，美国人在 20 世纪 60 年代称这种现象为“消失的疾病”，而 20 世纪 90 年代后期法国的蜂群损失被称为“神秘的蜜蜂损失”。所有的蜜蜂都去哪儿了？

15.1.1　经济价值

蜜蜂作为传粉者的巨大损失对农业传粉产生了巨大的影响。大约 130 种作物、坚果、水果和

---

① S. Giese，MSc · M. Giese，博士（✉）　（德国海德堡分子疫苗研究所，IMV。E-mail：info@imv-heidelberg.com）

蔬菜由意大利蜜蜂授粉，2005 年美国的总价值超过 150 亿美元，欧盟超过 140 亿欧元。一个蜂群每天生产大约 1 kg 或 2. 205lb 蜂蜜。作为回报，这些蜜蜂必须给 1 000万 ~ 1 500万朵花授粉。

人们应该记住，除了欧洲蜜蜂外，野生昆虫（其中有 30 000种野生蜜蜂）对授粉也有很大的影响，而且作为有管理的蜜蜂，似乎在授粉方面更有效率[1]。工业耕作也威胁着野生昆虫授粉者的自然生态环境。

15. 1. 2 生态价值

2005 年蜜蜂授粉的全球总经济价值超过 1 500亿欧元或 2 020亿美元。联合国粮农组织（FAO）估计，全世界有 6 500万个受管理的蜂群。除了这种专业的农业，蜜蜂对于生物多样性是不可替代的。如表 2. 5 所述，这种生物出现在第一批花卉植物的进化过程中，并且自 1 亿年前就已经存在。继猪和牛之后，蜜蜂在欧洲和北美是第三大农场动物。自 2007 年以来，在瑞士被正式列为农场动物。因此，CCD 不仅是一个经济问题，也是一个未来迫切需要解决的全球性生态问题。“蜜蜂不仅仅是蜂蜜”。

15. 2 是什么导致了 CCD?

过去几年的 CCD 似乎不同于以往的暴发：工蜂消失了，而不是死在原地，留下了蜂王和幼蜂。在剩余蜜蜂的肠道中，细菌、病毒和真菌的含量很高。瘫痪会在两天内发生。

15. 2. 1 一个复杂的问题

关于导致蜂群衰竭失调的原因，人们讨论了不同的理论，比如农药污染，激烈争论的是农药会干扰影响蜜蜂觅食行为的神经系统，导致蜜蜂放弃蜂巢；真菌疾病，如在西班牙以引起蜜蜂大量死亡而闻名的小芽孢虫属；单一栽培或基因改良作物；手机产生的电烟（无线电波）破坏了蜜蜂的指南针；以及在美国用卡车将蜜蜂从一种作物转运到另一种作物，非常艰辛。从 2 月开始，美国专业养蜂人带着他们的蜂群穿越美国，直到 12 月，因此蜜蜂必须迁移多达 15 次。欧洲蜂群在 9 月左右开始冬眠。除了气候变化，温度敏感性对作物授粉也有影响。CCD 可能是多种因素综合作用的结果[2-6]。

15. 2. 2 狄斯瓦螨

但是在所有 CCD 病例中，都能检测到吸血瓦螨数量过多，瓦螨目前被认为是养蜂业的主要威胁。这种感染和疾病被称为瓦螨病（Varoosis）。狄斯瓦螨是一种体外寄生虫，呈红棕色，扁平状，长 1~1. 8 mm，宽 1. 5~2 mm，有八条腿。狄斯瓦螨侵扰工蜂、雄蜂及其幼虫。螨虫在幼蜂细胞内发育。如图 15. 1 所示与蜜蜂的大小相比，瓦螨是一个真正的庞然大物。瓦螨属于蛛形纲，蜱螨亚纲。仅螨类就有 50 000种。一些螨虫更喜欢糖类作为食物如膳食或农作物。房子里的尘螨以脱落的人类皮肤薄片为食。瓦螨喜欢新鲜的“血液”，即蜜蜂的血淋巴，在 2 小时内可以喂养 0. 1 mg 或 0. 0000002205lb。

工蜂驮着瓦螨进入蜂巢。雌螨进入广泛的细胞，优先进入雄蜂细胞。一旦细胞被盖住，瓦螨就会在幼虫身上产卵。从卵到昆虫需要 7 天的时间。蜜蜂幼虫和螨虫大约在同一时间孵化，新生的瓦螨传播给其他蜜蜂[7-9]。夏螨的寿命为 3~6 周，秋螨的寿命可达数月。瓦螨只能在蜜蜂身上繁殖，因此被认为对其他昆虫无害。瓦螨不仅仅是一种疾病。瓦螨病是一种全球性害虫病，对蜜蜂有毁灭性影响。

15. 2. 3 世界性物种

1904 年在东南亚的爪哇岛发现了瓦螨的最初踪迹。瓦螨是亚洲蜜蜂中华蜜蜂的寄生虫，亚

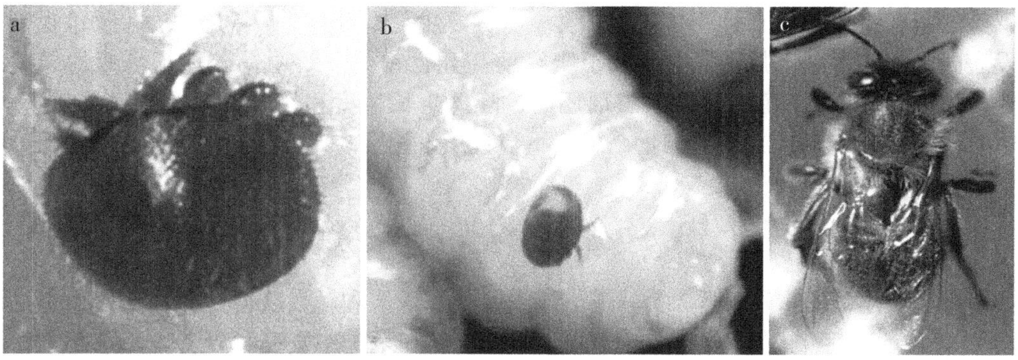

图 15.1 （a）瓦螨为红棕色，扁平状，长 1~1.8 mm，宽 1.5~2 mm，有 8 条腿。（b）蜜蜂幼虫上的狄斯瓦螨。（c）狄斯瓦螨侵害工蜂、雄蜂。瓦螨是病毒的重要载体，在蜂巢中传播这些病毒（图片由德国马格德堡 J. 穆勒博士提供）

洲蜜蜂对瓦螨有部分抵抗力。大约 45 年后，螨虫扩散到苏联东部地区。或许苏军士兵/苏联亚洲地区的业余养蜂人是使瓦螨向全球分布的先驱者。当他们被派往前德意志民主共和国（GDR）或苏联其他西部地区一个新的地方驻军时，他们带着受感染的蜂群重新定居并在西方世界自动分配瓦螨。值得注意的是，瓦螨仅在德意志民主共和国苏军驻扎的地区被发现。1971 年苏联发表了关于瓦螨的第一份报告[10]。几年后越来越多关于瓦螨感染的新文章发表[11-15]。仅在接下来的 20 年里，瓦螨已经蔓延到了除澳大利亚和中部非洲以外的每一个大陆。非洲蜜蜂对瓦螨完全有抵抗力，但由于它们的攻击行为，也称为杀手蜜蜂，因此不适合管理。今天瓦螨是一个令人信服的世界性物种。

15.2.4 瓦螨载体

不能认为瓦螨是该病的一种孤立病原体。在继发性病毒感染情况下，必须考虑成年蜜蜂及其后代的死亡率。至少有 18 种不同的病毒能够感染蜜蜂，主要是 ssRNA 病毒。已知 8 种病毒与瓦螨有关：急性蜜蜂麻痹病毒（ABPV）、黑皇后细胞病毒（BQCV）、慢性蜜蜂麻痹病毒（CBPV）、变形翼病毒（DWV）、克什米尔蜜蜂病毒（KBV）、囊状幼虫蜜蜂病毒（SBV）、云翼病毒（CWV）和慢蜜蜂麻痹病毒（SPV）[16-19]。瓦螨携带的病毒主动传播是水平传播路线的一部分，但请注意女王也可能向她的后代进行垂直传播。女王体内发现的病毒也存在于它们的卵中。但是这种垂直传播可能是雄性瓦螨感染的晚期后果。受感染的雄性精子中存在病毒，在交配期间将病毒传染给雌虫。因此对瓦螨的控制导致了对病毒的间接控制。

几个关键因素可以概括引起 CCD 的蜜蜂的一些重要健康问题：杀虫剂的长期污染，特别是滥用新烟碱类药物[20,21]；寄生虫和真菌（小芽孢虫属）太多；以及大量各种病毒的载体传播——对蜜蜂来说太多了。

15.2.5 瓦螨控制

许多天然和合成化学品在商业上可用于控制瓦螨感染。第一类化合物是溴螨酯、氟胺氰菊酯或其他拟除虫菊酯杀虫剂[15]。简而言之，瓦螨不仅对某一化学类别的一种产品有抵抗力，而且对整个类别的几种相关合成产品也有抵抗力。此外，使用天然产品如甲酸、矿物油或百里酚，只是部分和暂时有效并显示出副作用[22]。没有成功的化学处理，螨会很快对所有化学物质产生抗性。

15.2.6 蜜蜂和农作物杀虫剂

但最大的问题是，人们对杀虫剂和农药对蜂蜜的污染导致人类健康危害[23]和蜜蜂自身的健康问题深为担忧[24-26]。农药污染蜜蜂会导致小芽孢虫属增加，而且经过杀虫剂处理后，蜜蜂对病毒感染的敏感性似乎增加了[27]。新烟碱类全身性杀虫剂是世界上最常用的农作物杀虫剂之一，最近发表的文章认为由于归巢失败导致了蜜蜂的高死亡率[28,29]。中毒破坏了蜜蜂的导航系统。与未经处理的蜂群相比，新烟碱类杀虫剂还通过降低生长速度和显著减少新蜂王的产量来影响大黄蜂[30]。这些对蜜蜂友好的低剂量杀虫剂不会直接杀死蜜蜂，但会导致严重的行为困难吗？

15.3 昆虫的免疫系统

如第 2 章所述，动物和人类的第一道防线主要基于激活和触发获得性免疫的天然免疫应答。脊椎动物 TLRs 进化起源基于昆虫中的 Toll 受体，首次在果蝇中发现和分析[31,32]。为此 2011 年诺贝尔生理或医学奖授予了 J. A. Hoffmann。

昆虫和脊椎动物免疫的基本区别在于昆虫获得性免疫系统缺少高度特异性的抗原应答。尽管如此，在 4 亿年的进化过程中，昆虫发展了一种强大的防御策略来对抗细菌、真菌、病毒和寄生虫。只有受到这种“原始”免疫系统的保护，昆虫才能成功地在所有陆地生态系统中生存下来。

昆虫的天然免疫与脊椎动物和人类的天然免疫有许多相似之处，这是多方面的，涉及体液和细胞成分[33]。大多数对昆虫免疫的认识是通过黑腹果蝇研究提供的。在蜜蜂中也观察到了这一关键机制。

15.3.1 体液应答

对细菌和真菌感染的体液和全身应答由抗菌肽（AMP）控制。脂肪体细胞在刺激下分泌到血淋巴（昆虫血液）中：①防御素（对革兰氏阳性菌有活性）；②果蝇肽、双翅杀菌肽、抗菌肽和天蚕素（均对革兰氏阴性菌有活性）——蜜蜂的脂肪抑制素与果蝇肽相似；③碧蝽金属肽和果蝇抗真菌肽（对真菌有活性）[34]。显然体液应答可以区分不同种类的微生物[35]。AMP 基因的表达由两种不同的途径调节，Toll 通路控制对真菌和革兰氏阳性菌的反应基因，Imd 通路控制对革兰氏阴性菌的 AMP 基因[36]。循环受体感应危险信号并激活 Toll 通路，而膜结合受体激活 Imd 通路。两种途径都导致 NF-κB 样转录因子易位和产生 AMPs。NF-κB 反应元件可以在双翅杀菌肽基因启动子区域检测到。

15.3.2 细胞应答

细胞免疫应答由专门的血液细胞即血细胞、浆细胞、晶体细胞和片状细胞介导[37]。血浆细胞占大多数血细胞的 95%。这些细胞表达吞噬受体并在体内巡逻，清除微生物和细胞碎片，并向脂肪体细胞发出感染信号。昆虫的脂肪体细胞与脊椎动物的肝脏相似，因为它们储存营养，并合成蛋白质，也合成 AMP、脂类和糖类，这些蛋白质在体内循环。晶体细胞仅占血细胞的 5%，它们产生酶，通过这些入侵者的黑化杀死寄生虫或微生物。第三种血细胞类型是片状细胞，比其他血细胞大，仅在寄生时被诱导和测量。它们通过包裹和酶降解杀死入侵者[38]。

15.3.3 蜜蜂免疫

2006 年对蜜蜂全基因组进行了测序[39]。与果蝇相比，蜜蜂只有大约 1/3 的免疫基因[40]。免疫系统的减少仍然是一个推测问题。蜜蜂被认为是“一个”社会有机体，发展出一种特定的行为来降低感染寄生虫和疾病的风险。梳毛、卫生行为、对受感染蜜蜂的攻击以及营养在防御侵略者和感染中起着核心作用。营养对健康免疫系统的影响程度已在这本书第 2 章讨论过。这种进化的古老社会免疫力有明显局限性，这一点可以从瓦螨感染中看出，在个体水平上，免疫基因的低水平还受到杀虫剂和农药的影响，当然也受到病原的影响。

一个经典的蜜蜂科学家通常是一个保守的“群体思考者”，他谈到生物体“蜜蜂”时指的是一个有着成千上万个体的庞大群体。他管理健康问题的策略是只考虑群体。但从螨的角度来看，瓦螨一点也不在乎。瓦螨饿了，正在找下一顿饭。也许瓦螨对群体一无所知，忽视了社会免疫力和群体智慧。螨攻击个体蜜蜂，一步步地攻击幼虫，瓦螨意识到它已经进入了天堂——一个巨大的群体。

15.3.4　跛脚鸭

几年前我们应邀为控制德国瓦螨提出一项建议，这是对农业、耕作和蜜蜂健康创新的公开呼吁。在这之前，这里的大多数公共蜜蜂研究所都倾向于进行一项经典的育种研究，目的是创造一个能抵抗瓦螨的蜂王。10多年以来他们花费数百万欧元做这项研究，这是一项失败了的研究—— 跛脚鸭研究。这些经典的“群体思考者”不会接受与非群体昆虫相比，蜜蜂的免疫基因库只有其1/3，而且数量和功能绝对有限。

我们不是群体思考者，我们研究宿主-寄生虫的相互作用，着眼于现有基因及其规则，我们寻找如何干预的可能性。我们在个体层面上这样做，比如瓦螨。螨的目标是个体蜜蜂而不是群体。只有当个体蜜蜂受到保护时，才能防止瓦螨向蜂群扩散。

15.3.5　幻想世界

我们令人惊叹的项目是由精心挑选的和“特别”的评审员进行评估，特别是因为这些评审员也在同一个公开呼吁中申请，因此竞争对手也申请了相同的预算。这些评审员可以复制粘贴我们的科学项目供他们自己使用。好奇心的终结？值得注意的是，在这个充满创意的国度里，这样的创新是不受欢迎的。

15.4　蜜蜂疫苗

疫苗的特点可能是安全、无恢复致病性的风险、在田间条件下稳定和有效。生产应简单，符合GMP标准，价格低廉。最后但并非最不重要的是，在欧洲这种蜜蜂疫苗必须像所有其他动物疫苗一样符合EMA（欧洲药品管理局）的要求。

我们开发了一种DNA疫苗。基因疫苗需要直接原位接种表达载体，该载体编码抗原和/或免疫增强剂的序列[41]。图15.2给出了用于疫苗接种的表达质粒的示意图。

15.4.1　实验程序

用CMV启动子构建表达质粒。令人惊讶的是，没有蜜蜂或其他昆虫特异性启动子是必不可少的，以驱动蛋白质表达。增强型绿色荧光蛋白（EGFP）被选为报告基因，并与SV40增强子元件一起插入多个克隆位点。质粒构建体在大肠杆菌中生产，并通过标准技术高度纯化。

欧洲蜜蜂（意大利蜜蜂）是从当地养蜂人那里获得并在实验室条件下培育的。从受感染的蜜蜂身上采集到了瓦螨。

EGFP质粒的口服疫苗接种是通过给蜜蜂喂食糖和质粒DNA（疫苗糖）的混合溶液来进行的。养蜂人制作的标准糖溶液是冬季蜜蜂的正常食物。

15.4.2　结果

喂养开始后10天，我们用EGFP抗体通过免疫荧光和免疫印迹分析测量EGFP的表达。从第3天到第10天，在胸腔中尤其是在马氏管中，检测到清晰的EGFP信号。用缺少报告基因的DNA喂养的对照蜜蜂没有显示出任何信号。与此同时对转化大肠杆菌进行了对照实验，以研究EGFP在肠道细菌而不是蜜蜂细胞中表达的可能性。转化细菌中未检测到EGFP信号。

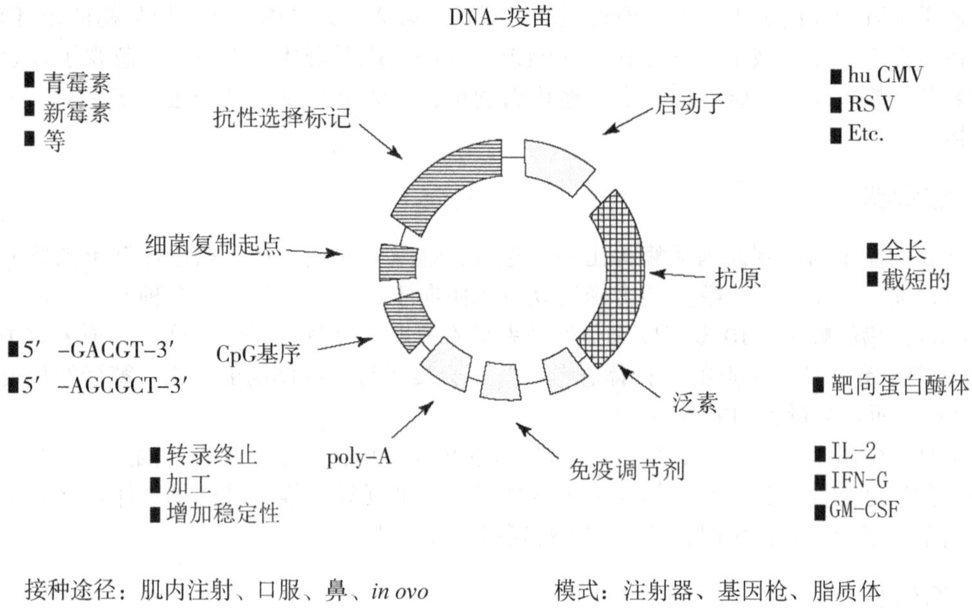

图 15.2　用于 DNA 疫苗接种的表达质粒模型

包含功能表达盒的单个元素。编码的抗原作为全长或截短的 cDNA，由一个强的启动子/增强子和多聚腺苷酸化序列控制。细胞因子的共表达可特异性增强免疫应答。免疫系统的非特异性激活可由 CpGs 启动。以蛋白酶体途径为靶点的泛素共表达可以增强聚焦于强细胞反应的疫苗。可以采用各种应用途径和管理模式（Giese[41]）。

最令人惊讶的是，在吸了蜜蜂（用疫苗糖溶液喂养的蜜蜂）血淋巴 5 天后的瓦螨身上发现了 EGFP 信号，而在对照蜜蜂（受感染的）的对照瓦螨身上没有检测到信号。质粒 DNA 的喂入导致报告基因在不同蜜蜂组织中的表达持续了几天，最终瓦螨通过吸血方式吸收了这种蛋白质。蜜蜂血液不是由动脉和静脉携带的，而是松散地在身体周围流动。蜂蜜、胃和粪便中都没有检测到 EGFP 信号。图 15.3 显示了 EGFP 通过蜜蜂身体并流向瓦螨。

15.5　展望

我们重组 DNA 疫苗的概念是基于瓦螨吸血和大分子活性生物通过血液转移到螨体内。这一概念首次在感染蜱的牛身上得到成功证明，即牛蜱属[42]。另见本书第 17 章。

蜜蜂具有体液和细胞免疫系统，如多种抗菌肽和免疫反应酶。瓦螨通过大量抑制这些免疫基因来克服蜜蜂的免疫反应[43]。我们正在研制一种能够完全恢复蜜蜂免疫系统功能的产品。还有很多工作要做。因此我们必须认识到，除了所有严重的科学瓶颈之外，抵抗力是由激进的人智学造成的：蜜蜂是具有灵魂纯洁性的“更高级生物”，不允许用基因技术治疗，而是要用化学杀虫剂和农药。

第一批兽用 DNA 疫苗是 2001 年在美国和 2002 年在德国由我们小组开发的[44,45]。世界上第一个获得批准的 DNA 疫苗是针对传染性造血坏死（IHN）病毒的鲑鱼 DNA 鱼疫苗（这是野生和养殖鲑鱼的常见病毒病原体），于 2005 年在加拿大由兽医生物制品科（VBS）、动物健康和生产部以及加拿大食品检验局（CFIA）注册。一种针对狗黑色素瘤的 DNA 疫苗最近在美国获得批准[46]。所有这些疫苗在生物学上绝对安全，耐受性好，没有副作用，而且是有效的。人类健康 DNA 疫苗正在临床试验中。

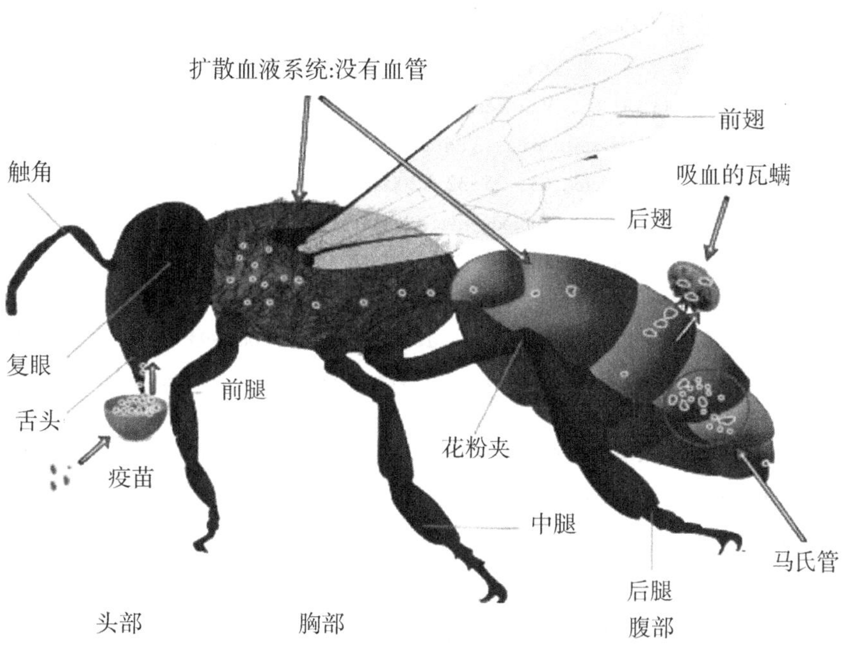

图 15.3　口服疫苗

将编码 EGFP 的质粒 DNA 作为疫苗抗原的载体，与溶于水中的标准糖溶液混合。这种糖溶液是冬季蜜蜂的营养，通常由养蜂人手工制作。实验 DNA 浓度为 500 μg DNA/mL 糖溶液。DNA 喂养 24 小时，没有加强喂养。

结论

生物学多迷人啊！我们从一个简单的想法开始，即不管是什么生物，真核细胞的生物化学都保持不变。免疫系统的结构有所不同。这意味着昆虫可以成功地对抗寄生虫和感染，但使用不同的武器——没有 T 细胞，没有 B 细胞，因此没有抗体和记忆。我们能够刺激蜜蜂的靶向免疫基因，并测量昆虫的典型免疫应答。首先为马开发的标准质粒 DNA 疫苗将鱼类、昆虫和哺乳动物的进化联系起来。没有其他疫苗能胜任这项工作。生物学是多么迷人啊！

除了进化论之外，生物学中没有任何东西是有意义的。

（Theodosius Dobzhansky，1900—1975，遗传学家和合成进化论领域的中心人物）

参考文献

[1] Garibaldi, L. A., et al.: Wild pollinators enhance fruit set of crops regardless of honey bee abundance. Science339, 1608-1611 (2013). doi: 1 0. 1126/science. 1230200.

[2] Cox-Foster, D. L., et al.: A meta-genomic survey of microbes in honey bee colony collapse disorder. Science 318, 283-287 (2007). doi: 10. 1126/ science. 1146498.

[3] Maori, E., et al.: IAPV, a bee-affecting virus associated with colony collapse disorder can be silenced by dsRNA ingestion. Insect Mol. Biol. 18, 55-60 (2009). doi: 10. 1111/j. 1365-2583. 2009. 00847. x.

[4] Vanengelsdorp, D., et al.: Colony collapse disorder: a descriptive study. PLoSOne4, e6481 (2009). doi: 10. 1371/journal. pone. 0006481.

[5] Stindl, R., Stindl Jr., W.: Vanishing honey bees: is the dying of adult worker bees a consequence of short telomeres and premature aging? Med. Hypotheses75, 387 - 390 (2010). doi: 10. 1016/j. mehy. 2010.

04. 003.

[6] Williams, G. R., et al.: Colony collapse disorder in context. Bioessays32, 845-846 (2010). doi: 10. 1002/ bies. 201000075.

[7] Calderon, R. A., Zamora, L. G., Van Veen, J. W., Quesada, M. V.: A comparison of the reproductive ability of Varroa destructor (Mesostigmata: Varroidae) in worker and drone brood of Africanized honey bees (Apismellifera). Exp. Appl. Acarol. 43, 25-32 (2007). doi: 10. 1007/s10493-007-9102-1.

[8] Garrido, C., Rosenkranz, P.: The reproductive program of female Varroa destructor mites is triggered by its host, Apis mellifera. Exp. Appl. Acarol. 31, 269-273 (2003).

[9] Maggi, M., et al.: Brood cell size of Apis mellifera modifies the reproductive behavior of Varroa destructor. Exp. Appl. Acarol. 50, 269-279 (2010). doi: 10. 1007/s10493-009-9314-7.

[10] Sal'chenko, V. L.: Diagnosis and treatment of Varroa mite infestations in bees. Veterinariia8, 62-64 (1971).

[11] Kamburov, G., K'Nchev, K., Stoimenov, V.: Testing of chemicals for control of Varroa jacobsoni infection in honeybees. Acta Microbiol. Virol. Immunol. (Sofiia) 2, 100-104 (1975).

[12] Betke, P., Ribbeck, R.: Varroatosis-a new mite pest of the honeybee in Europe. Angew. Parasitol. 20, 210-215 (1979).

[13] Dolejsky, W., Schley, P.: A mathematical simulation model to estimate the population development in a bee (Apis mellifera L.) colony following infestation with the mite Varro ajacobsoni Oud. Zentralbl. Veterinarmed. B 27, 798-805 (1980).

[14] Romaniuk, K., Bobrzecki, J., Kwiecien, S.: The course of Varroa infestation in bee families under treatment and the effect Varroa jacobsoni invasion on the body mass of bees. Wiad. Parazytol. 33, 185-192 (1987).

[15] Ritter, W., Perschil, F., Czarnecki, J. M.: Treatment of bee colonies with isopropyl-4, 4-dibromo-benzilate against varroa disease and acarine disease. Zentralbl. Veterinarmed. B30, 266-273 (1983).

[16] Bowen-Walker, P. L., Martin, S. J., Gunn, A.: The transmission of deformed wing virus between honeybees (Apismellifera L.) by the ectoparasitic mite varroa jacobsoni Oud. J. Invertebr. Pathol. 73, 101-106 (1999). doi: 10. 1006/jipa. 1998. 4807.

[17] Benjeddou, M., Leat, N., Allsopp, M., Davison, S.: Detection of acute bee paralysis virus and black queen cell virus from honeybees by reverse transcriptase pcr. Appl. Environ. Microbiol. 67, 2384-2387 (2001). doi: 10. 1128/AEM. 67. 5. 2384-2387. 2001.

[18] Di Prisco, G., et al.: Varroa destructor is an effective vector of Israeli acute paralysis virus in the honeybee, Apis mellifera. J. Gen. Virol. 92, 151-155 (2011). doi: 10. 1099/vir. 0. 023853-0.

[19] Chen, Y. P., Pettis, J. S., Collins, A., Feldlaufer, M. F.: Prevalence and transmission of honeybee viruses. Appl. Environ. Microbiol. 72, 606-611 (2006). doi: 10. 1128/AEM. 72. 1. 606-611. 2006.

[20] Blacquiere, T., Smagghe, G., van Gestel, C. A., Mommaerts, V.: Neonicotinoids in bees: a review on concentrations, side-effects and risk assessment. Ecotoxicology21, 973-992 (2012). doi: 10. 1007/ s10646-012-0863-x.

[21] Decourtye, A., Lacassie, E., Pham-Delegue, M. H.: Learning performances of honeybees (Apismellifera L) are differentially affected by imidacloprid according to the season. Pest Manag. Sci. 59, 269-278 (2003). doi: 10. 1002/ps. 631.

[22] Mondet, F., Goodwin, M., Mercer, A.: Age-related changes in the behavioural response of honeybees to Apiguard (R), a thymol-based treatment used to control the mite Varroa destructor. J. Comp. Physiol. A Neuroethol. Sens. Neural Behav. Physiol. 197, 1055-1062 (2011). doi: 10. 1007/s00359-011-0666-1.

[23] A l-Waili, N., Salom, K., Al-Ghamdi, A., Ansari, M. J.: Antibiotic, pesticide, and microbial contaminants of honey: human health hazards. Scientific World Journal2012, 930849 (2012). doi: 10. 1100/

2012/930849.

[24] Wu, J. Y., Anelli, C. M., Sheppard, W. S.: Sub-lethal effects of pesticide residues in brood comb on worker honey bee (Apis mellifera) development and longevity. PLoSOne 6, e14720 (2011). doi: 10. 1371/journal. pone. 0014720.

[25] Johnson, R. M., Dahlgren, L., Siegfried, B. D., Ellis, M. D.: Acaricide, fungicide and drug interactions in honey bees (Apis mellifera). PLoSOne 8, e54092 (2013). doi: 10. 1371/journal. pone. 0054092.

[26] Pettis, J. S., vanEngelsdorp, D., Johnson, J., Dively, G.: Pesticide exposure in honey bees results in increased levels of the gut pathogen Nosema. Naturwissenschaften99, 153-158 (2012).

[27] Locke, B., Forsgren, E., Fries, I., de Miranda, J. R.: Acaricide treatment affects viral dynamics in Varroa destructor-infested honey bee colonies via both host physiology and mite control. Appl. Environ. Microbiol. 78, 227-235 (2012). doi: 10. 1128/ AEM. 06094-11.

[28] Henry, M., et al.: A common pesticide decreases for - aging success and survival in honey bees. Science336, 348-350 (2012). doi: 10. 1126/science. 1215039.

[29] Guez, D.: A common pesticide decreases foraging success and survival in honey bees: questioning the ecological relevance. Front Physiol. 4, 37 (2013). doi: 10. 3389/fphys. 2013. 00037.

[30] Whitehorn, P. R., O'Connor, S., Wackers, F. L., Goulson, D.: Neonicotinoid pesticide reduces bumble bee colony growth and queen production. Science336, 351-352 (2012). doi: 10. 1126/science. 1215025.

[31] Lemaitre, B., et al.: Functional analysis and regulation of nuclear import of dorsal during the immune response in Drosophila. EMBO J. 14, 536-545 (1995).

[32] Hoffmann, J. A., Reichhart, J. M.: Drosophila innate immunity: an evolutionary perspective. Nat. Immunol. 3, 121-126 (2002). doi: 10. 1038/ni0202-121.

[33] Hoffmann, J. A.: The immune response of Drosophila. Nature426, 33-38 (2003). doi: 10. 1038/nature02021.

[34] Imler, J. L., Bulet, P.: Antimicrobial peptides in Drosophila: structures, activities and gene regulation. Chem. Immunol. Allergy86, 1-21 (2005). doi: 10. 1159/000086648.

[35] Lemaitre, B., Reichhart, J. M., Hoffmann, J. A.: Drosophila host defense: differential induction of antimicrobial peptide genes after infection by various classes of microorganisms. Proc. Natl. Acad. Sci. U. S. A. 94, 14614-14619 (1997).

[36] Lemaitre, B., et al.: A recessive mutation, immune deficiency (imd), defines two distinct control pathways in the Drosophila host defense. Proc. Natl. Acad. Sci. U. S. A. 92, 9465-9469 (1995).

[37] Williams, M. J.: Drosophila hemopoiesis and cellular immunity. J. Immunol. 178, 4711-4716 (2007).

[38] Meister, M., Ferrandon, D.: Immune cell trans-differentiation: a complex crosstalk between circulating immune cells and the haematopoietic niche. EMBO Rep. 13, 3 - 4 (2012). doi: 10. 1038/embor. 2011. 238.

[39] Honeybee Genome Sequencing, C.: Insights into social insects from the genome of the honeybee Apismellifera. Nature443, 931-949 (2006).

[40] Evans, J. D., et al.: Immune pathways and defence mechanisms in honey bees Apis mellifera. Insect Mol. Biol. 15, 645-656 (2006). doi: 10. 1111/j. 1365-2583. 2006. 00682. x.

[41] Giese, M.: DNA-antiviral vaccines: new developments and approache-a review. Virus Genes 17, 219-232 (1998).

[42] Willadsen, P., Bird, P., Cobon, G. S., Hungerford, J.: Commercialisation of a recombinant vaccine against Boophilus microplus. Parasitology 110 (Suppl), S43-S50 (1995).

[43] Navajas, M., et al.: Differential gene expression of the honey bee Apis mellifera associated with Varroa destructor infection. BMC Genomics 9, 301 (2008). doi: 10. 1186/1471-2164-9-301.

[44] Davis, B. S., et al.: West Nile virus recombinant DNA vaccine protects mouse and horse from virus challenge and expresses in vitro a noninfectious recombinant antigen that can be used in enzyme-linked immunosorbent assays. J. Virol. 75, 4040-4047 (2001). doi: 10. 1128/JVI. 75. 9. 4040-4047. 2001.

[45] Giese, M., et al.: Stable and long-lasting immune response in horses after DNA vaccination against equine arteritis virus. Virus Genes25, 159-167 (2002).

[46] Liao, J. C., et al.: Vaccination with human tyrosinase DNA induces antibody responses in dogs with advanced melanoma. Cancer Immun. 6, 8 (2006).

(何继军译)

第 16 章　莱姆病：宿主库靶向疫苗

Maria Gomes-Solecki①

摘要

在世界上许多地方，通过改善卫生习惯、监测、诊断、治疗、有效疫苗以及加强公众教育和对危险因素的认识，来控制细菌引起的疾病。对于病原体引起的疾病，控制策略尤其具有挑战性，这些病原体持续存在于哺乳类野生动物宿主库中，利用昆虫等媒介在该物种间循环传播。在这个群体中，对人类健康构成直接威胁的相关疾病是狂犬病、森林鼠疫和莱姆病[1]。

针对狂犬病和莱姆病的宿主库或传播媒介的疫苗接种策略，已开发出宿主库靶向疫苗。一个成功应用的例子是口服疫苗（Raboral™），目前美国地方政府使用该疫苗在受感染的野生动物和在人口稠密地区之间建立屏障，以防止狂犬病传播。

在本章中，我将讨论一种宿主库靶向口服疫苗的研究进展，以控制伯氏疏螺旋体在野生动物中的传播及其对降低人类莱姆病发病率的预期影响。

16.1　莱姆病和病原体

20 世纪 70 年代中期，莱姆疏螺旋体病或莱姆病的临床症状是在美国康涅狄格州莱姆镇附近的儿童中发现了一种影响多关节的不对称关节炎流行病[2]。这种疾病是由一组相关的螺旋体引起的，这些螺旋体通过特定的硬蜱属蜱传播到自然宿主和次级宿主[3]。

莱姆病螺旋体广泛分布在北半球温带地区，随着以前被砍伐用于农业的土地重新造林，为鹿、蜱和易感脊椎动物宿主创造了新的栖息地[4]，莱姆病螺旋体的范围不断扩大。

莱姆病占所有虫媒自然疫源性疾病 90%以上，美国每年报告近 3 万例确诊病例。在欧洲，每年在蓖子硬蜱的范围内发生多达 6 万例，是该区域主要的传播媒介[5]。

疏螺旋体属。分为两大类，一类含有莱姆病的病原体，另一类含有导致回归热的螺旋体[4]。根据系统发育分析已经把莱姆病螺旋体化分成许多种，广义上称为伯氏疏螺旋体。在伯氏疏螺旋体复合体中有 20 多种命名和未命名的种，其中 3 种基因种在人类病原体中占主导地位：狭义伯氏疏螺旋体（*B. burgdorferi sensustricto*），以美国和西欧为主；欧亚大陆的伽氏疏螺旋体（*B. garinii*）和阿弗西尼疏螺旋体（*B. afzelii*）[4]。

在蜱虫吸食过程中，螺旋体随蜱虫唾液一起沉积在咬伤处[6]。蜱虫附着 48 小时后感染的可能性越来越大[7]。由于伯氏疏螺旋体基因组不编码任何已知毒素，也不编码分泌毒素所需的机制[8]，因此组织损伤和疾病是由哺乳动物宿主感染后的炎症反应介导的[9]。游走性红斑是疏螺旋体感染最常见的临床表现（60%~80%[10]），潜伏期为 2~32 天[11]。大多数游走性红斑发生在 6—8 月。像大多数由蜱传播的感染一样，男性感染人数略多于女性。在美国存在双峰年龄分布，5~9 岁的儿童和 45~59 岁的成年人发病率最高，但所有年龄段的人都有感染风险[12]。低水平的螺旋体血症可能发生在大多数未经治疗的患者中[13]，偶尔会影响周围或中枢神经系统、关节或

---

① M. Gomes-Solecki，DVM（田纳西大学健康科学中心微生物、免疫和生物化学系，孟菲斯，田纳西州，美国，E-mail：mgomesso@ uthsc. edu）

心脏[11]。

在美国大多数感染是由狭义伯氏疏螺旋体引起的，其主要特征是关节炎，偶尔还有心脏炎和神经症状。在欧洲大多数感染是由阿弗西尼疏螺旋体引起的，该菌引起皮肤临床表现，如慢性萎缩性肢端皮炎、引起神经莱姆病的伽氏疏螺旋体和引起心脏炎和神经莱姆病的伯氏疏螺旋体（狭义上的）[14]。以缓慢进行性脑脊髓炎为特征的晚期莱姆病并不常见[15-17]。

欧洲少数患者的标本中已经分离出或通过 PCR 检测到了另外 5 种疏螺旋体（bavarensis 疏螺旋体、spielmanii 疏螺旋体、lusitania 疏螺旋体、valaisiana 疏螺旋体和 bissettii 疏螺旋体）[4,18]。

16.2 地方性循环和传播

在流行地区，莱姆病的传播发生在郊区或用于林业和娱乐活动的农村地区[19]。莱姆病传播的典型栖息地在这种疾病的整个地理范围内都是一样的，通常由落叶或混合林地组成，偶尔是针叶林，地上有大量下层植物和一层腐烂的植被，因此为蜱的发育和生存提供了足够的湿度，支持了一系列潜在的脊椎动物贮存宿主[3]。

伯氏疏螺旋体主要通过蓖麻属复合体内的 4 种硬蜱传播：北美东部的肩突硬蜱和西部的太平洋硬蜱，欧洲的蓖子硬蜱和亚洲的全沟硬蜱[2,3,20]。这些蜱经历了 4 个阶段的生命周期（卵、幼虫、若虫和成虫），每个活跃阶段只进食 1 次。雄蜱很少进食也从不暴饮暴食。当动物通过植被时，未进食的扁蜱附着在宿主的皮肤上。喂食几天后（幼虫 3 天，若虫 5 天，成年雌性 7 天）蜱离开宿主，在土壤表面附近冬眠，在那里它们需要 80% 的最小相对湿度才能生存。蜱需要几个月的时间才能发育到下一个发育阶段[3]。幼虫孵化时未受感染（不会经卵传播）[4]，伯氏疏螺旋体是在被感染的贮存宿主上进食后获得的。蜕皮至若虫阶段后，蜱会将病原体传播给动物或人类，为它们提供下一次血餐。莱姆病的传播是通过进食过程注射蜱唾液来实现的。肩突硬蜱或太平洋硬蜱传播伯氏疏螺旋体至少需要 36 小时的进食时间[21-23]。蓖子硬蜱传播阿弗西尼疏螺旋体的速度会更快。在沙鼠上进行的实验将这个数字定为 17 小时，但结果需要其他研究者重复，因此还没有确定最小喂食时间[24]。值得注意的是，莱姆病疏螺旋体向人类传播发生在肩突硬蜱、太平洋硬蜱和蓖子硬蜱的若虫期，但全沟硬蜱在若虫期则不会传播，全沟硬蜱在成年期才能将莱姆病疏螺旋体传染给人类（图 16.1）。

蜱生命周期的 4 种形态都有明显的季节性。肩突硬蜱若虫在初夏活跃，成虫在秋季活跃，一直持续到冬季和早春。至于蓖子硬蜱和全沟硬蜱，若虫和成虫在早春开始活跃，直到仲夏或一年中的晚些时候在潮湿的遮蔽环境中持续寻找寄主。蓖子硬蜱在秋季会出现第二个活动高峰。太平洋硬蜱的活动模式似乎更像蓖子硬蜱，而不是肩突硬蜱。在所有蜱类中，峰值活动通常发生在幼虫中，比若虫稍晚，特别是美国东部的肩突硬蜱。一组肩突硬蜱若虫活动与下一组幼虫活动之间的 3 个月差异允许有大量时间从受感染的宿主传播到幼虫，这可以解释美国东部的高传播率[23]。大部分传染给人类的病例表现为游走性红斑，发生在 5 月下旬至 9 月下旬，这与若虫的活动以及公众对蜱类栖息地的娱乐性利用日益增加相吻合[3]。

北美和欧亚大陆的田野研究已经确定了各种小型哺乳动物和鸟类是地方性动物病传播循环的贮存库[4]（表 16.1）。白足鼠被认为是美国东北部的主要贮存宿主，而在欧洲，啮齿类动物和候鸟则分别是阿弗西尼疏螺旋体和伽氏疏螺旋体的主要贮存宿主[2]。在大多数蜱虫栖息地，鹿是维持蜱虫种群的必要条件，因为它们是少数能够喂养足够数量成蜱并保持周期持续的野生宿主之一，但鹿不是螺旋体的合格贮存宿主。牛是不合格的宿主，绵羊也可能是不合格的贮存宿主[20,25,26]。伯氏疏螺旋体的不同致病基因种对某些脊椎动物贮存寄主表现出轻微偏好，尽管这种宿主特异性似乎不是绝对的。一个被认为与贮存能力有关的因素是，莱姆螺旋体的特定基因种对动物宿主的补充性介导杀伤的敏感性[27]。在蜱虫栖息地的鹿群数量少可以被认为是莱姆病风险的一个很好的指标，因为可能也会出现一系列其他宿主，包括贮存适应能力强的动物。然

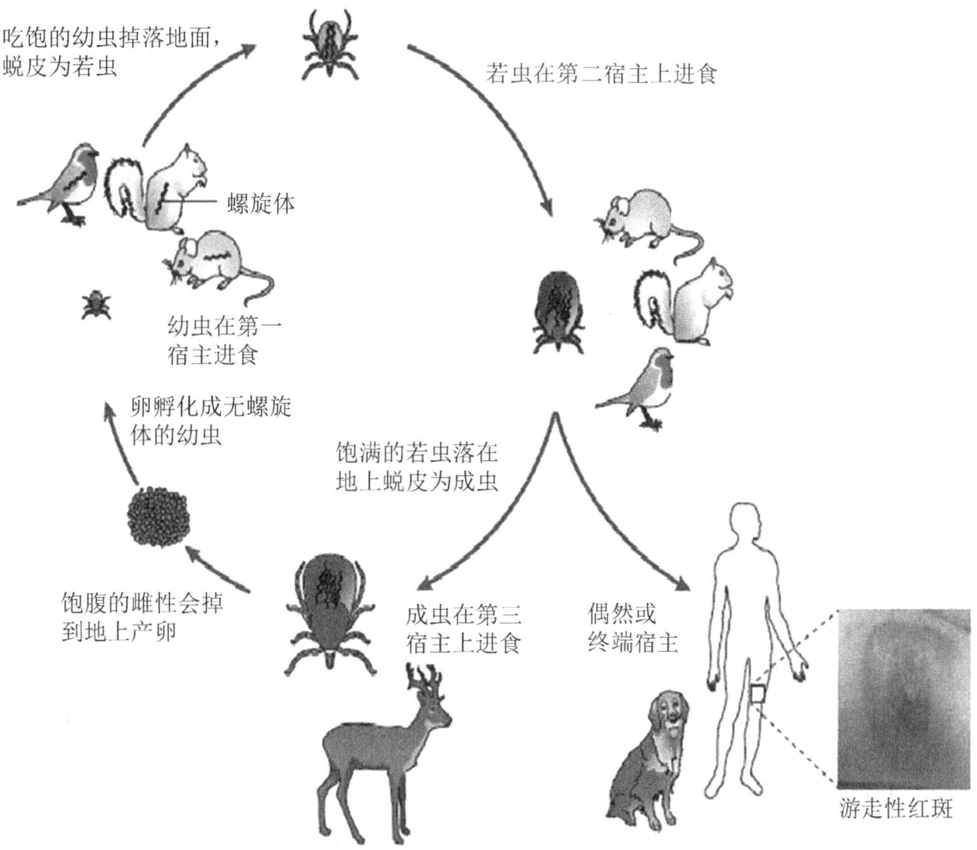

图 16.1　伯氏疏螺旋体的地方性循环（经麦克米伦出版有限公司［自然评论出版社许可印刷，鲁道夫，2012 年］）

而，如果一个栖息地中的大多数动物不是像鹿或牛这样的莱姆疏螺旋体的贮存宿主，那么莱姆病的风险就会降低，因为蜱主要以这些动物为食，因此不会受到感染[3,28]。

表 16.1　脊椎动物物种的贮存能力

贮存能力	物种	地理区域	参考文献
能够贮存	老鼠	美国/欧洲	[29-31]
	天鼠	欧洲	[31]
	金花鼠	美国	[29，30]
	鼩鼱	美国	[29，30，32]
	松鼠	美国/欧洲	[29，30，33]
	地面鸟类	美国/欧洲	[30，33]
	条纹臭鼬	美国	[30]
	蜥蜴	美国/欧洲	[23，36]
不能贮存	鹿	美国/欧洲	[30，35]
	牛	欧洲	[25，33，36，37]

（续表）

贮存能力	物种	地理区域	参考文献
	负鼠	美国	[30]
	浣熊	美国	[30]
不确定	绵羊	欧洲	[20]
	野兔	欧洲	[38]
	刺猬	欧洲	[39]
	獾	欧洲	[33]
	红狐	欧洲	[33]

贮存能力是指被感染的宿主将伯氏疏螺旋体传播给食蜱的概率。

16.3 诊断和经典疗法

16.3.1 临床表现和诊断

莱姆疏螺旋体病（LB）的临床表现从急性到慢性不等，由于感染所涉及不同的疏螺旋体基因种和基因型，LB 有很大的差异[16,18,40]。

简而言之，蜱叮咬后几天或几周，如果发生了疏螺旋体感染，在 60%~80%的病例中，这将表现为游走性红斑（蜱叮咬部位的皮疹，大约 10 cm 宽，可能向周边扩大，可能痒或不痒）[10,16,40]。其他早期症状包括流感样症状、发热、疲劳、头痛、肌肉或关节疼痛。然而早期感染可能完全没有症状。感染后几周或几个月（有或没有以往的游走性红斑病史），可能发生脑膜神经根炎、脑膜炎或脑膜脑炎形式的神经疏螺旋体病（在 10%~20%有症状的患者中发现）[17]、莱姆关节炎或疏螺旋体淋巴细胞瘤等[16]。很少有多发性红斑或心脏炎被诊断出来[16,41]。在疏螺旋体感染几个月甚至几年后，可以观察到慢性萎缩性肢端皮炎（在欧洲常见）、淋巴细胞瘤、慢性关节炎（在美国常见）、脑脊髓炎或慢性神经疏螺旋体病（在欧洲罕见）[16,18]。

除了典型的早期皮肤损伤外，该疾病的所有表现都需要微生物或血清学确认疏螺旋体感染。一些慢性 LB 诊断目前有争议[42]，也有人认为 LB 的过度诊断和过度治疗可能是一个重要问题[18]。

因此，诊断主要是临床上的；游走性红斑是一种病理征象，要考虑到蜱虫叮咬的风险。其他形式的 LB 诊断需要通过直接或指定的诊断试验进行确认[15]（表 16.2）。在临床组织标本中直接检测广义伯氏疏螺旋体的方法有很多，包括显微镜检查、伯氏疏螺旋体特异蛋白或核酸检测、培养等。未来的诊断方法可能包括基于 PCR 分子技术，可以快速确认 LB 临床诊断，并在组织标本或培养的分离株中鉴定出疏螺旋体基因种[43]。间接检测伯氏疏螺旋体最常用方法是血清间接免疫荧光抗体法（IFA）和酶联免疫吸附法（ELISA）[44]。在超过 50%的病例中，根据扩大的红斑可以作出 LB 诊断（经过 1 周的随访证实）。在没有游走性红斑情况下，必须注意至少 1 种其他临床表现，并使用血清学诊断血液和 CSF（脑脊液）中的疏螺旋体。然而，目前可用的检测方法往往无法在感染早期检测到特异性抗体。根据最新的欧洲和美国指南[45,46]，血清学诊断应遵循两步程序，从敏感性试验开始，如酶联免疫吸附法（ELISA）等。ELISA 阳性结果应通过免疫印迹等特异性检测加以证实。

表 16.2 莱姆螺旋体病的诊断测试，常规临床实践中莱姆螺旋体病的表现、简要临床病例定义和推荐诊断方法

疾病类型	初级诊断检测	辅助诊断检测	辅助临床研究结果
迁移性红斑 扩张红色或蓝红色斑块（≥5 cm 直径），有或没有中央清除	根据病史和皮肤损伤的目视检查进行诊断，不需要或不推荐实验室检测	皮肤活检标本的培养或 PCR 可用于研究，但不需要用于常规临床实践	蜱叮咬现场；北美患者区域性淋巴结病
前缘通常是明显的，通常颜色强烈，没有明显凸起	如果病变不典型，则建议进行急性期和恢复期血清学检测，因为急性期检测不敏感		
疏螺旋体淋巴细胞瘤（一种罕见的症状） 无痛的蓝红色结节或斑块，通常在耳垂、耳轮、乳头或阴囊上；儿童（尤其是耳朵）比成人更常见	血清学检测通常在症状出现时呈阳性；如果呈阴性，则检测恢复期血清（2~6 周后）	肿瘤活检可能是排除肿瘤所必需的皮肤活检标本的培养或 PCR 在研究中有用，但临床实践中不需要	蜱叮咬现场；近期或伴随的迁移性红斑
莱姆神经疏螺旋体病 成人以脑膜脊神经根炎、脑膜炎和周围面神经麻痹为主；很少发生脑炎、脊髓炎；脑血管炎非常罕见	脑脊液细胞增多和广义伯氏疏螺旋体鞘内抗体合成的实证	脑脊液培养或 PCR 检测莱姆螺旋体	近期或伴随的游走性红斑
儿童以脑膜炎和周围面神经麻痹为主（2~6 周后）	血清学检测通常在出现时呈阳性；如果是阴性，测试恢复期血清	鞘内合成总 IgM、IgG 或 IgA	
心脏莱姆病（一种罕见的症状） 急性发作房室传导紊乱（I-III），心律紊乱，有时并发心肌炎或心包炎应该排除其他解释	血清学检查通常阳性，但如果阴性和临床怀疑很强烈，则检测恢复期血清（2~6 周后）	不推荐（通过培养或 PCR 从心内膜心肌活检中检测莱姆螺旋体，仅限于研究）	近期或伴随的游走性红斑、神经失常或两者兼有
眼部病变（很罕见） 结膜炎、葡萄膜炎、视神经乳头炎、巩膜外层炎、角膜炎	血清学检查	培养或 PCR 检测眼液中广义的伯氏疏螺旋体	伴发或先前其他明确定义的莱姆病病变
莱姆关节炎 一个或多个大关节反复发作或持续客观关节肿胀	血清学检测结果通常存在高浓度的特异性血清 IgG 抗体	滑膜液分析。用 PCR 检测滑膜液或组织中广义的伯氏疏螺旋体	以前其他明确的莱姆病临床表现
慢性萎缩性肢端皮炎 长期存在红色或蓝红色病变，通常位于四肢伸肌表面	血清学检测	组织学检查	以前其他明确的莱姆病临床表现
最初的面团状肿胀病变最终会萎缩骨突上可能有皮肤硬结和纤维瘤结节	通常存在高浓度的特异性血清 IgG 抗体	皮肤活检组织培养或 PCR 检测伯氏疏螺旋体有助于研究，但不用于常规临床实践	

16.3.2 治疗

体外研究表明莱姆疏螺旋体对四环素、大多数青霉素、许多第二代和第三代头孢菌素以及大环内酯类药物敏感。莱姆疏螺旋体对特定的氟喹诺酮、利福平和第一代头孢菌素有耐药性[15,47,48]。

尽管不使用抗生素治疗，游走性红斑最终也会消失，但建议使用口服抗生素治疗以防止传播和发展到晚期后遗症（表16.3）。多西环素、阿莫西林、苯氧甲基青霉素和头孢呋辛非常有效，是这种表现的首选药物。多西环素是唯一一种前瞻性和大型回顾性临床试验都显示只有10天治疗有效的药物[49,50]。然而多西环素会引起光敏性，在8岁以下的儿童和怀孕或哺乳的妇女中禁用[15]。治疗莱姆病的首选胃肠外药物是头孢曲松，因为它在体外对莱姆病有很高的活性，能很好地穿越血脑屏障，并且有很长的血清半衰期，这意味着每天只能服用1次。非肠道给药抗生素的替代选择是头孢噻肟和静脉滴注青霉素。推荐胃肠外抗生素治疗晚期莱姆病患者并作为入院监测的心脏莱姆病患者的初始治疗。

表16.3 莱姆病的治疗

疾病类型	治疗方案	持续时间	评价
早期局部性和早期播散性莱姆病 游走性红斑	口服	14天	在美国用多西环素治疗10天是有效的，但其他一线口服抗生素10天疗程的疗效却没有得到充分证实。欧洲还没有做过研究 多西环素对嗜吞噬细胞无形体也有活性
脑膜炎或神经根病	肠胃外给药或多西环素	14天	欧洲的研究证据表明，口服多西环素治疗和胃肠外治疗一样有效，尽管这一发现尚未在北美得到系统测试
脑神经受累	口服	14天	虽然任何一线口服抗生素似乎对颅神经病变患者有效，但对除面神经麻痹以外的颅神经病变患者或除多西环素以外的药物有效的证据有限
心脏病	口服或胃肠外给药	14天	治疗信息受到限制 对于正在医院接受监测或正在医院放置临时起搏器的患者，肠胃外方案是首选方案 当心脏传导阻滞得到改善并且患者准备出院时，可以给予口服抗生素治疗方案 门诊病人可以用口服抗生素治疗
疏螺旋体淋巴细胞瘤	口服	14天	关于治疗的信息很少使用与治疗迁移性红斑相同的方法 北美没有记录
晚期莱姆病 无神经系统疾病的关节炎口服		28天	病人通常同时接受非甾体抗炎药
口服治疗一个疗程后复发性关节炎	口服 胃肠外给药	28天 14~28天	肠胃外治疗通常只保留给对口服治疗没有部分反应的患者
抗生素难治性关节炎	对症治疗	根据需要	抗生素难治性莱姆关节炎定义为静脉注射头孢曲松疗程结束后至少2个月（或口服抗生素疗程完成两个4周疗程后1个月）的持续性滑膜炎；此外，滑膜液（和滑膜组织，如果有）上的PCR方法对疏螺旋体核酸检测是阴性的

（续表）

疾病类型	治疗方案	持续时间	评价
中枢或周围神经系统疾病	胃肠外给药	14~28 天	没有将 14 天治疗和 28 天治疗进行比较研究，部分原因是病例很少
慢性萎缩性肢端皮炎	口服	21~28 天	没有将 21 天治疗与 28 天治疗进行比较研究，也没有研究将不同抗生素的治疗进行比较
莱姆病后综合征 莱姆病后综合征	对症治疗	根据需要	考虑和评估其他潜在的症状原因

16.4 疫苗

目前有多种行之有效的预防莱姆病传播的策略。避免蜱滋生的环境或覆盖裸露的皮肤，在这种环境下使用蜱驱虫剂是一种非常有效的方法。清除邻近森林草坪的木屑、施用杀螨剂和建造围栏以阻挡鹿也是有效的[49]，因为这些会扰乱寻找宿主的蜱密度高的栖息地。接触蜱后 2 小时内洗澡可降低莱姆病的风险[50]。由于蜱附着时间与莱姆病传播之间存在延迟，建议每天检查整个皮肤表面（包括头皮）以去除附着的蜱。临床研究表明，超过 96%的患者在没有任何其他干预的情况下，即使在高度流行的地理区域[51]，也不会感染莱姆病。如果蜱没有被发现或清除，感染的概率接近区域蜱群中的感染率（典型地是美国东北部和中西部高度流行地区感染了约 25%的若虫期肩突硬蜱，欧洲感染了约 10%的若虫期蓖子硬蜱）[52,53]。化学预防可以减少从皮肤上去除肩突硬蜱或全沟硬蜱后发生莱姆病的机会[3,51,54]。

目前还没有预防人类莱姆病的疫苗[3]。然而预防莱姆病的新战略包括新的人类疫苗、抗蜱疫苗、针对贮存宿主的疫苗和干预措施目前正在制定之中。本章将进一步讨论后一种针对贮存寄主的疫苗和干预措施。

肩突硬蜱广泛分布在美国各地，在许多地方，莱姆病不是地方病。这些地区少有莱姆病的一个原因是，蜥蜴是硬蜱的天然宿主，当暴露在蜥蜴血液中时，伯氏疏螺旋体会被杀死。因此，消灭宿主或降低生物体的载荷能力（即防止载体获得或传播生物体）可能是预防人类莱姆病的有效策略[55]。尽管从大自然中消灭莱姆病是不现实的，但减少莱姆病对人类的威胁是一个可以实现的目标[2]。

在美国，能够传播伯氏疏螺旋体的主要宿主是白足鼠（*Peromyscusleucpus*）[56,57]。然而花栗鼠（*Tamiasstriatus*）、松鼠、鼩鼱和其他小型脊椎动物正日益被认为是重要的宿主。此外，鸟类也可能在传播伯氏疏螺旋体到达的过程中发挥重要作用[58]。

研究人员一直在调查伯氏疏螺旋体贮存宿主的疫苗接种情况，以减少微生物的携带。Tsao 等进行了一项雄心勃勃的研究，他们捕获了白足鼠，在皮下接种伯氏疏螺旋体（OspA）的外表面蛋白 A 或对照疫苗。与对照相比[59]，接种疫苗显著降低了次年从使用 OspA 的场所采集的蜱中伯氏疏螺旋体的患病率。目前有几个小组正在开发一种口服 OspA 疫苗，将其作为诱饵分发给小鼠和其他宿主，最终目的是破坏这种螺旋体的地方性循环[6,61]。

开发诱饵疫苗作为减少病原体传播的策略是有先例的[62]。用于递送狂犬病疫苗[63-65]和鼠疫疫苗[66]的诱饵和诱饵系统已证明是成功的。已经探索了其他蜱控制系统来降低莱姆螺旋体病的风险。在一项研究中，白尾鹿的杀螨剂自我治疗导致蜱密度降低[67-70]。在另一项研究中，一种以啮齿动物为目标的杀螨剂（氟虫腈）在改良的商业诱饵盒中递送给白足鼠，也能有效地减少若虫和幼虫的蜱感染（蜱密度）[71]。在另一种方法中多西环素啮齿动物诱饵制剂可防止伯氏疏螺旋体向脊椎动物宿主的蜱传播，治愈小鼠中已确定的感染[72,73]。蜱密度或蜱感染降低以及脊椎动物宿主伯氏疏螺旋体感染的减少，预计将导致人类感染莱姆病风险的总体降低。

16.4.1 莱姆病动物模型

由于老鼠和其他啮齿类动物是伯氏疏螺旋体的天然宿主。因此，野生白足鼠属小鼠感染伯氏疏螺旋体期间无明显变化并不令人惊讶，因为症状性疾病会使受感染动物的生存处于不利地位，和/或限制螺旋体向新的载体群传播的机会[74]。然而当被伯氏疏螺旋体感染时，发现实验室小家鼠的特定近交系表现出与人类莱姆病相似的特征[9,75]。例如 C3H 和 BALB/c 小鼠感染伯氏疏螺旋体后，会出现踝关节关节炎和心脏炎，而 C57BL/6 和 DBA 小鼠对感染症状的发展更具抵抗力，并且通常心脏和关节的炎症程度最小。没有近交系小鼠发展成游走性红斑、脑膜炎或脑炎，因此都不是人类莱姆病的完美模型。恒河猴也可以感染伯氏疏螺旋体并被用作神经疏螺旋体病的模型，因为它们有中枢神经系统感染的倾向，特别是当用皮质类固醇免疫抑制时。恒河猴还会出现游走性红斑、多发性单神经炎和关节炎，这使它们成为与人类症状最相似的莱姆病动物模型[75]。然而出于成本和基因操作简便的原因，除了小鼠以外，其他动物模型的研究还很少。虽然白足鼠没有明显的疾病迹象，但是螺旋体的传播可以通过血液、心脏和膀胱组织中的伯氏疏螺旋体培养来确定，进一步通过伯氏疏螺旋体基因（如 FlaB、OspA 或 OspC 的 PCR）扩增来确认[76]。

16.4.2 OspA 疫苗

先前的研究表明，OspA 免疫可诱导小鼠的长期保护性免疫反应[77]。在 OspA 抗体存在的情况下，螺旋体在吸血蜱的中肠内被杀死[78]。OspA 疫苗成功完成了 I、II 和 III 期试验，于 1998 年获得美国联邦药品管理局的批准。接种疫苗个体在接受铝作为佐剂的三剂 OspA 疫苗后，对伯氏疏螺旋体感染显示出大约 80%的保护作用[79]。OspA 疫苗在人体中的一个缺点是保护性免疫与免疫后 OspA 抗体的高滴度相关，5%的疫苗接受者显示对 OspA 产生的抗体反应不足。疫苗失败与 Toll 样受体（TLR）-1 细胞表面表达降低有关[80]。因此，在接种疫苗后高抗体滴度不会持续很久，需要额外的增强剂来维持保护性滴度[79,81]。尽管有一些缺点，但 OspA 仍然是开发莱姆病疫苗的最佳免疫原。

两个小组使用 OspA 开发口服递送系统来给宿主接种疫苗，从而开发宿主靶向诱饵疫苗。其中一种递送系统是基于表达 OspA 的痘苗病毒[61]。另一种递送系统是基于表达 OspA 的大肠杆菌[60]。这两种系统在诱导近交系小家鼠和经口服接种疫苗的白足鼠产生保护性水平的抗 OspA 抗体方面一样有效[60,61,76,82]。此外，这两种疫苗在清除以接种小鼠为食的受感染蜱中的伯氏疏螺旋体方面同样有效。然而基于痘苗病毒的疫苗对免疫功能低下或患有湿疹的人具有传染性[83]。鉴于针对莱姆病的 RTV（宿主靶向疫苗）将部署在郊区附近，痘苗病毒是一种传染性病原体的事实将使监管批准变得非常复杂。因此基于细菌的递送系统被认为是最安全的方法。因此，所有疾病控制和预防中心现场测试宿主靶向疫苗预防莱姆病的努力以细菌递送系统为主。

细菌递送系统以大肠杆菌为基础，用一种编码伯氏疏螺旋体 OspA 全长序列的质粒进行转化[60,76]。诱导细菌表达蛋白质，分析细胞的生存能力，然后将活培养物冻干，形成一个递送系统，不需要纯化蛋白质。然后重新包装在可递送的微球中。在所报道的研究中，诱饵疫苗是在免疫前每天通过将 200 mg 冻干细菌与燕麦片（含有约 2 mg/mL OspA）混合制成的，并随意提供给白足鼠食用。测试 OspA-RTV 在白足鼠中诱导保护性免疫应答（白足鼠是螺旋体的天然宿主）。在优化 OspA-RTV 用于野外部署的研究中，测试了几种免疫计划（每天投放诱饵 4~16 周）和模拟诱饵在野外暴露的参数（高温和湿度）。

这些研究表明，用 OspA-RTV 免疫的小鼠产生了高滴度的 OspA 抗体，进行免疫后这种抗体持续了一整年，并且 OspA 抗体的高滴度与保护相关。此外研究人员确定白足鼠产生保护性免疫应答所需的最小单位数是 5 个。考虑到感染人类伯氏疏螺旋体的肩突硬蜱若虫在春季活跃，而将伯氏疏螺旋体感染转移到下一个蜱群（第二年将发育为若虫）的幼虫蜱在夏季活跃，将

OspA-RTV 部署到野生动物的可能计划包括每周 5 天，持续 4 周的诱饵递送。或者，可以通过部署 OspA-RTV 来实施替代计划，每周 2~3 次，为期 4 个月，从 4 月中旬至 8 月中旬。4 个月的部署而不是 1 个月的低劳动强度部署的基本原理可能同样有效，是为在这两个季节出生的年轻宿主接种疫苗。

16.4.3 临床前开发/田间应用

最近纽约州和宾夕法尼亚州与疾病控制和预防中心（CDC）合作，开展了使用基于大肠杆菌的 OspA-RTV 的田间试验。这些研究旨在评估 RTV 在降低野生贮存宿主物种（白足鼠、白足鼠属）以及蜱媒感染率方面的功效。疫苗的有效性将通过计算摄入 OspA-RTV 后出现 OspA 抗体的白足鼠数量以及通过与未部署 RTV 的配对对照场所相比，在接种了疫苗的场所感染了伯氏疏螺旋体的肩突硬蜱的比例降低来确定。

OspA-RTV 疫苗属于美国农业部的管辖范围。美国农业部颁发生物产品许可证的程序与联邦药品管理局（FDA）的程序相差甚远，根据技术的复杂程度和申请人的组织方式，时间范围从 2~3 年不等。大肠杆菌本身是公认的安全大肠杆菌 B2 的衍生物，在美国被批准用于生物制品。

16.5 优点和缺点

16.5.1 优点

基于大肠杆菌的 RTV（宿主靶向疫苗）技术的一个主要优势是在部署到田间目标区域后，它将卓越的功效与安全性相结合。使用的大肠杆菌菌株是 B2 衍生物，通常被认为是安全的（GRAS 状态）。

基于痘苗病毒的 RTV 技术的主要优势在于，它可能需要在田间进行一次应用以诱导对 OspA 的有效免疫应答。

16.5.2 缺点

基于大肠杆菌的 RTV（宿主靶向疫苗）技术的一个弱点是，这种疫苗的效力取决于漫长的免疫计划；考虑到在小鼠对 OspA 产生显著的免疫反应之前，必须接种至少 5 剂疫苗，因此在数周内多次部署 RTV 是有必要的。

痘苗病毒 RTV 技术的一个弱点是，在莱姆病流行的郊区广泛传播是不安全的，人类肯定会遇到和处理诱饵。

鉴于肩突硬蜱感染有 2 年的生命周期，一般来说这可能被视为该概念的一个弱点，即 OspA-RTV 在该领域的分布需要几年才能对降低若虫感染流行率产生影响。因此，要降低莱姆病的发病率，需要在田间应用几年时间。

16.5.3 生产和市场

制造和分配不受天气影响的基于大肠杆菌 OspA-RTV 疫苗是一个麻烦的过程，因为我们设想必须生产几种色调的材料才能部署在流行地区的目标区域。

目前，一种间接预防人类莱姆病的诱饵疫苗的市场将面向美国和国际的公共卫生组织。目前莱姆病流行地区遍布亚洲、欧洲和北美，感染率逐年上升。据我们所知，市场上没有针对伯氏疏螺旋体的诱饵疫苗。国家和城市卫生部门应形成国内的主要市场。

参考文献

[1] Cross, M. L., Buddle, B. M., Aldwell, F. E.: The potential of oral vaccines for disease control in wildlife

species. Vet. J. 174, 472–480 (2007). doi: 10. 1016/j. tvjl. 2006. 10. 005.

[2] Radolf, J. D., Caimano, M. J., Stevenson, B., Hu, L. T.: Of ticks, mice and men: understanding the dual–host lifestyle of Lyme disease spirochaetes. Nat. Rev. Microbiol. 10, 87–99 (2012). doi: 10. 1038/ nrmicro2714.

[3] Stanek, G., Wormser, G. P., Gray, J., Strle, F.: Lyme borreliosis. Lancet379, 461–473 (2012). doi: 10. 1016/ S0140-6736 (11) 60103-7.

[4] Kurtenbach, K., et al.: Fundamental processes in the evolutionary ecology of Lyme borreliosis. Nat. Rev. Microbiol. 4, 660–669 (2006). doi: 10. 1038/ nrmicro1475.

[5] O'Connell, S., Granstrom, M., Gray, J. S., Stanek, G.: Epidemiology of European Lyme borreliosis. Zentralbl. Bakteriol. 287, 229–240 (1998).

[6] Ribeiro, J. M., Mather, T. N., Piesman, J., Spielman, A.: Dissemination and salivary delivery of Lyme disease spirochetes in vector ticks (Acari: Ixodidae). J. Med. Entomol. 24, 201–205 (1987).

[7] Piesman, J., Mather, T. N., Sinsky, R. J., Spielman, A.: Duration of tick attachment and Borrelia burgdorferi transmission. J. Clin. Microbiol. 25, 557–558 (1987).

[8] Fraser, C. M., et al.: Genomic sequence of a Lyme disease spirochaete Borreliaburgdorferi. Nature 390, 580–586 (1997). doi: 10. 1038/37551.

[9] Weis, J. J., Bockenstedt, L. K.: In: Samuels, D. S., Radolf, J. D. (eds.) Lyme Disease in Humans. Borrelia: Molecular Biology, Host Interaction, and Pathogenesis, pp. 413–441. Caister Academic, Norfolk (2010).

[10] Cerar, D., Cerar, T., Ruzic–Sabljic, E., Wormser, G. P., Strle, F.: Subjective symptoms after treatment of early Lyme disease. Am. J. Med. 123, 79–86 (2010). doi: 10. 1016/j. amjmed. 2009. 05. 011.

[11] Radolf, J. D., Salazar, J. C., Dattwyler, R. J.: Lyme Disease in Humans. In: Samuels, D. S., Radolf, J. D. (eds.) Borrelia: Molecular Biology, Host Interaction, and Pathogenesis, pp. 487–533. Caister Academic, Norfolk (2010).

[12] Bacon, R. M., Kugeler, K. J., Mead, P. S.: Surveillance for Lyme disease–United States, 1992–2006. MMWR Surveill. Summ. 57, 1–9 (2008).

[13] Wormser, G. P., et al.: Brief communication: hematogenous dissemination in early Lyme disease. Ann. Intern. Med. 142, 751–755 (2005).

[14] Stanek, G., Strle, F.: Lyme disease: European perspective. Infect. Dis. Clin. North Am. 22, 327–339, vii (2008). doi: 10. 1016/j. idc. 2008. 01. 001.

[15] Wormser, G. P., et al.: The clinical assessment, treatment, and prevention of Lyme disease, human granulocytic anaplasmosis, and babesiosis: clinical practice guidelines by the Infectious Diseases Society of America. Clin. Infect. Dis. 43, 1089–1134 (2006). doi: 10. 1086/508667.

[16] Stanek, G., et al.: Lyme borreliosis: clinical case definitions for diagnosis and management in Europe. Clin. Microbiol. Infect. 17, 69–79 (2011). doi: 10. 1111/j. 1469-0691. 2010. 03175. x.

[17] Halperin, J. J.: Nervous system Lyme disease. Infect. Dis. Clin. North Am. 22, 261–274, vi (2008). doi: 10. 1016/j. idc. 2007. 12. 009 (2008).

[18] Rizzoli, A. et al.: Lyme borreliosis in Europe. Euro Surveill. 16 (27) (2011) p. 8, pii: 19906.

[19] Dennis, D. T., Hayes, E. B.: Epidemiology of lyme borreliosis. In: Gray, J. S., Kahl, O., Lane, R. S., Stanek, G. (eds.) Lyme Borreliosis: Biology, Epidemiology and Control, 1st edn, pp. 251–280. Cabi Publishing, New York (2002).

[20] Gray, J. S., Kahl, O., Janetzki, C., Stein, J., Guy, E.: The spatial distribution of Borrelia burgdorferi–infected Ixodes ricinus in the Connemara region of county Galway Ireland. Exp. Appl. Acarol. 19, 163–172 (1995).

[21] des Vignes, F., et al.: Effect of tick removal on transmission of Borrelia burgdorferi and Ehrlichia phagocytophila by Ixodes scapularis nymphs. J. Infect. Dis. 183, 773–778 (2001). doi: 10. 1086/318818.

[22] Peavey, C. A., Lane, R. S.: Transmission of Borrelia burgdorferi by Ixodes pacificus nymphs and reservoir competence of deer mice (Peromyscus maniculatus) infected by tick–bite. J. Parasitol. 81, 175–178 (1995).

[23] Einsen, L., Lane, R. S.: Vectors of Borrelia burgdorferi sensu lato. In: Gray, J. S., Kahl, O., Lane, R. S., Stanek, G. (eds.) Lyme Borreliosis: Biology, Epidemiology and Control, 1st edn, pp. 91-115. Cabi Publishing, New York (2002).

[24] Kahl, O., et al.: Risk of infection with Borrelia burgdorferi sensu lato for a host in relation to the duration of nymphal Ixodes ricinus feeding and the method of tick removal. ZentralblBakteriol287, 41-52 (1998).

[25] Ogden, N. H., Nuttall, P. A., Randolph, S. E.: Natural Lyme disease cycles maintained via sheep by co-feeding ticks. Parasitology 115 (Pt 6), 591-599 (1997).

[26] Matuschka, F. R., et al.: Diversionary role of hoofed game in the transmission of Lyme disease spirochetes. Am. J. Trop. Med. Hyg. 48, 693-699 (1993).

[27] Bykowski, T., et al.: Borrelia burgdorferi complement regulator-acquiring surface proteins (BbCRASPs): expression patterns during the mammal-tick infection cycle. Int. J. Med. Microbiol. 298 (Suppl 1), 249-256 (2008). doi: 10. 1016/j. ijmm. 2007. 10. 002.

[28] EUCALB: European Union Concerted Action on Lyme Borreliosis. An information resource of the ESCMID study group, ESGBOR. Accessed Aug 22, 2013. www. eucalb. com.

[29] Brisson, D., Dykhuizen, D. E.: OspC diversity in Borrelia burgdorferi: different hosts are different niches. Genetics 168, 713-722 (2004). doi: 10. 1534/ genetics. 104. 028738.

[30] LoGiudice, K., Ostfeld, R. S., Schmidt, K. A., Keesing, F.: The ecology of infectious disease: effects of host diversity and community composition on Lyme disease risk. Proc. Natl. Acad. Sci. U. S. A. 100, 567-571 (2003). doi: 10. 1073/pnas. 0233733100.

[31] De Boer, R., Hovius, K. E., Nohlmans, M. K., Gray, J. S.: The woodmouse (Apodemussylvaticus) as a reservoir of tick-transmitted spirochetes (Borrelia burgdorferi) in the Netherlands. Zentralbl. Bakteriol. 279, 404-416 (1993).

[32] Dykhuizen, D. E., et al.: The propensity of different Borrelia burgdorferi sensu stricto genotypes to cause disseminated infections in humans. Am. J. Trop. Med. Hyg. 78, 806-810 (2008).

[33] Mannelli, A., Bertolotti, L., Gern, L., Gray, J.: Ecology of Borrelia burgdorferi sensu lato in Europe: transmission dynamics in multi-host systems, influence of molecular processes and effects of climate change. FEMS Microbiol. Rev. 36, 837-861 (2012). doi: 10. 1111/j. 1574-6976. 2011. 00312. x.

[34] Dsouli, N., et al.: Reservoir role of lizard Psammodromus algirus in transmission cycle of Borrelia burgdorferi sensu lato (Spirochaetaceae) in Tunisia. J. Med. Entomol. 43, 737-742 (2006).

[35] Gray, J. S., Kahl, O., Janetzki, C., Stein, J.: Studies on the ecology of Lyme disease in a deer forest in county Galway Ireland. J. Med. Entomol. 29, 915-920 (1992).

[36] Kimura, K., et al.: Prevalence of antibodies against Borrelia species in patients with unclassified uveitis in regions in which Lyme disease is endemic and nonendemic. Clin. Diagn. Lab. Immunol. 2, 53-56 (1995).

[37] Pichon, B., Rogers, M., Egan, D., Gray, J.: Blood-meal analysis for the identification of reservoir hosts of tick-borne pathogens in Ireland. Vector Borne Zoonotic Dis. 5, 172-180 (2005). doi: 10. 1089/vbz. 2005. 5. 172.

[38] Jaenson, T. G., Talleklint, L.: Lyme borreliosis spirochetes in Ixodes ricinus (Acari: Ixodidae) and the varying hare on isolated islands in the Baltic Sea. J. Med. Entomol. 33, 339-343 (1996).

[39] Gern, L., Rouvinez, E., Toutoungi, L. N., Godfroid, E.: Transmission cycles of Borreliaburgdorferisen-sulato involving Ixodes ricinus and/or I. Hexagonus ticks and the European hedgehog, Erinaceus europaeus, in suburban and urban areas in Switzerland. Folia Parasitol. 44, 309-314 (1997).

[40] Wormser, G. P., et al.: Borrelia burgdorferi genotype predicts the capacity for hematogenous dissemination during early Lyme disease. J. Infect. Dis. 198, 1358-1364 (2008). doi: 10. 1086/592279.

[41] Fish, A. E., Pride, Y. B., Pinto, D. S.: Lyme carditis. Infect. Dis. Clin. North Am. 22, 275-288, vi (2008). doi: 10. 1016/j. idc. 2007. 12. 008.

[42] Marques, A.: Chronic Lyme disease: a review. Infect. Dis. Clin. North Am. 22, 341-360, vii-viii (2008). doi: 10. 1016/j. idc. 2007. 12. 011.

[43] Cerar, T., et al.: Validation of cultivation and PCR methods for diagnosis of Lyme neuroborreliosis. J. Clin. Microbiol. 46, 3375-3379 (2008). doi: 10. 1128/JCM. 00410-08.

[44] Aguero-Rosenfeld, M. E., Wang, G., Schwartz, I., Wormser, G. P.: Diagnosis of lyme borreliosis. Clin. Microbiol. Rev. 18, 484-509 (2005). doi: 10. 1128/CMR. 18. 3. 484-509. 2005.

[45] Wilske, B., Fingerle, V., Schulte-Spechtel, U.: Microbiological and serological diagnosis of Lyme borreliosis. FEMS Immunol. Med. Microbiol. 49, 13-21 (2007). doi: 1 0. 1111/j. 1574-695X. 2006. 00139. x.

[46] CDC. Lyme disease diagnosis and treatment. http: // www. cdc. gov/lyme/diagnosistreatment/index. html (2012).

[47] Hunfeld, K. P., Ruzic-Sabljic, E., Norris, D. E., Kraiczy, P., Strle, F.: In vitro susceptibility testing of Borrelia burgdorferi sensu lato isolates cultured from patients with erythema migrans before and after antimicrobial chemotherapy. Antimicrob. Agents Chemother. 49, 1294-1301 (2005). doi: 10. 1128/AAC. 49. 4. 1294-1301. 2005.

[48] Morgenstern, K., et al.: In vitro susceptibility of Borrelias pielmanii to antimicrobial agents commonly used for treatment of Lyme disease. Antimicrob. Agents Chemother. 53, 1281-1284 (2009). doi: 10. 1128/AAC. 01247-08.

[49] Stafford, K. I., Kitron, U.: Environmental management of Lyme borreliosis control. In: Gray, J. S., Kahl, O., Lane, R. S., Stanek, G. (eds.) Lyme Borreliosis: Biology, Epidemiology and Control, pp. 301-334. CABI Publishing, New York (2002).

[50] Connally, N. P., et al.: Peridomestic Lyme disease prevention: results of a population-based case-control study. Am. J. Prev. Med. 37, 201-206 (2009). doi: 10. 1016/j. amepre. 2009. 04. 026.

[51] Warshafsky, S., et al.: Efficacy of antibiotic prophylaxis for the prevention of Lyme disease: an updated systematic review and meta-analysis. J. Antimicrob. Chemother. 65, 1137-1144 (2010). doi: 10. 1093/jac/ dkq097.

[52] Piesman, J.: Lyme borreliosis in North America. In: Gray, J. S., Kahl, O., Lane, R. S., Stanek, G. (eds.) Lyme Borreliosis: Biology, Epidemiology and Control, pp. 223-249. CABI Publishing, New York (2002).

[53] Gern, L., Humair, P. F.: Lyme borreliosis in Europe. In: Gray, J. S., Kahl, O., Lane, R. S., Stanek, G. (eds.) Lyme Borreliosis: Biology, Epidemiology and Control, pp. 149-174. CABI Publishing, New York (2002).

[54] Korenberg, E. I., Horakova, M., Kovalevsky, J. V., Hubalek, Z., Karavanov, A. S.: Probability models of the rate of infection with tick-borne encephalitis virus in Ixodes persulcatus ticks. Folia Parasitol. 39, 85-92 (1992).

[55] Clark, R. P., Hu, L. T.: Prevention of Lyme disease and other tick-borne infections. Infect. Dis. Clin. North Am. 22, 381-396, vii (2008). doi: 10. 1016/j. idc. 2008. 03. 007.

[56] Anderson, J. F., Johnson, R. C., Magnarelli, L. A.: Seasonal prevalence of Borreliaburgdorferi in natural populations of white-footed mice, Peromyscus leucopus. J. Clin. Microbiol. 25, 1564-1566 (1987).

[57] Anderson, J. F.: Ecology of Lyme disease. Conn. Med. 53, 343-346 (1989).

[58] Comstedt, P., et al.: Migratory passerine birds as reservoirs of Lyme borreliosis in Europe. Emerg. Infect. Dis. 12, 1087-1095 (2006).

[59] Tsao, J. I., et al.: An ecological approach to preventing human infection: vaccinating wild mouse reservoirs intervenes in the Lyme disease cycle. Proc. Natl. Acad. Sci. U. S. A. 101, 18159-18164 (2004). doi: 10. 1073/pnas. 0405763102.

[60] Gomes-Solecki, M. J., Brisson, D. R., Dattwyler, R. J.: Oral vaccine that breaks the transmission cycle of the Lyme disease spirochete can be delivered via bait. Vaccine 24, 4440-4449 (2006). doi: 10. 1016/j. vaccine. 2005. 08. 089.

[61] Scheckelhoff, M. R., Telford, S. R., Hu, L. T.: Protective efficacy of an oral vaccine to reduce carriage of Borrelia burgdorferi (strain N40) in mouse and tick reservoirs. Vaccine 24, 1949-1957 (2006). doi: 10. 1016/j. vaccine. 2005. 10. 044.

[62] Piesman, J.: Strategies for reducing the risk of Lyme borreliosis in North America. Int. J. Med. Microbiol. 296 (Suppl40), 17-22 (2006). doi: 10. 1016/j. ijmm. 2005. 11. 007.

[63] Pastoret, P. P., et al.: First field trial of fox vaccination against rabies using a vaccinia-rabies recombinant

virus. Vet. Rec. 123, 481-483 (1988).

[64] Estrada, R., Vos, A., De Leon, R., Mueller, T.: Field trial with oral vaccination of dogs against rabies in the Philippines. BMC Infect. Dis. 1, 23 (2001).

[65] Knobel, D. L., du Toit, J. T., Bingham, J.: Development of a bait and baiting system for delivery of oral rabies vaccine to free-ranging African wild dogs (Lycaonpictus). J. Wildl. Dis. 38, 352-362 (2002).

[66] Creekmore, T. E., Rocke, T. E., Hurley, J.: A baiting system for delivery of an oral plague vaccine to black-tailed prairie dogs. J. Wildl. Dis. 38, 32-39 (2002).

[67] Daniels, T. J., et al.: Acaricidal treatment of white-tailed deer to control Ixodes scapularis (Acari: Ixodidae) in a New York Lyme disease-endemic community. Vector Borne Zoonotic Dis. 9, 381-387 (2009). doi: 10. 1089/vbz. 2008. 0197.

[68] Fish, D., Childs, J. E.: Community-based prevention of Lyme disease and other tick-borne diseases through topical application of acaricide to white-tailed deer: background and rationale. Vector Borne Zoonotic Dis. 9, 357-364 (2009). doi: 10. 1089/vbz. 2009. 0022.

[69] Hoen, A. G., et al.: Effects of tick control by acaricide self-treatment of white-tailed deer on host-seeking tick infection prevalence and entomologic risk for Ixodes scapularis-borne pathogens. Vector Borne Zoonotic Dis. 9, 431-438 (2009). doi: 10. 1089/ vbz. 2008. 0155.

[70] Stafford 3rd, K. C., Denicola, A. J., Pound, J. M., Miller, J. A., George, J. E.: Topical treatment of white-tailed deer with an acaricide for the control of Ixodesscapularis (Acari: Ixodidae) in a Connecticut Lyme borreliosis hyperendemic community. Vector Borne Zoonotic Dis. 9, 371-379 (2009). doi: 10. 1089/ vbz. 2008. 0161.

[71] Dolan, M. C., et al.: Control of immature Ixodes scapularis (Acari: Ixodidae) on rodent reservoirs of Borrelia burgdorferi in a residential community of southeastern Connecticut. J. Med. Entomol. 41, 1043-1054 (2004).

[72] Dolan, M. C., et al.: A doxycycline hyclate rodent bait formulation for prophylaxis and treatment of tick-transmitted Borrelia burgdorferi. Am. J. Trop. Med. Hyg. 78, 803-805 (2008).

[73] Zeidner, N. S., et al.: A sustained-release formulation of doxycycline hyclate (Atridox) prevents simultaneous infection of Anaplasma phagocytophilum and Borrelia burgdorferi transmitted by tick bite. J. Med. Microbiol. 57, 463-468 (2008). doi: 10. 1099/ jmm. 0. 47535-0.

[74] Oliver Jr., J. H., et al.: An enzootic transmission cycle of Lyme borreliosis spirochetes in the southeastern United States. Proc. Natl. Acad. Sci. U. S. A. 100, 11642 - 11645 (2003). doi: 10. 1073/ pnas. 1434553100.

[75] Embers ME., Hasenkampf NR., Jacobs MB., Philipp MT.: Dynamic longitudinal antibody responses during Borrelia burgdorferi infection and antibiotic treatment of rhesus macaques. Clin Vaccine Immunol. 19 (8), 1218-1226 (2012). doi: 10. 1128/CVI. 00228-12. Epub 2012 Jun 20. PMID: 22718128.

[76] Meirelles Richer, L., Aroso, M., Contente-Cuomo, T., Ivanova, L., Gomes-Solecki, M.: Reservoir targeted vaccine for lyme borreliosis induces a yearlong, neutralizing antibody response to OspA in white-footed mice. Clin. Vaccine Immunol. 18, 1809-1816 (2011). doi: 10. 1128/CVI. 05226-11.

[77] Fikrig, E., Barthold, S. W., Kantor, F. S., Flavell, R. A.: Long-term protection of mice from Lyme disease by vaccination with OspA. Infect. Immun. 60, 773-777 (1992).

[78] de Silva, A. M., Telford 3rd, S. R., Brunet, L. R., Barthold, S. W., Fikrig, E.: Borrelia burgdorferi OspA is an arthropod-specific transmission-blocking Lyme disease vaccine. J. Exp. Med. 183, 271-275 (1996).

[79] Steere, A. C., et al.: Vaccination against Lyme disease with recombinant Borreliaburgdorferi outer-surface lipoprotein a with adjuvant. Lyme Disease Vaccine Study Group. N. Engl. J. Med. 339, 209-215 (1998). doi: 10. 1056/NEJM199807233390401.

[80] Alexopoulou, L., et al.: Hyporesponsiveness to vaccination with BorreliaburgdorferiOspA in humans and in TLR1-and TLR2-deficient mice. Nat. Med. 8, 878-884 (2002). doi: 10. 1038/nm732.

[81] Schuijt, T. J., Hovius, J. W., van der Poll, T., van Dam, A. P., Fikrig, E.: Lyme borreliosis vaccination: the facts, the challenge, the future. Trends Parasitol. 27, 40-47 (2011). doi: 10. 1016/j. pt.

2010. 06. 006.

[82] Bhattacharya, D., et al.: Development of a baited oral vaccine for use in reservoir-targeted strategies against Lyme disease. Vaccine 29, 7818-7825 (2011). doi: 10. 1016/j. vaccine. 2011. 07. 100.

[83] Reed, J. L., Scott, D. E., Bray, M.: Eczema vaccinatum. Clin. Infect. Dis. 54, 832-840 (2012). doi: 10. 1093/ cid/cir952.

(马维民译)

第 17 章　控制蜱感染家畜的抗蜱疫苗

Cassandra Olds，Richard Bishop
和 Claudia Daubenberger①

摘要

蜱是专性吸血节肢动物寄生虫，影响大多数陆生脊椎动物。蜱作为疾病媒介的重要性在于它们向脊椎动物宿主传播生物体的丰富性和多样性。此外，蜱进食可能会导致附着部位继发性感染、直接中毒和瘫痪。蜱传疾病的影响在畜牧业中最为明显，由于蜱大量滋生和蜱传疾病的流行，畜牧业生产在许多地区受到限制。

目前，通过杀螨剂应用实现了对蜱大规模控制，但是持续使用已经导致对几种活性成分的抗性。作为一种替代方案，提出了通过接种疫苗对蜱进行控制。这种设想最终在控制微小扇头蜱的商业疫苗上得到了验证。尽管取得了初步成功，但是由于蜱和脊椎动物宿主之间复杂的相互作用，抗蜱疫苗的开发还面临着许多独特障碍。

然而，这种复杂性确实允许遵循疫苗开发的新领域，这些领域是其他虫媒病所不具备的。在蜱正常的摄食过程中，位于蜱肠内的蜱抗原隐藏在宿主的免疫反应中。从接种疫苗的宿主体内吸血后，肠壁会受到损伤，从而导致死亡或生殖能力下降。使用“隐蔽抗原”已成功有效地减少连续几代蜱的数量。此外研究人员还对唾液或胶结锥中存在的一些候选“暴露抗原”进行了研究，以确定它们是否有能力干扰蜱吸血和阻断寄生虫传播。随着蜱基因组测序项目的完成，作为进行候选疫苗评估的候选抗原数量不断增加。在这一章中概述了抗蜱疫苗开发历史和潜在的未来，特别提到了控制蜱和蜱传病对牲畜的影响。本章将讨论针对影响家畜的主要蜱类的抗蜱疫苗研究进展，包括扇头蜱、花蜱和璃眼蜱。

17.1　引言

蜱在世界各地广泛分布，影响世界上 80%的牛[1]。蜱和蜱传病（TBD）的经济重要性已经通过大量研究进行了评估；然而研究人员很可能低估了这些节肢动物媒介及其传播疾病的实际影响。部分因为研究只有一个焦点，如直接损失和控制成本；部分是因为他们通常只关注一种疾病或蜱种[2-4]。更复杂的估计是畜牧业损失与人类/伴侣动物的损失计量不同。

蜱吸血具有破坏性影响，包括疾病传播、瘫痪、中毒和蜱食地的继发性感染[4,5]。蜱和蜱媒病的影响在畜牧业中尤为显著，因为其对农民生计的影响一再被高度估价，特别是在严重依赖农业生产的南方国家[3,4]。牲畜中蜱的大量滋生会导致贫血、体重减轻、体况下降和皮革损伤，

---

① C. Olds，博士（✉）（国际家畜研究所生物技术系，内罗毕，肯尼亚，瑞士巴塞尔热带和公共卫生研究所医学寄生虫和感染生物学系，瑞士巴塞尔大学，ARC 安德斯波德兽医研究所寄生虫、媒介和媒传病，南非安德斯波德，E-mail：cassandra. leah. olds@ gmail. com）

R. Bishop，博士（肯尼亚内罗毕国际家畜研究所生物技术部）

C. Daubenberger，博士（瑞士热带和公共卫生研究所医学寄生虫和感染生物学，巴塞尔，瑞士，瑞士巴塞尔大学）

这是皮革生产行业的一个重要方面[2,6]。

尽管兽医上有重要的软蜱存在，但本综述将仅关注硬蜱科。硬蜱有 6 个重要的属，即花蜱属（钝眼蜱属）、革蜱属、血蜱属、璃眼蜱属、扇头蜱属和硬蜱属（总结于表 17. 1）。

硬蜱属中的硬蜱最著名的传播途径是疏螺旋体，这是欧洲、亚洲和北美常见的莱姆病媒介。由于这一物种在畜牧业中不受关注，因此不会进一步讨论。

如果蜱数量仍然很低，并且其分布与蜱媒疾病没有重叠，那么家畜可以忍受蜱。在某些情况下，蜱和蜱媒病已接近完全根除，从一个地理区域完全根除蜱种通常是不可行的[7,8]。从历史上看，蜱和蜱媒疾病控制重点是通过使用杀螨剂和适当的药物治疗，将蜱控制在可耐受水平。在某些情况下，基于杀螨剂的蜱控制通常是减少蜱数量而不牺牲生产力的唯一方法[8,9]。

杀螨剂在市场上能以多种配方买到，这些药剂可直接用于牲畜或浸泡池中，在浸泡池中可以定期让动物通过。杀螨剂的应用在很大程度上依赖于正确的配方和管理才能有效[9]。已发现大量化合物对蜱有效，包括砷（约 1983 年引入）、DDT（约 1946 年）、环二烯和毒杀芬（约 1947 年）、有机磷酸酯-氨基甲酸酯基（约 1955 年）、甲酰胺（约 1975 年）和大环内酯（约 1981 年）。许多上述化合物的效力和有用性随着许多蜱种（如扇头蜱、花蜱和璃眼蜱）的耐药性发展而逐渐减弱。

表 17. 1　家畜重要蜱类和蜱媒病概述[4]

属	分布	蜱传病
花蜱属	广泛分布在非洲（彩饰花蜱），非洲东南部（希伯来花蜱）	反刍动物考德里氏体病［反刍兽埃里克体（考德里氏体）（立克次体目）］ 低致病性泰勒虫病［变异泰勒虫（原生动物）］ 蜱在牛体吸食部位严重继发感染
革蜱属	欧洲、亚洲、北美、非洲[a]	牛无浆体病［边缘无浆体立克次体目］ 马巴贝斯虫病［驽巴贝斯虫（原生动物）］
血蜱属	亚洲、欧洲和澳大利亚范围较小	牛巴贝斯虫病［卵形巴贝斯虫（原生动物）］ 东亚牛泰勒虫病［水牛泰勒虫（原生动物）］ 小反刍动物巴贝斯虫病［莫氏巴贝斯虫（原生动物）］ 牛巴贝斯虫病［大巴贝斯虫（原生动物门）］
璃眼蜱属	亚洲、欧洲、非洲	牛热带泰勒虫病［牛环形泰勒虫（原生动物）］ 小反刍动物泰勒虫病［*lestoquardi* 泰勒虫（原生动物）］
扇头蜱属[b]	美洲（包括北美和南美）、非洲、亚洲、澳大利亚、欧洲	牛巴贝斯虫病［牛巴贝斯虫和双芽巴贝斯虫（原生动物）］ 牛泰勒虫病［小泰勒虫（原生动物）］ 低致病性牛泰勒虫病［陶罗特雷根氏泰勒虫（原生动物）］ 马焦虫病［马泰勒虫和驽巴贝斯虫（原生动物）］ 牛无浆体病［牛边缘无浆体，中央无浆体（原生动物）］ 小反刍动物巴贝斯虫病［羊巴贝斯虫（原生动物）］ 小反刍动物无浆体病［绵羊无浆体（原生动物）］

[a]虽然存在于非洲，但并不重要。

[b]扇头蜱属现在包括以前被称为牛蜱属的蜱类。最重要的蜱类包括牛的微小扇头蜱（牛巴贝斯虫病的主要媒介）以及具尾扇头蜱［牛东海岸热的媒介（牛泰勒虫病）］。

已经鉴定出多种杀螨剂耐药性蜱类，限制或完全排除了许多杀螨剂的使用[9-11]。除了耐药性之外，通过杀螨剂的应用进行化学控制还会导致环境污染和肉及奶制品残留污染。在经济薄弱的国家，使用杀螨剂控制蜱可能更不可行，因为基础设施维护和产品采购成本高[2]。对于小规模农民来说，成立合作社可能是允许杀螨剂控制蜱的一种替代方法[12]。在通过系统杀螨剂应用清除蜱的地方，由于政治动荡导致协议和基础设施崩溃，引起蜱种群的重建和与疾病暴发相关的高牲畜死亡率[6]。

图 17.1、图 17.2、图 17.3、图 17.4 和图 17.5 概述了在非洲感染牲畜的蜱。

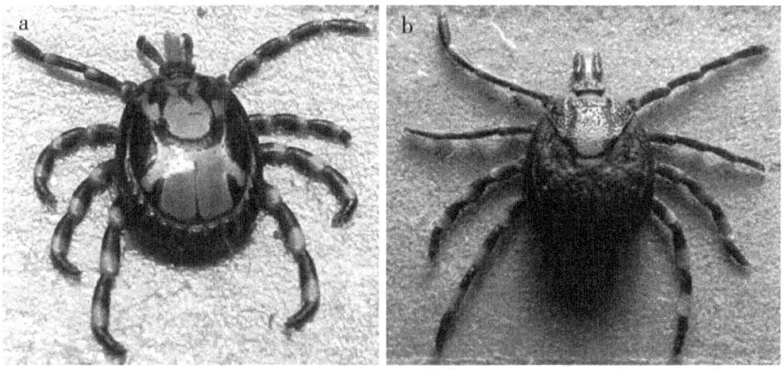

图 17.1 由于其大的口器，花蜱吸食可能导致蜱吸食部位严重继发感染

(a) 彩饰花蜱雄性成蜱；(b) 彩饰花蜱雌性成蜱。

图 17.2 具尾扇头蜱成蜱

以传播小泰勒虫为主，导致牛发病（东海岸热）。此外，该蜱还传播斑羚泰勒虫（低致病性牛泰勒虫病）、牛无浆体（牛埃立克体病）、康氏立克次体（蜱伤寒）和内罗毕绵羊病病毒等病原。

17.2 抗蜱疫苗接种

抗蜱疫苗接种的指导原则源于对获得性宿主抗蜱感染能力的早期研究。宿主反复暴露于蜱或蜱器官匀浆诱导对蜱再感染的抗性。虽然不同蜱类和宿主物种之间的抗性程度可能有所不同，但有证据强烈表明，对蜱侵扰的自然抗性是基于获得性免疫应答机制发展起来的[13-15]。从接种了蜱成分宿主中吸血的蜱在吸血过程中会吸收效应分子，从而介导对蜱的有害影响。这种效果表现为吸血时间减少、蜱的死亡率（在吸血期间或之后）、吸血重量减少和成年雌性生殖能力降

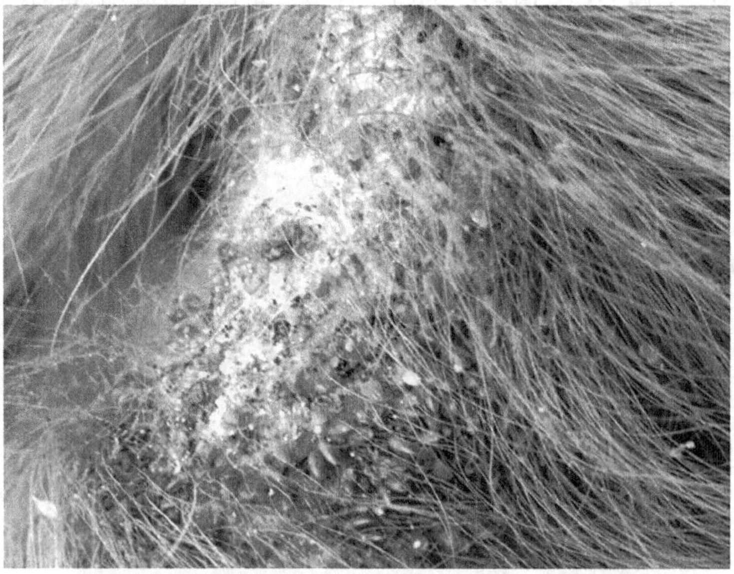

图 17.3　在牛耳朵上吸食的未成熟扇头蜱（由 S. Mwaura 提供）

低。从接种过的宿主吸血的蜱产卵也可能显示孵化率降低。总体结果导致蜱数量和蜱媒疾病减少。

与用于蜱控制的杀螨剂相比，抗蜱疫苗有几个优点。其中包括减少对动物产品和环境长期被有害残留物污染的担忧，提高所需基础设施的可持续性。此外，抗蜱疫苗的耐药性发展可能比杀螨剂慢，抗蜱疫苗开发、生产和注册成本被认为低于新型杀螨化合物（估计为 1 亿美元）[5,16,17]。尽管抗蜱疫苗接种通常被视为杀螨剂使用的替代方法，但也可被视为综合虫害治理方法框架内的一种补充方法。如果使用得当，抗蜱疫苗接种可减少杀螨剂的使用，并可能降低杀螨剂耐药性发展[5,17-20]。

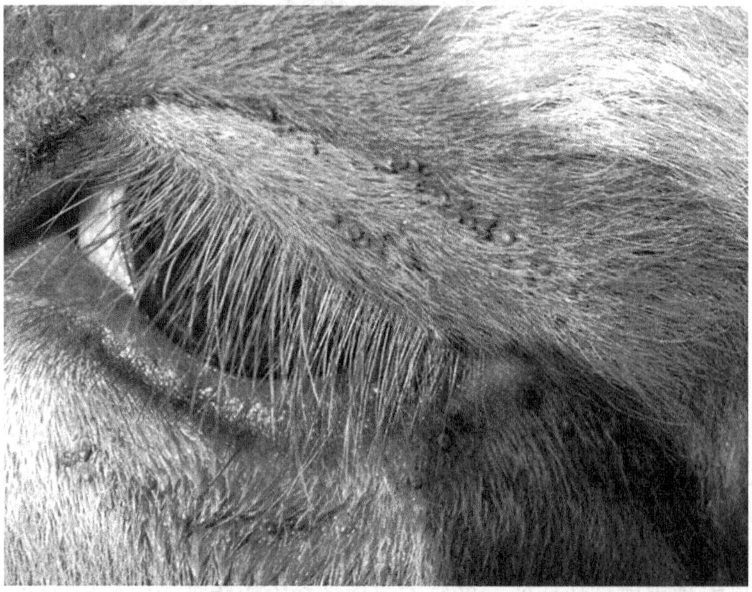

图 17.4　在牛眼睛周围吸食的未成熟扇头蜱（由 S. Mwaura 提供）

图 17.5　在肯利亚收集的杀螨剂抗药性扇头蜱

17.3　候选抗蜱疫苗

许多抗蜱疫苗的靶点已经用常规免疫筛选技术鉴定出来。用蜱匀浆或纯化的蜱提取物免疫脊椎动物宿主产生免疫血清，这些血清用于筛选宿主检测到的蜱抗原。

对蜱生存至关重要的蜱蛋白鉴定是更具针对性抗原发现的一种有用方法，随着蜱生物学信息的收集，这一方法越来越有可能实现。随着大量蜱类基因组序列的出现，通过反向疫苗学发现候选物的数量正在扩大。微小扇头蜱基因组数据是有价值的候选挖掘资源库[21]，希望其他重要蜱种的资源库也会随之出现。其他技术的使用（如 RNAi）在确认抗蜱候选疫苗的重要性方面非常有用，可能在未来发现抗蜱疫苗抗原方面发挥作用[22]。

候选抗蜱疫苗被分为两类：暴露或隐蔽抗原。暴露抗原在附着和摄食宿主期间分泌在蜱唾液中，而隐蔽抗原通常隐藏在宿主免疫反应中。虽然 Willadsen 提出抗原应该根据结构或功能进行分类[5]，但使用暴露和隐蔽区别抗原的主要优势在于对蜱潜在的免疫逃避机制有明显的重要影响。已经观察到宿主成分的蜱分子拟态，当使用暴露抗原时，接种疫苗可能诱发宿主敏感性和自身免疫应答[23]。

使用暴露抗原的一个优点是，通过蜱吸血可以自然地增强免疫反应。从机制上讲，暴露抗原的疫苗接种被认为会导致局部敌对环境，不支持蜱附着和取食。在自然蜱类吸血过程中，隐蔽抗原不会与宿主免疫反应接触。尽管某些唾液腺蛋白通常包含在蜱的胸腔内，但如果不分泌到蜱虫的吸血部位，就可以认为是隐蔽的。

隐蔽抗蜱疫苗研制的一个困难是抗原必须能够被诱导的体液疫苗应答所利用。这通常会限制候选抗原的数量，使它们与血粉长期直接接触，或体液反应可以通过肠道屏障进入血淋巴[24-26]。隐蔽抗原的第二个局限性与免疫反应的自然增强有关。由于抗原不与宿主内的免疫反应接触，必须通过反复接种诱导足够高的抗体水平。从积极的一面来看，这些抗原的自然选择压力较低，而耐药性的出现并不是主要关注的问题。

由于血粉作为效应免疫应答的媒介，与暴露的抗原相比，抗蜱作用可以在更长的时间内发生。这种影响甚至可能超出单纯的进食期，进入消化和蜕皮/产卵的非活跃阶段。

17.3.1 综述成功的候选抗蜱疫苗

基于Bm86的抗蜱疫苗仍然是唯一商业化生产的疫苗，已成为未来抗蜱疫苗开发和评估的基准。与肠道相关的Bm86糖蛋白首先在微小扇头蜱中鉴定出来，尽管其他蜱中的同源物已经被鉴定[27-33]。Bm86生物学功能仍然未知，但它被认为在血粉消化中起作用[34]。微小扇头蜱中Bm86的表达在胚胎形成过程中增加，在未吸血的幼虫中达到最高水平。在吸血和蜕皮过程中表达减少，在蜱的静止期检测到最低水平的表达[35]。Bm86有650个氨基酸的翻译编码序列，大小为71.7 kDa。Bm86蛋白含有4个潜在的N-连接糖基化位点和1个提示转运到细胞表面的前导肽[28]。定位研究表明，该分子主要位于肠上皮细胞的微绒毛上[29,36]。未经加工的蛋白质中存在单一的C-末端跨膜序列，该序列被成熟蛋白质中的糖基磷脂酰肌醇锚取代。Bm86蛋白还含有多个可预测的富含半胱氨酸残基的EGF（表皮生长因子）重复序列[28,34]。

疫苗接种主要是用整个分子进行的，Bm86保护表位尚未确定。已经确定了保护性B细胞表位位点，也可能存在额外的表位[37]。在Bm86和脱色扇头蜱同源物Bd86之间发现了重叠交叉免疫反应表位[36,38]。疫苗效力与抗Bm86抗体滴度直接相关，控制蜱数量的能力与获得强抗体应答直接相关[39-42]。

在产生抗Bm86抗体滴度的能力上观察到大量的动物间变异，这可能与表达的MHC Ⅱ类单倍型有关[42,43]。Bm86和牛补体系统抗体在血粉中被吸收。抗体结合导致肠上皮细胞溶解，最终导致血粉消化受损。强大的抗体反应可能导致蜱死亡，因为血液从肠道渗漏到血淋巴中，蜱可能会变成红色而不是灰色[34]。

Bm86的重组表达已在几种表达系统中进行尝试，包括大肠杆菌[34]、构巢曲霉黑曲霉[44]和毕赤酵母[45-48]。疫苗试验表明，Bm86疫苗主要针对微小扇头蜱成虫阶段，特别是成虫饱血和饱血后死亡率。雌性微小扇头蜱成蜱的繁殖能力受到产卵能力和卵孵化的影响[5,49]。

在田间条件下，给牛接种疫苗可使单代蜱数量减少56%，蜱繁殖能力下降72%[50]。在6个月的时间里，接种疫苗的动物体重平均增加18.6 kg，从而逆转了蜱虫吸血对活畜重量的负面影响[40,41]。在古巴、巴西、阿根廷和墨西哥进行的大量田间试验表明，在36周时间内微小扇头蜱控制率为55%~100%[42,51]。

重要的是，通过将Bm86疫苗接种与杀螨剂使用相结合，可以实现对杀螨剂耐药性蜱的完全控制[52]，表明综合控制系统在控制蜱数量方面是有效的。接种疫苗还减少了控制蜱虫所需的杀螨剂数量，延长了给牛消毒的时间间隔[5]。Bm86疫苗已被广泛评估其控制其他蜱类的能力。已经报道对具环牛蜱有几乎完全的交叉保护作用[53,54]；已观察到对小亚璃眼蜱、嗜驼璃眼蜱和脱色扇头蜱的显著保护作用；但是对具尾扇头蜱或彩饰花蜱没有交叉保护作用[38,55]。

Bm86氨基酸序列差异存在于不同的微小扇头蜱种群之间，这被认为是不同地理位置疫苗效力差异的原因之一。迄今为止，疫苗效力和Bm86序列同源性之间没有明确的相关性[5]。来自阿根廷的实验室微小扇头蜱株显示出对Bm86疫苗接种有抵抗力[47,51]，Bm86基因的核苷酸序列显示出显著差异，包括21个氨基酸被替换（成熟蛋白质中的610个氨基酸）[56]。这种蛋白质后来被重新命名为Bm95，尽管它是否应该单独分类仍有争议。

Bm95已在巴斯德毕赤酵母中成功生产[48]并已纳入多项疫苗试验。Bm95疫苗效果类似于Bm86疫苗，有蜱排斥、损伤和死亡，吸血重量、产卵和卵孵化也下降[57]。在牛上Bm95疫苗对印度镰性扇头蜱有高水平交叉保护性效力已得到证实[58]。利用一种新的生产技术，免疫原性Bm95肽和边缘无浆体MSP1 N-末端区域的融合对兔微小扇头蜱感染具有保护作用。蜱产卵和繁殖力降低与商业Bm86疫苗相当[59,60]。

17.3.2 Bm86衍生疫苗的商业化

澳大利亚联邦科学和工业研究组织（CSIRO）与澳大利亚生物技术公司（Biotech Australia

Pty）合作进行了名为TickGARD™的Bm86蜱疫苗商业化开发。1994年由赫斯特动物健康公司发布上市。在4年内，TickGARD™不断在商业上取得成功，直至成为澳大利亚价值最高的蜱治疗产品。不幸的是，TickGARD™生产和分销中涉及的商业合作伙伴的一系列变化导致生产中断。

Intervet澳大利亚有限公司暂时重新引入这种疫苗（本迪戈，澳大利亚），但目前没有出售疫苗[20]。基于Gavac Bm86的疫苗在拉丁美洲地区更加成功。它是由遗传工程和生物技术中心（哈瓦那，古巴）开发的，于1993年由南美洲的HeberBiotec供应商（哈瓦那，古巴）发行。该疫苗已在古巴广泛使用，可减少巴贝斯虫病和无浆体病，并显著减少所需的杀螨剂[20,42,61]。由于国家强制控制计划，古巴展示了抗蜱疫苗部署的最佳方案结果[20]。该疫苗分别于1994年、1995年和1997年在哥伦比亚、巴西和墨西哥注册，这些国家的报告显示达到了类似的控制水平[20]。

17.3.3 其他蜱类的Bm86同源物及其疫苗潜力

来自一系列其他蜱类的Bm86同源物已经显示出作为蜱虫控制工具的前景。牛Bm86疫苗接种诱导了与脱色扇头蜱肠道部分结合的交叉反应抗体，尽管其抗蜱作用尚未确定。合成的Bd86衍生肽诱导的小鼠单克隆抗体识别微小扇头蜱、脱色扇头蜱、小亚璃眼蜱和具尾扇头蜱的Bm86同源物[35]。长角血蜱Bm86同源物鉴定和RNAi沉默显示蜱吸血重量显著降低[62]。Bm86、Ba86的具环牛蜱同源物已在巴斯德毕赤酵母中成功表达，并在牛中进行了鉴定。

疫苗接种显著影响蜱的滋生、产卵和卵繁殖能力。此外，Ba86和具尾扇头蜱之间发生交叉保护反应。有趣的是，Ba86和Bm86对具环牛蜱的疗效高于微小扇头蜱，这表明Bm86样效应可能是蜱种依赖性的，而不是序列同源性依赖性的[60]。用小亚璃眼蜱同源物Haa86对牛进行接种，导致蜱排斥显著增加，蜱吸血减少和卵重量下降[32]。Bm86同源疫苗对蜱媒疾病传播的影响很少有人研究。在一项研究中，Haa86疫苗接种减少了环形泰勒虫向易感小牛传播[63]。具尾扇头蜱Bm86同源物Ra86表达模式明显不同于Bm86表达[35]。

在一个实验室的具尾扇头蜱种群中，Ra86同源物作为两个高度分化的等位基因变异体存在，Ra86和Bm86之间的核苷酸序列相似性为73%，两个变异体Ra85A和Ra92A核苷酸序列相似性为74%[33]。用杆状病毒表达的Ra86变异体接种兔子显著影响具尾扇头蜱的饱血重量和卵重量[64]。使用相同的杆状病毒构建体表达Ra86同源物，这些效应都不能在牛宿主中复制。然而在牛接种Ra86疫苗后，对蜱若虫蜕皮至成虫期有一个先前未报告的显著影响[65]。

17.3.4 其他候选抗蜱疫苗

许多其他候选抗蜱疫苗已经显示出控制同源和异源蜱物种的前景，尽管没有一种已经超出最初的概念验证研究。表17.2总结了这些抗原。

表17.2 评估控制家畜蜱的候选抗蜱疫苗

抗原	位置和功能	免疫效果	参考文献
BMA7	广泛分布于蜱组织中的63 kDa膜结合糖蛋白	用微小扇头蜱BMA7接种显示出对蜱感染的显著保护作用	[95]
	生物学功能未知	效果不如Bm86介导的疫苗接种	
SBm7462	源自Bm86的合成肽SBm7462	牛接种疫苗效果达到80%以上	[37]
Bm91	膜结合羧二肽酶 在微小扇头蜱唾液腺中鉴定	增加Bm86疫苗接种效果	[96，97]

（续表）

抗原	位置和功能	免疫效果	参考文献
丝氨酸蛋白酶抑制剂	丝氨酸蛋白酶抑制剂（serpins） 已经成为许多有前途的抗蜱疫苗研究焦点	用重组 HLS1 接种兔子分别导致若蜱和成蜱死亡 44%和 11% 用长角血蜱 serpin-2 接种兔子，若虫和成虫的死亡率分别为 45%和 43% 4 种丝氨酸蛋白酶抑制剂已在具尾扇头蜱中鉴定出来，命名为 RAS1-4 用重组 RAS1 和 RAS2 对牛进行接种后获得了保护性免疫，若蜱饱血重量降低了 61%，雌性成蜱和雄性成蜱死亡率分别降低了 28%和 43%	[98-101]
肌钙蛋白 I 样蛋白	从长角血蜱中分离并在兔和小鼠中诱导特异性抗体应答。位于肌肉、表皮、肠道和唾液腺	小鼠和兔子疫苗接种导致幼蜱和成年蜱吸血时间显著延长，幼蜱吸血率低，卵重减少	[102-104]
卵黄磷蛋白	微小扇头蜱卵中最丰富的卵黄蛋白。 针对卵黄蛋白升高的抗体也能识别成蜱血淋巴中的 200 kDa 多肽	以其天然形式诱导免疫应答保护绵羊防止微小扇头蜱侵染	[105]
Voraxin	充血因子	对兔进行抗希伯来花蜱 voraxin 免疫后，充血减少了 74%，而以对照兔为食的所有蜱都完全充血	[106]
谷胱甘肽硫转移酶	谷胱甘肽硫转移酶（GST）是一个参与异源和内源性化合物解毒的酶家族，已在许多蜱种中鉴定	重组长角血蜱 GST-HL 作为抗蜱候选疫苗已经在牛中进行了交叉保护抗微小扇头蜱评估，显示抗微小牛蜱的总体疫苗效力为 57%	[107-111]
RIM36	分离自具尾扇头蜱，存在于唾液腺（III 型唾液腺腺泡）和胶结锥中	在牛的自然感染疫苗接种下，免疫显性没有诱导明显的保护作用，尽管诱导了强抗体应答	[112-114]
HL34	蜱唾液免疫兔血清鉴定长角血蜱唾液蛋白，免疫筛选成蜱 cDNA 文库	兔子接种疫苗导致长角血蜱若虫期和成虫期发病率和死亡率增加，尽管效果没有达到统计学显著性	[115]
p29	来自长角血蜱的 29 kDa 细胞外基质样蛋白被认为是胶结锥的一个组成部分	兔子用重组 p29 免疫后导致长角血蜱幼蜱和若蜱的死亡率分别为 40%和 50%。观察到到成年雌性蜱的充血重量减少了 17%	[116]
组胺结合蛋白	在具尾扇头蜱中鉴定，并在瘙痒反应中起作用	多价疫苗接种无抗蜱作用。轻度阻断传播效应	[117]
雄性免疫球蛋白结合蛋白	在具尾扇头蜱中鉴定	豚鼠接种疫苗导致轻微的雌性吸血障碍	[24]
β-*N*-乙酰己糖胺酶	在微小扇头蜱中鉴定	将兔多克隆抗体直接注射到饱血雌性蜱体内，产卵量减少了 26%。不作为重组疫苗进行评估	[118，122]
5′-核苷酸酶	在微小扇头蜱中鉴定	非活性截短形式的蛋白质只在绵羊中诱导高抗体滴度，产生了显著效果。对牛没有效果	[119，120]
钙网蛋白	在美洲花蜱和微小扇头蜱中鉴定	牛免疫原性差，无效果	[121]

（续表）

抗原	位置和功能	免疫效果	参考文献
蛋黄前组织蛋白酶	在微小扇头蜱中鉴定	牛有抗原性，接种疫苗保护率为 25%	[123-126]

17.3.5 双重功能抗蜱疫苗

双重功能候选抗蜱疫苗代表了一个新概念，即暴露抗原与隐蔽抗原共享抗体结合表位。这组抗原的优势在于通过暴露的表位发生自然增强，而目标是隐蔽抗原。在这些双重作用蛋白中研究得最好的是 64P 蛋白，也称为 TRP64 和 64TP。具尾扇头蜱中鉴定出的蛋白质为 15 kDa，似乎是胶结锥的一部分[66]。交叉反应表位存在于雌性成蜱的唾液腺、血淋巴、中肠，以及具尾扇头蜱的幼虫和若虫的全身提取物中[66]。这种抗原的一个显著特征是截短型疫苗会影响蜱的不同阶段[66]。疫苗接种介导的影响表现为饱血重量和卵重量减少以及蜱的直接死亡率[66]。用 TRP64 疫苗接种诱导的炎症和适应性免疫反应会破坏蜱摄食。在发生吸血的地方，被摄取的抗体与交叉反应性中肠表位结合，导致蜱中肠破裂和饱血蜱死亡[66]。观察到对蓖子硬蜱、血红扇头蜱、彩饰花蜱、微小扇头蜱有广泛交叉保护作用。这种交叉反应可能基于不同蜱种中强表位的保护，以维持宿主胶原蛋白和角蛋白的分子模拟[66-68]，而抗体滴度的自然增强明显发生在蜱虫感染后[67]。

虽然已经在小鼠和豚鼠中证明了抗蜱作用[66,67]，但使用杆状病毒制备的抗原在家兔身上没有显示出保护效果[64]。在牛中使用细菌表达的 TRP64 作为多价疫苗的一部分，不能建立蜱效应。在蓖子硬蜱阻断了蜱传性脑炎病毒对易感小鼠的传播后，证实了 TRP64 作为候选抗蜱疫苗的潜力。64TRP 疫苗保护作用相当于单一剂量的商业蜱传脑炎疫苗[69]。

在摄取血粉的过程中，蜱肠道中铁离子的含量急剧增加，蜱铁蛋白作为铁储存蛋白。Fer2（铁蛋白 2）是一种分泌到蜱血淋巴中的肠道特异性蛋白质，在血淋巴中充当铁转运蛋白。在蜱所有发育阶段都有表达，Fer2RNAi 沉默对蜱摄食、产卵和幼虫孵化都有不利影响[70]。

用蓖子硬蜱和微小扇头蜱的 Fer2 重组同源物接种牛，显示出对微小扇头蜱同源物（如蓖子扇头蜱和微小扇头蜱及具环牛蜱）的良好控制[71]。值得注意的是，重组 Fer2 疫苗效力与针对微小扇头蜱的 Bm86 疫苗接种获得的基准相当[71]。

一种受到广泛关注的抗原是 subolesin 蛋白，最初被命名为 4D8，该抗原最初是在肩突硬蜱中发现的[72]，但在其他几个蜱种中也有其表征[73-75]。使用 RNAi 靶向降低 subolesin 蛋白的 mRNA 导致蜱肠、唾液腺、生殖组织和胚胎退化[74-78]。用重组 subolesin 蛋白对牛进行免疫，导致微小扇头蜱存活和繁殖能力下降[79,81]。此外发现，subolesin 蛋白可控制蜱基因表达，影响蜱对病原体的先天免疫应答，从而减少蜱媒疾病如边缘无浆体、嗜吞噬细胞无浆体和双芽巴贝斯虫感染[77,80-86]。最近 subolesin 免疫原性肽被融合到边缘无浆体 MSP1a 的 N 末端区域，命名为 SUB-MSP1a[30,59,77,87]。通过对蜱繁殖和卵繁殖力的负面影响，显示出牛疫苗接种对微小扇头蜱和具环牛蜱的高效控制[87]。

17.4 阻断传播的抗蜱疫苗

一种能够减少或阻断病原体从蜱媒向脊椎动物宿主传播的抗蜱疫苗将有助于蜱传疾病（TBD）控制。阻断传播的疫苗是通过减少已接种牛群的临床疾病发病率来实现地方病稳定发展的理想选择。众所周知，唾液有助于许多蜱传疾病的传播。唾液激活传播（SAT）被称为“通过脊椎动物宿主上具有生物活性的蜱唾液分子作用间接促进蜱传病原体传播”[88]。肩突硬蜱唾液腺蛋白 P11 促进嗜吞噬无浆体从蜱肠向唾液腺的迁移[89]。在具尾扇头蜱唾液腺提取物存在情况下，小泰勒虫在体外对淋巴细胞感染率增加[90]。以这些SAT组件为靶标似乎是理想的阻断传播选择；然而需要更多的研究来确定合适的非冗余候选蛋白。

抗蜱疫苗发展前景

正在进行的几个蜱种基因组项目将提供大量潜在新候选基因，作为抗蜱疫苗进行测试[17,21]。使用RNAi实验有助于在功能上优先考虑候选基因，这些候选基因将通过对蜱基因操纵而进一步研究[22]。理想情况下，未来抗蜱疫苗应该同时针对一系列蜱种，因为几个蜱种可能共同以牲畜为食。一种广泛作用的抗蜱疫苗可以通过鉴定几种蜱种中具有共同的、保守的和基本功能的抗原，或者将多种抗原合并到混合物中来研制。

在识别出新候选蛋白后，在合适的动物模型中进行评估就显得尤为重要。世界各地许多实验室都可以使用小型动物模型进行候选蛋白测试。然而处理家畜实验的大型动物设施既昂贵又稀有。此外，对于许多蜱和蜱传疾病（TBD）还没有开发出合适的体内攻毒模型。然而已有多次研究表明，通过反复的蜱感染，由蜱免疫动物产生的血清可以检测到不同的蜱抗原谱，这取决于所用的动物种类。蜱类成分免疫优势的这种差异很可能是长期宿主-病原体共同进化的结果[23]。

一个明显的例子是，血红扇头蜱以豚鼠等非自然宿主为食。即使少量的蜱叮咬（表现为蜱死亡率高、雌性蜱体重和生殖能力降低）也会产生对蜱感染的抵抗力[91]。相反当血红扇头蜱以狗（它们的天然宿主）为食时，反应显著降低，主要限于皮肤的即时炎症反应—迟发型超敏反应降低[91]。蜱唾液中存在的分子被认为抑制宿主免疫应答，在非宿主物种中反复暴露蜱会导致自然宿主系统中不存在的免疫应答[23]。当测试相同的候选疫苗在不同宿主-病原体相互作用中的效力时，经常观察到系统矛盾的结果。因此，对新的候选疫苗评价应该在相关自然宿主的早期发育阶段进行。

新型抗蜱疫苗的商业化还应考虑从Bm86研发中吸取经验教训[20]。尽管该疫苗可以在市场上买到，但不应视为商业上的成功。基于Bm86仅有两种商业产品，Gavac和TickGARD要么仅限于在拉丁美洲使用，要么分别从市场上撤下。这对于一种以全球分布的、具有高度经济重要性的蜱种为目标的产品来说是惊人的。有关抗蜱疫苗性能的期望应该向消费者详细解释。由于从全球根除不同蜱种是不现实的，抗蜱疫苗很可能在综合害虫控制策略中发挥作用。这些策略已经证明可以减少蜱种负荷和蜱传疾病传播，并有可能满足农民和政府对疾病控制的要求[20,92,93]。尽早将抗蜱疫苗测试与杀螨剂治疗计划结合起来，可能会为新产品的有效商业化以及更好的农民产品信息铺平道路。

综上所述，对蜱和蜱传疾病（TBD）控制的新方法需求越来越大，并且农民似乎已经接受使用抗蜱疫苗[94]。大多数学术机构都无法开发新型商业化抗蜱疫苗产品。新型公私伙伴关系可以大大加快在全球建立可持续、廉价和环境友好的蜱和蜱传疾病（TBD）控制战略[17]。

参考文献

[1] McCosker, P. J.: Global aspects of the management and control of ticks of veterinary importance. In: Rodriguez, J. (ed.) Recent Advances in Acarology. Academic, New York (1979).

[2] de Castro, J.: Sustainable tick and tick-borne disease control in livestock improvement in developing countries. Vet. Parasitol. 71, 77-97 (1997).

[3] Minjauw, B., McLeod, A.: Tick-borne diseases and poverty. The impact of ticks and tick-borne diseases on the livelihood of small scale and marginal livestock owners in India and eastern and southern Africa. Research report, DFID Animal Health Programme, Centre for Tropical Veterinary Medicine, University of Edinburgh (2003).

[4] Jongejan, F., Uilenberg, G.: The global importance of ticks. Parasitology129 (Suppl), S3-S14 (2004).

[5] Willadsen, P.: Anti-tick vaccines. Parasitology129 (Suppl), S367-S387 (2004).

[6] Norval, R. A. I., Perry, B. D., Young, A. S.: The Epidemiology of Theileriosis in Africa. Galliard (Printers) Ltd, Great Yarmouth (1992).

[7] Pérez de León, A. A., Teel, P. D., Auclair, A. N., Messenger, M. T., Guerrero, F. D., Schuster, G., Miller, R. J.: Integrated strategy for sustainable cattle fever tick eradication in USA is required to mitigate the impact of global change. Front Physiol. 3, 195 (2012).

[8] Kocan, K. M.: Targeting ticks for control of selected hemoparasitic diseases of cattle. Vet. Parasitol. 57, 121-151 (1995).

[9] George, J. E., Pound, J. M., Davey, R. B.: Chemical control of ticks on cattle and the resistance of these parasites to acaricides. Parasitology129, S353-S366 (2004).

[10] Nari, A., Hansen, J. W.: Resistance of Ecto-and Entoparasites: Current and Future Solutions. Office International des E'pizooties Technical Report 67 SG/10. OIE, Paris (1999).

[11] Foil, L. D., Coleman, P., Eisler, M., Fragoso-Sanchez, H., Garcia-Vazquez, Z., Guerrero, F. D., Jonsson, N. N., Langstaff, I. G., Li, A. Y., et al.: Factors that influence the prevalence of acaricide resistance and tick-borne diseases. Vet. Parasitol. 125, 163-181 (2004).

[12] Mbassa, G. K., Mgongo, F. O. K., Melau, L. S. B., Mlangwa, J. E. D., Silayo, R. S., Bimbita, E. N., Hayghaimo, A. A., Mbiha, E. R.: A financing system for the control of tick-borne diseases in pastoral herds: the Kambala (Tanzania) model. Livest. Res. Rural Dev. 21 (2012).

[13] Pruett, J. H.: Immunological control of arthropod ectoparasites-a review. Int. J. Parasitol. 29, 25-32 (1999).

[14] Willadsen, P., Jongejan, F.: Immunology of the tick-host interaction and the control of ticks and tick-borne diseases. Parasitol. Today 15 (7), 258-262 (1999).

[15] Willadsen, P.: The molecular revolution in the development of vaccines against ectoparasites. Vet. Parasitol. 101, 353-368 (2001).

[16] Graf, J. F., Gogolewski, R., Leach-Bing, N., Sabatini, G. A., Molento, M. B., Bordin, E. L., Arantes, G. J.: Tick control: an industry point of view. Parasitology 129 (Suppl), S427-S442 (2004).

[17] Guerrero, F. D., Miller, R. J., Pérez de León, A. A.: Cattle tick vaccines: many candidate antigens, but will a commercially viable product emerge? Int. J. Parasitol. 42, 421-427 (2012).

[18] Nuttall, P. A., Trimnell, A. R., Kazimirova, M., Labuda, M.: Exposed and concealed antigens as vaccine targets for controlling ticks and tick-borne diseases. Parasite Immunol. 28, 155-163 (2006).

[19] Willadsen, P.: Tick control: thoughts on a research agenda. Vet. Parasitol. 138, 161-168 (2006).

[20] de la Fuente, J., Almazán, C., Canales, M., Pérez de la Lastra, J. M., Kocan, K. M., Willadsen, P.: A ten-y ear review of commercial vaccine performance for control of tick infestations on cattle. Anim. Health Res. Rev. 8, 23-28 (2007).

[21] Bellgard, M. I., Moolhuijzen, P. M., Guerrero, F. D., Schibeci, D., Rodriguez-Valle, M., Peterson, D. G., Dowd, S. E., Barrero, R., Hunter, A., et al.: CattleTickBase: an integrated internet-based bioinformatics resource for *Rhipicephalus* (*Boophilus*) *microplus*. Int. J. Parasitol. 42, 161-169 (2012).

[22] de la Fuente, J., Kocan, K. M., Almazán, C., Blouin, E. F.: RNA interference for the study and genetic manipulation of ticks. Trends Parasitol. 23, 427-433 (2007).

[23] Steen, N. A., Barker, S. C., Alewood, P. F.: Proteins in the saliva of the Ixodida (ticks): pharmacological features and biological significance. Toxicon47, 1-20 (2006).

[24] Wang, H., Nuttall, P. A.: Comparison of the proteins in salivary glands, saliva and haemolymph of *Rhipicephalus appendiculatus* female ticks during feeding. Parasitology 109 (Pt 4), 517-523 (1994).

[25] Wang, H., Nuttall, P. A.: Immunoglobulin-binding proteins in ticks: new target for vaccine development against a blood-feeding parasite. Cell. Mol. Life Sci. 56, 286-295 (1999).

[26] Jasinskas, A., Jaworski, D. C., Barbour, A. G.: *Amblyomma americanum*: specific uptake of immuno-

globulins into tick hemolymph during feeding. Exp. Parasitol. 96, 213-221 (2000).

[27] Willadsen, P., Riding, G. A., McKenna, R. V., Kemp, D. H., Tellam, R. L., Nielsen, J. N., Lahnstein, J., Cobon, G. S., Gough, J. M.: Immunologic control of a parasitic arthropod. Identification of a protective antigen from *Boophilus microplus*. J. Immunol. 143, 1346-1351 (1989).

[28] Richardson, M. A., Smith, D. R., Kemp, D. H., Tellam, R. L.: Native and baculovirus-expressed forms of the immuno-protective protein BM86 from *Boophilus microplus* are anchored to the cell membrane by a glycosyl-phosphatidyl inositol linkage. Insect Mol. Biol. 1, 139-147 (1993).

[29] Gough, J. M., Kemp, D. H.: Localization of a low abundance membrane protein (Bm86) on the gut cells of the cattle tick *Boophilus microplus* by immunogold labeling. J. Parasitol. 79, 900-907 (1993).

[30] Canales, M., Labruna, M. B., Soares, J. F., Prudencio, C. R., de la Fuente, J.: Protective efficacy of bacterial membranes containing surface-exposed BM95 antigenic peptides for the control of cattle tick infestations. Vaccine 27, 7244-7248 (2009).

[31] Peconick, A. P., Sossai, S., Girão, F. A., Rodrigues, M. Q., Souza E Silva, C. H., Guzman, Q. F., Patarroyo, V. A. M., Vargas, M. I., Patarroyo, J. H.: Synthetic vaccine (SBm7462) against the cattle tick *Rhipicephalus* (*Boophilus*) *microplus*: preservation of immunogenic determinants in different strains from South America. Exp. Parasitol. 119, 37-43 (2008).

[32] Azhahianambi, P., De La Fuente, J., Suryanarayana, V. V., Ghosh, S.: Cloning, expression and immuno-protective efficacy of rHaa86, the homologue of the Bm86 tick vaccine antigen, from *Hyalomma anatolicumanatolicum*. Parasite Immunol. 31, 111-122 (2009).

[33] Kamau, L., Skilton, R. A., Odongo, D. O., Mwaura, S., Githaka, N., Kanduma, E., Obura, M., Kabiru, E., Orago, A., et al.: Differential transcription of two highly divergent gut-expressed Bm86 antigen gene homologues in the tick *Rhipicephalus appendiculatus* (Acari: Ixodida). Insect Mol. Biol. 20, 105-114 (2011).

[34] Rand, K. N., Moore, T., Sriskantha, A., Spring, K., Tellam, R., Willadsen, P., Cobon, G. S.: Cloning and expression of a protective antigen from the cattle tick *Boophilusmicroplus*. Proc. Natl. Acad. Sci. U. S. A. 86, 9657-9661 (1989).

[35] Nijhof, A. M., Balk, J. A., Postigo, M., Jongejan, F.: Selection of reference genes for quantitative RT-PCR studies in Rhipicephalus (Boophilus) microplus and Rhipicephalus appendiculatus ticks and determination of the expression profile of Bm86. BMC Mol. Biol. 10, 112 (2009).

[36] Kopp, N., Diaz, D., Amacker, M., Odongo, D. O., Beier, K., Nitsch, C., Bishop, R. P., Daubenberger, C. A.: Identifi-cation of a synthetic peptide inducing cross-reactive antibodies binding to *Rhipicephalus* (*Boophilus*) *decoloratus*, *Rhipicephalus* (*Boophilus*) *microplus*, *Hyalomma anatolicum anatolicum* and *Rhipicephalus appendiculatus* BM86 homologues. Vaccine 28, 261-269 (2009).

[37] Patarroyo, J. H., Portela, R. W., De Castro, R. O., Pimentel, J. C., Guzman, F., Patarroyo, M. E., Vargas, M. I., Prates, A. A., Mendes, M. A.: Immunization of cattle with synthetic peptides derived from the *Boophilus microplus* gut protein (Bm86). Vet. Immunol. Immunopathol. 88, 163-172 (2002).

[38] Odongo, D., Kamau, L., Skilton, R., Mwaura, S., Nitsch, C., Musoke, A., Taracha, E., Daubenberger, C., Bishop, R.: Vaccination of cattle with TickGARD induces cross-reactive antibodies binding to conserved linear peptides of Bm86 homologues in *Boophilus decoloratus*. Vaccine 25, 1287-1296 (2007).

[39] Willadsen, P., Bird, P., Cobon, G. S., Hungerford, J.: Commercialisation of a recombinant vaccine against *Boophilus microplus*. Parasitology 110 (Suppl), S43-S50 (1995).

[40] Rodríguez, M., Massard, C. L., da Fonseca, A. H., Ramos, N. F., Machado, H., Labarta, V., de la Fuente, J.: Effect of vaccination with a recombinant Bm86 antigen preparation on natural infestations of *Boophilus microplus* in grazing dairy and beef pure and cross-bred cattle in Brazil. Vaccine 13, 1804-1808 (1995).

[41] Rodríguez, M., Penichet, M. L., Mouris, A. E., Labarta, V., Luaces, L. L., Rubiera, R., Cordovés, C., Sánchez, P. A., Ramos, E., Soto, A.: Control of *Boophilus microplus* populations in grazing cattle vac-

cinated with a recombinant Bm86 antigen preparation. Vet. Parasitol. 57, 339-349 (1995).

[42] de la Fuente, J., Rodríguez, M., Redondo, M., Montero, C., García-García, J. C., Méndez, L., Serrano, E., Valdés, M., Enriquez, A., et al.: Field studies and cost - effectiveness analysis of vaccination with Gavac against the cattle tick *Boophilus microplus*. Vaccine 16, 366-373 (1998).

[43] Sitte, K., Brinkworth, R., East, I. J., Jazwinska, E. C.: A single amino acid deletion in the antigen binding site of BoLA-DRB3 is predicted to affect peptide binding. Vet. Immunol. Immunopathol. 85, 129-135 (2002).

[44] Turnbull, I. F., Smith, D. R., Sharp, P. J., Cobon, G. S., Hynes, M. J.: Expression and secretion in *Aspergillus nidulans* and *Aspergillus niger* of a cell surface glycoprotein from the cattle tick, *Boophilus microplus*, by using the fungal amdS promoter system. Appl. Environ. Microbiol. 56, 2847-2852 (1990).

[45] Rodriguez, M., Rubiera, R., Penichet, M., Montesinos, R., Cremata, J., Falcon, V., Sanchez, G., Bringas, R., Cordoves, C., Valdes, M.: High level expression of the *B. microplus* bm86 antigen in the yeast *Pichia pastoris* forming highly immunogenic particles for cattle. J. Biotechnol. 33, 135 - 146 (1994).

[46] Canales, M., Enríquez, A., Ramos, E., Cabrera, D., Dandie, H., Soto, A., Falcón, V., Rodríguez, M., de la Fuente, J.: Large-scale production in *Pichia pastoris* of the recombinant vaccine Gavac against cattle tick. Vaccine 15, 414-422 (1997).

[47] Garcia-Garcia, C. G., Montero, C. M., Rodriguez, M. A., Soto, A. S., Redondo, M. R., Valdes, M. V., Mendez, L. M., de la Fuente, J.: Effect of particulation on the immunogenic and protective properties of the recombinant Bm86 antigen expressed in *Pichia pastors*. Vaccine 16, 374-380 (1998).

[48] Boué, O., Farnós, O., González, A., Fernández, R., Acosta, J. A., Valdés, R., González, L. J., Guanche, Y., Izquierdo, G., et al.: Production and biochemical characterization of the recombinant *Boophilus microplus* Bm95 antigen from *Pichia pastoris*. Exp. Appl. Acarol. 32, 119-128 (2004).

[49] de la Fuente, J., Kocan, K. M., Blouin, E. F.: Tick vaccines and the transmission of tick-borne pathogens. Vet. Res. Commun. 31 (Suppl 1), 85-90 (2007).

[50] Jonsson, N. N., Mayer, D. G., Green, P. E.: Possible risk factors on Queensland dairy farms for acaricide resistance in cattle tick (*Boophilus microplus*). Vet. Parasitol. 88, 79-92 (2000).

[51] de la Fuente, J., Rodríguez, M., Montero, C., Redondo, M., García-García, J. C., Méndez, L., Serrano, E., Valdés, M., Enríquez, A., et al.: Vaccination against ticks (Boophilus spp.): the experience with the Bm86based vaccine Gavac. Genet. Anal. 15, 143-148 (1999).

[52] Redondo, M., Fragoso, H., Ortíz, M., Montero, C., Lona, J., Medellín, J. A., Fría, R., Hernández, V., Franco, R., et al.: Integrated control of acaricide-resistant *Boophilus microplus* populations on grazing cattle in Mexico using vaccination with Gavac and amidine treatments. Exp. Appl. Acarol. 23, 841-849 (1999).

[53] Fragoso, H., Hoshman Rad, P., Ortiz, M., Rodriguez, M., Redondo, M., Herrera, L., de la Fuente, J.: Protection against *Boophilus annulatus* infestations in cattle vaccinated with the *B. microplus* Bm86-containing vaccine Gavac. Vaccine 16 (20), 1990-1992 (1998).

[54] Pipano, E., Alekceev, E., Galker, F., Fish, L., Samish, M., Shkap, V.: Immunity against *Boophilus annulatus* induced by the Bm86 (Tick-GARD) vaccine. Exp. Appl. Acarol. 29, 141-149 (2003).

[55] de Vos, S., Zeinstra, L., Taoufi k, O., Willadsen, P., Jongejan, F.: Evidence for the utility of the Bm86 antigen from *Boophilus microplus* in vaccination against other tick species. Exp. Appl. Acarol. 25, 245-261 (2001).

[56] García-García, J. C., Gonzalez, I. L., González, D. M., Valdés, M., Méndez, L., Lamberti, J., D'Agostino, B., Citroni, D., Fragoso, H., et al.: Sequence variations in the *Boophilus microplus* Bm86 locus and implications for immuno-protection in cattle vaccinated with this antigen. Exp. Appl. Acarol. 23, 883-895 (1999).

[57] Kumar, A., Garg, R., Yadav, C. L., Vatsya, S., Kumar, R. R., Sugumar, P., Chandran, D., Mangamoorib, L. N., Bedarkar, S. N.: Immune responses against recombinant tick antigen, Bm95, for

the control of *Rhipicephalus* (*Boophilus*) *microplus* ticks in cattle. Vet. Parasitol. 165, 119-124 (2009).

[58] Sugumar, P., Chandran, D., Sudha Rani, G., Shahana, P. V., Maske, D. K., Rangarajan, P. N., Mangamoori, L. N., Srinivasan, V. A.: Recombinant mid gut antigen (Bm95) as a vaccine against Indian *Rhipicephalus haemaphysaloides* in *Bosindicus* cattle. Res. Vet. Sci. 90, 262-268 (2011).

[59] Canales, M., Almazán, C., Pérez de la Lastra, J. M., de la Fuente, J.: *Anaplas mamarginale* major surface protein 1a directs cell surface display of tick BM95 immunogenic peptides on *Escherichia coli*. J. Biotechnol. 135, 326-332 (2008).

[60] Canales, M., Almazán, C., Naranjo, V., Jongejan, F., de la Fuente, J.: Vaccination with recombinant *Boophilus annulatus* Bm86 ortholog protein, Ba86, protects cattle against *B. annulatus* and *B. microplus* infestations. BMC Biotechnol. 9, 29 (2009).

[61] Rodriguez, V. M., Méndez, L., Valdez, M., Redondo, M., Espinosa, C. M., Vargas, M., Cruz, R. L., Barrios, H. P., Seoane, G., Ramirez, E. S., Boue, O., Vigil, J. L., Machado, H., Nordelo, C. B., Pinēiro, M. J.: Integrated control of *Boophilus microplus* ticks in Cuba based on vaccination with the anti-tick vaccine Gavac. Exp. Appl. Acarol. 34, 375-382 (2004).

[62] Liao, M., Zhou, J., Hatta, T., Umemiya, R., Miyoshi, T., Tsuji, N., Xuan, X., Fujisaki, K.: Molecular characterization of *Rhipicephalus* (*Boophilus*) *microplus* Bm86 homologue from *Haemaphysalis longicornis* ticks. Vet. Parasitol. 146, 148-157 (2007).

[63] Jeyabal, L., Azhahianambi, P., Susitha, K., Ray, D. D., Chaudhuri, P., Vanlahmuaka, Ghosh, S.: Efficacy of rHaa86, an orthologue of Bm86, against challenge infestations of *Hyalomma anatolicum anatolicum*. Transbound. Emerg. Dis. 57, 96-102 (2010).

[64] Saimo, M., Odongo, D. O., Mwaura, S., Vlak, J. M., Musoke, A. J., Lubega, G. W., Bishop, R. P., van Oers, M. M.: Recombinant *Rhipicephalus appendiculatus* gut (ra86) and salivary gland cement (Trp64) proteins as candidate antigens for inclusion in tick vaccines: protective effects of Ra86 on infestation with adult *R. appendiculatus*. Vaccine: Dev. Therapy 1, 15-23 (2011).

[65] Olds, C., Mwaura, S., Crowder, D., Odongo, D., van Oers, M., Owen, J., Bishop, R., Daubenberger, C.: Immunization of cattle with Ra86 impedes *Rhipicephalus appendiculatus* nymphal-to-adult molting. Ticks Tick-borne Dis. 3, 170-178 (2012).

[66] Trimnell, A. R., Hails, R. S., Nuttall, P. A.: Dual action ectoparasite vaccine targeting 'exposed' and 'concealed' antigens. Vaccine 20, 3560-3568 (2002).

[67] Trimnell, A. R., Davies, G. M., Lissina, O., Hails, R. S., Nuttall, P. A.: A cross-reactive tick cement antigen is a candidate broad-spectrum tick vaccine. Vaccine 23, 4329-4341 (2005).

[68] Havlíková, S., Roller, L., Koci, J., Trimnell, A. R., Kazimírová, M., Klempa, B., Nuttall, P. A.: Functional role of 64P, the candidate transmission-b locking vaccine antigen from the tick. *Rhipicephalus appendiculatus*. Int. J. Parasitol. 39, 1485-1494 (2009).

[69] Labuda, M., Trimnell, A. R., Licková, M., Kazimírová, M., Davies, G. M., Lissina, O., Hails, R. S., Nuttall, P. A.: An antivector vaccine protects against a lethal vector-borne pathogen. PLoSPathog. 2, e27 (2006).

[70] Hajdusek, O., Sojka, D., Kopacek, P., Buresova, V., Franta, Z., Sauman, I., Winzerling, J., Grubhoffer, L.: Knockdown of proteins involved in iron metabolism limits tick reproduction and development. Proc. Natl. Acad. Sci. U. S. A. 106, 1033-1038 (2009).

[71] Hajdusek, O., Almazán, C., Loosova, G., Villar, M., Canales, M., Grubhoffer, L., Kopacek, P., de la Fuente, J.: Characterization of ferritin 2 for the control of tick infestations. Vaccine 28, 2993-2998 (2010).

[72] Almazán, C., Kocan, K. M., Bergman, D. K., Garcia-Garcia, J. C., Blouin, E. F., de la Fuente, J.: Identification of protective antigens for the control of *Ixodes scapularis* infestations using cDNA expression library immunization. Vaccine 21, 1492-1501 (2003).

[73] de la Fuente, J., Almazán, C., Blas-Machado, U., Naranjo, V., Mangold, A. J., Blouin, E. F.,

Gortazar, C., Kocan, K. M.: The tick protective antigen, 4D8, is a conserved protein involved in modulation of tick blood ingestion and reproduction. Vaccine 24, 4082-4095 (2006).

[74] Nijhof, A. M., Taoufi k, A., de la Fuente, J., Kocan, K. M., de Vries, E., Jongejan, F.: Gene silencing of the tick protective antigens, Bm86, Bm91 and subolesin, in the one-host tick *Boophilus microplus* by RNA interference. Int. J. Parasitol. 37, 653-662 (2007).

[75] Almazán, C., Kocan, K. M., Blouin, E. F., de la Fuente, J.: Vaccination with recombinant tick antigens for the control of *Ixodes scapularis* adult infestations. Vaccine 23, 5294-5298 (2005).

[76] de la Fuente, J., Almazán, C., Naranjo, V., Blouin, E. F., Kocan, K. M.: Synergistic effect of silencing the expression of tick protective antigens 4D8 and Rs86 in *Rhipicephalus sanguineus* by RNA interference. Parasitol. Res. 99, 108-113 (2006).

[77] Kocan, K. M., de la Fuente, J., Blouin, E. F.: Targeting the tick/pathogen interface for developing new anaplasmosis vaccine strategies. Vet. Res. Commun. 31 (Suppl 1), 91-96 (2007).

[78] de la Fuente, J., Manzano-Roman, R., Naranjo, V., Kocan, K. M., Zivkovic, Z., Blouin, E. F., Canales, M., Almazán, C., Galindo, R. C., et al.: Identification of protective antigens by RNA interference for control of the lone star tick, *Amblyomma americanum*. Vaccine 28, 1786-1795 (2010).

[79] Almazán, C., Lagunes, R., Villar, M., Canales, M., Rosario - Cruz, R., Jongejan, F., de la Fuente, J.: Identification and characterization of *Rhipicephalus* (*Boophilus*) *microplus* candidate protective antigens for the control of cattle tick infestations. Parasitol. Res. 106, 471-479 (2010).

[80] Merino, O., Almazán, C., Canales, M., Villar, M., Moreno-Cid, J. A., Galindo, R. C., de la Fuente, J.: Targeting the tick protective antigen subolesin reduces vector infestations and pathogen infection by *Anaplasma marginale* and *Babesiabigemina*. Vaccine 29, 8575-8579 (2011).

[81] de la Fuente, J., Almazán, C., Blouin, E. F., Naranjo, V., Kocan, K. M.: Reduction of tick infections with *Anaplasma marginale* and *A. phagocytophilum* by targeting the tick protective antigen subolesin. Parasitol. Res. 100, 85-91 (2006).

[82] de la Fuente, J., Maritz - Olivier, C., Naranjo, V., Ayoubi, P., Nijhof, A. M., Almazán, C., Canales, M., Pérez de la Lastra, J. M., Galindo, R. C., et al.: Evidence of the role of tick subolesin in gene expression. BMC Genomics 9, 372 (2008).

[83] Kocan, K. M., Zivkovic, Z., Blouin, E. F., Naranjo, V., Almazán, C., Mitra, R., de la Fuente, J.: Silencing of genes involved in Anaplasma marginale-tick interactions affects the pathogen developmental cycle in *Dermacento rvariabilis*. BMC Dev. Biol. 9, 42 (2009).

[84] Galindo, R. C., Doncel - Pérez, E., Zivkovic, Z., Naranjo, V., Gortazar, C., Mangold, A. J., Martín-Hernando, M. P., Kocan, K. M., de la Fuente, J.: Tick subolesin is an ortholog of the akirins described in insects and vertebrates. Dev. Comp. Immunol. 33, 612-617 (2009).

[85] Zivkovic, Z., Torina, A., Mitra, R., Alongi, A., Scimeca, S., Kocan, K. M., Galindo, R. C., et al.: Subolesin expression in response to pathogen infection in ticks. BMC Immunol. 11, 7 (2010). doi: 10.1186/1471-2172-11-7.

[86] Zivkovic, Z., Esteves, E., Almazán, C., Daffre, S., Nijhof, A. M., Kocan, K. M., Jongejan, F., de la Fuente, J.: Differential expression of genes in salivary glands of *Male Rhipicephalus* (*Boophilus*) *microplus* in response to infection with Anaplasma marginale. BMC Genomics 11, 186 (2010). doi: 10.1186/1471-2164-11-186.

[87] Almazán, C., Moreno-Cantú, O., Moreno-Cid, J. A., Galindo, R. C., Canales, M., Villar, M., de la Fuente, J.: Control of tick infestations in cattle vaccinated with bacterial membranes containing surface-exposed tick protective antigens. Vaccine 30, 265-272 (2012).

[88] Nuttall, P. A., Labuda, M.: Tick-host interactions: saliva-activated transmission. Parasitology 129, 177 (2004).

[89] Liu, L., Narasimhan, S., Dai, J., Zhang, L., Cheng, G., Fikrig, E.: *Ixodesscapularis* salivary gland protein P11 facilitates migration of *Anaplasma phagocytophilum* from the tick gut to salivary glands. EMBO

Rep. 12, 1196-1203 (2011).

[90] Shaw, M. K., Tilney, L. G., McKeever, D. J.: Tick salivary gland extract and interleukin-2 stimulation enhance susceptibility of lymphocytes to infection by *Theileria parva* sporozoites. Infect. Immun. 61, 1486-1495 (1993).

[91] Ferreira, B. R., Szabo, M. J. P., Cavassani, K. A., Bechara, G. H., Silva, J. S.: Antigens from *Rhipicephalus sanguineus ticks* elicit potent cell-mediated immune responses in resistant but not in susceptible animals. Vet. Parasitol. 115, 35-48 (2003).

[92] Ghosh, S., Azhahianambi, P., de la Fuente, J.: Control of ticks of ruminants, with special emphasis on livestock farming systems in India: present and future possibilities for integrated control-a review. Exp. Appl. Acarol. 40, 49-66 (2006).

[93] Miller, R., Estrada-Peña, A., Almazán, C., Allen, A., Jory, L., Yeater, K., Messenger, M., Ellis, D., Pérez de León, A. A.: Exploring the use of an anti-tick vaccine as a tool for the integrated eradication of the cattle fever tick, *Rhipicephalus* (*Boophilus*) *annulatus*. Vaccine 30 (38), 5682-5687 (2012).

[94] Pegram, R. G., Wilson, D. D., Hansen, J. W.: Past and present national tick control programs. Why they succeed or fail. Ann. N. Y. Acad. Sci. 916, 546-554 (2000).

[95] McKenna, R. V., Riding, G. A., Jarmey, J. M., Pearson, R. D., Willadsen, P.: Vaccination of cattle against the *Boophilus microplus* using a mucin-like membrane glycoprotein. Parasite Immunol. 20, 325-336 (1998).

[96] Riding, G. A., Jarmey, J. J., McKenna, R. V., Pearson, R., Cobon, G. S., Willadsen, P.: A protective 'concealed' antigen from *Boophilus microplus*. Purification, localization and possible function. J. Immunol. 153, 5158-5166 (1994).

[97] Willadsen, P., Smith, D., Cobon, G., McKenna, R. V.: Comparative vaccination of cattle against *Boophilus microplus* with recombinant antigen Bm86 alone or in combination with recombinant Bm91. Parasite Immunol. 18, 241-246 (1996).

[98] Sugino, M., Imamura, S., Mulenga, A., Nakajima, M., Tsuda, A., Ohashi, K., Onuma, M.: A serine proteinase inhibitor (serpin) from ixodid tick *Haemaphysalis longicornis*; cloning and preliminary assessment of its suitability as a candidate for a tick vaccine. Vaccine 21, 2844-2851 (2003).

[99] Imamura, S., da Silva Vaz Junior, I., Sugino, M., Ohashi, K., Onuma, M.: A serine protease inhibitor (serpin) from *Haemaphysalis longicornis* as an anti-tick vaccine. Vaccine 23, 1301-1311 (2005).

[100] Mulenga, A., Tsuda, A., Onuma, M., Sugimoto, C.: Four serine proteinase inhibitors (serpin) from the brown ear tick, Rhipicephalus appendiculatus; cDNA cloning and preliminary characterization. Insect Biochem. Mol. Biol. 33, 267-276 (2003).

[101] Imamura, S., Namangala, B., Tajima, T., Tembo, M. E., Yasuda, J., Ohashi, K., Onuma, M.: Two serine protease inhibitors (serpins) that induce a bovine protective immune response against *Rhipicephalus appendiculatus* ticks. Vaccine 24, 2230-2237 (2006).

[102] You, M., Xuan, X., Tsuji, N., Kamio, T., Igarashi, I., Nagasawa, H., Mikami, T., Fujisaki, K.: Molecular characterization of a troponin I-like protein from the hard tick *Haemaphysalis longicornis*. Insect Biochem. Mol. Biol. 32, 67-73 (2001).

[103] You, M. J.: Immunization effect of recombinant P27/30 protein expressed in *Escherichia coli* against the hard tick *Haemaphysalis longicornis* (Acari: Ixodidae) in rabbits. Korean J. Parasitol. 42, 195-200 (2004).

[104] You, M. J.: Immunization of mice with recombinant P27/30 protein confers protection against hard tick *Haemaphysalis longicornis* (Acari: Ixodidae) infestation. J. Vet. Sci. 6, 47-51 (2005).

[105] Tellam, R. L., Kemp, D., Riding, G., Briscoe, S., Smith, D., Sharp, P., Irving, D., Willadsen, P.: Reduced oviposition of *Boophilus microplus* feeding on sheep vaccinated with vitellin. Vet. Parasitol. 103,

141-156 (2002).

[106] Weiss, B. L., Kaufman, W. R.: Two feeding-induced proteins from the male gonad trigger engorgement of the female tick Amblyomma hebraeum. Proc. Natl. Acad. Sci. U. S. A. 101, 5874-5879 (2004).

[107] Agianian, B., Tucker, P. A., Schouten, A., Leonard, K., Bullard, B., Gros, P.: Structure of a Drosophila sigma class glutathione S-transferase reveals a novel active site topography suited for lipid peroxidation products. J. Mol. Biol. 326, 151-165 (2003).

[108] He, H., Chen, A. C., Davey, R. B., Ivie, G. W., George, J. E.: Characterization and molecular cloning of a glutathione S-transferase gene from the tick, *Boophilus microplus* (Acari: Ixodidae). Insect Biochem. Mol. Biol. 29, 737-743 (1999).

[109] Rosa de Lima, M. F., Sanchez Ferreira, C. A., Joaquim de Freitas, D. R., Valenzuela, J. G., Masuda, A.: Cloning and partial characterization of a *Boophilus microplus* (Acari: Ixodidae) glutathione S-transferase. Insect Biochem. Mol. Biol. 32, 747-754 (2002).

[110] da Silva Vaz Jnr, I., Imamura, S., Ohashi, K., Onuma, M.: Cloning, expression and partial characterization of a *Haemaphysalis longicornis* and a *Rhipicephalus appendiculatus* glutathione S-transferase. Insect Mol. Biol. 13, 329-335 (2004).

[111] Parizi, L. F., Utiumi, K. U., Imamura, S., Onuma, M., Ohashi, K., Masuda, A., da Silva Vaz, I.: Cross immunity with *Haemaphysalis longicornis* glutathione S-transferase reduces an experimental *Rhipicephalus* (*Boophilus*) *microplus* infestation. Exp. Parasitol. 127, 113-118 (2011).

[112] Bishop, R., Lambson, B., Wells, C., Pandit, P., Osaso, J., Nkonge, C., Morzaria, S., Musoke, A., Nene, V.: A cement protein of the tick *Rhipicephalus appendiculatus*, located in the secretory e cell granules of the type III salivary gland acini, induces strong antibody responses in cattle. Int. J. Parasitol. 32, 833-842 (2002).

[113] Konnai, S., Imamura, S., Nakajima, C., Witola, W. H., Yamada, S., Simuunza, M., Nambota, A., Yasuda, J., Ohashi, K., Onuma, M.: Acquisition and transmission of Theileri aparva by vector tick, *Rhipicephalus appendiculatus*. Acta Trop. 99, 34-41 (2006).

[114] Imamura, S., Konnai, S., Vaz Ida, S., Yamada, S., Nakajima, C., Ito, Y., Tajima, T., Yasuda, J., Simuunza, M., et al.: Effects of anti-tick cocktail vaccine against *Rhipicephalus appendiculatus*. Jpn. J. Vet. Res. 56, 85-98 (2008).

[115] Tsuda, A., Mulenga, A., Sugimoto, C., Nakajima, M., Ohashi, K., Onuma, M.: cDNA cloning, characterization and vaccine effect analysis of *Haemaphysalis longicornis* tick saliva proteins. Vaccine 19, 4287-4296 (2001).

[116] Mulenga, A., Sugimoto, C., Sako, Y., Ohashi, K., Musoke, A., Shubash, M., Onuma, M.: Molecular characterization of a *Haemaphysalis longicornis* tick salivary gland-associated 29-kilodalton protein and its effect as a vaccine against tick infestation in rabbits. Infect. Immun. 67, 1652-1658 (1999).

[117] Paesen, G. C., Adams, P. L., Harlos, K., Nuttall, P. A., Stuart, D. I.: Tick histamine-binding proteins: isolation, cloning, and three-dimensional structure. Mol. Cell 3, 661-671 (1999).

[118] Del Pino, F. A., Brandelli, A., Termignoni, C., Gonzales, J. C., Henriques, J. A., Dewes, H.: Purification and characterization of beta-N-acetylhexosaminidase from bovine tick *Boophilusmicroplus* (Ixodide) larvae. Comp. Biochem. Physiol. B Biochem. Mol. Biol. 123, 193-200 (1999).

[119] Liyou, M., Hamiltion, S., Elvin, C., Willadsen, P.: Cloning and expression of ecto 5′-nucleotidase from the cattle tick *Boophilus microplus*. Insect Mol. Biol. 8 (2), 257-266 (1999).

[120] Jaworski, D. C., Simmen, F. A., Lamoreaux, W., Coons, L. B., Muller, M. T., Needham, G. R.: A secreted calreticulin protein in Ixodid tick (*Amblyomma americanum*) saliva. J. Insect Physiol. 41 (4), 369-375 (1995).

[121] Ferreira, C. A., Da Silva Vaz, I., da Silva, S. S., Haag, K. L., Valenzuela, J. G., Masuda, A.: Cloning and partial characterization of a *Boophilus microplus* (Acari: Ixodidae) calreticulin. Exp. Parasitol. 101, 25-34 (2002).

[122] Del Pino, F. A. B., Brandelli, A., Gonzales, J. C., Henriques, J. A. P., Dewes, H.: Effect of antibodies against B-N-acetylhexosaminidase on reproductive efficiency of the bovine tick *Boophilus microplus*. Vet. Parasitol. 79, 247-255 (1998).

[123] da Silva Vaz, I., Logullo, C., Sorgine, M., Velloso, F. F., Rosa de Lima, M. F., Gonzales, J. C., Masuda, H., Oliveira, P. L., Masuda, A.: Immunization of bovines with an aspartic proteinase precursor isolated from *Boophilusmicroplus* eggs. Vet. Immunol. Immunopathol. 66, 331-341 (1998).

[124] Leal, A. T., Seixas, A., Pohl, P. C., Ferreira, C. A. S., Logullo, C., Oliveira, P. L., Farias, S. E., Termignoni, C., da Silva Vaz, I., Masuda, A.: Vaccination of bovines with recombinant Boophilus Yolk pro-Cathepsin. Vet. Immunol. Immunopathol. 114, 341-345 (2006). doi: 10. 1016/j. vetimm. 2006. 08. 011.

[125] Leal, A. T., Pohl, P. C., Ferreira, C. A. S., Nascimento-Silva, Maria, C. L., Sorgine, M. H. F., Logullo, C., Oliveira, P. L., Farias, S. E., da Silva Vaz, I., Masuda, A.: Purification and antigenicity of two recombinant forms of *Boophilus microplus* Yolk pro-Cathepsin expressed in inclusion bodies. Protein Expr. Purif. 45, 107-114 (2006). doi: 10. 1016/j. pep. 2005. 07. 009.

[126] Logullo, C., Vaz Ida, S., Sorgine, M. H., Paiva-Silva, G. O., Faria, F. S., Zingali, R. B., De Lima, M. F., Abreu, L., Oliveira, E. F., et al.: Isolation of an aspartic proteinase precursor from the egg of a hard tick, *Boophilus microplus*. Parasitology 116 (Pt 6), 525-532 (1998).

(马维民译)

第18章　开发安全有效的蓝舌病病毒疫苗

Polly Roy 和 Meredith Stewart①

摘要

蓝舌病病毒（BTV）感染是家畜的一种重要疾病。传统上BTV的暴发是通过使用减毒活疫苗来控制的。然而，最近在欧洲暴发的BTV已经被灭活疫苗控制住了。这些疫苗不符合DIVA标准（区分感染动物和免疫动物）。这篇综述总结了目前在开发符合DIVA亚单位疫苗方面的努力，重点是病毒样颗粒作为改良的蓝舌病疫苗。

尽管目前在欧洲用于控制病毒传播的商业疫苗主要基于灭活病毒，但这些疫苗并不符合DIVA要求。因此这导致了许多新的方法，以便开发一种符合DIVA的疫苗，并进行了相应的测试。本文介绍了不同的技术，重点介绍了病毒样颗粒作为蓝舌病疫苗的改进。

18.1　前言

蓝舌病（BT）是一种农业重要牲畜的病毒性疾病，已有100多年的历史。18世纪在非洲南部地区的国内反刍动物（主要是绵羊和不太常见的牛、山羊和水牛）以及野生反刍动物（例如大羚羊、鹿、麋鹿和羚羊）中观察到蓝舌病。蓝舌病局限于非洲几十年，直到1924年首次确认在非洲以外的塞浦路斯暴发。随后塞浦路斯在1943—1944年再次暴发疫情，约2 500只绵羊死亡，死亡率达到70%[1]。直到1956年欧洲才出现蓝舌病发病的报告，主要是从葡萄牙开始，沿着伊比利亚半岛一直延伸到西班牙，导致18万只以上的绵羊死亡[2]。20世纪40年代初和50年代初，在中东、东南亚、南欧和美国暴发了蓝舌病，才被描述为一种新兴疾病[3]。

蓝舌病病毒（BTV）是蓝舌病（BT）的病原体，在全世界范围内蓝舌病病毒（BTV）能够传播和发生的一个重要因素是是否有合适的媒介，通常是库蠓属的蠓。病毒分离株来自26种不同的血清型（BTV-1、BTV-2、BTV-3等），分布在热带、亚热带、温带，包括北美和南美、澳大利亚、南欧、以色列、非洲和东南亚。1998年BT在欧洲暴发，最初在希腊发生疫情，然后在许多其他地中海国家暴发，到2005年已造成100多万只羊死亡。在2006年和2007年BTV-8出现在西欧北部，最终扩展到英国、丹麦和捷克共和国。此次疫情暴发范围很广，共有超过180万只动物受到感染[5]，因为这种BTV分离物具有潜在的越冬能力。

该病毒的毒性如此之强，即使是通常不表现任何可见症状的受感染牛也表现出典型的BT疾病症状，导致动物死亡[6,7]。BTV菌株在不同动物宿主体内的毒力相关因素还有待进一步研究。控制BTV在欧洲的传播是通过接种灭活病毒疫苗进行的，而不是非洲和许多其他国家普遍使用的减毒活病毒疫苗（下文18.6）。由于最近在世界范围内暴发了各种BTV血清型，包括重组蛋白技术在内的许多新方法已被用于开发蓝舌病病毒疫苗。目前正在开发这些疫苗以提高所提供的安全性和保护范围。本文简要介绍了蓝舌病的发病、病毒和传播情况，并介绍了蓝舌病疫苗的一些改进技术，重点介绍了病毒样颗粒（VLP）的使用和开发。

---

①　P. Roy，博士，M. Stewart，博士（伦敦卫生和热带医学院传染病和热带疾病系，伦敦，英国，E-mail：polly. roy@ lshtm. ac. uk）

18.2 疾病和发病机制

BT 是家畜的非接触传染性疾病。传统上该病主要发生在绵羊身上，临床症状从亚临床感染到高发病率和死亡率不等。BTV 感染的临床结果取决于病毒株和宿主种类。特别是欧洲细毛羊和肉羊品种对 BTV 感染和传染的易感程度较高[8]。其他反刍动物包括牛、山羊、鹿和骆驼科动物（骆驼、美洲驼、羊驼、驼马和小羊驼）通常无症状，但可能表现出较温和的临床症状。通常 5%~10%的 BTV 感染牛会发展成轻度到重度疾病[9]，这在最近欧洲暴发的 BTV-8 疫情中也观察到了[10]。由于对感染的无症状或轻微反应，这些动物可能充当传染性病毒的贮存宿主。这部分是由于在这些动物中观察到的病毒血症延长，使得昆虫媒介能够在动物之间循环和传播传染性病毒。特别是初生的牛和小牛在感染后可出现长达 3 年的病毒血症[9]。

1905 年首次详细描述了 BT[11]，该病被称为疟疾卡他性热[12]。感染动物的口腔有独特的病变，舌头呈深蓝色是其特有的症状；然而肿胀发绀的舌头（即蓝舌病名字的来源）是一个非常罕见的临床症状。其他常见的临床症状包括发热、呼吸急促和窝地不动。与 BT 相关的病理生理学特征类似于许多其他病毒性出血热，包括在非洲马瘟病毒（AHSV）和流行性出血热病毒（EHDV）（两种相关的环状病毒）中观察到的那样[13-15]。在绵羊中病毒血症通常在 BTV 感染后 3~5 天可检测到，引起淋巴结、肺、心脏和骨骼肌的广泛水肿和出血[16]。此外，在感染过程中也可以观察到口鼻和消化系统黏膜表面坏死。肺对 BTV 引起的血管通透性紊乱尤其敏感[17,18]。BTV 和相关的 EHDV 均可穿过胎盘[19,20]，妊娠牛羊感染可导致胎儿感染、流产或先天性异常[21-23]。

反刍动物自然感染中 BTV 感染的细胞趋向性包括单核吞噬细胞、树突状细胞（DC）和内皮细胞（EC）[17,18,24]。传统树突状细胞是 BTV 感染的主要靶细胞，负责将病毒从皮肤传播到引流淋巴结，对 BT 的发病机制至关重要[18,24-26]。BTV 感染引起反刍动物出血性疾病并诱导细胞死亡，应主要归于凋亡而非坏死，可能与致细胞病变效应（CPE）、宿主细胞凋亡和发病机制有关[27]。在 BTV 感染的哺乳动物细胞中，感染后 24 小时内观察到凋亡的生化和形态学特征，包括 NF-κB、caspase-3 的激活、DNA 降解、细胞膜出泡和细胞收缩。这些变化在昆虫（库蠓属种）细胞中没有观察到，尽管这些细胞对 BTV 复制是有效的[28]。

18.3 病原

BTV 属于呼肠孤病毒科、环状病毒属。BTV 病毒粒子是由 7 个离散蛋白质组成的复杂结构（图 18.1）。该颗粒为无包膜的二十面体衣壳，由 3 个同心的蛋白质壳层即亚核心层、核心层和外壳层组成[29]。由 10 个线性 dsRNA 分子组成的基因组被包裹在病毒粒子中[30,31]（图 18.1）。片段分为大片段（S1-3）、中片段（S4-6）和小片段（S7-10）三大类，在 5’端都具有保守的五核苷酸序列，在 3’端具有不同的保守五核苷酸序列[30]。这些片段编码 12 种蛋白质和 7 种结构蛋白（VP1-VP7；图 18.1）和 4 种非结构蛋白（NS1、NS2、NS3/NS3A 和 NS4）[30,32,33]。

BTV 通过内吞途径进入易感细胞[34-36]（图 18.2）。外层衣壳由 VP2 和 VP5 组成，两者均附着于下面的核心表层 VP7[37,38]。VP2 和 VP5 在感染初期参与细胞黏附和细胞膜渗透[34,36]。黏附蛋白 VP2 是 BTV 中变异最大的病毒蛋白，具有血清型特异性并有中和表位[39-46]。通过高分辨率修正冷冻电子显微镜三维结构分析，发现 VP2 具有外露的受体结合尖端结构域和内部唾液酸结合位点；两者都是细胞表面结合活性所必需的[38]。VP5 在核内体的低 pH 下作为一种融合样蛋白，使病毒能够穿透内吞体膜[35,47]。

进入细胞后不久，病毒除去外层衣壳，释放一个转录活性核心粒子到细胞溶质中（图 18.2）。核心蛋白由 VP3 和 VP7 两种主要蛋白，VP1、VP4 和 VP6 3 种次要酶蛋白组成。此外，

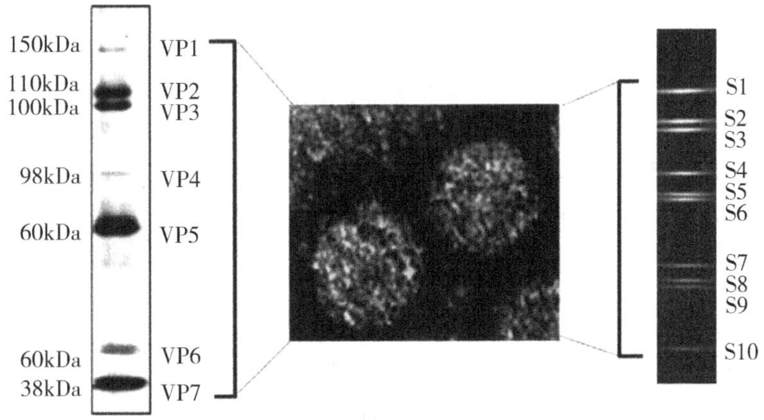

图 18.1 BTV 颗粒由 7 个蛋白质和 10 个离散双链 RNA 分子基因组组成

纯化病毒颗粒 SDS-PAGE（左）、BTV 颗粒电子显微图（中）、BTV dsRNA 基因组非变性 PAGE（右）。

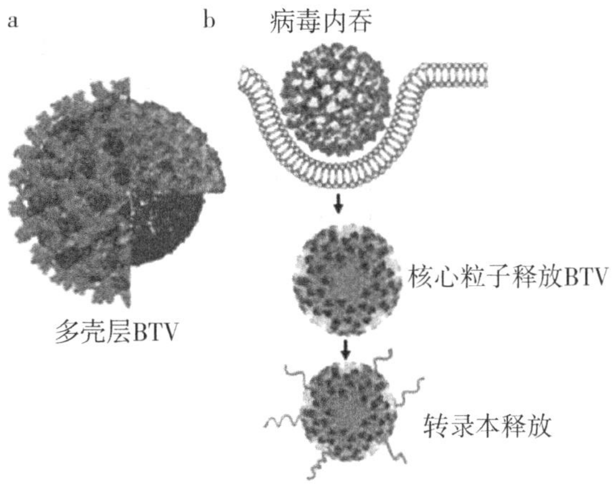

图 18.2 （a）BTV 病毒颗粒示意图。外衣壳由 VP2（绿色）和 VP5（洋红）、VP7（蓝色和绿色）内核和 VP3 亚核（红色）组成。（b）病毒颗粒通过内吞作用进入细胞质，释放核心。核心在进入细胞质释放病毒转录物时具有转录活性

还有 10 个 dsRNA 基因组片段。VP3 与位于五重轴的 3 种次要蛋白质一起形成病毒粒子的二十面体亚核心结构[48,49]。VP7 是主要的核心蛋白，是群体特异性抗原[39]，具有整合素受体结合位点，可进入昆虫媒介细胞[50]。转录活性核心粒子在释放到胞质中后，连续合成并向宿主细胞胞质中挤出多个拷贝的 10-帽子转录本。这些转录本（mRNAs）不仅负责病毒蛋白的合成，而且作为互补负链 RNA 合成的模板。然而在用作基因组 dsRNA 合成的模板之前，ssRNA 片段不是 dsRNA，是由正在装配的新生子代核心颗粒包装的[51]。在病毒感染细胞中，核心组装发生在由非结构 NS2 形成的病毒包涵体（VIBs）内[30,52-54]。所有 3 种次要的酶蛋白，病毒转录本以及两种主要的结构蛋白（VP3 和 VP7）都是由 NS2 召集并允许它们组装。这些新生成的核心蛋白随后从病毒包涵体中向外扩散，当它们沿着与 NS3 相关的脂筏运输到细胞膜时，外层衣壳蛋白 VP2 和 VP5 被组装到核心颗粒上[55,56]。NS3/NS3A 参与组装病毒粒子的转运和释放[57]。非结构蛋白 NS1 在感染细胞中形成管状结构，是表达最丰富的蛋白之一[58]。虽然管状结构的确切功能还不清楚，但最近已经证明它可以上调病毒蛋白的表达[59]。

18.4 传播和分布

蓝舌病在北美、南美、澳大利亚、东南亚和非洲大陆等许多热带和亚热带地区流行。然而在过去的15年中，新血清型的分布和出现都发生了变化。这一变化在欧洲大陆是典型的，地中海盆地沿岸国家不同血清型的暴发频率不断增加，北欧首次暴发BTV，最北至挪威。分子流行病学研究表明，在24个血清型中，8个不同的血清型（BTV-1、BTV-2、BTV-4、BTV-6、BTV-8、BTV-9、BTV-11和BTV-16）自1998年以来已传播到欧洲大陆[60]。其中BTV-8在北欧绵羊和牛中引起了最严重的疾病。此外，欧洲发现了一种新的血清型BTV-25[61]，科威特发现了另一种新的血清型BTV-26[62]。

BTV是通过吸血蠓（库蠓属物种）在动物间传播的[30,63]。气候变化（温度、降雨量）可能是导致库蠓种群分布和数量增加的原因，库蠓是促进病毒在非洲和亚洲传播的媒介。这与欧洲出现的BTV和美国出现的新血清型相吻合。此外，新型库蠓种类（Culicoideschiopterus，C. dewulffi、C. obsoletus、C. scotius和C. pulicaris）已在欧洲被确认能传播病毒[60,64]。媒介种群的扩大分布和新种库蠓病毒的传播（这在中欧和北欧是丰富的）使得BTV的地理范围保持扩展到欧洲[60,64]。

因此，BTV现在对所有欧洲国家的牲畜构成持续的威胁。在美洲，流行的血清型BTV-10、BTV-11、BTV-13是由C. sonorensis传播的。而BTV-2的传播仅限于C. insignis[65-67]。自1998年以来，美国东南部也分离到10种血清型（BTV-1、BTV-3、BTV-5、BTV-6、BTV-9、BTV-12、BTV-14、BTV-19、BTV-22和BTV-24），一般局限于加勒比/中美洲地区[67-70]。这些血清型的传播也强调了这种可能性，即其他库蠓属物种（C. pusillus、C. furens、C. filarifer和C. trilineatus）参与改变新血清型的分布和出现[67,69]。

18.5 BTV诊断

蓝舌病病毒和反刍动物感染是一个与国际贸易限制和动物迁徙相关的重大问题。2007年BTV-8病毒的暴发估计给法国造成高达14亿美元的经济损失，主要是由于感染和限制牲畜的活动[71]。这部分是由于病毒感染导致发病率0%~100%，绵羊[72]和牛[72]的平均发病率为20%，也是由于牛病毒血症持续时间长，能够感染库蠓属并将病毒传播给其他反刍动物，包括绵羊和山羊。在美国与牛的测试和限制移动相关的损失每年就高达1.3亿美元[71]。因此，世界动物卫生组织有许多国际规定的检测方法来分离、鉴定和血清学检测BTV感染动物。

为了鉴定BTV感染，采用了多种病毒分离方法，包括绵羊接种、昆虫细胞系（如C6/36）体外细胞培养分离以及最敏感的鸡胚接种（ECE；[73]）。病毒的血清型由病毒中和试验确定。血小板减少试验是确定血清型最可靠的方法，其结果是固定减少高达90%。其他方法包括微量滴定中和试验（50%的50-100 $TCID_{50}$）、免疫荧光抑制、使用特异性引物的逆转录聚合酶链反应（RT-PCR）和序列分析，由于易于使用正在成为普遍的做法。RT-PCR现已被公认为检测BTV感染的一种规定的国际鉴定程序[7]。

这些方法与血清学试验一起用于确定一个地区的BTV状态。确定血清中抗体血清型特异性的血清学方法通常更加复杂、昂贵和耗时，因为它们在中和试验中评估血清是否抑制已知病毒血清型板的传染性。因此，我们推荐采用竞争性酶联免疫吸附试验（ELISA）来检测BTV群特异性抗原VP7[7]。该方法利用BTV高度保守的核心蛋白VP7的特异性单克隆抗体，防止与之密切相关的流行性出血热病毒（EHDV）的VP7发生交叉反应。国际贸易的另一种检测方法是琼脂凝胶免疫扩散试验（AGID[74]），该试验取代了用于检测BTV抗体的补体结合（CF）试验，尽管补体结合试验仍在一些国家使用[7]。

18.6　BTV 疫苗

18.6.1　减毒病毒疫苗

不同国家的疫苗接种策略是根据各自的政策、发生 BTV 血清型的地理分布和适当疫苗的可获得性而制定的。在 BTV 来源国南非，减毒活疫苗已经使用了 40 多年，并且已知能诱导有效和持久的免疫[75-77]。这些减毒活疫苗也在意大利撒丁岛作为 BT 接种运动的一部分，与单价 BTV-2、BTV-4 和 BTV-16 联合使用[78]。由于循环血清型数量众多，改进的活疫苗通常作为多价疫苗施用，特别是在南非。这是由于外壳蛋白 VP2 上的中和表位提供了血清型特异性保护。

减毒活疫苗是通过在鸡胚中连续传代而研制的。尽管在流行地区取得了成功，但使用减毒活病毒疫苗也有一些缺点。例如用减毒 BTV 疫苗接种的致畸效果已得到很好的证明[79,80]。对减毒病毒的不良反应包括泌乳绵羊产奶量下降，在妊娠前半期接种疫苗的怀孕母羊的后代流产/胚胎死亡和畸形发生[81]。此外，这些减毒疫苗对强保护性抗体反应的刺激与它们在接种宿主中复制的能力直接相关[7]。

在实验室实验和田间接种疫苗后观察到的病毒血症足以使疫苗株传播给叮咬的蠓[80,82]。这将导致更大的通过媒介传播潜在的风险，最终导致毒性逆转和/或疫苗病毒基因与野生型病毒株的基因重组。BTV 基因组的分段性使得同时感染同一动物的毒株之间的基因重组成为可能[83]。2002 年在意大利的田间观察到这一点，当时流行的 BTV-16 毒株被发现是 BTV-2 和 BTV-16 减毒活疫苗的重组[84]。针对 BTV 的减毒病毒可能提供一种具有保护作用的控制疾病的途径，但它们不适用于在一个地区根除该疾病的规划。

18.6.2　灭活疫苗

最近在 BTV-8 和 BTV-1 疫情期间，许多北欧国家成功地将化学灭活 BTV 疫苗用于绵羊和牛的大规模接种。如果严格执行质量控制并实现完全减毒，灭活疫苗相比减毒病毒疫苗有重大的进步。这些灭活病毒疫苗已被证明是安全的，对动物具有免疫原性，但目前仅对有限数量的血清型有效[85]。然而对野外动物接种疫苗后，不良反应与其中一些灭活病毒制剂有关，其中包括轻度 BT 症状[85-87]。

值得注意的是，在 2007 年 BTV-8 暴发期间，德国用灭活疫苗（BLUEVAC 8）给绵羊和牛接种疫苗往往伴随着严重的副作用，包括流产、跛足、血便以及体温升高（Giese M，2012 年，个人交流）。这可能是由于用于灭活疫苗的佐剂包括福尔马林、β-丙内酯、二元乙烯亚胺和 γ 射线，或制剂中引起致病作用的不同佐剂[88-91]或病毒蛋白。这种疫苗的其他潜在缺点是，由于每次接种需要大量抗原以及需要加强免疫，生产成本很高，因为与减毒活疫苗相比，灭活疫苗通常诱导相对短暂的免疫。

重要的是，使用减毒活疫苗和灭活疫苗的最大限制是它们对农业贸易的重大经济影响。这些疫苗不能区分自然感染的动物与接种过疫苗的动物（DIVA），因此不符合 DIVA 的要求，对动物交易和活动施加限制。此外，由于缺乏可用于区分接种疫苗和自然感染动物的商业化检测，这些疫苗的 DIVA 依从性受到阻碍。这是因为灭活和活病毒疫苗制剂并非没有所有病毒成分，因此不允许在受感染动物和接种疫苗的动物之间进行血清学鉴别。此外，由于接种疫苗后基因组 RNA 的存在[92]，绵羊短期 BTV 病毒血症和牛长期病毒血症，基因 DIVA 并不总是可行的[7,23,93,94]。

18.6.3　基于重组蛋白质的疫苗

已经研究了许多基于重组蛋白的技术，包括基于杆状病毒[43,77,95,96]、基于痘病毒[97-99]和基于疱疹病毒[100]的表达系统，以符合 DIVA 并减少与灭活和减毒病毒疫苗相关的潜在副作用。在大多数情况下，重组蛋白已在绵羊实验性疫苗接种攻毒试验中进行了评估，并证明其具有保护作用，但尚未商业化[75,77,95]。BTV 重组蛋白疫苗是基于确定血清型的外层衣壳 VP2，能够在接种绵羊中获得保护性免疫[40,41,43]。这是由于 VP2 单独拥有多个中和表位[42,46,101]。这些结果清楚地表明，中和抗体对病毒的反应是针对这种蛋白质的[40,41,43,95]。后来的研究将两种外衣壳蛋白（VP2+VP5）结合起来，在接种绵羊身上比单独使用 VP2 有更好的保护作用（图 18.3）。此外，已使用其他活病毒（即痘病毒）载体系统开发了基于 VP2 的疫苗，该系统被设计成在接种疫苗后在动物中表达单个或多个蛋白质[98,102]。

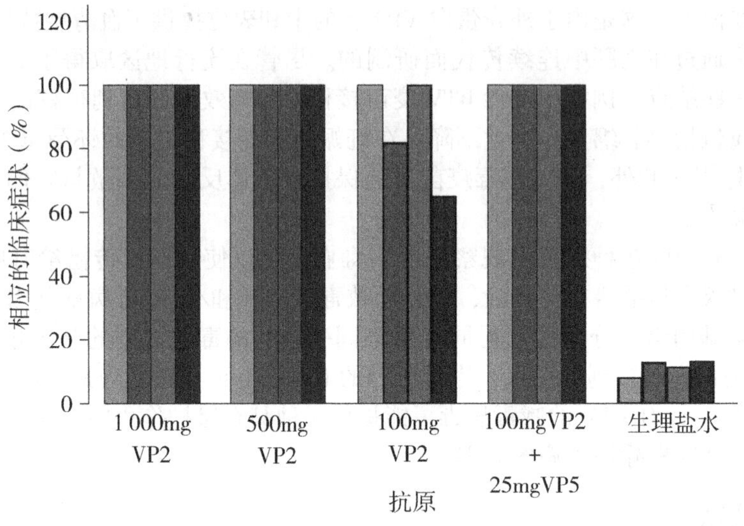

图 18.3　BTV 抗原剂量及组成对绵羊强毒攻击后临床症状的保护作用

采用不同浓度的 VP2 和 VP2+VP5 联合疫苗接种动物，测定对强毒攻击的临床反应。

18.6.4　蓝舌病病毒样颗粒（VLPs）

杆状病毒表达的重组蛋白 VP2 被证明在接种的绵羊中诱导血清型特异性，提供了针对同源强毒攻击的完全保护[103-105]。重组蛋白 VP2 和 VP5 的疫苗接种减少了抵抗病毒攻击所需的蛋白量[43]。这导致观察到以合适的构象呈现 VP2 将减少疫苗接种所需的蛋白量[96]。这是一个合乎逻辑的结论，两种 BTV 外层衣壳蛋白和支架核心蛋白 VP3 和 VP7 通过重组杆状病毒表达系统表达，以产生模仿病毒结构但不包含任何病毒基因组或酶蛋白[106]（图 18.4a）的双衣壳病毒样颗粒（VLPs）。这是第一个 BTV 颗粒组装不需要 NS1 或 NS2 的例子。

对 BTV 病毒样颗粒（VLP）的三维重建表明，其结构与病毒粒子完全相同[49]。VLPs 具有典型的二十面体结构（~90 nm），4 种重组蛋白的位置和比例与天然蛋白相同。进一步的研究表明，将不同血清型的外壳蛋白共表达到高度保守的内核上是可能的[107,108]。用不含 VP2 和 VP5 的具有核样颗粒（CLPs）的制剂对绵羊进行接种，对强毒攻击的保护水平较低，出现轻微发热和一些与 BT 相关的病理变化[76,105]。CLPs 的部分保护作用可能部分是由于细胞介导的免疫（CMI）应答直接针对内核蛋白上的抗原位点。事实上，每种结构和非结构 BTV 蛋白都被证明能诱导 CTL 应答[99]。已经证明，用 VP7 单独或与其他 BTV 蛋白联合接种动物可引发细胞介导的免疫应答，并且假设 CMI 应答可降低 BTV 病毒血症[99,109,110]，尽管 BTV 病毒血症的降低不太可能阻止病毒播[105]。

最近，CD8 和 CD4 表位都已在 VP7 序列中被定位[111]。动物表现出强烈的群体特异性 VP7 抗体反应，但没有中和抗体。

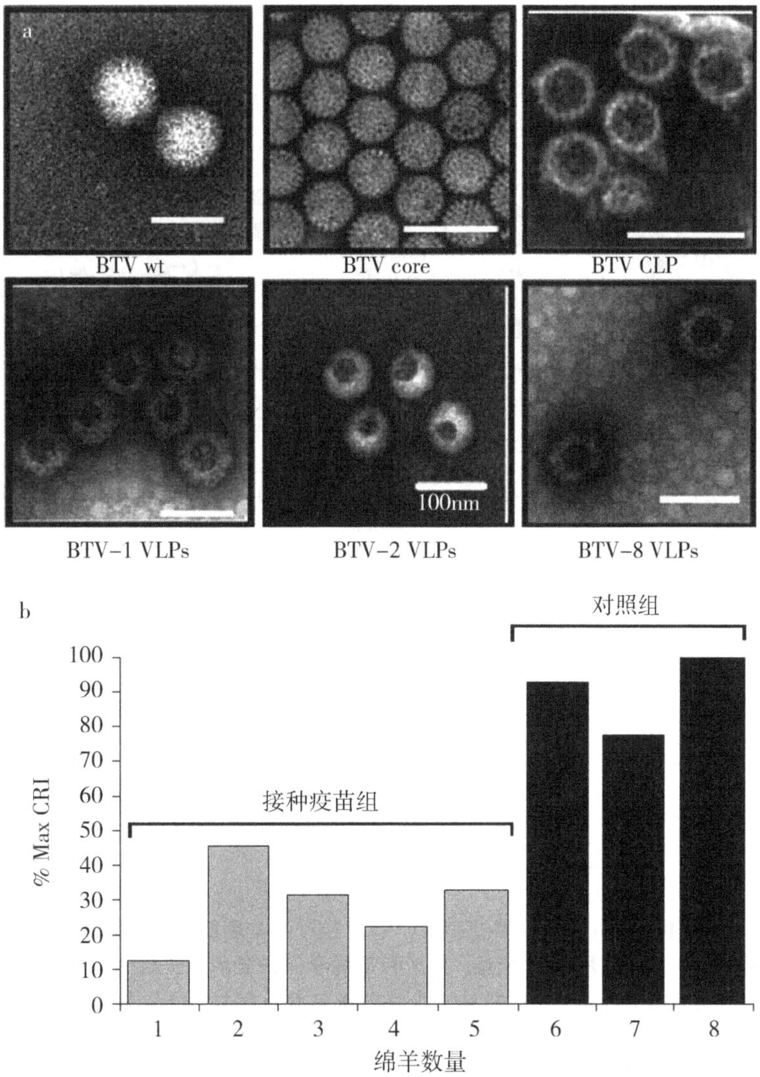

图 18.4　（a）BTV 病毒、核、核样颗粒和病毒样颗粒的电子显微照片。（b）与对照动物相比，绵羊接种 CLPs 疫苗可降低 BT 疾病的临床评分

在所有这些潜在的亚单位疫苗中，通过 4 种 BTV 结构蛋白的共表达形成的 BTV VLPs，以正确的构型显示每种蛋白，已经在动物中得到最广泛的测试，显示出最大的前景。在每项研究中，VLPs 都被证明是极好的免疫原，既能预防绵羊疾病，又能预防小规模试验中可检测到的病毒复制（图 18.5b）[77,96,103-105,112,113]。

BTV VLP 的成功是由于它能与细胞介导的免疫应答（CMI）结合产生强烈的中和抗体反应，CMI 在从感染中恢复和防止再感染方面也发挥了作用[109,110]。南非的早期研究表明，动物接种低至约 10 μg VLP（其中 VP2 仅为 1~2 μg；图 18.5a）[96]。近年来，欧洲利用易感绵羊品种（即美利奴羊、阿尔卑斯山前羊、卡拉古尼科杂交羊）进行了广泛的临床试验，其中 BTV VLPs 代表不同的欧洲血清型（表 18.1）[96,103-105,113]。这些试验还评估了单独或混合递送的病毒株以及病毒株的进化谱系，以评估它们商业化的潜力。

绵羊接种了由单一血清型或多血清型组成的 BTV VLP 疫苗。疫苗制备中每种血清型的中和抗体滴度在攻毒前测定。然后用强毒攻击动物，感染后 7 天检测复制 dsRNA 的 *Ct* 值。*Ct* 值 40 被认为是病毒复制的阴性值。

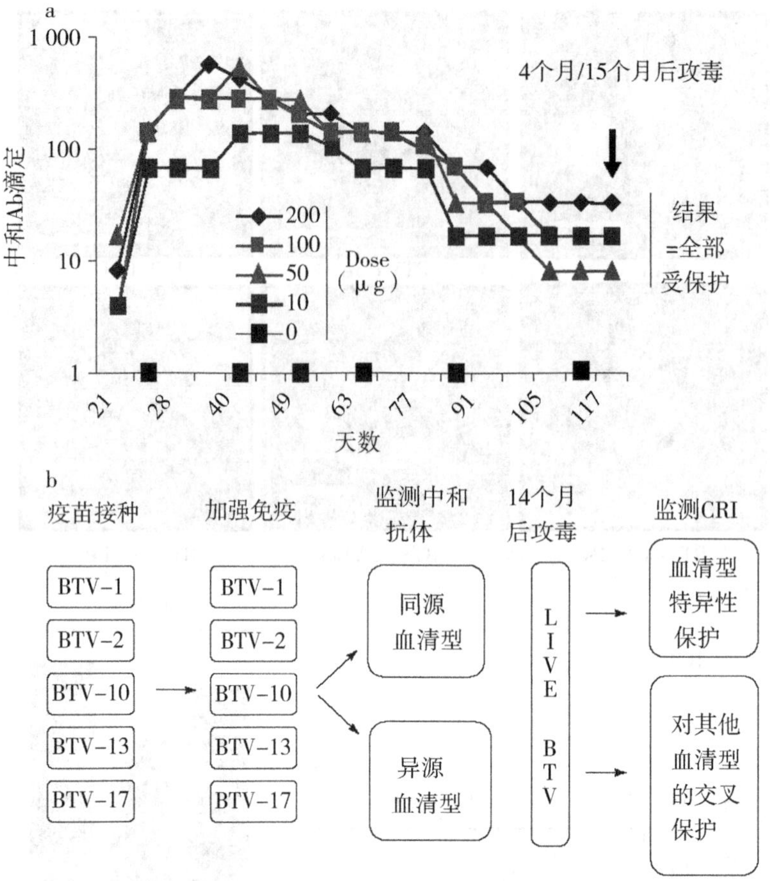

图 18.5 (a) 不同浓度 BTV VLP 接种绵羊后的中和抗体效价。在疫苗接种后 4 个月和 15 个月，临床和病毒学保护显示能长期预防 BTV 攻击。疫苗接种后即使中和抗体减少（改编自 Roy 等）。(b) 绵羊 VLP 疫苗接种攻毒试验示意图。绵羊接种了来自不同血清型的 VLP，用同源强毒和异源强毒攻击的结果（改编自罗伊等[103]）

进化趋异（东方和西方拓扑型[114-117]）对疫苗开发的影响清楚地表明，血清型特异性中和抗体的开发比循环毒株的拓扑型更为关键[105]。这表明，菌株的分离变异或进化距离不是 BTV 疫苗的关键因素[105]。此外，以不同血清型的混合物形式提供 VLPs 不会影响或影响每种血清型提高血清型特异性中和抗体反应的能力[43,96,112,113]或在动物受到攻毒时干扰疫苗的保护效力（表 18.1）[43,113]。这些发现强调可以使用已知的血清型开发多种血清型疫苗，而不是为每一次 BTV 暴发生产新疫苗。

表 18.1 欧洲绵羊品种 VLP 疫苗接种试验总结

疫苗		中和抗体滴度攻毒病毒（攻毒前测定）		Ct Day 7
单价苗	BTV-1	>128	BTV-1	40
单价苗	BTV-2	128	BTV-1	40
		32-64（10 μg）	BTV-2	40
		128（20 μg）	BTV-2	40

（续表）

疫苗		中和抗体滴度（攻毒前测定）	攻毒病毒	Ct Day 7
单价苗	BTV-8	>128	BTV-2	40
多价苗	BTV-1	64-128	BTV-1	40
	BTV-4	>16	BTV-4	40
多价苗	BTV-1	>128		
	BTV-2	32-128		
	BTV-8	64-128	BTV-8	40

VLPs 在绵羊试验中非常有效，由于它们不含任何病毒非结构蛋白，因此可以区分接种动物和病毒感染的动物，从而解决当前疫苗的一个主要问题。VLP 方法的另一个优点是，由于免疫原是完全基于蛋白质的，因此没有基因修饰的病毒成分需要在接种的动物中表达外源基因。

18.6.5 设计复制缺陷型 BTV 作为候选疫苗毒株

为 BTV 开发的反向遗传（RG）系统[118]允许直接引入病毒基因组的突变/缺失，从而利用特定的互补细胞系恢复具有复制缺陷的病毒株[119,120]。复制缺陷病毒缺乏必要的催化基因，在正常的非互补细胞中不具有传染性，但可以进入易感细胞。在易感细胞中，这些病毒粒子只能经历一个单一的复制周期；这是由于子代病毒无法包装 dsRNA（图 18.6）。这意味着这些有缺陷的病毒应该具有与强健的免疫反应相关的所有益处，类似于活病毒疫苗但没有感染的风险。此外，与 VLPs 类似的“核心”元素可以通过简单地替换外壳蛋白基因来生成新的血清型疫苗。例如使用 RG 系统，通过将编码 VP2 和 VP5 的两个 RNA 片段（如 BTV-8），连同来自 BTV-1 亲本复制缺陷病毒株的剩余 8 个片段转染细胞，可以产生重组体[120]。这些缺陷病毒已被证明对绵羊和牛具有高度的保护作用[120]。能够使用相同的骨架并仅对 VP2 和 VP5 进行重新排序，将使新疫苗能够迅速获得批准。RG 系统还将允许在病毒基因组中包含标记序列，从而将疫苗株与野生型病毒区分开来。

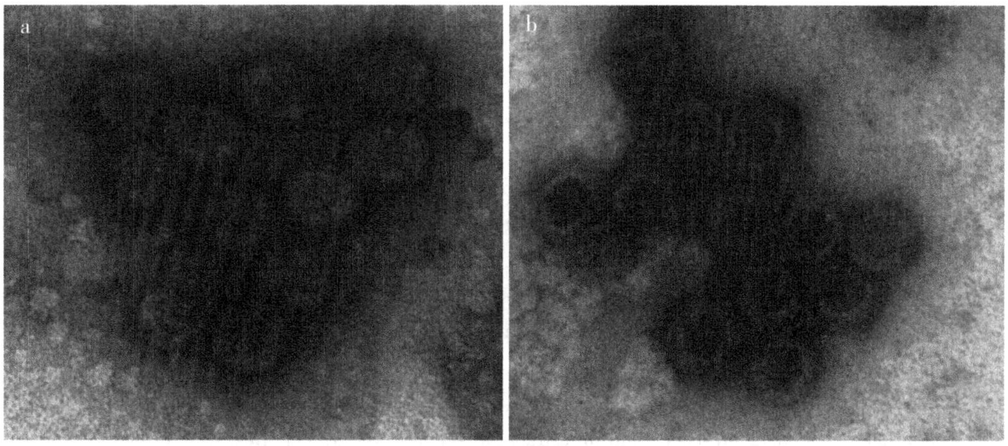

图 18.6 BTV-1 VP6 缺陷型病毒缺乏病毒基因组

BTV-1 复制缺陷纯化核心颗粒。(a) 来自反式表达 VP6 的 BSR 细胞的核。(b) 来自野生型 BSR 细胞的核

结论

蓝舌病是一种重要的农业病害，具有显著的经济影响。虽然有基于灭活或减毒病毒的有效疫苗，但接种疫苗的动物受到贸易限制和潜在的健康副作用。复制缺陷病毒株和 VLPs 的开发提供了两种新的潜在的符合 DIVA 要求的疫苗，可用于现有的制造技术中。这些技术的采用将取决于该国的具体贸易和农业需求，以及 BTV 状况。因此，疫苗的采用可以根据一个国家的需要进行调整。包括杆状病毒产生的 VLP 免疫原在内的 BTV 重组蛋白疫苗的研制是一个进步。这些疫苗很容易适应 DIVA 测试的开发和商业化，该测试可以针对任何小结构和非结构蛋白。这可以简单而廉价地实现，因为这些蛋白的重组蛋白已经被表达并用于基于 ELISA 的系统中[121]。与灭活疫苗或杀死的疫苗（约 12 个月）相比，VLP 免疫原可引起持久的免疫可检测反应[96]（大于 24 个月），因此，这些免疫 DIVA 试验是可行的。对于符合 DIVA 的灭活疫苗和减毒疫苗，必须采取许多不同的方法，包括去除非结构蛋白、加入特定免疫标记或去除表位，所有这些都将增加疫苗生产和 DIVA 检测成本。

参考文献

[1] Gambles, R. M.: Bluetongue of sheep in Cyprus. J. Comp. Pathol. 59, 176-190 (1949).

[2] Manso-Ribeiro, J., et al.: Fievre catarrhale du mouton (blue-tongue). Bull. Off. Int. Epizoot. 48, 350-367 (1957).

[3] Howell, P. G.: Emerging diseases of animals: Bluetongue. FAO Agric. Stud. 61, 109 (1963).

[4] Purse, B. V., et al.: Climate change and the recent emergence of bluetongue in Europe. Nat. Rev. Microbiol. 3, 171-181 (2005).

[5] Shaw, A. E., et al.: Development and initial evaluation of a real-time RT-PCR assay to detect bluetongue virus genome segment 1. J. Virol. Methods 145, 115- 126 (2007).

[6] Elbers, A. R., van der Spek, A. N., van Rijn, P. A.: Epidemiologic characteristics of bluetongue virus serotype 8 laboratory-confirmed outbreaks in The Netherlands in 2007 and a comparison with the situation in 2006. Prev. Vet. Med. 92, 1-8 (2009).

[7] OIE: Notifiable diseases list. http: //www. oie. int/eng/ maladies/en _ classification2009. htm? e1d7 (2009).

[8] MacLachlan, N. J.: The pathogenesis and immunology of bluetongue virus infection of ruminants. Comp. Immunol. Microbiol. Infect. Dis. 17, 197-206 (1994).

[9] Hourrigan, J. L., Klingsporn, A. L.: Bluetongue: the disease in cattle. Aust. Vet. J. 51, 170 - 174 (1975).

[10] Meroc, E., et al.: Bluetongue in Belgium: episode II. Transbound. Emerg. Dis. 56, 39-48 (2009).

[11] Spreull, J.: Malarial catarrhal fever (bluetongue) of sheep in South Africa. J. Comp. Pathol. Therap. 18, 321-337 (1905).

[12] Erasmus, B. J.: Bluetongue in sheep and goats. Aust. Vet. J. 51, 65-170 (1975).

[13] Laegreid, W. W., Burrage, T. G., Stone-Marschat, M., Skowronek, A.: Electron microscopic evidence for endothelial infection by African horse-sickness virus. Vet. Pathol. 29, 554-556 (1992).

[14] Mahrt, C. R., Osburn, B. I.: Experimental bluetongue virus infection of sheep; effect of vaccination: pathologic, immunofluorescent, and ultrastructural studies. Am. J. Vet. Res. 47, 1198-1203 (1986).

[15] Tsai, K. S., Karstad, L.: Epizootic hemorrhagic disease virus of deer: an electron microscopic study. Can. J. Microbiol. 16, 427-432 (1970).

[16] Pini, A.: Study on the pathogenesis of bluetongue: replication of the virus in the organs of infected sheep. Onderstepoort J. Vet. Res. 43, 159-164 (1976).

[17] DeMaula, C. D., Leutenegger, C. M., Bonneau, K. R., MacLachlan, N. J.: The role of endothelial

cell-derived inflammatory and vasoactive mediators in the pathogenesis of bluetongue. Virology 296, 330-337 (2002).

[18] DeMaula, C. D., Leutenegger, C. M., Jutila, M. A., MacLachlan, N. J.: Bluetongue virus-induced activation of primary bovine lung microvascular endothelial cells. Vet. Immunol. Immunopathol. 86, 147-157 (2002).

[19] De Clercq, K., et al.: Transplacental bluetongue infection in cattle. Vet. Rec. 162, 564 (2008).

[20] Gibbs, E. P., Lawman, M. J. P., Herniman, K. A. J.: Preliminary observations on transplacentar infection of bluetongue virus in sheep - a possible overwintering mechanism. Res. Vet. Sci. 27, 118-120 (1979).

[21] Darpel, K. E., et al.: Transplacental transmission of bluetongue virus 8 in cattle, UK. Emerg. Infect. Dis. 15, 2025-2028 (2009).

[22] De Clercq, K., et al.: Transplacental infection and apparently immunotolerance induced by a wild-type bluetongue virus serotype 8 natural infection. Transbound. Emerg. Dis. 55, 352-359 (2008).

[23] MacLachlan, N. J., et al.: Detection of bluetongue virus in the blood of inoculated calves: comparison of virus isolation, PCR assay, and in vitro feeding of Culicoides variipennis. Arch. Virol. 136, 1-8 (1994).

[24] Hemati, B., et al.: Bluetongue virus targets conventional dendritic cells in skin lymph. J. Virol. 83, 8789-8799 (2009).

[25] McLaughlin, B. E., DeMaula, C. D., Wilson, W. C., Boyce, W. M., MacLachlan, N. J.: Replication of bluetongue virus and epizootic hemorrhagic disease virus in pulmonary artery endothelial cells obtained from cattle, sheep, and deer. Am. J. Vet. Res. 64, 860-865 (2003).

[26] Whetter, L. E., Maclachlan, N. J., Gebhard, D. H., Heidner, H. W., Moore, P. F.: Bluetongue virus infection of bovine monocytes. J. Gen. Virol. 70 (Pt 7), 1663-1676 (1989).

[27] Stewart, M. E., Roy, P.: Role of cellular caspases, nuclear factor-kappa B and interferon regulatory factors in Bluetongue virus infection and cell fate. Virol. J. 7, 362 (2010).

[28] Mortola, E., Noad, R., Roy, P.: Bluetongue virus outer capsid proteins are sufficient to trigger apoptosis in mammalian cells. J. Virol. 78, 2875-2883 (2004).

[29] Verwoerd, D. W., Els, H. J., De Villiers, E. M., Huismans, H.: Structure of the bluetongue virus capsid. J. Virol. 10, 783-794 (1972).

[30] Roy, P.: Orbiviruses and their replication. In: Fields' Virology-Fifth Edition, (Knipe, D. M. et al., eds.). Lippincott Williams & Wilkins, Philadelphia, USA. (2007).

[31] Verwoerd, D. W., Louw, H., Oellermann, R. A.: Characterization of bluetongue virus ribonucleic acid. J. Virol. 5, 1-7 (1970).

[32] Firth, A. E.: Bioinformatic analysis suggests that the Orbivirus VP6 cistron encodes an overlapping gene. Virol. J. 5, 48 (2008).

[33] Ratinier, M., etal.: Identification and characterization of a novel non-structural protein of bluetongue virus. PLoSPathog. 7, e1002477 (2011).

[34] Forzan, M., Marsh, M., Roy, P.: Bluetongue virus entry into cells. J. Virol. 81, 4819-4827 (2007).

[35] Hassan, S. H., Wirblich, C., Forzan, M., Roy, P.: Expression and functional characterization of bluetongue virus VP5 protein: role in cellular permeabilization. J. Virol. 75, 8356-8367 (2001).

[36] Hassan, S. S., Roy, P.: Expression and functional characterization of bluetongue virus VP2 protein: role in cell entry. J. Virol. 73, 9832-9842 (1999).

[37] Nason, E. L., et al.: Interactions between the inner and outer capsids of bluetongue virus. J. Virol. 78, 8059- 8067 (2004).

[38] Zhang, X., et al.: Bluetongue virus coat protein VP2 contains sialic acid-binding domains, and VP5 resembles enveloped virus fusion proteins. Proc. Natl. Acad. Sci. U. S. A. 107, 6292-6297 (2010).

[39] Huismans, H., Erasmus, B. J.: Identification of the serotype-specific and group-specific antigens of bluetongue virus. Onderstepoort J. Vet. Res. 48, 51-58 (1981).

[40] Huismans, H., van der Walt, N. T., Cloete, M., Erasmus, B. J.: Isolation of a capsid protein of bluetongue virus that induces a protective immune response in sheep. Virology 157, 172-179 (1987).

[41] Huismans, H., van der Walt, N. T., Erasmus, B. J.: Immune response against the purified serotype specific antigen of bluetongue virus and initial attempts to clone the gene that codes for the synthesis of this protein. Prog. Clin. Biol. Res. 178, 347-353 (1985).

[42] Rossitto, P. V., MacLachlan, N. J.: Neutralizing epitopes of the serotypes of bluetongue virus present in the United States. J. Gen. Virol. 73 (Pt 8), 1947- 1952 (1992).

[43] Roy, P., Urakawa, T., Van Dijk, A. A., Erasmus, B. J.: Recombinant virus vaccine for bluetongue disease in sheep. J. Virol. 64, 1998-2003 (1990).

[44] Gould, A. R., Eaton, B. T.: The amino acid sequence of the outer coat protein VP2 of neutralizing monoclonal antibody-resistant, virulent and attenuated bluetongue viruses. Virus Res. 17, 161-172 (1990).

[45] Heidner, H. W., Rossitto, P. V., MacLachlan, N. J.: Identification of four distinct neutralizing epitopes on bluetongue virus serotype 10 using neutralizing monoclonal antibodies and neutralization-escape variants. Virology 176, 658-661 (1990).

[46] White, J. R., Eaton, B. T.: Conformation of the VP2 protein of bluetongue virus (BTV) determines the involvement in virus neutralization of highly conserved epitopes within the BTV serogroup. J. Gen. Virol. 71 (Pt 6), 1325-1332 (1990).

[47] Forzan, M., Wirblich, C., Roy, P.: A capsid protein of non-enveloped Bluetongue virus exhibits membrane fusion activity. Proc. Natl. Acad. Sci. U. S. A. 101, 2100-2105 (2004).

[48] Grimes, J. M., et al.: An atomic model of the outer layer of the bluetongue virus core derived from X-ray crystallography and electron cryomicroscopy. Structure 5, 885-893 (1997).

[49] Hewat, E. A., Booth, T. F., Roy, P.: Structure of correctly self-assembled bluetongue virus-like particles. J. Struct. Biol. 112, 183-191 (1994).

[50] Tan, B. H., et al.: RGD tripeptide of bluetongue virus VP7 protein is responsible for core attachment to Culicoides cells. J. Virol. 75, 3937-3947 (2001).

[51] Lourenco, S., Roy, P.: In vitro reconstitution of Bluetongue virus infectious cores. Proc. Natl. Acad. Sci. U. S. A. 108, 13746-13751 (2011).

[52] Browne, J. G., Jochim, M. M.: Cytopathologic changes and development of inclusion bodies in cultured cells infected with bluetongue virus. Am. J. Vet. Res. 28, 1091-1105 (1967).

[53] Modrof, J., Lymperopoulos, K., Roy, P.: Phosphorylation of bluetongue virus nonstructural protein 2 is essential for formation of viral inclusion bodies. J. Virol. 79, 10023-10031 (2005).

[54] Thomas, C. P., Booth, T. F., Roy, P.: Synthesis of blue-tongue virus-encoded phosphoprotein and formation of inclusion bodies by recombinant baculovirus in insect cells: it binds the single-stranded RNA species. J. Gen. Virol. 71 (Pt 9), 2073-2083 (1990).

[55] Bhattacharya, B., Noad, R. J., Roy, P.: Interaction between Bluetongue virus outer capsid protein VP2 and vimentin is necessary for virus egress. Virol. J. 4, 7 (2007).

[56] Bhattacharya, B., Roy, P.: Role of lipids on entry and exit of bluetongue virus, a complex non-enveloped virus. Viruses 2, 1218-1235 (2010).

[57] Celma, C. C., Roy, P.: A viral nonstructural protein regulates bluetongue virus trafficking and release. J. Virol. 83, 6806-6816 (2009).

[58] Urakawa, T., Roy, P.: Bluetongue virus tubules made in insect cells by recombinant baculoviruses: expression of the NS1 gene of bluetongue virus serotype 10. J. Virol. 62, 3919-3927 (1988).

[59] Boyce, M., Celma, C. C., Roy, P.: Bluetongue virus non-structural protein 1 is a positive regulator of viral protein synthesis. Virol. J. 9, 178 (2012).

[60] Wilson, A. J., Mellor, P. S.: Bluetongue in Europe: past, present and future. Philos. Trans. R. Soc. Lond. B Biol. Sci. 364, 2669-2681 (2009).

[61] Hofmann, M. A., et al.: Detection of Toggenburg Orbivirus by a segment 2-specific quantitative RT-PCR. J. Virol. Methods 165, 325-329 (2010).

[62] Maan, S., et al.: Novel bluetongue virus serotype from Kuwait. Emerg. Infect. Dis. 17, 886-889 (2011).

[63] Mellor, P. S.: The replication of bluetongue virus in Culicoides vectors. Curr. Top. Microbiol. Immunol. 162, 143-161 (1990).

[64] Wilson, A., Mellor, P.: Bluetongue in Europe: vectors, epidemiology and climate change. Parasitol. Res. 103 Suppl 1, S69-S77 (2008).

[65] Dargatz, D., et al.: Bluetongue surveillance methods in the United States of America. Vet. Ital. 40, 182-183 (2004).

[66] Mellor, P. S.: Infection of the vectors and bluetongue epidemiology in Europe. Vet. Ital. 40, 167-174 (2004).

[67] Walton, T. E.: The history of bluetongue and a current global overview. Vet. Ital. 40, 31-38 (2004).

[68] Becker, M. E., et al.: Detection of bluetongue virus RNA in field-collected Culicoides spp. (Diptera: Ceratopogonidae) following the discovery of bluetongue virus serotype 1 in white-tailed deer and cattle in Louisiana. J. Med. Entomol. 47, 269-273 (2010).

[69] Lager, I. A.: Bluetongue virus in South America: over-view of viruses, vectors, surveillance and unique features. Vet. Ital. 40, 89-93 (2004).

[70] Maclachlan, N. J., Guthrie, A. J.: Re-emergence of bluetongue, African horse sickness, and other orbivirus diseases. Vet. Res. 41, 35 (2010).

[71] Tabachnick, W. J., Smartt, C. T., Connelly, C. R.: Bluetongue. Report No. ENY-743 (IN768), Gainesville (2008).

[72] Elbers, A. R., et al.: Field observations during the Bluetongue serotype 8 epidemic in 2006. II. Morbidity and mortality rate, case fatality and clinical recovery in sheep and cattle in the Netherlands. Prev. Vet. Med. 87, 31-40 (2008).

[73] Goldsmit, L., Barzilai, E.: Isolation and propagation of bluetongue virus in embryonating chicken eggs. Prog. Clin. Biol. Res. 178, 307-318 (1985).

[74] Ward, M. P., Gardner, I. A., Flanagan, M.: Evaluation of an agar gel immunodiffusion test to detect infection of cattle with bluetongue viruses in Queensland, Australia. Vet. Microbiol. 45, 27-34 (1995).

[75] Roy, P.: From genes to complex structures of blue-tongue virus and their efficacy as vaccines. Vet. Microbiol. 33, 155-168 (1992).

[76] Roy, P.: Nature and duration of protective immunity to bluetongue virus infection. Dev. Biol. (Basel) 114, 169-183 (2003).

[77] Roy, P., Boyce, M., Noad, R.: Prospects for improved bluetongue vaccines. Nat. Rev. Microbiol. 7, 120-128 (2009).

[78] Savini, G., MacLachlan, N. J., Sanchez-Vizcaino, J. M., Zientara, S.: Vaccines against bluetongue in Europe. Comp. Immunol. Microbiol. Infect. Dis. 31, 101-120 (2008).

[79] Osburn, B. I., et al.: Experimental viral-induced congenital encephalopathies. II. The pathogenesis of bluetongue vaccine virus infection in fetal lambs. Lab. Invest. 25, 206-210 (1971).

[80] Schultz, G., Delay, P. D.: Losses in newborn lambs associated with bluetongue vaccination of pregnancy ewes. J. Am. Vet. Med. Assoc. 127, 224-226 (1955).

[81] Flanagan, M., Johnson, S. J.: The effects of vaccination of Merino ewes with an attenuated Australian bluetongue virus serotype 23 at different stages of gestation. Aust. Vet. J. 72, 455-457 (1995).

[82] Ferrari, G., et al.: Active circulation of bluetongue vaccine virus serotype-2 among unvaccinated cattle in central Italy. Prev. Vet. Med. 68, 103-113 (2005).

[83] Stott, J. L., Oberst, R. D., Channell, M. B., Osburn, B. I.: Genome segment reassortment between two serotypes of bluetongue virus in a natural host. J. Virol. 61, 2670-2674 (1987).

[84] Batten, C. A., Maan, S., Shaw, A. E., Maan, N. S., Mertens, P. P.: A European field strain of bluetongue virus derived from two parental vaccine strains by genome segment reassortment. Virus Res. 137, 56-63 (2008).

[85] Savini, G., et al.: Assessment of efficacy of a bivalent BTV-2 and BTV-4 inactivated vaccine by vaccination and challenge in cattle. Vet. Microbiol. 133, 1-8 (2009).

[86] Gethmann, J., et al.: Comparative safety study of three inactivated BTV-8 vaccines in sheep and cattle under field conditions. Vaccine 27 (31), 4118-4126 (2009).

[87] González, J. M., et al.: Possible adverse reactions in sheep after vaccination with inactivated BTV vaccines.

Vet. Rec. 166, 757-758 (2010).

[88] Campbell, C. H., Barber, T. L., Knudsen, R. C., Swaney, L. M.: Immune response of mice and sheep to bluetongue virus inactivated by gamma irradiation. Prog. Clin. Biol. Res. 178, 639-647 (1985).

[89] Parker, J., Herniman, K. A., Gibbs, E. P., Sellers, R. F.: An experimental inactivated vaccine against bluetongue. Vet. Rec. 96, 284-287 (1975).

[90] Stott, J. L., Barber, T. L., Osburn, B. I.: Immunologic response of sheep to inactivated and virulent bluetongue virus. Am. J. Vet. Res. 46, 1043-1049 (1985).

[91] Stott, J. L., Osburn, B. I., Barber, T. L.: The current status of research on an experimental inactivated bluetongue virus vaccine. Proc. Ann. Meet. U. S. Anim. Health Assoc. 83, 55-62 (1979).

[92] Steinrigl, A., Revilla-Fernandez, S., Eichinger, M., Koefer, J., Winter, P.: Bluetongue virus RNA detection by RT-qPCR in blood samples of sheep vaccinated with a commercially available inactivated BTV-8 vaccine. Vaccine 28, 5573-5581 (2010).

[93] Bonneau, K. R., DeMaula, C. D., Mullens, B. A., MacLachlan, N. J.: Duration of viraemia infectious to Culicoides sonorensis in bluetongue virus- infected cattle and sheep. Vet. Microbiol. 88, 115-125 (2002).

[94] Katz, J. B., Alstad, A. D., Gustafson, G. A., Moser, K. M.: Sensitive identification of bluetongue virus serogroup by a colorimetric dual oligonucleotide sorbent assay of amplified viral nucleic acid. J. Clin. Microbiol. 31, 3028-3030 (1993).

[95] Roy, P.: Use of baculovirus expression vectors: development of diagnostic reagents, vaccines and morphological counterparts of bluetongue virus. FEMS Microbiol. Immunol. 2, 223-234 (1990).

[96] Roy, P., Bishop, D. H., LeBlois, H., Erasmus, B. J.: Long-lasting protection of sheep against bluetongue challenge after vaccination with virus-like particles: evidence for homologous and partial heterologous protection. Vaccine12, 805-811 (1994).

[97] Boone, J. D., et al.: Recombinant canarypox virus vaccine co-expressing genes encoding the VP2 and VP5 outer capsid proteins of bluetongue virus induces high level protection in sheep. Vaccine 25, 672-678 (2007).

[98] Calvo-Pinilla, E., Navasa, N., Anguita, J., Ortego, J.: Multiserotype protection elicited by a combinatorial prime-boost vaccination strategy against bluetongue virus. PLoSOne7, e34735 (2012).

[99] Jones, L. D., Chuma, T., Hails, R., Williams, T., Roy, P.: The non - structural proteins of bluetongue virus are a dominant source of cytotoxic T cell peptide determinants. J. Gen. Virol. 77 (Pt 5), 997-1003 (1996).

[100] Ma, G., et al.: An equine herpesvirus type 1 (EHV - 1) expressing VP2 and VP5 of serotype 8 bluetongue virus (BTV-8) induces protection in a murine infection model. PLoSOne7, e34425 (2012).

[101] Inumaru, S., Roy, P.: Production and characterization of the neutralization antigen VP2 of bluetongue virus serotype 10 using a baculovirus expression vector. Virology 157, 472-479 (1987).

[102] Wade-Evans, A. M., et al.: Expression of the major core structural protein (VP7) of bluetongue virus, by a recombinant capripox virus, provides partial protection of sheep against a virulent heterotypic bluetongue virus challenge. Virology 220, 227-231 (1996).

[103] Roy, P., French, T., Erasmus, B. J.: Protective efficacy of virus-like particles for bluetongue disease. Vaccine 10, 28-32 (1992).

[104] Stewart, M., et al.: Validation of a novel approach for the rapid production of immunogenic virus-like particles for bluetongue virus. Vaccine 28, 3047-3054 (2010).

[105] Stewart, M., et al.: Protective efficacy of Bluetongue virus-like and subvirus-like particles in sheep: presence of the serotype-specific VP2, independent of its geographic lineage, is essential for protection. Vaccine 30, 2131-2139 (2012).

[106] French, T. J., Marshall, J. J., Roy, P.: Assembly of double-shelled, virus-like particles of bluetongue virus by the simultaneous expression of four structural proteins. J. Virol. 64, 5695-5700 (1990).

[107] French, T. J., Inumaru, S., Roy, P.: Expression of two related nonstructural proteins of bluetongue virus (BTV) type 10 in insect cells by a recombinant baculovirus: production of polyclonal ascetic fluid and characterization of the gene product in BTV-infected BHK cells. J. Virol. 63, 3270-3278 (1989).

[108] Liu, H. M., Booth, T. F., Roy, P.: Interactions between bluetongue virus core and capsid proteins

translated in vitro. J. Gen. Virol. 73 (Pt 10), 2577–2584 (1992).

[109] Ellis, J. A., et al.: T lymphocyte subset alterations following bluetongue virus infection in sheep and cattle. Vet. Immunol. Immunopathol. 24, 49–67 (1990).

[110] Jeggo, M. H., Wardley, R. C., Brownlie, J.: Importance of ovine cytotoxic T cells in protection against bluetongue virus infection. Prog. Clin. Biol. Res. 178, 477–487 (1985).

[111] Rojas, J. M., Rodriguez-Calvo, T., Pena, L., Sevilla, N.: T cell responses to bluetongue virus are directed against multiple and identical $CD4^{+}$and $CD8^{+}$T cell epitopes from the VP7 core protein in mouse and sheep. Vaccine 29, 6848–6857 (2011).

[112] de Perez de Diego, A. C., et al.: Characterization of protection afforded by a bivalent virus-like particle vaccine against bluetongue virus serotypes 1 and 4 in sheep. PLoSOne6, e26666 (2011).

[113] Stewart, M., et al.: Bluetongue virus serotype 8 virus-like particles protect sheep against virulent virus infection as a single or multi-serotype cocktail immunogen. Vaccine 31 (3), 553–558 (2012).

[114] Balasuriya, U. B., et al.: The NS3 proteins of global strains of bluetongue virus evolve into regionaltopotypes through negative (purifying) selection. Vet. Microbiol. 126, 91–100 (2008).

[115] Mertens, P. P., et al.: Design of primers and use of RT-PCR assays for typing European bluetongue virus isolates: differentiation of field and vaccine strains. J. Gen. Virol. 88, 2811–2823 (2007).

[116] Nomikou, K., et al.: Evolution and phylogenetic analysis of full-length VP3 genes of Eastern Mediterranean bluetongue virus isolates. PLoSOne4, e6437 (2009).

[117] Pritchard, L. I., et al.: Genetic diversity of bluetongue viruses in south east Asia. Virus Res. 101, 193–201 (2004).

[118] Boyce, M., Celma, C. C., Roy, P.: Development of reverse genetics systems for bluetongue virus: recovery of infectious virus from synthetic RNA transcripts. J. Virol. 82, 8339–8348 (2008).

[119] Matsuo, E., Roy, P.: Bluetongue virus VP6 acts early in the replication cycle and can form the basis of chimeric virus formation. J. Virol. 83, 8842–8848 (2009).

[120] Matsuo, E., et al.: Generation of replication-defective virus-based vaccines that confer full protection in sheep against virulent bluetongue virus challenge. J. Virol. 85, 10213–10221 (2011).

[121] Barros, S. C., et al.: A DIVA system based on the detection of antibodies to non-structural protein 3 (NS3) of bluetongue virus. Vet. Microbiol. 137, 252–259 (2009).

（何继军译）

第 19 章　非伤寒沙门氏菌病

Beatriz San Román，Victoria
Garrido 和 María-Jesús Grilló①

摘要

沙门氏菌是一种广泛存在的革兰氏阴性细菌，属于肠杆菌科，能感染动物和人类，具有中度到重度肠胃和全身症状。动物可以作为无症状载体，间歇地在粪便中排泄沙门氏菌并污染尸体。目前，家禽和猪被认为是人类感染的主要来源。人类沙门氏菌病的控制是基于“从农场到餐桌”的可持续生物安全和卫生措施，但有效疫苗将有助于避免动物感染。由于没有商用疫苗，因此在动物模型中进行了多种实验工作，以测试非活和活的减毒疫苗，要么使用添加佐剂的沙门氏菌亚细胞成分，或者使用缺乏结构元件、必需代谢物或毒力基因的活的转基因细菌。应特别努力设计有效的抗原标记疫苗，以便区分受感染动物和疫苗接种的动物。

19.1　病原

沙门氏菌是一种革兰氏阴性、兼性厌氧、能运动、非乳糖发酵细菌，属于肠杆菌科。这种病菌经常在动物和人粪便中排泄，因此普遍存在于污水、农场废水和任何受粪便污染的物质中。

由于血清型的不断增加，沙门氏菌的分类一直很复杂且有争议。现在，沙门氏菌分为两种类型：肠炎沙门氏菌和邦戈沙门氏菌[1]，而肠炎沙门氏菌又根据生化特性和对噬菌体的易感性被细分为 6 个亚种[2]。这 6 个亚种以前用罗马数字命名，现在替换如下：

亚种Ⅰ：肠炎亚种

亚种Ⅱ：萨拉姆亚种

亚种Ⅲa：亚利桑那亚种

亚种Ⅲb：双亚利桑那亚种

亚种Ⅳ：豪顿亚种

亚种Ⅴ：印度亚种

亚种Ⅰ菌株通常从人类和温血动物中分离，而亚种Ⅱ、Ⅲa、Ⅲb、Ⅳ和亚种Ⅴ菌株和邦戈沙门氏菌通常从冷血动物和环境（很少从人类）中分离。这 6 个亚种被细分为 51 个血清群（按字母和/或数字顺序命名）和 2 600多个血清型[3]（表 19.1）。

表 19.1　沙门氏菌属的分类和鉴定的血清型数量

种	亚种	血清群[a]	血清型数量
S. enterica 肠道沙门氏菌			2 587

① B. S. Román，博士 · V. Garrido，博士 · M. J. Grilló，博士（✉）（西班牙纳瓦拉公共大学阿罗萨迪亚校区科学研究高级理事会动物卫生研究组农业生物技术研究所（CSIC-UPNA），31006 潘普洛纳　E-mail：mariajesus. grillo@ unavarra. es）

（续表）

种	亚种	血清群[a]	血清型数量
	enterica（I） 肠炎	A-C4，D1，D2， E1-E4，F-Z， 51-54，57，67	1 547
	salamae（II） 萨拉姆	B-C2，C4， D1-D3，E1， E2，F-Z， 51-53，55-60， 65	513
	arizonae（IIIa） 亚利桑那	F，G，I-L，O，P， R-Z，51，53， 56，59，62，63	100
	diarizonae（IIIb） 双亚利桑那	C1，C3，C4， F-M，O，P， R-V，X-Z， 51-53，57-61， 63，65	341
	Houtenae（IV） 豪顿	C1，F，H-L，P， R-Z，51，53，57	73
	indica（VI） 印度	C1，F，H，K，S， W，Y，Z，59	13
S. bongori（*V*） 邦戈沙门氏菌		D1，G，H，R，V， Y，60，61，66	23

[a]按字母（A-Z）或数字（51~67）顺序命名。血清组C、D、Y和E细分为1~4。

根据抗原特征，应用考夫曼-怀特方案进行血清型分类：①由脂多糖（LPS）O-链表达的菌体抗原（O抗原）；②鞭毛蛋白表达的鞭毛抗原（H抗原），能运动，以及第1相和第2相抗原的表达；③荚膜抗原（Vi抗原）[2]。因此沙门氏菌菌株按种、亚种和血清型（如果存在）或完整抗原公式的名称顺序命名。例如肠炎沙门氏菌肠炎亚种鼠伤寒沙门氏菌血清型也可以被命名为肠炎沙门氏菌肠炎亚种血清型1，4，[5]，12：i：1，2。在实践中（和本文中），缩写可以是"*S*"（斜体和大写），后面跟着非大写的血清型名称，第一个字母是大写的，例如*S. Typhimurium*（鼠伤寒沙门氏菌）。

在人类沙门氏菌病临床病例中最常分离到的菌株属于血清组A、B、C1、C2、D和E以及血清型肠炎沙门氏菌和鼠伤寒沙门氏菌[4]。进一步分类，例如噬菌体分型和脉冲场凝胶电泳（PF-GE），以便对野外菌株分类进一步评估，特别是在暴发情况下。循环沙门氏菌菌株的其他特征，作为抗菌耐药性模式，对于在风险患者和/或患有复杂疾病的患者中应用有效的抗菌治疗具有重要临床意义（见下文）。

19.2　疾病

沙门氏菌病是世界范围内主要的人畜共患疾病之一，是发达国家重要的食源性疾病。事实上在美国，非伤寒肠炎沙门氏菌是急性肠胃炎的首要原因，2011年登记了1 027 561例人类病例，其中19 336例（1.9%）需要住院治疗，378例（1.95%）是致命的[5]。在欧盟（EU），沙门氏菌病是仅次于弯曲杆菌病的第二常见人畜共患病，2010年登记了99 020例人类病例[4]。在过去的10年中，据报告沙门氏菌病发病率有所下降[4]，这可能是由于从农场到餐桌的食物链中实施的控制措施以及公众对预防食源性感染的卫生控制措施重要性认识提高了。

在所描述的2 600多种沙门氏菌血清型中，临床表现和死亡率因细菌和宿主特征而不同。所有血清型都被认为对人类具有潜在的致病性，对宿主具有不同程度的适应性[6]（图19.1）。因此某些血清型，如伤寒沙门氏菌、副伤寒沙门氏菌和仙台沙门氏菌，会对人类造成严重的系统性疾病。然而，其他血清型特别易感动物，如猪霍乱沙门氏菌、牛都柏林沙门氏菌、绵羊流产沙门氏菌和家禽鸡沙门氏菌，仅偶尔感染人类，引起轻微症状[7,8]。无处不在的血清型，如鼠伤寒沙门氏菌、肠炎沙门氏菌和婴儿沙门氏菌，是最公认的人畜共患病因子，影响着广泛的动物物种[9-11]。

家禽和猪感染无症状，主要分别由肠炎沙门氏菌和鼠伤寒沙门氏菌感染，被认为是人类沙门氏菌感染的主要来源[4,9]，占欧盟报告的人类沙门氏菌病总血清型的75.6%。其他动物，如野生鸟类、啮齿动物、蜥蜴或家龟，无症状感染沙门氏菌可能有助于其传播并成为人类感染源[5,12,13]。

临床上，猪和家禽沙门氏菌病可产生败血症或小肠结肠炎（图19.1）。前者的特征是大量腹泻、全身感染症状（发热、虚脱等），且在没有抗菌药物治疗情况下死亡率很高。小肠结肠炎的特征是急性或慢性胃肠炎症状，腹泻是最常见的症状。接受治疗的动物可以从感染中恢复，在数月内通过粪便消除细菌，持续很长时间，并充当无症状携带者[14]。亚临床无症状沙门氏菌病是家禽和猪（人类感染的主要来源）中最常见的表现，由感染扁桃体、肠道和肠系膜淋巴结的多种血清型产生，并间歇排泄[14,15]。虽然临床形式易于识别，但在常规检查中无法检测到无症状携带者，这是人类面临的主要风险[16]。

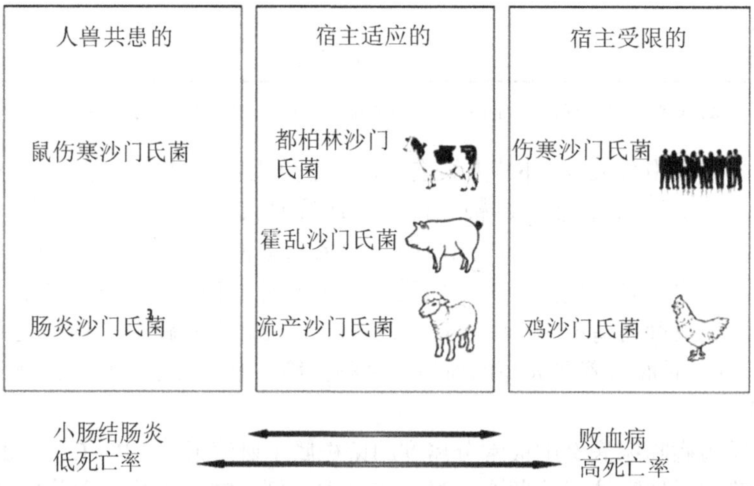

图19.1　沙门氏菌感染宿主的适应性及其对宿主临床形态的影响

人类沙门氏菌病主要是通过从动物源性摄取生的或未煮熟的受污染食物获得的，主要是从家禽（43.8%；鸡蛋和肉）和猪（26.9%；肉），以及通过摄取未经高温消毒的牛奶[11]。事实上，过去几年观察到的人类沙门氏菌病发病率下降与人类食用的鸡蛋和家禽产品中沙门氏菌控制行动的实施有关[6]。因此，猪源性感染变得更加相关，鼠伤寒沙门氏菌和德比沙门氏菌是欧盟最常见的血清型[17,18]。

人类感染的过程是在摄入受污染食物后12~72小时内出现急性胃肠炎，包括腹泻、呕吐、腹痛和发烧等症状。治疗以补液为基础，偶尔需要抗生素给药，只有当细菌到达血流时才会导致菌血症。这种并发症在免疫受损的患者中尤其危险，甚至是致命的，例如那些感染艾滋病病毒、患有癌症或正在接受免疫抑制治疗的患者，或者改变内源性肠道菌群的患者，尤其是儿童和老年人。在这些情况下，建立有效的抗生素治疗对控制感染至关重要，因此，应在感染早期评估病原体耐药性特性。与人类沙门氏菌病发生有关的另一个问题是沙门氏菌菌株对多种抗菌剂产生耐药性，通常与对动物施用抗生素有关。

19.3　发病机理

口服后，胃的酸性环境会破坏大部分沙门氏菌。存活的沙门氏菌到达回肠末端和盲肠，在肠细胞内复制并通过肠屏障，被覆盖在派伊尔结上的 M 细胞捕获，表达 CD18 分子的吞噬细胞和/或活跃进入非吞噬性肠细胞（图 19.2）[19]。一旦进入薄层介质，巨噬细胞（病原体靶细胞）入侵需要两个不同的Ⅲ型分泌系统（T3SSs），编码在单独沙门氏菌致病岛上（SPI-1 和 SPI-2）[20]。SPI-1 和 SPI-2 都提供了将细菌效应因子传递到宿主细胞、调节宿主细胞功能（SPI-1）所需的多种蛋白，如细胞骨架重组和细胞因子基因表达以及将含沙门氏菌液泡（SCV）转化为细胞内复制生态位（SPI-2）[21]。沙门氏菌局部感染后，细菌脂多糖与 Toll 样受体 4（TLR-4）的相互作用激活树突状细胞和巨噬细胞，进而触发宿主免疫应答[22]。病原体到达肠系膜淋巴结和一般淋巴系统，通过血液传播到脾脏、肝脏等器官，引起菌血症。

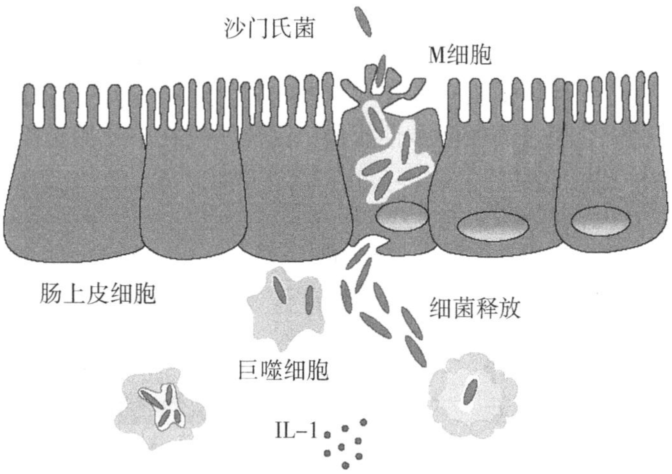

图 19.2　沙门氏菌在肠道内致病示意图

在肠道急性感染期间，病原体通过粪便大量排泄，最终可能导致持续的无症状感染，并通过粪便间歇性排泄较长时间。无症状感染的宿主作为健康宿主的传染源，通过直接接触或间接食物污染。建立持续性感染是一个复杂过程，包括宿主和病原体成分。

一方面宿主显示固有免疫反应，以消除来自生物体的感染（见下文），另一方面沙门氏菌与宿主微生物群竞争，以到达其目标细胞，从而发展或激活在极端环境中生存的策略[23]。事实上，有人认为菌毛和黏附素是一种试图维持在肠道内的表现[24]。同样对胃酸 pH 的抵抗和使用炎症相关代谢物的能力提供了对宿主微生物群的生长优势[25]。此外，外部因素（如最近的抗生素治疗）有利于通过竞争性微生物群抑制建立沙门氏菌感染，并延长病原体排泄[26]。因此，抗生素治疗仅推荐用于沙门氏菌感染的败血症临床症状，而不推荐用于轻度至中度健康患者（见下文）。

天然免疫系统首次识别沙门氏菌是由 TLR-4 通过 MD2 作为 LPS 特异性受体介导的，激活对细胞外和 SCV 空泡病原体的转录反应[27]。这种受体的刺激会触发巨噬细胞的细胞因子（如 IL-1β 和 TNF-α）和蛋白质（如蛋白水解酶和抗菌阳离子肽）表达。此后，多种其他机制有助于通过免疫系统控制沙门氏菌，如 SCV 酸化、防御素和活性氧中间产物的分泌[28]。初次感染后，在家畜和人类身上都能检测到抗体（主要针对沙门氏菌脂多糖和蛋白质）和特异性 T 细胞应答。体液应答不仅限于活细菌，因为灭活疫苗或亚单位疫苗能够诱发产生特异性抗体，但细胞免疫应答只有在主动感染后才能检测到。其他地方已经广泛指出，体液和细胞免疫应答水平并不总是与预防或消除病原体的先天免疫系统状态相关[29]。

19.4 沙门氏菌分离和分型

为了保护消费者的健康，欧盟当局制定了在所有成员国强制执行的法规，以控制食品和动物感染中的沙门氏菌属细菌。对于食品，微生物和卫生标准在 2005 年 11 月 15 日的委员会法规（EC）2073/2005 附录一中确定[30]，要求在 10 g/样品或 25 g/样品的 5 个样品中，根据产品类型，在屠宰后和冷藏前的整个产品保存期或动物尸体中，不存在沙门氏菌属细菌。为了控制动物沙门氏菌病，应沿着“从农场到餐桌”的生产链进行彻底控制[31,32]。目前，在对 27 个成员国进行了良好标准化参考研究后，控制猪沙门氏菌病正在所有欧盟地区实施[31]。

ISO（国际标准化组织）6579：2002/DAM 1：2007（国际标准化组织先行）[33]是国际上推荐的沙门氏菌分离的标准化方法，根据样本类型（粪便、淋巴结、食物、水）稍加修改。一般来说，样本是单独采集的，尽管动物和环境样本都可以在池中采集，用于流行病学目的。在亚临床感染中，沙门氏菌可能通过粪便间歇性排泄；因此，应在未使用抗生素的宿主中进行后续取样。

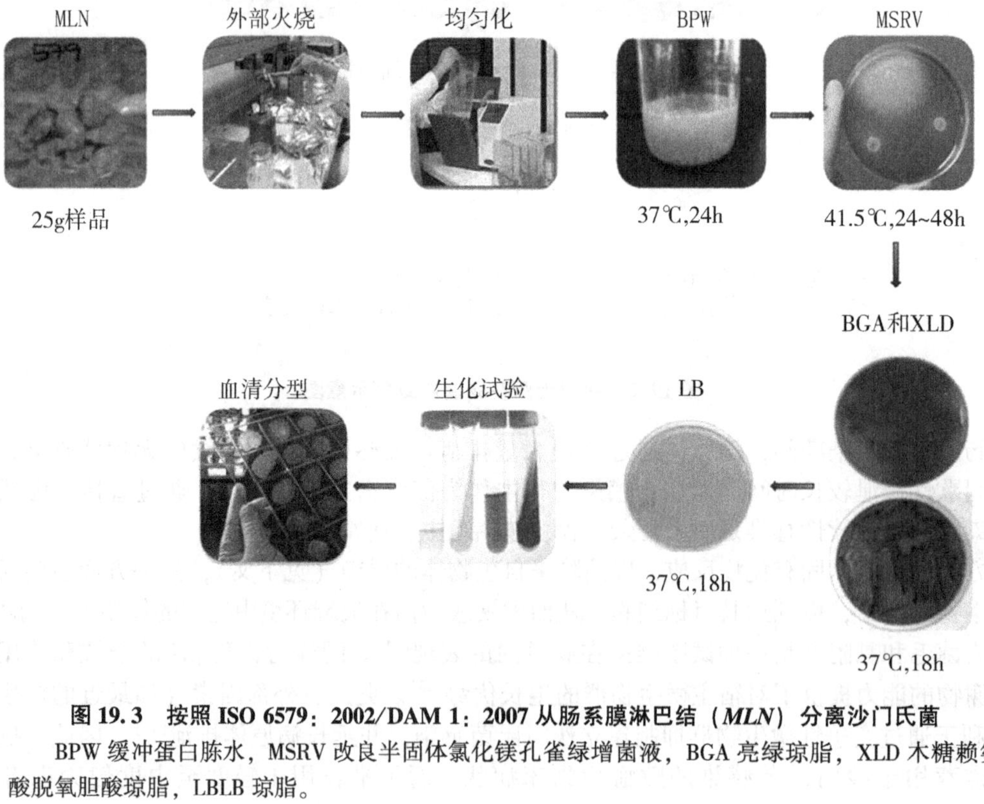

图 19.3 按照 ISO 6579：2002/DAM 1：2007 从肠系膜淋巴结（*MLN*）分离沙门氏菌

BPW 缓冲蛋白胨水，MSRV 改良半固体氯化镁孔雀绿增菌液，BGA 亮绿琼脂，XLD 木糖赖氨酸脱氧胆酸琼脂，LBLB 琼脂。

ISO 方法包括逐步使用培养基，从无培养基到高度选择性培养基，以成功分离沙门氏菌（图

19.3)。因此，可疑菌落应在缓冲蛋白胨水（BPW）中非选择性预富集后分离，随后在改良半固体氯化镁孔雀绿增菌液（MSRV）中半选择性富集，并在木糖赖氨酸脱氧胆酸琼脂（XLD）和亮绿琼脂（BGA）等两种固体选择性培养基中进行最终选择性培养。在琼脂平板中纯化单个菌落后，应进行生化试验，如尿素琼脂、赖氨酸、吲哚和三糖铁琼脂（TSI）或商业分析剖面指数（API），以确认可疑菌落的身份。最后，细菌应通过玻片凝集与特定的单克隆小鼠血清进行确认血清分型，以确定O、H和VI抗原的变体（见上文）。这项技术要求使用超过150种特定血清和经过精心培训的人员，因此应该在沙门氏菌参考实验中心进行。

其他技术（如噬菌体分型和分子分型）可用于更好地描述沙门氏菌菌株，特别适用于全球监测和疫情调查。噬菌体分型是通过特定噬菌体集合的溶解或溶原活性来确定的，如卫生保护局（HPA；科林达，英国）所述的17肠炎沙门氏菌分型噬菌体（SETP）[34]或34鼠伤寒沙门氏菌分型噬菌体[35]。最广泛使用的分子分型技术是多位点可变数目分析（MLVA）和脉冲场凝胶电泳（PFGE），后者是在暴发情况和系统发育研究中指纹菌株最广泛接受的技术。PFGE相对便宜，但费时费力，需要很好的标准化，对不同血清型表现出不同的敏感性。分子生物学发展导致了新诊断技术出现，例如通过使用多重PCR[36]或珠阵列[37]或ORF基因组比较[38,39]，分析编码O和H抗原的基因进行基因分型。

19.5 血清学诊断

在某些流行病学情况下，ELISA和其他血清学检查是有用的工具[40]。然而血清学在取样时并不代表沙门氏菌感染，因为体液应答在生物体内持续的时间比细菌长，相反沙门氏菌感染发生得很快，而血清转化需要很长的时间。

对于人类诊断，已经开发了4种用于检测伤寒沙门氏菌的ELISA试剂盒，而对于非伤寒沙门氏菌则很少。事实上，大多数实验室目前使用自己内部测试的试验方法，并取得了可接受的成功，标准化ELISA需求已经被搁置，或者在某些情况下被放弃，转而采用PCR或其他分子技术。在这种情况下，一些作者指出了属于不同血清群的脂多糖组合，以提高对大量血清型的检测[41]。

在兽医方面，有多种商业ELISA试剂盒可用于监测猪、家禽和食品感染。这些试剂盒可以使用动物健康监测研究收集的血清或肉汁进行快速诊断，但是血清阳性率并不总是与实际感染状况一致，限制了对预期沙门氏菌流行率较低的地区使用ELISA方法，但并不作为唯一的感染控制工具[42]。

19.6 沙门氏菌感染治疗

肠道细菌感染治疗的一般建议是水合作用和软性饮食，同时注意个人卫生。只有在特殊情况下，重病患者或高危患者（免疫抑制患者、婴儿等）应该用抗菌剂治疗。通常用于治疗人沙门氏菌病的药物是成人用氟喹诺酮类（环丙沙星）、甲氧苄啶-磺胺甲恶唑或阿莫西林-克拉维酸，儿童用第三代头孢菌素[43,44]。滥用抗生素治疗可能导致多重耐药菌株出现，正如20世纪80年代用氨苄西林、氯霉素和甲氧苄啶-磺胺甲恶唑进行大规模治疗后发生的那样。在这些情况下，抗菌剂不仅在进一步治疗中无效，而且会导致疾病严重程度和死亡率增加。此外已经证明，由于内源性微生物群的消除和整个胃肠道沙门氏菌的强化，用氨苄青霉素或阿莫西林治疗儿童经常会延长沙门氏菌排泄期和临床复发[45]。

动物被认为是多重耐药沙门氏菌菌株的主要来源，这是由于在饮食中系统使用抗生素作为生长促进剂以及在其生产期治疗多重感染过程而产生的选择性压力所致[46]。在欧盟，2008年猪和牛对四环素、氨苄青霉素和磺胺类药物耐药的沙门氏菌菌株数量最多[43]。还应该考虑到，使用抗菌剂进行治疗或促进生长也会破坏肠道菌群，这通常会增加猪对沙门氏菌感染的易感性[47]。因此，抗生素的使用可能会引发沙门氏菌在畜群中的传播。如果动物得不到治疗，这种情况就不会发生。除了选择性压力之外，获得耐药性还可能受到其他因素影响，例如某些沙门氏菌血清型获得和固定遗传因子的遗传倾向。

正如世界卫生组织早些时候所建议的那样[48]，可以得出结论，控制沙门氏菌感染不应该建立在使用抗菌剂基础上，耐药性的出现是非常谨慎使用抗菌剂的另一个严重原因，多重耐药性鼠伤寒沙门氏菌 DT104 的出现就是例证[49]。在 DT104 克隆中出现了经典的五抗性 ACSSuT（氨基青霉素、氯霉素、链霉素、磺胺类药物和四环素类药物）基因，这与动物中使用抗菌剂和受感染动物的国际贸易有关[50,51]。因此，多种抗生素（包括氟喹诺酮类和第三代头孢菌素类）耐药的沙门氏菌菌株出现是最近与猪源性相关的一个重要公共健康问题[43,52]。自 2006 年 1 月 1 日起，现行欧洲法规禁止使用抗生素作为生长促进剂，建议在动物中限制使用抗生素，并对所有沙门氏菌分离株进行抗菌监测[53]。

另外，正在实施基于改变内源性肠道菌群组成饮食的替代性新疗法，以利于从胃肠道中消除沙门氏菌[54]。

19.7 疫苗作用机理

在接种疫苗后的第一阶段诱导血清转化，其基础是 IgM 和 IgG 抗体。此外，在黏膜疫苗接种的情况下可以检测到产生 IgA。脂多糖、鞭毛蛋白、菌毛和其他蛋白质（脂蛋白、外膜蛋白、热休克蛋白）是刺激免疫系统的主要因素。体液免疫应答诱导已在广泛宿主中被引用，从实验室实验模型（小鼠）到家畜（即小牛）和人类，接受亚单位疫苗、灭活疫苗或减毒活疫苗。考虑到文献中描述的所有结果，抗体存在和对感染的抗性之间的相关性并不总是能够确定的。此外，基于体液免疫应答的保护作用仅在一些实验条件下描述过，例如低剂量或中等毒性菌株的挑战。

非亚单位疫苗和灭活疫苗的减毒活疫苗，已被描述为能够诱导 Th1 细胞免疫应答（CD4 和 CD8 淋巴细胞），同时在不同宿主中存在细胞因子（IL-12、IFN-γ）。事实上，据报道，一些亚单位疫苗和灭活疫苗是 Th2 免疫应答诱导剂。如上所述，关于体液应答、细胞免疫应答并不总是与沙门氏菌疫苗接种后的保护相关。因此，需要进一步研究来阐明沙门氏菌保护性免疫机制。

19.8 疫苗控制和动物模型

确保兽用疫苗纯度、安全性、效力和功效的方法可能因国家而异，这取决于当地的需求。然而，适当的标准和生产控制对于确保在动物健康计划中使用一致的、高质量产品的可用性是至关重要的[55]。应根据纯度（革兰氏染色、培养基生长均一性以及血清分型和噬菌体分型）、无害性（对小鼠和终末宿主的致死性、副作用、减毒活疫苗稳定性和无逆转性、向牛奶或鸡蛋的传播）、功效（小鼠和终末宿主的保护水平）和环境行为（疫苗在粪便和垃圾中的持久性、感染周围动物的能力）对设计的疫苗进行测试[55]。此外，所有这些最终控制测量应在不同批次的生产过程中应用，以确保制造方法的一致性。

灵长类沙门氏菌病与人类症状学非常相似，但由于实践、经济和伦理原因，对这些动物使用非常有限。此外，小鼠和小牛模型已成功地用于阐明毒性和免疫机制[56,57]。小鼠实验感染沙门氏菌可引起迅速的全身感染，表现为发热、寒冷引起的毛发直立、虚脱、饥饿等症状，继之以肝脾定殖最终导致死亡[58]。此外，小鼠已被用于评估沙门氏菌疫苗对致命性[59]或亚致死攻击的有效性[60,61]。

因此，人类肠炎的特征是腹泻，牛模型被认为是合适的，因为被都柏林沙门氏菌感染的小牛表现出菌血症、流产，并成为慢性携带者，而感染鼠伤寒沙门氏菌的小牛是无症状携带者[58]。结扎髂环模型在小牛、猪、兔和小鼠中得到了广泛使用，因为它具有研究细菌毒力因子和早期胃肠道感染步骤的潜力，尽管该模型不能预测疫苗的保护程度[58]。

19.9　疫苗和接种的基本原理

沙门氏菌作为人畜共患传染病的病原体，在人类中的流行与在动物中的流行直接相关。由于没有足够安全的沙门氏菌疫苗可用于人类，因此控制这种感染应以动物预防和卫生措施为基础，以避免病原体向其他动物和人类传播，后者主要通过家禽和猪源食物传播。为了保护消费者的健康，目前的立法涉及“从农场到餐桌”对沙门氏菌的完全控制[31,32]。在整个初级生产过程中的关键点，如饲料检测、屠宰时的健康和卫生控制以及处理和食用家禽、猪肉和衍生物时的卫生措施，被认为是控制沙门氏菌传播和感染的关键点。

19.9.1　非活疫苗

灭活疫苗和亚单位疫苗在过去已经在人类和动物中广泛使用，成功率参差不齐[62,63]。各种简单或组合的细菌组分，如 Vi 多糖、脂多糖、O 链、菌毛或孔蛋白已被用作非活疫苗。例如脂多糖与蛋白质载体的组合已证明能诱导兔和小鼠的抗体反应，但对攻毒感染的效力有限[64]。为了提高效力，非活疫苗需要与经典佐剂（如氢氧化铝、完全弗氏佐剂和不完全弗氏佐剂）或新型佐剂［如纤连蛋白（EDA）额外结构域 A］联合施用。一般来说，非活疫苗是安全的，但会诱导强烈的体液应答和差的 Th1 细胞介导的免疫应答，导致低效力[29,65]和对感染诊断的血清学不良干扰，因为脂多糖和菌毛是用于诊断试验的沙门氏菌免疫显性抗原。事实上，在设计动物沙门氏菌疫苗时，区分受感染动物和接种动物（DIVA）是一个优先事项。

目前，对动物沙门氏菌病疫苗的研究主要集中在活疫苗设计上，这种活疫苗允许一种减毒、安全和有效的沙门氏菌菌株，反过来又会引起血清反应，从而能够区分受感染动物和接种动物（DIVA）。然而，减毒概念因沙门氏菌血清型和涉及的动物种类而异，因为事实上大多数影响人类的沙门氏菌血清型对动物是“减毒”的，而不会诱发疾病。因此，一种“安全”的沙门氏菌疫苗应该被理解为不能被排泄并污染环境和食物链。

19.9.2　减毒活疫苗

一般来说，活疫苗被认为优于灭活疫苗，因为前者有以下优势：①同时诱导细胞介导的免疫应答和体液免疫应答；②单剂量给药后有效；③口服给药后诱导黏膜免疫应答；④用作递送其他重组抗原的载体并成为多价疫苗；⑤生产成本低，易于储存[29]。

早期开发的活沙门氏菌疫苗是体外培养后获得的自发突变体，如鸡沙门氏菌 9R[66]；温度处理，如伤寒沙门氏菌 TS[67]；化学选择性压力，如鼠伤寒沙门氏菌中的亚硝基胍-NTG[68]或流产沙门氏菌中的链霉素 Rv6[69]；或紫外线辐射[70]。这些疫苗已证明对小鼠、家禽和牛有效，其中一些已经获得使用许可。然而，这些活疫苗对动物有严重的副作用，如已经描述过的脓毒性关节炎或肝炎[71]。沙门氏菌致病性知识和分子生物学技术的进步为设计新的减毒活沙门氏菌株提供了可能性，这些减毒活沙门氏菌株在与毒力和/或免疫原性相关的基因中具有明确且不可逆的突变。沙门氏菌基因的功能鉴定使得选择和突变那些参与体内细菌存活和感染过程的基因成为可能，包括那些编码细菌结构成分和必需代谢物生物合成和毒力基因。所有这些描述如下。

19.9.2.1　细菌结构成分中的突变体

由于脂多糖是血清学诊断试验中的主要毒力因子和免疫显性抗原，开发粗糙脂多糖突变体已成为构建允许 DIVA 疫苗的有趣方法。总的来说，完整的脂多糖核心沙门氏菌突变体被认为比深度粗糙突变体更能有效抵抗恶性感染。事实上，一些缺乏不同脂多糖部分的沙门氏菌突变株

如 Δ*waaH* 和 Δ*waaL*，已经被提议作为活疫苗候选物[61,72,73]，但是其他粗糙突变株被认为太弱（例如鼠伤寒沙门氏菌 Δ*waaG*），不能提供足够的保护来防止强毒感染或毒力太强而不能作为安全疫苗，已经放弃了这些突变体作为候选疫苗的可能性[74]。

由于半乳糖是脂多糖核心的一个组成部分，无半乳糖差向异构酶或 *galE* 突变体不能合成尿苷二磷酸半乳糖（UDP-Gal）差向异构酶。因此，不能将尿苷二磷酸葡萄糖（UDP-Glu）转化为 UDP-Gal，反之亦然。这种类型的 Δ*galE* 突变体是在 20 世纪 80 年代开发的，在体外没有这种糖的情况下显示出不完整的 LPS（深度粗糙表型）。然而如果半乳糖是外源提供的，就像体内一样，Δ*galE* 突变体可以回复到光滑表型，也就是回复到毒性形式。这种现象已经在接种鼠伤寒沙门氏菌 Δ*galE* 突变株的小牛中描述过，它不仅不能保护小牛免受致命感染，还会引起小牛腹泻、发热甚至死亡[75,76]。在有伤寒沙门氏菌[77]和肠炎沙门氏菌[78]遗传背景的 Δ*galE* 突变体中获得了类似结果。为了试图避免从粗糙到光滑的表型转化，已经描述了肠炎沙门氏菌 Δ*gal* 操纵子（包括 *galM*、*galK*、*galT* 和 *galE* 基因）突变体[61]。尽管没有副作用，但是在用肠炎沙门氏菌 Δ*gal* 操纵子接种并通过腹膜内途径攻毒的小鼠中没有观察到保护作用，这表明深度粗糙突变体不是对抗平滑沙门氏菌感染的有效疫苗[6]。

外膜蛋白 OmpC 和 OmpF 合成受 *ompR* 基因调控。肠炎沙门氏菌 Δ*ompR* 基因高度减毒，能在口服后诱导适度保护[79]。此外，还描述了个体 Δ*ompC* 和 Δ*ompF* 减毒突变体。在 DIVA 背景下，Omp 突变体作为脂多糖突变体的替代物出现，导致接种但未感染动物中缺乏抗 Omp 的抗体。

19.9.2.2 细菌必需代谢产物中的突变体

编码芳香（*aro*）合成途径的基因突变已被描述为不同动物模型中的减毒和有效疫苗[29]。基因 *aroA*、*aroC* 和 *aroD* 已被广泛用于设计单突变体和双突变体疫苗，在都柏林沙门氏菌和鼠伤寒沙门氏菌中用于小牛[80]，在肠炎沙门氏菌和鸡沙门氏菌中用于鸡[81]。此外，Δ*aroC*Δ*aroD* 和 Δ*aroA*Δ*aroC* 双突变体保持免疫原性，回复到毒性表型机会最小，因此被提议作为候选疫苗[82]。

阻碍腺苷单磷酸合成的基因（*pur* 突变体）需要腺嘌呤的外部输入，导致体内急剧衰减。鼠伤寒沙门氏菌和都柏林沙门氏菌 *purA* 突变体显示在小鼠中定殖和存活的能力降低，刺激免疫反应不足，导致低水平的保护[83,84]。

最后，腺苷酸环化酶（*cya*）和环腺苷酸受体（*crp*）基因调节参与碳水化合物、氨基酸和细胞表面结构利用的其他基因的表达，如 Omps、菌毛和鞭毛。尽管 Δ*cya* 和 Δ*crp* 突变体高度减毒，它们在脾脏中的存活非常有限，但是口服免疫已经导致小鼠免受口服攻击。鼠伤寒沙门氏菌和霍乱沙门氏菌 Δ*cya*Δ*crpA* 双突变体分别在鸡和猪中对亲本或口服攻击有效[85,86]。

19.9.2.3 细菌毒力基因突变体

为了降低沙门氏菌在宿主体内生长的能力但能维持宿主免疫系统的刺激以抵抗毒性感染，研究了沙门氏菌的毒力基因。在此背景下，染色体（*invA*、*hilA*、*PhoP* / *PhoQ* 双组分调控系统）和质粒（*spvB*、*spvC*）的几个基因表现出不同程度的毒力。例如肠炎沙门氏菌 Δ*hilA*、Δ*spvB* 和 Δ*spvC* 突变体表现出中等至高毒力，而 Δ*invA* 和 Δ*phoP* 突变体表现出低毒力，但对强毒感染的保护也很低[78,87]。

结论

沙门氏菌病是一种主要通过食物获得的人畜共患病。人沙门氏菌病的发病率与家禽和猪的感染率直接相关，通常无症状。由于没有足够安全的疫苗可供人类使用，卫生当局建议在动物阶段"从农场到餐桌"控制感染。由于应避免在农场一级进行抗菌治疗，以避免沙门氏菌菌株出现多药耐药性，因此动物疫苗接种可以成功地加强（但不能替代）基于卫生和卫生措施的控制方案。减毒活疫苗通常比亚细胞疫苗更有效。然而活疫苗仍保留一些残留毒力，其大规模应用可能涉及将转基因新病原体引入食物链的风险。从实用的角度来看，未来发展动物疫苗的方向应该集中在新型沙门氏菌活疫苗上，这种疫苗能够区分感染动物和接种动物（DIVA）。

致谢

B. San Román 的合同由 CSIC（计划 JAE-Doc）提供资金，V. Garrido 的合同由 Gobierno de Navarra 提供资金（项目 IIQ14064. RI1）。

参考文献

[1] Tindall, B. J., Grimont, P. A., Garrity, G. M., Euzeby, J. P.: Nomenclature and taxonomy of the genus *Salmonella*. Int. J. Syst. Evol. Microbiol. 55, 521-524 (2005). doi: 10. 1099/ijs. 0. 63580-0.

[2] Grimont, P. A., Weill, F. X.: Antigenic formulae of the *Salmonella* serovars, WHO., I. P. a., ed. (Institute Pasteur and World Health Organization. Collaboration Centre for Reference Research on *Salmonella*), 9th edition. (2007).

[3] Guibourdenche, M., et al.: Supplement 2003-2007 (No. 47) to the White-Kauffmann-Le Minor scheme. Res. Microbiol. 161, 26-29 (2010).

[4] EFSA-ECDC: The European Union Summary report on trends and sources of Zoonoses, Zoonotic Agents and food-borne outbreaks in 2010. EFSA J. 10, 2597 (2012). doi: 10. 2903/j. efsa. 2012. 2597.

[5] CDC: In National Center for Emerging and Zoonotic Infectious Diseases 2. CDC: http: //www. cdc. gov/foodborneburden/2011-foodborne-estimates. html (2012).

[6] EFSA: Scientific Opinion on a quantitative estimation of the public health impact of setting a new target for the reduction of *Salmonella* in broilers. EFSA J. 9, 2106 (2011). doi: 10. 2903/j. efsa. 2011. 2106.

[7] Stevens, M. P., Humphrey, T. J., Maskell, D. J.: Molecular insights into farm animal and zoonotic *Salmonella* infections. Philos. Trans. R. Soc. Lond. B Biol. Sci. 364, 2709-2723 (2009). doi: 10. 1098/ rstb. 2009. 0094.

[8] Nielsen, L. R.: Review of pathogenesis and diagnostic methods of immediate relevance for epidemiology and control of *Salmonella* Dublin in cattle. Vet. Microbiol. (2012). doi: 10. 1016/j. vetmic. 2012. 08. 003.

[9] Hald, T., Vose, D., Wegener, H. C., Koupeev, T.: A Bayesian approach to quantify the contribution of animal food sources to human salmonellosis. Risk Anal. 24, 255-269 (2004). doi: 10. 1111/j. 0272-4332. 2004. 00427. x.

[10] Hald, T., Wingstrand, A., Brondsted, T., Lo Fo Wong, D. M.: Human health impact of *Salmonella* contamination in imported soybean products: a semi-quantitative risk assessment. Foodborne Pathog. Dis. 3, 422-431 (2006). doi: 10. 1089/fpd. 2006. 3. 422.

[11] Pires, S. M., de Knegt, L., Hald, T.: Estimation of the relative contribution of different food and animal sources to human *Salmonella* infections in the European Union. Scientific/technical report submitted to EFSA. http: //www. efsa. europa. eu/en/supporting/ doc/184e. pdf (2011).

[12] Hernandez, E. et al.: *Salmonella* Paratyphi B var Java infections associated with exposure to turtles in Bizkaia, Spain, September 2010 to October 2011. Euro Surveill. 17, 25 (2012) [pii: 20201].

[13] Lowther, S. A., et al.: Foodborne outbreak of *Salmonella* subspecies IV infections associated with contamination from bearded dragons. Zoonoses Public Health 58, 560-566 (2011). doi: 10. 1111/j. 1863-2378. 2011. 01403. x.

[14] Nollet, N., et al.: *Salmonella* in sows: a longitudinal study in farrow-to-finish pig herds. Vet. Res. 36, 645-656 (2005). doi: 10. 1051/vetres: 2005022.

[15] Merialdi, G., et al.: Longitudinal study of *Salmonella* infection in Italian farrow-to-finish swine herds. Zoonoses Public Health 55, 222-226 (2008). doi: 10. 1111/j. 1863-2378. 2008. 01111. x.

[16] Mousing, J., et al.: Nation-wide *Salmonella enterica* surveillance and control in Danish slaughter swine herds. Prev. Vet. Med. 29, 247-261 (1997), S0167587796010823 [pii].

[17] EFSA: Report of the Task Force on Zoonoses data collection on the analysis of the baseline survey on the

prevalence of *Salmonella* in slaughter pigs. Part A. EFSA J. 135, 1-111 (2008).

[18] EFSA: Analysis of the baseline survey on the prevalence of *Salmonella* in holdings with breeding pigs in the EU, 2008. Part A: *Salmonella* prevalence estimates. EFSA J. 7, 1-93 (2009).

[19] Carter, P. B., Collins, F. M.: The route of enteric infection in normal mice. J. Exp. Med. 139, 1189-1203 (1974).

[20] Stebbins, C. E., Galan, J. E.: Priming virulence factors for delivery into the host. Nat. Rev. Mol. Cell Biol. 4, 738-743 (2003). doi: 10. 1038/nrm1201.

[21] Galan, J. E.: Molecular genetic bases of *Salmonella* entry into host cells. Mol. Microbiol. 20, 263-271 (1996).

[22] Miller, S. I., Ernst, R. K., Bader, M. W.: LPS, TLR4 and infectious disease diversity. Nat. Rev. Microbiol. 3, 36-46 (2005). doi: 10. 1038/nrmicro1068.

[23] Bearson, B. L., Wilson, L., Foster, J. W.: A low pH-inducible, PhoPQ-dependent acid tolerance response protects *Salmonella typhimurium* against inorganic acid stress. J. Bacteriol. 180, 2409-2417 (1998).

[24] Wagner, C., Hensel, M.: Adhesive mechanisms of *Salmonella enterica*. Adv. Exp. Med. Biol. 715, 17-34 (2011). doi: 10. 1007/978-94-007-0940-9_2.

[25] Winter, S. E., et al.: Gut inflammation provides a respiratory electron acceptor for *Salmonella*. Nature 467, 426-429 (2010). doi: 10. 1038/nature09415.

[26] Hohmann, E. L.: Nontyphoidal salmonellosis. Clin. Infect. Dis. 32, 263-269 (2001). doi: 10. 1086/318457.

[27] Inohara, N., Chamaillard, M., McDonald, C., Nunez, G.: NOD-LRR proteins: role in host-microbial interactions and inflammatory disease. Annu. Rev. Biochem. 74, 355-383 (2005). doi: 10. 1146/annurev. biochem. 74. 082803. 133347.

[28] Prost, L. R., Sanowar, S., Miller, S. I.: *Salmonella* sensing of anti-microbial mechanisms to promote survival within macrophages. Immunol. Rev. 219, 55-65 (2007). doi: 10. 1111/j. 1600-065X. 2007. 00557. x.

[29] Mastroeni, P., Chabalgoity, J. A., Dunstan, S. J., Maskell, D. J., Dougan, G.: *Salmonella*: immune responses and vaccines. Vet. J. 161, 132-164 (2001). doi: 10. 1053/tvjl. 2000. 0502.

[30] EC: Reglamento (CE) no 2073/2005 de la Commission regulation (EC) No 2073/2005 of 15 November on 2005 microbiological criteria for foodstuffs. http: // eur-lex. europa. eu/LexUriServ/LexUriServ. do? uri=OJ: L: 2005: 338: 0001: 0026: EN: PDF (2005).

[31] DOUE: Commission regulation (EU) No 1086/2011 of 27 October 2011 amending Annex II to Regulation (EC) No 2160/2003 of the European Parliament and of the Council and Annex I to Commission Regulation (EC) No 2073/2005 as regards salmonella in fresh poultry meat. Official Journal of the European Union. L 281/7. (2006).

[32] DOUE: Commission regulation (EU) No 1086/2011 of 27 October 2011 amending Annex II to Regulation (EC) No 2160/2003 of the European Parliament and of the Council and Annex I to Commission Regulation (EC) No 2073/2005 as regards salmonella in fresh poultry meat. Official Journal of the European Union. L 281/7. (2011).

[33] ISO. International Organisation for Standardisation.: ISO 6579: 2002/DAM 1: 2007. Microbiology of food and animal feeding stuffs. Horizontal method for the detection of *Salmonella* spp. Annex D: detection of *Salmonella* spp. in animal faeces and in samples from the primary production stage. Geneva, Switzerland (2007).

[34] De Lappe, N., Doran, G., O'Connor, J., O'Hare, C., Cormican, M.: Characterization of bacteriophages used in the *Salmonella enterica* serovar Enteritidis phage-typing scheme. J. Med. Microbiol. 58, 86-93 (2009). doi: 10. 1099/jmm. 0. 000034-0.

[35] Baggesen, D. L., Wegener, H. C.: Phage types of *Salmonella enterica* ssp. *enterica* serovar *Typhimurium* isolated from production animals and humans in Denmark. Acta Vet. Scand. 35, 349-354 (1994).

[36] Herrera-Leon, S., et al.: Blind comparison of traditional serotyping with three multiplex PCRs for the identification of *Salmonella* serotypes. Res. Microbiol. 158, 122-127 (2007). doi: 10. 1016/j. resmic. 2006.

09. 009.

[37] McQuiston, J. R., Waters, R. J., Dinsmore, B. A., Mikoleit, M. L., Fields, P. I.: Molecular determination of H antigens of *Salmonella* by use of a microsphere-based liquid array. J. Clin. Microbiol. 49, 565-573 (2011). doi: 10. 1128/JCM. 01323-10.

[38] Arrach, N., et al.: *Salmonella* serovar identification using PCR-based detection of gene presence and absence. J. Clin. Microbiol. 46, 2581-2589 (2008). doi: 10. 1128/JCM. 02147-07.

[39] Porwollik, S., et al.: Characterization of *Salmonella enterica* subspecies I genovars by use of microarrays. J. Bacteriol. 186, 5883-5898 (2004). doi: 10. 1128/JB. 186. 17. 5883-5898. 2004.

[40] Narayanappa, D., Sripathi, R., Jagdishkumar, K., Rajani, H. S.: Comparative study of dot enzyme immunoassay (Typhidot-M) and Widal test in the diagnosis of typhoid fever. Indian Pediatr. 47, 331-333 (2010).

[41] Kuhn, K. G., et al.: Detecting non-typhoid *Salmonella* in humans by ELISAs: a literature review. J. Med. Microbiol. 61, 1-7 (2012). doi: 10. 1099/ jmm. 0. 034447-0.

[42] Vico, J. P., Engel, B., Buist, W. G., Mainar-Jaime, R. C.: Evaluation of three commercial enzyme-linked immunosorbent assays for the detection of antibodies against *Salmonella* spp. in meat juice from finishing pigs in Spain. Zoonoses Public Health 57 (Suppl 1), 107-114 (2010).

[43] EFSA-ECDC: European Union Summary Report Antimicrobial resistance in zoonotic and indicator bacteria from humans, animals and food in the European Union in 2010. EFSA J. 10, 233 (2012).

[44] Molbak, K.: Human health consequences of antimicrobial drug-resistant *Salmonella* and other food borne pathogens. Clin. Infect. Dis. 41, 1613-1620 (2005). doi: 10. 1086/497599.

[45] Nelson, J. D., Kusmiesz, H., Jackson, L. H., Woodman, E.: Treatment of *Salmonella* gastroenteritis with ampicillin, amoxicillin, or placebo. Pediatrics 65, 1125-1130 (1980).

[46] Tello, A., Austin, B., Telfer, T. C.: Selective pressure of antibiotic pollution on bacteria of importance to public health. Environ. Health Perspect. 120, 1100-1106 (2012). doi: 10. 1289/ehp. 1104650.

[47] van der Wolf, P. J., et al.: A longitudinal study of *Salmonella enterica* infections in high-and low-seroprevalence finishing swine herds in The Netherlands. Vet. Q. 23, 116-121 (2001). doi: 10. 1080/01652176. 2001. 9695096.

[48] Office international of epizooties (OIE) manual of diagnostic tests and vaccines for terrestrial animals, 7th ed. World health organization (WHO) -OIE, Paris, France (2012). http: //www. oie. int/fi leadmin/ Home/eng/Health_standards/tahm/2. 09. 09_SALMONELLOSIS. pdf.

[49] Threlfall, E. J.: Antimicrobial drug resistance in *Salmonella*: problems and perspectives in food-and water-borne infections. FEMS Microbiol. Rev. 26, 141-148 (2002).

[50] Besser, T. E., Goldoft, M., Pritchett, L. C., Khakhria, R., Hancock, D. D., Rice, D. H., Gay, J. M., Johnson, W., Gay, C. C.: Multiresistant *Salmonella* Typhimurium DT104 infections of humans and domestic animals in the Pacific Northwest of the United States. Epidemiol. Infect. 124, 193-200 (2000).

[51] Helms, M., Ethelberg, S., Molbak, K.: International *Salmonella* Typhimurium DT104 infections, 1992-2001. Emerg. Infect. Dis. 11, 859-867 (2005).

[52] Rodríguez, I., et al.: Extended-spectrum beta-lactamases and AmpC beta-lactamases in ceftiofur-resistant *Salmonella enterica* isolates from food and livestock obtained in Germany during 2003-07. J. Antimicrob. Chemother. 64, 301-309 (2009). doi: 10. 1093/jac/dkp195.

[53] DOUE: Commission Decision of 12 June 2007 on a Harmonised monitoring of antimicrobial resistance in *Salmonella* in poultry and pigs. 2007/407/EC. Official Journal of the European Union. OJ L 153 26-29 (2007).

[54] Sonnenburg, J. L., Chen, C. T., Gordon, J. I.: Genomic and metabolic studies of the impact of probiotics on a model gut symbiont and host. PLoS Biol. 4, e413 (2006). doi: 10. 1371/journal. pbio. 0040413.

[55] OIE: Manual of Diagnostic Tests & Vaccines for Terrestrial Animals. (2008).

[56] Costa, L. F., Paixao, T. A., Tsolis, R. M., Baumler, A. J., Santos, R. L.: Salmonellosis in cattle: advantages of being an experimental model. Res. Vet. Sci. 93, 1-6 (2012). doi: 10. 1016/j. rvsc. 2012. 03. 002.

[57] Kaiser, P., Diard, M., Stecher, B., Hardt, W. D.: The streptomycin mouse model for *Salmonella* diar-

rhea: functional analysis of the microbiota, the pathogen's virulence factors, and the host's mucosal immune response. Immunol. Rev. 245, 56–83 (2012). doi: 10. 1111/j. 1600–065X. 2011. 01070. x.

[58] Santos, R. L., et al.: Animal models of *Salmonella* infections: enteritis versus typhoid fever. Microbes Infect. 3, 1335–1344 (2001).

[59] Ochoa, J., et al.: Protective immunity of biodegradable nanoparticle–based vaccine against an experimental challenge with *Salmonella* Enteritidis in mice. Vaccine 25, 4410–4419 (2007). doi: 10. 1016/j. vaccine. 2007. 03. 025.

[60] Estevan, M., Irache, J. M., Grillo, M. J., Blasco, J. M., Gamazo, C.: Encapsulation of antigenic extracts of *Salmonella enterica* serovar. Abortusovis into polymeric systems and efficacy as vaccines in mice. Vet. Microbiol. 118, 124–132 (2006).

[61] San Román, B., et al.: The extradomain a of fibronectin enhances the efficacy of lipopolysaccharide defective *Salmonella* bacterins as vaccines in mice. Vet. Res. 43, 31 (2012). doi: 10. 1186/1297–9716–43–31.

[62] Collins, F. M.: Vaccines and cell–mediated immunity. Bacteriol. Rev. 38, 371–402 (1974).

[63] Thatte, J., Rath, S., Bal, V.: Immunization with live versus killed *Salmonella typhimurium* leads to the generation of an IFN–gamma–dominant versus an IL–4–dominant immune response. Int. Immunol. 5, 1431–1436 (1993).

[64] Watson, D. C., Robbins, J. B., Szu, S. C.: Protection of mice against *Salmonella typhimurium* with an O–specific polysaccharide–protein conjugate vaccine. Infect. Immun. 60, 4679–4686 (1992).

[65] Simon, R., Tennant, S. M., Galen, J. E., Levine, M. M.: Mouse models to assess the efficacy of nontyphoidal *Salmonella* vaccines: revisiting the role of host innate susceptibility and routes of challenge. Vaccine 29, 5094–5106 (2011). doi: 10. 1016/j. vaccine. 2011. 05. 022.

[66] Smith, H. W.: The use of live vaccines in experimental *Salmonella gallinarum* infection in chickens with observations on their interference effect. J. Hygiene 54, 419–432 (1956).

[67] Gherardi, M. M., Garcia, V. E., Brizzio, V., Sordelli, D. O., Cerquetti, M. C.: Differential persistence, immunogenicity and protective capacity of temperature–sensitive mutants of *Salmonella* Enteritidis after oral or intragastric administration to mice. FEMS Immunol. Med. Microbiol. 7, 161–168 (1993).

[68] Mitov, I., Denchev, V., Linde, K.: Humoral and cell–mediated immunity in mice after immunization with live oral vaccines of *Salmonella* Typhimurium: auxotrophic mutants with two attenuating markers. Vaccine 10, 61–66 (1992).

[69] Lantier, F., Pardon, P., Marly, J.: Vaccinal properties of *Salmonella* Abortusovis mutants for streptomycin: screening with a murine model. Infect. Immun. 34, 492–497 (1981).

[70] Tang, I. K., et al.: Characterization of a highly attenuated *Salmonella enterica* serovar Typhimurium mutant strain. J. Microbiol. Immunol. Infect. 35, 229–235 (2002).

[71] Hormaeche, C. E., Pettifor, R. A., Brock, J.: The fate of temperature–sensitive *salmonella* mutants *in vivo* in naturally resistant and susceptible mice. Immunology 42, 569–576 (1981).

[72] Germanier, R.: Immunity in experimental salmonellosis I. Protection induced by rough mutants of *Salmonella* Typhimurium. Infect. Immun. 2, 309–315 (1970).

[73] Nagy, G., et al.: Down–regulation of key virulence factors makes the *Salmonella enterica* serovar Typhimurium *rfaH* mutant a promising live–attenuated vaccine candidate. Infect. Immun. 74, 5914–5925 (2006). doi: 10. 1128/IAI. 00619–06.

[74] Nagy, G., et al.: "Gently rough": the vaccine potential of a *Salmonella enterica* regulatory lipopolysaccharide mutant. J. Infect. Dis. 198, 1699–1706 (2008). doi: 10. 1086/593069.

[75] Wray, C., Sojka, W. J., Pritchard, D. G., Morris, J. A.: Immunization of animals with *gal E* mutants of "*Salmonella* Typhimurium". Dev. Biol. Stand. 53, 41–46 (1983).

[76] Clarke, R. C., Gyles, C. L.: Galactose epimeraseless mutants of *Salmonella* Typhimurium as live vaccines for calves. Can. J. Vet. Res. 50, 165–173 (1986).

[77] Hone, D. M., et al.: A *galE* via (Vi antigen–negative) mutant of *Salmonella* typhi Ty2 retains virulence in humans. Infect. Immun. 56, 1326–1333 (1988).

[78] Karasova, D., et al.: Comparative analysis of *Salmonella enterica* serovar Enteritidis mutants with a vaccine potential. Vaccine 27, 5265-5270 (2009). doi: 10. 1016/j. vaccine. 2009. 06. 060.

[79] Dorman, C. J., Chatfield, S., Higgins, C. F., Hayward, C., Dougan, G.: Characterization of porin and *ompR* mutants of a virulent strain of *Salmonella* Typhimurium: *ompR* mutants are attenuated *in vivo*. Infect. Immun. 57, 2136-2140 (1989).

[80] Robertsson, J. A., Lindberg, A. A., Hoiseth, S., Stocker, B. A.: *Salmonella* Typhimurium infection in calves: protection and survival of virulent challenge bacteria after immunization with live or inactivated vaccines. Infect. Immun. 41, 742-750 (1983).

[81] Segall, T., Lindberg, A. A.: Oral vaccination of calves with an aromatic-dependent *Salmonella* Dublin (O9, 12) hybrid expressing O4, 12 protects against Ponercursiva. Dublin (O9, 12) but not against *Salmonella* Typhimurium (O4, 5, 12). Infect. Immun. 61, 1222-1231 (1993).

[82] Tacket, C. O., et al.: Comparison of the safety and immunogenicity of delta *aroC* delta *aroD* and delta *cya* delta *crp* *Salmonella*Typhi strains in adult volunteers. Infect. Immun. 60, 536-541 (1992).

[83] Sigwart, D. F., Stocker, B. A., Clements, J. D.: Effect of a *purA* mutation on efficacy of *Salmonella* live-vaccine vectors. Infect. Immun. 57, 1858-1861 (1989).

[84] McFarland, W. C., Stocker, B. A.: Effect of different purine auxotrophic mutations on mouse-virulence of a Vi-positive strain of *Salmonella* Dublin and of two strains of *Salmonella* Typhimurium. Microb. Pathog. 3, 129-141 (1987).

[85] Curtiss 3rd, R., Kelly, S. M.: *Salmonella* Typhimurium deletion mutants lacking adenylate cyclase and cyclic AMP receptor protein are avirulent and immunogenic. Infect. Immun. 55, 3035-3043 (1987).

[86] Kennedy, M. J., et al.: Attenuation and immunogenicity of Δ*cya*Δ*crp* derivatives of *Salmonella* Choleraesuis in pigs. Infect. Immun. 67, 4628-4636 (1999).

[87] Galan, J. E., Curtiss 3rd, R.: Distribution of the *invA*, *-B*, *-C*, and *-D* genes of *Salmonella* Typhimurium among other *Salmonella* serovars: *invA* mutants of *Salmonella* typhi are deficient for entry into mammalian cells. Infect. Immun. 59, 2901-2908 (1991).

（马维民，杨亚民译）

[78] [illegible]. Comparative analysis of Salmonella enterica serovar Typhimurium mutants with a vaccine potential. Vaccine 27, 5265-5277 (2009), doi: 10.1016/j.vaccine.2009.06.060.

[79] [illegible], Garfield, [illegible], Higgins, [illegible], [illegible], [illegible] [illegible] [illegible] [illegible] [illegible] [illegible] Infect. Immun. 57, 2136-2140 (1989).

[80] Robertsson, J. A., Lindberg, A. A., [illegible], Stocker, B. A. [illegible] Salmonella Typhimurium infection in calves: protection and survival [illegible] [illegible] [illegible] [illegible] [illegible] [illegible] [illegible] [illegible] vaccines. Infect. Immun. 41, 742-750 (1983).

[81] Sagall, [illegible], Chatfield, [illegible], A. A. [illegible] [illegible] [illegible] [illegible] [illegible] dependent Salmonella Dublin [illegible] [illegible] [illegible] [illegible] [illegible] [illegible] Dublin [illegible] [illegible] [illegible] against Salmonella Typhimurium [illegible] Infect. Immun. [illegible], 1222-1231 (1991).

[82] Tacket, C. O., et al. Comparison of the safety and immunogenicity of aroC aroD and cya crp deletion mutants of Salmonella Typhi strains in adult volunteers. Infect. Immun. 60, 536-541 (1992).

[83] [illegible], [illegible], [illegible], [illegible], [illegible] Effect of [illegible] [illegible] [illegible] [illegible] [illegible] [illegible] vaccine [illegible]. Infect. Immun. 57, 1858-1864 (1989).

[84] McFarland, W. C., Stocker, B. A. [illegible] [illegible] [illegible] [illegible] [illegible] [illegible] [illegible] Vi-positive strain of Salmonella Dublin and of two strains of Salmonella Typhimurium. Microb. Pathog. 3, 129-141 (1987).

[85] [illegible]

[86] [illegible]

[87] [illegible]. Immun. 57, 2136-2140 (1989).

第四部分

癌症疫苗

概　述

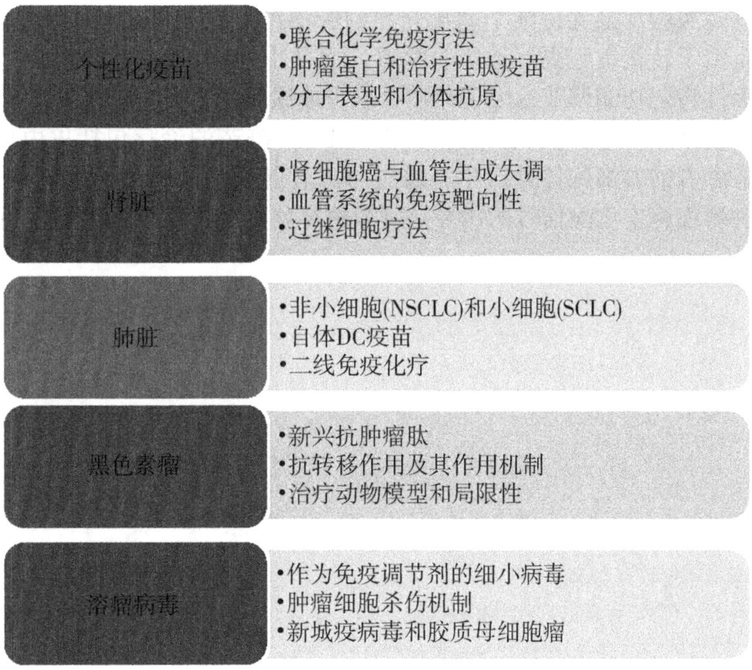

2009 年，第一个进入市场的个性化癌症疫苗是俄罗斯的肾细胞癌疫苗 Oncophage。美国和欧盟监管机构在后期研究失败后都拒绝了 Oncophage 疫苗。每种个性化疫苗的基础都是个体的基因图谱。这一新方法得到了科学证实，毫无疑问是一种有前景的治疗新策略。但分析的 DNA 与转录后的 mRNA 和翻译后的蛋白之间没有一一对应的关系，对个性化疫苗的功能反应主要取决于免疫系统的效率。肿瘤细胞过载和严格化疗将导致免疫能力丧失。因此对于个性化疫苗的成功，基于一组不同生物标志的治疗窗口（第 1 章）是必不可少的。

联合治疗　由于化疗药物的免疫抑制作用，化疗联合免疫治疗一直存在争议。在过去的 10 年中，数据积累表明，将这两种疗法结合起来可能更具有优势。为此，一些化疗药物导致癌细胞的免疫原性死亡和免疫调节细胞亚群的选择性消耗。

个性化肽疫苗　除了来自致癌传染因子的免疫原性抗原被宿主免疫系统识别为外源性外，用免疫原性较低的“自身”抗原肽进行疫苗接种也显示出实质性进展。特别是疫苗接种的改进策略，如针对“个性化”的抗原预筛选、多种肽联合使用以及与其他治疗方式的联合治疗增加了临床效益。

肾细胞癌及其血管系统　肿瘤血管系统不稳定会增加肿瘤内部间质压力，阻止致肿瘤细胞死亡药物和免疫效应细胞进入肿瘤微环境（TME）。与从健康组织分离的血管内皮细胞相比，肿瘤血管内皮细胞（VEC）表现出不同的基因表达，这种转录被认为是肿瘤血管失稳的基础。使用 siRNA 方法沉默这些基因可能阻止血管内皮细胞迁移和血管形成，从而支持肿瘤血管内皮细胞作为一个高度相关的治疗靶点。

肺癌免疫疗法　抗原没有理想的来源。自体与异体、单价与多价、全细胞与重组制剂之间的区别部分是理论性的，部分是实用性的。自体肿瘤抗原有针对患者产生特异性疫苗优势，在不了解患者癌症抗原表达的前提下，生产自体肿瘤疫苗通常所需的肿瘤组织体积将适于可手术切除的肿瘤。

黑色素瘤 已经从多种来源描述了具有抗肿瘤活性的天然肽。其中一些是游离分子，另一些是蛋白质水解释放或化学合成的蛋白质的内部序列。它们的抗菌和抗肿瘤活性类似于古老的先天免疫分子，在抗体和获得性 T 细胞免疫出现之前，这些分子在抵御威胁性疾病（感染或其他）方面是有效的。

细小病毒 许多实验观察支持细小病毒介导的肿瘤抑制是免疫原性的观点。如前所述，细小病毒嗜癌性不是由于转化细胞更好地吸收病毒，而是由于病毒生命周期细胞内步骤依赖于作为细胞周期和致癌性转化功能调节的因子存在、缺失或激活。

新城疫病毒 新城疫病毒-树突状细胞（NDV-DC）疫苗的治疗包括皮内应用树突状细胞，树突状细胞从癌症患者的血细胞体外产生，通过与患者肿瘤的病毒溶瘤物孵育而被激活。新城疫病毒感染时，肿瘤细胞中危险信号的诱导是刺激树突状细胞产生强抗肿瘤作用的关键。

第 20 章　癌症疫苗及其结合标准癌症疗法的潜在益处

Eva Ellebæk，Mads Hald Andersen
和 Inge Marie Svane①

摘要

许多不同种类的癌症疫苗正在开发中，包括广泛的不同抗原靶点和配方。尽管在过去的几十年中已经进行了许多癌症疫苗接种试验，但对于大多数患者，临床效果仍有待证实。免疫治疗的一个障碍可能是免疫抑制机制，如调节性 T 细胞和骨髓源性抑制细胞。

由于化疗药物的免疫抑制作用，化疗联合免疫治疗一直存在争议。在过去的 10 年中，累积数据表明，将这两种疗法结合起来可能更具有优势。为此，一些化疗药物导致癌细胞的免疫原性死亡和免疫调节细胞亚群的选择性消耗。

许多实验室和诊所正在研究免疫疗法与标准化疗方案或低剂量化疗相结合以提高免疫疗法的效果，目前结果似乎很有希望。

癌症疫苗方法也可能受益于与其他种类癌症治疗方法的结合，例如靶向治疗和免疫修饰抗体。

对选定的研究进行了综述，以解决目前关于结合癌症疫苗和现有癌症治疗的知识，还讨论了未来的展望。

20.1　引言/背景

癌症疫苗

利用癌症疫苗诱导治疗性宿主抗肿瘤免疫反应具有巨大的潜力，可以不重叠地补充传统癌症疗法。2010 年，首个治疗性癌症疫苗获得了 FDA 的批准（Sipuleucel-T，Provenge®）[1]。这是几十年来研究和临床试验高潮，研究癌症疫苗的潜在抗癌作用。尽管如此，疫苗接种试验的临床结果令人失望，但也有一些例外，新方法和策略正在进行深入调查。这包括疫苗本身，也包括疫苗接种时机以及最近疫苗与其他抗肿瘤疗法的结合。

一般来说，成功的疫苗接种需要两种成分：抗原和佐剂。"抗原"代表应该从体内清除的目标；佐剂是抗原本身诱导反应的放大器。癌症疫苗是不同于预防性疫苗的治疗性疫苗，因为它们的目的是克服患者已经存在的疾病。癌症疫苗接种的原则建立在针对所谓 TAA（肿瘤相关抗原）的免疫基础上。有核细胞在细胞核中加工细胞内蛋白质并在细胞表面与 HLA（人类白细胞抗原分子）结合呈现这些蛋白质的小部分。

在表面，肽可以被免疫细胞即 T 细胞识别，然后可以启动针对细胞的免疫反应。肿瘤细胞表面表达的一些抗原不同于正常细胞表面的抗原。TAA 可分为不同的亚型：过度表达的抗

① E. Ellebæk，医学博士 · I. M. Svane，医学博士（✉）（丹麦赫尔利夫哥本哈根大学医院血液科癌症免疫治疗中心（CCIT），丹麦赫尔利夫哥本哈根大学医院肿瘤科，E-mail：inge.marie.svane@regionh.dk）

M. H. Andersen，博士（丹麦赫尔利夫哥本哈根大学医院血液科癌症免疫治疗中心（CCIT））

原包括分化抗原，分化抗原在正常细胞上低表达，但在细胞发生恶性转化时高表达，癌症睾丸抗原仅在免疫专属睾丸细胞和肿瘤细胞尤其在突变抗原上表达（表 20.1）。肽应用代表了靶向单个或几个抗原的最简单的方法之一，但是实现这一点的其他方法也是可能的，例如通过使用全蛋白、RNA 或 DNA。因此癌症疫苗接种针对一种或多种这些 TAA，并且存在许多不同的疫苗方法。一般来说，佐剂可诱导抗体反应或细胞反应，这是因为该反应在某种程度上取决于佐剂。

表 20.1　肿瘤相关抗原分类

肿瘤相关抗原（TAA）	TAA 举例	TAA 典型表达的癌症	正常细胞上的表达	恶性细胞上的表达
分化抗原	MART - 1/Melan - A，gp100，tyrosinase	恶性黑色素瘤	低	高
过度表达的抗原	Survivin，hTERT HER2/neu oncogene	几乎所有癌症 乳腺癌和卵巢癌		
	CEA AFP	胃肠癌和肺癌 生殖细胞肿瘤，肝细胞癌	尤其是在胚胎发育期间	存在
癌症睾丸抗原	NY-ESO-1 MAGE，GAGE	各种癌症 恶性黑色素瘤	仅在睾丸细胞上	存在
突变抗原	p53，BRAF，KRAS	在几种肿瘤中有不同程度的表达	存在但没有突变	存在和突变
病毒抗原	HPV	宫颈癌、头颈癌	不存在	存在

由于该领域的大部分努力都致力于诱导细胞（T 细胞）反应，因此在癌症疫苗接种中对这一重点佐剂进行了最深入研究。在这方面，树突状细胞激活引起了广泛关注，因为树突状细胞被认为是先天免疫应答和获得性免疫应答之间相互作用的关键介质，被描述为自然界自身的佐剂。然而包括合成佐剂（例如油乳剂中的细菌提取物）、DNA 或病毒载体在内的其他方法正在研究中。佐剂可以增加疫苗的免疫原性（如卡介苗（BCG）、破伤风毒素，白介素（IL-2）、干扰素（IFN）、胸腺法新、粒细胞巨噬细胞集落刺激因子（GM-CSF））或降低免疫调节机制（如 CD25 抗体、化疗）。

20.2　免疫抑制机制

已经研究了许多不同的、有希望的疫苗策略，但迄今为止缺乏令人信服的临床疗效。免疫抑制机制可能是一个关键因素。为了生存，肿瘤发展了几种防御机制以避免来自免疫系统的攻击（知识点 20.1）。这些机制包括主要组织相容性复合物（MHC）分子的下调、免疫系统识别抗原的丢失，以及吸引和上调调节或抑制免疫细胞的分子分泌[2]，特别是已经描述了两种类型的调节性免疫细胞：调节性 T 细胞（Treg）和髓源性抑制细胞（MDSC）。

Treg 是由 CD4、CD25 和转录因子叉头盒 p3（Foxp3）表达所定义的 CD4 T 细胞的一个子集。Treg 通过多种不同的抑制机制工作；通过穿孔素和颗粒酶 B 依赖性途径杀死 T 细胞和抗原呈递细胞（APCs），释放 IL-10 和 TGF-β，从而抑制 T 细胞活化和抗原呈递细胞功能，在抗原呈递细胞上诱导吲哚胺 2，3-双加氧酶（IDO），导致在微环境中降低 T 细胞活化和耐受性[3]。

另一个免疫抑制细胞亚群是骨髓前体细胞的异质性群体 MDSC。定义为 HLA-DR 阴性和 CD33CD11b 阳性，分为单核细胞（$CD14^+$）和粒细胞亚型（$CD15^+$）。它们通过消耗半胱氨酸和精氨酸酶，直接抑制 CD4 和 CD8 T 细胞，或者分泌活性氧物质抑制 T 细胞功能[4]。此外，Treg 和 MDSC 都具有上调和吸引肿瘤环境中免疫抑制机制的能力。

知识点 20.1　抗免疫攻击的肿瘤防御机制

• 免疫识别能力降低/抗原表达丧失
-MHC 分子下调
-肿瘤内抗原处理缺失
• 抵抗免疫攻击
-诱导抗凋亡机制
• 在肿瘤环境中产生免疫抑制细胞因子，如 TGF-β、IDO 和 VEGF。
• 免疫抑制细胞亚群、Treg 和 MDSC 的吸引或诱导

IDO 酶是一种色氨酸分解代谢酶，近年来 IDO 被认为是肿瘤细胞微环境中免疫调节和免疫耐受发展的重要因素。在肿瘤环境和肿瘤引流淋巴结中，通过直接抑制 T 细胞或通过 Treg 诱导免疫抑制[5]，IDO 产生耐受性环境。在癌症患者中 IDO 在癌细胞和抗原呈递细胞中都有表达。因此，通过接种 IDO 诱导 IDO 特异性 T 细胞具有双重潜力，通过消除免疫抑制性 IDO 阳性免疫细胞和消除 IDO 阳性肿瘤细胞来阻止肿瘤免疫耐受的发展[6]。我们研究所正在对非小细胞肺癌（NSCLC）患者进行 I 期试验，以评估用 IDO 衍生肽进行疫苗接种的安全性和有效性（www. clinicaltrials. gov id：nct01219348）。

20.3　化学疗法

化疗免疫学优势

尽管由于化疗药物的免疫抑制潜力，将化疗与免疫疗法结合使用似乎有争议，但有证据表明，这种结合可能有更多益处，而不是负面影响。几个小组在临床前模型中检查了不同化疗药物对免疫细胞和免疫防御机制的效力[7-9]。

知识点 20.2　化疗药物的免疫调节作用

• 免疫原性细胞凋亡——肿瘤细胞凋亡后肿瘤相关抗原的释放是由化疗诱导的，其特征是：
-钙网蛋白在癌细胞表面的表达
-三磷酸腺苷的释放
• 脆弱的肿瘤细胞-化疗可诱导细胞膜渗透到颗粒酶 B→增加 T 细胞介导的杀伤的敏感性
• 淋巴耗竭后 T 细胞增殖增加→向增殖的 T 细胞克隆重新呈现抗原→T 细胞对肿瘤细胞的反应性增加
• 减少调节性 T 细胞的数量和功能

化疗药物通过多种机制发挥免疫调节功能（见知识点 20.2）。首先，当肿瘤细胞在用某些化疗药物治疗后发生凋亡时，会释放几种抗原。这些抗原被 APC 吸收，递呈给 T 细胞，并且可以启动针对在表面表达这些抗原的其他癌细胞的免疫反应。这被称为免疫原性凋亡，其特征在于细胞表面钙网蛋白的表达[9]和三磷酸腺苷的释放[10]。其次，肿瘤细胞在用某些化疗药物治疗后对 T 细胞介导的裂解更敏感，这是因为细胞膜对颗粒酶 B 更具渗透性[11]。这可能导致肿瘤细胞杀伤增加。再次，已经观察到化疗诱导的淋巴细胞减少症通常伴随着细胞因子的分泌，即 IL-7、IL-15 和 IL-21。这可能导致一段时间的 T 细胞增殖，在此期间，新的抗原被递呈给增殖的 T 细胞克隆，这种反应可以被用来重新定向免疫以抵抗肿瘤的敌对反应[7]。最后如上所述，Treg 在调节自然诱导的免疫反应以及免疫治疗中发挥重要作用，已经表明化疗药物通过直接杀死和降

低这些细胞的抑制功能来靶向 Treg[12,13]。

20.4 组合化学免疫疗法（表 20.2）

20.4.1 与小剂量免疫调节化疗联合的化学免疫疗法

环磷酰胺是近年来因其潜在的免疫调节作用而被广泛研究的化疗药物之一[8]。有研究表明，持续小剂量环磷酰胺可能会减少调节性 T 细胞数量，并通过增加成熟标志的表达和增强抗原递呈来影响树突状细胞的功能[7]。

表 20.2 组合化学免疫疗法举例

	试验示例	药物/剂量	疫苗	癌症	病人数量	反应/功效
低剂量化疗，环磷酰胺（Cy）	Ge 等[1]	Cy 50 mg/天	无	乳腺癌	12	Treg 短暂减少，抗肿瘤 T 细胞反应增强
	Emens 等[15]	Cy 200 mg/m^2+阿霉素 35 mg/m^2	HER2+同种异体 GM-CSF 疫苗	乳腺癌	28	HER2 特异性体液免疫增强
	Holtl 等[16]	接种疫苗前 3~4 天 Cy 为 300 mg/m^2，与无 Cy 相比	同种异体 DC 疫苗	肾细胞癌	10（+Cy）vs. 12（-Cy）	两名混合反应患者——也是免疫反应最高的患者；都接种了 Cy
	Ghiringhelli 等[17]	Cy 50 mg×2/天 每两周	无	晚期癌症	28	Treg 选择性降低
	Ellebaek 等[18]	Cy 50 mg×2/天 每两周	自体 DC 疫苗+IL-2	转移性黑色素瘤	28	58% 患者的 Treg 没有下降——疾病稳定
标准剂量化疗	Kyte 等[20]	替莫唑胺	端粒酶肽疫苗	转移性黑色素瘤	25	免疫应答者数量增加和长期临床应答
	Harrop 等[23]	5-FU，甲酰四氢叶酸，伊立替康	Trovax（5 种 T4 编码病毒疫苗）	转移性结直肠癌	12	免疫反应增强与临床活动和免疫反应的相关性
	Nistico 等[24]	达卡巴嗪±	Melan-A/MART-1 肽疫苗	II-IV 期无病黑色素瘤	5（+达卡巴嗪）5（-达卡巴嗪）	3 名接受无病联合治疗的患者的抗肿瘤 T 细胞数量增加
	Quoix 等[25]	顺铂和吉西他滨	±TG4010（MUC-1-和 IL-2-编码痘病毒）	非小细胞肺癌	每组 74 名患者（±疫苗）	使用该组合治疗的患者 ORR 改善和生存期延长

在一项初步研究中，对 12 名转移性乳腺癌患者进行了环磷酰胺持续剂量试验，每日 50 mg，为期 3 个月[14]。Treg 水平短暂降低，但在治疗期间完全恢复。与之相反，抗肿瘤 T 细胞反应随着 Treg 的减少而增加，并在整个治疗过程中维持在高水平。这是由于减少了对原有肿瘤特异性 T 细胞的抑制以及重新产生肿瘤特异性细胞。此外，肿瘤反应性 T 细胞的数量与临床疗效相关。

因此，即使这种剂量的环磷酰胺本质上不能降低Treg的水平和功能，仍然对抗肿瘤反应有影响。

Emens等在I期研究中[15]测试了环磷酰胺和阿霉素与HER2阳性同种异体GM-CSF分泌肿瘤疫苗联合使用时的最佳剂量，对每种药物的不同剂量进行测试，以确定最适合诱导HER2特异性免疫的化疗剂量。结果表明，200 mg/m^2环磷酰胺和35 mg/m^2阿霉素可提高HER2特异性体液免疫水平，维持DTH的发育。研究还表明，由于大剂量环磷酰胺会干扰免疫和DTH的发展，因此治疗窗口狭窄，强调在需要保持免疫调节作用时，化疗剂量很重要。

在另一项研究中，转移性肾细胞癌患者在Ⅰ期/Ⅱ期试验中接受同种异体树突状细胞疫苗接种治疗，10例患者与环磷酰胺联合接种，12例不接种环磷酰胺[16]。在接受治疗的22名患者中，所有患者均接受环磷酰胺治疗，2名患者有混合反应，1名患者病情稳定。免疫应答极弱或缺失，但2例混合应答患者具有最强的KLH特异性和非抗原依赖性增殖反应，提示环磷酰胺能增强免疫和临床反应。

最后，Ghiringhelli等用持续剂量的环磷酰胺治疗28例晚期癌症患者，每天两次，每次50 mg，疗程为1周，治疗后休息1周。这一方案导致Treg选择性降低。然而在临床试验中[18,19]，环磷酰胺方案与树突状细胞疫苗和IL-2联合使用，但在该试验中Treg的数量并未减少。这是否与IL-2联合治疗有关将在一项新的试验中进行评估，该试验使用持续剂量的环磷酰胺联合DC疫苗，但不使用IL-2（clinicaltrial. gov Identifier NCT00978913）。

20.4.2 标准剂量化疗联合化学免疫疗法

几项临床研究重点是在标准化疗中加入癌症疫苗接种策略的可行性。2011年在挪威开始的一项研究中[20]，Kyte等用替莫唑胺联合16个氨基酸的长端粒酶肽疫苗治疗25例转移性黑色素瘤患者。替莫唑胺的剂量为200 mg/m^2，每4周连续5天给药，在开始化疗后第2周和第3周给予前5次注射，间隔增加。临床结果显示，5例患者获得部分反应，另有6例患者病情稳定。所有5名临床应答者在疫苗接种期间均出现了疫苗特异性T细胞反应，这种反应持续整个肿瘤消退期。生存分析显示，与对照组相比，OS得到了扩展，应答者存活了5年以上。

我们做了两项有趣的观察：首先，该小组先前公布了仅用疫苗治疗黑色素瘤患者的临床试验结果，但是在将疫苗与化疗相结合的试验中，观察到对疫苗有免疫反应的患者数量增加。其次，观察到的临床反应是长期的（长达5年），并逐渐发展（长达近1年），而不是像化疗预期的那样在诱导治疗后不久。这些事实表明，标准剂量的替莫唑胺和疫苗接种的组合可以对免疫应答的诱导以及临床应答的持久性产生积极影响。

为此，Harrop等在三项连续研究[21-23]中使用TroVax疫苗治疗转移性结肠直肠癌患者；将肿瘤相关抗原5T4设计成病毒载体修饰安卡拉牛痘苗。在上述研究中，与单一治疗相比[21]，联合使用标准化疗（5-氟尿嘧啶、亚叶酸和伊立替康）时[23]，接种过程中诱导的免疫应答更强，寿命更长。此外，化疗结束后CEA水平稳定的临床活动迹象似乎与5T4特异性免疫应答的诱导有关。

10例II-IV期无病恶性黑色素瘤患者在I/II期试点研究中随机接受达卡巴嗪加黑色素-A/MART-1肽疫苗或单独接种疫苗[24]。免疫学分析显示，在接受达卡巴嗪和接种疫苗治疗组的患者中，可观察到肽特异性效应记忆$CD8^+T$细胞增加，这些细胞能够杀死黑色素-A^+肿瘤细胞系，但在仅接种疫苗的组中没有观察到这种情况。此外，在接受化疗后第一天，免疫调节因子基因表达谱出现上调，这可能会增强疫苗引起的免疫反应。在单独接种疫苗的组中，只有1例II期患者在随访29个月后仍然无病，而在联合用药组中，3例III/IV期患者没有发现疾病证据。有趣的是，这3名患者表现出显著的$CD8^+$抗肿瘤T细胞增加与效应记忆表型。然而这项研究还不能得出结论，因为每组只有5名患者。

在另一个有趣的IIB期试验中[25]，接受一线化疗的非小细胞肺癌（NSCLC）患者随机接受单独化疗或与编码MUC-1肿瘤相关抗原和IL-2的痘病毒TG4010联合化疗。这项研究的主要疗效指标是联合治疗组 6 个月的 PFS 达到 40%或更多。临床上他们发现联合治疗组的客观应答率更高，并且客观应答与该组的长期总生存率相关。此外，免疫表型分析显示，自然杀伤细胞 NK（$CD16^{+}$CD56+$CD69^{+}$）数量增加的患者具有更差的临床结果，表明对治疗患者免疫状态的影响。

总之，这些研究表明化疗具有调节免疫系统和改善疫苗接种环境的能力，以发挥其免疫刺激潜能（图 20.1）。然而同样清楚的是，化疗的最佳剂量和时间安排以及最佳化疗药物仍然未知，因此这一领域的研究非常有必要。

20.5 其他癌症疗法

20.5.1 癌症疫苗结合其他癌症疗法（表 20.3）

20.5.1.1 放射疗法

化疗不仅可以引起免疫系统变化，还可能有利于与癌症疫苗的结合。放射治疗（RT）可以诱导一些与化疗相同的调节，包括诱导免疫原性凋亡，导致树突状细胞抗原递呈增加[26]。此外，如果细胞死亡不是由辐射引起的，可以导致幸存的癌细胞增加 MHC Ⅰ类复合物和黏附分子的表达，从而增加 T 细胞的识别和杀死能力[27]。这可能会再次导致循环肿瘤特异性 T 细胞数量增加，这些 T 细胞有可能在身体其他部位介导肿瘤消退，而不是在受辐射部位。远离局部照射部位的肿瘤消退被称为远位效应，这种现象仅在少数情况下有所描述[28]。

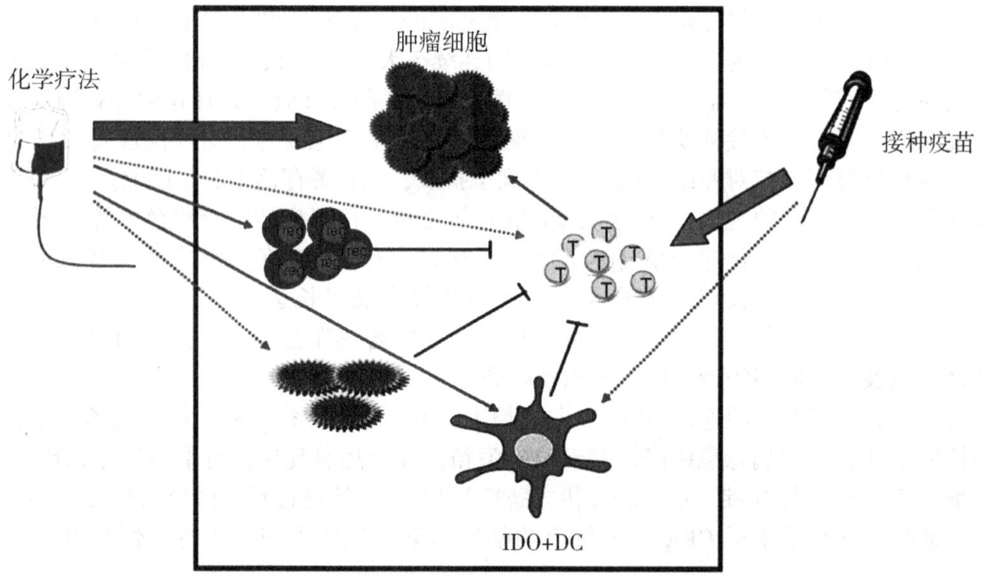

图 20.1 肿瘤细胞、效应 T 细胞与免疫调节细胞亚群的相互作用

在免疫抑制肿瘤环境中，效应 T 细胞（*T*）对肿瘤细胞的溶细胞活性较低。其功能受到调节性 T 细胞（Treg）、髓源性抑制细胞（MDSC）、IDO+树突状细胞（*DC*）以及肿瘤细胞通过下调 MHC 分子等逃避 T 细胞识别能力的抑制。化疗和癌症疫苗可以协同工作；疫苗产生大量肿瘤特异性 T 细胞，化疗直接杀死肿瘤细胞，启动肿瘤杀死 T 细胞，减少调节细胞亚群的数量和功能，诱导细胞因子释放，导致 T 细胞增殖。此外，通过对 IDO 肽的特异性免疫，选择性抑制 IDO+、DC 等可进一步提高 T 细胞活性。化疗/疫苗接种抑制活性用红色箭头表示，而增加活性用绿色箭头表示。

化疗或放化疗后病情或反应稳定的 NSCLC 患者纳入 IIB 期研究[29]，随机接种 BLP25 脂质体疫苗并辅以 BSC（最佳支持性护理）或单独使用 BSC。结果显示，接种疫苗组患者的生存时间显著延长，其中最显著的是 IIIB 期局部疾病亚组（30.6 个月 vs. 13.3 个月）。有人假设，除了免疫抑制较弱和侵袭性较低的患者有更多时间发展免疫应答外，这些患者更经常接受放射治疗预处理，这可能有助于免疫治疗效果。一项新的试验已经开始，只包括经放化疗预处理的IIIB期患者。

目前，只有很少的临床研究对放射治疗联合癌症疫苗疗效进行了研究。Gulley和他的同事[30]对接受局部放疗的前列腺癌患者进行了一项临床II期试验，随机将患者分为单独放疗组或放疗联合编码PSA的重组疫苗组。他们发现联合治疗组的PSA特异性T细胞增加了3倍，而单独治疗组没有发现增加。

因此，放射治疗似乎具有与免疫疗法结合的免疫调节潜力，未来临床试验正在进行或正在计划中[31]，这两项研究将放射治疗与易普利姆玛结合起来（ClinicalTrials.gov identifier：NCT00861614和NCT01557114）。

20.5.1.2 靶向治疗

另一种成熟的抗肿瘤治疗方式是靶向治疗，越来越多的被批准用于治疗。靶向治疗已被批准用于抗化疗癌症的一线治疗，如肾细胞癌[32,33]，在某些情况下也用于黑色素瘤患者[34]。这些疗法也可能通过诱导免疫原性细胞死亡和释放肿瘤抗原对免疫系统产生积极影响。此外，在这些疗法中，观察到快速、令人印象深刻但往往持续时间较短的客观反应，可能受益于与免疫疗法中观察到的较慢但更持久的反应相结合。

通过靶向治疗延长疾病进展时间，也可能延长从癌症疫苗等引起临床显著免疫反应的时间[35]。最近的研究发现，BRAF抑制剂不会对细胞毒性T细胞功能产生负面影响[36,37]，这些药物可能会增加黑色素瘤细胞表面的抗原表达，使其更容易杀死T细胞[38]。这表明BRAF抑制剂和免疫疗法结合是可行的，关于BRAF抑制剂和抗CTLA-4或MEK抑制剂结合的讨论正在进行中。

近年来我们发现，与经维莫非尼预处理的$BRAF^{V600}$突变自体黑色素瘤细胞系共培养时，用BRAF抑制剂维莫非尼阻断可增加肿瘤浸润淋巴细胞的抗肿瘤活性。肿瘤识别能力的提高与MHC Ⅰ类分子的上调和肿瘤细胞上的热休克蛋白相关[39]。这些数据表明，维莫非尼和过继细胞移植相结合可能有利于$BRAF^{V600}$突变型恶性黑色素瘤患者，但肿瘤特异性T细胞对肿瘤识别能力提高也可用于癌症疫苗接种。

表20.3 癌症疫苗联合其他疗法癌症疗法的例子

	试验示例	药物	疫苗	癌症	病人数量	反应/功效
放射疗法（RT）	Butts等[29]	化疗或放化疗	± Stimuvax（BLP25脂质体疫苗）	非小细胞肺癌	88（+疫苗）和83（-疫苗）	接受治疗的患者特别是联合组接受预先化学放射疗法的患者生存期延长
	Gulley等[30]	局部RT	± PSA编码的疫苗	前列腺癌	19（+疫苗）比11（-疫苗）	联合组的PSA特异性T细胞增加了3倍
靶向治疗	Comin-Anduix等[36]	BRAF抑制剂（PLX4032）	无	恶性黑色素瘤细胞BRAF V600E+（体外研究）		对T细胞的细胞毒性无负面影响
	Boni等[38]	BRAF抑制剂	无	恶性黑色素瘤细胞BRAF V600E+（体外研究）		抗原表达增加

（续表）

	试验示例	药物	疫苗	癌症	病人数量	反应/功效
免疫疗法	Schwartzen-truber 等[40]	高剂量 IL-2	±gp100 多肽疫苗	转移性黑色素瘤患者	94（仅 IL-2）比 91（IL-2+疫苗）	提高联合治疗组的临床反应和无进展生存率
	Hodi 等[42]	±易普利姆玛（Ipi）	±gp100 多肽疫苗	转移性黑色素瘤患者	403（Ipi+疫苗），137（仅 Ipi）。136（仅疫苗）	联合用药组无临床疗效

20.6 其他免疫疗法

通过癌症疫苗接种诱导肿瘤特异性细胞毒性 T 细胞在理论上，可以通过其他免疫治疗策略的协同作用来增强，例如用细胞因子治疗或阻断抑制性 T 细胞受体（如 CTLA-4 或 PD-1 抗体）。

在 2011 年的一项研究中[40]，Schwartzentruber 和他的同事报告了一项 III 期试验结果，该试验随机选取 185 名转移性黑色素瘤患者，在单独使用高剂量 IL-2 或联合使用 gp100 肽疫苗之间进行治疗。结果显示，联合组的临床反应有明显改善，客观反应率为 16%，而单独使用 IL-2 治疗组的客观反应率为 6%。此外，无进展生存期显著改善，总体生存率有利于联合治疗（17.8 个月 vs. 11.1 个月，$P=0.06$）。联合治疗组 37 例免疫评估患者中只有 7 例发展出可测量的肽特异性 T 细胞，免疫分析显示，抗肿瘤反应和临床反应之间无任何相关性。尽管如此，这项随机研究的结果强调了细胞因子治疗与癌症疫苗接种相结合的潜在优势。

Weber 等研究了 CTLA-4 抗体易普利姆玛（ipilimumab）（Yervoy，BMS）对免疫系统的影响[41]，他们发现在治疗期间活化的 $CD4^+$ 和 $CD8^+$T 细胞的百分比以及效应记忆 T 细胞的百分比均有所增加。与此同时，幼稚 T 细胞水平下降。这些数据表明，易普利姆玛增加了 T 细胞的活性，使其成为与癌症疫苗结合的一种明显制剂。在一个大规模随机 III 期试验中[42]，运用随机化策略分 3 组对易普利姆玛有效性进行了测试，即单独使用易普利姆玛、易普利姆玛与 gp100 肽疫苗联合使用或单独使用 gp100 疫苗。本研究表明，单独使用易普利姆玛或与 gp100 疫苗联合使用均优于单独使用 gp100 疫苗，但令人惊讶的是，将易普利姆玛与疫苗联合使用并无任何益处。在易普利姆玛联合 gp100 疫苗的其他研究中也观察到了这一点[43]。

IL-2 和 gp100 疫苗似乎具有协同临床效果，而 CTLA-4 抗体与同一疫苗联合使用并不能改善临床反应的原因尚不清楚，但很可能是由于不同的作用机制。这强调癌症疫苗与免疫激活剂结合的结果是不可预测的，可能不会导致协同或附加临床疗效。

20.7 展望

耐药性是癌症治疗中限制癌症化疗效力的主要问题。这种获得性耐药性的令人沮丧的特性之一是肿瘤不仅对正在使用的特定药物产生耐药性，而且还可能对具有不同作用机制的其他药物产生交叉耐药性[44]。细胞凋亡中的肿瘤相关缺陷在化疗和放疗的耐药性中起着重要作用。凋亡受损的一个重要原因是凋亡蛋白的抗凋亡调节因子（如肿瘤抗原生存素）过度表达[45]。

此外，肿瘤抗原 CYP1B1 药物失活可能是一种耐药机制，影响化疗临床疗效[46]。因此，以这些抗原为靶点的免疫疗法结合传统的化疗似乎特别有吸引力。在这种情况下，传统疗法会杀死大多数癌细胞，只留下表达高水平抗原的细胞，而这些细胞特别容易被疫苗诱导的T细胞杀死。这些措施协同作用可能比单独使用这两种方案提供更有效的治疗附加效果，从而加强已经描述的抗癌疫苗和化疗协同作用。因此，传统疗法和免疫疗法的协同效应需要重新考虑临床策略，不仅要考虑选择的化疗药物，而且要考虑所选择免疫疗法的设计。

毫无疑问，癌症疫苗与其他抗肿瘤治疗结合具有很高的潜力，可能性也很多。然而个体疗法之间的相互作用通常很难预测，因此彻底的临床试验是必不可少的。从某些化疗和放疗方案诱导的免疫原性细胞死亡到细胞因子和疫苗的协同作用，这种多样性指向许多潜在的新临床试验，这些试验应基于合理的科学论据和坚实的临床前研究。

我们研究所启动了两项结合标准化疗方案和癌症疫苗新试验。此外，还计划将过继细胞疗法与肽疫苗相结合进行试验（www. clinicaltrial. govIdentifier：NCT01446731 和 NCT01543464）。

癌症免疫治疗是一个发展迅速的领域；超过50个III期临床试验正在测试癌症治疗的免疫策略（www. clinicaltrial. gov）；其中几家公司正在将癌症疫苗与免疫调节剂结合起来，寻找提高疗效的方法。很有可能其中一些试验将达到临床意义，从而为组合癌症免疫治疗新时代铺平道路。

参考文献

[1] Higano, C. S., et al.: Sipuleucel-T. Nat. Rev. Drug Discov. 9, 513-514 (2010).

[2] Schreiber, R. D., Old, L. J., Smyth, M. J.: Cancer immunoediting: integrating immunity's roles in cancer suppression and promotion. Science 331, 1565-1570 (2011).

[3] Zou, W.: Regulatory T cells, tumour immunity and immunotherapy. Nat. Rev. Immunol. 6, 295-307 (2006).

[4] Srivastava, M. K., et al.: Myeloid-derived suppressor cells inhibit T-cell activation by depleting cystine and cysteine. Cancer Res. 70, 68-77 (2010).

[5] Munn, D. H., Mellor, A. L.: Indoleamine 2, 3-dioxygenase and tumor-induced tolerance. J. Clin. Invest. 117, 1147-1154 (2007).

[6] Andersen, M. H.: The specific targeting of immune regulation: T-cell responses against indoleamine 2, 3-dioxygenase. Cancer Immunol. Immunother. 61, 1289-1297 (2012), 1-9.

[7] Emens, L. A.: Chemoimmunotherapy. Cancer J. 16, 295-303 (2010).

[8] Sistigu, A., et al.: Immunomodulatory effects of cyclophosphamide and implementations for vaccine design. Semin. Immunopathol. 33, 369-383 (2011).

[9] Z itvogel, L., et al.: Immunogenic tumor cell death for optimal anticancer therapy: the calreticulin exposure pathway. Clin. Cancer Res. 16, 3100-3104 (2010).

[10] Aymeric, L., et al.: Tumor cell death and ATP release prime dendritic cells and efficient anticancer immunity. Cancer Res. 70, 855-858 (2010).

[11] Ramakrishnan, R., et al.: Chemotherapy enhances tumor cell susceptibility to CTL-mediated killing during cancer immunotherapy in mice. J. Clin. Invest. 120, 1111-1124 (2010).

[12] L iu, J. Y., et al.: Single administration of low dose cyclophosphamide augments the antitumor effect of dendritic cell vaccine. Cancer Immunol. Immunother. 56, 1597-1604 (2007).

[13] Lutsiak, M. E., et al.: Inhibition of CD4 (+) 25+Tregu-latory cell function implicated in enhanced immune response by low-dose cyclophosphamide. Blood 105, 2862-2868 (2005).

[14] Ge, Y., et al.: Metronomic cyclophosphamide treatment in metastasized breast cancer patients: immunological effects and clinical outcome. Cancer Immunol. Immunother. 61, 353-362 (2012).

[15] Emens, L. A., et al.: Timed sequential treatment with cyclophosphamide, doxorubicin, and an allogeneic granulocyte-macrophage colony-stimulating factor-secreting breast tumor vaccine: a chemotherapy dose-ranging factorial study of safety and immune activation. J. Clin. Oncol. 27, 5911-5918 (2009).

[16] Holtl, L., et al.: Allogeneic dendritic cell vaccination against metastatic renal cell carcinoma with or without

cyclophosphamide. Cancer Immunol. Immunother. 54, 663-670 (2005).

[17] Ghiringhelli, F., et al.: Metronomic cyclophosphamide regimen selectively depletes $CD4^+CD25+$regulatory T cells and restores T and NK effector functions in end stage cancer patients. Cancer Immunol. Immunother. 56, 641-648 (2007).

[18] Ellebaek, E., et al.: Metastatic melanoma patients treated with dendritic cell vaccination, interleukin-2 and metronomic cyclophosphamide: results from a phase II trial. Cancer Immunol. Immunother. 61, 1791-1804 (2012).

[19] Engell-Noerregaard, L., et al.: Influence of metronomic cyclophosphamide or interleukin-2 alone or combined on blood regulatory T cells in patients with advanced malignant melanoma treated with dendritic cell vaccines. J. Clin. Cell. Immunol 3, 1 (2012).

[20] Kyte, J. A., et al.: Telomerase peptide vaccination combined with temozolomide: a clinical trial in stage IV melanoma patients. Clin. Cancer Res. 17, 4568-4580 (2011).

[21] Harrop, R., et al.: Vaccination of colorectal cancer patients with modified vaccinia Ankara delivering the tumor antigen 5T4 (TroVax) induces immune responses which correlate with disease control: a phase I/II trial. Clin. Cancer Res. 12, 3416-3424 (2006).

[22] H arrop, R., et al.: Vaccination of colorectal cancer patients with modified vaccinia ankara encoding the tumor antigen 5T4 (TroVax) given alongside chemotherapy induces potent immune responses. Clin. Cancer Res. 13, 4487-4494 (2007).

[23] Harrop, R., et al.: Vaccination of colorectal cancer patients with TroVax given alongside chemotherapy (5-fluorouracil, leukovorin and irinotecan) is safe and induces potent immune responses. Cancer Immunol. Immunother. 57, 977-986 (2008).

[24] Nistico, P., et al.: Chemotherapy enhances vaccine-induced antitumor immunity in melanoma patients. Int. J. Cancer 124, 130-139 (2009).

[25] Quoix, E., et al.: Therapeutic vaccination with TG4010 and first-line chemotherapy in advanced non small-cell lung cancer: a controlled phase 2B trial. Lancet Oncol. 12, 1125-1133 (2011).

[26] Obeid, M., et al.: Calreticulin exposure dictates the immunogenicity of cancer cell death. Nat. Med. 13, 54-61 (2007).

[27] Chakraborty, M., et al.: External beam radiation of tumors alters phenotype of tumor cells to render them susceptible to vaccine-mediated T-cell killing. Cancer Res. 64, 4328-4337 (2004).

[28] Formenti, S. C., Demaria, S.: Systemic effects of local radiotherapy. Lancet Oncol. 10, 718-726 (2009).

[29] Butts, C., et al.: Updated survival analysis in patients with stage IIIB or IV non-small-cell lung cancer receiving BLP25 liposome vaccine (L-BLP25): phase IIB randomized, multicenter, open-label trial. J. Cancer Res. Clin. Oncol. 137, 1337-1342 (2011).

[30] Gulley, J. L., et al.: Combining a recombinant cancer vaccine with standard definitive radiotherapy in patients with localized prostate cancer. Clin. Cancer Res. 11, 3353-3362 (2005).

[31] Masucci, G., et al.: Stereotactic Ablative Radio Therapy (SABR) followed by immunotherapy a challenge for individualized treatment of metastatic solid tumours. J. Transl. Med. 10, 104 (2012).

[32] Coppin, C., et al.: Targeted therapy for advanced renal cell cancer (RCC): a Cochrane systematic review of published randomised trials. BJU Int. 108, 1556-1563 (2011).

[33] Escudier, B., Kataja, V.: Renal cell carcinoma: ESMO clinical practice guidelines for diagnosis, treatment and follow-up. Ann. Oncol. 21 (Suppl 5), v137-v139 (2010).

[34] Bhatia, S., Thompson, J. A.: Systemic therapy for metastatic melanoma in 2012: dawn of a new era. J. Natl. Compr. Canc. Netw. 10, 403-412 (2012).

[35] Blank, C. U., Hooijkaas, A. I., Haanen, J. B., Schumacher, T. N.: Combination of targeted therapy and immunotherapy in melanoma. Cancer Immunol. Immunother. 60, 1359-1371 (2011).

[36] Comin-Anduix, B., et al.: The oncogenic BRAF kinase inhibitor PLX4032/RG7204 does not affect the viability or function of human lymphocytes across a wide range of concentrations. Clin. Cancer Res. 16, 6040-6048 (2010).

[37] Hong, D. S., et al.: BRAF (V600) inhibitor GSK2118436 targeted inhibition of mutant BRAF in cancer patients does not impair overall immune competency. Clin. Cancer Res. 18, 2326–2335 (2012).

[38] B oni, A., et al.: Selective BRAFV600E inhibition enhances T-cell recognition of melanoma without affecting lymphocyte function. Cancer Res. 70, 5213–5219 (2010).

[39] Donia, M., et al.: Methods to improve adoptive T-cell therapy for melanoma: IFN-γ enhances anticancer responses of cell products for infusion. J. Invest. Dermatol. 133, 545–552 (2013).

[40] Schwartzentruber, D. J., et al.: gp100 peptide vaccine and interleukin-2 in patients with advanced melanoma. N. Engl. J. Med. 364, 2119–2127 (2011).

[41] Weber, J. S., et al.: Ipilimumab increases activated T cells and enhances humoral immunity in patients with advanced melanoma. J. Immunother. 35, 89–97 (2012).

[42] Hodi, F. S., et al.: Improved survival with ipilimumab in patients with metastatic melanoma. N. Engl. J. Med. 363, 711–723 (2010).

[43] Prieto, P. A., et al.: CTLA-4 blockade with ipilimumab: long-term follow-up of 177 patients with metastatic melanoma. Clin. Cancer Res. 18, 2039–2047 (2012).

[44] Helmbach, H., et al.: Drug-resistance in human melanoma. Int. J. Cancer. 93, 617–622 (2001).

[45] Wendel, H. G., Lowe, S. W.: Reversing drug resistance in vivo. Cell Cycle 3, 847–849 (2004).

[46] Rochat, B., et al.: Human CYP1B1 and anticancer agent metabolism: mechanism for tumor-specific drug inactivation? J. Pharmacol. Exp. Ther. 296, 537–541 (2001).

（吕律译）

第21章　个性化肽疫苗：一种新的晚期癌症免疫疗法

Nobukazu Komatsu，Satoko Matsueda，Masanori Noguchi，Akira Yamada，Kyogo Itoh 和 Tetsuro Sasada①

摘要

随着近几十年来基础免疫学和临床免疫学领域的最新研究进展，癌症免疫治疗作为一种新型的癌症疗法已经取得了一定的进展。我们开发了一种新的免疫治疗方法，即个性化肽疫苗接种（PPV），根据患者对疫苗抗原候选物的现有免疫应答选择2~4个合适的肽。最近PPV临床试验已经证明了这种新治疗方法在各种类型的晚期癌症中的可行性。炎症因子，如IL-6和CRP，以及免疫抑制细胞，如髓源性抑制细胞（MDSC），已被确定为接种癌症患者的预测性生物标志。在不久的将来，其他抑制炎症因子和MDSC治疗方法可以结合起来提高晚期癌症患者PPV临床疗效。

21.1　引言

随着过去几十年在基础和临床免疫学领域的研究进展，癌症免疫治疗作为一种新的癌症治疗方式已经取得了进展[1,2]。事实上，在过去的25年中，17种免疫治疗产品已经获得了美国食品和药物管理局（FDA）的许可[1]。包括非特异性免疫刺激剂、细胞因子、单克隆抗体、放射性标记抗体、免疫毒素和基于细胞的疗法。

值得注意的是，美国FDA最近批准了两种新型免疫治疗药物用于晚期癌症患者[3,4]。2010年一种自体树突状细胞疫苗sipuleucel-T（Provenge；Dendreon Corporation，Seattle，WA）被批准用于治疗去势抵抗型前列腺癌（CRPC）患者，其目的是刺激对人类前列腺酸性磷酸酶（PAP）的T细胞免疫反应，因为在影响研究中，这种疫苗显示改善了4.1个月的总体存活率，这是最大的3期sipuleucel-T随机对照试验[3]。

2011年，另一种药物易普利姆玛，即针对免疫检查点分子之一细胞毒性T淋巴细胞抗原4（CTLA-4）的阻断抗体药物，已被美国食品和药物管理局批准用于黑色素瘤患者。在关键的第3期试验中，这种药物导致总生存率提高了3个月，疾病控制率为28.5%，其中60%的应答患者疾病控制时间超过2年[4]。最近针对程序性死亡1（PD-1）分子（一种T细胞共抑制受体）及其配体之一PD-L1（在肿瘤细胞逃避宿主免疫系统能力中发挥关键作用的抗体）已经在治疗各种类型的癌症方面显示出有很好的结果，尽管还没有得到的正式批准[5,6]。Topalian等证明抗PD-1抗体在1/5~1/4的非小细胞肺癌、黑色素瘤或肾细胞癌患者中产生客

---

① N. Komatsu，博士 · K. Itoh，医学博士，博士 · T. Sasada，医学博士，博士（✉）（日本库鲁姆大学医学院免疫与免疫治疗系，E-mail：tsasada@med.kurume-u.ac.jp）

S. Matsueda，博士 · A. Yamada，博士（日本库鲁姆大学创新癌症治疗研究中心癌症疫苗开发室）

M. Noguchi，医学博士，博士（日本库鲁姆大学创新癌症治疗研究中心临床研究室）

观应答[5]。Brahmer 等表明，抗体介导的 PD-L1 阻断在晚期癌症患者（包括非小细胞肺癌、黑色素瘤和肾细胞癌）中诱导了持久肿瘤消退（客观有效率为 6%~17%）和延长疾病稳定期（24 周时为 12%~41%）[6]。目前这些有希望的进展为癌症疫苗的进一步发展带来了极大的乐观和热情。

21.2　基于肽的免疫疗法

通过几种不同的方法，已经鉴定出快速增加的肿瘤相关抗原（TAAs）和来源于 TAAs 的 HLA 限制性肽表位，例如 cDNA 表达克隆、重组 cDNA 表达文库的血清学分析（SEREX）和反向免疫学方法[7,8]。在过去 20 年中，源自 TAAs 的表位优先在肿瘤细胞中表达，但在正常组织中限制性表达，已被用作癌症的治疗性肽疫苗，但迄今为止进行的大多数临床试验的成功率很有限。[9,10]。

尽管如此，最近临床试验已经证明了治疗性肽疫苗的一些重大进展。例如，据报道治疗性人乳头状瘤病毒（HPV）疫苗对高危人群发展为 HPV 相关癌症有效。Melief 和他的同事们研究表明，一种由 HPV-16 E6/E7 癌蛋白衍生的合成长肽池组成的疫苗，成功诱导了 HPV 特异性免疫应答，并在大多数（79%）患者中导致了 HPV 感染的癌前生殖器病变可测量消退[11]。除了这些来源于致癌传染因子的免疫原性抗原被宿主免疫系统识别为外源性外，用免疫原性较低的“自身”抗原肽进行疫苗接种也显示出实质性进展。特别是疫苗接种的改进策略，如“个性化”抗原的预筛选、多种肽的联合使用以及与其他治疗方式的联合治疗，提高了临床效果[12,13]。在本综述中，我们讨论了我们的新疫苗策略“个性化肽疫苗（PPV）”，根据预先存在的对来自 TAAs 的候选疫苗抗原的免疫应答，选择适合于个体患者的疫苗抗原[12-16]。

21.3　个性化肽疫苗（PPV）的概念

一般来说，已知抗肿瘤免疫依赖于肿瘤细胞的抗原递呈和宿主免疫细胞库。考虑到宿主免疫细胞库的多样性和异质性，个体间的抗肿瘤免疫可能存在较大差异。因此，尽管在肿瘤细胞上高度优先表达的单个 TAAs 或 TAAs 组合通常被用作癌症疫苗接种的普通抗原，但它们可能并不总是适合于每个患者。在不考虑宿主免疫细胞库的情况下，选择和接种疫苗抗原可能不会有效地诱导有益的抗肿瘤免疫应答。因此，为了从癌症疫苗中产生更多的临床效益，应特别注意每个病人的免疫状况，在接种疫苗前确定其对疫苗抗原已有的免疫应答。

然而，在目前大多数癌症疫苗临床试验中，由于忽视了先前存在的个体免疫反应，普通抗原被用于疫苗接种。相比之下，我们开发了一种不同的方法即个性化肽疫苗接种（PPV），用于晚期癌症患者（图 21.1）[12-16]。在这种治疗中，根据患者对候选疫苗抗原已有的免疫应答，选择 2~4 个合适的肽。因此，PPV 有望有效地放大这些先前存在的免疫反应，从而促进抗肿瘤活性（图 21.2）

21.4　PPV 程序

目前，我们使用了 31 种 HLA（人类白细胞抗原）I 类限制性候选肽，这些候选肽是通过肿瘤浸润淋巴细胞克隆/系的 cDNA 表达克隆方法或反向免疫策略从多种 TAAs 中鉴定出来，包括 12 种 HLA-A2+患者的肽，14 种 HLA-A24+患者的肽，9 种 HLA-A 3 超型+（A3+，A11+，A31+或 A33+）患者的肽，4 种 HLA-A26+患者的肽（表 21.1）。这些候选疫苗的安全性和潜在免疫效应已在先前进行的临床研究中有所报道[12-16]。根据 HLA 分型结果和针对 31 种不同候选疫苗先前存在的免疫应答，选择最多 4 种适合单个患者的肽，每周或每两周皮下注射不完全弗氏佐剂乳化的肽。

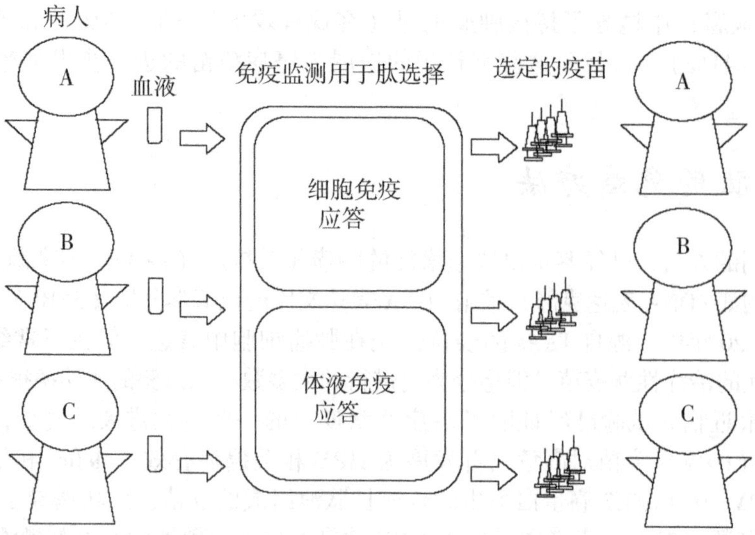

图 21.1 PPV 概念

31 个抗原被用作 HLA-A2+、HLAA24+、HLA-A3 超型+或 HLA-A26+癌症患者的候选疫苗抗原。通过筛选抗体反应和/或对候选疫苗的细胞免疫应答来评估患者外周血样本中先前存在的免疫应答。根据 HLA 分型结果和针对 31 种不同候选疫苗的现有免疫应答，最多选择 4 种肽，每周或每两周皮下注射不完全弗氏佐剂复合物。

对于 PPV，患者通过多种肽进行治疗，因为与仅单个表位相比，选择和接种多种 CTL 表位可以显著降低抗原阴性克隆逃避肽特异性免疫应答的风险。事实上，肿瘤细胞通过同时丢失所有选择用于疫苗接种的多种抗原来逃避抗原特异性免疫应答的情况相当罕见。为了防止接种部位的多种肽之间的相互作用/竞争，多种疫苗肽中的每一种都在不同部位单独接种，而不是在单一部位混合接种。因为先前可行性研究表明，每次接种 4 个肽似乎是可以接受的，但是每次接种超过 5 个肽是不可容忍的，因为皮肤不良反应有时会导致不舒服的症状，例如瘙痒和疼痛（未公布的数据），所以目前患者每次接种最多可接受 4 个肽的治疗。

尽管临床效果可能主要归因于对疫苗抗原的细胞免疫应答，但我们目前将体液免疫应答作为疫苗接种前选择疫苗抗原的一种独特方法进行评估。当为 PPV 选择疫苗抗原时，对候选抗原特异性细胞免疫应答的评估有时是困难的，因为抗原特异性效应 T 细胞在大多数癌症患者中通常以非常低的浓度存在，并且用于确定细胞免疫应答的分析可能会受到血样处理条件的很大影响，例如外周血单核细胞（PBMC）的纯化和储存条件[17-19]。这就是为什么在选择用于评估抗原特异性细胞免疫反应分析方法时，必须考虑诸如灵敏度、特异性和再现性等实际问题。

另一方面，用于评估体液免疫应答的测定，如抗原特异性抗体测定，往往比用于测量细胞免疫应答的测定更稳定[19,20]。特别是我们开发的基于 Luminex ® 微悬浮阵列技术用于监测体液免疫应答，允许简单、快速和高度可重复性的高通量筛选，仅针对少量血浆或血清的大量肽抗原 IgG 反应[20]。由于该检测的动态范围比传统的酶联免疫吸附检测宽得多，因此不仅适用于疫苗接种后，也适用于疫苗接种前监测患者血浆或血清中抗原特异性体液免疫反应。

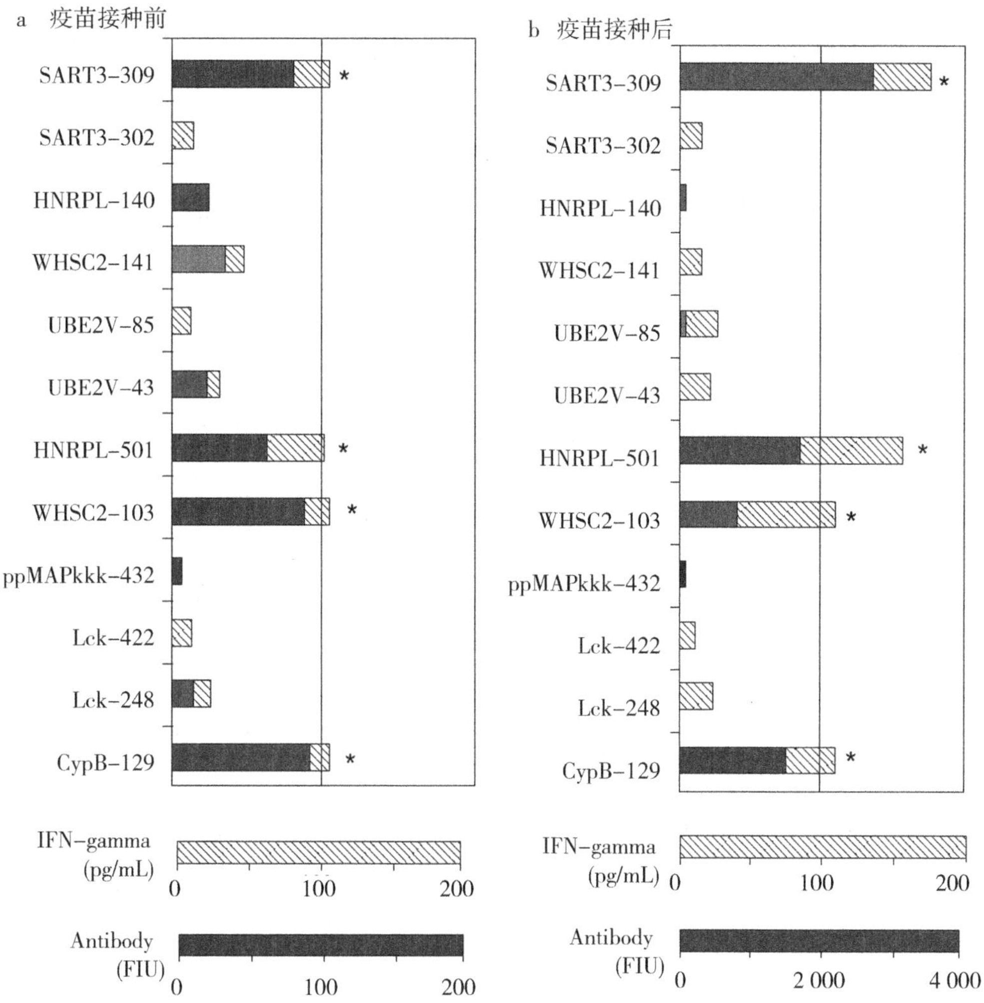

图 21.2 PPV 免疫增强的代表性数据

显示（a）疫苗接种前和（b）疫苗接种后采集的血液样本免疫监测的代表性结果。抗原特异性抗体（IgG）反应通过微悬浮阵列（Luminex® 分析）测定，抗原特异性细胞免疫反应通过 IFN-γ 分泌（Elisa）测定。4 种疫苗肽（SART3-309、HNRPL-501、WHSC2-103 和 CypB-129，以“*”表示）通过先前存在的候选抗原特异性 IgG 滴度进行选择。在接种后（第 6 次）疫苗样本中，对 SART3-309 肽的 IgG 应答从 80 FIU 增加到 2 700 FIU。其他 3 种肽的体液应答也有类似的增强。所有 4 种免疫肽的抗原特异性细胞免疫应答也增强。

表 21.1 用于 PPV 候选肽的信息

肽名称	原始蛋白质	位置	序列	HLA-IA 限制	参考文献
Lck-246	p56 lck	246-254	KLVERLGAA	A2+	*Int. J. Cancer*94，237 - 242（2001）
CypB-129	Cyclophilin B	129-138	KLKHYGPGWV	A2+/A3+超类型	*Jpn. J. Cancer Res.*92，762 - 767（2001）
Lck-422	p56 lck	422-430	DVWSFGILL	A2+/A3+超类型	*Int. J. Cancer*94，237 - 242（2001）

（续表）

肽名称	原始蛋白质	位置	序列	HLA-IA 限制	参考文献
ppMAPkkk-432	ppMAPkkk	432-440	DLLSHAFFA	A2+/A26+	*Cancer Res.* 61, 2038 - 2046 (2001)
WHSC2-103	WHSC2	103-111	ASLDSDPWV	A2+/A26+/ A3+超类型	*Cancer Res.* 61, 2038 - 2046 (2001)
HNRPL-501	HNRPL	501-510	NVLHFFNAPL	A2+/A26+	*Cancer Res.* 61, 2038 - 2046 (2001)
UBE2V-43	UBE2V	43-51	RLQEWCSVI	A2+	*Cancer Res.* 61, 2038 - 2046 (2001)
UBE2V-85	UBE2V	85-93	LIADFLSGL	A2+	*Cancer Res.* 61, 2038 - 2046 (2001)
WHSC2-141	WHSC2	141-149	ILGELREKV	A2+	*Cancer Res.* 61, 2038 - 2046 (2001)
HNRPL-140	HNRPL	140-148	ALVEFEDVL	A2+	*Cancer Res.* 61, 2038 - 2046 (2001)
SART3-302	SART3	302-310	LLQAEAPRL	A2+	*Int. J. Cancer*88, 633 - 639 (2000)
SART3-309	SART3	309-317	RLAEYQAYI	A2+	*Int. J. Cancer*88, 633 - 639 (2000)
SART2-93	SART2	93-101	DYSARWNEI	A24+	*J. Immunol.* 164, 2565 - 2574 (2000)
SART3-109	SART3	109-118	VYDYNCHVDL	A24+/A24+/ A3+超类型	*Cancer Res.* 59, 4056 - 4063 (1999)
Lck-208	p56 lck	208-216	HYTNASDGL	A24+	*Eur. J. Immunol.* 31, 323 - 332 (2001)
PAP-213	PAP	213-221	LYCESVHNF	A24+	*J. Urol.* 166, 1508 - 1513 (2001)
PSA-248	PSA	248-257	HYRKWIKDTI	A24+	*Prostate*57, 152-159 (2003)
EGF-R-800	EGF-R	800-809	DYVREHKDNI	A24+	*Eur. J. Cancer*40, 1776 - 1786 (2004)
MRP3-503	MRP3	503-511	LYAWEPSFL	A24+	*Cancer Res.* 61, 6459 - 6466 (2001)
MRP3-1293	MRP3	1293-1302	NYSVRYRPGL	A24+	*Cancer Res.* 61, 6459 - 6466 (2001)
SART2-161	SART2	161-169	AYDFLYNYL	A24+	*J. Immunol.* 164, 2565 - 2574 (2000)
Lck-486	p56 lck	486-494	TFDYLRSVL	A24+	*Eur. J. Immunol.* 31, 323 - 332 (2001)
Lck-488	p56 lck	488-497	DYLRSVLEDF	A24+	*Eur. J. Immunol.* 31, 323 - 332 (2001)
PSMA-624	PSMA	624-632	TYSVSFDSL	A24+	*Cancer Sci.* 94, 622 - 627 (2003)

（续表）

肽名称	原始蛋白质	位置	序列	HLA-IA 限制	参考文献
EZH2-735	EZH2	735-743	KYVGIEREM	A24+	*Prostate*60，273-281（2004）
PTHrP-102	PTHrP	102-111	RYLTQETNKV	A24+	*Br. J. Cancer*91，287－296（2004）
SART3-511	SART3	511-519	WLEYYNLER	A3+超类型	*Cancer Immunol. Immunother.* 56，689-698（2007）
SART3-734	SART3	734-742	QIRPIFSNR	A3+超类型	*Cancer Immunol. Immunother.* 56，689-698（2007）
Lck-90	p56 lck	90-99	ILEQSGEWWK	A3+超类型	*Br. J. Cancer*97，1648－1654（2007）
Lck-449	p56 lck	449-458	VIQNLERGYR	A3+超类型	*Br. J. Cancer*97，1648－1654（2007）
PAP-248	PAP	248-257	GIHKQKEKSR	A3+超类型	*Clin. Cancer Res.* 11，6933－6943（2005）

除了来自前列腺相关抗原（PSA、PAP 和 PSMA）的前列腺癌肽外，所有列出的肽目前都用于各种癌症的疫苗接种。根据 HLA 分型结果和针对 31 种不同候选疫苗的预先存在的免疫应答，最多选择和给予 4 种肽。

尽管 TAA 衍生肽特异性抗体的生物学功能尚不清楚，但可能包括促进抗肿瘤免疫的功能。例如抗原特异性抗体的功能之一可能是促进肿瘤细胞死亡。事实上，Vaughan 等报道了一种模仿 T 细胞受体（TCR）的单克隆抗体，通过识别并结合到 MHC 肽表位来特异性地靶向并杀死肿瘤细胞[21]。同样我们也可以检测到特定于人血清中结合在 MHC Ⅰ类分子上肽的抗体（未发表的观察）。另一种可能性是肽特异性抗体诱导抗体依赖性细胞介导的细胞毒性（ADCC），如 Frank 等在乳腺癌患者上所报道的[22]。进一步的研究将需要澄清哪些抗原特异性抗体的生物学功能与免疫治疗的临床反应最相关。

21.5 PPV 临床效果

在过去 10 年中，500 多名不同类型的癌症患者参加了 PPV 临床试验，结果令人鼓舞（表 21.2）[12-15,23]。436 名可评估患者中，观察到最佳临床反应是 43 名患者（10%）表现部分反应（PR），144 名患者表现稳定疾病（SD）（33%），249 名患者表现为进展性疾病（PD）（57%），总生存期中位数为 9.9 个月[14]。例如最近的一项随机 II 期临床试验表明，PPV 对晚期去势抵抗型前列腺癌（CRPC）患者有潜在的临床效果[15]。在这项研究中，57 名 CRPC 患者被纳入并随机分为两组：接受 PPV 联合小剂量雌莫司汀磷酸盐（EMP）治疗的患者或仅接受标准剂量 EMP 治疗的患者。接受 PPV 联合低剂量 EMP 治疗的患者无进展时间明显延长［中位生存时间（MST）8.5 个月 vs. 2.8 个月；危险比（HR），0.28（95%置信区间（CI），0.14～0.61）；P= 0.0012］和总生存率［MST，未定义 vs. 16.1 个月；HR，0.30（95%置信区间，0.1～0.91）；P=0.0328］与单纯用标准剂量 EMP 治疗的患者相比，提示联合治疗的可行性。

表 21.2　PPV 治疗晚期癌症患者的临床反应

项目	最佳临床反应（n）[a]				反应率（%）	疾病控制率（%）
	患者（n）	PR	SD	PD		
总计	436	43	144	249	9.9	42.9

（续表）

项目	最佳临床反应（n）[a]				反应率（%）	疾病控制率（%）
	患者（n）	PR	SD	PD		
前列腺癌	155	29	36	90	18.7	41.9
结肠直肠癌	68	1	23	44	1.5	35.3
胰腺癌	41	4	23	14	9.8	65.9
胃癌	35	0	8	27	0	22.9
脑癌	30	5	11	14	16.7	53.3
宫颈癌	23	3	7	13	13.0	43.5
非小细胞肺癌	21	0	11	10	0	52.4
肾细胞癌	12	0	9	3	0	75.0
黑色素瘤	11	0	5	6	0	45.5
乳腺癌	10	0	1	9	0	10.0
尿路上皮癌	7	1	2	4	14.3	42.9
其他	23	0	8	15	0	34.8

PR 部分反应，*SD* 稳定的疾病，*PD* 进行性疾病；

[a]通过 RECIST 标准（或前列腺癌的 PSA 值）评估最佳临床反应。

21.6 PPV 生物标志

只有一部分患者表现出癌症免疫疗法的临床效果，包括 PPV[14]。因此对于 PPV 的进一步发展，识别准确描述抗肿瘤免疫反应和预测免疫患者预后的生物标志将是至关重要的。关于接种疫苗后的生物标志，在一些癌症疫苗临床试验中，一些因素如 CTL 反应、Th1 反应、迟发型超敏反应（DTH）和自身免疫性，已报道与临床反应有关[24-26]。此外，我们还报告了疫苗接种后血浆中肽反应性 IgG 水平升高患者的总生存期比未增加的患者明显延长（$P=0.0003$）。因此，接种疫苗后免疫反应的评估可能有助于预测癌症疫苗的有效性。

然而关于生物标志的可用信息很少，这些生物标志可用于预测肽疫苗使用前的临床疗效。最近为了在接种疫苗前确定有助于选择合适患者的生物标志，我们评估了接受 PPV 治疗的几种不同类型晚期癌症患者接种疫苗前的预后标志（表 21.3）。在接受 PPV 治疗的 CRPC 患者中（$n=40$），通过微阵列分析对可溶性因子和基因表达谱综合研究表明，IL-6 水平较高，髓源抑制性细胞（MDSC）频率较高，其表达抑制分子，如 ARG1 和诱导型一氧化氮合酶[27,28]，免疫前外周血 T 细胞和其他免疫细胞的免疫功能受损与预后差密切相关[29]。Yoshiyama 等报道，在难治性非小细胞肺癌（$n=41$）患者中，接种疫苗前 C 反应蛋白（CRP）水平是不良总生存率的一个重要预测因子（HR=10.115，95%CI=2.447-41.806，$P=0.001$）[30]。此外，在难治性胆管癌患者中（$n=25$），多变量 Cox 回归分析显示，接种前血浆中较高的 IL-6 和较低的白蛋白水平，以及较少的所选疫苗肽是影响总生存率的显著不利因素［分别为：HR=1.123，95% CI=1.008-1.252，$P=0.035$；HR=0.158，95%置信区间=0.029-0.860，$P=0.033$；HR=0.258，95%置信区间=0.098-0.682，$P=0.006$］[31]。此外用 PPV 对化疗耐药的晚期胰腺癌患者（$n=36$）治疗中，血清淀粉样蛋白 A（SAA）水平与总生存率相关［HR=1.21，95% CI=1.05~1.40，$P=0.006$］[32]。总的来说，这些发现表明较少的炎症可能有助于对 PPV 更好地反应，表明在接种疫苗前对炎症因素的评估可能有助于选择合适的癌症患者进行 PPV 治疗。

基于这些发现，在接种疫苗之前，可以使用外周血中的基因表达谱和/或可溶性因子的测量来监测炎症因子和免疫抑制细胞，这可以帮助预测接受癌症疫苗治疗患者的预后，并导致发展更好的治疗策略。

表 21.3　PPV 的预后生物标志

癌症类型	影响因素	统计分析（HR，95% CI，*P* value）[a]	参考文献
杂癌（$n=500$）	身体状况	HR=2.295；95% CI，1.653-3.188；$P<0.0001$	*Cancer Biol. Ther.* 10，1266-1279（2011）
	淋巴细胞计数	HR=1.472；95% CI，1.099-1.972，$P=0.0095$	
	IgG 对抗原的反应（接种疫苗后）	HR=1.455；95% CI，1.087-1.948，$P=0.0116$	
非小细胞肺癌（$n=41$）	C 反应蛋白（CRP）	HR=10.115；95% CI，2.447-41.806；$P=0.001$	*Int. J. Oncol.* 40，1492-1500（2012）
	IL-6	HR=1.123；95% CI，1.008-1.252；$P=0.035$	
胆管癌（$n=25$）	白蛋白	HR=0.158；95% CI，0.029-0.860；$P=0.033$	*Exp. Ther. Med.* 3，463-469（2012）
	疫苗肽数量	HR=0.258；95% CI，0.098-0.682；$P=0.006$	
胰腺癌（$n=36$）	血清淀粉样蛋白 A（SAA）	HR=1.21；95% CI，1.05-1.40；$P=0.006$	*Oncol. Rep. in press*（2013）
前列腺癌（$n=40$）	IL-6	（未确定）	*Cancer*118，3208-3221（2012）
	MDSC	（未确定）	

HR 危险比率，*CI* 置信区间，*MDSC* 髓源性抑制细胞；

[a]通过多变量 Cox 回归分析确定 PPV 潜在的预后生物标志。

21.7　小结

我们开发了一种新的癌症疫苗免疫方法，即 PPV，旨在提高个体患者对 TAA 衍生肽表位的现有免疫反应。在疫苗接种前，通过测量每个抗原的体液免疫应答来选择适合每个患者的疫苗抗原。最近，PPV 临床试验已经证明了这种新的治疗方法在各种类型的晚期癌症中的可行性。为了提高 PPV 临床疗效，我们已经确定了参与 PPV 试验的各种癌症患者的预测性生物标志。炎症因子，如 IL-6、CRP、SAA 和免疫抑制细胞（MDSC），与接种患者的临床效果密切相关。在不久的将来，PPV 可与其他治疗方法结合使用，以抑制晚期癌症患者中常见的炎症因子和 MD-SC。

致谢

这项研究得到了日本教育、文化、体育、科技省区域创新集群项目的资助，以及日本教育、文化、体育、科技省癌症治疗创新研究发展项目（P-Direct）的资助。

参考文献

[1] Mellman, I., et al.: Cancer immunotherapy comes of age. Nature480, 480-489 (2011).

[2] Topalian, S. L., et al.: Cancer immunotherapy comes of age. J. Clin. Oncol. 29, 4828-4836 (2011).

[3] Kantoff, P. W., et al.: Sipuleucel-T immunotherapy for castration-resistant prostate cancer. N. Engl. J. Med. 363, 411-422 (2010).

[4] Hodi, F. S., et al.: Improved survival with ipilimumab in patients with metastatic melanoma. N. Engl. J. Med. 363, 711-723 (2010).

[5] Topalian, S. L., et al.: Safety, activity, and immune correlates of anti-PD-1 antibody in cancer. N. Engl. J. Med. 366, 2443-2454 (2012).

[6] Brahmer, J. R., et al.: Safety and activity of anti-PD L1 antibody in patients with advanced cancer. N. Engl. J. Med. 366, 2455-2465 (2012).

[7] van der Bruggen, P., et al.: A gene encoding an antigen recognized by cytolytic T lymphocytes on a human melanoma. Science 254, 1643-1647 (1991).

[8] Chen, Y. T., et al.: A testicular antigen aberrantly expressed in human cancers detected by autologous antibody screening. Proc. Natl. Acad. Sci. U. S. A. 94, 1914-1918 (1997).

[9] Purcell, A. W., et al.: More than one reason to rethink the use of peptides in vaccine design. Nat. Rev. Drug Discov. 6, 404-414 (2007).

[10] Sasada, T., et al.: Overcoming the hurdles of randomised clinical trials of therapeutic cancer vaccines. Eur. J. Cancer46, 1514-1519 (2010).

[11] Kenter, G. G., et al.: Vaccination against HPV-16 oncoproteins for vulvar intraepithelial neoplasia. N. Engl. J. Med. 361, 1838-1847 (2009).

[12] Itoh, K., et al.: Personalized peptide vaccines: a new therapeutic modality for cancer. Cancer Sci. 97, 970-976 (2006).

[13] Itoh, K., et al.: Recent advances in cancer vaccines: an overview. Jpn. J. Clin. Oncol. 39, 73-80 (2009).

[14] Noguchi, M., et al.: Assessment of immunological biomarkers in patients with advanced cancer treated by personalized peptide vaccination. Cancer Biol. Ther. 10, 1266-1279 (2011).

[15] Noguchi, M., et al.: A randomized phase II trial of personalized peptide vaccine plus low dose estramustine phosphate (EMP) versus standard dose EMP in patients with castration resistant prostate cancer. Cancer Immunol. Immunother. 59, 1001-1009 (2010).

[16] Yoshida, K., et al.: Characteristics of severe adverse events after peptide vaccination for advanced cancer patients: analysis of 500 cases. Oncol. Rep. 25, 57-62 (2011).

[17] Asai, T., et al.: Evaluation of the modified ELISPOT assay for gamma interferon production in cancer patients receiving antitumor vaccines. Clin. Diagn. Lab. Immunol. 7, 145-154 (2000).

[18] Sharma, P., et al.: Novel cancer immunotherapy agents with survival benefit: recent successes and next steps. Nat. Rev. Cancer11, 805-812 (2011).

[19] Whiteside, T. L.: Immune monitoring of clinical trials with biotherapies. Adv. Clin. Chem. 45, 75-97 (2008).

[20] Komatsu, N., et al.: New multiplexed flow cytometric assay to measure anti-peptide antibody: a novel tool for monitoring immune responses to peptides used for immunization. Scand. J. Clin. Lab. Invest. 64, 535-545 (2004).

[21] Vaughan, P. W., et al.: Antibody targeting to a class I MHC-peptide epitope promotes tumor cell death. J. Immunol. 177, 4187-4195 (2006).

[22] Frank, G. M. S., et al.: Antibody-dependent cell-mediated cytotoxicity can be induced by MUC1 peptide vaccination of breast cancer patients. Int. J. Cancer93, 97-106 (2001).

[23] Terasaki, M., et al.: Phase I trial of a personalized pep-tide vaccine for patients positive for human leuko-

cyte antigen - A24 with recurrent or progressive glioblastoma multiforme. J. Clin. Oncol. 29, 337 - 344 (2011).

[24] Disis, M. L.: Immunologic biomarkers as correlates of clinical response to cancer immunotherapy. Cancer Immunol. Immunother. 60, 433-442 (2011).

[25] Hoos, A., et al.: Improved endpoints for cancer immunotherapy trials. J. Natl. Cancer Inst. 102, 1388-1397 (2010).

[26] Amos, S. M., et al.: Autoimmunity associated with immunotherapy of cancer. Blood118, 499 - 509 (2011).

[27] Gabrilovich, D. I., et al.: Myeloid-derived suppressor cells as regulators of the immune system. Nat. Rev. Immunol. 9, 162-174 (2009).

[28] Ostrand-Rosenberg, S., et al.: Myeloid-derived sup-pressor cells: linking inflammation and cancer. J. Immunol. 182, 4499-4506 (2009).

[29] Komatsu, N., et al.: Gene expression profiles in peripheral blood as a biomarker in cancer patients receiving peptide vaccination. Cancer118, 3208-3221 (2012).

[30] Yoshiyama, K., et al.: Personalized peptide vaccination in patients with refractory non-small cell lung cancer. Int. J. Oncol. 40, 1492-1500 (2012).

[31] Yoshitomi, M., et al.: Personalized peptide vaccination for advanced biliary tract cancer: IL-6, nutritional status, and pre-existing antigen-specific immunity as possible biomarkers for patient prognosis. Exp. Ther. Med. 3, 463-469 (2012).

[32] Yutani, S., et al.: A phase II study of a personalized pep-tide vaccination for chemotherapy-resistant advanced pancreatic cancer patients. Oncol. Rep. in press (2013).

（马维民，何继军译）

第 22 章　肾细胞癌及其血管的分子免疫治疗和疫苗

Nina Chi Sabins，Jennifer L. Taylor，
Devin B. Lowe 和 Walter J. Storkus①

摘要

肾细胞癌（RCC）仍然是一个重要的健康问题，因为许多患者在初次诊断时就出现了转移性疾病。肾细胞癌肿瘤通常由于血管生成失调而高度血管化，导致血管渗漏、混乱，从而限制免疫效应细胞接近肿瘤，这可能是导致癌症免疫治疗方法取得适度临床成功的部分原因。目前，晚期癌症一线治疗方法包括抗血管生成小分子药物，这些药物产生较高的客观临床反应率，尽管这些药物在本质上往往是暂时性的，最终大多数患者用药物难以治疗。

实施以肿瘤血管为靶标、基于基因治疗的疫苗是一种有希望的治疗选项，无论作为一种独立疗法还是与现有疗法选项相结合，都可以延长无病间隔和患者的整体生存率。这种分子疗法范围包括体外基因修饰抗原呈递细胞的递送到直接注射重组病毒疫苗。在这里我们将讨论当前和即将出现的与 RCC 患者密切相关的新型分子治疗方法。

22.1　肾细胞癌：改进分子疗法的需要

肾细胞癌（RCC）约占成人所有癌症的 3%，初次诊断时有 20%～30%的患者发现已经转移。如果不治疗，转移性肾细胞癌患者 5 年无病生存率只有 2%～11%[1]。肾切除手术后，使用标准化疗药物、激素和放射疗法的常规治疗收效甚微。这促使人们对佐剂和晚期癌症环境中的替代治疗策略（包括免疫疗法）进行广泛评估。

22.1.1　肿瘤血管异常为治疗干预提供了靶点

各种组织类型实体瘤，包括肾、卵巢和肺，通常具有高度的血管化。肿瘤微环境（TME）中血管和淋巴管支持肿瘤恶化，包括：①输送氧气和营养物质；②提供原发性肿瘤可以转移的管道；③募集/支持促进肿瘤新生血管形成的肿瘤干细胞和血管细胞前体；④募集限制保护性抗肿瘤免疫应答的免疫调节细胞群（Treg，MDSC）。

肿瘤微环境（TME）血管典型表现为“异常”，定义为弯曲的、不规则的血管网络，具有从主支血管床分支的高度不稳定和可渗透的毛细血管。肿瘤血管的不稳定性增加了肿瘤内的间质压力，阻止了杀死肿瘤的药物和免疫效应细胞进入 TME[2]。与从健康组织中分离的血管内皮细

① N. C. Sabins，博士（匹兹堡大学医学院免疫系，美国宾夕法尼亚州匹兹堡市 邮编：15213）

J. L. Taylor，博士 · D. B. Lowe，博士（匹兹堡大学医学院皮肤系，美国宾夕法尼亚州匹兹堡市 邮编：15213）

W. J. Storkus，博士（✉）（匹兹堡大学医学院免疫系，美国宾夕法尼亚州匹兹堡市 邮编：15213，匹兹堡大学医学院皮肤系，美国宾夕法尼亚州匹兹堡市 邮编：15213，匹兹堡大学癌症研究所皮肤病和免疫学系，美国宾夕法尼亚州匹兹堡市 邮编：15213，E-mail：storkuswj@ upmc. edu）

胞（VEC）相比，肿瘤血管内皮细胞显示不同的基因表达，这种转录物被认为是肿瘤血管不稳定的基础[3]。使用 siRNA 方法沉默这些基因可能阻断 VEC 迁移和血管形成[4]，从而支持肿瘤 VEC 作为高度相关的治疗靶点。肿瘤血管的另一个特征是血管壁细胞覆盖异常。VEC 通过分泌 PDGF（血小板衍生生长因子）募集血小板衍生生长因子受体（PDGFR）-β 表达的壁细胞[5]。

壁细胞支持 VEC 生存和分化，可以细分为平滑肌细胞和周细胞群体。已发现正常成熟周细胞在血管稳定、成熟、重塑和修复中起主要作用。肿瘤周细胞通常不成熟，松散地附着在毛细血管上，甚至在某些区域可能不存在，导致血管不稳定[6]。这种周细胞的表型特征是在一系列肿瘤模型中 SMA、RGS5 和 TEM1（肿瘤内皮标志-1）的过度表达[7-9]。有趣的是，最近发现 RGS5 缺陷小鼠的肿瘤显示正常的血管结构模式，其特征是与野生型宿主中生长的肿瘤相比，平均血管密度（MVD）和成熟的周细胞覆盖减少，缺氧和血管渗漏减少。这些血管变化与过继转移肿瘤特异性 T 细胞向 TME 渗透增加有关，被治疗动物的存活时间显著延长[10]。

22.1.2　TME 免疫功能障碍

实体肿瘤经常缺氧，导致 HIFs（缺氧诱导因子）表达增加，效应 T 细胞和 DC 功能受损[11]。上调 HIFs 表达也引起 VEGF（血管内皮生长因子）的表达增强，能够减弱 DC 分化以及通过 VEGFR2（血管内皮生长因子受体 2）介导的信号增加 STAT3 激活，从而促进免疫抑制因子（包括 IL-10 和 TGF-β）在这些调节性抗原呈递细胞中的内在表达[12]。许多癌症患者由于 Treg 或 MDSC（髓源性抑制细胞）和/或 2 型炎症分子的抑制作用，肿瘤破坏所需的 1 型促炎反应被抑制或功能失调[13-16]。

此外在癌症患者中，由于同源肿瘤抗原的慢性刺激，1 型肿瘤特异性 T 细胞可能功能衰竭并经历过早（凋亡）死亡[17,18]。这种由肿瘤假定的免疫逃避策略用于限制保护性宿主免疫反应，并允许肿瘤存活和发展。此外，由于肿瘤逐渐生长过程中血管不稳定，循环 1 型抗肿瘤 T 效应细胞可能无法有效地运输到肿瘤部位[10]。

22.1.3　RCC 免疫靶向

像黑色素瘤一样，过去的报告显示肾细胞癌进展和消退可以在免疫学上得到控制，这为生物反应调节剂和疫苗的使用带来了乐观情绪[19]。值得注意的是，少数但频率显著的肾细胞癌患者表现出明显的自发肿瘤消退[20]，而采用慢性免疫抑制方案保留同种异体肾移植的患者表现出增加患肾细胞癌的风险[21]。此外，肿瘤浸润淋巴细胞（TIL）的大小被认为是患者生存的预后指标[22]。在这方面，1 型 TIL（即产生 IFN-γ 并能介导细胞溶解功能的 TIL）[23]似乎代表客观临床反应的主要免疫介质。

尽管施用大剂量白介素-2（IL-2；又名 T 细胞生长因子）在少数转移性肾细胞癌患者中产生了持久的完全应答，但这种方法也观察到了严重毒性[24]，表明需要确定更特异/集中的免疫治疗方法。

22.1.4　在 TME 使用抗血管生成小分子药物实现血管正常化

Judah Folkman 提出了把血管生成作为癌症治疗靶点的想法[25]。他的理论促进了几种抗血管生成疗法的发展，包括抗血管生成生长因子（及其同源受体）的单克隆抗体和促血管生成受体酪氨酸激酶小分子抑制剂。事实上在 2005 年有人提出，作为一种新的癌症疗法，可以通过使用抗血管生成剂来实现肿瘤血管正常化[2,10]。这种方法协调允许改善联合应用化疗药物到 TME 的递送。

人源化单克隆抗体贝伐单抗（Avastin）已被证明可诱导治疗患者肿瘤血管短暂正常化，当与化疗结合时，可促进增强抗肿瘤效果[26]。以可溶性 VEGF（血管内皮生长因子）为靶点可增加 TME 中 PDGF（血小板衍生生长因子）的优势，从而诱导周细胞募集、活化和毛细血管稳定[27]。黄等最近报道，在给予一定剂量的抗 VEGFR2（血管内皮生长因子受体 2）抗体后，小鼠的肿瘤血管恢复正常。此外，这种治疗将 TME 重新编程为更具促炎性并越来越容易接受 T 效应细胞浸润[28]。另外，药物制剂（如舒尼替尼）以 VEGF 受体（即 VEGFR1-3）为靶点，以及其他促血管生成的酪氨酸激酶受体（即 PDGFR-β），也可以抑制失调的血管生成，导致血管暂时正常化。

由于肾细胞癌具有高度血管化特性，一些临床试验已经利用拮抗血管生成的酪氨酸激酶抑制剂（TKI）作为治疗这种疾病的手段。其中一种抑制剂是苹果酸舒尼替尼[29]，在 I/II 期临床试验中显示出明显（尽管短暂）疗效，并被批准为肾细胞癌患者的一线治疗方案[30-32]。然而一项 III 期试验显示，虽然接受舒尼替尼（11 个月）的患者早期无进展，生存率高于干扰素治疗（5 个月），但两组患者之间与此方法相关的总生存效益并无显著差异（26.4 个月对 21.8 个月）[30]。

尚不确定舒尼替尼的直接抗肿瘤作用，因为在人肾细胞癌中未发现 RTK（受体酪氨酸激酶）的体细胞突变[33]。有趣的是体外研究表明，舒尼替尼的作用机制包括诱导肿瘤相关内皮细胞而不是人肾细胞癌细胞自身的凋亡。与贝伐单抗一样，体内研究显示舒尼替尼可在 TME 中诱导血管正常化的临时窗口[34]。

虽然舒尼替尼最初是作为血管抑制剂开发的，但最近的报道表明，这种 TKI 实际上通过选择性修剪未成熟和脆弱的血管使肿瘤血管“正常化”，使分化较好的血管完好无损并覆盖成熟的周细胞。这种正常化导致间质压力降低，改善化疗药物和效应性 T 细胞向 TME 递送[35]。这与 Ganss 等的研究结果一致，表明由局部放射治疗诱导的 TME 炎症连同肿瘤特异性细胞过继转移，可以诱导微血管重构至类似正常组织的表型[36]。标准化内皮显示 1 型趋化因子 CXCL9 和 CXCL10 及其同源受体 CXCR3 表达增加。CXCL10 是一种内源性血管生成抑制剂，可作为 1 型 T 细胞的化学引诱剂，已证明可诱导人 CXCR3b+内皮细胞凋亡，但不能诱导肿瘤细胞凋亡[37]。

此外，舒尼替尼治疗的患者在体外表现出 MDSC 和 Treg 群体减少和标准化 1 型 T 细胞反应[31,32,38,39]。小鼠肿瘤模型表明舒尼替尼抑制 STAT3 激活[40]，并通过促进肿瘤特异性效应子 T 细胞提高免疫治疗功效，同时还抑制体内 MDSC 和 Treg[32,38,41]。舒尼替尼治疗已显示趋化因子及其受体的表达偏向 1 型分布[42]，以及 VCAM1 和 CXCL9（MIG）表达增加，TME 和 TDLN（肿瘤引流淋巴结）内产生 IFN-γ 的 T 细胞增加，这意味着 TKI 可能是一种有效的免疫佐剂。我们预计，肿瘤相关血管及其细胞亚群的特异性免疫靶向，可能会对血管正常化产生类似且潜在的更持久影响。

22.1.5 当前的 RCC 治疗疫苗

设计用于促进针对 RCC 的特异性获得性免疫的疫苗传统上涉及肿瘤细胞本身的使用[43]。自体肿瘤细胞疫苗通过共同使用或基因修饰，来表达炎症细胞因子，如 GM-CSF、IFN-γ 和 IL-2，来改进疫苗效力[44,45]。另一种肿瘤疫苗配方是以 RCC 抗原呈递细胞（APC）融合杂交种为代表，该杂交种产生能够表达 RCC 基因产物，并向 T 细胞递呈其衍生肽表位的杂交抗原呈递细胞[46]。

另一种基于细胞的疫苗（即改良的肿瘤细胞、APC、T 细胞）方法是使用重组病毒的递送系统进行基因疫苗接种。各种报告显示，基于病毒的疫苗比基于蛋白质抗原/佐剂的疫苗具有治疗优势[47,48]，这可能是由于病毒固有的促炎特性（通过激活由 APC 表达的 Toll 样受体（TLR））以及它们感染专业 APC 的能力，允许患者在 DC 中异位表达疫苗抗原[49-51]。此外，高滴度重组病毒很容易产生，与因其患者特异性而需要昂贵的耗时方法的细胞疗法相比，病毒载体可以作为现成治疗方式应用于任何特定患者。因此，尽管对复制能力强的污染物病毒或逆转录病毒插入突变的关注较少[52]，但在癌症（RCC）环境中，基因疫苗仍然是有吸引力的治疗选择。

重组改良疫苗病毒 Ankara（MVA）编码 RCC 抗原 5T4（Trovax），Trovax 疫苗已经在一些临床试验中进行测试。RCC 患者初步试验显示，接种 Trovax 疫苗后有一些客观的临床反应[53,54]；然而在第 III 期试验中，使用添加或不添加细胞因子（IFN-α 和 IL-2）的 Trovax 疫苗和舒尼替尼联合使用，实验组和对照组之间生存率没有显著差异[55]。第二种 MVA 疫苗是含有重组 MUC1 和 IL-2 的转基因疫苗，使用或不使用细胞因子。尽管一些 RCC 患者表现出抗 MUC1T 细胞应答，但根据 RECIST 标准，用这种疫苗进行治疗并未产生客观的临床反应[56]。

值得注意的是，也有 RCC 非病毒分子疫苗正在研发中。另一种方法包括 DC 致敏或转染 RCC 衍生的总 mRNA（编码完整的 RCC 相关抗原库）[57-59]。最近在一项涉及 30 名 RCC 患者的 I/II 期临床试验中，对直接皮内注射编码 MUC1、CEA、HER-2/neu、端粒酶、生存素和 MAGE-A1 的 mRNA 疫苗进行了评估。报告的结果是有希望的，一些接种疫苗的患者病情稳定，协同增加肿瘤抗原特异性 T 细胞应答[60]。

表 22.1 概述了靶向分子定义 RCC 相关抗原的治疗性疫苗。总的来说，这些方法已经证明能产生免疫原性，接种疫苗后 1 型 T 细胞应答往往与临床结果改善有关。

22.1.6 编码肿瘤基质细胞相关抗原基因疫苗

在免疫选择压力作用下，基因畸变包括逐步发展和异质性 RCC 病变，可能发展数月至数年[72]。基于全肿瘤细胞、肿瘤 APC 杂交和/或肿瘤源性 mRNA 或 cDNA 的疫苗，以及来自突变或过度表达蛋白质的肿瘤抗原，目前作为治疗药物疗效远不如预期。这些疫苗配方可能仅仅加强已有但失败的免疫库，这些免疫库提供了特定肿瘤抗原高于其他抗原的免疫优势，以及由异质肿瘤块假设的免疫逃避模式。大量肽表位在体内竞争通过（交叉递呈）APCs 表达的主要组织相容性复合体（MHC）分子，也可能限制对广泛的治疗性肿瘤抗原靶标的有效免疫激活。

最近一些研究小组研究表明[73,74]，尽管用抗血管生成剂治疗可以导致肿瘤血管短暂正常化，并至少导致短暂的肿瘤消退，但最终在治疗停止后，肿瘤经常复发，甚至侵袭性和转移潜力方面表现出更具侵略性的行为。由于这些药物仅限制血管生成［即通过受体酪氨酸激酶（RTK）信号拮抗作用］，而不是彻底根除血管，因此肿瘤相关血管细胞很可能最终适应耐药性状态。规避这一问题的一种可能方法是特定地激发肿瘤周细胞和/或血管内皮细胞的免疫靶向性，以提供持久的维持力，能够优先破坏/调节肿瘤相关基质细胞，同时保留患者正常的、与肿瘤无关器官的血管。通过聚焦肿瘤血管生态位的 1 型免疫反应，还可以激活 TME（即调节其细胞因子环境和表达黏附分子的模式/水平/激活状态），从而增强抗肾细胞癌 T 效应细胞的募集和（多聚）功能。

我们还有和其他研究人员最近倡导实施促进肿瘤血管细胞（即周细胞、VEC）群体特异性 1 型 T 细胞识别的疫苗[75-77]；然而这些疫苗制剂是基于细胞的，利用过继转移 DC 递呈基质抗原衍生肽。Rosenberg 研究小组最近发现，T 细胞被设计成表达 VEGFR2 特异性嵌合抗原受体（CAR），在许多实体癌（包括肾细胞癌）血管中过度表达，能够限制 5 种不同类型的已建立的血管化肿瘤在小鼠体内生长，并协同诱导 VEGFR2 特异性宿主 T 细胞应答[78]。这种治疗策略目前正在 RCC 和黑色素瘤患者 I 期试验中进行研究。也有研究表明，在转移性 RCC 小鼠模型中，用基因改变以表达内皮抑素（另一种对肿瘤血管生成至关重要的分子）的成纤维细胞治疗，当与 rIL-2 给药结合时，随着 $CD4^+$、$CD8^+$和 CD49b/ VLA-4^+淋巴细胞的浸润增加，导致肿瘤负担减轻，这表明疫苗具有额外的免疫调节作用[79]。

目前，对肾细胞癌的基因免疫策略仅限于那些使用编码肿瘤抗原的重组痘苗病毒，迄今为止，这些策略在临床上取得了一定的成功。不幸的是，痘苗载体只能应用 1 次或 2 次，因为宿主中和抗体的快速发展限制了以这种方式进一步给药的增强能力[80]。鉴于慢病毒在缺乏中和抗病毒免疫情况下，在体内转导内源性 DC 能力以及向患者提供重复给药的潜力[81]，我们最近开发了一种基于慢病毒的疫苗，其设计用于促进以肿瘤血管为靶标的特异性免疫，更具体地说以血

表 22.1 含有确定分子的肾细胞癌相关抗原临床疫苗

疫苗配方	疫苗靶向抗原	试验期	接受治疗的肾细胞癌患者人数	疫苗免疫原性	临床效果	参考文献
多肽疫苗+/-IL-2或IFN-α	基于患者体外反应预处理选择4种确定的TAA肽	Ⅰ	10	8/10的患者对疫苗DTH反应阳性，2/10的患者接种疫苗后对肽1型T细胞反应增强	6 SD	[61]
AutoDC+MUC1肽+PA-DRE肽+IL-2	MUC1	Ⅰ	20	10/20患者对肽有1型T细胞反应	1CR，2 PR，5 SD，2MR；T细胞反应与OCR相关	[62]
MKC1106-PP基因初免/肽加强免疫	PRAME，PSMA	Ⅰ	2	2/2患者对多肽有1型T细胞反应	2/2患者SD>11个月	[63]
CA-IX（G250）肽+IFA	CA-IX	Ⅰ	23	5/10可评估的患者对肽有1型$CD8^{+}$T细胞反应	3 PR，6 SD	[64]
成熟AutoDC+CA-IX（G250）多肽	CA-IX	Ⅰ	6	对CA-IX无特异性免疫	No OCR	[65]
肿瘤抗原RNA+GM-CSF	MUC1、CEA，HER2/Neu、端粒酶、生存素、MAGE-A1	Ⅰ/Ⅱ	30	8/10可评估的患者对疫苗肽有1型T细胞反应	16 SD，1 PR	[60]
端粒酶肽+Montanide佐剂	端粒酶	Ⅰ/Ⅱ	4	无描述，1例患者的进展期肿瘤为MHC Ⅰ型丢失变异体	No OCR	[66]
AutoDC+端粒酶肽	端粒酶	Ⅰ/Ⅱ	10	2/8的可评估患者有DTH反应，只有3名有OCR的患者有CTL反应	1SD，2 MR与最强的T细胞IFN-γ对肽的反应相关	[67]
突变VHL肽+Montanide佐剂	突变VHL	Ⅰ/Ⅱ	6	4/5可评估的患者对抗原有$CD8^{+}$T细胞反应	没有OCR，如果病人对疫苗有反应，延长OS	[68]
IM901+GM-CSF+/-CY	9个天然MHC Ⅰ类分子递呈TAA（PLIN2、APOL1（2）、CCND1、GUCY1A3、PRUNE2、MET、MUC1、RGS5）+1个天然MHC Ⅱ类分子递呈TAA（MMP7）肽	Ⅰ+Ⅱ	96	64%~74%的可评估患者对1种或多种疫苗肽表现出$CD8^{+}$T细胞应答	如果患者对多种多肽产生免疫反应，就可获得更好的OS	[69]

（续表）

疫苗配方	疫苗靶向抗原	试验期	接受治疗的肾细胞癌患者人数	疫苗免疫原性	临床效果	参考文献
TG4010（重组痘苗病毒）+/-IFN-α/IL-2	MUC1	Ⅱ	37	6/28 和 6/23 可评估患者对疫苗分别产生 $CD4^+$ 和 $CD8^+$T 细胞反应，免疫应答者有较长的 OS	6/20 患者 SD 持续 6 个月以上	[56]
MVA - 5T4（TroVAX；改良的安卡拉痘苗）+/- 舒尼替尼、IL-2 或 IFN-α	5T4	Ⅲ	288	56%的可评估患者在接种疫苗后出现抗 5T4 抗体反应，未评估 T 细胞反应	抗体反应的强度与较长的 OS 相关，对较小肿瘤的影响最大	[53-55，70，71]
IMA901+/-舒尼替尼	9 个天然 MHC Ⅱ类分子递呈 TAA（PLIN2、APOL1（2）、CCND1、GUCY1A3、PRUNE2、MET、MUC1、RGS5）+ 1 个天然 MHC Ⅱ类分子递呈 TAA（MMP7）多肽	Ⅲ	330	进展中，结果预计在 2014 年	进展中，结果预计在 2014 年	进展中，结果预计在 2014 年

缩略词：*AlloDC* 异体树突状细胞，*APOL*1 载脂蛋白 L1 抗原，*AutoDC* 自体树突状细胞，*CA-IX* 碳酸酐酶-IX 抗原（也称为 G250），*CCND*1 细胞周期蛋白 D1 抗原，*CEA* 癌胚抗原，*CR* 完全应答，*CTL* 细胞毒性 T 淋巴细胞，*CY* 环磷酰胺，*DTH* 延迟型超敏反应，*GUCY*1A3 可溶性鸟苷酸环化酶亚基 alpha-3 抗原，*HER*2 人表皮生长因子受体 2（亦称 Neu），*IFA* 不完全弗氏佐剂，*MAGE-A*1 黑素瘤相关抗原 1，*MET*c-MET 原癌基因抗原（亦称肝细胞生长因子受体），*MHC* 主要组织相容性复合体，*MMP*7 基质金属蛋白酶 7 抗原，*MR* 混合反应，*MUC*1 黏蛋白 1 抗原，*OCR* 客观临床反应，*OS* 总生存期，*PADRE*Pan-DR 递呈抗原表位，*PR* 部分反应，*PRAME* 在黑色素瘤中优先表达抗原，*PRUNE*2 蛋白质修剪同源物 2 抗原，*PSMA* 前列腺特异膜抗原，*RGS*5G 蛋白信号 5 抗原调节因子，*SD* 疾病稳定，*TAA* 肿瘤相关抗原，*VHL*Von Hippel-Lindau 蛋白，*5T4* 滋养细胞糖蛋白抗原（亦称 TPBG）。

管周细胞为靶标。从人和小鼠肾细胞癌肿瘤中分离的周细胞差异过度表达一组抗原（与正常肾中的同源细胞相比），包括delta样1同源物（DLK1）。据观察，治疗性DLK1肽和（慢病毒）基因疫苗均具有免疫原性，并能有效调节体内肾细胞癌（RCC）肿瘤生长。有效治疗的肾细胞癌在减少血管渗漏和组织缺氧的基础上表现出血管正常化，在血管周围空间被$CD8^+$TIL高度浸润。TME残留周细胞与血管紧密接近，缺乏DLK1表达，支持疫苗诱导体内成熟壁细胞群的免疫选择。我们认为针对肿瘤相关血管抗原（如DLK1）的疫苗在肾细胞癌和其他血管化实体癌的治疗，可能具有重要的临床前景。

22.1.7 现有的疫苗平台可能不太理想，需要发展联合免疫治疗方法

目前单一RCC治疗方法，无论是针对癌细胞免疫靶向治疗还是抗血管生成药物治疗，迄今为止取得的临床成功有限。虽然功能性肿瘤特异性T细胞可通过疫苗接种激活，但由于缺乏补充信号和作为淋巴细胞外渗物理屏障的高间质压力，这些细胞通常不能有效地进入，表现出不稳定血管的肿瘤病变[82]。如果与疫苗疗法结合使用，抗血管生成疗法治疗的肿瘤中出现短暂血管正常化，可能会绕过T细胞运输限制。

促进癌细胞免疫靶向疫苗可以选择耐药性（即抗原或MHC丢失）变异体[83]，从而维持隐性疾病或使获得性免疫系统“看不见”肿瘤进展。预期由血管抗原特异性$CD8^+$T细胞库介导的抗血管生成作用将删除特定的血管细胞群，从而补充受体信号的特异性抑制，这是可选药理学抗血管生成治疗方式的标志，例如抗VEGF抗体（即贝伐单抗）和小分子酪氨酸激酶抑制剂（即舒尼替尼）[29,84,85]。通常用这些药物治疗的血管化肿瘤由于采用代偿性生长/进展途径而迅速成为药物难治性肿瘤。鉴于这种局限性，靶向血管基质抗原分子疫苗在许多对贝伐单抗、舒尼替尼或类似抗血管生成药物产生耐药性的病例中，可能是一种合乎逻辑的二线方法。这些疫苗也可能代表有效的一线治疗剂（图22.1），因为预计TME抗肿瘤基质相关抗原T效应细胞特异性激活、募集和发挥作用，将通过联合施用抗血管生成TKI减少抑制细胞人群（最显著的是肾细胞癌患者）和体内促炎性TME的激活而得到改善[39-42]。

22.2 过继细胞疗法（ACT）

用基因修饰的重组T细胞受体（TCR）治疗，近期证实能促进血液系统恶性肿瘤的癌症消退[86]。实施类似的策略，肿瘤抗原特异性T细胞可以在体外从肾细胞癌肿瘤浸润淋巴细胞中分离，随后肾细胞癌抗原特异性T细胞受体被分离、克隆和转导到患者T细胞中，以赋予抗肿瘤特异性[87]。一旦体外扩增，这种抗肿瘤效应细胞作为治疗剂可以被过继转移。

此外，考虑到最近在利用基因工程表达CARS（嵌合型抗原受体）T细胞过继转移治疗方面取得的成功[88]，在实体肿瘤治疗中应用这一策略的可能性越来越大。例如正在进行的临床试验旨在测试转移性结肠癌中T细胞工程化表达癌胚抗原（CEA）的反应性[89]，和晚期卵巢癌中叶酸受体-α（FR-α）反应性CARs的安全性和有效性[90]。在针对肾细胞癌（RCC）患者的1期临床试验中，许多接受基因工程改造以表达对碳酸酐酶-IX（CA-IX，又名G250）肾细胞癌相关蛋白个体产生了相应的抗G250抗体应答和T细胞介导的免疫[91]。在肾细胞癌患者中，使用携带对肿瘤相关血管抗原具有反应性CARs的工程化T细胞，来监测ACT发展和未来临床实施将是非常有意义的。虽然此类治疗仍处于患者评估的初步阶段，但我们可以预测，用抗血管性CAR^+T细胞治疗肾细胞癌患者，将极大地改善首选临床结果的幅度和持续时间，尤其是与贝伐单抗和舒尼替尼等其他TME正常化药物联合使用时。

22.3 结束语

肾细胞癌的异质性以及大多数其他实体形式的癌症，使得很难开发出针对肿瘤相关抗原一

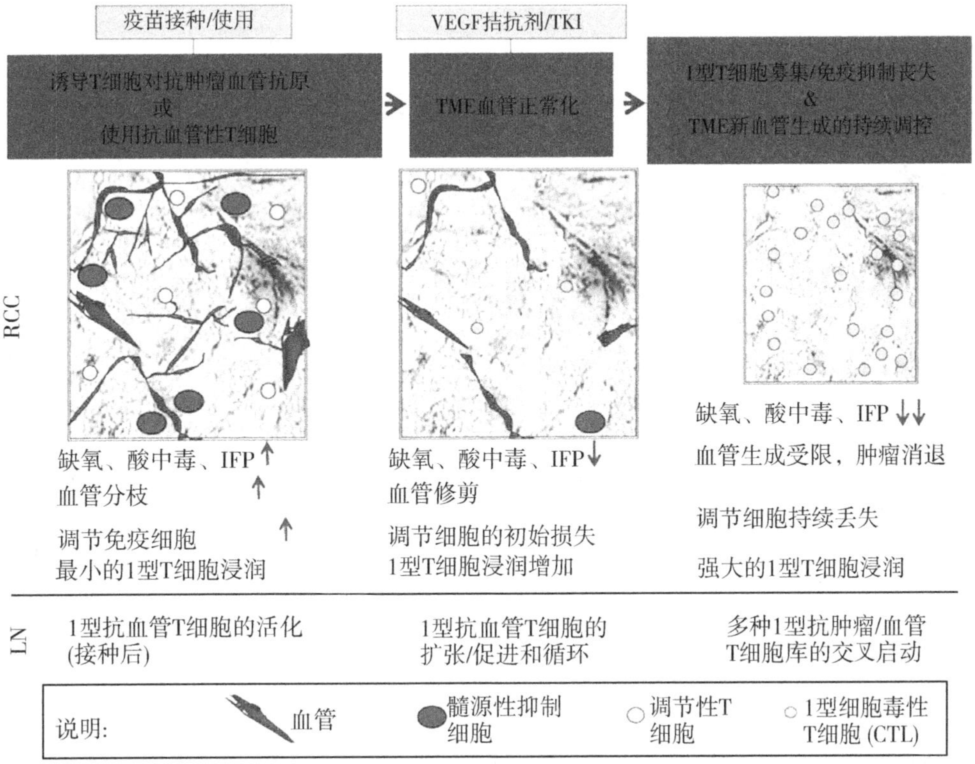

图 22.1 针对 RCC 血管系统联合疫苗治疗效益改进范例

未治疗的 RCC 肿瘤（左图）由调节性免疫细胞（髓源性抑制细胞（MDSC）和调节性 T 细胞（Treg））组成，仅有很少的 1 型 T 效应细胞。进展期病变中不稳定、无组织的血管系统导致缺氧、酸中毒和间质液压力高。针对肿瘤血管抗原的主动疫苗接种以及随后在疫苗引流淋巴结中特异性 T 细胞的激活，或血管抗原特异性 T 细胞的过继转移（即过继细胞疗法，ACT）导致循环水平的 1 型 T 效应细胞在向 RCC 肿瘤的转运中效率低下。与抗血管生成药物（如 VEGF 拮抗剂或 TKI）的交错联合治疗至少暂时使肿瘤血管系统正常化，导致低氧/酸中毒/干扰素减少并去除调节细胞群（中间图），允许增加 1 型疫苗诱导或 ACT T 细胞的募集和效应功能（右图）。随后免疫介导的肿瘤细胞根除和局部 DC 人群肿瘤（和替代间质抗原）的出现，可能导致肿瘤引流淋巴结中治疗性 1 型 T 细胞多样化库的交叉启动，从而扩大临床效益（右图）。

致的成功疗法。事实上，虽然已经发展出几种分子策略来诱导对肿瘤抗原的免疫，但这种方法的临床成功率很低。因此有必要确定肿瘤的共同（不变）特征，例如不规则的血管，这是宏观实体肿瘤的标志。在这方面，TME 中未成熟血管细胞群的亚群过度表达一系列抗原，例如受体酪氨酸激酶，这些抗原可能是能导致短暂疗效的药理学药物或免疫治疗方法的靶标，这些药物提供了以肿瘤相关血管为靶标的持续获得性免疫的希望。

这些策略包括结合肿瘤相关血管蛋白/肽抗原的疫苗、编码全长或亚单位血管抗原的病毒疫苗以及对肿瘤相关血管靶标具有反应性的离体扩增或分子工程化 T 细胞。我们相信，虽然这些以肿瘤血管为靶标的分子疗法可以为肾细胞癌患者提供治疗益处，但是如果与现有的“血管正常化”药物（包括单克隆抗体（贝伐单抗）和受体酪氨酸激酶抑制剂（舒尼替尼））联合施用，它们的效果可能会进一步增强。

致谢

这项工作得到了美国国立卫生研究院 R01 拨款 CA140375（W. J. S.）和美国癌症学会（D. B. L.）博士后奖学金（PF-11-151-01-LIB）的支持。作者声明没有利益冲突。

参考文献

[1] Rohrmann, K., et al.: Immunotherapy in metastatic renal cell carcinoma. World J. Urol. 23, 196-201 (2005).

[2] Jain, R. K.: Normalization of tumor vasculature: an emerging concept in antiangiogenic therapy. Science307, 58-62 (2005).

[3] Gerhardt, H., Semb, H.: Pericytes: gatekeepers in tumour cell metastasis? J. Mol. Med. 86, 135-144 (2008).

[4] Lu, C., et al.: Gene alterations identified by expression profiling in tumor-associated endothelial cells from invasive ovarian carcinoma. Cancer Res. 67, 1757-1768 (2007).

[5] Hellberg, C., Ostman, A., Heldin, C. H.: PDGF and vessel maturation. Recent Results Cancer Res. 180, 103-114 (2009).

[6] Morikawa, S., et al.: Abnormalities in pericytes on blood vessels and endothelial sprouts in tumors. Am. J. Pathol. 160, 985-1000 (2002).

[7] Heath, V. L., Bicknell, R.: Anticancer strategies involving the vasculature. Nat. Rev. Clin. Oncol. 6, 395-404 (2009).

[8] Berger, M., Bergers, G., Arnold, B., Hammerling, G. J., Ganss, R.: Regulator of G-protein signaling-5 induction in pericytes coincides with active vessel remodeling during neovascularization. Blood105, 1094-1101 (2005).

[9] Christian, S., et al.: Endosialin (Tem1) is a marker of tumor-associated myofibroblasts and tumor vessel-associated mural cells. Am. J. Pathol. 172, 486-494 (2008).

[10] Hamzah, J., et al.: Vascular normalization in Rgs5-deficienttumours promotes immune destruction. Nature 453, 410-414 (2008).

[11] Mancino, A., et al.: Divergent effects of hypoxia on dendritic cell functions. Blood 112, 3723-3734 (2008).

[12] Yu, H., Kortylewski, M., Pardoll, D.: Crosstalk between cancer and immune cells: role of STAT3 in the tumour microenvironment. Nat. Rev. Immunol. 7, 41-51 (2007).

[13] Tatsumi, T., et al.: Disease stage variation in $CD4^+$ and $CD8^+$ T-cell reactivity to the receptor tyrosine kinase EphA2 in patients with renal cell carcinoma. Cancer Res. 63, 4481-4489 (2003).

[14] Tatsumi, T., et al.: Disease-associated bias in T helper type 1 (Th1) /Th2 CD4 (+) T cell responses against MAGE-6 in HLA-DRB10401 (+) patients with renal cell carcinoma or melanoma. J. Exp. Med. 196, 619-628 (2002).

[15] Onishi, T., Ohishi, Y., Imagawa, K., Ohmoto, Y., Murata, K.: An assessment of the immunological environment based on intratumoral cytokine production in renal cell carcinoma. BJU Int. 83, 488-492 (1999).

[16] Griffiths, R. W., et al.: Frequency of regulatory T cells in renal cell carcinoma patients and investigation of correlation with survival. Cancer Immunol. Immunother. 56, 1743-1753 (2007).

[17] Saff, R. R., Spanjaard, E. S., Hohlbaum, A. M., Marshak-Rothstein, A.: Activation-induced cell death limits effector function of CD4 tumor-specific T cells. J. Immunol. 172, 6598-6606 (2004).

[18] Lu, B., Finn, O. J.: T-cell death and cancer immune tolerance. Cell Death Differ. 15, 70-79 (2008).

[19] Kolbeck, P. C., Kaveggia, F. F., Johansson, S. L., Grune, M. T., Taylor, R. J.: The relationships among tumor-infiltratinglymphocytes, histopathologic findings, and long-term clinical follow-up in renal cell

carcinoma. Mod. Pathol. 5, 420-425 (1992).

[20] Ritchie, A. W., deKernion, J. B.: The natural history of renal carcinoma. Semin. Nephrol. 7, 131-139 (1987).

[21] Penn, I.: Primary kidney tumors before and after renal transplantation. Transplantation 59, 480-485 (1995).

[22] Hernberg, M.: Lymphocyte subsets as prognostic markers for cancer patients receiving immunomodulative therapy. Med. Oncol. 16, 145-153 (1999).

[23] Kondo, T., et al.: Favorable prognosis of renal cell carcinoma with increased expression of chemokines associated with a Th1-type immune response. Cancer Sci. 97, 780-786 (2006).

[24] Turcotte, S., Rosenberg, S. A.: Immunotherapy for metastatic solid cancers. Adv. Surg. 45, 341-360 (2011).

[25] Folkman, J.: Role of angiogenesis in tumor growth and metastasis. Semin. Oncol. 29, 15-18 (2002).

[26] Ma, J., Waxman, D. J.: Combination of anti-angiogenesis with chemotherapy for more effective cancer treatment. Mol. Cancer Ther. 7, 3670-3684 (2008).

[27] Greenberg, J. I., Cheresh, D. A.: VEGF as an inhibitor of tumor vessel maturation: implications for cancer therapy. Expert Opin. Biol. Ther. 9, 1347-1356 (2009).

[28] Huang, Y., et al.: Vascular normalizing doses of antiangiogenic treatment reprogram the immunosuppressive tumor microenvironment and enhance immunotherapy. Proc. Natl. Acad. Sci. U. S. A. 109, 17561-17566 (2012).

[29] Faivre, S., Demetri, G., Sargent, W., Raymond, E.: Molecular basis for sunitinib efficacy and future clinical development. Nat. Rev. Drug Discov. 6, 734-745 (2007).

[30] Motzer, R. J., et al.: Overall survival and updated results for sunitinib compared with interferon alfa in patients with metastatic renal cell carcinoma. J. Clin. Oncol. 27, 3584-3590 (2009).

[31] Gore, M. E., et al.: Safety and efficacy of sunitinib for metastatic renal-cell carcinoma: an expanded-access trial. Lancet Oncol. 10, 757-763 (2009).

[32] Escudier, B., et al.: Phase II study of sunitinib administered in a continuous once-daily dosing regimen in patients with cytokine-refractory metastatic renal cell carcinoma. J. Clin. Oncol. 27, 4068-4075 (2009).

[33] Huang, D., et al.: Sunitinib acts primarily on tumor endothelium rather than tumor cells to inhibit the growth of renal cell carcinoma. Cancer Res. 70, 1053-1062 (2010).

[34] Czabanka, M., Vinci, M., Heppner, F., Ullrich, A., Vajkoczy, P.: Effects of sunitinib on tumor hemodynamics and delivery of chemotherapy. Int. J. Cancer124, 1293-1300 (2009).

[35] Zhou, Q., Guo, P., Gallo, J. M.: Impact of angiogenesis inhibition by sunitinib on tumor distribution of temozolomide. Clin. Cancer Res. 14, 1540-1549 (2008).

[36] Ganss, R., Ryschich, E., Klar, E., Arnold, B., Hammerling, G. J.: Combination of T-cell therapy and trigger of inflammation induces remodeling of the vasculature and tumor eradication. Cancer Res. 62, 1462-1470 (2002).

[37] Feldman, E. D., et al.: Interferon gamma-inducible protein 10 selectively inhibits proliferation and induces apoptosis in endothelial cells. Ann. Surg. Oncol. 13, 125-133 (2006).

[38] van Cruijsen, H., et al.: Sunitinib-induced myeloid lineage redistribution in renal cell cancer patients: CD1c+dendritic cell frequency predicts progression-free survival. Clin. Cancer Res. 14, 5884-5892 (2008).

[39] Finke, J. H., et al.: Sunitinib reverses type-1 immune suppression and decreases T-regulatory cells in renal cell carcinoma patients. Clin. Cancer Res. 14, 6674-6682 (2008).

[40] Ko, J. S., et al.: Sunitinib mediates reversal of myeloid-derived suppressor cell accumulation in renal cell carcinoma patients. Clin. Cancer Res. 15, 2148-2157 (2009).

[41] Ozao-Choy, J., et al.: The novel role of tyrosine kinase inhibitor in the reversal of immune suppression and modulation of tumor microenvironment for immune-based cancer therapies. Cancer Res. 69, 2514-2522 (2009).

[42] Bose, A., et al.: Sunitinib facilitates the activation and recruitment of therapeutic anti-tumor immunity in

concert with specific vaccination. Int. J. Cancer129, 2158-2170 (2011).

[43] Chi, N., et al.: Update on vaccine development for renal cell cancer. Res. Rep. Urol. 2010, 125-141 (2010).

[44] Tani, K., et al.: Progress reports on immune gene therapy for stage IV renal cell cancer using lethally irradiated granulocyte-macrophage colony-stimulating factor-transduced autologous renal cancer cells. Cancer Chemother. Pharmacol. 46 (Suppl), S73-S76 (2000).

[45] Maini, A., et al.: Combination of radiation and vaccination with autologous tumor cells expressing IL-2, IFN-γ and GM-CSF for treatment of murine renal carcinoma. In Vivo 17, 119-123 (2003).

[46] Avigan, D. E., et al.: Phase I/II study of vaccination with electro-fused allogeneic dendritic cells/autologous tumor-derived cells in patients with stage IV renal cell carcinoma. J. Immunother. 30, 749-761 (2007).

[47] Bonini, C., Lee, S. P., Riddell, S. R., Greenberg, P. D.: Targeting antigen in mature dendritic cells for simultaneous stimulation of $CD4^+$ and $CD8^+$ T cells. J. Immunol. 166, 5250-5257 (2001).

[48] Hodge, J. W., Chakraborty, M., Kudo-Saito, C., Garnett, C. T., Schlom, J.: Multiple costimulatory modalities enhance CTL avidity. J. Immunol. 174, 5994-6004 (2005).

[49] Kass, E., et al.: Induction of protective host immunity to carcinoembryonic antigen (CEA), a self-antigen in CEA transgenic mice, by immunizing with a recombinant vaccinia-CEA virus. Cancer Res. 59, 676-683 (1999).

[50] Yang, S., Tsang, K. Y., Schlom, J.: Induction of higher-avidity human CTLs by vector-mediated enhanced co-stimulation of antigen-presenting cells. Clin. Cancer Res. 11, 5603-5615 (2005).

[51] He, Y., Zhang, J., Mi, Z., Robbins, P., Falo Jr., L. D.: Immunization with lentiviral vector-transduced dendritic cells induces strong and long-lasting T cell responses and therapeutic immunity. J. Immunol. 174, 3808-3817 (2005).

[52] Auman, J. T.: Gene therapy: have the risks associated with viral vectors beensolved? Curr. Opin. Mol. Ther. 12, 637-638 (2010).

[53] Amato, R. J., et al.: Vaccination of renal cell cancer patients with modified vaccinia Ankara delivering tumor antigen 5T4 (TroVax) administered with interleukin 2: a phase II trial. Clin. Cancer Res. 14, 7504-7510 (2008).

[54] Amato, R. J., et al.: Vaccination of renal cell cancer patients with modified vaccinia Ankara delivering the tumor antigen 5T4 (TroVax) alone or administered in combination with interferon-alpha (IFN-alpha): a phase 2 trial. J. Immunother. 32, 765-772 (2009).

[55] Amato, R. J., et al.: Vaccination of metastatic renal cancer patients with MVA-5T4: a randomized, double-blind, placebo-controlled phase III study. Clin. Cancer Res. 16, 5539-5547 (2010).

[56] Oudard, S., et al.: A phase II study of the cancer vaccine TG4010 alone and in combination with cytokines in patients with metastatic renal clear-cell carcinoma: clinical and immunological findings. Cancer Immunol. Immunother. 60, 261-271 (2011).

[57] Heiser, A., et al.: Human dendritic cells transfected with renal tumor RNA stimulate polyclonal T-cell responses against antigens expressed by primary and metastatic tumors. Cancer Res. 61, 3388-3393 (2001).

[58] Su, Z., et al.: Immunological and clinical responses in metastatic renal cancer patients vaccinated with tumor RNA-transfected dendritic cells. Cancer Res. 63, 2127-2133 (2003).

[59] Geiger, C., Regn, S., Weinzierl, A., Noessner, E., Schendel, D. J.: A generic RNA-pulsed dendritic cell vaccine strategy for renal cell carcinoma. J. Transl. Med. 3, 29 (2005).

[60] Rittig, S. M., et al.: Intradermal vaccinations with RNA coding for TAA generate $CD8^+$ and $CD4^+$ immune responses and induce clinical benefit in vaccinated patients. Mol. Ther. 19, 990-999 (2011).

[61] Suekane, S., et al.: Phase I trial of personalized pep-tide vaccination for cytokine-refractory metastatic renal cell carcinoma patients. Cancer Sci. 98, 1965-1968 (2007).

[62] Wierecky, J., et al.: Immunologic and clinical responses after vaccinations with peptide-pulsed dendritic cells in metastatic renal cancer patients. Cancer Res. 66, 5910-5918 (2006).

[63] Weber, J. S., et al.: A phase 1 study of a vaccine targeting preferentially expressed antigen in melanoma

and prostate-specific membrane antigen in patients with advanced solid tumors. J. Immunother. 34, 556-567 (2011).

[64] Uemura, H., et al.: A phase I trial of vaccination of CA9-derived peptides for HLA-A24-positive patients with cytokine-refractory metastatic renal cell carcinoma. Clin. Cancer Res. 12, 1768-1775 (2006).

[65] Bleumer, I., et al.: Preliminary analysis of patients with progressive renal cell carcinoma vaccinated with CA9-peptide-pulsed mature dendritic cells. J. Immunother. 30, 116-122 (2007).

[66] Yasukawa, M., Ochi, T., Fujiwara, H.: Relapse of renal cell carcinoma with disappearance of HLA class I following hTERT peptide vaccination. Ann. Oncol. 21, 2122-2124 (2010).

[67] Märten, A., et al.: Telomerase-pulsed dendritic cells: preclinical results and outcome of a clinical phase I/II trial in patients with metastatic renal cell carcinoma. Ger. Med. Sci. 4, Doc02 (2006).

[68] Rahma, O. E., et al.: A pilot clinical trial testing mutant von Hippel-Lindau peptide as a novel immune therapy in metastatic renal cell carcinoma. J. Transl. Med. 8, 8 (2010).

[69] Walter, S., et al.: Multi-peptide immune response to cancer vaccine IMA901 after single-dose cyclophosphamide associates with longer patient survival. Nat. Med. 18, 1254-1261 (2012).

[70] Harrop, R., et al.: MVA-5T4-induced immune responses are an early marker of efficacy in renal cancer patients. Cancer Immunol. Immunother. 60, 829-837 (2011).

[71] Harrop, R., et al.: Analysis of pre-treatment markers predictive of treatment benefit for the therapeutic cancer vaccine MVA-5T4 (TroVax). Cancer Immunol. Immunother. 61, 2283-2294 (2012).

[72] Khong, H. T., Restifo, N. P.: Natural selection of tumor variants in the generation of "tumor escape" phenotypes. Nat. Immunol. 3, 999-1005 (2002).

[73] Ebos, J. M., et al.: Accelerated metastasis after short-term treatment with a potent inhibitor of tumor angiogenesis. Cancer Cell15, 232-239 (2009).

[74] Paez-Ribes, M., et al.: Antiangiogenic therapy elicits malignant progression of tumors to increased local invasion and distant metastasis. Cancer Cell15, 220-231 (2009).

[75] Ahmed, F., Steele, J. C., Herbert, J. M., Steven, N. M., Bicknell, R.: Tumor stroma as a target in cancer. Curr. Cancer Drug Targets8, 447-453 (2008).

[76] Zhao, X., et al.: Vaccines targeting tumor blood vessel antigens promote $CD8^+$T cell-dependent tumor eradication or dormancy in HLA-A2 transgenic mice. J. Immunol. 188, 1782-1788 (2012).

[77] Komita, H., et al.: $CD8^+$T-cell responses against hemoglobin-beta prevent solid tumor growth. Cancer Res. 68, 8076-8084 (2008).

[78] Chinnasamy, D., et al.: Gene therapy using genetically modified lymphocytes targeting VEGFR-2 inhibits the growth of vascularized syngenic tumors in mice. J. Clin. Invest. 120, 3953-3968 (2010).

[79] Rocha, F. G., et al.: Endostatin gene therapy enhances the efficacy of IL-2 in suppressing metastatic renal cell carcinoma in mice. Cancer Immunol. Immunother. 59, 1357-1365 (2010).

[80] Kundig, T. M., Kalberer, C. P., Hengartner, H., Zinkernagel, R. M.: Vaccination with two different vaccinia recombinant viruses: long-term inhibition of secondary vaccination. Vaccine 11, 1154-1158 (1993).

[81] He, Y., Zhang, J., Donahue, C., Falo Jr., L. D.: Skin-derived dendritic cells induce potent $CD8^+$T cell immunity in recombinant lentivector-mediated genetic immunization. Immunity24, 643-656 (2006).

[82] Ganss, R., Hanahan, D.: Tumor microenvironment can restrict the effectiveness of activated antitumor lymphocytes. Cancer Res. 58, 4673-4681 (1998).

[83] Ganss, R., Arnold, B., Hammerling, G. J.: Mini-review: overcoming tumor-intrinsic resistance to immune effector function. Eur. J. Immunol. 34, 2635-2641 (2004).

[84] Helfrich, I., et al.: Resistance to antiangiogenic therapy is directed by vascular phenotype, vessel stabilization, and maturation in malignant melanoma. J. Exp. Med. 207, 491-503 (2010).

[85] Rini, B. I., Atkins, M. B.: Resistance to targeted therapy in renal-cell carcinoma. Lancet Oncol. 10, 992-1000 (2009).

[86] Morgan, R. A., et al.: Cancer regression in patients after transfer of genetically engineered lymphocytes. Science314, 126-129 (2006).

[87] Leisegang, M., et al.: T-cell receptor gene-modified T cells with shared renal cell carcinoma specificity for adoptive T-cell therapy. Clin. Cancer Res. 16, 2333-2343 (2010).

[88] Porter, D. L., Levine, B. L., Kalos, M., Bagg, A., June, C. H.: Chimeric antigen receptor-modified T cells in chronic lymphoid leukemia. N. Engl. J. Med. 365, 725-733 (2011).

[89] Parkhurst, M. R., et al.: T cells targeting carcinoembryonic antigen can mediate regression of metastatic colorectal cancer but induce severe transient colitis. Mol. Ther. 19, 620-626 (2011).

[90] Kandalaft, L. E., Powell Jr., D. J., Coukos, G.: A phase I clinical trial of adoptive transfer of folate receptor-alpha redirected autologous T cells for recurrent ovarian cancer. J. Transl. Med. 10, 157 (2012).

[91] Lamers, C. H., et al.: Immune responses to transgene and retroviral vector in patients treated with ex vivo-engineered T cells. Blood117, 72-82 (2011).

（马维民译）

第23章　肺癌免疫治疗：规划发展、进展和展望

Edward A. Hirschowitz，Terry H. Foody
和 John R. Yannelli①

摘要

本章介绍了一个机构开发肺癌疫苗的10年经验。回顾要素包括基本原理、发展战略与历史观点相关的科学方法和临床情况。

23.1　肺癌

肺癌是全世界最常见的癌症，每年死亡130万人，比结肠癌、乳腺癌和前列腺癌这3种最常见的癌症加起来还要多。肺癌的隐蔽性使绝大多数病例在诊断时进入晚期。由于分期与预后关系最为密切，因此晚期诊断是15% 5年生存率的主要原因[1,2]。

近一半可接受有潜在疗效的手术切除患者最终死于疾病[1,2]。每年第一次癌症诊断后的第二次原发性肺癌发病率为1%~2%[1,2]。早期检测和预防措施可能有助于降低整体疾病死亡率，但改善诊断结果的负担更直接地落在治疗上。

23.2　常规疗法

肺癌主要分为非小细胞肺癌（NSCLC）和小细胞肺癌（SCLC）两种组织学亚型，它们描述了不同的体细胞起源、行为、表型和预后。NSCLC∶SCLC相对发生率为4∶1（80% vs 20%）。传统治疗方法是针对分期和组织学细胞类型（NSCLC与SCLC）量身定制的；最近分子表型已被用于指导最有效的治疗方案[1-4]。

23.2.1　小细胞肺癌

在实际应用中，所有这种侵袭性肿瘤类型的病例在诊断时均假定为转移。化疗是治疗的主要手段，手术没有任何治疗优势。在有限的胸部疾病中，增加放疗以进行局部控制。SCLC死亡率约为95%[2,3]。

23.2.2　非小细胞肺癌

手术是早期癌症的主要治疗方式，化疗和放疗是晚期疾病的主要治疗方式。在过去10年中，外科和非外科治疗的结合增加了生存率[4]。

---

① E. A. Hirschowitz，MD（✉）（肯塔基大学钱德勒医疗中心内科肺和重症监护医学室，美国肯塔基州列克星敦市，列克星敦退伍军人管理医疗中心，美国肯塔基州列克星敦市，E-mail：eahirs2@ uky. edu）

T. H. Foody，RN（肯塔基大学钱德勒医疗中心内科肺和重症监护医学室，美国肯塔基州列克星敦市）

J. R. Yannelli，博士（肯塔基大学微生物、免疫和人类遗传学系，美国肯塔基州列克星敦）

23.2.3 早期疾病

手术切除仍然为少数早期诊断为非小细胞肺癌的患者提供最好的治愈机会。小肿瘤治愈率接近80%，并随着肿瘤负担的增加而稳步下降。辅助化疗现在通常提供给疾病潜在治愈性切除术后复发机会大于35%的个体（Ⅱ期和Ⅲ期），提供了4%~15%的5年生存率优势。除了病理阶段，没有临床验证的复发预测措施来帮助选择合适的候选辅助化疗[4]。

23.2.4 晚期疾病

80%的非小细胞肺癌（NSCLC）患者出现局部晚期或转移性癌症，另外10%的患者在可能治愈的外科切除术后出现致命复发。治疗后的中位生存期为7~12个月。只有60%的晚期非小细胞肺癌患者对一线化疗有临床反应，通常持续3~6个月；新药物降低了毒性，但对产生持久反应作用微乎其微。维持性化疗在概念上很有吸引力，但只有培美曲塞能有效延缓腺癌患者的生存时间。在进展过程中，二线化疗比最佳支持治疗提高了8个月的中位生存率[4]。

23.3 癌症免疫疗法

这一不断进化的领域是基于免疫系统能够识别癌细胞为外来细胞的知识[5]。选择性靶向对正常组织影响最小的播散性疾病能力是一个基本原则。免疫生物学知识的不断积累已经产生了无数治疗策略，每一种策略都有共同的目标，即改善不可接受的不良治疗结果。对适应性免疫、抗原发现以及体外和动物研究中免疫佐剂的各种作用的深入了解，催生了许多肺癌和其他癌症的早期研究[5]。尽管有几十年的预期前景，一些Ⅱ期试验已经让位给令人失望的Ⅲ期结果[5,6]。

2010年，美国食品和药物管理局（FDA）批准了由前列腺酸性磷酸酶（PAP）和粒细胞-巨噬细胞集落刺激因子（GMCSF）组成的融合蛋白自体树突状细胞疫苗（Provenge），用于转移性激素难治性前列腺癌。可能开创了一个新纪元[6,7]。疗效标准是在3个随机Ⅲ期试验中增加4个月的中位生存率[6]。其他几种癌症疫苗和免疫调节剂正在研制中；一些正在进行的肺癌随机试验值得谨慎乐观。大多数积极的癌症免疫疗法已经并继续指导产生抗原特异性T细胞应答。最近对癌症功能性免疫调节的认识，为调节免疫系统以获得临床效果提供了新途径[8,9]。尽管近年来这种方法的复兴可能会改变这一历史先例，但肿瘤抗原特异性抗体治疗效果证据却少得多。

23.4 肺癌免疫疗法

疾病特征确实影响了肺癌的发展策略。抗原异质性在一定程度上依赖于肿瘤组织学，是抗原特异性治疗的决定性特征和重大挑战。其他考虑因素包括分期分布的显著差异、常规治疗范围和不良预后。针对近半数手术切除的非小细胞肺癌（NSCLC）患者复发的微小残留疾病是一个合理的应用。风险收益比与这些患者特别相关，因为还不能超出统计概率预测复发。相反地，疗效试验受到不同复发率和延长临床终点阻碍。晚期疾病患者尚未被广泛认为可能受益于免疫疗法，尽管以最小的附加风险减缓疾病进展是一个有价值的目标。晚期非小细胞肺癌（NSCLC）最终预后和准确确定肿瘤进展和存活的能力使这一人群成为临床试验的合理选择。由于这些病例中的大部分是通过很少的细胞学取样进行诊断的，因此进入胸腔内疾病部位的有限途径会带来第二个障碍。除了是原位方法的明显障碍外，肺肿瘤组织可用性也是自体疫苗生产、抗原鉴定和免疫监测的障碍；通过侵入性程序重复取样的风险几乎是不合理的。小细胞肺癌（SCLC）的免疫治疗尚未得到广泛研究。这种疾病往往具有很强的侵袭性，免疫疗法并不被认为具有显著的影响预后潜力。由于小细胞肺癌不能用于体外研究或异种肿瘤建模，也没有合适的小细胞肺癌动物模型，病理标本缺乏也延缓了临床前研究（同样地，容易获得的非小细胞肺癌标本也促进了研究）。然而，与 SCLC 相关的不良预后使其成为进一步调查的合理选择。

23.5　方案发展

我们基于大学的项目根植于临床医生、免疫学家和专门的护士协调员之间的跨学科合作，他们对 NSCLC 新疗法有共同的兴趣。从一开始，目标就是开发肺癌通用疫苗，检验免疫治疗的中心规则并提供信息，从而合理实施肺癌疫苗。尽管我们受益于多种药物在不同癌症中的大量临床前和可行性试验，但我们也面临着一些令人失望的负面疗效研究和对免疫治疗基本原则的大量猜测[5]。NSCLC 进展需要对有形终点连续确认，并取决于制剂的免疫可靠性。疫苗制备的选择多种多样，最合理的选择将提供治疗活动潜力，同时也是探索免疫治疗相关问题的工具。

23.6　选择免疫佐剂

从非特异性到高度选择性的多种药剂通过各种交叉机制有效促进抗原摄取、递呈和/或识别[11]。树突状细胞完全协调适应性反应的先天能力导致它们被描述为“自然佐剂”[12]。当时已经建立了从外周血中产生大量树突状细胞（DCs）的技术，通过培养 DC 体外携带抗原来产生有效疫苗的能力刚刚从实验室转移到临床。一些早期人类研究中有前景的临床前数据和免疫诱导的明显可靠性表明，这种方法是癌症疫苗的新标准[12,13]。由于非小细胞肺癌集体经验仅限于少数携带与其他癌症相同抗原的患者，因此非小细胞肺癌 DC 疫苗的开放生态位似乎是临床研究的合理入口[12,13]。与其他更复杂的、重组或药理学替代品相比，对细胞生物学的历史熟悉程度也保证了一个更简化调节过程。

23.7　抗原

除了抗原相关性的限制外，没有理想的抗原来源。自体制剂与同种异体制剂、单价制剂与多价制剂、全细胞制剂与重组制剂之间的区别是部分理论、部分实用的[11]。自体肿瘤抗原将提供一种患者特异性疫苗优势，这种疫苗可以避免了解患者癌症所表达的抗原需要，但通常生产自体肿瘤疫苗所需的肿瘤组织体积限制了这种方法对于具有外科可切除肿瘤的个体应用。利用同种异体重组蛋白（包括基因疗法衍生的）或合成肽单价方法将限制对携带相关抗原个体人群的使用，并在有或无 HLA 限制情况下引入抗原特征第二个障碍。同种异体全细胞制剂提供了一系列描述良好的抗原，以及一个次要决定因素和不匹配抗原的补充以增强交叉启动[14]。细胞系是肿瘤蛋白的可再生来源，可提供标准化多价抗原制剂[14]。受体 APCs 通过Ⅰ类和Ⅱ类途径处理并递呈最相关的表位，从而产生免疫效应物和免疫记忆应答潜力。

23.8　免疫监测

免疫学实验是对疫苗效力和免疫能力的测量。IFN-γ 酶联免疫斑点测定法（ELISPOT）是免疫监测的主要手段。这种高度敏感和可靠的检测方法可以定量混合淋巴细胞或分级群体中识别特定抗原的 $CD4^+$ 和 $CD8^+$ T 细胞数量。相比之下，细胞毒性 T 淋巴细胞（CTL）检测是一种高度特异性适应性免疫功能测量方法，但其灵敏度仅为中等，需要的细胞数量大于常规临床抽血的细胞数量。四聚体分析定量携带选定抗原受体 HLA 匹配 T 细胞，是最广泛用于监测 HLA 限制性肽疫苗方法。迟发型超敏反应（DTH）对皮内抗原的应答通常用于监测同种异体方法，其中不同个体疫苗的抗原特征不同[15]。

人们普遍认为，免疫后可测量反应性的统计增加，反映了 T 细胞对疫苗提供的 1 种或多种抗原的动员[15]。“免疫应答”的临床相关性仍然是推测性的[15-17]。有问题的是，检测是多方面、多样化的，不是各实验室标准化的[15,18]。此外，由于个体患者缺乏用于检测的活肿瘤，大多数免疫监测推测同种异体替代物的抗原相关性和特异性。尽管抗原与个体肿瘤相关，但有多种下游元件可在诱导后调节效应物功能。尽管如此，在该领域目前的成熟水平上，免疫学检测是疫苗发展阶段比较效力和成功指标的唯一实际措施。尽管必须强调的是，免疫应答并不能直接转化为治疗效果，但多种免疫治疗研究已经观察到临床反应和诱导免疫之间的联系。大多数疫苗诱导可测量免疫应答能力表明，许多疫苗可能具有类似疗效。一旦在 NSCLC 中定义了疗效，并且监测分析与结果的相关性得到更好的定义，标准化监测可以促进发展，提供比较分析，促进最佳研究设计，支持提高应答率的策略[15-18]（图 23.1）。

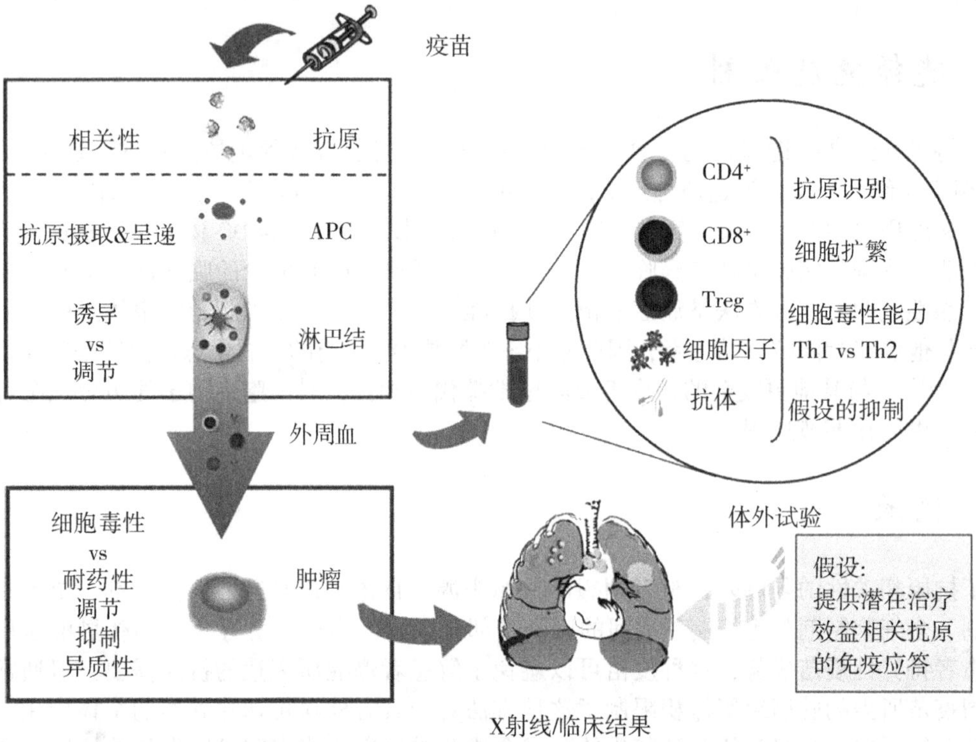

图 23.1 免疫监测

随附的示意图描述了免疫监测的无数已知、可测量和未知要素，包括可测量免疫标准与临床结果之间的某种推测性联系。适应性免疫的假定机制如图左侧所示：抗肿瘤免疫诱导的顺序位点和竞争途径的作用显示在黑框中。以假定或已知的抗原相关性生产的疫苗被送入生物坩埚中。右侧显示了疫苗递送的生物学和治疗效果窗口；免疫学和临床终点提供了可量化的指标。免疫状态是从具有外周血成分特征的一连串体外试验推断出来的。一系列措施的改变是免疫“应答”的基本基础。影像学改变和临床结果反映了肿瘤微环境中细胞毒性、抑制和耐药性的总和。免疫监测试验与治疗效果之间的关系在提供潜在治疗效益相关抗原的免疫应答的假设中得到了概括。

23.9　自体树突状疫苗

23.9.1　原理和试验设计

我们开始了一项关于 DC 疫苗的临床试验，以确定可行性，获取建立未来研究的信息，并致力于优化非小细胞肺癌的疫苗[19,20]。疫苗的生产和作用机制是相当清楚的（图 23.2）。试验设计的重点是在手术治疗和医药治疗以及多模式治疗的非小细胞肺癌患者对抗原刺激 DC 疫苗的免疫应答。任何经组织学证实的 I-IIIB 期非小细胞肺癌患者，无论之前的治疗如何，均符合资格。受试者将接受两次免疫，间隔 1 个月。在免疫前和免疫后连续抽取血液 1 年，持续监测安全参数。主要终点是用 IFN-γ ELISPOT（酶联免疫斑点测定法），测定间隔 1 个月的初始疫苗接种和强化疫苗免疫接种的个体患者的免疫诱导率。受试者之间的比较分析将确定肿瘤负担（手术切除与可测量的持续性疾病）的相对重要性，以及既往放疗或化疗对免疫诱导的影响。数据将根据可能影响疫苗效力的宿主因素进行分析，包括负责可变诱导率的免疫调节因素。在这种情况下，直接目标和纵向目标取决于免疫诱导的可靠性，而动量将建立在这一可定义的度量上。

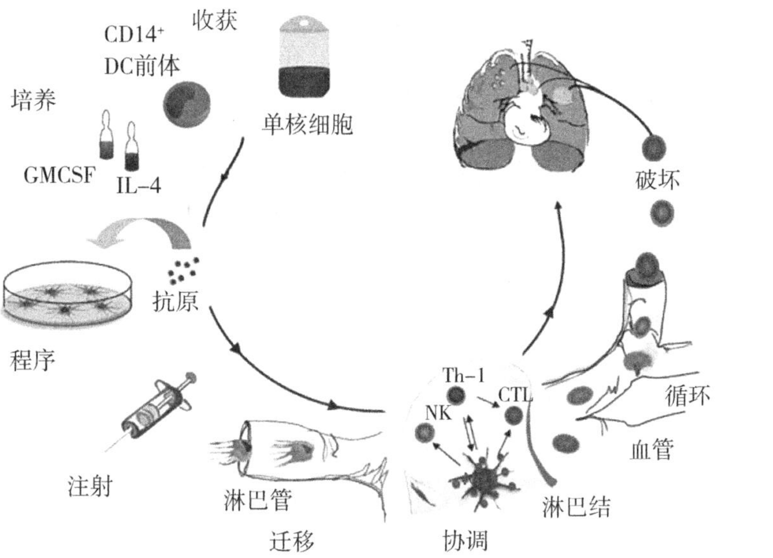

图 23.2　树突状细胞（DCs）疫苗

来自患者的外周血单核细胞在巨噬细胞集落刺激因子和 IL-4 中培养，以驱动 CD14⁻DC 前体向未成熟髓样树突状细胞分化。未成熟树突状细胞具有吞噬作用，容易在培养物中吸收颗粒抗原。树突状细胞将抗原加工成较小的肽，以 MHC Ⅰ类分子或Ⅱ类分子的形式呈现在细胞表面。当 DCs 在体外通过添加生物活性分子或通过与体内驻留 T 细胞的同源相互作用而被驱动成熟时，完全抗原递呈能力得以实现。皮下注射的树突状细胞迁移到局部淋巴结，并与肿瘤特异性 T 细胞前体相互作用。效应细胞被释放到外周循环中，并迁移到遇到并破坏抗原表达细胞（例如肿瘤）的组织中。

23.9.2　配方和剂量

疫苗是根据美国食品和药物管理局细胞治疗指南制备的。患者接受了 2 小时白细胞介导术，以采集足够数量的单核细胞用于疫苗生产。自体树突状细胞由外周血单核细胞在富含生长因子培养基中培养产生。两种不同DC制剂在类似免疫计划中依次进行了试验[19,20]。在第一组18名患者中，DC疫苗在抗原刺激作用下使用生物活性因子混合物在体外成熟。第二组18名受试者接受未成熟DC疫苗。所有疫苗都是用来自非小细胞肺癌肿瘤细胞系TC1650抗原制备的。该细胞系过度表达CEA、Her2/neu、WT1、生存素和NY-ESO-1，这意味着至少有1种抗原与95%以上的非小细胞肺癌共享。建立了由确认不含污染物和外膜病毒的肿瘤细胞系组成的主细胞库，作为抗原的可再生来源。

基于文献表明，凋亡体为简单坏死细胞碎片提供了更高免疫原性的替代物，细胞在致死性照射之前用紫外线照射凋亡。疫苗生产时间超过8天。最终疫苗产品（经处理的TC1650体外致敏突状细胞）被等分，用于两次免疫接种（预接种和加强接种），以及几个无菌检验和质量保证试剂瓶。冷冻保存24小时后，解冻1个测试瓶，检测内毒素，并送至大学临床微生物实验室进行细菌和真菌的扩展培养测试。在无菌检验至少两周之后，疫苗可以进行注射。每名受试者在大腿间隔1个月接受两次连续皮内注射10^8个致敏树突状细胞。对受试者进行临床副反应监测。

23.9.3 结果

对36名确诊为I-IIIB期非小细胞肺癌的患者用两种DC疫苗制剂（成熟和未成熟树突状细胞）中的1种免疫接种。观察到的主要副反应是免疫接种后24~48小时注射部位出现局部水泡和潮红反应。实验治疗没有严重的副反应。免疫反应用IFN-γ ELISPOT测定，比较免疫前和免疫后一系列时间点的抗原反应性T细胞的相对数值。在67%的受试者中，通过IFN-γ ELISPOT法测定发现了可测量的免疫反应，与疾病期或既往治疗无关。诱导率与树突状细胞在递送时的表型成熟状态无关，两种制剂的诱导率与文献中的其他癌症疫苗相当。免疫诱导与免疫时循环的T调节细胞的相对数量成反比，免疫后任何受试者的T调节细胞没有明显增加。由于数据太小，无法得出关于疗效的结论，尽管生存率尤其是那些晚期疾病患者的生存率比基于历史对照预测要好得多[19,20]。

23.10 结论和展望

肺癌患者可能从免疫治疗中获益。虽然DC疫苗在大多数患者中明显诱导免疫应答，可能在更大的临床试验中显示临床疗效，但DC疫苗生产是长期的、资源密集型的和昂贵的；此外，DC疫苗是单独生产的，相对不可运输，并且要求每个患者接受一种白血病去除术程序，这限制在主要医疗中心为选定的患者生产和输送疫苗。这些因素使得DC疫苗不适合进行高级阶段的研究，也不适合广泛的临床应用。尽管如此，DC疫苗仍然是固体肿瘤疫苗的一个有价值的生物学标准。这些经验和数据将为进一步开发疫苗提供一个有价值的标准。更具体地说，我们很容易认识到，没有标准的疫苗方法和比较来表明某种癌症疫苗类型的优越性，这提供了以下工作假设：任何能够诱导可测量的免疫应答，同时利用相关肿瘤抗原宿主的药物都具有类似的治疗效果潜力。即使如此，在进行比较治疗研究之前，这无疑是一个漫长的过程。

23.11 1650-G疫苗

23.11.1 原理和试验设计

我们认为，疗效更多地取决于诱导免疫应答的能力，而不是任何特定的疫苗配方或佐剂，我们生产了1650-G，一种DC疫苗的简化衍生物[21]。该疫苗将我们的DC疫苗（肿瘤细胞系1650）的加工抗原成分与重组GMCSF（作为免疫佐剂）结合。GMCSF已知能够刺激体内抗原摄取和递呈，将被用来代替自体DCs体外的抗原装载；同种异体肿瘤细胞系1650的抗原谱将继续提供多价覆盖；使用类似的抗原、免疫程序和监测有助于与先前测试的 DC 疫苗进行比较，该疫苗将用作免疫活性的基准。与 DC 疫苗相比，1650-G 将是一种经济和可运输的替代品，可继续用于更高级阶段的研究。

初步可行性试验是对 I/II 期非小细胞肺癌患者进行开放性非随机多位点研究，手术切除后无疾病迹象。这项研究是在肯塔基州的 4 个地方进行的，代表了学者和地区社区癌症中心之间的合作关系。初免疫苗和单一的加强免疫间隔 1 个月，类似于自体 DC 疫苗的规定方案。每次免疫接种 1.0×10^8 个处理过的 1650 肿瘤细胞，加 100 μg GMCSF，体积为 0.6 mL；在大腿的两处皮内注射疫苗（0.3 mL/次，总量为 0.6 mL）；另外一只大腿进行初免和加强疫苗接种。以 IFN-γ ELISPOT 法测量免疫应答为主要终点。与文献和以往的经验一致，只有一部分研究患者预期会出现免疫应答。相对于先前测试的 DC 疫苗的免疫诱导率是疫苗效力的主要指标；1650-G 到 DC 疫苗的免疫学可比性将为进一步研究提供基础。为疫苗合作小组建立基础设施，并测试疫苗在社区癌症诊所中运输和交付给癌症患者的潜力，也实现了一个重要研究目标。

23.11.2　配方和剂量

疫苗是从主细胞库中批量生产的。每个剂量由 10^8 个处理过的 1650 细胞（凋亡和致死性照射）加上 100 μg 临床级 GMCSF（0.6 mL 无菌盐水）组成。小瓶制备好的疫苗直到输送前要储存在液氮气相中。如前所述，要对试验小瓶进行无菌检验。

23.11.3　结果

11 个人接受了两次连续免疫接种。主要副反应是自限注射部位反应，没有严重的意外不良事件。在 6/11 的免疫患者中观察到 ELISPOT 测量的免疫反应，这与资源更密集的 DC 疫苗相当（图 23.3）。观察到的生物活性加上易于生产、成本降低和便携性，使 1650-G 成为随机临床研究和随后临床实施有吸引力的候选者[21]。

23.11.4　方向

疫苗 1650-G 的早期结果得到了适当的鼓励，目前的兴趣是确定在非小细胞肺癌中实施免疫治疗的参数。更具体地说，已经设计了两个随机临床试验来评估晚期非小细胞肺癌免疫化疗的应急概念；每种方法的主要终点是临床反应。第一项试验即一线化疗后维持 1650-G 加培美曲塞的随机Ⅱ/Ⅲ期研究，于 2010 年开始累积。由于缺乏外部资金，该试验在临床上被搁置了。第二项研究是二线 1650-G 加多西他赛治疗进展性疾病的随机Ⅱ/Ⅲ期研究，由于缺乏资金，尚未开始累积。该试验描述如下。正在进行 I 期试验以检查膳食补充剂 β-葡聚糖的免疫增强作用。

23.12　二线免疫化疗随机试验（SLIC）用于晚期 NSCLC

23.12.1　原理和试验设计

认识到疫苗作为单一疗法在肿瘤负担较大的患者中所面临的挑战，我们被临床效果和临床前研究所吸引，这些研究表明化疗和癌症疫苗之间存在免疫和治疗协同作用[22,23]。多西他赛免疫增强作用的深入了解尤其与晚期非小细胞肺癌患者相关[24,25]。主要目的是证明在常规化疗可预测结果的人群中，免疫系统可以被动员起来以获得治疗效果。

23.12.2　研究描述

本研究是一项非盲 2∶1 随机试验，旨在评估 1650-G 联合多西他赛（相同的治疗周期）作

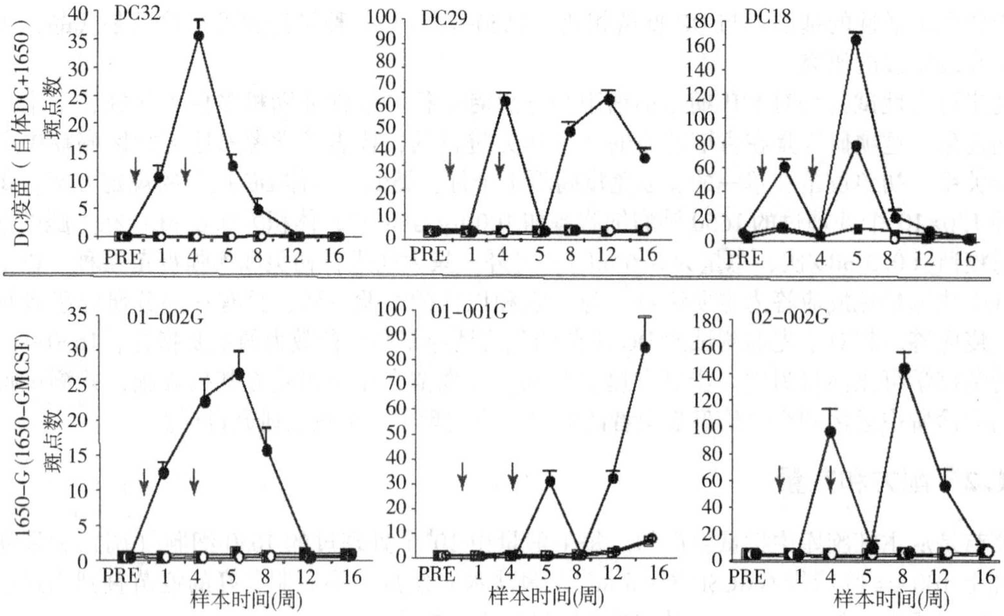

图 23.3 IFN-γ ELISPOT：DC 疫苗与 1650-G 疫苗的比较

代表性免疫反应来自参与 DC 疫苗研究的个体（上图）和接受 1650-G 疫苗的患者（下图）。两种疫苗都使用经过加工的 1650 细胞系作为抗原来源。在两项研究中，免疫计划是相同的。图表显示了从可用时间点（疫苗接种前到接种后 16 周）对（●）1650 致敏自体树突状细胞，（■）单独树突状细胞，（○）淋巴细胞单独反应的淋巴细胞（“斑点”）的测量结果。箭头表示疫苗递送的时间。在整个研究人群中，免疫诱导率和反应模式的相似性掩盖了免疫佐剂在动力学上的相似性，表明两种方法都有潜在的治疗益处。这些数据是使用 DC 疫苗的简化衍生物（1650-G）进行高级阶段研究的基础。

为二线疗法治疗进展期 IIIB 或 IV 期非小细胞肺癌的疗效，不适于外科手术或放疗。根据 II/III 期综合试验设计，最低研究人数为 120 名患者，最高研究人数为 174 名患者。患者将从门诊诊所、社区自我转诊或初级治疗医师转诊的方式在肯塔基州的 8 个地点登记。两个研究组的化疗（多西他赛）和治疗标准相关程序相同。所有受试者将经历 4 个治疗周期；每个治疗周期为 3 周。在每个周期的第 1 天注射疫苗（实验组），在每个周期的第 5 天注射多西他赛（图 23.4）。

23.12.3 终点

这项研究专门研究了实验组和对照组在结果上的差异。主要终点包括进展时间（TTP）、生存率和免疫相关应答标准（irRC）。后一项措施包括允许肿瘤负荷适度增加，以适应长期免疫介导的反应和炎症变化。与文献和以往使用该疫苗的经验相一致，50%～70%的个体在免疫后预计会表现出免疫反应。第二个假设是临床反应有利于免疫反应组。相关研究旨在解决治疗成功或治疗失败的最可理解的原因。

23.13 优点和缺点

DC 疫苗在非小细胞肺癌治疗中仍有潜在的疗效，但价格昂贵且资源密集[12,19,20]。1650-G 是一种免疫上可比较的替代物。除了低成本的明显吸引力，1650-G 是便携式的、统一的、多价的[21]。我们继续被这种疫苗的理论、实践和实用优势所吸引，并受到非小细胞肺癌早期试验结果的适当鼓励。为了满足更高级研究的标准，疫苗 1650-G 面临着新的挑战。

毫无疑问，资金的可获得是进一步发展的最大障碍。简单的非专利战略一般不会从行业支持中受益，而且在目前外部资助水平上，用这种药剂完成 III 期研究的前景似乎很渺茫。尽管文献中描述的癌症疫苗数量持续增长，排列继续增加，但对疗效试验的热情似乎矛盾地减少了[26]。可以理解的是，疗效试验，特别是在晚期患者中，承担了阴性结果的风险。考虑到药物准备情况、肺癌严重程度以及 sipuleucel-T 治疗前列腺癌的显著疗效，解决非小细胞肺癌的疗效问题远比详尽的生物学研究更有说服力。如果没有任何证据表明一种疫苗比另一种疫苗更具治疗活性，我们也无法证明开发替代配方的时间和费用是合理的。

虽然复杂性不是疗效的必要条件，但 1650-G 疫苗的简单性引起了人们对其治疗潜力的怀疑。对全细胞疫苗有不同的看法，但没有任何特征妨碍疗效。作为一种免疫佐剂，GMCSF 并没有取得持久的成功，但是在疫苗配方、肿瘤类型、时间表、监测化验以及 GMCSF 接种剂量和途径上的差异使得比较很困难，无法得出结论[14,27]。尽管存在这些现实，历史偏见仍然是一个明显的障碍，特别是在资金方面。疫苗治疗成功和 GVAX 疫苗的新希望（两者都是用 GMCSF 配制的）可能会改变人们的看法。

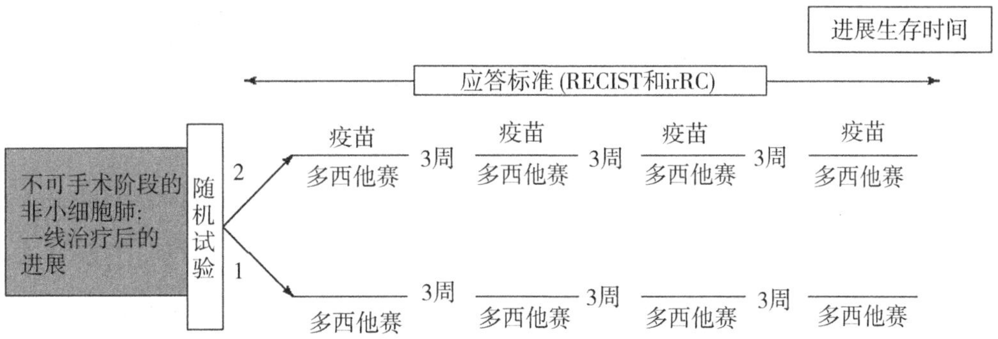

图 23.4　疫苗 1650-G 试验方案：晚期非小细胞肺癌二线免疫化疗随机试验

在追求功效的过程中，这种疫苗与癌症疫苗有一系列共同的障碍。获得最佳潜能最公认的生物障碍包括含有调节性 T 细胞和异常成熟髓细胞受损肿瘤微环境、易于免疫衰竭和衰老的肿瘤特异性 T 细胞库，以及能够丢失抗原和逃避免疫的高度可变肿瘤靶位。继续影响开发和实施的临床因素包括非标准的检测和免疫监测的不确定相关性、治疗终点的长期疗程以及缺乏评估临床疗效的相关生物标志。

对癌症疫苗的认知是这个领域的一个含蓄但非常现实的挑战。关于免疫疗法治疗潜力的观点基于少数具有统计学意义的阳性和阴性临床研究，涵盖了所有领域[5,6]。尽管大多数研究者不认为免疫疗法是治疗癌症的灵丹妙药，但医学界不切实际地期望继续产生怀疑和偏见。然而重要的是要记住，在过去几十年中，晚期肺癌的预后改善是在生存期增加数周内进行测量的，在手术切除的患者中，增加辅助化疗只能略微降低复发的相对风险。在没有达到任何全球疗效标准的情况下，为一定比例的治疗选择有限的患者提供参与肺癌研究的机会，即使不能从肺癌研究中获益，也为其作出了贡献，这是实实在在的成功。

致谢

我们的项目很幸运地得到了机构、癌症中心、慈善家和基金会的支持，以及国家支持的倡议和国家卫生研究院的外部支持。

参考文献

[1] Tarver, T.: Cancer facts & figures 2012. American Cancer Society (ACS). J. Consum. Health Internet16, 366-367 (2012).

[2] Howlader, N., et al. (eds.): SEER cancer statistics review, 1975-2009 (Vintage 2009 populations), National Cancer Institute. Bethesda, MD. http://seer. cancer. gov/csr/1975_2009_pops09/, based on Nov 2011 SEER data submission, posted to the SEER web site, Apr 2012.

[3] van Meerbeeck, J. P., Fennell, D. A., De Ruysscher, D. K.: Small-cell lung cancer. Lancet378, 1741-1755 (2011). doi: 10. 1016/S0140-6736 (11) 60165-7.

[4] Goldstraw, P., et al.: Non-small-cell lung cancer. Lancet378, 1727-1740 (2011). doi: 10. 1016/S0140-6736 (10) 62101-0.

[5] Dalgleish, A. G.: Therapeutic cancer vaccines: why so few randomised phase III studies reflect the initial optimism of phase II studies. Vaccine29, 8501-8505 (2011). doi: 10. 1016/j. vaccine. 2011. 09. 012.

[6] Perez, C. A., Santos, E. S., Raez, L. E.: Active immunotherapy for non-small-cell lung cancer: moving toward a reality. Expert Rev. Anticancer Ther. 11, 1599-1605 (2011). doi: 10. 1586/era. 11. 155.

[7] Kantoff, P. W., et al.: Sipuleucel-T immunotherapy for castration-resistant prostate cancer. N. Engl. J. Med. 363, 411-422 (2010). doi: 10. 1056/NEJMoa1001294.

[8] Disis, M. L.: Immune regulation of cancer. J. Clin. Oncol. 28, 4531-4538 (2010). doi: 10. 1200/JCO. 2009. 27. 2146.

[9] Shepherd, F. A., Douillard, J. Y., Blumenschein Jr., G. R.: Immunotherapy for non-small cell lung cancer: novel approaches to improve patient outcome. J. Thorac. Oncol. 6, 1763-1773 (2011). doi: 10. 1097/JTO. 0b013e31822e28fc.

[10] Scott, A. M., Wolchok, J. D., Old, L. J.: Antibody therapy of cancer. Nat. Rev. Cancer 12, 278-287 (2012). doi: 10. 1038/nrc3236.

[11] Hirschowitz, E. A., Hiestand, D. M., Yannelli, J. R.: Vaccines for lung cancer. J. Thorac. Oncol. 1, 93-104 (2006).

[12] Palucka, K., Banchereau, J.: Cancer immunotherapy via dendritic cells. Nat. Rev. Cancer 12, 265-277 (2012). doi: 10. 1038/nrc3258.

[13] Cranmer, L. D., Trevor, K. T., Hersh, E. M.: Clinical applications of dendritic cell vaccination in the treatment of cancer. Cancer Immunol. Immunother. CII 53, 275-306 (2004). doi: 10. 1007/s00262-003-0432-5.

[14] Keenan, B. P., Jaffee, E. M.: Whole cell vaccines-past progress and future strategies. Semin. Oncol. 39, 276-286 (2012). doi: 10. 1053/j. seminoncol. 2012. 02. 007.

[15] Keilholz, U., et al.: Immunologic monitoring of cancer vaccine therapy: results of a workshop sponsored by the Society for Biological Therapy. J. Immunother. 25, 97-138 (2002).

[16] Wolchok, J. D., et al.: Guidelines for the evaluation of immune therapy activity in solid tumors: immune-related response criteria. Clin. Cancer Res. 15, 7412-7420 (2009). doi: 10. 1158/1078-0432. CCR-09-1624.

[17] Disis, M. L.: Immunologic biomarkers as correlates of clinical response to cancer immunotherapy. Cancer Immunol. Immunother. CII 60, 433-442 (2011). doi: 10. 1007/s00262-010-0960-8.

[18] Moodie, Z., et al.: Response definition criteria for ELISPOT assays revisited. Cancer Immunol. Immunother. CII 59, 1489-1501 (2010). doi: 10. 1007/ s00262-010-0875-4.

[19] Hirschowitz, E. A., et al.: Autologous dendritic cell vaccines for non-small-cell lung cancer. J. Clin. Oncol. 22, 2808-2815 (2004). doi: 1 0. 1200/JCO. 2004. 01. 074.

[20] Hirschowitz, E. A., Foody, T., Hidalgo, G. E., Yannelli, J. R.: Immunization of NSCLC patients with antigen-pulsed immature autologous dendritic cells. Lung Cancer 57, 365-372 (2007). doi: 10. 1016/j. lungcan. 2007. 04. 002.

[21] Hirschowitz, E. A., et al.: Pilot study of 1650-G: a simplified cellular vaccine for lung cancer. J. Thorac. Oncol. 6, 169-173 (2011). doi: 10. 1097/JTO. 0b013e3181fb5c22.

[22] Emens, L. A., Jaffee, E. M.: Leveraging the activity of tumor vaccines with cytotoxic chemotherapy. Cancer Res. 65, 8059-8064 (2005). doi: 10. 1158/0008-5472. CAN-05-1797.

[23] Zitvogel, L., Apetoh, L., Ghiringhelli, F., Kroemer, G.: Immunological aspects of cancer chemotherapy. Nat. Rev. Immunol. 8, 59-73 (2008). doi: 10. 1038/nri2216.

[24] Garnett, C. T., Schlom, J., Hodge, J. W.: Combination of docetaxel and recombinant vaccine enhances T-cell responses and antitumor activity: effects of docetaxel on immune enhancement. Clin. Cancer Res. 14, 3536-3544 (2008). doi: 10. 1158/1078-0432. CCR-07-4025.

[25] Kodumudi, K. N., et al.: A novel chemoimmunomodulating property of docetaxel: suppression of myeloid-derived suppressor cells in tumor bearers. Clin. Cancer Res. 16, 4583-4594 (2010). doi: 10. 1158/1078-0432. CCR-10-0733.

[26] Kudrin, A., Hanna Jr., M. G.: Overview of the cancer vaccine field: are we moving forward? Hum. Vaccines Immunother. 8, 1135-1140 (2012).

[27] Parmiani, G., et al.: Opposite immune functions of GM-CSF administered as vaccine adjuvant in cancer patients. Ann. Oncol. 18, 226-232 (2007). doi: 10. 1093/annonc/mdl158.

（马维民译）

第 24 章　黑色素瘤：基于肽疫苗的展望

Mariana H. Massaoka，Alisson L. Matsuo，Jorge A. B. Scutti，
Denise C. Arruda，Aline N. Rabaça，Carlos R. Figueiredo，
Camyla F. Farias，Natalia Girola 和 Luiz R. Travassos①

黑色素瘤患者的标准治疗包括手术、化疗、放射治疗、生物和靶向治疗。肿瘤手术切除是否成功取决于疾病的早期诊断。转移性黑色素瘤预后最差，已尝试多种新药物，包括细胞因子、单克隆抗体、信号转导抑制剂、溶瘤病毒和血管生成抑制剂，一般阳性反应率较低。迄今为止，疫苗在治疗转移性黑色素瘤方面尚未取得成功，但却是一个合理的研究领域。

人源化抗 CTLA-4 抗体易普利姆玛最近在转移性黑色素瘤患者中展示了广泛的应用前景。在所有情况下，癌症疗法的毒性都很高。许多中心正在积极研究新药物和方案。目前我们发现，易获得的肽可能为抗黑色素瘤治疗的发展提供基础。有些肽凋亡或影响肿瘤细胞在体外的迁移，但在黑色素瘤实验模型中，在体内，肽对树突状细胞和抗转移活性的激活正在鼓励和刺激进一步的研究。

24.1　疾病描述

恶性黑色素瘤是最具侵袭性和抗治疗性的皮肤癌类型，几十年来，它在世界范围内皮肤白皙人群中发病率一直在上升[1]。

尽管在黑素瘤发生和发展的生物学和分子遗传学方面取得了进展，但可用的治疗方案尚未转化为显著提高转移性疾病患者的生存率，据估计 2012 年美国有 9 180人死亡[2]。

皮肤黑色素瘤发病率的迅速增加似乎是与日晒有关的行为模式改变造成的，但其他因素可能与黑色素瘤发病机制有关，如遗传和化学致癌[3]。

黑色素瘤患者最重要的预后因素是原发性病变的侵袭深度。病变直径与存活率之间的大致相反关系也得到证实[4]。早期黑色素瘤的治疗包括完全切除原发肿瘤，这通常足以治愈该病。淋巴结定位和前哨淋巴结活检显示癌细胞扩散的程度。一旦黑色素瘤扩散到远处和内部，目前的治疗方法不能可靠地限制疾病的侵袭性进程[5]。早期发现、正确诊断和新的治疗方法是必要的，以抑制这种潜在的致命恶性肿瘤。

24.2　简史

黑色素瘤是人类已知最古老的恶性肿瘤之一，尽管其在古代发生的历史证据很少。20 世纪 60 年代，一份关于 9 具秘鲁印加木乃伊的检查报告显示，这些木乃伊大约是 2 400年之前的，有

① M. H. Massaoka，博士 · J. A. B. Scutti
D. C. Arruda，博士 · A. N. Rabaça
C. R. Figueiredo，博士 · C. F. Farias（巴西圣保罗联邦大学微生物、免疫和寄生虫学系肿瘤实验组）
A. L. Matsuo，博士 · N. Girola（巴西圣保罗生物制药有限公司）
L. R. Travassos，MD，博士（✉）（巴西圣保罗联邦大学微生物、免疫和寄生虫学系细胞生物学室。E-mail：travassos@ unifesp. br）

明显的黑素瘤迹象，如皮肤上的黑素瘤和扩散性骨转移[6]。

据报道，第一位对转移性黑色素瘤进行手术的外科医生是1787年的John Hunter。他切除了一个35岁男性下颚角后面的复发性肿块，并描述它包含两个不同的部分：一个白色坚硬的部分，另一个海绵状和黑色的外观[7,8]。1804年法国内科医生René Laennec在巴黎医学院（the Faculté de Médecine de Paris）的一次演讲中将黑色素瘤描述为一种疾病。1840年Samuel Cooper认为晚期黑色素瘤患者无法治疗，并指出生存的机会取决于早期手术切除。大约20年后，全科医生William Norris报告称，黑色素瘤的发生有家族性倾向[10]。

自从这些初步观察以来，黑色素瘤生物学和治疗研究已经加强，并且一直鼓励人们努力控制疾病和延长患者的存活时间。

24.3 皮肤黑色素瘤的起源和病因

皮肤黑色素瘤是一种神经外胚层来源的肿瘤，由产生色素的细胞（生黑色素细胞）增殖和恶性转化产生，黑色素细胞来源于神经嵴祖细胞，在胚胎发生过程中迁移到皮肤和毛囊中[11]。

黑色素细胞位于表皮、毛球、眼睛、耳朵和脑膜的基底层以及存于其他功能中，通过黑色素生成促进光保护作用和体温调节[12]。黑色素细胞的稳态是通过旁分泌生长因子、黏附分子和缝隙连接的复杂系统与表皮角质形成细胞的相互作用而受到严格调节的[13]。细胞增殖控制、自分泌生长因子的产生和黏附受体的丧失相关的关键基因突变，与黑色素细胞中细胞内信号的解除调节协同作用。这种突变使细胞能够逃脱角质细胞的控制，为黑色素细胞不受控制和独立增殖创造了有利的微环境[14,15]。

尽管恶性黑色素瘤的确切病因尚不清楚，但有证据表明遗传和环境的相互作用在黑色素瘤发展中起着重要作用。黑色素瘤发病的重要宿主因子列表很长，包括黑色素瘤家族史、黑色素瘤易感基因、痣的数量和类型、皮肤类型和色素沉着[16]。

阳光和紫外线照射被认为是黑色素瘤最重要的危险因素。事实上，流行病学数据表明，黑色素瘤发病率的增长与太阳照射模式的变化密切相关。高度遗传毒性的紫外线辐射可能导致DNA损伤，如果不修复，这种损伤在黑色素瘤的发病和发展中起着核心作用[17]。

流行病学研究已经证实，金发或红发、皮肤苍白的人更容易受到晒伤和雀斑的影响，他们比深色肤色的人更容易患上黑色素瘤。此外研究表明，儿童时期暴露在紫外线辐射下尤其会增加患黑素瘤的风险[18,19]。

多发性黑色素细胞痣、发育异常痣和非典型痣综合征（AMS）是表明黑色素瘤易感性的主要临床表现型[20]。因此，在黑色素瘤发展之前，已发现多个常见（普通、获得性）痣、5个以上发育异常（非典型）痣、大型先天性痣和直径大于6 mm的病变[21,22]。

家族性黑色素瘤是指两个或两个以上一级亲属被诊断出患有这种肿瘤性疾病的频率。一般来说，患者在早期就有了第一次黑色素瘤诊断，病变较薄，分布模式不同，多发性原发性黑色素瘤的频率较高。家族成员的病变在组织学上与散发性黑色素瘤不可区分，预后相似。

迄今为止，已经发现了两种黑色素瘤易感基因CDKN2A和CDK4。CDKN2A编码细胞周期蛋白依赖性激酶抑制剂2A（CDKN2A，$p16^{Ink4A}$），也称为多重肿瘤抑制因子1（MTS-1）。P16在调节细胞周期中起着重要作用。这种基因的突变增加了各种癌症的风险，主要是黑色素瘤。CDK4基因编码细胞周期蛋白依赖性激酶4，这是蛋白激酶复合物的一个组成部分，对细胞周期G1期的进展很重要。3种编码不同蛋白质的选择性剪接变异体被认为是由CDKN2A转录产生的。其中两种编码作为CDK4抑制剂的亚型。第三个包括另一个外显子1，它包含一个交替的开放阅读框（ARF），编码一种通过与MDM2相互作用来稳定p53的蛋白质。因此，这些基因的突变可能影响控制细胞分裂的相关信号通路，并可能导致患黑色素瘤的高风险[23,24]。

黑色素瘤中经常突变的其他基因包括有丝分裂原激活蛋白激酶途径上的*BRAF*和蛋白激酶

B/Akt 途径上的 *PTEN*。然而据报道，这些突变是遗传破坏的结果，而不是导致黑色素瘤引发事件的种系易感性改变[25,26]。

24.4 流行病学

过去几十年来，皮肤黑色素瘤发病率和黑色素瘤相关死亡人数急剧增加，因此黑色素瘤已成为许多地区的公共健康问题。尽管恶性黑色素瘤在皮肤癌中所占的比例不到 5%，但却导致了 80%的死亡，这主要是因为它具有很高的转移潜能[20]。

皮肤黑色素瘤主要发生在白人人群中，而在肤色较深的非洲或亚洲裔人群中则少得多。美国癌症统计数据显示，在 20 世纪 70 年代初，10 万人中有 6 人被诊断出患有黑色素瘤，而 2000 年代末，10 万人中有 21 人被诊断出患有黑素瘤[27]，这是所有癌症中增长最快的。澳大利亚和新西兰是世界上黑色素瘤发病率和死亡率最高的国家[18]。在欧洲，黑素瘤发病率和死亡率在人群中存在相当大的差异，但总体而言，该地区黑色素瘤发病率和死亡率也有所增加[28]。

尽管不同国家的数据库中男性/女性黑色素瘤发病率有所不同，但女性通常比男性患者的生存期要长得多[29]。黑色素瘤诊断的中位年龄为 55 岁[18]。

24.5 体征和症状

黑色素瘤主要症状和体征是皮肤病变，其直径或厚度增加、颜色变化、出血、瘙痒、溃疡和压痛[30]。

1985 年提出了黑色素瘤识别的 ABCD 标准，并已用于评估潜在的黑色素瘤病变[31]。最近，临床数据支持 ABCDE 扩展以强调黑色素瘤进展中色素病变的演变意义[32]。以 ABCDE 首字母缩略词的临床表现为：A，不对称，即病变的一半不同于另一半；B，表示边缘不规则，即边缘有缺口、不均匀或模糊；C，表示颜色斑驳，表示存在褐色、棕褐色和黑色病变的阴影；D，表示直径大于 6 mm（尽管有些黑色素瘤可能较小）；以及 E，病变的时间演变过程，是表示恶性黑色素瘤的主要特征。对表现出这些特征的病变进行调查和监测，提高了医生在常规皮肤检查早期识别黑色素瘤的能力，从而大大促进了最终的外科治疗。

24.6 分期和分类

基于黑色素瘤进展过程中的变化，Clark 等[33]提出了一种微分级系统，该系统反映了痣形成过程中黑色素细胞的增殖、痣发育不良、增生、侵袭和转移[34,35]。因此，根据该模型，黑色素瘤的开始和进展涉及一系列组织学改变，可分为 5 个不同阶段：①良性痣的发展，这是一种以黑色素细胞生长受限为特征的病变；②发育不良痣，以黑色素细胞的不连续和随机排列为特征；③放射状生长阶段（RGP），以表皮内细胞增殖为特征；④垂直生长阶段（VGP），其特征为黑色素细胞穿过基膜向真皮和皮下组织渗透；⑤黑色素瘤转移，其特征为扩散到皮肤和/或器官的其他区域，常见的是肝、肺、骨和脑[36]。Breslow 深度标准是评估黑色素瘤微分级的另一种方法[37]。通过使用眼用千分尺测量原发性肿瘤的垂直厚度，以评估黑色素瘤的侵袭程度，并推断疾病预后。

目前，Breslow 和 Clark 的分期参数都已纳入美国癌症联合委员会（AJCC）黑色素瘤分期系统[38]。广泛使用 TNM（肿瘤、淋巴结、转移）参数：T，根据 Breslow 的深度、细胞分裂和溃疡程度判断原发性肿瘤的特征；N，表示累及区域淋巴结；M，表示远处转移。此外，血清乳酸脱氢酶（LDH）已被选为评价疾病转移潜能的重要标志[39]。根据修订的 AJCC 指南[40]，皮肤黑色素瘤可分为以下几个阶段。

•0期，特征是存在局限于真皮的异常黑色素细胞，也被称为原位黑色素瘤。定义为Tis、N0、M0。

•Ⅰ期，以肿瘤厚度（<1 mm）、有丝分裂指数和溃疡状态为特征。定义为T1a、N0、M0；T1b、N0、M0；或T2a、N0、M0。

•Ⅱ期，也以肿瘤厚度（>1 mm）和溃疡状态为特征。没有淋巴结受累或远处转移的证据。定义为T2b、N0、M0；T3a、N0、M0；T3b、N0、M0；T4a、N0、M0；或T4b、N0、M0。

•Ⅲ期，以局部淋巴结受累和小转移或大转移为特征。定义为任意T、N1、M0；任意T、N2、M0；或任意T、N3、M0。

•Ⅳ期，其特征是存在远处转移和血清乳酸脱氢酶（LDH）水平。定义为任意T、任意N、M1。

黑色素瘤分期系统根据临床结果准确反映了黑色素瘤的生物学特性，预测了这种具有挑战性的肿瘤行为，为临床决策提供了宝贵的工具。

结合组织形态学特征对日照程度和相关分子变化的分析已经确定了不同组原发性黑色素瘤基因组突变的不同模式[41]。对比研究表明，肢端（手掌、脚底和指甲下部位）或黏膜黑色素瘤比其他有或没有长期日光诱导损伤的黑色素瘤组表现出更高频率的染色体畸变（例如，局部放大和丢失）[42,43]。此外，BRAF突变在间歇性暴露于太阳下的黑色素瘤中常见，而在长期暴露于太阳下的皮肤黑色素瘤中很少发现。在V^{600}突变的BRAF黑色素瘤中，已检测到PTEN失活或缺失，并与病灶Breslow厚度增加有关[45]。基于分子生物标志的分类可以重新定义诊断和预后类别，从而提供关于恶性黑色素瘤治疗和管理的额外信息。

24.7 诊断和经典疗法

在常规皮肤或身体检查期间，医生应根据ABCDE标准警惕潜在黑色素瘤病变的迹象。由于黑色素瘤在早期发现时是可以治愈的，所以任何可疑的病变都应进行手术切除，并提交组织病理学评估[46]。前哨淋巴结活检被推荐为原发性黑色素瘤患者的淋巴结分期程序，厚度为1～4 mm[47]。黑色素瘤的诊断基于一般形态、表皮、真皮和细胞学特征，这些特征与预后信息密切相关[48]。

大多数原发性黑色素瘤在早期（0、Ⅰ、Ⅱ）被发现时可通过手术切除成功治疗[49]。尽管进行了几十年的临床研究，但晚期黑色素瘤患者的预后仍然非常差，现有的治疗方案通常无效。对于累及区域性淋巴结的Ⅲ期黑色素瘤患者，通常建议进行完全淋巴结切除术。手术切除后，IFN-α辅助治疗是一种选择[46]。最近，抗CTLA-4单克隆抗体（mAb）治疗不可切除的Ⅲ期黑色素瘤患者，在临床试验中取得了统计学上的显著改善[50]，因此认为是一线或二线治疗[51]。此外，在BRAF V^{600}突变阳性和不可切除的黑色素瘤患者中使用特定的BRAF突变抑制剂维莫非尼已显示出部分反应，可能是Ⅲ期疾病的替代治疗方法[52]。

在IV期转移性黑色素瘤患者中，估计中位生存时间约为8个月，5年生存率小于10%[53]。对于转移性黑色素瘤的标准治疗尚无共识[54]，但系统治疗是唯一选择[55]。达卡巴嗪是最广泛使用的单一化疗药物，联合免疫治疗（如易普利姆玛、IL-2）可略微提高应答率[48]。此外，维莫非尼最近获得了美国食品和药物管理局（FDA）的许可，对BRAF $V^{600}E$突变患者进行单药治疗试验[52]，但临床研究表明，无进展生存率和总生存率只有谨慎的改善[56]。其他BRAF抑制剂如PLX4032（Plexxikon/Roche）、RAF265（Novartis）、XL281（Exelixis）和GSK2118436（GSK）正在进行转移性黑色素瘤临床试验。尤其是PLX4032，在携带V^{600} BRAF突变的IV期黑色素瘤中取得了令人鼓舞的效果，大多数患者的肿瘤完全或部分消退[57]。然而据报道，肿瘤对PLX4032治疗有抵抗力[58]。最近的研究表明，慢性BRAF抑制通过增强IGF-1R（胰岛素样生长因子受体1）信号传导介导黑色素瘤的生存，该信号传导途径涉及调节细胞增殖、防止细胞凋亡和肿瘤对治疗的抵抗[59]。总之，这些观察结果可能表明靶向一个单一的途径可能不足以根除黑色素瘤。

24.8 新兴抗肿瘤肽

已经从多种来源描述了具有抗肿瘤活性的天然肽。其中一些是游离分子，另一些是蛋白质水解释放或化学合成的蛋白质内部序列。它们的抗菌和抗肿瘤活性类似于古老的先天免疫分子，在抗体和适应性 T 细胞免疫出现之前，这些分子在抵御威胁性情况（感染或其他）方面是有效的。

Polonelli 等[60]在抗独特型抗体中鉴定出具有抗感染活性的内部序列，在一项合作研究中，我们发现来自不同单克隆抗体的 CDRs 对念珠菌和 HIV 有细胞毒性活性，对高度侵袭性鼠黑色素瘤 B16F10 也有细胞毒性活性[61]。进一步研究证实，免疫球蛋白的内部序列，即使是来自恒定（Fc）区域，独立于抗体特异性，也能显示抗感染和抗肿瘤活性，因此是生物活性肽的来源[62,63]。具体而言，来自单克隆抗体 C7（C7H2）的 V_HCDR 2（H2）定向到来自白色念珠菌的甘露糖蛋白。来自单克隆抗体 HuA（HuAL1）的 V_L CDR 1（L1）直接导入人类 A 型血，作为 C-酰胺化物合成肽进行检测，导致黑色素瘤细胞在体外凋亡。此外，在体内使用经静脉注射攻击的 C57BL6 小鼠，转移性同源基因模型中具有保护作用。C7H2 肽（YISCYNGATSYNQKFK）不仅在 B16F10 小鼠黑色素瘤细胞中发挥其凋亡活性，而且在几个 EC50 浓度相近的人肿瘤细胞系中也发挥其凋亡活性。最近 Arruda 等[64]研究表明，C7H2 结合并引起 G 肌动蛋白的聚合，同时作用于 F 肌动蛋白以对其进行稳定，从而改变肌动蛋白的动力学，在大量产生超氧化物阴离子后导致细胞凋亡。一个典型的半胱天冬酶依赖性细胞凋亡随之而来，记录了许多细胞改变和细胞器破坏。HuAL1（RASQSVSSYLA）也发现了类似的作用，其明显结合不同的受体，导致肿瘤细胞死亡，具有坏死特征（Arruda D. C.，2013 年未发表数据）。

B16F10 小鼠黑色素瘤（A4 和 A4M）单克隆抗体也提供了具有抗肿瘤活性的 CDR 肽[63]。有趣的是，两种单克隆抗体的 V_H CDR 3（H3）都是与单克隆抗体竞争结合黑色素瘤细胞的微体。A4H3 肽呈线性或循环延伸形式，其细胞毒性与 mAb A4 相同，靶向小鼠黑色素瘤细胞的原钙黏蛋白 β-13（与人类原钙黏蛋白 β-6 高度相似）。线性 A4 H3 肽（IRDGHYGSTSHWYF）在 EC50 0.06 mM 时能够抑制黑色素瘤细胞。与 A4 相似，它能诱导 B16F10-Nex2 细胞的 DNA 降解，并且在体内对转移性黑色素瘤也有活性。来自 mAb A4M 的 CDRs L1（RASGNIHNYLA）和 L2（NVKTLA）抑制了 B16F10-Nex2 细胞的生长，并诱导黑色素瘤和 H L-60 细胞的 DNA 降解，后者被 Bcr-A bl、Bcl-2 和 Bcl-X_L高表达所抵消。L1 和 L2 的凋亡效应伴随着两种肽的抗血管生成活性，这两种肽均以 HUVEC 为靶细胞[63]。

来自单克隆抗体 REB 200（人源化单克隆抗体 MX35，针对钠依赖性磷酸转运蛋白 2b，NaPi2b）的 CDR H3 肽，命名为 Rb10，其环状延伸衍生物 Rb9 不具有凋亡作用，但对肿瘤细胞具有显著的诱导作用。它们引起培养细胞的高度黏附，从而通过基底膜基质抑制迁移和入侵（图 24.1）。使用鼠黑色素瘤模型，Rb10 在体内具有抗转移性，并且通过 HSP90 结合和抑制细胞运动而明显作用于肿瘤细胞（PTC/1B2011/03053）。

抗血型 A 鼠单克隆抗体（Ac1001）CDR（H3）的另一个作用是巨噬细胞的免疫调节[65]。H3 肽刺激细胞因子的产生，激活 PI3K-Akt，并增强 TLR-4 的表达。CDR 肽的不同性质进一步增强了它们作为药物的潜在用途[66]。表 24.1 列出了针对几种肿瘤细胞系的凋亡/坏死肽。

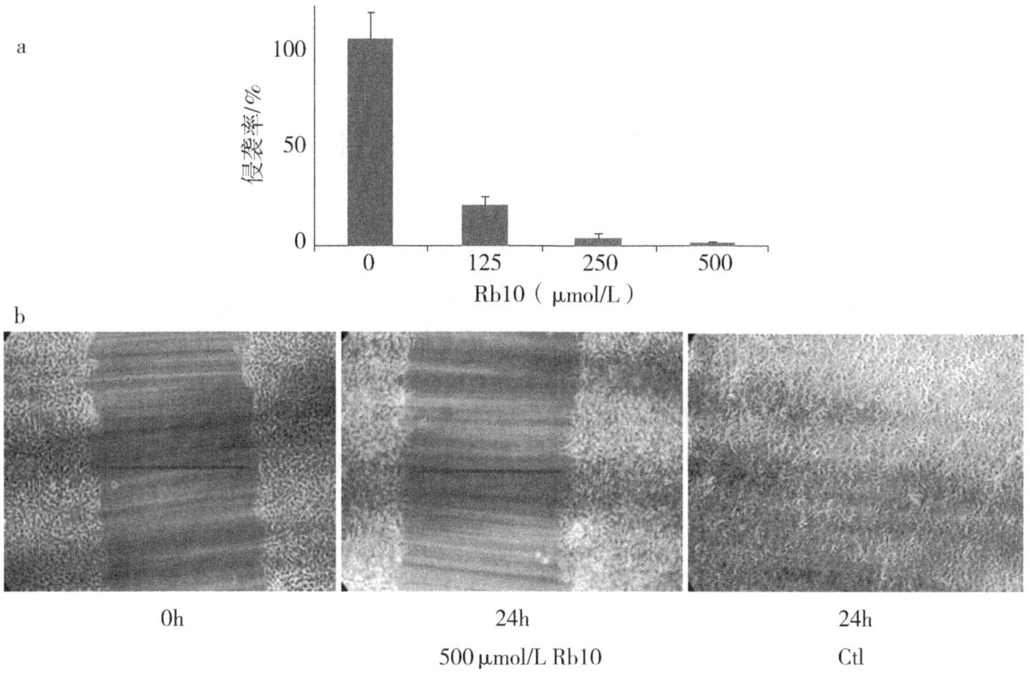

图 24.1　Rb10 肽（H3）抑制黑色素瘤细胞侵袭（基质凝胶）和迁移

在通透性隔离腔中测定 B16F10-Nex2 黑色素瘤细胞对基质凝胶的侵袭，在顶部培养相中加入 2×10^5 肿瘤细胞悬浮的无血清培养基，在底部培养相中加入 10% FBS 作为细胞引诱剂。迁移细胞被固定，用姬姆萨染色并计数。(a) 24 小时后，Rb10 在指定浓度下抑制细胞入侵。数值来自 3 次测定。(b) 使用划痕伤口愈合试验，Rb10 抑制 B16F10-Nex2 黑色素瘤细胞迁移（在 12 孔板中允许 3×10^5 细胞达到 70%~80%汇合）。与未处理的对照组相比，孵育 24 小时后观察到完全抑制。

表 24.1　显示凋亡/坏死抗肿瘤活性的各种特异性单克隆抗体的 CDR 肽

单抗	免疫原	免疫球蛋白	CDR（肽）	参考文献
A4	黑色素瘤 B16F10-Nex4	IgG	A4 CDR H3 (mic. mAb)	[64]
A4M	黑色素瘤 B16F10-Nex4	IgM	A4M CDR L1 和 L2	[64]
C7	念珠菌甘露糖蛋白	IgM	C7 CDR H2	[62, 65]
Pc42	乙型肝炎/恶性疟原虫杂交种	IgM	pc42 CDR H2	[62]
HuA	人类血型 A（Fuc_2）	IgM	HuA CDR L1	[62]
AC1001	小鼠抗血型 A（Fuc_2）	IgM	Ac1001 CDR H3	[66]
C36	牛痘	IgG	C36 CDR L1	—

H3 可以作为一种微量（mic）抗体。

24.9　肽的体内活性和抗肿瘤疫苗前景

免疫球蛋白 CDRs 在体外凋亡作用的发现，是以多个肿瘤细胞系为靶标而不是非致瘤细胞，我们不禁要问，它们在体内的保护性抗肿瘤活性是否可能涉及相同的机制。游离肽，即使有酰胺化保护的羧基末端，在作为血浆肽酶底物的循环中通常也不是很稳定的，根据大小，可以通过肾过滤清除。然而在转移性模型中，通过腹腔注射和隔日注射的 16（C7H2）和 11

（HuAL1）氨基酸肽对 B16F10 肿瘤细胞具有显著的保护作用[61]。

这两种肽都有免疫应答能力，在 C7H2 中，在 N 末端（Y1A 和 C4A）丙氨酸替换的肽在体内缺乏活性，这与特定的氨基酸识别相符[64]。C7H2 的凋亡效应与 β 肌动蛋白结合有关，但是 C7H2 的 C 末端序列而非 N 末端参与了这种反应。

C7H2 在 NOD/ Scid/IL-2rγ^{null}小鼠的转移性模型中对黑色素瘤的保护作用无效，这表明免疫系统的重要性，意味着免疫系统细胞参与其中。

为了探讨免疫应答的参与，我们检测了受体外凋亡 CDRs 刺激与否的树突状细胞的保护活性。采用相同的 B16F10 黑色素瘤细胞转移模型。当用黑色素瘤抗原初免时，来自 C57BL6 小鼠的同源基因树突状细胞具有部分保护作用，但这些细胞在用 CDRs 预刺激后显示出非常显著的抗肿瘤作用（图 24.2）。因此，体内的抗肿瘤作用似乎不依赖于肽直接靶向肿瘤细胞，但树突状细胞可能会在 CDRs 自然清除前放大抗肿瘤作用。我们假设 CDR 肽在降解前被树突状细胞有效吸收，激活的树突状细胞最有效地递呈肿瘤抗原，以产生有能力的 T 淋巴细胞 CTL 反应。更广泛的研究必须遵循这些观察，包括其他凋亡/坏死肽，以确定效应细胞和保护性免疫反应的其他特征。

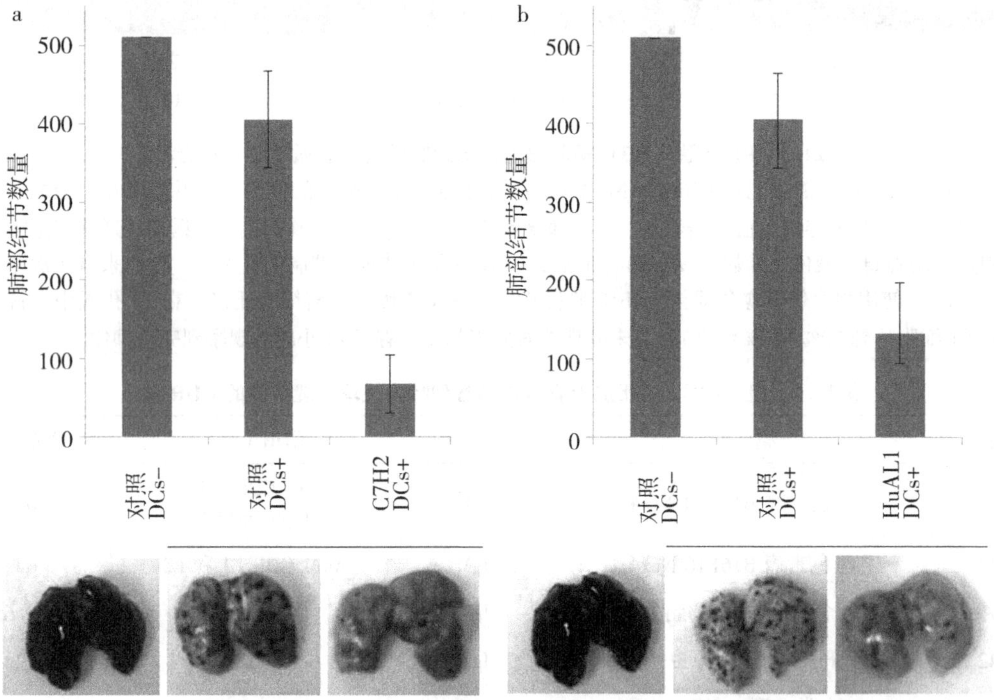

图 24.2　预防转移性黑色素瘤的保护性肽激活树突状细胞疫苗

同源 C57Bl/6 小鼠骨髓树突状细胞用黑色素瘤抗原致敏（5×10^5 B16F10-Nex2 细胞裂解物与 3×10^{5}树突状细胞孵育 1 小时）。将致敏树突状细胞与 20 μg/mL C7H2 或 HuAL1 孵育 3 小时。在第 1 天和第 4 天将肽刺激树突状细胞皮下注射到小鼠体内。动物在第 11 天接受 5×10^{5}黑色素瘤细胞静脉注射。(a) 用黑色素瘤抗原致敏 C7H2-DCs（+）的保护作用。与未致敏、未活化的树突状细胞相比，未用肽活化的对照致敏（+）树突状细胞效果差。(b) 与（a）中的相同，但有 HuAL1 肽激活的树突状细胞。如图所示，肺部有黑色素瘤结节。

A4H3（凋亡单抗 A4 的微抗体序列）和 Rb10（和 Rb9）（抑制肿瘤转移的 CDRs）的体内保护作用是否也依赖于树突状细胞（图 24.3），还没有确定。

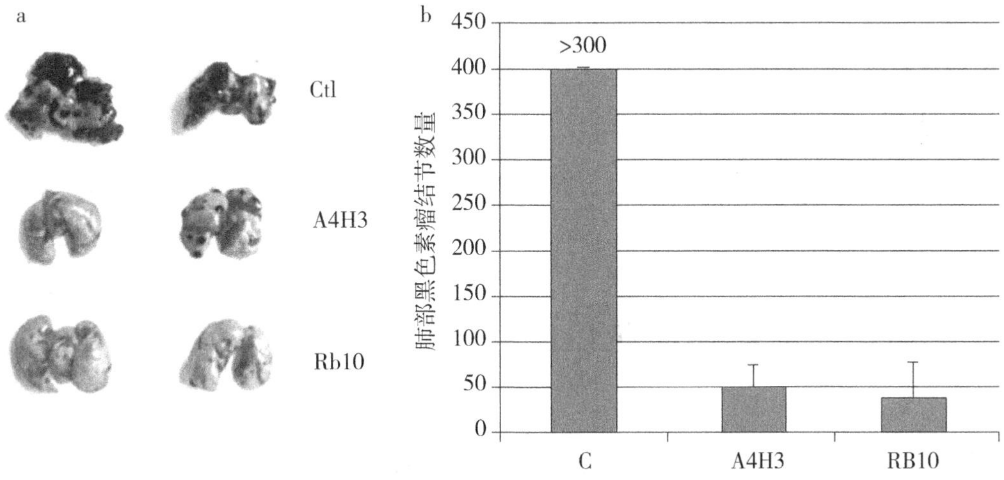

图 24.3 肽 A4H3 和 Rb10 的抗转移作用

用 100 μL 的 2×10^5B16F10-Nex2 黑色素瘤细胞对同源小鼠（C57Bl/6）进行体内攻击。从攻击后第 1 天开始，以 6 剂 250 μg/小鼠/天腹腔内注射肽。小鼠（每组 5 只）在攻击后 22 天处死，未用肽治疗的小鼠第 10 天处死。（a）未治疗动物和用肽治疗的小鼠的肺。（b）对照组（未治疗）和肽治疗动物肺中的黑色素瘤结节。

24.10 来源于信号蛋白的抗肿瘤肽

由于 CDRs 氨基酸序列的高度多样性和可变性，CDRs 一部分可能集中其生物活性，来自信号蛋白的其他肽已经成为潜在的抗癌剂。其中一些在体内是有活性的，而在体外就像 SOCS1 一样没有活性（Scutti J，2013 年未发表的结果），另外一些还有几个目标和复杂的作用机制仍在研究中。目前，我们提到肽 pTj 来源于 WT1 的 C 末端锌指结构域[67]。WT1 是与黑色素瘤发生和增殖相关的转录因子。pTj 肽显示出与不可逆 G2/M 细胞周期停滞和衰老诱导相关的抗增殖活性。在体内，pTj 显示出显著的抗肿瘤作用，减少了 B16F10 转移性黑色素瘤同源基因模型中肺结节的数量，延长了人黑色素瘤 A2058 皮下攻击裸鼠的存活时间（图 24.4）。

来自免疫球蛋白和信号蛋白域的天然肽可能是早期先天免疫分子肽序列的系统发育对应物，在防御和调节机制中起着重要作用，在我们的时代，一些回忆性分子仍发挥着重要作用，当作为分离的肽进行测试时，它们可以自己或通过诸如免疫系统的效应细胞激活而起作用。

24.11 抗原特性

传统上，免疫肽在抗原蛋白中定位后很容易被鉴定。在人类黑色素瘤中，来自 TRP2、NY-ESO-1、MelanA/MART-I、酪氨酸酶和 gp100/ Pmel17 的 MHC 限制性肽被用作免疫原。由于抗原表达的异质性，几个研究小组尝试用全黑色素瘤细胞通过基因转导的方法表达细胞因子。最有希望的是使用与 NY-ESO-1 反应的基因工程淋巴细胞[68]。NY-ESO-1 癌症/睾丸抗原在 80% 的滑膜细胞肉瘤患者和大约 25%的黑色素瘤患者中表达。阻断 CTLA-4 可将 CD8（$^+$）T 细胞增加到更广泛的预免疫黑色素瘤抗原阵列中[69]。

就本综述中的肽而言，它们不会与肿瘤抗原发生交叉反应以引起特定的抗肿瘤反应。它们直接作用于与不同受体结合的癌细胞，或间接激活免疫系统细胞。

如上所述，CDRs C7H2（pc42H2）和 A4H3 分别识别 β 肌动蛋白和原钙黏蛋白 β-13（人 β-16）。Rb10M3 是 Rb10 的衍生物，在第二个位置具有 dA，与 HSP90 反应，抑制其蛋白抗聚集

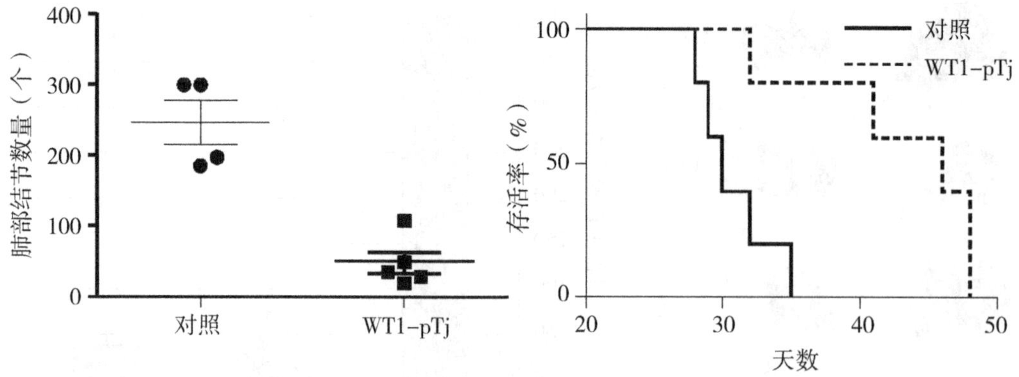

图 24.4　WT1 pTj 在体内的保护作用

（a）B16F10-Nex2 细胞在同源系统中的肺定殖用于测试 WT1 衍生肽的保护活性。动物连续几天接受 5 次腹腔注射 300 μg WT1 pTj 或 PBS。WT1 pTj 治疗后肺转移结节的数量显著减少了 81%。（b）用 WT1 衍生肽治疗的携带人黑色素瘤裸鼠的存活曲线。动物连续几天注射 300 μg 的 WT1 pTj 或 PBS。瘤周治疗从接种黑素瘤细胞后的第二天开始，并延长 5 天。未治疗肿瘤对照组的平均存活时间为（31±2.7）天，WT1 pTj 治疗组为（43±6.7）天。

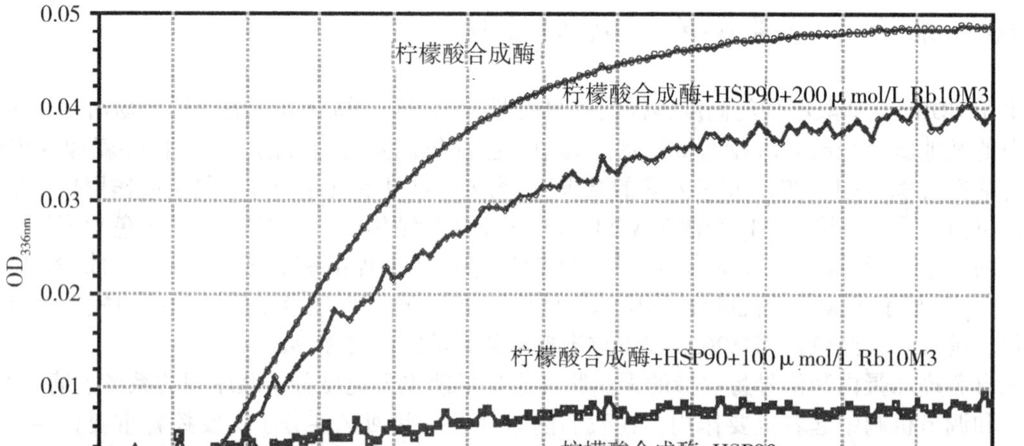

图 24.5　使用柠檬酸合酶聚集试验分析 Rb10M3 对 HSP90 伴侣功能的剂量依赖性抑制

Rb10M3 是具有 dA（A2dA）的 Rb10 的衍生物，用于提高稳定性。柠檬酸合酶（0.115 μmol/L），在 40mmol/L HEPES 中，pH 7.5，在 43℃温育变性和聚集。聚集在 $OD_{336\ nm}$ 处测量。

活性（图 24.5）。在肿瘤细胞中，高度表达的 HSP90 分子伴侣特性的干扰，影响细胞中许多信号通路。根据 Rb10 对肿瘤细胞转移的抑制，表明肌动蛋白动力学介导的负责细胞迁移和基质侵入的细胞突起与 HSP90 抑制相关[70]。HuAL1 结合组蛋白 3。WT1 pTj 增强 p53 活性，并与 WT1 蛋白竞争结合 p53[67]。WT1 pTj 可能通过锌指基序阻止 WT1 蛋白与其伴侣之间的重要相互作用。因此，该肽是对抗表达 WT1 恶性肿瘤的有希望药物（例如，MCF-7，但不是 SK-BR-3 乳腺癌或正常人成纤维细胞以及 HL-60 急性髓细胞白血病）。

24.12　动物模型

C57Bl/6 小鼠和同源基因小鼠黑色素瘤 B16F10（亚系 Nex2，由圣保罗联邦大学实验性肿瘤科开发，UNONEX-UNIFESP）通常用于体内实验。黑色素瘤细胞皮下注射（4×10^4 ~ 1×10^5个细胞），并且用公式测量每 mm^3中的肿瘤生长：$V=(d)^2\times D\times0.52$（其中 V 是肿瘤体积，D 是最大直径，d 是最短直径）。对于转移性模型（肺定殖），小鼠静脉内注射 B16F10-Nex2 黑色素瘤 1×10^5 ~ 5×10^5个细胞，15 ~ 20 天后处死动物，在肺中计数黑色结节。保护实验中使用的肽按 150 ~ 350 μg/只·天腹膜内注射。根据方案，肽可隔天给药，治疗可在 11 ~ 15 天后中断以绘制存活曲线。对于 B16F10 和人类肿瘤细胞的某些实验，使用免疫缺陷动物。$Nude^{-/-}$、RAG $1/2^{-/-}$ 和 NOD/Scid/IL-$2r\gamma^{null}$鼠已经与皮下注射的肿瘤细胞（A2058，SKMel28 人黑色素瘤细胞）或静脉内注射的肿瘤细胞（B16F10-Nex2 鼠黑色素瘤细胞）一起使用。癌周注射用于 WT1 pTj 肽的保护实验。

这些研究中使用了预防和治疗方案。对于 SOCS-1 肽，在皮下移植肿瘤达到 200 mm^3后开始治疗方案（图 24.6）。在该模型中，将 5×10^4B16F10-Nex2 细胞注射入 C57BL/6 小鼠（$n=5$）的右侧，15 天后用肽 300 μg/只腹腔内处理动物 5 天。如图所示，用 SOCS-1 P5 治疗抑制了肿瘤生长。加扰肽用作阴性对照。

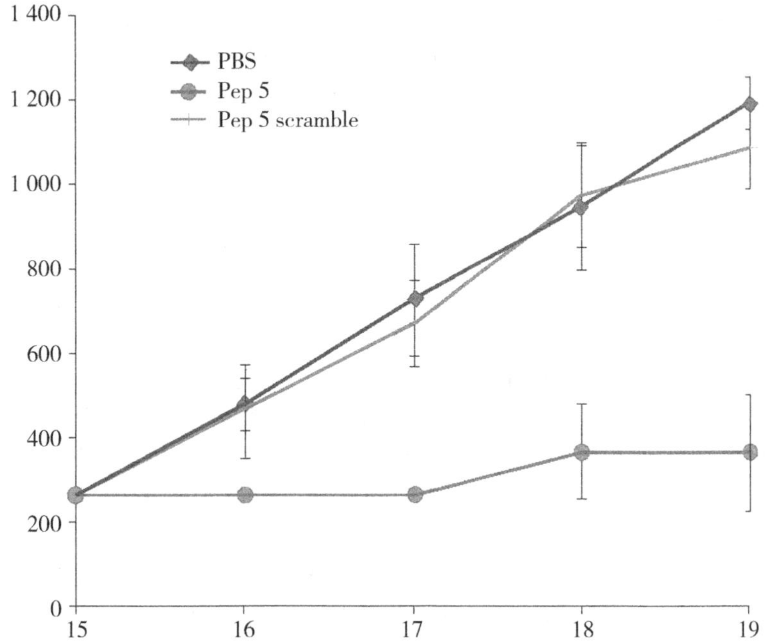

图 24.6　SOCS-1 衍生肽（P5）抑制皮下肿瘤发展

肿瘤是通过将 5×10^4　B16F10Nex2 细胞注射入 C57BL/6 小鼠的右侧而产生的（每组 5 个）。15 天后（肿瘤大小为 200 mm^3），小鼠用肽 300 μg/只对动物进行连续 5 天腹腔注射治疗（* $P<0.01$）。

24.13　可行性和临床前开发

表 24.1 所列的肽以及来源于 SOCS-1 和 WT1 的肽易于获得，并且可以在高纯度和再现性条件下制备。这些肽衍生物将取决于化学合成的复杂性。如果从抗肿瘤活性方面验证，与纳米颗

粒或大分子的结合物更容易获得。

本章所述的几种肽依赖于免疫细胞的体内保护作用，目前我们发现，其中两种通过树突状细胞（DCs）激活而发挥作用。在体内通过肽激活树突状细胞以及这些细胞作为治疗药的活性有一种可能性，在基于DC疫苗试验中有许多先例，这些试验使用不同的繁殖方法、抗原负荷和细胞给药[71]。在20世纪90年代，肿瘤肽或全细胞抗原已被用于树突状细胞的离体致敏DCs[72,73]。相反地，本章综述的肽与肿瘤细胞无关，显然直接作用于树突状细胞，使这些APCs效力增加。它们似乎在体内激活树突状细胞，树突状细胞是触发保护性免疫应答的必要条件。在转移性黑色素瘤中，用负载黑色素瘤抗原的树突状细胞接种导致9.5%的肿瘤消退，相比之下，用肿瘤细胞接种的肿瘤消退率最高为4.6%，用免疫肽疫苗接种的肿瘤消退率为2.7%，用病毒载体接种肿瘤消退率为1.9%[74]。在我们的体内实验中，在体内简单注射肽足以保护小鼠免受转移性黑色素瘤侵袭，这意味着激活树突状细胞的肿瘤抗原递呈是有效的。如图所示，通过肽处理的肿瘤抗原致敏树突状细胞在体外再现保护作用（图24.2）。此外，已经表明在黑色素瘤小鼠模型中，用肿瘤肽致敏树突状细胞进行静脉或皮下免疫诱导记忆性T细胞，从而控制肺部转移[75]。

Bonifaz等（2004）利用B16黑色素瘤对体内以DC为靶标和体外抗原负载的树突状细胞进行了比较。用表达的B16攻击小鼠，然后在抗CD40存在的情况下用抗DEC-205-OVA进行免疫，与体外负载OVA的成熟树突状细胞进行了比较。后者免疫并不抑制肿瘤生长，用抗DEC-205-OVA接种在控制肿瘤生长方面是有效的。就本综述中描述的抗肿瘤肽而言，迄今为止尚未鉴定出靶向受体，但有直接证据表明树突状细胞被激活。另一方面除了来自SOCS-1的肽之外，CDR衍生肽和来自WT1的肽似乎识别不同的配体，尽管明显发挥相似的刺激作用。

如果树突状细胞确实是抗肿瘤肽的主要靶向细胞，小鼠临床前数据必须转化为患者的成功结果。在这些研究中，识别小鼠和人类细胞中的保守靶点是了解免疫调节与疫苗接种之间关系的重要步骤。

24.14 优点和缺点

由于蛋白水解降解和肾过滤，短肽在体内的稳定性通常较差。然而本文综述的抗肿瘤肽隔天腹腔接种，对转移性黑色素瘤有保护作用。这种保护作用在免疫缺陷小鼠身上未观察到，这意味着免疫系统的细胞是必要的。骨髓树突状细胞激活的直接证据是，当在体外用黑色素瘤抗原致敏时，骨髓树突状细胞可有效防止转移性B16F10黑色素瘤。我们假设注射的肽在从血液循环中清除之前被树突细胞迅速吸收。肽激活树突状细胞在递呈肿瘤抗原的任务中是非常有效的，即使没有用抗体在体内靶向这些细胞，正如卡明斯基等所讨论的那样[76]。此外，由这些肽激活的树突状细胞类型能够诱导控制肿瘤生长的T细胞。最近我们实验室的Scutti（Scutti，JAB，2013年未发表的结果）显示，在患有B16F10黑色素瘤的小鼠中注射SOCS-1肽产生脾脏T-$CD8^+$细胞，其对$RAG^{-/-}$小鼠的转移性黑色素瘤生长具有保护作用。

目前研究的主要弱点是肽早期发展阶段。许多问题仍有待回答，主要涉及肽的作用机制、生物分布以及为抗肿瘤作用而激活的免疫细胞的类型。肽受体和如何将小鼠黑色素瘤模型中的实验转化为临床有效的程序是其需要深入研究的重要问题。然而这些简单存在的肽在纯化状态下如何容易生产，以及其抗癌作用值得进一步地研究。

致谢

目前的工作是由圣保罗州研究资助基金会（FAPESP）和巴西国家研究与发展委员会（CNPq）通过研究赠款和研究金资助的。

参考文献

[1] Godar, D. E.: Worldwide increasing incidence of cutaneous malignant melanoma. J. Skin Cancer2011, 858425 (2011).

[2] American Cancer Society: Cancer Facts & Figures 2013. Atlanta: American Cancer Society (2013). Retrieved from: http: //www. cancer. gov/cancertopics/ types/melanoma. Last accessed 5 Sept 2013.

[3] Houghton, A. N., Viola, M. V.: Solar radiation and malignant melanoma of the skin. J. Am. Acad. Dermatol. 5, 477-483 (1981).

[4] Lee, Y. T.: Diagnosis, treatment and prognosis of early melanoma. The importance of depth of micro-invasion. Ann. Surg. 191, 87-97 (1980).

[5] Dummer, R., Hauschild, A., Guggenheim, M., Jost, L., Pentheroudakis, G.: Melanoma: ESMO Clinical Practice Guidelines for diagnosis, treatment and follow-up. Ann. Oncol. 21 (Suppl 5), v194-v197 (2010).

[6] Urteaga, O., Pack, G. T.: On the antiquity of melanoma. Cancer19, 607-610 (1966).

[7] Bodenham, D. C.: A study of 650 observed malignant melanomas in the South-West region. Ann. R. Coll. Surg. Engl. 43, 218-239 (1968).

[8] Kuno, Y., Ishihara, K., Yamazaki, N., Mukai, K.: Clinical and pathological features of cutaneous malignant melanoma: a retrospective analysis of 124 Japanese patients. Jpn. J. Clin. Oncol. 26, 144-151 (1996).

[9] Laennec, R. T. H.: Sur les melanoses. Bull de la Faculté de Méd. de Paris1, 24-26 (1806).

[10] Norris, W.: Eight Cases of Melanosis with Pathological and Therapeutical Remarks on That Disease. Longman, Brown, Green, Longman, and Roberts, London (1857).

[11] Bandarchi, B., Ma, L., Navab, R., Seth, A., Rasty, G.: From melanocyte to metastatic malignant melanoma. Dermatol. Res. Pract. (2010).

[12] Kawakami, A., Fisher, D. E.: Key discoveries in melanocyte development. J. Invest. Dermatol. 131, E2-E4 (2011).

[13] Haass, N. K., Smalley, K. S., Herlyn, M.: The role of altered cell-cell communication in melanoma progression. J. Mol. Histol. 35, 309-318 (2004).

[14] Haass, N. K., Smalley, K. S., Li, L., Herlyn, M.: Adhesion, migration and communication in melanocytes and melanoma. Pigment Cell Res. 18, 150-159 (2005).

[15] Gray-Schopfer, V., Wellbrock, C., Marais, R.: Melanoma biology and new targeted therapy. Nature 445, 851-857 (2007).

[16] Newton-Bishop, J. A., et al.: Melanocytic nevi, nevus genes, and melanoma risk in a large case-control study in the United Kingdom. Cancer Epidemiol. Biomarkers Prev. 19, 2043-2054 (2010).

[17] de Gruijl, F. R.: Skin cancer and solar UV radiation. Eur. J. Cancer 35, 2003-2009 (1999).

[18] Garbe, C., Leiter, U.: Melanoma epidemiology and trends. Clin. Dermatol. 27, 3-9 (2009).

[19] Gandini, S., et al.: Meta-analysis of risk factors for cutaneous melanoma: II Sun exposure. Eur. J. Cancer 41, 45-60 (2005).

[20] Silva, J. H., Sa, B. C., Avila, A. L., Landman, G., DupratNeto, J. P.: Atypical mole syndrome and dysplastic nevi: identification of populations at risk for developing melanoma-review article. Clinics (Sao Paulo) 66, 493-499 (2011).

[21] Schneider, J. S., Moore 2nd, D. H., Sagebiel, R. W.: Risk factors for melanoma incidence in prospective follow-up. The importance of atypical (dysplastic) nevi. Arch. Dermatol. 130, 1002-1007 (1994).

[22] Gogotov, I. N., Kosiak, A. V.: Hydrogen metabolism in Anabaena variabilis in the dark. Mikrobiologiia45, 586-591 (1976).

[23] Tucker, M. A., Goldstein, A. M.: Melanoma etiology: where are we? Oncogene 22, 3042-3052 (2003).

[24] Udayakumar, D., Mahato, B., Gabree, M., Tsao, H.: Genetic determinants of cutaneous melanoma p

redisposition. Semin. Cutan. Med. Surg. 29, 190-195 (2010).

[25] Haluska, F. G., et al.: Genetic alterations in signaling pathways in melanoma. Clin. Cancer Res. 12, 2301s-2307s (2006).

[26] Casula, M., et al.: BRAF gene is somatically mutated but does not make a major contribution to malignant melanoma susceptibility: the Italian Melanoma Intergroup Study. J. Clin. Oncol. 22, 286-292 (2004).

[27] Howlader, N., et al.: SEER Cancer Statistics Review, 1975-2010. Bethesda: National Cancer Institute. http://seer.cancer.gov/csr/1975_2010/, based on November 2012 SEER data submission, posted to the SEER web site, 2013. Last accessed 7 Sept 2013.

[28] de Vries, E., Coebergh, J. W.: Cutaneous malignant melanoma in Europe. Eur. J. Cancer 40, 2355-2366 (2004).

[29] Joosse, A., et al.: Gender differences in melanoma survival: female patients have a decreased risk of metastasis. J. Invest. Dermatol. 131, 719-726 (2011).

[30] Shashanka, R., Smitha, B. R.: Head and neck melanoma. ISRN Surg. 2012, 948302 (2012).

[31] Friedman, R. J., Rigel, D. S., Kopf, A. W.: Early detection of malignant melanoma: the role of physician examination and self-examination of the skin. CA Cancer J. Clin. 35, 130-151 (1985).

[32] Abbasi, N. R., et al.: Early diagnosis of cutaneous melanoma: revisiting the ABCD criteria. JAMA 292, 2771-2776 (2004).

[33] Clark, W. H. J., et al.: A study of tumor progression: the precursor lesions of superficial spreading and nodular melanoma. Hum. Pathol. 15, 1147-1165 (1984).

[34] Takata, M., Murata, H., Saida, T.: Molecular pathogenesis of malignant melanoma: a different perspective from the studies of melanocytic nevus and acral melanoma. Pigment Cell Melanoma Res. 23, 64-71 (2010).

[35] Miller, A. J., Mihm, M. C. J.: Melanoma. N. Engl. J. Med. 355, 51-65 (2006).

[36] Patel, J. K., Didolkar, M. S., Pickren, J. W., Moore, R. H.: Metastatic pattern of malignant melanoma. A study of 216 autopsy cases. Am. J. Surg. 135, 807-810 (1978).

[37] Breslow, A.: Thickness, cross-sectional areas and depth of invasion in the prognosis of cutaneous melanoma. Ann. Surg. 172, 902-908 (1970).

[38] Balch, C. M., et al.: Final version of 2009 AJCC melanoma staging and classification. J. Clin. Oncol. 27, 6199-6206 (2009).

[39] Gogas, H., et al.: Biomarkers in melanoma. Ann. Oncol. 20 (Suppl 6), vi8-vi13 (2009).

[40] Piris, A., Mihm, M. C. J., Duncan, L. M.: AJCC melanoma staging update: impact on dermatopathology practice and patient management. J. Cutan. Pathol. 38, 394-400 (2011).

[41] Kyrgidis, A., Tzellos, T. G., Triaridis, S.: Melanoma: stem cells, sun exposure and hallmarks for carcinogenesis, molecular concepts and future clinical implications. J. Carcinog. 9, 3 (2010).

[42] Whiteman, D. C., et al.: Melanocytic nevi, solar keratoses, and divergent pathways to cutaneous melanoma. J. Natl. Cancer Inst. 95, 806-812 (2003).

[43] Bastian, B. C., et al.: Gene amplifications characterize acral melanoma and permit the detection of occult tumor cells in the surrounding skin. Cancer Res. 60, 1968-1973 (2000).

[44] Curtin, J. A., et al.: Distinct sets of genetic alterations in melanoma. N. Engl. J. Med. 353, 2135-2147 (2005).

[45] Goel, V. K., Lazar, A. J., Warneke, C. L., Redston, M. S., Haluska, F. G.: Examination of mutations in BRAF, NRAS, and PTEN in primary cutaneous melanoma. J. Invest. Dermatol. 126, 154-160 (2006).

[46] Dummer, R., Hauschild, A., Pentheroudakis, G.: Cutaneous malignant melanoma: ESMO clinical recommendations for diagnosis, treatment and follow up. Ann. Oncol. 20 (Suppl 4), 129-131 (2009).

[47] Boland, G. M., Gershenwald, J. E.: Sentinel lymph node biopsy in melanoma. Cancer J. 18, 185-191 (2012).

[48] Smoller, B. R.: Histologic criteria for diagnosing primary cutaneous malignant melanoma. Mod. Pathol. 19 (Suppl 2), S34-S40 (2006).

[49] National Cancer Institute: Melanoma Treatment - Treatment Option Overview. National Cancer Institute (2013). Retrieved from: http: //www. cancer. gov/cancertopics/pdq/treatment/melanoma/HealthProfessional/page4. Last accessed 5 Sept 2013.

[50] O'Day, S. J., et al.: Efficacy and safety of ipilimumab monotherapy in patients with pretreated advanced melanoma: a multicenter single-arm phase II study. Ann. Oncol. 21, 1712-1717 (2010).

[51] Ebert, S. N., Shtrom, S. S., Muller, M. T.: Topoisomerase II cleavage of herpes simplex virus type 1 DNA in vivo is replication dependent. J. Virol. 64, 4059-4066 (1990).

[52] Sinha, R., et al.: Cutaneous adverse events associated with vemurafenib in patients with metastatic melanoma: practical advice on diagnosis, prevention and management of the main treatment-related skin toxicities. Br. J. Dermatol. 167, 987-994 (2012).

[53] Garbe, C., Eigentler, T. K., Keilholz, U., Hauschild, A., Kirkwood, J. M.: Systematic review of medical treatment in melanoma: current status and future prospects. Oncologist 16, 5-24 (2011).

[54] Bermejo-Perez, M. J., Villar-Chamorro, E., Lemos-Simosomo, M., Garcia-Gonzalez, C., Galeote-Miguel, A. M.: Complete response in a patient with a metastatic cutaneous melanoma. An. Sist. Sanit. Navar. 34, 307-310 (2011).

[55] Bhatia, S., Tykodi, S. S., Thompson, J. A.: Treatment of metastatic melanoma: an overview. Oncology (Williston Park) 23, 488-496 (2009).

[56] Luke, J. J., Hodi, F. S.: Vemurafenib and BRAF inhibition: a new class of treatment for metastatic melanoma. Clin. Cancer Res. 18, 9-14 (2012).

[57] Flaherty, K. T., et al.: Inhibition of mutated, activated BRAF in metastatic melanoma. N. Engl. J. Med. 363, 809-819 (2010).

[58] Kim, T., Kim, J., Lee, M. G.: Inhibition of mutated BRAF in melanoma. N. Engl. J. Med. 363, 2261 (2010); author reply 2261-2262.

[59] Villanueva, J., et al.: Acquired resistance to BRAF inhibitors mediated by a RAF kinase switch in melanoma can be overcome by cotargeting MEK and IGF-1R/PI3K. Cancer Cell 18, 683-695 (2010).

[60] Polonelli, L., et al.: Therapeutic activity of an engineered synthetic killer anti-idiotypic antibody fragment against experimental mucosal and systemic candidiasis. Infect. Immun. 71, 6205-6212 (2003).

[61] Polonelli, L., et al.: Antibody complementarity-determining regions (CDRs) can display differential antimicrobial, antiviral and antitumor activities. PLoS One 3, e2371 (2008).

[62] Polonelli, L., et al.: Peptides of the constant region of antibodies display fungicidal activity. PLoS One 7, e34105 (2012).

[63] Dobroff, A. S., et al.: Differential antitumor effects of IgG and IgM monoclonal antibodies and their synthetic complementarity-determining regions directed to new targets of B16F10-Nex2 melanoma cells. Transl. Oncol. 3, 204-217 (2010).

[64] Arruda, D. C., et al.: β-Actin-binding complementarity-determining region 2 of variable heavy chain from monoclonal antibody C7 induces apoptosis in several human tumor cells and is protective against metastatic melanoma. J. Biol. Chem. 287, 14912-14922 (2012).

[65] Gabrielli, E., et al.: Antibody complementarity-determining regions (CDRs): a bridge between adaptive and innate immunity. PLoS One 4, e8187 (2009).

[66] Magliani, W., Conti, S., Cunha, R. L., Travassos, L. R., Polonelli, L.: Antibodies as crypts of anti-infective and antitumor peptides. Curr. Med. Chem. 16, 2305-2323 (2009).

[67] Massaoka, M. H., et al.: A novel WT1-derived peptide induces cellular senescence and inhibits tumor growth in a human melanoma cell line and xenograft model. Cancer Res. 2 (8 Suppl), 2867 (2012).

[68] Robbins, P. F., et al.: Tumor regression in patients with metastatic synovial cell sarcoma and melanoma using genetically engineered lymphocytes reactive with NY-ESO-1. J. Clin. Oncol. 29, 917-924 (2011).

[69] Yuan, J., et al.: CTLA-4 blockade increases antigen-specific CD8 (+) T cells in pre-vaccinated patients with melanoma: three cases. Cancer Immunol. Immunother. 60, 1137-1146 (2011).

[70] Taiyab, A., Rao, C.: HSP90 modulates actin dynamics: inhibition of HSP90 leads to decreased cell motility and impairs invasion. Biochim. Biophys. Acta1813, 213-221 (2011).

[71] Tacken, P. J., de Vries, I. J., Torensma, R., Figdor, C. G.: Dendritic-cell immunotherapy: from ex vivo loading to in vivo targeting. Nat. Rev. Immunol. 7, 790-802 (2007).

[72] Mayordomo, J. I., et al.: Bone marrow-derived dendritic cells pulsed with synthetic tumour peptides elicit protective and therapeutic anti-tumour immunity. Nat. Med. 1, 1297-1302 (1995).

[73] Nestle, F. O., et al.: Vaccination of melanoma patients with peptide-or tumorlysate-pulsed dendritic cells. Nat. Med. 4, 328-332 (1998).

[74] Banchereau, J., Palucka, K.: Dendritic cells as therapeutic vaccines against cancer. Nat. Rev. Immunol. 5, 296-306 (2005).

[75] Mullins, D. W., et al.: Route of immunization with peptide-pulsed dendritic cells controls the distribution of memory and effector T cells in lymphoid tissues and determines the pattern of regional tumor control. J. Exp. Med. 198, 1-13 (2003).

[76] Caminschi, I., Maraskovsky, E., Ross Heath, W.: Targeting dendritic cells in vivo for cancer therapy. Front Immunol. 3 (13), 1-13 (2012).

（马维民译）

第 25 章　细小病毒：友好的抗癌免疫调节剂

Zahari Raykov，Svitlana P. Grekova，
Assia L. Angelova 和 Jean Rommelaere①

摘要

溶瘤病毒是 1 种有多用途的工具，通过自然机制或基因操作，可以特异性地靶向并杀死肿瘤细胞。在过去 10 年里，这些药物的主要作用方式之一是作为原位（肿瘤内）抗癌疫苗。

最近批准细小病毒（PVs）作为一种用于临床治疗胶质瘤的溶瘤药物。本章论述了溶瘤性 PVs 免疫调节机制的几个要点，例如间接（通过免疫原性杀死肿瘤细胞）或直接（流产感染）激活人类免疫细胞。

此外，还讨论了细胞因子修饰或富含 CpG DNA 的细小病毒和免疫调节组合治疗策略。

这一主题的最新研究表明，PVs 是一种沉默的抗癌免疫调节剂。

25.1　癌症作为免疫治疗的挑战

1909 年，Paul Ehrlich 第一个提出免疫系统在保护宿主患癌方面发挥关键作用的概念。从那时起，免疫系统对有效控制癌症发展的作用多年来一直是一个有争议的话题。Galon 等在 2006 年提供了一个主要的支持证据，表明肿瘤免疫反应程度与提高结直肠癌患者存活率呈正相关[1]。然而免疫系统和进展中的肿瘤通过耐受性和免疫编辑过程相互塑造。癌细胞利用基质细胞作为它们的帮凶，通过特异性消除肿瘤反应效应物或使癌细胞处于无反应状态来改变免疫细胞环境，使其更加“自由”。另一方面，先天免疫应答和获得性免疫应答都会编辑肿瘤，选择免疫原性最小和抗性最强的克隆。此外，每个个体免疫系统和肿瘤都有一套独特的基因组和表观基因组特征，这些特征可以影响前者对后者的控制。这解释了开发通用抗癌疫苗的困难，除了少数肿瘤实体（例如宫颈癌）表达由转化病毒蛋白衍生的共同表位，免疫系统将识别这些表位[2]。

25.2　溶瘤病毒（OVs）作为癌症疫苗

病毒疗法最近成为传统癌症治疗的替代方法。由于对病毒和癌症生物学的更好理解以及对 DNA/RNA 操作技术的掌握，这一新研究领域正在爆炸式发展，这源于对与病毒感染相关的肿瘤退化的一些历史观察。在过去 20 年里，十几种病毒的溶瘤活性已经被鉴定出来，有几种已经进入临床试验，其中大多数来源于改良的人类致病病毒，如单纯疱疹病毒、腺病毒和痘病毒，具有很大的克隆能力。这为设计多功能基于病毒的产品（治疗和成像分子）在人类肿瘤中以高浓度选择性表达辟了可能性。溶瘤病毒（OVs）感染肿瘤，并通过连续几轮感染-复制-溶瘤和再感染，导致其根除的最初想法已经变得相当理想化，因为肿瘤内病毒复制和在人体内传播的效率受到限制，并且大多数肿瘤内存在大部分正常基质细胞。然而，这些限制并不一定反映在溶瘤病毒不良抑瘤效果上。事实上，除了它们的直接裂解活性，这些病毒还有能力触发肿瘤内免

---

① Z. Raykov，MD（✉）· S. P. Grekova，博士 A. L. Angelova，博士 · J. Rommelaere，博士（德国癌症研究中心（DKFZ）肿瘤病毒学研究室，海德堡，德国，E-mail：z. raykov@ dkfz. de）

疫系统的细胞和体液分支，以这种方式代表了理想的疫苗佐剂。

免疫系统在癌症病毒治疗中起着双重作用。

一方面，它有助于保护机体免受感染，从而对抗 OV 活性。大多数 OVs 会触发一种涉及 I 型干扰素诱导的细胞防御机制，并限制病毒复制和传播。这可能不一定会妨碍 OVs，因为肿瘤细胞中的 IFN 信号经常受损。这种损伤实际上被用来设计不能阻断干扰素通路 OVs，因此在肿瘤中优先复制。此外，由 NK 细胞、单核细胞、巨噬细胞和中性粒细胞介导的早期先天性免疫应答提供了宿主的初始防御线以限制 OV 感染、复制和传播，同时促进病毒抗原呈递细胞成熟，为随后的获得性免疫应答奠定了基础[3]。尤其对受感染肿瘤的 NK 细胞募集与 IFN-γ 产生有关，IFN-γ 在巨噬细胞激活中起着重要作用，巨噬细胞有助于 OV 清除和随后诱导 T 细胞和 B 细胞介导的获得性应答。

另一方面，这种由肿瘤感染 OVs 引起的免疫前微环境也有利于激活抗癌天然应答、交叉递呈肿瘤相关抗原以及诱导肿瘤特异性获得性免疫[4]。在这方面，溶瘤病毒可通过向免疫系统揭示患者自身的肿瘤抗原来充当原位个性化疫苗佐剂。这种 OV 活性的免疫成分尤其重要，因为它有助于破坏非感染性肿瘤细胞，这可能是 OV 治疗肿瘤中大多数恶性细胞的表现。还应指出，通过给 OVs 配备适当的调节基序或转基因，可以增强 OVs 的免疫刺激活性。这一点可以通过以基于疱疹和痘苗为基础的 OncoVEX 和 JX594 载体，并辅以 GM-CSF 编码的转基因来证明，可以假设这些载体在最近报告的静脉接种后，这些 OVs 影响转移性癌能力中起作用[5]。

25.3 细小病毒（PVs）作为溶瘤药物

细小病毒是小的（25 nm）二十面体粒子，包含一个约 5kb 的单链 DNA 基因组。源于啮齿动物的细小病毒属成员，因其溶瘤特性而备受关注。在溶瘤病毒中，啮齿类 PVs 与新城疫病毒（NDV）（鸟）、黏液瘤（兔）Semliki 森林病毒和其他病毒属于非工程（自然溶瘤）动物病毒[6-8]。这些病毒可以在其自然宿主中引起疾病，但对人类没有致病性。此外，它们在动物和人类癌细胞中都显示出致瘤性和溶瘤性特征[9]。PVs 自然肿瘤选择性似乎至少部分地由病毒 DNA 复制[10]基因表达[11]和非结构蛋白细胞毒性活性[12]对宿主细胞增殖和转化的依赖性引起。特别是我们实验室研究已经表明，大鼠细小病毒 H-1PV 选择性地杀死一组人类肿瘤源性细胞系[13-16]。此外，在动物模型中 H-1PV 全身或肿瘤内注射可抑制各种人类癌、淋巴瘤和神经系统肿瘤[17-21]。重要的是，H-1PV 具有传染性，但对人类没有明显的致病性[22]。

通过动物模型中不同途径进行的 PV 感染研究表明，病毒血症是短暂的，随后在感染后 7~10 天内产生病毒特异性抗体。在人类实验性感染 H-1PV 中观察到类似现象[23]。抗 PV 抗体至少部分是中和的，因此预期会阻碍病毒性肿瘤抑制。事实上，当动物在治疗前几周预先感染 H-1PV，在肿瘤细胞注射后不久施用病毒，H-1PV 抑制成年 ACI 大鼠肝癌转移的能力受损[24]。然而在人类群体中甚至日常使用这些药物的实验室人员中，没有或很少发现抗 PVs 的天然免疫力[22]（未发表的观察结果）。这些数据表明，PVs 对人体的自然感染力较低，适合临床应用。

25.4 通过 PVs 激活免疫系统

McKisic 等对小鼠细小病毒 MPV-1 的开创性研究提高了 PVs 引发抗癌疫苗接种的可能性。这些作者研究表明，病毒感染后治愈肿瘤的动物对随后相同肿瘤细胞的攻击有保护作用。这种病毒被证明能持续感染小鼠淋巴组织，加速同种肉瘤移植物的排斥反应，而不会直接感染肿瘤细胞[25]。

通过比较免疫状态不同的小鼠对细小病毒 MVMp 诱导移植同源基因胶质瘤抑制的反应，给出了免疫系统参与 PV 肿瘤抑制的另一个线索。免疫活性小鼠完全不受 MVMp 感染 GL261 胶质瘤细胞体外肿瘤生长的影响。相反地，免疫缺陷（$RAG2^{-/-}$）动物对 MVMp 依赖的肿瘤抑制能力较低，只有 20%的受体受到保护，因此，需要额外的获得性免疫成分来完全抑制肿瘤[26]。

在胰腺导管腺癌（PDAC）同基因免疫活性大鼠模型中，细小病毒 H-1PV 可以在只有相对较小部分的肿瘤细胞接受直接病毒诱导裂解的条件下，实现显著的肿瘤抑制[19]。在这个系统中，病毒主要起到抗癌免疫应答的触发作用。事实上，H-1PV 治疗荷瘤供体的脾细胞过继转移，导致携带有预先定制的原发性原位胰腺癌患者肿瘤消退[27]。在另一个研究中，发现受感染胰腺肿瘤 H-1PV 诱导的溶瘤作用导致同一动物中未感染的远处肿瘤消退。在这两种类型的实验中，受体或远处肿瘤消退分别发生在没有病毒转移的情况下，大概是通过病毒触发的抗肿瘤免疫反应。在 PV 抑瘤中，免疫系统作用的另外一系列证据来源于免疫耗竭实验，该实验表明，从免疫活性大鼠中选择性地去除 $CD8^+$细胞，与 H-1PV 根除移植性胶质瘤（未发表的观察）和肝癌转移有效性降低有关[24]。

25.4.1　PVs 通过杀死肿瘤间接激活免疫系统

许多实验观察支持 PV 介导的肿瘤抑制具有免疫原性的观点（参见图 25.1，左上图）。如上所述，PVs 嗜癌性不是由于转化细胞更好地吸收病毒，而是由于病毒生命周期的细胞内步骤，依赖于作为细胞周期和致癌转化功能调节因子的存在、缺失或激活[28]。由于其基因组体积小，编码能力有限，PVs 利用其非结构蛋白来实现多种功能。特别是主要非结构蛋白 NS1 活性，通过翻译后修饰和/或亚细胞分布得到及时调节。不同 NS1 域的磷酸化控制病毒产物的酶（ATP 酶、螺旋酶）和非酶（DNA 和蛋白质结合）功能，在感染期间发生变化[29]。细胞死亡和最终溶解是允许性恶性细胞感染 PVs 的最终结果。根据细胞类型和代谢状态，PV 感染的细胞可以通过不同的机制死亡，即凋亡、坏死或溶酶体中的组织蛋白酶溢出[13,16,19]。

几项研究表明，在体外系统中，受 PV 感染的肿瘤细胞与未受感染的肿瘤细胞相比，其激活人和啮齿动物体内 APC 的能力增强。这种效应不仅仅是 PV 诱导的肿瘤细胞杀伤结果。事实上，H-1PV 感染的人类黑色素瘤细胞比通过冷冻/解冻获得的黑色素瘤细胞裂解物能更有效地激活体外成熟的树突状细胞（DCs）。当 DCs 与 PV 感染的 SK-29MEL 细胞共培养时，DCs 吞噬作用和肿瘤相关抗原的交叉递呈，导致黑色素瘤特异性 CTLs 有效激活[30]。还测试了感染细小病毒 MVMp（小鼠微小病毒）的小鼠神经胶质瘤细胞（GL261）激活 DCs 和小胶质细胞（MG）的能力，DCs 和 MG 是两种不同的细胞类型，它们可以充当 APCs。MG 和离散 DC 亚群在与 MVMp 感染的胶质瘤 GL261 细胞共培养后被激活，特异性激活标志（CD80、CD86）上调和促炎细胞因子（TNF-α 和 IL-6）释放证明了这一点[26]。

近年来，细胞凋亡被认为是免疫“沉默”，坏死被认为是细胞死亡的“报警”方式，这一教条在免疫原性细胞死亡（ICD）方面得到了重新审视和表述。死亡细胞释放的钙网蛋白（CRT）高迁移率族蛋白 B（HMGB1）和 ATP 等危险相关分子模式（DAMPs），是某些化疗药物（如蒽环类药物）诱导 ICD 的基础[31,32]。细胞死亡模式似乎并不决定 ICD 特征，尽管 CRT 暴露与内质网应激、HMGB1 释放对创伤性死亡的影响以及 ATP 释放对细胞焦亡的影响更为密切。

在此背景下，细胞凋亡和坏死分别作为抗炎和促炎性死亡方式的观点已被修正，以将细胞死亡的免疫原性与上述 3 个关键 ICD 决定因素的诱导联系起来[33]。CRT 显示、HMGB1 释放和 ATP 足以使旁分泌成熟和树突状细胞激活以及随后 CTL 的引发——这是获得性免疫应答的基础事件。CRT 表面暴露介导树突状细胞对死亡肿瘤细胞的吸收，而 HMGB 和 ATP 的释放分别参与 TLR2/TLR4/RAGE 介导信号的激活和 P2X7/炎症介导的半胱天冬酶 1（CASP1）激活，导致成熟白细胞介素-1（IL1）加工和分泌[34]。

我们实验室最近研究表明，H-1PV 是胰腺癌细胞释放 HMGB1 的有效诱导因子。在这些细胞中，H-1PV 与标准 PDAC 化疗药物吉西他滨联合使用，可以刺激 ICD 的所有决定因素，甚至包括 IL-1。在早期研究中，报道称 H1 诱导的凋亡杀伤与 SK29Mel-1 黑色素瘤细胞的免疫原性

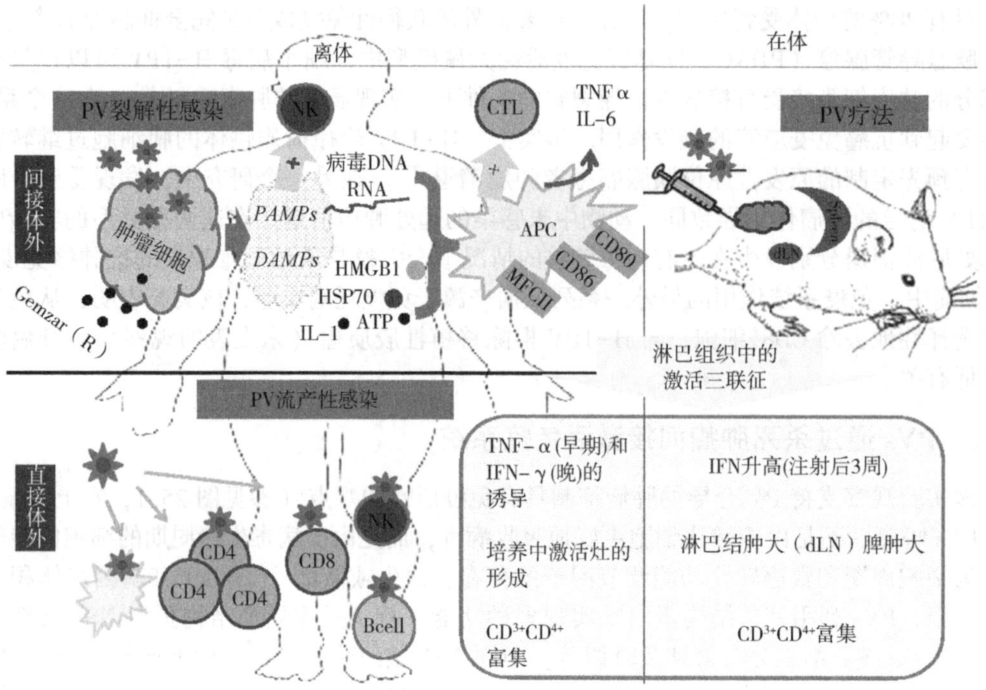

图 25.1　PVs 的免疫调节作用

图上半部分-PVs 间接体外（左面）免疫调节结合标准胰腺癌化疗。肿瘤细胞经历免疫原细胞死亡（ICD），指出了由 H-1PV 而来的 PAMPs（*DNA*、*RNA*）和 H-1PV-健择（抗肿瘤药）诱导 DAMPS（HMGB1、HSP70、ATP、IL-1）激活抗原呈递细胞（APC）的贡献。免疫调节 PDAC 模型体内实验（右面）；肿瘤引流淋巴结（dLN）。图下半部分-直接体外（人 PBMCs）的免疫调节。在两个系统中都观察到了 PV 激活的 3 个变化；CD4 T 辅助细胞 CD8 细胞毒性 T 细胞，NK 自然杀伤细胞。

信号热休克蛋白（HSP72）强释放有关[35]。在本系统中，H-1PV 感染与细胞毒性制剂或舒尼替尼联合治疗可增强黑素瘤细胞的免疫原性，这是由于与单独治疗的肿瘤细胞相比[36]，联合培养的树突状细胞具有更高的诱导成熟能力。综上所述，这些研究强调联合病毒化疗的力量不仅可以在其细胞减压术的背景下使用，而且可以作为局部肿瘤内佐剂。

除细胞衍生的危险信号外，含有双链 RNA 或非甲基化 CpG 等病毒元素也被证明起病原体相关分子模式（PAMPs）作用，并通过病原体识别受体（PRRs）（PKR、TLR3、TLR9）触发免疫激活。在自主性 PVs 情况下，这一问题仍在研究中，同时考虑到病毒基因组的独特单链性质，这为增加这些病毒的真正佐剂效应提供了独特的机会。

除树突状细胞外，先天免疫的重要效应物，即 NK 细胞也似乎“感觉”到 PV 介导的对 PDAC 细胞的杀伤，导致 IFN-γ、TNF-α 和 MIP-1α/β 产生。最近在细胞培养物中观察到这种激活，尤其在免疫应答早期可能起重要作用。NK 细胞依赖性杀伤体外受感染的肿瘤细胞可追溯到 DNAM-1 配体 CD155 上调和 MHCI 类分子表达下调[37]。

总之，用各种人类癌症模型进行的研究支持这样的观点，即 PVs（特别是 H-1PV 和 MVMp）可以通过免疫原性途径杀死人类肿瘤细胞，从而激活先天免疫系统的邻近细胞，充当间接免疫刺激剂。这些方法为使用 PV 感染或化学病毒疗法治疗肿瘤细胞，进行同种异体接种提供了诱人的策略。

25.4.2 PVs 对免疫系统（IS）的直接影响

有证据表明，除了能够间接激活免疫细胞外，PVs 还可能对免疫系统产生直接影响，如人类 PBMCs 体外实验所示（图 25.1，左下方）。早期研究表明，长期培养的淋巴细胞、单核细胞以及未成熟和成熟的树突状细胞不易受 H-1PV 的细胞毒性作用影响[35]。此外，将 H-1PV 直接接种到人 PBMC 培养物中只会导致流产感染[27,38]。病毒可以进入细胞，根据细胞性质和激活状态实现不同程度的复制，但不能检测到病毒产生。有丝分裂激活后，感染模式（病毒倾向于特定细胞类型）从单核细胞、B 细胞和 NK 细胞转移到 T 细胞，特别是 CD4 淋巴细胞[27]。

然而，CD4 细胞是相当稳定的，在感染后没有对增殖和生存能力造成损害，与表达细胞毒性 PV 蛋白 NS1 的能力低一致。在感染的非刺激 PBMCs 中，CD4 细胞的比例甚至增加，这与培养物中激活灶的形成有关。这些变化伴随着 PBMC 感染后早期 TNF-α（24 小时）和晚期 IFN-γ（72 小时）的产生。单核细胞被鉴定为 TNFα 的来源，这是通过分离和分离培养物的黏附部分而建立的[27]。$CD4^+T$ 细胞在感染的 PBMC 培养中也被激活，这是由激活标志表达和 Th1 和 Th2 细胞因子分泌所揭示的[27,38]。此外，由于 H-1PV 感染，Tregs 免疫抑制特性降低，而病毒不影响这些细胞活性[38]。

除这些体外数据外，一些观察表明，在体内条件下 PVs 也可以与免疫细胞直接相互作用。无论是否患有肿瘤（黑色素瘤、淋巴瘤、血管肉瘤），在感染小鼠的淋巴组织中均检测到显著的 MVMp 基因表达[39]。这种表达在引流病毒注射肿瘤（DLN）的淋巴结中尤为明显。在血管肉瘤模型中，利用转化型多瘤中间 T 癌基因（PymT）作为特异性肿瘤标志，可以很容易地识别肿瘤细胞。携带血管肉瘤的小鼠其肿瘤显示有很强的 PymT 积累，而这个标志在脾脏中几乎没有检测到，在 DLN 中也没有检测到。因此，MVMp 在淋巴结中的表达不太可能是由于它们侵入了被感染的转移瘤细胞，而是可以被分配到感染的淋巴细胞中。事实上，髓性树突状细胞和 B-1 淋巴细胞是 DLN 中病毒 mRNA 表达的主要来源，为细小病毒与免疫细胞的直接相互作用提供了证据。

在大鼠模型（见上文）中对 H-1PV 治疗的胰腺癌进行分析证实，在肿瘤内接种后第一天内，PV 表达在肿瘤中占主导地位，但也可以在淋巴组织和肿瘤引流淋巴结中检测到[19]。在胶质瘤和胰腺癌模型中，用 PVs 治疗的免疫活性动物表现为脾脏和淋巴结肿大，伴有 T 细胞室增加，CD4/CD8 比率升高以及 DLN 和脾细胞中诱导 IFN-γ[27]。在直接感染 PBMC 培养物（见上文）时也观察到了相同的变化，这一事实表明在整个生物体中，发生免疫细胞 PV 感染并导致 Th1 偏倚。

总之，这些观察结果使我们提出，尽管免疫细胞 PV 感染具有流产特征，但是用 PV 治疗癌症生物并不是免疫沉默的，而是导致直接和肿瘤细胞介导的调节，这可能赋予这些病毒佐剂性质。值得注意的是，PV 诱导的免疫系统对肿瘤启动似乎不需要诱导强烈的局部炎症反应。在接受了基于 PV 的肿瘤抑制荷瘤动物的脑或胰腺中，没有检测到炎性细胞因子大暴发。PVs 的这种低促炎性特征与其几乎不存在的致病性一致，表明 PVs 相当有助于微调免疫系统[40]。我们实验室目前正在研究这种调节对上述诱导的脾细胞，对既定肿瘤反应性的确切影响。

25.5 PVs 作为新型免疫治疗组合的一部分

事实上，瘤内注射溶瘤病毒产生了以病毒作为佐剂的自体原位疫苗，可以为设计更有效的病毒提供基础。H-1V 的沉默免疫调节特性要求其免疫原性有一定的提高，以便产生一种更“危险”的病毒，能够更有效地逆转肿瘤中普遍存在的耐受性条件。

25.5.1 用免疫刺激转基因培育繁殖缺陷型细小病毒载体

历史上，首次尝试增加 PVs 的佐剂性质是通过从细小病毒 MVMp、H-1PV 和 LuIII 中产生衣壳替代载体。这些载体补充了编码毒性（凋亡素）或免疫刺激性（细胞因子/趋化因子）产物的转基因[41-43]。使用合适的辅助系统，产生转导这些异源基因的重组病毒。中试动物实验表明，与空载体相比，携带白细胞介素 2（IL-2）细胞因子基因的重组 PVs 具有增强的抗癌能力[44]。表达 INF-γ 诱导蛋白 10（IP-10）和单核细胞趋化蛋白 3（MCP-3）的重组细小病毒也分别对免疫活性小鼠中已建立的血管肉瘤和黑色素瘤有很强的抗癌作用[42,43]。有趣的是，发现转导 TNF-α 和 IP-10 的 MVMp 载体在抑制小鼠胶质瘤中发挥协同作用[45]。应该指出的是，由于包装的 DNA 大小限制，部分衣壳基因必须从载体上除去，以容纳上述转基因。结果重组病毒粒子不能产生后代粒子，因此不能引起继发性感染。

25.5.2 利用具有复制能力的细小病毒载体作为疫苗佐剂

上述携带转基因的 PV 载体局限性促使我们设计了另一类重组体，在没有同时删除病毒序列的情况下，将一个短调控元件插入 PV 基因组非编码区。这种新型具有复制能力的 PVs 利用了病毒衣壳（aprx 150bp）包装所提供的最小基因组扩增，而不影响后代的生产效率。位于病毒衣壳基因的终止密码子和多聚 A 尾部之间的基因组区域被证明最适合这种插入。细菌和其他微生物的 DNA 中非甲基化 CpG 基序长期以来通过 TLR-9 受体被认为是抗原呈递细胞的有效危险信号刺激因子[46]。

我们知道，CpG 基序在包括 PVs 在内的几乎所有小型真核病毒基因组中都没有得到充分表达。因此，我们假设用免疫刺激 CpG 模式来丰富 H-1PV 基因组可以改善其免疫调节特性[47]。H-1PV 肿瘤特异性复制应导致肿瘤细胞内这些基序的选择性扩增，然后在肿瘤抗原的作用下释放这些基序。通过 TLR-9 受体将导致更高水平的树突状细胞活化，病毒和肿瘤抗原交叉递呈给主要 CTLs，用于杀死受感染和未受感染的肿瘤细胞。在典型的治疗性疫苗接种方案中（图 25.2，上半部分），制备完全感染携带 CpG 的 H-1PV 载体，测试其佐剂能力。将具有肺转移倾向的大鼠肝癌细胞静脉注射到具有同源免疫功能的 ACI 大鼠体内。已确定的肺转移瘤通过皮下注射用野生型或修饰的 H-1PVs 病毒感染辐照自体肿瘤细胞进行治疗。在这些条件下，病毒不会到达靶转移瘤以发挥直接的溶瘤作用，而是在自体疫苗中复制并用作佐剂。

与上述细小病毒免疫调节证据一致，感染野生型 H-1PV（或添加了 GpC 的对照载体）提高了细胞疫苗疗效。携带 CpG 基序的 H-1PV 显著增强了这种病毒佐剂效应，从而抑制了 H-1PV-CG 感染自体疫苗的大多数转移。这种治疗性疫苗接种效果伴随着对细胞介导免疫应答（IFN-γ）标志的强烈诱导，尤其是在引流含肺转移纵隔淋巴结中 DC 活化（CD80 和 CD86）的强烈诱导。这是一个令人兴奋的结果，首次表明无缺陷细小病毒载体可以被工程化，并用于产生富含免疫刺激性 CpG 基序的溶菌剂，甚至在没有病毒传播的情况下也可以增强疫苗接种对远处转移的癌细胞的疗效。最近采用相同的策略（CpG 基因组整合）来修饰溶瘤腺病毒，显著地提高了治疗效果[49]。

25.5.3 PVs 和 IF-γ 结合

我们对 H-1PV 抑制 PDAC 的上述研究强烈表明，免疫系统在该系统中对细小病毒的治疗效果起着重要作用。大多数 PDAC 患者在诊断时已经有远处（肝、腹膜）转移，现阶段需要积极化疗。我们假设，在这些条件下腹腔局部应用 PVs 是有效的[50]。发现肿瘤 DLN 中 IFNγ 的产生在系统性上伴随着 PVs 的治疗效果，意味着这种有效的细胞因子在介导 PV 免疫调节中的重要性。在卵巢癌随机 III 期临床试验中，腹腔内应用 IFN-γ 已被证明可获得手术证明的一线和二线治疗效果。

综上所述，这些数据使我们有理由认为，在治疗晚期不可治愈的 PDAC（如腹膜癌）时，IFN-γ 可以提高 H-1PV 的治疗性能。通过单次或腹腔内注射 H-1PV 治疗腹膜转移的原位胰腺癌大鼠，同时施用或不施用 IFN-γ（图 25.2，下半部分）。与单纯腹腔内注射 H-1PV 相比，这

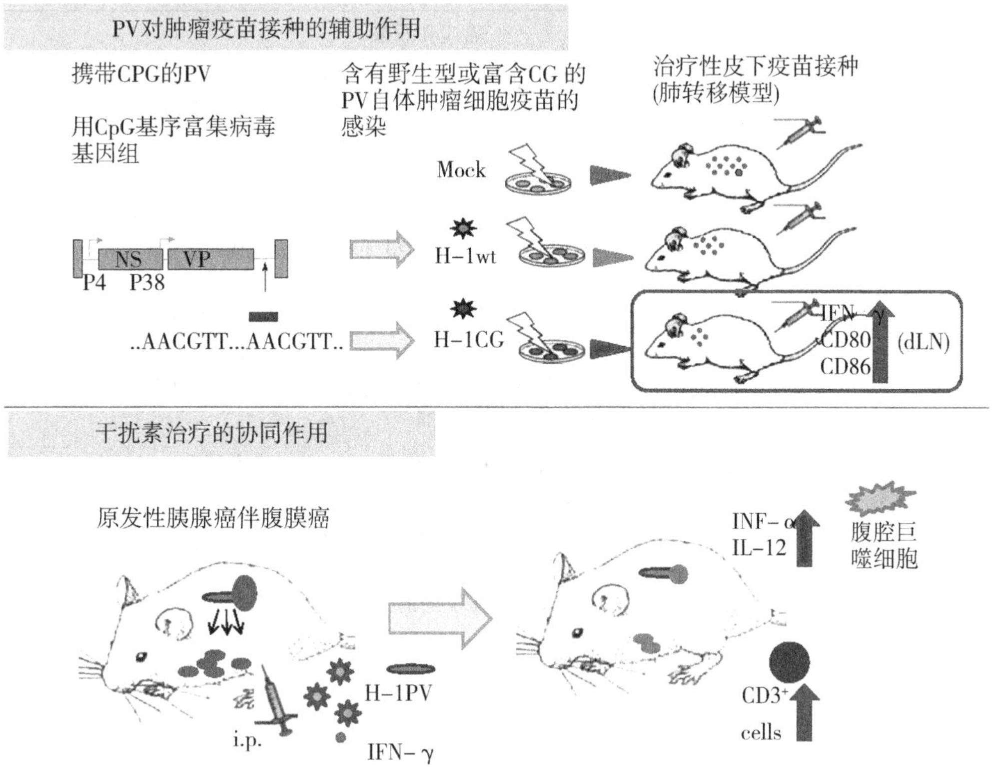

图 25.2　使用 PVs 的免疫治疗

改进的 TLR-9 参与的疫苗接种方法（图上半部分）。皮下应用自体 PV 感染疫苗治疗已确定的肝癌肺转移。用一种基因工程化的 H-1PV 病毒感染疫苗，这种病毒富含免疫刺激性 C_pG 序列，通过减少转移来提高疫苗接种效果。接种 H-1PV-C_pG 疫苗后，纵隔淋巴结中 IFN-γ、CD80 和 CD86 的表达增加。联合免疫治疗（H-1PV 与 IFN-γ）（图下半部分）。采用 H-1PV+IFN-γ 腹腔注射，治疗原位胰腺肿瘤腹膜转移。IFN-γ 增强脾脏腹膜巨噬细胞活化和 CD^{3+} 淋巴细胞增殖，导致原发性肿瘤生长和癌变减慢，延长生存期。

种 H-1PV 和 recIFN-γ 联合治疗提高了治疗效果（延长了动物存活时间）。有趣的是，联合治疗组也可以通过其免疫学参数来区分，即①分离的腹膜巨噬细胞易于产生 TNF-α；②在体外条件下，在肿瘤细胞存在的情况下脾脏 T 细胞的增殖增强；③腹水中 H-1PV 中和抗体的滴度显著降低。这些结果证实了我们的论点，即 PV 抑癌作用诱导免疫决定因素，这些决定因素可以通过适当的联合治疗加以控制。据我们所知，我们的研究是第一次利用 IFN-γ 与溶瘤病毒的组合。

结论

总之溶瘤病毒特别是 PVs，代表了多功能的抗癌工具，既能特异性杀死肿瘤细胞，又可作为佐剂，将进一步的肿瘤根除交给免疫系统。与为了变成溶瘤病毒（例如痘病毒和疱疹病毒）而必须被修饰的人类致病性病毒和触发明显免疫反应的动物病毒（例如 NDV）相比，PVs 代表了一种安全且不引人注目的入侵者角色，它更能使免疫系统精确调节到抗癌状态。然而这一事实可能允许以免疫原性方式暴露和重新发现隐伏的肿瘤抗原，而不会强烈地“吸引”免疫系统对固有病毒表位的注意。当前和计划中的细小病毒疗法临床试验（表 25.1）应首先指出，即 PV 溶瘤作用能否转化为免疫系统的长期保护，从而提高肿瘤根除率和患者存活率[51]。

表 25.1　细小病毒疗法的临床前数据和正在进行的临床试验

模型	描述和效果	
神经胶质瘤	单次鼻内灌注 H-1PV 足以到达大脑，诱导大鼠胶质瘤的有效消退。在具有人胶质瘤异种移植的大鼠中，H-1PV 溶瘤活性不太明显，但导致存活率显著延长	[21]
胰腺癌（PDAC）	H-1PV 单药治疗在体外（人 PDAC 细胞）和体内对大鼠和严重联合免疫缺陷小鼠（人异种移植）均有很强的抗癌作用，可降低肿瘤生长，延长动物存活时间，CT 扫描无转移。H-1PV+健择组合具有协同作用	[19]
伯基特淋巴瘤	H-1PV 通过坏死、源于伯基特淋巴瘤（BL）的细胞培养物（包括对利妥昔单抗细胞诱导凋亡的抗性）有效感染和杀死。基于细小病毒单一疗法有效抑制了在严重联合免疫缺陷（SCID）小鼠疾病模型中晚期建立的人伯基特淋巴瘤	[20]
宫颈癌	H-1PV 引起从人宫颈癌（HeLa）细胞到免疫缺陷大鼠和 SCID 小鼠皮下肿瘤的剂量依赖性消退。	[17]
乳腺癌	在成年裸鼠体内人肿瘤细胞植入部位注射细小病毒 H-1PV，可显著抑制肿瘤生长（大于 80%）。当对携带预制瘤的动物静脉注射时，H-1 病毒能够减缓甚至在某些情况下能够恢复肿瘤生长	[18]
转移性癌（肺、腹膜）	通过皮下注射病毒感染（野生型 H-1PV 或浓缩 CpG）自体疫苗成功治疗大鼠肝癌肺转移 同时用 recIFN-γ 腹膜内免疫治疗 H-1PV 可提高大鼠由胰腺癌引起的腹膜癌治疗效果	[48, 50]
胶质瘤 I 期临床试验（GBM）	（正在进行）Parvoryx01 是一项开放的、非控制的、分两组、组内剂量递增、单中心、I/IIA 期试验。18 例复发性 GBM 患者将分为 2 组，每组 9 例。治疗组 1 将首先接受肿瘤内注射 H-1PV，其次在肿瘤切除过程中通过肿瘤腔壁注射。在第 2 组中，病毒最初通过静脉注射，随后在肿瘤切除过程中（与第 1 组相同）注入周围的脑组织。试验的主要目标是局部和全身安全性和耐受性，并确定最大耐受剂量（MTD）。次要目标是概念验证（POC）和长达 6 个月的无进展生存期（PFS）	[51]

参考文献

[1] Galon, J., et al.: Type, density, and location of immune cells within human colorectal tumors predict clinical outcome. Science 313, 1960-1964 (2006). doi: 10. 1126/ science. 1129139, 313/5795/1960 [pii].

[2] Ogino, S., Galon, J., Fuchs, C. S., Dranoff, G.: Cancer immunology-analysis of host and tumor factors for personalized medicine. Nat. Rev. Clin. Oncol. 8, 711-719 (2011). doi: 10. 1038/nrclinonc. 2011. 122.

[3] Cooper, M. A., et al.: Human natural killer cells: a unique innate immuno-regulatory role for the CD56 (bright) subset. Blood 97, 3146-3151 (2001).

[4] Gauvrit, A., et al.: Measles virus induces oncolysis of mesothelioma cells and allows dendritic cells to cross-prime tumor-specific CD8 response. Cancer Res. 68, 4882-4892 (2008). doi: 10. 1158/0008-5472. CAN-07-6265, 68/12/4882 [pii].

[5] Breitbach, C. J., et al.: Intravenous delivery of a multi-mechanistic cancer-targeted oncolytic poxvirus in humans. Nature 477, 99-102 (2011). doi: 10. 1038/ nature10358.

[6] Zamarin, D., Palese, P.: Oncolytic Newcastle disease virus for cancer therapy: old challenges and new directions. Future Microbiol. 7, 347-367 (2012). doi: 10. 2217/fmb. 12. 4.

[7] Liu, J., Wennier, S., McFadden, G.: The immuno-regulatory properties of oncolytic myxoma virus and

their implications in therapeutics. Microbes Infect. 12, 1144–1152 (2010). doi: 10. 1016/j. micinf. 2010. 08. 012.

[8] Maatta, A. M., et al.: Evaluation of cancer viro–therapy with attenuated replicative Semliki forest virus in different rodent tumor models. Int. J. Cancer 121, 863–870 (2007). doi: 10. 1002/ijc. 22758.

[9] Rommelaere, J., et al.: Oncolytic parvoviruses as cancer therapeutics. Cytokine Growth Factor Rev. 21, 185–195 (2010). doi: 10. 1016/j. cytogfr. 2010. 02. 011.

[10] Bashir, T., Horlein, R., Rommelaere, J., Willwand, K.: Cyclin A activates the DNA polymerase delta–dependent elongation machinery in vitro: a parvovirus DNA replication model. Proc. Natl. Acad. Sci. U. S. A. 97, 5522–5527 (2000). doi: 10. 1073/pnas. 090485297090485297, [pii].

[11] Perros, M., et al.: Upstream CREs participate in the basal activity of minute virus of mice promoter P4 and in its stimulation in ras–transformed cells. J. Virol. 69, 5506–5515 (1995).

[12] Mousset, S., Ouadrhiri, Y., Caillet–Fauquet, P., Rommelaere, J.: The cytotoxicity of the autonomous parvovirus minute virus of mice nonstructural proteins in FR3T3 rat cells depends on oncogene expression. J. Virol. 68, 6446–6453 (1994).

[13] Rayet, B., Lopez–Guerrero, J. A., Rommelaere, J., Dinsart, C.: Induction of programmed cell death by parvovirus H–1 in U937 cells: connection with the tumor necrosis factor alpha signalling pathway. J. Virol. 72, 8893–8903 (1998).

[14] Ran, Z., Rayet, B., Rommelaere, J., Faisst, S.: Parvovirus H–1–induced cell death: influence of intracellular NAD consumption on the regulation of necrosis and apoptosis. Virus Res. 65, 161–174 (1999), S016817029900115X [pii].

[15] Moehler, M., et al.: Effective infection, apoptotic cell killing and gene transfer of human hepatoma cells but not primary hepatocytes by parvovirus H1 and derived vectors. Cancer Gene Ther. 8, 158–167 (2001). doi: 10. 1038/sj. cgt. 7700288.

[16] DiPiazza, M., et al.: Cytosolic activation of cathepsins mediates parvovirus H–1–induced killing of cisplatin and TRAIL–resistant glioma cells. J. Virol. 81, 4186–4198 (2007). doi: 10. 1128/JVI. 02601–06, JVI. 02601–06 [pii].

[17] Faisst, S., et al.: Dose–dependent regression of HeLa cell–derived tumours in SCID mice after parvovirus H–1 infection. Int. J. Cancer 75, 584–589 (1998). doi: 10. 1002/ (SICI) 1097–0215 (19980209) 75: 4<584:: AID-IJC15>3. 0. CO; 2–9, [pii].

[18] Dupressoir, T., Vanacker, J. M., Cornelis, J. J., Duponchel, N., Rommelaere, J.: Inhibition by parvovirus H–1 of the formation of tumors in nude mice and colonies in vitro by transformed human mammary epithelial cells. Cancer Res. 49, 3203–3208 (1989).

[19] Angelova, A. L., et al.: Improvement of gemcitabine–based therapy of pancreatic carcinoma by means of oncolytic parvovirus H–1PV. Clin. Cancer Res. 15, 511–519 (2009). doi: 10. 1158/1078–0432. CCR–08–1088, 15/2/511 [pii].

[20] Angelova, A. L., et al.: Oncolytic rat parvovirus H–1PV, a candidate for the treatment of human lymphoma: in vitro and in vivo studies. Mol. Ther. 17, 1164 – 1172 (2009). doi: 10. 1038/mt. 2009. 78, mt200978 [pii].

[21] Kiprianova, I., et al.: Regression of glioma in rat models by intranasal application of parvovirus h–1. Clin. Cancer Res. 17, 5333–5342 (2011). doi: 10. 1158/1078–0432. CCR–10–3124, 1078–0432. CCR–10–3124 [pii].

[22] Rommelaere, J., Cornelis, J. J.: Antineoplastic activity of parvoviruses. J. Virol. Methods 33, 233–251 (1991).

[23] Toolan, H. W., Saunders, E. L., Southam, C. M., Moore, A. E., Levin, A. G.: H–1 virus viremia in the human. Proc. Soc. Exp. Biol. Med. 119, 711–715 (1965).

[24] Raykov, Z., et al.: Combined oncolytic and vaccination activities of parvovirus H–1 in a metastatic tumor model. Oncol. Rep. 17, 1493–1499 (2007).

[25] McKisic, M. D., Paturzo, F. X., Smith, A. L.: Mouse parvovirus infection potentiates rejection of tumor allografts and modulates T cell effector functions. Transplantation 61, 292–299 (1996).

[26] Grekova, S. P., Raykov, Z., Zawatzky, R., Rommelaere, J., Koch, U.: Activationof a glioma-specific immune response by oncolytic parvovirus Minute Virus of Mice infection. Cancer Gene Ther. 19, 468-475 (2012). doi: 10. 1038/cgt. 2012. 20.

[27] Grekova, S., et al.: Immune cells participate in the oncosuppressive activity of parvovirus H-1PV and are activated as a result of their abortive infection with this agent. Cancer Biol. Ther. 10, 1280-1289 (2011), 13455 [pii].

[28] Nuesch, J. P., Lacroix, J., Marchini, A., Rommelaere, J.: Molecular pathways: rodent parvoviruses-mechanisms of oncolysis and prospects for clinical cancer treatment. Clin. Cancer Res. 18, 3516-3523 (2012). doi: 10. 1158/1078-0432. CCR-11-2325.

[29] Cotmore, S. F., Tattersall, P.: Parvoviral host range and cell entry mechanisms. Adv. Virus Res. 70, 183-232 (2007). doi: 10. 1016/S0065-3527 (07) 70005-2, S0065-3527 (07) 70005-2 [pii].

[30] Moehler, M. H., et al.: Parvovirus H-1-induced tumor cell death enhances human immune response in vitro via increased phagocytosis, maturation, and cross-presentation by dendritic cells. Hum. Gene Ther. 16, 996-1005 (2005). doi: 10. 1089/hum. 2005. 16. 996.

[31] Obeid, M., et al.: Calreticulin exposure dictates the immunogenicity of cancer cell death. Nat. Med. 13, 54-61 (2007). doi: 10. 1038/nm1523, nm1523 [pii].

[32] Michaud, M., et al.: Autophagy-dependent anticancer immune responses induced by chemotherapeutic agents in mice. Science 334, 1573-1577 (2011). doi: 10. 1126/science. 1208347.

[33] Kepp, O., et al.: Molecular determinants of immuno-genic cell death elicited by anticancer chemotherapy. Cancer Metastasis Rev. 30, 61-69 (2011). doi: 10. 1007/ s10555-011-9273-4.

[34] Zitvogel, L., Kepp, O., Galluzzi, L., Kroemer, G.: Inflammasomes in carcinogenesis and anticancer immune responses. Nat. Immunol. 13, 343-351 (2012). doi: 10. 1038/ni. 2224.

[35] Moehler, M., et al.: Oncolytic parvovirus H1 induces release of heat-shock protein HSP72 in susceptible human tumor cells but may not affect primary immune cells. Cancer Gene Ther. 10, 477-480 (2003). doi: 10. 1038/sj. cgt. 7700591.

[36] Moehler, M., et al.: Activation of the human immune system by chemotherapeutic or targeted agents combined with the oncolytic parvovirus H-1. BMC Cancer 11, 464 (2011). doi: 10. 1186/1471-2407-11-464.

[37] Bhat, R., Dempe, S., Dinsart, C., Rommelaere, J.: Enhancement of NK cell antitumor responses using an oncolytic parvovirus. Int. J. Cancer 128, 908-919 (2011). doi: 10. 1002/ijc. 25415.

[38] Morales, O., et al.: Activation of a helper and not regulatory human CD4+T cell response by oncolytic H-1 parvovirus. PLoS One 7, e32197 (2012). doi: 10. 1371/ journal. pone. 0032197PONE-D-12-00550, [pii].

[39] Raykov, Z., et al.: B1 lymphocytes and myeloid dendritic cells in lymphoid organs are preferential extra-tumoral sites of parvovirus minute virus of mice prototype strain expression. J. Virol. 79, 3517-3524 (2005). doi: 10. 1128/JVI. 79. 6. 3517-3524. 2005, 79/6/3517 [pii].

[40] Lang, S. I., Giese, N. A., Rommelaere, J., Dinsart, C., Cornelis, J. J.: Humoral immune responses against minute virus of mice vectors. J. Gene Med. 8, 1141-1150 (2006). doi: 10. 1002/jgm. 940.

[41] Olijslagers, S., et al.: Potentiation of a recombinant oncolytic parvovirus by expression of Apoptin. Cancer Gene Ther. 8, 958-965 (2001). doi: 10. 1038/ sj. cgt. 7700392.

[42] Giese, N. A., et al.: Suppression of metastatic hemangiosarcoma by a parvovirus MVMp vector transducingthe IP-10 chemokine into immunocompetent mice. Cancer Gene Ther. 9, 432-442 (2002). doi: 10. 1038/sj. cgt. 7700457.

[43] Wetzel, K., et al.: MCP-3 (CCL7) delivered by parvo-virus MVMp reduces tumorigenicity of mouse melanoma cells through activation of T lymphocytes and NK cells. Int. J. Cancer 120, 1364-1371 (2007). doi: 10. 1002/ijc. 22421.

[44] Haag, A., et al.: Highly efficient transduction and expression of cytokine genes in human tumor cells by means of autonomous parvovirus vectors; generation of antitumor responses in recipient mice. Hum. Gene Ther. 11, 597-609 (2000). doi: 10. 1089/10430340050015789.

[45] Enderlin, M., et al.: TNF-alpha and the IFN-gamma-inducible protein 10 (IP-10/CXCL-10) delivered by parvoviral vectors act in synergy to induce antitumor effects in mouse glioblastoma. Cancer Gene Ther. 16, 149-160 (2009). doi: 10. 1038/cgt. 2008. 62, cgt200862 [pii].

[46] Krieg, A. M.: Development of TLR9 agonists for cancer therapy. J. Clin. Invest. 117, 1184-1194 (2007). doi: 10. 1172/JCI31414.

[47] Karlin, S., Doerfler, W., Cardon, L. R.: Why is CpG suppressed in the genomes of virtually all small eukaryotic viruses but not in those of large eukaryotic viruses? J. Virol. 68, 2889-2897 (1994).

[48] Raykov, Z., Grekova, S., Leuchs, B., Aprahamian, M., Rommelaere, J.: Arming parvoviruses with CpG motifs to improve their onco-suppressive capacity. Int. J. Cancer122, 2880-2884 (2008). doi: 10. 1002/ijc. 23472.

[49] Cerullo, V., et al.: An oncolytic adenovirus enhanced for toll-like receptor 9 stimulation increases antitumor immune responses and tumor clearance. Mol. Ther. 20, 2076-2086 (2012). doi: 10. 1038/mt. 2012. 137, mt2012137 [pii].

[50] Grekova, S. P., et al.: Interferon gamma improves the vaccination potential of oncolytic parvovirus H-1PV for the treatment of peritoneal carcinomatosis in pancreatic cancer. Cancer Biol. Ther. 12, 888-895 (2011). doi: 10. 4161/cbt. 12. 10. 17678, 17678 [pii].

[51] Geletneky, K., et al.: Phase I/IIa study of intratumoral/ intracerebral or intravenous/intracerebral administration of Parvovirus H-1 (ParvOryx) in patients with progressive primary or recurrent glioblastoma multiforme: ParvOryx01 protocol. BMC Cancer 12, 99 (2012). doi: 1 0. 1186/1471-2407-12-99.

（马维民译）

第 26 章　用病毒溶瘤物致敏树突状细胞

Philippe Fournier 和 Volker Schirrmacher[①]

摘要

多形性成胶质细胞瘤（GBM）是一种侵袭性中枢神经系统（CNS）恶性肿瘤。尽管使用替莫唑胺（TMZ）通过放射治疗和辅助化疗改善了手术管理和治疗，但这种类型的癌症预后较差，平均生存期不到 15 个月。利用树突状细胞（DC）为基础的主动免疫接种，来增强胶质瘤患者在其他方面受损的抗肿瘤免疫反应，已受到越来越多的关注。早期临床试验证明了外源性加载多种抗原的自体树突状细胞的安全性和免疫原性。然而不幸的是，这种树突状细胞疫苗接种很少能转化为强的临床疗效。主要原因似乎是诱导的强大细胞抗肿瘤免疫反应以抵消胶质瘤诱导的免疫抑制存在局限性。

这可能是由于：①所用树突状细胞不成熟/中间状态可能诱导耐受性；②缺乏危险信号和 1 型极化信号，这创造了树突状细胞向 T 细胞递呈肿瘤抗原的适当环境。在本章中，我们介绍了一种新的 DC 疫苗，它具有降低耐受性和增强免疫原性的潜力。改进的疫苗（NDV-DC）将树突状细胞与自体肿瘤细胞以及由新城疫病毒（NDV）溶瘤株感染提供的危险信号相结合。这种方法的独特性与 NDV 作为禽类副黏病毒的 3 个重要特性有关：肿瘤选择性复制、溶瘤潜力和免疫刺激能力。胶质细胞瘤患者术后将用 NDV-DC 疫苗通过皮内进行免疫接种，该疫苗由体外培养的来自患者的树突状细胞组成，树突状细胞载有来自 NDV 感染的自体肿瘤细胞的病毒肿瘤溶解剂。

当这种肿瘤抗原递呈树突状细胞应用于患者时，它们的肿瘤抗原特异性 T 细胞将接收几个激活信号：①肿瘤抗原将提供 T 细胞激活信号 1；②同时 NDV-DC 将提供所谓的危险信号，其对于信号 2（T 细胞共刺激）和信号 3（T 细胞向 Th1 极化）的诱导具有重要意义。与经典 DC 疫苗相比，NDV-DC 疫苗提供了诱导 $CD4^+$和 $CD8^+$T 细胞介导的免疫应答所需改善的 T 细胞激活信号，从而刺激强大的系统性杀瘤 T 细胞免疫。

通过 NDV 诱导 IFN-α 和 IFN-β 的能力，NDV 提供了先天免疫系统和获得性免疫系统之间的联系，从而进一步增强抗癌免疫反应。NDV 诱导的分子危险信号（病毒 RNA 和 HN 蛋白）驱动树突状细胞成为 Th1 导向的免疫原性抗原呈递细胞，能够克服免疫抑制和耐受机制。溶瘤 NDV 与患者肿瘤细胞和自体树突状细胞的这种组合将首先对胶质细胞瘤进行评估，但实际上适用于所有类型的癌症。

26.1　引言

多形性成胶质细胞瘤（GBM）是成人最常见的原发性恶性脑肿瘤，是一种极具挑战性的癌症。目前的护理标准包括最安全的手术切除，然后用替莫唑胺（TMZ）进行放疗和化疗。尽管

---

① P. Fournier，博士（德国海德堡癌症研究中心（DKFZ））

V. Schirrmacher，博士（✉）（德国海德堡癌症研究中心（DKFZ），IOZK 科隆，科隆，德国，E-mail：v. schirrmacher@ dkfz. de）

如此，复发还是很常见。这解释了寻找更有效的治疗方法来更好预防复发的原因。

与上述针对肿瘤细胞的标准治疗相比，免疫学涉及宿主对抗肿瘤，旨在激活患者自身的免疫系统，通过攻击和排斥恶性肿瘤细胞来帮助消除癌症。T细胞介导的抗肿瘤免疫涉及持久的记忆应答和抗肿瘤免疫监测机制，以防止肿瘤复发。近年来，这方面研究取得了很大进展。2010年FDA批准了一种来自美国Provenge（由Denderon公司销售）基于树突状细胞的疫苗，来治疗难治性前列腺癌。第一种针对非病毒性癌症的疫苗[1-3]为T细胞免疫疗法打开了大门，加入标准癌症疗法的武器库。

虽然对中枢神经系统（CNS）肿瘤进行早期树突状细胞免疫治疗试验显示了令人鼓舞的结果，但人们认为，使用树突状细胞对于足够的抗肿瘤免疫反应是次优的，只可以在原位调节有临床意义的变化。这可能主要是由于多种机制的存在，这些机制导致胶质瘤肿瘤的高度免疫抑制微环境[4,5]。

我们首先介绍了多形性成胶质细胞瘤的特点，它是神经胶质瘤中最常见和最具侵袭性的肿瘤之一。我们还综述了这类肿瘤的诊断进展以及作为金标准的经典治疗方法。

然后我们将介绍中枢神经系统及其肿瘤免疫生物学的最新进展。这些知识对于改善目前利用免疫系统治疗脑肿瘤的治疗策略至关重要。

然后我们将提出一种新的抗癌疫苗（NDV-DC），基于树突状细胞、肿瘤细胞和病毒的组合，旨在诱导神经胶质瘤特异性Th1反应和细胞毒性T淋巴细胞（CTLs）。所选择的病毒是一种禽类副黏病毒，具有特别有趣的特性：肿瘤选择性复制、溶瘤能力和免疫刺激特性。

最后，我们将讨论这种新型免疫治疗策略的利弊。这将在成胶质细胞瘤患者的临床I/II期研究中进行评估，该研究正在计划中。

26.2　多形性成胶质细胞瘤

原发性脑肿瘤总年发病率为5/100 000~6/100 000[6]，是指任何涉及脑实质或周围结构（如脑膜、垂体或松果体）的颅内肿瘤。在这些肿瘤中，我们可以区分非胶质瘤（脑膜瘤、垂体瘤和髓母细胞瘤）和胶质瘤。这些是由非神经胶质细胞引起的肿瘤，其功能是维持体内平衡和产生髓磷脂，从而为神经元提供支持和保护。胶质细胞由星形胶质细胞、少突胶质细胞和室管膜组成，在癌变发生时可分别形成星形细胞瘤、少突胶质细胞瘤和室管膜瘤[7]。这些占所有原发性脑肿瘤的36%和恶性脑肿瘤的80%[8]。

在星形胶质瘤中，GBM是最常见的脑瘤，占所有胶质瘤的大多数[9]。占所有颅内肿瘤的12%~15%，占星形细胞肿瘤的60%~75%。世界卫生组织（WHO）提出的分级方案区分了4种不同级别的胶质瘤，其中GBM被归类为世卫组织4级，是所有胶质瘤中最恶性的变体[11]。这种分级与细胞恶性、有丝分裂活跃、易坏死的肿瘤相关，且疾病发展迅速[12,13]。GBM最常见于成人，通常出现在脑叶白质或大脑深灰质其特点是弥漫性渗透生长[10,14]。

这种恶性肿瘤发展涉及一个多方面过程，导致遗传或表观遗传基因控制丧失，细胞生长不受调控，免疫耐受丧失。它由高度增殖和异常迁移的肿瘤细胞组成：分化较差的圆形或多形性细胞，偶尔多核，核不典型，具有多种获得性遗传改变。原发性肿瘤的病因尚不清楚，危险因素也很难确定。此外，GBM肿瘤细胞起源仍然是个谜。

它可能从胶质细胞发展为新生细胞，通常临床病史小于6个月，最常见于老年患者（原发性GBM），也可能是由低级别胶质瘤（世卫组织II级和III级）从先前存在的低级别星形细胞瘤恶性进展的结果，主要影响年轻患者（继发性GBM）。

与其他恶性肿瘤一样，GBMs由肿瘤细胞和非肿瘤细胞的异质混合组成，包括原生细胞和募集细胞。GBM干细胞（GBMSCs）可定义为一种缓慢循环但高度致瘤的GBM细胞亚群，能够选择性地无限期自我更新和分化为大量无致瘤性癌细胞。这两个基本特性使GBMSC人群成为肿瘤维持和复发的主要候选者。它们也与化疗和放射耐药性的发展以及肿瘤复发有关[15-20]。这些肿瘤的扩散性使得目前的治疗干预措施在根除所有残留的颅内肿瘤库方面效果不佳，几乎导致普遍的肿瘤复发，反过来又导致这种疾病的致命性。

26.3 多形性成胶质细胞瘤的诊断和经典疗法

GBM 患者的症状取决于肿瘤的位置、大小和生长速度。它们可以立即出现，但在某些情况下，只有当肿瘤达到相当大的尺寸时才会出现。包括不同强度的头痛；癫痫、恶心和呕吐；偏瘫或身体一侧瘫痪；运动无力和感觉丧失；头盖骨内压力增加的感觉；认知障碍；情绪、个性、注意力和精神能力的变化；以及失明或失语症。神经胶质瘤的存在主要通过非侵入性方法诊断，如磁共振成像（MRI）和磁共振波谱（MRS）或基于计算机断层扫描（计算机断层扫描（CT）、单光子发射计算机断层扫描/X 射线计算机断层扫描（SPECT/CT）和正电子发射断层扫描（PET））方法。这些技术可以对患者体内组织学变化进行成像，检测分子功能变化。因为此时也可能被误诊为良性脑损伤，通常需要对肿瘤活检样本进行组织学分析，以确定肿瘤的确切类型和等级。原发性 GBM 可通过坏死、微血管增生和可能的血栓形成从组织病理学上区别于间变性星形细胞瘤[21]。

从 20 世纪 70 年代末开始，增加放射疗法治疗 GBMs 使患者存活率首次显著提高。最近 Stupp 等已经表明[22]，化疗剂 TMZ 加入可以增加 GBM 患者存活率。尽管外科技术、影像学和辅助放疗或化疗取得了巨大进步，恶性神经胶质肿瘤患者预后仍然很差。目前对高级别胶质瘤患者的护理标准是最大限度的手术切除、联合放射治疗以及伴随和辅助 TMZ 治疗[22]。这种治疗的中位生存期为 14.6 个月，2 年生存率为 27.2%，5 年生存率为 9.8%[22]。主要原因是 GBMs 高度侵入性，这使得几乎不可能完全切除，导致高复发率和低总生存率。

根据世卫组织的说法，有 5 个主要的原因解释了对侵袭性治疗的抵抗力。在对抗 GBMs 方面，最重要的问题之一是巨大的基因型和表型异质性。在临床上这意味着遗传异质性是获得性耐药的一个主要原因，只有极少数癌细胞亚群的存活仍可能导致肿瘤的再生或转移。另一个导致细胞异质性的重要因素是高度未分化细胞以及癌干细胞的存在，这些细胞很可能具有抵抗机制。此外，恶性细胞具有高度侵袭性，使它们能够渗透到健康组织中，最远可达距离大体积肿瘤几厘米。另一个导致对不同治疗反应不佳的障碍是血脑屏障（BBB），众所周知，血脑屏障可以大大减少药物进入大脑的机会。最后但并非最不重要的是，DNA 修复系统的退出取消了化疗和放疗的有效性[12]。

恶性脑肿瘤成功治疗方式发展将取决于设计一种手段来消除原发肿瘤肿块手术切除后遗留的所有颅内肿瘤灶。这是一个艰巨的任务，由于高度传播的疾病过程，我们目前无法充分可视化和治疗每一个剩余肿瘤细胞。然而旨在利用患者免疫系统对抗颅内肿瘤的治疗方法有望实现这一目标。免疫治疗方法旨在加强免疫系统的武器，这已受到肿瘤诱导抑制因素的影响，似乎非常有希望。可以想象，如果有效，与治疗相关的免疫介导的神经胶质瘤治疗策略，应该能成功地克服与肿瘤相关的免疫抑制，使免疫系统能够充分识别和清除手术切除原发肿瘤后大脑内所有残留肿瘤病灶。GBM 不良结果促使人们寻求新的实验疗法，以解决和克服与这种恶性肿瘤致命性相关的基本生物学现象。利用免疫疗法，提高胶质瘤患者抗肿瘤免疫反应受到越来越多的关注。产生针对癌症免疫反应的主要优点之一是免疫效应细胞能够区分肿瘤和非肿瘤细胞并破坏肿瘤细胞。这些可能位于传统手术、放疗或化疗药物无法到达的地方。

26.4 大脑免疫学特性及对癌症免疫治疗的意义

大脑是由细胞组成的器官，对外来物质的毒性反应极为敏感。有许多因素限制中枢神经系统炎症。血脑屏障（BBB）是毛细血管内皮细胞的一种特殊结构，通过它们之间紧密连接的选择性来保护大脑[23]。中枢神经系统具有一些独特特征，多年来一直被认为是免疫豁免部位[24]，与免疫系统没有直接联系。这是由于缺乏淋巴管[25]、血脑屏障存在以及中枢神经系统内缺乏常驻专门抗原呈递细胞（APC）[26,27]。然而目前的数据表明，大脑不是一个免疫无活性的器官，并且可能发生神经炎症反应。

脑脊液（CSF）经魏-罗二氏隙（血管周隙）经血管周围鞘和鼻黏膜下引流至颈深淋巴管[28-31]。现在认识到，虽然在中枢神经系统内未发现幼稚 T 细胞，但在病理条件下，如炎症或肿瘤生长，淋巴细胞可能浸润大脑，并可能引起全身免疫反应[32,33]。此外，小胶质细胞、巨噬细胞以及血管内皮细胞、平滑肌细胞、星形胶质细胞、血管周围巨噬细胞、脉络丛上皮细胞、神经元和树突状细胞被证明是中枢神经系统内的常驻 APCs[34]。小胶质细胞作为中枢神经系统 APCs，具有巨噬细胞和树突状细胞的表型和功能特征[35,36]，表达Ⅱ类抗原和 T 细胞共刺激分子[36-38]，并具有抗原（Ag）递呈能力[39]。DCs 存在于脉络丛[40]和脑膜[41]中，但不存在于实质中。所有这些观察结果都解释了中枢神经系统内的抗原通过这些途径进入颈淋巴结（CLN），可以诱导免疫应答。但这种免疫激活的特点是向非肿瘤 Th2 表型倾斜[30,42,43]，导致强烈的抗体反应和刺激细胞毒性 T 细胞，但没有延迟型超敏反应（DTH）。

在胶质瘤组织中，免疫系统是细胞多样化的驱动力，引起遗传和表观遗传变化的连续积累。这意味着神经胶质瘤细胞产生新的蛋白质和过度表达的分子，这些分子通常只在低水平上存在[44]。肿瘤异质性的增加也包括抗原异质性。虽然免疫系统可能施加压力以消除某些胶质瘤细胞，但其他肿瘤细胞可能通过修饰来逃避免疫反应。这种“免疫编辑”导致肿瘤发展出各种逃避免疫攻击的策略。

恶性胶质瘤表现出局部免疫抑制特征（图 26.1；［31，45-48］）。肿瘤微环境中某些类型的免疫细胞似乎积极参与肿瘤的发展和进程。

除了观察到大脑不包含或仅包含少量自然杀伤细胞、显示出极少量的 MHC-1 类分子以及缺乏常规淋巴管之外[49,50]，大脑肿瘤细胞发展出许多免疫逃逸和免疫抑制的被动和主动机制（图 26.1）。例如，它们释放/表达几种具有免疫抑制特性分子，如转化生长因子 β（TGF-β）、白细胞介素 10（IL-10）和血管内皮生长因子（VEGF）。众所周知，TGF-β 抑制 T 细胞中 INF-γ 和 Fas 配体（FasL）的产生，这是细胞活化所必需的[51,52]。此外，已知 TGF-β 下调 NKs 和细胞毒性 T 细胞上的激活受体。另一种机制在于几种免疫抑制细胞类型的募集或转化，如调节性 T 细胞（Tregs）和髓源性抑制细胞（MDSC）。这导致对 T 细胞、DCs 和 NK 的抑制[53]。

这种肿瘤微环境通过积极预防对肿瘤表达“外来抗原”的免疫反应来维持肿瘤内的免疫耐受性[54,55]。这有助于肿瘤部位存在的 T 细胞衰竭，并诱导对肿瘤细胞的免疫耐受。

癌症免疫疗法旨在指导免疫系统正确排斥肿瘤，为肿瘤切除后清除所有残留胶质瘤细胞提供了前提。已经表明，在鼠胶质瘤模型中[56-58]，肿瘤特异性 T 细胞可以在 CLNs 中激活。由于中枢神经系统和中枢神经系统肿瘤的免疫微环境对于有效抗肿瘤免疫反应在原位介导具有临床意义的变化似乎不太理想[4,5]，因此使用基于在大脑外产生针对 GBM 的强细胞免疫反应的免疫治疗策略，例如，在外周皮肤，似乎是一个好的策略。已经表明，树突状细胞在癌症患者中表现出主要的功能障碍。当从癌症患者中分离时，树突状细胞在刺激免疫反应方面的效果明显低于从健康供体中分离的，正如乳腺癌所显示的那样[59,60]。

为了克服患者树突状细胞的低效率，体外制备和激活树突状细胞似乎是合适的。为此，最有希望的免疫治疗方法之一，是对 GBM 患者用肿瘤抗原结合微生物危险信号在体外致敏树突状细胞，进行外周接种。

26.5 NDV-DC 疫苗

用 NDV-DC 疫苗进行治疗包括皮内接种树突状细胞，树突状细胞是从癌症患者的血细胞体

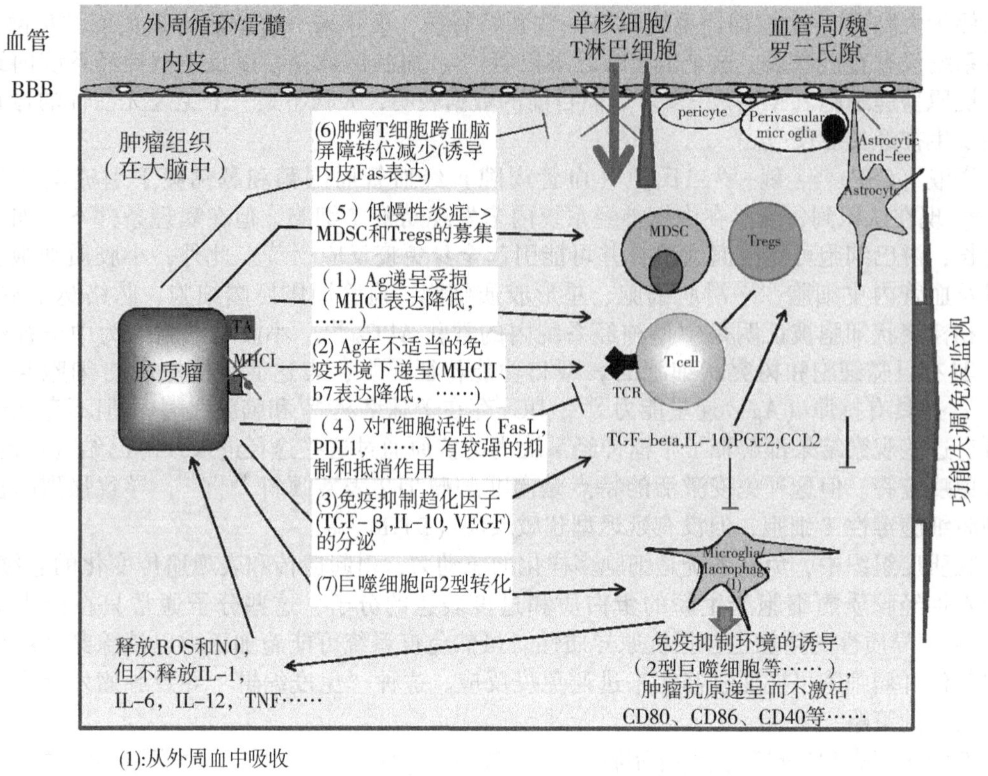

图 26.1 胶质瘤和免疫抑制-胶质瘤利用复杂的免疫抑制机制来抑制内源性免疫识别和清除恶性胶质瘤，避免免疫系统通过多种途径识别和清除

其中一些如图所示。(1) 由于缺乏抗原递呈，通过显著降低 MHC Ⅰ的表达，T 细胞介导的反应得以逃避。(2) MHCII 和 B7 表达下降（共刺激信号）损害了免疫反应的诱导。(3) 肿瘤可通过免疫抑制细胞因子及其他因子（包括 TGF-β、IL-10、VEGF）和免疫抑制细胞（如 $CD4^+CD25^+$ Tregs 和 MDSCs）抑制 T 细胞活性。(4) 肿瘤也可通过上调 FasL 和程序性细胞死亡 1 配体 1（PDL-1）表达来抑制或抵消 T 细胞活性。(5) 低慢性肿瘤内炎症导致 MDSC 和 Tregs 的募集。(6) 胶质瘤可以减少魏-罗二氏隙内单核细胞和肿瘤性 T 细胞跨 BBB 的易位（即周细胞、血管周围小胶质细胞和星形胶质细胞）。(7) 最终吸收的巨噬细胞将转化为 2 型巨噬细胞，通过向 T 细胞提供肿瘤抗原而不激活 CD80、CD86 和 CD40 等信号，促进肿瘤进展。

外产生的，并通过与患者肿瘤的病毒溶瘤剂孵育而被激活（图 26.2）。

目的是通过激活来自针对肿瘤细胞 T 细胞库的 T 细胞，并通过重新刺激潜在的肿瘤特异性记忆 T 细胞，在外周诱导新的抗肿瘤免疫应答，即使这些细胞可能部分消失。由于胶质瘤诱导局部免疫抑制和全身免疫抑制，树突状细胞适当的体外制备、信息化和激活是治疗效果的关键。为了抵消神经胶质瘤诱导的免疫调节系统效应，癌症患者可以在 NDV-DC 免疫疗法之前接受系统的 NDV 注射。由于强烈的 I 型干扰素应答激活抵消了系统肿瘤诱导的 Th2 极化效应，这可能对免疫系统有调节作用。

有效 T 细胞免疫所需信号首先是抗原递呈和 T 细胞相互作用（“信号 1”），这决定了 T 细胞应答的特异性。共刺激和共刺激分子介导的扩增需要进一步的信号（“信号 2”），这决定了所选肿瘤特异性 T 细胞反应大小[61-63]。对树突状细胞诱导有效的癌症免疫至关重要的附加特征印记是强制性的。这导致极化“信号 3”（决定 $CD4^+$和 $CD8^+$T 细胞中[64]的效应器功能（反应性或抑制性）和免疫类型（Th1 或 Th2）），从而选择性地增强 Th1-、CTL-和 NK 细胞介导的 1 型免疫。

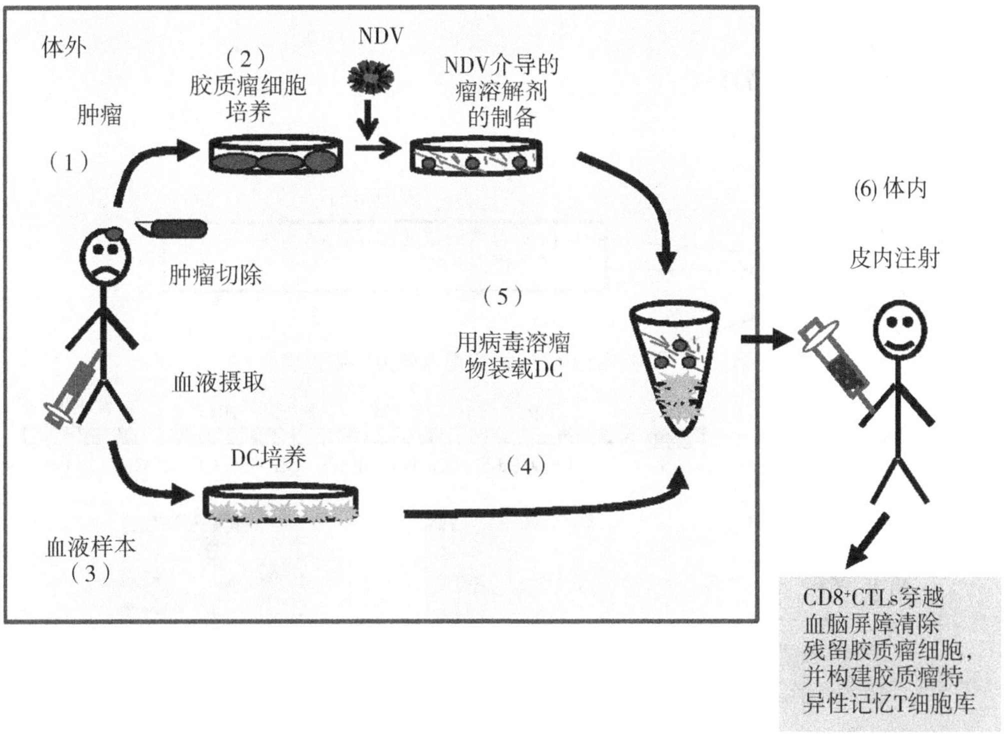

图 26.2　基于患者源性的来自 NDV 感染的自体肿瘤细胞裂解物致敏 DCs 的 NDV-DC 疗法示意图

（1）为了制备肿瘤溶解剂，从新手术的肿瘤样品中分离肿瘤细胞。（2）肿瘤细胞随后在体外生长。当细胞呈指数增长时，NDV 感染是由病毒对肿瘤细胞的 1-h 吸附引起的。然后将细胞留在培养物中直到溶解。随后，肿瘤溶解剂作为冷冻的小份储存。（3）每次接种时，从血样中获得单核细胞。（4）这些细胞用于在 IL-4 和 GM-CSF 存在下产生树突状细胞。（5）树突状细胞随后装载病毒肿瘤溶解剂。（6）然后通过皮内注射将激活的树突状细胞注射到患者体内。

为此，NDV-DC 疫苗的理论基础如下。

（1）使用自体 DCs。自体 DCs 是最有效的 APCs，在 T 细胞介导免疫应答启动和调节中起核心作用。可以在体外通过使用重组细胞因子分化外周血祖细胞而获得[65]。

（2）使用自体肿瘤细胞。自体肿瘤细胞使得疫苗肿瘤抗原与病人的肿瘤抗原能够紧密匹配。这些抗原包括常见肿瘤抗原和个别独特的抗原，这些抗原来源于肿瘤细胞表达的许多不同蛋白质中体细胞点突变[66]。后者是唯一不被正常组织表达的肿瘤特异性抗原。独特（或突变）抗原被假定为免疫治疗的几个潜在优势。

（a）独特 T 细胞库不应被耐受或删除，应被免疫系统识别为非自身细胞，就像病毒 Ags 情况一样[67]。

（b）在突变蛋白对细胞存活至关重要的情况下，对阴性选择的潜在抗性[67]。

最近对乳腺肿瘤基因分析揭示了它们的个体特征[68]。肿瘤抗原表达也存在异质性。正如用于制备致敏 DCs 的病毒溶瘤剂，100 万个肿瘤细胞可能比在其他疫苗试验中使用的明确定义的普通肿瘤抗原更能代表这种肿瘤抗原异质性[69,70]。从这些观察中得出的逻辑推断是，一种用患者来源的肿瘤材料对 DCs 进行指导（“启动”）的个性化方法。

（3）使用新城疫病毒。这是一种禽类 RNA 副黏病毒见图 26.3a)。在人体中会诱导一种强烈的 I 型干扰素反应，阻止任何病毒在正常细胞中复制。在肿瘤细胞中，NDV 感染诱导较弱的干扰素应答，因为这些细胞的 I 型干扰素应答存在各种缺陷见图 26.3b；[72，73]）。由于这些缺陷，NDV 在肿瘤细胞中复制没有受到抑制[74]。这就解释了该病毒的 3 个特性，这些特性可以赋予该病毒[75]。

（a）肿瘤选择性复制；

（b）某些菌株的溶瘤特性；

（c）免疫刺激性质。

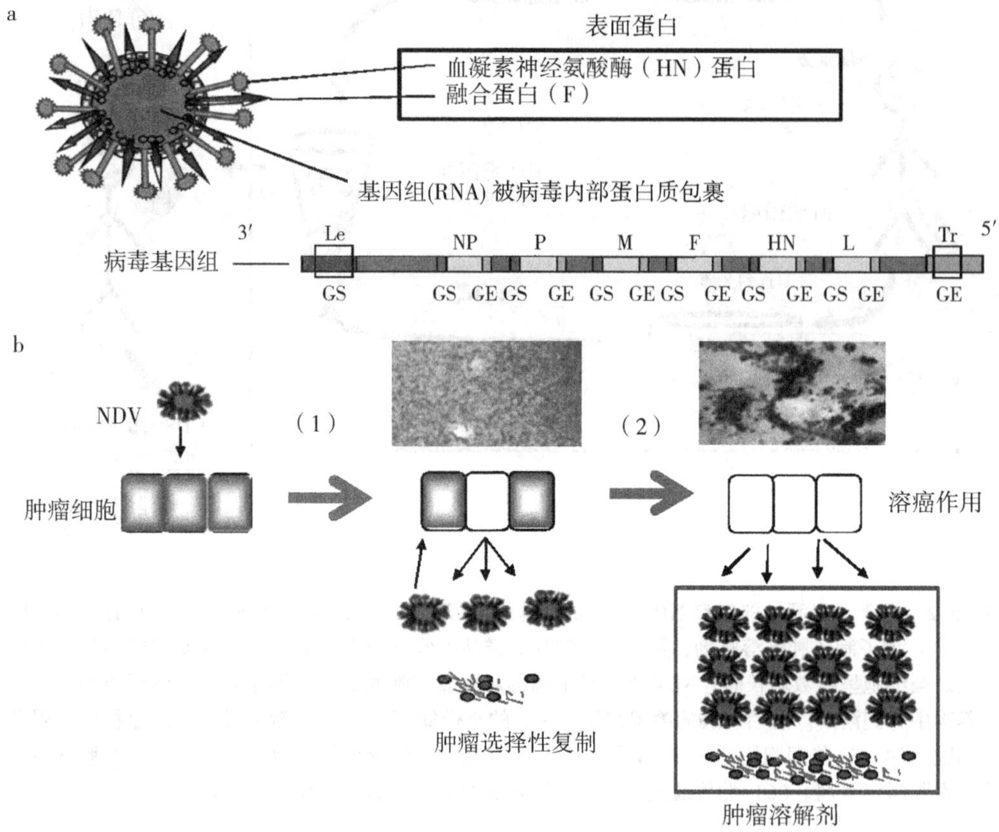

图 26.3 NDV——一种通过肿瘤选择性复制诱导肿瘤分解的禽副黏病毒

（a）病毒结构：NDV 是一种直径为 150~300 nm 的禽副黏病毒，其包膜含有病毒蛋白，核衣壳含有负链 RNA 基因组。15kb 非节段基因组在每个末端都有一个前导序列（*Le*）和一个尾部序列（*Tr*），允许结合和去除细胞核糖体。6 个病毒基因（从 3’ 末端到 5’ 末端：核蛋白（NP）、磷蛋白（P）、基质蛋白（M）、融合蛋白（F）、血凝素神经氨酸酶蛋白（HN）和大蛋白（L））的转录从基因起始（GS）序列开始，终止于基因结束（GE）序列。（b）肿瘤细胞中 NDV 的复制：（1）第一轮结合/复制：病毒周期由两个步骤组成：结合和复制。（i）结合：第一种涉及病毒通过 HN 分子的凝集素样细胞结合结构域与表达独特糖类侧链的普遍表达的宿主细胞表面受体结合[71]。随后是融合蛋白 F 激活。HN 和融合蛋白 F 协同作用导致病毒和宿主细胞膜融合。这种膜融合允许病毒基因组进入宿主细胞细胞质。在那里，负链 RNA 基因组被转录成 mRNAs，翻译成病毒蛋白。核衣壳组装需要蛋白质 NP、P 和 L。（ii）复制（第二步）：核衣壳作为“反基因组”，然后被用作病毒复制的模板。在翻译后修饰后，M 蛋白和包膜蛋白 HN 和 F 转移到发生病毒组装和出芽的膜上。在这个过程中，NDV 基因组单个拷贝被包裹在由宿主细胞质膜制成的外壳中。（2）一连串的病毒结合和复制产生肿瘤溶解产物（包含病毒颗粒和肿瘤细胞片段）。所用 NDV 株毒性很强，在它们的 F 蛋白中有一个弗林蛋白酶切割位点，允许它在蛋白水解培养基中活化。然后可以通过感染相邻细胞，并随后通过相同机制（多周期复制）破坏来重复周期结合/复制。病毒复制的这一特性提供了病毒的产生，只有当所有的肿瘤细胞都被破坏时，病毒才会停止产生。裂解性 NDV 毒株杀灭潜力是显著的。这种毒株已被证明具有很高的杀死肿瘤细胞能力。一个传染性粒子在体外导致大约 10 000个癌细胞在 2~3 天内死亡。

NDV 感染时肿瘤细胞中危险信号的诱导是刺激树突状细胞产生强抗肿瘤效应的关键（图 26.4）。

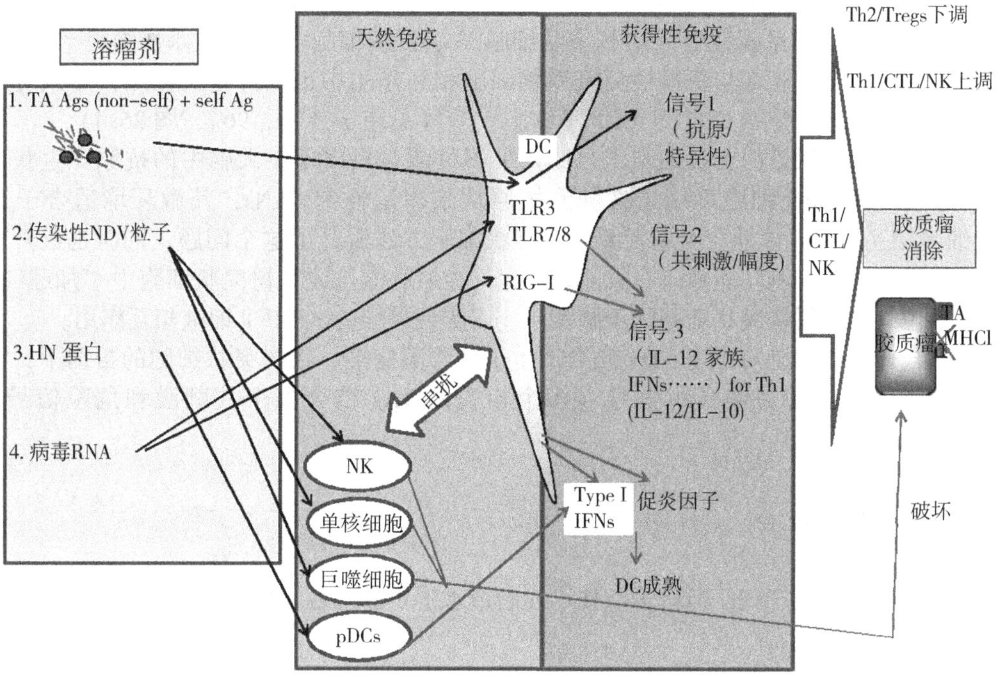

图 26.4 NDV 感染的自体肿瘤细胞溶瘤产物对 DCs 的体外印记用于 Th1-/ CTL-和 NK 细胞介导的 1 型免疫

为了激活 DCs，将 NDV 感染的自体肿瘤细胞溶瘤产物外源性添加到来自患者的 DCs 中。这种肿瘤溶解剂包含：(1) 肿瘤细胞片段［带有肿瘤抗原（*TAs*）和非肿瘤抗原（*self Ags*）］；(2) 传染性 NDV 粒子；(3) NDV 的 HN 蛋白；(4) 病毒 RNA。(1) TAs 被加工成 MHC Ⅰ类通路并在 DCs 表面交叉递呈为 TA 肽 MHC（pMHC）复合物。(2) 肿瘤溶解剂中存在的传染性 NDV 粒子激活自然杀伤细胞[76]、单核细胞[77,78]和巨噬细胞[79]。这些活化细胞有助于破坏肿瘤。NK 和 DCs 之间的串扰[80]导致 1 型 DC 激活。NDV 还刺激 pDCs[81,82]和单核细胞[77]释放对免疫系统有很大影响的 I 型干扰素（IFN）。1 型 IFN 在 CTL 活性产生中起着重要作用，因为在混合淋巴细胞肿瘤细胞培养（MLTC）试验中，这种诱导可以被 1 型 IFN 抗血清特异性阻断[83]。在 CTL 启动过程中在体内观察到类似的效应，表明 IFN-α/IFN-β 不仅增加了 CTL 活性，而且对 CTL 活性的产生也是必不可少的[83]。据报道，1 型 IFN 在细胞中诱导 IL-12 受体-α 链[84]。IFN-α 与 IL-12 一起，使 T 细胞向以 DTH 和 CTL 活性为特征的细胞介导的 Th1 反应极化。此外，IFN-α 诱导分子上调，对于抗原识别［例如人类白细胞抗原（HLA）］和细胞间相互作用（例如细胞黏附分子[85]）很重要。最近观察到 NDV 将 DCs 极化为 DC1 表型[86]。(3) HN 蛋白通过 pDCs 诱导产生 1 型 IFNs[81,82]。(4) 病毒 RNA 通过 TLR3[87]和 RIG-1[88]激活 DCs，有助于 DCs 产生强危险信号。虽然 TAs 激活 DCs 以诱导信号 1（抗原/特异性），但病毒元件（NDV、HN 和 RNA）诱导 DCs 信号 2（共刺激）和信号 3（极化）。这解释了为什么在 DCs 中添加肿瘤溶解剂会赋予它们 3 种信号——信号 1（抗原）、信号 2（共刺激）和信号 3（极化），它们的同时表达是产生强肿瘤特异性 Th1-/CTL-和 NK 细胞介导型免疫所必需的。

NDV 通过在肿瘤细胞中复制，导致病毒血凝素-神经氨酸酶（HN）蛋白产生，这是一种跨膜蛋白，在被感染的肿瘤细胞表面表达[89]。这为肿瘤细胞引入了新的细胞黏附特性，用于淋巴细胞相互作用[90,91]和用于 T 细胞共刺激[92]。HN 蛋白已被证明通过与 pDCs[81,82]的旁分泌相互作用诱导 IFN-α 产生。此外，NDV 对人类肿瘤细胞感染导致 HLA 和 ICAM-1 类分子上调并诱导 IFNs、趋化因子，最终导致细胞凋亡[85]。NDV 对人类肿瘤细胞的感染也导致病毒复制的副产品，如 dsRNA。这些 RNA 衍生物可以激活细胞质 PKR[72]以及细胞质 RNA 依赖的解旋酶 RIG-

1[88]和核内体 TLR3[87]。

所有危险信号都来自 NDV 感染（dsRNA 和 HN 细胞表面蛋白[93]以及 IFN-α[81,82]），允许激活多种先天免疫应答（单核细胞[77]、树突状细胞[94]、巨噬细胞[79]和自然杀伤细胞[76]）。

这些活化的天然细胞在与获得性免疫系统细胞相互作用中也发挥重要作用，导致有效的获得性抗肿瘤免疫反应，包括 $CD4^+$和 $CD8^+T$ 细胞[83,95]（总结于［90，96］，图 26.4）。

由于正常脑实质缺乏树突状细胞，因此几乎不可能对只出现在大脑中的抗原产生获得性免疫反应。没有专业 APC 细胞可用于将肿瘤抗原从脑瘤运输到 CLNs，其微环境适合于最佳的 APC-T 细胞相互作用。体外产生患者源性树突状细胞方法避开了这个问题，同时也避开了与癌症患者内源性树突状细胞功能障碍有关问题。允许控制肿瘤抗原在树突状细胞中“加载”，并且在真皮等部位外周注射树突状细胞优于脑实质，以实现最佳的 APC-T 细胞相互作用。

最后，自体肿瘤细胞与溶瘤 NDV 结合产生了病毒溶瘤剂，在树突状细胞的帮助下，能传递有效抗肿瘤反应所需的关键指示信号（多个相关的 TAs 结合多个共刺激和危险信号）（图 26.5）。

26.6 优点和缺点

尽管这一战略前景广阔，但科学、技术或制造水平仍面临各种挑战[100]。

26.6.1 科学

首先，皮质类固醇的使用（几乎总是脑肿瘤患者标准护理的一部分）被认为会抑制免疫反应。因此，在主动免疫治疗期间禁止使用。免疫接种应该尽早进行，而不是在治疗方案中推迟。此外，从皮质类固醇治疗时分离的血液单核细胞中生产高质量的树突状细胞可能很困难。

其次，肿瘤裂解物致敏树突状细胞诱导自身免疫反应的问题没有得到解决。总的来说，这种治疗耐受性良好。迄今为止，在临床前体内模型中用基于自体肿瘤细胞的前一代 NDV 疫苗和 NDV 治疗的患者中[90,96]，均未观察到自身免疫反应，该疫苗已在成胶质细胞瘤患者中得到积极评价[101]。其原因已在其他地方讨论过[102]。

26.6.2 技术

对 GBM 患者临床研究的挑战包括免疫治疗后获得肿瘤组织相对困难。与其他癌症不同，颅内胶质瘤组织在疫苗治疗后很容易获得。

此外，由于 GBM 很少，大型随机研究通常很难进行。由于需要对患者进行个性化高度选择性治疗，在临床试验中，患者积累速度可能比通常要慢。获得批准的产品其潜在市场更小，而且 FDA 或 EMA 批准的开发时间更长。

此外，用疫苗设计新辅助设置并不总是可行的，因为复发性恶性胶质瘤（临床上需要手术切除）不允许在手术前等待数周，而且常常需要使用大剂量皮质类固醇治疗。

26.6.3 制造

这里介绍治疗的临床发展面临着重大的制造挑战。用肿瘤裂解物致敏树突状细胞需要自体树突状细胞和胶质瘤细胞的培养和耗时程序，而质量控制/质量保证比率并不总是可行的。此外，目前良好生 GMP 要求细胞在转移到人体前需要培养数周，这使得治疗极其费力和昂贵。为了实现治疗个性化，疫苗的生产任务需要严格开发。所有这些要求特别是 GMP，尽管是必要的，但需要大量投资。还需要一个集中制造设施。疫苗作为培养和活化步骤的成品，必须尽快以适当方式储存。此外，这种自体疗法需要自动化生产。标准化也有潜在困难。总的来说，这减缓了转化研究步伐。

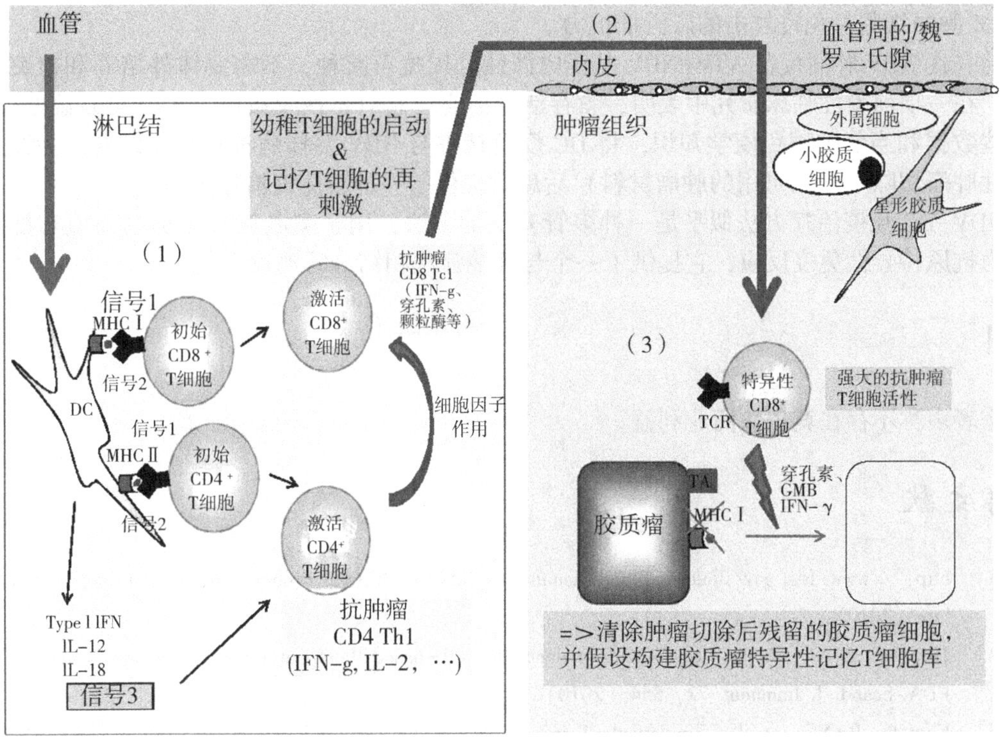

图 26.5 外周淋巴结内 T 细胞外周活化和肿瘤特异性 CD8 T 细胞迁移，用于杀死淋巴结内

(1) 脑内残留的肿瘤细胞，活化 DCs 将 3 个信号 1、2 和 3 传递给 $CD4^+$ 和 $CD8^+$T 细胞。这导致了初始 T 细胞启动和肿瘤特异性记忆 T 细胞再刺激。$CD4^+$T 细胞有助于肿瘤抗原特异性 CTLs 扩增[97]，同时也有助于肿瘤部位巨噬细胞活化[98]和杀伤肿瘤细胞[99]。(2) $CD8^+$T 细胞成为肿瘤特异性 CTLs，并通过外周循环和血管周魏-罗二氏隙（血管周隙）迁移到肿瘤部位。(3) 在大脑中，肿瘤内 $CD8^+$T 细胞通过穿孔素和颗粒酶 B（GMB）杀死残留肿瘤细胞，并通过局部产生 IFN-γ 促进抗肿瘤活性。

随着 DC 疫苗效率障碍消除和疫苗标准化推进，NDV-DC 似乎有可能成为 GBM 患者的可行选择。

结论

多形性成胶质细胞瘤是一种预后极差的脑癌。具有充分治疗性抗肿瘤免疫反应的关键障碍是胶质瘤诱导的免疫抑制。然而免疫疗法似乎是新治疗干预的未来选择。随着技术进步和新颖、个性化、合理的治疗，患者存活率和生活质量有望得到提高。

体外成熟的树突状细胞对抑制因子有明显抵抗力，已成为治疗的选择。NDV-DC 似乎是解决这一挑战非常好的候选疫苗。在该疫苗中，DCs 通过与自体肿瘤细胞和 NDV 结合，合理地接受关键和有效的指令指导，同时传递 3 种信号（信号 1（抗原）、信号 2（共刺激）和信号 3（极化）），3 种信号是激活和产生强抗肿瘤活性所必需的。

未来基于 DC 的免疫疗法需要多管齐下的策略，来刺激对肿瘤抑制环境具有抗性和强有力的抗原特异性免疫反应。NDV 作为具有肿瘤选择性、溶瘤性和强免疫刺激性的多价药物，是一种与 DC 联合使用的优秀药物。这篇综述中的科学数据为 NDV-DC 肿瘤疫苗的原理和设计提供了新见解。这种疫苗作为一种有效 Th1 反应介质，有利于诱导 DC 成熟、释放促炎细胞因子和改善抗原交叉递呈，这对于 $CD8^+$T 细胞介导的免疫反应启动和激活至关重要，从而产生抗肿瘤临床疗效。非常重要的是，NDV溶解剂用于DC活化所提供的机制存在很大多样性，这可以确保克服GBM肿瘤细胞对免疫攻击的广泛抵抗力。

前一代NDV肿瘤疫苗ATV-NDV基于内源性外周疫苗接种，不需要体外培养和树突状细胞激活，在一项前瞻性临床研究中表明，就存活率而言，它对恶性胶质瘤患者有明显益处[101]。基于这些数据和当前肿瘤免疫学知识，将DC疫苗接种与NDV溶瘤病毒疗法相结合，为原发性或继发性肿瘤切除（使用可用的肿瘤材料）后患者提供了一种新免疫治疗策略。

NDV-DC免疫治疗方法似乎是一种多管齐下的方法，用于刺激对肿瘤抑制环境有抵抗力的强有力抗原特异性免疫反应，它提供了一个与其他疗法相结合的视角[103]。

声明

作者声称不存在竞争性商业利益。

参考文献

[1] http://www.fda.gov/BiologicsBloodVaccines/CellularGeneTherapyProducts/ApprovedProducts/ ucm210012.htm (2010).

[2] Hovden, A. O., Appel, S.: The first dendritic cell-based therapeutic cancer vaccine is approved by the FDA. Scand. J. Immunol. 72, 554 (2010).

[3] Kantoff, P. W., et al.: Sipuleucel-T immunotherapy for castration-resistant prostate cancer. N. Engl. J. Med. 363, 411-422 (2010).

[4] Okada, H., et al.: Immunotherapeutic approaches for glioma. Crit. Rev. Immunol. 29, 1-42 (2009).

[5] Walker, P. R., Calzascia, T., de Tribolet, N., Dietrich, P. Y.: T-cell immune responses in the brain and their relevance for cerebral malignancies. Brain Res. Rev. 42, 97-122 (2003).

[6] Central Brain Tumor Registry of the United States (CBTRUS), www.cbtrus.org.

[7] Buckner, J. C., et al.: Central nervous system tumors. Mayo Clin. Proc. 82, 1271-1286 (2007).

[8] Dolecek, T. A., Propp, J. M., Stroup, N. E., Kruchko, C.: CBTRUS Statistical Report: Primary brain and central nervous system tumors diagnosed in the United States in 2005-2009. Neuro-Oncol14, 1-49 (2012).

[9] Porter, K. R., McCarthy, B. J., Freels, S., Kim, Y., Davis, F. G.: Prevalence estimates for primary brain tumors in the United States by Age, Gender, Behavior, and Histology. Neuro-Oncology, 12, 520-527 (2010).

[10] Ohgaki, H., Kleihues, P.: Population-based studies on incidence, survival rates, and genetic alterations in astrocytic and oligodendroglial gliomas. J. Neuropathol. Exp. Neurol. 6, 479-489 (2005).

[11] Kleihues, P., et al.: The WHO classification of tumors of the nervous system. J. Neuropathol. Exp. Neurol. 61, 215-229 (2002).

[12] Louis, D. N., et al.: The 2007 WHO classification of tumors of the central nervous system. ActaNeuropathol. 114, 97-109 (2007).

[13] Wrensch, M., Minn, Y., Chew, T., Bondy, M., Berger, M. S.: Epidemiology of primary brain tumors: current concepts and review of the literature. Neuro Oncol. 4, 278-299 (2002).

[14] Swanson, K. R., Alvord Jr., E. C., Murray, J. D.: Virtual brain tumors (gliomas) enhance the reality of medical imaging and highlight inadequacies of current therapy. Br. J. Cancer 86, 14-18 (2002).

[15] Singh, S. K., et al.: Identification of a cancer stem cell in human brain tumors. Cancer Res. 63, 5821-5828 (2003).

[16] Singh, S. K., et al.: Identification of human brain tumor initiating cells. Nature432, 396-401 (2004).

[17] Sanai, N., Alvarez-Buylla, A., Berger, M. S.: Neural stem cells and the origin of gliomas. N. Engl. J. Med. 353, 811-822 (2005).

[18] Galli, R., Binda, E., Orfanelli, U., Cipelletti, B., Gritti, A., De Vitis, S., Fiocco, R., Foroni, C., Dimeco, F., Vescovi, A.: Isolation and characterization of tumorigenic, stem-like neural precursors from human glioblastoma. Cancer Res. 64, 7011-7021 (2004).

[19] Bao, S., et al.: Glioma stem cells promote radio-resistance by preferential activation of the DNA damage response. Nature 444, 756-760 (2006).

[20] Eramo, A., et al.: Chemotherapy resistance of glioblastoma stem cells. Cell Death Differ. 13, 1238-1241 (2006).

[21] Tehrani, M., Friedman, T. M., Olson, J. J., Brat, D. J.: Intravascular thrombosis in central nervous system malignancies: a potential role in astrocytoma progression to glioblastoma. Brain Pathol. 18, 164-171 (2008).

[22] Stupp, R., European Organisation for Research and Treatment of Cancer Brain Tumor and Radiation Oncology Groups, National Cancer Institute of Canada Clinical Trials Group, et al.: Efficacy of radiotherapy with concomitant and adjuvant temozolomide versus radiotherapy alone on survival in glioblastoma in a randomized phase III study: 5-analysis of the EORTC-NCIC trial. Lancet Oncol. 10, 459-466 (2009).

[23] Abbott, N. J., Rönnbäck, L., Hansson, E.: Astrocyte-endothelial interactions at the blood-brain barrier. Nat. Rev. Neurosci. 7, 41-53 (2006).

[24] Medawar, P. B.: Immunity to homologous grafted skin. III. The fate of skin homo-grafts transplanted to the brain, to subcutaneous tissue, and to the anterior chamber of the eye. Br. J. Exp. Pathol. 29, 58-69 (1948).

[25] Ransohoff, R. M., Kivisakk, P., Kidd, G.: Three or more routes for leukocyte migration into the central nervous system. Nat. Rev. Immunol. 3, 569-581 (2003).

[26] Fabry, Z., Raine, C. S., Hart, M. N.: Nervous tissue as an immune compartment: the dialect of the immune response in the CNS. Immunol. Today15, 218-224 (1994).

[27] Parney, I. F., Hao, C., Petruk, K. C.: Glioma immunology and immunotherapy. Neurosurgery46, 778-792 (2000).

[28] Cserr, H. F., Knopf, P. M.: Cervical lymphatics, the blood-brain barrier and the immunoreactivity of the brain: a new view. Immunol. Today13, 507-512 (1992).

[29] Cserr, H. F., Harling-Berg, C. J., Knopf, P. M.: Drainage of brain extracellular fluid into blood and deep cervical lymph and its immunological significance. Brain Pathol. 2, 269-276 (1992).

[30] Harling-Berg, C. J., Park, T. J., Knopf, P. M.: Role of the cervical lymphatics in the Th2-type hierarchy of CNS immune regulation. J. Neuroimmunol. 101, 111-127 (1999).

[31] Weller, R. O., Engelhardt, B., Phillips, M. J.: Lymphocyte targeting of the central nervous system: a review of afferent and efferent CNS-immune pathways. Brain Pathol. 6, 275-288 (1996).

[32] Perrin, G., et al.: Astrocytoma infiltrating lymphocytes include major T cell clonal expansions confined to the CD8 subset. Int. Immunol. 11, 1337-1349 (1999).

[33] Walker, P. R., et al.: The brain parenchyma is permissive for full antitumor CTL effector function, even in the absence of CD4 T cells. J. Immunol. 165, 3128-3135 (2000).

[34] Gehrmann, J., Banati, R. B., Wiessner, C., Hossmann, K. A., Kreutzberg, G. W.: Reactive microglia in cerebral ischaemia: an early mediator of tissue damage? Neuropathol. Appl. Neurobiol. 21, 277-289 (1995).

[35] Lowe, J., MacLennan, K. A., Powe, D. G., Pound, J. D., Palmer, J. B.: Microglial cells in human brain have phenotypic characteristics related to possible function as dendritic antigen presenting cells. J. Pathol. 159, 143-149 (1989).

[36] Ulvestad, E., et al.: Human microglial cells have phenotypic and functional characteristics in common with both macrophages and dendritic antigen-presenting cells. J. Leukoc. Biol. 56, 732-740 (1994).

[37] Gehrmann, J., Banati, R. B., Kreutzberg, G. W.: Microglia in the immune surveillance of the brain: human microglia constitutively express HLA-DR molecules. J. Neuroimmunol. 48, 189-198 (1993).

[38] Williams Jr., K., Ulvestad, E., Cragg, L., Blain, M., Antel, J. P.: Induction of primary T cell responses by human glial cells. J. Neurosci. Res. 36, 382-390 (1993).

[39] Hickey, W. F., Kimura, H.: Perivascular microglial cells of the CNS are bone marrow-derived and present antigen in vivo. Science239, 290-292 (1988).

[40] S erot, J. M., Foliguet, B., Bene, M. C., Faure, G. C.: Ultrastructural and immuno-histological evidence for dendritic-like cells within human choroid plexus epithelium. Neuroreport8, 1995-1998 (1997).

[41] McMenamin, P. G., Forrester, J. V.: Dendritic cells in the central nervous system and eye and their associated supporting tissues. In: Dendritic cells: biology and clinical applications, pp. 205-248. Academic Press, New York (1999).

[42] Harling-Berg, C. J., Hallett, J. J., Park, J. T., Knopf, P. M.: Hierarchy of immune responses to antigen in the normal brain. Curr. Top. Microbiol. Immunol. 265, 1-22 (2002).

[43] Mosmann, T. R., Coffman, R. L.: TH1 and TH2 cells: different patterns of lymphokine secretion lead to different functional properties. Annu. Rev. Immunol. 7, 145-173 (1989).

[44] P ardoll, D.: Does the immune system see tumors as foreign or self? Annu. Rev. Immunol. 21, 807-839 (2003).

[45] Hussain, S. F., Heimberger, A. B.: Immunotherapy for human glioma: innovative approaches and recent results. Expert Rev. Anticancer Ther. 5, 777-790 (2005).

[46] Bodey, B., Bodey Jr., B., Siegel, S. E., Kaiser, H. E.: Failure of cancer vaccines: the significant limitations of this approach to immunotherapy. Anticancer Res. 20, 2665-2676 (2000).

[47] Pawelec, G., Engel, A., Adibzadeh, M.: Prerequisites for the immunotherapy of cancer. Cancer Immunol. Immunother. 48, 214-217 (1999).

[48] Roszman, T., Elliott, L., Brooks, W.: Modulation of T-cell function by gliomas. Immunol. Today12, 370-374 (1991).

[49] Vauleon, E., Avril, T., Collet, B., Mosser, J., Quillien, V.: Overview of cellular immunotherapy for patients with glioblastoma. Clin. Dev. Immunol. 2010, (2010).

[50] Yamasaki, T., Moritake, K., Klein, G.: Experimental appraisal of the lack of antitumor natural killer cell mediated immuno-surveillance in response to lymphomas growing in the mouse brain. J. Neurosurg. 98, 599-606 (2003).

[51] Platten, M., et al.: Transforming growth factors beta (1) (TGF beta (1)) and TGF beta (2) promote glioma cell migration via upregulation of alpha (V) beta (3) integrin expression. Biochem. Biophys. Res. Commun. 268, 607-611 (2000).

[52] Weller, M., Fontana, A.: The failure of current immunotherapy for malignant glioma. Tumor derived TGF beta, T cell apoptosis, and the immune privilege of the brain. Brain Res. Rev. 21, 128-151 (1995).

[53] Albesiano, E., Han, J. E., Lim, M.: Mechanisms of local immuno-resistance in glioma. Neurosurg. Clin. N. Am. 21, 17-29 (2010).

[54] Gregori, S., Goudy, K. S., Roncarolo, M. G.: The cellular and molecular mechanisms of immuno-suppression by human type 1 regulatory T cells. Front. Immunol. 3, 30 (2012).

[55] Roncarolo, M. G., Bacchetta, R., Bordignon, C., Narula, S., Levings, M. K.: Type 1 T regulatory cells. Immunol. Rev. 182, 68-79 (2001).

[56] Fujita, M., et al.: Effective immunotherapy against murine gliomas using type 1 polarizing dendritic cells-significant roles of CXCL10. Cancer Res. 69, 1587-1595 (2009).

[57] Kuwashima, N., et al.: Delivery of dendritic cells engineered to secrete IFN-α into central nervous system tumors enhances the efficacy of peripheral tumor cell vaccines: dependence on apoptotic pathways. J. Immunol. 175, 2730-2740 (2005).

[58] Okada, N., et al.: Augmentation of the migratory ability of DC-based vaccine into regional lymph nodes by efficient CCR7 gene transduction. Gene Ther. 12, 129-139 (2005).

[59] Gabrilovich, D. I., Corak, J., Ciernik, I. F., Kavanaugh, D., Carbone, D. P.: Decreased antigen presentation by dendritic cells in patients with breast cancer. Clin. Cancer Res. 3, 90-97 (1997).

[60] Satthaporn, S., et al.: Dendritic cells are dysfunctional in patients with operable breast cancer. Cancer Immunol. Immunother. 53, 510-518 (2004).

[61] Banchereau, J., Steinman, R. M.: Dendritic cells and the control of immunity. Nature392, 245-252 (1998).

[62] Schuler, G., Schuler-Thurner, B., Steinman, R. M.: The use of dendritic cells in cancer immunotherapy. Curr. Opin. Immunol. 15, 138-147 (2003).

[63] Schuler, G., Steinman, R. M.: Dendritic cells as adjuvants for immune-mediated resistance to tumors. J. Exp. Med. 186, 1183-1187 (1997).

[64] Kalinski, P., Hilkens, C. M., Wierenga, E. A., Kapsenberg, M. L.: T-cell priming by type-1 and type-2 polarized dendritic cells: the concept of a third signal. Immunol. Today20, 561-567 (1999).

[65] Markowicz, S., Engleman, E. G.: Granulocyte - macrophage colony stimulating factor promotes differentiation and survival of human peripheral blood dendritic cells in vitro. J. Clin. Invest. 85, 955-961 (1990).

[66] Gilboa, E.: The makings of a tumor rejection antigen. Immunity11, 263-270 (1999).

[67] Parmiani, G., De Filippo, A., Novellino, L., Castelli, C.: Unique human tumor antigens: immuno-biology and use in clinical trials. J. Immunol. 178, 1975-1979 (2007).

[68] Sabatier, R., et al.: Kinome expression profiling and prognosis of basal breast cancers. Mol. Cancer10, 86 (2011).

[69] Horvath, J. C., et al.: Cancer vaccines with emphasis on a viral oncolysate melanoma vaccine. ActaMicrobiol. Immunol. Hung. 46, 1-20 (1999).

[70] Fournier, P., Schirrmacher, V.: Randomized clinical studies of antitumor vaccination: state of the art in 2008. Expert Rev. Vaccines8, 51-66 (2009).

[71] Ferreira, L., Villar, E., Muñoz-Barroso, I.: Gangliosides and N-glycoproteins function as Newcastle disease virus receptors. Int. J. Biochem. Cell Biol. 36, 2344-2356 (2004).

[72] Fiola, C., et al.: Tumor-selective replication of Newcastle disease virus: association with defects of tumor cells defence. Int. J. Cancer119, 328-338 (2006).

[73] Wilden, H., Fournier, P., Zawatzky, R., Schirrmacher, V.: Expression of RIG-I, IRF3, IFN-beta and IRF7 determines resistance or susceptibility of cells to infection by Newcastle disease virus. Int. J. Oncol. 34, 971-982 (2009).

[74] Critchley-Thorne, R. J., et al.: Impaired interferon signaling is a common immune defect in human cancer. Proc. Natl. Acad. Sci. U. S. A. 106, 9010-9015 (2009).

[75] Fournier, P., Bian, H., Szeberényi, J., Schirrmacher, V.: Analysis of three properties of Newcastle disease virus for fighting cancer: tumor-selective replication, antitumor cytotoxicity, and immuno-stimulation. Methods Mol. Biol. 797, 177-204 (2012).

[76] Jarahian, M., et al.: Activation of natural killer cells by Newcastle disease virus hemagglutinin-neuraminidase. J. Virol. 83, 8108-8121 (2009).

[77] Washburn, B., et al.: TNF-related apoptosis-inducing ligand mediates tumoricidal activity of human monocytes stimulated by Newcastle disease virus. J. Immunol. 170, 1814-1821 (2003).

[78] Janke, M., et al.: Recombinant Newcastle disease virus (NDV) with inserted gene coding for GM-CSF as a new vector for cancer immuno-gene therapy. Gene Ther. 14, 1639-1649 (2007).

[79] Schirrmacher, V., et al.: Newcastle disease virus activates macrophages for antitumor activity. Int. J. Oncol. 16, 363-373 (2000).

[80] Gerosa, F., et al.: Reciprocal activating interaction between natural killer cells and dendritic cells. J. Exp. Med. 195, 327-333 (2002).

[81] Zeng, J., Fournier, P., Schirrmacher, V.: Induction of interferon-alpha and tumor necrosis factor-related apoptosis-inducing ligand in human blood mononuclear cells by hemagglutinin-neuraminidase but not F protein of Newcastle disease virus. Virology297, 19-30 (2002).

[82] Zeng, J., Fournier, P., Schirrmacher, V.: Stimulation of human natural interferon-alpha response via paramyxovirus hemagglutinin lectin-cell interaction. J. Mol. Med. 80, 443-451 (2002).

[83] von Hoegen, P., Zawatzky, R., Schirrmacher, V.: Modification of tumor cells by a low dose of Newcastle disease virus. III. Potentiation of tumor-specific cytolytic T cell activity via induction of interferon-α/β. Cell. Immunol. 126, 80-90 (1990).

[84] Rogge, L., et al.: Selective expression of an interleukin-12 receptor component by human T helper 1 cells. J. Exp. Med. 185, 825-831 (1997).

[85] Washburn, B., Schirrmacher, V.: Human tumor cell infection by Newcastle disease virus leads to up-regulation of HLA and cell adhesion molecules and to induction of interferons, chemokines and finally apoptosis. Int. J. Oncol. 21, 85-93 (2002).

[86] Fournier, P., Arnold, A., Schirrmacher, V.: Polarization of human monocyte-derived dendritic cells to DC1 by in vitro stimulation with Newcastle disease virus. J. BUON14, 111-122 (2009).

[87] Alexopoulou, L., Holt, A. C., Medzhitov, R., Flavell, R. A.: Recognition of double-stranded RNA and activation of NF-kappaB by Toll-like receptor 3. Nature413, 732-738 (2001).

[88] Kato, H., et al.: Cell type-specific involvement of RIG-I in antiviral response. Immunity23, 19-28 (2005).

[89] Schirrmacher, V., et al.: Human tumor cell modify-cation by virus infection: an efficient and safe way to produce cancer vaccine with pleiotropic immune stimulatory properties when using Newcastle disease virus. Gene Ther. 6, 63-73 (1999).

[90] Schirrmacher, V., et al.: Immunization with virus-modified tumor cells. Semin. Oncol. 25, 677-696 (1998).

[91] Ertel, C., Millar, N. S., Emmerson, P. T., Schirrmacher, V., von Hoegen, P.: Viral hemagglutinin augments peptide-specific cytotoxic T cell responses. Eur. J. Immunol. 23, 2592-2596 (1993).

[92] Termeer, C. C., Schirrmacher, V., Bröcker, E. B., Becker, J. C.: Newcastle-disease-virus infection induces a B7-1/ B7-2 independent T cell-co-stimulatory activity in human melanoma cells. Cancer Gene Ther. 7, 316-323 (2000).

[93] Fournier, P., Zeng, J., Schirrmacher, V.: Two ways to induce innate immune responses in human PBMCs: paracrine stimulation of IFN-α responses by viral protein or dsRNA. Int. J. Oncol. 23, 673-680 (2003).

[94] Bai, L., Koopmann, J., Fiola, C., Fournier, P., Schirrmacher, V.: Dendritic cells pulsed with viral oncolysates potently stimulate autologous T cells from cancer patients. Int. J. Oncol. 21, 685-694 (2002).

[95] Schild, H., von Hoegen, P., Schirrmacher, V.: Modification of tumor cells by a low dose of Newcastle disease virus. II. Augmented tumor-specific T cell response as a result of $CD4^+$ and $CD8^+$ immune T cell cooperation. Cancer Immunol. Immunother. 28, 22-28 (1989).

[96] Schirrmacher, V.: Antitumor immune memory and its activation for control of residual tumor cells and improvement of patient survival. In: Sinkovics, J., Horvath, J. (eds.) Virus therapy of human cancers, pp.

481-574. Marcel Decker, New York (2005).

[97] Antony, P. A., et al.: CD8$^+$T cell immunity against a tumor/self-antigen is augmented by CD4$^+$T helper cells and hindered by naturally occurring T regulatory cells. J. Immunol. 174, 2591-2601 (2005).

[98] Corthay, A., et al.: Primary antitumor immune response mediated by CD4$^+$T cells. Immunity22, 371-383 (2005).

[99] Quezada, S. A., et al.: Tumor-reactive CD4 (+) T cells develop cytotoxic activity and eradicate large e stablished melanoma after transfer into lymphopenic hosts. J. Exp. Med. 207, 637-650 (2010).

[100] Yang, M. Y., Zetler, P. M., Prins, R. M., Khan-Farooqi, H., Liau, L. M.: Immunotherapy for patients with malignant glioma: from theoretical principles to clinical applications. Expert Rev. Neurother. 6, 1481-1494 (2006).

[101] Steiner, H. H., et al.: Antitumor vaccination of patients with glioblastoma multiforme: a pilot study to assess feasibility, safety, and clinical benefit. J. Clin. Oncol. 22, 4272-4281 (2004).

[102] Schirrmacher, V., Fournier, P.: Danger signals in tumor cells: a risk factor for autoimmune disease? Expert Rev. Vaccines9, 347-350 (2010).

[103] Moschella, F., Proietti, E., Capone, I., Belardelli, F.: Combination strategies for enhancing the efficacy of immunotherapy in cancer patients. Ann. N. Y. Acad. Sci. 1194, 169-178 (2010).

(马维民译)

第五部分

非传染性疾病疫苗和非癌症疫苗

概　述

传统疫苗开发的重点是病毒性和细菌性传染病。与此同时，治疗性抗癌疫苗也正在申请获得市场授权。根据各种非传染性和非癌症（NINC）疾病的分子研究及其途径和网络的分析，提出了创制非传染性和非癌症疾病疫苗接种策略的不同目标。非传染性和非癌症（NINC）疾病疫苗最突出的例子是阿尔茨海默病。痴呆症已成为增长最快的全球卫生流行病（世界卫生组织）。以淀粉样 β 肽和单磷酸脂酶 A（第 33 章和第 34 章）为佐剂的疫苗能够刺激阿尔茨海默病患者大脑的自然防御机制。在这一部分中，您将了解到专家学者们对疫苗开发中一些最紧迫问题和最具前景方法的观点。非传染性和非癌症（NINC）疾病疫苗将极大地改善公共卫生。

高血压和动脉粥样硬化　全身血压由心输出量和总外周阻力的乘积决定，主要由肾、内分泌系统、中枢神经系统和血管系统等多种机制控制。肾素-血管紧张素系统（RAS）是血压的重要内分泌调节因子之一。虽然高血压疫苗主要集中于 RAS，但就动脉粥样硬化疫苗而言，不同细胞中的多个靶标已成为动脉粥样硬化疫苗和免疫调节治疗的候选对象。

肥胖治疗　胃饥饿素是一种促进食物摄入和减少能量消耗的胃肠道激素。胃饥饿素主要在胃底产生，并向下丘脑传递食欲信号。分别检测了以用钉形贝血蓝蛋白作为载体蛋白或牛血清白蛋白的抗胃饥饿素疫苗对小鼠和猪内源性胃饥饿素生物活性的抑制作用。这些疫苗能够减少体重增加和脂肪量，但在人类中应用时存在一些限制，例如免疫应答的风险。

I 型糖尿病　致病性免疫应答被认为是由对胰岛 β 细胞自身抗原（自反应性 T 细胞）有反应的 T 淋巴细胞介导的，而保护性免疫应答可能是由抑制自身反应性 T 细胞（调节性 T 细胞）的

T 细胞介导的。谷氨酸脱羧酶（GAD）存在于胰岛细胞中。GAD 是自身免疫性糖尿病的主要自身抗原，也是疫苗研制的靶标。在临床前研究和 I 期临床试验中，使用或不使用佐剂的重组人谷氨酸脱羧酶均未在人和小鼠中诱发不良副作用或加重 1 型糖尿病。

I 型过敏反应 新一代过敏诊断和治疗将基于重组蛋白，允许根据患者的致敏概况进行定制治疗。除了产品具有高度标准化的优势外，重组变应原的生产也允许对目的变应原进行修饰。这种具有低免疫球蛋白 E 结合潜能的修饰性变应原被称为低变应原。由于基因疫苗是设计先进类型疫苗的一个高度通用平台，各种创新方法已经在过敏动物模型中进行了试验。

抗过敏大米 日本柳杉花粉病是日本最主要的季节性变态反应性疾病，由 2 月至 4 月初在日本大部分地区传播的花粉引起。屋尘螨（HDM）也是吸入性变应原的主要来源，吸入性变应原可引起慢性变应性疾病，如支气管哮喘。45%~80%的过敏性哮喘患者对 HDM 中的变应原敏感。作为一种新型的口服抗过敏疫苗，水稻种子中积累了低过敏性柳杉花粉变应原和 HDM 变应原衍生物，可以作为一种载体传递给肠道相关淋巴样组织（GALT）。

第 27 章　高血压和动脉粥样硬化疫苗

Hiroyuki Sasamura，Tasuhiko
Azegami 和 Hiroshi Itoh①

摘要

高血压和动脉粥样硬化都是需要持续治疗的慢性疾病。最近的基础和临床研究表明，针对肾素-血管紧张素系统进行疫苗接种可有效降低血压。已在动物模型中证明，在动脉粥样硬化的情况下，靶向内皮、巨噬细胞、免疫系统和脂质代谢靶标的疫苗能减少动脉粥样硬化。研发高血压和动脉粥样硬化疫苗可能是预防和治疗这些疾病的重要策略。

27.1　高血压疫苗

27.1.1　疾病

高血压是一种以动脉系统血压升高为特征的慢性疾病。虽然高血压通常是无症状的，但它是心血管疾病的潜在危险因素，如中风、心肌梗塞、心力衰竭和终末期肾病[1]。

据估计，2000 年全世界成年人口中约有 26.4%患有高血压[2]。到 2025 年，患病率预计将达到 29.2%。近年来大量新的、有效的抗高血压药物已经广泛使用，并且在过去 30 年中，药物种类的选择急剧增加的情况下，此数据让人非常吃惊[3]。

尽管在高血压病理生理学和治疗方面取得了许多重要进展，但高血压及其并发症的患病率仍在继续上升，这一事实被称为“高血压悖论”[4]。造成这一矛盾的原因之一是，药物治疗依从性低所致的高血压的治疗率和控制率低。美国是控制高血压成功率最高的国家之一；即便如此，最近的数据表明，大约 20%的高血压患者不清楚自己的病情，28%的人没有接受治疗，大约一半的人没有将血压控制在建议的水平[5]。其他国家的数据通常要低得多[3]。基于这些原因，需要有新的策略来对付日益蔓延的高血压。

27.1.2　病理生理学和经典疗法

全身血压由心输出量和总外周阻力的乘积决定。心输出量和总外周阻力主要由肾脏、内分泌系统、中枢神经系统和血管系统等多种机制控制。肾素-血管紧张素系统（renin-angiotensin system，RAS）是血压的重要内分泌调节因子之一[6]。

肾素是在近肾小球细胞（JGA）中合成并储存在细胞内颗粒中的一种蛋白水解酶。在低血压状态下，肾素从 JGA 细胞分泌并进入循环。血管紧张素原是肾素的生理底物，是一种在肝脏合成并进入血液的糖蛋白。在肾素蛋白水解后，形成一个 10 氨基酸肽（血管紧张素 I）。然后在与主要位于内皮细胞上的血管紧张素转换酶（ACE）相互作用后转化为 8 肽血管紧张素 II。

血管紧张素 II 作用于多个靶器官，包括血管系统、心脏、肾脏、肾上腺和中枢神经系统。

---

① H. Sasamura，MD，PhD（✉）· T. Azegami，MD，PhD · H. Itoh，MD，PhD（日本东京庆应大学医学院内科，E-mail：sasamura@ a8. keio. jp）

升压作用几乎完全由 G 蛋白偶联血管紧张素 1 型（AT1）受体介导。相反地，AT2 受体对 AT1 受体具有主要的拮抗作用，并可起到降低血压的作用[7]。

RAS 在血压控制中的重要性已经被 RAS 抑制剂的有效性所证实，RAS 抑制剂不仅可以控制血压，而且可以减少高血压并发症，包括心血管疾病和中风[8]。虽然除 RAS 外的许多其他调节系统对高血压的病理生理学有重要作用[1]，但 RAS 之所以重要是因为抑制 RAS 是一种降低人类高血压的安全有效方法。因此，很容易理解为什么 RAS 是生产有效高血压疫苗的主要靶标（图 27.1）。

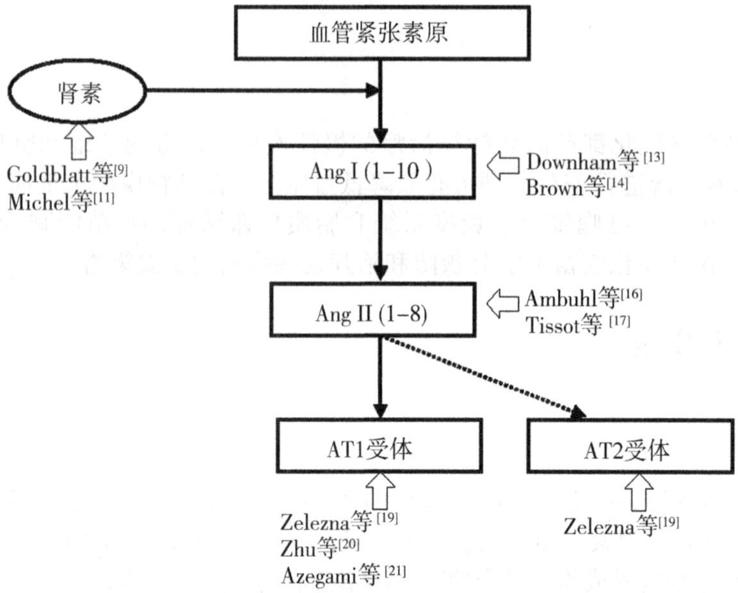

图 27.1　肾素–血管紧张素系统及高血压疫苗主要靶标

27.1.3　疫苗

Goldblatt 等在 1951 年报道了首次尝试用 RAS 免疫治疗高血压的临床研究[9]。在该研究中，作者将异源性（猪）肾素注射到严重高血压患者体内，并观察其对抗胰岛素滴度和血压的影响。他们的论文描述了 8 名患者的结果，这些患者在 3.5～33 周内接受了 7～66 次猪肾素注射(28 000～142 000单位)。几乎所有患者血清抗肾素抗体均有明显升高。然而，在这些患者中没有检测到明显的血压下降，作者推断这可能是因为所用肾素来源于猪而不是人。由于肾素的作用具有种属特异性，因此异源肾素抗体的产生可能不影响人体内源性肾素的作用。

随后还利用动物模型进行了进一步的研究。1987 年，米歇尔等报道了用纯人肾素免疫绒猴的结果。3 次皮下注射 30 μg 纯肾素蛋白均可产生较高滴度的肾素抗体。

此外，血压从（125±13）mmHg 降至（87±8）mmHg，同时血浆肾素活性和血浆醛固酮显著降低。然而，作者发现绒猴也会发展成免疫性肾病，其特征为免疫球蛋白和巨噬细胞浸润与肾素共定位于肾脏[10]。同样地，纯化的小鼠颌下腺肾素免疫自发性高血压大鼠，收缩压明显降低；然而，作者再次发现肾素免疫 SHR 的肾脏表现出慢性自身免疫性间质性肾炎，与肾素生成的 JGA 细胞有关[11]。

以整个肾素蛋白为免疫原，进行了肾素免疫实验。已经有人提出，肾素的巨大体积可能促进了自身免疫性疾病的发展，而事实上，肾素的产生只发生在一个解剖部位，并可能导致抗体的局部积累[12]。

因此，针对 RAS 的疫苗接种靶标转移到血管紧张素肽本身。在对大鼠和人类进行初步研究后[13]，Brown 等对原发性高血压患者血管紧张素 I（PMD3117）疫苗进行了临床随机双盲安慰剂对照研究。在他们的方案中，患者每隔 21 天接受 3 剂 100 μg 肽当量疫苗或每隔 14 天接受 4 剂肽当量疫苗，并且根据活性疫苗和安慰剂在停用 ACE 抑制剂或 ARB2 周后 24 小时动态血压升高的差异来估计血压变化。尽管作者发现血浆肾素和尿醛固酮有显著变化，但并未发现主动疫苗组和安慰剂治疗组之间的血压有显著差异[14]。

血管紧张素 II 疫苗的下一个重要突破是苏黎世 Cytos 公司开发的新类病毒颗粒（VLP）技术[15]。病毒样颗粒是由噬菌体外壳蛋白单体形成的大分子组装体，其自发地组装在衣壳结构中。血管紧张素肽在其 N 末端与 CGG 间隔序列融合，并与 VLP 结合。使用这种疫苗（AngQb），Ambuhl 等首先在大鼠模型（SHR）和健康的人体志愿者中，检查了皮下注射血管紧张素 II 疫苗的效果[16]。

随后，Tissot 等对 72 例轻中度高血压患者进行了多中心、双盲、安慰剂对照的 IIA 期 AngQB 疫苗疗效试验。患者在第 0 周、4 周和 12 周随机接受 100 μg、300 μg 疫苗或安慰剂。在治疗前和治疗 14 周时测量 24 小时动态血压。当比较平均日间动态血压时，作者发现收缩压显著降低 9 mmHg（$P=0.015$），300 μg 组的舒张压也有类似的趋势（降低 4 mmHg，$P=0.064$）。此外，与安慰剂相比（收缩压降低 25 mmHg（$P<0.0001$）和舒张压降低 13 mmHg（$P=0.0035$））[17]（图 27.2），300 μg 疫苗降低了清晨血压激增。发现注射部位的轻度短暂反应显著增加，共有 10 名受试者报告了轻度短暂流感样症状。Tissot 等的研究可以认为是一个里程碑式的研究，因为这是第一个表明疫苗接种可以降低人类血压的研究[18]。

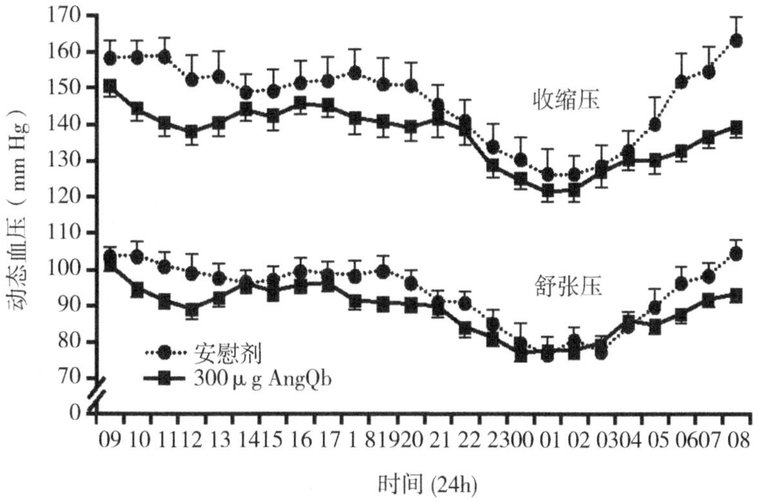

图 27.2 注射血管紧张素 II 疫苗（AngQb）或安慰剂对轻中度高血压患者 24 小时动态血压的影响（经许可摘自 Tissot 等[17]）

包括我们在内的几个研究小组，均已将血管紧张素 1 型受体（AT1 受体）作为未来开发高血压和高血压并发症疫苗的潜在靶标[19-21]。与血管紧张素 II 疫苗相比，AT1 受体疫苗的理论优势之一是直接抑制受体水平可能导致更完全抑制 RAS，并增强疫苗的有效性。此外，与阻断血管紧张素 II 在 AT1 和 AT2 受体上作用的血管紧张素 II 疫苗不同，特异性阻断 AT1 受体不会影响血管紧张素 II 在 AT2 受体上的作用，这可能对器官保护具有优势[7]

我们在实验室中合成了一个七肽氨基酸序列，对应于 Zhu 等[20]首次报道的大鼠 AT1a 受体（AFHYESR）胞外第二环 181~187 位氨基酸，并将其与载体蛋白钉形贝血蓝蛋白（KLH）结合（图 27.3）。在初步实验中，我们比较了注射疫苗 1 次、3 次和 6 次对自发性高血压大鼠（一种原发性高血压的啮齿动物模型）血压和 AT1 受体抗体滴度的影响[21]。我们发现，单次注射 AT1 疫苗导致抗体滴度升高，但这些水平明显低于 3 次或 6 次注射产生的滴度。相比之下，3 次和 6

次注射AT1疫苗都能使AT1抗体滴度升高，血压显著下降。

a

AFHYESR-C-$H_2H_2(CO)_2NH$-NH_2-KLH

b

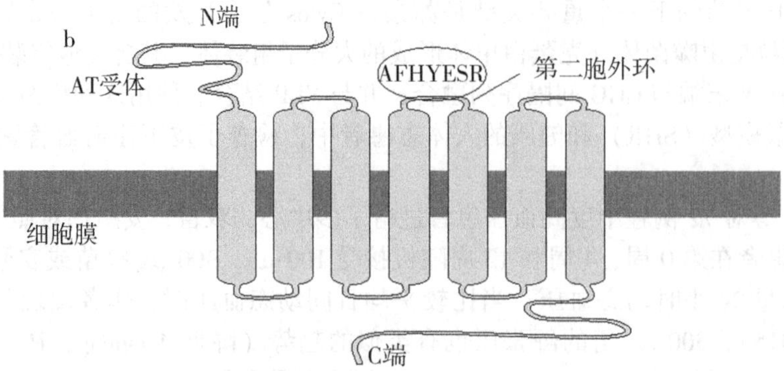

图27.3　（a）在高血压和动脉粥样硬化研究中使用的KLH结合AT1受体肽序列的结构；（b）血管紧张素1型（AT1）受体的结构和肽序列在受体第二胞外环的位置

我们还用一氧化氮合酶抑制剂L-NAME研究了AT1受体疫苗对肾损伤的影响。我们和其他实验室先前的研究表明，使用L-NAME会导致肾损伤，包括蛋白尿和肾小球及肾血管损伤[22]。我们发现，给大鼠注射3次AT1受体疫苗后，L-NAME对蛋白尿有明显的抑制作用。此外，AT1受体疫苗组肾损伤的组织学指标明显受到抑制，与持续用ARB坎地沙坦治疗的大鼠的结果相似，提示RAS疫苗不仅可以有效地减轻高血压，而且可以有效地预防高血压性肾损伤[21]。

27.1.4　优点与缺点

Tissot等的研究证明，高血压患者可以通过注射抗血管活性物质的疫苗来治疗的概念。3次注射血管紧张素疫苗后，抗体滴度在几个月内增加，第三次注射后平均半衰期约为4个月。这种半衰期可能与每年几次注射的治疗方案相一致[17]。这类治疗对于增加高血压治疗的依从性非常有吸引力，目前高血压治疗需要每日和终身的药物治疗。

尽管这些结果令人鼓舞，但在将疫苗作为治疗高血压的策略时，有许多重要的注意事项。首先，已有多种抗高血压药物可供使用；因此，有可能质疑开发高血压疫苗的重要性[23]。另一方面，尽管有这些药物，高血压和高血压并发症的患病率仍在增加，这一事实表明，应当考虑高血压治疗的新方法，并与其他策略相比较，考虑疫苗方法的风险效益比[12,24-26]。

一个重要的问题是，疫苗治疗的效果可能是不可逆的，并导致低血压，特别是在盐和容量耗尽的状态下，活性RAS有助于存活。其他重要的安全性问题包括形成刺激性和抑制性抗体的潜力，以及发展自身免疫性疾病的风险，如肾素免疫动物模型中所报道的情况[18,25]。

尽管Tissot等在研究中没有报告与疫苗接种相关的严重不良反应，但在包括安慰剂组在内的所有组中均发现局部注射部位有相关红斑。可以通过使用替代疫苗接种途径（例如口服或鼻用疫苗）来避免这种局部反应。因为高血压本身是无症状的，所以如果要开发用于常规临床使用的高血压疫苗，非常重要的一点是其副作用应该最小。

27.2　动脉粥样硬化疫苗

27.2.1　疾病

动脉粥样硬化主要是一种中、大动脉疾病，其特征是血管壁中存在含脂斑块，最终导致管腔狭窄，易发生斑块破裂和管腔血栓形成[27]。动脉粥样硬化直接导致心血管疾病的主要综合征，包括中风、心绞痛、心肌梗死、动脉瘤病和周围动脉粥样硬化，因此是全世界发病率和死亡率的主要原因[27]。

已知多种危险因素会增加动脉粥样硬化的易发性。其中最重要因素之一的是年龄，动脉粥样硬化在某种程度上是衰老过程的一部分。然而，风险增加也与高胆固醇血症、高血压、糖尿病、肥胖和男性性别有关。

动脉粥样硬化可能被认为是高血压以及其他生活方式相关疾病（包括糖尿病、代谢综合征和血脂异常）有害影响的“最终共同途径”，因此是减少中风、心肌梗塞和其他心血管疾病发病率治疗策略的重要靶标。

27.2.2 病理生理学和经典疗法

虽然动脉粥样硬化与高胆固醇水平的关系已被公认，但最近的研究表明，动脉粥样硬化本质上是一种慢性炎症性疾病，由血管壁内皮细胞和平滑肌细胞与多种免疫调节细胞的复杂相互作用介导[28,29]。

胆固醇与载脂蛋白家族一起在血液中转运，形成脂蛋白。低密度脂蛋白（LDL）是动脉粥样硬化发病机制中的一种重要脂蛋白，可在动脉内膜中积聚，易被氧化修饰为氧化低密度脂蛋白。氧化低密度脂蛋白可引起动脉壁损伤，引发一系列复杂的反应，导致血管壁炎症和免疫细胞活化。

氧化低密度脂蛋白相关血管壁损伤免疫应答的重要组成部分包括巨噬细胞、单核细胞、白细胞、树突状细胞和 T 细胞。特别是，巨噬细胞摄入氧化低密度脂蛋白并转化为泡沫细胞，体积扩大直至最终破裂，并导致氧化低密度脂蛋白进一步沉积，免疫调节细胞进一步聚集的恶性循环。

动脉中发生的炎症过程导致多种细胞因子、生长因子和其他血管活性化合物的激活，均有助于平滑肌细胞的增殖和侵袭，胶原和其他细胞外基质蛋白的沉积，以及纤维帽的形成，纤维帽可以破裂并导致血栓形成。

目前对动脉粥样硬化的治疗始于旨在改变生活方式的非药物干预。这些措施均旨在降低动脉粥样硬化发展的风险因素，包括改变饮食习惯、戒烟、控制体重和增加运动。

药物干预是一级和二级预防的最大临床证据，使用 HMG-CoA 还原酶抑制剂（他汀类）通过抑制肝脏胆固醇生成途径，降低血清胆固醇水平。尽管他汀类药物的大多数效果可能是通过其对脂质代谢的影响来介导的，但有人认为他汀类药物的抗炎作用也可能与这些药物的效果有关[27]。

其他有证据表明可以二级预防心血管病的药物包括烟酸（niacin）和抗血小板药物。此外，用 RAS 抑制剂治疗可能有一个超出其降压效果的有益效果。此外，用 RAS 抑制剂治疗除了降压作用可能还有一个有益的效果。

27.2.3 疫苗

与高血压疫苗相似，动脉粥样硬化疫苗的发展也有很长的历史。虽然高血压疫苗主要集中于 RAS，但就动脉粥样硬化疫苗而言，不同细胞中的多个靶标已成为动脉粥样硬化疫苗和免疫调节治疗的候选对象（图 27.4）。

最早抑制动脉粥样硬化的方法之一是使用由 Gero 等 1959 年报道的 β-脂蛋白免疫[30]。随后的研究检验了低密度脂蛋白及其衍生物的免疫效果。Palinski 等用丙二醛修饰的低密度脂蛋白免疫高胆固醇血症家兔，发现动脉粥样硬化病变程度明显减轻[31]。Ameli 等报道了用同源低密度脂蛋白进行免疫的类似效果，但用氧化低密度脂蛋白免疫的效果较小（不显著）[32]。这些结果已在其他研究中得到了验证[33,34]。

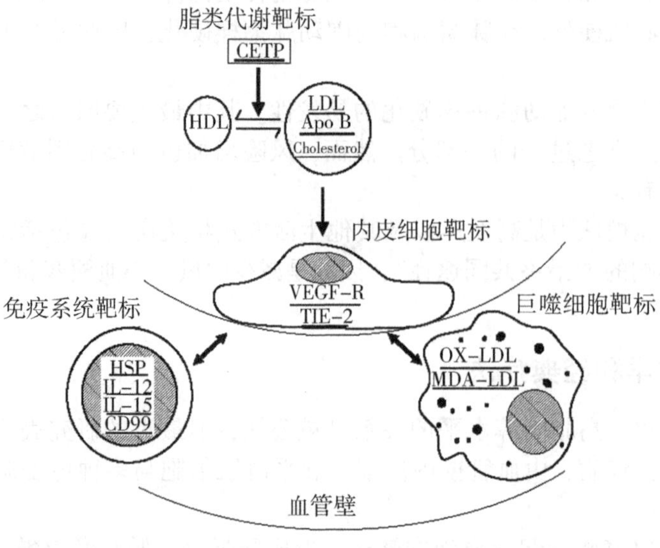

图 27.4　动脉粥样硬化疫苗/免疫疗法主要靶点的简要概述（带下划线的为疫苗靶点）

载脂蛋白 B-100 是低密度脂蛋白的主要蛋白质组分，低密度脂蛋白的氧化导致载脂蛋白降解成更小的片段，并被活性醛进一步修饰。Nilsson 等提出，基于载脂蛋白 B 的肽抗原疫苗可能是开发动脉粥样硬化疫苗的重要候选抗原[35]。Fredrikson 等检测了载脂蛋白 B100 肽序列免疫对高胆固醇血症小鼠动脉粥样硬化病变发展的影响，发现动脉粥样硬化减少了 60%[36]。有趣的是，后来发现在肽特异性 IgG 不增加的情况下，可以发现抗动脉粥样硬化作用，这表明细胞免疫反应在动脉粥样硬化保护效应中具有作用[37]。

另一种方法是开发抗胆固醇酯转移蛋白的疫苗，该疫苗参与调节高密度脂蛋白和低密度脂蛋白的平衡[38]。Rittershaus 等用含有中性脂质转移功能所需的 CETP 区域的肽免疫家兔。他们发现，在喂食高胆固醇饮食的接种家兔中，高密度脂蛋白水平升高，低密度脂蛋白水平降低。他们还发现主动脉粥样硬化区域减少了 39.6%[39]。在第一阶段的人体试验中，同一组发现 15 名受试者中有 8 名（53%）在两次注射后出现抗 CETP 抗体，没有明显副作用[40]。随后对 HDL 水平低的患者进行的第二阶段临床试验证实，该疫苗耐受性好，90%以上的患者产生抗 CETP 抗体，且 HDL 增加的百分比与抗体滴度峰值相关[38]。应该指出的是，这些结果在 CETP 抑制剂托彻普（torcetrapib）的发病率和死亡率增加数据发表之前，就已经有报道[41]。

开发动脉粥样硬化疫苗的新方法包括使用 DNA 疫苗和树突状细胞策略。在 Mao 等的研究中，高胆固醇血症家兔肌肉内用含有 CETP 羧基末端片段表位的质粒以及免疫调节性 CPG 序列进行免疫，发现动脉粥样硬化斑块病变显著减少（80%）[42]。针对细胞因子和生长因子如 IL-15[43]、CD99[44]、TIE2[45]和 VEGFR-2[46,47]的 DNA 疫苗策略也报道了有益的结果。另一种实验方法是利用树突状细胞诱导特异性体液免疫反应。氧化低密度脂蛋白[48]或载脂蛋白 B-100[49]对树突状细胞有良好的作用。（关于其他脂质靶标、炎性细胞因子和内皮细胞标记物研究的更多细节，可以查找最近的几篇综述[33,34,50]。）

本实验室最近研究了 AT1 受体疫苗对高脂高盐饮食诱导的载脂蛋白 E 缺陷小鼠动脉粥样硬化的影响。我们的初步结果表明，使用这种以 AT1 受体为靶标的疫苗，不仅会导致该模型中主动脉苏丹红阳性动脉粥样斑块面积的减少，而且会导致蛋白尿的减少（Azegami 等，未发表的研究结果）。我们和其他实验室的研究表明，药物抑制 RAS 可通过多种机制减轻动脉粥样硬化，包括降低血管的前胰岛素反应、降低动脉粥样硬化脂蛋白的滞留和增加脂质释放[51]，我们正计划进一步研究 AT1 受体疫苗是否具有抗动脉粥样硬化和抗高血压的作用。

其他研究小组已经研究了口服和鼻黏膜免疫策略的使用，作为诱导对促动脉粥样硬化抗原的耐受性和免疫无应答性的方法。热休克蛋白（HSPs）是研究最多的蛋白之一，其被认为与动脉粥样硬化前免疫介导的反应有关。Harats 和 Maron 等在小鼠模型中检测了口服或鼻内免疫 HSP65 的效果，两组均发现动脉粥样硬化减轻[52,53]。同样地，据 van Puijvelde 等报道，对动脉粥样硬化小鼠口服 HSP60 或 HSP60 肽可诱导口服耐受，并显著减少动脉粥样硬化斑块的大小[54]。另一方面，Yuan 等报告称，针对 CETP 的 DNA 疫苗经鼻内免疫可导致持续 28 周的显著抗 CETP-IgG 反应，并显著降低动脉粥样硬化[55,56]。需要进一步的研究来阐明不同方法减少动脉粥样硬化的优点。

在一项小型临床研究中，Bourinbaiar 等检查了 13 名志愿者在 3 个月内服用含有来自猪脂肪组织的混合抗原的片剂的效果[57]。他们报告 13 例患者中有 12 例 HDL-C 水平升高，腰围、中臂围和大腿围明显减少。由于实验设计规模较小，因此很难描述这些初步观察的机制和临床意义。

27.2.4　优点与缺点

开发广泛临床应用的动脉粥样硬化疫苗的主要挑战是，难以进行能够得出关于疫苗治疗安全性和有效性结论的且足够时间和足够数量的患者临床研究。

在高血压的情况下，可以设计一个时间尺度为几周或几个月的临床方案。而相比之下，动脉粥样硬化的发展需要数年甚至数十年的时间，准确评估动脉粥样硬化可能需要昂贵的介入性方法，如血管造影或血管内超声。生物标记物作为替代标记物是解决这一问题的方法之一，但目前其对硬终点标记物的预测能力仍然有限[27]。

虽然不应低估设计和进行临床研究以测试人类动脉粥样硬化疫苗效力的困难，但也应认识到，已经出现令人鼓舞的临床结果，另一种形式的疫苗，即流感疫苗，可能与心血管病的减少有关。

2005 年，Nichol 等在两组至少 65 岁的社区居民中评估了流感疫苗接种，对心脏病、中风、肺炎、流感和各种原因死亡住院风险的影响[58]。除了流感或肺炎住院风险预期下降 29%~32% 外，作者还注意到心脏病住院风险下降 19%，脑血管病住院风险下降 16%~23%，各种原因死亡风险下降 48%~50%。

在 Fluvacs 的前瞻性研究中，古芬克尔等选择了有心肌梗死或计划冠状动脉介入治疗的患者，并随机分配他们接受流感疫苗接种或不接种。作者在 1 年的随访中发现疫苗治疗组的心血管死亡和缺血性疾病均有减少[59]。同样，Phrommintikul 等研究了 439 例急性冠脉综合征患者接种流感疫苗的效果。他们发现与未接种疫苗组相比，主要心血管病显著减少[60]。

总之，这些前瞻性临床研究的结果，鼓励人们相信疫苗治疗可能成为未来治疗动脉粥样硬化并发症的可行策略。然而，将动物实验的结果转化为临床试验具有许多新的挑战，需要克服这些挑战才能阐明预防人类动脉粥样硬化疫苗的有效性和安全性。

参考文献

[1] Kaplan, N. M., Victor, R. G.: Clinical Hypertension, 10th edn. Lippincott Williams and Wilkins, Philadelphia (2010).

[2] Kearney, P. M., et al.: Global burden of hypertension: analysis of worldwide data. Lancet 365, 217-223 (2005).

[3] Israili, Z. H., Hernandez-Hernandez, R., Valasco, M.: The future of antihypertensive treatment. Am. J. Ther. 14, 121-134 (2007).

[4] Chobanian, A., Shattuck Lecture, V.: The hypertension paradox - more uncontrolled disease despite improved therapy. N. Engl. J. Med. 361, 878-887 (2009).

[5] Egan, B. M., Zhao, Y., Axon, R. N.: US trends in prevalence, awareness, treatment, and control of hypertension, 1988-2008. JAMA 303, 2043-2050 (2010).

[6] Nguyen, D. C. A., Touyz, R.: M. A new look at the renin-angiotensin system-focusing on the vascular system. Peptides 32, 2141-2150 (2011).

[7] Wright, J. W., Yamamoto, B. J., Harding, J. W.: Angiotensin receptor subtype mediated physiologies and behaviors: new discoveries and clinical targets. Prog. Neurobiol. 84, 157-181 (2008).

[8] Weber, M.: Achieving blood pressure goals: should angiotensin II receptor blockers become fi rst-line treatment in hypertension? J. Hypertens. Suppl. 27, S9-S14 (2009).

[9] Goldblatt, H., Haas, E., Lamfrom, H.: Antirenin in man and animals. Trans. Assoc. Am. Physicians 64, 122-125 (1951).

[10] Michel, J. B., et al.: Active immunization against renin in normotensive marmoset. Proc. Natl. Acad. Sci. U. S. A. 84, 4346-4350 (1987).

[11] Michel, J. B., et al.: Physiological and immunopatho-logical consequences of active immunization of spontaneously hypertensive and normotensive rats against murine renin. Circulation 81, 1899-1910 (1990).

[12] Gradman, A. H., Pinto, R.: Vaccination: a novel strat - egy for inhibiting the renin - angiotensin - aldosterone system. Curr. Hypertens. Rep. 10, 473-479 (2008).

[13] Downham, M. R., et al.: Evaluation of two carrier protein-angiotensin I conjugate vaccines to assess their future potential to control high blood pressure (hypertension) in man. Br. J. Clin. Pharmacol. 56, 505-512 (2003).

[14] Brown, M. J., et al.: Randomized double-blind placebo-c ontrolled study of an angiotensin immunotherapeutic vaccine (PMD3117) in hypertensive subjects. Clin. Sci. (Lond.) 107, 167-173 (2004).

[15] Jegerlehner, A., et al.: A molecular assembly system that renders antigens of choice highly repetitive for induction of protective B cell responses. Vaccine 20, 3104-3112 (2002).

[16] Ambuhl, P. M., et al.: A vaccine for hypertension based on virus-like particles: preclinical effi cacy and phase I safety and immunogenicity. J. Hypertens. 25, 63-72 (2007).

[17] Tissot, A. C., et al.: Effect of immunisation against angiotensin II with CYT006-AngQb on ambulatory blood pressure: a double-blind, randomised, placebo-controlled phase IIa study. Lancet 371, 821-827 (2008).

[18] Samuelsson, O., Herlitz, H.: Vaccination against high blood pressure: a new strategy. Lancet 371, 788-789 (2008).

[19] Zelezna, B., et al.: Infl uence of active immunization against angiotensin AT1 or AT2 receptor on hypertension development in young and adult SHR. Physiol. Res. 48, 259-265 (1999).

[20] Zhu, F., et al.: Target organ protection from a novel angiotensin II receptor (AT1) vaccine ATR12181 in spontaneously hypertensive rats. Cell. Mol. Immunol. 3, 107-114 (2006).

[21] Azegami, T., Sasamura, H., Hayashi, K., Itoh, H.: Vaccination against the angiotensin type 1 receptor for the prevention of L-NAME-induced nephropathy. Hypertens. Res. 35, 492-499 (2012).

[22] Ishiguro, K., Sasamura, H., Sakamaki, Y., Itoh, H., Saruta, T.: Developmental activity of the renin-angiotensin system during the "critical period" modulates later L-NAME-induced hypertension and renal injury. Hypertens. Res. 30, 63-75 (2007).

[23] Menard, J.: A vaccine for hypertension. J. Hypertens. 25, 41-46 (2007).

[24] Brown, M. J.: Therapeutic potential of vaccines in the management of hypertension. Drugs 68, 2557-2560 (2008).

[25] Campbell, D. J.: Angiotensin vaccination: what is the prospect of success? Curr. Hypertens. Rep. 11, 63-68 (2009).

[26] Pandey, R., Quan, W. Y., Hong, F., Jie, S. L.: Vaccine for hypertension: modulating the renin-angiotensin system. Int. J. Cardiol. 134, 160-168 (2009).

[27] Weber, C., Noels, H.: Atherosclerosis: currentpatho - genesis and therapeutic options. Nat. Med. 17, 1410-1422 (2011).

[28] Ross, R.: Atherosclerosis-an infl ammatory disease. N. Engl. J. Med. 340, 115-126 (1999).

[29] Libby, P.: Infl ammation in atherosclerosis. Nature 420, 868-874 (2002).

[30] Gero, S., et al.: Inhibition of cholesterol atheroslero-sis by immunisation with beta-lipoprotein. Lancet 2,

6-7 (1959).

[31] Palinski, W., Miller, E., Witztum, J. L.: Immunization of low density lipoprotein (LDL) receptor-defi cient rabbits with homologous malondialdehyde-modifi ed LDL reduces atherogenesis. Proc. Natl. Acad. Sci. U. S. A. 92, 821-825 (1995).

[32] Ameli, S., et al.: Effect of immunization with homolo-gous LDL and oxidized LDL on early atherosclerosis in hypercholesterolemic rabbits. Arterioscler. Thromb. Vasc. Biol. 16, 1074-1079 (1996).

[33] de Carvalho, J. F., Pereira, R. M., Shoenfeld, Y.: Vaccination for atherosclerosis. Clin. Rev. Allergy Immunol. 38, 135-140 (2010).

[34] de Jager, S. C., Kuiper, J.: Vaccination strategies in atherosclerosis. Thromb. Haemost. 106, 796-803 (2011).

[35] Nilsson, J., Fredrikson, G. N., Bjorkbacka, H., Chyu, K. Y., Shah, P. K.: Vaccines modulating lipoprotein autoimmunity as a possible future therapy for cardiovascular disease. J. Intern. Med. 266, 221-231 (2009).

[36] Fredrikson, G. N., et al.: Inhibition of atherosclerosis in apoE-null mice by immunization with apoB-100 peptide sequences. Arterioscler. Thromb. Vasc. Biol. 23, 879-884 (2003).

[37] Fredrikson, G. N., Bjorkbacka, H., Soderberg, I., Ljungcrantz, I., Nilsson, J.: Treatment with apo B peptide vaccines inhibits atherosclerosis in human apo B-100 transgenic mice without inducing an increase in peptide-specifi c antibodies. J. Intern. Med. 264, 563-570 (2008).

[38] Ryan, U. S., Rittershaus, C. W.: Vaccines for the pre-vention of cardiovascular disease. Vascul. Pharmacol. 45, 253-257 (2006).

[39] Rittershaus, C. W., et al.: Vaccine-induced antibodies inhibit CETP activity in vivo and reduce aortic lesions in a rabbit model of atherosclerosis. Arterioscler. Thromb. Vasc. Biol. 20, 2106-2112 (2000).

[40] Davidson, M. H., et al.: The safety and immunogenic-ity of a CETP vaccine in healthy adults. Atherosclerosis 169, 113-120 (2003).

[41] Barter, P. J., et al.: Effects of torcetrapib in patients at high risk for coronary events. N. Engl. J. Med. 357, 2109-2122 (2007).

[42] Mao, D., et al.: Intramuscular immunization with a DNA vaccine encoding a 26-amino acid CETP epitope displayed by HBc protein and containing CpG DNA inhibits atherosclerosis in a rabbit model of atherosclerosis. Vaccine 24, 4942-4950 (2006).

[43] van Es, T., et al.: IL-15 aggravates atherosclerotic lesion development in LDL receptor deficient mice. Vaccine 29, 976-983 (2011).

[44] van Wanrooij, E. J., et al.: Vaccination against CD99 inhibits atherogenesis in low-density lipoprotein receptor-deficient mice. Cardiovasc. Res. 78, 590-596 (2008).

[45] Hauer, A. D., et al.: Vaccination against TIE2 reduces atherosclerosis. Atherosclerosis 204, 365-371 (2009).

[46] Hauer, A. D., et al.: Vaccination against VEGFR2 attenuates initiation and progression of atherosclerosis. Arterioscler. Thromb. Vasc. Biol. 27, 2050-2057 (2007).

[47] Petrovan, R. J., Kaplan, C. D., Reisfeld, R. A., Curtiss, L. K.: DNA vaccination against VEGF receptor 2 reduces atherosclerosis in LDL receptor-defi cient mice. Arterioscler. Thromb. Vasc. Biol. 27, 1095-1100 (2007).

[48] Habets, K. L., et al.: Vaccination using oxidized low-density lipoprotein-pulsed dendritic cells reduces atherosclerosis in LDL receptor-defi cient mice. Cardiovasc. Res. 85, 622-630 (2010).

[49] Hermansson, A., et al.: Immunotherapy with tolero-genic apolipoprotein B-100-loaded dendritic cells attenuates atherosclerosis in hypercholesterolemic mice. Circulation 123, 1083-1091 (2011).

[50] Chyu, K. Y., Nilsson, J., Shah, P. K.: Immune mecha-nisms in atherosclerosis and potential for an atherosclerosis vaccine. Discov. Med. 11, 403-412 (2011).

[51] Hayashi, K., Sasamura, H., Azegami, T., Itoh, H.: Regression of atherosclerosis in apolipoprotein E-defi cient mice is feasible using high-dose angiotensin receptor blocker, candesartan. J. Atheroscler. Thromb. 19 (8), 736-746 (2012).

[52] Harats, D., Yacov, N., Gilburd, B., Shoenfeld, Y., George, J.: Oral tolerance with heat shock protein 65 attenuates Mycobacterium tuberculosis-induced and high-fat-diet-driven atherosclerotic lesions. J. Am. Coll. Cardiol. 40, 1333-1338 (2002).

[53] Maron, R., et al.: Mucosal administration of heat shock protein-65 decreases atherosclerosis and infl ammation in aortic arch of low-density lipoprotein receptor-defi cient mice. Circulation 106, 1708-1715 (2002).

[54] van Puijvelde, G. H., et al.: Induction of oral tolerance to HSP60 or an HSP60-peptide activates T cell regulation and reduces atherosclerosis. Arterioscler. Thromb. Vasc. Biol. 27, 2677-2683 (2007).

[55] Yuan, X., et al.: Intranasal immunization with chito-san/pCETP nanoparticles inhibits atherosclerosis in a rabbit model of atherosclerosis. Vaccine 26, 3727-3734 (2008).

[56] Jun, L., et al.: Effects of nasal immunization of multi-target preventive vaccines on atherosclerosis. Vaccine 30, 1029-1037 (2012).

[57] Bourinbaiar, A. S., Jirathitikal, V.: Effect of oral immunization with pooled antigens derived from adipose tissue on atherosclerosis and obesity indices. Vaccine 28, 2763-2768 (2010).

[58] Nichol, K. L., et al.: Infl uenza vaccination and reduc-tion in hospitalizations for cardiac disease and stroke among the elderly. N. Engl. J. Med. 348, 1322-1332 (2003).

[59] Gurfi nkel, E. P., Leon de la Fuente, R., Mendiz, O., Mautner, B.: Flu vaccination in acute coronary syndromes and planned percutaneous coronary interventions (FLUVACS) study. Eur. Heart J. 25, 25-31 (2004).

[60] Phrommintikul, A., et al.: Infl uenza vaccination reduces cardiovascular events in patients with acute coronary syndrome. Eur. Heart J. 32, 1730-1735 (2011).

（傅松波，杨欣燐，吴国华译）

第 28 章 抗胃饥饿素治疗性疫苗：一种治疗肥胖症的新途径

Sara Andrade，Marcos Carreira，Felipe F. Casanueva，Polly Roy 和 Mariana P. Monteiro①

摘要

肥胖是当前重大公共卫生问题之一，由于全世界肥胖症的发病率不断上升，其相关并发症，如 2 型糖尿病、心血管疾病和癌症的负担也不断加重。尽管肥胖症的临床相关性越来越强，但治疗肥胖症的手段仍然很少。肥胖治疗的基石仍然是节食和运动；引起厌食或营养吸收不良的抗肥胖药物可作为辅助治疗，但由于体重恢复，只能实现适度的减肥，而且往往是短期有效。对于重度肥胖症，唯一被证明有效的治疗方法是减肥手术，这是一种具有内在风险的侵入性手术，仅推荐用于特定患者。

胃饥饿素（Ghrelin）是唯一已知的刺激食物摄取的激素。在生理条件下，胃饥饿素水平随禁食而升高，餐后则随之下降。大多数肥胖者的空腹胃饥饿素水平较低，在节食和减肥后则会升高，这也是减肥难以维持的原因之一。在体重大幅度下降的情况下，一些减肥手术可以阻止胃饥饿素水平的增加，这对于维持体重下降可能会有所帮助。

由于胃饥饿素是唯一已知的促食欲激素，人们一直假设阻断反应性胃饥饿素增加，可以诱导持续的体重控制。

以往中和胃饥饿素食欲作用的尝试包括被动接种抗胃饥饿素单克隆抗体和针对不同胃饥饿素半抗原的单克隆抗体混合物，这些抗体混合物能够减少胃饥饿素介导的和剥夺性诱导的食物摄入，同时促进能量消耗的增加，但其仅局限于急性作用；已证明使用胃饥饿素受体拮抗剂可改善糖耐量、抑制食欲和促进体重减轻；并且使用钉形贝血蓝蛋白和牛血清白蛋白作为载体蛋白对胃饥饿素进行主动免疫，需要使用佐剂，而这些佐剂可能会引起炎症反应，限制了在人体中的应用。

一种新的分子方法是使用病毒样颗粒作为免疫原性载体的抗胃饥饿素疫苗，该疫苗在正常体重和饮食诱导肥胖（DIO）的小鼠中，表现出良好的耐受性、食物摄入减少和能量消耗增加。接种 DIO 小鼠下丘脑基底核 NPY 基因表达也显著降低，反应出了中枢驱动食欲减弱。总之，数据表明，这种新的治疗性抗胃饥饿素疫苗是安全的，对能量平衡有积极影响，可作为治疗肥胖的有效方法。

---

① S. Andrade，MBSc · M. P. Monteiro，MD，PhD（✉）（波尔图大学解剖学系和多学科生物医学研究部，葡萄牙波尔图，E-mail：mpmonteiro@ icbas. up. pt）

M. Carreira，PhD（卡洛斯三世研究所营养保健中心，西班牙圣地亚哥德孔波斯特拉）

F. F. Casanueva，MD，PhD（卡洛斯三世研究所营养保健中心，西班牙圣地亚哥德孔波斯特拉，圣地亚哥·德·孔波斯特拉大学医药系，西班牙圣地亚哥德孔波斯特拉）

P. Roy，PhD（伦敦卫生与热带医学院病原分子生物学系，英国伦敦）

28.1 引言

在过去几十年里，超重、肥胖和极度肥胖的患病率在全世界范围内不断增加，不仅影响到成年人，而且儿童和青少年人群患病率亦呈上升趋势[1,2]。最新的评估表明，体重增加趋势将继续上升，尤其是年轻人群[3]。

众所周知，肥胖是许多慢性病（包括 2 型糖尿病、高血压、代谢综合征、心血管疾病和癌症）的危险因素[4]。

不利的健康后果不仅发生在超重人群身上，而且在正常体重指数上限（体重指数为 22～24.9 kg/m^2）也开始增加，而体重减轻可改善或消除与肥胖相关几种疾病的共存状态[4]。

据估计，由于直接损失（个人保健、医院护理、医生服务、相关保健服务和药品）和间接损失（因发病率或死亡率导致生产力下降或停止而造成的产出损失），肥胖症对保健系统有着巨大的经济影响[5,6]。这不仅归因于肥胖本身，如肥胖造成的过度就医、工作日减少、活动受限和卧床天数，更多的是与之相关的共病所致，如 2 型糖尿病、冠心病、乳腺癌、子宫内膜癌、结肠癌以及骨关节炎[7]。此外，肥胖还与拒绝就业、限制职业发展和较高的保险费有关[7]。

肥胖和超重与预期寿命的大幅度降低有关，与高血压和糖尿病等主要的、潜在可预防的早发性发病和死亡的共病无关[8]。与肥胖相关的预期寿命缩短和早期死亡率上升与吸烟者的情况相似[8]。

鉴于肥胖是世界范围内可预防性致死的主要因素，各国政府现已将该病视为本世纪最严重的公共卫生问题之一[9]。

28.2 肥胖的诊断和经典疗法

肥胖被定义为一种医学病症，其特征是体内积聚过多脂肪，并可能对健康产生不利影响[2]。

根据身高平方（m^2）的体重比（kg）计算得出的体重指数（BMI），是临床上诊断超重和肥胖的常规测量工具。尽管 BMI 没有提供体脂分布的相关信息，但它与体脂百分比有很好的相关性。体重指数将体重指数在 18.5～24.9 kg/m^2 的人群定义为正常体重，在 25～29.9 kg/m^2 的人群定义为超重，大于 30 kg/m^2 的人群定义为肥胖[10,11]。

肥胖通常是遗传易感人群摄入过多食物能量和缺乏体育锻炼的综合结果。只有少数病例是由于单基因引起；内分泌失调，如库欣综合征和甲状腺功能减退；或以前使用过导致体重增加的药物所致[4]。

肥胖症是一种慢性疾病，医学治疗减肥后体重恢复的可能性很大，因此需要对这种疾病采取长期治疗办法[12]。但临床医生几乎没有治疗肥胖的手段。节食和运动仍然是肥胖治疗的基石，目前的抗肥胖药物只能达到相对短期的减肥效果，而且往往会导致体重回升[10,12]。

现有的减肥治疗包括节食、运动、行为改变和药物治疗等不同组合。尽管目前还没有明显的证据支持单一膳食方法优于其他用于减肥的膳食，许多不同微量营养素组成的饮食已显示出减肥功效。坚持按规定减少卡路里摄入量水平，似乎是成功的最重要决定因素[12]。体育锻炼也是一种有效的减肥方法；而且，一旦减肥成功，对维持体重更为重要[12]。肥胖的药物治疗包括旨在诱发体重减轻的抑制食欲或改变营养吸收药物。

能够引起 5%～10% 的体重减轻，这是美国食品和药物管理局（FDA）和欧洲药品管理局（EMEA）等监管机构批准减肥药物的最低要求。除了减肥之外，这些药物还应提供良好的安全性，并对几种心血管危险因素起到有益的作用[13]。即使只有轻微的体重减轻，且患者并未达到正常体重，这些亦能够给患者带来健康益处，并改善肥胖症合并症[13]。肥胖症患者的体重减轻与低密度脂蛋白胆固醇、总胆固醇和血压的降低相关，并可降低患 2 型糖尿病的风险，从长远来看可能对心血管疾病有益[14]。

目前，市场上有两类经美国食品和药物管理局批准的抗肥胖药物：食欲抑制剂和脂肪酶抑制剂奥利司他，前者是欧洲药品管理局批准的唯一在欧洲上市的抗肥胖药物。抑制食欲的药物包括中枢神经系统兴奋剂，如芬特明、苯二甲吗啉和二乙胺苯丙酮。芬特明因其已在美国上市几十年且价格低廉，是目前使用最广泛的减肥药物。盐酸芬特明是一种被批准用于肥胖症短期治疗的去甲肾上腺素能交感神经胺。芬特明属于一类能刺激中枢神经系统并传递“战斗或逃跑”反应的药物，这种反应会阻碍进食动力并诱发厌食症。与芬特明相关的最常见副作用包括失眠、易怒和血压升高[12]。另外，两种抑制食欲的药物是西布曲明和利莫那班，但它们在分别因心血管和精神副作用而建议终止使用，最近出于安全考虑已从市场上撤出。

奥利司他是一种脂肪酶抑制剂，可阻止膳食甘油三酯的水解，从而防止吸收不会被胃肠道排出的膳食脂肪。该药物可广泛获得并批准长期使用；尽管如此，它只产生轻微的体重减轻，并伴有高比率的胃肠道副作用，如脂肪泻、大便失禁和胃肠胀气[15]。

美国食品和药物管理局最近批准了两种近期上市的新药，即选择性5-羟色胺受体2C激动剂洛卡色林[16]和芬特明与托吡酯的联合用药[17]。托吡酯是一种氨基磺酸盐取代的单糖，自1996年开始上市，先前被美国食品和药物管理局批准用于癫痫发作和偏头痛的预防。托吡酯以及此前临床试验数据表明可促进体重减轻的其他药物，连同抗抑郁药氟西汀、舍曲林、安非他酮和抗糖尿病药物二甲双胍一起作为肥胖治疗的辅助治疗。

对于病态肥胖症，外科手术（被称为减肥手术）是提供可持续减肥的唯一疗法[19]。1992年，美国国立卫生研究院的共识发展会议在一份立场声明中肯定了在这种情况下，外科手术相对于非外科手术的优越性[20,21]。减肥手术是为那些已知药物减肥治疗失败且未显示长期疗效的患者所保留，即体重指数>40 kg/m^2 或>35 kg/m^2，并伴有高危合并症的患者[22]。

在严重肥胖患者中，通过降低新的肥胖相关并发症的发生风险并改善现有并发症的外科治疗，在生产和减肥保持方面亦更具成本效益[23]。肥胖症外科手术大大降低了新的健康相关疾病风险，即心血管疾病、癌症、内分泌疾病（包括糖尿病和高血压）、呼吸系统疾病、肌肉骨骼疾病、感染性疾病、精神病和精神障碍[24]。减肥手术后的体重减轻，通常会改善或消除多种疾病，从而消除以前病态肥胖患者的药物使用和旷工现象。与相关控制措施相比，接受手术患者的医疗保健使用率和直接医疗保健总成本均显著降低。减肥手术也显著降低了5年内的总死亡率，与相关控制措施相比，相对死亡风险降低了89%[24]。

因此，为了应对严重肥胖症药物治疗的相对无效性，近年来对减肥手术的需求大幅增加[25]。

28.3　抗胃饥饿素治疗性疫苗

28.3.1　食物摄入量的管理以及能量平衡

调节能量稳态的生理系统包括脑中枢，如下丘脑、脑干和边缘系统的奖赏中枢，它们通过神经肽的分泌来调节食物的摄取和能量消耗。这些中枢受到周围神经和激素信号的调节。由脂肪组织合成的激素，如瘦素，反映了身体的长期营养状况，并能够影响长期体重调节，而胃肠激素，如胃饥饿素、肽酪氨酸-酪氨酸（PYY）和胰高血糖素样肽1（GLP-1）等，以及其他几种激素，能快速调节这些通路，并能够调节食物摄取和能量消耗[26]（图28.1）。在下丘脑中有最重要的食物摄取调节胞核，弓状核（ARC）和室旁核（PVN）。基底下丘脑弓状核接受外周信号，并在食欲调节中发挥综合作用。弓状核向参与内脏传出活动调节的室旁核，投射二级神经元。在弓状核（ARC）中，涉及两类典型的神经元集群，一类是共表达神经肽Y和刺鼠色蛋白相关蛋白（NPY/AGRP）的食欲刺激，另一类是共表达阿皮素原和可卡因-苯丙胺调节转录肽（POMC/CART）的食欲抑制[27]（图28.2）。

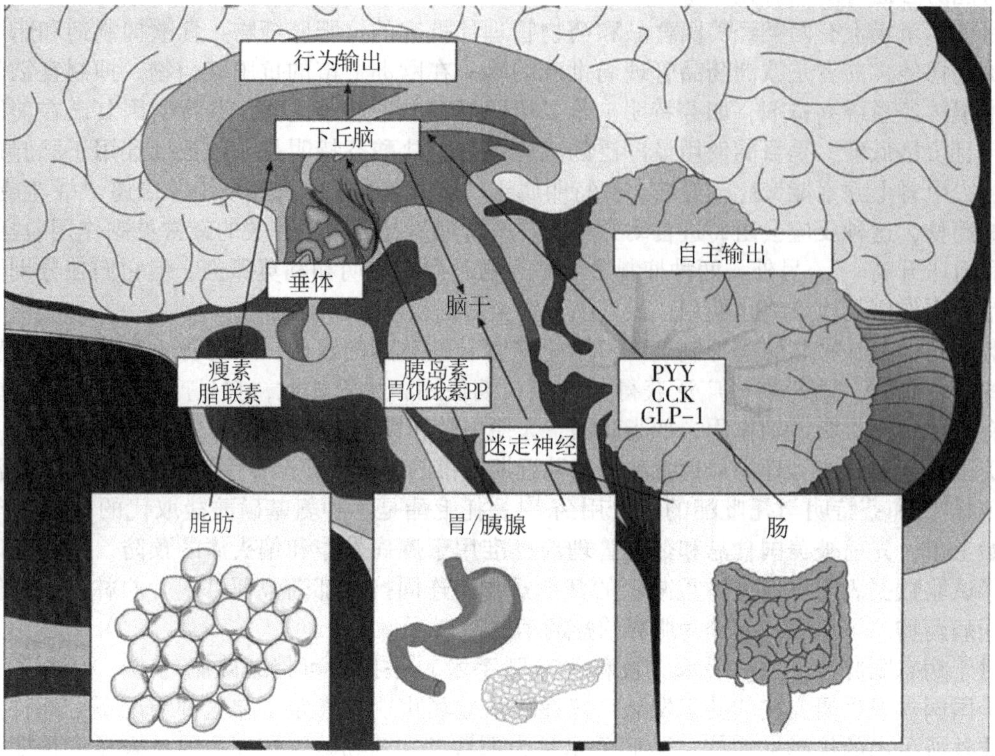

图 28.1 调节能量平衡的生理系统

大脑中枢，如下丘脑、脑干和边缘系统的奖赏中枢，通过分泌神经肽调节食物摄取和能量消耗。这些中枢受到周围神经和激素信号的调节。由脂肪组织合成的激素，如瘦素和胃肠激素，如胃肠激素、肽酪氨酸-酪氨酸（PYY）和胰高血糖素样肽 1（GLP-1），可调节这些通路，并能够调节食物摄取和能量消耗。

28.3.2 胃饥饿素

胃饥饿素（Ghrelin）是一种促进食物摄入和减少能量消耗的胃肠道激素[28]。胃饥饿素主要在胃底产生[29]，并向下丘脑传递食欲信号[30]。

胃饥饿素作用于基底下丘脑弓状核，可刺激 NPY 的产生和释放，并抑制 POMC[31]。NPY 是中枢神经系统中最强的刺激进食和减少能量消耗的信号，而 POMC 是一种前体蛋白，通过蛋白水解裂解产生多种多肽，其中就有降低食欲和增加能量消耗的 α-MSH[32,33]。

胃饥饿素血浆水平在饭前升高，进食后则受到抑制[34]，但肥胖患者不受此影响[35]。空腹胃饥饿素水平与体重指数呈负相关。肥胖受试者的空腹血清胃饥饿素水平通常低于对照组[36]，而饮食诱导体重减轻[34,37]后和神经性厌食患者的空腹血浆水平则升高[38]。因此，除了隐睾-侏儒-肥胖-低智能综合征（Prader-Willi syndrome）患者[39]外，总的来说，胃饥饿素似乎在肥胖中不起致病作用，其浓度降低被认为是对正能量平衡的生理适应。

最近的研究表明，减肥手术后体重减轻也是由于手术的内分泌效应所致，这种效应能够通过抑制通常在卡路里缺乏后观察到的胃饥饿素水平的升高，而干扰食欲途径[34,40]。

28.3.3 使用抗胃饥饿素疫苗的合理性

鉴于胃饥饿素是迄今为止唯一确定的促食欲激素，它被认为是开发新肥胖治疗方法的一个有希望靶标。

作为这一理念的证明，已证明小鼠接种抗胃饥饿素单克隆抗体，可抑制急性胃饥饿素介导

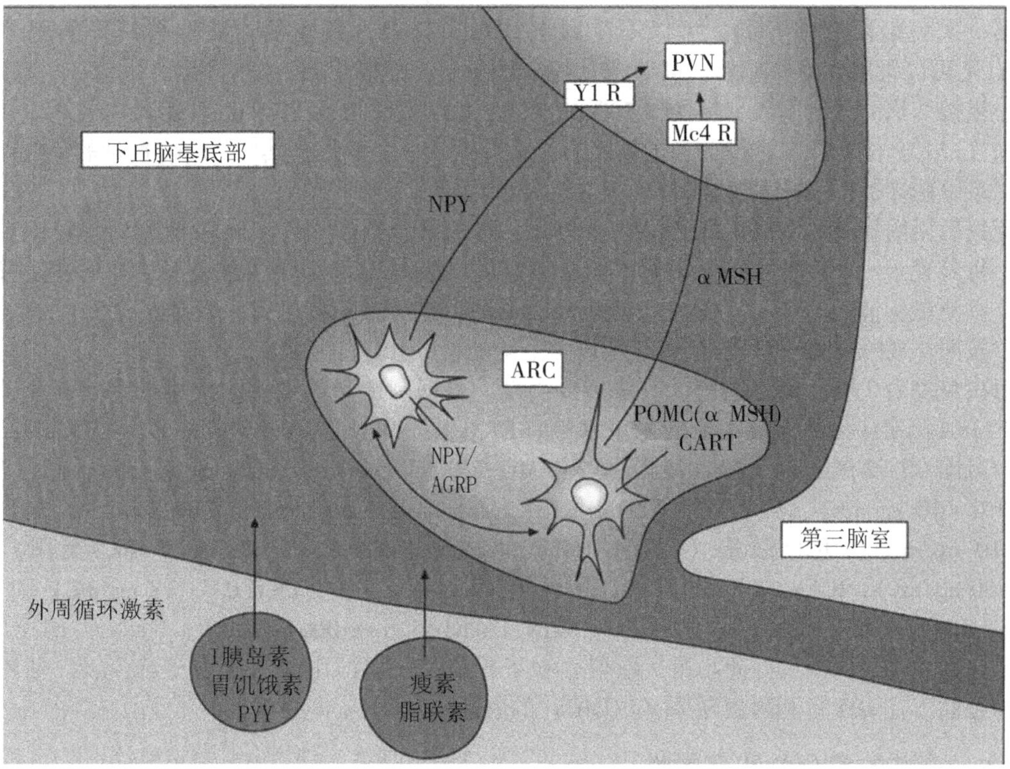

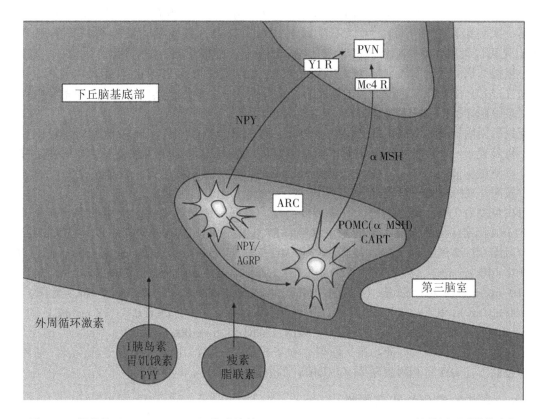

图 28.2　弓状核（arcuate，ARC）和室旁核（paravencular nuclei，PVN）均位于下丘脑基底部

弓状核接收外周的信号，并在食欲调节中发挥综合作用。弓状核向参与内脏传出活动调节的室旁核投射二级神经元。在弓状核（ARC）中，涉及两类典型的神经元集群，一类是共表达神经肽 Y 和刺鼠色蛋白相关蛋白（NPY/AGRP）的食欲刺激，另一类是共表达阿皮素原和可卡因-苯丙胺调节转录肽（POMC/CART）。

的食欲效应，但不能改变长期的食物摄入量[41]。最近，另一项研究表明，使用针对不同半抗原的单克隆抗体混合物（而不是单独使用抗体）不仅能促进能量消耗的增加，还能减少剥夺诱导的食物摄入[42]。已证明胃饥饿素受体拮抗剂 GSH-R1 具有改善糖耐量、抑制食欲和促进体重减轻的作用[43]，从而证实了胃饥饿素阻断作为肥胖潜在治疗靶标可能性。还分别检测了以用钉形贝血蓝蛋白（KLH）作为载体蛋白[44]或牛血清白蛋白（BSA）[45]的抗胃饥饿素疫苗，对小鼠和猪内源性胃饥饿素生物活性的抑制作用。这些疫苗能够诱导胃饥饿素（Ghrelin）活性形式抗体的产生[44]，并减少体重增加和脂肪量[45]。然而，这些疫苗需要使用佐剂，如明矾和弗氏不完全佐剂，这些佐剂可能会引起炎症反应或限制在人体中的应用。

28.3.4　使用病毒样颗粒的抗胃饥饿素疫苗

近年来，病毒样颗粒（Virus-like particles，VLPs）作为免疫原性分子被用于多种重组疫苗中，以诱导产生针对内源性分子的特异性抗体，这种抗体在慢性疾病中发挥着重要作用[46]，如为治疗动脉高血压而开发的抗血管紧张素疫苗[47]。

我们研究的主要目的是利用蓝舌病病毒（BTV）NS1 蛋白小管与活性胃饥饿素的化学偶联物，研制一种有效的抗胃饥饿素疫苗[48]。虽然该蛋白不是病毒衣壳的一部分，但 NS1 小管具有与经典 VLPs 相同的免疫原性特征[49]。

28.3.5　动物模型及可行性研究

将雄性成年 C57BL6/J 小鼠（西班牙、巴塞罗那、查尔斯河），正常体重小鼠和饮食诱导的肥胖小鼠（DIO）（n=18/组）随机分成3个体重匹配组（n=6/组）。正常体重的小鼠在断奶后直到第一次免疫研究前一周，可以不受限制饮水和常规大鼠食物，而饮食诱导的肥胖小鼠（DIO）则可以食用含60%脂肪的高热量食物（西班牙、巴塞罗那、查尔斯河）。

小鼠按2周的注射间隔，接受含500 μL 75 μg免疫偶联物、仅75 μg NS1蛋白或PBS的3次腹腔内（i. p.）注射。在进行了剂量确定研究后选择了一个剂量，其中75 μg剂量的免疫偶联物已证明足以诱导抗胃饥饿素抗体的产生和减少食物摄入。

免疫后用间接量热法测定能量消耗。为此，将小鼠分别置于一个小格栅笼中以限制其运动活性，将其置于一个装有氢氧化钠受体的密封室内以吸附二氧化碳。密封室盖并用容量移液管刺穿，以测量所消耗的氧气体积。记录消耗1 mL的时间，并重复进行，直至获得5个一致的值。然后，按照4.82 kcal是每升消耗氧气释放的平均能量，计算能量消耗。

每次免疫后2周，测定抗胃饥饿素抗体滴度。根据制造商的说明，使用特定的商业试剂盒通过ELISA测定，活性胃饥饿素的血浆水平（EZRGRA-90K，Linco Research，St. Charles，Mo，美国，范围25~2 000 pg/mL）、瘦蛋白（EZML-82K，Linco Research，St. Charles，Mo，美国，范围0.2~30 ng/mL）、胰岛素（EZRMI-13K，Linco Research，St. Charles，Mo，美国，范围0.2~10 ng/mL）、生长激素（EZRMGH45K，Linco Research，St. Charles，Mo，美国，范围0.07~50 ng/mL），IGF-1（E25，Mediagnost，Reutlingen，德国，范围0.5~18 ng/mL）和TNF-α（Quantikine，R&D Systems，Abingdon，英国，范围15.6~1 000 pg/mL）。

第三次免疫后两周，处死小鼠，取胃底和下丘脑，并立即用液氮浸泡冷冻，以评估胃饥饿素、神经肽Y（NPY）和阿皮素原（POMC）的表达。

28.3.6 疫苗的安全性和有效性

用该免疫偶联物治疗的正常体重小鼠，每日食物摄入量显著减少（0.44g NS1-Ghr vs 0.14g PBS vs 0.16g NS1，$P<0.001$）（图28.3）。此外，尽管没有达到统计学意义，在前两次接种后，与PBS对照相比，接受免疫偶联物的小鼠组食物摄入量也急剧减少，分别相当于PBS对照的95.3%和94.8%。在研究期间，各组小鼠体重增长没有显著差异（3.83g±0.40g NS1-Ghr vs 5.00g±0.26g PBS vs 5.33g±0.49g NS1，P=NS）[50]

在饮食诱导的肥胖小鼠（DIO）中，从高热量饮食转变为标准饮食后，每日食物摄入量增加，随后快速稳定。尽管在每次接种免疫偶联物后的24小时内食物摄入量显著减少，分别相当于接种3次后PBS组食物摄入量（$P=0.036$），82.22%（$P=0.008$）和50.09%（$P=0.039$）的66.16%，但与对照组相比（（147.64±2.46）g NS1-Ghr，（147.80±5.89）g NS1，（150.37±3.65）g PBS，P=NS），接种免疫偶联物的DIO小鼠累积食物摄入量并无显著减少。尽管接种后不同实验组的体重并无显著差异（32.17±0.872g NS1-Ghr vs（31.33±1.282）g NS1 vs（31.83±0.833）g PBS，P=NS），但DIO小鼠体重确实随着高热量饮食到标准饮食的变化而下降（与基线相比为13.84%）[50]。

用该免疫偶联物免疫的正常体重小鼠产生了特异性的抗胃饥饿素抗体，每次免疫后抗体滴度增加，最后一次免疫后2周达到最大值1 265±492。单独接受NS1蛋白或PBS的对照组的基础滴度分别为332±114和324±143，$P=0.035$，这些滴度在整个研究过程中保持不变，并且没有被免疫改变，这表明存在与检测方法相关的非特异性结合。与维持其基础滴度的对照组相比，用免疫偶联物接种的DIO小鼠也产生了滴度递增的特异性抗胃饥饿素抗体，直到第三次接种后达到最大值（2 680±1 197 NS1-Ghr，458±31 NS1和257±78 PBS组，分别为$P=0.03$[50]。

与对照组相比，接种免疫偶联物的正常体重小鼠的能量消耗显著增加（（0.0146±0.001）kcal（h·kg）NS1-Ghr，（0.0138±0.001）kcal/（h·kg）NS1，（0.0129±0.001）kcal/（h·kg）PBS，$P=0.038$）（图28.4）。与对照组相比，接种免疫偶联物的DIO小鼠的能量消耗也更高（0.0207±0.01）kcal/（h·kg）NS1-Ghr，（0.0140±0.002）kcal/（h·kg）NS1，（0.0159±0.002）kcal/（h·

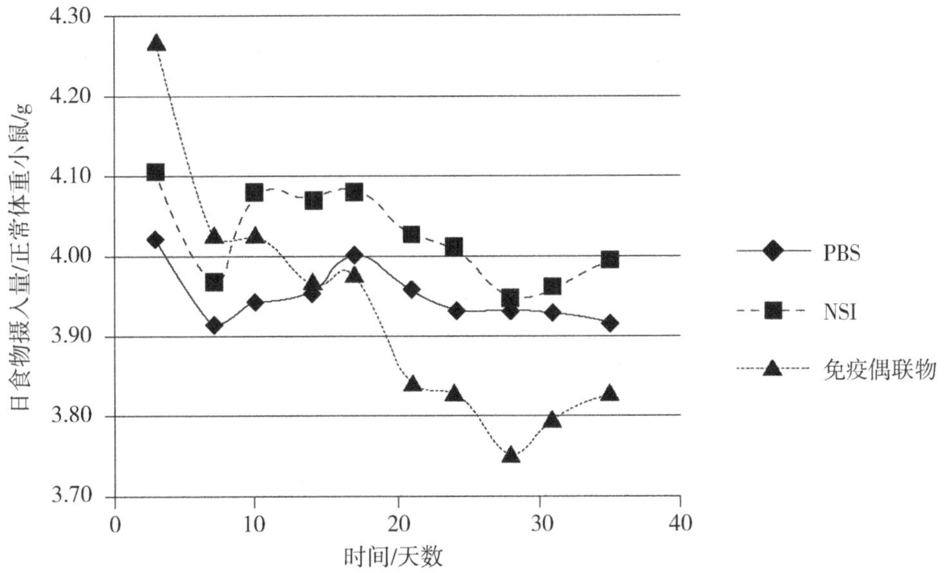

图 28.3　与对照组小鼠相比，用免疫偶联物治疗的正常体重小鼠，每日食物摄入量显著减少（0.44g NS1-Ghr vs 0.14g PBS vs 0.16g NS1，P<0.001）。

kg）PBS；P=0.044，NS1-Ghr vs PBS 和 P=0.008，NS1-Ghr vs NS1）[50]。

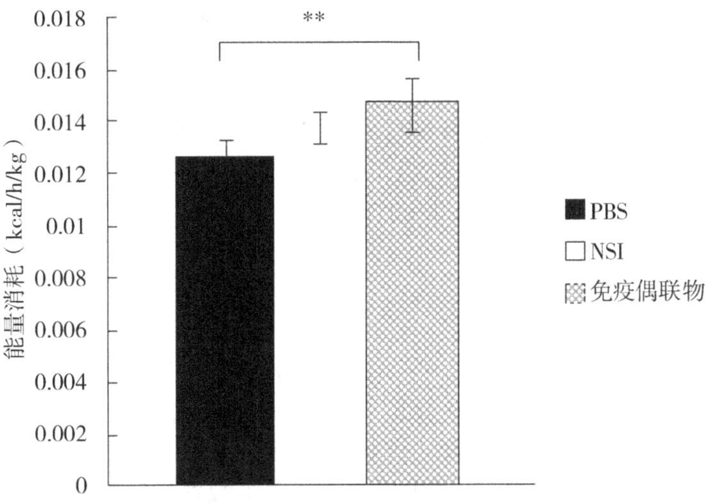

图 28.4　与对照组相比，接种免疫偶联物的正常体重小鼠的能量消耗显著增加

（（0.0146±0.001）kcal/h/kg　NS1，（0.0129±0.001）kcal/h/kg PBS，P=0.038）

正常体重小鼠空腹血浆活性胃饥饿素水平（（361.3±79.9）pg/mL）显著高于对照组（（186.9±14.8）pg/mL NS1 和（114.1±27.9）pg/mL PBS，P=0.009）。免疫偶联物免疫的 DIO 小鼠的空腹血浆胃饥饿素水平也高于对照组（（429.63±179.27）pg/mL NS1-Ghr，（147.29±53.17）pg/mL NS1，（105.88±27.76）pg/ mL PBS，P=NS），但无统计学意义。各组间血清瘦素、胰岛素、血糖、生长激素、胰岛素样生长因子-1 和肿瘤坏死因子-α 并无显著性差异。酶联免疫吸附试验结果证实，用该免疫偶联物免疫的正常体重小鼠血浆中存在胃饥饿素-抗胃饥饿素抗体的循环免疫复合物。胃饥饿素血浆水平与循环免疫复合物滴度之间也呈正相关（r=0.846）（图 28.5）。用免疫组织化学方法在肾脏中寻找免疫球蛋白沉积，未能发现肾小球基底

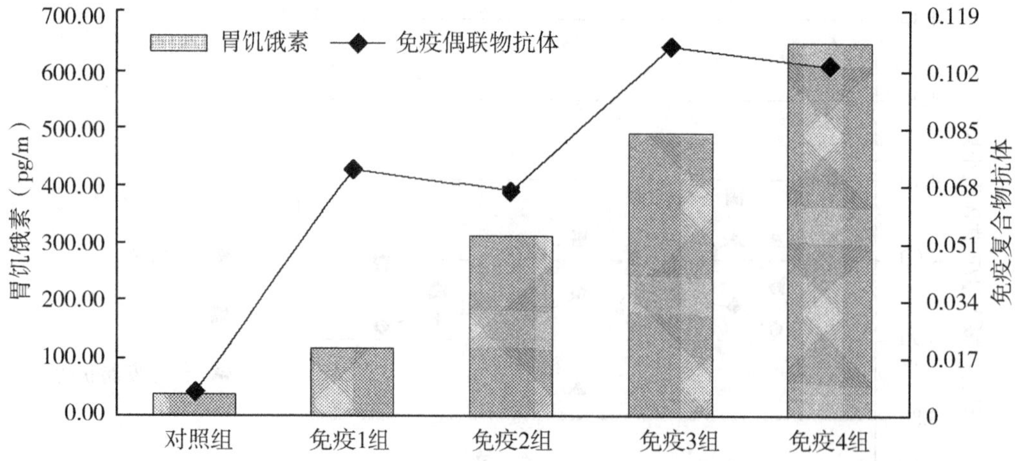

图 28.5 循环胃饥饿素-抗胃饥饿素抗体、免疫复合物、滴度和血浆胃饥饿素水平

酶联免疫吸附试验结果证实，免疫偶联物免疫正常小鼠血浆中存在胃饥饿素抗体循环免疫复合物。血浆胃饥饿素水平与循环免疫复合物滴度呈正相关（$r=0.846$）。

膜上沉积免疫复合物的任何证据[50]。

正常体重小鼠胃底胃饥饿素的表达，在 3 个实验组间无显著性差异（0.94±0.17 NS1-Ghr，1.79±0.35 NS1，1.00±0.30 PBS，P=NS）。NPY 在下丘脑基底核区的表达亦无显著性差异（1.32±0.17 NS1Ghr，0.94±0.10 NS1，1.00±0.20 PBS，P=NS）。与对照组相比，免疫偶联物组小鼠 POMC mRNA 表达明显降低（0.20±0.17 NS1-Ghr，0.93±0.17 NS1，1.00±0.10 PBS，$P<0.05$）。在 DIO 小鼠中，不同研究组胃细胞中 GAPDH 正常化后的胃饥饿素表达，也没有显著差异（1.27±0.30 NS1-Ghr，0.38±0.13 NS1，1.00±0.12 PBS，P=NS）。然而，与对照组相比，接种免疫偶联物的 DIO 小鼠下丘脑基底核 NPY 表达较低（0.59±0.09 NS1 GHR，1.03±0.12 NS1，1.00±0.13 PBS，$P<0.05$）。不同研究组下丘脑基底核的 POMC 表达，并无显著性差异（1.04±0.14 NS1-Ghr，1.32±0.25 NS1，1.00±0.12 PBS，P=NS）[50]。

28.4 优点与缺点

肥胖是当今一个主要的公共卫生问题[11,51]，且缺乏相关医疗资源[12,18]。由于胃饥饿素是迄今为止发现的唯一一种促食欲激素，它被认为是治疗肥胖症的一个有希望靶标[52]。多个研究小组先前曾尝试过胃饥饿素中和。被动转移单克隆抗胃饥饿素抗体，不能改变小鼠的长期食物摄入量[41]。靶向水解胃饥饿素的辛酰基部分形成的去酰基胃饥饿素抗体，没有生物活性，导致小鼠新陈代谢率增加，并在 24 小时食物匮乏之后抑制 6 小时再进食，但这种方法意味着需要定期注射抗体[53]。

最近，另一项研究得出结论，在禁食和剥夺诱导的食物摄入期间，需要一种寡克隆反应来维持增加的能量消耗，并在重新进食后减少总体食物摄入[42]。还测试了胃饥饿素受体拮抗剂，并且由于葡萄糖依赖性胰岛素分泌增加，GSH-R1a 降低了食物摄入和体重，并改善了葡萄糖耐量[43]。使用 KLH 或 BSA 作为免疫原性物质的抗胃饥饿素疫苗，通过降低大鼠的饲料效率[44]、猪的食物摄入量和体重[45]来降低体重增长。

然而，这些抗胃饥饿素疫苗接种和中和策略在应用于人类时，因为需要使用佐剂而存在几个局限性，即针对内源性物质的免疫反应加剧的风险，以及在被动免疫的情况下，获得耐受性和缺乏长期有效性。与传统的免疫技术相比，由于缺乏遗传物质，VLPs 仅由病毒蛋白组成，并能有效地诱导 B 细胞活化，因此 VLPs 是安全的。由于表位在分子表面的有序呈现和高度的免疫原性，这些结构的高度重复性具有允许 B 细胞受体交联的优点，且不管免疫途径如何，均可使用低剂量的免疫接种和较少数量的疫苗，使这类疫苗接种方案更有效，且更具成本效益[54]。

当前疫苗方法的主要目标是开发一种可用于人类治疗的更安全和更有效的抗胃饥饿素疫苗。为此，我们研制了一种由胃饥饿素和 BTV NS1 蛋白组成的免疫偶联物。之所以选择 NS1 小管作为 VLP 样载体蛋白，是由于它以前被用作针对人类常见传染病的预防性疫苗分子的分配系统，如口蹄疫和甲型流感病毒的蛋白质[49,55]。

在正常体重和饮食诱导的肥胖小鼠（DIO）雄性小鼠中测试了该疫苗触发免疫应答的能力，这些小鼠均产生了滴度递增的特异性抗胃饥饿素抗体，证实了仅由免疫偶联物组成的疫苗能够在不需要佐剂的情况下触发免疫应答的假设。此外，与常见传染病后的抗体滴度相比，免疫方案后获得的抗体滴度并不很高，这在安全性方面也令人放心，因胃饥饿素除了食欲调节之外，还干预了几个关键的生物过程，如生长激素分泌以及胃肠和心血管功能，完全中和胃饥饿素并不是用于肥胖症治疗的抗胃饥饿素疫苗接种战略的目的所在[52]。

接种小鼠的能量消耗明显高于单独接受 NS1 蛋白或 PBS 的分组。更高的能量消耗通常意味着更容易的体重减轻和维持。众所周知，胃饥饿素能抑制能量代谢，并且胃饥饿素替代部分逆转了胃切除小鼠体重和体脂的减少[56]。胃饥饿素已被证明通过增加呼吸商并减少脂肪作为能量的利用，对能量平衡具有长期影响[30]。此外，与野生型小鼠相比，胃饥饿素基因敲除小鼠的食物摄入量没有变化，但在喂食高脂肪食物时呼吸商降低，这表明内源性胃饥饿素在决定用于维持能量平衡的代谢底物类型方面，比在调节食物摄入量方面发挥了更突出的作用[57]。虽然接种疫苗动物的体重与对照动物相比增加较少，但这种差异并未达到统计学意义，可能是由于随访时间较短或能量稳态途径代偿机制被激活所致。

但相矛盾的是，与对照组相比，接种疫苗的小鼠具有更高的胃饥饿素水平。鉴于胃饥饿素水平的升高似乎没有生物学效应，我们假设循环中的胃饥饿素，可能是胃饥饿素-抗胃饥饿素抗体的免疫复合物形式。尽管循环免疫复合物的存在尚未有记录，以前关于抗胃饥饿素疫苗的报告也记录了免疫动物体内胃饥饿素的增加[58]。可能是由于清除率较低而存在的循环免疫复合物，引起了人们对已排除在外的肾小球基底膜沉积所致肾毒性的关注。因为胃饥饿素在胃中的表达并无差异，所以胃饥饿素在免疫动物中的合成与对照组相似，并且在胃饥饿素生物活性中和后，并未上调胃饥饿素的表达以维持体内平衡。

在接种疫苗的正常体重小鼠中，小鼠下丘脑基底核 NPY 基因表达与对照组相比，并无显著性差异。但在接种疫苗的饮食诱导的肥胖小鼠中，与对照组相比，其下丘脑基底核 NPY 基因表达显著降低，反映了中枢食欲信号减弱[59]。阿皮素原在正常体重小鼠的下丘脑基底区的表达明显低于对照组，这可能是为了防止正常体重小鼠的摄食阈值降低，而对降低的外周食欲信号进行的代偿机制，这也可以解释为什么这些发现只出现在正常体重小鼠中而不出现在饮食诱导的肥胖小鼠中。

胃饥饿素是生长激素促分泌剂[52]，且胃饥饿素中和可引起生长激素/胰岛素样生长因子-1 轴的改变。由于这种疫苗似乎对生长激素和胰岛素样生长因子-1 水平没有影响，表明我们的疫苗不太可能对生长激素轴造成内分泌不良影响。

能量动态平衡和食欲控制的调节机制是一个非常复杂的过程，包括高度冗余的信号通路[31]。因此，在某些生物学参数（如食物摄入量和体重）上缺乏显著差异，可能是由于激活了使可用活性胃饥饿素减少的代偿机制，与胃饥饿素基因敲除小鼠中发生的情况相似[57,60]。

28.5 结语

这种抗胃饥饿素疫苗似乎对动物有良好的耐受性，并且没有炎症反应或毒性迹象。与对照组动物相比，抗胃饥饿素抗体的产生可有效减少接种动物的快速食物摄入，并增加能量消耗，对建立负能量平衡、促进体重减轻具有重要作用。

大多数肥胖患者的胃饥饿素水平较低，因此，在没有饮食诱导的胃饥饿素升高的情况下，预计疫苗不会有效，故抗胃饥饿素疫苗将作为减肥和预防体重恢复的辅助疗法，对纳入饮食和锻炼计划的患者有益[34]。此外，具有高胃饥饿素水平的肥胖患者可以受益于通过这种抗胃饥饿素疫苗进行的胃饥饿素阻断，例如患有 Prader-Willi 综合征患者[39]。

总之，这些结果表明，这种抗胃饥饿素疫苗对能量动态平衡有积极的影响，可能是治疗肥胖的有效手段。

参考文献

[1] d o Carmo, I., et al.: Overweight and obesity in Portugal: national prevalence in 2003-2005. Obes. Rev. 9, 11-19 (2008). doi: 10. 1111/j. 1467-789X. 2007. 00422. x. OBR422 [pii].

[2] WHO: Obesity and overweight fact sheet N°311. http: //www. who. int/mediacentre/factsheets/fs311/en/print. html (2006). Accessed 3 July 2009.

[3] Ogden, C. L., Carroll, M. D., Kit, B. K., Flegal, K. M.: Prevalence of obesity and trends in body mass index among US children and adolescents, 1999-2010. JAMA 307, 483-490 (2012). doi: 10. 1001/jama. 2012. 40. jama. 2012. 40 [pii].

[4] Pi-Sunyer, F. X.: The obesity epidemic: pathophysiol-ogy and consequences of obesity. Obes. Res. 10 (Suppl 2), 97S-104S (2002). doi: 10. 1038/oby. 2002. 202.

[5] Finkelstein, E. A., Trogdon, J. G., Cohen, J. W., Dietz, W.: Annual medical spending attributable to obesity: payer-and service-specifi c estimates. Health Aff. 28, w822-w831 (2009). doi: 10. 1377/hlthaff. 28. 5. w822.

[6] Rappange, D. R., Brouwer, W. B., Brouwer, W. B., Hoogenveen, R. T., Van Baal, P. H.: Healthcare costs and obesity prevention: drug costs and other sector-s pecifi c consequences. Pharmacoeconomics 27, 1031-1044 (2009). doi: 10. 2165/11319900-000000000-00000.

[7] Wolf, A. M., Colditz, G. A.: Current estimates of the economic cost of obesity in the United States. Obes. Res. 6, 97-106 (1998).

[8] Peeters, A., et al.: Obesity in adulthood and its conse-quences for life expectancy: a life-table analysis. Ann. Intern. Med. 138, 24-32 (2003). doi: 200301070 00008 [pii].

[9] WHO: Diet, nutrition and the prevention of chronic dis-eases. http: //www. who. int/mediacentre/news/releases/2003/pr20/en/ (2003). Accessed 3 July 2009.

[10] NHLBI Obesity Education Initiative Expert Panel on the Identifi cation, Evaluation, and Treatment of Obesity in Adults (US). Clinical Guidelines on the Identifi cation, Evaluation, and Treatment of Overweight and Obesity in Adults: The Evidence Report. Bethesda (MD): National Heart, Lung, and Blood Institute; Sep 1998 Available from: http: // www. ncbi. nlm. nih. gov/books/NBK2003/

[11] Puska, P., Nishida, C., Porter, D., World Health Organization: Obesity and overweight. http: // www. who. int/dietphysicalactivity/publications/ facts/obesity/en/print. html (2006). Accessed 3 July 2009.

[12] Bray, G. A.: Lifestyle and pharmacological approaches to weight loss: effi cacy and safety. J. Clin. Endocrinol. Metab. 93, S81-S88 (2008). doi: 10. 1210/jc. 2008 1294. jc. 2008-1294 [pii].

[13] Fujioka, K.: Management of obesity as a chronic dis-ease: nonpharmacologic, pharmacologic, and surgical options. Obes. Res. 10 (Suppl 2), 116S-123S (2002). doi: 10. 1038/oby. 2002. 204.

[14] Avenell, A., et al.: Systematic review of the long-term effects and economic consequences of treatments for obesity and implications for health improvement. Health Technol. Assess. 8 (iii-iv), 1-182 (2004). doi: 99-02-02 [pii].

[15] Rucker, D., Padwal, R., Li, S. K., Curioni, C., Lau, D. C.: Long term pharmacotherapy for obesity and overweight: updated meta-analysis. BMJ 335, 1194-1199 (2007). doi: 10. 1136/bmj. 39385. 413113. 25. bmj. 39385. 413113. 25 [pii].

[16] Bays, H. E.: Lorcaserin: drug profi le and illustrative model of the regulatory challenges of weight-loss drug

development. Expert Rev. Cardiovasc. Ther. 9, 265-277 (2011). doi: 10. 1586/erc. 10. 22.

[17] B ays, H. E., Gadde, K. M.: Phentermine/topiramate for weight reduction and treatment of adverse metabolic consequences in obesity. Drugs Today (Barc.) 47, 903-914 (2011). doi: 1 0. 1358/dot. 2011. 47. 12. 1718738. 1718738 [pii].

[18] Snow, V., Barry, P., Fitterman, N., Qaseem, A., Weiss, K.: Pharmacologic and surgical management of obesity in primary care: a clinical practice guideline from the American College of Physicians. Ann. Intern. Med. 142, 525-531 (2005). doi: 142/7/525 [pii].

[19] Bult, M. J., van Dalen, T., Muller, A. F.: Surgical treat-ment of obesity. Eur. J. Endocrinol. 158, 135-145 (2008). doi: 10. 1530/EJE-07-0145. 158/2/135 [pii].

[20] NIH: The Practical Guide: Identifi cation, Evaluation, and Treatment of Overweight and Obesity in Adults. National Institutes of Health, National Heart, Lung, and Blood Institute, and North American Association for the Study of Obesity, Bethesda (2000) (NIH Publication Number 00-4084).

[21] Fisher, B. L., Schauer, P.: Medical and surgical options in the treatment of severe obesity. Am. J. Surg. 184, 9S-16S (2002). doi: S000296100201173X [pii].

[22] Schneider, B. E., Mun, E. C.: Surgical management of morbid obesity. Diabetes Care 28, 475-480 (2005). doi: 28/2/475 [pii].

[23] Picot, J., et al.: The clinical effectiveness and cost-effectiveness of bariatric (weight loss) surgery for obesity: a systematic review and economic evaluation. Health Technol. Assess. 13, 1-190, 215-357, iii-iv (2009). doi: 10. 3310/hta13410.

[24] Christou, N. V. et al.: Surgery decreases long-term mor-tality, morbidity, and health care use in morbidly obese patients. Ann. Surg. 240, 416-423; discussion 423-414, (2004). doi: 00000658-200409000-00003 [pii].

[25] Buchwald, H., Williams, S. E.: Bariatric surgery worldwide. Obes. Surg. 14, 1157-1164 (2003). doi: 10. 1381/0960892042387057 (2004).

[26] Field, B. C., Chaudhri, O. B., Bloom, S. R.: Obesity treat-ment: novel peripheral targets. Br. J. Clin. Pharmacol. 68, 830-843 (2009). doi: 10. 1111/j. 1365-2125. 2009. 03522. x. BCP3522 [pii].

[27] Schwartz, M. W., Woods, S. C., Porte Jr., D., Seeley, R. J., Baskin, D. G.: Central nervous system control of food intake. Nature 404, 661-671 (2000). doi: 10. 1038/ 35007534.

[28] De Vriese, C., Delporte, C.: Ghrelin: a new peptide reg-ulating growth hormone release and food intake. Int. J. Biochem. Cell Biol. 40, 1420-1424 (2008). doi: 10. 1016/j. biocel. 2007. 04. 020. S1357-2725 (07) 00138-0 [pii].

[29] Kojima, M., et al.: Ghrelin is a growth-hormone-releasing acylated peptide fromstomach. Nature 402, 656-660 (1999). doi: 10. 1038/45230.

[30] Tschop, M., Smiley, D. L., Heiman, M. L.: Ghrelin induces adiposity in rodents. Nature 407, 908-913 (2000). doi: 10. 1038/35038090.

[31] Cone, R. D., et al.: The arcuate nucleus as a conduit for diverse signals relevant to energy homeostasis. Int. J. Obes. Relat. Metab. Disord. 25 (Suppl 5), S63-S67 (2001). doi: 10. 1038/sj. ijo. 0801913.

[32] Millington, G. W.: The role of proopiomelanocortin (POMC) neurones in feeding behaviour. Nutr. Metab. (Lond.) 4, 18 (2007). doi: 10. 1186/1743-7075-4-18. 1743-7075-4-18 [pii].

[33] Williams, G., Harrold, J. A., Cutler, D. J.: The hypo-thalamus and the regulation of energy homeostasis: lifting the lid on a black box. Proc. Nutr. Soc. 59, 385-396 (2000). doi: S0029665100000434 [pii].

[34] Cummings, D. E., et al.: Plasma ghrelin levels after diet-induced weight loss or gastric bypass surgery. N. Engl. J. Med. 346, 1623-1630 (2002). doi: 10. 1056/NEJMoa012908346/21/1623 [pii].

[35] English, P. J., Ghatei, M. A., Malik, I. A., Bloom, S. R., Wilding, J. P.: Food fails to suppress ghrelin levels in obese humans. J. Clin. Endocrinol. Metab. 87, 2984 (2002).

[36] Stock, S., et al.: Ghrelin, peptide YY, glucose-dependent insulinotropic polypeptide, and hunger responses to a mixed meal in anorexic, obese, and control female adolescents. J. Clin. Endocrinol. Metab. 90, 2161-2168 (2005). doi: 10. 1210/jc. 2004-1251. jc. 2004-1251 [pii].

[37] Hansen, T. K., et al.: Weight loss increases circulating levels of ghrelin in human obesity. Clin. Endocrinol. (Oxf) 56, 203-206 (2002). doi: 1456 [pii].

[38] J anas-Kozik, M., Krupka-Matuszczyk, I., MalinowskaKolodziej, I., Lewin-Kowalik, J.: Total ghrelin plasma level in patients with the restrictive type of anorexia nervosa. Regul. Pept. 140, 43-46 (2007). doi: 1 0. 1016/j. regpep. 2006. 11. 005. S0167-0115 (06) 00223-0 [pii].

[39] Goldstone, A. P., et al.: Elevated fasting plasma ghre-lin in prader-willi syndrome adults is not solely explained by their reduced visceral adiposity and insulin resistance. J. Clin. Endocrinol. Metab. 89, 1718-1726 (2004).

[40] Monteiro, M. P., et al.: Increase in ghrelin levels after weight loss in obese Zucker rats is prevented by gastric banding. Obes. Surg. 17, 1599-1607 (2007). doi: 10. 1007/s11695-007-9324-7.

[41] Lu, S. C., et al.: An acyl-ghrelin-specifi c neutralizing antibody inhibits the acute ghrelin-mediated orexigenic effects in mice. Mol. Pharmacol. 75, 901-907 (2009). doi: 1 0. 1124/mol. 108. 052852. mol. 108. 052852 [pii].

[42] Zakhari, J. S., Zorrilla, E. P., Zhou, B., Mayorov, A. V., Janda, K. D.: Oligoclonal antibody targeting ghrelin increases energy expenditure and reduces food intake in fasted mice. Mol. Pharm. 9, 281-289 (2012). doi: 10. 1021/mp200376c.

[43] Esler, W. P., et al.: Small-molecule ghrelin receptor antagonists improve glucose tolerance, suppress appetite, and promote weight loss. Endocrinology 148, 5175-5185 (2007). doi: 10. 1210/en. 2007-0239. en. 2007-0239 [pii].

[44] Zorrilla, E. P., et al.: Vaccination against weight gain. Proc. Natl. Acad. Sci. U. S. A. 103, 13226-13231 (2006). doi: 10. 1073/pnas. 0605376103. 0605376103 [pii].

[45] Vizcarra, J. A., Kirby, J. D., Kim, S. K., Galyean, M. L.: Active immunization against ghrelin decreases weightgain and alters plasma concentrations of growth hormone in growing pigs. Domest. Anim. Endocrinol. 33, 176-189 (2007). doi: 10. 1016/j. domaniend. 2006. 05. 005. S0739-7240 (06) 00112-3 [pii].

[46] Jennings, G. T., Bachmann, M. F.: Immunodrugs: ther-apeutic VLP-based vaccines for chronic diseases. Annu. Rev. Pharmacol. Toxicol. 49, 303-326 (2009). doi: 10. 1146/annurev-pharmtox-061008-103129.

[47] Brown, M. J.: Therapeutic potential of vaccines in the management of hypertension. Drugs 68, 2557-2560 (2008). doi: 68182 [pii].

[48] Hewat, E. A., Booth, T. F., Wade, R. H., Roy, P.: 3-D reconstruction of bluetongue virus tubules using cryoelectron microscopy. J. Struct. Biol. 108, 35-48 (1992). doi: 1047-8477 (92) 90005-U [pii].

[49] Ghosh, M. K., Borca, M. V., Roy, P.: Virus-derived tubular structure displaying foreign sequences on the surface elicit $CD4^+$ Th cell and protective humoral responses. Virology 302, 383-392 (2002). doi: S004268220291648X [pii].

[50] Andrade, S., Carreira, M., Ribeiro, A.: Development of an Anti-Ghrelin Vaccine for Obesity Treatment Endocr. Rev. 32, P2-305 (2011). (The Endocrine Society).

[51] Moayyedi, P.: The epidemiology of obesity and gas-trointestinal and other diseases: an overview. Dig. Dis. Sci. 53, 2293-2299 (2008). doi: 10. 1007/ s10620-008-0410-z.

[52] Kojima, M., Kangawa, K.: Ghrelin: structure and function. Physiol. Rev. 85, 495-522 (2005). doi: 10. 1152/physrev. 00012. 2004. 85/2/495 [pii].

[53] Mayorov, A. V., et al.: Catalytic antibody degradation of ghrelin increases whole-body metabolic rate and reduces refeeding in fasting mice. Proc. Natl. Acad. Sci. U. S. A. 105, 17487-17492 (2008). doi: 10. 1073/ pnas. 0711808105.

[54] Lechner, F., et al.: Virus-like particles as a modular system for novel vaccines. Intervirology 45, 212-217 (2002). doi: 10. 1159/000067912int45212 [pii].

[55] Mikhailov, M., Monastyrskaya, K., Bakker, T., Roy, P.: A new form of particulate single and multiple immunogen delivery system based on recombinant bluetongue virus-derived tubules. Virology 217, 323-331

(1996). doi: 10. 1006/viro. 1996. 0119. S0042-6822 (96) 90119-1 [pii].

[56] Dornonville de la Cour, C., et al.: Ghrelin treatment reverses the reduction in weight gain and body fat in gastrectomised mice. Gut 54, 907-913 (2005). doi: 10. 1136/gut. 2004. 058578.

[57] Wortley, K. E., et al.: Genetic deletion of ghrelin does not decrease food intake but infl uences metabolic fuel preference. Proc. Natl. Acad. Sci. U. S. A. 101, 8227 - 8232 (2004). doi: 10. 1073/pnas. 04027631010402763101 [pii].

[58] Kellokoski, E., et al.: Ghrelin vaccination decreases plasma MCP - 1 level in LDLR (-/-) - mice. Peptides 30, 2292-2300 (2009). doi: 10. 1016/j. peptides. 2009. 09. 008. S0196-9781 (09) 00366-0 [pii].

[59] Kalra, S. P., Dube, M. G., Sahu, A., Phelps, C. P., Kalra, P. S.: Neuropeptide Y secretion increases in the paraventricular nucleus in association with increased appetite for food. Proc. Natl. Acad. Sci. U. S. A. 88, 10931-10935 (1991).

[60] De Smet, B., et al.: Energy homeostasis and gastric emptying in ghrelin knockout mice. J. Pharmacol. Exp. Ther. 316, 431-439 (2006). doi: 10. 1124/ jpet. 105. 091504. jpet. 105. 091504 [pii].

（傅松波，吴国华译）

第29章　1型糖尿病疫苗

Sandeep Kumar Gupta①

摘要

通过以避免或防止会促进遗传易感个体疾病的环境因素的一级预防，理想情况下可以根除1型糖尿病，但这些因素尚未确定，并且可能普遍存在。自20世纪80年代初期以来，二级预防在疾病开始后的作用，一直是备受人们关注的焦点，并在临床糖尿病发作后，试用了许多候选药物（主要是免疫抑制药物）。然而，预防更适用于早期、临床前疾病，而不是近期发病的临床疾病，后者的β细胞破坏更为严重。传统的接种方法是，通过使免疫系统暴露于致弱或灭活的传染因子来预防传染病。另一种称为“逆向接种”（免疫反应的抑制）的方法，通过操纵免疫系统的先天和适应性功能来抑制自身免疫。逆向接种可特异性降低病理适应性自身免疫反应。通过逆向接种，可以有针对性地减少不需要的抗体和T细胞对自身抗原的反应，同时保持免疫系统的其余部分完好无损。当前治疗自身免疫性的选择，如免疫抑制药物（如环孢素）和抗T细胞抗体（如抗CD3抗体），在NOD小鼠中不同程度显示了其在抑制β细胞自身免疫性方面的成功。但这些方法的缺点是需要反复给药，并可能导致非特异性的有害影响，如干扰正常的免疫系统功能。而抗原特异性免疫治疗（ASI）则是针对特定的自身抗原进行逆向免疫。ASI的优点是选择性灭活自身反应性T细胞，而不干扰正常免疫功能。在临床试验各个阶段的抗原特异性免疫治疗剂的主要实例是明矾配制的谷氨酸脱羧酶、热休克蛋白和肽277。

29.1　引言

1型糖尿病（T1DM）是一种自身免疫性疾病。在过去的几十年中，发病率总体每年增加约3%，据估计，15岁以下儿童每年新增病例约为65 000例[1]。1型糖尿病患者胰岛先天性和适应性免疫细胞逐渐破坏胰岛中产生胰岛素的β细胞[2]。

自身免疫性糖尿病不仅包括T1DM，还包括成人隐匿性自身免疫性糖尿病（LADA）。成人隐匿性自身免疫性糖尿病与2型糖尿病相似，常与2型糖尿病混淆。但成人隐匿性自身免疫性糖尿病（LADA）以自身抗体为主，主要是谷氨酸脱羧酶（GAD）自身抗体（GADA）。与典型的2型糖尿病相比，成人隐匿性自身免疫性糖尿病（LADA）患者通常以更快的速度依赖胰岛素。有些人认为，成人隐匿性自身免疫性糖尿病（LADA）是糖尿病的一个独特的实体和类型；另一些人则认为，成人隐匿性自身免疫性糖尿病（LADA）只是1型糖尿病（T1DM）的一个轻微变体，应该同样治疗[3]。表29.1总结了1型糖尿病、2型糖尿病和成人隐匿性自身免疫性糖尿病（LADA）的特征。

---

① S. K. Gupta，MD（✉）（Department of Pharmacology，DhanalakshmiSrinivasan Medical College and Hospital，Siruvachur，Perambalur，Tamil Nadu，India，达娜拉希米斯里尼瓦桑医学院药理学系及附属医院，印度泰米尔纳德邦贝伦伯卢尔西鲁瓦赫，E-mail：drsandeep_gupta@ rediffmail. com）

表29.1　1型糖尿病、2型糖尿病和成人隐匿性自身免疫性糖尿病（LADA）的特征[4]

项目	1型糖尿病	2型糖尿病	成人隐匿性自身免疫性糖尿病
发病年龄	青年或成人（<35岁）	成人（>35岁）	成人（>35岁）
发展为胰岛素依赖	快速（天/周）	缓慢（年）	潜伏（月/年）
存在自身抗体	是	否	是
胰岛素依赖	在诊断时	如有，随时间推移发展	6年内
胰岛素耐受	否	是	部分
对生活方式改变或口服药物的反应	差	良好	先混合后恶化
DKA频率	高	低	低
糖尿病家族史	罕见	常见	罕见
身型	健康或清瘦	超重到肥胖	正常到超重
黑色棘皮症	否	是	否
代谢综合征	否	是	否
C肽水平	检测不到	正常	低/正常

29.2　1型糖尿病：一种免疫调节紊乱疾病

胰岛中分泌胰岛素的β细胞选择性破坏，导致了1型糖尿病。由对1种或多种β细胞蛋白（自身抗原）有特异性反应的T淋巴细胞（T细胞）介导的自身免疫反应，破坏了胰岛β细胞。遗传因素和环境因素相互作用，根据个体所拥有的基因/等位基因以及个体所接触的环境因素，赋予个体对疾病的易感性或抗性。疾病易感性导致病原性免疫应答，而疾病抗性导致保护性免疫应答。

致病性免疫应答被认为是由对胰岛β细胞自身抗原（自反应性T细胞）有反应的T淋巴细胞（T细胞）所介导，而保护性免疫应答可能是由抑制自反应性T细胞（调节性T细胞）的T细胞所介导。病原性免疫应答占优势会导致胰岛炎症（胰岛炎）。其特征是胰岛被巨噬细胞和T细胞浸润，这些细胞具有细胞毒性，直接或间接产生损伤β细胞的细胞因子（如IL-1、TNF-α、TNF-β和IFN）和自由基。胰岛的主要部分在临床症状出现时缺乏β细胞，并且与对照组相比，胰岛的总体积明显减少。遗传和环境因素也可能直接增加或降低β细胞修复损伤和预防不可逆β细胞死亡、胰岛素缺乏和糖尿病的能力[2]。

29.2.1　细胞因子在免疫反应中的作用

胸腺中的T细胞前体，在阳性和阴性选择过程中成熟。具有识别和结合由人类白细胞抗原（HLA）分子所递呈抗原的T细胞受体（TCR）细胞，将在阳性选择中存活，而其他细胞将死亡。T细胞被认为仅识别人类白细胞抗原递呈的外源肽[5]。

在选择过程中存活下来的T细胞离开胸腺，并不断地从血液循环到外周淋巴组织（淋巴结、脾脏和黏膜组织）。当抗原呈递细胞（APC）识别外周的抗原时，它们会将抗原转移到将其递呈给T细胞的淋巴结。具有抗原特异性受体的T细胞将结合，然后增殖并分化为效应T细胞。识别由人类白细胞抗原I类分子所递呈的细胞内病原体肽的细胞毒性T细胞，杀死病原体感染的细胞。靶细胞上的Fas受体和浸润细胞上的Fas配体之间的相互作用，以及穿孔素和颗粒酶的细胞毒性作用，均可以诱导细胞凋亡[5]。

T细胞可根据其对CD4和CD8的表达，进一步分为不同的亚群，分别定义为T辅助细胞（Th）和细胞毒性T细胞（Tc）。根据其细胞因子谱，T辅助细胞（Th）可分为Th1和Th2细胞。Th1和Th2相关细胞因子之间的失衡，被认为在介导β细胞的破坏中具有重要作用，见于1型糖尿病[5]。

Th1分泌的细胞因子［例如干扰素（IFN）］被认为在早期糖尿病中启动和传播炎症过程，而Th2分泌的抗炎细胞因子［例如白介素（IL）-4、IL-10］则被认为抑制Th1所分泌的细胞因子。这就出现了一个假设，即使用Th2分泌的细胞因子可以预防糖尿病。另一种可以改变从Th1到Th2反应的细胞因子级联的方法是，使用非消耗性抗CD3抗体[6]。

29.3 1型糖尿病的预测

1型糖尿病的发展可结合遗传、胰岛自身抗体和代谢检测进行预测。

29.3.1 1型糖尿病的遗传预测

人类白细胞抗原（human leucocyte antigen，HLA）基因在主要组织相容性复合体（major histocompatibility complex，MHC）中，特别是HLA DR和DQ位点的等位基因，是1型糖尿病的最重要遗传决定因素。人类白细胞抗原DR3、4-DQ2、8单倍型的风险最高，而人类白细胞抗原DR2-DQ6单倍型则可以预防1型糖尿病。人类白细胞抗原分子通过与T细胞受体识别的特异性抗原肽结合，而形成良好或不良的免疫应答。人类白细胞抗原遗传学在很大程度上解释了为什么我们大多数人不会发展成1型糖尿病。非人类白细胞抗原基因、微生物等环境因素和膳食成分也可能促进或延缓1型糖尿病的发展[2,7]。

29.3.2 用于预测1型糖尿病的自身抗体

我们可以在1型糖尿病临床发病前检测到β细胞抗原自身抗体，但其在人类疾病中的作用尚不清楚。胰岛细胞自身抗体（islet cell autoantibodies，ICA）是通过免疫荧光技术，首次在人胰腺组织切片中检测到的1型糖尿病相关自身抗体。目前，用于预测1型糖尿病的主要自身抗体是针对胰岛素（即抗胰岛素抗体（IAA）、酪氨酸磷酸酶胰岛素瘤抗原（IA-2和IA-2β）以及谷氨酸脱羧酶（GAD）的自身抗体。最近，人们发现了抗锌转运蛋白（ZNT8）的抗体，并将其用于糖尿病的预测和诊断。我们可以通过多种抗体的存在来识别1型糖尿病的高危人群[7]。

29.3.2.1 胰岛素

在初诊1型糖尿病患者中检测到的第一个β细胞抗原是胰岛素。在发展为1型糖尿病的个体中出现的第一种自身抗体，就是胰岛素自身抗体（IAA）。40%~70%的新诊断患者均有胰岛素自身抗体[5]。

29.3.2.2 谷氨酸脱羧酶

谷氨酸脱羧酶存在于胰岛细胞、中枢神经系统和睾丸中。其最初是在1型糖尿病患者血浆中检测到的一种64 kDa蛋白。进一步研究表明，初诊1型糖尿病患者血清中的抗体，系针对这种胰岛细胞蛋白，该胰岛细胞蛋白后来被鉴定为GAD65酶。谷氨酸脱羧酶抗体在自身免疫性胰岛炎早期就出现了，并且在70%的诊断患者中均发现了该抗体。谷氨酸脱羧酶是一种参与谷氨酸向抑制性神经递质γ-氨基丁酸（GABA）转化的酶。谷氨酸脱羧酶在胰岛中的生理作用尚不清楚，但已有研究表明，谷氨酸脱羧酶可能是胰岛素分泌的负调节因子[5,8]。

29.3.2.3　酪氨酸磷酸酶样蛋白

在对64 kDa蛋白的进一步分析中，发现了另外两个结合与T1DM进展密切相关抗体的40 kDa和37 kDa蛋白抗原靶标，并鉴定为GAD65。37 kDa抗原被认为是与IA-2结构相似的另一种蛋白，而40 kDa抗原被鉴定为酪氨酸磷酸酶样蛋白IA-2（ICA512）。IA-2位于胰岛和其他神经内分泌细胞的分泌颗粒膜中。抗IA-2（IA-2A）的自身抗体直接作用于蛋白质的细胞内部分。表29.2汇总了预测1型糖尿病的自身抗体。

表29.2　用于预测1型糖尿病的自身抗体

抗体类型	自身抗体产生部位与作用
胰岛素	在发展为1型糖尿病的个体中出现的第一种自身抗体，就是胰岛素自身抗体（IAA） 40%~70%的新诊断患者均有胰岛素自身抗体
谷氨酸脱羧酶	谷氨酸脱羧酶存在于胰岛细胞、中枢神经系统和睾丸中 谷氨酸脱羧酶抗体在自身免疫性胰岛炎早期就出现了，并且在70%的诊断患者中均发现了该抗体 谷氨酸脱羧酶是一种参与谷氨酸向抑制性神经递质γ-氨基丁酸（GABA）转化的酶 谷氨酸脱羧酶在胰岛中的生理作用尚不清楚，但已有研究表明，谷氨酸脱羧酶可能是胰岛素分泌的负调节因子
酪氨酸磷酸酶样蛋白	IA-2位于胰岛和其他神经内分泌细胞的分泌颗粒膜中 抗IA-2（IA-2A）的自身抗体直接作用于蛋白质的细胞内部分

29.4　1型糖尿病疫苗的基本原理

通过旨在避免或防止会促进基因高危人群疾病的环境因素的一级预防，理想情况下可以根除1型糖尿病，但这些因素尚未确定，并且可能普遍存在。因不知道它们是什么，亦不能够改变其遗传易感性，一级预防的前景仍然无法确定。自20世纪80年代初期以来，疾病进程开始后的二级预防一直是备受人们关注的焦点，并且许多候选药物，主要是免疫抑制药物，通常在临床糖尿病发病后进行试验。然而，预防更适用于早期、临床前疾病，而不是近期发病的临床疾病，后者的β细胞破坏更为严重[2]。

1型糖尿病临床前阶段的个体，可通过特定胰岛抗原的循环自身抗体胰岛素原、谷氨酸脱羧酶的分子量65 000亚型（GAD65）和酪氨酸磷酸酶样胰岛素瘤抗原2（IA2）来鉴定。早期干预这些无症状、高危个体的决定，不仅取决于更大疗效的可能性，而且需要仔细考虑安全性。在临床前阶段，促进保护性免疫稳态的疫苗应该是相对安全和有效的，而潜在毒性免疫抑制药物则需要在无症状个体，特别是儿童中得到强有力的证明。对于新近发病的个体，免疫抑制药物可用于减轻病原免疫的负担，并允许出现或主动诱导发病机制[2,9]。

传统的接种方法是，通过使免疫系统暴露于致弱或灭活的传染因子来预防传染病。另一种称为“逆向接种”（免疫反应的抑制）的方法，通过操纵免疫系统的先天性和适应性功能来抑制自身免疫。逆向接种可特异性降低病理适应性自身免疫反应。通过逆向接种，可以有针对性地减少不需要的抗体和T细胞对自身抗原的反应，同时保持免疫系统的其余部分完好无损[10,11]。

当前治疗自身免疫性的选择，如免疫抑制药物（如环孢素）和抗T细胞抗体（如抗CD3抗体），在NOD小鼠中显示了其在抑制β细胞自身免疫性方面的不同程度成功。但这些方法的缺点是需要反复给药，并可能导致非特异性的有害影响，如干扰正常的免疫系统功能。而抗原特异性免疫治疗（ASI）则是针对特定的自身抗原进行逆向免疫。抗原特异性免疫治疗（ASI）的优点是选择性灭活自身反应性T细胞，而不干扰正常免疫功能[10,11]。

糖尿病疫苗可通过各种机制发挥作用：（1）将免疫应答从破坏性（例如Th1）转变为良性（例如Th2）应答；（2）诱导抗原特异性调节性T细胞；（3）删除自身反应性T细胞；（4）防止免疫细胞相互作用[12]。通过调节性T细胞释放的细胞因子信号分子类型的改变，而发生Th1-Th2转变。调节性T细胞开始释放抑制炎症的细胞因子，而不是促炎性细胞因子[13]。

药物开发的基本要求是在动物模型中证明其有效性和安全性。1型糖尿病最常用的动物模型是NOD小鼠，其在我们对1型糖尿病发病机制和防治理论的认识中发挥了重要作用。人类1型糖尿病与NOD小鼠的自身免疫性糖尿病具有一些共同特征，如在对胰岛素原和GAD65的主要组织相容性复合体（MHC）自身免疫应答中以抗原递呈分子基因为主的多基因遗传、通过骨髓转移疾病、临床前阶段延长。但与人类不同的是，NOD小鼠是近亲繁殖的，对许多免疫和其他干预有反应。而且，大多数干预措施仅在一部分NOD小鼠中能预防疾病[2]。

在最近的研究中，有报道称NOD小鼠和易感人群的免疫发病机制可能有很大不同，即人类胰岛炎的程度要轻得多，并且在任何给定的时间内，最多涉及所有胰岛的15%~35%。此外，与NOD小鼠相比，很难想象炎症可以由人类胰岛新形成的淋巴结构自主驱动。并且在人胰岛炎中，CD8淋巴细胞占主导地位，而在NOD小鼠中，则主要发现的是CD4 T细胞。已建议采用淋巴细胞减少的BB大鼠和几种抗原驱动的小鼠模型作为替代模型，其可用于突出在NOD小鼠中未准确反映的糖尿病发生方面的信息。另一个有吸引力的替代品是“人源化”的小鼠，这些小鼠部分地用人类免疫系统的成分进行了重组[9]。

29.5 疫苗接种对象是谁？

新生儿高危人类白细胞抗原-II类易感基因筛查，可确定大多数注定发展为1型糖尿病的个体。对1个或多个胰岛自身抗原（即胰岛素原、谷氨酸脱羧酶和胰岛素瘤抗原2）呈阳性的，自身抗体呈一级相对阳性的年轻人，患1型糖尿病的风险较高。在这些人群中，如果他们分别对1个、两个和3个自身抗原有自身抗体，则发展为1型糖尿病的5年风险为≤25%、26%~50%和>50%。此外，静脉血糖低于第10百分位数的第一阶段胰岛素反应（FPIR），提示预后不良。在自身抗体阳性且FPIR正常的亲属中，胰岛素抗性的风险最高[2]。

29.6 免疫预防的各种干预措施

29.6.1 抗体免疫疗法

29.6.1.1 抗CD3单克隆抗体

针对CD3的单克隆抗体OKT3，抑制了细胞毒性T细胞介导的靶细胞溶解。但是，OKT3具有强烈的促有丝分裂活性，并可诱导大量细胞因子导致不良反应[14]。大多数患者出现了一定程度的细胞因子释放综合征。并出现了许多副作用，如寒颤、恶心、低血压、呼吸困难、发热、肌肉疼痛、血小板减少伴出血风险、白细胞减少伴感染频率增加和贫血[3]。

经修饰的抗人CD3单克隆抗体，被认为是预防NOD小鼠1型糖尿病OKT3抗体的下一个替代物。已经在临床试验中，对经修饰的非Fc受体（FcR）结合CD3抗体进行了测试。已发现与Fc CD3抗体的功能相比，它们的促有丝分裂作用较弱，并且具有同等的耐受性[14]。

在使用两种不同人源化抗CD3（Teplizumab和Otelixizumab）的两项第二阶段试验中，报道了在新发病1型糖尿病患者治疗组中，通过C-肽反应评估的内源性胰岛素分泌的持续性，同时降低血红蛋白A1c水平和胰岛素需求量超过2年。在最近的随访研究中，已经显示了一剂后持续达 5 年的良好效果。目前正在进行人源化抗 CD3 单克隆抗体的第三阶段试验[1]。

29.6.1.2　抗 CD20 单克隆抗体（利妥昔单抗）

最初认为，B 细胞在引发 T 细胞中起重要作用。但是，最近研究表明，B 细胞可以促进胰岛中 $CD8^{+}$T 细胞的存活，从而促进疾病的发生。所有成熟 B 细胞均表达细胞表面标志物 CD20。利妥昔单抗是一种人源化的抗 CD20 单克隆抗体。其最初用于治疗 B 细胞淋巴瘤，后来被证明是治疗类风湿性关节炎的有效药物。已证明利妥昔单抗通过 Fc 和补体介导的细胞毒性机制以及促凋亡信号，成功从外周循环中清除人 B 细胞。现正在进行临床试验，测试利妥昔单抗在新患 1 型糖尿病患者中的疗效和安全性[14]。

在新诊断 1 型糖尿病患者的第二阶段试验中，为期 1 年的 4 剂量利妥昔单抗（抗 CD20）疗程，部分保留了 β 细胞功能[1]。

29.6.2　抗原免疫治疗

29.6.2.1　外源抗原免疫

用卡介苗（BCG）等“非特异性”免疫刺激剂接种 NOD 小鼠，可抑制糖尿病的发展。这些药物刺激了先天免疫途径，并重置了免疫稳态。但是卡介苗接种的最初人体试验，并未显示出证明残余 β 细胞功能的益处[2]。

最近，Faustman DL 等[15]使用卡介苗对 6 名长期（平均 15.3 年）患有 1 型糖尿病（平均年龄 35 岁）的受试者进行了一项小型概念验证研究。之所以选择卡介苗，是因为它通过诱导宿主产生肿瘤坏死因子（TNF）来刺激先天免疫，而肿瘤坏死因子又能杀死致病的自身免疫细胞，并恢复胰岛 β 细胞功能。这 6 名受试者随机分为卡介苗组和安慰剂组，并与自身、健康配对对照（$n=6$）或与患有（$n=57$）或无（$n=16$）1 型糖尿病参考受试者进行比较，具体取决于结果测量。疫苗和安慰剂注射间隔 4 周，共 2 次。每周监测血液中胰岛素自身反应性 T 细胞、调节性 T 细胞（Tregs）、谷氨酸脱羧酶（GAD）和其他自身抗体以及胰岛素分泌标志物 C 肽，共为期 20 周。研究发现，如 3 个疫苗接种者中有两个受试者的 C 肽水平短暂但显著增加所证实的那样，接种卡介苗的受试者显示出了胰岛素自身活性 T 细胞增加，并与配对健康对照组相比，其调节性 T 细胞数量增加、胰岛素敏感性获改善[15]。

病毒可导致 1 型糖尿病，如果明确指出某种特定病毒是致病因素，应在儿童生命早期进行疫苗接种，但该疫苗须具有良好的安全性。怀孕初期感染风疹的母亲所生的先天性风疹患儿，有脑、胰腺和其他组织感染的证据，其大约 20% 患有 1 型糖尿病。据报道，这类先天性风疹儿童中，约有两倍的比例会产生胰岛细胞抗体。此外，据报道先天性风疹患儿随后发生糖尿病的 1 型糖尿病易感性单倍型人类白细胞抗原-A1-B8-（DR3-DQ2）频率较高[2]。

怀孕期间感染了一些肠道病毒［如柯萨奇病毒 B（CVB）和埃可病毒］的母亲所生婴儿，在生命早期患糖尿病的可能性更大，但最近的研究并未证实这一点[2,9]。

流行性腮腺炎疫苗接种对预防 1 型糖尿病尚无价值。1 型糖尿病中巨细胞病毒（CMV）感染的证据尚不充分。已发现轮状病毒感染与胰岛自身抗体的增加在时间上相关。轮状病毒能感染小鼠、猪和猴胰岛 β 细胞[2]。

29.6.2.2　内源性抗原免疫

（1）明矾配制的谷氨酸脱羧酶（GAD）

谷氨酸脱羧酶是一种存在于胰岛细胞、中枢神经系统和睾丸中的 65 kDa 蛋白。谷氨酸脱羧酶是自身免疫性糖尿病的主要自身抗原。尚不清楚谷氨酸脱羧酶为自身免疫性糖尿病主要自身抗原的原因。谷氨酸脱羧酶抗体在自身免疫性胰岛炎早期就出现了，并且在 70% 的诊断患者中均发现了该抗体。谷氨酸脱羧酶自身抗体（GADA）可能是糖尿病自身免疫过程的早期征兆，并且 GADA 已成为 1 型糖尿病风险的重要预测指标之一。谷氨酸脱羧酶疫苗可以调节免疫系统，从而防止 β 细胞的破坏。在非肥胖糖尿病小鼠的研究中，已证明 GAD65 亚型可以防止 β 细胞的自身免疫性破坏[3,6]。

以氢氧化铝（Alum）为佐剂，制备了增强抗原呈递细胞抗原递呈的谷氨酸脱羧酶疫苗。抗原呈递细胞处理该注射的GAD65，以提供T细胞识别的肽片段。并促成了包括GAD65特异性调节性T细胞亚群诱导和增殖的Th1/Th2转换。这些特异性T细胞下调会攻击β细胞的抗原特异性杀伤性T细胞[3]。

在临床前研究和I期临床试验中，给予含有或不含有佐剂的重组人谷氨酸脱羧酶均未在人和小鼠中诱发不良反应或加重1型糖尿病。随后，在LADA（成人隐匿性自身免疫性糖尿病）患者中进行的第二阶段试验，进一步支持了明矾制剂谷氨酸脱羧酶疫苗的临床开发[8,16]。

受试者皮下注射安慰剂或谷氨酸脱羧酶/明矾（4 μg、20 μg、100 μg或500 μg），在谷氨酸脱羧酶/明矾疫苗的临床试验中进行了两次，第一次是在患有成人隐匿性自身免疫性糖尿病（LADA）的个体中进行的（LADA描述了成人缓慢进展型1型糖尿病）。成人隐匿性自身免疫性糖尿病（LADA）的诊断系基于，(1) 成人糖尿病发作；(2) 循环胰岛自身抗体；(3) 诊断时胰岛素的独立性。约10%的非胰岛素依赖型糖尿病成人患有成人隐匿性自身免疫性糖尿病（LADA）。$CD4^+CD25^+$/ $CD4^+CD25^-$细胞比率以及血清C肽水平，6个月后仅在20 μg剂量组中从基线增加，并且只有500 μg剂量提高了谷氨酸脱羧酶自身抗体水平。显然，至少在LADA患者中，抗原治疗（ABT）可在不改变对所给抗原的体液反应情况下，具有良好的免疫调节作用。在5年随访中没有显著的研究相关副作用报道，且C肽水平仅在20 μg剂量组显著升高[3,8]。

随后，对新患糖尿病儿童进行的一项更大规模临床试验进一步证实了20 μg谷氨酸脱羧酶/明矾剂量的良好作用。在1型糖尿病发病6个月内接种谷氨酸脱羧酶/明矾（20 μg）疫苗的患者β细胞功能得到了保护，而发病6个月后接种的患者β细胞功能则未得到保护。在最近诊断患者中的治疗有效性，可能反映出更佳的剩余β细胞质量。这与抗CD3治疗相似，在治疗时，抗CD3治疗对残留β细胞功能最高的患者最为有效。即使在15个月后，治疗也能诱导较高水平的IL-5、IL-10和IL-13GAD特异性反应，并伴有较高频率的$Foxp3^+$和TGF-β分泌T细胞[8]。

已证明对1型糖尿病和成人隐匿性自身免疫性糖尿病有效的唯一基于抗原的候选疫苗是明矾配制的谷氨酸脱羧酶[13]。然而，在最近由Diamyd和TrialNet进行的两项II / III期研究中，明矾配制的谷氨酸脱羧酶疫苗未能达到保持胰岛素生成的主要疗效临界点[16]。

(2) 热休克蛋白和肽277

热休克蛋白（heat shock proteins，Hsp）在原核生物和真核生物中都具有高度的保守性，并且在细胞内具有作为伴侣分子的作用。人Hsp60也在先天免疫系统的调节中发挥作用。这种作用通过刺激巨噬细胞上的Toll样受体4（TLR-4）（刺激性炎症效应）和T细胞上的TLR-2（免疫调节抗炎效应）来传递[6]。

热休克蛋白（HsP）是糖尿病发病机制中的另一个重要自身抗原。在发展为胰岛素炎的NOD小鼠中，发现了针对分枝杆菌Hsp65及其人类变体Hsp60的抗体。这些自身抗体随着糖尿病的发展而消失，并且在没有发展成糖尿病的NOD小鼠中并不存在。将抗Hsp60 T细胞克隆移植到健康小鼠诱导胰岛炎。T细胞识别的Hsp60表位被鉴定为24个氨基酸的肽，称为肽277（p277）[6]。

NOD小鼠或暴露于低剂量链脲佐菌素（p277）不完全弗氏佐剂的小鼠，均可预防胰岛素炎和糖尿病。Th1细胞的脾群转移至Th2免疫调节表型，伴随接种后胰岛中白细胞数量和Th1产生细胞因子的减少。从p277处理的小鼠胰岛中恢复的T细胞，降低了将糖尿病转移到NOD受体小鼠的能力。即使在疾病发作后，该疫苗也能防止NOD小鼠糖尿病的恶化[6]。

根据p277在动物模型中的保护和治疗作用，以及在人类疾病中的研究结果，研制了人用p277疫苗DiaPep277。在最近诊断为1型糖尿病的患者中进行的一项双盲研究中，对这种疫苗的有效性进行了测试。35 名患者分别在 0、1、6 和 12 个月时，接受皮下注射 DiaPep277 或安慰剂。研究的主要终点是，通过停止 C 肽生成，来检测 β 细胞功能的保留；次要终点是对外源性胰岛素需求的减少、血红蛋白 A1c（HbA1c）水平的降低和 T 细胞细胞因子表型的改变。C 肽水平在对照组迅速下降，但在对 DiaPep277 组（$n=15$）10 个月的随访中保持不变。随访 7 个月、10 个月及 18 个月后的差异，均有统计学意义。此外，在试验结束时，DiaPep277 干预组需要显著减少外源性胰岛素，以达到相同水平的 HbA1c（7%）。在试验结束时，暴露于 Hsp60 的 T 细胞的细胞因子应答，从 Th1-IFN 应答显著转变为 Th2、IL-10 和 IL-13 应答[6]。尚不确定这种疫苗在成人隐匿性自身免疫性糖尿病患者中的临床试验结果[16]。

（3）胰岛素

因其潜在的治疗价值，胰岛素是 β 细胞自身抗原的主要选择。胰岛素的生物活性形式是由其前体胰岛素原通过连续的酶裂解，来释放前导肽和 C 肽[16]。

在口服猪胰岛素后，Zhang 等首次报告了 NOD 小鼠对糖尿病的保护作用[17]。许多后续研究报道，口服胰岛素原/胰岛素或胰岛素表位可部分保护 NOD 小鼠不发生糖尿病。口服人胰岛素可诱导糖尿病前期小鼠的 $CD4^+$ 调节性 T 细胞增殖，从而保护糖尿病前期小鼠，免受糖尿病的侵袭。胰岛素在树突细胞（DCs）中促进抗炎状态的能力，最终导致认为 Th2 细胞群扩增的原因为 T 细胞功能的免疫抑制。而且，当通过不同进入体内的途径给予胰岛素时，即鼻腔内接种的胰岛素原或皮下递送的胰岛素 B 链肽 B：9-23，据报道在 NOD 小鼠中部分抑制糖尿病发作是有效的[10,17]。

迄今为止，在使用胰岛素作为治疗或预防性免疫疗法的人类临床试验中，结果一直令人失望。在一项安全性评估试验中，对有患 1 型糖尿病风险的个体，每天给予 1.6 mg 雾化胰岛素治疗，连续 10 天，随后每周 2 天，每次 1.6 mg，共持续 6 个月。未见不良反应或加速 β-细胞丢失，但见胰岛素特异性 T 细胞反应降低。在一项针对新发糖尿病患者的研究中，鼻内胰岛素降低了血清胰岛素自身抗体（IAAs），并在一组患者中降低了胰岛素特异性 T 细胞（IFN-γ）的应答；但并未延迟 β-细胞的破坏。

在芬兰 1 型糖尿病预测和预防研究（DIPP）中，当鼻内胰岛素（每天 1 单位/千克）施予携带高危人类白细胞抗原（HLA）单倍型的婴儿、或两种或两种以上 1 型糖尿病相关自身抗体阳性的同胞兄妹时，其并不能延迟或预防 1 型糖尿病。在为了研究特慢胰岛素（0.25 单位皮下注射，每年 4 天连续静脉输注）作为高危人群（5 年预测风险大于 50%）预防性疫苗的效果，而进行的 1 型糖尿病预防试验（DPT-1）结果表明，在中位数为 3.7 年的随访期内，该方案并不能预防或延缓 1 型糖尿病的发展。在另一个 DPT-1 试验中，具有 5 年预测风险 26%～50%（由代谢、免疫和遗传分期决定）的一级或二级亲属接受口服胰岛素（7.5 mg/d）或安慰剂，在中位数为 4.3 年的随访期内，既未延迟也未预防 1 型糖尿病。但在亚组分析中显示，口服胰岛素对胰岛素自身抗体水平高的患者有益。TrialNet 正在基于这些结果进行口服胰岛素预防试验，并招募具有与上述亚组相似特征的受试者——具有正常糖耐量的亲属，其血清中携带至少两种自身抗体特异性，其中一种必须是抗胰岛素，并在 5 年内呈现 35% 的 1 型糖尿病风险[16]。

在免疫治疗 1 型糖尿病（IMDIAB）试验中，对初诊糖尿病患者在确诊后 4 周内，给予口服胰岛素（5 mg/d）联合强化皮下胰岛素治疗 12 个月。在 12 个月的随访中，口服胰岛素未能维持 C 肽水平并降低胰岛素需求。此外，15 岁以下患者的 β-细胞丢失明显加快。在 1 型糖尿病胰岛素口服研究中，每日口服胰岛素（2.5 mg/d 或 7.5 mg/d）也不能抑制新发糖尿病患者的 T1DM。在一期临床试验中，对新近发病的糖尿病患者进行单次肌内注射胰岛素 B 链/IFA，结果表明，这种方法可诱导对胰岛素的强大抗体和 T 细胞反应，且不会引起不良反应。然而，在 C 肽水平上并未检测到显著差异[16]。表 29.3 总结了自身免疫性糖尿病免疫预防的各种干预措施。

表 29.3　预防自身免疫性糖尿病的各种干预措施总结

制剂	作用靶点/机制	递送方式	动物实验结果	人体实验结果	安全性
抗 CD3 单克隆抗体	对致病性 T 细胞的直接影响，调节性细胞群的诱导，或两者兼而有之[18]	肠外	诱导 NOD 小鼠长期缓解显性自身免疫[19]	能更好地维持 C 肽水平和降低胰岛素用量[1]	修饰的非 FC 受体(FcR)结合 CD3 抗体引起有丝分裂减弱，但具有同等的致耐受性
抗 CD20 单克隆抗体(利妥昔单抗)	通过 Fc 和补体介导的细胞毒性和促凋亡信号机制从外周循环中消耗人类 B 细胞[14]	肠外	由于缺乏能导致小鼠 B 细胞耗竭的抗 CD20 试剂而受到限制[20]	在 1 年的时间里，四个剂量的过程中部分地保持了 β 细胞的功能[21]	在第一次输注后主要是一级或二级不良反应。在随后的注射中，反应似乎很轻微[21]
BCG 接种	可刺激先天免疫途径并重置免疫内稳态[2]	皮内	注射卡介苗可抑制 NOD 小鼠的糖尿病发展[2]	增加胰岛素活性 T 细胞和 Tregs 数量，提高胰岛素敏感性[15,22]	低剂量重复接种卡介苗安全且耐受性好[22]
明矾配伍谷氨酸脱羧酶	Th1/Th2 转换，包括 GAD65 特异性调节性 T 细胞亚群的诱导和增殖[3]	GAD65 注射	对 NOD 小鼠没有不良反应或加重 T1D[8,16]	唯一被证明在 T1D 和 LADA 中都有效的抗原疫苗候选物[14]	临床结果支持 GAD65 免疫调节的安全性[3]
热休克蛋白和肽 277	人 Hsp60 在天然免疫系统的调节中起作用[6]	肠外	NOD 小鼠胰岛炎和糖尿病均得到预防[6]	在 10 个月的随访中，DiaPep277 组的 C 肽水平保持不变[6]	临床研究结果支持 DiaPep277 免疫调节的安全性[23]
胰岛素	胰岛素通过诱导调节性 T 细胞诱导自身抗原特异性耐受[24]	皮下、口服、鼻内	通过不同的途径给药时，部分保护 NOD 小鼠不患糖尿病[10,17]	迄今为止，人类临床试验的结果令人失望[16]	在安全性评估试验中，胰岛素没有产生明显的不良反应[16]

参考文献

[1] Cernea, S., Dobreanu, M., Raz, I.: Prevention of type 1 diabetes: today and tomorrow. Diabetes Metab. Res. Rev. 26, 602-605 (2010).

[2] Harrison, L. C.: The prospect of vaccination to prevent type 1 diabetes. Hum. Vaccin. 1, 143-150 (2005).

[3] Ludvigsson, J.: The role of immunomodulation t herapy in autoimmune diabetes. J. Diabetes Sci. Technol. 3, 320-330 (2009).

[4] Appel, S. J., et al.: Latent autoimmune diabetes of adulthood (LADA): an often misdiagnosed type of diabetes mellitus. J. Am. Acad. Nurse Pract. 21, 156-159 (2009).

[5] Hjorth, M. Immunological profi le and aspects of immunotherapy in type 1 diabetes. Linköping University Medical dissertations, vol. 1161 (2010). l iu. diva-portal. org/smash/get/diva2: 279876/ FULLTEXT01. Accessed 26 Aug 2012.

[6] Raz, I., Eldor, R., Naparstek, Y.: Immune modulation for prevention of type 1 diabetes mellitus. Trends Biotechnol. 23, 128-134 (2005).

[7] Zhang, L., Eisenbarth, G. S.: Prediction and prevention of type 1 diabetes mellitus. J. Diabetes 3, 48-57 (2011).

[8] Tian, J., Kaufman, D. L.: Antigen-based therapy for the treatment of type 1 diabetes. Diabetes 58, 1939-1946 (2009).

[9] Boettler, T., von Herrath, M.: Type 1 diabetes vaccine development: animal models vs. humans. Hum. Vaccin. 7, 19-26 (2011).

[10] Nicholas, D., Odumosu, O., Langridge, W. H.: Autoantigen based vaccines for type 1 diabetes. Discov. Med. 11, 293-301 (2011).

[11] Steinman, L.: Inverse vaccination, the opposite of Jenner's concept, for therapy of autoimmunity. J. Intern. Med. 67, 441-451 (2010).

[12] Petrovsky, N., Silva, D., Schatz, D. A.: Vaccine thera-pies for the prevention of type 1 diabetes mellitus. Paediatr. Drugs 5, 575-582 (2003).

[13] Gupta, S. K.: Vaccines for type 1 diabetes in the late stage of clinical development. Indian J. Pharmacol. 43, 485 (2011).

[14] Sanjeevi, C. B.: Type 1 diabetes research: Newer approaches and exciting developments. Int. J. Diabetes Dev Ctries 29, 49-51 (2009).

[15] Faustman, D. L., et al.: Proof-of-concept, randomized, controlled clinical trial of Bacillus-Calmette-Guerin for treatment of long-term type 1 diabetes. PLoS One 7, e41756 (2012).

[16] Clemente-Casares, X., Tsai, S., Huang, C., Santamaria, P.: Antigen-specifi c therapeutic approaches in type 1 diabetes. Cold Spring Harb. Perspect. Med. 2, a007773 (2012).

[17] Zhang, Z. J., Davidson, L., Eisenbarth, G., Weiner, H. L.: Suppression of diabetes in nonobese diabetic mice by oral administration of porcine insulin. Proc. Natl. Acad. Sci. U. S. A. 88, 10252-10256 (1991).

[18] Herold, K. C., et al.: Anti-CD3 monoclonal antibody in new-onset type 1 diabetes mellitus. N. Engl. J. Med. 346, 1692-1698 (2002).

[19] Chatenoud, L., Thervet, E., Primo, J., Bach, J. F.: Anti-CD3 antibody induces long-term remission of overt autoimmunity in nonobese diabetic mice. Proc. Natl. Acad. Sci. U. S. A. 91, 123-127 (1994).

[20] Bour-Jordan, H., Bluestone, J. A.: B cell depletion: a novel therapy for autoimmune diabetes? J. Clin. Invest. 117, 3642-3645 (2007).

[21] Pescovitz, M. D., et al.: Rituximab, B-lymphocyte depletion, and preservation of beta-cell function. N. Engl. J. Med. 361, 2143-2152 (2009).

[22] Huppmann, M., Baumgarten, A., Ziegler, A. G., Bonifacio, E.: Neonatal Bacille Calmette-Guerin vaccination and type 1 diabetes. Diabetes Care 28, 1204-1206 (2005).

[23] Fischer, B., Elias, D., Bretzel, R. G., Linn, T.: Immunomodulation with heat shock protein DiaPep277 to preserve beta cell function in type 1 diabetes–an update. Expert Opin. Biol. Ther. 10, 265–272 (2010).

[24] Rewers, M., Gottlieb, P.: Immunotherapy for the pre–vention and treatment of type 1 diabetes. Diabetes Care 32, 1769–1782 (2009).

（傅松波译）

第 30 章　I 型过敏症新型疫苗

Sandra Scheiblhofer，Josef Thalhamer
和 Richard Weiss①

摘要

如今在西方工业化国家，I 型过敏症影响了超过 30%的人口，给公共卫生系统带来了越来越重的负担。在本章中，我们对过敏原的分子特征和过敏致敏机制进行了综述。讨论了致敏的危险因素，如遗传易感性或环境因素（“卫生假说”），并总结了当前的诊断和药物治疗标准。由于传统的过敏原-特异性免疫疗法存在副作用大、患者依从性低、疗效差等缺点，本章重点介绍了克服这些缺点的新治疗方法，包括新的分子，如重组（低）过敏原或肽，也包括有先进的载体疫苗和基因疫苗。这种疫苗类型针对先天免疫系统的特异性受体，可导致免疫原性增加和调节不需要的 TH2 型应答。最后，提出了针对高免疫活性组织如皮肤或淋巴结的标准皮下注射或舌下应用替代途径。

30.1　疾病

目前，在发达国家，高达 1/3 的人口受到了 I 型过敏性疾病的影响，给公共卫生系统带来了巨大的经济负担。这种情况促进了对阐明 I 型过敏性疾病潜在机制、识别危险因素和开发新治疗干预措施方面的深入研究。

在某些条件下，树木和草花粉、尘螨、蟑螂、动物毛屑、食物、带刺昆虫的毒液、霉菌或乳胶等来源的其他无害的无处不在的蛋白质可作为过敏原，引发由 TH2 细胞驱动的过度免疫反应称为 I 型（或速发型）过敏反应或超敏反应，特应性一词，有时也用来表示这种疾病，仅用于指定过敏性疾病发展的遗传易感性。其特点是，血清 IgE 水平升高导致血液中肥大细胞和嗜碱性粒细胞活化。由此产生的炎症反应导致了诸如湿疹（特应性皮炎）、鼻结膜炎（“花粉热”）、胃肠道症状、哮喘和全身过敏在内的临床症状。

在儿童早期，过敏患者往往首先出现主要由牛奶和小麦的食物过敏原引起的胃肠道症状和过敏性皮炎。在以后的生活中，过敏性疾病可以消退或继续发展，如若为后者，则成年期会导致一系列病症，称之为“自然进程”[1]。

Prausnitz 和 Küstner 于 1921 年首次描述了过敏的 3 个主要临床表现成分：疾病诱发抗原（过敏原）、可区分过敏和健康个体的可转移血清因子（IgE）以及组织成分（肥大细胞）。

过敏性致敏发生在过敏原通过上皮屏障进入之时（图 30.1）。被激活之后，专业抗原呈递细胞将迁移至引流淋巴结，并促进静息性 T 淋巴细胞发展为涉及关键细胞因子白细胞介素-4（IL-4）的过敏原特异性 TH2 细胞[2]。TH2 细胞将迁移到淋巴结 B 细胞区，诱导亲和力成熟，并通过 B 细胞产生过敏原特异性免疫球蛋白 E（IgE），或将进入过敏原接触部位充当 TH2 效应细胞。

IgE 抗体与嗜碱性粒细胞和肥大细胞上的高亲和力受体 FcεRI 结合。在与相同的过敏原反复

---

① S. Scheiblhofer，PhD · J. Thalhamer，PhD · R. Weiss，PhD（✉）（萨尔茨堡大学分子生物学系，奥地利萨尔茨堡，E-mail：richard. weiss@ sbg. ac. at）

接触后，这些细胞表面的 IgE 受体可通过过敏原与结合的 IgE 分子的结合而交联[3]。这种交联引发细胞内信号的级联反应，最终激活嗜碱细胞、嗜酸性粒细胞和肥大细胞，使得这些细胞从其颗粒中释放炎症介质，包括组胺、白三烯和前列腺素，从而导致急性疾病的局部或全身症状。包括平滑肌收缩，血管扩张，黏液分泌，水肿，甚至危及生命的过敏性休克。

晚期反应通常发生在急性反应后 24 小时，起源于中性粒细胞、淋巴细胞、嗜酸性粒细胞和巨噬细胞，而且 TH2 淋巴细胞也会迁移到最初接触过敏原的部位，引起过敏性炎症。

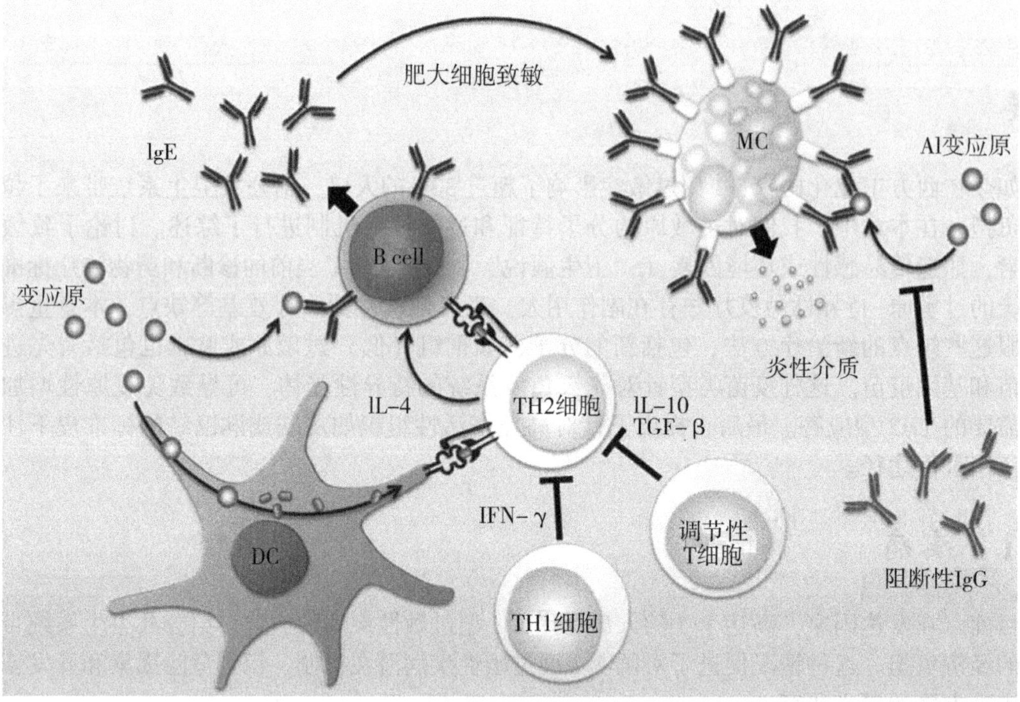

图 30.1 过敏性致敏

过敏原通过皮肤或呼吸道和肠道黏膜等上皮表面吸收。树突状细胞（dendritic cells，DC）吞噬并将过敏原加工成肽，然后递呈到 MHC Ⅱ类分子上，诱导 TH2 细胞产生。分泌 IFN-γ 的 TH1 细胞或分泌 IL-10 和/或 TGF-β 的调节性 T 细胞（Treg）可抑制这一过程。TH2 细胞产生促进过敏原特异性 B 细胞向 IgE 转换的 IL-4。IgE 抗体结合肥大细胞上的 FcεRI 受体，并在再次遇到过敏原时诱导释放炎症介质（如组胺）。高水平阻断性 IgG 的存在可以抑制过敏原与肥大细胞表面 IgE 结合。

30.2 过敏原

虽然人类暴露于大量的蛋白质和化合物中，但其中只有一小部分具有过敏原性，即诱导可产生特异性 IgE 的不当 TH2 偏倚免疫反应的能力，此过程称为过敏性致敏。已提出内源性以及外源性因素通过影响过敏原通过上皮屏障的进入、过敏原对 T 细胞的摄取、加工和递呈，以及与抗原呈递细胞有关的过程，来促进过敏性的说法。

与过敏原同时存在的影响致敏作用的外来因素，既包括来自过敏原来源的物质，如花粉相关脂质介质[4]、几丁质[5]，也包括环境污染物[6]。空气污染、香烟烟雾或臭氧可影响花粉过敏原的释放[7]，并可导致吸入性过敏原硝化反应，从而产生新的抗原表位[8]，或可直接作用于抗原呈递细胞[9]。从革兰氏阴性细菌的细胞外膜中提取的脂多糖（LPS），是一种具有促进或减少（取决于其在过敏原接触时的浓度）过敏性致敏作用的特殊性质的试剂。而极低剂量的脂多糖和模型过敏原卵白蛋白可导致耐受性，低剂量可诱导TH2反应，高剂量甚至可促进TH1免疫反应[10,11]。脂多糖通过与CD14/Toll样受体4/MD-2复合物结合发挥作用，导致促炎性细胞因子的分泌，特别是B细胞和巨噬细胞的分泌。该脂多糖识别Toll样受体4（TLR4）属于先天免疫系统的Toll样受体家族，其特异性识别高度保守的微生物结构[12]。MD-2，也称为淋巴细胞抗原96，通过与细胞表面的Toll样受体4结合而起到接头分子的作用，实现LPS诱导的信号传导[13]。

过敏原的内在特性包括折叠稳定性、化学修饰、寡聚、分子动力学和配体结合。蛋白质过敏原促进致敏的功能之一是蛋白酶活性。研究表明，屋尘螨过敏原Der P1作为半胱氨酸蛋白酶，在上皮屏障上分裂紧密连接蛋白，导致过敏原的渗透增强。此外，这种过敏原可蛋白水解T细胞和B细胞、以及树突状细胞（DC）上与Th2反应启动或增强有关的几个受体[14]。过敏原木瓜蛋白酶也是一种半胱氨酸蛋白酶，通过激活嗜碱性粒细胞可明显诱导Th2反应，并产生IgE[15]。

一些室内尘螨和霉菌过敏原以及蟑螂提取物中的丝氨酸蛋白酶活性可诱导上皮细胞的分子变化，从而促使树突状细胞极化T细胞，形成th2表型[16]。脂质结合和随后激活TLR的能力，是引起过敏性致敏的另一个内在因素。研究表明，屋尘螨过敏原Der p2通过与脂多糖结合而取代MD-2，从而通过Toll样受体4信号传导促进Th2反应[17]。树突状细胞上的C型凝集素受体结合碳水化合物，并被鉴定为针对病原体的免疫反应[18]。甘露糖受体介导了树突状细胞对屋尘螨、狗毛屑、蟑螂和花生过敏原的吸收，并导致了Th2极化[19]。因此，作者认为甘露糖受体在识别糖过敏原和形成树突状细胞的Th2反应中起着主要作用。

30.3 过敏性致敏的危险因素

特应性的个体具有遗传介导的过度免疫球蛋白E（IgE）介导反应的倾向。通过遗传和双胞胎研究发现，如双亲均是特应性个体，其后代患IgE介导疾病的风险将增加到60%，而健康个体的后代则为5%~10%[20]。已经确定了一些可以促进或阻止特应性发生的遗传多态性。其中，导致皮肤屏障功能减弱的丝聚蛋白基因编码突变，与湿疹向哮喘的易化转变有关。这些多态性可作为疾病症状发生前预测哮喘发展的标志物[21]。相比之下，已发现发生特应性的患病率降低，与编码先天免疫系统受体基因（如Toll样受体4）中的某些多态性有关[22]。

在高度发达的国家，已对多种生活方式相关因素展开密切调查，以确定其与Ⅰ型过敏性疾病的相关性，包括饮食习惯、空气污染、被动吸烟以及儿童期传染病数量的减少。其中，儿童期传染病数量的减少是通过提高卫生标准和大规模接种疫苗而实现，已被发现是特应性疾病的重要危险因素之一[23]。

20世纪90年代末，发表了第一份关于农场儿童与来自同一地区但在成长过程中未与马厩和牲畜密切接触的儿童相比，其过敏性致敏显著降低的报告。并将此类现象命名为“卫生假说”，解释了过去几十年来特应性疾病发病率不断上升的原因（图30.2）。由此可见，儿童早期接触微生物化合物的减少，可能导致对先天免疫系统的刺激减少，同时亦导致对过敏原的TH2偏倚适应性免疫反应的转变。同时，已经确定了导致农场儿童过敏和哮喘风险降低的一些关键因素。

包括接触牲畜（牛、猪、家禽）、接触动物饲料（干草、谷物、稻草、青贮饲料）和食用未经加工的牛奶[24]。吸收的主要途径是吸入和摄入。最好的保护措施是在子宫内和出生的最初几年内进行接触[25,26]。在分子水平上，起保护作用的化合物来源于细菌和真菌。其中包括革兰氏阳性细菌细胞壁的一种成分胞壁酸，和青霉菌属及曲霉属真菌的胞外多糖[27,28]。有趣的是，脂多糖似乎可以预防过敏性致敏，但不能预防儿童哮喘和喘息[29]。

TLR2、TLR4和CD14mRNA水平显著高于非农场儿童[25,30]。此外，在母亲未接触动物棚和饲料的新生儿脐带血中观察到的，季节性过敏原特异性IgE抗体水平显著升高。免疫球蛋白E（IgE）反应与TH1相关细胞因子IFN-γ和TNF的减少有关[31]。这些发现支持这样的假设：先天免疫系统感知了在农业环境中的微生物化合物，并随后形成了适应性免疫系统。

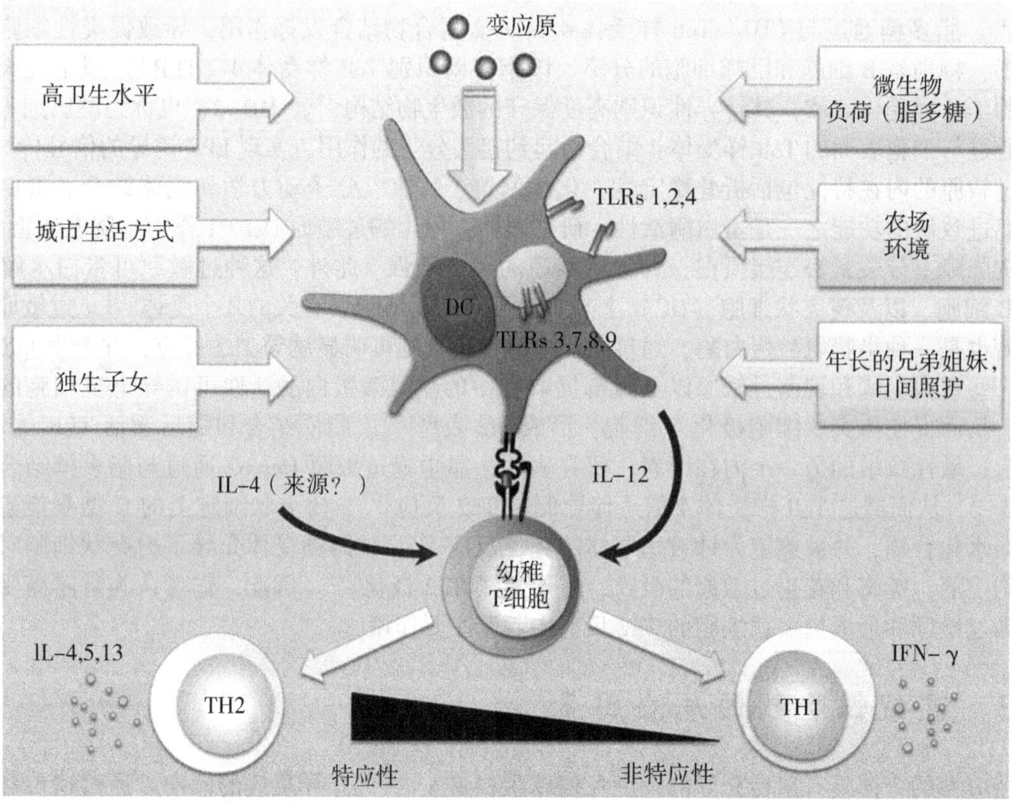

图 30.2　卫生假说

通过刺激促进 IL-12 分泌的先天免疫受体，导致保护性 Th1 免疫产生的外源性因素。高卫生水平导致免疫刺激不足，促进了 TH2 反应和特应性的产生。在这一过程中，IL-4 的作用及其细胞来源仍具有很大争议

对 TH1 的“缺失免疫偏离”是否是导致 I 型过敏症患病率持续上升的唯一机制，尚待阐明。动物模型研究和流行病学证据表明，调节性 T 细胞活性降低引起的免疫抑制缺乏也可能参与了这一过程。最有可能的是，两种机制（可能还有其他机制）的结合可能是主要原因[32]。

30.4　诊断和经典治疗（临床“黄金标准”）

30.4.1　诊断

两种具有相似特异性和敏感性的方法可用于评估过敏原特异性 IgE，即皮肤点刺试验和过敏性血液试验。

皮肤试验，也称为点刺或穿刺试验，是使用一个小的金属或塑料装置，在患者前臂内侧或背部皮肤上进行一系列小点刺或穿刺。之前，在皮肤区域上应用疑似过敏原或过敏原提取物，并用染料或笔在这些区域作标记。通过穿刺将相应的过敏原或提取物引入皮肤之中。或者，也可以使用针头和注射器进行皮内注射。如果患者对 1 种或多种测试过敏原或提取物过敏，通常会在 30 分钟内出现炎症反应。皮肤点刺试验的结果是从临界反应（±）到强烈反应（4+）进行评分，反映了皮肤从轻微发红，到所谓的让人联想起蚊虫叮咬的风团和潮红的反应。

同时，还可以确定风团和潮红的直径。如果病人患有广泛的皮肤病，最近服用了抗组胺药，或已经受到危及生命的过敏反应的困扰，通常血液测试将是首选方法。与皮肤点刺试验相比，

血液测试有几个优点：患者的年龄、皮肤状况、药物、症状或疾病严重程度不是排除标准。这种方法对于非常年幼的儿童和婴儿来说也更容易被接受，因为一根针头足以采集一份检测一大组过敏原的血液样本。患者不接触潜在的致敏物质，也提供了额外的安全性。通过血液检测，确定过敏原特异性 IgE 的浓度。最初在 20 世纪 70 年代由瑞典 Pharmacia Diagnostics AB 公司开发，用于使用放射性同位素标记的抗 IgE 抗体，并作为 RAST（放射性过敏原吸附试验）销售，如今，利用荧光标记的 IgE 结合抗体的所谓免疫 CAP 试验，已使 RAST（作为几种体外过敏试验的口语称谓）一词过时。

除皮肤点刺试验和过敏性血液试验外，还可采用其他方法。诱发试验特别适用于食物和药物过敏的情况。通过口服或吸入引入少量疑似过敏原，并由过敏医生对患者进行严密监护。在斑贴试验中，使用粘性贴片贴在患者背部，作为确定皮肤接触过敏或接触性皮炎原因的方法。随后，检查皮肤的局部反应。

30.4.2 药物治疗

虽然在食品过敏的情况下，避免过敏原可减轻症状，甚至可预防危及生命的过敏反应，是一种简单有效治疗方法，但这项措施对于患有空气过敏的患者来说，却很难实现。

对于标准的药物治疗，可选择几种免疫抑制和抗炎药物，包括抗组胺药、糖皮质激素和 β 受体激动剂。抗组胺药可阻断肥大细胞和嗜碱性粒细胞在脱粒过程中释放介质的作用。鼻内或全身应用包括强的松在内的皮质类固醇作为消炎药来治疗过敏性鼻炎。β-2-肾上腺素受体激动剂，如沙丁胺醇（短效）或沙美特罗（长效），是快速缓解哮喘症状的有效支气管扩张剂。

虽然所有这些药物均能缓解过敏症状，但它们并不针对潜在的免疫性疾病。

另一种已被批准用于中重度哮喘的治疗方法是皮下注射人源化单克隆抗 IgE 抗体[33]，如奥马珠单抗（商品名为 Xolair）。由于其高成本，奥马珠单抗主要用于严重持续性哮喘患者。单克隆抗体选择性地结合血液和间质液中的游离 IgE 和 B 淋巴细胞表面的膜结合 IgE，但不与肥大细胞、嗜碱性粒细胞和抗原递呈树突状细胞表面的高亲和力 IgE 受体结合。奥马珠单抗通过部分掩盖与免疫球蛋白 E 结合的位点，来阻断免疫球蛋白 E 与肥大细胞和嗜碱性粒细胞上的高亲和力受体结合。一旦 IgE 与该受体结合，由于空间位阻，奥马珠单抗就不能再结合，从而避免了 IgE 分子交联后肥大细胞和嗜碱细胞释放介质所产生的过敏反应。最重要的是，奥马珠单抗还会耗尽患者体内的游离免疫球蛋白 E（IgE），导致嗜碱性粒细胞、肥大细胞和树突状细胞上的 IgE 受体逐渐下调。由此，这些细胞对过敏原的刺激变得不那么敏感[34]。

30.4.3 特异性免疫治疗

努恩和弗里曼在 100 多年前就已引入过敏原特异性免疫疗法（SIT）[35]，并且其仍然是过敏性患者的唯一因果治疗，该治疗法更改了针对过敏原的不适当和夸大的 TH2 驱动免疫反应（图 30.3）。这些免疫变化的特征是促进 Th1 细胞因子，如 IFN-γ，并通过血液和炎症气道中的 T 调节细胞诱导 IL-10/TGF-β 分泌。过敏原特异性免疫治疗（SIT）还与抑制过敏原特异性 IgE、诱导 IgG4，以及抑制肥大细胞、嗜碱性粒细胞和嗜酸性粒细胞有关。自 SIT 首次应用以来，其临床实践并未实质性改变或改进。这种疗法主要是通过 50~80 次皮下注射（SCIT），在 3~5 年内逐渐增加过敏原剂量[36]。已有这种疗法引起的局部或全身性副作用相关报道[37]。由于几乎不使用特征性的过敏原提取物，因此存在治疗诱导新致敏的潜在风险。

作为一种对患者更友好的替代方案，使用滴剂或片剂的舌下免疫疗法（SLIT）已获得批准（SLIT）[38]。这种方法避免了使用针头和注射器，并提供了自我给药的可能性。舌下免疫疗法需要在与皮下注射相当的时间间隔内，每天摄入大量过敏原，并且经常伴随口腔和胃肠道副作用[39]。此外，由于与口腔黏膜接触时间较短，导致的过敏原吸收不良，使得舌下免疫疗法的疗效不如皮下注射[40]。

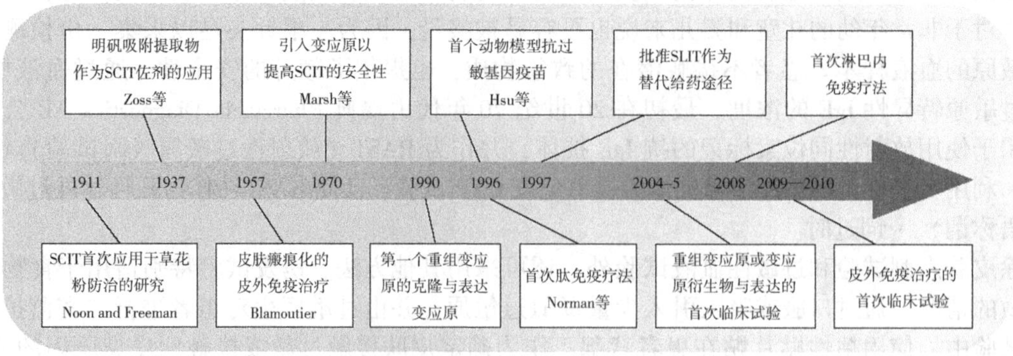

图 30.3　过敏原特异性免疫治疗（SIT）的里程碑

30.5　新型过敏疫苗

经典皮下注射相关副作用以及过敏原特异性舌下含服免疫治疗的低依从性和缺乏效力，促使人们努力开发新的治疗和（最近）预防性疫苗接种方法（图 30.4）。

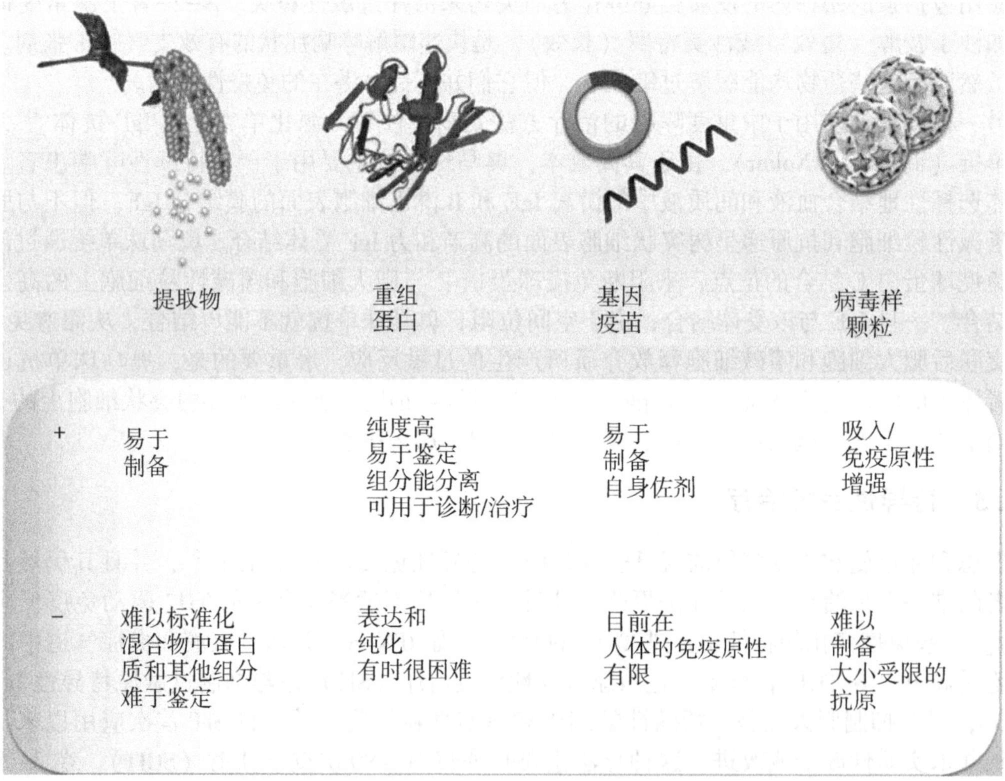

图 30.4　新型过敏原特异性免疫治疗疫苗及其与经典过敏原提取物的优缺点比较

30.5.1　重组（低）过敏原

目前，过敏原特异性免疫治疗（SIT）是基于天然来源的过敏原提取物，通常定义不清，难以标准化。此外，这种提取物中含有大量不同的蛋白质，患者有可能会对治疗前未致敏的成分产生新的致敏作用[41]。因此，下一代过敏诊断和治疗将基于重组蛋白，允许根据患者的致敏概况进行定制治疗[42]。除了高度标准化产品的益处之外，重组过敏原的生产还允许对目的过敏原进行修改。通过改变蛋白B细胞表位的三维结构，可以破坏B细胞表位，减少过敏原与患者体内原有免疫球蛋白E（IgE）的结合，从而避免肥大细胞介导物的释放，该介导物是常规过敏原特异性免疫治疗（SIT）过程中副作用的主要来源。这种具有低IgE结合潜能的修饰过敏原被称为低过敏原[43-45]。低过敏原也可以通过电子模拟突变系统地产生。突变对编码分子折叠稳定性的影响可以通过z-score计算来预测。我们最近已经证明，显示出降低IgE结合能力的基于折叠稳定性降低的新型低过敏原[45]，可以通过z-score计算产生。

30.5.2　肽免疫治疗

肽免疫疗法将这一概念向前推进了一步，因为它只应用合成肽。由于B细胞表位通常是不连续的（它们由蛋白质折叠而成的连续分离的氨基酸片段组成），这些疫苗不能再与患者的IgE结合，但能保持刺激T细胞反应的能力。虽然这是一个令人信服的概念，但肽疫苗常常具有低免疫原性。临床试验表明，虽然可以避免立即反应，但在治疗的早期阶段，已经报告了独立于IgE的晚期哮喘反应[46]。

30.5.3　病毒和细菌载体疫苗

除了使用纯化的重组过敏原或过敏原肽外，还可以通过重组病毒或细菌载体施用治疗性蛋白。这些系统目前仅在动物的临床前研究中进行了测试，并显示了改进的递送方式以及源自微生物载体成分共同刺激的益处。例如，改良型安卡拉痘苗病毒（MVA）[47]以及表达（模型）过敏原的重组卡介苗（BCG）[48]或植物乳杆菌[49]，通过刺激先天免疫受体建立TH1偏向的免疫应答，来保护免受过敏性敏化。由于使用活的或致弱病毒活载体时的安全性问题以及与抗载体免疫原性相关的问题，一种简化的方法是使用目的抗原的病毒样颗粒，其已在最近针对主要的猫过敏原Fel d1被证实。在本研究中认为阻断IgG抗体是主要的效应机制[50]。

30.5.4　基因疫苗

20世纪90年代，一种结合病毒感染的免疫刺激性和重组蛋白安全性的简化方法开始流行，即使用基因疫苗。基因疫苗接种仅适用于抗原的遗传信息，可作为含有真核启动子（DNA疫苗接种）的质粒DNA[51]，或合成的信使RNA（mRNA疫苗接种）[52]。遗传物质被宿主细胞吸收，例如角质形成细胞或成纤维细胞（当皮内施用时），并翻译成相应的蛋白质。除了利用体细胞作为蛋白工厂外，引入的DNA或mRNA还可以通过先天免疫受体如TLR9（DNA）和TLR3、TLR7、TLR8（RNA）刺激免疫系统。因为它们不能整合到基因组中，只能瞬时表达，mRNA疫苗比DNA疫苗更具优势[53]。由于基因疫苗是设计先进类型疫苗的一个高度通用平台，各种创新方法已经在过敏动物模型中进行了试验。通过泛素与其N-末端的共价连接，将质粒编码的过敏原表达。我们已经使基因产物低过敏原性，即虽然T细胞反应基本上不受影响，但天然过敏原的完全降解阻止了抗体的形成[43]。

自我复制的DNA和RNA疫苗代表了另一类基因疫苗，我们已经将其用于小鼠过敏性疾病的预防和治疗。编码可驱动自身扩增以及编码过敏原的转录和翻译的α病毒衍生复制酶分子、自复制DNA以及RNA疫苗，已经证明了在比常规载体低100倍的剂量下，具有相同功效[52,54,55]。它们优越的免疫原性均基于通过双链RNA中间体提供危险信号，这些双链RNA中间体被几种固有的抗病毒受体所识别[56]。

30.5.5　过敏原特异性免疫治疗的新途径

虽然已证实皮下注射和舌下免疫疗法（SLIT）的临床疗效[57]，但仅有一小部分过敏患者决定接受这种治疗，来取代服用缓解症状的药物。传统过敏原特异性免疫治疗接受率低、退出率高的主要原因是，治疗时间长和治疗引起的副作用。因此，创新治疗 I 型过敏症的方法（图 30.5）旨在通过避免过敏原与大循环的接触来提升疗效、缩短时间，并提高安全性。

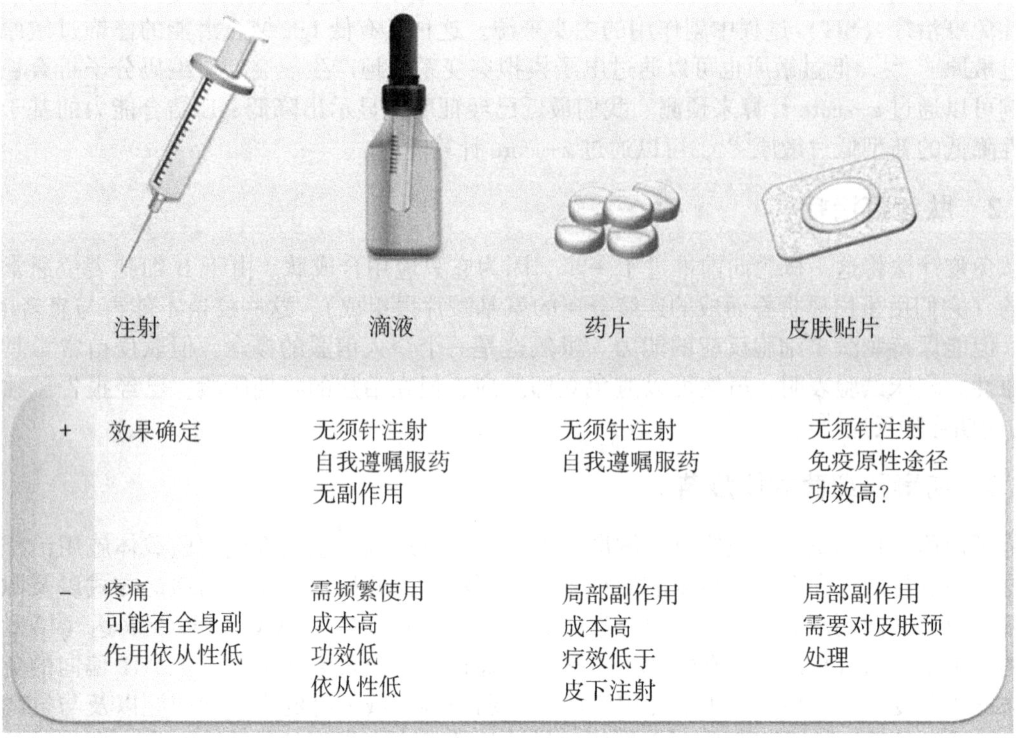

图 30.5　过敏原特异性免疫治疗不同免疫途径的优缺点

皮肤是一种很具治疗意义的靶向组织，它易于接近，富含抗原呈递细胞，并且在表层没有血管[58]。皮肤的真皮和表皮层含有高度免疫活性的细胞类型，如真皮树突状细胞和朗格汉斯细胞，但也含有有助于皮肤免疫功能的其他类型细胞，如肥大细胞、天然杀伤细胞和角质形成细胞[59]。另外，皮肤被局部淋巴结有效地引流，这是皮下组织的另一个优势。在 20 世纪 50 年代[60]就已引入了皮肤免疫疗法，最近又被重新发现，并称为经皮免疫疗法（TCIT）或表皮免疫疗法（EPIT）。

临床研究主要采用两种不同的方法来打破皮肤的最外层，即角质层，以增加过敏原的吸收。一种简单的方法是采用胶带，多次剥离胶带后，在贴剂中加入含过敏原的溶液[61,62]。另一种方法是将含有过敏原的贴片作为干粉末涂在完好的皮肤上。长期贴敷会产生一种湿润腔，削弱皮肤的屏障功能，导致粉剂的水合作用，促进吸收[63]。基于激光的微孔技术，是一种高度可重复性且适应性强的方式，以绕过皮肤屏障的创新方法。

我们已经证明，通过使用发射远红外激光束的装置，可以产生数量、密度和深度可变的水性微孔，并且通过这些微孔进行经皮免疫，是诱导特异性免疫反应的合适技术[64]。在一个过敏性哮喘小鼠模型中，我们比较了激光微孔经皮免疫治疗和标准皮下免疫治疗，发现这些治疗方法在减少气道高反应性和白细胞浸润肺方面同样有效。经皮应用避免了皮下注射观察到的，TH2 细胞因子治疗相关的全身性增加[65]。

过敏原特异性免疫治疗的另一个有趣途径是淋巴管内应用，称为 ILIT，通过针直接将过敏原传递到皮下淋巴结。并对草花粉过敏患者和猫皮屑过敏患者进行了临床研究，均取得了很好的效果[66,67]。

参考文献

[1] Locksley, R. M.: Asthma and allergic infl ammation. Cell 140, 777-783 (2010). doi: 10. 1016/j. cell. 2010. 03. 004.

[2] Mowen, K. A., Glimcher, L. H.: Signaling pathways in Th2 development. Immunol. Rev. 202, 203-222 (2004). doi: 10. 1111/j. 0105-2896. 2004. 00209. x.

[3] Turner, H., Kinet, J. P.: Signalling through the high-affi nity IgE receptor Fc epsilonRI. Nature 402, B24-B30 (1999).

[4] Gilles, S., et al.: Pollen allergens do not come alone: pollen associated lipid mediators (PALMS) shift the human immune systems towards a T (H) 2-dominated response. Allergy Asthma Clin. Immunol. 5, 3 (2009). doi: 10. 1186/1710-1492-5-3.

[5] Burton, O. T., Zaccone, P.: The potential role of chitin in allergic reactions. Trends Immunol. 28, 419-422 (2007). doi: 10. 1016/j. it. 2007. 08. 005.

[6] Morgenstern, V., et al.: Atopic diseases, allergic sensitization, and exposure totraffi c-related air pollution in children. Am. J. Respir. Crit. Care Med. 177, 1331-1337 (2008). doi: 10. 1164/rccm. 200701-036OC.

[7] Behrendt, H., Becker, W. M.: Localization, release and bioavailability of pollen allergens: the infl uence of environmental factors. Curr. Opin. Immunol. 13, 709-715 (2001). doi: S0952-7915 (01) 00283-7 [pii].

[8] Gruijthuijsen, Y. K., et al.: Nitration enhances the allergenic potential of proteins. Int. Arch. Allergy Immunol. 141, 265-275 (2006). doi: 10. 1159/000095296.

[9] Williams, M. A., et al.: Disruption of the transcription factor Nrf2 promotes pro-oxidative dendritic cells that stimulate Th2-likeimmunoresponsiveness upon activation by ambient particulate matter. J. Immunol. 181, 4545-4559 (2008).

[10] Eisenbarth, S. C., et al.: Lipopolysaccharide-enhanced, toll-like receptor 4-dependent T helper cell type 2 responses to inhaled antigen. J. Exp. Med. 196, 1645-1651 (2002).

[11] Herrick, C. A., Bottomly, K.: To respond or not to respond: T cells in allergic asthma. Nat. Rev. Immunol. 3, 405-412 (2003). doi: 10. 1038/nri1084.

[12] Poltorak, A., et al.: Defective LPS signaling in C3H/HeJ and C57BL/10ScCr mice: mutations in Tlr4 gene. Science 282, 2085-2088 (1998).

[13] Shimazu, R., et al.: MD-2, a molecule that confers lipopolysaccharide responsiveness on Toll-like receptor 4. J. Exp. Med. 189, 1777-1782 (1999).

[14] Furmonaviciene, R., et al.: The protease aller-gen Der p 1 cleaves cell surface DC-SIGN and DC-SIGNR: experimental analysis of in silico substrate identifi cation and implications in allergic responses. Clin. Exp. Allergy 37, 231-242 (2007). doi: 10. 1111/j. 1365-2222. 2007. 02651. x.

[15] Sokol, C. L., Barton, G. M., Farr, A. G., Medzhitov, R.: A mechanism for the initiationof allergen-induced T helper type 2 responses. Nat. Immunol. 9, 310-318 (2008). doi: 10. 1038/ni1558.

[16] Comeau, M. R., Ziegler, S. F.: The infl uence of TSLP on the allergic response. Mucosal Immunol. 3, 138-147 (2010). doi: 10. 1038/mi. 2009. 134.

[17] Trompette, A., et al.: Allergenicity resulting from functional mimicry of a Toll-like receptor complex protein. Nature 457, 585-588 (2009). doi: 10. 1038/ nature07548.

[18] van Kooyk, Y.: C-type lectins on dendritic cells: key modulators for the induction of immune responses. Biochem. Soc. Trans. 36, 1478-1481 (2008). doi: 10. 1042/BST0361478.

[19] Royer, P. J., et al.: The mannose receptor mediates the uptake of diverse native allergens by dendritic cells and determines allergen-induced T cell polarization through modulation of IDO activity. J. Immunol. 185, 1522-1531 (2010). doi: 10. 4049/jimmunol. 1000774.

[20] Kjellman, N. I.: Atopic disease in seven-year-old chil-dren. Incidence in relation to family history. Acta Paediatr. Scand. 66, 465-471 (1977).

[21] Marenholz, I., et al.: An interaction between fi laggrin mutations and early food sensitization improves the prediction of childhood asthma. J. Allergy Clin. Immunol. 123, 911-916 (2009). doi: 10. 1016/j. jaci. 2009. 01. 051.

[22] Senthilselvan, A., et al.: Association of polymor-phisms of toll-like receptor 4 with a reduced prevalence of hay fever and atopy. Ann. Allergy Asthma Immunol. 100, 463-468 (2008). doi: 10. 1016/S1081-1206 (10) 60472-3.

[23] Floistrup, H., et al.: Allergic disease and sensitization in Steiner school children. J. Allergy Clin. Immunol. 117, 59-66 (2006). doi: 10. 1016/j. jaci. 2005. 09. 039.

[24] von Mutius, E., Vercelli, D.: Farm living: effects on childhood asthma and allergy. Nat. Rev. Immunol. 10, 861-868 (2010). doi: 10. 1038/nri2871.

[25] Ege, M. J., et al.: Prenatal farm exposure is related to the expression of receptors of the innate immunity and to atopic sensitization in school-age children. J. Allergy Clin. Immunol. 117, 817-823 (2006). doi: 10. 1016/j. jaci. 2005. 12. 1307.

[26] Riedler, J., et al.: Exposure to farming in early life and development of asthma and allergy: a cross-sectional survey. Lancet 358, 1129-1133 (2001). doi: 10. 1016/ S0140-6736 (01) 06252-3.

[27] Ege, M. J., et al.: Not all farming environments protect against the development of asthma and wheeze in children. J. Allergy Clin. Immunol. 119, 1140-1147 (2007). doi: 10. 1016/j. jaci. 2007. 01. 037.

[28] vanStrien, R. T., et al.: Microbial exposure of rural school children, as assessed by levels of N-acetyl-m uramic acid in mattress dust, and its association with respiratory health. J. Allergy Clin. Immunol. 113, 860-867 (2004). doi: 10. 1016/j. jaci. 2004. 01. 783.

[29] Vogel, K., et al.: Animal shed Bacillus licheniformis spores possess allergy-protective as well as infl ammatory properties. J. Allergy Clin. Immunol. 122, 307-312, 312 e301-e308 (2008). doi: 10. 1016/j. jaci. 2008. 05. 016.

[30] Lauener, R. P., et al.: Expression of CD14 and Toll-like receptor 2 in farmers' and non-farmers' children. Lancet 360, 465-466 (2002). doi: 10. 1016/ S0140-6736 (02) 09641-1.

[31] Pfefferle, P. I., et al.: Cord blood allergen-specifi c IgE is associated with reduced IFN-gamma production by cord blood cells: the Protection against Allergy-Study in Rural Environments (PASTURE) Study. J. Allergy Clin. Immunol. 122, 711-716 (2008). doi: 10. 1016/j. jaci. 2008. 06. 035.

[32] Romagnani, S.: The increased prevalence of allergy and the hygiene hypothesis: missing immune deviation, reduced immune suppression, or both? Immunology 112, 352-363 (2004). doi: 10. 1111/j. 1365-2567. 2004. 01925. x.

[33] Chang, T. W., Wu, P. C., Hsu, C. L., Hung, A. F.: Anti-IgE antibodies for the treatment of IgE-mediated allergic diseases. Adv. Immunol. 93, 63-119 (2007). doi: 10. 1016/S0065-2776 (06) 93002-8.

[34] Scheinfeld, N.: Omalizumab: a recombinant human-ized monoclonal IgE-blocking antibody. Dermatol. Online J. 11, 2 (2005).

[35] Calderon, M., Cardona, V., Demoly, P.: One hundred years of allergen immunotherapy European Academy of Allergy and Clinical Immunology celebration: review of unanswered questions. Allergy 67, 462-476 (2012). doi: 10. 1111/j. 1398-9995. 2012. 02785. x.

[36] Cox, L., Calderon, M. A.: Subcutaneous specifi c immunotherapy for seasonal allergic rhinitis: a review of treatment practices in the US and Europe. Curr. Med. Res. Opin. 26, 2723-2733 (2010). doi: 10. 1185/ 03007995. 2010. 528647.

[37] Frew, A. J.: Allergen immunotherapy. J. Allergy Clin. Immunol. 125, S306-S313 (2010). doi: 10. 1016/ j. jaci. 2009. 10. 064.

[38] Canonica, G. W., et al.: Sub-lingual immuno-therapy: World Allergy Organization Position Paper 2009. Allergy 64 (Suppl 91), 1-59 (2009). doi: 10. 1111/j. 1398-9995. 2009. 02309. x.

[39] Frew, A. J.: Sublingual immunotherapy. N. Engl. J. Med. 358, 2259-2264 (2008). doi: 10. 1056/NEJMct0708337.

[40] Razafi ndratsita, A., et al.: Improvement of sublingual immunotherapy effi cacy with a mucoadhesive allergen formulation. J. Allergy Clin. Immunol. 120, 278-285 (2007). doi: 10. 1016/j. jaci. 2007.

04. 009.

[41] Ball, T., et al.: Induction of antibody responses to new B cell epitopes indicates vaccination character of allergen immunotherapy. Eur. J. Immunol. 29, 2026-2036 (1999). doi: 10. 1002/ (SICI) 1521-4141 (199906) 29: 06< 2026:: AID-IMMU2026> 3. 0. CO; 2-2.

[42] Pauli, G., Malling, H. J.: The current state of recombinant allergens for immunotherapy. Curr. Opin. Allergy Clin. Immunol. 10, 575-581 (2010). doi: 10. 1097/ACI. 0b013e32833fd6c5.

[43] Bauer, R., et al.: Generation of hypoallergenic DNA vaccines by forced ubiquitination: preventive and therapeutic effectsin a mouse model of allergy. J. Allergy Clin. Immunol. 118, 269-276 (2006). doi: 10. 1016/j. jaci. 2006. 03. 033.

[44] Purohit, A., et al.: Clinical effects of immunotherapy with genetically modifi ed recombinant birch pollen Bet v1 derivatives. Clin. Exp. Allergy 38, 1514-1525 (2008). doi: 10. 1111/j. 1365-2222. 2008. 03042. x.

[45] Thalhamer, T., et al.: Designing hypoallergenic deriv-atives for allergy treatment by means of in silico mutation and screening. J. Allergy Clin. Immunol. 125, 926-934, e910 (2010). doi: 10. 1016/j. jaci. 2010. 01. 031.

[46] Haselden, B. M., Kay, A. B., Larche, M.: Immunoglobulin E-independent major histocompatibility complex-restricted T cell peptide epitope-induced late asthmatic reactions. J. Exp. Med. 189, 1885-1894 (1999).

[47] Albrecht, M., et al.: Vaccination with a Modifi ed Vaccinia Virus Ankara-based vaccine protects mice from allergic sensitization. J. Gene Med. 10, 1324-1333 (2008). doi: 10. 1002/jgm. 1256.

[48] Kumar, M., Behera, A. K., Matsuse, H., Lockey, R. F., Mohapatra, S. S.: A recombinant BCG vaccine generates a Th1-like response and inhibits IgE synthesis in BALB/c mice. Immunology 97, 515-521 (1999).

[49] Rigaux, P., et al.: Immunomodulatory properties of Lactobacillus plantarum and itsuse as a recombinant vaccine against mite allergy. Allergy 64, 406-414 (2009). doi: 10. 1111/j. 1398-9995. 2008. 01825. x.

[50] Schmitz, N., et al.: Displaying Fel d1 on virus-like particles prevents reactogenicity despite greatly enhanced immunogenicity: a novel therapy for cat allergy. J. Exp. Med. 206, 1941-1955 (2009). doi: 10. 1084/jem. 20090199.

[51] Weiss, R., et al.: Is genetic vaccination against allergy possible? Int. Arch. Allergy Immunol. 139, 332-345 (2006). doi: 10. 1159/000091946.

[52] Roesler, E., et al.: Immunize and disappear-safety-optimized mRNA vaccination with a panel of 29 allergens. J. Allergy Clin. Immunol. 124, 1070-1077, e1071-1011, doi: 10. 1016/j. jaci. 2009. 06. 036 (2009).

[53] Weiss, R., Scheiblhofer, S., Roesler, E., Weinberger, E., Thalhamer, J.: mRNA vaccination as a safe approach for specifi c protection from type I allergy. Expert Rev. Vaccines 11, 55-67 (2012). doi: 10. 1586/ erv. 11. 168.

[54] Gabler, M., et al.: Immunization with a low-dose rep-licon DNA vaccine encoding Phl p 5 effectively prevents allergic sensitization. J. Allergy Clin. Immunol. 118, 734-741 (2006). doi: 10. 1016/j. jaci. 2006. 04. 048. S0091-6749 (06) 00943-2 [pii].

[55] Scheiblhofer, S., et al.: Inhibition of type I allergic responses with nanogram doses of replicon-based DNA vaccines. Allergy 61, 828-835 (2006). doi: 10. 1111/ j. 1398-9995. 2006. 01142. x. ALL1142 [pii].

[56] Leitner, W. W., Bergmann-Leitner, E. S., Hwang, L. N., Restifo, N. P.: Type I Interferons are essential for the effi cacy of replicase-based DNA vaccines. Vaccine 24, 5110-5118 (2006). doi: 10. 1016/j. vaccine. 2006. 04. 059.

[57] Eifan, A. O., Shamji, M. H., Durham, S. R.: Long-term clinical and immunological effects of allergen immunotherapy. Curr. Opin. Allergy Clin. Immunol. 11, 586 - 593 (2011). doi: 10. 1097/ ACI. 0b013e32834cb994.

[58] Bal, S. M., Ding, Z., Jiskoot, W., Bouwstra, J. A.: Advances in transcutaneous vaccine delivery: do all ways lead to Rome? J. Control. Release 148, 266-282 (2010). doi: 10. 1016/j. jconrel. 2010. 09. 018.

[59] Gutowska-Owsiak, D., Ogg, G. S.: The epidermis as an adjuvant. J. Invest. Dermatol. 132, 940-948 (2012). doi: 10. 1038/jid. 2011. 398.

[60] Blamoutier, P., Blamoutier, J., Guibert, L.: Treatment of pollinosis with pollen extracts by the method of cutaneous quadrille ruling. Presse Med. 67, 2299-2301 (1959).

[61] Senti, G., et al.: Epicutaneous allergen administration as a novel method of allergen-specifi c immunotherapy. J. Allergy Clin. Immunol. 124, 997-1002 (2009). doi: 10. 1016/j. jaci. 2009. 07. 019.

[62] Senti, G., et al.: Epicutaneous allergen-specifi c immunotherapy ameliorates grass pollen-induced rhinoconjunctivitis: a double-blind, placebo-controlled dose escalation study. J. Allergy Clin. Immunol. 129, 128-135 (2012). doi: 10. 1016/j. jaci. 2011. 08. 036.

[63] Dupont, C., et al.: Cow's milk epicutaneous immuno-therapy in children: a pilot trial of safety, acceptability, and impact on allergic reactivity. J. Allergy Clin. Immunol. 125, 1165-1167 (2010). doi: 10. 1016/j. jaci. 2010. 02. 029.

[64] Weiss, R., et al.: Transcutaneous vaccination via laser microporation. J. Control. Release (2012). doi: 10. 1016/j. jconrel. 2012. 06. 031.

[65] Bach, D., et al.: Transcutaneous immunotherapy via laser-generated micropores effi ciently alleviates allergic asthma in Phl p 5-sensitized mice. Allergy (2012). doi: 10. 1111/all. 12005.

[66] Senti, G., et al.: Intralymphatic immunotherapy for cat allergy induces tolerance after only 3 injections. J. Allergy Clin. Immunol. 129, 1290-1296 (2012). doi: 10. 1016/j. jaci. 2012. 02. 026.

[67] Senti, G., et al.: Intralymphatic allergen administra-tion renders specifi c immunotherapy faster and safer: a randomized controlled trial. Proc. Natl. Acad. Sci. U. S. A. 105, 17908-17912 (2008). doi: 10. 1073/pnas. 0803725105.

（吴国华译）

第31章　基于水稻种子的过敏疫苗：针对柳杉花粉和屋尘螨过敏的过敏原特异性口服耐受性的诱导

Fumio Takaiwa 和 Takachika Hiroi①

摘要

日本柳杉花粉和屋尘螨过敏原是免疫球蛋白（IG）E 介导 I 型变态反应的主要原因。针对这些过敏症，使用粗过敏原提取物进行全身免疫的过敏原特异性免疫疗法已经成为唯一的治疗方法。在这里，我们介绍一种新的过敏原特异性免疫疗法，使用基于水稻种子的口服疫苗作为理想的替代方案，其包含基因改良的低过敏性耐受原或来源于过敏原的 T 细胞表位肽。

31.1　日本主要过敏性疾病

在工业国家中，有30%~40%的人口患有某种 IgE 介导的I型过敏性疾病，如哮喘、过敏性鼻炎、结膜炎和特应性皮炎[1,2]。日本柳杉花粉病（*Cryptomeria japonica*：Cry j）是日本最主要的季节性变态反应性疾病，由2月至4月初在日本大部分地区传播的花粉引起。流行病研究表明，大约27%的日本人均患有这种过敏性疾病[3]。

此外，超过半数的普通人群具有针对柳杉花粉过敏原的循环免疫球蛋白 E（IgE）。预计柳杉花粉症患者的数量和与之相关的经济成本将稳步增加，导致社会对开发控制这种花粉症可靠有效方法的强烈需求。

屋尘螨（HDM）也是吸入性过敏原的主要来源，在世界许多地区，多达10%的人口患有支气管哮喘等慢性过敏性疾病。过敏性哮喘患者中有45%~80%对 HDM 中的过敏原敏感，尤其是粉尘螨（Der f）和屋尘螨（Der p）[4]。因此，这些患者患有对屋尘螨的过敏症状或反应，和/或具有升高的过敏原特异性血清免疫球蛋白 E，表明屋尘螨过敏原对于支气管哮喘的发展至关重要[5,6]。

31.2　过敏性病原体

已经分离和鉴定了日本柳杉花粉的两种主要过敏原 Cry j 1 和 Cry j 2。超过90%的日本柳杉花粉症患者均具有针对这两种过敏原的免疫球蛋白 E[6]。Cry j 1 是碱性糖蛋白果胶酸裂合酶，其表观分子量为41~45 kDa，pI 为8.9~9.2；Cry j 2 也是具有多聚半乳糖醛酸活性的碱性蛋白质，其分子量为37 kDa，pI 为8.6~8.8。它们分别特异性地定位于柳杉花粉的乳突和淀粉体细胞壁中。Cry j 1 和 Cry j 2 均具有几种在一级结构上不同且由翻译后修饰产生的亚型。已经确定了 Cry

---

① F. Takaiwa，PhD（✉）（日本国家农业生物科学研究所转基因作物功能研究室，日本茨城县筑波，E-mail：takaiwa@nias.affrc.go.jp）

T. Hiroi，DDS，PhD（东京都医学科学研究所过敏与免疫系，日本东京濑户区，E-mail：hiroi-tk@igakuken.or.jp） j 1 和 Cry j 2 cDNA 的核苷酸序列，并推导出了它们的氨基酸序列[7-10]。

这些来源于屋尘螨的过敏原，是一些与支气管哮喘、鼻炎和特应性皮炎相关的最常见室内过敏原。屋尘螨是 70%以上的儿童支气管哮喘的病因，到目前为止，已鉴定出 20 多种屋尘螨过敏原，并对其进行了特征分析[5,11]。主要的屋尘螨过敏原分为 1 组（Derf1 和 Derp1，分子量为 25 kDa）和 2 组（Derf2 和 Derp2，分子量为 14 kDa）。50%~70%的螨敏感患者血清，均对这些过敏原中的 1 种或 2 种过敏原有反应。第 1 组过敏原是一种不耐热的酸性糖蛋白，主要存在于屋尘螨粪便中。这种蛋白具有木瓜蛋白酶样半胱氨酸蛋白酶活性，通过从免疫细胞表面分裂 CD23 和 CD25，参与过敏的发病机制。第 1 组过敏原被合成为一种无酶活性的前第 1 组蛋白，然后加工成具有 82%序列同源性的 222（Der p1）和 223（Der f1）氨基酸的，更易过敏原成熟和酶活性蛋白[12,13]。免疫球蛋白 E 与 I 类过敏原的结合，高度依赖于它们的三级结构。第 2 组过敏原包含 129 个氨基酸残基，并具有 87%的同源性[14-17]。超过 80%的螨过敏患者对第 2 组过敏原敏感，这些过敏原在螨类粪便中以高浓度存在[18]。虽然尚不清楚第 2 组过敏原的生物学功能，但它们在半胱氨酸残基的序列、大小和分布上与附睾蛋白家族有相似之处[19]。

31.3 常规治疗

变态反应性疾病的特征是，产生过敏原特异性免疫球蛋白 E 和激活包括嗜酸性粒细胞、肥大细胞和嗜碱性粒细胞在内的效应细胞。这些事件受 Th2 细胞调节，后者优先产生 IL-4、IL-5 和 IL-13。因此，变态反应性疾病被定义为，外周对过敏原特异性 T 细胞的调节不足。

这些变态反应性疾病的治疗策略通常包括使用抗组胺药、皮质类固醇等进行药物治疗。然而，尽管这些方法通过阻断变态反应的关键介质释放或通过抑制变态反应性炎症来减少临床症状，但它们并不是治愈性的，并且有时由于副作用而导致性能受损。众所周知，过敏原特异性免疫治疗是调节这些免疫反应的唯一途径，产生的效果可持续多年而无需进一步治疗，并可降低鼻炎患儿对进一步过敏原过敏或发生哮喘的风险[20,21]。传统的过敏原特异性免疫疗法已经实践了将近 1 个世纪[22]。通过反复皮下注射增加剂量的天然过敏原提取物，在至少 3~5 年的时间内诱导免疫耐受（脱敏），即可以获得成功。

这种治疗有时伴随着严重的副作用，如过敏原捕获引起的过敏反应，以及肥大细胞和嗜碱性粒细胞表面的特异性抗过敏原免疫球蛋白 E。最近已开发出了舌下免疫疗法，并提供了一种更安全、更有益的途径[23,24]。过敏原特异性免疫治疗的基本原理是，通过多种细胞和分子机制[25]诱导对过敏原的免疫耐受，从而减少炎症细胞的募集和活化，以及介质的分泌。诱导外周 T 细胞的耐受状态，是过敏原特异性免疫治疗的重要步骤[26-28]。外周 T 细胞耐受的主要特征是产生过敏原特异性调节性 T 细胞（Tregs）（Foxp3+$CD4^+CD25^-$，诱导 Foxp3+$CD4^+$、Tr1 和 Th3），导致抑制 T 细胞增殖和抗过敏原 Th1 和 Th2 细胞因子的反应。这伴随着 IgG 抗体和 IgA 的过敏原特异性 IgG4 和 IgG1 亚类的显著增加，以及疾病晚期的 IgE 减少。

31.4 新型过敏性疾病疫苗开发

长期以来，安全和无创口服免疫疗法被认为是一种替代传统皮下应用的方法，具有降低免疫球蛋白 E 结合活性（低过敏原性）的修饰过敏原，是在过敏原特异性免疫治疗中规避过敏反应风险的最有前途和简单的方法。为了满足这些需求，最近开发了基于水稻种子的口服过敏原疫苗，该疫苗含有过敏原性降低或无过敏原性的修饰重组过敏原或过敏原衍生的 T 细胞表位肽。这些口服疫苗是一种简单、安全、经济有效的免疫疗法。

31.4.1 致敏性降低的重组低过敏原抗体表达耐受原表达

为了产生低过敏原性耐受原，需要以过敏原性降低的重组蛋白取代用于诱导免疫耐受（脱敏）的天然过敏原提取物，该天然过敏原提取物可结合肥大细胞和嗜碱细胞上的特异性免疫球蛋白 E，并导致过敏性副作用[29]。因此，需要使用这种修饰低过敏原耐受原的过敏原特异性免疫治疗新方法[30-33]。据报道，过敏原与特异性免疫球蛋白 E（IgE）的结合是由氨基酸的连续延伸或构象结构所决定；因此，寻找具有低 IgE 结合活性的理想低过敏性耐受原，涉及通过诸如缺失、定点突变、片段化、寡聚体形成或分子重组等手段来实现的，形成各种镶嵌结构的过敏原片段重组研究[30-33]。因此，将去除与 IgE 结合有关的氨基酸和肽，并且过敏原折叠将因序列改变而改变。

尤其是，半胱氨酸残基突变破坏二硫键依赖性构象可以显著改变三级结构。重组低过敏原衍生物表现出降低免疫球蛋白 E（IgE）的反应性，但保留 T 细胞反应性和免疫原性，因此即使对过敏患者给予相对高剂量的耐受原，亦不会诱导 IgE 介导的副作用。降低的过敏原性使得能够施用更高的剂量，促成了过敏原特异性免疫治疗的更大效力。有趣的是，重组寡聚体保留了 IgE 反应性，但由于 IgE 表位的递呈改变而失去了过敏原活性。已在过敏患者的临床试验（目前处于 III 期）中证实了主要桦木花粉过敏原 BET V1 的重组低过敏原性过敏原衍生物的功效[34,35]。皮下和舌下免疫治疗可诱导产生具有阻断活性的过敏原特异性 IgG 抗体（特别是 IgG4），该抗体不仅抑制了过敏原诱导的肥大细胞和嗜碱细胞释放炎性介质，而且抑制了 IgE 促进的抗原对 T 细胞的递呈。

31.4.2 T 细胞表位肽作为最小耐受原

过敏原通过与其 T 细胞表位的相互作用，被 T 细胞受体（TCR）和主要组织相容性复合物（MHC Ⅱ）识别。因其不包含可诱导过敏反应的 IgE 交联表位，利用 T 细胞表位肽进行免疫治疗，是一种很有吸引力的安全过敏原特异性免疫治疗方法[36,37]。但由过敏原衍生的 T 细胞表位被特异性 T 细胞所识别，并根据 MHC Ⅱ类单倍型而变化，这严重障碍了利用 T 细胞表位肽的肽免疫治疗的发展。为了克服这个问题，已经产生了由衍生自一种或几种过敏原的，完整 T 细胞表位库组成的人工杂交表位肽。

从许多变态反应患者中鉴定出几个主要的 T 细胞表位，并将其包含在人工杂交多肽中。在日本柳杉花粉过敏的情况下，将来自两种主要过敏原分子（Cryj1 和 Cryj2）的 5 个或 7 个候选 T 细胞杂交表位肽[38,39]。这些杂交种在用作免疫治疗耐受原时有许多好处，它们由于缺乏 B 细胞表位和构象变化而表现出很少或没有免疫球蛋白 E 反应。例如，在 48 例有日本柳杉花粉症症状的患者血清中，杂交 7Crp 肽（7 个连接表位）不与特异性免疫球蛋白 E 结合[39]。值得注意的是，7Crp 肽诱导 T 细胞增殖的免疫原性，比用于构建杂交肽的 T 细胞表位混合物高 100 倍，主要采用来自猫和蜂毒过敏原 Feld1 和 Apim1 的肽进行了临床试验。众所周知，用 Feld1 或磷脂酶 A2 Apim1 肽进行的肽免疫治疗，诱导了 Tr1 型过敏原特异性免疫反应，并伴有 Th2 型细胞因子减少，IL-10 和过敏原特异性 IgG4 产生增加[40]。低剂量的肽也可以改善临床症状而不诱发不良反应。

31.4.3 种子口服疫苗

种子为重组过敏疫苗提供了一个合适的生产平台，当口服时，可作为肠道相关淋巴样组织（GALT）［包括淋巴滤泡集结（pps）和黏膜免疫组织的一大簇淋巴滤泡］的运载工具。种子疫苗的生产有几个优点，包括不需要冷藏和储存、无针给药、不存在被人类病原体污染的风险、低成本以及生产工艺可以根据需要轻松扩大规模[41,42]。

对于实际使用植物制成的口服变态反应疫苗而言，连续施用诱导免疫耐受所需的高剂量耐受原（脱敏），意味着高度且一致的表达对于疗效至关重要。为了在这个系统中最大限度地提高产量，需要使用为在种子中表达而优化的密码子来合成编码这些耐受原的转移基因[43,44]。此外，由于重组蛋白的积累在很大程度上也依赖于细胞内定位，因此通过细胞内靶向信号将其稳定、高浓度地靶向到合适的位置非常重要[45,46]。在种子的情况下，针对分泌途径的信号肽和针对靶蛋白体（PBS）的分类信号，是产品稳定积累所必需的。相比于组成性启动子，强组织特异性启动子更适合在所需组织中高水平表达抗原[47]。

我们已经报道，因外源产品有充足和稳定的储存空间，水稻种子胚乳是一个很好的重组蛋白生产平台[48]。对于转基因水稻种子中抗原的表达，强胚乳特异性启动子，如主要储存蛋白谷蛋白（GluB）、26-kDa 球蛋白或 10-kDa 和 16-kDa 醇溶谷蛋白，已用于在水稻胚乳中的表达[49]。利用这些启动子，靶基因产物在 CBB 染色的 SDS-PAGE 凝胶中高度积累，占种子总蛋白（3~5 mg/g 干种子）的 4%~6%。到目前为止，使用该系统，来自日本柳杉花粉过敏原 Cry j 1 和 Cry j 2 的杂交 T 细胞表位肽（7Crp）及其结构破坏的全长形式、主要螨过敏原 Der p 1 的亚基抗原和修饰的螨过敏原 Der f 2［其中与二硫键相关的 Cys 残基突变为 Ser（ALa）］，均被发现在转基因水稻的胚乳中高度积累[50,51]。它们主要沉积在 ER 衍生的 PBs 或独特的 ER 衍生隔室中，即 Der f 体中。这些重组蛋白在随后的几代中稳定遗传，且无沉默现象。此外，这些抗原即使在室温下储存数年也可保持稳定，而不丧失免疫原性。

31.4.4 以口服种子过敏疫苗向肠道相关淋巴样组织（GALT）递送

口服植物抗原不仅能诱导黏膜免疫反应，还能诱导全身免疫反应。反复经口给药，尤其是当抗原作为种子的贮存细胞器贮存在颗粒状 PBs 中时，更易诱导对所服抗原的免疫耐受，而不是致敏。小于 10 nm 的 PBs 主要被小肠 GALT 中 PPs 滤泡相关上皮中的 M 细胞吸收，然后递呈给邻近黏膜 T 细胞区的抗原呈递细胞，如树突状细胞[52,53]。此外，固有层（LP）中的树突状细胞有可能向管腔和标本 PBs 延伸。肠道相关淋巴样组织包含 B 淋巴细胞和 T 淋巴细胞对树突状细胞递呈的抗原作出反应区域的有组织的宏观结构，该结构可以诱导记忆 B 细胞和 T 细胞。在 PPS 中至少存在 3 个具有不同组织分布的 DC 亚群[54,55]，在上皮下结构域（SED）区域中存在 $CD11b^+DCs$，在富含 T 细胞的滤泡间区域（IFR）中存在 $CD8^+DCs$，在 SED 和 IFRs 两者中均存在双阴性 $CD4^-CD8\alpha^-DCs$（图 31.1）。

产生 IL-10 的未成熟树突状细胞，通过诱导 Tr1-调节细胞而促进免疫耐受。应注意，肠道相关淋巴样组织是 Foxp3+Tregs 外周诱导的优先位点（图 31.2）。小肠 LP 的 $CD103^+DCs$（黏膜树突状细胞）尤其能通过 TGF-β 和维甲酸（RA）诱导 Foxp3+T 细胞（图 31.2）。递呈膳食抗原的浆细胞样树突状细胞（pDCs）负责诱导口服耐受和免疫抑制（图 31.2）。

肠黏膜的环境特别容易产生可诱导的调节性 T 细胞。$CD103^+DCs$ 可在缺乏 TGF-β 时，在小肠固有层产生维生素 A 代谢物维甲酸的情况下，将 $CD4^+$静息性 T 细胞转化为 $Foxp3^+$诱导调节性 T 细胞。

已经证明，为了达到同样的疗效水平，口服黏膜免疫疗法所需的抗原>皮下注射的 100~1 000倍[56]。这是因为纯化的抗原在到达肠道相关淋巴样组织中的黏膜免疫细胞之前，会被胃肠道中的消化酶降解，并且暴露于胃中的恶劣环境。相比之下，通过稻谷等谷类种子口服时，利用 PBs 的双重屏障和植物细胞的细胞壁特性进行抗原的生物包埋，具有保护抗原不被蛋白质水解的优点[57]。当抗原由其他植物营养细胞递送时，情况并非如此。值得注意的是，从 ER 衍生的 PBs 对胃肠酶消化的抗性比在蛋白质贮存空泡（PSVs）中的抗性更强，如比较 ER 衍生的 PB（PB-I）和 PSV（PB-II）定位抗原之间的消化率和免疫耐受诱导能力所示[58]。这种物理差异可能与在 ER 衍生的 PBs 中观察到的半胱氨酸醇溶蛋白形成的抗原聚合或聚集排列有关。事实上，当将含有沉积在 ER 衍生的 PB 或 PSV 中的几种过敏原衍生物的转基因水稻种子喂养给小鼠时，PB-1 定位的过敏原衍生物表现出比 PB-II 定位或合成的裸露形式更大的酶促消化抗性。抑制小鼠过敏原特异性免疫球蛋白 E 水平所需的 Pb-I 定位 T 细胞表位的剂量，分别是合成肽和 Pb-II 定位形式所需剂量的 20 倍和 3 倍[58]。

大米过敏疫苗具有热稳定性，能耐受常温贮存和煮沸煮食过程。已证实蒸制或煮熟的转基

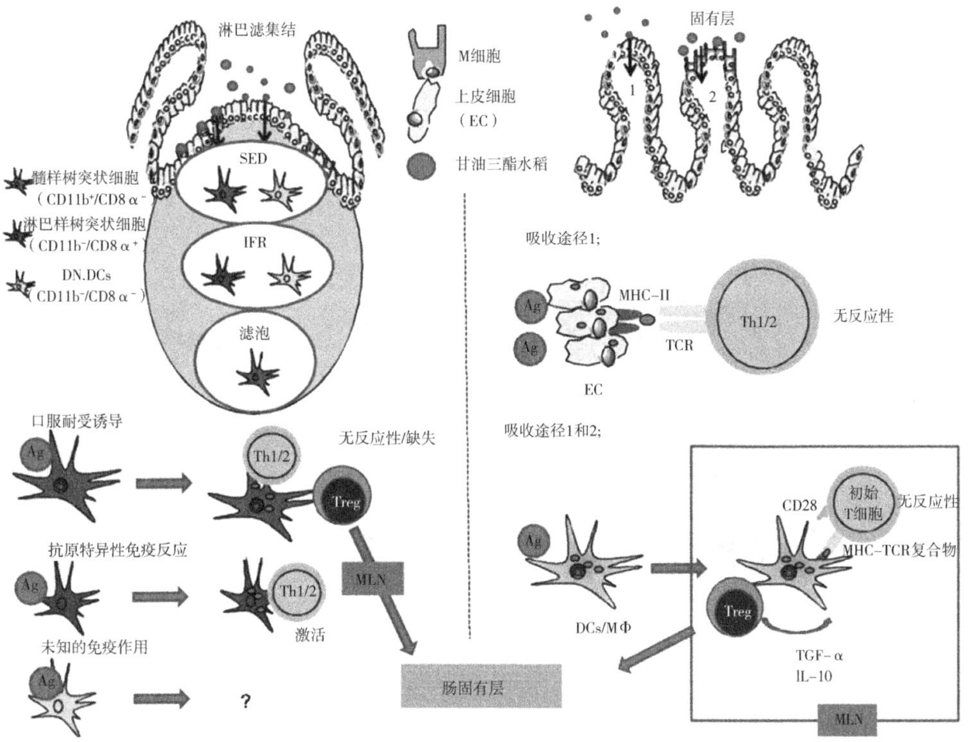

图 31.1　口服的过敏原被带到淋巴滤泡集结和肠固有层中的黏膜免疫系统

通过口服疫苗接种的抗原，穿过固有层（LP）中的淋巴滤泡集结（PPs）和肠上皮细胞的 M 细胞。淋巴滤泡集结（PPs）中的 $CD11b^+CD8\alpha^-$髓系树突状细胞，在 MHC Ⅱ分子上处理和递呈抗原，以诱导口服耐受，但不诱导 $CD11b^-CD8\alpha^+$淋巴树突状细胞。肠固有层中的树突状细胞 /巨噬细胞（Mϕ）捕获的抗原，也在 MHC Ⅱ分子上处理和递呈，然后作为肠黏膜中的主淋巴结转移至肠系膜淋巴结（MLN）。此外，肠上皮细胞也加工抗原，并将其直接递呈在主要组织相容性复合体（MHC）分子上，以诱导肠固有层的口服耐受。

因大米仍具有诱导口服免疫耐受的功效。事实上，处理后的种子提取物即使在高压灭菌 20 分钟后，仍能诱导 T 细胞的增殖活性。

31.4.5　免疫诱导肠道相关淋巴样组织中的耐受性

诱导耐受的免疫机制，取决于给予抗原的频率、持续时间和剂量。重复低剂量抗原给药有利于由调节性 T 细胞驱动的耐受性，而高剂量给药有利于无反应或细胞缺失驱动的耐受性，这类耐受性被认为是诱导耐受性作为静止免疫状态的原因[59-61]。已有研究表明，对于口服耐受性，Fas（CD95）依赖性细胞凋亡导致了效应器 T 细胞的缺失[62]（图 31.3a）。在没有共刺激信号的情况下，将通过 T 细胞受体连接发生 T 淋巴细胞无反应，并且可通过 APC 上的 CD80 / 86（B7）和效应 T 细胞上的 CD28（CTLA-4）之间的同源相互作用来预防 T 淋巴细胞无反应（图 31.3b）。众所周知，CTLA-4 在控制调节性 T 细胞的抑制功能方面起着关键作用。共刺激分子的相互作用在调节 T 细胞活化和耐受中起着关键作用[63]。还发现了其他 B7 和 CD28 家族成员，作为共刺激因子参与免疫调节（图 31.3b）。

相比之下，口服低剂量抗原可通过诱导调节性 T 细胞介导的抑制性免疫反应，来诱导主动抑制。这些细胞包括天然存在的 $Foxp3^+CD4^+CD25^+$Treg 细胞和抗原诱导的 Treg 细胞，所述抗原诱导的 Treg 包括 $Foxp3^+$诱导的 $CD4^+$细胞、产生 IL-10 的 TR1 细胞、产生 TGF-β 的 Th3 细胞和调节性 $CD8^+$T 细胞[64]。天然存在的 Treg 在胸腺中产生，而诱导的 Treg 细胞由静息性 T 细胞在

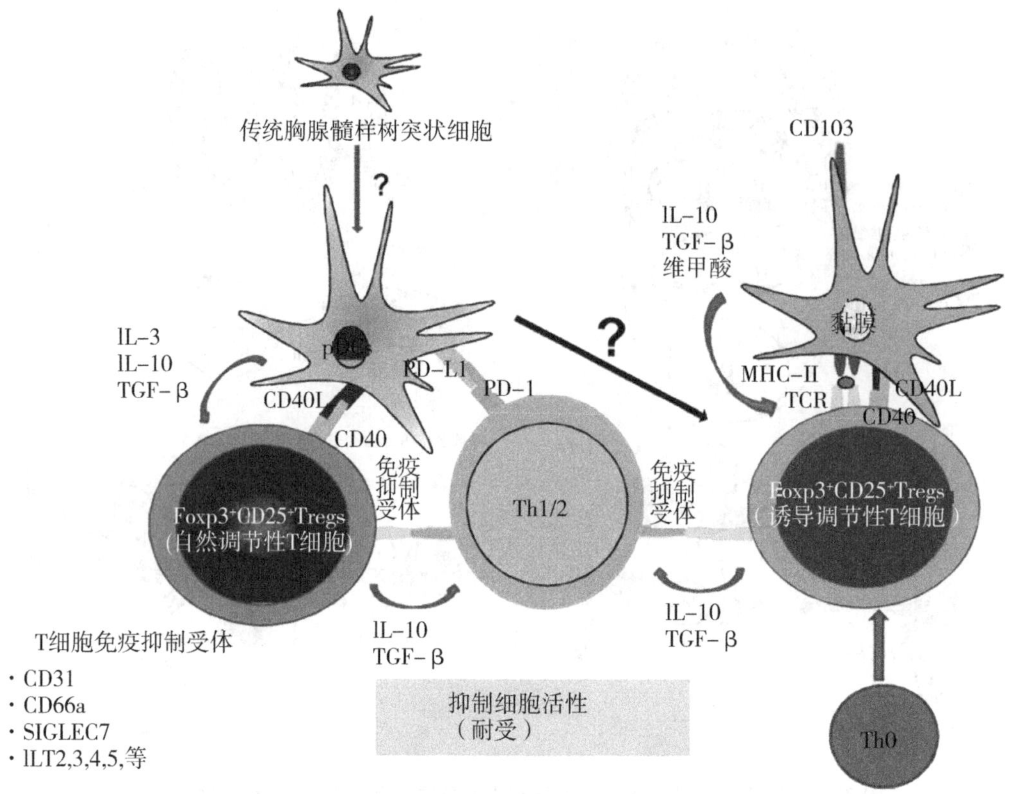

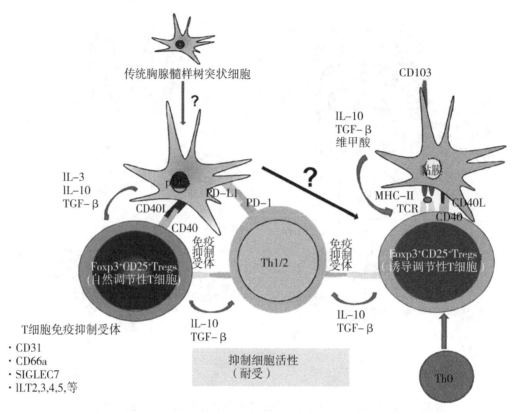

图 31.2　诱导抗原特异性耐受的两个 Foxp3$^+$CD25$^+$CD4$^+$Tregs 亚群

源自传统胸腺髓样树突状细胞的浆细胞样树突细胞（pDCs）产生 IL-10 和 TGF-β，并诱导更多产生 IL-10 的自然调节性 T 细胞（Tregs），从而抑制免疫反应。

外周产生。

已有研究表明，CD4$^+$T 细胞向 Treg 的外周转化主要发生在 GALT，因为肠道是许多抗原（如食物和共生细菌）的进入部位。在 TGF-β 和 IL-6 存在的情况下，Treg 修饰树突状细胞产生的 IL-27，介导了静息性 CD4$^+$T 细胞分化为产生 IL-10 的 Tr1 细胞[65]。肠道 CD103$^+$树突状细胞在肠道固有层中产生的 TGF-β 和 RA，诱导产生了 Foxp3$^+$诱导 Treg 细胞[65]。RA 是 TGF-β 依赖性免疫反应的关键调节因子。IL-6 抑制 TGF-β 诱导的 Foxp3 介导，产生了 Th17 细胞，而 RA 可以抵消这种抑制作用[66,67]。调节性 T 细胞抑制 Th1-和 Th17-细胞反应，与预防炎症有关。

31.4.6　水稻过敏疫苗的疗效

在日本柳杉花粉症的小鼠模型中，口服施用 Cry j 2 的显性 T 细胞表位，抑制了 Cry j 2 致 Cry j 2 对 Cry j 2 致敏小鼠的特异性 T 细胞反应具有抑制作用[68]。喷嚏频率是一种可测量的临床症状，其不仅可通过全身注射降低，也可通过口服 CRY j 2 的显性 T 细胞表位降低[69,70]。这种转基因水稻种子积有源自 Cry j 1 和 Cry j 2 的主要小鼠 T 细胞表位，它们作为融合蛋白和大豆种子蛋白甘氨酸一起，通过经口营养的方式检测口服转基因水稻种子的功效[71]（图 31.4a）。连续 4 周每日给药后，用柳杉花粉总蛋白进行致敏，相比用非转基因水稻种子饲喂的小鼠而言，过敏原特异性血清 IgE 和 IgG 抗体的产生受抑制（图 31.4b）。过敏原特异性 CD4$^+$T 细胞增殖反应和 Th2 细胞因子（如 IL-4、IL-5 和 IL-13）水平均受到抑制（图 31.4c）。作为过敏临床症状的测量，肥大细胞释放组胺和打喷嚏也被抑制（图 31.4d 和图 31.4e）。这些结果表明，基于水稻来表达 T 细胞表位肽的口服疫苗可诱导柳杉花粉过敏原口服免疫耐受性，且可预防性地缓解临床症状。

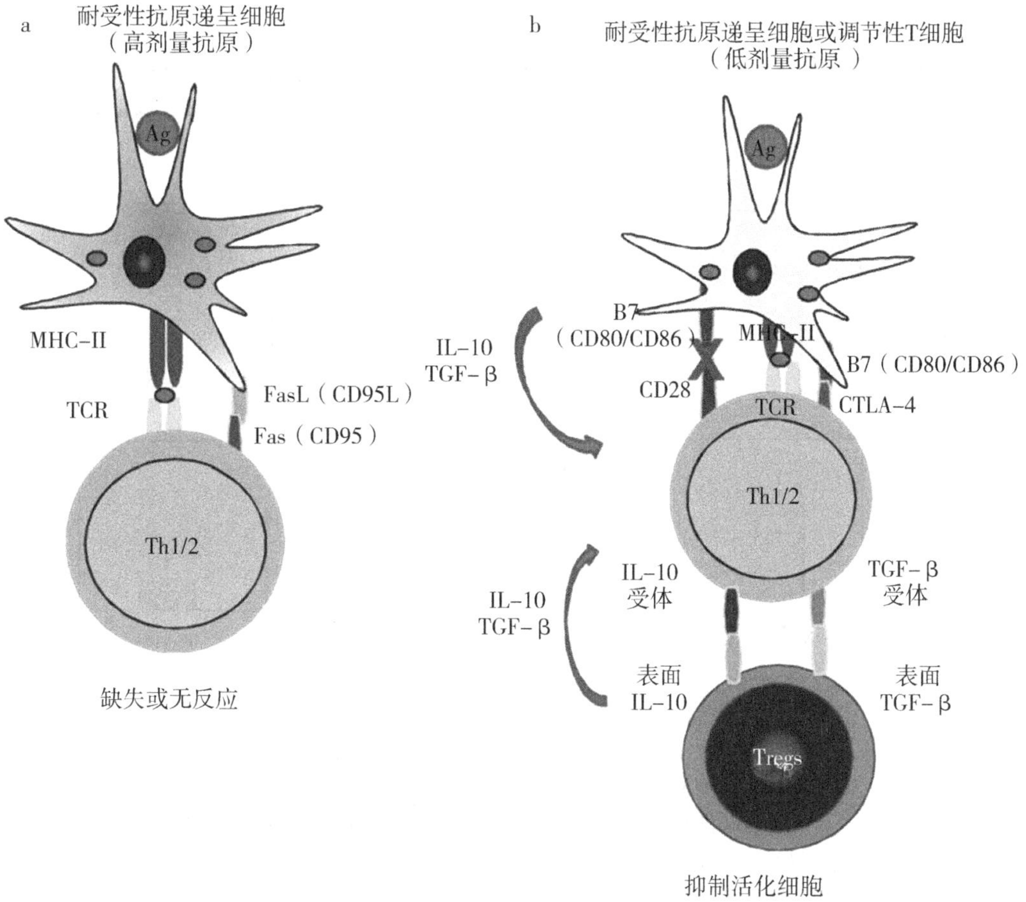

图 31.3　与免疫耐受有关的免疫机制取决于抗原的剂量

无论是高剂量抗原还是低剂量无调节性 T 细胞（Tregs），免疫反应的产生均需要将 T 细胞受体（TCR）与肽 MHC Ⅱ 复合物结合。（a）在通过高剂量抗原诱导的口服耐受中，抑制性连接（FasFas-L）导致免疫调节反应，例如过敏原特异性 CD4⁺ Th2 细胞的缺失和无反应性。（b）在重复低剂量抗原的口服耐受诱导中，诱导了调节性 T 细胞和/或耐受性抗原呈递细胞（APCs）。通过调节性 T 细胞连接细胞表面 T 细胞抑制性细胞因子 IL-10 和 TGF-β，来抑制通过产生这些细胞因子的可溶性形式维持的免疫反应。在没有 B7-CD28 连接的情况下，耐受性抗原呈递细胞导致抗原特异性免疫反应的抑制。

基于该证据，将积累了由来自 Cry j 1 和 Cry j 2 的人类主要 T 细胞表位组成的 7Crp 肽的转基因水稻种子开发成用于治疗人类柳杉花粉过敏的基于水稻的肽疫苗[72]。7Crp 肽在转基因水稻种子（约 3 mg/g 干种子）中高度积累，并沉积在 ER 衍生的 PBs 中，适用于输送至肠道相关淋巴样组织。为了检验其安全性，我们将转基因水稻种子喂给猴子和大鼠，以检测其慢性毒性和生殖毒性；未检测到任何不良反应[73]。

尽管其安全，但是由于人类具有不同的遗传背景，所以对各种 T 细胞表位的反应不尽相同，使用 T 细胞表位的肽免疫疗法并不适用于所有患者。为了能覆盖到较为广泛的过敏患者群体，最近已测试了另一种策略。通过分子改装或片段化产生镶嵌结构，为 Cry j 1 和 Cry j 2 生成带有降低的 IgE 结合的重组全过敏原衍生物，以影响需用于识别过敏原特异性 IgE 的三级结构[74]。连续 3 周每天将转基因水稻种子喂给小鼠，然后用原柳杉花粉过敏原攻击，在这种情况下，与喂食非转基因水稻种子的小鼠相比，其过敏原特异性 $CD4^+$T 细胞增殖和 IgE 以及 IgG 水平明显受到抑制。花粉症的临床症状之一即打喷嚏频率，以及鼻组织中的嗜酸性粒细胞和嗜中性粒细

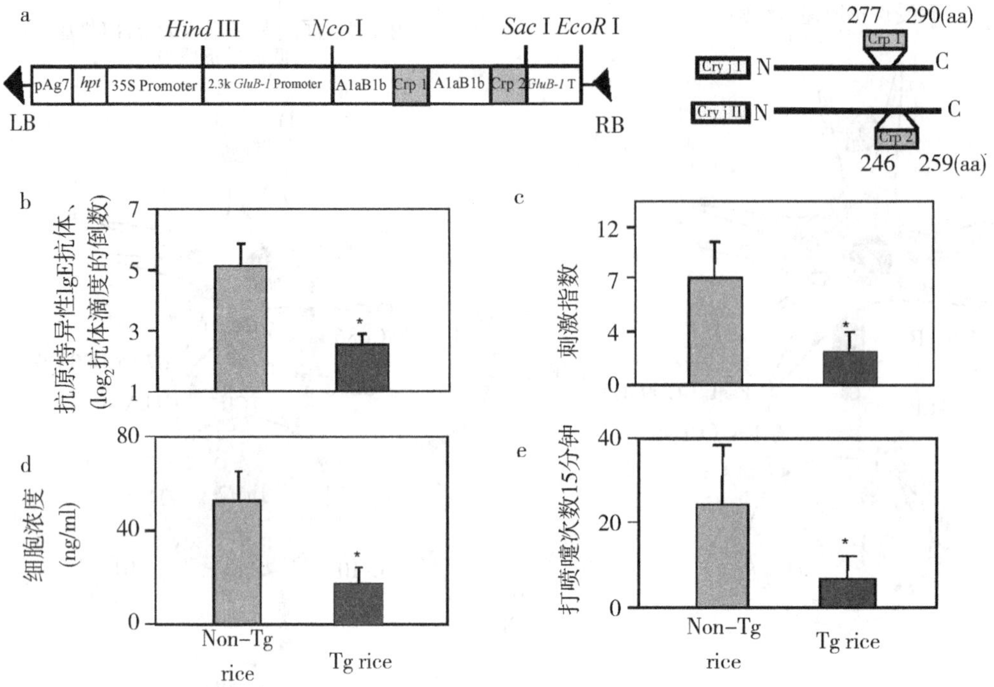

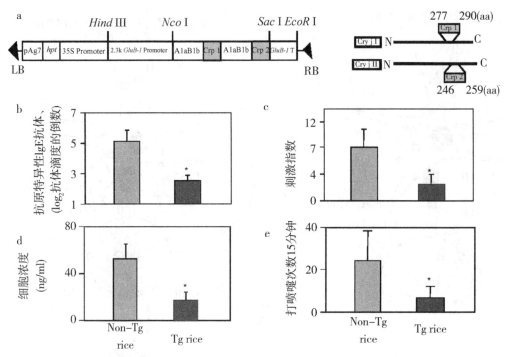

图 31.4 表达 Cry j 1 和 Cry j 2 的 T 细胞表位的转基因水稻，通过抑制 Th2 介导的免疫球蛋白 E 反应，诱导口服耐受

质粒转化示意图和 Cry j 1 及 Cry j 2 两个选定的主要 T 细胞表位（分别是 Crp 1 和 Crp 2）（a）示意图。通过口服抑制过敏原特异性血清 IgE 水平（b）和 $CD4^{+}T$ 细胞反应（c）。服用转基因（甘油三酯）水稻种子。在喂养甘油三酯水稻种子的小鼠组中，血清组胺水平（d）和打喷嚏次数（e）受到抑制。数据表示为平均值±SD. *，与喂食非甘油三酯水稻种子（品红色）的小鼠组相比，喂食甘油三酯水稻种子（蓝色）的小鼠组的$P<0.01$。

胞一样的炎性细胞浸润都显著降低。结果表明，含有结构已破坏的 Cry j 1 和 Cry j 2 抗原的转基因水稻种子作为一种口服的治疗性过敏疫苗前景可观，适合所有日本柳杉花粉症患者。

为应对屋尘螨过敏，研制成了转基因水稻谷粒，其积累成熟的 Der p 1，覆盖大多数人和小鼠主要 T 细胞表位的 Der p 1 片段（p45145），以及由分子内二硫键破坏导致的具有降低免疫球蛋白 E 反应性的 Der f 2 衍生物[51,75,76]。植物源性的 Der p 1 以高甘露糖型聚糖结构后转译修饰。与其未糖基化的对应物相比，糖基化的 Der p 1 显示出降低的免疫球蛋白 E 结合能力[75]（图 31.5a）。

将含有 Der p 1 片段（p45~145）的螨过敏原转基因水稻种子喂给小鼠，随后用重组 Der p 1 免疫，在这种情况下，与喂食非转基因水稻种子的小鼠相比，其过敏原特异性 $CD4^{+}T$ 细胞增殖（图 31.5b）和过敏原特异性免疫球蛋白 E 和免疫球蛋白 G（图 31.5c）的产生受抑制[76]。口服免疫转基因水稻种子后，Th2 细胞因子 IL-4、IL-5 和 IL-13 的生成也显著减少（图 31.5d）。预防接种转基因水稻种子能抑制过敏性气道炎症的发展以及支气管高反应性（图 31.5f）。两者皆为可测量的临床症状，前者由嗜酸性粒细胞和淋巴细胞渗入气道（图 31.5e）导致。

当将含有 Der f 2 衍生物的转基因水稻种子喂给小鼠时，用类似于其他过敏原的方式，诱导了对 Der f 2 抗原的免疫耐受性[51]。有趣的是，对 Der f 2 特异性免疫球蛋白 E 和免疫球蛋白 G 的产生进行抑制，表明了这些衍生物的功效与含有 Der f 2 的 PBs 对胃肠道消化酶的耐药性有关。

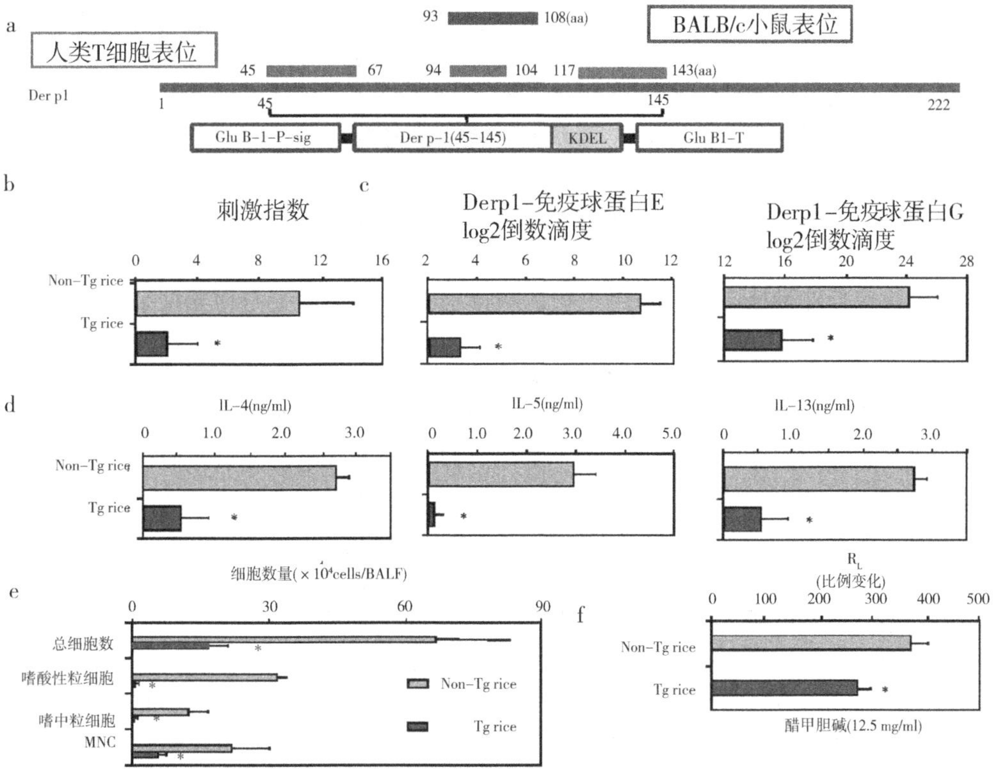

图 31.5 表达 Der p 1T 细胞表位的转基因水稻通过抑制 Der p 1 特异性免疫应答诱导口服耐受性

转化质粒及所选 Der p 1 主要 T 细胞表位图示表征（a）。将由阴性对照小鼠（数据未显示）制备而得的脾脏 $CD4^+T$ 细胞、喂食非转基因（甘油三酯）水稻（品红）的小鼠，或喂食甘油三酯水稻（蓝色）的小鼠与辐射后的正常脾细胞以及过敏原放在一起培养共计 7 天。口服甘油三酯水稻种子使过敏原特异性 $CD4^+T$ 细胞增殖反应受抑制（b），且血清免疫球蛋白 E 和免疫球蛋白 G 水平也受到抑制（c）。测定释放至培养上清液的细胞因子（IL-4、IL-5 和 IL-13）的浓度（d）。预防性口服 Der p 1-甘油三酯水稻后，过敏原诱导的炎症细胞浸润现象受到抑制。测量支气管肺泡灌洗液（*BALF*）中的细胞总数、嗜酸性粒细胞、嗜中性粒细胞和单核细胞（MNC）的数量（e）。第二次过敏原激发 24 小时后测量支气管对雾化乙酰甲胆碱（MCh）的高反应性（f）。与喂食非甘油三酯水稻种子的小鼠组相比，喂食甘油三酯水稻种子的小鼠组的数据表示为平均值±SD. *，$P< 0.01$。

31.5 基于水稻的过敏疫苗的实际应用

作为一种疫苗抗原运作系统，与其他作物相比，水稻种子具有加工工艺成熟、产量高、自花传粉转基因逃逸风险低等优点。此外，水稻转化体系业已建立，并且关于全基因组序列和表达的遗传信息亦可获取。作为一种新型的口服过敏疫苗，水稻种子中积累了低过敏性柳杉花粉过敏原和屋尘螨过敏原衍生物，可以作为一种载体递送给肠道相关淋巴样组织。当将这些基于水稻的疫苗口服给小鼠时，与这些气敏原相关的免疫和生化反应，包括抗原特异性免疫球蛋白 E 抗体的产生、$CD4^+T$ 细胞的增殖，以及组胺的释放都受到抑制。先让小鼠接触柳杉花粉和屋尘螨过敏原，然后接种转基因水稻，发现包括打喷嚏、过敏性气道炎症和支气管高反应性在内的几种过敏临床症状皆有所缓解。

人类通常使用的疫苗大多为用于肠外递送配制，需高度纯化[77]。疫苗抗原的生产必须符合特定的药品生产质量管理规范（GMP）条件下的药品指南和法规[78]。因此，为每种产品开发可靠的加工步骤（纯化）非常有必要。迄今为止，运用病毒和农杆菌载体的瞬时表达系统，主要开发了能进入临床试验的用于肠外递送的植物制疫苗候选株。通常情况下，生产的下游加工过程比较昂贵，其费用占生产总成本的80%[77]。需要注意的是，从其稳定性及向肠道相关淋巴样组织递送的有效性来看，直接口服未进一步加工的转基因作物产品疫苗非常有利，并且由于作物种子含有抗原，可考虑作为带天然包装的口服制剂，它们适合作为载体进行黏膜递送。验证制备商业药物试剂的植物生产系统可能遭遇阻碍。此外，由于生物系统的生理差异，产品积累可能存在差异。

此外，如果在野外种植药用转基因植物，需特别注意与野生型作物异型杂交的情况。消费者和环保组织对食物链污染风险的态度是零容忍。因此，建议指出药用作物的生产需在具体法规和指南的控制下来进行[78]。理想条件下，采用严格控制环境条件的封闭式分子农业设施（封闭式温室），并且应减少目标组织中产品浓度的变化，同时防止异型杂交的发生。但是，假如使用过敏疫苗，其商业化需要大量使用种子。例如，若要使柳杉花粉过敏原脱敏，需每日口服烹调的转基因水稻（约70g）至少3个月，并且持续数年。必须为约两千万患有日本柳杉花粉症的患者制造该转基因水稻。

因此，须制定一种替代策略，因为在封闭的温室中培养所需量的转基因水稻，正如大多数的植物制药一样，不太可能。我们必须再考虑其他的生产系统，例如效仿欧盟商业作物和生物防范采用的做法，在物理隔离地或在用带细孔的网构成的温室中种植，并制定具体的指导方针和规定[79]。此外，在运输及装卸过程中需严格控制避免进入食物链。基于水稻的过敏疫苗的种植地仍面临严峻挑战，包括药品生产质量管理规范中产品一致性和药物作物控制及生产方面的问题[78]。

相比预防传染病疫苗的免疫诱导疗法，采用免疫球蛋白E介导的I型过敏原的免疫耐受诱导的免疫疗法可控且简单，因为受过医学训练的人员不需要为了避免免疫耐受性而进行治疗。然而，口服基于水稻种子的含有低过敏性衍生物的疫苗是否有引起新过敏（致敏）的风险还不确定，所以对这种风险的检测尤为重要。

参考文献

[1] Gerth van Wijk, R.: Allergy: a global problem. Allergy 57, 1097-1110 (2002).

[2] Bauchau, V., Durham, S. R.: Prevalence and rate of diagnosis of allergic rhinitis in Europe. Eur. Respir. J. 24, 758-764 (2004).

[3] Okamoto, Y., et al.: Present situation of cedar pollinosis in Japan and its immune responses. Allergol. Int. 58, 155-162 (2009).

[4] Tovey, E. R., Chapman, M. D., Platts-Mills, T. A.: Mite faces are a major source of house dust allergens. Nature 289, 592-593 (1981).

[5] Thomas, W. R., et al.: Characterization and immuno-biology of house dust mite allergens. Int. Arch. Allergy Immunol. 129, 1-18 (2002).

[6] Hashimoto, M., et al.: Sensitivity to two major aller-gens (Cry j 1 and Cry j 2) in patients with Japanese cedar (Cryptomeria japonica) pollinosis. Clin. Exp. Allergy 25, 848-852 (1995).

[7] Yasueda, H., et al.: Isolation and partial characteriza-tion of the major allergen from Japanese cedar (Cryptomeria japonica) pollen. J. Allergy Clin. Immunol. 71, 77-86 (1993).

[8] Sone, T., et al.: Cloning and sequencing of cDNA coding for Cry j 1, a major allergen of Japanese cedar pollen. Biochem. Biophys. Res. Commun. 199, 619-625 (1994).

[9] Sakaguchi, M., et al.: Identifi cation of the second major allergen of Japanese cedar pollen. Allergy 45,

309-312 (1990).

[10] Namba, M., et al.: Molecular cloning of the second allergen, Cry j II, from Japanese cedar pollen. FEBS Lett. 353, 124-128 (1994).

[11] Kawamoto, S., et al.: Toward elucidating the full spectrum of mite allergens-state of the art. J. Biosci. Bioeng. 94, 285-298 (2002).

[12] Chua, K. Y., et al.: Sequence analysis of cDNA coding for a major house dust mite allergen, Der p 1. Homology with cysteine proteases. J. Exp. Med. 167, 175-182 (1988).

[13] Dilworth, R. J., Chua, K. Y., Thomas, W. R.: Sequence analysis of cDNA coding for a major house dust mite allergen, Der f I. Clin. Exp. Allergy 21, 25-32 (1991).

[14] Thomas, W. R., Smith, W.: House-dust-mite allergens. Allergy 53, 821-832 (1998).

[15] Haide, M., et al.: Allergens of the house dust mite Dermatophagoides farinae-immunochemical studies of four allergenic fractions. J. Allergy Clin. Immunol. 75, 686-692 (1985).

[16] Chua, K. Y., et al.: Analysis of sequence polymorphism of a major allergen, Der p 2. Clin. Exp. Allergy 26, 829-837 (1996).

[17] Yuuki, T., et al.: Cloning and sequencing of cDNA corresponding to mite major allergen Der f II. Jpn. J. Allergy 39, 557-561 (1990).

[18] Park, G. M., et al.: Localization of a major allergen, Der p 2, in the gut and faecal pellets of Dermatophagoides pteronyssinus. Clin. Exp. Allergy 30, 1293-1297 (2000).

[19] Thomas, W. R., Chua, K. Y.: The major mite allergen Der p 2-a secretion of the male mite reproductive tract? Clin. Exp. Allergy 25, 667-669 (1995).

[20] Bousquet, J., Lockey, R. F., Malling, H. U.: Allergen immunotherapy: therapeutic vaccines for allergic diseases. WHO position paper. Allergy 53, 1-42 (1998).

[21] Frew, A. J.: Immunotherapy of allergic diseases. J. Allergy Clin. Immunol. 111, S712-S719 (2003).

[22] Noon, L.: Prophylactic inoculation against hay fever. Lancet 2, 1572-1573 (1911).

[23] Cox, L. S., et al.: Sublingual immunotherapy: a com-prehensive review. J. Allergy Clin. Immunol. 117, 1021-1035 (2006).

[24] Moingeon, P., et al.: Immune mechanisms of allergen-specifi c sublingual immunotherapy. Allergy 61, 151-165 (2006).

[25] Larche, M., Akids, C. A., Valenta, R.: Immunological mechanisms of allergen-specifi c immunotherapy. Nat. Rev. Immunol. 6, 761-766 (2006).

[26] Verhagen, J., et al.: Mechanisms of allergen-specifi c immunotherapy: T-regulatory cells and more. Immunol. Allergy Clin. N. Am. 26, 207-231 (2006).

[27] Till, S. J., et al.: Mechanisms of immunotherapy. J. Allergy Clin. Immunol. 113, 1024-1034 (2004).

[28] Akdis, M., Akids, C. A.: Mechanisms of allergen-specifi c immunotherapy. J. Allergy Clin. Immunol. 119, 780-791 (2007).

[29] Linhart, B., Valenta, R.: Molecular design of allergy vaccines. Curr. Opin. Immunol. 17, 646-655 (2005).

[30] Valenta, R., Kraft, D.: From allergen structure to new forms of allergen-specifi c immunotherapy. Curr. Opin. Immunol. 14, 718-727 (2002).

[31] Cromwell, O., Hafner, D., Nandy, A.: Recombinant allergens for specifi c immunotherapy. J. Allergy Clin. Immunol. 127, 865-872 (2011).

[32] Valenta R. et al. From allergen genes to allergy vaccines. Ann Rev Immunol28, 211-241 (2010).

[33] Valenta, R., et al.: Recombinant allergens: what does the future hold? J. Allergy Clin. Immunol. 127, 860-864 (2011).

[34] Purohit, A., et al.: Clinical effects of immunotherapy with genetically modifi ed recombinant birth pollen

Bet v1 derivatives. Clin. Exp. Allergy 38, 1514-1525 (2008).

[35] Gronlund, H., Gafvelin, G.: Recombinant Bet v1 vaccine for treatment of allergy tobirch pollen. Hum. Vaccin. 6, 970-977 (2010).

[36] Larche, M.: Peptide immunotherapy for allergic diseases. Allergy 62, 325-331 (2007).

[37] Ali, F. R., Larche, M.: Peptide-based immunotherapy: a novel strategy for allergic disease. Exp. Rev. Vaccine 4, 881-889 (2005).

[38] Sone, T., et al.: T cell epitopes in Japanese cedar (Cryptomeria japonica) pollen allergens: choice of major T cell epitopes in Cry j 1 and Cry j 2 toward design of the peptide-based immunotherapeutics for management of Japanese cedar pollinosis. J. Immunol. 161, 448-457 (1998).

[39] Hirahara, K., et al.: Preclinical evaluation of an immunotherapeutic peptide comprising 7 T-cell determinants of Cry j1 and Cry j2, the major Japanese cedar pollen allergens. J. Allergy Clin. Immunol. 108, 94-100 (2001).

[40] Larche, M.: Update on the current status of peptide immunotherapy. J. Allergy Clin. Immunol. 119, 906-909 (2007).

[41] Streatfi eld, S. J., et al.: Plant-based vaccines: unique advantages. Vaccine 19, 2742-2748 (2001).

[42] Nochi, T., et al.: Rice-based mucosal vaccine as a global strategy for cold-chain-and needle-free vaccination. Proc. Natl. Acad. Sci. U. S. A. 104, 10986-10991 (2007).

[43] Stoger, E., et al.: Sowing the seeds of success: phar - maceutical proteins from proteins. Curr. Opin. Biotech. 16, 167-173 (2005).

[44] Lau, O. S., Sun, S. S. M.: Plant seeds as bioreactors for recombinant protein production. Biotechnol. Adv. 27, 1015-1022 (2009).

[45] Vitale, A., Hinz, G.: Sorting of proteins to storage vacuoles: how many mechanisms? Trends Plant Sci. 10, 315-323 (2005).

[46] Benchabane, M., et al.: Preventing unintended prote-olysis in protein biofactories. Plant Biotech. J. 6, 633-648 (2008).

[47] Kawakatsu, T., Takaiwa, F.: Cereal seed storage pro - tein synthesis: fundamental processes for recombinant protein production in cereal grains. Plant Biotech. J. 8, 939-953 (2010).

[48] Takaiwa, F., et al.: Endosperm tissue is good production platform for artifi cial recombinant proteins in transgenic rice. Plant Biotech. J. 5, 84-92 (2007).

[49] Qu, L. Q., Takaiwa, F.: Tissue specifi c expression and quantitative potential evaluate of seed storage component gene promoters in transgenic rice. Plant Biotech. J. 2, 113-125 (2004).

[50] Takaiwa, F., et al.: Deposition of a recombinant peptide in ER-derived protein bodies by retention with cysteine-rich prolamins in transgenic rice seed. Planta 229, 1147-1158 (2009).

[51] Yang, L., et al.: Expression of hypoallergenic Der f 2 derivatives with altered intramolecular disulphide bonds induces the formation of novel ER-derived protein bodies in transgenic rice seeds. J. Exp. Bot. 63, 2947-2959 (2012).

[52] Kunisawa, J., Kurashima, Y., Kiyono, H.: Gut-associated lymphoid tissues for the development of oral vaccines. Adv. Drug Deliv. Rev. 64, 523-530 (2012).

[53] Neutra, M. R., Kozlowski, P. A.: Mucosal vaccines: the promise and the challenge. Nat. Rev. Immunol. 6, 148-158 (2006).

[54] Kelsall, B. L., Leon, F.: Involvement of intestinal den-dritic cells in oral tolerance, immunity to pathogens, and infl ammatory bowel disease. Immunol. Rev. 206, 132-148 (2005).

[55] Tsuji, N. M., Kosaka, A.: Oral tolerance: intestinal homeostasis and antigen-specifi c regulatory T cells. Trends Immunol. 29, 532-540 (2008).

[56] Streatfi led, S. J.: Mucosal immunization using recom-binant plant-based oral vaccines. Methods 38, 150-

157 (2006).

[57] Takaiwa, F.: Seed-based oral vaccines as allergen-specifi c immunotherapy. Hum. Vaccin. 7, 357-366 (2011).

[58] Takagi, H., et al.: Rice seed ER-derived protein body as an effi cient delivery vehicle for oral tolerogenic peptides. Peptides 31, 421-425 (2010).

[59] Weiner, H. L.: Oral tolerance: immune mechanisms and treatment of autoimmune diseases. Immunol. Today 18, 335-343 (1997).

[60] Maer, L., Shao, L.: Therapeutic potential of oral tolerance. Nat. Rev. Immunol. 4, 407-419 (2004).

[61] Burks, A. W., Laubach, S., Jones, S. M.: Oral toler-ance, food allergy, and immunotherapy: Implications for future treatment. J. Allergy Clin. Immunol. 121, 1344-1350 (2008).

[62] Appeman, L. J., Boussiotis, V. A.: Tcell anergy and co-stimulation. Immunol. Rev. 192, 161-180 (2003).

[63] Greenwald, R. J., Freeman, G. J., Sharpe, A. H.: B7 family revised. Annu. Rev. Immunol. 23, 515-548 (2005).

[64] Ozdemir, C., Akids, M., Akids, C. A.: T regulatory cells and their counterparts: masters of immune regulation. Clin. Exp. Allergy 39, 626-639 (2009).

[65] Awasthi, A., et al.: A dominant function for interleu-kin 27 in generating interleukin 10-producing anti-infl ammatory T cells. Nat. Immunol. 12, 1380-1390 (2007).

[66] Bettelli, E., et al.: Reciprocal developmental path-ways for generation of pathogenic effector Th17 and regulatory T cells. Nature 441, 235-238 (2006).

[67] Mucida, D., Salek-Ardakani, S.: Regulation of Th17 cells in the mucosal surfaces. J. Allergy Clin. Immunol. 123, 997-1003 (2009).

[68] Hirahara, K., et al.: Oral administration of a dominant T-cell determinant peptide inhibits allergen-specifi c TH1 and TH2 cell response in Cry j 2-primed mice. J. Allergy Clin. Immunol. 102, 961-967 (1998).

[69] Murasugi, T., et al.: Oral administration of a T cell epitope inhibits symptoms and reactions of allergic rhinitis in Japanese cedar pollen allergen-sensitized mice. Eur. J. Pharm. 510, 143-148 (2005).

[70] Tsunematsu, M., et al.: Effect of Cry-consensus pep-tide, a novel recombinant peptide for immunotherapy of Japanese cedar pollinosis, on an experimental allergic rhinitis model in B10. S mice. Allegol. Int. 56, 465-472 (2007).

[71] Takagi, H., et al.: A rice-based edible vaccine expressing multiple epitopes induces oral tolerance for inhibition of Th2-mediated IgE response. Proc. Natl. Acad. Sci. U. S. A. 102, 17525-17530 (2005).

[72] Takagi, H., et al.: Oral immunotherapy against a pollen allergy using a seed-based peptide vaccine. Plant Biotech. J. 3, 521-533 (2005).

[73] Domon, E., et al.: 26-week oral safety study in macaques for transgenic rice containing major human T-cell epitope peptides from Japanese cedar pollen allergens. Agric. Food Chem. 57, 5633-5638 (2009).

[74] Wakasa, Y., et al.: Oral immunotherapy with trans-genic rice seed containing destructed Japanese cedar pollen allergens, Cry j 1 and Cry j 2, against Japanese cedar pollinosis. Plant Biotechnol. J. 11, 66-76 (2013).

[75] Yang, L., et al.: Generation of a transgenic rice seed-based edible vaccine against house dust mite allergy. Biochem. Biophys. Res. Commun. 365, 334-338 (2008).

[76] Suzuki, K., et al.: Prevention of allergic asthma by vaccination with transgenic rice seed expressing mite allergen: induction of allergen-specifi c oral tolerance without bystander suppression. Plant Biotech. J. 9, 982-990 (2011).

[77] Wilken, L. R., Nikolow, Z. L.: Recovery and purifi ca-tion of plant-made recombinant proteins. Biotechnol. Adv. 30, 419-433 (2012).

[78] Fisher, R., et al.: GMP issues for recombinant plant-derived pharmaceutical proteins. Biotechnol. Adv. 30, 434-439 (2012).

[79] Spok, A., et al.: Evolution of a regulatory framework for pharmaceuticals derived from genetically modifi ed plants. Trends Biotechnol. 26, 506-517 (2008).

(杨欣燐译)

第六部分

佐剂和纳米技术

概　述

激光
- 疫苗接种前的激光疫苗佐剂
- 激光促进皮肤抗原呈递细胞迁移
- 贴片疫苗递送

细菌佐剂
- 源自外毒素和内毒素的佐剂
- 基于脂多糖的佐剂和脂质A类似物
- 用于抗原递送的细菌毒素

热休克蛋白
- 植物生物活性物质
- 自身佐剂抗原
- 植物热休克蛋白及其免疫特性

纳米颗粒
- 基于脂质体的疫苗
- 肺疫苗递送纳米技术
- 纳米颗粒口服佐剂

明矾和壳聚糖
- 含铝佐剂配方疫苗
- 分子通路和作用机制
- 一种碳水化合物聚合物佐剂：壳聚糖

高聚物
- 从胶体至树枝状聚合物的发展历程
- 纳米粒子的突出特点
- 控制释放策略

完整的疫苗开发过程并不是止步于抗原。大多数抗原的免疫原性较差，需要强佐剂的支持。这不仅适用于基于蛋白质抗原的重组疫苗，也适用于 DNA 疫苗。纯蛋白质和裸 DNA 都不起作用。天然免疫应答的佐剂、激活剂和增强剂触发了必需的促炎反应。佐剂和抗原必须形成完美的混合物。到目前为止，全球只有两种佐剂被授权用于人类疫苗。不幸的是，疫苗佐剂的研究至今很少受到关注。开发创新的无毒佐剂，特别是黏膜疫苗的输送是必要的。

激光佐剂　激光疫苗佐剂（LVA）是在皮内疫苗接种前对注射部位进行短暂（2 min）的激光照射，使机体对疫苗作出更好的反应。

脂多糖佐剂　新一代去毒脂多糖佐剂比明矾具有更高的佐剂特性和可接受的毒性，这为疫苗接种开辟了新的前景。

细菌毒素 内毒素和外毒素已开发用作佐剂，主要是酶活性消失或降低、或修饰化学结构的突变毒素以减少其细胞毒性效应。

植物热休克蛋白 这些数据证明了植物热休克蛋白具有刺激免疫反应的一般能力，同时也揭示了从表达重组抗原的植物组织中提取的热休克蛋白多肽复合物更特殊的用途。

功能性纳米脂质体 显然，脂质体疫苗的物理化学特性（抗原附着方法、脂质成分、双层流动性、粒子电荷和其他特性）对产生的免疫反应有着强烈的影响。

新兴纳米技术 研制干粉疫苗可以潜在地减少或可能消除冷链需求，促进无菌，提高抗原的整体稳定性，从而降低产品的整体成本。

口服佐剂 关于口服疫苗，免疫系统面临的一大挑战是在黏膜水平上达到耐受和炎症反应之间的适当平衡。肠道是哺乳动物最大的淋巴器官，比包括脾脏和肝脏在内的任何其他器官都含有更多的免疫细胞和最大浓度的抗体。

壳聚糖 在探索新型佐剂的过程中，糖基聚合物壳聚糖因其具有免疫刺激作用而受到越来越多的关注。壳聚糖来源于天然产物甲壳素。

铝佐剂疫苗 目前，明矾依然是唯一在美国获得许可的疫苗佐剂。然而，在欧洲，自 20 世纪 90 年代以来，水包油乳液 MF59、MPL（单磷脂 A，脂多糖类似物）+明矾和 AS03 也被批准使用。

新聚合物 纳米系统可以是无机胶体、有机胶体（通过乳液聚合或微/纳米乳液技术合成）、聚合物聚集体（胶束或聚合物体）、核心交联聚集体（纳米水凝胶、交联胶束或多聚体）、多功能聚合物线圈、树枝状聚合物或完美树枝状聚合物。

第 32 章　用于皮肤疫苗递送的激光和佐剂

Xinyuan Chen 和 Mei X. Wu①

摘要

过去 10 年间，因富含抗原呈递细胞（耐受性抗原呈递细胞）且免疫应答优于肌肉，接种选择上皮肤更受青睐。疫苗经由皮肤而非肌肉有可能研发出更安全的佐剂和非注射接种系统。过去 5 年我们开发了一种基于激光的新型疫苗佐剂（LVA），采用安全激光短暂照射接种部位，以增强耐受性抗原呈递细胞的活性和抗原吸收。短暂照射还能促进耐受性抗原呈递细胞从皮肤向引流淋巴结迁移。激光疫苗佐剂使疫苗接种效率大为提高，除抗原本身外，无外来或自身物质注入身体，所以局部副作用最小，也无长期副作用。传统疫苗佐剂改变抗原形式增强免疫原性，与之相比，这种新策略使接种部位做好准备作出更好的免疫应答。因此，激光疫苗佐剂不需要与抗原结合，并可充当通用疫苗佐剂。实际上，这种物理类型的激光疫苗佐剂可以有效增强卵清蛋白（OVA）、流感疫苗、尼古丁疫苗等诱导的免疫应答。除激光疫苗佐剂外，我们还探索了一种激光辅助贴片接种系统，无针、无痛经皮免疫。该系统以烧蚀分数激光（AFL）治疗为基础在皮肤中生成大量可自更新微通道（MCs），局部应用疫苗通过该微通道轻松进入皮肤。激光治疗主动将耐受性抗原呈递细胞募集到每个微通道附近，后耐受性抗原呈递细胞更有效地吸收微通道附近周围或内部的抗原，致使对卵清蛋白免疫应答比基于胶带剥离贴片递送的免疫应答高出 100 倍。本章，我们将介绍这些技术及其临床前研究并讨论它们的优缺点。

32.1　背景

到目前为止，大多数疫苗为肌内注射，相对方便且易于管理，但肌肉组织含有较少耐受性抗原呈递细胞（APCs），并且肌内非疫苗接种有效靶点。相比之下，皮肤富含固有的耐受性抗原呈递细胞和淋巴管网。越来越多证据表明，与肌内疫苗接种相比，直接进入皮肤中的疫苗接种能产生更强效的免疫反应[1-4]。约两世纪前划痕接种天花疫苗取得巨大成功，这种高传染性疾病于 1979 年最终被消灭，使皮肤免疫效用大放异彩[5]。因损伤皮肤、结疤，且有感染风险，现代不再使用划痕[6,7]。今天的结核病皮肤试验采用曼图克斯法，是一种皮内（ID）注射技术，但由于需要特殊的培训人员，可靠皮肤靶点有技术难度，因此不适用于常规免疫接种[7]。为增进患者依从性和疫苗免疫原性，我们正研发一种激光辅助贴片接种系统，这种皮肤免疫实用、无针、无痛。

除可靠的皮肤接种疫苗外，加入佐剂以进一步加强皮肤免疫还面临着其他挑战，缘于皮内注射要求小注射量，且皮肤对炎症敏感性高[2-4]。许多化学物质和生物分子能够增强由疫苗诱导的免疫反应，但它们也会引起皮肤高度炎症，因此不适用于皮肤免疫[8]。这里，我们将介绍一种新型、安全的方法来增强皮肤接种疫苗免疫反应。

---

① X. Chen，PhD. M. X. Wu，MD，PhD（✉）（马萨诸塞州总医院（MGH）威尔曼医学中心，美国马萨诸塞州波士顿；哈佛医学院（HMS）皮肤病学系，美国马萨诸塞州波士顿，哈佛-麻省理工学院医疗科技（HST）学院，美国马萨诸塞州剑桥，E-mail：mwu2@ partners. org）

32.2 激光介导的佐剂及疫苗的递送

32.2.1 激光疫苗佐剂

激光疫苗佐剂（LVA）先于皮内疫苗接种，它以注射部位短暂（2 分钟）激光照射为基础，使我们的身体对疫苗有更好的反应[9]。照射用 Q 开关 532-纳米 Nd：YAG 激光，脉冲宽度 5~7 纳秒、光束直径 7 mm、频率 10 Hz（光谱物理有限公司，山景城，美国加州）[9]。这种无创激光照射对皮肤损伤微小或可见，并提升局部皮肤耐受性抗原呈递细胞运动与功能[9]。当注入激光照射部位时，抗原可被高度移动的耐受性抗原呈递细胞充分吸收，并从皮肤转移到引流淋巴结，增强免疫反应[9]。

与传统疫苗佐剂相比，使用激光疫苗佐剂助推皮肤免疫，安全方便[8]。如图 32.1 所示，用激光治疗的皮肤未发现明显的组织学炎症、损伤或发红现象[8]。相比之下，肌内免疫佐剂使用最多的是基于铝盐（明矾）的佐剂，它在皮肤中停留数月，皮内接种后（高割面）引发皮肤反应强烈持久，有大量炎性细胞浸润到佐剂治疗部位（低割面）[8]。同样地，几种佐剂像油包水乳剂 montanideISA 720 和 Toll 样受体-7（TLR7）激动剂咪喹莫特（R837），每种都诱发严重局部皮肤刺激，并伴随长达数周皮肤溃疡和炎症[8]。注意到结合使用 Toll 样受体 4 激动剂单磷酰脂质 A 也伴有严重持久的皮肤刺激（膜磷脂）和 Toll 样受体 9 激动剂非甲基化 CpG 寡核苷酸（CpG）、R837、或明矾[8]。这些佐剂中，尤其是乳状佐剂，导致持续皮肤炎症可破坏皮肤完整性，使皮肤易受到各种感染，因此排除其皮肤接种疫苗临床使用[8]。在测试的佐剂中，可溶性膜磷脂或 CpG 仅引起轻微局部反应，2 周内完全分解，因此可用于皮肤免疫[8]。有趣的是，激光疫苗佐剂与膜磷脂或 CpG 相结合不会扩大局部反应，由此我们可以探索激光疫苗佐剂和膜磷脂或 CpG 对皮肤免疫的协同作用。

激光介导免疫增强机制还不完全清楚。激光照射看似显著增强了耐受性抗原呈递细胞运动性：伪足移动得更快，与对照皮肤相比，它们在激光处理皮肤中的分布更分散[9]。除增强皮肤耐受性抗原呈递细胞的运动性外，透射电镜显示激光照射后皮肤胶原蛋白纤维紊乱，皮肤细胞与周围组织支架之间的相互作用被扰乱，据推测这可以降低组织对耐受性抗原呈递细胞迁移的免疫力[8,10]。这种松散的皮肤结缔组织与未经处理的皮肤组织形成对比，后者充满致密且排列整齐的胶原蛋白纤维，与皮肤细胞密切相关[8,10]。耐受性抗原呈递细胞的运动性增强和组织抵抗力降低共同促进耐受性抗原呈递细胞从皮肤通行至引流淋巴结，激光治疗后立即形成许多皮肤线来反映这一点，这由大量耐受性抗原呈递细胞迁移到淋巴管中所致（图 32.2）[10]。未经治疗的

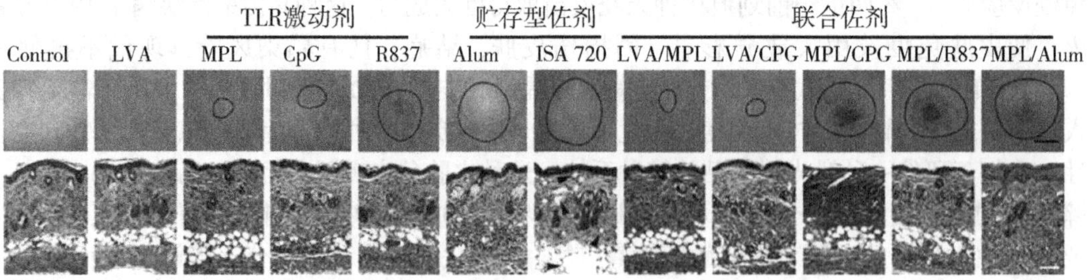

图 32.1 皮内注射后各种佐剂的反应原性

用激光或皮内照射 BALB/c 小鼠，后每个注射指定量佐剂 20 μL，有或无激光照射。5 天后分析局部皮肤反应。上半部、皮肤照片（比例尺，2 mm）和下半部、组织学检查（比例尺，100 μm）。用 Montanide ISA 720 注射皮肤点上半部上的圆圈和下半部上的箭头勾勒病变大小，以去除油包水佐剂占据的鳞茎。

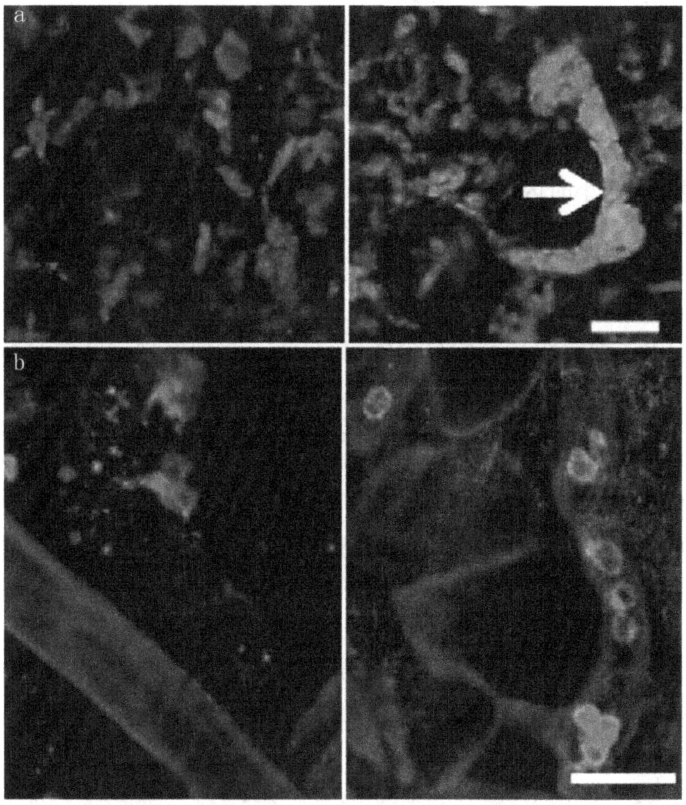

图 32.2　激光增强真皮抗原呈递细胞控制激光迁移

（a）主要组织相容性复合体 II-EGFP 转基因小鼠皮肤激光照射后 1 小时内形成真皮索（箭头）。标尺，50 μm。

（b）典型整体标本耳像，显示 GFP+细胞 30 分钟内进入激光照射耳朵淋巴管内。红色胶原蛋白 IV，绿色主要组织相容性复合体 II+细胞。标尺，25 μm。

皮肤中未发现真皮索。由于它们的活动性增加，激光照射也增强了耐受性抗原呈递细胞对皮肤和引流淋巴结的抗原吸收[9]。增加耐受性抗原呈递细胞运动性时，激光照射不会引起炎症或显著增加炎症细胞因子生成[8,9]。它对局部皮肤树突状细胞（DCs）上 CD86、CD80 和 CD40 表达也无影响[8-10]。因此，激光疫苗佐剂被认定为非炎症性疫苗佐剂。简言之，特定环境下皮肤激光治疗会扰乱致密的皮肤结缔组织，增强皮肤耐受性抗原呈递细胞运动性和抗原吸收，促进疫苗诱导免疫反应。

32.2.2　激光辅助贴片疫苗接种

浅层角质层（SC）表层是阻止经皮疫苗有效接种的主要障碍物[11]。我们探索各种各样物理（如热、超声波）、机械（如胶带剥离、微针）和化学方法（如二甲基亚砜、酒精），破坏角质层以便于经皮疫苗接种[12]。然而，这些方法要么是效率低下，要么有引起皮肤刺激或其他局部或全身反应的高风险[12]。

还评估了 SC 激光介导烧蚀促进经皮疫苗接种潜力，但传统全表面烧蚀需要一段相对较长（数周）的恢复时间，并且对于疫苗接种不安全[13]。过去几年，烧蚀分数激光（AFL）已被用于取代传统皮肤表面重修，明显缩短恢复时间[14]。烧蚀分数激光发射微激光束，并在皮肤表面生成大量自我更新微通道，直径和深度为 100～300 μm，具体取决于微激光束的能量[12]。事实证明，这些微通道可使敷于局部的疫苗通过[12]。烧蚀分数激光治疗后，局部应用疫苗涂层纱布贴片，可通过这些微通道将疫苗直接接种至皮肤，由此疫苗进一步扩散到周围圆柱形尺寸的组织，然后被微通道周围募集的耐受性抗原呈递细胞吸收[12]。

研究发现，在经烧蚀分数激光处理的皮肤上局部应用卵清蛋白（OVA）涂层的纱布贴片，其免疫反应比同样在胶带剥离皮肤上应用卵清蛋白涂层纱布贴片的免疫反应增加了99倍[12]。此外，激光治疗后皮肤在24小时内迅速表皮再生，而胶带剥离皮肤至少需要2天才能恢复[12]。快速表皮再生可确保受影响皮肤完整，且可确保无针技术经皮疫苗接种的安全。

为直观观察烧蚀分数激光对疫苗递送和耐受性抗原呈递细胞募集，我们将荧光标记的OVA局部应用于经烧蚀分数激光处理的基因修饰小鼠皮肤上，以表达框架内GFP，它被注入大多数耐受性抗原呈递细胞上主要组织相容性复合体-II分子。我们发现，烧蚀分数激光产生的微通道不仅为OVA进入皮肤提供了自由路径（图32.3，红点、激光），而且还吸引大量GFP标记耐受性抗原呈递细胞到微通道邻近（图32.3，第1组、第2组），这样每个微通道邻近可直接有效的摄取抗原（图32.3，第3组、第4组）。募集耐受性抗原呈递细胞的增强抗原摄取发挥功效成为“内置”佐剂，并且有望增强疫苗诱导免疫应答。激光疫苗佐剂仅施压力但不杀死皮肤细胞，与上述激光疫苗佐剂使用的激光不同，烧蚀分数激光引入光热效应，引起组织不可逆损伤，在每个微通道周围形成多个微通道和微热区。在微热区，凋亡或坏死细胞物质释放可能会向免疫系统发出“危险信号”，并诱导局部巨噬细胞或浸润中性粒细胞分泌趋化因子[15]，将吸引大量耐受性抗原呈递细胞。据我们所知，耐受性抗原呈递细胞募集是烧蚀分数激光技术的一个特性，不能用其他SC剥离法进行概括。

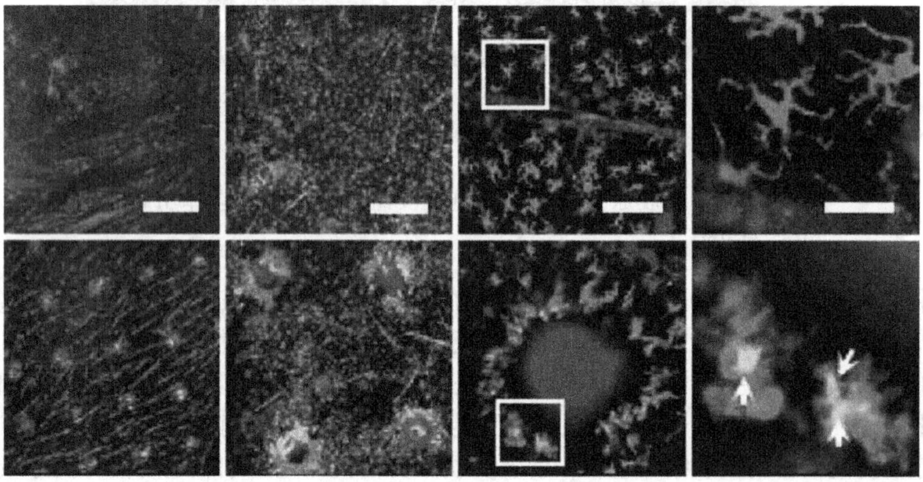

图32.3 抗原呈递细胞的有效抗原递送和募集

用烧蚀分数激光治疗主要组织相容性复合体II-EGFP小鼠耳朵或根本不接受治疗，然后局部用Alexa Fluor 647（AF647）-联合OVA包衣纱布贴片30分钟。24小时后对表皮层进行活体共聚焦成像。显示了低（第1组，比例：750 μm）、中（第2组，比例：300 μm）放大率和高放大率（第3组，比例：75 μm）的代表性图片。在第4组中，比例：25 μm显示单个耐受性抗原呈递细胞（箭头）抗原提取，第3组的矩形中有突出显示。未治疗的对照耳中未发生抗原摄取。绿色GFP标记耐受性抗原呈递细胞，红色标记AF647-OVA，黄色则标记耐受性抗原呈递细胞摄取AF647-OVA。

32.3　临床前开发和功效

激光照射可增强耐受性抗原呈递细胞一般功能，因此，激光疫苗佐剂可能增强所有现有的或新的、基于蛋白质的疫苗的免疫反应。由这项技术我们获得了类似的疫苗佐剂，包括模型抗原 OVA[9]、尼古丁疫苗、季节性和大流行性流感疫苗或乙型肝炎表面抗原（HBsAg）。抗尼古丁免疫疗法是一种新型的有希望的疗法，通过诱导抗尼古丁抗体（NicAb），结合并阻断尼古丁进入大脑，来治疗尼古丁成瘾[16-18]。然而，只有 30%的吸烟者产生相对较高的 NicAb 滴度，与使用安慰剂且有明矾佐剂存在的情况下 5~7 次肌内（im）免疫后的吸烟者相比，该亚组的戒烟率明显增加[16,17]。血清 NicAb 滴度处于低水平或中位水平时，戒烟率无明显提高[16,17]。NicAb 滴度与戒断率正相关近期 NicVAX（一种与重组外蛋白 A 结合的尼古丁疫苗）III 期临床试验失败，提升了改善尼古丁疫苗免疫原性的必要性[19]。因此，我们评估了皮内免疫与激光疫苗佐剂组合是否可以显著增强小鼠中尼古丁疫苗免疫原性。

尼古丁结合钉形贝血蓝蛋白（Nic-KLH）评估激光佐剂功效。如表 32.1 所示，每 2 周对小鼠进行多次免疫。我们发现，3-皮内免疫诱导 NicAb 滴度比 7-肌内免疫高 60%，并且在峰值水平下明矾佐剂 4-肌内免疫接种，有相当的 NicAb 滴度（表 32.1），这证实尼古丁疫苗的皮内免疫接种优于肌内免疫接种。将激光疫苗佐剂加入原发性皮内免疫后，NicAb 滴度较皮内免疫提高 140%，较肌内免疫提高 280%（表 32.1）。此外，激光疫苗佐剂/皮内免疫接种能维持 NicAb 滴度峰值超 4 个半月或更久，若无佐剂，肌内或皮内免疫约 3 个月后，NicAb 峰值产量将逐渐明显减少（$P<0.05$）。与非免疫对照组相比，尼古丁激发后，7-肌内免疫接种或 3-皮内免疫接种显著抑制尼古丁进入大脑的程度分别达到 37%和 40%，而往 3-皮内免疫接种方案的初级免疫中加入激光疫苗佐剂可减少 60%的脑尼古丁进入大脑。

表 32.1　激光疫苗佐剂/皮内免疫优于明矾佐剂肌内免疫

免疫方式	免疫次数	NicAb 滴度[a]峰值（$\times10^3$）	增加百分比	*P* 值
肌内	7	10.1±2.7	—	—
明矾+肌内	4	20.3±5.0	100	0.035 2
皮内	3	15.7±3.1	60	0.283
激光疫苗佐剂+皮内	3	38.2±5.1	280	0.002 8

NicAb 滴度值表示为平均值±SEM（平均值标准误差），并用 T-检验计算 *P* 值，然后与 7-肌内疫苗接种方案相比，*P* 值小于 0.05 则以粗体 a 标识。

[a]尼古丁抗体滴度通常在最后一次免疫后 2 周达到峰值。

与尼古丁疫苗的相当弱的免疫原性不同，流感疫苗显示出强免疫原性，也可以与激光疫苗佐剂类似地加强免疫原性。我们的研究表明，激光疫苗佐剂与皮内免疫的组合显著增强了血细胞凝集抑制（HAI）抗体滴度，在皮内免疫中对成年小鼠中的小鼠适应的 PR8 疫苗提供了 5~10 倍的剂量保留效应。此外，激光疫苗佐剂与皮内免疫（而不是单独的皮内免疫）相结合，可显著提高致死性病毒攻击后老年小鼠的 HAI 抗体滴度和存活率。这些临床前研究证明了激光疫苗佐剂在成年和老年小鼠中具有弱或强免疫原性的疫苗免疫应答的能力。

32.4　优势和缺点

与大多数化学或生物佐剂相比，这种非炎性激光治疗的疫苗佐剂相对较弱。然而，与传统佐剂相比，激光疫苗佐剂具有许多优势。首先，激光疫苗佐剂不会引起局部反应或长期副作用，因为除抗原本身外，没有外来物质注入体内。相反地，除了少数例外，几乎所有传统的佐剂都是外来的，并且可能通过分子模拟诱导自毁性免疫交叉反应而潜在地引起长期副作用。其次，激光疫苗佐剂不需要修改疫苗生产程序或开发特定的疫苗/佐剂配方，可以随时立即重复使用。当我们在遇到流感暴发，新病毒株暴发或生物恐怖袭击事件中遇到疫苗短缺时，这比传统佐剂具有很大的优势。再次，激光疫苗佐剂可以方便地与新开发的皮内或皮肤免疫策略相结合，以进一步增强疫苗诱导的免疫应答。激光疫苗佐剂也可以被批准作为各种临床疫苗的独立佐剂。

为了将疫苗输送到皮肤中，过去10年中已经开发了几种策略，包括喷射注射器，微针和微型阵列贴片[20-23]。AFL具有优于这些策略的优点。首先，它是无针和无痛的，并且不会引起任何风疹或皮肤形状改变，因此更符合患者的要求。其次，AFL引导疫苗直接输送到各个微通道，最大限度地减少局部反应，同时与将疫苗输送到单一比较大的局部相比，提高了疫苗接种效率。再次，AFL不仅向皮肤提供足够的疫苗，而且还可能增强疫苗诱导的免疫应答，这是由于“内置”的适应性，如发生在每个微通道周围的APC募集所示，并可能从微热区向免疫系统释放危险信号。我们正在制造一种小型手持装置，将AFL和微通道引导的疫苗递送结合在一起，这将加速这种新技术在疫苗输送和佐剂方面的广泛应用。

参考文献

[1] Barraclough, K. A., et al.: Intradermal versus intra-muscular hepatitis B vaccination in hemodialysis patients: a prospective open-label randomized controlled trial in nonresponders to primary vaccination. Am. J. Kidney Dis. 54, 95-103 (2009).

[2] Belshe, R. B., et al.: Serum antibody responses after intradermal vaccination against infl uenza. N. Engl. J. Med. 351, 2286-2294 (2004).

[3] Beran, J., et al.: Intradermal infl uenza vaccination of healthy adults using a new microinjection system: a 3-year randomised controlled safety and immunogenicity trial. BMC Med. 7, 13 (2009).

[4] Kenney, R. T., Frech, S. A., Muenz, L. R., Villar, C. P., Glenn, G. M.: Dose sparing with intradermal injection of infl uenza vaccine. N. Engl. J. Med. 351, 2295-2301 (2004).

[5] Lofquist, J. M., Weimert, N. A., Hayney, M. S.: Smallpox: a review of clinical disease and vaccination. Am. J. Health Syst. Pharm. 60, 749-756 (2003).

[6] Kim, Y. C., Jarrahian, C., Zehrung, D., Mitragotri, S., Prausnitz, M. R.: Delivery systems for intra-dermal vaccination. Curr. Top. Microbiol. Immunol. 351, 77-112 (2012).

[7] Liu, L., et al.: Epidermal injury and infection during poxvirus immunization is crucial for the generation of highly protective T cell-mediated immunity. Nat. Med. 16, 224-227 (2010).

[8] Chen, X., Wu, M. X.: Laser vaccine adjuvant for cuta-neous immunization. Expert Rev. Vaccines 10, 1397-1403 (2011).

[9] Chen, X., et al.: A novel laser vaccine adjuvant increases the motility of antigen presenting cells. PLoS One 5, e13776 (2010).

[10] Chen, X., Zeng, Q., Wu, M. X.: Improved effi cacy of dendritic cell-based immunotherapy by cutaneous laser illumination. Clin. Cancer Res. 18, 2240-2249 (2012).

[11] Li, N., Peng, L. H., Chen, X., Nakagawa, S., Gao, J. Q.: Transcutaneous vaccines: novel advances in technology and delivery for overcoming the barriers. Vaccine 29, 6179-6190 (2011).

[12] Chen, X., et al.: Facilitation of transcutaneous drug delivery and vaccine immunization by a safe laser technology. J. Control. Release 159 (1), 43-51 (2012).

[13] Lee, W. R., et al.: Erbium: YAG laser enhances trans-dermal peptide delivery and skin vaccination. J. Control. Release 128, 200-208 (2008).

[14] Manstein, D., Herron, G. S., Sink, R. K., Tanner, H., Anderson, R. R.: Fractional photothermolysis: a new concept for cutaneous remodeling using microscopic patterns of thermal injury. Lasers Surg. Med. 34,

426-438 (2004).

[15] Chen, G. Y., Nunez, G.: Sterile infl ammation: sensing and reacting to damage. Nat. Rev. Immunol. 10, 826-837 (2010).

[16] Cornuz, J., et al.: A vaccine against nicotine for smok-ing cessation: a randomized controlled trial. PLoS One 3, e2547 (2008).

[17] Hatsukami, D. K., et al.: Immunogenicity and smoking-cessation outcomes for a novel nicotine immunotherapeutic. Clin. Pharmacol. Ther. 89, 392-399 (2011).

[18] Pentel, P. R., et al.: A nicotine conjugate vaccine reduces nicotine distribution to brain and attenuates its behavioral and cardiovascular effects in rats. Pharmacol. Biochem. Behav. 65, 191-198 (2000).

[19] Fahim, R. E., Kessler, P. D., Fuller, S. A., Kalnik, M. W.: Nicotine vaccines. CNS Neurol. Disord. Drug Targets 10, 905-915 (2011).

[20] Prausnitz, M. R., Mikszta, J. A., Cormier, M., Andrianov, A. K.: Microneedle-based vaccines. Curr. Top. Microbiol. Immunol. 333, 369-393 (2009).

[21] Matriano, J. A., et al.: Macrofl ux microprojection array patch technology: a new and effi cient approach for intracutaneous immunization. Pharm. Res. 19, 63-70 (2002).

[22] Taberner, A. J., Ball, N. B., Hogan, N. C., Hunter, I. W.: A portable needle-free jet injector based on a custom high power-density voice-coil actuator. Conf. Proc. IEEE Eng. Med. Biol. Soc. 1, 5001-5004 (2006).

[23] Hingson, R. A., Davis, H. S., Bloomfi eld, R. A., Brailey, R. F.: Mass inoculation of the Salk polio vaccine with the multiple dose jet injector. GP 15, 94-96 (1957).

（吴国华译）

第 33 章　作为佐剂的细菌脂多糖

Jesús Arenas①

摘要

脂多糖（LPS，内毒素）是革兰氏阴性细菌外膜的主要成分。它是人类体液和细胞反应的激活剂，有可能在疫苗技术中用作佐剂。重要的是，脂多糖具有诱导 Th1 型反应和刺激细胞毒性 T 淋巴细胞的能力，这是通过标准佐剂很难获得的，而且需要特异性免疫刺激疗法。相比之下，脂多糖具有极强毒性，限制了其在人体中的临床应用。其化学结构的改变导致产生毒性降低，但保留佐剂性质的基于脂多糖的衍生物。单磷酰脂质 A（MPLA）是最成功的基于脂多糖的佐剂，目前纳入批准的疫苗制剂中，并广泛用于疫苗试验和临床前研究。新型设计的结构类似于脂多糖，并通过化学合成产生，与 MPLA 相比，可以提供更低的生产成本和更少的异源配方。此外，很最适合特定的免疫疗法。因此，基于脂多糖的结构在人类疫苗学中作为佐剂很有价值，并为现有的特定治疗需求提供了新的可能性。

亚单位疫苗的研制提高了人类疫苗预防的安全性，但需要强有力的佐剂。明矾（各种铝盐）是最常用的佐剂，具有可接受的副作用和诱导对许多人类病原体的最佳保护。然而，明矾不适合某些病原体或新的疫苗疗法，如癌症或过敏。脂多糖是革兰氏阴性细菌外膜的组分，主要通过其刺激免疫应答的固有能力作为佐剂进行研究。但是，脂多糖在人体中会引起不可接受的毒性作用。

新一代脱毒脂多糖品种佐剂特性高于明矾，其可接受毒性也为疫苗接种开辟了新的前景。在本章节中，首先介绍了脂多糖结构会毒性和免疫系统激活的基础知识。接下来，简要地引用明矾的佐剂特征和相应的缺点。随后，进一步讨论最有希望的脱毒脂多糖分子和佐剂特征，并特别介绍当前的进展。最后的概述部分总结了最相关的要点。

33.1　脂多糖结构和生物活性

脂多糖是革兰氏阴性细菌外膜的外部小叶的组分。它是由 3 个结构域形成的复合糖脂，富含脂肪酸的结构域（脂质 A）、寡糖结构域（核心）和重复寡糖结构域（O-抗原）[1]。图 33.1 表示典型的脂多糖组织。

脂质 A 结构域是与可变数量的酯-和酰胺-连接的 3-羟基脂肪酸和磷酸基团连接的 β-1，6-连接的 d-葡糖胺二糖（图 33.1）。其结构高度保守，但不同的微生物可能存在脂肪酸侧链的数量和长度，末端磷酸酯残基以及相关的变化和修饰。核心结构域是由 9 或 10 个糖形成的支链寡糖区域，其组成在物种之间比脂质 A 更易变。最后，O-抗原（如果存在的话）是树结构域中变化最大的，并且由多达 50 个由 2～8m 的糖基部分形成的重复寡糖单元组成。此外，某些修饰酶可以改变脂多糖的组成，有助于增加脂多糖异质性[2-5]。

---

① J. Arenas，PhD（乌得勒支大学生物系分子微生物学研究室，荷兰乌得勒支帕多瓦拉安，邮编：83584，E-mail：jesusarenasbust@ yahoo. es，j. a. arenasbusto@ uu. nl）

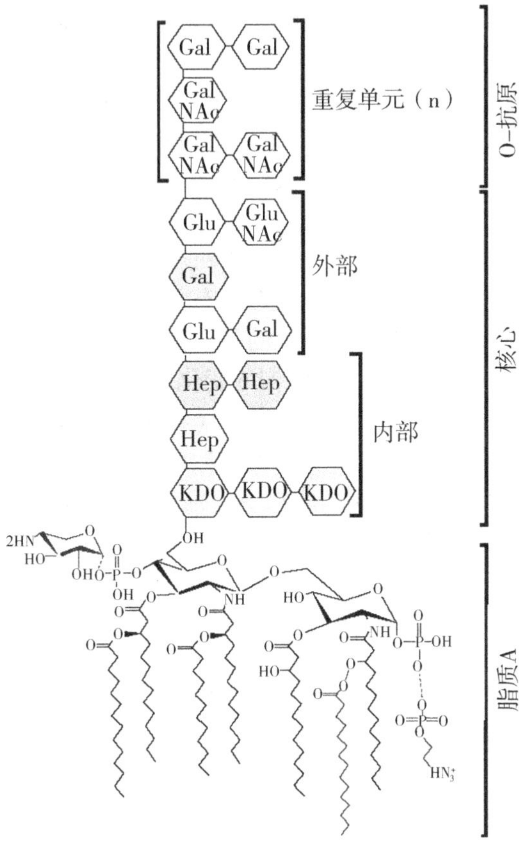

图 33.1 革兰氏阴性菌脂多糖的一般化学结构

描述了大肠杆菌的脂多糖结构。脂多糖由 3 个区域组成：脂质 A、核心和 O-抗原。显示人体最大免疫刺激或内毒素活性的大肠杆菌脂质 A 的化学结构（本工作的主题）进一步详述（黑色）。另外，具有用于获得 MPL 的明尼苏达沙门氏菌的脂质 A 的那些取代基是蓝色的。核心和 O-抗原区域的残基被示意性地缩写为 *KDO* 2-酮-3-脱氧辛酸、*Hep* d-甘油-甘露-庚糖、*Glu* d-葡萄糖、*Gal* d-半乳糖、*GluNAc* N-乙酰基-葡糖胺和 *GalNAc* N-乙酰半乳糖胺。

表 33.1 脂多糖对人体的相关影响

有利		有害	
影响	机制	影响	机制
抗菌免疫力	激活 DC、启动 B 细胞和 T 细胞、调理、刺激自然杀伤细胞，激活巨噬细胞	高热	IL-1 释放
抗病毒免疫	刺激 CD8（+）T 细胞免疫	强烈的炎症反应	大量分泌各种炎症介质
抗肿瘤免疫力	刺激 CD8（+）T 细胞免疫	弥漫性血管内凝血	血液凝固中元素的减少
减轻对 Th1 的 Th2 反应	增强 Th1 表型	休克，低血压，淋巴细胞减少	血流量减少

Toll 样受体（TLR）s 属于可识别多种特定但保守的病原微生物结构的一个受体家族[6]。受

刺激后，TLRs 立即启动免疫防御机制的激活。Toll 样受体 4（TLR4）与糖蛋白 MD-2 结合，构成脂多糖受体[7,8]。TLR4 是存在于抗原呈递细胞（APC）（巨噬细胞和树突细胞）和人类上皮细胞上的跨膜蛋白。其刺激需要相关分子的配合，如脂多糖结合蛋白和促进脂多糖转移至受体的 CD14。TLR4 刺激诱导细胞内蛋白质复合物的形成，导致细胞内信号级联的激活[9,10]。这些反应引发各种促炎细胞因子（IL-1、IL-8、IL-12、TNFα 和 IFNγ）的生物合成和分泌以及共刺激分子的产生[11]，最终激活体液和细胞反应，包括激活补体系统[12,13]、巨噬细胞[14,15] B 和 T 细胞的激活，以及细胞协同的增强[11]。因此，这种反应有利于控制局部感染。相反地，特别是在脓毒症期间释放到血液系统的高脂多糖剂量导致细胞因子和炎症介质的大量分泌，具有严重的[16,17]和/或致命后果[18,19]。表 33.1 总结了脂多糖对人类的有益和有害影响，有关详细信息，请参见修订版[20,21]。总之，脂多糖是免疫系统（佐剂）的强激活剂，但也是一种剧毒物质（内毒素）。

脂质 A 区域是刺激 TLR4 的主导因素。其结构的变化，主要是关于酰基链的数量和长度，以及电荷方面是至关重要的[22]。六酰化大肠杆菌脂质 A（经典的脂多糖结构，如图 33.1 所示）含 12~14 个碳的脂肪酸以及两个磷酸残基是人 Toll 样受体 4（hTLR4）的最大刺激物[23,24]。相比之下，脂肪酸含量为 18~16 个碳和磷酸残基（脂质 A 生物合成途径的中间产物）的四酰化脂质 IVa 不刺激 hTLR4（典型的 hTLR4 拮抗剂）[25]。

33.2 脂质 A 的类似结构及其作为佐剂的作用

基于传染性致弱或灭活的全病原体的疫苗含有多种用于 TLRs 的靶结构，从而促进强大和长期的保护。然而，它们也会产生多种副作用，甚至危及生命[26,27]。基于一种或某些纯化组分（亚单位疫苗）的疫苗显示出可接受的安全性，但免疫原性差并且需要额外的免疫刺激剂（佐剂）。明矾是指几种铝盐，是最常用的佐剂。它是安全的，主要引发 Th2 型抗体反应，显示在多种疫苗中有效[28]。然而，明矾几乎不会促进 Th1 型抗体反应[29]。有利于 Th1 或更平衡的 Th1 / Th2 反应的佐剂需要诱导针对某些病原体[30]、癌症[31]或过敏[32]的疾病的最佳免疫保护。除此之外，明矾对黏膜免疫刺激很差。黏膜组织是抵御许多病原体的第一道防线，也是共生和机会性微生物的生态位，例如脑膜炎奈瑟菌。

因此，黏膜免疫被认为是逃避病原体定植和提供群体免疫特定病原体的黄金疗法。针对黏膜免疫的疫苗佐剂必须促进一系列复杂的生物学活性，如 Th17 细胞发育、APC 增殖和 IgA 产生[33,34]。在这方面，已经广泛研究了几种物质作为细菌毒素或 CpG 等[34]，但迄今为止，还没有可用的批准佐剂（MPL 除外，将在下面讨论）。除明矾外，另外有 3 种佐剂获得许可：MF59、AS03 和 RC-529。MF59 是一种低油含量的水包油乳剂，纳入在批准的流感疫苗中[35]。虽然 MF59 诱导的 Th1/Th2 反应比明矾更平衡，但它通常需要联合使用 Th1 增强剂才显示出部分疗效。AS03 和 RC-529 含有基于脂多糖的物质，稍后将讨论。

脂多糖作为一种佐剂，以其对共给药抗原的高诱导 Th1 型反应的能力而备受关注。IL-12 是影响该反应发展的最相关因素之一。值得注意的是，脂多糖是白细胞介素的刺激物。有趣的是，TLR4 受体存在于黏膜表面；因此，预期 TLR4 激动剂可以促进局部和远端黏膜部位的免疫应答。在过去的几十年中，人们采取了几种策略，在不改变这种固有能力的情况下，降低其极端毒性；通过化学处理和脂质 A 类似物的化学合成来改变脂多糖的组成，特别是脂质 A，是一个很好的例子（见详细修订[36]）。因此，尽管只有少数具有所需的性质，也产生了多样的脂质 A 种类。接下来，进一步讨论最相关的物质和临床应用。

沙门氏菌脂多糖的化学水解，其含有具有 7 个酰基链和 3 个磷酸基团的脂质 A，如图 33.1 所示（具有蓝色取代基），产生了最成功的基于脂多糖的佐剂之一，单磷酰脂质 A（MPL）。该衍生物结构是具有 1 个磷酰基的六酰基侧链脂质 A[37]（参见图 33.2）。MPL 的毒性低于母体（0.1%的毒性）[38]，其毒性副作用与明矾[39,40]相当，同时保留了脂多糖的刺激特性。目前，

MPL是欧洲批准的人类疫苗制剂（人乳头瘤病毒（Cervarix）[33,34]），花粉过敏（Pollinexquattro）[41,42]）和澳大利亚［乙型肝炎病毒（Fendrix）[43]］的佐剂，它被广泛应用于人类疫苗试验中，用于多种传染病，如疟疾[44]、结核病[45,46]或肿瘤生长[47]。

因为MPL是高度疏水的并且在水溶液中产生可能显著影响TLR4活化的聚集体，所以它通常与明矾或其他递送系统组合配制[48]。这些组合与其他因子（伴随的抗原或给药途径）一起可以改变其佐剂作用。例如，在含水制剂中，MPL促进抗体产生，而在水包油乳剂中，它能更好地刺激T细胞应答。相反地，MPL与其他递送系统组合是细胞毒性T淋巴细胞增殖的强刺激物。输送系统也可以改变它们的生物特性。脂质体是由磷脂双层膜形成的球形囊泡，广泛用于以其自身结构传递抗原。将MPL掺入脂质体中减少了残留物的毒性但保留完整，其佐剂脂质体MPL制剂具有广泛的潜力[49]；在人类试验中，对其他脱毒脂多糖物种的不同适应症也观察到了这种作用[50]。因此，如疟疾[49]、肺炎球菌病、[51]或生殖器疱疹2型[52]和实验动物用于化脓性链球菌感染[53]或大肠杆菌毒素中和[54]。最后，值得注意的是，MPL通过黏膜[55]和肌内[56,57]给药后，成功提供黏膜免疫功能；黏膜疫苗的MPL配方已被广泛探索用于治疗不同疾病；生殖器疱疹[58]和黏膜利什曼病[59]是很好的例子。

MPL的主要缺点之一是其生产过程中产生的脂质A同系物的不均一性以及随后的纯化成本。解决这个问题的办法可能是生成合成脂质A类似物。化学合成产生了纯净和明确的结构，降低了生产成本。与天然脂多糖物种一样，这些结构与hTLR4相互作用。因此，产生了许多具有酰基链长度和位置，磷酸基团或骨架单元变化的类似物，并进一步分析了它们的生物活性。迄今为止，最适合疫苗开发的分子是RC-529、E6200、GLA和ONO-4007（化学结构见图33.2）。

RC-529是一种MPL的合成类似物，由一个单糖主链和6个脂肪酰链[38]组成。它是一种非常有吸引力的佐剂。与MPL一样，它在临床试验中具有良好的耐受性和有效性[38]，但生产成本较低。事实上，它在阿根廷被批准用于乙肝疫苗，通常与不同的递送系统结合以增强其溶解性或提高其递送性而不影响免疫刺激能力[60]。此外，在实验动物身上进行的几项研究表明，RC-529是一种有效的黏膜佐剂，可以对抗缺乏有效疫苗治疗的病原体。例如，它在用肺炎链球菌蛋白PppA[61]和脑膜蛋白P2086[60]进行鼻内免疫后引发了杀菌抗体，并且它促进了鼻腔和生殖器黏膜中的猕猴抗HIV肽免疫原的高抗体滴度[62]。同样地，它减少了小鼠非典型流感嗜血杆菌和卡他莫拉菌的鼻腔定植，通过鼻腔用重组蛋白免疫[63]，并且提供针对致命性流感攻击的显著保护[64]。

E6200是一种六酰基无环骨架[65,66]，其生物活性高于明矾[65,66]或MPL[65]，且毒性较低[67]。其简单的结构允许生产比其他合成TLR4激动剂高纯度的材料[65]。各种研究表明，当与传统疫苗结合时，其将免疫反应转化为Th1特征的能力很强[65,67,68]。这种免疫谱的产生在癌症疫苗中尤其相关。事实上，在动物模型中，E6200与单克隆抗体（曲妥珠单抗）联合使用可显著增强对肿瘤生长的保护[69]。

GLA是一种六酰基合成脂质，是由一个磷酸基双糖主链组成的衍生物。结果表明，它比MPL[70,71]具有更强的辅助能力，在动物和第一阶段试验中表现出良好的安全性[72]。MPL有很强但不是压倒性的能力来促进TH1反应。有趣的是，GLA显示出将抗原特异性免疫反应转化为Th1型[73,74]的强大能力，因此，我们建议将其作为一种更好的替代品，以提供对某些病原体的充分保护。事实上，实验动物对弓形虫[75]、结核分枝杆菌[76,77]或流感病毒[72]有显著的保护作用。

ONO-4007是一种三酰化无环磺化骨架。这种分子在动物模型中诱导了肿瘤和转移的退化[78,79]。这一特性是由于它具有很强的刺激巨噬细胞分泌肿瘤坏死因子（TNF-α）的能力[80,81]。对啮齿类动物的研究显示了对肿瘤坏死因子-α敏感性肿瘤的显著但有选择性的疗效，与其他抗肿瘤治疗相结合，这种疗效得到了改善[82]。不幸的是，在人类细胞中只检测到一种初始状态的TNF-α诱导[83]。第一阶段临床研究显示能力有限，抗肿瘤研究没有继续进行[79]，但

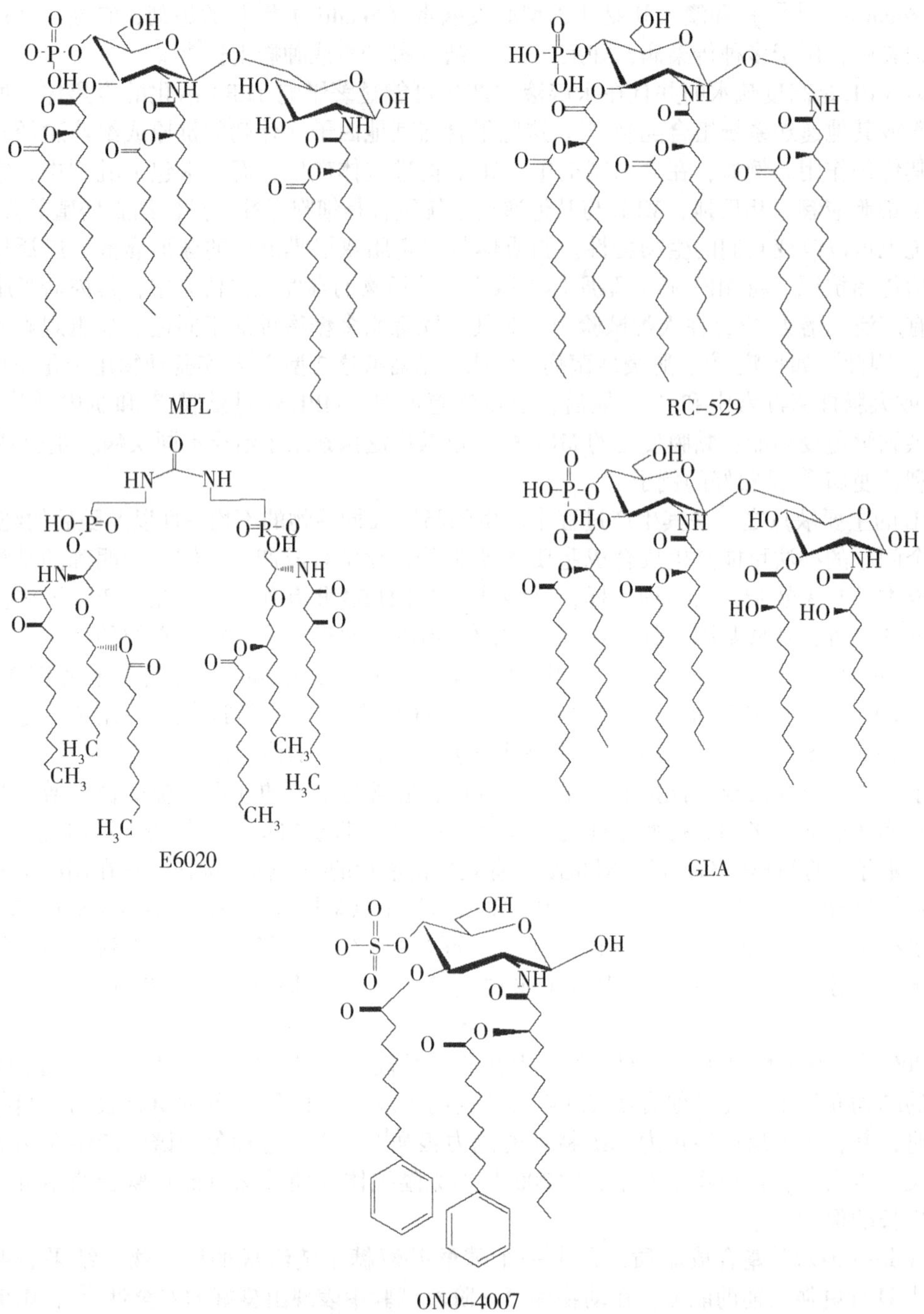

图 33.2 相关基于脂多糖的佐剂的化学结构

该分子表现出抗利什曼原虫[84]和抗过敏活性[85]。

结论

虽然脂多糖长期以来被认为是具有潜在辅助用途的免疫刺激物质，但是大量不可接受的毒性作用极大地限制了其临床应用。然而，发现 MPL 是安全的且保留了理想的佐剂性质的脂多糖开辟了治疗病原体疾病新的可能性。与以前的佐剂相比，基于脂多糖的佐剂具有增强 Th1 型反应和刺激细胞毒性 T 淋巴细胞的能力，从而提供了新的益处。这项活动对于保护人体免受许多病原体的侵害以及开发预防性治疗癌症或过敏症等其他疾病至关重要。事实上，MPL 在现有疫苗中的功效支持了这一点，而标准佐剂未能提供保护。此外，基于脂多糖的佐剂具有较高的佐剂能力，在黏膜保护、更快地激活保护、减少辅助剂剂量、老年人的功能性免疫或制备多价疫苗制剂方面具有显著和明显的优势。某些归因于 MPL 的缺点，例如提高生产成本和 TLR4 相关自身免疫疾病的可能激活/增强。合成脂 A 类似物具有类似的生物活性，如 MPL，已证明生产成本大大降低。关于激活 TLR4 自身免疫疾病，迄今为止从人免疫接种的累积数据提供了进一步的安全性证据。总之，基于脂多糖的佐剂改善了当前的疫苗接种疗法，并为解决现有的挑战提供了可能性。

参考文献

[1] Rietschel, E. T., et al.: Bacterial endotoxins: chemical structure, biological activity and role in septicaemia. Scand. J. Infect. Dis. Suppl. 31, 8-21 (1982).

[2] Guo, L., et al.: Lipid A acylation and bacterial resistance against vertebrate antimicrobial peptides. Cell 95, 189-198 (1998).

[3] Gibbons, H. S., Lin, S., Cotter, R. J., Raetz, C. R.: Oxygen requirement for the biosynthesis of the S-2-hydroxymyristate moiety in Salmonella typhimurium lipid A. Function of LpxO, A new Fe^{2+}/alphaketoglutarate-dependent dioxygenase homologue. J. Biol. Chem. 275, 32940-32949 (2000).

[4] Reynolds, C. M., et al.: An outer membrane enzyme encoded by Salmonella typhimurium lpxR that removes the 3' -acyloxyacyl moiety of lipid A. J. Biol. Chem. 281, 21974-21987 (2006).

[5] Trent, M. S., Pabich, W., Raetz, C. R., Miller, S. I.: A PhoP/PhoQ-induced Lipase (PagL) that catalyzes 3-O-deacylation of lipid A precursors in membranes of Salmonella typhmurium. J. Biol. Chem. 276, 9083-9092 (2001).

[6] Kawai, T., Akira, S.: Toll-like receptors and their crosstalk with other innate receptors in infection and immunity. Immunity 34, 637-650 (2011).

[7] Palsson-McDermott, E. M., O'Neill, L. A.: Signal transduction by the lipopolysaccharide receptor, Toll-like receptor-4. Immunology 113, 153-162 (2004).

[8] Trinchieri, G., Sher, A.: Cooperation of Toll-like receptor signals in innate immune defence. Nat. Rev. Immunol. 7, 179-190 (2007).

[9] Kawai, T., Akira, S.: TLR signaling. Cell Death Differ. 13, 816-825 (2006).

[10] Miggin, S. M., O'Neill, L. A.: New insights into the regulation of TLR signaling. J. Leukoc. Biol. 80, 220-226 (2006).

[11] Alexander, C., Rietschel, E. T.: Bacterial lipopolysac-charides and innate immunity. J. Endotoxin Res. 7, 167-202 (2001).

[12] Morrison, D. C., Kline, L. F.: Activation of the classi-cal and properdin pathways of complement by bacterial lipopolysaccharides (LPS). J. Immunol. 118, 362-368 (1977).

[13] Cooper, N. R., Morrison, D. C.: Binding and activation of the first component of human complement by the lipid A region of lipopolysaccharides. J. Immunol. 120, 1862-1868 (1978).

[14] Conti, P., et al.: Activation of human natural killer cells by lipopolysaccharide and generation of i nterleukin-1 alpha, beta, tumour necrosis factor and interleukin-6. Effect of IL-1 receptor antagonist. Immunology 73, 450-456 (1991).

[15] Kobayashi, M., et al.: Identifi cation and purifi cation of natural killer cell stimulatory factor (NKSF), a cytokine with multiple biologic effects on human lymphocytes. J. Exp. Med. 170, 827-845 (1989).

[16] Cinel, I., Dellinger, R. P.: Advances in pathogenesis and management of sepsis. Curr. Opin. Infect. Dis. 20, 345-352 (2007).

[17] Annane, D., Bellissant, E., Cavaillon, J. M.: Septic shock. Lancet 365 (63-78) (2005).

[18] Vincent, J. L., et al.: Sepsis in European intensive care units: results of the SOAP study. Crit. Care Med. 34, 344-353 (2006).

[19] Martin, G. S., Mannino, D. M., Eaton, S., Moss, M.: The epidemiology of sepsis in the United States from 1979 through 2000. N. Engl. J. Med. 348, 1546-1554 (2003).

[20] Kotani, S., Takada, H.: Structural requirements of lipid A for endotoxicity and other biological activities-an overview. Adv. Exp. Med. Biol. 256, 13-43 (1990).

[21] Nowotny, A.: Molecular aspects of endotoxic reac-tions. Bacteriol. Rev. 33, 72-98 (1969).

[22] Miller, S. I., Ernst, R. K., Bader, M. W.: LPS, TLR4 and infectious disease diversity. Nat. Rev. Microbiol. 3, 36-46 (2005).

[23] Golenbock, D. T., Hampton, R. Y., Qureshi, N., Takayama, K., Raetz, C. R.: Lipid A-like molecules that antagonize the effects of endotoxins on human monocytes. J. Biol. Chem. 266, 19490-19498 (1991).

[24] Kotani, S., et al.: Synthetic lipid A with endotoxic and related biological activities comparable to those of a natural lipid A from an *Escherichia coli* re-mutant. Infect. Immun. 49, 225-237 (1985).

[25] Saitoh, S., et al.: Lipid A antagonist, lipid IVa, is dis-tinct from lipid A in interaction with Toll-like receptor 4 (TLR4) -MD-2 and ligand-induced TLR4 oligomerization. Int. Immunol. 16, 961-969 (2004).

[26] David, S., Vermeer-de Bondt, P. E., van der Maas, N. A.: Reactogenicity of infant whole cell pertussis combination vaccine compared with acellular pertussis vaccines with or without simultaneous pneumococcal vaccine in the Netherlands. Vaccine 26, 5883-5887 (2008).

[27] Gustafsson, L., Hallander, H. O., Olin, P., Reizenstein, E., Storsaeter, J.: A controlled trial of a two-component acellular, afive-component acellular, and a whole-cell pertussis vaccine. N. Engl. J. Med. 334, 349-355 (1996).

[28] Brewer, J. M., et al.: Aluminium hydroxide adjuvant initiates strong antigen-specifi c Th2 responses in the absence of IL-4-or IL-13-mediated signaling. J. Immunol. 163, 6448-6454 (1999).

[29] Dubensky Jr., T. W., Reed, S. G.: Adjuvants for cancer vaccines. Semin. Immunol. 22, 155-161 (2010).

[30] Soghoian, D. Z., Streeck, H.: Cytolytic CD4 (+) T cells in viral immunity. Expert Rev. Vaccines 9, 1453-1463 (2010).

[31] Ziegler, A., et al.: EpCAM, a human tumor-associated antigen promotes Th2 development and tumor immune evasion. Blood 113, 3494-3502 (2009).

[32] Akkoc, T., Akdis, M., Akdis, C. A.: Update in the mechanisms of allergen-specifi c immunotheraphy. Allergy Asthma Immunol. Res. 3, 11-20 (2011).

[33] Chen, K., Cerutti, A.: Vaccination strategies to pro-mote mucosal antibody responses. Immunity 33, 479-491 (2010).

[34] Lawson, L. B., Norton, E. B., Clements, J. D.: Defending the mucosa: adjuvant and carrier formulations for mucosal immunity. Curr. Opin. Immunol. 23, 414-420 (2011).

[35] O'Hagan, D. T.: MF59 is a safe and potent vaccine adjuvant that enhances protection against infl uenza virus infection. Expert Rev. Vaccines 6, 699-710 (2007).

[36] Arenas, J.: The role of bacterial lipopolysaccharides as immune modulator in vaccine and drug development. Endocr. Metab. Immune Disord. Drug Targets 12 (3), 221-235 (2012).

[37] Baldridge, J. R., Crane, R. T.: Monophosphoryl lipid A (MPL) formulations for the next generation of vaccines. Methods 19, 103-107 (1999).

[38] Evans, J. T., et al.: Enhancement of antigen-specifi c immunity via the TLR4 ligands MPL adjuvant and Ribi. 529. Expert Rev. Vaccines 2, 219-229 (2003).

[39] Ulrich, J. T., Myers, K. R.: Monophosphoryl lipid A as an adjuvant. Past experiences and new directions. Pharm. Biotechnol. 6, 495-524 (1995).

[40] Vajdy, M., Singh, M.: The role of adjuvants in the development of mucosal vaccines. Expert Opin. Biol. Ther. 5, 953-965 (2005).

[41] Drachenberg, K. J., Wheeler, A. W., Stuebner, P., Horak, F.: A well-tolerated grass pollen-specifi c

allergy vaccine containing a novel adjuvant, monophosphoryl lipid A, reduces allergic symptoms after only four preseasonal injections. Allergy 56, 498-505 (2001).

[42] Drachenberg, K. J., Heinzkill, M., Urban, E., Woroniecki, S. R.: Effi cacy and tolerability of short-term specifi c immunotherapy with pollen allergoids adjuvanted by monophosphoryl lipid A (MPL) for children and adolescents. Allergol. Immunopathol. (Madr.) 31, 270-277 (2003).

[43] Kundi, M.: New hepatitis B vaccine formulated with an improved adjuvant system. Expert Rev. Vaccines 6, 133-140 (2007).

[44] Aide, P., et al.: Four year immunogenicity of the RTS, S/AS02 (A) malaria vaccine in Mozambican children during a phase IIb trial. Vaccine 29, 6059-6067 (2011).

[45] Von, E. K., et al.: The candidate tuberculosis vaccine Mtb72F/AS02A: Tolerability and immunogenicity in humans. Hum. Vaccin. 5, 475-482 (2009).

[46] Polhemus, M. E., et al.: Evaluation of RTS, S/AS02A and RTS, S/AS01B in adults in a high malaria transmission area. PLoS One 4, e6465 (2009).

[47] Brichard, V. G., Lejeune, D.: Cancer immunotherapy targeting tumour-specifi c antigens: towards a new therapy for minimal residual disease. Expert Opin. Biol. Ther. 8, 951-968 (2008).

[48] Garcon, N., Chomez, P., Van, M. M.: GlaxoSmithKline Adjuvant Systems in vaccines: concepts, achievements and perspectives. Expert Rev. Vaccines 6, 723-739 (2007).

[49] Fries, L. F., et al.: Liposomal malaria vaccine in humans: a safe and potent adjuvant strategy. Proc. Natl. Acad. Sci. U. S. A. 89, 358-362 (1992).

[50] Arenas, J., et al.: Coincorporation of LpxL1 and PagL mutant lipopolysaccharides into liposomes with Neisseria meningitidis opacity protein: infl uence on endotoxic and adjuvant activity. Clin. Vaccine Immunol. 17, 487-495 (2010).

[51] Vernacchio, L., et al.: Effect of monophosphoryl lipid A (MPL) on T-helper cells when administered as an adjuvant with pneumocococcal-CRM197 conjugate vaccine in healthy toddlers. Vaccine 20, 3658-3667 (2002).

[52] Olson, K., et al.: Liposomal gD ectodomain (gD1-306) vaccine protects against HSV2 genital or rectal infection of female and male mice. Vaccine 28, 548-560 (2009).

[53] Hall, M. A., et al.: Intranasal immunization with mul-tivalent group A streptococcal vaccines protects mice against intranasal challenge infections. Infect. Immun. 72, 2507-2512 (2004).

[54] Tana, W. S., Isogai, E., Oguma, K.: Induction of intestinal IgA and IgG antibodies preventing adhesion ofverotoxin-producing Escherichia coli to Caco-2 cells by oral immunization with liposomes. Lett. Appl. Microbiol. 36, 135-139 (2003).

[55] Baldridge, J. R., Yorgensen, Y., Ward, J. R., Ulrich, J. T.: Monophosphoryl lipid A enhances mucosal and systemic immunity to vaccine antigens following intranasal administration. Vaccine 18, 2416-2425 (2000).

[56] Kidon, M. I., Shechter, E., Toubi, E.: Vaccination against human papilloma virus and cervical cancer. Harefuah 150, 33-36 (2011). 68.

[57] Labadie, J.: Postlicensure safety evaluation of human papilloma virus vaccines. Int. J. Risk Saf. Med. 23, 103-112 (2011).

[58] Morello, C. S., Levinson, M. S., Kraynyak, K. A., Spector, D. H.: Immunization with herpes simplex virus 2 (HSV-2) genes plus inactivated HSV-2 is highly protective against acute and recurrent HSV-2 disease. J. Virol. 85, 3461-3472 (2011).

[59] Llanos-Cuentas, A., et al.: A clinical trial to evaluate the safety and immunogenicity of the LEISH-F1+MPL-SE vaccine when used in combination with sodium stibogluconate for the treatment of mucosal leishmaniasis. Vaccine 28, 7427-7435 (2010).

[60] Zhu, D., Barniak, V., Zhang, Y., Green, B., Zlotnick, G.: Intranasal immunization of mice with recombinant lipidated P2086 protein reduces nasal colonization of group B Neisseria meningitidis. Vaccine 24, 5420-5425 (2006).

[61] Green, B. A., et al.: PppA, a surface-exposed protein of Streptococcus pneumoniae, elicits cross-reactive

antibodies that reduce colonization in a murine intranasal immunization and challenge model. Infect. Immun. 73, 981-989 (2005).

[62] Egan, M. A., et al.: A comparative evaluation of nasal and parenteral vaccine adjuvants to elicit systemic and mucosal HIV-1 peptide-specifi c humoral immune responses in cynomolgus macaques. Vaccine 22, 3774-3788 (2004).

[63] Mason, K. W., et al.: Reduction of nasal colonization of nontypeable Haemophilus infl uenzae following intranasal immunization with rLP4/rLP6/UspA2 proteins combined with aqueous formulation of RC529. Vaccine 22, 3449-3456 (2004).

[64] Baldridge, J. R., et al.: Immunostimulatory activity of aminoalkyl glucosaminide 4-phosphates (AGPs): induction of protective innate immune responses by RC-524 and RC-529. J. Endotoxin Res. 8, 453-458 (2002).

[65] Ishizaka, S. T., Hawkins, L. D.: E6020: a synthetic Toll-like receptor 4 agonist as a vaccine adjuvant. Expert Rev. Vaccines 6, 773-784 (2007).

[66] Morefi eld, G. L., Hawkins, L. D., Ishizaka, S. T., Kissner, T. L., Ulrich, R. G.: Synthetic Toll-like receptor 4 agonist enhances vaccine effi cacy in an experimental model of toxic shock syndrome. Clin. Vaccine Immunol. 14, 1499-1504 (2007).

[67] Przetak, M., et al.: Novel synthetic LPS receptor ago-nists boost systemic and mucosal antibody responses in mice. Vaccine 21, 961-970 (2003).

[68] Baudner, B. C., et al.: MF59 emulsion is an effective delivery system for a synthetic TLR4 agonist (E6020). Pharm. Res. 26, 1477-1485 (2009).

[69] Wang, S., et al.: Effective antibody therapy induces host-protective antitumor immunity that is augmented by TLR4 agonist treatment. Cancer Immunol. Immunother. 61, 49-61 (2012).

[70] Baldwin, S. L., et al.: Intradermal immunization improves protective effi cacy of a novel TB vaccine candidate. Vaccine 27, 3063-3071 (2009).

[71] Bertholet, S., et al.: Optimized subunit vaccine pro-tects against experimental leishmaniasis. Vaccine 27, 7036-7045 (2009).

[72] Coler, R. N., et al.: A synthetic adjuvant to enhance and expand immune responses to infl uenza vaccines. PLoS One 5, e13677 (2010).

[73] Xiao, L., et al.: A TLR4 agonist synergizes with den-dritic cell-directed lentiviral vectors for inducing antigen-specifi c immune responses. Vaccine 30, 2570-2581 (2012).

[74] Pantel, A., et al.: A new synthetic TLR4 agonist, GLA, allows dendritic cells targeted with antigen to elicit Th1 T-cell immunity in vivo. Eur. J. Immunol. 42, 101-109 (2012).

[75] Cong, H., et al.: Toxoplasma gondii HLA-B * 0702-restricted GRA7 (20-28) peptidewith adjuvants and a universal helper T cell epitope elicits CD8 (+) T cells producing interferon-gamma and reduces parasite burden in HLA-B * 0702 mice. Hum. Immunol. 73, 1-10 (2012).

[76] Baldwin, S. L., et al.: The importance of adjuvant for-mulation in the development of a tuberculosis vaccine. J. Immunol. 188, 2189-2197 (2012).

[77] Windish, H. P., et al.: Protection of mice from Mycobacterium tuberculosis by ID87/GLA-SE, a novel tuberculosis subunit vaccine candidate. Vaccine 29, 7842-7848 (2011).

[78] Ueda, H., Yamazaki, M.: Induction of tumor necrosis factor in a murine tumor by systemic administration of a novel synthetic lipid A analogue, ONO-4007. J. Immunother. 20, 65-69 (1997).

[79] Matsumoto, N., Aze, Y., Akimoto, A., Fujita, T.: Restoration of immune responses in tumor-bearing mice by ONO-4007, an antitumor lipid A derivative. Immunopharmacology 36, 69-78 (1997).

[80] Kuramitsu, Y., et al.: A new synthetic lipid A analog, ONO-4007, stimulates the production of tumor necrosis factor-alpha in tumor tissues, resulting in the rejection of transplanted rat hepatoma cells. Anticancer Drugs 8, 500-508 (1997).

[81] Matsumoto, N., Oida, H., Aze, Y., Akimoto, A., Fujita, T.: Intratumoral tumor necrosis factor induction and tumor growth suppression by ONO-4007, a low-toxicity lipid A analog. Anticancer Res. 18, 4283-4289 (1998).

[82] Inagawa, H., et al.: Mechanisms by which chemother-apeutic agents augment the antitumor effects of tumor necrosis factor: involvement of the pattern shift of cytokines from Th2 to Th1 in tumor lesions. Anticancer Res. 18, 3957-3964 (1998).

[83] Matsumoto, N., Aze, Y., Akimoto, A., Fujita, T.: ONO-4007, an antitumor lipid A analog, induces tumor necrosis factor-alpha production by human monocytes only under primed state: different effects of ONO-4007 and lipopolysaccharide on cytokine production. J. Pharmacol. Exp. Ther. 284, 189-195 (1998).

[84] Khan, M. A., et al.: Inhibition of intracellular prolif-eration of Leishmania parasites in vitro and suppression of skin lesion development in BALB/c mice by a novel lipid A analog (ONO-4007). Am. J. Trop. Med. Hyg. 67, 184-190 (2002).

[85] Iio, J., et al.: Lipid A analogue, ONO-4007, inhibits IgE response and antigen-induced eosinophilic recruitment into airways in BALB/c mice. Int. Arch. Allergy Immunol. 127, 217-225 (2002).

(吴国华译)

第 34 章　细菌毒素是成功的免疫治疗佐剂和免疫毒素

Irena Adkins①

摘要

病原菌的细菌蛋白毒素在感染性疾病中起着重要作用，可以劫持哺乳动物细胞，操纵宿主免疫反应。在过去的几十年中，人们对几种细菌毒素及其无毒突变体进行了深入研究，以利用它们进入宿主细胞的能力来刺激适应性 T 细胞反应、直接清除癌细胞或作为佐剂提高免疫力。此外，另一组称为效应器的细菌毒素被专门的分泌系统转运到宿主细胞中，被用来促进异源抗原向抗原呈递细胞的传递。一些最强大的细菌神经毒素已应用于治疗各种神经疾病以及化妆品行业。虽然目前很少有细菌毒素应用于临床癌症试验的各个阶段，但还需要进一步确定其他细菌毒素的免疫治疗潜力。本章将介绍用于免疫治疗的细菌毒素，并介绍这一领域的主要成就。同样地，在人类免疫治疗中使用细菌毒素的障碍和局限性将与其他诊断应用一起讨论。

34.1　引言

通过进化完善细菌毒素，增强细菌的竞争潜力和防御能力。它们是致病菌劫持哺乳动物细胞的主要毒力因子。许多细菌毒素是由染色体、质粒或噬菌体编码的，可以大致分为内毒素和外毒素两类。内毒素是细菌细胞包膜的细胞相关结构成分，由位于革兰氏阴性细菌外膜的肽聚糖组分胞壁二肽（MDP）和胞壁肽或脂多糖（脂多糖）或脂寡糖（LOS）表示。内毒素通常作用于细菌细胞附近。外毒素主要是蛋白质毒素，属于已知最强大的人体毒素。它们通过宿主防御机制或抗生素的作用从细菌中主动运输或通过细胞裂解释放。有些细菌产生许多毒素，如葡萄球菌或链球菌，而其他细菌（如白喉棒状杆菌）只产生一种毒素。一些毒素，如造孔溶血素或磷脂酶，对各种细胞类型表现出广泛的细胞毒性活性，而其他毒素，如梭状芽孢杆菌神经毒素，只对神经元起作用。有些细菌毒素可作为入侵蛋白；另一些细菌毒素通过酶活性或孔形成破坏细胞完整性，导致细胞死亡。

在过去的几十年中，细菌毒素在真核细胞信号通路的阐明中发挥了重要作用。毒素的细胞毒性活性已被用于治疗各种神经系统疾病以及产生免疫毒素以杀死恶性或病毒感染的细胞。此外，通过单点突变、插入特定序列或删除部分毒素分子以改变其结合能力和/或去除其用于疫苗接种、抗原/药物输送或最重要的佐剂的酶活性，各种细菌毒素已被不同地修饰。根据表 34.1 中的作用模式列出了用于免疫疗法的研究最多的细菌毒素。细菌毒素的特殊免疫治疗用途列见表 34.2。

① I. Adkins，PhD（ASCR 微生物研究所，捷克布拉格，E-mail：irena. adkins@ seznam. cz）

表34.1　在免疫疗法中常用细菌毒素的特点

毒素/来源/英文缩写	疾病	作用方式	目标
激活免疫反应 胞壁二肽（MDP）/细菌细胞壁肽聚糖成分	—	TLR配体	NOD受体
肠毒素A、B、C/*S.*金黄色葡萄球菌（SEA，SEB，SEC）	食物中毒[b]	超抗原	TCR和MHC Ⅱ
抑制蛋白质合成志贺毒素/志贺氏痢疾杆菌（Stx）	HC	N-糖苷酶	28S rRNA
志贺样毒素/大肠杆菌（Stx1）	HUS	N-糖苷酶	28S rRNA
外毒素A/铜绿假单胞菌（PE）	肺炎[b]	ADP核糖转移酶	延伸因子2
白喉毒素/白喉棒状杆菌（DT）	白喉	ADP核糖转移酶	延伸因子2
蛋白酶致死因子/炭疽芽孢杆菌	炭疽病	锌金属蛋白酶	MAPKKs
破伤风毒素/破伤风梭菌（TeNT）	破伤风	锌金属蛋白酶	SNARE蛋白
神经毒素A-G/肉毒梭菌（BoNTs）	肉毒杆菌中毒	锌金属蛋白酶	SNARE蛋白
激活第二信使通路 腺苷酸环化酶毒素/百日咳博德特氏菌（CyaA）	百日咳腺苷酸环化酶	三磷酸腺苷	
百日咳毒素/百日咳博德特氏菌（PT）	百日咳形成孔道	G蛋白	
皮肤坏死毒素/百日咳博德特氏菌（DNT）	百日咳，ADP-核糖基转移酶鼻炎		Rho G蛋白
水肿毒素/炭疽杆菌	炭疽病	脱氨酶	三磷酸腺苷
霍乱毒素/霍乱弧菌（CT）	霍乱	腺苷酸环化酶	G蛋白
封闭带毒素/霍乱弧菌（ZOT）	霍乱	ADP核糖转移酶	肌动蛋白 通过蛋白激酶C信号传导聚合
C2毒素/肉毒杆菌	肉毒毒素[C]	G-肌动蛋白细胞间紧密连接的可逆性破坏	
C3毒素/肉毒杆菌	肉毒毒素[C]	模拟连蛋白（zonulin）	Rho G蛋白
毒素A/艰难梭菌	腹泻/PC	ADP核糖转移酶	Rho G蛋白
细胞毒性坏死因子1/大肠杆菌（CNF1）	尿路感染[c]	ADP核糖转移酶	胆固醇
热不稳定毒素/大肠杆菌（LT）	腹泻	UDP葡萄糖基转移酶	Rho G蛋白
		形成孔道 脱氨酶 ADP核糖转移酶	G蛋白
损坏膜 溶血素[a]/大肠杆菌（HlyA）	尿路感染	形成孔道	质膜
李斯特菌溶血素O/单核细胞增生李斯特氏菌（LLO）	食源性系统性疾病，脑膜炎	形成孔道	胆固醇

（续表）

毒素/来源/英文缩写	疾病	作用方式	目标
产气荚膜梭菌溶素 O/ C 产气荚膜炎	气性坏疽[b,c]	形成孔道	胆固醇
气单胞菌溶素 /嗜水气单胞菌	腹泻	形成孔道	GPI-AP
肺炎球菌溶血素/肺炎链球菌	肺炎[b]	形成孔道	胆固醇
中间型链球菌溶素/中间葡萄球菌	脓肿	形成孔道	胆固醇
晶体蛋白/苏云金芽孢杆菌（Cry）	—	形成孔道	质膜

摘自 Schmitt 等[1]。

缩写：*TLR* Toll 样受体；*NOD* 核苷酸寡聚化结构域；TCR T 细胞受体；*MHC* Ⅱ主要组织相容性复合体 *II* 类分子；*HC* 出血性结肠炎；*HUS* 溶血性尿毒症综合征；*MAPKKs* 丝裂原活化蛋白激酶激酶；*PC* 抗生素相关伪膜性结肠炎；*UTIs* 尿路感染；*GPI-AP* 糖基磷脂酰肌醇锚定蛋白。

[a]由其他细菌属产生的毒素。

[b]其他疾病也与生物体有关。

[c]毒素与疾病之间无明确因果关系[1]。

34.2 细菌毒素进入宿主细胞的结构和方式

表 34.3 中描述了细菌毒素的结构及其对免疫疗法的特异性修饰。大多数毒素被称为 AB 毒素，其中 A 部分通常具有酶活性，单体或多聚体 B 部分将毒素与细胞表面受体结合。B 部分可以在 A 部分易位到胞质溶胶中起作用[65]。受体通过 B 部分结合后，大多数蛋白毒素进入宿主细胞主要通过受体介导的内吞作用。

毒素通过酸性内体或以逆行方式通过高尔基体和内质网转运到宿主细胞胞质溶胶中。在此，易位的酶 A 部分干扰蛋白质合成机制的组分、各种细胞信号传导途径、肌动蛋白聚合和囊泡的细胞内运输，在大多数情况下引起中毒细胞的死亡（表 34.3，第 1 和第 2 部分）。相反地，百日咳博德特氏菌腺苷酸环化酶毒素（CyaA）与其受体 CD11b 结合后，将其酶促腺苷酸环化酶结构域直接转移到细胞膜上，进入宿主细胞胞质，产生非生理水平的 cAMP，从而麻痹骨髓吞噬细胞的先天免疫功能，并诱导细胞死亡。受体结合后，成孔毒素直接作用于质膜中形成孔，而无须内吞作用，导致靶细胞溶解[5]。

表 34.2 细菌毒素的实验性免疫治疗应用

毒素	佐剂	实验性免疫治疗应用						
		抗原递送[a]	药物/DNA递送	IT	抗病毒制剂	诊断/成像	癌症治疗	神经系统疾病或损伤的治疗
胞壁二肽/黏肽	x		x		x		x	
肠毒素 A、B、C	x						x	
志贺毒素	x	x	x	x	x	x	x	
志贺样毒素	x	x		x	x	x	x	
外毒素 A	x	x		x	x		x	
白喉毒素	x	x		x	x		x	
致死因子		x		x		x	x	

（续表）

毒素	佐剂	实验性免疫治疗应用						
		抗原递送[a]	药物/DNA递送	IT	抗病毒制剂	诊断/成像	癌症治疗	神经系统疾病或损伤的治疗
破伤风毒素		x	x				x	x
神经毒素			x				x	x
腺苷酸环化酶毒素	x	x					x	
百日咳毒素	x	x					x	
皮肤坏死毒素	x				x			
水肿毒素	x	x						
霍乱毒素	x						x	
封闭带毒素	x	x	x				x	
C2 毒素			x				x	
C3 毒素								x
毒素 A							x	
细胞毒性坏死因子 1	x							x
热不稳定毒素	x						x	
溶血素		x						
李斯特菌溶血素 O	x	x	x				x	
产气荚膜梭菌溶素 O	x	x	x					
气单胞菌溶素				x		x	x	
肺炎球菌溶血素	x	x						
中间型链球菌溶素			x		x			
晶体蛋白	x	x					x	

缩写：IT 免疫毒素。

[a]与毒素分子基因融合或化学偶联的抗原。

细菌毒素结合其特定细胞受体的能力已被用作体内人类肿瘤概况分析的诊断工具。与许多可检测部分（包括荧光色素、放射性核素、荧光蛋白，甚至磁共振图像对比剂）的毒素结合可提供特定细胞或细胞相关酶活性的实时、无创成像[28-30]。

表 34.3　细菌毒素的结构及其在免疫疗法中的修饰

毒素	结构/受体/宿主细胞进入	免疫疗法的修改
皮肤坏死毒素	160.6 kDa，单链多肽，由氨基末端结合域（B）和羧基末端催化结构域（A）/未知受体/发动蛋白依赖性内吞作用，由蛋白酶弗林蛋白酶切割，从酸性内体转运到宿主细胞胞质中	未经修改[14]

（续表）

毒素	结构/受体/宿主细胞进入	免疫疗法的修改
致命毒素 水肿毒素	蛋白酶弗林蛋白酶对保护性抗原（PA，83 kDa）（B 部分）的切割暴露了竞争性结合的致死因子（LF，90 kDa）或水肿因子（EF，89 kDa）（A 部分）的结合位点。具有 LF 或 EF 的 PA 七聚体结合 TEM8 和 CMG2 受体/ RME，LF 和 EF 随后从酸性内体转运至宿主细胞胞质溶胶中	LFn（1～255aa），EFn（1～260aa），带有插入或偶联的 Ag；具有改变的切割位点的 PA 未经修饰的综述[5]
霍乱毒素 热不稳定的毒素	～85 kDa，两种毒素形成由单个 A 亚基（27 kDa，CTA 或 LTA）组成的寡聚体（AB5），其含有通过二硫桥连接在一起的两个结构域（A1 和 A2）和 5 个相同的 B 亚基（每个 11.5 kDa，CTB 或 LTB）/ GM1 神经节苷脂受体/ RME，通过 GA 和 ER 逆行转运，其中酶 A 部分易位至宿主细胞胞质溶胶中	各种单点突变，例如 LTR72、R192G 综述[15] CTA1-DD[16] CTB-CpG[17]
Zonula 封闭毒素	单个 45 kDa 的多肽，其经历切割。12 kDa（aa 288～293）的羧基末端片段从细菌中排出并通过可逆地分解模拟连蛋白 / 连蛋白受体的细胞间紧密连接发挥其生物学效应	未修饰［18，19］ AT1001[20] AT1002[21]
C2 毒素	二元 AB 型毒素，由结合成分 C2II（80.5～100 kDa，B）和酶成分 C2I（49.4 kDa，A）/天冬酰胺连接组成 碳水化合物结构/ RME，A 部分从酸性内体转移到宿主细胞胞质溶胶中	C2IN-链霉抗生物素蛋白融合用于药物输送[22]
C3 毒素	25 kDa，没有鉴定易位结构域的单一多肽。毒素到达宿主细胞胞质溶胶；然而，具体的细胞进入机制尚不清楚	未修饰或 $C_{154-182}$ 片段[23,24]
毒素 A	308 kDa，大型梭菌毒素家族，由氨基末端催化结构域（A）组成的单链多肽，其后是疏水性易位结构域和羧基末端结合结构域（B）/各种类型的碳水化合物结构/ RME，A 部分从酸性内体转移到宿主细胞胞质溶胶中；胆固醇依赖性孔隙形成	福尔马林治疗的毒素[25]
细胞毒性坏死因子 1	113 kDa，由氨基末端结合结构域（B）和羧基末端催化结构域（A）/层粘连蛋白/ RME 组成的单链多肽，A 部分从酸性内体转运到宿主细胞胞质溶胶中	未经修改[14]
溶血素	107 kDa，属于 RTX 毒素家族的单一多肽。HlyA 被合成为一种无毒的原溶血素（proHlyA），通过共同合成的脂肪酸酰基转移酶 HlyC/HlyA 进一步激活为一种成熟的毒素，靶向内皮细胞和免疫系统的细胞/HlyA 通过在质膜中产生阳离子选择性孔引起细胞溶解	与 HlyAs（羧基末端 50～60 aa）连接的 Ag 作为分泌信号进行了综述[5]
李斯特菌溶血素 O	58 kDa，胆固醇依赖性溶细胞素家族成员。它在具有吞噬的单核细胞增生李斯特菌的细胞的酸性吞噬体内被选择性激活。它溶解吞噬体和艾滋病细菌逃逸到细胞质中	LLO 与肿瘤 Ag 融合或未经修改参考文献[5]

（续表）

毒素	结构/受体/宿主细胞进入	免疫疗法的修改
全环水解酶 间质溶解素 溶血素	53~54 kDa 单链多肽，硫醇激活的胆固醇依赖性溶细胞素，在细胞质膜中形成大的同源异构孔（~50个单体亚基）	未修改，域4 中间溶血素，解毒 用 Ag（C428G、W433F、ΔA146） 评价肺炎球菌溶血素[5]
气溶素	在通过弗林蛋白酶和其他蛋白酶蛋白水解去除羧基末端片段之前或之后，非活性的溶血素（52 kDa）结合细胞表面上的 GPI-AP；它被转化为气溶素（50 kDa）。气溶素寡聚形成七聚体，插入质膜形成导致靶细胞死亡的通道	PRX302 含有 PSA 可切割序列[26] R336A 不结合 GPI-AC[27]
晶体蛋白	~130 kDa，球状分子含有3个结构 Cry1Aa8 或未修饰的[5]结构域通过单个连接子连接。氨基末端结构域Ⅰ（膜插入和孔形成），结构域Ⅱ（受体结合结构域）和羧基末端结构域Ⅲ（毒性片段的稳定性和成孔活性的调节）。由肠道蛋白酶切割产生 60~70 kDa 蛋白质/钙黏蛋白样蛋白质、GPI 锚定氨肽酶 N、GPI 锚定碱性磷酸酶和 270 kDa 糖缀合物受体/中肠上皮细胞孔形成	Cry1Aa8 或未经修改

缩写：*aa* 氨基酸；*Ag* 抗原；*CMG*2 毛细管形态发生蛋白 2；*DAP* 二氨基庚二酸；*ER* 内质网；*GA* 高尔基体；*GPI-AP* 糖基磷脂酰肌醇锚定蛋白；*RTX* 重复毒素；*TEM*8 肿瘤内皮标记物 8；*RME* 受体介导内吞作用。

34.3 用作佐剂的细菌毒素

佐剂通常增强对疫苗或实验抗原的免疫力。它们作用于先天免疫系统的细胞，而不是直接作用于淋巴细胞，但有佐剂活性的超抗原除外[31]。主要在抗原呈递细胞（如巨噬细胞和树突细胞）中，佐剂诱导共刺激和细胞黏附分子以及抗原加工机制的蛋白质的表达，从而增加它们对T细胞的抗原递呈能力。该过程称为树突细胞的成熟。此外，佐剂诱导细胞因子和趋化因子的产生，刺激T细胞的迁移、增殖和分化。它们进一步诱导吞噬和抗微生物活性，并促进抗体介导的细胞毒性。大多数细菌毒素对宿主免疫细胞具有佐剂活性；然而，它们同时作为免疫原起作用，引发对自身的免疫反应。内毒素以及外毒素已被用作佐剂，主要是具有消除或降低的酶活性或改变的化学结构的突变毒素以降低其细胞毒性作用（表34.2和表34.3）。

研究最多的细菌内毒素佐剂是脂多糖（第33章）、胞壁酰二肽（MDP）和胞壁肽。发现MDP是弗氏完全佐剂（FCA）中佐剂活性的最小结构（表34.3），然而，缺乏FCA中存在的抗原，MDP的效率不高[32]。胞壁肽是革兰氏阴性菌和革兰氏阳性菌肽聚糖的其他分解产物，与其他Toll样受体（TLR）配体如脂多糖具有很强的协同作用。大量的胞壁肽和衍生物已被化学合成，以探索其辅助活性，并发现各种临床用途。莫拉布胺（Murabutide）是一种来自MDP的合成免疫调节剂，已被证明可增强对细菌和病毒感染的抵抗力，并在小鼠中显示出抗肿瘤作用[2]。其他MDP衍生药物，如与MDP结合的polyG（10-聚鸟苷酸）或紫杉醇（Taxol®），已显示出抗肿瘤活性和免疫增强作用[33,34]。最重要的是，米伐木肽（mifamurtide），一种脂质体胞壁酰三肽磷脂酰乙醇胺（L-MTP-PE）激活巨噬细胞和单核细胞，最近在欧洲被批准用于化疗治疗非转移性骨肉瘤[35]。有趣的是，最近制备了一种新的抗炎MDP衍生物[2]。

在细菌外毒素中，霍乱弧菌霍乱毒素（CT）和大肠杆菌热不稳定毒素（LT）代表了最有效的口腔黏膜免疫原和多种共同施用抗原的佐剂。它们固有的免疫调节性质取决于ADP-核糖基化活性和A亚基的结构特性以及五聚体B亚基与广泛表达的GM1神经节苷脂结合的活性。由于其毒性，主要是无毒的 B 亚单位已广泛用作人类黏膜免疫原，例如霍乱疫苗。由于单独的 B 亚单位是较差的黏膜佐剂，因此引入 CT 和 LT 突变体，其中酶促 A 亚单位发生突变。当口服、鼻内或阴道给予所有抗原时，这些突变体表现为强黏膜佐剂，诱导强烈的全身和黏膜抗原特异性抗体应答[15,36]。其中一些突变体（如 LTK63 或 LTR192G）在 I 期临床试验中作为佐剂进行了测试[37]。一些基于 CT 的其他修饰佐剂已有介绍，如 CTA1-DD，其中完全活性的 CTA1 和名为 DD 的金黄色葡萄球菌蛋白 A 衍生物是基因工程融合的[16]或 CTB 与 CpG 寡核苷酸偶联[17]。

另一方面，含有结合抗原的 CTB 和 LTB 能有效地诱导口腔耐受[38]。在实验模型中，CTB-抗原结合物显示出抑制几种自身免疫疾病、I 型过敏和同种异体移植物排斥的发展[5,36]。在概念验证临床试验中，CTB 的治疗耐受性已扩展到白塞病患者[39]。CT 和 LT 的一个有趣的替代品可能是昆虫病原性苏云金芽孢杆菌的 Cry 蛋白，其表现出与共同施用的抗原相似的有效黏膜免疫和全身免疫原性和佐剂活性[5,40]。尽管 Cry 蛋白似乎对哺乳动物细胞无害，表现出稳定的特性，并且可以大规模廉价生产，但缺乏它们对哺乳动物细胞影响的详细评估。

另一组被用作佐剂的毒素以热原毒素超级抗原，如金黄色葡萄球菌的肠毒素 A、B 和 C（SEA、SEB 和 SEC）为代表。这些毒素结合抗原呈递细胞上表达的主要组织相容性 II 类（主要组织相容性复合体 II）分子的肽结合裂缝外的不同区域和 T 细胞受体上的特异性 Vβ 元件，诱导大量 T 细胞增殖和细胞因子释放。修饰的肠毒素已被用于癌症免疫疗法。对肠毒素进行基因工程改造以降低 MHC Ⅱ结合，并将其与肿瘤定向单克隆抗体的 Fab 部分融合。这些肿瘤靶向的超级抗原能够在体外和动物模型中介导各种肿瘤细胞的裂解[4,31]。此外，C215Fab-SEA 与多西紫杉醇联合免疫治疗可产生协同抗肿瘤作用[3]。

许多其他细菌毒素已被证明可增强对融合或共同施用抗原的抗体反应和/或对增强 T 细胞适应性反应的免疫细胞发挥佐剂活性[5]（表 34.2）。然而，必须仔细评估它们作为佐剂的治疗潜力，例如百日咳博德特氏菌百日咳毒素加剧了多发性硬化小鼠模型中的实验性自身免疫性脑脊髓炎[41]。

34.4 免疫毒素类

除蛋白毒素作为佐剂外，免疫毒素在肿瘤免疫治疗中也发挥着重要作用[6]。免疫毒素是由蛋白质毒素和配体（通常是抗体或其衍生物、生长因子或细胞因子）组成的分子。在靶细胞表面，配体结合肿瘤相关抗原，然后将毒素传递到细胞胞质溶胶中，从而导致细胞死亡。目前研究最多的肿瘤免疫毒素是以铜绿假单胞菌外毒素 A（PE）和白喉杆菌白喉毒素（DT）为基础的。有趣的是，最近已经报道了基于嗜水气单胞菌改良的气单胞菌溶素的新型免疫毒素[26,27]。除细菌来源的免疫毒素外（表 34.2），植物来源的毒素，如蓖麻毒素、皂草素和商陆抗病毒蛋白也用于制备免疫毒素[6]。

铜绿假单胞菌外毒素 A（PE）和白喉杆菌白喉毒素（DT）为基础的免疫毒素[42]，主要在血液恶性肿瘤中比实体肿瘤有效，对标准化疗失败后的免疫反应有显著影响，提示它们可以成为治疗微小残留疾病的有用工具。构建了大量针对各种肿瘤相关抗原的 PE 和 DT 免疫毒素，并在临床前和/或 I/II 期临床试验中进行测试[6-8]。美国食品和药物管理局批准用于皮肤 T 细胞淋巴瘤的基于重组 DT 免疫毒素 DAB（389）IL-2（地尼白介素；ONTAK）是免疫毒素的成功的标志。然而，需要进一步改善免疫毒素，以限制其副作用，如血管渗漏综合征和肝毒性，增加其分子特异性和通过生理屏障的转运，并增强其抵抗免疫系统失活的能力[7,8]。

蛋白酶激活毒素的产生和毒素自杀基因疗法是毒素癌症免疫疗法的新途径[6]。蛋白酶激活毒素被设计为在输送到靶细胞后通过分解疾病相关蛋白酶。临床前研究表明，炭疽毒素的主要修饰保护性抗原与致死因子 PE 或 DT 结合，可靶向基质金属蛋白酶或尿激酶纤溶酶原激活系统过度表达的肿瘤细胞[6,43,44]。同样地，将前列腺特异性抗原可诱导序列引入到原溶血素中，以调节其对前列腺癌细胞的毒性[26]。基于毒素的自杀基因治疗涉及一个 DNA 结构，编码毒性部分，其表达受癌症特异性启动子的控制。携带 DT 亚基 A 的许多治疗载体在小鼠中成功地用于治疗各种癌症类型；但是，只有 DTA-H19[45] 目前在 I/II 期临床试验[6] 中进行测试。

34.5　用作抗病毒制剂的细菌毒素

很少有细菌毒素显示有潜力用作主要对抗艾滋病毒感染的抗病毒药物[42]（表 34.2）。以痢疾杆菌志贺毒素、铜绿假单胞菌 PE 和白喉杆菌 DT 为基础的免疫毒素[42,46] 通过杀死受感染的细胞而直接起作用。特别是，基于 PE 的 Env 靶向免疫毒素作为 HAART 的互补可能会被重新考虑用于临床研究，以消耗持久存在的 HIV 感染的细胞[46]。百日咳毒素及其受体结合五聚体 B 亚单位单独在体外抑制了艾滋病毒感染，并在实验动物中对其他病毒介导了部分抗病毒作用[5,42]。同样地，大肠杆菌产生的志贺样毒素被证明可以减轻实验感染绵羊的牛白血病病毒感染[47]。中间体中间溶素增加宿主细胞对补体介导的裂解敏感性的固有能力已被用于增强 HIV 调理作用[48]。

34.6　细菌毒素用作抗原和药物传递制剂

一些毒素及其衍生物被证明是治疗病毒、细菌和恶性疾病的抗原传递载体。抗原表位或多表位可以通过基因工程或化学方式与毒素分子结合。同样地，蛋白质毒素也可以作为药物或 DNA 传递剂（表 34.2）。研究最多的毒素是百日咳博德特氏菌 CyaA，特别是其遗传解毒变体 CyaA-AC⁻其中催化细胞溶质 ATP 转化为 cAMP 的能力被破坏[49,50]。与修饰的炭疽杆菌炭疽毒素类似，CyaA-AC⁻被证明能够在体内以非常低的浓度容纳大的多表位，并刺激抗原特异性 T 细胞反应，而不需要佐剂[12,51-53]。最重要的是，已经证明 CyaA-AC⁻类毒素可以在小鼠中引发针对 HPV-16 诱导的肿瘤和黑素瘤的保护性和治疗性免疫[54,55]。用于宫颈肿瘤和转移性黑色素瘤免疫治疗的 CyaA-AC⁻疫苗已进入 I / II 期临床试验[5]。携带结核分枝杆菌抗原的 CyaA-AC⁻进一步被证明是诊断潜伏性结核的有用诊断工具[56]。此外，大肠杆菌 α-溶血素、单核细胞增生李斯特菌蛋白酶 O（LLO）或由 III 型分泌系统分泌的耶尔森氏菌和沙门氏菌的效应蛋白被用来促进融合异源蛋白或肽的传递以供抗原呈现[5,57-60]。I 期研究表明，使用基于 LLO 的抗原递送系统的致弱单核细胞增生李斯特菌疫苗是安全的，并且具有良好的耐受性，在终末期癌症患者中观察到疗效信号[61]。

34.7　细菌毒素用于治疗神经系统疾病紊乱和脊髓损伤

肉毒杆菌神经毒素代表了一种治疗范例，因为 BoNT / A 和 BoNT / B 在商业上用于治疗各种神经和肌张力障碍，或用于化妆品行业[10]。然而，使用肉毒杆菌神经毒素有一些限制，如抗体的形成和对 A 型毒素反应的消除，毒素向邻近肌肉的扩散，或者需要在慢性疾病中反复注射毒素[10,11]。对破伤风杆菌破伤风毒素（TeNT）及其衍生物无毒 C-片段作为外周神经的载体分子也提出了治疗用途[9]。肉毒杆菌 C3 毒素，称为 BA-210（Centhrin®），达到治疗脊髓损伤的临床研究 IIb 期[62]。有趣的是，最近已显示酶不活跃

C3154-182 肽在体内创伤后神经再生中是有效的[23,24]。此外，据报道，大肠杆菌的细胞毒性坏死因子 1（CNF1）可改善学习和记忆[63]。

34.8　细菌毒素在商业上用于疫苗接种

分子工程的科学进展使基于突变细菌毒素的疫苗得以开发；然而，目前只有很少的疫苗用

于治疗和/或预防人类疾病。解毒的白喉杆菌白喉类毒素或破伤风破伤风类毒素分别用于针对白喉或破伤风的通用疫苗接种。霍乱弧菌霍乱毒素（CTB）的无毒B亚基在广泛许可的口服霍乱疫苗中与灭活的弧菌一起用作保护性抗原。同样地，百日咳博德特氏菌百日咳（PT9K/129G）[13]已被批准用于人类，用作针对百日咳博德特氏菌感染的疫苗成分[64]。虽然白喉类毒素成功地作为糖结合疫苗的载体，对抗多种细菌病原体，但4项临床试验未能证明铜绿假单胞菌外毒素A在这类疫苗中的益处。

结论

肉毒杆菌神经毒素、米伐木肽为代表的MDP衍生物用作佐剂，地尼白介素为代表的免疫毒素用于治疗皮肤T细胞淋巴瘤，是细菌毒素免疫治疗领域最成功的成果。在临床前和/或临床研究的早期阶段正在评估的许多其他细菌毒素显示出有希望的结果。然而，我们需要进一步的研究来提高它们的效率和安全性，同时也要加深对细菌毒素作用方式的认识，以利用其巨大潜力为人类谋福利。

参考文献

[1] Schmitt, C. K., Meysick, K. C., O'Brien, A. D.: Bacterial toxins: friends or foes? Emerg. Infect. Dis. 5, 224-234 (1999).

[2] Ogawa, C., Liu, Y. J., Kobayashi, K. S.: Muramyl dipeptide and its derivatives: peptide adjuvant in immunological disorders and cancer therapy. Curr. Bioact. Compd. 7, 180 - 197 (2011). doi: 1 0. 2174/157340711796817913.

[3] Sundstedt, A., Celander, M., Ohman, M. W., Forsberg, G., Hedlund, G.: Immunotherapy with tumor-targeted superantigens (TTS) in combination with docetaxel results in synergistic anti-tumor effects. Int. Immunopharmacol. 9, 1063-1070 (2009). doi: 1 0. 1016/j. intimp. 2009. 04. 013.

[4] Xu, M., et al.: An engineered superantigen SEC2 exhibits promising antitumor activity and low toxicity. Cancer Immunol. Immunother. 60, 705-713 (2011). doi: 10. 1007/s00262-011-0986-6.

[5] Adkins, I., Holubova, J., Kosova, M., Sadilkova, L.: Bacteria and their toxins tamed for immunotherapy. Curr. Pharm. Biotechnol 13 (8), 1446-1473 (2012).

[6] Shapira, A., Benhar, I.: Toxin-based therapeutic approaches. Toxins (Basel) 2, 2519-2583 (2010). doi: 10. 3390/toxins2112519.

[7] Kreitman, R. J.: Recombinant immunotoxins contain-ing truncated bacterial toxins for the treatment of hematologic malignancies. BioDrugs 23, 1-13 (2009).

[8] Wolf, P., Elsasser-Beile, U.: Pseudomonas exotoxin A: from virulence factor to anti-cancer agent. Int. J. Med. Microbiol. 299, 161-176 (2009).

[9] Toivonen, J. M., Olivan, S., Osta, R.: Tetanus toxin C-fragment: the courier and the cure? Toxins (Basel) 2, 2622-2644 (2010). doi: 10. 3390/toxins2112622.

[10] Truong, D. D., Jost, W. H.: Botulinum toxin: clinical use. Parkinsonism Relat. Disord. 12, 331-355 (2006). doi: 10. 1016/j. parkreldis. 2006. 06. 002.

[11] Johnson, E. A.: Clostridial toxins as therapeutic agents: benefi ts of nature's most toxic proteins. Annu. Rev. Microbiol. 53, 551-575 (1999). doi: 10. 1146/ annurev. micro. 53. 1. 551.

[12] Fayolle, C., et al.: Delivery of multiple epitopes by recombinant detoxifi ed adenylate cyclase of Bordetella pertussis induces protective antiviral immunity. J. Virol. 75, 7330-7338 (2001).

[13] Pizza, M., et al.: Mutants of pertussis toxin suitable for vaccine development. Science 246, 497-500 (1989).

[14] Munro, P., et al.: The Rho GTPase activators CNF1 and DNT bacterial toxins have mucosal adjuvant properties. Vaccine 23, 2551-2556 (2005). doi: 10. 1016/j. vaccine. 2004. 11. 042.

[15] Pizza, M., et al.: Mucosal vaccines: non toxic deriva-tives of LT and CT as mucosal adjuvants. Vaccine 19, 2534-2541 (2001).

[16] Lycke, N.: Targeted vaccine adjuvants based on modi-fi ed cholera toxin. Curr. Mol. Med. 5, 591-597 (2005).

[17] Adamsson, J., et al.: Novel immunostimulatory agent based on CpG oligodeoxynucleotide linked to the non-toxic B subunit of cholera toxin. J. Immunol. 176, 4902-4913 (2006).

[18] Fasano, A., Uzzau, S.: Modulation of intestinal tight junctions by Zonula occludens toxin permits enteral administration of insulin and other macromolecules in an animal model. J. Clin. Investig. 99, 1158-1164 (1997). doi: 10. 1172/JCI119271.

[19] Marinaro, M., Di Tommaso, A., Uzzau, S., Fasano, A., De Magistris, M. T.: Zonula occludens toxin is a powerful mucosal adjuvant for intranasally delivered antigens. Infect. Immun. 67, 1287-1291 (1999).

[20] Paterson, B. M., Lammers, K. M., Arrieta, M. C., Fasano, A., Meddings, J. B.: The safety, tolerance, pharmacokinetic and pharmacodynamic effects of single doses of AT-1001 in coeliac disease subjects: a proof of concept study. Aliment. Pharmacol. Ther. 26, 757-766 (2007). doi: 10. 1111/j. 1365-2036. 2007. 03413. x.

[21] Song, K. H., Fasano, A., Eddington, N. D.: Effect of the six-mer synthetic peptide (AT1002) fragment of zonula occludens toxin on the intestinal absorption of cyclosporin A. Int. J. Pharm. 351, 8-14 (2008). doi: 10. 1016/j. ijpharm. 2007. 09. 011.

[22] Fahrer, J., Rieger, J., van Zandbergen, G., Barth, H.: The C2-streptavidin delivery system promotes the uptake of biotinylated molecules in macrophages and T-leukemia cells. Biol. Chem. 391, 1315-1325 (2010). doi: 10. 1515/BC. 2010. 132.

[23] Boato, F., et al.: C3 peptide enhances recovery from spinal cord injury by improved regenerative growth of descending fi ber tracts. J. Cell Sci. 123, 1652-1662 (2010). doi: 10. 1242/jcs. 066050.

[24] Holtje, M., Just, I., Ahnert-Hilger, G.: Clostridial C3 proteins: recent approaches to improve neuronal growth and regeneration. Ann. Anat. 193, 314-320 (2011). doi: 10. 1016/j. aanat. 2011. 01. 008.

[25] Ghose, C., et al.: Transcutaneous immunization with Clostridium diffi cile toxoid A induces systemic and mucosal immune responses and toxin A-neutralizing antibodies in mice. Infect. Immun. 75, 2826-2832 (2007). doi: 10. 1128/IAI. 00127-07.

[26] Williams, S. A., et al.: A prostate-specifi c antigen-activated channel-forming toxin as therapy for prostatic disease. J. Natl. Cancer Inst. 99, 376-385 (2007). doi: 10. 1093/jnci/djk065.

[27] Osusky, M., Teschke, L., Wang, X., Buckley, J. T.: A chimera of interleukin 2 and a binding variant of aerolysin is selectively toxic to cells displaying the interleukin 2 receptor. J. Biol. Chem. 283, 1572-1579 (2008). doi: 10. 1074/jbc. M706424200.

[28] Janssen, K. P., et al.: In vivo tumor targeting using a novel intestinal pathogen-based delivery approach. Cancer Res. 66, 7230-7236 (2006).

[29] Engedal, N., Skotland, T., Torgersen, M. L., Sandvig, K.: Shiga toxin and its use in targeted cancer therapy and imaging. Microb. Biotechnol. 4, 32-46 (2011). doi: 10. 1111/j. 1751-7915. 2010. 00180. x.

[30] Battiwalla, M., et al.: Multiparameter fl ow cytometry for the diagnosis and monitoring of small GPI-defi cient cellular populations. Cytometry B Clin. Cytom. 78, 348-356 (2010). doi: 10. 1002/cyto. b. 20519.

[31] Totterman, T. H., et al.: Targeted superantigens for immunotherapy of haematopoietic tumours. Vox Sang. 74 (Suppl 2), 483-487 (1998).

[32] Ellouz, F., Adam, A., Ciorbaru, R., Lederer, E.: Minimal structural requirements for adjuvant activity of bacterial peptidoglycan derivatives. Biochem. Biophys. Res. Commun. 59, 1317-1325 (1974).

[33] Killion, J. J., Fidler, I. J.: Therapy of cancer metastasis by tumoricidal activation of tissue macrophages using liposome-encapsulated immunomodulators. Pharmacol. Ther. 78, 141-154 (1998).

[34] Li, X., et al.: Chemical conjugation of muramyl dipeptide and paclitaxel to explore the combination of immunotherapy and chemotherapy for cancer. Glycoconj. J. 25, 415-425 (2008). doi: 10. 1007/s10719-007-9095-3.

[35] Ando, K., Mori, K., Corradini, N., Redini, F., Heymann, D.: Mifamurtide for the treatment of non-

metastatic osteosarcoma. Expert Opin. Pharmacother. 12, 285 - 292 (2011). doi: 10. 1517/14656566. 2011. 543129.

[36] Sanchez, J., Holmgren, J.: Cholera toxin-a foe & a friend. Indian J. Med. Res. 133, 153-163 (2011).

[37] Lapa, J. A., et al.: Randomized clinical trial assessing the safety and immunogenicity of oral microencapsulated enterotoxigenic Escherichia coli surface antigen 6 with or without heat-labile enterotoxin with mutation R192G. Clin. Vaccine Immunol. 15, 1222-1228 (2008).

[38] Sun, J. B., Holmgren, J., Czerkinsky, C.: Cholera toxin B subunit: an effi cient transmucosal carrier-delivery system for induction of peripheral immunological tolerance. Proc. Natl. Acad. Sci. U. S. A. 91, 10795-10799 (1994).

[39] Stanford, M., et al.: Oral tolerization with peptide 336-351 linked to cholera toxin B subunit in preventing relapses of uveitis in Behcet's disease. Clin. Exp. Immunol. 137, 201-208 (2004).

[40] Vazquez, R. I., Moreno - Fierros, L., Neri - Bazan, L., De La Riva, G. A., Lopez - Revilla, R.: Bacillus thuringiensis Cry1Ac protoxin is a potent systemic and mucosal adjuvant. Scand. J. Immunol. 49, 578-584 (1999).

[41] Munoz, J. J., Bernard, C. C., Mackay, I. R.: Elicitation of experimental allergic encephalomyelitis (EAE) in mice with the aid of pertussigen. Cell. Immunol. 83, 92-100 (1984).

[42] Alfano, M., Rizzi, C., Corti, D., Adduce, L., Poli, G.: Bacterial toxins: potential weapons against HIV infection. Curr. Pharm. Des. 11, 2909-2926 (2005).

[43] Liu, S., Bugge, T. H., Leppla, S. H.: Targeting of tumor cells by cell surface urokinase plasminogen activator-dependent anthrax toxin. J. Biol. Chem. 276, 17976-17984 (2001).

[44] Liu, S., Netzel-Arnett, S., Birkedal-Hansen, H., Leppla, S. H.: Tumor cell-selective cytotoxicity of matrix metalloproteinase-activated anthrax toxin. Cancer Res. 60, 6061-6067 (2000).

[45] Ohana, P., et al.: Use of H19 regulatory sequences for targeted gene therapy in cancer. Int. J. Cancer 98, 645-650 (2002).

[46] Berger, E. A., Pastan, I.: Immunotoxin complementa-tion of HAART to deplete persisting HIV-infected cell reservoirs. PLoS Pathog. 6, e1000803 (2010). doi: 10. 1371/journal. ppat. 1000803.

[47] Ferens, W. A., Hovde, C. J.: The non-toxic A subunit of Shiga toxin type 1 prevents replication of bovine immunodefi ciency virus in infected cells. Virus Res. 125, 29-41 (2007).

[48] Nagamune, H., et al.: The human-specifi c action of intermedilysin, a homolog of streptolysin O, is dictated by domain 4 of the protein. Microbiol. Immunol. 48, 677-692 (2004).

[49] Sebo, P., et al.: Cell-invasive activity of epitope-tagged adenylate cyclase of Bordetella pertussis allows in vitro presentation of a foreign epitope to $CD8^+$ cytotoxic T cells. Infect. Immun. 63, 3851-3857 (1995).

[50] Osicka, R., et al.: Delivery of CD8 (+) T-cell epitopes into major histocompatibility complex class I antigen presentation pathway by Bordetella pertussis adenylate cyclase: delineation of cell invasive structures and permissive insertion sites. Infect. Immun. 68, 247-256 (2000).

[51] Ballard, J. D., Collier, R. J., Starnbach, M. N.: Anthrax toxin-mediated delivery of a cytotoxic T-cell epitope in vivo. Proc. Natl. Acad. Sci. U. S. A. 93, 12531-12534 (1996).

[52] Fayolle, C., Sebo, P., Ladant, D., Ullmann, A., Leclerc, C.: In vivo induction of CTL responses by recombinant adenylate cyclase of Bordetella pertussis carrying viral $CD8^+$ T cell epitopes. J. Immunol. 156, 4697-4706 (1996).

[53] Loucka, J., Schlecht, G., Vodolanova, J., Leclerc, C., Sebo, P.: Delivery of a MalE CD4 (+) -T-cell epitope into the major histocompatibility complex class Ⅱ antigen presentation pathway by Bordetella pertussis adenylate cyclase. Infect. Immun. 70, 1002-1005 (2002).

[54] Preville, X., Ladant, D., Timmerman, B., Leclerc, C.: Eradication of established tumors by vaccination with recombinant Bordetella pertussis adenylate cyclase carrying the human papillomavirus 16 E7 oncoprotein. Cancer Res. 65, 641-649 (2005).

[55] Fayolle, C., Ladant, D., Karimova, G., Ullmann, A., Leclerc, C.: Therapy of murine tumors with recombinant Bordetella pertussis adenylate cyclase carrying a cytotoxic T cell epitope. J. Immunol. 162, 4157-4162 (1999).

[56] Vordermeier, H. M., et al.: Recognition of mycobacte-rial antigens delivered by genetically detoxifi ed Bordetella pertussis adenylate cyclase by T cells from cattle with bovine tuberculosis. Infect. Immun. 72, 6255-6261 (2004).

[57] Blight, M. A., Holland, I. B.: Heterologous protein secretion and the versatile Escherichia coli haemolysin translocator. Trends Biotechnol. 12, 450-455 (1994).

[58] Lee, K. D., Oh, Y. K., Portnoy, D. A., Swanson, J. A.: Delivery of macromolecules into cytosol using liposomes containing hemolysin from Listeria monocytogenes. J. Biol. Chem. 271, 7249-7252 (1996).

[59] Russmann, H., et al.: Protection against murine listeriosis by oral vaccination with recombinant Salmonellaexpressing hybrid Yersinia type Ⅲ proteins. J. Immunol. 167, 357-365 (2001).

[60] Russmann, H., et al.: Delivery of epitopes by the Salmonella type III secretion system for vaccine development. Science 281, 565-568 (1998).

[61] Maciag, P. C., Radulovic, S., Rothman, J.: The fi rst clinical use of a live-attenuated Listeria monocytogenes vaccine: a Phase Ⅰ safety study of Lm-LLO-E7 in patients with advanced carcinoma of the cervix. Vaccine 27, 3975-3983 (2009).

[62] Fehlings, M. G., et al.: A phase I/IIa clinical trial of a recombinant Rho protein antagonist in acute spinal cord injury. J. Neurotrauma 28, 787-796 (2011). doi: 10. 1089/neu. 2011. 1765.

[63] Diana, G., et al.: Enhancement of learning and memory after activation of cerebral Rho GTPases. Proc. Natl. Acad. Sci. U. S. A. 104, 636-641 (2007). doi: 10. 1073/pnas. 0610059104.

[64] Podda, A., et al.: Metabolic, humoral, and cellular responses in adult volunteers immunized with the genetically inactivated pertussis toxin mutant PT-9K/129G. J. Exp. Med. 172, 861-868 (1990).

[65] Sandvig, K., van Deurs, B.: Delivery into cells: lessons learned from plant and bacterial toxins. Gene Ther. 12, 865-872 (2005). doi: 10. 1038/sj. gt. 3302525.

(吴国华译)

第 35 章　基于植物热休克蛋白的自身佐剂免疫原

Selene Baschieri①

摘要

亚单位疫苗基于分离的纯或半纯微生物组分（抗原）。与基于全病原体（灭活或致弱）的传统制剂相比，这些疫苗除非添加佐剂，否则不能增强免疫应答，但更安全。如今，由于高效基因工程和生化工艺的发展，市场上出现的基于亚单位的疫苗是用产自细菌、酵母或动物细胞的重组抗原配制的。基于植物的表达系统也被证明是重组抗原的非常有吸引力的“生物工厂”，因为它们确保了成本低、速度快和制造规模易扩大以及最终产品的内在生物安全性。

然而，植物产生的重组抗原本身免疫原性较差。已经有人尝试建立从植物中获得自体佐剂免疫原的策略。本章主要讨论利用植物热休克蛋白的可能性。

35.1　引言

理想情况下，疫苗应能够尽可能模拟病原体感染，从而激活持久的保护性免疫反应，但不会产生不良反应[1]。传统的疫苗配方基于完全灭活或致弱的病原体，尽管非常有效，但偶尔与不良反应有关[2]。为了克服这些问题，人们已经开始尝试仅基于那些被认为能够启动保护性免疫反应的病原体成分（抗原）来定义更安全的制剂。在过去，这些亚单位疫苗是通过直接从病原体中纯化关键抗原而制备的。现在，抗原开始用细菌、酵母或动物细胞在重组方式中产生[3]。

利用植物来实现这一目标是一个机会，这是由于稳定和短暂的细胞转化方法的进展而产生的，以过度表达外源基因[4]。尽管如此，与所有重组抗原相似，纯化后的植物源性蛋白质也需要保存冷链，并提供佐剂和注射剂。已经有人尝试从植物中获得固有免疫原性抗原，从而充分利用这些表达系统的优势。为此目的，合理设计所表达的免疫原以实现直接传递到抗原呈递细胞（APCs），尤其是树突状细胞（DCs），并获得先天性和病原体特异性适应性免疫反应的激活[5]。在简要介绍了将植物转化为抗原“生物工厂”的常用方法后，本文将考虑使用植物的热休克蛋白（HSPs）作为疫苗成分的可能性。

35.2　植物作为“生物工厂”

通过稳定或短暂的转化方法，可以获得植物中表达重组蛋白的基因转移[6]。

为了获得稳定的转化（和表达），需要在培养的叶片外植体细胞的核或叶绿体基因组中插入一个编码目的蛋白的 DNA 片段和一个可选择的标记物[7]。利用用双元载体（即人工载体，包括来源于天然肿瘤诱导质粒的元素，最初在根瘤病菌中发现，需要将异源 DNA 转移到植物细胞中）[8]变换的植物病原菌肿瘤农杆菌可以有效地将异源序列转移到核基因组中（图 35.1），虽然

① S. Baschieri，PhD（意大利能源与可持续经济发展署卡萨西亚研究中心生物技术实验室，意大利罗马，E-mail：selene. baschieri@ enea. it）

转移到质体基因组需要使用直接的 DNA 传递方法，如粒子轰击或聚乙二醇处理[9]。

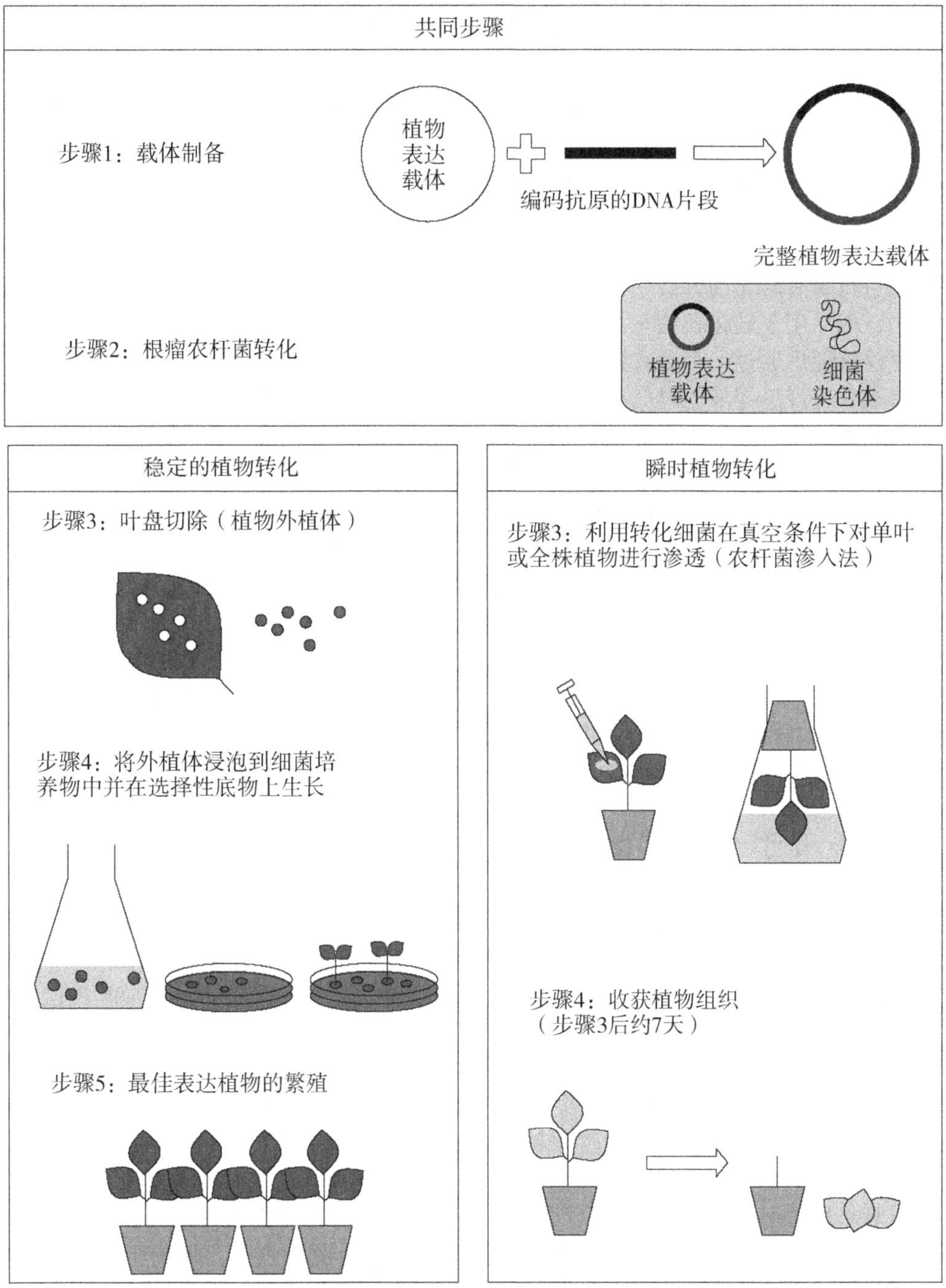

图 35.1 植物转化过程中的主要基因工程和细菌转化步骤示意图

在上框中，初始基因工程和细菌转化这两个步骤是植物利用根瘤农杆菌进行稳定或短暂转化的共同步骤。

在这两种情况下，经过转化后，外植体被用来再生整个植物，依靠植物细胞的全能性，通过在一种选择性培养基上生长，使存活（和植物再生）仅限于表达可选择标记基因的细胞。为了获得基因的纯合性和稳定的遗传，转化为核基因组的植物（以及它们通过自受精产生的种子）在选择性条件下生长数代，而在质体转化的情况下，实现纯质体所需的选择过程（转基因整合到每个质体基因组中的条件，即每个叶细胞多达10 000个）需要更长的时间[10]。尽管如此，叶绿体中产生的重组蛋白在翻译后修饰（如糖基化模式的差异）方面可能有类似于细菌中产生的“缺陷”，但质体转化具有一些有趣的特征。事实上，外源基因通过同源重组被插入基因组的特定位置，这一机会消除了随机基因插入核基因组的突出问题（“位置效应”）。因此，由于叶绿体中不存在基因沉默，通过这种转化方法可以获得很高的表达水平。此外，由于质体不是通过花粉传播的，所以转基因植物是“天然的”[11]。

与稳定的转化方法相比，瞬时转化是一种更快速、更容易在植物中产生大量重组蛋白的高通量过程，但正如其名称所暗示的那样，异源序列只是暂时表达的，不能传递给植物后代，因为是在完全发育的植物体内进行转化[12]。

可用于进行瞬时转化的方法之一具有与稳定核转化相同的初始步骤，即将表达的序列插入二元载体中，然后用于转化根瘤土壤杆菌细胞[13]（图 35.1）。然而，在短暂的转化过程中，转化细菌的悬浮液随后通过一个没有针头的注射器渗透到植物叶片的背面（即底面）。这个过程被称为农业渗透，允许悬浮液通过气孔或叶片上的一个小切口渗透到叶片的空隙中。或者，可以将整个植物倒置浸入含有细菌悬浮液的烧杯中，并放置在真空室中。当应用真空时，释放真空时，空气从叶肉中被细菌悬浮液取代。农杆菌介导的瞬时表达仅限于植物浸润组织。

通过将异源序列插入植物病毒基因组（或插入编码完整病毒基因组的表达载体）并利用转基因植物病原体感染植物，而不是农业渗透，可以获得瞬时表达[14]。为此，含有重组植物病毒颗粒或编码重组病毒基因组的表达载体的溶液中的几微升分布在叶片的近轴（即上部）表面，异源蛋白的表达与病毒感染一起在整个植物中进行。

最近，一些瞬时的过表达系统已经开发出来，将农杆菌介导的转化效率与病毒的表达率结合起来[15]。这些载体编码对控制基因表达起重要作用的植物病毒基因组中的少数元素，并被重组农杆菌导入植物细胞。这些模块化的载体系统除在植物组织中有很高的表达水平外，本质上也是安全的，因为植物病毒永远不会组装。

35.3 植物用于生产重组抗原

所有这些策略都被广泛应用于植物体内表达重组抗原[16]。虽然稳定的转化方法与瞬时转化相比，具有非常耗时的主要缺点，但当目的是将植物本身用作（可食用）疫苗时，它们是生产重组抗原的首选方法[17]。这种植物“生物活性”应用的巨大潜力已经通过几项临床试验得到证明，在这些试验中，人类口服各种表达不同抗原的植物[18]。例如，植物科学家正在努力改进这项技术，试图通过同源重组将异源基因插入核基因组的特定位置，以获得植物组织中编码抗原的更高和可重复表达水平，以确保在一次服药中摄入足以提供保护的抗原剂量[19]。其目的是通过降低生产和保存成本，可以在消除西方和发展中国家之间仍然存在的差距方面取得真正的突破，从而提高全球卫生公平性[20]。

如果打算将植物用作“生物工厂”，则采用瞬时转化方法可能更为合适，因为这些方法可以在更短的时间内确保更高的表达水平。在这种情况下，一个关键点是，作为生产步骤效率的纯化阶段可能是极其可变的[21]。如果植物被用来表达产生病毒样颗粒（VLPS）的自行组装抗原，问题则不那么明显[22]。该方法已经成功实施，通过在本塞姆氏烟草植物中瞬时表达血凝素来产生流感 VLP[23]。利用这个平台生产流感疫苗在生产时间、安全性和扩大规模方面具有显著优势。特别是植物“生物工厂”可确保在新型病毒序列释放后 3 周内从约 1 kg 农业灌溉叶片中实现 1 500剂量，而供应基于鸡胚的全球季节性流感疫苗则需要 4~6 个月[24]。

35.4 植物作为自身佐剂抗原的“生物工厂”

亚单位疫苗的一个主要局限性是纯化的抗原本身不能引起保护性的广谱免疫反应以刺激先

天和适应性免疫系统，激活 **B** 淋巴细胞和 **T** 淋巴细胞[25]。因此，免疫刺激分子（佐剂）必须增强这些疫苗的效力。目前可用于人类和兽医的佐剂很少，因为有效的免疫刺激可导致不良反应[26]。研究正在进行，以确定新的传递策略或分子作为对抗原免疫反应的真正增强剂，并且没有不必要的副作用[27]。这项工作也由植物生物技术学家提出，他们正试图鉴定可用于运载重组抗原的植物产品/结构。在这方面的尝试中，包括构建嵌合植物病毒[28]或展示肽或蛋白质的嵌合植物细胞器[29]，以及在植物中产生由抗原和抗体组成的免疫复合物[30]。另一种方法考虑了植物热休克蛋白在疫苗配方中的可能用途。

35.5 热休克蛋白和疫苗开发

Hsp 是一组异质性的普遍存在的蛋白质，分为六大家族，根据分子量（钙网蛋白、钙调蛋白和泛素除外）命名[31]。除了小的热休克蛋白外，它们是活生物体中最具系统保护性的蛋白质[32]。最初，Hsp 表达被认为与压力紧密相关，因为它在压力条件下（如温度变化）非常明显，但现在很清楚 Hsp 是组成型表达的管家蛋白。事实上，它们在控制细胞蛋白质组稳态中是必不可少的，如参与蛋白质折叠和展开、细胞内和细胞外靶向、降解和多聚体组装[33]。每个 Hsp 家族包括几个高度同源的亚型，与特定的细胞室和/或组织相关，并在正常或应激条件下表达[34]。Hsp90 和 Hsp70 是研究最广泛的 Hsp 家族[35]。

Hsp90 是一组动态蛋白质（共伴侣）的核心组成部分，其数量和作用尚未完全阐明。在真核生物中，参与关键信号蛋白的成熟和活化，例如荷尔蒙、激素受体、转录因子和蛋白激酶。它起到二聚体的作用，其结构特征是 3 个结构域[36]。氨基末端结构域具有 ATP 酶活性，是分子伴侣生物学功能的基础。中间区域参与与客户蛋白的相互作用。羧基末端域对于二聚化是必不可少的。在不存在 ATP 的情况下，蛋白质与 Hsp90 二聚体缔合。在 ATP 结合时，两个 Hsp90 组分的氨基末端结构域旋转，彼此接触并通过仍未定义的机制产生客户蛋白的活化、稳定化或组装。

Hsp70 控制蛋白质的构象状态，其在蛋白质生物合成，转运和降解中是必不可少的[37]。对来自不同生物的 Hsp70 同种型结构的分析表明，由于一级结构中的高度同源性，该伴侣分子的折叠是相似的[38-40]。所有的亚型都有两个主结构域。氨基末端结构域具有 ATP 酶活性（核苷酸结合结构域，NBD），结合 ATP 并将其水解为 ADP。羧基末端域可分为两个子域：①由两个 β 折叠并参与多肽结合的口袋状底物结合域（SBD）[41]；②由一对 α-螺旋形成的“盖子域”，其构象影响底物结合[42]。两个结构域之间的通信是双向的，Hsp70 存在于两个确定的和稳定的分子构象中。当 NBD 被 ATP 占据时，“盖子”打开并且 SBD 是自由的，而当 NBD 被 ADP 占据时，盖子关闭并且 SBD 被多肽占据（未折叠、天然折叠或聚集）[43]。

除了参与控制蛋白质整形外，Hsp90 和 Hsp70 还具有免疫刺激特性[44]。一项研究首次揭示了这一特征，该研究表明，从小鼠肉瘤中分离出的、后来被鉴定为 Hsp 的选定蛋白质组分在全身注射时能够抗肿瘤，这种保护是由肿瘤特异性的适应性免疫反应激活引起的[45]。基于肿瘤源性热休克蛋白的癌症疫苗目前正在临床试验中进行评估[46,47]。

据报道，从病毒感染细胞中提取的热休克蛋白也能激活病毒肽特异性的 $CD8^+$ T 细胞克隆[48]，并且从表达异源蛋白的哺乳动物 P815 细胞中纯化的 Hsp 能够激活能够在 H-2d 小鼠（即 P815 细胞的相同单倍型）以及不同单倍型的小鼠中[49]激活对重组蛋白特异的主要组织相容性复合物（MHC）I 类限制性 T 淋巴细胞。

一旦纯化，热休克蛋白也可以在体外卸载自然结合的肽，随后装载一个选择的肽。这些复合物在没有佐剂的情况下传递给小鼠，能够引起抗体和细胞介导的针对携带肽的免疫反应[50]。与肽相比，载有肽的分枝杆菌 Hsp70 在产生肽特异性细胞毒性 T 细胞反应方面效率更高[51]。

热休克蛋白具有免疫学特性的主要原因[52]：①Hsp 结合的肽是给定时间内细胞内表达蛋白

质的指纹图谱；②Hsp 和肽形成的复合物是稳定的并且可以纯化；③Hsp-肽复合物可以被抗原呈递细胞摄取，通过与特定受体（如 CD91，在 DCs 上表达）的相互作用[53,54]；④在 APCS 内部，肽从热休克蛋白转移至 MHC 分子，转运至细胞膜递呈。

除了通过受体结合和内化将抗原传递给 APC（特别是 DCs）的能力外，热休克蛋白还被证明具有佐剂性质，因为它们单独在诱导 DCs 释放促炎性细胞因子方面具有活性[55]。这两种功能（抗原传递与树突状细胞的刺激）在结构上是分离的，因为一个能够阻止肽结合到 SBD 的突变不会影响刺激先天免疫的能力[51]。

总的来说，这些观察和来自系统发育远缘生物的热休克蛋白能够激活小鼠免疫反应的证据[56]表明，植物热休克蛋白也可以作为简单的佐剂或抗原传递载体在疫苗学中进行应用。

35.6 植物热休克蛋白及其免疫特性

植物是固着的有机体，环境和生物胁迫是蛋白质稳态的主要挑战。它们通过产生热休克蛋白来应对不同的非生物胁迫，如低温和高温、干旱、盐度或洪水。此外，植物病毒的感染和农业渗透是很强的热休克蛋白诱导物[57]，因为这些生物应激导致大量错误折叠的蛋白质积累，严重影响细胞新陈代谢[58]。此外，在复制和/或移动过程中，已证明热休克蛋白积极参与病毒蛋白成熟和转换[59]。

在不同的植物细胞室（叶绿体、线粒体、内质网）中发现特定的 Hsp90 和 Hsp70 蛋白。这些植物伴侣分子在结构和功能上都与动物细胞中存在的植物伴侣分子极为相似。首先，这一假设得到了证据的支持，证明植物和其他真核生物的热休克蛋白之间的氨基酸序列同源性很高。烟草细胞溶质 Hsp70 和相应蛋白在智人、牛磺酸和肌肉中的表达率均超过 75%。序列的比对表明，对 ATP/ADP 或肽结合关键的基序是完全保守的。通过将哺乳动物 Hsp70 的解析结构与植物 Hsp70 的模型（采用同源建模方法设计）进行比较，确定了结构的优越性[57]。

由于它涉及功能方面，研究表明，与哺乳动物细胞的情况类似，拟南芥植物中细胞溶质 Hsp70 的下调降低了耐热性[60]。此外，对动物 Hsp90 有特异作用的药物（格尔德霉素）也对植物 Hsp90 有效，而由植物 Hsp90 和 Hsp70 形成的复合物能够正确组装动物类固醇受体和致癌蛋白激酶[61]。

在此基础上，探讨了植物和哺乳动物热休克蛋白的相似性也可以扩展到免疫特性的可能性。

一项研究旨在确定来自植物的 Hsp90 是否具有与其哺乳动物对应物相同的佐剂特性。在该研究中，克隆了拟南芥和本塞姆氏烟草中编码 Hsp90 的序列，在大肠杆菌中独立表达，融合组氨酸标记，通过镍亲和层析纯化。纯化的蛋白质（与人 Hsp90 中具有约 68%的氨基酸序列同一性）在体外刺激鼠脾细胞增殖的能力进行了测试。这些实验的结果表明，植物来源的 Hsp90 能够与 B 淋巴细胞质膜上表达的 Toll 样受体 4 结合，并且对这些细胞具有选择性促有丝分裂作用，表明这种植物伴侣分子具有免疫刺激特性[62]。

另外两项研究旨在确定植物 Hsp70 是否可用于有效传递植物表达的蛋白质/抗原[57,63]。为此目的，使用本塞姆氏烟草植物瞬时表达报告蛋白［植物病毒马铃薯病毒 X（PVX）的外壳蛋白（CP）］或高度保守的甲型流感病毒核蛋白（NP）。该抗原的选择是为了利用现有的数据，对 T 淋巴细胞介导的免疫反应进行精细地表征，该免疫反应在基因不同的小鼠株中触发[64,65]。NP 对疫苗设计也有启发，因为它可以诱导针对不同病毒变体的交叉保护免疫反应的激活，因此在“新”流感病毒暴发的情况下，对先前存在的体液免疫无效（或效果不佳）的流感病毒可能有用[66]。

通过用编码完整病毒基因组的表达载体接种植物诱导 PVX 的 CP 的瞬时表达，而通过用携带包含 NP 编码盒的二元载体的根癌土壤杆菌细胞浸润植物叶来获得 NP 的瞬时表达。在病毒感染或农杆菌渗入后 7 天（在两种情况下获得异源蛋白质和内源植物 Hsp70 的最大表达水平所需的时间间隔），取样本氏烟草叶子并用于纯化植物 Hsp70。用于该目的的色谱方法基本上与用于纯化哺乳动物Hsp70的方法相同，并且基于该伴侣蛋白对ATP／ADP的亲和力。用ADP-琼脂糖树脂进行蛋白纯化，可防止伴随肽的SBD解离，使pHsp70与多肽结合。

每克新鲜叶中回收高纯度植物Hsp70伴肽约23 μg植物Hsp70是可重复的。对纯化的植物Hsp70多肽复合物激活植物组织中异源蛋白特异性免疫反应的能力进行了测试。这些实验的结果清楚地证明，无佐剂的小鼠免疫能够诱导CP特异性机体反应和NP特异性抗体反应。

正如所预测的，血清中的抗体也能对野生植物粗提物产生反应（植物Hsp70耗尽）。这是因为植物Hsp70不仅可能携带重组蛋白产生的肽，而且可能携带内源性植物细胞蛋白产生的肽。即便如此，与CP或NP特异性反应相比，这些植物靶点的抗体反应更低，可能是因为在取样时，与内源性蛋白质相比，植物细胞中的CP和NP的表达过高。小鼠体内的抗体反应也针对植物Hsp70，这也是预期的结果。尽管如此，已经证明，尽管哺乳动物和植物Hsp70之间的高度相似，抗植物Hsp70抗体不能与小鼠或人类Hsp70交叉反应，因此不能尽可能地触发自身免疫反应，这是一个不能排除从自体肿瘤细胞制备的基于Hsp的癌症疫苗的问题，其可以诱导针对自身抗原的不期望的免疫应答的激活。

纯化的植物Hsp70-多肽复合物也证实了其对NP抗原特异性T淋巴细胞介导免疫应答的能力。在这种情况下，免疫应答通过干扰素-γ酶联免疫吸附斑点试验（IFN-γ ELISPOT）测定来评价。这种方法可以检测和列举脾细胞体外培养中的抗原特异性$CD8^+T$细胞，因为这些细胞在用主要组织相容性复合体I类限制性合成肽刺激时分泌IFN-γ，从序列外推抗原。简言之，用从表达NP的植物组织中纯化的pHSP70-多肽复合物对C57BL/6小鼠进行免疫，用5种不同的NP衍生肽（单个或不同组合）刺激小鼠的体外脾细胞。已知所有肽都能特异性激活$H-2D^b$-限制性$CD8^+T$细胞应答。其中一个肽的序列是C57BL/6J小鼠的NP免疫优势表位序列[64]，而另一个是该菌株的次优势表位序列[65]。

总的来说，结果清楚地表明，植物Hsp70多肽复合物能够启动T细胞介导的对NP的免疫应答，而无需辅助性共传递，应答是多克隆的，针对优势和次优势表位。用完整的肽库（免疫优势+次优势）刺激脾细胞获得较高的反应。通过实验证实了该反应的多克隆性，证明用于免疫C57BL/6J小鼠的同一植物Hsp70制剂也能有效地诱导具有不同遗传背景（$H-2K^d$）的BALB/C小鼠$CD8^+T$细胞的活化，但在这种情况下，T细胞的反应是针对不同的肽。在两种小鼠中，即使在去除内毒素后，pHSP70制剂仍能保持诱导$CD8^+T$细胞活化的作用。

这些数据证明了植物热休克蛋白具有刺激免疫反应的一般能力，同时也揭示了表达重组抗原的植物组织衍生的Hsp多肽复合物的更具体用途。携带植物Hsp70的肽似乎诱导抗体产生和$CD8^+T$细胞活化。这两种情况下的反应都是多克隆的，有趣的是，不需要辅助性输送。这些数据表明，来自表达重组抗原的植物的Hsp70可用作多表位疫苗，以刺激任何给定MHC类型的个体中对抗原特异的免疫应答。以植物Hsp70复合物为基础的免疫策略对于开发抗复杂病原体的疫苗也特别有吸引力，在复杂病原体中，保护性抗原很难实现或未知。事实上，表达的外源多肽并不代表终端提取过程是这一技术的主要优势，因为通过这种技术，植物表达系统的所有优点都是通过克服纯化过程中的困难来体现的，而纯化的效率通常取决于重组蛋白的内在特征。

结论

已经设想在疫苗开发领域使用植物，这可能对一些尚未解决的问题是一个机会，如效力、安全性，以及疫苗学研究的伦理和社会方面。

由于特别关注植物热休克蛋白的可能用途，因此迄今为止获得的结果非常具有前景，并表明需要进一步探索来完善这些天然植物产品的免疫特性。不计其数的动物传染病需要合适且价格低廉的保护性疫苗，我们通过为其中的一种传染病开发疫苗，以在兽医领域有条件测试它们的功效。

免疫学中植物的非传统开发表明，对当前与疫苗接种有关的问题采用多学科“交叉培育”方法可以成为创新的捷径。

参考文献

[1] Germain, R. N.: Vaccines and the future of human immunology. Immunity 33, 441-450 (2010).

[2] Plotkin, S. A., Plotkin, S. L.: The development of vaccines: how the past let to the future. Nat. Rev. Immunol. 9, 889-893 (2011).

[3] Ulmer, J. B., Valley, U., Rappuoli, R.: Vaccine manufacturing: challenges and solutions. Nat. Biotechnol. 24, 1377-1383 (2006).

[4] Streatfi eld, S. J.: Approaches to achieve high-level heterologous protein production in plants. Plant Biotechnol. J. 5, 2-15 (2007).

[5] Lico, C., et al.: Plant-based vaccine delivery strategy. In: Baschieri, S. (ed.) Innovation in Vaccinology. From Design, Through to delivery and Testing. Springer, New York (2012).

[6] Fischer, R., et al.: Plant-based production of biophar-maceuticals. Curr. Opin. Plant Biol. 7, 152-158 (2004).

[7] Newell, C. A.: Plant transformation technology. Mol. Biotechnol. 16, 53-65 (2000).

[8] Lee, L. -Y., Gelvin, S. B.: T-DNA binary vectors and systems. Plant Physiol. 146, 325-332 (2008).

[9] Bock, R., Khan, M. S.: Taming plastids for a green future. Trends Biotechnol. 22, 311-318 (2004).

[10] Meyers, B., et al.: Nuclear and plastid genetic engineering of plants: comparison of opportunities and challenges. Biotechnol. Adv. 28, 747-756 (2010).

[11] Bock, R.: Plastid biotechnology: prospects for herbicide and insect resistance, metabolic engineering and molecular farming. Curr. Opin. Biotechnol. 18, 100-106 (2007).

[12] Komarova, T. V., et al.: Transient expression systems for plant-derived biopharmaceuticals. Expert Rev. Vaccines 9, 859-876 (2010).

[13] Sheludko, Y. V.: Agrobacterium-mediated transient expression as an approach to production of recombinant proteins in plants. Recent Pat. Biotechnol. 2, 198-208 (2008).

[14] Pogue, G. P., et al.: Making an ally from an enemy: plant virology and the new agriculture. Annu. Rev. Phytopathol. 40, 45-74 (2002).

[15] Lico, C., Chen, Q., Santi, L.: Viral vectors for produc-tion of recombinant proteins in plants. J. Cell. Physiol. 216, 366-377 (2008).

[16] Rybicki, E. P.: Plant-made vaccines for humans and animals. Plant Biotechnol. J. 8, 620-637 (2010).

[17] Walmsley, A. M., Arntzen, C. J.: Plants for delivery of edi-ble vaccines. Curr. Opin. Biotechnol. 11, 126-129 (2000).

[18] Tacket, C. O.: Plant-based oral vaccines: results of human trials. Curr. Top. Microbiol. Immunol. 332, 103-117 (2009).

[19] Butaye, K. M. J., et al.: Approaches to minimize varia-tion of transgene expression in plants. Mol. Breed. 16, 79-91 (2005).

[20] Penney, C. A., et al.: Plant-made vaccines in support of the Millennium Development Goals. Plant Cell Rep. 30, 789-798 (2011).

[21] Wilken, L. R., Nikolov, Z. L.: Recovery and purifi ca-tion of plant-made recombinant proteins. Biotechnol. Adv. 30, 419-433 (2012).

[22] Santi, L., Huang, Z., Mason, H.: Virus-like particles production in green plants. Methods 40, 66-76 (2006).

[23] D'Aoust, M. A., et al.: Infl uenza virus-like particles produced by transient expression in Nicotiana benthamiana induce a protective immune response against a lethal viral challenge in mice. Plant Biotechnol. J. 6, 930-940 (2008).

[24] D'Aoust, M. A., et al.: The production of hemagglutinin-based virus-like particles in plants: a rapid, effi

cient and safe response to pandemic infl uenza. Plant Biotechnol. J. 8, 607-619 (2010).
[25] Plotkin, S. A.: Vaccines: the fourth century. Clin. Vaccine Immunol. 16, 1709-1719 (2009).
[26] Guy, B.: The perfect mix: recent progress in adjuvant research. Nat. Rev. Microbiol. 5, 505-517 (2007).
[27] Bachmann, M. F., Jennings, G. T.: Vaccine delivery: a matter of size, geometry, kinetics and molecular patterns. Nat. Rev. Immunol. 10, 787-796 (2010).
[28] Lico, C., et al.: Plant-produced potato virus X chime-ric particles displaying an infl uenza virus-derived peptide activate specifi c $CD8^+$T cells in mice. Vaccine 27, 5069-5076 (2009).
[29] Capuano, F., et al.: LC-MS/MS methods for absolute quantifi cation and identifi cation of proteins associated with chimeric plant oil bodies. Anal. Chem. 83, 9267-9272 (2011).
[30] Chargelegue, D., et al.: Highly immunogenic and pro-tective recombinant vaccine candidate expressed in transgenic plants. Infect. Immun. 73, 5915-5922 (2005).
[31] Feder, M. E., Hoffmann, G. E.: Heat-shock proteins, molecular chaperones, and the stress response: evolutionary and ecological physiology. Annu. Rev. Physiol. 61, 243-282 (1999).
[32] Lindquist, S., Craig, E. A.: The heat-shock proteins. Annu. Rev. Genet. 22, 631-677 (1988).
[33] Christis, C., Lubsen, N. H., Braakman, I.: Protein folding includes oligomerization-examples from the endoplasmic reticulum and cytosol. FEBS J. 275, 4700-4727 (2008).
[34] Rutherford, S. L.: Between genotype and phenotype: protein chaperones and evolvability. Nat. Rev. Genet. 4, 263-274 (2003).
[35] Wegele, H., Muller, L., Buchner, J.: Hsp70 and Hsp90-a relay team for protein folding. Rev. Physiol. Biochem. Pharmacol. 151, 1-44 (2004).
[36] Pearl, L. H., Prodromou, C.: Structure and mechanism of the HSP90 molecular chaperone machinery. Annu. Rev. Biochem. 75, 271-294 (2006).
[37] Morano, K. A.: New tricks for an old dog. The evolving world of Hsp70. Ann. N. Y. Acad. Sci. 1113, 1-14 (2007).
[38] Zhu, X., et al.: Structural analysis of substrate binding by the molecular chaperone DnaK. Science 72, 1606-1614 (1996).
[39] Worrall, L. J., Walkinshaw, M. D.: Crystal structure of the C-terminal three-helix bundle subdomain of C. elegans HSP70. Biochem. Biophys. Res. Commun. 357, 105-110 (2007).
[40] Jiang, J., Lafer, E. M., Sousa, R.: Crystallization of a functionally intact HSC70 chaperone. Acta Crystallogr. Sect. F Struct. Biol. Cryst. Commun. 62, 39-43 (2006).
[41] Rudiger, S., Buchberger, A., Bukau, B.: Interaction of Hsp70 chaperones with substrates. Nat. Struct. Biol. 4, 342-349 (1997).
[42] Schlecht, R., et al.: Mechanics of Hsp70 chaperones enables differential interaction with client proteins. Nat. Struct. Mol. Biol. 18, 345-351 (2011).
[43] Nicolai, A., et al.: Human inducible Hsp70: structures, dynamics, and interdomain communication from all-atom molecular dynamics simulations. J. Chem. Theory Comput. 6, 2501-2519 (2010).
[44] Srivastava, P. K.: Roles of heat-shock proteins in innate and adaptive immunity. Nat. Rev. Immunol. 2, 185-194 (2002).
[45] Srivastava, P. K., Deleo, A. B., Old, L. J.: Tumor rejection antigens of chemically induced sarcomas of inbred mice. Proc. Natl. Acad. Sci. U. S. A. 83, 3407-3411 (1986).
[46] Murshid, A., Gong, J., Calderwood, S. K.: Heat-shock proteins in cancer vaccines: agents of antigen cross-presentation. Expert Rev. Vaccines 7, 1019-1030 (2008).
[47] Srivastava, R. M., Khar, A.: Dendritic cells and their receptors in antitumor immune response. Curr. Mol. Med. 6, 708-724 (2009).
[48] Nieland, T. J. F., et al.: Isolation of an immunodomi-nant viral peptide that is endogenously bound to the stress protein GP96/GRP94. Proc. Natl. Acad. Sci. U. S. A. 93, 6135-6139 (1996).
[49] Arnold, D., et al.: Cross-priming of minor histocom-patibility antigen-specifi c cytotoxic T cells upon immunization with the heat shock protein gp96. J. Exp. Med. 182, 885-889 (1995).
[50] Blachere, N. E., et al.: Heat shock protein-peptide c omplexes, reconstituted in vitro, elicit p eptide-s

pecifi c cytotoxic T lymphocyte response and tumor i mmunity. J. Exp. Med. 186, 1315-1323 (1997).
[51] MacAry, P. A., et al.: HSP70 peptide binding mutants separate antigen delivery from dendritic cell stimulation. Immunity 20, 95-106 (2004).
[52] Javid, B., MacAry, P. A., Lehner, P. J.: Structure and function: heat shock proteins and adaptive immunity. J. Immunol. 179, 2035-2040 (2007).
[53] Basu, S., et al.: CD91 is a common receptor for heat shock proteins gp96, hsp90, hsp70, and calreticulin. Immunity 14, 303-313 (2001).
[54] Basu, S., Matsutake, T.: Heat shock protein-antigen presenting cell interactions. Methods 32, 38-41 (2004).
[55] Asea, A., et al.: HSP70 stimulates cytokine production through a CD14-dependantpathway, demonstrating its dual role as a chaperone and cytokine. Nat. Med. 6, 435-442 (2000).
[56] Kumaraguru, U., et al.: Antigenic peptides complexed to phylogenically diverse Hsp70s induce differential immune responses. Cell Stress Chaperones 8, 134-143 (2003).
[57] Buriani, G., et al.: Plant heat shock protein 70 as carrier for immunization against a plant-expressed reporter antigen. Transgenic Res. 20, 331-344 (2011).
[58] Sugio, A., et al.: The cytosolic protein response as a subcomponent of the wider heat shock response in Arabidopsis. Plant Cell 21, 642-654 (2009).
[59] Aparicio, F., et al.: Virus induction of heat shock pro-tein 70 refl ects a general response to protein accumulation in the plant cytosol. Plant Physiol. 138, 529-536 (2005).
[60] Lee, J. H., Schöffl, F.: An Hsp70 antisense gene affects the expression of HSP70/HSC70, the regulation of HSF, and the acquisition of thermotolerance in transgenic Arabidopsis thaliana. Mol. Gen. Genet. 252, 11-19 (1996).
[61] Kadota, Y., Shirasu, K.: The HSP90 complex of plants. Biochim. Biophys. Acta 1823, 689-697 (2012).
[62] Corigliano, M. G., et al.: Plant Hsp90 proteins interact with B-cells and stimulate their proliferation. PLoS One 6, e21231 (2011).
[63] Buriani, G., et al.: Heat-shock protein 70 from plant biofactories of recombinant antigens activate multiepitope-t argeted immune responses. Plant Biotechnol. J. 10, 363-371 (2012).
[64] Townsend, A. R., et al.: The epitopes of infl uenza nucleoprotein recognized by cytotoxic T lymphocytes can be defi ned with short synthetic peptides. Cell 44, 959-968 (1986).
[65] Oukka, M., et al.: Protection against lethal viral infec-tion by vaccination with non immunodominant peptides. J. Immunol. 157, 3039-3045 (1996).
[66] Doherty, P. C., Kelso, A.: Toward a broadly protective infl uenza vaccine. J. Clin. Invest. 118, 3273-3275 (2008).

（赵志荀，吴国华译）

第 36 章　用于构建重组疫苗的功能化纳米脂质体：以莱姆病作为实例

Jaroslav Turánek，Josef Mašek，Michal Křupka 和 Milan Raška①

摘要

脂质体（磷脂双层囊泡）代表了用于制备合成疫苗的近乎完美的载体系统，因为它们具有生物降解性及保护和转运不同理化性质分子（包括大小、亲水性、疏水性和电荷）的能力。脂质体载体可以通过侵入性（例如，im、sc、id）以及非侵入性（透皮和黏膜）途径施用。在过去的 15 年中，脂质体疫苗技术已经成熟，并且几种含有基于脂质体的佐剂的疫苗已被批准用于人类和兽医用途或已经达到临床评估的后期阶段。

鉴于人们对基于脂质体的疫苗越来越感兴趣，准确理解脂质体如何与免疫系统相互作用以及如何刺激免疫显得尤为重要。显然，脂质体疫苗的物理化学性质（抗原附着方法、脂质组成、脂质双层流动性、颗粒电荷和其他性质）对产生的免疫反应有着强烈的影响。在本章中，我们将讨论脂质体疫苗的以下几个方面问题，包括新型和新兴免疫调节剂组合的作用，以及将金属螯合纳米脂质体用于开发针对莱姆病的重组疫苗的应用。

36.1　引言

疫苗学作为一个科学领域正在经历着巨大的发展。先进的技术和免疫机制方面知识的快速增长，随时可以充分开发通过疫苗接种来预防和治愈疾病的潜力。尽管在 20 世纪 70 年代根除天花和在较大程度上根除脊髓灰质炎（医学史上两个重要里程碑）等方面取得了巨大成功，但仍面临新的挑战。迅速变化的生态系统和人类行为，人类和养殖动物种群的密度不断增加，高度的流动性导致病原体在受感染的人和动物中快速传播，第三世界的贫困和战争冲突以及许多其他因素促成新的以及一些旧传染病的发生和迅速传播。

影响全球健康的 3 种最严重的传染病是艾滋病、肺结核和疟疾。埃博拉病毒、SARS 冠状病毒及新流感病毒株，作为新型病毒病的病原体经常会被提及[1]。病原体基因组的快速测序促进了敏感分子诊断工具的发展，并增强了对未来疫苗的重组抗原靶点的识别和表达[2]。在癌症的免疫治疗这一特殊领域，抗癌疫苗可能是长期有效治疗癌症的有力武器。

疫苗开发的进展不仅与免疫学的新发现密切相关，而且与分子生物学和生物技术也密切相关。拉波利提出了新术语“反向疫苗学”，以描述一种复杂的基于基因组的疫苗设计方法[3]。与传统方法相比，传统方法需要费力的减毒或灭活病原体，或选择对诱导免疫反应很重要的个体成分，反向疫苗学提供了使用来自计算机分析的基因组信息的可能性，用于使用重组技术直接设计和产生保护性抗原。

---

① J. Turánek，Dr. Sc（✉）· J. Mašek，PharmDr（捷克兽医研究所药理及学免疫治疗系，捷克布尔诺，E-mail：turanek@ vri. cz）

M. Křupka，PhD · M. Raška，MD，PhD（帕拉基大学医学和牙科学院免疫学系，捷克奥洛莫克）

这种方法可以显著减少用在开发候选疫苗的抗原鉴定上的时间，并且能够系统地鉴定所有潜在抗原，即使是目前难以培养的病原体也可应用。当然，这种方法仅限于鉴定蛋白质或糖蛋白抗原，省略了诸如多糖和糖脂之类的重要疫苗组分。反向疫苗学的关键问题在于鉴定保护性抗原，这是该方法的主要障碍。然而，一旦确定了保护性抗原，科学家就能对这些抗原进行系统分类，并开发出对任何确定基因组序列的病原体有效的制剂。

亚单位疫苗具有优异的安全性，制造过程的污染风险最低[4,5]。当与适当的佐剂偶联时，它们还可以将免疫应答集中在保护性或高度保守的抗原决定簇上，这些抗原决定簇在自然感染期间或接种灭活或减毒病原体后可能不会引起强烈反应[6,7]。

从诱导适应性免疫反应的过程中常见的现象是，抗原本身不是一种刺激剂，换句话说，绝对纯的重组蛋白抗原和合成肽抗原的施用通常不会诱导特异性免疫应答。因此，需要通过共同施用适当的佐剂、生物相容性载体系统和用于由高度纯化的抗原组成的疫苗的施用装置进行有效的共刺激。这些微粒系统被认为能够有效地传递给抗原呈递细胞，并可能通过激活先天免疫来引发炎症[8-10]。

36.2 莱姆病

莱姆病是欧洲和美国最常见的人畜共患病。疾病可能发展为慢性形式，并导致神经或心血管系统、关节、皮肤或眼睛的损害。感染患者可能长期不能工作甚至永久丧失劳动能力。因此，疾病的预防和治疗成为医学研究长期的重点领域。

36.3 历史

20 世纪初，人们就了解了一些典型的莱姆病症状——慢性肢端萎缩性皮炎、游走性红斑性、淋巴细胞瘤和脑膜多神经根神经炎[11]。但是，最近有人发现地方性青少年的伴有游走性红斑的关节炎病的病原为伯氏疏螺旋体。[12,13]

后来研究表明，广义上的伯氏疏螺旋体是几个亲缘物种的复合体。目前，已区分出 12 个物种，并且不断发现新变种，因此已知疏螺旋体物种的数量可能不是最终数量。已知至少有 3 种物种（广义上的伯氏疏螺旋体、埃氏疏螺旋体和伽氏疏螺旋体）对人类具有致病性。在许多地理区域，每年报告的莱姆病病例数不断增加。这可能是由于疾病的实际传播或诊断方法的改进[14]。

36.4 病原

1907 年包柔氏旋体属与钩端螺旋体属、密螺旋体属同属于螺旋体科。在美国，病原体仅伯氏疏螺旋体一种。在欧洲，最常见的有两种：埃氏疏螺旋体和伽氏疏螺旋体。

疏螺旋体是典型的螺旋形细胞，长 10~30 μm，直径 0. 2~0. 3 μm。这种形状是由周质鞭毛引起的，鞭毛在结缔组织等黏性环境中具有很高的运动性。疏螺旋体缺乏坚硬的细胞壁，表面主要由脂蛋白组成，具有严格控制的表达模式，对适应外界条件至关重要[15]。

36.5 流行病学、临床疾病和治疗

这种疾病具有媒介特征。大多数时候，它经由 Ixodes 属感染扁虱传播给人类，但是在蚊子和其他吸血昆虫的中肠中也发现了疏螺旋体。噬血昆虫在疾病传播中的作用还不清楚[16,17]。大多数野生动物都能成为本病的储存宿主，如啮齿动物（无节肢动物属、节肢动物属、田鼠属、

大鼠等）、松鼠（松鼠属）、野兔属（兔属）和鸟（鸫属、雉属、金翅雀属、燕雀属等）。人类是本病的终末宿主，不能进一步传播感染[18]。

莱姆病可分为 3 个阶段——早期局部感染、早期传播感染和晚期持续感染，但并非每个个体感染后都会出现这 3 个阶段。似乎很大一部分感染都无临床症状，但抗伯氏疏螺旋体特异性抗体水平升高。第一阶段的特征是非特异性症状，包括发热、寒战、头痛、嗜睡和/或肌肉和关节疼痛，约 70%伴有细菌进入部位的红斑性偏头痛。如果宿主免疫系统或抗生素治疗不能消除感染，它可能会影响中枢神经系统（脑膜炎、神经根病变、第七颅神经麻痹）或心血管系统（房室传导阻滞、心包炎）。晚期持续性感染可能在感染后数月至数年内出现，通常影响神经系统、皮肤、关节，或心脏、眼睛或其他器官的频率较低[15]。

莱姆病的治疗主要依赖于抗生素。建议成人服用强力霉素，儿童建议使用 β-内酰胺类抗生素。对于晚期患者的治疗，推荐使用头孢菌素类或青霉素 G。长期使用抗生素治疗莱姆病的持续性感染效果有限[19]。

在发现疏螺旋体是莱姆病的致病因素后不久，人们便尝试研制莱姆病疫苗。首次实验证明了细菌全细胞裂解物的免疫原性。随后，鉴定出能被宿主免疫系统识别的特异性蛋白质[20]。

36.6　抗原和免疫反应

螺旋体表面抗原的变异性很强。在蜱宿主中，疏螺旋体表达外表面蛋白 A（OspA）和外表面蛋白 B（OspB）。编码两种蛋白质的基因位于一个操纵子内，这些蛋白质的表达依赖于管家西格玛因子 RpoD（σ70）。据推测，OspA 和 OspB 的功能是粘附在钩状肠上皮上。当蜱开始寄生在脊椎动物宿主上时，两种蛋白质都被下调。同时，诱导依赖于替代西格玛因子 RpoS（σ38）的蛋白质的表达，例如，（OspC）和 OspF 和核心蛋白聚糖结合蛋白 DbpA 和 DbpB。这些蛋白是将疏螺旋体传播到脊椎动物以及脊椎动物感染的初始阶段所必需的。在后期阶段，OspC 被下调并且诱导 VlsE 蛋白（表达的 Vmp 样序列）的表达。由于高 VlsE 变异性，使疏螺旋体逃避特定的体液反应。

对疏螺旋体的早期体液反应的特征为产生对 OspC、鞭毛蛋白（p39 和 p41）和 BmpA（p39）特异的 IgM 抗体。随后出现针对 p39、p41、p83/100、DbpA 和 VlsE 的 IgG 抗体。在某些情况下，可以在疾病的后期检测出抗 OspA 和 OspB 的抗体[15]。

第一代亚单位疫苗基于 OspA 抗原，其在体外培养莱姆病螺旋体期间高度表达。基于 OSPA 抗原的疫苗的临床前试验证明了其对动物模型的保护作用。OspA 抗原以重组蛋白或脂蛋白的形式产生，或者在大肠杆菌、鼠伤寒沙门氏菌或牛分枝杆菌的表面上表达。

现已开发出两种基于 OspA 的候选疫苗并进行了测试。基于在大肠杆菌中表达的纯化的重组 OspA 抗原的 ImuLyme 疫苗（Pasteur Merieux-Connaught）和含有来自 *B. burgdorferi* s. s 的纯化的重组 OspA 的 LYMErix（GlaxoSmithKline）。[21]。在两种疫苗中，蛋白质都被氢氧化铝吸附。在涉及10 000多人的临床试验中，发现 LYMErix 疫苗在 76%的成人和 100%仅有轻度或中度和短暂不良反应的儿童中能够产生保护性免疫[20]。基于这些结果，LYMErix 于 1998 年 12 月 21 日获得美国食品和药物管理局（FDA）的批准。

随后，数百名接种者报告了自身免疫副作用的进展。在一些患者权益团体的支持下，针对葛兰素史克公司提起了一系列集体诉讼，指控该疫苗已经引发了这些健康问题。美国食品和药物管理局和美国疾病控制中心（CDC）对这些说法进行了调查，他们发现疫苗和自身免疫投诉之间没有任何联系[22,23]。尽管缺乏证据，但在负面媒体报道和对疫苗副作用的担忧的背景下，2002 年 2 月葛兰素史克公司的销量暴跌，LYMErix 退出美国市场。

莱姆病疫苗研发缓慢的原因有很多，包括制药公司不愿意像 LYMErix 生产商一样受到同样的影响。另一些人则认为，药物制造商通过对症治疗莱姆病的症状、给予消炎药、止痛药和其他抗生素无法治疗患者疾病的药物，会更有利可图。持续供应狗用莱姆病疫苗导致人们质疑制药公司的动机，并且很多人出现心理上的疾病也归咎于莱姆病。

36.7 兽用莱姆病疫苗

与人用药物不同，一些获得许可的兽用莱姆病疫苗可用于宠物。其中 3 种是伯氏杆菌素（Merilym，Merial，德国；Galaxy Lyme，美国先灵葆雅；Lymevax，美国道奇堡）。基于来自 *B. garinii* 和 *B. afzelii* 的菌苗组合的欧洲疫苗是为狗生产的（Biocan，Bioveta，捷克共和国）[24]。还有基于重组 OspC 抗原的兽医亚单位疫苗可用（ProLyme，Intervet，美国；Recombitek Lyme，Merial，美国）。所有兽用疫苗均采用二次免疫方案，推荐每年使用增效剂。疫苗的安全性数据有限，有报道称不同种的疏螺旋体之间有轻微交叉反应[25]。因此，低交叉保护覆盖率是目前可用疫苗的一个问题。

第二代 OspA 亚单位疫苗是基于重组 OspA，其基因去除了潜在的交叉反应性 T 细胞表位，之前曾被怀疑诱导自身免疫[26]。开发广泛交叉保护的 rOspA 疫苗必须解决与序列变异有关的问题，因为至少存在 7 种 OspA 血清型。多价嵌合 OspA 蛋白是通过分子克隆结合几个 OspA 血清型的保护性表位而开发出来的[27]。然而，基于 OspA 抗原的疫苗需要重复接种疫苗以在接种者血清中保持高滴度的特异性抗体，以防止细菌从蜱转移到人宿主中。

36.8 铁蛋白

用于疫苗开发的新的有趣抗原是蜱铁蛋白 2，其负责维持铁稳态。铁是一种基本元素但具有潜在毒性；因此，在喂养期间，蜱必须应对血粉中大量铁供应的挑战。铁蛋白是储存铁的蛋白质，在这个过程中起着关键作用。结果表明，重组 FER2 疫苗能显著降低感染蓖子硬蜱的家兔和感染微小牛蜱和扇头蜱属软蛴扁角跳小蜂的蜱虫感染。这些结果有希望使 FER2 成为开发新型抗蜱疫苗的候选抗原[28]。尽管铁蛋白 2 不是源自疏螺旋体的抗原，但用该抗原接种可防止长期饲养的蜱将疏螺旋体以及其他蜱传播的病原体传播到宿主中。在这方面，铁蛋白 2 优于 OspA，并且将其包括在开发组合的多抗原疫苗中是有意义的。

36.9 OspC 抗原

如上所述，人们发现了一些新的抗原并将其作为新疫苗的可能靶点进行了研究。我们专注于 OspC 抗原，该抗原在开发广泛保护性莱姆病疫苗中受到了大量关注。OspC 是一种重要的毒力因子，对于在哺乳动物中建立早期感染至关重要[29]，并且该蛋白质的序列在感染期间不会发生突变[30]。OspC 是一种 22 kDa 免疫优势脂蛋白，通过 N-末端三棕榈酰-S-甘油基-半胱氨酸锚定在螺旋体外膜中[31]。

使 OspC 作为疫苗候选物的利用复杂化的因素是 OspC 变异的存在。分子系统发育分析揭示，OspC 序列形成至少 38 种不同的 OspC 类型。这种 OspC 变异主要通过遗传交换和重组而不是伴随免疫选择的超突变而产生[32]。由于这种序列变异，单一基于 OspC 的疫苗的保护范围很窄。解决 OspC 高变异性的潜在方法有 3 个：（1）制备含有几个 OspC 变异的多价疫苗[33]；（2）从与疾病相关的 OSPC 类型中产生由保护性表位组成的重组嵌合蛋白[34]；（3）联合两种或两种以上不同重组免疫原，如 OSPC、DBPA 和 FI-支气管连接蛋白（BBK32）[35]。

从生物技术的观点来看，全长 OspC 难以作为重组蛋白进行大量并高纯度表达[36]。去除脂质化信号显著增加了重组 OspC 蛋白的产率（28 mg/L 的细菌培养物）和纯度（93%）[37]。另一方面，脱脂 rOspC 具有非常低的免疫原性[38,39]，并且疫苗制剂的强佐剂性对于诱导特异性抗体应答是必需的，尤其是在 IgG2 亚类（在实验小鼠中）对于有效补体激活和调理作用非常重要。

为了遵守 OspC 疫苗的所有上述要求，我们开发了基于功能化金属螯合纳米脂质体和衍生自胞壁酰二肽的合成无热原佐剂的实验疫苗。该制剂在实验小鼠中诱导强烈的免疫应答，其优于氢氧化铝制剂。由于作为佐剂的多功能性和安全性，脂质体因此代表了开发临床上可接受的针对莱姆病疫苗的有希望的平台。本章的下一部分重点介绍脂质体疫苗。

36.10　脂质体作为生物相容性和多功能载体用于构建重组疫苗

在迄今为止开发的众多颗粒递送系统中，磷脂双层囊泡（脂质体）是最有前景的。Gregoriadis 和 Allison 在 1974 年首次报道了使用脂质体作为免疫佐剂[40,41]。自此，脂质体和相关的囊泡载体已被确立为用于诱导体液和细胞介导的广谱传染病和癌症免疫的强大系统[42]。

脂质体代表了最古老的纳米颗粒系统，其描述用于作为模型膜和医学的生物学研究中的应用。超过 44 年，脂质体已被证明是适用于从化妆品和皮肤病到医疗应用的给药系统，如感染治疗、抗癌治疗和兽医疫苗接种[43]。

基于脂质体的疫苗已经存在大约 30 年，并且已经开发出许多脂质体变体，其中一些具有明显的免疫刺激特性和吸引人的安全特性，从而使一些产品在市场上注册，而另一些则发展到临床试验的高级阶段。Epaxal ®（Crucell、甲型肝炎、福尔马林灭活甲型肝炎病毒吸附至病毒体）和 Inflexal ® V（Crucell、流感病毒、病毒体-重组流感病毒膜）是用于人类的市售脂质体疫苗的两个实例。Inflexal ® V 在 43 个国家获得许可，使用的剂量超过 6 000万[44,45]。

此外，目前至少有 8 种基于脂质体的佐剂系统被批准用于人类或正在进行临床评估[46]。

36.11　影响脂质体疫苗活性的理化和结构参数

脂质体代表了合成疫苗成分的几乎理想的载体/递送系统，因为它们具有生物降解性，并且能够同时保留/掺入各种必需的疫苗成分，即使这些成分具有完全不同的物理化学性质（不同的大小、疏水性、电荷等）也可应用。不同的合成疫苗组分可以包封在脂质体的水性腔内（如果是亲水的）或与脂质体双层结合（如果在性质上至少部分是疏水的）。此外，基本成分可以通过静电、共价或金属螯合作用附着在内部或外部小叶膜上。生物化学上大多数不同的疫苗组分是有效激活先天免疫细胞所需的佐剂，包括抗原呈递细胞（APC）（如单磷酰脂质 A、CpG 寡核苷酸、胞壁酰二肽和类似物）。此外，佐剂可与可溶性或膜蛋白等抗原结合，以引起强烈的特异性免疫反应。最后，脂质体可作为抗原靶向载体和抗原呈递细胞佐剂[47]。已经建立了脂质体制备的实验室和工业工艺，并且脂质体已经被美国食品和药物管理局批准用于生物医学应用。基于脂质体的重组疫苗进入人类和兽医疫苗市场的潜力很大[48]。

脂质体载体的有益作用在于其能够确保抗原的保护和稳定（包括恢复其自身构象）、通过被动或主动靶向增强 APC 的摄取、摄取后增强或控制抗原加工。

脂质体疫苗和脂质体免疫调节剂与免疫细胞的相互作用以及免疫系统固有和适应性分支之间的合作，如图 36.1 所示。

用于疫苗接种的脂质体纳米颗粒制剂可大致分为以下几类。

传统的由中性脂质和胆固醇组成的脂质体由于其最大的通用性而得到了广泛的研究——通过改变脂质组成或囊泡制备方法可以获得所需的配方参数。在脂质体疫苗研究的早期阶段，这些脂质体被用于蛋白质抗原的空间包埋。

这些经典脂质体在增强抗原免疫原性方面相对无效，因为磷脂不起辅助作用。这些脂质体的佐剂效应是基于抗原的多聚体呈现，其佐剂效应可通过脂质组成的变化、多种佐剂分子（如 MLP-A、MDP、CpG 寡核苷酸、海藻糖二元酸）的共掺以及抗原的附着来优化。

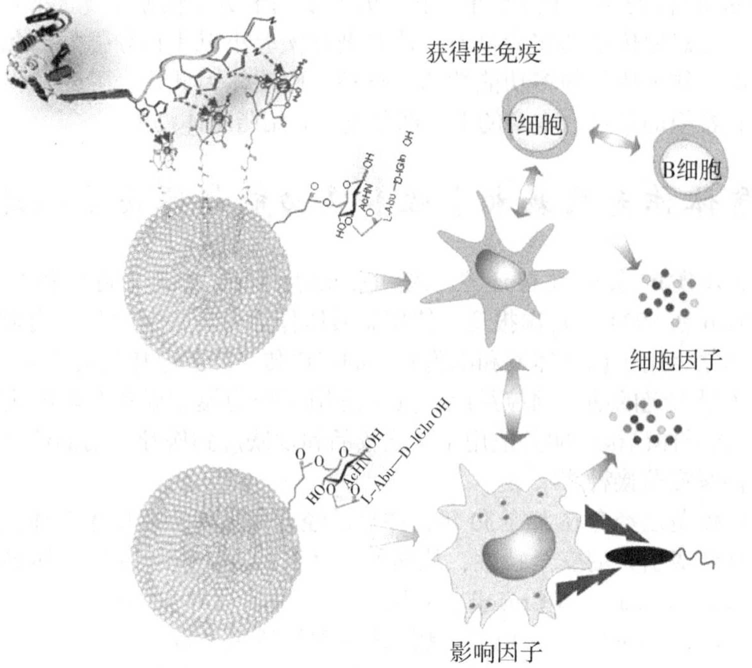

图 36.1　基于脂质体的免疫调节剂和蛋白脂质体疫苗与免疫细胞的相互作用

功能化的纳米脂质体促进抗原与免疫细胞（树突细胞、巨噬细胞和 B 细胞）的共递送。脂质体表面的高定向蛋白抗原能够与 B 细胞上的膜结合免疫球蛋白相互作用，然后将整个复杂的脂质体抗原内化。合成亲脂性胞壁酰二肽（norAbuMDP）的衍生物通过细胞内受体（如 NOD2 和 NALP-3）提供危险信号，以诱导先天性和适应性免疫应答。此外，暴露在脂质体表面的分子的糖肽部分形成免疫细胞识别的分子模式，并增强脂质体的内化（脂质体和免疫细胞不在规模内）。

转移体是具有皮肤渗透特性的超变形脂质体。由磷脂酰胆碱和胆酸盐组成（比例为9：2 mol/L）。胆酸盐的存在赋予这些囊泡高弹性，使它们能够通过角质层中的孔挤压。由于双层的高弹性，200~300 nm 大小的囊泡可以通过高度致密的角质层，并将其内容物输送到皮肤的深层。转移体被用作疫苗局部应用的载体系统[49]。据测试，阳离子转移体可用于针对乙型肝炎的 pDNA 疫苗的非侵入性局部疫苗接种[50]。

36.12　阳离子脂质体

通过静电相互作用在阳离子脂质体上的表面吸附是另一种将抗原（主要是蛋白质）附着在脂质体上的有效方法。Christensen、Perrie 和其他蛋白质抗原吸附到阳离子 DDA / TDB（二甲基二（十八烷基）溴化铵/海藻糖-6，6-二苯甲酸乙酯）脂质体上的研究表明，具有表面吸附抗原的制剂可以高度稳定，并在小鼠和雪貂体内引起强大的抗体和细胞介导的免疫反应[42,51-53]。将合成的免疫刺激分子（如亲脂性 TDB）掺入阳离子脂质体中能够显著增强其佐剂效应[52]。据观察，某些阳离子脂质，最初合成为转染试剂，在小鼠肺转移和纤维肉瘤模型中诱导血清 IFN-γ 和 TNF-α 的应答和抗肿瘤免疫[54]。DOTAP 是一个阳离子脂质的例子，作为脂质体基因传递系统的组成部分，多年来在树突状细胞中发挥了对映体特异性辅助作用[55]。

病毒体是由重组流感病毒膜辅以 PC[56] 制备的一类特殊蛋白脂质体。由于病毒体有明确的结构及制备方法，所以它的物理化学特征受限，但这些小泡从流感病毒内在递送特性（细胞结合、内化和胞质释放效率高）和免疫原性中受益巨大。

四醚脂质体是由从古细菌中分离的特殊脂质制备的脂质体，这些古细菌是极端微生物，并且它们的细胞膜含有对极端温度、酸碱度和盐浓度具有抗性的脂质和蛋白质。这些脂质对氧化反应不太敏感，并且古细菌的膜对胆汁酸盐的作用具有抵抗力，因此适用于肠内应用。四醚脂质体诱导的免疫反应与完全弗氏佐剂诱导的免疫反应相当[57]。

功能化脂质体通常代表各种脂质体结构，其表面被各种配体（生物素、低聚糖、肽）、具有反应性头基的脂质、pH或温度敏感聚合物、单克隆抗体或其片段、凝集素等修饰，调整其与目标细胞外或细胞内结构（如病原微生物、肿瘤细胞、免疫细胞、血衣）的选择性相互作用的功能。抗原靶向抗原呈递细胞对于提高亚单位重组疫苗的疗效具有重要意义。在本章中，我们将介绍一种新的脂质体疫苗的构建系统，该系统使用金属螯合脂质将HIS标记的重组抗原结合到脂质体表面上。

36.13 抗原与脂质体的结合

影响脂质体疫苗免疫原性的关键参数之一是抗原与制剂物理或化学结合的方法。最常见的结合方式包括共价脂质结合（囊泡形成前或囊泡形成后）、非共价表面附着（通过生物素、NTA-Ni-His或抗体-表位相互作用）、包封和表面吸附（图36.2）。许多关于小鼠中脂质体肽和蛋白质抗原性的早期研究将包封的抗原与缀合到预先形成的脂质体表面的抗原进行比较。Alving、Gregoriadis、Therien和其他人的早期研究证实，这两种方法通常可有效诱导抗体和T细胞对相关蛋白抗原（如白蛋白和破伤风类毒素）的反应[58-60]。

在某些情况下，共价抗原结合可诱导更高的抗体水平，这并不奇怪，因为B细胞受体可以识别脂质体表面上的完整抗原[61]。随着抗原的大小和复杂度的降低，其表面结合诱导抗体的效果变得更加明显，正如表面结合提供优于包囊的免疫应答的合成肽所发现的那样[61,62]。

36.14 金属螯合键和金属螯合脂质体

关于金属螯合脂质体在疫苗构建中的潜在应用，体外特别是体内稳定性问题具有重要意义。这一问题可分为两个领域：第一，脂质体本身的稳定性；第二，生物流体中的成分（如蛋白质和离子）对金属螯合键稳定性的影响。在OspC蛋白脂质体的实例中，凝胶渗透色谱数据表明OspC蛋白脂质体在色谱过程中具有良好的体外稳定性，在此过程中它们经历剪切应力和稀释。此外，在37℃的血清中孵育OspC蛋白脂质体后，我们证明了连接蛋白质与脂质体表面的金属螯合键的稳定性[37]。事实上，皮内注射后脂质体的体内命运与静脉注射后不同。首先，皮内注射后蛋白脂质体的稀释速度不太快，其次，由于组织液蛋白与蛋白脂质体在注射部位的浓度相对较高，因此组织液蛋白与蛋白脂质体的比例对蛋白脂质体更有利。在另一种应用途径（皮内）与肌肉组织或血管相比，皮内细胞外基质内组织液的流速甚至更低。这一事实经常被忽视。金属螯合键的稳定性可能也取决于特定蛋白质的性质。结果表明，具有结合到脂质体表面上的单链Fv片段（抗CD11c）的Ni-NTA3-DTDA脂质体（含有3种功能性螯合脂质）能够在体外和体内靶向树突细胞。Ni-NTA3-DTDA的应用可能赋予金属螯合键具有更高的体内稳定性[63]，但这种改善的稳定性不会影响更高的免疫原性[64]。通常，金属离子、金属螯合脂质的物理化学特性和它们的表面密度属于可以被优化以获得所需的体内稳定性，并因此具有强免疫应答的因子。

通过TEM、GPC、SDS PAGE和动态光散射法证实重组OspC（rOspC）与金属硫化脂质体的结合。通过rOspC脂质体与人血清的孵育，研究了模型生物流体中金属螯合键的稳定性。该研究显示，在37℃温育1小时后，超过60%的rOspC仍然与脂质体相关。因此，基于该数据，估计血清中rOspC蛋白脂质体的半衰期为至少1小时。脂质体上表面暴露的抗原的稳定性的增加可以

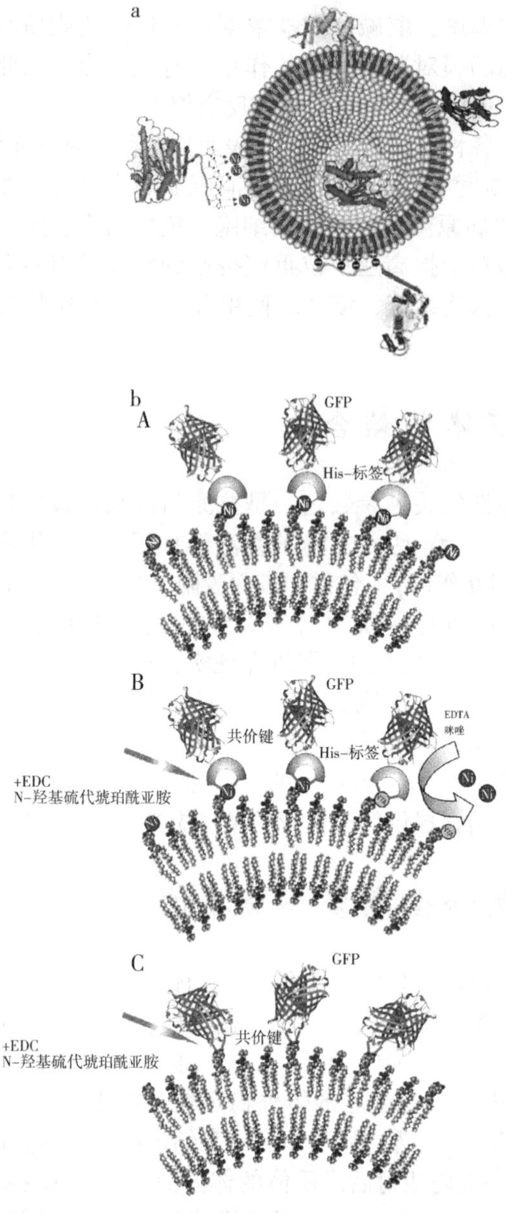

图 36.2

（a）蛋白质抗原与脂质体的结合。蛋白质抗原可以通过各种方式与脂质体结合。膜蛋白可在脂质体膜（黄色）中以其天然构象重组；亲脂蛋白（如棕榈酰化蛋白）可被脂质残基（红色）锚定在脂质膜中。可制备功能化脂质体以促进蛋白质与脂质体表面的共价或非共价结合。选择性金属螯合键适用于His-标签的重组蛋白（绿色），并且该非共价键可通过碳二亚胺缀合转化为共价键。以上所有例子都代表了蛋白质在脂质体表面的高度定向结合。蛋白质与脂质体表面的共价偶联（例如反应性马来酰亚胺磷脂基）导致结合蛋白的随机定向。静电相互作用代表蛋白质与脂质体结合的另一种方式。这种关联是非特异性的，并且在生物环境（粉红色）中相对不稳定。可溶性蛋白分子也可以在脂质体（蓝色）的水室中进行空间包埋。这对于肽抗原和蛋白质抗原尤其重要，它们直接传递到免疫细胞的细胞质中，以模拟病毒样途径进行抗原加工。（b）蛋白质与脂质体表面的结合。A. 定向 HIS 标签的 GFP（绿色荧光蛋白）与脂质体表面的非共价结合。B. 通过碳二亚胺化学将金属螯合键转化为共价酰胺键，保持绿色荧光蛋白的取向结合。通过 EDTA 除去 Ni^{+2} 离子作为金属去除剂。C. 无金属螯合键预定向碳二亚胺化学在脂质体表面结合的绿色荧光蛋白随机定向实例通过化学结合到适当官能化的外部脂质体表面来实现。关于构建疫苗的潜在应用，体外和特别是体内稳定性的问题非常重要。

只有少数参考文献报道了在超分子结构构建中作为疫苗载体实施的金属螯合键[65-69]。蛋白质抗原与脂质体结合的各种方法如图 36.2a 所示，各种显微镜方法揭示的真实结构如图 36.3 所示。

36.15 脂质体疫苗的佐剂

先天性和适应性免疫应答之间的相互作用对于产生抗原特异性免疫应答是至关重要的。先天免疫反应始于病原体侧的病原体相关分子模式（PAMPs）与参与先天免疫的宿主细胞（例如树突细胞）上的模式识别受体（PRR），如 Toll 样受体（TLRs）的相互作用。通常用于评估各种新佐剂的主要功能标准涉及其刺激先天免疫细胞的能力。这将包括参与其他 PRRs 和与其相关的共受体和细胞内衔接子信号传导蛋白。佐剂显影剂利用 PAMPs 及其衍生物来利用先天免疫的力量，引导免疫反应向所需的方向发展。

根据鉴定出的几种 TLRs 和 PAMPs，将各种 PAMP 激动剂作为佐剂进行试验。TLR-PAMP 特异性相互作用的实例包括与 TLR9 和脂糖相互作用的细菌或病毒非甲基化免疫刺激性 CpG 寡核苷酸及其与 TRL4 相互作用的组分单磷酰脂质 A（MPLA）。这两种类型的佐剂处于临床试验的后期测试阶段，一些已经许可的疫苗含有脂质体形式的 MPLA[70,71]。

适用于构建基于脂质体的疫苗的新型亲脂性佐剂（参见图 36.4）是分枝杆菌因子海藻糖-6，6-二霉菌酸酯（TDM），其合成类似物海藻糖-6，6-二山嵛酸酯（TDB）是有效的佐剂，用于在耐受性抗原呈递细胞中激活 SykCard9 信号传导的 Th1 / Th17 疫苗接种[72]。

最后一组有效的合成或半合成佐剂衍生自胞壁酰二肽（MDP）。

胞壁酰二肽和其他鼠肽衍生自肽聚糖（PGN）代表的非常特异的 PAMPs 组。革兰氏阳性和革兰氏阴性细菌都含有 PGN，其由许多通过寡肽交联的聚糖链组成。这些聚糖链由交替的 N-乙酰氨基葡萄糖（GlcNAc）和 N-乙酰壁酸（MurNAc）与壁酸偶联的氨基酸组成。Muropeptides 是 PGN 的分解产物，至少带有 MurNAc 部分和 1 个氨基酸[73]。其中一种突出的 Muropeptide 是胞壁酰二肽（MDP），自 20 世纪 70 年代以来已为人所知，我们将更详细地讨论这个化合物。

最近，已发现 MDP 识别和随后刺激宿主免疫系统的分子基础。骨髓免疫细胞（单核细胞、粒细胞、中性粒细胞和 DC）具有两种类型的 MDP / MDP 类似物的细胞内受体，即 NOD2 和 cryopyrin（炎性体-NALP-3 复合物）[74-76]。这两种受体识别 MDP / MDP 类似物对细菌细胞壁肽聚糖的最小识别基序[77]。另一个最近报道的 MDP 传感器是 cryopyrin（也称 CIAS1 和 NALP3），它是 NOD-LRR 家族的成员[74]。Cryopyrin 是炎性体复合物的一部分，负责将胱天蛋白酶-1 加工成其活性形式。Caspase-1 切割白细胞介素 IL-1β 和 IL-18 的前体，从而激活这些促炎细胞因子，并促进其分泌。已知 IL-1β 是由 MDP 诱导的强内源性热原。我们发现 norAbu-MDP 和 norAbu-GMDP 类似物即使在高浓度下也不会产生高温，远高于用于疫苗接种的浓度[15,65,66]。

NOD2 在树突细胞中的表达对于 MDP 类似物作为佐剂的应用至关重要。纳米颗粒样脂质体能够与重组抗原直接共同传递危险信号（例如，MDP），从而诱导免疫反应而不是免疫耐受性。这对于弱重组抗原或肽抗原尤其重要。显然，DCs 对 MDP 的识别对于 MDP 类似物作为佐剂的应用是至关重要的。在细胞内，MDP/MDP 类似物触发细胞内信号级联，最终激活炎症介质，如核转录因子 NF-κB 的转录激活。脂质体可能在 MDP 及其类似物从细胞外环境进入胞质溶胶的途径中起到有效载体的作用。在那里，它们触发细胞内信号级联，最终在炎性介质的转录激活，如核转录因子 NF-κB 途径。自 MDP 的发现和首次合成以来，已经设计、合成和测试了大约 1 000 种 MDP 衍生物，以开发出一种适合免疫治疗应用的药物，该药物不会产生 MDP 的副作用。MDP 的主要副作用是发热、僵硬、头痛、流感样症状、高血压等。只有几种制剂达到临床试验

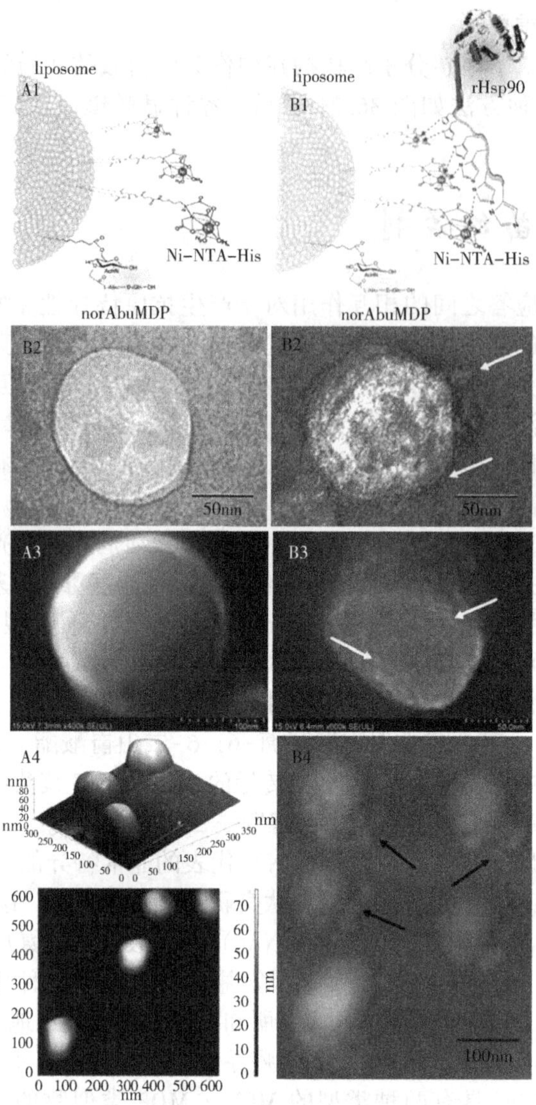

图 36.3 金属螯合脂质体和蛋白脂质体的的结构，通过 TEM、SEM 和 AFM 显示

来自白色念珠菌的重组热休克蛋白用作说明性实例。A1：具有金属螯合脂质和掺入的脂质化 norAbuMDP 分子的金属螯合脂质体的示意图。A2：具有金属螯合脂质和掺入的脂质化 norAbuMDP 分子的金属螯合脂质体的 TEM 照片。A3：具有金属螯合脂质并掺入脂质化 norAbuMDP 分子的金属螯合脂质体的 SEM 照片。A4：具有金属螯合脂质并掺入脂质化 norAbuMDP 分子的金属螯合脂质体的 AFM 照片。B1：具有通过金属螯合键结合的 rHsp90 的金属螯合脂质体（A1）的示意图。B2：具有通过金属螯合键结合的 rHsp90 的金属螯合脂质体的 TEM 照片。B3：具有通过金属螯合键结合的 rHsp90 的金属螯合脂质体的 SEM 照片。B4：具有通过金属螯合键结合的 rHsp90 的金属螯合脂质体的 AFM 照片

阶段，只有米非那肽被批准用于治疗骨肉瘤[77]。

参考文献

[1] Wack, A., Rappuoli, R.: Vaccinology at the beginning of the 21st century. Curr. Opin. Immunol. 17, 411-

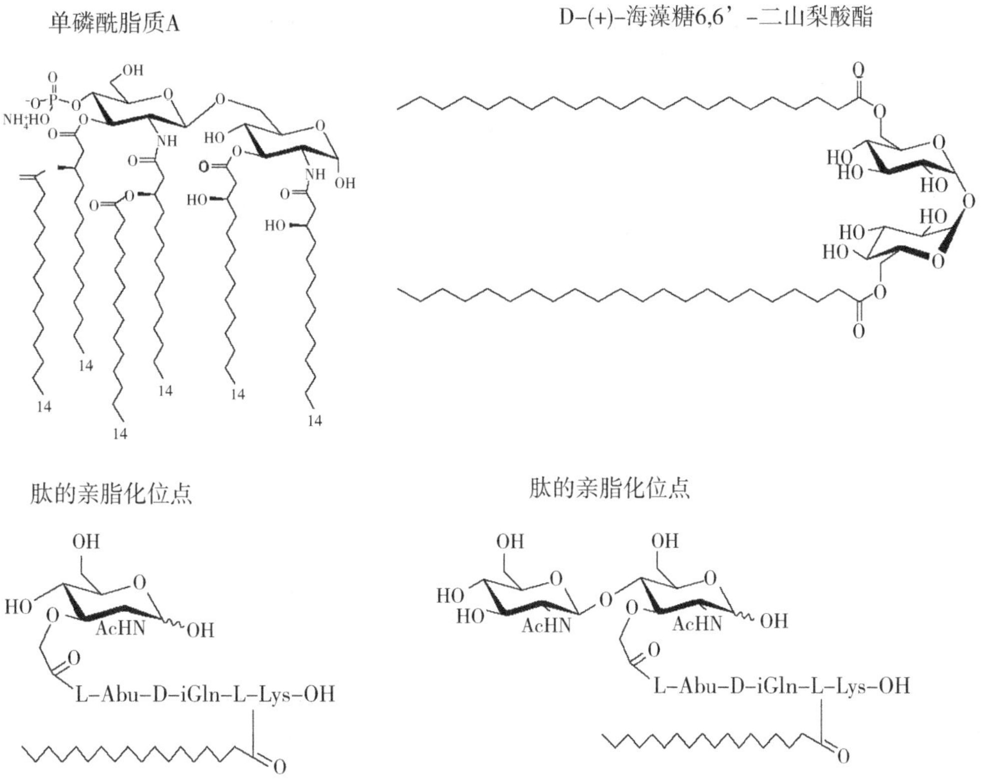

图 36.4 脂质体疫苗的新型亲油性佐剂

418 (2005).

[2] Stadler, K., et al.: SARS-beginning to understand a new virus. Nat. Rev. Microbiol. 1, 209-218 (2003).

[3] Rappuoli, R.: Reverse vaccinology. Curr. Opin. Microbiol. 3, 445-450 (2000).

[4] Zanetti, A. R., Van Damme, P., Shouval, D.: The global impact of vaccination against hepatitis B: a historical overview. Vaccine 26, 6266-6273 (2008).

[5] Munoz, N., et al.: Safety, immunogenicity, and effi-cacy of quadrivalent human papillomavirus (types 6, 11, 16, 18) recombinant vaccine in women aged 24-45 years: a randomised, double-blind trial. Lancet 373, 1949-1957 (2009).

[6] O'Hagan, D. T., Valiante, N. M.: Recent advances in the discovery and delivery of vaccine adjuvants. Nat. Rev. Drug Discov. 2, 727-735 (2003).

[7] Perrie, Y., Mohammed, A. R., Kirby, D. J., McNeil, S. E., Bramwell, V. W.: Vaccine adjuvant systems: enhancing the effi cacy of sub-unit protein antigens. Int. J. Pharm. 364, 272-280 (2008).

[8] Storni, T., Kundig, T. M., Senti, G., Johansen, P.: Immunity in response to particulate antigen-delivery systems. Adv. Drug Deliv. Rev. 57, 333-355 (2005).

[9] Harris, J., Sharp, F. A., Lavelle, E. C.: The role of infl ammasomes in the immunostimulatory effects of particulate vaccine adjuvants. Eur. J. Immunol. 40, 634-638 (2010).

[10] Marrack, P., McKee, A. S., Munks, M. W.: Towards an understanding of the adjuvant action of aluminium. Nat. Rev. Immunol. 9, 287-293 (2009).

[11] Stanek, G., Strle, F., Gray, J., Wormser, G. P.: History and characteristics of Lyme borreliosis. In: Gray, J. S., Kahl, O., Lane, R. S., Stanek, G. (eds.) Lyme Borreliosis-Biology, Epidemiology and Control, vol. 1, Ch. 1, pp. 1-28. CABI Publishing, Oxford (2002).

[12] Steere, A. C., et al.: Erythema chronicum migrans and Lyme arthritis. The enlarging clinical spectrum. Ann. Intern. Med. 86, 685-698 (1977).

[13] Burgdorfer, W., et al.: Lyme disease-a tick-borne spi-rochetosis? Science 216, 1317-1319 (1982).

[14] Krupka, M., et al.: Biological aspects of Lyme disease spirochetes: Unique bacteria of the Borrelia burgdorferi species group. Biomed. Pap. Med. Fac. Univ. Palacky Olomouc Czech Repub. 151, 175 - 186 (2007).

[15] Krupka, M., Zachova, K., Weigl, E., Raska, M.: Prevention of lyme disease: promising research or sisyphean task? Arch. Immunol. Ther. Exp. (Warsz.) 59, 261-275 (2011).

[16] Zakovska, A., Nejedla, P., Holikova, A., Dendis, M.: Positive fi ndings of Borrelia burgdorferi in Culex (Culex) pipiens pipiens larvae in the surrounding of Brno city determined by the PCR method. Ann. Agric. Environ. Med. 9, 257-259 (2002).

[17] Nejedla, P., Norek, A., Vostal, K., Zakovska, A.: What is the percentage of pathogenic borreliae in spirochaetal fi ndings of mosquito larvae? Ann. Agric. Environ. Med. 16, 273-276 (2009).

[18] Gern, L., Humair, P. F.: Ecology of Borrelia burgdorferi sensu lato in Europe. In: Gray, J. S., Kahl, O., Lane, R. S., Stanek, G. (eds.) Lyme Borreliosis-Biology, Epidemiology and Control, vol. 1, Ch. 1, pp 149-174. CABI Publishing, Oxford (2002).

[19] Klempner, M. S., et al.: Two controlled trials of antibiotic treatment in patients with persistent symptoms and a history of Lyme disease. N. Engl. J. Med. 345, 85-92 (2001).

[20] Poland, G. A., Jacobson, R. M.: The prevention of Lyme disease with vaccine. Vaccine 19, 2303-2308 (2001).

[21] Sigal, L. H., et al.: A vaccine consisting of recombi-nant Borrelia burgdorferi outer-surface protein A to prevent Lyme disease. Recombinant Outer-Surface Protein A Lyme Disease Vaccine Study Consortium. N. Engl. J. Med. 339, 216-222 (1998).

[22] Abbott, A.: Lyme disease: uphill struggle. Nature 439, 524-525 (2006).

[23] Nigrovic, L. E., Thompson, K. M.: The Lyme vaccine: a cautionary tale. Epidemiol. Infect. 135, 1-8 (2007).

[24] Tuhackova, J., et al.: Testing of the Biocan B inj. ad us. vet. vaccine and development of the new recombinant vaccine against canine borreliosis. Biomed. Pap. Med. Fac. Univ. Palacky Olomouc Czech Repub. 149, 297-302 (2005).

[25] Topfer, K. H., Straubinger, R. K.: Characterization of the humoral immune response in dogs after vaccination against the Lyme borreliosis agent A study with fi ve commercial vaccines using two different vaccination schedules. Vaccine 25, 314-326 (2007).

[26] Trollmo, C., Meyer, A. L., Steere, A. C., Hafl er, D. A., Huber, B. T.: Molecular mimicry in Lyme arthritis demonstrated at the single cell level: LFA-1 alpha L is a partial agonist for outer surface protein A-reactive T cells. J. Immunol. 166, 5286-5291 (2001).

[27] Gern, L., Hu, C. M., Voet, P., Hauser, P., Lobet, Y.: Immunization with a polyvalent OspA vaccine protects mice against Ixodes ricinus tick bites infected by Borrelia burgdorferi ss, Borrelia garinii and Borrelia afzelii. Vaccine 15, 1551-1557 (1997).

[28] Hajdusek, O., et al.: Characterization of ferritin 2 for the control of tick infestations. Vaccine 28, 2993-2998 (2010).

[29] Stewart, P. E., et al.: Delineating the requirement for the Borrelia burgdorferi virulence factor OspC in the mammalian host. Infect. Immun. 74, 3547-3553 (2006).

[30] Hodzic, E., Feng, S., Barthold, S. W.: Stability of Borrelia burgdorferi outer surface protein Cunder immune selection pressure. J. Infect. Dis. 181, 750-753 (2000).

[31] Brooks, C. S., Vuppala, S. R., Jett, A. M., Akins, D. R.: Identifi cation of Borrelia burgdorferi outer surface proteins. Infect. Immun. 74, 296-304 (2006).

[32] Earnhart, C. G., Marconi, R. T.: OspC phylogenetic analyses support the feasibility of a broadly protective polyvalent chimeric Lyme disease vaccine. Clin. Vaccine Immunol. 14, 628-634 (2007).

[33] Earnhart, C. G., Marconi, R. T.: An octavalent lyme disease vaccine induces antibodies that recognize all incorporated OspC type-specifi c sequences. Hum. Vaccin. 3, 281-289 (2007).

[34] Buckles, E. L., Earnhart, C. G., Marconi, R. T.: Analysis of antibody response in humans to the type A OspC loop 5 domain and assessment of the potential utility of the loop 5 epitope in Lyme disease vaccine devel-

opment. Clin. Vaccine Immunol. 13, 1162-1165 (2006).

[35] Brown, E. L., Kim, J. H., Reisenbichler, E. S., Hook, M.: Multicomponent Lyme vaccine: three is not a crowd. Vaccine 23, 3687-3696 (2005).

[36] Krupka, M., et al.: Isolation and purifi cation of recom-binant outer surface protein C (rOspC) of Borrelia burgdorferi sensu lato. Biomed. Pap. Med. Fac. Univ. Palacky Olomouc Czech Repub. 149, 261 - 264 (2005).

[37] Krupka, M., et al.: Enhancement of immune response towards non - lipidized Borrelia burgdorferi recombinant OspC antigen by binding onto the surface of metallochelating nanoliposomes with entrapped lipophilic derivatives of norAbuMDP. J. Control. Release 160, 374-381 (2012).

[38] Weis, J. J., Ma, Y., Erdile, L. F.: Biological activities of native and recombinant Borrelia burgdorferi outer surface protein A: dependence on lipid modifi cation. Infect. Immun. 62, 4632-4636 (1994).

[39] Gilmore Jr., R. D., et al.: Inability of outer - surface pro - tein C (OspC) - primed mice to elicit a protective anamnestic immune response to a tick-transmitted challenge of Borrelia burgdorferi. J. Med. Microbiol. 52, 551-556 (2003).

[40] Allison, A. G., Gregoriadis, G.: Liposomes as immu-nological adjuvants. Nature 252, 252 (1974).

[41] Gregoriadis, G., Allison, A. C.: Entrapment of proteins in liposomes prevents allergic reactions in pre-i mmunised mice. FEBS Lett. 45, 71-74 (1974).

[42] Henriksen - Lacey, M., Korsholm, K. S., Andersen, P., Perrie, Y., Christensen, D.: Liposomal vaccine delivery systems. Expert Opin. Drug Deliv. 8, 505-519 (2011).

[43] Gregoriadis, G.: Engineering liposomes for drug delivery: progress and problems. Trends Biotechnol. 13, 527-537 (1995).

[44] Bovier, P. A.: Epaxal: a virosomal vaccine to prevent hepatitis A infection. Expert Rev. Vaccines 7, 1141-1150 (2008).

[45] Herzog, C., et al.: Eleven years of Infl exal V-a viro-somal adjuvanted infl uenza vaccine. Vaccine 27, 4381-4387 (2009).

[46] Watson, D. S., Endsley, A. N., Huang, L.: Design con-siderations for liposomal vaccines: infl uence of formulation parameters on antibody and cell-mediated immune responses to liposome associated antigens. Vaccine 30, 2256-2272 (2012).

[47] Altin, J. G., Parish, C. R.: Liposomal vaccines-target-ing the delivery of antigen. Methods 40, 39-52 (2006).

[48] Adu-Bobie, J., Capecchi, B., Serruto, D., Rappuoli, R., Pizza, M.: Two years into reverse vaccinology. Vaccine 21, 605-610 (2003).

[49] Gupta, P. N., et al.: Non-invasive vaccine delivery in transfersomes, niosomes and liposomes: a comparative study. Int. J. Pharm. 293, 73-82 (2005).

[50] Mahor, S., et al.: Cationic transfersomes based topical genetic vaccine against hepatitis B. Int. J. Pharm. 340, 13-19 (2007).

[51] Henriksen-Lacey, M., et al.: Comparison of the depot effect and immunogenicity of liposomes based on dimethyldioctadecylammonium (DDA), 3beta - [N (N', N' - Dimethylaminoethane) carbomyl] cholesterol (DC - Chol), and 1, 2 - Dioleoyl - 3 - trimethylammonium propane (DOTAP): prolonged liposome retention mediates stronger Th1 responses. Mol. Pharm. 8, 153-161 (2011).

[52] Henriksen - Lacey, M., Devitt, A., Perrie, Y.: The vesi - cle size of DDA: TDB liposomal adjuvants plays a role in the cell-mediated immune response but has no signifi cant effect on antibody production. J. Control. Release 154, 131-137 (2011).

[53] Henriksen-Lacey, M., et al.: Liposomal cationic charge and antigen adsorption are important properties for the effi cient deposition of antigen at the injection site and ability of the vaccine to induce a CMI response. J. Control. Release 145, 102-108 (2010).

[54] Whitmore, M., Li, S., Huang, L.: LPD lipopolyplex initiates a potent cytokine response and inhibits tumor growth. Gene Ther. 6, 1867-1875 (1999).

[55] Vasievich, E. A., Chen, W., Huang, L.: Enantiospecifi c adjuvant activity of cationic lipid DOTAP in

cancer vaccine. Cancer Immunol. Immunother. 60, 629-638 (2011).

[56] Daemen, T., et al.: Virosomes for antigen and DNA delivery. Adv. Drug Deliv. Rev. 57, 451-463 (2005).

[57] Conlan, J. W., Krishnan, L., Willick, G. E., Patel, G. B., Sprott, G. D.: Immunization of mice with lipopeptide antigens encapsulated in novel liposomes prepared from the polar lipids of various Archaeobacteria elicits rapid and prolonged specifi c protective immunity against infection with the facultative intracellular pathogen. Listeria monocytogenes. Vaccine 19, 3509-3517 (2001).

[58] Davis, D., Gregoriadis, G.: Liposomes as adjuvants with immunopurifi ed tetanus toxoid: infl uence of liposomal characteristics. Immunology 61, 229-234 (1987).

[59] Shahum, E., Therien, H. M.: Liposomal adjuvanticity: effect of encapsulation and surface-linkage on antibody production and proliferative response. Int. J. Immunopharmacol. 17, 9-20 (1995).

[60] Shahum, E., Therien, H. M.: Immunopotentiation of the humoral response by liposomes: encapsulation versus covalent linkage. Immunology 65, 315-317 (1988).

[61] White, W. I., et al.: Antibody and cytotoxic T-lymphocyte responses to a single liposome-associated peptide antigen. Vaccine13, 1111-1122 (1995).

[62] Frisch, B., Muller, S., Briand, J. P., Van Regenmortel, M. H., Schuber, F.: Parameters affecting the immunogenicity of a liposome-associated synthetic hexapeptide antigen. Eur. J. Immunol. 21, 185-193 (1991).

[63] van Broekhoven, C. L., Parish, C. R., Demangel, C., Britton, W. J., Altin, J. G.: Targeting dendritic cells with antigen-containing liposomes: a highly effective procedure for induction of antitumor immunity and for tumor immunotherapy. Cancer Res. 64, 4357-4365 (2004).

[64] Watson, D. S., Platt, V. M., Cao, L., Venditto, V. J., Szoka Jr., F. C.: Antibody response to polyhistidine-tagged peptide and protein antigens attached to liposomes via lipid-linked nitrilotriacetic acid in mice. Clin. Vaccine Immunol. 18, 289-297 (2011).

[65] Masek, J., et al.: Metallochelating liposomes with associated lipophilised norAbuMDP as biocompatible platform for construction of vaccines with recombinant His-tagged antigens: preparation, structural study and immune response towards rHsp90. J. Control. Release151, 193-201 (2011).

[66] Masek, J., et al.: Immobilization of histidine-tagged proteins on monodisperse metallochelation liposomes: preparation and study of their structure. Anal. Biochem. 408, 95-104 (2011).

[67] Chikh, G. G., Li, W. M., Schutze-Redelmeier, M. P., Meunier, J. C., Bally, M. B.: Attaching histidine-tagged peptides and proteins to lipid-based carriers through use of metal-ion-chelating lipids. Biochim. Biophys. Acta 1567, 204-212 (2002).

[68] Malliaros, J., et al.: Association of antigens to ISCOMATRIX adjuvant using metal chelation leads to improved CTL responses. Vaccine 22, 3968-3975 (2004).

[69] Patel, J. D., O'Carra, R., Jones, J., Woodward, J. G., Mumper, R. J.: Preparation and characterization of nickel nanoparticles for binding to his-tag proteins and antigens. Pharm. Res. 24, 343-352 (2007).

[70] Alving, C. R., Rao, M., Steers, N. J., Matyas, G. R., Mayorov, A. V.: Liposomes containing lipid A: an effective, safe, generic adjuvant system for synthetic vaccines. Expert Rev. Vaccines 11, 733-744 (2012).

[71] Kim, D., Kwon, H. J., Lee, Y.: Activation of Toll-like receptor 9 and production of epitope specifi c antibody by liposome-encapsulated CpG-DNA. BMB Rep. 44, 607-612 (2011).

[72] Schoenen, H., et al.: Cutting edge: Mincle is essential for recognition and adjuvanticity of the mycobacterial cord factor and its synthetic analog trehalose-dibehenate. J. Immunol. 184, 2756-2760 (2010).

[73] Traub, S., von Aulock, S., Hartung, T., Hermann, C.: MDP and other muropeptides-direct and synergistic effects on the immune system. J. Endotoxin Res. 12, 69-85 (2006).

[74] Agostini, L., et al.: NALP3 forms an IL-1beta-processing infl ammasome with increased activity in Muckle-Wells autoinfl ammatory disorder. Immunity 20, 319-325 (2004).

[75] Girardin, S. E., Philpott, D. J.: Mini-review: the role of peptidoglycan recognition in innate immunity.

Eur. J. Immunol. 34，1777-1782（2004）.
[76] McDonald，C.，Inohara，N.，Nunez，G.：Peptidog lycan signaling in innate immunity and infl am matory disease. J. Biol. Chem. 280，20177-20180（2005）.
[77] Ando，K.，Mori，K.，Corradini，N.，Redini，F.，Heymann，D.：Mifamurtide for the treatment of non-metastatic osteosarcoma. Expert Opin. Pharmacother. 12，285-292（2011）.

（吴国华译）

第 37 章　新兴纳米技术在肺部疫苗递送中的应用

Amit K. Goyal，GoutamRath 和 Basant Malik①

摘要

肺免疫是近年来研究的一种替代肠外免疫的方法。通过肺部途径注射疫苗可以诱导全身和局部黏膜免疫反应。在巨噬细胞（间质和肺泡）的呼吸上皮内层和中枢中大量的树突状细胞（DC）在诱导强免疫反应中起着重要作用。有几个因素限制了肺部免疫的有效性，包括肺泡区抗原沉积不良、肺和外周气道有上皮屏障导致吸收率低、肺中上部黏膜中的纤毛可自动排出异物，这些可迅速清除中央呼吸道的抗原或颗粒。在过去的几年中，许多高效的新型纳米载体已经被开发出来，用于通过肺部途径安全有效地疫苗接种。本章概述了疫苗肺部给药的各个方面。

37.1　引言

随着抗生素、疫苗接种计划、卫生条件的改善以及公众对健康的认识的提高，人们的健康情况日益发生变化。现在公众对健康的意识越来越高，世界各地由传染病引起的死亡率正在显著降低。在发达国家，针对天花、白喉、百日咳、破伤风、脊髓灰质炎、麻疹、腮腺炎、风疹和乙型流感等多种传染病的广泛免疫计划已显著降低了这些传染病的发病率[1]。然而，惊人数量的致命感染仍然在世界上严重流行，一些突变株甚至对新一代抗生素有耐药性，这是科学社会面临的重大挑战。这些疾病通常没有有效的疫苗，并且不能用抗生素治疗。

大多数这样的疾病沿着黏膜屏障本身的路径侵入身体[1,2]。许多感染性微生物，包括霍乱弧菌、肠毒性和肠致病性大肠杆菌、沙门氏菌和各种志贺氏菌菌株，以及涉及性传播疾病的病毒，如艾滋病病毒，破坏黏膜屏障，感染我们的身体。这种情况在发展中国家或公众不太关心卫生的地区更为严重[3]。

为了解决这些黏膜缺损，强大的黏膜屏障对于在初始阶段预防问题至关重要。这是对成功黏膜疫苗开发的巨大研究的强大背景。但不幸的是，只有少数黏膜疫苗处于商业水平；然而，其中一些疫苗正在创造奇迹，例如脊髓灰质炎疫苗。在本章中，我们将详细讨论黏膜免疫的各个部位，以及肺部免疫的意义。此外，我们还将讨论纳米技术在通过肺部途径进行靶向抗原输送以获得最佳免疫反应方面的作用。

37.2　黏膜免疫反应

所有黏膜室是相互连接的，这被称为共同的黏膜免疫系统（CIMS）（图 37.1）。因此，一个黏膜部位的免疫也保护其他黏膜室。每个黏膜室都有一些共同的特征，可以被细分为两个一般的室，通常被称为感应和效应部位，它们在免疫中起着特殊的作用。免疫诱导位点是抗原取样

---

① A. K. Goyal（✉）· G. Rath · B. Malik（加尔克兰 ISF 药剂学院药剂系，印度旁遮普省莫加，E-mail：amitkumargoyal1979@ yahoo. com）

激活免疫细胞的主要位点，包括有组织的黏膜相关淋巴组织（MALT）和局部黏膜引流淋巴结（LNs）（图 37.1）。

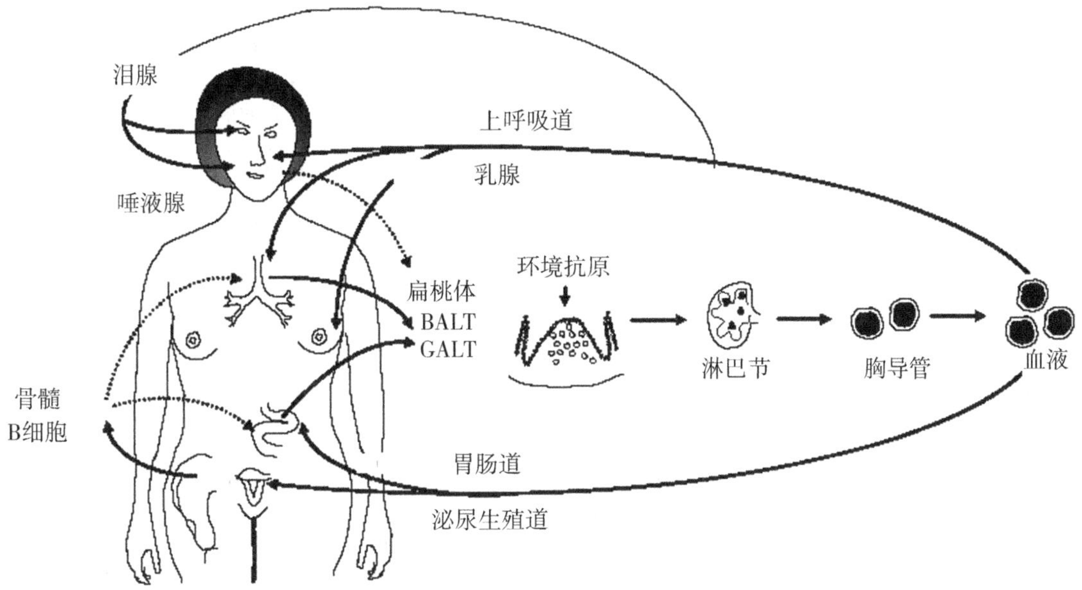

图 37.1　常见黏膜免疫系统的部位

效应部位是免疫系统的抗体和细胞在人体防御系统中实际发挥其功能的场所。效应部位包括各种黏膜相关淋巴组织的固有层（LP）、外分泌腺基质、表面表皮膜-以及小肠中的 Peyer's 斑（PPs）（表 37.1）[4-6]。此外，MALT 结构富含 T 细胞区、B 细胞滤泡和各种抗原呈递细胞，包括 DCs 和巨噬细胞，在这些地区不断循环。所有 MALT 结构都被一个特征性的滤泡相关上皮（FAE）所覆盖，该上皮（FAE）含有微血管（M）细胞，这些微血管（M）细胞积极参与直接从黏膜表面摄取外源性抗原[7]。这些特化的薄上皮细胞也有效地参与任何颗粒物质向免疫诱导位点的转运[8,9]。我们之前的文章对此进行了深入解释，为什么针对这些特定的细胞即 M 细胞和树突状细胞[2,9]进行靶向抗原传递进行高级的研究。

表 37.1　MALT 的各个部分

人体解剖系统	诱导位点	效应位点	淋巴结构
呼吸道	鼻咽相关淋巴组织（NALT）	鼻咽黏膜	咽淋巴环 腺样体（咽扁桃体） 腭扁桃体 舌扁桃体 咽鼓管扁桃体
	支气管相关淋巴组织（BALT）	支气管黏膜 下呼吸道	派尔集合淋巴结 淋巴集结 分离的淋巴滤泡
	喉相关淋巴组织（LALT）	喉	喉扁桃体 淋巴滤泡 具有生发中心的淋巴滤泡

（续表）

人体解剖系统	诱导位点	效应位点	淋巴结构
消化系统	唾液管相关淋巴组织（DALT）	唾液腺	淋巴滤泡
	肠道相关淋巴组织（GALT）	胃肠黏液	派尔集合淋巴结 淋巴腺复合物 分离的淋巴滤泡 隐窝结节 囊尾
皮肤	皮肤相关淋巴组织（SALT）	角蛋白细胞 朗格罕细胞	皮肤营养性T细胞 淋巴内皮细胞
眼部系统	结膜相关淋巴组织（CALT）	结膜	淋巴上皮 淋巴滤泡与B细胞区和T细胞区 邻近血管内皮增厚 淋巴管
	泪管相关淋巴组织（LDALT）	眼组织	淋巴滤泡
生殖系统	外阴阴道相关淋巴组织（VALT）	泌尿生殖道	淋巴滤泡
排泄系统	直肠淋巴上皮组织	胃肠黏液	淋巴滤泡

37.3 疫苗递呈的肺通路

如今，大多数商用疫苗都是通过注射给药的，这种方法伴随疼痛，注射可能引起感染，还需要受过训练的人操作，并且引起的黏膜免疫反应低，因此人们通常不愿接受注射给药。为了避免这些限制，在过去几年中已经广泛探索了非侵入性免疫策略，但成功率有限。近年来，黏膜免疫受到了科学界的广泛关注。这可能是由于许多原因造成的，例如对黏膜疫苗的科学优势的认识、有关黏膜参与各种疾病发病机制的知识的提高、易于扩展的生产以及轻松疫苗递送。在各种途径中，由于呼吸道的某些特殊特点，肺部免疫受到了特别的重视。

呼吸上皮具有很强的渗透性，吸入的微粒很容易进入。呼吸道上皮含有高度活跃的免疫系统，据报道，正常呼吸道上皮存在数千种病原体，具有强大的先天免疫力，包括黏膜自动排出异物，具有抗菌特性的分泌物、趋化因子、细胞因子和黏液覆盖物。此外，它还含有大量的肺巨噬细胞和树突状细胞，这些细胞既参与先天免疫，也参与适应性免疫。肺周围和间质中存在数十亿巨噬细胞，而DCs位于导气管的上皮内层、黏膜下层、肺泡隔膜内和肺泡表面上。这些细胞处理并递呈抗原以激活免疫系统（图37.2）[10-12]。这些特殊特征使肺部途径成为疫苗接种的潜在场所。

37.4 淋巴组织和呼吸道免疫反应

根据位置的不同，整个呼吸道分为上呼吸道（URT）和下呼吸道（LRT）两部分。上呼吸道的免疫系统由附着于上咽的腺样体（鼻咽扁桃体）、咽鼓管开口处的成对管状扁桃体、口咽处的成对腭扁桃体和舌根后部的舌扁桃体构成。总的来说，这些结构构成Waldeyer环，也称为鼻相关淋巴组织（NALT）。

传出淋巴管从NALT排出到上胸部浅表颈部淋巴结，这些淋巴结又由颈后淋巴结引流传入淋

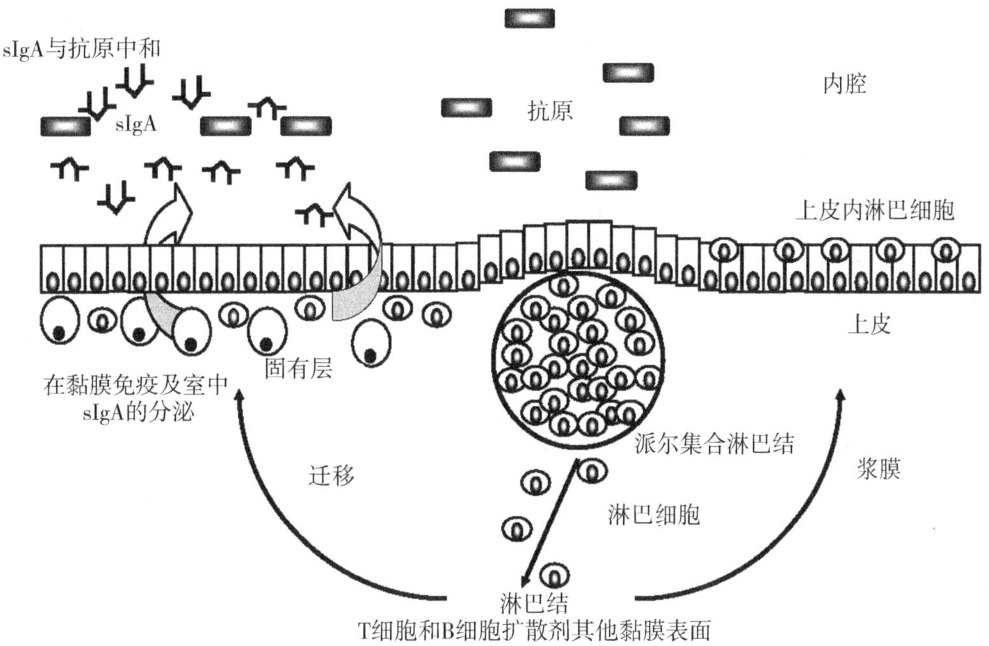

图 37.2 黏膜 S-IgA 反应的诱导

巴管。NALT 还涉及淋巴滤泡（B 细胞区域）、滤泡间区域（T 细胞区域）、巨噬细胞和树突细胞（DCs）的聚集体。然而，与 NALT 相反，下呼吸道（LRT）中的淋巴组织，也称为支气管相关淋巴组织（BALT），并非如此组织。BALT 免疫系统的结构在外部刺激作用之前和之后完全不同。在刺激前，BALT 在松散的基质网络中充满未分化的淋巴细胞和巨噬细胞，由独特的“淋巴上皮”保护，该“淋巴上皮”没有杯状细胞和纤毛，这是相邻呼吸黏膜上皮的特征[13]。淋巴上皮含有特殊细胞，类似于淋巴滤泡集结中的“M”或微细胞，这些细胞在从气道内腔采集抗原或颗粒物质时具有的特征（图 37.2）。

刺激后，BALT 改变其组织并完全转变为圆顶形状，稍微膨胀到上皮下气道的腔内，以促进入侵病原体与免疫细胞的相遇[13]。由于大多数呼吸道感染是在上呼吸道中引发的，因此 NALT 应该作为抗原识别的第一个位点似乎是合乎逻辑的。也有强有力的证据表明，它是免疫反应的一个重要的诱导部位，可以清除上呼吸道的感染。但是，在没有免疫诱导的情况下，上呼吸道中的黏膜自动排出异物作用，能迅速从黏膜表面去除微粒抗原，并清除抗原。这就是为什么我们更关注针对深层气道的抗原。

具有黏附特性的颗粒可能被覆盖在 NALT 上的上皮中的微囊（M）细胞所吸收。M 细胞的基底外侧表面折叠并内陷，形成充满淋巴细胞和巨噬细胞的囊袋。在细胞质中经过长时间处理后，抗原由抗原呈递细胞（APCs）呈现[14]。可溶性抗原被鼻上皮吸收，并通过流出淋巴管直接引流到淋巴结。病毒抗原和颗粒载体被口袋区域中的巨噬细胞和（DCs）吸收，导致 IgA 前体的引发，其离开 NALT 并进入颈淋巴结以进一步扩增和分化成产生免疫球蛋白的细胞，然后这些细胞迁移到黏膜效应部位和呼吸道黏膜。

分泌 IgA（sIgA）通过聚合 Ig 受体（pIgR）[15-17]主动转运到黏膜表面。这可能是黏膜表面感染导致全身（血清 IgG）和黏膜（IgA）抗体应答的原因。致敏 T 细胞和 B 细胞也通过淋巴结迁移到效应部位，提供长期的 B 细胞和 T 细胞记忆，在继发感染后迅速分化为抗体生产细胞。各种研究表明，自然病毒感染和肺疫苗接种都会产生抗病毒细胞毒性 T 淋巴细胞（CTL）[18]。研究表明，鼻腔内感染后，NALT 是病毒特异性 CTLs 的有效诱导位点[19]。除病毒特异性 IgA 之外，强抗病毒 CTL 反应诱导也是通过肺途径进行疫苗接种的主要目标。BALT 中的黏膜反应通过类似的机制在 BALT 进行诱导。因此，我们可以同时针对 URT 和 LRT。但针对 LRT 可能会跳过黏膜自动排出作用，颗粒会停留更长时间，从而延长抗原接种时间。

37.5 肺部免疫

由于其独特的特征，肺部给药成为儿童和成人接种疫苗的一种有吸引力的替代途径。肺部免疫不使用注射器和针头，进而简化了给药程序，缓解注射痛苦，提高了免疫接受性，降低受污染针头感染的风险，并可减少严重的废物处理问题[20]。由于几乎所有的呼吸道病原体都通过黏膜进入，在入口建立免疫能更好地建立一级防线[21,22]。

如上所述，呼吸道具有一个由相关淋巴组织组成的巨大且相互关联的网络，能容易地共享抗原信息，并启动快速免疫反应[1,23]。肺深层的一些独特特征，如生理 pH 和黏膜纤毛作用减少，可能会避免其他非侵入性途径的许多问题，如快速清除、吸收不良、暴露于消化酶和抗原耐受性[2,24]，几乎相当于单个网球场面积的吸入式颗粒沉积表面。因此，与常规肠胃外给药相比，有效的系统能够在较低抗原剂量下诱导良好的免疫反应。这不仅对于少数呼吸系统疾病是一种选择，而且能够探索更多的不同抗原[12]。

气溶胶疫苗的另一优势是易于配制成干粉，其使用具有自身优势的材料科学[25,26]。研制以干粉为基础的疫苗可以潜在地减少或消除冷链需求，促进无菌性，提高抗原的整体稳定性，从而降低产品的总体成本[27,28]。因此，肺部疫苗接种引起了广泛的研究关注度，并且为成功获得研制流感[29]和麻疹[30]的干粉疫苗带来了新的希望。

37.6 肺部免疫面临的挑战

一些抗原在肺部给药后具有诱导免疫作用，尽管已证明其有效性，但目前还没有商用肺疫苗[31,32]。根据文献，成功研制肺部疫苗的主要挑战是开发合适的工具以有效地将抗原输送到更深的呼吸道。然而，呼吸道的薄上皮屏障对高分子具有很高的渗透性。常规的肠胃外疫苗无法通过呼吸道给药，因为这些制剂的安全性已确定用于其他途径，并且仅批准用于口服给药。使用渗透增强剂（包括表面活性剂和胆盐）等赋形剂可能会提高抗原的渗透和吸收，但可能会对上皮细胞造成不可逆损伤。更严重的是，其会导致短期内丧失肺上皮细胞的完整性，造成严重的影响，进而更加难以进行治疗[33]。目前，需要进行长期的毒性研究，才能清楚地确定向肺部给药的新分子的毒性。近年来，以皂苷为基础的 ISCOMATRIX™和用于肺部给药的无毒生物可降解聚合物的研究取得了可喜的进展，但仍需要很多努力[34-36]。

使用皂苷造成的毒性仍然就健康问题引起诸多争议。在体外研究中，还筛选了胞壁酰二肽（MDP）和海藻糖二山嵛酸酯（TDB），以用于肺结核疫苗的潜在用途[37]。呼吸道中疫苗接种的最佳部位仍有待确定。最近在小鼠身上的研究表明，需要将靶向抗原接种至下呼吸道以产生强烈的免疫反应，但临床阶段的结果可能差异很大[34,38]。人类和啮齿动物的呼吸系统结构有许多不同之处。此外，目前尚不清楚经肺途径抗原接种后免疫诱导的准确机制。

37.7 呼吸道中的颗粒沉积

吸入颗粒物在呼吸系统不同部位的沉积非常复杂，可能会因多种因素而有所不同。影响呼吸沉积的因素有以下几个方面。

- 呼吸率
- 口鼻呼吸

- 肺活量
- 呼吸量
- 个人健康状况
- 气道的分叉导致了流体力学流场的持续变化。

制剂类型也可能对经肺途径接种的分子生物活性造成明显影响。此外，诸如颗粒大小、颗粒密度、形态和空气流量等因素也会影响给药效率。有几种机制（如嵌塞、沉积、截留和扩散）（图37.3）都与呼吸途径中的颗粒沉降有关。

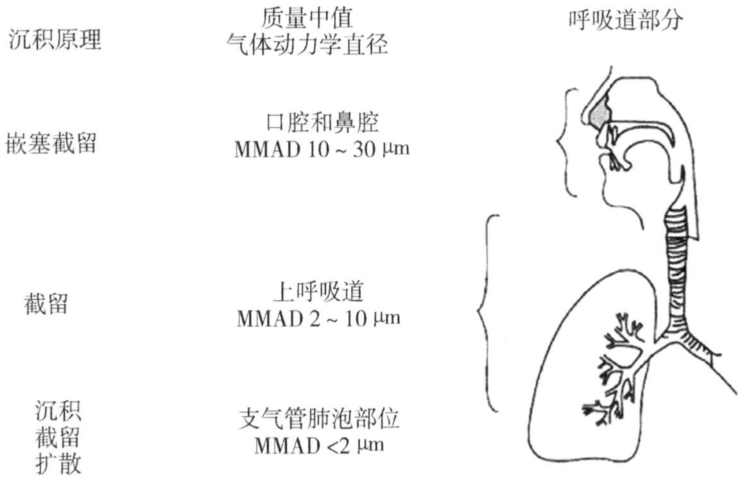

图37.3 呼吸道中的颗粒沉积及其机制

37.7.1 嵌塞

以质量为基础的颗粒沉积主要由嵌塞作用引起。较大颗粒通常会造成嵌塞，这些颗粒非常靠近气道壁，并且主要靠近第一气道分叉处。嵌塞与气体动力学直径息息相关，且在支气管区[39]颗粒嵌塞程度最高。因此，较大颗粒（MMAD>10 μm）不适合肺免疫，更具体而言，是为防止疾病的抗原出现在更深的呼吸道，如结核病。当颗粒较大时，会由上呼吸道的免疫细胞呈现抗原，包括颈部淋巴结、扁桃体、咽部和喉部的免疫细胞。

37.7.2 沉积

沉积是细支气管和肺泡小气道中的颗粒沉积，其气道尺寸小、吸气量低。颗粒沉积机理取决于气体动力学直径较大的颗粒的末端沉降速度[40]。吸湿颗粒在通过温暖潮湿的气道时，其尺寸可能会增大，从而逐渐增加通过沉积作用进行沉降的可能性。因此，即使初始吸湿颗粒很小，也不适合肺抗原接种，因为吸湿颗粒在吸入过程中容易聚集。

37.7.3 截留

当微粒接触气道表面时就会出现截留现象。当空气流线靠近修复壁时，截留就会影响颗粒沉积过程，且其多发生于较小的气道中。截留所沉积的颗粒不像嵌塞时偏离空气流线。然而，也有报告指出截留对颗粒沉积的影响很小，但其对于形状不规则的颗粒（细长颗粒，如纤维）作用非常明显，因为其长度易接触气道表面[41]。此外，纤维相对其自身大小具有较小的气体动力学直径，所以往往可以到达最小的气道。

37.7.4 扩散

扩散是纳米范围（直径<0.5 μm）的颗粒沉积的主要机制。这主要取决于几何大小，而不是气体动力学直径。扩散是通过布朗运动，颗粒从高浓度部位向低浓度部位的移动。这是由于空气分子持续轰击而造成的粒子随机运动。当颗粒刚进入鼻咽部时会出现扩散沉积，最可能发生在空气流量较低的肺（肺泡）部位的较小气道。纳米载体可在较深气道内扩散，在疫苗接种过程中发挥重要作用[40]。

根据以上所述因素和呼吸道的颗粒沉积原理，我们可以根据需要设计给药工具。我们甚至无法想象患者的解剖学和生理学会出现任何变化，但可以根据要求研制出相应的处方。

37.8 纳米技术在肺疫苗接种中的应用

尽管用于其他途径的基于纳米颗粒的制剂已经获得成功并达到商业规模，但吸入纳米颗粒的使用仍然有限。虽然有一些成功的临床研究和新技术，但目前尚未有吸入纳米颗粒疫苗配方被批准用于人类。研究表明，纳米颗粒在肺部的沉积是无效且不可预测的[42]。在雾化期间，颗粒间相互作用超过惯性分离，会导致无法预测的聚集。干燥纳米材料的均匀填充本身就很难，这是因为颗粒内聚导致较差的流动性能。

此外，在被动干粉吸入器中，这些黏性聚集体难以按正常流速再分散。此外，由于缺乏沉积所需的惯性力和重力，高达80%的粒径小于1 μm的吸入式纳米材料会分散[42]。纳米颗粒在肺部的沉积是基于布朗运动的扩散力，而布朗运动是一种时间依赖性过程。与惯性和沉积相反，扩散是一种不太有效的肺沉积方式。数项研究指出，由于更快的扩散，小于1 μm的粒径减少，肺沉积效率增加[42]。

使用计算机建模的研究表明，下气道沉积开始上升，50%以上吸入超细（小于100 nm）颗粒沉积在肺泡区[43,44]。有趣的是，临床试验中也出现了相同的结果，40%～60%的平均粒径为40 nm的吸入颗粒物沉积于肺部。这些研究表明，布朗运动可能是小颗粒扩散的一种可能机制。然而，由于呼吸模式的偏差、昂贵的制造过程以及处理超精细材料的差异，一些限制因素与纳米颗粒有关，如沉积变异性。

此外，纳米颗粒在体内的命运也是可变的。带有包封抗原的生物可降解纳米颗粒（如蛋白质、多肽或DNA），释放并以受控的方式递呈抗原。此外，所研发的纳米颗粒可以选择性地靶向抗原呈递细胞（APC），如树突状细胞（DC），以获得所需的免疫反应。这些细胞在人体防御系统中起着重要的作用。诱导免疫反应的关键步骤包括通过抗原呈递细胞（APC），捕获、处理并加载外源抗原至MHC Ⅰ类和Ⅱ类分子上[45]。研究表明，可溶性抗原很少出现交叉递呈，而颗粒状抗原会视为疫苗接种后诱导$CD4^+$和$CD8^+$细胞活化的关键[46]。因此，抗原纳米载体的接种将导致体液和细胞免疫反应的诱导，而可溶性抗原仅诱导优势体液反应。

除有效结果之外，小于1 μm的纳米颗粒无法有效地沉积于肺上皮细胞，且难以全面制造也是当前面临的困境。

为解决与纳米颗粒相关的问题，制药科学家研究了许多方法，包括使用液体和固体载体、粒子工程来改善雾化（图37.4）。由于离散的纳米颗粒不适合吸入，因此在保留纳米颗粒对肺有效接种的基础上，针对可吸入纳米颗粒配方进行了很多改进。

37.8.1 雾化液体

水凝胶分散体或悬浮液的雾化是一种有效的接种策略，可使纳米颗粒沉积在肺深部。由于药物不能作为水溶液传递，分散纳米颗粒的雾化比药物的微分散具有更多优势。纳米颗粒可以在1～5 μm的液滴中更均匀地分布，实现更高剂量的均匀性和更有效的深肺沉积[47,48]。此外，

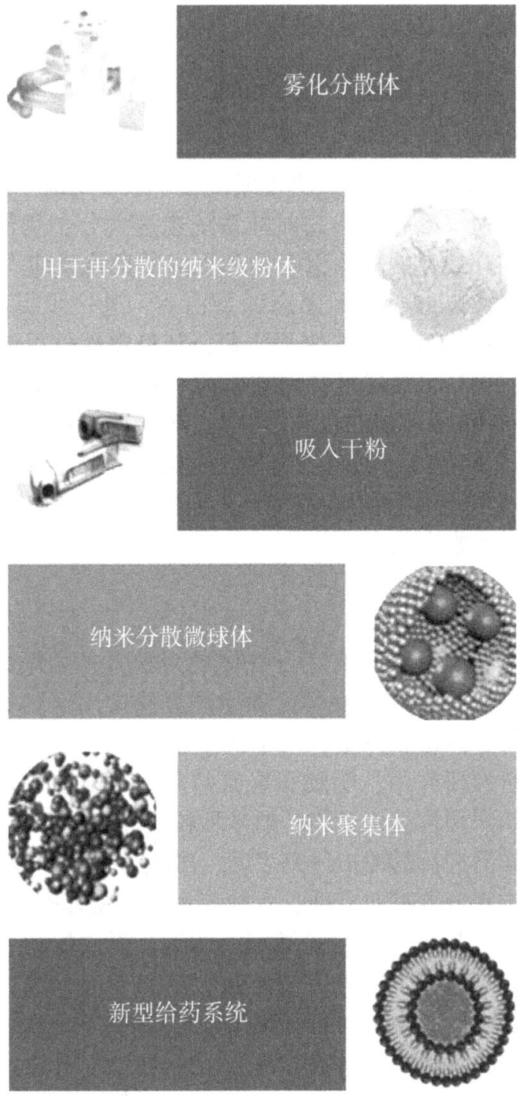

图 37.4　肺部给药的纳米技术途径

由于气流变性几乎与溶液相同，与微颗粒相比，雾化颗粒分散体对雾化器功能和气溶胶液滴尺寸的负面影响最小[49]。稳定材料的选择非常重要，尤其是含有蛋白质药物的治疗性纳米颗粒的水分散体的输送。许多通常用于口服和静脉注射的合成表面活性剂和稳定剂都不适合疫苗接种，尤其对于高浓度。此外，这种方式的缺点是具有毒性、在雾化过程中有发泡倾向，以及表面活性剂可能改变蛋白质的结构完整性。干粉制剂的提升限制了液体制剂的应用。

37.8.2　用于再分散的纳米粉剂

由于水解、颗粒沉降或聚集，大多数分散体可能会出现化学和物理不稳定，或需要使用稳定剂（必须在人类使用前证明其在肺部是安全无毒的）。另一方面，用于再分散的纳米颗粒具有更高的稳定性，但通常需要临床医生给药，因此不便于强化免疫。

在一系列的临床前和动物实验中，已证明分散到水性介质中的纳米粒子是合适的载体，但在临床阶段失败了。大多数抗原是大分子，如多糖、蛋白质和多肽，其在液体制剂中容易发生化学和物理降解[50]。由于蛋白质在液相中的稳定性问题，一些研究人员使用了干燥制剂，目的是在给药前进行分散。纳米干粉配方不需要稳定的表面活性剂，同时减少了蛋白质与水相的接触时间。许多研究表明，乳糖、甘露醇、葡萄糖和蔗糖等保护剂可以增加蛋白质的稳定性，而不会明显改变粒径[51]。

37.8.3 用于吸入的干粉

干粉吸入器（DPI）是一种简单、廉价、高度紧凑且单次使用的一次性设备，对于通过肺部途径疫苗接种非常有效。以干粉气溶胶的形式向肺系统递送大分子是替代肠外免疫最成功的方法，与传统的液体制剂相比，其具有更高的稳定性和更好的免疫反应[52]。通过添加甘露醇、海藻糖、蔗糖和菊粉等添加剂，可以形成干燥固态的蛋白质，进一步提高蛋白质的稳定性[53,54]。最近关于干粉麻疹疫苗配方的报告显示，即使不冷藏，其稳定性也更好[53,55]。多项研究已经证明了干粉气溶胶疫苗的效用，例如，与通过肺部或肌内（IM）途径接种的液体疫苗相比，通过喷雾冷冻干燥制备的干粉流感亚单位疫苗显示在肺部给药后诱导小鼠全身和黏膜的体液和细胞介导的免疫反应[56]。此外，卡介苗（BCG）的喷雾干燥纳米颗粒显示出更好的免疫反应[57]。研究还证明，APC 能更有效地吸收干燥颗粒抗原，导致更强的免疫反应[58]。因此，可吸入干粉疫苗的研制似乎对肺部疫苗接种非常有效。

37.8.4 纳米分散微球

对于肺部给药，微囊化可用于形成具有适合气体动力学直径的颗粒，以用于深肺沉积。生产适合吸入粉末的一种常用方法是喷雾干燥。这种生产方法常用于制备可吸入颗粒物。一些研究还探索了微胶囊化，以制备分散在载体微粒子中的纳米颗粒[59]。在研究喷雾干燥作为将纳米颗粒掺入载体微粒技术的可行性研究中，明胶和聚氰基丙烯酸酯纳米颗粒分散在乳糖基质微粒中。在水介质中再分散后，纳米颗粒仍未聚集，纳米颗粒的优势就在于存在于体内。

此外，利用阶式碰撞取样器进行的冲击研究表明，可吸入乳糖基质颗粒可产生 38%～42%的细颗粒组分。由于最终配方中水分含量非常低，且加工温度相对较低（40～45℃），这种方法更有利于多肽和蛋白类药物。纳米胶囊为肺巨噬细胞、酯酶、蛋白酶和各种上皮代谢途径在体内降解提供了解决方案。此外，这些纳米颗粒在固体微粒基质中的分散可提供比水分散体雾化更合适的替代方案，而水分散体由于缺乏水而具有更高的稳定性。在一项研究中，采用含脂/壳聚糖纳米复合物的微球对肺内的大分子进行给药。喷雾干燥时，以甘露醇为微囊化辅料，胰岛素为模型蛋白。其报告称，所开发的系统可以成功地用于通过肺途径递送治疗性大分子[60]。

37.8.5 纳米聚集体

如前所述，由于惯性和沉积力的影响可以忽略不计，所以很难在较低气道中进行离散的气溶胶化纳米颗粒沉积。形成多个纳米颗粒的低密度聚集体是非常有效的策略之一。这些纳米集料的密度较低（$<0.1\ g/cm^3$），可能具有不同的形状，如空心球[61,62]、球形团聚体[63]、非球形聚体[64,65]或聚集板[66]。为制成纳米聚集体，人们已经发明了各种技术，如喷雾干燥、盐絮凝和快速冷冻工艺。喷雾干燥常用于生产可吸入的纳米聚集体，如空心球和实心球。最初，Tsapis 等提出了由聚集的纳米颗粒组成的空心球体以用于肺部给药[61]。Kho 等利用喷雾干燥技术研制了生物相容性二氧化硅纳米颗粒的空球形聚集体，以促进有效的肺沉积。由于纳米聚集体的大几何尺寸和低密度，可实现高雾化效率和有效的肺沉积。基于以上结果，他们得出结论，空心球形二氧化硅纳米聚集体是吸入给药的潜在候选材料[67]。

37.9 用于肺部给药聚合物的选择

已制定不同的策略和专利，以促进和加强肺部给药。目前已研制出不同的脂质递送系统，即脂质体、免疫刺激复合物、固体脂质纳米颗粒以及使用多种聚合物的聚合物纳米颗粒。影响制剂性能的因素包括：降解速率和机理、副产品、抗原附着或包封的易用性、热稳定性、成本和可用性以及安全性。表 37.2 总结了聚合物的理想属性。

这些因素都是制备安全、能够刺激免疫系统并适合制成气溶胶的纳米颗粒的重要标准。目前已探索了给药的多种聚合物[68,69]。

表 37.2 选择适合肺部免疫聚合物的理想特性

序号	性能	理想的要求
1	物理安全性能 颗粒大小	聚合物应是无毒的，具有可生物降解和生物相容性 具有较小气动直径（最佳 2~5 μm）的肺泡定位颗粒 纳米范围内的颗粒可适当设计以促进特定细胞的吸收
2	多功能性	此载体体系和聚合物在不改变抗原结构完整性的前提下，可广泛应用于多种抗原
3	水分含量	最终配方不应具有吸湿性［水分含量<5%］，并应在较长时间内保持稳定
4	给药	应与患者相容，应易于吸入，不得刺激呼吸道黏膜
5	免疫反应	必须能够诱导黏膜和全身免疫反应 应该没有耐受性 应消除加强剂量的要求（可控的抗原输送） 在低抗原剂量下有效
6	制造	应具有成本效益，能够消毒，批量供应，技术可扩展，必须消除冷链要求，并且符合监管要求

37.10 基于脂质的载体系统

尽管有许多可用的更先进的给药系统，如聚合物微粒、纳米颗粒和金属纳米颗粒，但由于具有许多鲜明的特征，脂质载体仍被视为极具希望的肺给药备选方案（图 37.5）[70]。一些早期研究已经证明脂质体、固体脂质纳米颗粒和 ISCOM 等脂质载体的安全性和有效性，可以进行更深的肺部给药[71,72]。

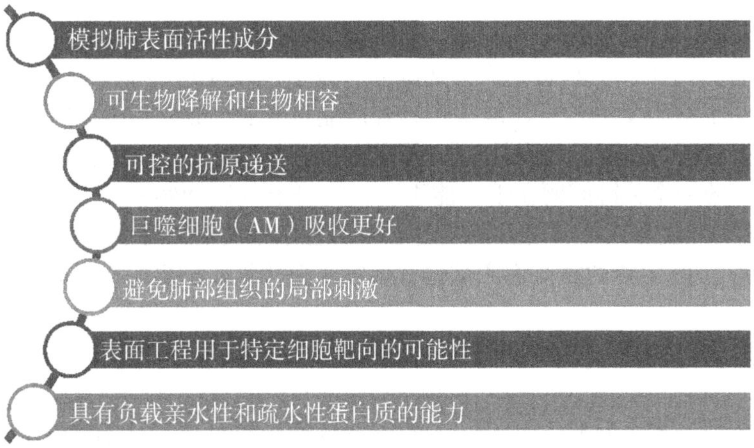

图 37.5 脂质载体系统的优点

37.10.1 脂质体

脂质体是一种具有1个或多个封闭同心圆双分子层的囊泡载体，其与水性腔室和包膜药物交替，且由于储存作用可实现延长的治疗反应。目前已证实，脂质体大小和电荷对巨噬细胞摄取具有重要影响。Chono等研究了粒径对AM吸收的影响。他们使用氢化大豆磷脂酰胆碱（HSPC）、胆固醇和磷酸双十六烷基酯（DCP）制备了粒径为0.1～2 μm的脂质体。脂质体对AM的递送效率随着粒径增加（最多1 μm）而增加，其没有明显变化，并且其在纳米范围内提供了超过24小时的药物缓释[73]。

此外，脂质体表面可锚定多个路径导航分子进行靶向给药。据报道，甘露糖修饰后的脂质体比未修饰脂质体更易被AM吸收[74]。同样地，我们小组也观察到甘露醇修饰脂质体比未修饰脂质体具有更好的AM处理[75]。这些研究证实了脂质体作为靶向给药至肺巨噬细胞的可能，这些巨噬细胞通常积极参与肺免疫。因此，脂质体可能是更好的肺部免疫工具。

37.10.2 固体脂质纳米粒子

固体脂质纳米颗粒是亚微米级的脂质纳米载体。由于具有包覆磷脂单层的固体疏水核，其在结构上不同于其他脂基囊泡。SLN还证明了其通过肺部途径对全身或局部作用有效给药的潜力。许多研究，如人类肺泡上皮细胞系（A549）和小鼠精确切割肺切片（PCLS）细胞系，已经证明了一种蛋白（乳糖脱氢酶、LDH）的控释，并证明了SLN对肺部给药的适用性[76]。Liu等评估了SLN作为一种通过荧光标记胰岛素经肺途径有效递送蛋白的给药工具的潜力。研究表明，SLN在肺泡内有效且均匀分布[77]。此外，SLN还可以通过表面修饰来提高包膜药物的稳定性、选择性摄取和生物利用度，如SLN聚乙二醇化可以很容易地改变表面性质，并对特定的肺泡位点表现出增强的生物利用度。

37.10.3 ISCOM

通常情况下，ISCOM为球形、中空、刚性、笼状结构，粒径约为40 nm，带强负电荷[78,79]。ISCOM由磷脂、胆固醇、皂苷和免疫原（抗原）组成，通常是一种蛋白质。这些系统实际上为了有效地疫苗接种而设计的。这些系统模仿病毒颗粒，例如大小、表面蛋白方向（这些表面蛋白由于皂苷而具有强大的免疫刺激活性）[80]。ISCOM的这些特征使其能够对许多物种的多种抗原产生强烈的免疫反应。与许多其他载体系统相比，ISCOM通过同时诱导高水平的抗体滴度和强T细胞反应，包括增强细胞因子分泌和激活细胞毒性T淋巴细胞（CTL）反应，促进广泛的免疫反应[81]。

刺激CTL的能力对于产生对病毒感染细胞的有效免疫反应可能非常重要，因为在病毒感染细胞中需要强烈的T细胞反应。与其他基于皂苷的制剂相比，ISCOM的另一个优势在于其毒性和反应原性较低。目前，市场上已有针对马流感的基于ISCOM的兽医用疫苗。有关使用ISCOM作为一般疫苗佐剂的一个关键问题在于，能否在人类身上复制动物模型中的结果（如CTL反应的有效诱导）。ISCOM诱导人类明显且持续的CTL活性水平的潜力可能对病毒疫苗特别重要[82]。目前正在进行的基于ISCOM的疫苗临床试验有望在未来几年显出更好的结果。

37.11 聚合物基纳米颗粒递送系统

聚合物纳米颗粒是大小为10～1 000 nm的胶体载体，较小的尺寸有助于靶向和维护被包裹的颗粒，这是一种有效的递送蛋白质的方法。聚合物材料的广泛选择会影响载体结构的各种物理化学特性，如药物释放行为、ζ电位和疏水性等。所选的聚合物必须是安全、生物相容且可降解的。生物降解是抗原释放的必要条件，可避免通过手术来治愈衰竭系统。制备纳米颗粒所用的聚合物包括壳聚糖、海藻酸盐、PLGA、PLA 等。天然聚合物（如壳聚糖、明胶、白蛋白和海藻酸钠）（图 37.6）似乎比合成聚合物更安全。以下将简要说明一些持续用于肺部给药的聚合物。

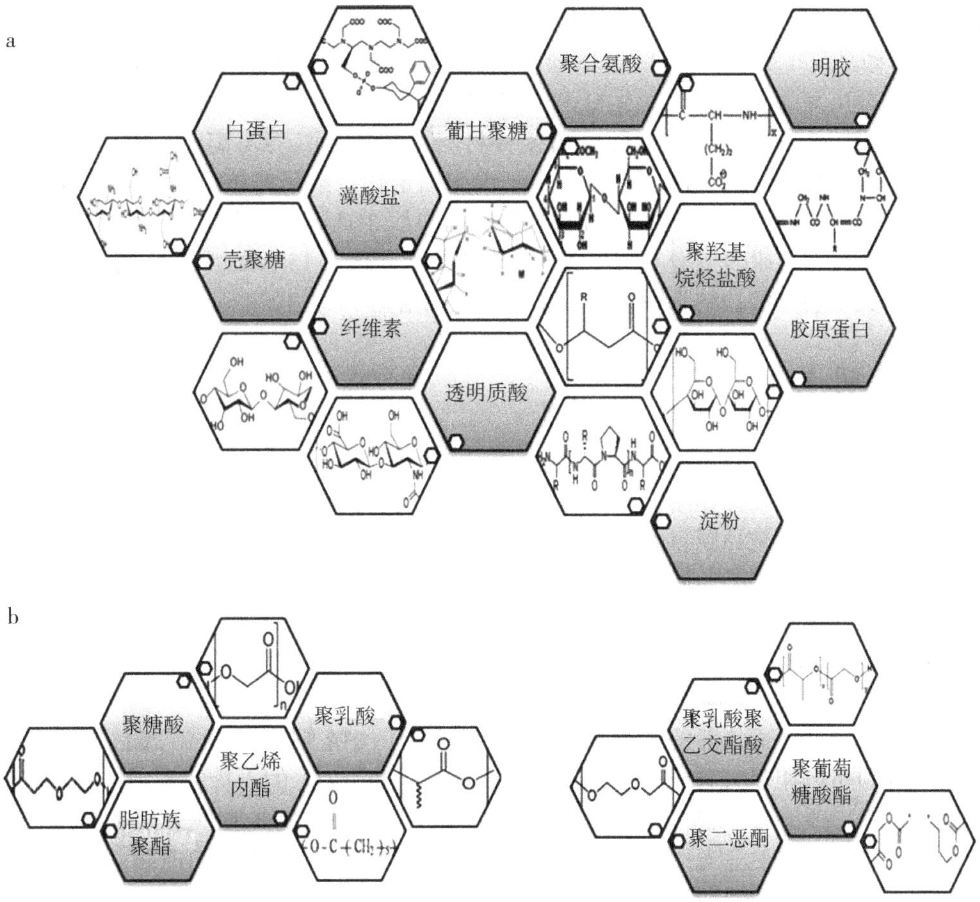

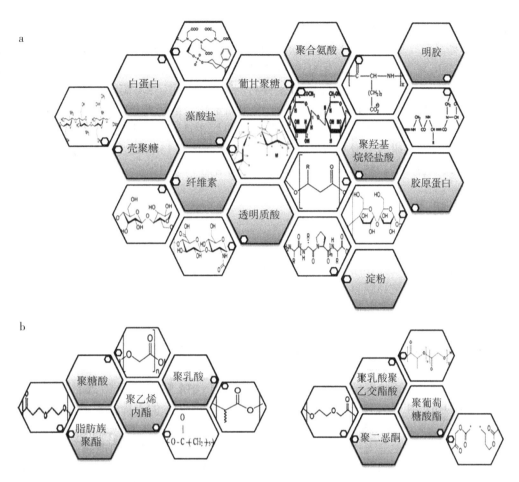

图 37.6　常用的肺部给药聚合物（a）天然来源（b）合成来源
（a）天然的聚合物（b）合成的聚合物

37.12　壳聚糖

壳聚糖是一种具有生物相容性的生物材料，在生物医学工程和药物/疫苗接种领域有着广泛的应用。壳聚糖是氨基葡萄糖与 n-乙酰-d-氨基葡萄糖的共聚物，由于其多阳离子性质，具有粘附性能。由于壳聚糖容易形成微粒子和纳米粒子，可以包封大量抗原，因此可以更好地用于受控疫苗接种，诱导免疫反应，包括体液免疫反应和细胞免疫反应[9]。壳聚糖的阳离子性质赋予其固有的免疫刺激活性和生物粘附特性，从而改善细胞对抗原的吸收和渗透。其还能在体内和体外保护抗原[83,84]。

近年来，壳聚糖在肺蛋白给药中的应用已受到广泛的关注。根据不同的工艺变量，可以在 50 nm 至几微米的广泛尺寸范围内生成壳聚糖颗粒[85]。在一项研究中，使用涂有 N，N，N-三甲基壳聚糖（TMC）的全灭活流感病毒（WIV）可提高流感肺部疫苗的接种效果和免疫原性[86]。Benita 等的研究也证实了壳聚糖纳米颗粒对于肺部免疫的潜力。他们发现与溶液形式的质粒肺部给药或更频繁使用的肌内免疫相比，加载在壳聚糖纳米颗粒中的 DNA 质粒的肺部给药会导致

IFN-γ 分泌水平增加[36]。

37.13 海藻酸盐

海藻酸盐是从褐藻中提取的一种天然高分子。这是一种线性无支链多糖，含有 1，4′-连接的 β-D-甘露糖醛酸（M）和 α-1-古洛糖醛酸（G）的交替残基。类藻酸盐分子量和化学结构的特征取决于分离藻酸盐的来源以及 M 和 G 残基的比例和序列。与壳聚糖类型相似，海藻酸盐具有较强的黏附性，是一种适宜的肺疫苗接种材料。其在 Ca^{++} 等反离子存在下迅速形成凝胶，由于其无毒、可生物降解的特性，通常会用于细胞固定化、给药和抗原[87]。已经制备了海藻酸钠纳米粒，用于经肺途径进行抗结核药物给药。与口服游离药物相比，藻酸盐中药物的相对生物利用度明显更高，结果表明，3 剂海藻酸盐负载纳米雾化器 15 天的疗效相当于 45 剂口服免费药物。这些研究表明藻酸盐纳米颗粒对肺靶向给药的潜力[88]。

37.14 透明质酸

透明质酸（HA）是一种优良的肺部免疫聚合物，其对肺环境具有内源性，在多种炎症介质和肺泡巨噬细胞凝集中发挥作用，并抑制吞噬作用。此外，它是一种高分子量生物黏附聚合物，药物释放时间较长，不被黏膜自动排出作用主动排出[89]。透明质酸是由 N-乙酰-D-葡萄糖胺和葡萄糖醛酸组成的聚阴离子非硫酸盐多糖。HA 是许多细菌细胞壁的重要成分。透明质酸具有良好的生物相容性和黏附性，目前受到研究者的广泛关注。相对于壳聚糖盐酸盐，最低分子量（202kD）的透明质酸具有更好的渗透增强性能[90,91]。已经通过喷雾干燥技术制备了 HA 颗粒，用于靶向肺部给药，尤其是肺泡巨噬细胞。

研究表明，空气表面培养的肺泡巨噬细胞会有效吸收 HA 颗粒（RAW 264.7）[92]。喷雾干燥的透明质酸颗粒也已用于大分子的肺给药。研究表明，基于 HA 的干粉给药系统可用于控制肺部胰岛素递送[93,94]。此外，由透明质酸和氧化铁组成的新型混合纳米颗粒表明，铁络合物透明质酸颗粒可有效地用于多肽递送[95]。带正电荷的溶菌酶、带负电荷的透明质酸盐、环糊精衍生物为致孔剂制备了低密度的透明质酸溶菌酶微粒子，溶菌酶与透明质酸盐的相互作用会提高蛋白质包封效率，并使溶菌酶对变性有机溶剂保持稳定。结果表明，多孔微粒可应用于蛋白质或肽类药物的长期肺部给药，尤其是肺深层上皮细胞给药[96]。

37.15 羧甲基纤维素

羧甲基纤维素（CMC）是带负电荷的、可生物降解的线性聚合物纤维素醚。它具有许多迷人的特征，包括渗透性增强、更好的转染、黏膜细胞内治疗药物的内化，以及负载药物/生物分子的延迟释放。因此，CMC 广泛应用于黏膜途径给药，也广泛用于制备可吸入的蛋白质制剂的喷雾干燥赋形剂。含有蛋白质和多肽的 NaCMC 喷雾干燥颗粒可提供其他方法来制备用于肺部给药的稳定制剂[97]。最近的研究表明，羧甲基纤维素钠喷雾干燥蛋白质可保护蛋白质的结构完整性[97]。研究表明，所得的喷雾干粉可用于研制一种包膜蛋白稳定性高、可呼吸成分高的气溶胶制剂[98]。此外，研究表明，CMC 还能抑制黏液纤毛清除；因此，将有越来越多的抗原用于免疫诱导[99]。

37.16 环糊精

环糊精是 α-d-葡聚糖的环状聚合物，由糖分子结合在环中，能够与疏水分子形成主-客体复合物。通过形成包合物，环糊精能够改变客体分子的物理、化学和生物学特征。环糊精具有较高的渗透增强性能，是理想的渗透增强剂。此外，环糊精可通过减少与外界环境的接触以稳定脆弱的药物分子。研究表明，基于环糊精的多孔颗粒可以在不改变其活性的情况下，将胰岛素输送到更深的气道[100]。使用二甲基-β-环糊精制备了一种可吸入的重组人生长激素干粉，用于蛋白的全身输送，结果表明喷雾干粉气溶胶性能得到改善[101]。

37.17 羧乙烯聚合物

羧乙烯聚合物是丙烯酸与聚烯烃醚或二乙烯基乙二醇交联的聚合物。研究表明，聚羧乙烯具有良好的佐剂性能，已应用于某些已上市动物疫苗[102,103]。Suvaxyn® M. Hyo 猪疫苗含有羧乙烯聚合物作为佐剂。更有趣的是，这种疫苗可以预防呼吸道感染。这种疫苗专用于为猪肺炎支原体引起的临床症状的健康猪疫苗接种，有助于减少支原体引起的呼吸系统疾病和生长抑制。这也证明，由于一些重要的特征，包括允许生物惰性的三维结构、不溶于水和其在水中的溶胀能力，可以控制抗原的释放并且将延长免疫反应。目前正进行许多研究，以研制成功的疫苗，不仅针对肺途径，而且针对其他黏膜途径，即鼻、眼、阴道和直肠。

37.18 聚 ε-己内酯（PCL）

合成的可生物降解的 PCL 是线性的、疏水的、部分结晶的聚酯，可慢慢被微生物用于酶生产[104]。有报告称，这种脂肪族聚酯在生活环境中降解可能是由于酶促作用或酯键的化学水解，或两者兼有[105]。PCL 是一种成本低廉的常用合成聚酯，已被用于治疗性递送载体[106]。与聚乳酸和聚乳酸聚乙交酯酸聚合物相比，其良好的特征使其成为合适的选择，包括在生理液体中的长期稳定性、可忽略的毒性和体内缓慢降解[107]。PCL 的疏水性也使其成为黏膜疫苗递送的理想选择[108]。已经在一些研究中探索了 PCL 的喷雾干燥法[108,109]。

37.19 聚乳酸（PLA）

PLA 是一种具有生物相容性的可降解聚合物，可转化为单体乳酸。乳酸是一种天然的无氧呼吸的中间体/副产物，在柯里氏循环中促进肝脏产生葡萄糖。葡萄糖被用作体内的能量来源。因此，使用聚乳酸纳米颗粒非常安全，且没有任何严重不良反应。PLA 具有广泛的生物医学应用，如缝合线、支架、透析介质和给药装置。阴离子 PEG-PLA 纳米颗粒毒性较低，被证明为治疗肺部疾病的高效药物载体[110]。已证明 PLA 的喷雾干燥配方具有令人满意的气溶胶特性[111]。吸入后的溶液中巨噬细胞的药物浓度比其他溶液高 20 倍。由于肺巨噬细胞吸收较好，因此 PLA 纳米颗粒似乎是肺部免疫的一种合适选择。

37.20 聚乳酸-共-乙醇酸（PLGA）

PLGA 由于其良好的生物相容性、生物降解性和无毒性，近年来得到了广泛的应用。含有 PLC（具有甘露醇）的纳米颗粒吸入显示出良好的气溶胶性能。此外，所制备的纳米颗粒很容易被肺泡巨噬细胞识别，并在肺的肺泡区域表现出浓聚[112]。PLGA 纳米颗粒中含有乳糖和二棕榈酰磷脂酰胆碱（DPPC），且含有抗体（人体 IgG、模型抗体），可延长蛋白释放时间（>35 天）[113]。经凝胶电泳（SDS-PAGE）、场流分离（FFF）、酶联免疫吸附试验（ELISA）证明，释放的抗体稳定、活性好，且无结构变化。同样地，我们制备了胰岛素负载的大孔 PLGA 颗粒用于肺内给药，体内数据显示大孔 PLGA 颗粒达到肺泡，且释放的胰岛素呈生物活性形式[100]。此外，表面修饰的PLGA纳米球比平面PLGA纳米球在肺中的清除速度慢。这可能是因为它们黏附在支气管黏液和肺组织上，并在黏附部位持续保持药物释放[114]。

37.21 金属纳米颗粒递送

近年来，金属纳米粒子在肺疫苗接种方面受到了广泛的关注。它们具有一些独特的特征(图37.7)，其能将这些载体与其他探索用于肺部给药的系统区分开来。然而，未来可以继续深入研究有关金属纳米颗粒安全性的文献，以获得有关安全性的结果。

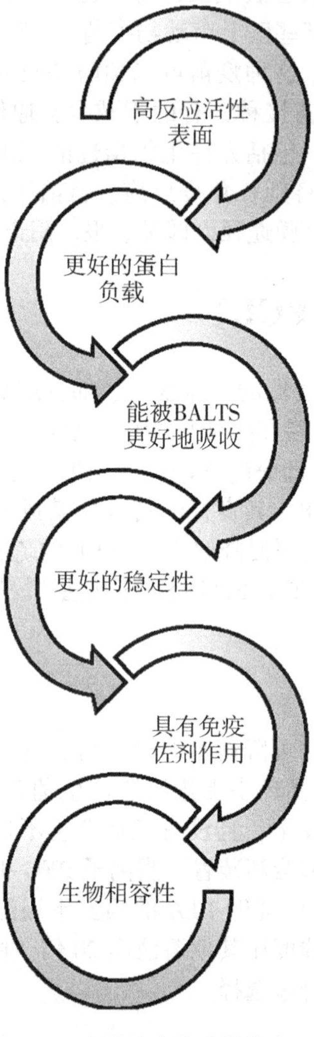

图37.7 金属纳米粒子的优点

37.21.1 铁

铁纳米粒子广泛应用于医学、诊断和实验室应用。由于铁纳米粒子的表面积大，且当存在氧气和水时会氧化，因此具有高反应活性。目前人们预测，铁由于高度氧化还原活性而具有高毒性。然而，在体外细胞系研究中，铁纳米颗粒既不影响细胞活力，也不影响细胞系的形态学参数，同时可通过大胞吞作用促进纳米颗粒内化入细胞[115]。超顺磁性氧化铁纳米颗粒导致肺内吸入气溶胶的磁性靶向。这些颗粒也被用作磁共振成像（MRI）的造影剂，研究表明，超顺磁性氧化铁纳米颗粒能更好地定位于肺部[116]。基于乳酸脱氢酶法研究纳米颗粒对细胞膜的损伤，也有研究报道超磁性铁纳米颗粒具有浓度和时间依赖性损伤[117]。毫无疑问，铁纳米颗粒可以向肺部给药，但其安全性仍有待商榷，在这一领域还需要大量研究。

37.21.2　金

金纳米粒子具有独特的光学、电子和分子识别特性，广泛用于医学领域。金纳米粒子具有化学稳定性、电子密度高、与生物分子的亲和性等特性，使其成为药物载体和诊断工具。除了相关优势，纳米金颗粒的这些特性也对纳米金颗粒在体内的安全性提出了实质性疑问。在过去的几十年内，已经研究了金纳米粒子与肺结构相互作用的机制。利用金纳米颗粒对肺泡 II 型细胞株 A549 和 NCIH441 进行的细胞毒性初步研究表明，金纳米颗粒以与大小无关的方式被细胞内化，可以引起轻微的细胞毒性[118]。在一项研究中，Sadauskas 等研究了金纳米粒子经气管内灌注后在小鼠肺内的生物分布，发现惰性金纳米粒子被肺巨噬细胞吞噬，只有小部分金颗粒根据颗粒大小进入体循环，最小的颗粒（2 nm）迁移的距离最大。研究发现，单次灌注后 1 小时内，注入的纳米颗粒被肺巨噬细胞内化[119]，证明了气管内施用的惰性金纳米颗粒被肺巨噬细胞吞噬。因此，金纳米颗粒可以作为合适的肺部免疫给药工具，但需要对其安全性进行全面的评估。

37.21.3　锌

锌纳米颗粒似乎也是一种合适的肺给药选择。研究开发了体外筛选方法，以确定纳米级（NZO）以及细小氧化锌（FZO）的肺潜在危险性。在大鼠中也评估了灌注与吸入的效果。使用基于细胞毒性生物标志物（LDH）和促炎细胞因子（MIP-2 和 TNF-α）的大鼠肺上皮细胞（L2）、初级肺泡巨噬细胞（AM）和肺泡巨噬细胞-肺上皮细胞共培养（AM-L2）以评估暴露后几个时间点的肺炎症、细胞毒性和组织病理学端点。NZO 或 FZO 颗粒会生成强烈但可逆的炎症反应，其会在滴注后 1 个月消退。此外，根据 LDH 结果，发现 L2 细胞对纳米或细粒度 ZnO 的暴露敏感。然而，巨噬细胞呈剂量依赖性耐药[120]。本研究所涉及的不同研究结果并不相关，所得的结果也不相同。因此，在得出有关锌的安全性结论之前，需要这一领域纳米颗粒大量研究工作，以研究用于肺部给药的工具。

37.21.4　二氧化硅

已证明，暴露于二氧化硅中会造成肺部炎症和肺组织损伤。研究表明，活性氧和氮在二氧化硅急性期肺反应中起一定作用[121]。但在过去几年里，介孔二氧化硅纳米颗粒（MSN）在疫苗中的应用已越来越引起人们的关注。MSN 具有无毒、生物相容性好、孔径可调、表面积大、内部结构呈六边形、结构良好、热化学稳定性高、可控降解速率等独特特征[122,123]。介孔二氧化硅纳米颗粒由于其中空的特性、较大的表面积和较低密度，具有理想的气动特性。这些特征使 MSN 成为理想的肺疫苗接种载体系统。此外，在我们的小组研究中，制备了二氧化硅纳米颗粒原细胞，喷雾干燥后的颗粒具有较好的气动性能，可用于肺免疫[124]。

结论

由于具有高度反应性免疫反应，肺部免疫似乎非常有效。然而，将大分子/抗原输送至更深的呼吸道是一项具有挑战性的任务。许多障碍，包括蛋白水解酶的存在、黏膜自动排出作用的存在以及上呼吸道中的颗粒沉积，让我们的研究面临更多的困难。用于肺部疫苗接种的纳米载体的引入带来了新的希望。为了在肺部免疫后产生较强的免疫反应，目前已发明了许多脂质、聚合物和金属纳米载体。这些系统通过控制抗原向特定免疫细胞（即树突状细胞和肺巨噬细胞）的递呈来改善免疫反应。然而，需要大量研究以证明纳米载体的实用性，尤其是临床阶段的金属纳米颗粒。

参考文献

[1] Ogra, P. L., Faden, H., Welliver, R. C.: Vaccination strategies for mucosal immune responses. Clin. Microbiol. Rev. 14, 430-45 (2001). doi: 10. 1128/CMR. 14. 2. 430-445. 2001.

[2] Malik, B., Goyal, A. K., Mangal, S., Zakir, F., Vyas, S. P.: Implication of gut immunology in the design of oral vaccines. Curr. Mol. Med. 10, 47-70 (2010).

[3] Russell-Jones, G. J.: Oral vaccine delivery. J. Control. Release 65, 49-54 (2000).

[4] Cesta, M. F.: Normal structure, function, and histology of mucosa-associated lymphoid tissue. Toxicol. Pathol. 34, 599-608 (2006). doi: 1 0. 1080/01926230600865531.

[5] Chadwick, S., Kriegel, C., Amiji, M.: Nanotechnology solutions for mucosal immunization. Adv. Drug Deliv. Rev. 62, 394-407 (2010). doi: 10. 1016/j. addr. 2009. 11. 012.

[6] McKenzie, B. S., Brady, J. L., Lew, A. M.: Mucosal immunity: overcoming the barrier for induction of proximal responses. Immunol. Res. 30, 35-71 (2004). doi: 10. 1385/IR: 30: 1: 035.

[7] Brandtzaeg, P., Pabst, R.: Let's go mucosal: commu-nication on slippery ground. Trends Immunol. 25, 570-7 (2004). doi: 10. 1016/j. it. 2004. 09. 005.

[8] Moghaddami, M., Cummins, A., Mayrhofer, G.: Lymphocyte-fi lled villi: comparison with other lymphoid aggregations in the mucosa of the human small intestine. Gastroenterology 115, 1414-25 (1998).

[9] Malik, B., et al.: Microfold-cell targeted surface engineered polymeric nanoparticles for oral immunization. J. Drug Target. 20, 76-84 (2012). doi: 10. 3 109/1061186X. 2011. 611516.

[10] McCray Jr., P. B., Bentley, L.: Human airway epithe-lia express a beta-defensin. Am. J. Respir. Cell Mol. Biol. 16, 343-9 (1997).

[11] Larrick, J. W., et al.: Human CAP18: a novel antimi-crobial lipopolysaccharide-binding protein. Infect. Immun. 63, 1291-7 (1995).

[12] Bivas-Benita, M., Ottenhoff, T. H., Junginger, H. E., Borchard, G.: Pulmonary DNA vaccination: concepts, possibilities and perspectives. J. Control. Release 107, 1-29 (2005). doi: 10. 1016/j. jconrel. 2005. 05. 028.

[13] Bienenstock, J.: Gut and bronchus associated lym-phoid tissue: an overview. Adv. Exp. Med. Biol. 149, 471-7 (1982).

[14] Kuper, C. F., et al.: The role of nasopharyngeal lym-phoid tissue. Immunol. Today 13, 219-24 (1992).

[15] Tamura, S., Kurata, T.: Defense mechanisms against infl uenza virus infection in the respiratory tract mucosa. Jpn. J. Infect. Dis. 57, 236-47 (2004).

[16] Zuercher, A. W.: Upper respiratory tract immunity. Viral Immunol. 16, 279-89 (2003). doi: 1 0. 1089/088282403322396091.

[17] Perry, M., Whyte, A.: Immunology of the tonsils. Immunol. Today 19, 414-21 (1998).

[18] Cerwenka, A., Morgan, T. M., Dutton, R. W.: Naive, effector, and memory CD8 T cells in protection against pulmonary infl uenza virus infection: homing properties rather than initial frequencies are crucial. J. Immunol. 163, 5535-43 (1999).

[19] Zuercher, A. W., Coffi n, S. E., Thurnheer, M. C., Fundova, P., Cebra, J. J.: Nasal-associated lymphoid tissue is a mucosal inductive site for virus-specifi c humoral and cellular immune responses. J. Immunol. 168, 1796-803 (2002).

[20] Giudice, E. L., Campbell, J. D.: Needle-free vaccine delivery. Adv. DrugDeliv. Rev. 58, 68-89

(2006). doi: 10. 1016/j. addr. 2005. 12. 003.

[21] Bouvet, J. P., Decroix, N., Pamonsinlapatham, P.: Stimulation of local antibody production: parenteral or mucosal vaccination? Trends Immunol. 23, 209-13 (2002).

[22] Stevceva, L., Abimiku, A. G., Franchini, G.: Targeting the mucosa: genetically engineered vaccines and mucosal immune responses. Genes Immun. 1, 308-15 (2000). doi: 10. 1038/sj. gene. 6363680.

[23] McNeela, E. A., Mills, K. H.: Manipulating the immune system: humoral versus cell-mediated immunity. Adv. Drug Deliv. Rev. 51, 43-54 (2001).

[24] Moyle, P. M., McGeary, R. P., Blanchfi eld, J. T., Toth, I.: Mucosal immunisation: adjuvants and delivery systems. Curr. Drug Deliv. 1, 385-96 (2004).

[25] Valente, A. X., Langer, R., Stone, H. A., Edwards, D. A.: Recent advances in the development of an inhaled insulin product. BioDrugs 17, 9-17 (2003).

[26] Bosquillon, C., Preat, V., Vanbever, R.: Pulmonary delivery of growth hormone using dry powders and visualization of its local fate in rats. J. Control. Release 96, 233-44 (2004). doi: 1 0. 1016/j. jconrel. 2004. 01. 027.

[27] Huang, J., et al.: A novel dry powderinfl uenza vaccine and intranasal delivery technology: induction of systemic and mucosal immune responses in rats. Vaccine 23, 794-801 (2004). doi: 10. 1016/j. vaccine. 2004. 06. 049.

[28] LiCalsi, C., Christensen, T., Bennett, J. V., Phillips, E., Witham, C.: Dry powder inhalation as a potential delivery method for vaccines. Vaccine 17, 1796-803 (1999).

[29] Smith, D. J., Bot, S., Dellamary, L., Bot, A.: Evaluation of novel aerosol formulations designed for mucosal vaccination against infl uenza virus. Vaccine 21, 2805-12 (2003).

[30] LiCalsi, C., et al.: A powder formulation of measles vaccine for aerosol delivery. Vaccine 19, 2629-36 (2001).

[31] Amorij, J. P., et al.: Pulmonary delivery of an inulin-stabilized infl uenza subunit vaccine prepared by spray-freeze drying induces systemic, mucosal humoral as well ascell-mediated immune responses in BALB/c mice. Vaccine 25, 8707-17 (2007). doi: 10. 1016/j. vaccine. 2007.10. 035.

[32] Wee, J. L., et al.: Pulmonary delivery of ISCOMATRIXinfl uenza vaccine induces both systemic and mucosal immunity with antigen dose sparing. Mucosal Immunol. 1, 489-96 (2008). doi: 10. 1038/mi. 2008. 59.

[33] Weers, J. G., Tarara, T. E., Clark, A. R.: Design of fi ne particles for pulmonary drug delivery. Expert Opin. Drug Deliv. 4, 297-313 (2007). doi: 10. 1517/17425247. 4. 3. 297.

[34] Sanders, M. T., Deliyannis, G., Pearse, M. J., McNamara, M. K., Brown, L. E.: Single dose intranasal immunization with ISCOMATRIX vaccines to elicit antibody-mediated clearance of infl uenza virus requires delivery to the lower respiratory tract. Vaccine 27, 2475-82 (2009). doi: 10. 1016/j. vaccine. 2009. 02. 054.

[35] Vujanic, A., et al.: Combined mucosal and systemic immunity following pulmonary delivery of ISCOMATRIX adjuvanted recombinant antigens. Vaccine 28, 2593-7 (2010). doi: 10. 1016/j. vaccine. 2010. 01. 018.

[36] Bivas-Benita, M., et al.: Pulmonary delivery of chitosan-DNA nanoparticles enhances the immunogenicity of a DNA vaccine encoding HLA-A *0201-restricted T-cell epitopes of Mycobacterium tuberculosis. Vaccine 22, 1609-15 (2004). doi: 10. 1016/j. vaccine. 2003. 09. 044.

[37] Wang, C., et al.: Screening for potential adjuvants administered by the pulmonary route for tuberculosis vaccines. AAPS J. 11, 139-47 (2009). doi: 10. 1208/ s12248-009-9089-0.

[38] Minne, A., et al.: The delivery site of a monovalent infl uenza vaccine within the respiratory tract impacts on the immune response. Immunology 122, 316–25 (2007). doi: 10. 1111/j. 1365-2567. 2007. 02641. x.

[39] Heyder, J.: Deposition of inhaled particles in the human respiratory tract and consequences for regional targeting in respiratory drug delivery. Proc. Am. Thorac. Soc. 1, 315 – 20 (2004). doi: 10. 1513/ pats. 200409-046TA.

[40] Rogueda, P. G., Traini, D.: The nanoscale in pulmo-nary delivery. Part 1: deposition, fate, toxicology and effects. Expert Opin. Drug Deliv. 4, 595–606 (2007). doi: 10. 1517/17425247. 4. 6. 595.

[41] Hinds, W. C.: Aerosol Technology: Properties, Behavior, and Measurement of Airborne Particles, 2nd edn. Wiley, New York (1999).

[42] Watts, A. B., Williams, R. O., 3rd: Chapter 15: Nanoparticles for Pulmonary Delivery. In: Smyth, H. D., Hickey, A. J. (eds.) Controlled Pulmonary Drug Delivery, Springer, New York (2011) p335.

[43] Byron, P. R.: Prediction of drug residence times in regions of the human respiratory tract following aerosol inhalation. J. Pharm. Sci. 75, 433–8 (1986).

[44] Yang, W., Peters, J. I., Williams Ⅲ, R. O.: Inhaled nanoparticles – a current review. Int. J. Pharm. 356, 239–47 (2008). doi: 10. 1016/j. ijpharm. 2008. 02. 011.

[45] Kurts, C., Robinson, B. W., Knolle, P. A.: Cross-priming in health and disease. Nat. Rev. Immunol. 10, 403–14 (2010). doi: 10. 1038/nri2780.

[46] Shen, H., et al.: Enhanced and prolonged cross-presentation following endosomal escape of exogenous antigens encapsulated in biodegradable nanoparticles. Immunology 117, 78–88 (2006). doi: 10. 1111/j. 1365-2567. 2005. 02268. x.

[47] Ostrander, K. D., Bosch, H. W., Bondanza, D. M.: An in – vitro assessment of a NanoCrystal beclomethasone dipropionate colloidal dispersion via ultrasonic nebulization. Eur. J. Pharm. Biopharm. 48, 207–15 (1999).

[48] Wiedmann, T. S., DeCastro, L., Wood, R. W.: Nebulization of NanoCrystals: production of a respirable solid-in-liquid-in-air colloidal dispersion. Pharm. Res. 14, 112–6 (1997).

[49] Dailey, L. A., et al.: Nebulization of biodegradable nanoparticles: impact of nebulizer technology and nanoparticle characteristics on aerosol features. J. Control. Release 86, 131–44 (2003).

[50] Mahler, H. C., Muller, R., Friess, W., Delille, A., Matheus, S.: Induction and analysis of aggregates in a liquid IgG1 – antibody formulation. Eur. J. Pharm. Biopharm. 59, 407 – 17 (2005). doi: 10. 1016/j. ejpb. 2004. 12. 004.

[51] Packhaeuser, C. B., et al.: Stabilization of aerosoliz-able nano-carriers by freeze-drying. Pharm. Res. 26, 129–38 (2009). doi: 10. 1007/s11095-008-9714-0.

[52] Schule, S., Schulz – Fademrecht, T., Garidel, P., Bechtold – Peters, K., Frieb, W.: Stabilization of IgG1 in spray – dried powders for inhalation. Eur. J. Pharm. Biopharm. 69, 793 – 807 (2008). doi: 10. 1016/j. ejpb. 2008. 02. 010.

[53] Geeraedts, F., et al.: Preservation of the immunogenicity of dry-powder infl uenza H5N1 whole inactivated virus vaccine at elevated storage temperatures. AAPS J. 12, 215–22 (2010). doi: 1 0. 1208/s12248-010-9179-z.

[54] Maury, M., Murphy, K., Kumar, S., Mauerer, A., Lee, G.: Spray-drying of proteins: effects of sorbitol and trehalose on aggregation and FT-IR amide Ⅰ spectrum of an immunoglobulin G. Eur. J. Pharm. Biopharm. 59, 251–61 (2005). doi: 10. 1016/j. ejpb. 2004. 07. 010.

[55] Ohtake, S., et al.: Heat – stable measles vaccine pro – duced by spray drying. Vaccine 28, 1275 – 84

(2010). doi: 10. 1016/j. vaccine. 2009. 11. 024.

[56] Saluja, V., et al.: A comparison between spray dry-ing and spray freeze drying to produce an infl uenza subunit vaccine powder for inhalation. J. Control. Release 144, 127-33 (2010). doi: 10. 1016/j. jconrel. 2010. 02. 025.

[57] Garcia-Contreras, L., et al.: Immunization by a bacterial aerosol. Proc. Natl. Acad. Sci. U. S. A. 105, 4656-60 (2008). doi: 10. 1073/pnas. 0800043105.

[58] Thomas, C., Gupta, V., Ahsan, F.: Particle size infl uences the immune response produced by hepatitis B vaccine formulated in inhalable particles. Pharm. Res. 27, 905-19 (2010). doi: 10. 1007/s11095-010-0094-x.

[59] Sham, J. O., Zhang, Y., Finlay, W. H., Roa, W. H., Lobenberg, R.: Formulation and characterization of spray-dried powders containing nanoparticles for aerosol delivery to the lung. Int. J. Pharm. 269, 457-67 (2004).

[60] Grenha, A., Remunan-Lopez, C., Carvalho, E. L., Seijo, B.: Microspheres containing lipid/chitosan nanoparticles complexes for pulmonary delivery of therapeutic proteins. Eur. J. Pharm. Biopharm. 69, 83-93 (2008). doi: 10. 1016/j. ejpb. 2007. 10. 017.

[61] Tsapis, N., Bennett, D., Jackson, B., Weitz, D. A., Edwards, D. A.: Trojan particles: large porous carriers of nanoparticles for drug delivery. Proc. Natl. Acad. Sci. U. S. A. 99, 12001-5 (2002). doi: 10. 1073/ pnas. 182233999.

[62] Hadinoto, K., Phanapavudhikul, P., Kewu, Z., Tan, R. B.: Dry powder aerosol delivery of large hollow nanoparticulate aggregates as prospective carriers of nanoparticulatedrugs: effects of phospholipids. Int. J. Pharm. 333, 187-98 (2007). doi: 10. 1016/j. ijpharm. 2006. 10. 009.

[63] Hu, T., Chiou, H., Chan, H. K., Chen, J. F., Yun, J.: Preparation of inhalable salbutamol sulphate using reactive high gravity controlled precipitation. J. Pharm. Sci. 97, 944-9 (2008).

[64] McConville, J. T., et al.: Targeted high lung concentrations of itraconazole using nebulized dispersions in a murine model. Pharm. Res. 23, 901-11 (2006). doi: 10. 1007/s11095-006-9904-6.

[65] Plumley, C., et al.: Nifedipine nanoparticle agglom-eration as a dry powder aerosol formulation strategy. Int. J. Pharm. 369, 136-43 (2009). doi: 10. 1016/j. ijpharm. 2008. 10. 016.

[66] Richardson, P. C., Boss, A. H.: Technosphere insulin technology. J. Diabetes Sci. Technol. 9 (Suppl 1), S65-72 (2007). doi: 10. 1089/dia. 2007. 0212.

[67] Katherine, K., Kunn, H.: Aqueous re-dispersibility characterization of spray-dried hollow spherical silica nano-aggregates. Powder Technol. 198, 354-63 (2010).

[68] Nair, L. S., Laurencin, C. T.: Polymers as biomaterials for tissue engineering and controlled drug delivery. Adv. Biochem. Eng. Biotechnol. 102, 47-90 (2006).

[69] Langer, R.: New methods of drug delivery. Science 249, 1527-33 (1990).

[70] Gaspar, M. M., Bakowsky, U., Ehrhardt, C.: Inhaled liposomes-current strategies and future challenges. J. Biomed. Nanotechnol. 4, 1-13 (2008).

[71] Henderson, A., Propst, K., Kedl, R., Dow, S.: ucosal immunization with liposome-nucleic acid adjuvants generates effective humoral and cellular immunity. Vaccine 29, 5304-12 (2011). doi: 10. 1016/j. vaccine. 2011. 05. 009.

[72] Videira, M. A., et al.: Lymphatic uptake of pulmonary delivered radiolabelled solid lipid nanoparticles. J. Drug Target. 10, 607-13 (2002). doi: 10. 1080/1061186021000054933.

[73] Chono, S., Tanino, T., Seki, T., Morimoto, K.: Infl uence of particle size on drug delivery to rat

alveolar macrophages following pulmonary administration of ciprofl oxacin incorporated into liposomes. J. Drug Target. 14, 557-66 (2006). doi: 10. 1080/10611860600834375.

[74] Chono, S., Tanino, T., Seki, T., Morimoto, K.: Uptake characteristics of liposomes by rat alveolar macrophages: infl uence of particle size and surface mannose modifi cation. J. Pharm. Pharmacol. 59, 75-80 (2007). doi: 10. 1211/jpp. 59. 1. 0010.

[75] Vyas, S. P., Quraishi, S., Gupta, S., Jaganathan, K. S.: Aerosolized liposome-based delivery of amphotericin B to alveolar macrophages. Int. J. Pharm. 296, 12-25 (2005). doi: 10. 1016/j. ijpharm. 2005. 02. 003.

[76] Nassimi, M., et al.: Low cytotoxicity of solid lipid nanoparticles in in vitro and ex vivo lung models. Inhal. Toxicol. 21 (Suppl 1), 104-9 (2009). doi: 10. 1080/08958370903005769.

[77] Liu, J., et al.: Solid lipid nanoparticles for pulmonary delivery of insulin. Int. J. Pharm. 356, 333-44 (2008). doi: 10. 1016/j. ijpharm. 2008. 01. 008.

[78] Rimmelzwaan, G. F., Osterhaus, A. D.: A novel generation of viral vaccines based on the ISCOM matrix. Pharm. Biotechnol. 6, 543-58 (1995).

[79] Morein, B.: Iscom-an immunostimulating com-plex. Arzneimittelforschung 37, 1418 (1987).

[80] Morein, B., Sundquist, B., Hoglund, S., Dalsgaard, K., Osterhaus, A.: Iscom, a novel structure for antigenic presentation of membrane proteins from enveloped viruses. Nature 308, 457-60 (1984).

[81] Barr, I. G., Mitchell, G. F.: ISCOMs (immunostimu-lating complexes): the fi rst decade. Immunol. Cell Biol. 74, 8-25 (1996). doi: 10. 1038/icb. 1996. 2.

[82] Sjolander, A., Cox, J. C., Barr, I. G.: ISCOMs: an adjuvant with multiple functions. J. Leukoc. Biol. 64, 713-23 (1998).

[83] Khatri, K., Goyal, A. K., Gupta, P. N., Mishra, N., Vyas, S. P.: Plasmid DNA loaded chitosan nanoparticles for nasal mucosal immunization against hepatitis B. Int. J. Pharm. 354, 235-41 (2008). doi: 10. 1016/j. ijpharm. 2007. 11. 027.

[84] Marcinkiewicz, J., Polewska, A., Knapczyk, J.: Immunoadjuvant properties of chitosan. Arch. Immunol. Ther. Exp. (Warsz.) 39 (127-132) (1991).

[85] Wright, I. K., Higginbotham, A., Baker, S. M., Donnelly, T. D.: Generation of nanoparticles of controlled size using ultrasonic piezoelectric oscillators in solution. ACS Appl. Mater. Interfaces 2, 2360-4 (2010). doi: 10. 1021/am100375w.

[86] Hagenaars, N., et al.: Physicochemical and immuno-logical characterization of N, N, N-trimethyl chitosan-coated whole inactivated infl uenza virus vaccine for intranasal administration. Pharm. Res. 26, 1353-64 (2009). doi: 1 0. 1007/s11095-009-9845-y.

[87] Jain, A., Gupta, Y., Jain, S. K.: Perspectives of biode-gradable natural polysaccharides for site-specifi c drug delivery to the colon. J. Pharm. Pharm. Sci. 10, 86-128 (2007).

[88] Ahmad, Z., Khuller, G. K.: Alginate-based sustained release drug delivery systems for tuberculosis. Expert Opin. Drug Deliv. 5, 1323-34 (2008). doi: 10. 1517/17425240802600662.

[89] Rouse, J. J., Whateley, T. L., Thomas, M., Eccleston, G. M.: Controlled drug delivery to the lung: Infl uence of hyaluronic acid solution conformation on its adsorption to hydrophobic drug particles. Int. J. Pharm. 330, 175-182 (2007). doi: 1 0. 1016/j. ijpharm. 2006. 11. 066.

[90] Fraser, J. R., Laurent, T. C., Laurent, U. B.: Hyaluronan: its nature, distribution, functions and turnover. J. Intern. Med. 242, 27-33 (1997).

[91] Liao, Y. H., Jones, S. A., Forbes, B., Martin, G. P., Brown, M. B.: Hyaluronan: pharmaceutical

characterization and drug delivery. Drug Deliv. 12, 327-42 (2005). doi: 10. 1080/10717540590952555.

[92] Hwang, S. M., Kim, D. D., Chung, S. J., Shim, C. K.: Delivery ofofl oxacin to the lung and alveolar macrophages via hyaluronan microspheres for the treatment of tuberculosis. J. Control. Release 129, 100-6 (2008). doi: 10. 1016/j. jconrel. 2008. 04. 009.

[93] Surendrakumar, K., Martyn, G. P., Hodgers, E. C., Jansen, M., Blair, J. A.: Sustained release of insulin from sodium hyaluronate based dry powder formulations after pulmonary delivery to beagle dogs. J. Control. Release 91, 385-94 (2003).

[94] Morimoto, K., Metsugi, K., Katsumata, H., Iwanaga, K., Kakemi, M.: Effects of low-viscosity sodium hyaluronate preparation on the pulmonary absorption of rh-insulin in rats. Drug Dev. Ind. Pharm. 27, 365-71 (2001). doi: 10. 1081/DDC-100103737.

[95] Kumar, A., et al.: Development of hyaluronic acid-Fe2O3 hybrid magnetic nanoparticles for targeted delivery of peptides. Nanomedicine 3, 132-7 (2007). doi: 10. 1016/j. nano. 2007. 03. 001.

[96] Lee, E. S., Kwon, M. J.: Protein release behavior from porous microparticle with lysozyme/hyaluronate ionic complex. Colloids Surf. B Biointerfaces 55, 125-30 (2007). doi: 10. 1016/j. colsurfb. 2006. 11. 024.

[97] Li, H. Y., Song, X., Seville, P. C.: The use of sodium carboxymethylcellulose in the preparation of spray-dried proteins for pulmonary drug delivery. Eur. J. Pharm. Sci. 40, 56-61 (2010). doi: 10. 1016/j. ejps. 2010. 02. 007.

[98] Li, H. Y., Seville, P. C.: Novel pMDI formulations for pulmonary delivery of proteins. Int. J. Pharm. 385, 73-8 (2010). doi: 10. 1016/j. ijpharm. 2009. 10. 032.

[99] Dailey, L. A., et al.: Surfactant-free, biodegradable nanoparticles for aerosol therapy based on the branched polyesters. DEAPA-PVAL-g-PLGA. Pharm. Res. 20, 2011-20 (2003).

[100] Ungaro, F., et al.: Insulin-loaded PLGA/cyclodextrin large porous particles with improved aerosolization properties: in vivo deposition and hypoglycaemic activity after delivery to rat lungs. J. Control. Release 135, 25-34 (2009). doi: 1 0. 1016/j. jconrel. 2008. 12. 011.

[101] Jalalipour, M., Najafabadi, A. R., Gilani, K., Esmaily, H., Tajerzadeh, H.: Effect of dimethyl-beta-cyclodextrin concentrations on the pulmonary delivery of recombinant human growth hormone dry powder in rats. J. Pharm. Sci. 97, 5176-85 (2008). doi: 10. 1002/jps. 21353.

[102] Krashias, G., et al.: Potent adaptive immune responses induced against HIV-1 gp140 and infl uenza virus HA by a polyanionic carbomer. Vaccine 28, 2482-9 (2010). doi: 10. 1016/j. vaccine. 2010. 01. 046.

[103] Robert, E.: Chapter 1: An Overview of Adjuvant Use. In: O'Hagan, D. T. (ed.). Vaccine Adjuvants: Preparation Methods and Research Protocols. Humana Press, Totowa (2000) p 5.

[104] Singh, R. P., Pandey, J. K., Rutot, D., Degee, P., Dubois, P.: Biodegradation of poly (epsilon-caprolactone) /starch blends and composites in composting and culture environments: the effect of compatibilization on the inherent biodegradability of the host polymer. Carbohydr. Res. 338, 1759-69 (2003).

[105] Khatiwala, V., Shekhar, N., Aggarwal, S., Mandal, U.: Biodegradation of Poly (ε-caprolactone) (PCL) Film by Alcaligenes faecalis. J Polym Environ 16, 61-7 (2008). doi: 10. 1007/s10924-008-0104-9.

[106] Balmayor, E. R., Tuzlakoglu, K., Azevedo, H. S., Reis, R. L.: Preparation and characterization of starch-poly-epsilon-caprolactone microparticles incorporating bioactive agents for drug delivery and tissue engineering applications. Acta Biomater. 5, 1035-45 (2009). doi: 10. 1016/j. actbio. 2008. 11. 006.

[107] Sinha, V. R., Trehan, A.: Formulation, characteriza-tion, and evaluation of ketorolac tromethamine-loaded biodegradable microspheres. Drug Deliv. 12, 133-9 (2005).

[108] Baras, B., Benoit, M. A., Gillard, J.: Infl uence of various technological parameters on the preparation of spray-dried poly (epsilon-caprolactone) microparticles containing a model antigen. J. Microencapsul. 17, 485-98 (2000). doi: 10. 1080/026520400405732.

[109] Zalfen, A. M., et al.: Controlled release of drugs from multi-component biomaterials. Acta Biomater. 4, 1788-96 (2008). doi: 10. 1016/j. actbio. 2008. 05. 021.

[110] Harush-Frenkel, O., et al.: A safety and tolerability study of differently-charged nanoparticles for local pulmonary drug delivery. Toxicol. Appl. Pharmacol. 246, 83-90 (2010). doi: 1 0. 1016/j. taap. 2010. 04. 011.

[111] Muttil, P., et al.: Inhalable microparticles containing large payload of anti-tuberculosis drugs. Eur. J. Pharm. Sci. 32, 140-50 (2007). doi: 10. 1016/j. ejps. 2007. 06. 006.

[112] Ohashi, K., Kabasawa, T., Ozeki, T.: One-step preparation of rifampicin/poly (lactic-co-glycolic acid) nanoparticle-containing mannitol microspheres using a four-fl uid nozzle spray drier for inhalation therapy of tuberculosis. J. Control. Release 135, 19-24 (2009). doi: 10. 1016/j. jconrel. 2008. 11. 027.

[113] Kaye, R. S., Purewal, T. S., Alpar, H. O.: Simultaneously manufactured nano-in-micro (SIMANIM) particles for dry-powder modifi ed-r elease delivery of antibodies. J. Pharm. Sci. 98, 4055-68 (2009). doi: 1 0. 1002/jps. 21673.

[114] Yamamoto, H., Kuno, Y., Sugimoto, S., Takeuchi, H., Kawashima, Y.: Surface-modifi ed PLGA nanosphere with chitosan improved pulmonary delivery of calcitonin by mucoadhesion and opening of the intercellular tight junctions. J. Control. Release 102, 373-81 (2005). doi: 10. 1016/j. jconrel. 2004. 10. 010.

[115] Canete, M., et al.: The endocytic penetration mechanism of iron oxide magnetic nanoparticles with positively charged cover: a morphological approach. Int. J. Mol. Med. 26, 533-9 (2010).

[116] Hasenpusch, G., et al.: Magnetized aerosols com-prising superparamagnetic iron oxide nanoparticles improve targeted drug and gene delivery to the lung. Pharm. Res. 29, 1308-18 (2012). doi: 10. 1007/s11095-012-0682-z.

[117] Naqvi, S., et al.: Concentration-dependent toxicity of iron oxide nanoparticles mediated by increased oxidative stress. Int. J. Nanomed. 5, 983-9 (2010). doi: 10. 2147/IJN. S13244.

[118] Uboldi, C., et al.: Gold nanoparticles induce cytotox-icity in the alveolar type-Ⅱ cell lines A549 and NCIH441. Part. Fibre Toxicol. 6, 18 (2009). doi: 10. 1186/1743-8977-6-18.

[119] Sadauskas, E., et al.: Biodistribution of gold nanoparticles in mouse lung following intratracheal instillation. Chem. Cent. J. 3, 16 (2009). doi: 10. 1186/1752-153X-3-16.

[120] Sayes, C. M., Reed, K. L., Warheit, D. B.: Assessing toxicity of fi ne and nanoparticles: comparing in vitro measurements to in vivo pulmonary toxicity profi les. Toxicol. Sci. 97, 163-80 (2007). doi: 10. 1093/toxsci/ kfm018.

[121] DiMatteo, M., Antonini, J. M., Van Dyke, K., Reasor, M. J.: Characteristics of the acute-phase pulmonary response to silica in rats. J. Toxicol. Environ. Health A 47, 93-108 (1996). doi: 10. 1080/009841096161951.

[122] Balas, F., Manzano, M., Horcajada, P., Vallet-Regi, M.: Confi nement and controlled release of bisphosphonates on ordered mesoporous silica-based materials. J. Am. Chem. Soc. 128, 8116-7 (2006). doi: 10. 1021/ja062286z.

[123] Lai, C. Y., et al.: A mesoporous silica nanosphere-based carrier system with chemically removable CdS nanoparticle caps for stimuli-responsive controlled release of neurotransmitters and drug molecules. J. Am.

Chem. Soc. 125, 4451-9 (2003). doi: 10. 1021/ja028650l.

[124] Kaur, G., Rath, G., Heer, H., Goyal, A. K.: Optimization of protocell of silica nanoparticles using 3 (2) factorial designs. AAPS Pharm. Sci. Tech. 13, 167-73 (2012). doi: 10. 1208/s12249-011-9741-8.

(吴国华译)

第 38 章　用于抗原递送系统的口服佐剂

Carlos Gamazo 和 Juan M. Irache①

摘要

动物（包括人类）已经进化出一套复杂的黏膜免疫系统。通过对黏膜免疫激活的了解，我们可以合理地设计佐剂，从而引起持久的保护性免疫反应。在本章中，我们将讨论已知的黏膜免疫和口服佐剂。尤其是，我们将特别关注作为黏膜佐剂的聚合物纳米颗粒，其已被研制用于刺激疫苗产生终身免疫记忆。

38.1　引言

通常情况下，含有活微生物的疫苗非常有效且低成本，但其具有毒力返强的风险，并对免疫功能低下的个体诱导疾病；相比较而言，基于亚单位病原体的疫苗更安全，但需要使用佐剂进行诱导保护。面对如此形势，业界（尤其是兽医和制药公司）更愿意趋向于疫苗的使用便利性和利润，而不是疫苗的安全性。然而，找到成本合理的抗原复合物佐剂组合，甚至易于口服途经使用的疫苗是许多研究人员的强大动力，这是一项巨大的挑战[1]。

在大多数疫苗学家看来，关键在于佐剂。本章主要介绍口服佐剂。从目前市场来看，尽管佐剂的历史已近百年，但很少有佐剂符合这一要求。1925—1926 年，Ramon、Glenny 及 col. 发表了他们的研究结果，证明了通过吸附“惰性”物质，如木炭、火棉胶颗粒、葡聚糖或硫酸铝钾，可以提高类毒素的免疫力[2,3]。

这些物质可称为佐剂（拉丁语，佐剂：“辅助”）。根据 Glenny 的研究成果，铝盐被用于破伤风和白喉类毒素的疫苗制剂中。整体来看，从那时起就开始实验性地使用佐剂。虽然已研制了佐剂，但不溶性铝盐仍然是世界上最常用的方式。此外，如果我们还需面临口服的挑战，则更难实现目标。使用疫苗接种诱导黏膜免疫是一项相当困难的任务。为说明黏膜疫苗研制的复杂性，我们可以查看已批准用于人类使用的有限数量的口服疫苗：减毒脊髓灰质炎疫苗，减毒霍乱疫苗、减毒伤寒疫苗、减毒结核杆菌（BCG），减毒轮状病毒疫苗，以及唯一的非生物口服疫苗——含有霍乱毒素 B 亚基的霍乱弧菌菌苗。很明显，活体——一种在宿主体内复制的减毒疫苗是我们需要关注的重点；实际上，当前的趋势是制备佐剂，以帮助非生物疫苗模拟减毒活疫苗，但没有活疫苗的本身的危险性。面对黏膜免疫系统的特殊性，成功的口服疫苗接种需要有效的黏膜佐剂（包括特殊的给药系统）[4]。

38.2　黏膜免疫系统

宿主对外部世界开放，具有广泛的内部膜层，即所谓的黏膜。因此，免疫防御集中在其上，而且，大约 60%的全身淋巴细胞在沿着黏膜组织的囊泡中，这种黏膜组织通常泛称为黏膜相关

---

① C. Gamazo，PhD（✉）（纳瓦拉大学微生物系，西班牙潘普洛纳，E-mail：cgamazo@ unav. es）
J. M. Irache，PhD（纳瓦拉大学药学与制药技术系，西班牙潘普洛纳）

淋巴组织（MALT）[5]。根据其物理情况，主要系统包括内脏相关淋巴组织（GALT）、鼻咽相关淋巴组织（NALT）、支气管相关淋巴组织（BALT）或结膜相关淋巴组织（CALT）。无论解剖位置在哪儿，MALT在功能上都是相连的，被视为“共同的黏膜免疫系统”。因此，通过口服接种疫苗后，我们有望发现特定记忆细胞和效应反应，如肠道、阴道和其他黏膜道的特异性IgA。

对口服免疫而言，免疫系统面临的一大挑战是黏膜水平耐受性和炎症反应之间能否达到恰当平衡。这是多种因素作用的结果，首先取决于GALT的特殊解剖结构。肠道是哺乳动物最大的淋巴器官，比任何其他器官（包括脾脏和肝脏）都含有更多的免疫细胞和最大浓度的抗体[6]。抗原可通过不同的途径穿过上皮细胞屏障。在任何情况下，树突状细胞（DC）充当哨兵，吸收抗原，然后迁移至派尔集派尔集合淋巴结（PP）中的可迁移至上皮下区域（SED）或肠系膜淋巴结（MLN）以激活初始T细胞（图38.1）。实际上，肠道树突状细胞被视为决定耐受性与炎症反应的关键细胞[7]。树突状细胞有许多不同的亚群，其“自然”或诱导性倾向程度不同，可递呈并释放特定的细胞因子，并随后产生特定的免疫反应。我们可以发现不同成熟程度下固有层（LP）和PP处的DC。有些是“常驻”，而有些则是“移民”。一般而言，常驻黏膜DC在微环境条件下有利于致耐受性反应，而新出现的炎症性树突状细胞可能会避免这种情况，并引起炎症性反应（图38.2）。这是黏膜接种中要考虑的一个主要问题。

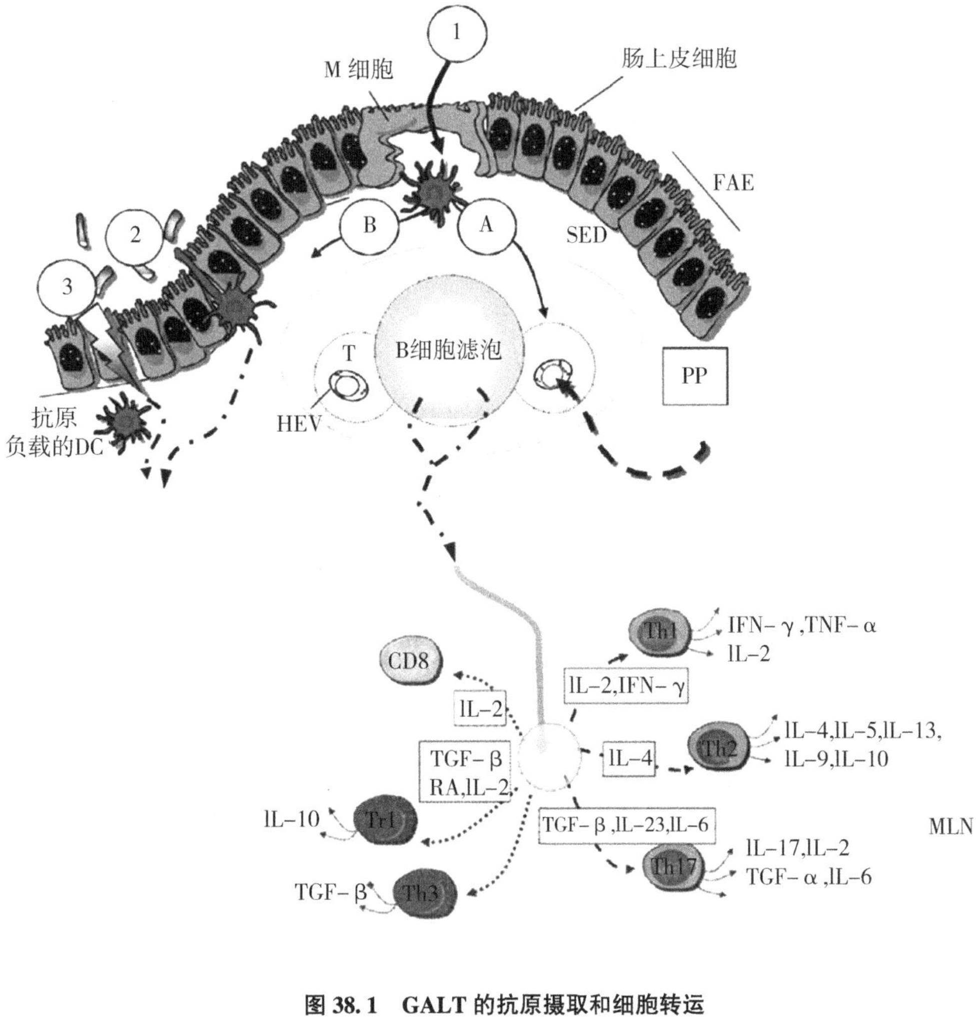

图38.1 GALT的抗原摄取和细胞转运

基本上，未成熟的树突状细胞存在于周围组织，主要是吞噬细胞；成熟的树突状细胞存在于PP和MLN中，因为它们的特征在于T细胞活化所需的协调刺激分子的表面递呈。成熟的DC

是由炎症刺激引发的成熟过程后的未成熟细胞产生的，这将导致这些 DC 大量迁移至引流淋巴结。在这种情况下，会诱导炎症反应。然而，当未成熟或半成熟树突状细胞从周边组织到达引流淋巴结时，则诱导免疫耐受（图 38.2 和图 38.3）。

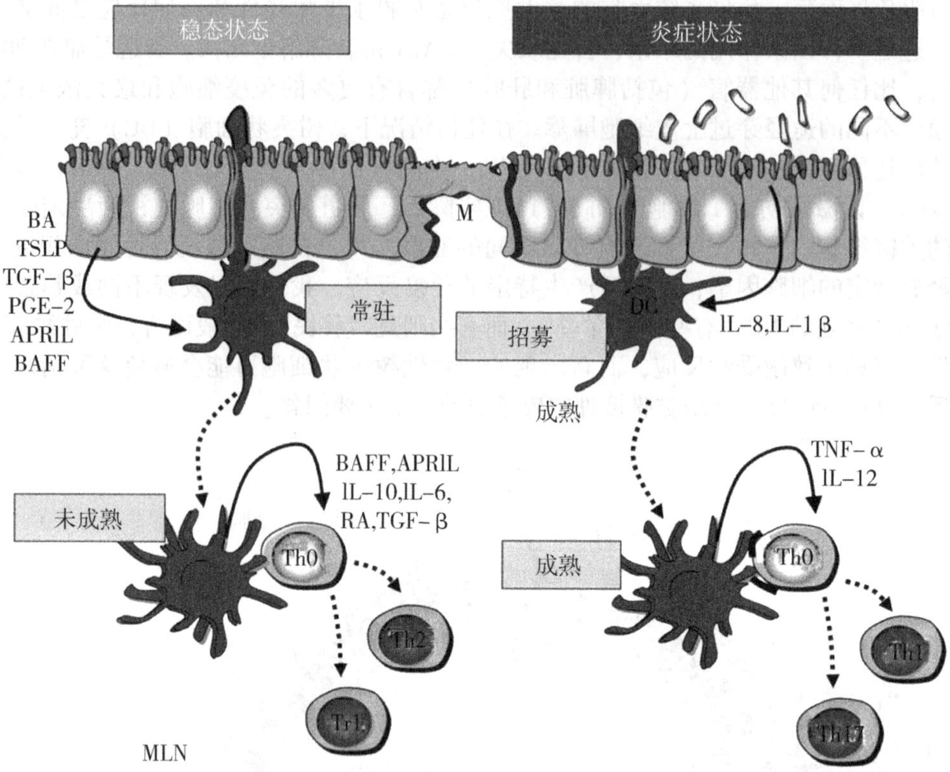

图 38.2　肠相关淋巴组织

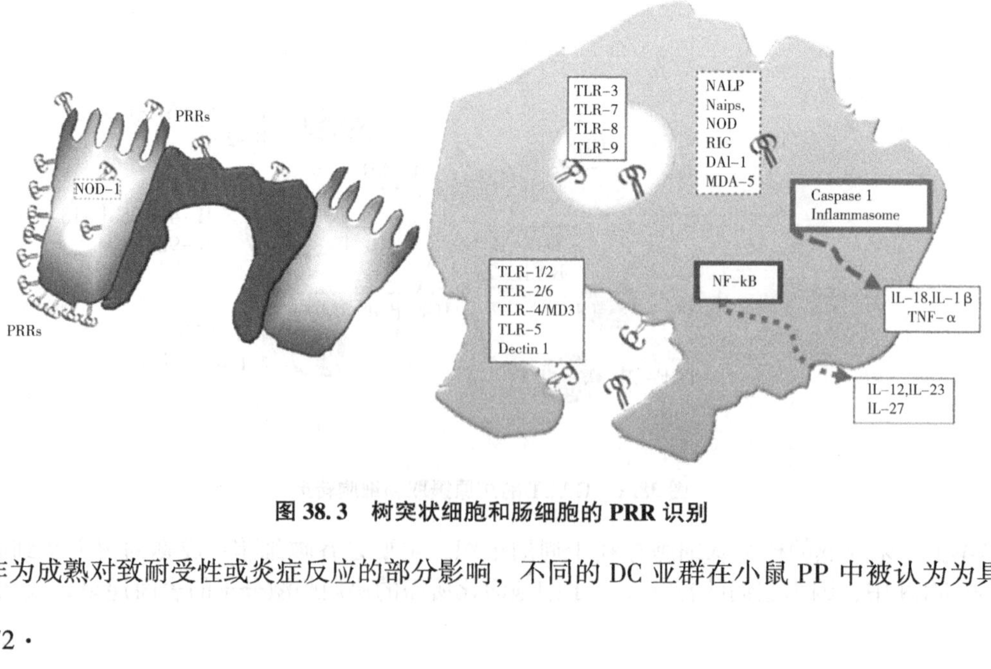

图 38.3　树突状细胞和肠细胞的 PRR 识别

作为成熟对致耐受性或炎症反应的部分影响，不同的 DC 亚群在小鼠 PP 中被认为为具有特殊功能[8,9]。例如，表型$CD11b^{+}CD8\alpha^{-}CCR6^{+}$主要发现于SED内的表面滤泡相关上皮，因此最先摄取M细胞转运的腔内抗原，优先诱导Th2分化和IgA浆母细胞[10]。黏膜$CXCR1^{+}CD11b^{+}DC$是适应性免疫最重要的激活因子反应和疫苗诱导的保护性免疫。然而，固有层$CD103^{+}DC$诱导调节性T细胞，对诱导和维持免疫耐受非常重要[11]。由专职的APC（如$CXCR1^{+}CD11b^{+}DC$）和特定的微折叠（M）完成黏膜表面抗原摄取。

可以推断，不同的条件DC亚群或非条件DC亚群将根据穿过肠上皮的途径处理抗原，从而引起不同的免疫反应[12]。所使用的佐剂在迁移路线中发挥主要作用，我们将在以下章节进行讨论。

GALT的诱导部位包括阑尾、孤立的淋巴滤泡集结、派尔集合淋巴结（PP）（小肠）和淋巴腺复合体（大肠）。孤立的淋巴滤泡虽较小，但与淋巴腺复合物与PP相似（如图38.1所示）。整体而言，它们的组织结构类似于淋巴结，包含APC网络、B细胞和T细胞区，并通过传出淋巴管与宿主系统连接。特异性免疫反应的成功激活依赖于这种组织的进化，以调节T细胞和B细胞的活化。

滤泡相关上皮（FAE）是由肠上皮细胞和微折叠细胞（M细胞）组成的单层。在小鼠和人体中，M细胞分别占FAE的10%和1%，其由正常的肠细胞进化而来，用于捕获和转运未修饰的抗原。因此，M细胞含有低微绒毛、少量溶酶体，缺乏与膜有关的水解酶，并且与B细胞、T细胞和APC所在的开放囊一致，以便于捕获经细胞抗药的抗原。其下是由常驻DC构成的网状结构，以及包含由T细胞区分隔的B细胞囊泡的上皮下区域（SED），称为滤泡间区域，也富含APC。T区还包括用于免疫细胞循环的高内皮小静脉（HEV）。B细胞囊泡面积是T细胞的5倍，这与黏膜部位IgA反应的重要性一致。

固有层是黏膜上皮下的一层高度血管化的疏松结缔组织，被视为GALT内的效应部位。

图中所示为惰性颗粒从内腔进入固有层的3种已知途径：①通过M细胞；②通过DC将树突扩张至内腔；③通过受损或破坏的上皮细胞。

主要途径是通过FAE中的M细胞。受体（PRR和pIgR）对抗原识别和转运非常重要。转移至DC后，抗原可以直接呈现至派尔集合淋巴结（A）中的T细胞上，或抗原或抗原负载DC通过随后的T细胞识别（B）进入肠系膜淋巴结（MLN）。在这两种情况下，DC通过传入淋巴区进入滤泡间T细胞富集区，而在此处DC发现通过HEV到达的淋巴细胞。随后，DC特异性T细胞离开此部位成为效应细胞或记忆T细胞，或者可以迁移到边缘区，并帮助初始和记忆B细胞。然后，B细胞迁移至生发中心，在生发中心会激活体细胞超突变和转换。最后，B细胞分化为记忆B细胞或浆细胞，并通过HEV离开生发中心。反之，活化的T细胞获得整联蛋白和趋化因子受体递呈，以离开MLN，并在进入血流后进入绒毛固有层，到达不同的MALT效应点。如果抗原通过效应部位（绒毛）进入，可能会出现类似抗原或抗原负载DC转运至MLN的过程。

目前，我们假设（为进行简化）$CD4^{+}T$细胞被细分为辅助T细胞（Th1、Th2、Th17）和调节性T细胞（包括Tr1和Th3）。已有研究表明，在T细胞活化过程中，信号细胞因子的存在会指引产生的反应。因此，Th1细胞是释放IFN-γ作为关键特征的炎性细胞，并且主要涉及炎症和针对细胞内病原体的保护性免疫。Th2细胞主要涉及B细胞辅助，因为它们会释放诸如IL-4的B-细胞生长因子。Th17细胞可加重炎症反应，在宿主对多种细菌和真菌的防御中发挥重要作用。另外，有几个亚群的T调节细胞抑制调节性T细胞的功能，有助于维持黏膜免疫和控制炎症性疾病。

38.3　口服佐剂

黏膜免疫后，肠道的反应是驱动耐受性或非炎症反应（图38.2）；然而，通过使用合适的佐剂，有可能将其消除，从而获得通常所需的平衡Th1/Th2/Th17反应。简而言之，在黏膜耐受性压力下，口服佐剂采用适当的方式激活 DC。然而，在此之前，口服佐剂需具有强大的能力来面临酸性条件、疏水环境、多样的水解酶、黏液蛋白网络和覆盖其靶标的密集微生物群，以及推动其前进的蠕动力（表 38.1）。因此，目前市场上口服佐剂的数量很少也就不足为奇了。

肠上皮黏膜基本上可分为两种状态：稳态状态和炎症状态。这两种情况主要取决于肠上皮细胞决定的微环境以及 DC 的子集类型和活化状态。在稳定的非活化黏膜中，腔内抗原的摄取主要由位于固有层的未成熟 DC 进行介导（图 38.1）。DC 通过分泌细胞因子来调节 T 细胞亚群的活化和分化。例如，IL-12（p70）促进 Th1 细胞；IL-10 有利于 Th2 或 Treg 反应及诱导 DC 释放 TGF-β 的刺激；IL-6 和 IL-23 促进 Th17 分化。

此外，肠细胞会分泌一些重要的细胞因子（图 38.3），“推动”肠 DC 诱导 Treg 分化。肠细胞可能通过释放这些所谓的“调节因子”来抑制炎症反应，这些“调节因子”驱动分化为致耐受性未成熟树突状细胞。因此，维甲酸（RA）、胸腺基质淋巴细胞生成素（TSLP）、PGE-2 和 TGF-β 促进致耐受性 DC，并有利于释放细胞因子［包括 BAFF（B 细胞激活因子）和 APRIL（细胞因子增殖-诱导配体）］，驱动 Treg 或 Th2 反应和 IgA 生成细胞。此外，从肠上皮细胞或这些致耐受性 DC 释放的自身调节细胞因子可以与 T 细胞无关的“原始”方式促进 IgA 转为 IgB 细胞。当出现刺激或上皮屏障损坏，上皮细胞和新 DC 释放 IL-1 和 IL-8，以吸引 DC 和巨噬细胞来解决这些问题。免疫细胞向受损区域的特征性补充部分依赖于上皮细胞释放的趋化因子 IL-8。随后，抗原由树突状细胞转运至 MLN，并在此处引发炎症免疫反应。

在微生物学教学中，我们经常使用模拟术语描述肠道病原体在被摄入后传播过程中可能面临的恶劣条件。已经不断改进调整了病原体工具以发挥作用，进而不断突破。因此，大多数成功的口服佐剂即使是明显模仿了肠道病原体的佐剂也就不足为奇了。目前也有人正作为口服佐剂来研究来自微生物的外毒素、凝集素和其他结构成分。

外毒素

外毒素是一种蛋白质，其中许多具有催化活性，由一些细菌分泌，通过宿主进入体内。由霍乱弧菌（霍乱毒素，CT）和大肠杆菌（不耐热毒素，LT）分泌的肠毒素、影响胃肠道细胞的外毒素是目前研究最多的黏膜佐剂[13]。亚基 A，具有酶促 ADP 核糖基转移酶活性；亚基 B，与哺乳动物细胞上的 GM1 神经节苷脂受体结合（包括 APC）[14]。其可通过黏膜或全身途径给药，对抗体和细胞免疫反应有影响，包括细胞毒性 CD8 T 细胞反应。具体而言，APC 的 CT 暴露对 IL-1、IL-6 和 IL-10 的生成，以及 IL-12 和 TNF-α 的下调具有增强作用，表明促炎症和抗炎症功能[15,16]。此外，CT 佐剂抗原在口服免疫后产生长期特异性记忆 B 细胞[17-19]，因为其具有促进大量生发中心和 CD86 上调的强大能力。这一点非常重要，因为免疫后 GC 的形成与 B 记忆细胞的分化以及对高亲和力和类别转换抗体的成熟有关[20]。

然而，外毒素主要缺点之一是具有毒性。由于所有有核细胞（包括上皮细胞和神经细胞）均含有 GM1 神经节苷脂受体，CT 和 LT 可诱发腹泻和神经系统疾病[21]。

宿主的一个基本特征是区分潜在病原体和非致病性共生微生物群。这种特征仅存在于一系列模式识别受体（PRR）中，这些蛋白质能够识别病原体中共有但在宿主中并不存在的保守分子标记。这些危险信号可能以病原体相关分子模式（PAMP）的形式存在于病原体中，或可能是其作为损伤相关分子模式（DAMP）作用的产物。PAMP 和 DAMP 由树突状细胞（DC）和肠上皮细胞递呈的 PRR 直接进行检测，然后这些细胞或释放致耐受性或促炎性细胞因子（图 38.2）。因此，递呈破坏宿主体内平衡的毒力因子的微生物会触发多个 PRR 系统的快速动员，从而导致炎症反应。与此相反，正常的微生物群会导致致耐受性反应[22]。因此，DC 抗原递呈的结果是成熟水平的作用：不成熟的 DC 更有可能产生耐受性，而完全成熟的 DC 将引发强大的 T 细胞免疫[23]。

Toll 样受体（TLR）是跨膜 PRR，而 Nods、Naips、Nalps、RIG-I、MDA-5 和 DAI-1 位于细

胞溶质中。TLR的连接激活信号级联，其导致转录因子NF-κB的激活，以及随后的共激分子以及炎性细胞因子和趋化因子和激素因子的上调。另一方面，Nalp和Naip PRR成分控制炎性小体的激活，而炎性小体是一种负责处理和分泌IL-1b的多蛋白复合物。

目前，已经报道了哺乳动物中的13种TLR，分别具有不同的性能参与引发的免疫反应[24]。因此，对相应PAMP的特异识别将直接导致先天差异和进一步的适应性免疫反应。TLR3（识别来自病毒的双链RNA）、TLR4（细菌脂多糖）、TLR5（细菌鞭毛蛋白）、TLR7（来自病毒的单链RNA）和TLR9（含细菌cpg DNA）偏向于Th1反应。相比之下，TLR2（识别细菌肽聚糖和脂肽）可引起Th2反应，尽管与TLR6组合可引起调节性反应（Th3/Treg）[25]。

肠细胞能转录所有类型TLR的mRNA[26]。实际上，TLR在肠上皮细胞中的递呈要高于肝脏等其他主要器官。然而，为避免永久性炎症，有几种机制可以维持对正常微生物群的耐受性。因此，肠上皮细胞中的TLR数量和位置有很大的影响。例如，TLR-4仅在高尔基体内部递呈，这意味着LPS仅在能够穿透这些细胞时才会激活肠上皮细胞。此外，大多数TLR在基底侧递呈，且只有少数在根尖区递呈。胞质PRR，如Nod样受体（NLR），也对肠上皮细胞非常重要。M细胞也递呈几种PRR，如TLR-4、PAF-R（血小板活化因子受体）和α5β1整合蛋白[27]。TLR和NLR在肠道中的递呈有所差异。这一重要演变结果使沿着肠上皮面涵盖了特定病原体的识别信号，并且对疫苗学具有直接作用，因为肠不同部位可能对相同的抗原刺激反应不同[7]。

结合基和酶亚基都有助于免疫调节，尽管酶活性似乎非常重要[28]。CTA1-DD是一种缺乏B亚基的无毒突变体，酶促A亚基与金黄色葡萄球菌（抗体结合域）蛋白A连接[29]，在鼻腔给药后仍保持佐剂活性[30,31]，但口服后无活性。目前，已有人提出A1亚基中的其他CT和LT突变体（TK63、LT196/211、LTR72或CT112[32]通过口服途径保留了佐剂功能，尽管LTK63和LTR196/211仍然能够结合GM1神经节苷脂，因此可能有毒）[33]。

知识点38.1　研制优良的口服疫苗要点

要避免

耐受性：口服疫苗接种过程中，剂量和剂量方案是主要的难题。单次高剂量或重复低剂量抗原的口服给药可诱导以调节Tr1和Th3亚群为特征的耐受性。另外，在强毒攻击后，例如由于致病性病原体破坏上皮细胞，或是使用强效佐剂，可引起高炎症反应（Th1/Th2/Th17）（图38.1）[7]。

宿主中的微量营养素缺乏状态：如维甲酸（维生素A）或锌缺乏，尤其在发展中国家这种微量营养素缺乏的人群较多，可影响肠道树突状细胞和T细胞的分散亚群以降低免疫反应[34]。

要做到

疫苗最终产品：安全、无毒、无污染。

疫苗制备工艺：标准化、重复性好、成本相对低廉、易于推广。

对胃低pH和水解消化酶具有抗性。

能抵抗母源中和抗体的干扰[35]。

增加肠细胞的吸收，即用配基修饰特定受体，因为已观察到抗原穿过肠上皮细胞到达GALT的不良转运导致免疫无应答[36,37]。

诱导长期保护性免疫：APC通过MHC Ⅰ类和Ⅱ类途径[38]交叉递呈抗原，诱导Th1、Th2、Th17、T_{FH}或Treg细胞的发育[39,40]。

38.4　微生物定植因子

另一种策略是模仿微生物的定植过程。肠道病原体利用不同的结构通过M细胞和肠细胞进行附着和侵袭，导致它们在肠上皮细胞间的转运[41]。这种仿生方法已导致呼肠孤病毒衣壳蛋白[42]、来自不同细菌的FI菌毛[43-45]或来自耶尔森菌的外膜蛋白[46]的应用。这种方法的主要局限性在于，疗效取决于特异性C受体的宿主表达，并且微生物黏附物是免疫原性的，因此，预先存在的抗体可以对其进行中和。

微生物相关分子模式

通过类似模拟的方法使其他结构——称为微生物相关分子模式（MAMPs），其中包括PAMPs（病原体相关分子模式）得以获得应用。这些可能是微生物结构或代谢物的标记物，被认为是对宿主的危险分子，引起细胞因子和趋化因子的释放，从而引发炎症防御反应。MAMPS在细胞、免疫细胞和上皮细胞上的特殊受体称为PRRS，包括TLR和NLR（非TLR受体）。因此，用携带PRR激动剂的佐剂修饰疫苗可能是面对自然黏膜耐受性倾向的一种出路[47]。

主要的PAMP为来自革兰氏阴性细菌的脂多糖（LPS）[48]。TLR4/MD-2异二聚体可识别脂多糖，通过肠细胞和M细胞介导细菌移位[49]，使其适合作为口服佐剂。然而，称为内毒素的LPS，可能具有有害的生物活性和有害的副作用。目前正研究具有较低毒性的LPS的天然或合成衍生物[48]。布鲁氏菌LPS因其毒性低，但仍能诱导Th1对携带抗原产生反应而被广泛应用[50]。来自肠沙门氏菌的单磷酸脂类A（MPL）。Minnesota被FDA批准用于人乳头瘤病毒疫苗的AS04佐药Cervarix™，与明矾结合使用[51]。AS02佐剂也含有MPL，但在水包油乳剂中与QS-21结合。在一项针对非人类灵长类动物的HIV疫苗研究中，已证明含有MPL和皂苷脂质体的AS01可以增强全身免疫和黏膜免疫[52]。

鞭毛蛋白是一种主要的危险分子，因为其不含细菌。鞭毛蛋白是组成细菌鞭毛的单体蛋白，因此，其作为一种PAMP被进行广泛研究，因为肠上皮细胞递呈的鞭毛蛋白结合TLR5[53]适合用于口服佐剂[54,55]。鞭毛蛋白诱导肠DC成熟，激活体内$CD4^+T$细胞，促进混合效应细胞反应的发育[56]。鞭毛蛋白作为黏膜佐剂，几乎与CT或LT具有同样的效果，但更安全[53]。

微生物富含未甲基化的CpG寡脱氧核苷酸（CpG），因此宿主将其视为危险信号。CpG结构域与TLR9相互作用，形成强烈的Th1反应。当在黏膜递送时，其已在动物疫苗研究中显示有效[57]。人体临床试验证明，CpG结构域具有良好的安全性，为扩大CpG在黏膜疫苗中的应用提供了科学依据[58]。

其他有效作为口服佐剂的PRR激动剂包括α-半乳糖酰基鞘氨醇、CD1d配体和NKT细胞激活剂[59]，以及结合APC上非TLR PRR的真菌或细菌结构，如树突状细胞相关性C型植物血凝素1[60]、甘露糖[61-63]或细菌第二信使3′，5′-环二鸟苷酸被视作为PAMP和有效黏膜佐剂宿主的危险信号[64]。

38.5 非微生物配体

凝集素是与特定碳水化合物结合的蛋白质。植物凝集素是常用的凝集素之一，因其对肠道降解更具抗性，因此也易于用作M细胞和肠细胞的靶向配体，其表面由糖蛋白和糖脂覆盖。在几种二聚体植物凝集素碳水化合物中，最常研究的凝集素是荆豆凝集素1（UEA-1），其与M细胞、杯状细胞和潘氏细胞上的α-胰岛素结合[65,66]。然而，毒性和抗营养特性是限制其用于口服疫苗接种的主要缺点[67]。

已研制了针对细胞表面抗原的抗体，如M细胞的单克隆抗体[68]。

已经研制了几种具有免疫刺激特性细胞因子。这些包括重组IL-1家族细胞因子（IL-1α/β、IL-18和IL-33）或胸腺基质淋巴细胞生成素（TSLP）[69,70]。然而，成本与功效比并不理想。

上述策略提出了优点和需要解决的不足之处。如下文所述，抗原递送系统已准备克服所有口服疫苗接种的种种障碍。

38.6　用于口服的纳米颗粒系统

制药技术专家正研制许多粒子系统，用于将抗原递送至宿主。其主要特征具体表现为：保护抗原不受消化酶的影响；如果其主要特性用上皮受体特异性配体修饰，尤其会促进肠上皮细胞吸收[11]；提高 APC 对抗原的吸收，或通过增加专业 APC 流入注射部位，并通过 MHC Ⅰ类和Ⅱ类途径的抗原交叉递呈[38]。不同类型的颗粒递送系统中有脂质体、原生脂质体，其由原核古细菌的脂质生成[71]。Iscomatrix 由 Quil A（植物皂甙）、胆固醇和磷脂组成，但不适用于亲水性抗原；VLP 由黏膜感染病毒的衣壳蛋白、脂质体、免疫刺激复合物（ISCOM）、酵母细胞膜和聚合物微或纳米颗粒、螺旋形和聚合物微粒给药系统制备而成[72-78]。

38.6.1　纳米颗粒

基于保护抗原免受极端 pH 条件、胆汁和胰腺分泌物的影响[79,80]，使用抗原装载或包裹在颗粒，特别是纳米颗粒的口服疫苗似乎具有良好的科学基础，并具有持久性效应[81]。同时，利用黏膜 APCs 自然捕获颗粒的固有倾向，作为其触发黏膜免疫对抗病原体的预警作用的一部分。[82-85]。事实上，已清楚地证明，纳米颗粒可以与黏膜的不同成分相互作用。

38.6.2　生物黏附

为增加纳米颗粒与黏膜上皮表面的紧密接触，进而提高其到达特异性免疫感受性细胞和 GALT 的能力，已经制定出不同的策略。其中一种策略可能是通过改造或修饰这些负载表面的配体来增强其生物黏附性能，以穿过保护黏膜表面的黏液层和/或靶向黏膜免疫功能细胞上的特异性受体。

聚乙二醇化是纳米颗粒表面修饰的一种常用技术。用聚乙二醇将纳米颗粒完全包被可减少纳米颗粒与黏蛋白的相互作用，并通过黏液层到达黏膜上皮[86,87]。此外，聚（酸酐）纳米颗粒的聚乙二醇化产生的载体可尽量减少与胃黏膜的黏附作用，并将其汇集于动物小肠黏膜[88]。

另一种可能是使用仿生方法，如涉及微生物定植过程或宿主免疫系统活化的纳米颗粒化合物或分子的修饰。微生物可通过多种不同的特异性粘附因子侵袭并定植于宿主组织，包括外膜蛋白[89]、鞭毛[90]、菌毛[91]、凝集素[92]和糖蛋白[55]。这些黏附因子大多数也被视为免疫调制剂，通常包含于 PAMP（病原体相关分子模式）的通用名称中。

最近，我们的研究小组证明，肠炎沙门氏菌或甘露糖胺中的鞭毛蛋白与 Gantrez AN 纳米颗粒的结合可以增强纳米颗粒的生物粘附能力。这些纳米粒子在动物肠道内具有很强的定植能力，具有特征性的回肠趋向性和对淋巴滤泡集结的高亲和力[62,93]（图 38.4）。

38.6.3　免疫细胞的直接相互作用

此外，纳米颗粒的使用还具有其他优势，比如可以将上述口服佐剂（如 CT、凝集素、PRR 激动剂）与抗原结合，进而增加并诱导更合适的免疫反应[94,95]。CTA1-DD 是一种无毒突变体，其能产生一种缺乏 B 亚基的霍乱毒素，在口服给药时并不稳定，但在微粒给药系统中受到保护[96-98]。

通过将肠道沙门氏菌鞭毛蛋白与 PVM/MA 纳米颗粒结合，获得了“类沙门氏菌”纳米颗粒[99]。这些载体显示出回肠的重要趋性，其在肠道内的分布与细菌的定植特性密切相关，包括在淋巴滤泡集结中广泛的浓度[62]。利用卵清蛋白作为模型抗原，“类沙门氏菌”纳米颗粒诱导强且平衡的 $IgG2\alpha^-$（Th1）和 IgG1（Th2）特异性抗体的分泌。此外，这些纳米颗粒能够诱导比对照组更强的黏膜 IgA 反应[99-103]。

表 38.1 总结了一些与纳米颗粒作为黏膜疫苗佐剂研制有关的其他示例。有几个因素会影响

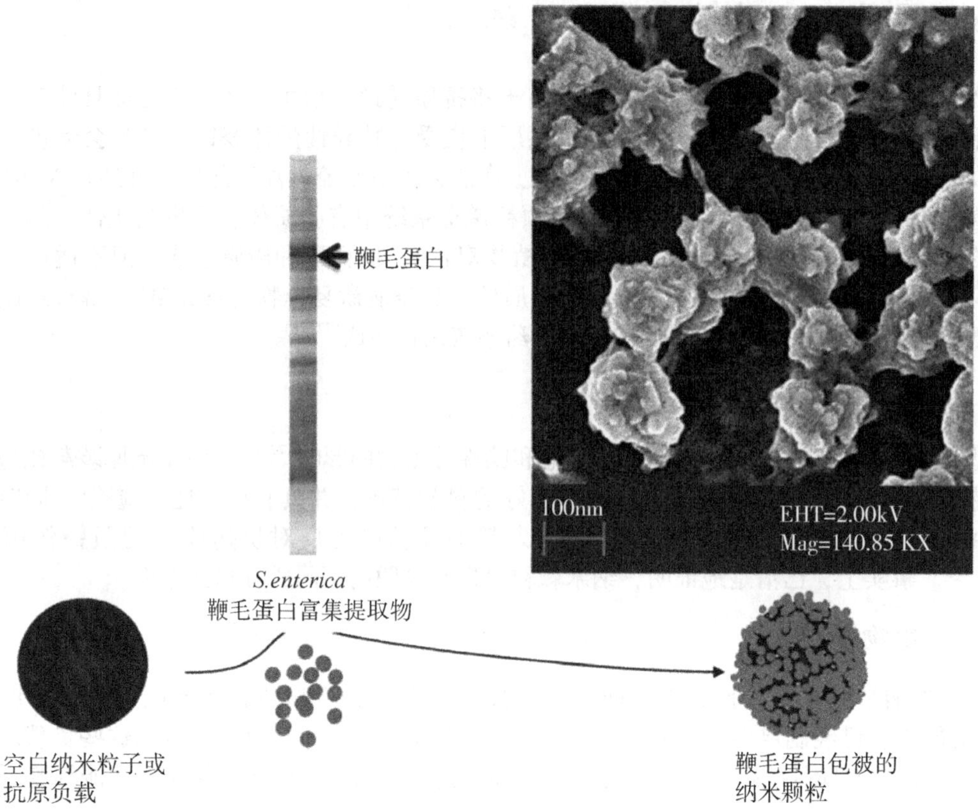

图 38.4 含鞭毛蛋白纳米颗粒的图形结构

微粒作为口服佐剂的有效性。聚合物与抗原的适当结合以及抗原负载的制备方法对微粒的物理化学性质、粒径和表面性质的测定有很大的影响。[113]。虽然颗粒大小与其佐剂作用之间的相关性一直存在争议[111,114-121]，但纳米级颗粒而非微尺度的颗粒似乎更适合胃肠道中的细胞摄取和免疫刺激[122-124]。

表 38.1 使用含抗原的可生物降解ᵃ纳米颗粒进行口服免疫示例

抗原	聚合物	观察结果	参考
BSA	疏水性碳纳米颗粒	血清 IgG 滴度和 BSA 与 FCA 乳化后肌内接种的结果接近	[104]
*艾美球虫*重组抑制蛋白	Montanide IMS 纳米粒子	减少粪便中寄生虫的排出减少受感染动物（鸡）的肠道病变	[105]
大肠杆菌 F4 菌毛抗原	Gantrez AN 纳米颗粒	脾脏中 F4-特异性抗体分泌细胞水平升高	[106]
卵清蛋白	含有 MPLA 的纳米颗粒	强烈的 IgG 免疫反应 IgA 滴度高	[107]
DNA 编码的 LCDV 主要衣壳蛋白	PLGA 纳米颗粒	降低鱼类感染率	[108]
HBsAg	翅荚百脉根植物凝集素包被的 PLGA 纳米颗粒	强烈的黏膜和全身免疫反应	[109]

（续表）

抗原	聚合物	观察结果	参考
HBsAg	聚乙二醇化 PLA 纳米颗粒	有效的细胞 Th-1 反应与黏膜体液免疫（IgA）	[110]
甲型流感抗原	Bilosomes（脂质囊泡）	当口服给药时，产生的抗体比肌内途径高	[111]
DNA 编码的屋尘螨过敏原 Der p2	壳聚糖纳米粒	高血清水平的 IFN-g 诱导 Th-1 反应	[112]

BSA 牛血清白蛋白、FCA 弗氏完全佐剂、MPLA 单磷酸脂类 A、LCDV 淋巴细胞病病毒、HBsAg 乙肝表面抗原、PLGA 聚（D，L-丙交酯-丙二醇酯）、PLA 聚（D，L-乳酸）。

[a]在生理条件下，从 PLGA、壳聚糖、脂质或甘曲聚糖中提取的纳米颗粒受到生物降解或生物侵蚀。从二氧化硅中提取的疏水碳纳米颗粒，目前还没有关于其降解性的信息，我们认为其在生理条件下不会降解。在任何情况下，还需考虑所有这些纳米颗粒的拟议给药途径是口服/黏膜途径。在口服时，蠕动、肠上皮细胞/黏膜、黏液更新等生理过程非常重要，对肠道内容物的清除起着重要作用。因此，纳米颗粒的降解没有负载抗原的充分释放重要。

另外，所用聚合物不仅决定了所制备纳米颗粒的抗原负载，而且决定了所制备纳米颗粒在胃肠道中的稳定性及其与黏膜成分的相互作用。值得注意的是，稳定性可能是影响负载抗原释放的一个重要因素。

细菌鞭毛蛋白除是特异免疫系统的合理靶点外，还可直接激活先天免疫细胞，进而分泌炎症细胞因子。这种免疫刺激能力由单核/巨噬细胞和树突状细胞递呈的表面受体 TLR-5 介导，也由肠道内的肠细胞介导。这种先天能力使鞭毛蛋白成为有效的佐剂。此外，鞭毛蛋白是黏膜细菌病原体用于侵入宿主表面，并因此移植肠道的黏附因子。考虑到这两种特性，鞭毛蛋白可以用作很好的黏膜辅助剂。图中显示了鞭毛蛋白包覆的聚（酸酐）纳米颗粒的制备过程。采用脱溶法制备聚（酸酐）纳米粒子，随后进行冷冻干燥。简言之，将来自肠炎沙门氏菌的鞭毛蛋白与聚合物（甲基乙烯基醚和马来酸酐的共聚物，Gantrez AN）在丙酮中一起温育。然后，通过加入乙醇和水的混合物获得纳米颗粒。在减压下除去有机溶剂，并通过与 100 μg 1，3-二氨基丙烷一起温育 5 分钟，使所得纳米颗粒交联。然后，通过离心纯化悬浮液，最后，使用蔗糖（5%）作为冷冻保护剂冻干纳米颗粒。SEM 显微照片显示了所得纳米颗粒的形态和形状，鞭毛蛋白在其表面直接递呈给肠上皮细胞，然后递呈给免疫功能细胞。

另外，聚 D，L-丙交酯聚乙交酯（PLGA）、壳聚糖、脂类、淀粉、磷腈、聚 ε-已内酯、聚酸酐和阳离子交联多糖也被视为口服疫苗的抗原递送纳米颗粒。

以这种方式，大量抗原成功地包裹于 PLGA 颗粒中，即使在临床试验[127]中，口服后仍能充分保持结构、抗原完整性和免疫原性[125,126]。

壳聚糖也可能具有免疫调节作用，因为已证明其在体外刺激免疫细胞产生细胞因子[128]，并在没有抗原的情况下增强黏膜水平的天然 Th2/Th3 偏移微环境[129]。

已证明由癸二酸、1，6-双（p-羧基苯氧基）己烷和 1，8-双（p-羧基苯氧基）-3，6-二氧硅氧烷共聚物制成的聚酸酐纳米颗粒体系在体内和体外均表现出生物相容性，其浓度预计可供人类使用[63,130-133]。

据报道，一些两亲性聚酸酐也具有佐剂特性[134]。尤其是甲基乙烯醚和马来酸酐（PVM/MA）共聚物制备的聚酸酐纳米颗粒体系，已证明其作为佐剂诱导 Th1 免疫反应的有效性[50,135-138]。此外，由于 PVM/MA 纳米颗粒具有独特的生物黏附性，可增强载抗原向肠道淋巴细胞的递送，因此可作为口服佐剂使用[79]。

颗粒与 DC 之间的相互作用取决于颗粒大小和形状、电荷和疏水性等特性，但 DC 成熟的机理可能主要与 DC 中的 TLR 识别有关。许多研究表明，TLR 激动剂刺激诱导共刺激分子的表面递呈，从而对成熟 DC 的典型特征进行表型调控。研究表明，PMVA 的聚酸酐纳米颗粒可诱导 TLR-2 和 TLR-4 依赖介导的先天免疫反应[128,138]。

这些纳米颗粒诱导 DC 成熟，CD40、CD80 和 CD86 明显上调，在动物模型中出现偏向 Th1 的反应。这是一项重要发现，因为最近有研究表明，使用纳米颗粒携带的多种 TLR 激动剂会影响长期记忆细胞的诱导，而刺激长期保护性免疫记忆是任何疫苗的最终目标[139,140]。

一般来说，关于佐剂效应和记忆形成的信息很少，关于黏膜佐剂效应和记忆反应的信息更少。通常由关键成功示例提供长期记忆范例，包括天花、麻疹、腮腺炎、脊髓灰质炎、风疹和典型的黄热病[141,142]。使用减毒活疫苗免疫人，细胞介导的免疫可维持数十年。最后一种疫苗（YF-17D）含有几种 Toll 样受体的激动剂。对先天免疫激活的影响可以解释为何这种疫苗能有效地刺激长期记忆。实际上，成功研制疫苗佐剂的新方法是以 TLR 激动剂的使用为基础[143]。

其他聚合物经化学式配基修饰后潜在效果各不相同（上文）。例如，在体外实验中发现，聚乳酸和聚苯乙烯微粒可以像明矾一样有效地通过 NLRP3 激活 caspase-1[144]。其他实验佐剂包括植物皂甙，一种从皂树皮中提取的皂苷以及壳聚糖[145]也证明可以介导 NLRP-依赖性 IL-1 释放。

结论

疫苗接种覆盖率，是当今社会尤其在发展中国家面临的主要挑战[146]。与疫苗研制相关的若干关键问题如下：安全性、低成本生产、大规模接种和便利的黏膜给药等。亚单位疫苗的黏膜给药既安全又方便，但其需要强有力的佐剂来克服上皮屏障。纳米颗粒作为一组具有佐剂功能的递送系统迅速兴起。然而，不管纳米技术抗原递送的前景如何（表 38.1），需要更好地了解人体黏膜免疫反应以确定这些佐剂的有效性，目前，还没有制备且许可用于黏膜给药的纯亚单位疫苗。这主要是由于很少有佐剂被批准用于人类[147,148]。毒素作为黏膜佐剂的主要缺点已引起对新黏膜佐剂研制的攻关研究[149,150]。因此，这是我们面临的主要挑战，即开发佐剂，以增强未来通过不同黏膜途径接种的亚单位疫苗的效力。

亚群疫苗的主要挑战是不能诱导终生免疫记忆，而减毒活疫苗，如牛痘病毒疫苗、麻疹、腮腺炎、脊髓灰质炎、风疹或黄热病能诱导终生免疫记忆。因此，未来的佐剂设计可关注减毒活疫苗的作用机制。黏膜免疫后，肠道的默认反应是引起致耐受性或非炎症反应，这是食物摄取的进化结果。未成熟的树突状细胞没有太多的共刺激分子，对肠道内容物进行连续采样，将产生抗炎反应。然而，病原体和减毒疫苗激活多种途径并呈现“危险分子”（PAMP），其包括 Toll 样受体、C 型凝集素受体、RIG-1 样受体和 NOD-样受体[151]，其与黏膜上皮细胞结合，导致上皮细胞释放细胞因子和趋化因子，吸引免疫细胞诱导炎症反应。因此，用携带 PAMP 或其他 PRR 激动剂的免疫佐剂进行疫苗修饰，可能是一种解决黏膜天然致耐受倾向的方法。综上所述，确定调节 Th1、Th2、Th17、T_{FH}或 T_{reg}的方法将是成功研制新型疫苗佐剂的关键。

因此，纳米颗粒能够同时诱导外周和黏膜免疫，已被证明是良好的口服佐剂。通过使用纳米微粒递送系统，有可能消除耐受性，并在黏膜和系统部位产生所需的平衡 Th1/Th2/Th17 型反应，为黏膜疫苗佐剂领域开辟新的空间。

参考文献

［1］ Nepom，G. T.：Mucosal matters. Foreword. Nat. Rev. Immunol. 8，409（2008）.

［2］ Ramon，G.：Sur l’augmentation anormale de l’antitoxine chez les chevaux producteurs de serum antidiphterique. Bull. Soc. Centr. Med. Vet. 101，227-234（1925）.

[3] Glenny, A., Pope, C., Waddington, H., Wallace, V.: The antigenic value of toxoid precipitated by potassium-alum. J. Path. Bacteriol. 29, 38-45 (1926).

[4] Holmgren, J., Czerkinsky, C.: Mucosal immunity and vaccines. Nat. Med. 11, S45-S53 (2005).

[5] Czerkinsky, C., Holmgren, J.: Mucosal delivery routes for optimal immunization: targeting immunity to the right tissues. Curr. Top. Microbiol. Immunol. 354, 1-18 (2012).

[6] Mayer, L.: Mucosal immunity and gastrointestinal antigen processing. J. Pediatr. Gastroenterol. Nutr. 30, S4-S12 (2000).

[7] Ng, S. C., Kamm, M. A., Stagg, A. J., Knight, S. C.: Intestinal dendritic cells: their role in bacterial recognition, lymphocyte homing, and intestinal infl ammation. Infl amm. Bowel Dis. 16, 1787 - 1807 (2010).

[8] Persson, K. E., Jaensson, E., Agace, W. W.: The diverse ontogeny and function of murine small intestinal dendritic cell/macrophage subsets. Immunobiology 215, 692-697 (2010).

[9] Rescigno, M., Di Sabatino, A.: Dendritic cells in intestinal homeostasis and disease. J. Clin. Invest. 119, 2441-2450 (2009).

[10] Sato, A., Hashiguchi, M., Toda, E., Iwasaki, A., Hachimura, S., Kaminogawa, S.: CD11b^{+}Peyer's patch dendritic cells secrete IL-6 and induce IgA secretion from naive B cells. J. Immunol. 171, 3684-3690 (2003).

[11] Devriendt, B., Devriendt, B., De Geest, B. G., Goddeeris, B. M., Cox, E.: Crossing the barrier: targeting epithelial receptors for enhanced oral vaccine delivery. J. Control. Release 160, 431-439 (2012).

[12] Dudziak, D., Kamphorst, A. O., Heidkamp, G. F., Buchholz, V. R., Trumpfheller, C., Yamazaki, S., Cheong, C., Liu, K., Lee, H. W., Park, C. G., Steinman, R. M., Nussenzweig, M. C.: Differential antigen processing by dendritic cell subsets in vivo. Science 315, 107-111 (2007).

[13] Freytag, L. C., Clements, J. D., Eliasson, D. G., Lycke, N.: Use of genetically or chemically detoxifi ed mutants of cholera and Escherichia coli heat-labile enterotoxins as mucosal adjuvants. In: Levine, M. M. (ed.) New Generation Vaccines, 4th edn, pp. 273-283. Informa Healthcare USA, New York (2010).

[14] Fan, E., O'Neal, C. J., Mitchell, D. D., Robien, M. A., Zhang, Z., Pickens, J. C., Tan, X. J., Korotkov, K., Roach, C., Krumm, B., Verlinde, C. L., Merritt, E. A., Hol, W. G.: Structural biology and structure-based inhibitor design of cholera toxin and heat-labile enterotoxin. Int. J. Med. Microbiol. 294, 217-223 (2004).

[15] Lavelle, E. C., Jarnicki, A., McNeela, E., Armstrong, M. E., Higgins, S. C., Leavy, O., Mills, K. H.: Effects ofcholera toxin on innate and adaptive immunity and its application as an immunomodulatory agent. J. Leukoc. Biol. 75, 756-763 (2004).

[16] Lavelle, E. C., McNeela, E., Armstrong, M. E., Leavy, O., Higgins, S. C., Mills, K. H.: Cholera toxin promotes the induction of regulatory T cells specifi c for bystander antigens by modulating dendritic cell activation. J. Immunol. 171, 2384-2392 (2003).

[17] Lycke, N.: Targeted vaccine adjuvants based on modifi ed cholera toxin. Curr. Mol. Med. 5, 591 - 597 (2005).

[18] Vajdy, M., Lycke, N. Y.: Cholera toxin adjuvant promotes long-term immunological memory in the gut mucosa to unrelated immunogens after oral immunization. Immunology 75, 488-492 (1992).

[19] Soenawan, E., Srivastava, I., Gupta, S., Kan, E., Janani, R., Kazzaz, J., Singh, M., Shreedhar, V., Vajdy, M.: Maintenance of long-term immunological memory by low avidity IgM-secreting cells in bone marrow after mucosal immunizations with cholera toxin adjuvant. Vaccine 22, 1553-1563 (2004).

[20] Schwickert, T. A., Alabyev, B., Manser, T., Nussenzweig, M. C.: Germinal center reutilization by newly activated B cells. J. Exp. Med. 206, 2907-2914 (2009).

[21] Fujihashi, K., Koga, T., van Ginkel, F. W., Hagiwara, Y., McGhee, J. R.: A dilemma for mucosal vaccination: effi cacy versus toxicity using enterotoxin-based adjuvants. Vaccine 20, 2431-2438 (2002).

[22] Miron, N., Cristea, V.: Enterocytes: active cells in tolerance to food and microbial antigens in the gut.

Clin. Exp. Immunol. 167, 405-412 (2011).

[23] Blander, M. J., Sander, L. E.: Beyond pattern recogni-tion: fi ve immune checkpoints for scaling the microbial threat. Nat. Rev. Immunol. 12, 215-225 (2012).

[24] Takeda, K., Kaisho, T., Akira, S.: Toll-like receptors. Annu. Rev. Immunol. 21, 335-376 (2003).

[25] Yamamoto, M., Sato, S., Mori, K., Hoshino, K., Takeuchi, O., Takeda, K., Akira, S.: Cutting edge: role of Toll-like receptor 1 in mediating immune response to microbial lipoproteins. J. Immunol. 169, 10-14 (2002).

[26] Gewirtz, T.: Intestinal epithelial toll-like receptors: to protect. And serve? Curr. Pharm. Des. 9, 1-5 (2003).

[27] Tyrer, P., Foxwell, A. R., Cripps, A. W., Apicella, M. A., Kyd, J. M.: Microbial pattern recognition receptors mediate M-Cell uptake of a gram-negative bacterium. Infect. Immun. 74 (1), 625-631 (2006).

[28] Kawamura, Y. I., Kawashima, R., Shirai, Y., Kato, R., Hamabata, T., Yamamoto, M., Furukawa, K., Fujihashi, K., McGhee, J. R., Hayashi, H., Dohi, T.: Cholera toxin activates dendritic cells through dependence on GM1-ganglioside which is mediated by NF-kappaB translocation. Eur. J. Immunol. 33, 3205-3212 (2003).

[29] Agren, L., Löwenadler, B., Lycke, N.: A novel con-cept in mucosal adjuvanticity: the CTA1-DD adjuvant is a B cell-targeted fusion protein that incorporates the enzymatically active cholera toxin A1 subunit. Immunol. Cell Biol. 76, 280-287 (1998).

[30] van Ginkel, F. W., Jackson, R. J., Yuki, Y., McGhee, J. R.: Cutting edge: the mucosal adjuvant cholera toxin redirects vaccine proteins into olfactory tissues. J. Immunol. 165, 4778-4782 (2000).

[31] Eliasson, D. G., El Bakkouri, K., Schön, K., Ramne, A., Festjens, E., Löwenadler, B., Fiers, W., Saelens, X., Lycke, N.: CTA1-M2e-DD: a novel mucosal adjuvant targeted infl uenza vaccine. Vaccine 26, 1243-1252 (2008).

[32] Giuliani, M. M., Del Giudice, G., Giannelli, V., Dougan, G., Douce, G., Rappuoli, R., Pizza, M.: Mucosal adjuvanticity and immunogenicity of LTR72, a novel mutant of Escherichia coli heat-labile enterotoxin with partial knockout of ADP-ribosyltransferase activity. J. Exp. Med. 187, 1123-1132 (1998).

[33] Summerton, N. A., Welch, R. W., Bondoc, L., Yang, H. H., Pleune, B., Ramachandran, N., Harris, A. M., Bland, D., Jackson, W. J., Park, S., Clements, J. D., Nabors, G. S.: Toward the development of a stable, freeze-dried formulation of Helicobacter pylori killed whole cell vaccine adjuvanted with a novel mutant of Escherichia coli heat-labile toxin. Vaccine 28, 1404-1411 (2010).

[34] Spencer, S. P., Belkaid, Y.: Dietary and commensal derived nutrients: shaping mucosal and systemic immunity. Curr. Opin. Immunol. 24, 379-384 (2012).

[35] Holmgren, J., Svennerholm, A. M.: Vaccines against mucosal infections. Curr. Opin. Immunol. 24, 343-353 (2012).

[36] Dunne, A., Marshall, N. A., Mills, K. H.: TLR based therapeutics. Curr. Opin. Pharmacol. 11, 404-411 (2011).

[37] Higgins, S. C., Mills, K. H.: TLR, NLR agonists, and other immune modulators as infectious disease vaccine adjuvants. Curr. Infect. Dis. Rep. 12, 4-12 (2010).

[38] Men, Y., Audran, R., Thomasin, C., Eberl, G., Demotz, S., Merkle, H. P., Gander, B., Corradin, G.: MHC class Ⅰ-and class Ⅱ-restricted processing and presentation of microencapsulated antigens. Vaccine 17, 1047-1056 (1999).

[39] Lefrancois, L.: Development, traffi cking, and func-tion of memory T-cell subsets. Immunol. Rev. 211, 93-103 (2006).

[40] Sallusto, F., Mackay, C. R., Lanzavecchia, A.: The role of chemokine receptors in primary, effector, and memory immune responses. Annu. Rev. Immunol. 18, 593-620 (2000).

[41] Khader, S. A., Gaffen, S. L., Kolls, J. K.: Th17 cells at the crossroads of innate and adaptive immunity against infectious diseases at the mucosa. Mucosal Immunol. 2, 403-411 (2009).

[42] Rubas, W., Banerjea, A. C., Gallati, H., Speiser, P. P., Joklik, W. K.: Incorporation of the reovirus M cell attachment protein into small unilamellar vesicles: incorporation effi ciency and binding capability to L929 cells in vitro. J. Microencapsul. 7, 385–395 (1990).

[43] Hase, K., Kawano, K., Nochi, T., Pontes, G. S., Fukuda, S., Ebisawa, M., Kadokura, K., Tobe, T., Fujimura, Y., Kawano, S., Yabashi, A., Waguri, S., Nakato, G., Kimura, S., Murakami, T., Iimura, M., Hamura, K., Fukuoka, S., Lowe, A. W., Itoh, K., Kiyono, H., Ohno, H.: Uptake through glycoprotein 2 of FimH+bacteria by M cells initiates mucosal immune response. Nature 462, 226–230 (2009).

[44] Verdonck, F., De Hauwere, V., Bouckaert, J., Goddeeris, B. M., Cox, E.: Fimbriae of enterotoxigenic Escherichia coli function as a mucosal carrier for a coupled heterologous antigen. J. Control. Release 104, 243–258 (2005).

[45] Melkebeek, V., Rasschaert, K., Bellot, P., Tilleman, K., Favoreel, H., Deforce, D., De Geest, B. G., Goddeeris, B. M., Cox, E.: Targeting aminopeptidase N, a newly identifi ed receptor for F4ac fi mbriae, enhances the intestinal mucosal immune response. Mucosal Immunol. 5 (6), 635–645 (2012).

[46] Suzuki, T., Yoshikawa, Y., Ashida, H., Iwai, H., Toyotome, T., Matsui, H., Sasakawa, C.: High vaccine effi cacy against shigellosis of recombinant non–invasive Shigella mutant that expresses Yersinia invasion. J. Immunol. 177, 4709–4717 (2006).

[47] Chadwick, S., Kriegel, C., Amiji, M.: Delivery strat–egies to enhance mucosal vaccination. Expert Opin. Biol. Ther. 9, 427–440 (2009).

[48] Fox, C. B., Friede, M., Reed, S. G., Ireton, G. C.: Synthetic and natural TLR4 agonists as safe and effective vaccine adjuvants. Subcel. Biochem. 53, 303–321 (2010).

[49] Neal, M. D., Leaphart, C., Levy, R., Prince, J., Billiar, T. R., Watkins, S., Li, J., Cetin, S., Ford, H., Schreiber, A., Hackam, D. J.: Enterocyte TLR4 mediates phagocytosis and transcytosis of bacteria across the intestinal barrier. J. Immunol. 176, 3070–3079 (2006).

[50] Gómez, S., Gamazo, C., Roman, B. S., Ferrer, M., Sanz, M. L., Irache, J. M.: Gantrez © AN nanoparticles as an adjuvant for oral immunotherapy with allergens. Vaccine 25, 5263–5271 (2007).

[51] Schwarz, T.: Clinical update of the AS04–adjuvanted human papillomavirus–16/18 cervical cancer vaccine, cervarix ©. Adv. Ther. 26, 983–998 (2009).

[52] Cranage, M. P., Fraser, C. A., Cope, A., McKay, P. F., Seaman, M. S., Cole, T., Mahmoud, A. N., Hall, J., Giles, E., Voss, G., Page, M., Almond, N., Shattock, R. J.: Antibody responses after intravaginal immunisation with trimeric HIV–1CN54 clade C gp140 in Carbopol gel are augmented by systemic priming or boosting with an adjuvanted formulation. Vaccine 4, 1421–1430 (2011).

[53] Mizel, S. B., Bates, J. T.: Flagellin as an adjuvant: cel–lular mechanisms and potential. J. Immunol. 185, 5677–5682 (2010).

[54] Salman, H. H., Irache, J. M., Gamazo, C.: Immunoadjuvant capacity of fl agellin and mannosamine–c oated poly (anhydride) nanoparticles in oral vaccination. Vaccine 27, 4784–4790 (2009).

[55] Salman, H. H., Gamazo, C., Campanero, M. A., Irache, J. M.: Salmonella–like bioadhesive nanoparticles. J. Control. Release 106, 1–13 (2005).

[56] Uematsu, S., Fujimoto, K., Jang, M. H., Yang, B. G., Jung, Y. J., Nishiyama, M., Sato, S., Tsujimura, T., Yamamoto, M., Yokota, Y., Kiyono, H., Miyasaka, M., Ishii, K. J., Akira, S.: Regulation of humoral and cellular gut immunity by lamina propria dendritic cells expressing Toll–like receptor 5. Nat. Immunol. 9, 769–776 (2008).

[57] Harandi, A. M. J., Holmgren, J.: CpG DNA as a potent inducer of mucosal immunity: implications for immunoprophylaxis and immunotherapy of mucosal infections. Curr. Opin. Invest. Drugs 5, 141 – 145 (2004).

[58] Lawson, L. B., Norton, E. B., Clements, J. D.: Defending the mucosa: adjuvant and carrier formulations for mucosal immunity. Curr. Opin. Immunol. 23, 414–420 (2011).

[59] Courtney, A. N., Nehete, P. N., Nehete, B. P., Thapa, P., Zhou, D., Sastry, K. J.: Alpha-galactosylceramide is an effective mucosal adjuvant for repeated intranasal or oral delivery of HIV peptide antigens. Vaccine 27, 3335-3341 (2009).

[60] Agrawal, S., Gupta, S., Agrawal, A.: Human dendritic cells activated via dectin-1 are effi cient at priming Th17, cytotoxic CD8T and B cell responses. PLoS One 5 (10), e13418 (2010).

[61] Da Costa Martins, R., Gamazo, C., Sánchez-Martínez, M., Barberán, M., Peñuelas, I., Irache, J. M.: Conjunctival vaccination against Brucella ovis in mice with mannosylated nanoparticles. J. Control. Release 162, 553-560 (2012).

[62] Salman, H. H., Gamazo, C., Campanero, M. A., Irache, J. M.: Bioadhesive mannosylated nanoparticles for oral drug delivery. J. Nanosci. Nanotechnol. 6, 3203-3209 (2006).

[63] Carrillo-Conde, B., Song, E. H., Chavez-Santoscoy, A., Phanse, Y., Ramer-Tait, A. E., Pohl, N. L., Wannemuehler, M. J., Bellaire, B. H., Narasimhan, B.: Mannose-functionalized "pathogen-like" polyanhydride nanoparticles target C-type lectin receptors on dendritic cells. Mol. Pharm. 8, 1877-1886 (2011).

[64] Chen, W., Kuolee, R., Yan, H.: The potential of 3′, 5′-cyclic diguanylic acid (c-di-GMP) as an effective vaccine adjuvant. Vaccine 28, 3080-3085 (2010).

[65] Clark, M. A., Hirst, B. H., Jepson, M. A.: Lectin - mediated mucosal delivery of drugs and microparticles. Adv. DrugDeliv. Rev. 43, 207-223 (2000).

[66] Jepson, M. A., Clark, M. A., Hirst, B. H.: M cell targeting by lectins: a strategy for mucosal vaccination and drug delivery. Adv. Drug Deliv. Rev. 56, 511-525 (2004).

[67] Rajapaksa, T. E., Lo, D. D.: Microencapsulation of vaccine antigens and adjuvants for mucosal targeting. Curr. Immunol. Rev. 6, 29-37 (2010).

[68] Nochi, T., Yuki, Y., Matsumura, A., Mejima, M., Terahara, K., Kim, D. Y., Fukuyama, S., Iwatsuki-Horimoto, K., Kawaoka, Y., Kohda, T., Kozaki, S., Igarashi, O., Kiyono, H.: A novel M cell-specifi c carbohydrate-targeted mucosal vaccine effectively induces antigen-specifi c immune responses. J. Exp. Med. 204, 2789-2796 (2007).

[69] Kayamuro, H., Yoshioka, Y., Abe, Y., Arita, S., Katayama, K., Nomura, T., Yoshikawa, T., Kubota-Koketsu, R., Ikuta, K., Okamoto, S., Mori, Y., Kunisawa, J., Kiyono, H., Itoh, N., Nagano, K., Kamada, H., Tsutsumi, Y., Tsunoda, S.: Interleukin-1 family cytokines as mucosal vaccine adjuvants for induction of protective immunity against infl uenza virus. J. Virol. 84, 12703-12712 (2010).

[70] Tovey, M. G., Lallemand, C.: Adjuvant activity of cytokines. Methods Mol. Biol. 626, 287-309 (2010).

[71] Gmajner, D., Ota, A., Sentjurc, M., Ulrih, N. P.: Stability of diether C25, 25liposomes from the hyperthermophilic archaeon Aeropyrum pernix K1. Chem. Phys. Lipids 164, 236-245 (2011).

[72] Nordly, P., Madsen, H. B., Nielsen, H. M., Foged, C.: Status and future prospects of lipid-based particulate delivery systems as vaccine adjuvants and their combination with immunostimulators. Expert Opin. Drug Deliv. 6, 657-672 (2009).

[73] Kersten, G. F., Crommelin, D. J.: Liposomes and ISCOMs. Vaccine 21, 915-920 (2003).

[74] Rao, R., Squillante, E. I. I. I., Kim, K. H.: Lipid-based cochleates: a promising formulation platform for oral and parenteral delivery of therapeutic agents. Crit. Rev. Ther. Drug Carrier Syst. 24, 41-61 (2007).

[75] McDermott, M. R., Heritage, P. L., Bartzoka, V., Brook, M. A.: Polymer-grafted starch microparticles for oral and nasal immunization. Immunol. Cell Biol. 76, 256-262 (1998).

[76] Eldridge, J. H., Gilley, R. M., Staas, J. K., Moldoveanu, Z., Meulbroek, J. A., Tice, T. R.: Biodegradable microspheres: vaccine delivery system for oral immunization. Curr. Top. Microbiol. Immunol. 146, 59-66 (1989).

[77] Payne, L. G., Jenkins, S. A., Andrianov, A., Roberts, B. E.: Water-soluble phosphazene polymers for parenteral and mucosal vaccine delivery. Pharm. Biotechnol. 6, 473-493 (1995).

[78] Payne, L. G., Jenkins, S. A., Woods, A. L., Grund, E. M., Geribo, W. E., Loebelenz, J. R., An-

drianov, A. K., Roberts, B. E.: Poly [di (carboxylatophenoxy) phosphazene] (PCPP) is a potent immunoadjuvant for an infl uenza vaccine. Vaccine 16, 92–98 (1998).

[79] Arbós, P., Campanero, M. A., Arangoa, M. A., Renedo, M. J., Irache, J. M.: Infl uence of the surface characteristics of PVM/MA nanoparticles on their bioadhesive properties. J. Control. Release 89, 19–30 (2003).

[80] Olbrich, C., Müller, R. H., Tabatt, K., Kayser, O., Schulze, C., Schade, R.: Stable biocompatible adjuvants–a new type of adjuvant based on solid lipid nanoparticles: a study on cytotoxicity, compatibility and effi cacy in chicken. Altern. Lab. Anim. 30, 443–458 (2002).

[81] Storni, T., Kündig, T. M., Senti, G., Johansen, P.: Immunity in response to particulate antigen-delivery systems. Adv. Drug Deliv. Rev. 57, 333–355 (2005).

[82] Krahenbuhl, J. P., Neutra, M. R.: Epithelial M cells: differentiation and function. Annu. Rev. Cell Dev. Biol. 16, 301–332 (2000).

[83] Espuelas, S., Irache, J. M., Gamazo, C.: Synthetic particulate antigen delivery systems for vaccination. Inmunologia 24, 207–223 (2005).

[84] O'Hagan, D. T., Jeffery, H., Maloy, K. J., Mowat, A. M., Rahman, D., Challacombe, S. J.: Biodegradable microparticles as oral vaccines. Adv. Exp. Med. Biol. 371B, 1463–1467 (1995).

[85] Vila, A., Sánchez, A., Tobío, M., Calvo, P., Alonso, M. J.: Design of biodegradable particles A. for protein delivery. J. Control. Release 78, 15–24 (2002).

[86] Yoncheva, K., Gómez, S., Campanero, M. A., Gamazo, C., Irache, J. M.: Bioadhesive properties of pegylated nanoparticles. Expert Opin. Drug Deliv. 2 (2), 205–218 (2005).

[87] Yoncheva, K., Lizarraga, E., Irache, J. M.: Pegylated nanoparticles based on poly (methyl vinyl ether-co-maleic anhydride): preparation and evaluation of their bioadhesive properties. Eur. J. Pharm. Sci. 24, 411–419 (2005).

[88] Fadl, A. A., Venkitanarayanan, K. S., Khan, M. I.: Identifi cation of Salmonella enteritidis outer membrane proteins expressed during attachment to human intestinal epithelial cells. J. Appl. Microbiol. 92, 180–186 (2002).

[89] Allen-Vercoe, E., Woodward, M. J.: The role of flagella, but not fimbriae, in the adherence of Salmonella enterica serotype Enteritidis to chick gut explant. J. Med. Microbiol. 48, 771–780 (1999).

[90] Humphries, A. D., Townsend, S. M., Kingsley, R. A., Nicholson, T. L., Tsolis, R. M., Bäumler, A. J.: Role of fi mbriae as antigens and intestinal colonization factors of Salmonella serovars. FEMS Microbiol. Lett. 201, 121–125 (2001).

[91] Kaltner, H., Stierstorfer, B.: Animal lectins as cell adhesion molecules. Acta Anat. (Basel) 161, 162–179 (1998).

[92] Lloyd, D. H., Viac, J., Werling, D., Rème, C. A., Gatto, H.: Role of sugars in surface microbe-host interactions and immune reaction modulation. Vet. Dermatol. 18, 197–204 (2007).

[93] Conway, M. A., Madrigal-Estebas, L., McClean, S., Brayden, D. J., Mills, K. H.: Protection against Bordetella pertussis infection following parenteral or oral immunization with antigens entrapped in biodegradable particles: effect of formulation and route of immunization on induction of Th1 and Th2 cells. Vaccine 19, 1940–1950 (2001).

[94] Carcaboso, A. M., Hernández, R. M., Igartua, M., Gascón, A. R., Rosas, J. E., Patarroyo, M. E., Pedraz, J. L.: Immune response after oral administration of the encapsulated malaria synthetic peptide SPf66. Int. J. Pharm. 260, 273–282 (2003).

[95] Sundling, C., Schön, K., Mörner, A., Forsell, M. N., Wyatt, R. T., Thorstensson, R., Karlsson Hedestam, G. B., Lycke, N. Y.: CTA1-DD adjuvant promotes strong immunity against human immunodefi ciency virus type 1 envelope glycoproteins following mucosal immunization. J. Gen. Virol. 89, 2954–2964 (2008).

[96] Helgeby, A., Robson, N. C., Donachie, A. M., Beackock-Sharp, H., Lövgren, K., Schön, K., Mo-

wat, A., Lycke, N. Y.: The combined CTA1-DD/ ISCOM adjuvant vector promotes priming of mucosal and systemic immunity to incorporated antigens by specifi c targeting of B cells. J. Immunol. 176, 3697-3706 (2006).

[97] Smith, R. E., Donachie, A. M., Grdic, D., Lycke, N., Mowat, A. M.: Immune - stimulating complexes induce an IL-12-dependent cascade of innate immune responses. J. Immunol. 162, 5536-5546 (1999).

[98] Dalle, F., Jouault, T., Trinel, P. A., Esnault, J., Mallet, J. M., d'Athis, P., Poulain, D., Bonnin, A.: Beta-1, 2-and alpha-1, 2-linked oligomannosides mediate adherence of Candida albicans blastospores to human enterocytes in vitro. Infect. Immun. 71, 7061-7068 (2003).

[99] Irache, J. M., Salman, H. H., Gamazo, C., Espuelas, S.: Mannose-targeted systems for the delivery of therapeutics. Expert Opin. Drug Deliv. 5, 703-724 (2008).

[100] Jack, D. L., Turner, M. W.: Anti-microbial activities of mannose-binding lectin. Biochem. Soc. Trans. 31, 753-775 (2003).

[101] Uemura, K., Hiromatsu, K., Xiong, X., Sugita, M., Buhlmann, J. E., Dodge, I. L., Lee, S. Y., Roura- Mir, C., Watts, G. F., Roy, C. J., Behar, S. M., Clemens, D. L., Porcelli, S. A., Brenner, M. B.: Conservation of CD1 intracellular traffi cking patterns between mammalian species. J. Immunol. 169, 6945-6958 (2002).

[102] Wagner, S., Lynch, N. J., Walter, W., Schwaeble, W. J., Loos, M.: Differential expression of the murine mannose-binding lectins A and C in lymphoid and nonlymphoid organs and tissues. J. Immunol. 170, 1462-1465 (2003).

[103] Tamayo, I., Irache, J. M., Mansilla, C., Ochoa-Repáraz, J., Lasarte, J. J., Gamazo, C.: Poly (anhydride) nanoparticles as adjuvants for mucosal vaccination. Front. Biosci. (Schol. Ed.) 2, 876-890 (2010).

[104] Wang, T., Zou, M., Jiang, H., Ji, Z., Gao, P., Cheng, G.: Synthesis of a novel kind of carbon nanoparticle with large mesopores and macropores and its application as an oral vaccine adjuvant. Eur. J. Pharm. Sci. 44 (5), 653-659 (2011).

[105] Jang, S. I., Lillehoj, H. S., Lee, S. H., Lee, K. W., Lillehoj, E. P., Bertrand, F., Dupuis, L., Deville, S.: TMIMS 1313 N VG PR nanoparticle adjuvant enhances antigen-specifi c immune responses to profi lin following mucosal vaccination against Eimeria acervulina. Vet. Parasitol. 182 (2-4), 163-170 (2011).

[106] Vandamme, K., Melkebeek, V., Cox, E., Adriaensens, P., Van Vlierberghe, S., Dubruel, P., Vervaet, C., Remon, J. P.: Infl uence of polymer hydrolysis on adjuvant effect of Gantrez © AN nanoparticles: implications for oral vaccination. Eur. J. Pharm. Biopharm. 79 (2), 392-398 (2011).

[107] Sarti, F., Perera, G., Hintzen, F., Kotti, K., Karageorgiou, V., Kammona, O., Kiparissides, C., Bernkop-Schnürch, A.: In vivo evidence of oral vaccination with PLGA nanoparticles containing the immunostimulant monophosphoryl lipid A. Biomaterials 32 (16), 4052-4057 (2011).

[108] Tian, J., Yu, J.: Poly (lactic-co-glycolic acid) nanoparticles as candidate DNA vaccine carrier for oral immunization of Japanese fl ounder (Paralichthys olivaceus) against lymphocystis disease virus. Fish Shellfi sh Immunol. 30 (1), 109-117 (2011).

[109] Mishra, N., Tiwari, S., Vaidya, B., Agrawal, G. P., Vyas, S. P.: Lectin anchored PLGA nanoparticles for oral mucosal immunization against hepatitis B. J. Drug Target. 19 (1), 67-78 (2011).

[110] Jain, A. K., Goyal, A. K., Mishra, N., Vaidya, B., Mangal, S.: Vyas SP PEG-PLA-PEG block copolymeric nanoparticles for oral immunization against hepatitis B. Int. J. Pharm. 387 (1-2), 253-262 (2010).

[111] Mann, J. F., Shakir, E., Carter, K. C., Mullen, A. B., Alexander, J., Ferro, V. A.: Lipid vesicle size of an oralinfl uenza vaccine delivery vehicle infl uences the Th1/Th2 bias in the immune response and protection against infection. Vaccine 27, 3643-3649 (2009).

[112] Li, G. P., Liu, Z. G., Liao, B., Zhong, N. S.: Induction of Th1-type immune response by chitosan nanoparticles containing plasmid DNA encoding house dust mite allergen Der p 2 for oral vaccination in mice. Cell. Mol. Immunol. 6 (1), 45-50 (2009).

[113] Oyewumi, M. O., Kumar, A., Cui, Z.: Nano-microparticles as immune adjuvants: correlating particle sizes and the resultant immune responses. Expert Rev. Vaccines 9, 1095-1107 (2010).

[114] Gutierro, I., Hernández, R. M., Igartua, M., Gascón, A. R., Pedraz, J. L.: Size dependent immune response after subcutaneous, oral and intranasal administration of BSA loaded nanospheres. Vaccine 21, 67-77 (2002).

[115] Kanchan, V., Panda, A. K.: Interactions of antigen-loaded polylactide particles with macrophages and their correlation with the immune response. Biomaterials 28, 5344-5357 (2007).

[116] Jung, T., Kamm, W., Breitenbach, A., Hungerer, K. D., Hundt, E., Kissel, T.: Tetanus toxoid loaded nanoparticles from sulfobutylated poly (vinyl alcohol) -graft-poly (lactide-co-glycolide): evaluation of antibody response after oral and nasal application in mice. Pharm. Res. 18, 352-360 (2001).

[117] Wendorf, J., Chesko, J., Kazzaz, J., Ugozzoli, M., Vajdy, M., O'Hagan, D., Singh, M.: A comparison of anionic nanoparticles and microparticles as vaccine delivery systems. Hum. Vaccin. 4, 44-49 (2008).

[118] Fifi s, T., Gamvrellis, A., Crimeen - Irwin, B., Pietersz, G. A., Li, J., Mottram, P. L., McKenzie, I. F., Plebanski, M.: Size-dependent immunogenicity: therapeutic and protective properties of nano-v accines against tumors. J. Immunol. 173, 3148-3154 (2004).

[119] Kalkanidis, M., Pietersz, G. A., Xiang, S. D., Mottram, P. L., Crimeen-Irwin, B., Ardipradja, K., Plebanski, M.: Methods for nano-particle based vaccine formulation and evaluation of their immunogenicity. Methods 40, 20-29 (2006).

[120] Mottram, P. L., Leong, D., Crimeen-Irwin, B., Gloster, S., Xiang, S. D., Meanger, J., Ghildyal, R., Vardaxis, N., Plebanski, M.: Type 1 and 2 immunity following vaccination is infl uenced by nanoparticle size: formulation of a model vaccine for respiratory syncytial virus. Mol. Pharm. 4, 73-84 (2007).

[121] Caputo, A., Castaldello, A., Brocca-Cofano, E., Voltan, R., Bortolazzi, F., Altavilla, G., Sparnacci, K., Laus, M., Tondelli, L., Gavioli, R., Ensoli, B.: Induction of humoral and enhanced cellular immune responses by novel core-shell nanosphere-and microsphere-based vaccine formulations following systemic and mucosal administration. Vaccine 27, 3605-3615 (2009).

[122] Estevan, M., Irache, J. M., Grilló, M. J., Blasco, J. M., Gamazo, C.: Encapsulation of antigenic extracts of Salmonella enterica serovar. Abortus-ovis into polymeric systems and effi cacy as vaccines in mice. Vet. Microbiol. 118, 124-132 (2006).

[123] Shakweh, M., Ponchel, G., Fattal, E.: Particle uptake by Peyer's patches: a pathway for drug and vaccine delivery. Expert Opin. Drug Deliv. 1, 141-163 (2004).

[124] Peppoloni, S., Ruggiero, P., Contorni, M., Morandi, M., Pizza, M., Rappuoli, R., Podda, A., Del Giudice, G.: Mutants of the Escherichia coli heat-labile enterotoxin as safe and strong adjuvants for intranasal delivery of vaccines. Expert Rev. Vaccines 2, 285-293 (2003).

[125] Maloy, K. J., Donachie, A. M., O'Hagan, D. T., Mowat, A. M.: Induction of mucosal and systemic immune responses by immunization with ovalbumin entrapped in poly (lactide - co - glycolide) microparticles. Immunology 81, 661-667 (1994).

[126] Kim, S. Y., Doh, H. J., Jang, M. H., Ha, Y. J., Chung, S. I., Park, H. J.: Oral immunization with Helicobacter pylori-loaded poly (D, L-lactide-co-glycolide) nanoparticles. Helicobacter 4, 33-39 (1999).

[127] Katz, D. E., DeLorimier, A. J., Wolf, M. K., Hall, E. R., Cassels, F. J., van Hamont, J. E., Newcomer, R. L., Davachi, M. A., Taylor, D. N., McQueen, C. E.: Oral immunization of adult volunteers with microencapsulated enterotoxigenic Escherichia coli (ETEC) CS6 antigen. Vaccine 21, 341-346 (2003).

[128] Otterlei, M., Vårum, K. M., Ryan, L., Espevik, T.: Characterization of binding and TNF-alpha-inducing ability of chitosans on monocytes: the involvement of CD14. Vaccine 12, 825-832 (1994).

[129] Porporatto, C., Bianco, I. D., Correa, S. G.: Local and systemic activity of the polysaccharide chitosan at lymphoid tissues after oral administration. J. Leukoc. Biol. 78, 62-69 (2005).

[130] Kumar, N., Langer, R. S., Domb, A. J.: Polyanhydrides: an overview. Adv. Drug Deliv. Rev. 54, 889-910 (2002).

[131] Torres, M. P., Wilson-Welder, J. H., Lopac, S. K., Phanse, Y., Carrillo-Conde, B., Ramer-Tait, A. E., Bellaire, B. H., Wannemuehler, M. J., Narasimhan, B.: Polyanhydride microparticles enhance dendritic cell antigen presentation and activation. Acta Biomater. 7, 2857-2864 (2011).

[132] Petersen, L. K., Ramer-Tait, A. E., Broderick, S. R., Kong, C. S., Ulery, B. D., Rajan, K., Wannemuehler, M. J., Narasimhan, B.: Activation of innate immune responses in a pathogen-mimicking manner by amphiphilic polyanhydride nanoparticle adjuvants. Biomaterials 32, 6815-6822 (2011).

[133] Ulery, B. D., Phanse, Y., Sinha, A., Wannemuehler, M. J., Narasimhan, B., Bellaire, B. H.: Polymer chemistry infl uences monocytic uptake of polyanhydride nanospheres. Pharm. Res. 26, 683-690 (2009).

[134] Ulery, B. D., Kumar, D., Ramer-Tait, A. E., Metzger, D. W., Wannemuehler, M. J., Narasimhan, B.: Design ofa protective single-dose intranasal nanoparticle-based vaccine platform for respiratory infectious diseases. PLoSOne6 (3), e17642 (2011). doi: 10. 1371/ journal. pone. 0017642.

[135] Ochoa, J., Irache, J. M., Tamayo, I., Walz, A., DelVecchio, V. G., Gamazo, C.: Protective immunity of biodegradable nanoparticle-based vaccine against an experimental challenge with Salmonella enteritidis in mice. Vaccine 25, 4410-4419 (2007).

[136] Gómez, S., Gamazo, C., San Roman, B., Vauthier, C., Ferrer, M., Irachel, J. M.: Development of a novel vaccine delivery system based on Gantrez nanoparticles. J. Nanosci. Nanotechnol. 6, 3283-3289 (2006).

[137] Tamayo, I., Irache, J. M., Mansilla, C., Ochoa-Repáraz, J., Lasarte, J. J., Gamazo, C.: Poly (anhydride) nanoparticles act as active Th1 adjuvants through toll-like receptor exploitation. Clin. Vaccine Immunol. 17, 1356-1362 (2010).

[138] Camacho, A. I., Da Costa Martins, Tamayo, I., de Souza, J., Lasarte, J. J., Mansilla, C., Esparza, I., Irache, J. M., Gamazo, C.: Poly (methyl vinyl ether-co-maleic anhydride) nanoparticles as innate immune system activators. Vaccine 22, 7130-7135 (2011).

[139] Lycke, N., Bemark, M.: Mucosal adjuvants and long-term memory development with special focus on CTA1-DD and other ADP-ribosylating toxins. Mucosal Immunol. 3, 556-566 (2010).

[140] Kasturi, S. P., Skountzou, I., Albrecht, R. A., Koutsonanos, D., Hua, T., Nakaya, H. I., Ravindran, R., Stewart, S., Alam, M., Kwissa, M., Villinger, F., Murthy, N., Steel, J., Jacob, J., Hogan, R. J., García-Sastre, A., Compans, R., Pulendran, B.: Programming the magnitude and persistence of antibody responses with innate immunity. Nature 470, 543-547 (2011).

[141] Amanna, I. J., Carlson, N. E., Slifka, M. K.: Duration of humoral immunity to common viral and vaccine antigens. N. Engl. J. Med. 357, 1903-1915 (2007).

[142] Hammarlund, E., Lewis, M. W., Hansen, S. G., Strelow, L. I., Nelson, J. A., Sexton, G. J., Hanifi n, J. M., Slifka, M. K.: Duration of antiviral immunity after smallpox vaccination. Nat. Med. 9, 1131-1137 (2003).

[143] Querec, T. D., Akondy, R. S., Lee, E. K., Cao, W., Nakaya, H. I., Teuwen, D., Pirani, A., Gernert, K., Deng, J., Marzolf, B., Kennedy, K., Wu, H., Bennouna, S., Oluoch, H., Miller, J., Vencio, R. Z., Mulligan, M., Aderem, A., Ahmed, R., Pulendran, B.: Systems biology approach predicts immunogenicity of the yellow fever vaccine in humans. Nat. Immunol. 10, 116-125 (2009).

[144] Sharp, F. A., Ruane, D., Claass, B., Creagh, E., Harris, J., Malyala, P., Singh, M., O'Hagan,

D. T., Pétrilli, V., Tschopp, J., O'Neill, L. A., Lavelle, E. C.: Uptake of particulate vaccine adjuvants by dendritic cells activates the NALP3 infl ammasome. Proc. Natl. Acad. Sci. U. S. A. 106, 870-875 (2009).

[145] Li, H., Willingham, S. B., Ting, J. P., Re, F.: Cutting edge: infl ammasome activation by alum and alum's adjuvant effect are mediated by NLRP3. J. Immunol. 181, 17-21 (2008).

[146] Correia-Pinto, J. F., Csaba, N., Alonso, M. J.: Vaccine delivery carriers: insights and future perspectives. Int. J. Pharm. 440 (1), 27-38 (2013). doi: 10. 1016/j. bbr. 2011. 03. 031.

[147] Lambrecht, B. N., Kool, M., Willart, M. A., Hammad, H.: Mechanism of action of clinically approved adjuvants. Curr. Opin. Immunol. 21, 23-29 (2009).

[148] Tagliabue, A., Rappuoli, R.: Vaccine adjuvants: the dream becomes real. Hum. Vaccin. 4, 347-349 (2008).

[149] Chen, W., Patel, G. B., Yan, H., Zhang, J.: Recent advances in the development of novel mucosal adjuvants and antigen delivery systems. Hum. Vaccin. 6, 706-714 (2010).

[150] Holmgren, J., Czerkinsky, C., Eriksson, K., Mharandi, A.: Mucosal immunisation and adjuvants: a brief overview of recent advances and challenges. Vaccine 21, S89-S95 (2003).

[151] Pulendran, B.: Learning immunology from the yel-low fever vaccine: innate immunity to systems vaccinology. Nat. Rev. Immunol. 9, 741-747 (2009).

(吴国华译)

第39章　壳聚糖基佐剂

Guro Gafvelin 和 Hans Grönlund①

摘要

一些传染病和与免疫有关的疾病缺乏安全有效的预防和治疗疫苗。新型佐剂能够增强和改变免疫反应，因此能够改进预防和治疗疫苗。在寻找新型佐剂的过程中，碳水化合物聚合物壳聚糖受到越来越多的关注。壳聚糖来源于天然产物甲壳素，化学成分为N-乙酰氨基-葡糖胺和氨基葡萄糖。壳聚糖具有天然来源、完全生物降解、无毒、低成本等优点，是一种具有广阔应用前景的新型佐剂。壳聚糖具有刺激先天免疫的作用，但其作用机制尚不明确。其佐剂效应也可能通过改善抗原对免疫细胞的递呈和增强抗原摄取进行介导。由于其阳离子性质，壳聚糖粘附在黏膜表面。目前已研究了壳聚糖的各种化学衍生物和颗粒变体在疫苗黏膜给药中的应用。临床试验的数据支持壳聚糖作为鼻内疫苗的有效佐剂和给药系统。为研制适合不同疫苗应用和给药途径的壳聚糖基佐剂，高质量的医用级壳聚糖至关重要。在本章中，我们介绍了壳聚糖作为佐剂的研究进展，并重点关注最近研制的壳聚糖佐剂黏性凝胶。

39.1　疾病/应用领域

疫苗和疫苗规划在与许多传染性疾病的斗争中取得了巨大的成功，这表明，新疫苗战略可能是解决各种免疫相关疾病的方法。然而，对于研制治疗性疫苗（如癌症、自身免疫疾病和过敏反应）以及传染病，仍存在很多挑战。迫切需要有效疫苗的疾病包括疟疾、结核病和艾滋病病毒感染/艾滋病，这些疾病每年会造成数百万人死亡[1,2]。尤其需要有效疫苗的患者是全球范围内不断增长的老年人口。65岁以上的人往往无法从目前的疫苗接种得到充足的保护，例如季节性流感和肺炎球菌感染[2-4]。因此，需要研制能够引起有效和适当免疫反应的疫苗。

改进次优疫苗的一种方法是使用更有效的佐剂，即免疫反应增强剂。佐剂还提供将反应引导至特定方向的方式，例如体液或细胞免疫反应。已知给药途径会影响受刺激的免疫反应，此外，可将佐剂设计成疫苗给药的载体或给药系统。

铝盐（明矾）是应用最广泛的佐剂，也是迄今为止唯一被批准用于人类的佐剂。明矾主要增强免疫系统的体液反应，尤其是对小鼠，其强烈促进TH2型免疫反应[5]。为满足对能够引起细胞反应的疫苗的广泛需求，已经研制或正在研究几种新型佐剂[6,7]。设计用于非注射给药途径的疫苗中也需要新型佐剂。

在寻找新型佐剂的过程中，碳水化合物聚合物壳聚糖因其具有良好的免疫刺激作用而受到越来越多的关注[8,9]。壳聚糖来源于天然产物甲壳素。甲壳素是自然界最丰富的聚合物之一，是甲壳类动物外壳、昆虫外骨骼和微生物细胞壁的结构元素。化学上，壳聚糖由N-乙酰氨基葡糖

---

① G. Gafvelin，PhD（卡罗琳斯卡医学院分子医学中心、临床神经科学系治疗免疫设计研究室。瑞典斯德哥尔摩）

H. Grönlund，PhD（✉）（卡罗琳斯卡医学院分子医学中心、临床神经科学系治疗免疫设计研究室。瑞典斯德哥尔摩，E-mail：hans. gronlund@ ki. se）

和葡萄糖胺组成。由于其阳离子性质，壳聚糖能有效粘附在黏膜表面[10]。壳聚糖具有天然来源、完全生物降解、无毒、低成本等优点，是一种具有广阔应用前景的新型佐剂。在本章节中，我们将介绍壳聚糖作为佐剂的研究进展，并重点关注最近研制的壳聚糖佐剂黏性凝胶。

39.2 疫苗接种

39.2.1 制剂/化学

壳聚糖由甲壳类动物的外骨骼制成，作为海产品工业的副产品回收利用。制备方法包括对几丁质（一种由 1-4-β-连接的 N-乙酰葡糖胺单元组成的均聚物）进行强碱处理，以除去乙酰基并形成水溶性壳聚糖。因此，壳聚糖以各种比例由 N-乙酰葡糖胺和葡糖胺（图 39.1）组成。壳聚糖中葡萄糖胺的百分比称为脱乙酰度（DD）。标准制备过程通常 DD 为 80%~95%，但是 DD 可能具有很大的差异。目前已发表的研究中应用的壳聚糖质量往往没有进行充分定义，因此很难将壳聚糖的物理和化学成分与某些免疫反应刺激联系起来。现有数据大多是在 80%~95%的 DD 范围内的壳聚糖，因为市售脱乙酰壳多糖通常高度脱乙酰化。

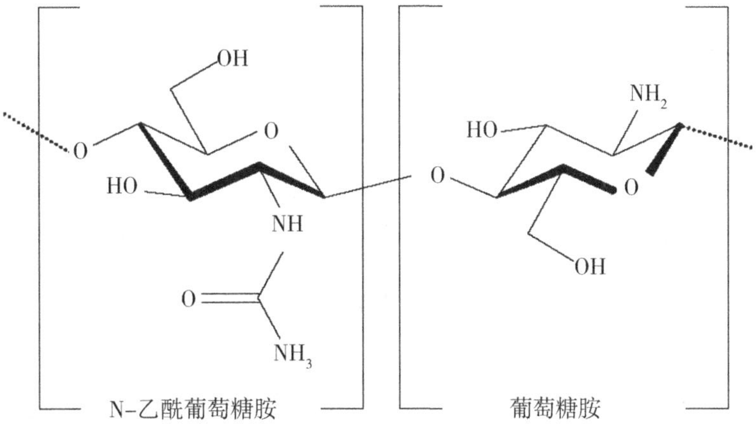

图 39.1 壳聚糖是由 N-乙酰氨基葡萄糖和氨基葡萄糖单元组成的碳水化合物聚合物

壳聚糖的聚合物长度、N-乙酰氨基葡萄糖和氨基葡萄糖单元的分布和总体组成各不相同。聚合物中葡萄糖胺单元的百分比表示脱乙酰度（DD）

壳聚糖由于其高含量的葡萄糖胺，是带正电[11]和 muco 黏着性的[10,12]，还具有渗透增强性能[10,13]。这些特征对黏膜疫苗的研制具有重要意义。然而，壳聚糖在医学上应用有一个常见的限制，其只能在 pH 低于 6~6.5 的情况下以质子化形式溶解。

为将壳聚糖保持在生理条件下（中性 pH、盐等）的溶液中，目前已经尝试了各种化学修饰[10,14]。黏膜应用中最常见的可溶性壳聚糖衍生物是 N-三甲基化壳聚糖和羧甲基化壳聚糖[15,16]。目前已经以多种不同的物理形式制备壳聚糖及其衍生物，以满足疫苗递送系统的各种需求，如凝胶、粉末和纳米颗粒等[16,17]。

目前已研制了一种生理条件下可溶的壳聚糖，并以 Viscosan（黏胶原）的名义制作。乙酰基和 DD 的随机分布促进了水溶度接近 50%。此外，Viscosan 在生物环境中降解速度快、纯度高。Viscosan 可加工成黏性水凝胶 ViscoGel，其具有良好的佐剂和疫苗接种性能。此凝胶由壳聚糖链交联而成，以 99%的水和 1%的粘胶原组成物理稳定的水凝胶。黏性凝胶进一步加工成各种尺寸的颗粒。最后，黏性凝胶可以与不同的抗原或免疫调节剂混合或共价键连接，进而制成疫苗佐剂配方。具有共价连接抗原的制剂可能非常有利，因为已有研究表明，与非相关抗原/佐剂相比，与佐剂物理连接的抗原能产生更好的免疫反应[18,19]。

39.2.2 微技术/纳米技术

颗粒系统为疫苗技术提供了几个好处。这种颗粒状结构能够暴露大表面积，并有效展示抗原。微粒大小与微生物相似，免疫系统的进化使其能与之对抗。因此，它们被有效地吞噬，并激活抗原呈递细胞（APC）[20]。黏性凝胶提供微粒系统，可加工成几微米至 200 μm 的凝胶颗粒。值得注意的是，当受到生理 pH 的影响时，黏性凝胶不会沉淀，而是以透明的凝胶颗粒的形式存在。

目前，已对纳米壳聚糖体系在黏膜疫苗的应用进行了大量研究。例如，壳聚糖纳米球（约 350 nm）已被用作疫苗载体，用于临床前测试破伤风类毒素的鼻腔给药[21,22]。与液体疫苗相比，壳聚糖纳米颗粒制剂破伤风类毒素能引起较好的全身免疫反应和局部免疫反应。在大鼠鼻腔给药后，负载有荧光标记白蛋白（OVA）的壳聚糖纳米颗粒（350 nm）被鼻腔上皮细胞内化并转运至黏膜下层[23]。在用于舌下过敏原特异性免疫疗法的小鼠模型中，另一方面表明，用壳聚糖微粒（1~3 μm）配制的模型过敏原 OVA 处理优于纳米颗粒（300~800 nm）配制的疫苗。采用微颗粒疫苗进行治疗后，舌下 APC 对 OVA 的吸收和 OVA 特异性 T 细胞反应的激活均增加，气道高反应性和肺部炎症反应降低[24]。

39.2.3 作用机制

壳聚糖经鼻[8,21]、口服[25]、皮下[9,26]和肌内抗原给药后，已被证明具有免疫激活作用[9,27]。口服途径能够诱导致耐受性免疫反应。在没有任何蛋白质抗原的情况下，口服壳聚糖刺激肠黏膜中的 Th2/调节性 T 细胞可促进环境，其特征在于 IL-10 的分泌以及 IL-4 和 TGF-β 的递呈[25]。当通过皮下或肌内注射途径给药时，壳聚糖能刺激体液和细胞对共同给药抗原的强烈反应。Th1/Th2 混合反应以及 CTL 活化已在小鼠中得到证实[9,27]。有人提出，几丁质是壳聚糖的前体，其通过病原体模式识别受体（PRR）和几丁质酶两种机制的激活来激活先天性免疫反应[28]。被几丁质激活的 PRR 包括甘露糖受体、TLR-2 和 dectin-1[29-32]。

壳聚糖被定义为部分去脱乙酰几丁质，可发挥一些几丁质显示的免疫激活作用。然而，关于壳聚糖如何作为佐剂的报道却很少。早期研究表明，壳聚糖可以激活巨噬细胞[33]。后来证明壳聚糖具有抗原贮存库效应，可以增强佐剂作用[9]。有报告称壳聚糖以 NLRP3 依赖的方式诱导炎性体活化，而这种机制可以在颗粒佐剂（包括明矾）之间共享[34]。有趣的是，在对小鼠骨髓来源的树突状细胞（DC）进行的体外研究结果表明，与明矾相比，壳聚糖不抑制从激活的 DC 刺激产生 Th1 极化细胞因子 IL-12，为壳聚糖比明矾[35]更容易激发 Th1 反应提供了一种可能的解释[35]。这些研究和其他研究表明，壳聚糖激活先天免疫信号，但迄今为止尚未确定与壳聚糖特异性相互作用的候选 PRR。

黏性凝胶含有低脱乙酰化壳聚糖，具有随机分布的乙酰基[26]。其特殊性表明，尽管某些免疫激活机制与其他壳聚糖相同，但黏性凝胶将以一种独特的方式激活免疫系统。黏性凝胶刺激的免疫反应取决于黏性凝胶制剂的性质，例如其颗粒大小、DD 以及混合或将抗原与黏性凝胶连接的方法。应针对每种疫苗制剂对这些因素进行评估，以获得对任何特定疾病的最佳反应。重要的是，黏性凝胶系统的多功能性提供了一种可能性，即可以为特定应用设计具有特性的疫苗。

39.2.4 临床前开发、安全和功效

许多基于壳聚糖的候选佐剂已通过各种疫苗和给药途径进行了临床前评估，但主要用于黏膜疫苗应用[16]。此处，我们从小鼠的概念验证到批准用于人体临床试验的产品，来描述黏性凝胶的临床前发展。最初在小鼠身上进行了一项研究，以证明黏性凝胶作为佐剂的概念。将 B 型

流感嗜血杆菌（Act-HIB）的糖结合疫苗用作模型抗原[26]。之所以选择模型疫苗，是因为其已批准用于人体，且疫苗接种的结果可以作为标准试验的抗体滴度进行测量[36]。将疫苗配制为 Act-HIB 与 200 μm ViscoGel 颗粒的混合物。用佐剂配制的疫苗免疫各组小鼠，并与未加佐剂的 Act-HIB 进行比较[26]。在用黏性凝胶配制的 Act-HIB 免疫的小鼠血清中发现明显增强的 IgG1 和 IgG2a 滴度。事实上，当黏性凝胶用作佐剂时，抗原剂量可以减少 10 倍。此外，根据细胞因子谱，Act-HIB-特异性细胞反应明显更强，且为混合 Th1/Th2/Th17 型。当在皮下或肌内施用佐剂时，也可观察到类似的效果。佐剂特性在小鼠体内的成功演示，为临床概念验证做了准备。在经验证的工艺条件下，制备了医用级与 Act-HIB 混合的黏性凝胶。然后进行一系列质量控制分析，以记录物理和化学特性，并评估纯度和稳定性。

安全对佐剂应用于人类疫苗至关重要。佐剂应增强抗原特异性反应，而不促进对佐剂的强非特异性炎症或免疫反应。以小鼠和家兔为实验对象，研究了黏性凝胶的毒性，以及 Act-HIB 与黏性凝胶混合研制疫苗的毒性。此研究仔细考察了 3 次肌内注射黏性凝胶后的局部和全身毒性作用，未记录全身毒性反应。观察到的局部反应是温和的，并与免疫刺激作用有关，这已由接种疫苗对 Act-HIB 产生的抗体测量得到证实。对壳聚糖产生免疫反应的可能性，以及可能的过敏反应，IgE 介导的反应，已在前述予以讨论[27]。仅用壳聚糖肌内注射或与流感疫苗或添加佐剂（费氏不完全佐剂）联合注射的小鼠，未检测到明显的壳聚糖抗体反应。此外，注射壳聚糖或壳聚糖疫苗的大鼠血清中未记录到壳聚糖或流感抗原的 IgE 增加[27]。目前有人已提出海鲜过敏的人是否会对壳聚糖产生反应的问题[37]。这一忧虑也适用于壳聚糖制剂中可能存在的微量过敏蛋白。具有良好特性的高纯度壳聚糖，例如黏性凝胶，却不具有这种风险。

39.2.5　临床研究

虽然已经研制几种壳聚糖佐剂进行临床前试验[15,16]，但很少有研究评估壳聚糖作为人体内的佐剂（表 39.1）。到目前为止，已有研究免疫刺激效应的临床试验涉及黏膜疫苗接种。目前已设计了几项研究来比较壳聚糖对无佐剂或其他佐剂的疫苗接种的佐剂效果。与不加佐剂的鼻内接种疫苗和肌内明矾配制的 CRM 197 相比，鼻内注射用壳聚糖谷氨酸配制的脱毒白喉毒素疫苗（CRM 197）可产生明显更高水平的中和抗毒素血清抗体[38]。只有壳聚糖制剂疫苗能诱导抗毒素 IgA 反应[38]，鼻内 CRM197/壳聚糖疫苗对白喉毒素的 Th2 型细胞反应也比单独接种或肌内注射含矾的疫苗[39]更强。在另一项研究中，用壳聚糖谷氨酸配制的三价流感疫苗鼻内接种与常规肌内途径疫苗接种进行了比较[40]。鼻内免疫生成了令人满意的保护反应，其与肌内疫苗产生的保护反应无明显差异。第三项研究调查了与壳聚糖鼻内给药的脑膜炎奈瑟球菌糖缀合物疫苗，与用明矾配制并与肌内注射的相同疫苗相比较[41]。鼻内壳聚糖疫苗耐受性好，并产生与肠胃外疫苗相当的保护性抗体反应。此外，目前已在诺瓦克病毒样颗粒上进行利用壳聚糖作为佐剂和疫苗载体的相关研究，但没有与其他给药途径或非佐剂疫苗进行比较的试验用佐剂单磷酰脂质 A（MLP）和壳聚糖配制的壳聚糖给药激光免疫治疗疫苗用于鼻内递送。安全性和剂量发现研究[42]中对此疫苗进行了评估，随后进行了随机、双盲、安慰剂对照、多中心研究[43]。结果显示，鼻用疫苗耐受性好，免疫原性强，对实验诱导的诺瓦克病毒性胃肠炎有较好的保护作用。壳聚糖也被应用于鼻内疫苗的 I 期试验（MUCOVAC2），评估通过鼻内、肌内和阴道途径接种的 HIV 病毒涂层疫苗[44]。最后，壳聚糖被用作癌症免疫治疗的免疫激活剂。在晚期转移性乳腺癌患者的安全性-功效性研究中，糖化壳聚糖联合激光免疫治疗[45]。本研究的临床结果前景广阔，显示出全身抗肿瘤反应，副作用比常规治疗更温和。

表 39.1 壳聚糖佐剂的临床试验

抗原	接种途径	受试者	研究说明	参考文献
灭活的白喉类毒素 CRM_{197} 突变体	鼻内（肌内对照组）	健康成人	一组鼻内接种 CRM_{197}（$n=10$） 一组鼻内接种 CRM_{197}+壳聚糖（$n=10$） 一组肌内接种 CRM_{197}+明矾（$n=5$）	Mills 等[38] McNeela 等[39]
三价流感亚单位疫苗	鼻内（肌内对照组）	健康成人	壳聚糖鼻内接种两个不同剂量组（$n=23$），一组疫苗肌内接种（$n=22$）	Read 等[40]
基于 CRM_{197} 的脑膜炎球菌 C 寡糖偶联物	鼻内（肌内对照组）	健康成人	两组壳聚糖鼻内接种（分成亚组，总共 $n=30$），一组肌内接种疫苗+明矾（$n=6$）	Huo 等[41]
诺瓦克病毒样颗粒疫苗，与 MPL 和壳聚糖共同配制	鼻内	健康成人	随机分组，安慰剂分两步控制： 1. 3 种疫苗剂量（$n=5+5+10$，剂佐对照分别为 2+2+4），两组鼻内接种 2. 两种疫苗剂量（$n=20+20$，10 个佐剂对照，11 个安慰剂），两组鼻内接种	El-Kamary 等[42] Atmar 等[43]
HIV 病毒外壳蛋白	鼻内、肌内和阴道内接种	健康成年妇女	共 3 组疫苗接种，其中鼻内组接种壳聚糖佐剂疫苗（$n=36$）	Mucovac2[44]
转移性乳腺癌的激光免疫治疗	局部瘤内接种	成人晚期乳腺癌患者	激光免疫疗后进行壳聚糖接种（$n=10$），治疗 3 次	Li 等[45]

总之，临床试验的数据支持壳聚糖作为鼻内疫苗的有效佐剂和给药系统。使用基于壳聚糖的佐剂（包括黏性凝胶）进行其他疫苗和给药途径的临床试验正在进行中。

39.2.6 优势和缺点

壳聚糖作为佐剂和疫苗载体具有许多优点，但也具有一些局限性。表 39.2 列出了与应用相关的优点和限制壳聚糖在疫苗研制中的应用。获得高品质壳聚糖是阐明免疫刺激机制和临床应用进展的关键。如 ViscoGel 系统所示，可以解决与提供高纯度、特性良好的医用级壳聚糖相关的问题。因此，通过受控生产工艺制作的新型壳聚糖及其衍生物的研制，为针对各种疾病靶点和给药途径而设计的疫苗打开了大门。

表 39.2 壳聚糖作为佐剂的优点和缺点

优点	缺点
天然产品；无毒，可生物降解	天然产品；配方和鉴定方法不规范
在高品质壳聚糖中，受控制造工艺可确保其纯度高，不含任何致敏成分	劣质壳聚糖含有杂质，结果不可靠
其可制成多种形式，例如凝胶、粉末、纳米/微粒、化学衍生物	大多数未修饰壳聚糖在生理 pH 下不溶

（续表）

优点	缺点
阳离子和黏膜黏合剂；作为黏膜疫苗接种的载体/佐剂，可用于多种给药途径	除了鼻内给药途径外，很少有可用的其他给药途径研究
商品成本低，可无限制获取原材料	很少有可用的医用级、高品质壳聚糖
不同的化学性质（如DD）和物理性质（如颗粒大小）决定免疫反应，多功能性允许特定设计的灭活疫苗	决定免疫反应的化学和物理特性不明确
刺激体液和细胞免疫反应	作用机制知之甚少

参考文献

[1] Nossal，G. J.：Vaccines of the future. Vaccine 29 Suppl 4，D111-D115（2011）. doi：1 0. 1016/j. vaccine. 2011. 06. 089. S0264-410X（11）00983-2［pii］.

[2] Leroux-Roels，G.：Unmet needs in modern vaccinol-ogy：adjuvants to improve the immune response. Vaccine 28 Suppl 3，C25-C36（2010）. doi：10. 1016/j. vaccine. 2010. 07. 021. S0264-410X（10）01004-2［pii］.

[3] McElhaney，J. E.：Prevention of infectious diseases in older adults through immunization：the challenge of the senescent immune response. Expert Rev. Vaccines 8，593-606（2009）. doi：10. 1586/erv. 09. 12.

[4] Schubert，C.：New vaccine tailored to the weakened elderly immune system. Nat. Med. 16，137（2010）. doi：10. 1038/nm0210-137a. nm0210-137a［pii］.

[5] Brewer，J. M.，et al.：Aluminium hydroxide adjuvant initi-ates strong antigen-specifi c Th2 responses in the absence of IL-4-or IL-13-mediated signaling. J. Immunol. 163，6448-6454（1999）. ji_v163n12p6448［pii］.

[6] Mbow，M. L.，De Gregorio，E.，Valiante，N. M.，Rappuoli，R.：New adjuvants for human vaccines. Curr. Opin. Immunol. 22，411-416（2010）. doi：10. 1016/j. coi. 2010. 04. 004. 0952-7915（10）00068-3［pii］.

[7] Schijns，V. E.，Lavelle，E. C.：Trends in vaccine adjuvants. Expert Rev. Vaccines 10，539-550（2011）. doi：10. 1586/erv. 11. 21.

[8] Jabbal-Gill，I.，Fisher，A. N.，Rappuoli，R.，Davis，S. S.，Illum，L.：Stimulation of mucosal and systemic antibody responses against Bordetella pertussis fi lamentous haemagglutinin and recombinant pertussis toxin after nasal administration with chitosan in mice. Vaccine 16，2039-2046（1998）. S0264410X98000772［pii］.

[9] Zaharoff，D. A.，Rogers，C. J.，Hance，K. W.，Schlom，J.，Greiner，J. W.：Chitosan solution enhances both humoral and cell-mediated immune responses to subcutaneous vaccination. Vaccine 25，2085-2094（2007）. doi：10. 1016/j. vaccine. 2006. 11. 034. S0264-410X（06）01225-4［pii］.

[10] Bonferoni，M. C.，Sandri，G.，Rossi，S.，Ferrari，F.，Caramella，C.：Chitosan and its salts for mucosal and transmucosal delivery. Expert Opin. Drug Deliv. 6，923-939（2009）. doi：10. 1517/17425240903114142.

[11] Sorlier，P.，Denuziere，A.，Viton，C.，Domard，A.：Relation between the degree of acetylation and the electrostatic properties of chitin and chitosan. Biomacromolecules 2，765-772（2001）.

[12] Kumar，M. N.，Muzzarelli，R. A.，Muzzarelli，C.，Sashiwa，H.，Domb，A. J.：Chitosan chemistry and pharmaceutical perspectives. Chem. Rev. 104，6017-6084（2004）. doi：10. 1021/cr030441b.

[13] I llum，L.，Farraj，N. F.，Davis，S. S.：Chitosan as a novel nasal delivery system for peptide drugs. Pharm. Res. 11，1186-1189（1994）.

[14] Alves，N. M.，Mano，J. F.：Chitosan derivatives obtained by chemical modifi cations for biomedical and

environmental applications. Int. J. Biol. Macromol. 43, 401-414 (2008). doi: 1 0. 1016/j. ijbiomac. 2008. 09. 007. S0141-8130 (08) 00208-0 [pii].

[15] Arca, H. C., Gunbeyaz, M., Senel, S.: Chitosan-based systems for the delivery of vaccine antigens. Expert Rev. Vaccines 8, 937 (2009). doi: 10. 1586/erv. 09. 47.

[16] Jabbal-Gill, I., Watts, P., Smith, A.: Chitosan-based delivery systems for mucosal vaccines. Expert Opin. Drug Deliv. 9, 1051-1067 (2012). doi: 10. 1517/17425247. 2012. 697455.

[17] Luppi, B., Bigucci, F., Cerchiara, T., Zecchi, V.: Chitosan-based hydrogels for nasal drug delivery: from inserts to nanoparticles. Expert Opin. Drug Deliv. 7, 811-828 (2010). doi: 10. 1517/17425247. 2010. 495981.

[18] Gronlund, H., et al.: Carbohydrate-based particles: a new adjuvant for allergen-specifi c immunotherapy. Immunology 107, 523-529 (2002). 1535 [pii].

[19] Kamath, A. T., et al.: Synchronization of dendritic cell activation and antigen exposure is required for the induction of Th1/Th17 responses. J. Immunol. 188, 4828-4837 (2012). doi: 10. 4049/jimmunol. 1103183. jimmunol. 1103183 [pii].

[20] Kovacsovics-Bankowski, M., Clark, K., Benacerraf, B., Rock, K. L.: Effi cient major histocompatibility complex class I presentation of exogenous antigen upon phagocytosis by macrophages. Proc. Natl. Acad. Sci. U. S. A. 90, 4942-4946 (1993).

[21] Vila, A., Sanchez, A., Tobio, M., Calvo, P., Alonso, M. J.: Design of biodegradable particles for protein delivery. J. Control. Release 78, 15-24 (2002). S0168365901004862 [pii].

[22] Vila, A., et al.: Low molecular weight chitosan nanoparticles as new carriers for nasal vaccine delivery in mice. Eur. J. Pharm. Biopharm. 57, 123-131 (2004). S0939641103001620 [pii].

[23] Amidi, M., et al.: Preparation and characterization of protein-loaded N-trimethyl chitosan nanoparticles as nasal delivery system. J. Control. Release 111, 107-116 (2006). doi: 10. 1016/j. jconrel. 2005. 11. 014. S0168-3659 (05) 00652-8 [pii].

[24] Saint-Lu, N., et al.: Targeting the allergen to oral dendritic cells with mucoadhesive chitosan particles enhances tolerance induction. Allergy 64, 1003-1013 (2009). doi: 10. 1111/j. 1398-9995. 2009. 01945. x. ALL1945 [pii].

[25] Porporatto, C., Bianco, I. D., Correa, S. G.: Local and systemic activity of the polysaccharide chitosan at lymphoid tissues after oral administration. J. Leukoc. Biol. 78, 62-69 (2005). doi: 10. 1189/jlb. 0904541. jlb. 0904541 [pii].

[26] Neimert-Andersson, T., et al.: Improved immune responses in mice using the novel chitosan adjuvant ViscoGel, with a haemophilus infl uenzae type b glycoconjugate vaccine. Vaccine 29, 8965-8973 (2011). doi: 1 0. 1016/j. vaccine. 2011. 09. 041. S0264-410X (11) 01447-2 [pii].

[27] Ghendon, Y., et al.: Evaluation of properties of chitosan as an adjuvant for inactivated infl uenza vaccines administered parenterally. J. Med. Virol. 81, 494-506 (2009). doi: 10. 1002/jmv. 21415.

[28] Lee, C. G., Da Silva, C. A., Lee, J. Y., Hartl, D., Elias, J. A.: Chitin regulation of immune responses: an old molecule with new roles. Curr. Opin. Immunol. 20, 684-689 (2008). doi: 10. 1016/j. coi. 2008. 10. 002. S0952-7915 (08) 00185-4 [pii].

[29] Shibata, Y., Metzger, W. J., Myrvik, Q. N.: Chitin particle-induced cell-mediated immunity is inhibited by soluble mannan: mannose receptor-mediated phagocytosis initiates IL-12 production. J. Immunol. 159, 2462-2467 (1997).

[30] Da Silva, C. A., Hartl, D., Liu, W., Lee, C. G., Elias, J. A.: TLR-2 and IL-17A in chitin-induced macrophage activation and acute infl ammation. J. Immunol. 181, 4279-4286 (2008). 181/6/4279 [pii].

[31] Da Silva, C. A., Pochard, P., Lee, C. G., Elias, J. A.: Chitin particles are multifaceted immune adjuvants. Am. J. Respir. Crit. Care Med. 182, 1482-1491 (2010). doi: 1 0. 1164/rccm. 200912-1877OC. 200912-1 877OC [pii].

[32] Mora-Montes, H. M., et al.: Recognition and blocking of innate immunity cells by Candida albicans chitin. Infect. Immun. 79, 1961-1970 (2011). doi: 10. 1128/IAI. 01282-10. IAI. 01282-10 [pii].

[33] Peluso, G., et al.: Chitosan-mediated stimulation of macrophage function. Biomaterials 15, 1215-1220 (1994).

[34] Li, H., Willingham, S. B., Ting, J. P., Re, F.: Cutting edge: infl ammasome activation by alum and alum's adjuvant effect are mediated by NLRP3. J. Immunol. 181, 17-21 (2008). 181/1/17 [pii].

[35] Mori, A., et al.: The vaccine adjuvant alum inhibits IL-12 by promoting PI3 kinase signaling while chitosan does not inhibit IL-12 and enhances Th1 and Th17 responses. Eur. J. Immunol. (2012). doi: 10. 1002/ eji. 201242372.

[36] Hallander, H. O., Lepp, T., Ljungman, M., Netterlid, E., Andersson, M.: Do we need a booster of Hib vaccine after primary vaccination? A study on antiHib seroprevalencein Sweden 5 and 15 years after the introduction of universal Hib vaccination related to notifi cations of invasive disease. APMIS 118, 878-887 (2010). doi: 10. 1111/j. 1600-0463. 2010. 02674. x.

[37] Muzzarelli, R. A.: Chitins and chitosans as immuno-adjuvants and non-allergenic drug carriers. Mar. Drugs 8, 292-312 (2010). doi: 10. 3390/md8020292.

[38] Mills, K. H., et al.: Protective levels of diphtheria-neutralizing antibody induced in healthy volunteers by unilateral priming-boosting intranasal immunization associated with restricted ipsilateral mucosal secretory immunoglobulin a. Infect. Immun. 71, 726-732 (2003).

[39] McNeela, E. A., et al.: Intranasal immunization with genetically detoxifi ed diphtheria toxin induces T cell responses in humans: enhancement of Th2 responses and toxin-neutralizing antibodies by formulation with chitosan. Vaccine 22, 909-914 (2004). doi: 10. 1016/j. vaccine. 2003. 09. 012. S0264410X03006728 [pii].

[40] Read, R. C., et al.: Effective nasal infl uenza vaccine delivery using chitosan. Vaccine 23, 4367-4374 (2005). doi: 10. 1016/j. vaccine. 2005. 04. 021. S0264-410X (05) 00463-9 [pii].

[41] Huo, Z., et al.: Induction of protective serum meningococcal bactericidal and diphtheria-neutralizing antibodies and mucosal immunoglobulin A in volunteers by nasal insuffl ations of the Neisseria meningitidis serogroup C polysaccharide-CRM197 conjugate vaccine mixed with chitosan. Infect. Immun. 73, 8256-8265 (2005). doi: 1 0. 1128/IAI. 73. 12. 8256-8 265. 2005. 73/12/8256 [pii].

[42] El-Kamary, S. S., et al.: Adjuvanted intranasal Norwalk virus-like particle vaccine elicits antibodies and antibody-secreting cells that express homing receptors for mucosal and peripheral lymphoid tissues. J. Infect. Dis. 202, 1649-1658 (2010). doi: 10. 1086/657087.

[43] Atmar, R. L., et al.: Norovirus vaccine against experimental human Norwalk Virus illness. N. Engl. J. Med. 365, 2178-2187 (2011). doi: 10. 1056/NEJMoa1101245.

[44] Mucovac2. http: //public. ukcrn. org. uk/Search/ StudyDetail. aspx? StudyID=11679 (2012).

[45] Li, X., et al.: Preliminary safety andeffi cacy results of laser immunotherapy for the treatment of metastatic breast cancer patients. Photochem. Photobiol. Sci. 10, 817-821 (2011). doi: 10. 1039/c0pp00306a.

(吴国华译)

第 40 章　含铝配方佐剂的作用机理

Mirjam Kool 和 Bart N. Lambrecht①

80 多年来，明矾是应用最为广泛的佐剂，人们一直在探索替代佐剂的应用，但铝佐剂在未来多年内仍将继续使用。这是因为它们具有良好的安全性、低成本和多种抗原的佐剂性。令人惊讶的是，它的作用机制仍不清楚。

在本章中，我们将介绍不同的明矾配方以及我们目前对其作用机制的理解，尽管明矾的最终作用方式尚不确定。

40.1　引言

目前接种疫苗已有 100 多年的历史。Edward Jenner 于 1796 年完成[1,2]首次报告的疫苗接种。他给一个小男孩接种了牛痘病毒，进而导致这个小男孩对随后的天花病毒的感染产生抵抗力，而这一实验如今一定不会得到监管机构的批准。通过接种病毒颗粒的灭活微生物、致弱活病毒或亚单位疫苗可实现疫苗接种的保护。然而，亚单位疫苗接种不会引起强烈的免疫反应，而这可通过施用佐剂予以实现。在免疫学中，佐剂是一种可以刺激免疫系统并增强疫苗反应的制剂，而不会对其产生任何特异性抗原作用。

1926 年，Alexander Glenny 及其同事报道，用硫酸铝钾（称为明矾）沉淀的类毒素在注射豚鼠时比单独使用可溶性类毒素诱导更强的抗体产生[3,4]。目前，明矾凭借出色的安全性和增强保护性体液免疫反应的能力，已获批成为预防性疫苗中最常用的佐剂。然而明矾作为佐剂使用已有很长一段时间，但我们对它的作用机制却不甚了解。直到现在，我们才开始探索明矾的作用机制。

目前，明矾仍然是美国唯一获得许可的疫苗佐剂。但在欧洲，自 20 世纪 90 年代以来，MF59、AS03（均为水包油乳剂）和 MPL（单磷脂 A；LPS 类似物）也被批准使用[5-7]。除用于人类疫苗接种外，铝佐剂还用于实验免疫学[8,9]，并用于生产鼠单克隆抗体以及多克隆抗血清。在兽医学中，铝佐剂已被大量用于抗病毒和细菌性疾病的疫苗配方中[10,11]。

由于疫苗用于包括婴儿和儿童在内的健康个体，而且新型佐剂存在潜在的安全性问题，因此需要进行广泛的临床前安全性研究，包括局部反应原性和全身毒性试验。虽然大量佐剂明显比明矾更有效，但其大多具有更高的毒性水平，这是将其排除作为人疫苗佐剂的主要原因。

40.2　疫苗接种

40.2.1　铝佐剂疫苗

很多文献在提及“明矾”一词时，有很多不准确之处。最常用的“明矾”一词仅适用于

---

① M. Kool，PhD（✉）（伊拉斯谟医学中心肺内科，荷兰鹿特丹，根特大学免疫调节与黏膜免疫学实验室，比利时根特，E-mail：m. kool@ erasmusmc. nl）

B. N. Lambrecht，MD，PhD（伊拉斯谟医学中心肺内科，荷兰鹿特丹，比利时根特 VIB 研究所分子生物医学研究部，比利时根特，根特大学免疫调节与黏膜免疫学实验室，比利时根特）

硫酸铝钾[12-15]（表40.1）。由于在制造过程中出现了问题，因此不再使用明矾用于疫苗。相反生产中使用了其他几种不溶性铝盐，如氢氧化铝和磷酸铝。这是因为这些不溶性铝盐可以以更标准化的方式进行制备，并通过直接吸附捕获抗原。两种不同的吸附力主导抗原与佐剂之间的吸附作用[8,13,16,17]。首先是静电相互作用，这种作用在带负电荷的蛋白质和氢氧化铝之间，以及带正电荷的蛋白质和磷酸铝之间最为强烈。制备过程中出现的阴离子可能会共沉淀，并改变“纯”氢氧化铝的特性。具有这种影响的阴离子的一个关键示例是磷酸盐。应尽量减少氢氧化铝佐剂和磷酸化抗原对配方中磷酸盐离子的暴露，以最大程度地吸附抗原[7,18,19]。第二种吸附力是阴离子配体交换，是一种共价相互作用，当抗原含有磷酸基时，磷酸基可以取代佐剂表面上的羟基。其与铝（共价键的无机等价物）形成了一个内球面复合体。配体交换是最强的吸附力，即使存在静电斥力也会进行吸附[7,20,21]。重要的是，在小鼠体内，吸附力的强弱与抗体滴度和T细胞活化成反比[1,22-24]。通过静电吸引吸附至含铝佐剂上的抗原比通过配体交换吸附到含铝佐剂上的抗原更容易在肌内或皮下注射时洗脱[3]。树突状细胞可适当提取从佐剂中释放的抗原、在佐剂内空隙中捕获的抗原，以及吸附到含铝佐剂中的抗原[5,7]。值得注意的是，氢氧化铝或磷酸铝佐剂吸附抗原分别可在4小时或15分钟内完全洗脱[8]。

表40.1　不同含铝佐剂剂配方概述

商业名称	化学配方	化学名称	应用
Alum	$AlK(SO_4)_2 \cdot 12H_2O$	硫酸铝钾	不再使用
Imject-alum	$Al(OH)_3+Mg(OH)_2$	氢氧化铝和氢氧化镁	实验免疫学
Alhydrogel	$Al(OH)_3$	氢氧化铝	人用和兽用疫苗
Adju-Phos	$Al(PO_4)_3$	磷酸铝	人用和兽用疫苗

40.2.2　作用机制

经常引用3种可能机制来解释佐剂如何增加体液免疫，尽管现在还没有实验证据来支持这3种机制说法。首先，需要形成缓慢释放抗原的贮库，以增强抗体的产生[10]；其次，诱导炎症，进而引进和激活捕获抗原的抗原呈递细胞[12,14,15]；再次，将可溶性抗原转化为颗粒形式，使其被抗原呈递细胞，如巨噬细胞、DC和B细胞吞噬。主要理论就是，贮存形式和相关抗原缓慢释放[8,16]是明矾增强抗原递呈和随后T细胞和B细胞反应的原因。然而，最近有几篇论文显示，明矾的佐剂活性与贮存形式无关[18,19]。虽然从腹腔中析出的铝可在注射后1个月的时间内将佐剂效应从一只小鼠转移至另一只小鼠，但这并不是明矾的主要作用机制。据观察，明矾结节周围有中性粒细胞，$CD11c^+$DC和抗原特异性T细胞的积累，表明贮存形式可能与记忆库维持有关（MK和BNL 2008，[20,21]）。

40.2.2.1　铝佐剂激活的细胞通路

免疫系统包括两个主要的分支，即先天或非特异性免疫系统和适应性或特异性免疫系统。先天免疫系统是我们抵御生物入侵的第一道防线，而适应性免疫系统则是第二道防线，并且能保护我们免受同一病原体的再次侵袭。快速作用的先天免疫反应提供了必要的第一道防御。相反，适应性免疫利用识别来自病原体抗原的免疫细胞选择和克隆扩增，从而获取特异性和持久的免疫记忆。虽然免疫系统的两个分支具有不同的功能，但它们之间相互作用（即先天免疫系统的组成部分影响适应性免疫系统，反之亦然）。DC在先天免疫反应和适应性免疫反应之间形成桥梁[22-24]，并在佐剂介导的疫苗抗原免疫原性增强中发挥重要作用。

不同佐剂功效可通过DC中信号通路的差异进行解释，以进行启动T细胞分化和激活所需的复杂成熟事件[22]。明矾注射在注射部位引起炎症，即皮下、肌内或腹腔内的实验环境中（图

40.1)。注射后数小时内，常规和浆细胞样 DC、中性粒细胞、NK 细胞、嗜酸性粒细胞和炎症性单核细胞数量明显增加，肥大细胞和巨噬细胞消失[25-28]。这些固有细胞的增进伴随着几种必要趋化因子的分泌，如单核细胞趋化性 MCP1（CCL2）、MIP-1αCCL2、MIP-1α（CCL2）、MIP-1β（CCL4）、中性粒细胞趋化因子 KC/IL-8（CXCL8），嗜酸性粒细胞趋化因子 1（CCL11），IL-5 和 RANTES（CCL5）[25,27,29]。

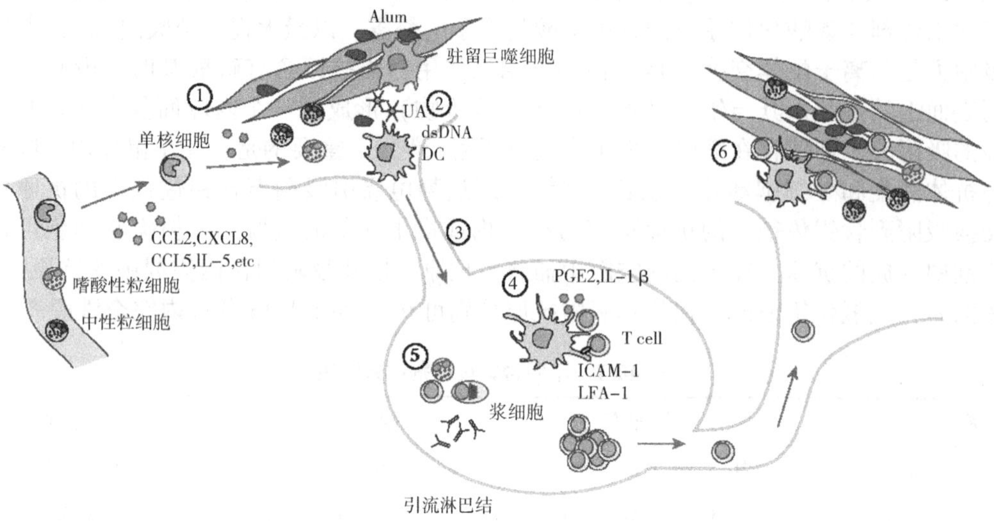

图 40.1 明矾佐剂诱导的细胞免疫反应综述

（1）肌内注射明矾配制的抗原后，肌肉细胞立即响应释放趋化因子和细胞因子，如 CCL2、CCL5、IL-5 和 CXCL8，其会吸引先天免疫系统的细胞，如单核细胞、嗜酸性粒细胞和中性粒细胞至注射部位。（2）此外，通过局部组织损伤或巨噬细胞的活性产生，会形成尿酸（UA）和 dsDNA（至少在一些模型中）的局部释放。（3）新进的单核细胞分化为树突状细胞（DC），吸收抗原，在 MHCI 和 MHCII 分子上处理抗原，并将其运送到引流淋巴结。（4）在淋巴结中，成熟的 DC 选择分化成 Th 效应细胞的抗原特异性 T 细胞。尤其在小鼠中，由于 PGE2 和 IL1β 的生成以及 ICAM-1 和 LFA-1 的递呈，这种反应是 Th2 型反应。（5）在脾脏中，并且可能在引流淋巴结中，还存在刺激 B 细胞反应的 $Gr1^+$、$IL-4^-$ 嗜酸性粒细胞的补充。明矾主要诱导长寿的 B 细胞反应。（6）将效应 T 细胞补充至注射部位，并发现于明矾和抗原库周围也有树突状细胞（DC）、中性粒细胞和嗜酸性粒细胞。可能，储库的形成与刺激长期记忆有关。

目前尚不清楚含铝佐剂是否以及如何诱导 DC 调动和成熟。正如体外数据所证明，明矾在体内的作用模式很可能是 DC 直接激活的上游[30]。至少在体外，明矾并未增强共刺激分子递呈和 DC 的成熟，因此有人提出，明矾只有在结构细胞或炎症细胞释放内源性危险信号时，才可能在体外诱导 DC 活化[31-33]。已有研究表明，当注射明矾时，尿酸和 dsDNA 都作为危险信号释放[25,34]。鉴于 DC 对活化适应性免疫起着重要的作用，已有多项研究关表明矾在体内腹腔及肌内注射抗原后对 DC 及其单核细胞前体的影响[25,35]。因此，单核细胞吸收抗原并迁移至引流淋巴结，变成单核细胞衍生的 DC，表达高水平的 MHC Ⅱ类和 CD86，并诱导抗原特异性 T 细胞的剧烈增殖。

重要的是，在有条件地耗尽 CD11c 树突状细胞的小鼠中，明矾对 T 细胞反应和体液免疫的所有佐剂效应均被消除，但分选的 Ly6C 高单核细胞的授受性转移恢复了这些效应。在人体中，明矾配制疫苗主要作用于单核细胞水平，诱导表型和功能成熟[36,37]。总之，这些实验表明，单核细胞衍生的 DC 是必需的，且足以介导明矾的佐剂作用。此外，伴随着 MHCII 表达的增加[38]，明矾还增强了 DC 表面上肽对 MHCII 复合物递呈的幅度和持续时间。这与增加的 ICAM-1 递呈一起将导致与 T 细胞接触时间的增加，从而引起 $CD4^+T$ 细胞[39]产生强烈的反应[39]。

DC与带抗原的明矾接触不会导致颗粒抗原的吞噬；DC只通过内吞作用获得可溶性抗原[39]。这种抗原摄取途径不会导致交叉递呈（如MHC Ⅰ中外源性抗原的递呈）[40]。众所周知，含铝佐剂主要诱导体液免疫，而不影响细胞免疫。明矾通过脾Gr-1$^+$骨髓IL-4$^-$生成细胞类型诱导B细胞致敏，这很可能代表嗜酸性粒细胞[41-43]，这一事实进一步支持了这一观点[41-43]。明矾缺乏交叉致敏，通过DTH反应和诱导CD8$^+$CTL对一系列多肽和蛋白抗原的反应来测量经典的细胞介导免疫，但明矾对蛋白抗原的诱导效果较差[13,44,45]。尽管如此，已发现CD4$^+$T细胞和Th2细胞因子产生的增殖反应在许多小鼠和人类研究中得到增强，这表明明矾通过为滤泡B细胞提供Th2帮助来增强体液免疫[12,46-48]。

虽然动物研究对揭示控制免疫反应的基本机制非常重要，但在理解人类免疫系统的复杂性、遗传异质性及揭示人类病原体的复杂逃逸机制方面，我们仍然面临着种种挑战。这些知识将有助于制定新的疫苗策略。

40.2.2.2　含铝佐剂在DC中激活的分子途径

含铝佐剂如何被抗原呈递细胞识别一直是一个难题（图40.2）。过去几年的几篇论文表明，通过MyD88或TRIF衔接途径的Toll样受体（TLR）和TLR信号传导，先天免疫的经典激活剂和体内DC网络对于明矾而言并不总是必须作为体液免疫佐剂[49-52]。然而，先天免疫反应依赖于关键的衔接分子MyD88[25]。在这种情况下，有必要说明TLR信号和白细胞介素-1受体（IL-1R）信号都要依赖MyD88来进行有效的细胞内信号传导。

此外，已清楚得知其他病原体检测系统的存在，其依赖于称为NOD样受体（NLR）的细胞内受体族[53,54]。NLR具有多种调节炎症反应和凋亡反应的功能。尽管TLR能感知感染性生物的细胞外非自身基序，而NLR能感知微生物来源的刺激以及内源性信号（DAMP），而不是微生物模式[55]。其可以识别压力、异常自我或危险信号，如DNA、RNA和尿酸[31]。NLRP3（NALP3）是NLR族成分之一，与ASC和caspase-1共同形成一个称为炎性体的分子平台[56]。NLRP3可由几种激动剂激活，如内源性ATP、尿酸、二氧化硅和明矾佐剂，导致IL-1β和IL-18的加工和释放[57]。

这些促炎细胞因子的释放需要两个信号。首个信号来自Toll样受体激动剂，例如脂多糖（LPS）和炎性细胞因子，激活NF-κB途径[58]，进而激活前体蛋白的转录和积累。其次，炎性小体会介导细胞因子的分裂和分泌。反之，IL-1β会触发另一系列导致炎症的分子事件[59]。IL-1β是非常有效的炎性细胞因子，与急性和慢性炎症性疾病有关。

Fabio Re的开创性论文显示，在人体外DC和巨噬细胞通过明矾和TLR激动剂的组合刺激生成IL-1β和IL-18。这种细胞因子的产生依赖于caspase-1和NLRP3[33]。随后的研究表明，小鼠DC在明矾刺激下也表现出类似的结果（图40.2）[27,60-63]。明矾激活NLRP3炎性小体的体外实验结果相比，明矾诱导的体液免疫应答是否需要炎性小体-IL-1β途径存在争议。在体外实验中，先天性和获得性免疫反应在NLRP3炎性小体的必要性方面似乎存在差异。在NLRP3缺陷小鼠中，抗原摄取、免疫细胞向注射部位的募集、树突状细胞向引流淋巴结的迁移以及CD4$^+$T细胞的增殖减少，但仍然存在[60]。明矾的佐剂活性的发挥和体液免疫的诱导是否需要NLRP3还存在矛盾。结果显示，如果没有NLRP3，佐剂活性可能被终止[61,62]，而有些研究则显示佐剂活性无须NLRP3的参与[27,63]，而我们的研究[60]表明需要NLRP3炎性体来生成免疫球蛋白E。目前尚无法解释不同组别的结果有何不同。然而，不同类型的明矾和免疫方案可能导致不同的表型。更令人困惑的是，对NLRP3的需求受到了质疑，因为铝诱导的体液反应不需要TLR4和IL-1R途径的衔接分子MyD88和TRIF[49,51]。

最近已有研究显示，明矾与鞘磷脂、胆固醇等膜脂相互作用，通过脂筏形成激活DC[39]。通过脂质分选，含有免疫受体信号传导基序（ITAM）的受体聚集，进而导致ITAM磷酸化。磷酸化的ITAM基序补充Syk，随后激活PI3激酶。这一途径最终将导致吞噬和细胞因子的产生[64-66]。与诱导脂筏形成和Syk活化的尿酸结晶不同[67]，明矾不会被DC吞噬。然而，用明矾

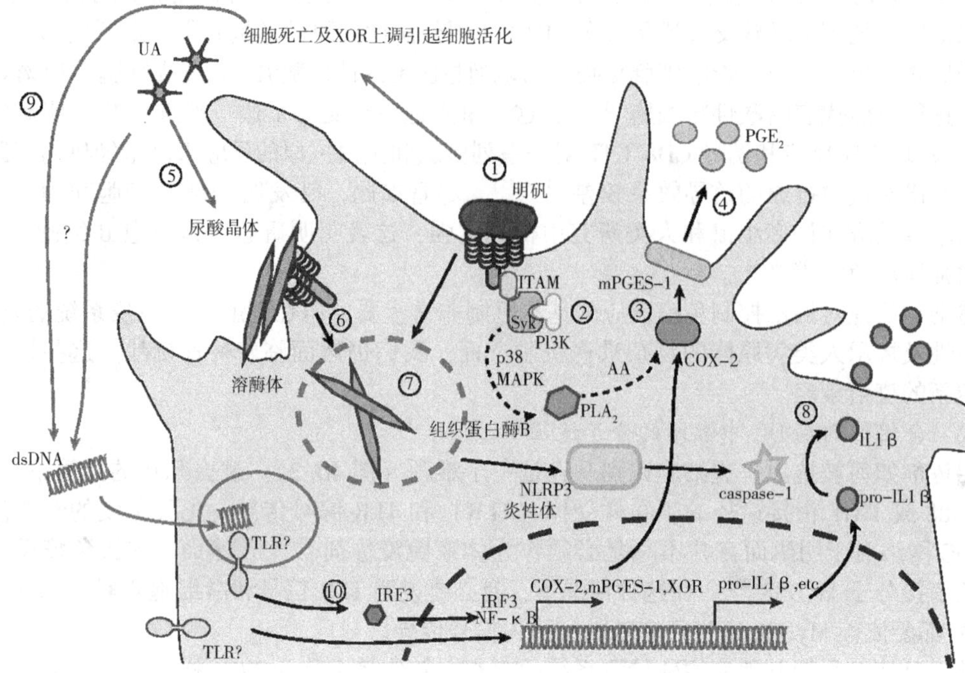

图 40.2 明矾在 DC 和巨噬细胞中诱导的分子路径

①明矾与细胞膜结合，进而诱导脂筏形成。②在脂筏中，含有 ITAM 的受体聚集并激活 Syk-PI3Kδ 路径。③这一路径反过来激活细胞磷脂酶 A2（cPLA2），可能通过 p38 MAP 激酶，导致膜脂质稀释花生四烯酸。COX-2 和膜相关的 PGE 合酶-1（mPGES-1）将花生四烯酸转化为前列腺素 E2（PGE2）。④从细胞中释放 PGE 2 并指示 Th2 反应。⑤此外，明矾通过诱导 XOR 诱导尿酸的释放。⑥在明矾或尿酸晶体的吞噬作用下诱导溶酶体损伤。⑦这激活了由组织蛋白酶 B 等酶释放到细胞质中而产生的 NLRP3 炎性体。⑧活化的 caspase-1 介导 IL-1β 前体的溶蛋白性裂解产生生物活性 IL-1β。⑨明矾除释放尿酸外，还可能通过细胞死亡诱导 dsDNA 的释放。⑩dsDNA 通过 Irf3 激活免疫系统。

引发的 DC 与 $CD4^+T$ 细胞相互作用更强，$CD4^+T$ 细胞不依赖于抗原，但依赖于 ICAM-1 和 LFA-1 递呈[39,65]。此外，巨噬细胞释放明矾诱导的前列腺素 E2（PGE 2）对 Th2 极化至关重要[68]。PGE 2 是一种特性良好的促炎脂质介质。其作为一种花生四烯酸代谢物，通过提高 DC、巨噬细胞和 Th1 细胞的细胞内 cAMP 浓度来抑制 Th1 反应。PGE 2 的产生与炎性体激活无关，但依赖于 Syk 和 P38 MAP 激酶途径。同样地，Syk 和 PI3 激酶信号在 DC 活化和 Th2 极化过程中发挥了重要作用[69]。总之，明矾是 LPS 诱导 DC 产生 IL-12 的有效抑制剂，这是由于明矾诱导 PI3 激酶信号转导[70]。

40.2.2.3 内源性危险信号的介入

除直接效应外，明矾佐剂还会产生一定程度的细胞毒性[15]。在人体中，可在注射部位发现坏死细胞的增加、内源性危险信号的潜在来源或损伤相关分子模式（DAMPs）[71]。濒死细胞释放大量的分子作为 DAMP（阻尼），可以通过激活各种 PRRR 信号传导途径警示先天免疫系统。含铝佐剂诱导尿酸（UA）和宿主 dsDNA 的内源性释放，间接激活免疫系统[25,34]。UA 是嘌呤分解代谢过程中产生的一种危险信号，是哺乳动物中嘌呤代谢的终产物。另外，在 RNA 和 DNA 降解以及释放的嘌呤置换成 UA 后，可以从受损细胞中释放 UA。

宿主 dsDNA 通过从坏死细胞中释放而暴露于免疫系统。这两种 DAMP 均会对细胞死亡向免疫系统示警。为小鼠注射明矾后，注射部位 UA 和 dsDNA 水平均迅速升高[25,34]。UA 和 dsDNA 都能作为佐剂，促进 Th2 相关的体液反应[25,34,69]。UA 晶体通过激活 Syk/PI3 激酶途径诱导 Th2

免疫[69]，而dsDNA信号通过Tbk1和Irf3[34]。研究这两种途径之间的相互作用机制，将有助于深入了解明矾佐剂的作用机理。

40.3　结语

如本章所示，多个大系统之间具有复杂的相互作用；而免疫学与复杂性和控制力有关。然而，疫苗研制与复杂性协调利用有关。现代疫苗研制努力将复杂简单化，而实现最佳疫苗研制的免疫杠杆的实际实现通常会受到疫苗佐剂的影响[72]。因此，了解最古老、最常用的佐剂明矾的作用机理，对于研制新型佐剂具有重要意义。

参考文献

[1] Hansen, B., Sokolovska, A., HogenEsch, H., Hem, S. L.: Relationship between the strength of antigen adsorption to an aluminum-containing adjuvant and the immune response. Vaccine 25, 6618-6624 (2007).

[2] Jenner, E.: The Three Original Publications on Vaccination Against Smallpox. Harvard Classics. P. F. Collier & Son, New York (1909).

[3] Jiang, D., Morefi eld, G. L., HogenEsch, H., Hem, S. L.: Relationship of adsorption mechanism of antigensby aluminum-containing adjuvants to in vitro elution in interstitial fl uid. Vaccine 24, 1665-1669 (2006).

[4] Glenny, A.: Insoluble precipitates in diphtheria and tetanus immunization. Br. Med. J. 2, 244-245 (1930).

[5] Romero Méndez, I. Z., Shi, Y., HogenEsch, H., Hem, S. L.: Potentiation of the immune response to non-a dsorbed antigens by aluminum-containing adjuvants. Vaccine 25, 825-833 (2007).

[6] Tritto, E., Mosca, F., De Gregorio, E.: Mechanism of action of licensed vaccine adjuvants. Vaccine 27, 3331-3334 (2009).

[7] Iyer, S., HogenEsch, H., Hem, S. L.: Effect of the degree of phosphate substitution in aluminum hydroxide adjuvant on the adsorption of phosphorylated proteins. Pharm. Dev. Technol. 8, 81-86 (2003).

[8] Shi, Y., HogenEsch, H., Hem, S. L.: Change in the degree of adsorption of proteins by aluminum-containing adjuvants following exposure to interstitial fl uid: freshly prepared and aged model vaccines. Vaccine 20, 80-85 (2001).

[9] Kool, M., et al.: Alum adjuvant boosts adaptive immunity by inducing uric acid and activating infl ammatory dendritic cells. J. Exp. Med. 205, 869-882 (2008).

[10] Glenny, A., Pope, C., Waddington, H., Wallace, U.: Immunological notes. XVII. The antigenic value of toxoid precipitated by potassium alum. J. Path. and Bact 29, 31-40 (1926).

[11] Heegaard, P. M. H., et al.: Adjuvants and delivery systems in veterinary vaccinology: current state and future developments. Arch. Virol. 156, 183-202 (2011).

[12] Mannhalter, J. W., Neychev, H. O., Zlabinger, G. J., Ahmad, R., Eibl, M. M.: Modulation of the human immune response by the non-toxic and non-pyrogenic adjuvant aluminium hydroxide: effect on antigen uptake and antigen presentation. Clin. Exp. Immunol. 61, 143-151 (1985).

[13] Hem, S. L., HogenEsch, H.: Relationship between physical and chemical properties of aluminum-containing adjuvants and immunopotentiation. Expert Rev. Vaccines 6, 685-698 (2007).

[14] Goto, N., Akama, K.: Histopathological studies of reactions in mice injected with aluminum-adsorbed tetanus toxoid. Microbiol. Immunol. 26, 1121-1132 (1982).

[15] Goto, N., et al.: Local tissue irritating effects and adjuvant activities of calcium phosphate and aluminium hydroxide with different physical properties. Vaccine 15, 1364-1371 (1997).

[16] Gupta, R. K., Chang, A. C., Griffi n, P., Rivera, R., Siber, G. R.: In vivo distribution of radioactivity in mice after injection of biodegradable polymer microspheres containing 14C-labeled tetanus toxoid. Vaccine 14, 1412-1416 (1996).

[17] Hem, S. L., HogenEsch, H., Middaugh, C. R., Volkin, D. B.: Preformulation studies-the next advance in aluminum adjuvant-containing vaccines. Vaccine 28, 4868-4870 (2010).

[18] Hutchison, S., et al.: Antigen depot is not required for alum adjuvanticity. FASEB J. 26, 1272-1279 (2011).

[19] Noe, S. M., Green, M. A., HogenEsch, H., Hem, S. L.: Mechanism of immunopotentiation by aluminum-containing adjuvants elucidated by the relationship between antigen retention at the inoculation site and the immune response. Vaccine 28, 3588-3594 (2010).

[20] Munks, M. W., et al.: Aluminum adjuvants elicit fi brin-dependent extracellular traps in vivo. Blood 116, 5191-5199 (2010).

[21] Lambrecht, B. N., Kool, M., Willart, M. A. M., Hammad, H.: Mechanism of action of clinically approved adjuvants. Curr. Opin. Immunol. 21, 23-29 (2009).

[22] Steinman, R. M., Pope, M.: Exploiting dendritic cells to improve vaccine effi cacy. J. Clin. Invest. 109, 1519-1526 (2002).

[23] Pashine, A., Valiante, N. M., Ulmer, J. B.: Targeting the innate immune response with improved vaccine adjuvants. Nat. Med. 11, S63-S68 (2005).

[24] Bendelac, A., Medzhitov, R.: Adjuvants of immunity: harnessing innate immunity to promote adaptive immunity. J. Exp. Med. 195, F19-F23 (2002).

[25] Kool M., et al.: Cutting edge: Alum adjuvant stimu-lates infl ammatory dendritic cells through activation of the NALP3 infl ammasome. J. Immunol. 181, 3755-3759 (2008).

[26] Seubert, A., et al.: Adjuvanticity of the oil-in-water emulsion MF59 is independent of Nlrp3 infl ammasome but requires the adaptor protein MyD88. Proc. Natl. Acad. Sci. 108, 11169-11174 (2011).

[27] McKee, A. S., et al.: Alum induces innate immune responses through macrophage and mast cell sensors, but these sensors are not required for alum to act as an adjuvant for specifi c immunity. J. Immunol. 183, 4403-4414 (2009).

[28] Calabro, S., et al.: Vaccine adjuvants alum and MF59 induce rapid recruitment of neutrophils and monocytes that participate in antigen transport to draining lymph nodes. Vaccine 29, 1812-1823 (2011).

[29] Didierlaurent, A. M., et al.: AS04, an aluminum salt-and TLR4 agonist-based adjuvant system, induces a transient localized innate immune response leading to enhanced adaptive immunity. J. Immunol. 183, 6186-6197 (2009).

[30] Mosca, F., et al.: Molecular and cellular signatures of human vaccine adjuvants. Proc. Natl. Acad. Sci. 105, 10501-10506 (2008).

[31] Shi, Y., Evans, J. E., Rock, K. L.: Molecular identifi ca-tion of a danger signal that alerts the immune system to dying cells. Nature 425, 516-521 (2003).

[32] Sun, H., Pollock, K. G. J., Brewer, J. M.: Analysis of the role of vaccine adjuvants in modulating dendritic cell activation and antigen presentation in vitro. Vaccine 21, 849-855 (2003).

[33] Li, H., Nookala, S., Re, F.: Aluminum hydroxide adjuvants activate caspase-1 and induce IL-1beta and IL-18 release. J. Immunol. 178, 5271-5276 (2007).

[34] Marichal, T., et al.: DNA released from dying host cells mediates aluminum adjuvant activity. Nat. Med. 17, 996-1002 (2011).

[35] Langlet, C., et al.: CD64 expression distinguishes monocyte-derived and conventional dendritic cells and reveals their distinct role during intramuscular immunization. J. Immunol. 188, 1751-1760 (2012).

[36] Ulanova, M., Tarkowski, A., Hahn-Zoric, M., Hanson, L. A.: The common vaccine adjuvant aluminum hydroxide up-regulates accessory properties of human monocytes via an interleukin-4-dependent mechanism. Infect. Immun. 69, 1151-1159 (2001).

[37] Seubert, A., Monaci, E., Pizza, M., O'Hagan, D. T., Wack, A.: The adjuvants aluminum hydroxide and MF59 induce monocyte and granulocyte chemoattractants and enhance monocyte differentiation toward dendritic cells. J. Immunol. 180, 5402-5412 (2008).

[38] Ghimire, T. R., Benson, R. A., Garside, P., Brewer, J. M.: Alum increases antigen uptake, reduces antigen degradation and sustains antigen presentation by DCs in vitro. Immunol. Lett. 147, 55-62 (2012).

[39] Flach, T. L., et al.: Alum interaction with dendritic cell membrane lipids is essential for its adjuvanticity. Nat. Med. 17, 479–487 (2011).

[40] Burgdorf, S., Kautz, A., Böhnert, V., Knolle, P. A., Kurts, C.: Distinct pathways of antigen uptake and intracellular routing in CD4 and CD8 T cell activation. Science 316, 612–616 (2007).

[41] Jordan, M. B., Mills, D. M., Kappler, J., Marrack, P., Cambier, J. C.: Promotion of B cell immune responses via an alum–induced myeloid cell population. Science 304, 1808–1810 (2004).

[42] Wang, H. –B., Weller, P. F.: Pivotal advance: eosino–phils mediate early alum adjuvant–elicited B cell priming and IgM production. J. Leukoc. Biol. 83, 817–821 (2008).

[43] McKee, A. S., et al.: Gr1 + IL – 4 – producing innate cells are induced in response to Th2 stimuli and suppress Th1–dependent antibody responses. Int. Immunol. 20, 659–669 (2008).

[44] Wijburg, O. L., et al.: The role of macrophages in the induction and regulation of immunity elicited by exogenous antigens. Eur. J. Immunol. 28, 479–487 (1998).

[45] Bomford, R.: The comparative selectivity of adju–vants for humoral and cell–mediated immunity. II. Effect on delayed – type hypersensitivity in the mouse and guinea pig, and cell – mediated immunity to tumour antigens in the mouse of Freund's incomplete and complete adjuvants, alhydrogel, Corynebacterium parvum, Bordetella pertussis, muramyl dipeptide and saponin. Clin. Exp. Immunol. 39, 435–441 (1980).

[46] Brewer, J. M., et al.: Aluminium hydroxide adjuvant initiates strong antigen–specifi c Th2 responses in the absence of IL–4–or IL–13–mediated signaling. J. Immunol. 163, 6448–6454 (1999).

[47] Grun, J. L., Maurer, P. H.: Different T helper cell subsets elicited in mice utilizing two different adjuvant vehicles: the role of endogenous interleukin 1 in proliferative responses. Cell. Immunol. 121, 134 – 145 (1989).

[48] Serre, K., et al.: IL–4 directs both CD4 and CD8 T cells to produce Th2 cytokines in vitro, but only CD4 T cells produce these cytokines in response to alum–precipitated protein in vivo. Mol. Immunol. 47, 1914–1922 (2010).

[49] Gavin, A. L., et al.: Adjuvant–enhanced antibody responses in the absence of toll–like receptor signaling. Science 314, 1936–1938 (2006).

[50] Schnare, M., et al.: Toll–like receptors control activation of adaptive immune responses. Nat. Immunol. 2, 947–950 (2001).

[51] Nemazee, D., Gavin, A., Hoebe, K., Beutler, B.: Immunology: toll – like receptors and antibody responses. Nature 441, (2006).

[52] Palm, N. W., Medzhitov, R.: Immunostimulatory activity of haptenated proteins. Proc. Natl. Acad. Sci. 106, 4782–4787 (2009).

[53] Martinon, F., Tschopp, J.: Infl ammatory caspases and infl ammasomes: master switches of infl ammation. Cell Death Differ. 14, 10–22 (2007).

[54] Ting, J. P. Y., Willingham, S. B., Bergstralh, D. T.: NLRs at the intersection of cell death and immunity. Nat. Rev. Immunol. 8, 372–379 (2008).

[55] Mariathasan, S., Monack, D. M.: Infl ammasome adap – tors and sensors: intracellular regulators of infection and infl ammation. Nat. Rev. Immunol. 7, 31–40 (2007).

[56] Martinon, F., Gaide, O., Pétrilli, V., Mayor, A., Tschopp, J.: NALP infl ammasomes: a central role in innate immunity. Semin. Immunopathol. 29, 213–229 (2007).

[57] Fritz, J. H., Ferrero, R. L., Philpott, D. J., Girardin, S. E.: Nod–like proteins in immunity, infl ammation and disease. Nat. Immunol. 7, 1250–1257 (2006).

[58] Kawai, T., Akira, S.: Signaling to NF – κB by Toll – like receptors. Trends Mol. Med. 13, 460 – 469 (2007).

[59] Arend, W. P., Palmer, G., Gabay, C.: IL–1, IL–18, and IL–33 families of cytokines. Immunol. Rev. 223, 20–38 (2008).

[60] Kool, M., et al.: Cutting edge: alum adjuvant stimulatesinfl ammatory dendritic cells through activation of the NALP3 infl ammasome. J. Immunol. 181, 3755–3759 (2008).

[61] Eisenbarth, S. C., Colegio, O. R., O'Connor, W., Sutterwala, F. S., Flavell, R. A.: Crucial role for

the Nalp3 infl ammasome in the immunostimulatory p roperties of aluminium adjuvants. Nature 453, 1122–1126 (2008).

[62] Li, H., Willingham, S. B., Ting, J. P. Y., Re, F.: Cutting edge: infl ammasome activation by alum and alum's adjuvant effect are mediated by NLRP3. J. Immunol. 181, 17–21 (2008).

[63] Franchi, L., Núñez, G.: The Nlrp3 infl ammasome is critical for aluminium hydroxide–mediated IL–1β secretion but dispensable for adjuvant activity. Eur. J. Immunol. 38, 2085–2089 (2008).

[64] Nakashima, K., et al.: A novel Syk kinase–selective inhibitor blocks antigen presentation of immune complexes in dendritic cells. Eur. J. Pharmacol. 505, 223–228 (2004).

[65] Greenberg, S., Chang, P., Wang, D. C., Xavier, R., Seed, B.: Clustered syk tyrosine kinase domains trigger phagocytosis. Proc. Natl. Acad. Sci. U. S. A. 93, 1103–1107 (1996).

[66] Turner, M., Schweighoffer, E., Colucci, F., Di Santo, J. P., Tybulewicz, V. L.: Tyrosine kinase SYK: essential functions for immunoreceptor signalling. Immunol. Today 21, 148–154 (2000).

[67] Ng, G., et al.: Receptor–independent, direct mem–brane binding leads to cell–surface lipid sorting and Syk kinase activation in dendritic cells. Immunity 29, 807–818 (2008).

[68] Kuroda, E., et al. Silica crystals and aluminum salts regulate the production of prostaglandin in macrophages via NALP3 infl ammasome–independent mechanisms. Immunity. 34, 1–13 (2011).

[69] Kool, M., et al.: An unexpected role for uric acid as an inducer of T helper 2 cell immunity to inhaled antigens and infl ammatory mediator of allergic asthma. Immunity 34, 527–540 (2011).

[70] Mori, A., et al.: The vaccine adjuvant alum inhibits IL–12 by promoting PI3 kinase signaling while chitosan does not inhibit IL–12 and enhances Th1 and Th17 responses. Eur. J. Immunol. 42, 2709–2719 (2012).

[71] Goto, N., Akama, K.: Local histopathological reac–tions to aluminum–adsorbed tetanus toxoid. Naturwissenschaften 71, 427–428 (1984).

[72] Alving, C. R., Peachman, K. K., Rao, M., Reed, S. G.: Adjuvants for human vaccines. Curr. Opin. Immunol. 24, 310–315 (2012).

（何继军，吴国华译）

第41章　从聚合物到纳米药物的发展：未来疫苗的新材料

Philipp Heller，David Huesmann，
Martin Scherer 和 Matthias Barz①

摘要

纳米医学是纳米技术的医学应用，因此涵盖各种纳米颗粒。在本章中，我们将简要介绍和概述纳米颗粒对免疫系统的调节作用。整体来看，这些纳米物体可以是无机胶体、有机胶体（通过乳液聚合或微/纳米乳液技术合成）、聚合物聚集体（胶束或聚合物囊泡）、核心交联聚集体（纳米水凝胶、交联胶束或聚合物复合物）、多功能聚合物线圈、树枝状聚合物或完美的树枝状聚合物。聚合物材料受到极大的关注，因为粒子电晕的化学成分将通过提供位阻稳定作用来重塑粒子性质，避免蛋白质在体内的吸附和粒子聚集。除新材料合成之外，颗粒表征也同样重要，其可能是更详细地了解纳米级系统行为的关键。此外，我们想强调针对基于纳米颗粒的免疫疗法的方法。

41.1　引言

顾名思义，纳米颗粒是大小从1到几百纳米的物体[1]。其可由各种材料制成，由于其表面积与体积比大，通常具有不同于其所组成的基体材料的特性。可以通过自下向上或自上向下的方法合成纳米颗粒。整体来看，这些纳米系统可以是无机胶体、有机胶体（通过乳液聚合或微/纳米乳液技术合成）、聚合物聚集体（胶束或聚合物囊泡）、核心交联聚集体（纳米水凝胶、交联胶束或聚合物复合物）、多功能聚合物线圈、树枝状聚合物或完美的树枝状聚合物。

例如，顺磁性氧化铁核可用于磁共振成像（MRI）或可生物降解的核，降解时通过扩散释放生物活性。由于纳米颗粒大小不同，其可用来在一个物体之间结合不同的功能，从而实现多功能性或多价性。此外，这些系统能够包裹生物活性化合物。保护药物要求采用核壳结构；当核心包裹负载物时，外壳需要具有溶解性，并抑制蛋白质吸附或聚集。

当纳米尺寸物体用于诊断或治疗疾病时，其就成为所谓的纳米药物，这是一个包含纳米药物、纳米显影剂和治疗诊断学的总称[2]。

本文开头我们就想提及一个非常重要的观点，而令人惊讶的是，这一点常常被人忽视。纳米颗粒与环境的相互作用发生于颗粒与周围介质的交界处——颗粒表面。因此，稳定在微粒表面朝向其周围环境的分子决定了微粒与生物物质的相互作用，进而决定了微粒在体内的命运。在许多情况下，很难说明纳米颗粒的特性。就这一事实而言，我们认为有必要仔细考量、理解和设计纳米颗粒，尤其是当其用于复杂的体内应用时。

---

① P. Heller，Dipl. Chemist · D. Huesmann，Dipl. Chemist · M. Scherer，Dipl. Chemist · M. Barz，PhD（✉）（美因茨约翰内斯古腾堡大学有机化学研究所，德国美因茨，E-mail：barz@uni-mainz.de）

纳米颗粒合成的术语

聚合物 由称为单体的基本结构重复单元组成的化合物。可在称为聚合作用的过程中由一种或多种单体形成聚合物。材料性能不仅取决于单体的化学组成，还取决于彼此连接形成大分子的单体分子的数量（聚合度）和空间结构。

分散性：除一些示例外（蛋白质、DNA或完美树枝状聚合物），聚合物通常由分子量/聚合度不同的大分子混合组成，这是由于单个聚合物分子所构成单体数量的差异。分散度（正式称为多分散指数，PDI）定义为 $Đ=Mw/Mn$

其中，Mw 是质量平均摩尔质量，

$$Mw=\sum M_i^2\cdot N_i/(\sum M_i\cdot N_i)$$

Mn 是平均摩尔质量，

$$Mn=\sum M_i\cdot N_i/\sum N_i$$

低分散度（接近1）意味着大部分分子具有相似的聚合度。

嵌段共聚物 嵌段共聚物是一种重要的共聚物（由多个单体组成的聚合物）。当一种单体（均聚物）序列后跟着另一种单体形成的序列时，就可使用此术语。嵌段共聚物可由两种（二嵌段共聚物）、3种（三嵌段共聚物）或多种嵌段（多嵌段共聚物）组成。共价高分子中不同嵌段的化学性质可以形成独特的性质（如亲水/疏水嵌段共聚物的表面活性剂类两亲性），也可以使嵌段特异性功能化（使用染料、受体分子）成为可能。

胶体 胶体显微镜下分散在连续介质中的颗粒，大小为1～1 000 nm。胶体溶液是介于真溶液和悬浮液之间的级间，不具有依数性。胶体系统可以由有机和无机颗粒组成。为防止胶体凝固（这在热力学上是有利的），颗粒必须通过静电和位阻排斥力保持彼此分离。

两亲分子 两亲分子是同时具有亲水（极性）和疏水（非极性）的分子。当两亲分子与水接触时，其会自组装成不同的上层结构，从而将相间能降至最低。通过这种方法，可以形成各种聚集体，如胶束、聚合体（聚合物基脂质体状结构）或双层膜。精确的结构取决于制备和储存过程中亲疏水结构域的比例以及环境条件。

树突状大分子/树突 树突状大分子是一种具有重复分支的聚合物分子，并且有高度对称和清晰的结构。它们在重复循环中合成，新一代单体添加至具有两个或更多反应位点的核中。然后，新添加的外层的反应基（同样，每单位1个以上）作为新添加下一代的基础。或者，使用收敛法合成树突（具有单个反应基团的支链分子，即所谓的焦点），然后在最后一步耦合至多功能核上。在这两种情况下，均可合成单分散树枝状大分子，提供大量的功能性边缘团体。

胶束 胶束是由两亲性分子（如低分子量表面活性剂或两亲性聚合物）形成的一种重要的胶束。在临界胶束浓度（CMC）以上，表面活性剂分子会自发地排列成极性头基团指向水相（或其他极性溶剂），而尾部形成疏水内核，可用来包裹疏水负载物。

表面活性剂 表面活性剂（表面活化剂）是降低液体表面张力或两种不混溶相之间张力的分子。它们是两亲性化合物，吸附在亲疏水相界面，极性头部分别排列在亲水环境中，非极性部分延伸至疏水相。表面活性剂广泛应用于相间科学技术中，作为洗涤剂或分散剂、润湿剂、乳化稳定剂等。

除天然和合成的小分子表面活性剂外，聚合物基表面活性剂也引起了人们的广泛关注。由于嵌段材料的可变性和聚合后改性的可能性，这种两亲分子可以专门用于预期的应用。与小分子两亲分子相比，其高分子量使其具有更强的吸附性能和更优越的稳定性能，因此空间稳定成为一个很重要的因素。

乳状液 乳状液是一种分散体系，其分散在第二不混溶的液相（分散相）中。通常，稳定的乳液需要添加表面活性剂和溶解在分散相中的共稳定剂。通过不同的方法，可以得到大约30

到几百纳米的液滴粒度。这种所谓的微乳液制备已成为纳米颗粒合成中有趣的工具。这种颗粒既可通过微乳液聚合获得，也可通过（聚合物）包覆和预成型液滴的稳定（微乳液技术）获得。

41.2　纳米颗粒的优良性能

在过去的几十年中，纳米颗粒在给药和疫苗接种[3]方面显示出巨大的应用潜力，有望为当前此领域的各种问题提供解决方案。第一个问题是溶解度的根本问题：根据托尔钦林[4]，通过高通量筛选技术（包括活性最高的技术）确定的潜在有价值候选药物中，约有一半在水中溶解度较差，因此未得到进一步发展，通过包封或与纳米载体结合可以解决这一问题。

此外，敏感试剂如DNA、RNA或肽包封可以防止酶或蛋白的过早降解。

与载体的“隐形”表面相结合，可以防止通过位阻排斥力[5]的调理素作用，实现长时间循环。此外，通过化学改性，可以确定识别和位点特异性释放。

使用纳米载体的一个主要目的是提高对某些细胞或组织的特异性。这可通过被动靶向实现，正如在血管化良好的肿瘤中，由EPR（增强渗透性和保留率）效应[6]驱动的积累，也可通过颗粒与某些细胞递呈的特定受体结合予以实现。这种靶向定位是通过配体分子与受体结合的颗粒表面功能化予以实现的，例如，已经使用抗体[7]、糖基[8]、转铁蛋白[9]或叶酸[10]予以实现。

整体而言，包封和靶向可以提高各自治疗药物的生物利用度。此外，如果活性成分与环境隔离，并且仅在指定的作用部位释放，那么就可以减少与许多药物相关的毒副作用。以阿霉素为例，采用HPMA共聚物偶联PK1/PK2，使药物的最大耐受剂量（MTD）提高了3倍[11]。可由pH[12]、氧化还原电位[13]、酶活性[14]、温度[15]或磁性[16]、电信号、超声信号等因素触发[17,18]控制释放。

如Little所指，纳米颗粒在疫苗研制中的另一个很有前景的方面是亚单位抗原的多重表达[19]。合成颗粒可设计成抗原在表面的重复定向，从而具有生成B细胞受体交联和增强活化的潜力[20]。

41.2.1　纳米颗粒的稳定性

当使用纳米颗粒作为药物传递载体时，需要考虑几个因素：不仅载体本身必须保持胶体形式，而且必须避免或至少控制体内的非特异性聚集[21]。

静电稳定和空间稳定是分散和乳化体系稳定的两种主要机理。静电相互作用由以Derjaguin、Landau、Verwey和Overbeek命名的DLVO理论进行说明。聚合物吸附或共价键连接至粒子表面，形成空间位阻稳定，主要是由于熵效应和溶剂化能的少量焓[21]。

对于体内应用，似乎很难实现静电稳定，因为血清蛋白中既有阴离子大分子，也有阳离子大分子，其可以吸附于带相反电荷的胶体上。人体内存在一系列针对外来物质的保护机制，远远超出了本章的讨论范围。其中之一是从血液中去除大于肾阈值的异物。然而，这包括去除纳米级药物载体，而这是由单核吞噬细胞系统（MPS）的调理作用激发的过程。MPS的巨噬细胞可在静脉注射后几秒钟内清除颗粒，由于过早被身体清除阻碍了其作为药物载体的应用[22,23]。

尽管尚未完全了解这一复杂过程的具体机制，但人们普遍认为，调理素作用始于一系列蛋白吸收，包括免疫球蛋白、补体系统成分和其他血清蛋白。目前确认的相关蛋白（不完整）列表包括C3、C4、C5、层粘连蛋白、纤连蛋白、C反应蛋白、I型胶原蛋白等[24,25]。这些蛋白质通过随机的布朗运动接近纳米载体，一旦接近表面，就可通过不同的力（范德瓦尔斯或离子力，仅举两例）与颗粒结合[23]。调理素结合之后，吞噬细胞附于纳米颗粒上。吞噬细胞本身无法识别载体，但会通过表面结合的调理素与微粒结合。两种替代机制是通过非特异性黏附于调理素和补体激活来刺激吞噬作用，从而导致单核吞噬细胞与异物结合[23]。

在最后一步，颗粒被吞噬细胞吸收，通常通过吞噬作用。一旦微粒被内吞，其就会被各种酶和化学物质降解，或者（如果是不可降解的物质）会储存并富集在体内。

迄今为止，还没有能够完全防止颗粒不均匀化的通用解决方案。然而，观测结果显示，总体趋势是亲水性但中性的配体阻碍了蛋白质的吸附，从而减缓了其从血液中的清除[26]。

因此，空间稳定似乎是体内应用的方法。使用商业聚合物表面活性剂（例如，Tween 20/40/60/80、Pluronics 及 Cremophor）或其他两亲性聚合物可以实现空间稳定。据 Scheibe[27] 和 Kelsch[28] 等报告，嵌段共聚物在颗粒和水溶液接口似乎形成了更稳定的层，并能抑制血清中的聚集。此外，聚合物还可用功能群进行改性，而这些随后可用于进一步功能化。

41.3 纳米连接载体的类型和制备方法

如上所述，生物活性化合物的位置确实对其作用有重要影响。例如，在疫苗接种中，可以用抗原位置调节免疫反应。根据具体目的和试剂性质，将抗原包埋以防止其过早降解，并将其递呈于内体途径（细胞内受体）中可能非常有利。在其他情况下，可优选在载体外部（细胞外受体）优化的免疫反应上以重复方向递呈。

因此，可以使用各种不同的胶体载体。最常用的系统包括脂质体、固体脂质纳米颗粒（SLN）、免疫刺激复合物（ISCOM）、胶束、水凝胶、微乳液制备的聚合物胶体，以及聚合物纳米颗粒，如在非润湿模板中复制的聚合物纳米颗粒（PRINT）（图 41.1）。

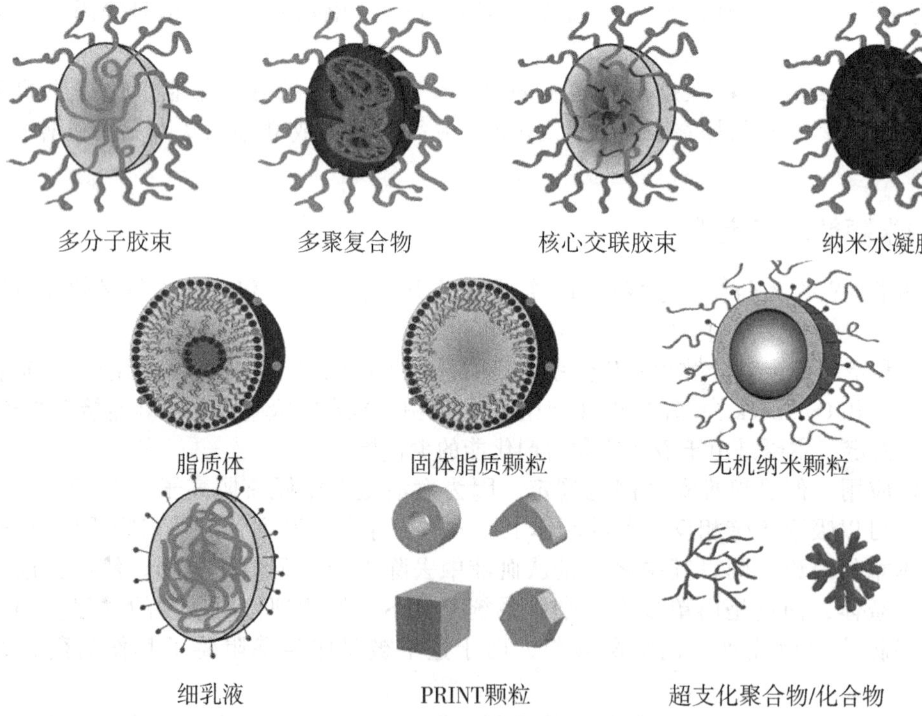

图 41.1　纳米颗粒载体

聚合物胶束[29]、核心交联胶束[30]、纳米水凝胶[31]和聚合物[32,33]主要是通过嵌段共聚物组装成超结构，主要受疏水、静电、金属络合和氢键 4 种力的驱动[34]。通常，其涉及疏水（胶束和核心交联胶束）和静电（多聚物）相互作用，而金属络合和 H 键形成仅适用于特定情况。对于疏水相互作用，不同极性块导致相分离，从而形成聚合物胶束，其中疏水部分聚集以避免与水接触。在核-壳结构中，疏水核可用于输送疏水有效载荷。在这种情况下，胶囊化合物的释放基于胶束的被动扩散或解体。另外，静电相互作用依赖于带相反电荷的分子之间的相互作用以形成上部结构。形成所谓聚合物的一个典型示例是带正电荷的聚合物（如聚赖氨酸）和带负电荷的核酸（DNA或RNA）的相互作用。驱动力主要是熵的增加，这是由于在复杂的形成过程中反荷离子的释放。

聚合物胶束形成于临界胶束浓度（CMC）以上，但始终与游离单基缔合物处于平衡状态。CMC可以视为溶液中游离单体的最大浓度，强调了这些系统的动力学特征。注射入血液后，胶束溶液迅速稀释，可能导致胶束分解[35]。虽然聚合物胶束的CMC远低于表面活性剂胶束的CMC[36]，但这仍然存在一定的问题。

因此，稳定胶束进一步引发关注。进一步稳定胶束最简单的方法是通过化学交联核心[37]。这些核心交联胶束发现自身既要稳定地封装负载物，又要在囊泡到达目的地后进行释放。要实现这一点，交联必须是可逆的，而目标部位的刺激必须打破稳定黏合剂。这可通过各种刺激响应性交联剂实现。最常用的刺激是pH或氧化还原电位，但也实现了温度、酶、光等因素（表41.1）[38]。pH-不稳定键在内胚体内吞过程中被裂解，而氧化还原不稳定键在细胞质还原环境中或抗原呈递细胞内吞过程中实现裂解。从而导致负载物在被细胞吸收后释放。

表41.1　控制释放策略

触发	链接示例	释放点
pH[12]	乙酰基、腙、酯中共价键的裂解	酸性环境（如内体/溶酶体室）
还原电位[13]	二硫化物或苯醌的裂解	还原环境（胞质）
酶活力[14]	肽键的裂解	特定的细胞/组织/疾病
温度[15]	相变或溶解度的变化	高温组织/应用过高热
磁信号[16]	通过磁铁运动形成孔隙	应用信号
电讯号[17]	通过电刺激聚合物膨胀或压电元件形成孔隙	应用信号
超声信号[18]	聚合物降解	应用信号

脂质体是一种球形结构，由1个或多个磷脂双层围绕在水核周围组成，模仿细胞膜的基本结构。根据双层数量不同，其大小范围从20 nm至几微米[39]。其结构组成使脂质体在负载药物活性剂和药剂特性方面具有很强的通用性。抗原或免疫增强剂可包封在核内，嵌入脂质双层膜，或与外表面连接[40]。脂质体通常由薄膜水化、溶剂注入或反相蒸发生成。为形成更小的囊泡和缩小尺寸分布，可以采用超声和挤压技术[41]。脂质体常用聚合物修饰以屏蔽脂质体，减少蛋白质吸附（如聚乙二醇化）[42]或整合pH反应元件[43]和分子，以特异性靶向免疫系统细胞，如甘露糖衍生物[44]。脂质体不仅可携带抗原，还可以携带佐剂来刺激先天免疫反应[45,46]。目前临床使用的脂质体制剂包括病毒体流感和甲型肝炎疫苗[47,48]。尽管具有上述优势，脂质体仍然受到几个问题的限制。这些因素包括长期的物理不稳定性和低分子包封率，但也难以大规模生产，进而增加成本[49]。其中一些问题可通过聚合体解决[50,51]，通过嵌段共聚物两亲体而不是脂质构建类似于脂质体的囊泡载体。由于聚合物的分子量比脂类的分子量大两个数量级，因此具有优越的材料性能和储存能力[50]。与脂类相比，大多数聚合物不具有FDA的GRAS（通常被认为是安全的）状态。

另一种脂质纳米颗粒是免疫刺激复合物（ISCOM）。除抗原外，其还有球形的开放晶格结构，由磷脂、胆固醇和强佐剂奎尔A皂苷形成。这类ISCOMs可携带多种抗原，据报告称，由于其含有皂苷，其免疫原性比同类胶体颗粒高10倍[52]。必须指出，由于Quil A（植物皂甙）的溶血活性，人们非常关注其毒性[53]。

固体脂质纳米颗粒（SLN）也以脂质为基础，但其结构与脂质体不同。其包括由脂质形成的核，脂质在生理条件下处于固态，被乳化剂外壳包围。这里的方法是将常规脂质载体（良好的生物相容性、易于生产/规模化）与聚合物纳米颗粒（用于物理和化学稳定的固体基质）的优势相结合[54,55]。主要通过高压均质化和细乳液方法制备 SLN。对于固体脂类加工，需要提高温度，但可以避免使用有机溶剂。SLN 的主要问题是由于脂质堆积（例如结晶）的改变而导致的多态性转变，影响颗粒形，进而影响与生理环境的相互作用[56]。

乳剂技术是一种多用途的颗粒形成方法，在疫苗接种和给药中产生不同种类的胶体载体。包括直接给药的纳米和微乳液颗粒[57-59]，以及通过乳液聚合或对先前形成的乳液滴进行处理而产生的聚合物或无机纳米颗粒[3,53,60,61]。

（微型）乳剂（图 41.2）由一种液体（分散相）滴入另一种不混相（连续相）液滴组成。根据其成分和制备方法，微乳液的尺寸范围为 30~500 nm[62]。

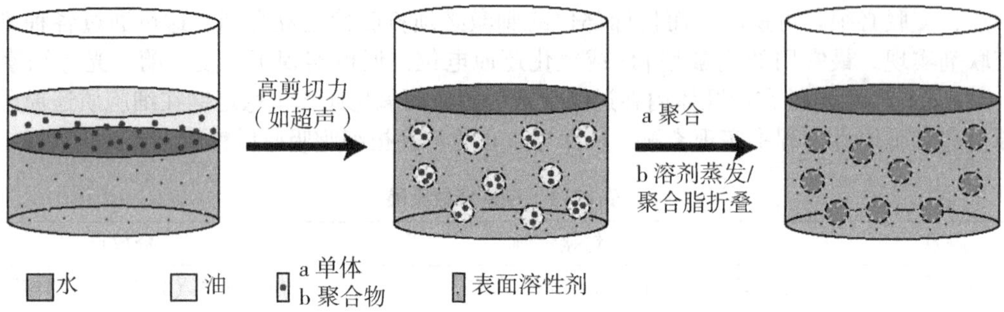

图 41.2　细乳液聚合（a）和细乳液技术（b）

使用表面活性剂后，导致静电或空间排斥，需要稳定液滴以防止聚集。由于不同尺寸颗粒的拉氏压力不同，可以通过混合到分散相中的第二种添加剂来防止分散液滴的净质量通量不稳定。这些添加剂具有高度疏水性，可以防止其从一个液滴扩散到另一个液滴，从而形成抵消拉普拉斯压力的渗透压[63]。此方法的有效性取决于添加剂的疏水性。许多不同类型的分子被用作疏水分子，包括烷烃[64]、染料[65]和用于聚合的疏水引发剂[66]。鉴于上述机理及其纳米尺寸可确保沉积的动力学稳定性，微型乳液被认为具有热力学稳定性[46,58]，而根据报道，这种乳液的稳定性的时间量程为几个月[62]。

一种物质在另一种物质中的纳米或微乳化（以及分散过程）包括两个主要方面：导致高比表面积液滴的待乳化相的破坏以及用表面活性剂稳定进行界面稳定。在乳化过程中，由待乳化分子和疏水物组成的油相通过适当的表面活性剂分散在水相中，然后同质化。由超声、转子-定子分散器、微流化器或高压均质器提供所需的高剪切力[62]。均质化后，加入第二批表面活性剂，使系统稳定有效[67]。

除上述之外，乳液技术的一个重要优势是相对容易规模化。尽管这是任何一种高度有序、自组装结构的临界点，但乳液配方的批量生产在短时间内提供了均匀且高度稳定的结果。

抗原包裹最常见的乳化方法是溶剂蒸发和凝聚。两种方法都包括形成双重乳液（水相 1/油相/水相 2，图 41.3），并共用第一步，其中在有机聚合物溶液中乳化各试剂的水溶液。在溶剂蒸发过程中，第一个 W/O 乳液随后进一步分散至含有适当乳化剂的水溶液中——聚乙烯醇经常用于此目的[68]。最后，有机溶剂在对比压力下蒸发，导致聚合物硬化，并形成所需颗粒。

凝聚是将聚合物（和抗原）的非溶剂加入之前形成的 W/O 乳液中的相分离过程。这导致相分离以形成包覆抗原的凝聚层相。

纳米载体形成的另一种方法是纳米沉淀法，也称为溶剂置换法。此处，通过预先形成的聚合物在乳液滴上的界面沉积形成稳定的纳米颗粒。反相乳液是在含有表面活性剂和用于外壳形成的混合溶剂/非溶剂介质中形成的。添加非溶剂会沉淀聚合物，并沉积于较大的表面液滴上。

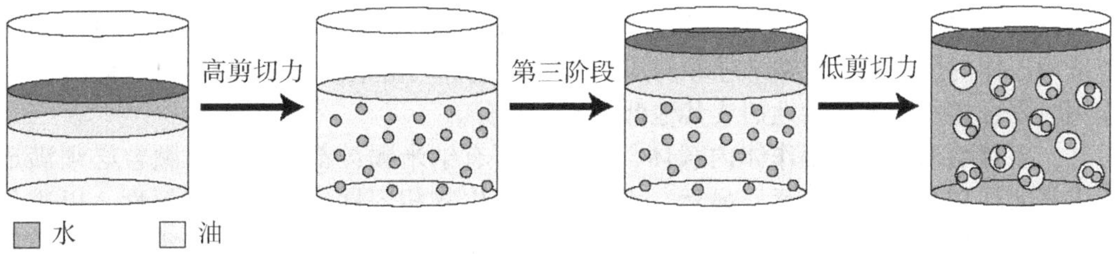

图 41.3 双重乳液的制备

沉淀后，颗粒可转换为水。

然而，与微米级的类似物体相比，纳米载体的形成需要更严苛的条件[69]。因此，诸如DNA这样的敏感分子可能在结构上受到破坏，在设计形成这种颗粒的条件时，需要特别小心。这些措施可能包括药剂与阳离子两亲分子的络合（疏水离子配对，HIP）[70]、合理的溶剂选择以及抗原在预成型颗粒表面的吸附[71]。

聚乳酸（PLA）和聚乳酸-丙二酸（PLGA）等聚酯是一类广泛用于药物载体形成的聚合材料。其可通过水解进行生物降解，具有良好的生物相容性和缓释性能[72]。为了在微乳液或纳米乳液过程中稳定这种聚合物，PEG-b-PLA共聚物似乎是有利的，因为PLA嵌段与聚酯核心充分混合，并在溶剂蒸发过程中瓦解。而PEG块则形成亲水、抗蛋白的电晕。当需要具有更高功能的电晕时，就可应用由Barz及其同事研制的多功能PHPMA-b-PLA共聚物[73,74]。

然而，天然或半合成聚合物也得到了广泛应用。这些聚合物通常以多糖或蛋白质为基础，对其的综述超出了本章的内容范围。因此，此处仅提及葡聚糖和壳聚糖这两个重要示例。

葡聚糖是主要包括α-1，6-糖苷和导致支化的小比例1，3-交联的聚合物[75]。它们极易溶于水，并且因为糖苷键而相对稳定，仅在极酸性或碱性条件下水解。羟基允许衍生化，例如通过部分疏水作用获得两亲分子[76]。壳聚糖在生理pH下不溶于水，因此在胺或羟基上进行化学改性以获得可溶性衍生物。壳聚糖由于其黏附特性，显示出强大的黏膜递送潜力[77,78]。

Kundu等对用于生物聚合物纳米颗粒的材料进行了很好的综述[79]。目前对胶态纳米粒的研究包括研制针对各种传染病的疫苗[80]，如乙型肝炎[81]、艾滋病[82]以及癌症[83]。

一种形成类似单分散和形状特异性聚合物纳米颗粒的有趣方法是由DeSimone团队研制的一种名为PRINT（非润湿模板中的颗粒复制）的制备方法[84]（图41.1）。在这种“自上而下”的方法中，会在由全氟聚醚制成的模具中形成颗粒。由聚乙二醇三丙烯酸酯或对羟基苯乙烯等组成的分子作为溶液或熔体充装到模具中，然后通过紫外线或化学方法进行固化。这可通过全氟聚醚的低表面能予以实现，允许通过毛细管作用填充空腔，但需防止材料在空腔之间的沉积[85]。用治疗剂负载颗粒可通过包埋到颗粒基质中或通过制备后功能化予以实现。例如，最近将用于基因沉默的siRNA整合至PRINT颗粒中[86]。

除了由有机材料制成的颗粒外，还研究了由无机物质制成的胶体颗粒。此处，研究主要集中于由金和磁铁矿（Fe_3O_4）组成的颗粒。文献中已说明了多种粒子合成途径[87]。金属纳米颗粒通常使用合适的还原剂还原金属盐制备而成，而金属氧化物纳米颗粒通常使用分子前体的碱水解进行制备[61]。胶体的稳定性通过适当的配体实现，例如能够提供空间稳定作用的螯合聚合物，可以在初始合成期间将其直接加入或者在合成后期通过交换或聚合物涂层修饰配体球[88,89]。

到目前为止，金纳米颗粒已基本用于肿瘤免疫疗法研究。基于金纳米颗粒的近红外（NIR）光吸收，当肿瘤特异性抗体靶向肿瘤时，金纳米颗粒能够局部增加温度，因此在光热癌症治疗中有广阔的应用前景[90]。最近，金纳米颗粒-抗原决定簇-Fc复合物也成功地用作抗原载体，并且与脂质体相比表现出更好的活性和免疫应答[91]。

磁纳米颗粒也用于热疗研究，主要影响抗肿瘤 T 淋巴细胞介导的免疫中热休克蛋白的表达[92]。此外，还进行了将它们作为磁共振对比剂 MRI 的研究[93]。在上述所有情况中需要注意，在制备适用于体内使用的无机纳米颗粒时应使用稳定剂和增溶剂。基于与有机胶体相同的原因，发挥作用时聚合物优于表面活性剂或其他低分子量组分。

总之，多种纳米颗粒可用作体内载体，但不是所有纳米颗粒都可应用。颗粒必须满足一些基本要求来实现在人体内的转运。显然，它们应具有生物相容性或生物可降解性，以避免出现类似聚乙二醇出现过的急性毒性或贮积疾病[94-96]。它们应具有水溶性，并且很少与蛋白质发生作用或者不与蛋白质发生相互作用，以确保血液循环顺畅。不与蛋白质相互作用的材料应具有亲水性和电中性，并且具有氢键受体，但没有氢键供体[26,97,98]。颗粒的外部材料会与身体接触，因此可以通过 MPS 的摄取确定身体对它的反应（或不反应）。在许多情况下，PEG 可以屏蔽颗粒，但有一些有趣的替代物可供选择，我们将在下面的章节中介绍。

由于许多纳米颗粒由聚合物组成或用聚合物在颗粒周围形成壳，因此有必要对可使用的

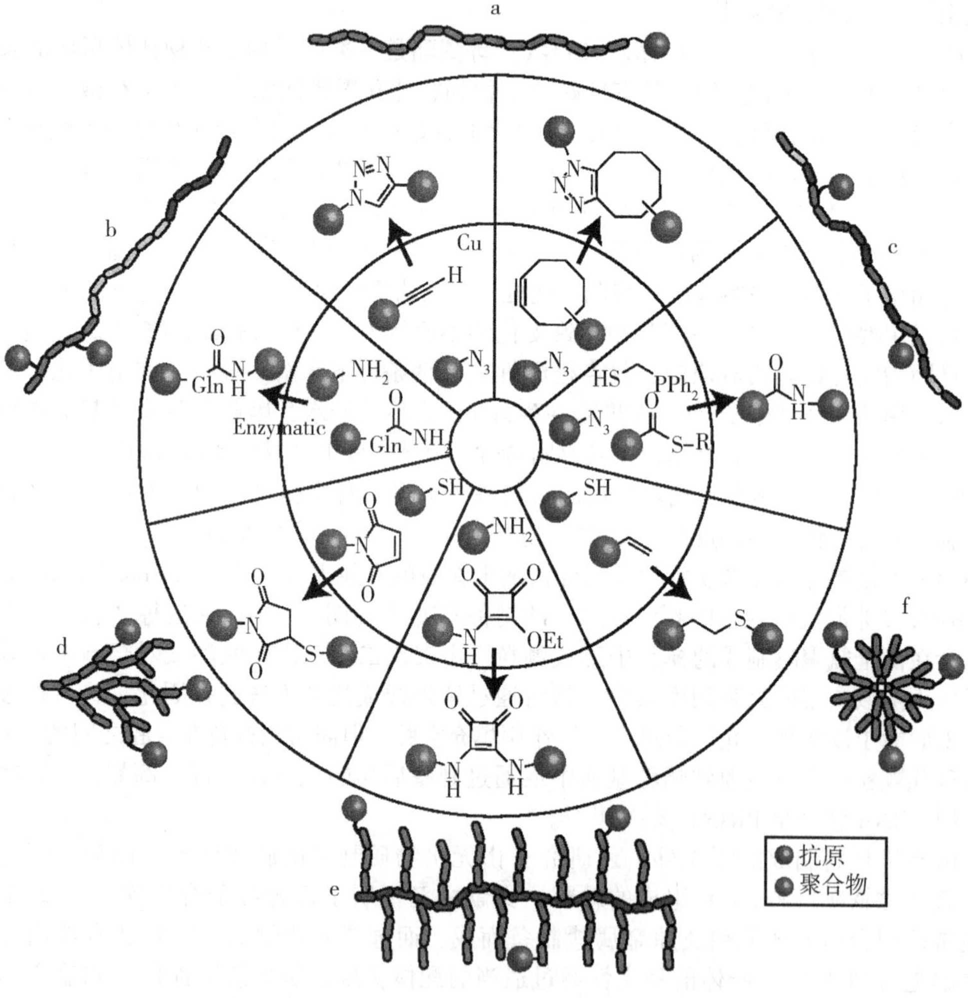

图 41.4 聚合物结构以及可能的抗原结合

（a）均聚物；（b）嵌段共聚物；（c）无规共聚物；（d）超支化聚合物；（e）聚合物刷；（f）树枝状聚合物。抗原连接可以使用（左上方顺时针方向）铜催化的叠氮-烷基环加成[101]、非铜催化叠氮-链烷-环加成[101]、无痕施陶丁格连接反应[102]、硫醇-烯化学[101]、方酸[103]、马来酰亚胺[101]、酶促共轭[104]以及其他化学方法[105]。

聚合物的结构进行简要说明（图41.4）。高分子由多个重复单元组成，具有不同的复杂程度。最简单的结构为线性均聚物，其仅具有一种重复单元。嵌段共聚物和无规共聚物由两种或更多种重复单元组成。但是，在第一种情况中重复单元以不同段排序，在第二种情况中它们沿着链随机分布。使用多功能单体时，也可以使用分支结构和刷结构。特殊情况为完美状态的树枝状大分子，这是一种分支结构，其每一代分支都是在不同的步骤中合成的。相比经典聚合物的每分子重复单元的分布（分子量分布），将树枝状聚合物定义为蛋白质类似物。它们表面的高密度官能团使它们参与多价相互作用[99]。例如，已将糖肽树枝状聚合物用于免疫疗法[100]。但是，完美状态的树枝状聚合物的合成要求很高，并且可能难以进行更大规模的合成。

但除了聚合物结构的变化之外，聚合物组成的材料也可以变化。在下一段中，我们将要介绍聚合物材料，其能够在纳米颗粒周围建立亲水性电晕，因此可以使其适用于体内，例如，药物递送、成像或疫苗接种。

41.4　高分子材料

41.4.1　聚（甲基）丙烯酸酯和丙烯酰胺

受控自由基聚合（CRP）技术的发展对合成高分子化学产生了巨大影响，因为它能够控制（甲基）丙烯酸酯和丙烯酰胺的聚合（表41.2）。开发CRP技术能减少终止，可分为3个小组：稳定的自由基聚合（如NMP）、变性转移聚合（例如RAFT、MADIX）和过渡金属介导的可控自由基聚合（例如ATRP）。其中，原子转移自由基聚合（ATRP）和RAFT（可逆加成-断裂链转移）聚合是最常用和最通用的方法。已经有一些综述报道了RAFT[106,107]或ATRP[108,109]的机制以及最新进展。

表41.2　传统自由基聚合（FRP）与可控自由基聚合（CRP）的比较

	自由基聚合	可控自由基聚合
优点	与各种单体相容	低分散性/窄分子量分布
	对杂质不敏感 已用于工业几十年	对聚合物架构的良好控制（链拓扑、端基功能）
缺点	缺乏对链拓扑的控制（没有分段体系结构，没有明确的端基！）	由于活性物质数量有限，反应速率较低
	分散性控制不足/宽分子量分布	对杂质的敏感性更高

CRP技术可用于合成复杂的聚合物结构，例如（多）嵌段共聚物、支化聚合物或混合体系[110-113]。

有趣的（甲基）丙烯酸酯组是低聚乙二醇侧链（OEGMA）的体系，例如甲基丙烯酸二乙二醇酯（DEGMA）或聚乙二醇甲基丙烯酸酯（PEGMA）。最近，人们对这些体系的兴趣不断增长。这些单体的聚合物具有良好的特性，例如在水中的高溶解性、非免疫原性和低毒性、较低的临界溶解温度（LCST）、提高了血液循环时间[114-116]。通过两种单体的共聚可以很好地调节LCST。据Lutz和Hoth报道，通过改变共聚物中OEGMA与DEGMA单元的比例，可以将LCST从26℃调节到90℃[117]。

在各种细胞系中进行了细胞毒性的研究，例如Caco-2、HT29-MTX-E12或HepG2，确定浓度达到5 mg/mL，暴露72小时没有毒性作用[118,119]。

另一种有趣的聚合物是聚（2-（甲基）丙烯酰氧基乙基磷酰胆碱）（PMPC）。由于侧链含有天然 PC-磷脂的头部基团，因此单体结构具有高度生物启发性，确保了高生物相容性[120,121]。

已经合成了 PMPC 的聚合物囊泡并应用于研究口腔上皮的扩散[122]。另外，Battaglia 等已将 PMPC 的嵌段共聚物用于转染[123]。PMPC 聚合物囊泡显示出明显的细胞摄取，并且高达 3 mg/mL无细胞毒性。此外，Lewis 和同事将 PMPC 聚合物用于蛋白质结合[124]。并且发现，与相同流体力学体积的 PEG-蛋白质结合物相比，组织迁移减少。因此，组织中增强的贮积效应以及随后更长的消除半衰期提高了药代动力学性能。这些结果显示了 PMPC 聚合物体系的潜力。

最后，也可以通过 CRP 方法以受控方式制备成熟的聚合物，例如 2-（羟丙基）甲基丙烯酰胺（HPMA）。在 20 世纪 90 年代，HPMA 是第一个进入临床试验的聚合物药物结合物[125]，并且 HPMA 的共聚物可以直接合成或使用后聚合修饰[126,127]。

这 3 种类型的单体可能是未来纳米药物的材料。然而，必须注意聚（甲基）丙烯酸酯的聚合物主链和丙烯酰胺都不能在体内降解。需要将聚合物排出，因此限制了单一聚合物的分子量以及潜在的应用。

41.4.2 多肽

多肽是自然界中的主要构建物之一，是一种非常通用的材料。它们本身具有多种功能，使其能够响应 pH 或氧化电位的变化等刺激。

有很多种合成多肽的方法。可以使用在细胞中合成肽或固相肽合成的生物技术方法合成特定序列的多肽，但是氨基酸有限，价格昂贵，并且只能以相对小的规模进行，因此我们不在这里进行讨论。当需要大量同源或嵌段共聚肽时，合适的制备方法是进行 α-氨基酸-N-羧酸酐（NCAs）的聚合（图 41.5）。NCAs 的聚合自 20 世纪初就已为人所知[128]，并且在经典的书籍和综述中总结了之后的发展[129,130]。但仅在最近几年，已经开发出控制这种聚合的不同方法[131]。

由于合成多肽与天然蛋白质一样由氨基酸组成，因此它们在体内是无毒、生物相容和可降解的，同时它们在水溶液中很稳定。由于它们与天然蛋白质的相似性，必须解决免疫原性的问题。尽管尚未发现同源和嵌段共聚肽的免疫原性[132,133]，但在这方面必须研究新的多肽，特别是当混入各种氨基酸时。

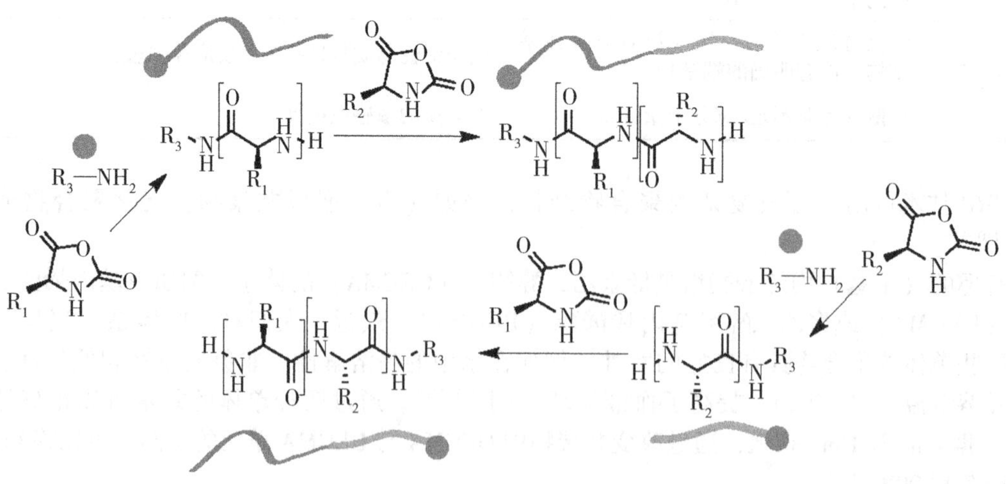

图 41.5 嵌段共聚肽的合成

根据多种不同的侧链能够设计肽的超结构，如聚合物复合物[33,134]、聚合物胶束[135,136]、聚合物囊泡[137,138]、纳米纤维或纳米管[139]和水凝胶[140]。

这些不同的结构使其具有广泛的生物医学应用。在DNA和RNA转运领域，肽聚合复合物已从简单的聚赖氨酸演变为高度复杂的结构，这些复杂结构中含有可去除的亲水壳[141]，并能够对pH[142]、氧化电位[143]或ATP浓度变化等刺激产生反应[144]。

多肽也用于包封疏水性药物。可能最典型的例子是PEG-嵌段-聚天冬氨酸，其中Asp被4-苯基-1-丁醇修饰，使其更加疏水[145]。该胶束制剂含有紫杉醇，目前正在进行名为NK105的乳腺癌Ⅲ期临床试验。

多肽也已进入诊断领域。例如，将聚乙二醇化聚赖氨酸与NIR染料用作体内成像探针，以显示裸鼠中的肿瘤相关蛋白酶[146]。

多肽本身也可以作为生物活性剂。免疫调节剂药物Copaxone是丙氨酸、赖氨酸、谷氨酸和酪氨酸的无规共聚物，已批准其用于治疗多发性硬化[147]。Copaxone的具体作用方式仍在探讨中，但据报道由于Copaxone和髓鞘碱性蛋白（MBP）的相似性，该药物可以转移自身免疫反应使其远离神经元髓鞘。

药物与多肽的共价连接也很有发展前景。在Opaxio（以前称作Xyotax或CT-2103）中，紫杉醇通过酯键与聚谷氨酸（PGA）结合，并且该结合物目前用于卵巢和非小细胞肺癌的Ⅲ期临床试验[133]。其中紫杉醇仅在聚合物被肿瘤相关蛋白酶降解时释放，使得化学治疗剂可以选择性释放[148]。

Deming使用赖氨酸-嵌段-亮氨酸共聚肽制备首个低于100 nm的稳定的双乳剂[149]。这些双重乳液由水相中油滴包裹水滴组成，能够携载疏水性和亲水性药物，这使它们在药物递送中的应用引人注目。

表41.3总结了在临床试验中用作载体或活性剂的多肽[150-152]。

尽管在临床中大多数多肽用于治疗癌症，但多肽也有用作载体的潜力。多肽的多功能性、生物相容性和生物降解性使其成为在体内转运中理想的生物活性材料。

表41.3 临床试验中的多肽

名称	多肽	药物	应用	阶段
NK105	PEG-b-Asp（buPh）	紫杉醇	胃癌	Ⅲ
			乳腺癌	Ⅱ Ⅲ
NK012	PEG-b-Glu	SN-38	小细胞肺癌	Ⅱ Ⅲ
			乳腺癌	Ⅱ Ⅲ
NC-6004	PEG-b-Glu	顺铂	胰腺癌	Ⅱ Ⅲ
			实体癌	Ⅰ
NC-4016	PEG-b-Glu	DACH-platin	实体癌	Ⅰ
Opaxio	p（Glu）	紫杉醇	卵巢癌	Ⅲ
			非小细胞肺癌	Ⅲ
CT-2106	p（Glu）	喜树碱	卵巢癌	Ⅱ Ⅲ
			实体癌	Ⅰ

（续表）

名称	多肽	药物	应用	阶段
Copaxone	Ala-Lys-Glu-Tyr copolymer Ala-Lys-Glu-Tyr 共聚物	肽	多发性硬化	美国食品和药物管理局/欧洲药品局/医药品医疗器械综合机构-认证
Vivagel	改性聚（赖氨酸）树枝状聚合物	肽	细菌性阴道病	Ⅲ

41.4.3 聚甘油

聚甘油（PG）是一类重要的生物相容性支化聚合物，可以将其合成为树枝状分子、树枝状聚合物或者超支化结构[153-155]。

与相似分子量的线性聚合物相比，树枝状聚合物和超支化聚合物具有有趣的性质，如球状结构和低黏性[156]。PG 的化学结构与 PEG 相似，因此 PG 与蛋白质的亲和力较低[157]。但是 PEG 仅提供两个修饰位点，而 PG 具有许多游离 OH 基团，它们可用于连接生物活性分子或靶向部分。

多步法[158,159]能够合成树枝状分子和树枝状聚合物，但也可以通过缩水甘油的阴离子聚合制备明确定义的超支化聚甘油[160,161]。可以使用电晕处理的醇基在核[162]或壳中修饰聚合物[163]。

关于线性和超支化 PG（分子量 6.4 kg/mol）补体激活的能力，未观察到显著延长的红细胞聚集和凝固。PG 的细胞毒性与 PEG 和羟乙基淀粉相当（浓度为 10 mg/mL 时，细胞存活率为 80%；L-929 成纤维细胞）。当注射剂量高达 62.5 mg/kg 时，发现小鼠对超支化聚甘油具有良好的耐受性[164]。

在体外[165]和体内[166]进行了较大的超支化 PGs（160 kg/mol 和 540 kg/mol）的生物相容性研究。体外试验与用低分子量 PG 进行的试验相似，并得出类似的结论，即 PGs 具有生物相容性。在小鼠体内进行这些聚合物的药代动力学试验。血浆半衰期分别为 32 小时和 57 小时。使用 1 g/kg 的剂量没有观察到毒性作用。但由于 PG 的分子量很高，尿排泄非常低，观察到单核吞噬细胞系统的显著积累。在肝脏和脾脏中至少持续 30 天可以检测到 PGs。这再次表明需要使用可生物降解材料。

由于其良好的生物相容性，PGs 有很多生物医学应用[38,155,167]。PG 可以通过共价连接或包封来携载生物活性化合物。

对于共价连接，非常有趣的是醇末端 PG 的功能性添加。例如，引入胺基团后[163]，可以通过酰胺连接酸，并且使用可生物降解的肽接头，PG 表面的硫醇已经用于连接药物[168]。多柔比星通过酸可切割接头与 PG 连接。在体外和卵巢癌 A2780 异种移植模型中长达 30 天检测结合物，显示出耐受性明显增加以及缓解肿瘤的抗肿瘤疗效[169]。此外，在位点选择性点击化学反应中已引入炔基，并且已将其用于合成对所选蛋白具有高结合亲和力的糖末端 PG[170]。发现带负电荷的 PG 硫酸盐具有相同的结果[171,172]。此类体系使用其多价性来实现高结合亲和力。

采用反相细乳液技术通过含有多元醇和聚环氧化物的二硫化物的酸催化开环加成反应合成 PG，部分改善了 PG 的不可降解性[173]。这些纳米颗粒可以在还原环境中裂解成能够从体内排出的较小碎片。

PG 的应用还包括用于组织工程的水凝胶支架，这里将不再讨论。

由于树枝状聚合物非常小（1~5 nm），超支化结构也没有大很多（2~20 nm），因此需要组装成超结构以达到 20~100 nm 的大小范围。

使用反向细乳技术制备微凝胶[174]。颗粒直径为 23~80 nm，可能在人肺癌细胞的核周区域积聚。MTT 试验表明低于 0.5 mg/mL 的微凝胶对人造血细胞 U-937 细胞的代谢活性无影响。

使用细乳液技术[175]，用叠氮化物和炔烃修饰的PG制备交联的超支化PG，即所谓的巨型聚合物。制备直径为25~90 nm的亲水性和疏水性聚合物。

聚甘油具有较低的蛋白质吸附性和良好的生物相容性，是一类很有前景的材料，已经在各种生物医学应用中证明了它们的优势。聚甘油具有较低的蛋白质吸附性和良好的生物相容性，是一类很有前景的材料，已经在各种生物医学应用中证明了它们的优势。未来，他们希望研发出更先进、功能更强的生物活性分子转运体系。

41.4.4 聚噁唑啉

聚（2-噁唑啉）（POx）也有望成为即将出现的纳米药物的聚合物之一[176,177]。POx满足研发下一代纳米药物所需的特定要求，例如生物相容性、溶解度的高调节性、大小的变化、结构以及化学功能。POx由2-噁唑啉阳离子开环聚合合成（图41.6）。该方法可精确控制分子量和链端功能（Đ≤1.2）[178]，因此科研人员可以调整聚合物以满足必要的要求。

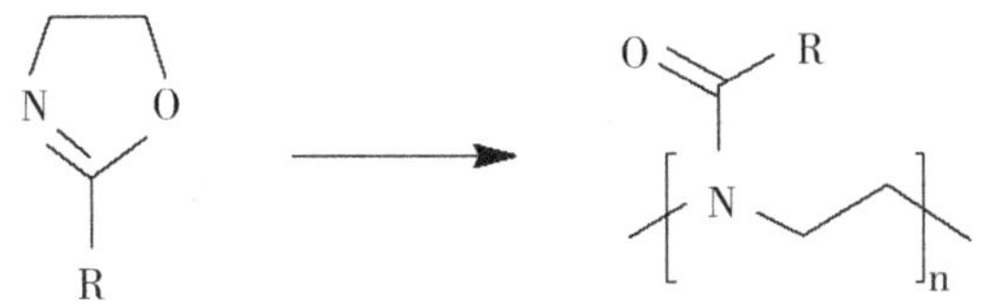

图41.6 多噁唑啉的合成

POx具有LCST，低于该温度的聚合物完全可溶于水溶液中。高于该温度时它们不溶，产生聚合物沉淀。POx的LCST取决于聚合物组成和结构，并且可以通过单体nPrOx与iPrOx的共聚调节到26~38.7℃，使用更加亲水的EtOx，温度范围调节至35~75.1℃。在较高的体外温度（高温）下，这些特性引起聚合物在体内部位沉淀，实现特定积累。此外，POx满足将聚合物用于体内的几个重要要求。考虑到体外细胞毒性，发现不同的POx的聚合物基本无毒害作用[179-182]。除此之外，PEtOx在20 g/L时没有溶血作用[183]。含有POx部分的混合体系的结果是可变的。发现三嵌段共聚物聚L-丙交酯-b-PEtOx-b-聚L-丙交酯和PEtOx-b-PCL-b-PEtOx分别在高达50 g/L和10 g/L的浓度下无细胞毒性[184,185]。与此相反，PEtOx-b-聚ε-己内酯共聚物在低于1 g/L的浓度下显示出明显的细胞毒性，但在同一研究的另一个实验中，浓度10 g/L的PEtOx-b-PCL没有溶血作用[186]。

以前关于聚（2-噁唑啉）的体内毒性的研究主要集中于PEtOx和PMeOx。Viegas等静脉注射给予大鼠重复剂量的10 kg/mol PEtOx（2 g/kg）和50 kg/mol（50 mg/kg）。没有出现不良反应，肾脏、肝脏和脾脏的组织学无异常[183]。

与其他聚合物体系类似，聚2-噁唑啉与蛋白质相互作用的敏感性取决于它们的亲水性或疏水性。总体而言，亲水性聚2-噁唑啉，如PMeOx和PEtOx，对蛋白质吸附的亲和力较低[187-190]。因此，它们显示隐性性质，可使用其介导作用，例如作为脂质体的包衣部分[191,192]。PMeOx和PEtOx的体内分布特征与其隐性特性一致，即低器官摄取和根据分子量的快速或缓慢的肾脏消除[193,194]。

对于其潜在的免疫原性，根据激活补体系统的能力筛选了PMeOx、PEtOx、PBuOx和PPheOx的不同嵌段共聚物。与阴性对照相比，补体蛋白水平略有升高[181,195]。尽管已经进行了大量研究，但还需要进一步的临床数据来证明POx的潜力，使其成为PEG的替代物。

41.5 纳米颗粒的特征

除了开发不同种类的新材料纳米颗粒外，还需要关注纳米药物的特征，尤其是在这些体系应用于药物靶向递送或疫苗接种时。

使用（低温）透射电子显微镜（TEM）、扫描电子显微镜（SEM）或原子显微镜（AFM）在溶液中观察合成的颗粒。此外，可以采用光散射技术和荧光相关光谱确定颗粒的流体动力学直径。然而，这些仪器限于人工操作，因此通常不能提供关于蛋白质吸附或诱导蛋白质聚集的信息。

特别是在疫苗接种中，当纳米颗粒将抗原和佐剂递送至抗原呈递细胞时，蛋白质电晕会引起免疫反应，对抗诱导自身免疫疾病的不能更好吸收的血清蛋白。为了解决这些重要问题，需要采用新方法，简要介绍其中一些方法。

如上所述，电荷稳定的纳米颗粒可能不影响蛋白质吸附，因为蛋白质是含有阴离子或阳离子基团的聚电解质。在含盐溶液中，电荷由抗衡离子补偿。当存在多价结合配体时，这些离子就被释放，因为自由扩散的离子增加了系统熵。虽然系统能量的焓贡献也起作用，但对于大分子通常熵贡献是主要因素。因此，纳米颗粒的空间稳定性可能是合理的，但在这些情况下需要仔细研究蛋白质吸附。

检测和解释电动学现象时，通常在低离子强度和不存在蛋白质的水溶液中表征颗粒。很多情况下都使用商业系统进行 zeta（ζ）电位测量。ζ-电位本身无法测量，必须使用复杂的模型从电动信号中得出，测量时要考虑表面特性，如表面性质、粗糙度、电荷、电解质浓度、电解质的性质以及电解质中使用的溶剂的性质。由于从一个粒子到另一个粒子测量，一些参数很容易改变，因此 IUPAC 技术报告中强烈提出应谨慎使用这种方法，不能过度解释所获得的结果[196]。但是，ζ-电位是颗粒合成的质量控制中的有用工具。

据 Rausch 及其同事报道的血清中的动态光散射以确定流体动力学半径的变化可能是更合适的方法[28,197]。该方法可以研究纳米颗粒在血清中的聚集行为，甚至可以识别微小的聚集颗粒痕迹。该方法的缺点在于，由于流体动力学半径的变化低于测量精度，因此无法检测到单个或少量蛋白质的吸附。

一个更人工的方法，即使用等温滴定量热法（ITC）测量热流，从而量化所有颗粒的平均吸附能[198-200]。与无标记定量的液相色谱和质谱（MS）相结合，可以识别和定量吸附的蛋白质[201]。

此外，纳米颗粒体内分布的体内成像对于监测靶位置的局部积累或确定机体的排泄是必要的。需要用近红外染料[202]或放射性同位素[203]标记纳米粒子，然后使用单光子发射计算机断层扫描（SPECT）或正电子发射断层扫描（PET）成像。在过去 10 年中，已经开发了多种方法标记纳米颗粒。已经建立了使用 18F 和 72/74Ar 修饰基团的聚合物标记方法[204-206]。

总之，已开发新的方法，并可以表明纳米颗粒在体内的行为，同时有助于下一代纳米药物的开发。

41.6 纳米颗粒疫苗制剂

41.6.1 纳米颗粒疫苗的优点

接下来的几年将会有越来越多的合成的亚单位抗原疫苗，因为相比减毒活疫苗、灭活苗或灭活的病原体等传统疫苗，它们具有更多优势[207]。然而，众所周知，合成肽通常缺乏诱导强免疫应答的免疫原性。因此，进行有效的疫苗接种需要使用佐剂[208]。在肿瘤免疫疗法领域，佐剂对于逆转由调节性 T 细胞或髓源性抑制细胞介导的肿瘤免疫抑制尤其重要。

其中，佐剂通过与先天免疫系统的受体（例如树突细胞）的相互作用促进抗原呈递细胞的成熟。只有成熟的树突细胞能够产生共刺激信号，它们与所递呈的抗原一起引起效应 T 细胞应答。因此，抗原和佐剂同时作用于树突细胞很重要。如今大多数人用疫苗制剂使用铝化合物作为佐剂。然而，近年来人们对更有效、更具体的替代物进行了大量研究[209]，需要新的共同递送药物的方式。

亚单位抗原疫苗的另一个问题是它们对酶促降解的敏感性。需要保护性载体分子来增加它们在体内的时间和生物活性。纳米颗粒很有趣，是保护抗原和佐剂免受血清蛋白质影响，以及确保其安全并同时递送至树突细胞的工具。

除了它们的保护能力[210]和共同递送抗原和佐剂外，纳米颗粒还具有其他几个优点。

- 它们可以使抗原和佐剂组合，对抗调节性T细胞的免疫抑制[211-213]。
- 它们包裹药物，防止药物全身分布和毒性作用[214]。
- 它们可以用作载体，在其表面呈现多个抗原拷贝，尤其可以改善B细胞活化[215]。
- 它们能促进树突细胞细胞质中抗原的积累。这对于交叉递呈很重要，是$CD8^+$T细胞激活的关键机制。
- 用惰性物质如聚乙二醇修饰纳米颗粒，介导隐性性质[216]可以最大限度地减少蛋白质吸附。这增加了它们的血液循环时间[217]。
- 纳米颗粒与受体配体或抗体的表面功能化能够靶向特定组织或细胞[217]。因此，抗原和佐剂保留在作用部位，可以减少非特异性分布，使其仅需要较低剂量，并且最小化疫苗接种的不良反应。
- N纳米颗粒提供佐剂的持续释放，从而延长免疫系统暴露，这对树突细胞的激活很重要[218]。

41.6.2　方法

纳米微粒疫苗制剂在癌症免疫疗法方面很可能有巨大潜力。除此之外，还进行了抗各种传染病的疫苗接种研究。例如，抗流感[219]、西尼罗河脑炎[220]和疟疾[221]的纳米颗粒疫苗。

代表性的纳米颗粒疫苗制剂将抗原（T细胞和/或B细胞表位），佐剂和靶向部分组合在一个颗粒上[214]。检测多种抗原与纳米颗粒载体的组合，例如肽/蛋白[222]、脂蛋白[223]、糖肽[224]、肿瘤相关碳水化合物抗原（TACA）[225]和碳水化合物模拟肽[226]。

抗原靶向树突细胞有两种基本方式：离体脉冲然后给予成熟树突细胞或直接体内靶向未成熟的树突细胞。纳米颗粒可能对体内治疗有意义[227]。

在纳米颗粒疫苗的背景下主要研究的佐剂是几种Toll样受体配体。LPS是革兰氏阴性细菌外膜的主要成分，是病原体来源的TLR4配体。LPS作为经典疫苗的免疫增强剂对于全身应用而言毒性很大，但在体外和小鼠模型中给予在PLGA纳米颗粒中的LPS时显示出很好的结果[228]。在小鼠中进行了将7-酰基脂质A（LPS的组分）作为PLGA疫苗的添加剂的研究[229]。除病原体的TLR配体外，还评估了模拟病原体相关分子模式（PAMP）的合成分子。该组中代表性的是与TLR 9结合的CpG寡脱氧核苷酸（CpG ODN）和聚核糖核酸：肌胞苷酸（Poly I：C），即TLR 3配体[211]。除了树突状细胞的活化外，将某些TLR配体（如CpG寡脱氧核苷酸和LPS）与纳米颗粒载体一起应用很有趣，因为它们与明矾相反，激活偏向Th1的免疫应答[230]。

Demento等开发了PLGA纳米颗粒，其携载西尼罗河病毒的重组包膜蛋白抗原，并用CpG ODN表面功能化。在几个实验中，他们将该系统与未修饰的纳米颗粒和氢氧化铝疫苗制剂进行了比较。在小鼠中，CpG ODN修饰的纳米颗粒产生偏向Th1免疫应答的抗体谱，与由氢氧化铝激活的Th2谱相反。此外，与对照组相比，给予修饰纳米颗粒的免疫小鼠诱导的循环效应T细胞数量更多，抗原特异性IL-2和IFN-γ的产生增多。在西尼罗河脑炎小鼠模型中，与氢氧化铝相比，修饰纳米颗粒显示出更好的保护作用[220]。

研究表明，有些纳米颗粒可能通过与Toll样受体结合，可以作为佐剂[231]。其中包括高分子量γ-PGA颗粒[232]、PLGA纳米和微粒[233]、聚酸酐纳米颗粒[234]、脂质体[235]、聚苯乙烯纳米和微粒[236]和酸-可降解的阳离子聚丙烯酰胺纳米颗粒[237]。虽然粒径和浓度似乎对树突细胞的激活至关重要，但在其他实验中，PLGA 颗粒不能诱导树突细胞成熟[238]。

在一项试验中检测了小鼠抗高致病性甲型流感 H5N1 病毒的黏膜免疫。该疫苗由重组流感血凝素（rHA）抗原和 γ-PGA／壳聚糖纳米颗粒组成。将该体系与另外两种制剂进行比较：含有 γ-PGA／壳聚糖纳米颗粒的灭活病毒和含有重组流感血凝素抗原的霍乱毒素。γ-PGA／壳聚糖纳米颗粒与 rHA 抗原组合诱导有效的病毒特异性体液和细胞免疫应答。与霍乱毒素相比，γ-PGA／壳聚糖纳米颗粒的佐剂性质良好。给予致死剂量的高致病性甲型流感 H5N1 病毒保护免疫小鼠对抗刺激[239]。

有多项研究[240,241]报道了纳米颗粒疫苗的另一个概念，将编码特定抗原的核酸（mRNA 或 pDNA）递送到树突细胞中。与肽抗原相反，mRNA 和 pDNA 必须在内吞途径中保持完整和维持其功能性，直到它们分别到达胞质或细胞核中。在内吞后，用缓冲基团或细胞穿透肽功能化的纳米颗粒可以保护核酸免受水解酶的影响，并促进内体逃逸[242]。一旦核酸到达细胞内的目的地，就可以促进编码抗原肽的表达。

pDNA 的一个优点是它可以编码各种抗原肽。因此，转运 pDNA 的纳米颗粒体系可用于广泛的疫苗接种。此外，pDNA 含有 TLR 9 识别的 CpG 序列，因此可以用作免疫刺激剂[243]。

Sudowe 等研究了体内 pDNA 疫苗接种作为模型抗原产生抗 β-Gal 的免疫应答的潜力。编码 β-Gal 的质粒在树突状特异性启动子、角化细胞特异性启动子或非特异性启动子的控制下，吸附在金纳米颗粒上，并通过基因枪递送至小鼠皮肤中。观察到偏向 Th1 的免疫应答结果。此外，发现转染的皮肤树突细胞产生的内源性抗原的递呈与外源性抗原的交叉递呈对 $CD8^+$T 细胞激活效果相同[244]。

如上所述，纳米载体的表面可以用多种不同的配体进行功能化，包括蛋白质、肽、抗体、抗体片段、适体和小分子[217]。用附着于纳米颗粒的受体配体或抗体靶向确定的细胞表面结构，可以实现增加的和/或更特异性的细胞摄取。特别考虑到疫苗制剂，具有模式识别受体配体的载体系统的构型，例如 Toll 样受体或 C 型凝集素受体，可能增强其免疫原性，并促进树突细胞活化[241]。已经用各种颗粒载体实现了不同的靶向概念，并且与非靶向载体相比显示出更强的免疫应答[245]。甘露糖是在树突细胞上表达的 C 型凝集素受体的配体。已有多个研究项目研究了其靶向树突状细胞和巨噬细胞的能力[44,235,246-251]。比较甘露糖修饰的 PLGA 颗粒与普通 PLGA 颗粒的试验显示，甘露糖仅引起细胞摄取的轻微增加。据推测，根据试验条件，非特异性摄取通常控制特异性摄取[252]。然而，在另一项研究中，证实了甘露糖，一种来自酿酒酵母细胞壁的天然多聚糖，与 PLGA 颗粒结合时，引起抗原特异性 $CD4^+$和 $CD8^+$T 细胞应答增加[253]。

文献中报道的处理树突细胞的其他方法是使用 Fc 受体配体或抗 DEC－205 抗体。Raghuwanshi 等评价了一种双组分疫苗制剂，其中一个组分是生物素化的 PLGA 纳米颗粒，其携载有卵清蛋白作为模型抗原，另一个组分是双功能融合蛋白，由截短的核心-链霉抗生物素蛋白和抗 DEC-205 单链抗体组成。他们观察到与非靶向纳米颗粒相比，受体介导的功能化纳米颗粒的摄入增加了两倍。小鼠使用抗 DEC-205 修饰的纳米颗粒和用作佐剂的 αCD40 联合诱导免疫后，其 OVA 特异性 IgG 反应显著高于对照组[254]。

为了诱导抗体反应，不仅需要树突细胞也需要 B 细胞识别和处理抗原。因此，除 T 细胞表位外，将 B 细胞表位包含在疫苗制剂中是一种重要方法。如前所述，纳米颗粒可以在表面上呈现多个表位[215]。这使得能够诱导 T 细胞非依赖性 I 型抗体应答，如病毒样颗粒所示[255]。T 细胞非依赖性 II 型抗体应答可以通过共同递送适合 B 细胞活化的佐剂来实现[256]。

特别是在肿瘤免疫疗法中，糖肽、TACA 或碳水化合物模拟肽的 B 细胞表位的递送是非常有趣的。在 Kunz 等的几项研究中研究了 B 细胞激活不同合成方法的肿瘤相关 MUC1 糖肽的潜力，糖肽包含 Thomsen-Friedenreich 抗原、它的前体或唾液酸衍生物。它们将糖肽与牛血清白蛋白（BSA）或破伤风类毒素化学连接作为载体分子。将这些疫苗体系应用于 Balb/c 小鼠。在 BSA 制剂存在的情况下，共同给予完全弗氏佐剂。观察到所有疫苗抗 MUC1 抗原的强特异性抗体反应[257-259]。

Brinãs等使用金纳米颗粒和补体蛋白C3d的肽作为B细胞活化佐剂，共同递送含有Thomsen-Friedenreich抗原的MUC4糖肽。给予这些纳米颗粒免疫的小鼠显示出统计学上显著的免疫应答[256]。

41.7　总结与结论

纳米颗粒在医学应用上具有很大的潜力。它们的颗粒特性在免疫系统的刺激中可能特别有吸引力。可以预测纳米颗粒疫苗对未来疫苗具有重大影响，因为它们具有向一个部位定向递送敏感性抗原、抗原或抗原/佐剂组合的潜力。因此，能够增强免疫调节。然而，纳米颗粒的合成仍然是困难的，但是新工具增加了对颗粒性质的控制，可以将合成最佳的免疫调节纳米药物用于有效的疫苗接种。

参考文献

[1] Duffus, J. H., Nordberg, M., Templeton, D. M.: Glossary of terms used in toxicology, 2nd edition (IUPAC Recommendations 2007). Pure Appl. Chem. 79, 1153-1344 (2007).

[2] Duncan, R., Gaspar, R.: Nanomedicine (s) under the microscope. Mol. Pharm. 8, 2101-2141 (2011).

[3] Johnston, A. P. R., Such, G. K., Ng, S. L., Caruso, F.: Challenges facing colloidal delivery systems: from synthesis to the clinic. Curr. Opin. ColloidIn. 16, 171-181 (2011).

[4] Torchilin, V. P.: Micellar nanocarriers: pharmaceutical perspectives. Pharm. Res. 24, 1-16 (2007).

[5] Lasic, D. D., Martin, F. J.: Stealth Liposomes. CRC Press, Boca Raton (1995).

[6] Hobbs, S. K., et al.: Regulation of transport pathways in tumor vessels: role of tumor type and microenvironment. Proc. Natl. Acad. Sci. U. S. A. 95, 4607-4612 (1998).

[7] Torchilin, V. P.: Targeted polymeric micelles for delivery of poorly soluble drugs. Cell. Mol. Life Sci. 61, 2549-2559 (2004).

[8] Nagasaki, Y., Yasugi, K., Yamamoto, Y., Harada, A., Kataoka, K.: Sugar-installed block copolymer micelles: their preparation and specific interaction with lectin molecules. Biomacromolecules 2, 1067-1070 (2001).

[9] Ogris, M., Brunner, S., Schuller, S., Kircheis, R., Wagner, E.: PEGylated DNA/transferrin-PEI complexes: reduced interaction with blood components, extended circulation in blood and potential for systemic gene delivery. Gene Ther. 6, 595-605 (1999).

[10] Leamon, C. P., Weigl, D., Hendren, R. W.: Folate copolymer-mediated transfection of cultured cells. Bioconjug. Chem. 10, 947-957 (1999).

[11] Hopewell, J. W., Duncan, R., Wilding, D., Chakrabarti, K.: Preclinical evaluation of the cardiotoxicity of PK2: a novel HPMA copolymer-doxorubicin-galactosamine conjugate antitumour agent. Hum. Exp. Toxicol. 20, 461-470 (2001).

[12] Ahmed, F., et al.: Shrinkage of a rapidly growing tumor by drug-loaded polymersomes: pH-triggered release through copolymer degradation. Mol. Pharm. 3, 340-350 (2006).

[13] Saito, G., Swanson, J. A., Lee, K. -D.: Drug delivery strategy utilizing conjugation via reversible disulfide linkages: role and site of cellular reducing activities. Adv. Drug Deliv. Rev. 55, 199-215 (2003).

[14] Thornton, P. D., Mart, R. J., Webb, S. J., Ulijn, R. V.: Enzyme-responsive hydrogel particles for the controlled release of proteins: designing peptide actuators to match payload. Soft Matter 4, 821-827 (2008).

[15] Ishida, O., Maruyama, K., Yanagie, H., Iwatsuru, M., Eriguchi, M.: Targeting chemotherapy to solid tumors with long circulating thermosensitive liposomes and local hyperthermia. Jpn. J. Cancer Res. 91,

118-126 (2000).

[16] Edelman, E. R., Kost, J., Bobeck, H., Langer, R.: Regulation of drug release from polymer matrices by oscillating magnetic fields. J. Biomed. Mater. Res. 19, 67-83 (1985).

[17] Langer, R.: New methods of drug delivery. Science 249, 1527-1533 (1990).

[18] Uhrich, K. E., Cannizzaro, S. M., Langer, R. S., Shakesheff, K. M.: Polymeric systems for controlled drug release. Chem. Rev. 99, 3181-3198 (1999).

[19] Little, S. R.: Reorienting our view of particle-based adjuvants for subunit vaccines. Proc. Natl. Acad. Sci. 109, 999-1000 (2012).

[20] Moon, J. J., et al.: Enhancing humoral responses to a malaria antigen with nanoparticle vaccines that expand Tfh cells and promote germinal center induction. Proc. Natl. Acad. Sci. U. S. A. 109, 1080-1085 (2012).

[21] Prokop, A., Davidson, J. M.: Nanovehicular intracellular delivery systems. J. Pharm. Sci. 97, 3518-3590 (2008).

[22] Gref, R., et al.: Biodegradable long-circulating polymeric nanospheres. Science 263, 1600-1603 (1994).

[23] Owens, D. E., Peppas, N. A.: Opsonization, biodistribution, and pharmacokinetics of polymeric nanoparticles. Int. J. Pharm. 307, 93-102 (2006).

[24] Frank, M., Fries, L.: The role of complement in inflammation and phagocytosis. Immunol. Today 12, 322-326 (1991).

[25] Johnson, R. J.: The complement system. In: Ratner, B. D., Hoffman, A. S., Schoen, F. J., Lemons, J. E. (eds.) Biomaterials Science: An Introduction to Materials in Medicine, pp. 318-328. Elsevier/Academic, Amsterdam (2004).

[26] Ostuni, E., Chapman, R. G., Holmlin, R. E., Takayama, S., Whitesides, G. M.: A survey of structure-property relationships of surfaces that resist the adsorption of protein. Langmuir 17, 5605-5620 (2001).

[27] Scheibe, P., Barz, M., Hemmelmann, M., Zentel, R.: Langmuir-Blodgett films of biocompatible poly (HPMA) -block-poly (lauryl methacrylate) and poly (HPMA) -random-poly (lauryl methacrylate): influence of polymer structure on membrane formation and stability. Langmuir 26, 5661-5669 (2010).

[28] Kelsch, A., et al.: HPMA copolymers as surfactants in the preparation of biocompatible nanoparticles for biomedical application. Biomacromolecules 13, 4179-4187 (2012).

[29] Riess, G.: Micellization of block copolymers. Prog. Polym. Sci. 28, 1107-1170 (2003).

[30] O'Reilly, R. K., Hawker, C. J., Wooley, K. L.: Cross-linked block copolymer micelles: functional nanostructures of great potential and versatility. Chem. Soc. Rev. 35, 1068-1083 (2006).

[31] Kabanov, A. V., Vinogradov, S. V.: Nanogels as pharmaceutical carriers: finite networks of infinite capabilities. Angew. Chem. Int. Ed. Engl. 48, 5418-5429 (2009).

[32] Christie, R. J., Nishiyama, N., Kataoka, K.: Delivering the code: polyplex carriers for deoxyribonucleic acid and ribonucleic acid interference therapies. Endocrinology 151, 466-473 (2010).

[33] Miyata, K., Nishiyama, N., Kataoka, K.: Rational design of smart supramolecular assemblies for gene delivery: chemical challenges in the creation of artificial viruses. Chem. Soc. Rev. 41, 2562-2574 (2012).

[34] Kataoka, K., Harada, A., Nagasaki, Y.: Block copolymer micelles for drug delivery: design, characterization and biological significance. Adv. Drug Deliv. Rev. 47, 113-131 (2001).

[35] Torchilin, V. P.: Structure and design of polymeric surfactant-based drug delivery systems. J. Control. Release 73, 137-172 (2001).

[36] Gaucher, G., et al.: Block copolymer micelles: preparation, characterization and application in drug delivery. J. Control. Release 109, 169-188 (2005).

[37] Nuhn, L., et al.: Cationic nanohydrogel particles as potential siRNA carriers for cellular delivery. ACS Nano 6, 2198-2214 (2012).

[38] Fleige, E., Quadir, M. A., Haag, R.: Stimuli-responsive polymeric nanocarriers for the controlled transport of active compounds: concepts and applications. Adv. Drug Deliv. Rev. 64, 866-884 (2012).

[39] Jesorka, A., Orwar, O.: Liposomes: technologies and analytical applications. Annu. Rev. Anal. Chem. 1, 801-832 (2008).

[40] Torchilin, V. P.: Recent advances with liposomes as pharmaceutical carriers. Nat. Rev. Drug Discov. 4, 145-160 (2005).

[41] Szoka, F. C.: Comparative properties and methods of preparation of lipid vesicles (liposomes). Ann. Rev. Biophys. Bioeng. 9, 467-508 (1980).

[42] Lasic, D. D.: Sterically stabilized vesicles. Angew. Chem. Int. Ed. Engl. 33, 1685-1698 (1994).

[43] Allen, T. M., Cullis, P. R.: Drug delivery systems: entering the mainstream. Science 303, 1818-1822 (2004).

[44] White, K. L., Rades, T., Furneaux, R. H., Tyler, P. C., Hook, S.: Mannosylated liposomes as antigen deliv-ery vehicles for targeting to dendritic cells. J. Pharm. Pharmacol. 58, 729-737 (2006).

[45] Drummond, D. C., Meyer, O., Hong, K., Kirpotin, D. B., Papahadjopoulos, D.: Optimizing liposomes for delivery of chemotherapeutic agents to solid tumors. Pharmacol. Rev. 51, 691-744 (1999).

[46] Krishnamachari, Y., Geary, S. M., Lemke, C. D., Salem, A. K.: Nanoparticle delivery systems in cancer vaccines. Pharm. Res. 28, 215-236 (2011).

[47] Gluck, R.: Immunopotentiating reconstituted influenza virosomes (IRIVs) and other adjuvants for improved presentation of small antigens. Vaccine 10, 915-919 (1992).

[48] Moser, C., et al.: Influenza virosomes as a combined vaccine carrier and adjuvant system for prophylactic and therapeutic immunizations. Expert Rev. Vaccines 6, 711-721 (2007).

[49] Kayser, O., Olbrich, C., Croft, S. L., Kiderlein, A. F.: Formulation and biopharmaceutical issues in the development of drug delivery systems for antiparasitic drugs. Parasitol. Res. 90, S63-S70 (2003).

[50] Discher, D. E., Ahmed, F.: Polymersomes. Annu. Rev. Biomed. Eng. 8, 323-341 (2006).

[51] Levine, D. H., et al.: Polymersomes: a new multi-functional tool for cancer diagnosis and therapy. Methods 46, 25-32 (2008).

[52] Osterhaus, A., Rimmelzwaan, G. F.: Induction of virus-specific immunity by ISCOMs. Dev. Biol. Stand. 92, 49-58 (1998).

[53] Saupe, A., McBurney, W., Rades, T., Hook, S.: Immunostimulatory colloidal delivery systems for cancer vaccines. Expert Opin. Drug Deliv. 3, 345-354 (2006).

[54] Westesen, K., Siekmann, B.: Biodegradable colloidal drug carrier systems based on solid lipids. In: Benita, S. (ed.) Microencapsulation, pp. 213-258. Marcel Dekker, New York (1996).

[55] Bunjes, H.: Lipid nanoparticles for the delivery of poorly water-soluble drugs. J. Pharm. Pharmacol. 62, 1637-1645 (2010).

[56] Petersen, S., Steiniger, F., Fischer, D., Fahr, A., Bunjes, H.: The physical state of lipid nanoparticles affects their in vitro cell viability. Eur. J. Pharm. Biopharm. 79, 150-161 (2011).

[57] Shi, R., et al.: Enhanced immune response to gastric cancer specific antigen peptide by coencapsulation with CpG oligodeoxynucleotides in nanoemulsion. Cancer Biol. Ther. 4, 218-242 (2005).

[58] Gupta, S., Moulik, S. P.: Biocompatible microemulsions and their prospective uses in drug delivery. J. Pharm. Sci. 97, 22-45 (2008).

[59] Fanun, M.: Microemulsions as delivery systems. Curr. Opin. Colloid In. 17, 306-313 (2012).

[60] Sailaja, A. K., Amareshwar, P., Chakravarty, P.: Chitosan nanoparticles as a drug delivery system. Res. J. Pharm. Biol. Chem. Sci. 1, 474-484 (2010).

[61] Lohse, S. E., Murphy, C. J.: Applications of colloidal inorganic nanoparticles: from medicine to energy. J. Am. Chem. Soc. 134, 15607-15620 (2012).

[62] Landfester, K.: Synthesis of colloidal particles in miniemulsions. Annu. Rev. Mater. Res. 36, 231-279 (2006).

[63] Klinger, D., Landfester, K.: Stimuli-responsive microgels for the loading and release of functional compounds: Fundamental concepts and applications. Polymer 53, 5209-5231 (2012).

[64] Ugelstad, J., Mork, P. C., Kaggerud, K. H., Ellingsen, T., Berge, A.: Swelling of oligomer-polymer particles: new method of preparation of emulsions and polymer dispersions. Adv. Colloid Interface Sci. 13, 101-140 (1980).

[65] Chern, C. S., Chen, T. J., Liou, Y. C.: Miniemulsion polymerization of styrene in the presence of a water-insoluble blue dye. Polymer 37, 3767-3777 (1998).

[66] Reimers, J. L., Schork, F. J.: Lauroyl peroxide as a cosurfactant in miniemulsion polymerization. Ind. Eng. Chem. Res. 36, 1085-1087 (1997).

[67] Landfester, K.: Recent developments in miniemulsions - formation and stability mechanisms. Macromol. Symp. 150, 171-178 (2000).

[68] Tamber, H., Johansen, P., Merkle, H. P., Gander, B.: Formulation aspects of biodegradable polymeric microspheres for antigen delivery. Adv. Drug Deliv. Rev. 57, 357-376 (2005).

[69] Mok, H., Park, T. G.: Direct plasmid DNA encapsulation within PLGA nanospheres by single oil-in-water emulsion method. Eur. J. Pharm. Biopharm. 68, 105-111 (2008).

[70] Meyer, J. D., Manning, M. C.: Hydrophobic ion pairing: altering the solubility properties of biomolecules. Pharm. Res. 15, 188-193 (1998).

[71] Kazzaz, J., Neidleman, J., Singh, M., Ott, G., O'Hagan, D. T.: Novel anionic microparticles are a potent adjuvant for the induction of cytotoxic T lymphocytes against recombinant p55 gag from HIV-1. J. Control. Release 67, 347-356 (2000).

[72] Schwendeman, S. P.: Recent advances in the stabilization of proteins encapsulated in injectable PLGA delivery systems. Crit. Rev. Ther. Drug Carr. Syst. 19, 73-98 (2002).

[73] Barz, M., et al.: Synthesis, characterization and preliminary biological evaluation of P (HPMA) -b-P (LLA) copolymers: a new type of functional biocompatible block copolymer. Macromol. Rapid Comm. 31, 1492-1500 (2010).

[74] Barz, M., et al.: P (HPMA) -block-P (LA) copolymers in paclitaxel formulations: polylactide stereochemistry controls micellization, cellular uptake kinetics, intracellular localization and drug efficiency. J. Control. Release 163, 63-74 (2012).

[75] Aspinall, G. O.: The Polysaccharides 35. Academic, New York (1982).

[76] Leonard, M., et al.: Preparation of polysaccharide-covered polymeric nanoparticles by several processes involving amphiphilic polysaccharides. ACS Symp. Ser. 996, 322-340 (2008).

[77] Artursson, P., Lindmark, T., Davis, S., Illum, L.: Effect of chitosan on the permeability of monolayers of intestinal epithelial-cells (Caco-2). Pharm. Res. 11, 1358-1361 (1994).

[78] Domard, A., Gey, C., Rinaudo, M., Terrassin, C., et al.: C-13 and H-1-NMR spectroscopy of chitosan and Ntrimethyl chloride derivates. Int. J. Biol. Macromol. 9, 233-237 (1987).

[79] Sundar, S., Kundu, J., Kundu, S. C.: Biopolymeric nanoparticles. Sci. Technol. Adv. Mater. (11) (2010).

[80] Schultze, V., et al.: Safety of MF59 (TM) adjuvant. Vaccine 26, 3209-3222 (2008).

[81] Makidon, P. E., et al.: Pre-clinical evaluation of a novel nanoemulsion-based hepatitis B mucosal vaccine. PLoS One 3, e2954 (2008).

[82] Bielinska, A. U., et al.: Nasal immunization with a recombinant HIV gp120 and nanoemulsion adjuvant produces Th1 polarized responses and neutralizing antibodies to primary HIV type 1 isolates. AIDS Res. Hum. Retrov. 24, 271-281 (2008).

[83] Ge, W., et al.: The antitumor immune responses induced by nanoemulsion encapsulated MAGE1-H SP70/SEA complex protein vaccine following different administration routes. Oncol. Rep. 22, 915-920 (2009).

[84] Rolland, J. P., et al.: Direct fabrication and harvesting of monodisperse. Shape-specific nanobiomaterials. J. Am. Chem. Soc. 127, 10096-10100 (2005).

[85] Gratton, S. E., et al.: Nanofabricated particles for engineered drug therapies: a preliminary biodistribution study of PRINT nanoparticles. J. Control. Release 121, 10-18 (2007).

[86] Dunn, S. S., et al.: Reductively responsive siRNA-conjugated hydrogel nanoparticles for gene silencing. J. Am. Chem. Soc. 134, 7423-7430 (2012).

[87] Laurent, S., et al.: Magnetic iron oxide nanoparticles: synthesis, stabilization, vectorization, physico-chemical characterizations, and biological applications. Chem. Rev. 108, 2064-2110 (2008).

[88] Dahl, J. A., Maddux, B. L. S., Hutchison, J. E.: Toward greener nanosynthesis. Chem. Rev. 107, 2228-2269 (2007).

[89] Caragheorgheopol, A., Chechik, V.: Mechanistic aspects of ligand exchange in Au nanoparticles. Phys. Chem. Chem. Phys. 10, 5029-5041 (2008).

[90] Bernardi, R. J., Lowery, A. R., Thompson, P. A., Blaney, S. M., West, J. L.: Immunonanoshells for targeted photothermal ablation in medulloblastoma and glioma: an in vitro evaluation using human cell lines. J. Neurooncol 86, 165-172 (2008).

[91] Cruz, L. J., et al.: Targeting nanosystems to human DCs via Fc receptor as an effective strategy to deliver antigen for immunotherapy. Mol. Pharm. 8, 104-116 (2011).

[92] Ito, A., Honda, H., Kobayashi, T.: Cancer immunotherapy based on intracellular hyperthermia using magnetite nanoparticles: a novel concept of "heat-controlled necrosis" with heat shock protein expression. Cancer Immunol. Immunother. 55, 320-328 (2006).

[93] Masoudi, A., Madaah Hosseini, H. R., Shokrgozar, M. A., Ahmadi, R., Oghabian, M. A.: The effect of poly (ethylene glycol) coating on colloidal stability of superparamagnetic iron oxide nanoparticles as potential MRI contrast agent. Int. J. Pharm. 433, 129-141 (2012).

[94] Webster, R. et al.: PEG and PEG conjugates toxicity: towards an understanding of the toxicity of PEG and its relevance to PEGylated biologicals. In: PEGylated Protein Drugs: Basic Science and Clinical Applications. Birkhäuser Verlag, Basel (2009) pp. 127-146.

[95] Bendele, A., Seely, J., Richey, C., Sennello, G., Shopp, G.: Short communication: renal tubular vacuolation in animals treated with polyethyleneglycol-conjugated proteins. Toxicol. Sci. 42, 152-157 (1998).

[96] Young, M. A., Malavalli, A., Winslow, N., Vandegriff, K. D., Winslow, R. M.: Toxicity and hemodynamic effects after single dose administration of MalPEG-hemoglobin (MP4) in rhesus monkeys. Transl. Res. 149, 333-342 (2007).

[97] Chapman, R. G., et al.: Surveying for surfaces that resist the adsorption of proteins. J. Am. Chem. Soc. 122, 8303-8304 (2000).

[98] Zhou, M., et al.: High throughput discovery of new fouling-resistant surfaces. J. Mater. Chem. 21, 693 (2011).

[99] Fasting, C., et al.: Multivalency as a chemical organization and action principle. Angew. Chem. Int. Ed. Engl. 51, 10472-10498 (2012).

[100] Niederhafner, P., Reinis, M., Sebestík, J., Jezek, J.: Glycopeptide dendrimers, part III: a review. Use of glycopeptide dendrimers in immunotherapy and diagnosis of cancer and viral diseases. J. Pept. Sci. 14, 556-587 (2008).

[101] Günay, K. A., Theato, P., Klok, H. A.: Standing on the shoulders of Hermann Staudinger: post-polymerization modification from past to present. J. Polym. Sci. A1 51, 1-28 (2013).

[102] Grandjean, C., Boutonnier, A., Guerreiro, C., Fournier, J. -M., Mulard, L. A.: On the preparation of carbohydrate-protein conjugates using the traceless Staudinger ligation. J. Org. Chem. 70, 7123-7132 (2005).

[103] Xu, P., et al.: Simple, direct conjugation of bacterial O-SP-core antigens to proteins: development of cholera conjugate vaccines. Bioconjugate Chem. 22, 2179-2185 (2011).

[104] Scaramuzza, S., et al.: A new site-specific monoPEGylated filgrastim derivative prepared by enzymatic

conjugation: production and physicochemical characterization. J. Control. Release 164, 355-363 (2012).

[105] Jung, B., Theato, P.: Chemical strategies for the syn-thesis of protein-polymer conjugates. Bio-synth. Polym. Conjugates 253, 37-70 (2013).

[106] Moad, G., Rizzardo, E., Thang, S. H.: Living radical polymerization by the RAFT process. Aust. J. Chem. 58, 379 (2005).

[107] Moad, G., Rizzardo, E., Thang, S. H.: Radical addition-fragmentation chemistry in polymer synthesis. Polymer 49, 1079-1131 (2008).

[108] Braunecker, W. A., Matyjaszewski, K.: Controlled/living radical polymerization: features, developments, and perspectives. Prog. Polym. Sci. 32, 93-146 (2007).

[109] Matyjaszewski, K., Xia, J.: Atom transfer radical polymerization. Chem. Rev. 101, 2921-2990 (2001).

[110] York, A. W., Kirkland, S. E., McCormick, C. L.: Advances in the synthesis of amphiphilic block copolymers via RAFT polymerization: stimuli-responsive drug and gene delivery. Adv. Drug Deliv. Rev. 60, 1018-1036 (2008).

[111] Gao, H., Matyjaszewski, K.: Synthesis of functional polymers with controlled architecture by CRP of monomers in the presence of cross-linkers: from stars to gels. Prog. Polym. Sci. 34, 317-350 (2009).

[112] Marsden, H. R., Kros, A.: Polymer-peptide block copolymers-an overview and assessment of synthesis methods. Macromol. Biosci. 9, 939-951 (2009).

[113] Tizzotti, M., Charlot, A., Fleury, E., Stenzel, M., Bernard, J.: Modification of polysaccharides through controlled/living radical polymerization grafting-towards the generation of high performance hybrids. Macromol. Rapid Comm. 31, 1751-1772 (2010).

[114] Lutz, J. F.: Polymerization of oligo (ethylene glycol) (meth) acrylates: toward new generations of smart biocompatible materials. J. Polym. Sci. A1 46, 3459-3470 (2008).

[115] Lutz, J. -F., Akdemir, O., Hoth, A.: Point by point comparison of two thermosensitive polymers exhibiting a similar LCST: is the age of poly (NIPAM) over? J. Am. Chem. Soc. 128, 13046-13047 (2006).

[116] Tao, L., Mantovani, G., Lecolley, F., Haddleton, D. M.: Alpha-aldehyde terminally functional methacrylic polymers from living radical polymerization: application in protein conjugation "pegylation". J. Am. Chem. Soc. 126, 13220-13221 (2004).

[117] Lutz, J. -F., Hoth, A.: Preparation of Ideal PEG ana-logues with a tunable thermosensitivity by controlled radical copolymerization of 2- (2-Methoxyethoxy) ethyl methacrylate and oligo (ethylene glycol) methacrylate. Macromolecules 39, 893-896 (2006).

[118] Ryan, S. M., et al.: Conjugation of salmon calcitonin to a combed-shaped end functionalized poly (poly (ethylene glycol) methyl ether methacrylate) yields a bioactive stable conjugate. J. Control. Release 135, 51-59 (2009).

[119] Lutz, J. -F., Andrieu, J., Üzgün, S., Rudolph, C., Agarwal, S.: Biocompatible, thermoresponsive, and biodegradable: simple preparation of "all-in-one" biorelevant polymers. Macromolecules 40, 8540-8543 (2007).

[120] Ishihara, K., Ziats, N. P., Tierney, B. P., Nakabayashi, N., Anderson, J. M.: Protein adsorption from human plasma is reduced on phospholipid polymers. J. Biomed. Mater. Res. A 25, 1397-1407 (1991).

[121] Salvage, J. P., et al.: Novel biocompatible phosphorylcholine-based self-assembled nanoparticles for drug delivery. J. Control. Release 104, 259-270 (2005).

[122] Murdoch, C., et al.: Internalization and biodistribution of polymersomes into oral squamous cell carcinoma cells in vitro and in vivo. Nanomedicine 5, 1025-1036 (2010).

[123] Lomas, H., et al.: Non-cytotoxic polymer vesicles for rapid and efficient intracellular delivery. Faraday Discuss. 139, 143-159 (2008).

[124] Lewis, A., Tang, Y., Brocchini, S., Choi, J. -W., Godwin, A.: Poly (2-methacryloyloxyethyl phosphorylcholine) for protein conjugation. Bioconjugate Chem. 19, 2144-2155 (2008).

[125] Kopecek, J., Kopecková, P.: HPMA copolymers: origins, early developments, present, and future. Adv. Drug Deliv. Rev. 62, 122–149 (2010).

[126] Barz, M., et al.: From defined reactive diblock copolymers to functional HPMA-based self-assembled nanoaggregates. Biomacromolecules 9, 3114–3118 (2008).

[127] Barz, M., Canal, F., Koynov, K., Zentel, R., Vicent, M. J.: Synthesis and in vitro evaluation of defined HPMA folate conjugates: influence of aggregation on folate receptor (FR) mediated cellular uptake. Biomacromolecules 11, 2274–2282 (2010).

[128] Leuchs, H.: Über die Glycin-carbonsäure. Ber. Dtsch. Chem. Ges. 39, 857–861 (1906).

[129] Kricheldorf, H. R.: α-Amino acid-N-Carboxy-Anhydrides and Related Heterocycles: Syntheses, Properties, Peptide Synthesis, Polymerization. Springer, Berlin/Heidelberg/New York (1987).

[130] Kricheldorf, H. R.: Polypeptides and 100 years of chemistry of alpha-amino acid N-carboxyanhydrides. Angew. Chem. Int. Ed. Engl. 45, 5752–5784 (2006).

[131] Hadjichristidis, N., Iatrou, H., Pitsikalis, M., Sakellariou, G.: Synthesis of well - defined polypeptide-based materials via the ring-opening polymerization of alpha-amino acid N-carboxyanhydrides. Chem. Rev. 109, 5528–5578 (2009).

[132] Bogdanov, A. A., et al.: A new macromolecule as a contrast agent for MR angiography: preparation, properties, and animal studies. Radiology 187, 701–706 (1993).

[133] Singer, J. W., et al.: Paclitaxel poliglumex (XYOTAX; CT-2103): an intracellularly targeted taxane. Anti-cancer Drug. 16, 243–254 (2005).

[134] Harada, A., Kataoka, K.: Formation of polyion com-plex micelles in an aqueous milieu from a pair of oppositely-charged block copolymers with poly (ethylene glycol) segments. Macromolecules 28, 5294–5299 (1995).

[135] Carlsen, A., Lecommandoux, S.: Self-assembly of polypeptide-based block copolymer amphiphiles. Curr. Opin. Colloid In. 14, 329–339 (2009).

[136] Deng, J., et al.: Self-assembled cationic micelles based on PEG-PLL-PLLeu hybrid polypeptides as highly effective gene vectors. Biomacromolecules 13, 3795–3804 (2012).

[137] Bellomo, E. G., Wyrsta, M. D., Pakstis, L., Pochan, D. J., Deming, T. J.: Stimuli-responsive polypeptide vesicles by conformation-specific assembly. Nat. Mater. 3, 244–248 (2004).

[138] Holowka, E. P., Sun, V. Z., Kamei, D. T., Deming, T. J.: Polyarginine segments in block copolypeptides drive both vesicular assembly and intracellular delivery. Nat. Mater. 6, 52–57 (2007).

[139] Kanzaki, T., Horikawa, Y., Makino, A., Sugiyama, J., Kimura, S.: Nanotube and three-way nanotube formation with nonionic amphiphilic block peptides. Macromol. Biosci. 8, 1026–1033 (2008).

[140] Nowak, A. P., et al.: Rapidly recovering hydrogel scaffolds from self-assembling diblock copolypeptide amphiphiles. Nature 417, 424–428 (2002).

[141] Takae, S., et al.: PEG-detachable polyplex micelles based on disulfide-linked block catiomers as bioresponsive nonviral gene vectors. J. Am. Chem. Soc. 130, 6001–6009 (2008).

[142] Uchida, H., et al.: Odd-even effect of repeating ami-noethylene units in the side chain of N-substituted polyaspartamides on gene transfection profiles. J. Am. Chem. Soc. 133, 15524–15532 (2011).

[143] Sanjoh, M., et al.: Dual environment-responsive polyplex carriers for enhanced intracellular delivery of plasmid DNA. Biomacromolecules 13, 3641–3649 (2012).

[144] Naito, M., et al.: A phenylboronate-functionalized polyion complex micelle for ATP-triggered release of siRNA. Angew. Chem. Int. Ed. Engl. 124, 10909–10913 (2012).

[145] Hamaguchi, T., et al.: NK105, a paclitaxel-incorporating micellar nanoparticle formulation, can extend in vivo antitumour activity and reduce the neurotoxicity of paclitaxel. Brit. J. Cancer 92, 1240 – 1246 (2005).

[146] Weissleder, R., Tung, C. H., Mahmood, U., Bogdanov, A.: In vivo imaging of tumors with protease-activated near-infrared fluorescent probes. Nat. Biotechnol. 17, 375–378 (1999).

[147] Arnon, R.: The development of Cop 1 (Copaxone), an innovative drug for the treatment of multiple sclerosis: personal reflections. Immunol. Lett. 50, 1-15 (1996).

[148] Shaffer, S. A., et al.: In vitro and in vivo metabolism of paclitaxel poliglumex: identification of metabolites and active proteases. Cancer Chemother. Pharmacol. 59, 537-548 (2007).

[149] Hanson, J. A., et al.: Nanoscale double emulsions stabilized by single-component block copolypeptides. Nature 455, 85-88 (2008).

[150] Matsumura, Y.: Poly (amino acid) micelle nanocarriers in preclinical and clinical studies. Adv. Drug Deliv. Rev. 60, 899-914 (2008).

[151] http: //www. clinicaltrials. gov.

[152] Li, C., Wallace, S.: Polymer-drug conjugates: recent development in clinical oncology. Adv. Drug Deliv. Rev. 60, 886-898 (2008).

[153] Wilms, D., Stiriba, S. -E., Frey, H.: Hyperbranched polyglycerols: from the controlled synthesis of biocompatible polyether polyols to multipurpose applications. Acc. Chem. Res. 43, 129-141 (2010).

[154] Quadir, M. A., Haag, R.: Biofunctional nanosystems based on dendritic polymers. J. Control. Release 161, 484-495 (2012).

[155] Khandare, J., Calderón, M., Dagia, N. M., Haag, R.: Multifunctional dendritic polymers in nanomedi-cine: opportunities and challenges. Chem. Soc. Rev. 41, 2824-2848 (2012).

[156] Mourey, T. H., et al.: Unique behavior of dendritic macromolecules: intrinsic viscosity of polyether dendrimers. Macromolecules 25, 2401-2406 (1992).

[157] Wyszogrodzka, M., Haag, R.: Study of single protein adsorption onto monoamino oligoglycerol derivatives: a structure-activity relationship. Langmuir 25, 5703-5712 (2009).

[158] Wyszogrodzka, M., et al.: New approaches towards monoamino polyglycerol dendrons and dendritic triblock amphiphiles. Eur. J. Org. Chem. 2008, 53-63 (2008).

[159] Haag, R., Sunder, A., Stumbé, J. -F.: An approach to glycerol dendrimers and pseudo-dendritic polyglycerols. J. Am. Chem. Soc. 122, 2954-2955 (2000).

[160] Sunder, A., Krämer, M., Hanselmann, R., Mülhaupt, R., Frey, H.: Molecular nanocapsules based on amphiphilic hyperbranched polyglycerols. Angew. Chem. Int. Ed. Engl. 38, 3552-3555 (1999).

[161] Wilms, D., et al.: Hyperbranched polyglycerols with elevated molecular weights: a facile two-step synthesis protocol based on polyglycerol macroinitiators. Macromolecules 42, 3230-3236 (2009).

[162] Barriau, E., et al.: Systematic investigation of functional core variation within hyperbranched polyglycerols. J. Polym. Sci. A1 46, 2049-2061 (2008).

[163] Roller, S., Zhou, H., Haag, R.: High-loading polyglycerol supported reagents for Mitsunobu-and acylation-reactions and other useful polyglycerol derivatives. Mol. Divers. 9, 305-316 (2005).

[164] Kainthan, R. K., Janzen, J., Levin, E., Devine, D. V., Brooks, D. E.: Biocompatibility testing of branched and linear polyglycidol. Biomacromolecules 7, 703-709 (2006).

[165] Kainthan, R. K., Hester, S. R., Levin, E., Devine, D. V., Brooks, D. E.: In vitro biological evaluation of high molecular weight hyperbranched polyglycerols. Biomaterials 28, 4581-4590 (2007).

[166] Kainthan, R. K., Brooks, D. E.: In vivo biological evaluation of high molecular weight hyperbranched polyglycerols. Biomaterials 28, 4779-4787 (2007).

[167] Calderón, M., Quadir, M. A., Sharma, S. K., Haag, R.: Dendritic polyglycerols for biomedical applications. Adv. Mater. 22, 190-218 (2010).

[168] Calderón, M., Graeser, R., Kratz, F., Haag, R.: Development of enzymatically cleavable prodrugs derived from dendritic polyglycerol. Bioorg. Med. Chem. Lett. 19, 3725-3728 (2009).

[169] Calderón, M., et al.: Development of efficient acid cleavable multifunctional prodrugs derived from dendritic polyglycerol with a poly (ethylene glycol) shell. J. Control. Release 151, 295-301 (2011).

[170] Papp, I., Dernedde, J., Enders, S., Haag, R.: Modular synthesis of multivalent glycoarchitectures and their unique selectin binding behavior. Chem. Commun. 4, 5851-5853 (2008).

[171] Türk, H., Haag, R., Alban, S.: Dendritic polyglycerol sulfates as new heparin analogues and potent inhibitors of the complement system. Bioconjug. Chem. 15, 162-167 (2003).

[172] Dernedde, J., et al.: Dendritic polyglycerol sulfates as multivalent inhibitors of inflammation. Proc. Natl. Acad. Sci. U. S. A. 44, 19679-19684 (2010).

[173] Steinhilber, D., et al.: Synthesis, reductive cleavage, and cellular interaction studies of biodegradable polyglycerol nanogels. Adv. Funct. Mater. 20, 4133-4138 (2010).

[174] Sisson, A. L., et al.: Biocompatible functionalized polyglycerol microgels with cell penetrating properties. Angew. Chem. Int. Ed. Engl. 48, 7540-7545 (2009).

[175] Sisson, A. L., Papp, I., Landfester, K., Haag, R.: Functional nanoparticles from dendritic precursors: hierarchical assembly in miniemulsion. Macromolecules 42, 556-559 (2009).

[176] Luxenhofer, R., et al.: Poly (2-oxazoline) s as polymer therapeutics. Macromol. Rapid Comm. 33, 1613-1631 (2012).

[177] Viegas, T. X., et al.: Polyoxazoline: chemistry, properties, and applications in drug delivery. Bioconjugate Chem. 22, 976-986 (2011).

[178] Knop, K., Hoogenboom, R., Fischer, D., Schubert, U. S.: Poly (ethylene glycol) in drug delivery: pros and cons as well as potential alternatives. Angew. Chem. Int. Ed. Engl. 49, 6288-6308 (2010).

[179] Kempe, K., et al.: Multifunctional poly (2-oxazoline) nanoparticles for biological applications. Macromol. Rapid Comm. 31, 1869-1873 (2010).

[180] Luxenhofer, R., et al.: Structure-property relationship in cytotoxicity and cell uptake of poly (2-oxazoline) amphiphiles. J. Control. Release 153, 73-82 (2011).

[181] Donev, R., Koseva, N., Petrov, P., Kowalczuk, A., Thome, J.: Characterisation of different nanoparticles with a potential use for drug delivery in neuropsychiatric disorders. World J. Biol. Psychiatry 12, 44-51 (2011).

[182] Tong, J., et al.: Neuronal uptake and intracellular superoxide scavenging of a fullerene (C60) -poly (2-oxazoline) s nanoformulation. Biomaterials 32, 3654-3665 (2011).

[183] Viegas, T. X., et al.: Polyoxazoline: chemistry, properties, and applications in drug delivery. Bioconjug. Chem. 22, 976-986 (2011).

[184] Wang, X., et al.: Synthesis, characterization and biocompatibility of poly (2-ethyl-2-oxazoline) -poly (D, L-lactide) -poly (2-ethyl-2-oxazoline) hydrogels. Acta Biomater. 7, 4149-4159 (2011).

[185] Wang, C. -H., et al.: Extended release of bevaci-zumab by thermosensitive biodegradable and biocompatible hydrogel. Biomacromolecules 13, 40-48 (2012).

[186] Cheon Lee, S., Kim, C., Chan Kwon, I., Chung, H., Young Jeong, S.: Polymeric micelles of poly (2-ethyl-2-oxazoline) -block-poly (epsilon-caprolactone) copolymer as a carrier for paclitaxel. J. Control. Release 89, 437-446 (2003).

[187] Konradi, R., Pidhatika, B., Mühlebach, A., Textor, M.: Poly-2-methyl-2-oxazoline: a peptide-like polymer for protein-repellent surfaces. Langmuir 24, 613-616 (2008).

[188] Pidhatika, B., et al.: The role of the interplay between polymer architecture and bacterial surface properties on the microbial adhesion to polyoxazoline-based ultrathin films. Biomaterials 31, 9462-9472 (2010).

[189] Zhang, N., et al.: Tailored poly (2-oxazoline) polymer brushes to control protein adsorption and cell adhesion. Macromol. Biosci. 12, 926-936 (2012).

[190] Wang, H., Li, L., Tong, Q., Yan, M.: Evaluation of photochemically immobilized poly (2-ethyl-2-oxazoline) thin films as protein-resistant surfaces. ACS Appl. Mater. Interfaces 3, 3463-3471 (2011).

[191] Woodle, M. C., Engbers, C. M., Zalipsky, S.: New amphipathic polymer-lipid conjugates forming long-circulating reticuloendothelial system-evading liposomes. Bioconjug. Chem. 5, 493-496 (1994).

[192] Zalipsky, S., Hansen, C. B., Oaks, J. M., Allen, T. M.: Evaluation of blood clearance rates and biodistribution of poly (2-oxazoline) -grafted liposomes. J. Pharm. Sci. 85, 133-137 (1996).

[193] Gaertner, F. C., Luxenhofer, R., Blechert, B., Jordan, R., Essler, M.: Synthesis, biodistribution and excretion of radiolabeled poly (2-alkyl-2-oxazoline) s. J. Control. Release 119, 291-300 (2007).

[194] Goddard, P., Hutchinson, L.: Soluble polymeric carriers for drug delivery. Part 2. Preparation and in vivo behaviour of N-acylethylenimine copolymers. J. Control. Release 10, 5-16 (1989).

[195] Luxenhofer, R., et al.: Doubly amphiphilic poly (2-oxazoline) s as high-capacity delivery systems for hydrophobic drugs. Biomaterials 31, 4972-4979 (2010).

[196] Delgado, A. V., González-Caballero, F., Hunter, R. J., Koopal, L. K., Lyklema, J.: Measurement and interpretation of electrokinetic phenomena. J. Colloid Interface Sci. 309, 194-224 (2007).

[197] Rausch, K., Reuter, A., Fischer, K., Schmidt, M.: Evaluation of nanoparticle aggregation in human blood serum. Biomacromolecules 11, 2836-2839 (2010).

[198] Olsen, S. N.: Applications of isothermal titration calorimetry to measure enzyme kinetics and activity in complex solutions. Thermochim. Acta 448, 12-18 (2006).

[199] Gourishankar, A., Shukla, S., Ganesh, K. N., Sastry, M.: Isothermal titration calorimetry studies on the binding of DNA bases and PNA base monomers to gold nanoparticles. J. Am. Chem. Soc. 126, 13186-13187 (2004).

[200] Cedervall, T., et al.: Understanding the nanoparticle-protein corona using methods to quantify exchange rates and affinities of proteins for nanoparticles. Proc. Natl. Acad. Sci. U. S. A. 104, 2050-2055 (2007).

[201] Tenzer, S., et al.: Nanoparticle size is a critical physicochemical determinant of the human blood plasma corona: a comprehensive quantitative proteomic analysis. ACS Nano 5, 7155-7167 (2011).

[202] Ghoroghchian, P. P., Therien, M. J., Hammer, D. A.: In vivo fluorescence imaging: a personal perspective. Wiley Interdiscip. Rev. Nanomed. Nanobiotechnol. 1, 156-167 (2009).

[203] Herzog, H., Rösch, F.: PET - und SPECT - Technik: Chemie und Physik der Bildgebung. Pharm. Unserer Zeit 34, 468-473 (2005).

[204] Herth, M. M., et al.: Radioactive labeling of defined HPMA-based polymeric structures using [18F] FETos for in vivo imaging by positron emission tomography. Biomacromolecules 10 (4), 1697 - 1703 (2009).

[205] Devaraj, N. K., Keliher, E. J., Thurber, G. M., Nahrendorf, M., Weissleder, R.: 18F labeled nanoparticles for in vivo PET-CT imaging. Bioconjugate Chem. 20, 397-401 (2009).

[206] Herth, M. M., Barz, M., Jahn, M., Zentel, R., Rösch, F.: 72/74As-labeling of HPMA based polymers for long-term in vivo PET imaging. Bioorg. Med. Chem. Lett. 20, 5454-5458 (2010).

[207] Fujita, Y., Taguchi, H.: Current status of multiple antigen-presenting peptide vaccine systems: application of organic and inorganic nanoparticles. Chem. Cent. J. 5, 1-8 (2011).

[208] Vogel, F. R.: Immunologic adjuvants for modern vaccine formulations. Ann. N. Y. Acad. Sci. 754, 153-160 (1995).

[209] Petrovsky, N., Aguilar, J.: Vaccine adjuvants: current state and future trends. Immunol. Cell Biol. 82, 488-496 (2004).

[210] Panyam, J., Labhasetwar, V.: Biodegradable nanoparticles for drug and gene delivery to cells and tissue. Adv. Drug Deliv. Rev. 55, 329-347 (2003).

[211] Lee, Y. -R., Lee, Y. -H., et al.: Biodegradable nanoparticles containing TLR3 or TLR9 agonists together with antigen enhance MHC-restricted presentation of the antigen. Arch. Pharm. Res. 33, 1859-1866 (2010).

[212] Stone, G. W., et al.: Nanoparticle-delivered multimeric soluble CD40L DNA combined with Toll-like receptor agonists as a treatment for melanoma. PLoS One 4, e7334 (2009).

[213] Kasturi, S. P., et al.: Programming the magnitude and persistence of antibody responses with innate immunity. Nature 470, 543-547 (2011).

[214] Malyala, P., O'Hagan, D. T., Singh, M.: Enhancing the therapeutic efficacy of CpG oligonucleotides using biodegradable microparticles. Adv. Drug Deliv. Rev. 61, 218-225 (2009).

[215] O'Hagan, D. T., Singh, M., Ulmer, J. B.: Microparticle-based technologies for vaccines. Methods 40, 10-19 (2006).

[216] Sherman, M. R. et al.: Conjugation of high-m olecular weight poly (ethylene glycol) to cytokines: granulocyte-macrophage colony-stimulating factors as model substrates. ACS Symposium Series, Vol. 680, pp. 155-169. (1997).

[217] Alexis, F., Pridgen, E., Molnar, L. K., Farokhzad, O. C.: Factors affecting the clearance and biodistribution of polymeric nanoparticles. Mol. Pharm. 5, 505-515 (2008).

[218] Yang, Y., Huang, C. -T., Huang, X., Pardoll, D. M.: Persistent Toll-like receptor signals are required for reversal of regulatory T cell-mediated CD8 tolerance. Nat. Immunol. 5, 508-515 (2004).

[219] Galloway, A. L., et al.: Development of a nanoparticle-based influenza vaccine using the PRINT © technology. Nanomedicine 9 (4), 523-531 (2013).

[220] Demento, S. L., et al.: TLR9-targeted biodegradable nanoparticles as immunization vectors protect against West Nile encephalitis. J. Immunol. 185, 2989-2997 (2010).

[221] Tyagi, R. K., Garg, N. K., Sahu, T.: Vaccination strategies against malaria: novel carrier (s) more than a tour de force. J. Control. Release 162, 242-254 (2012).

[222] Brandt, E. R., et al.: New multi-determinant strategy for a group A streptococcal vaccine designed for the Australian Aboriginal population. Nat. Med. 6, 455-459 (2000).

[223] Shi, L., et al.: Pharmaceutical and immunological evaluation of a single-shot hepatitis B vaccine formulated with PLGA microspheres. J. Pharm. Sci. 91, 1019-1035 (2002).

[224] Pejawar-Gaddy, S., et al.: Generation of a tumor vaccine candidate based on conjugation of a MUC1 peptide to polyionic papillomavirus virus-like particles. Cancer Immunol. Immunother. 59, 1685-1696 (2010).

[225] Sundgren, A., Barchi, J.: Varied presentation of the Thomsen-Friedenreich disaccharide tumor-associated carbohydrate antigen on gold nanoparticles. Carbohydr. Res. 343, 1594-1604 (2008).

[226] Monzavi-Karbassi, B., Pashov, A., Jousheghany, F., Artaud, C., Kieber-Emmons, T.: Evaluating strategies to enhance the anti-tumor immune response to a carbohydrate mimetic peptide vaccine. Int. J. Mol. Med. 17, 1045-1052 (2006).

[227] Reddy, S., Swartz, M., Hubbell, J.: Targeting dendritic cells with biomaterials: developing the next generation of vaccines. Trends Immunol. 27, 573-579 (2006).

[228] Demento, S. L., et al.: Inflammasome-activating nanoparticles as modular systems for optimizing vaccine efficacy. Vaccine 27, 3013-3021 (2009).

[229] Hamdy, S., et al.: Co-delivery of cancer-associated antigen and Toll-like receptor 4 ligand in PLGA nanoparticles induces potent $CD8^{+}$T cell-mediated anti-tumor immunity. Vaccine 26, 5046-5057 (2008).

[230] Barton, G. M., Medzhitov, R.: Control of adaptive immune responses by Toll-like receptors. Curr. Opin. Immunol. 14, 380-383 (2002).

[231] Akagi, T., Baba, M., Akashi, M.: Biodegradable nanoparticles as vaccine adjuvants and delivery systems: regulation of immune responses by nanoparticle-based vaccine. Adv. Polym. Sci. 247, 31-64 (2011).

[232] Lee, T. Y., et al.: Oral administration of poly-gamma-glutamate induces TLR4-and dendritic cell-dependent antitumor effect. Cancer Immunol Imm. 58, 1781-1794 (2009).

[233] Yoshida, M., Babensee, J. E.: Poly (lactic-co-glycolic acid) enhances maturation of human monocyte-derived dendritic cells. J. Biomed. Mater. Res. A 71, 45-54 (2004).

[234] Tamayo, I., et al.: Poly (anhydride) nanoparticles act as active Th1 adjuvants through Toll-like receptor exploitation. Clin. Vaccine Immunol. 17, 1356-1362 (2010).

[235] Copland, M. J., et al.: Liposomal delivery of antigen to human dendritic cells. Vaccine 21, 883-890 (2003).

[236] Matsusaki, M., et al.: Nanosphere induced gene expression in human dendritic cells. Nano Lett. 5, 2168-

2173 (2005).

[237] Kwon, Y. J., Standley, S. M., Goh, S. L., Fréchet, J. M. J.: Enhanced antigen presentation and immunostimulation of dendritic cells using acid-degradable cationic nanoparticles. J. Control. Release 105, 199-212 (2005).

[238] Sun, H., Pollock, K. G. J., Brewer, J. M.: Analysis of the role of vaccine adjuvants in modulating dendritic cell activation and antigen presentation in vitro. Vaccine 21, 849-855 (2003).

[239] Moon, H. -J., et al.: Mucosal immunization with recombinant influenza hemagglutinin protein and poly gamma-glutamate/chitosan nanoparticles induces protection against highly pathogenic influenza A virus. Vet. Microbiol. 160, 277-289 (2012).

[240] Geall, A., Verma, A.: Nonviral delivery of self-amplifying RNA vaccines. Proc. Natl. Acad. Sci. U. S. A. 109, 14604-14609 (2012).

[241] Van den Berg, J. H., et al.: Shielding the cationic charge of nanoparticle-formulated dermal DNA vaccines is essential for antigen expression and immunogenicity. J. Control. Release 141, 234-240 (2010).

[242] Varkouhi, A. K., Scholte, M., Storm, G., Haisma, H. J.: Endosomal escape pathways for delivery of biologicals. J. Control. Release 151, 220-228 (2011).

[243] Krieg, A. M.: CpG motifs in bacterial DNA and their immune effects. Annu. Rev. Immunol. 20, 709-760 (2002).

[244] Sudowe, S., et al.: Uptake and presentation of exogenous antigen and presentation of endogenously produced antigen by skin dendritic cells represent equivalent pathways for the priming of cellular immune responses following biolistic DNA immunization. Immunology 128, e193-e205 (2009).

[245] Joshi, M. D., Unger, W. J., Storm, G., Van Kooyk, Y., Mastrobattista, E.: Targeting tumor antigens to dendritic cells using particulate carriers. J. Control. Release 161, 25-37 (2012).

[246] Arigita, C., et al.: Liposomal meningococcal B vac-cination: role of dendritic cell targeting in the development of a protective immune response. Infect. Immun. 71, 5210-5218 (2003).

[247] Espuelas, S., Haller, P., Schuber, F., Frisch, B.: Synthesis of an amphiphilic tetraantennary mannosyl conjugate and incorporation into liposome carriers. Bioorg. Med. Chem. Lett. 13, 2557-2560 (2003).

[248] Espuelas, S., Thumann, C., Heurtault, B., Schuber, F., Frisch, B.: Influence of ligand valency on the targeting of immature human dendritic cells by mannosylated liposomes. Bioconjug. Chem. 19, 2385-2393 (2008).

[249] Sheng, K. - C., et al.: Delivery of antigen using a novel mannosylated dendrimer potentiates immunogenicity in vitro and in vivo. Eur. J. Immunol. 38, 424-436 (2008).

[250] Chenevier, P., et al.: Grafting of synthetic mannose receptor-ligands onto onion vectors for human dendritic cells targeting. Chem. Commun. 20, 2446-2447 (2002).

[251] Saraogi, G. K., et al.: Mannosylated gelatin nanoparticles bearing isoniazid for effective management of tuberculosis. J. Drug Target. 19, 219-227 (2011).

[252] Brandhonneur, N., et al.: Specific and non-specific phagocytosis of ligand-grafted PLGA microspheres by macrophages. Eur. J. Pharm. Sci. 36, 474-485 (2009).

[253] Hamdy, S., Haddadi, A., Shayeganpour, A., Samuel, J., Lavasanifar, A.: Activation of antigenspecific T cell-responses by mannan-decorated PLGA nanoparticles. Pharm. Res. 28, 2288-2301 (2011).

[254] Raghuwanshi, D., Mishra, V., Suresh, M. R., Kaur, K.: A simple approach for enhanced immune response using engineered dendritic cell targeted nanoparticles. Vaccine 30, 7292-7299 (2012).

[255] Fehr, T., Skrastina, D., Pumpens, P., Zinkernagel, R. M.: T cell-independent type I antibody response against B cell epitopes expressed repetitively on recombinant virus particles. Proc. Natl. Acad. Sci. U. S. A. 95, 9477-9481 (1998).

[256] Brinãs, R. P., et al.: Design and synthesis of multifunctional gold nanoparticles bearing tumor-associated glycopeptide antigens as potential cancer vaccines. Bioconjug. Chem. 23, 1513-1523 (2012).

[257] Hoffmann-Röder, A., et al.: Synthetic antitumor vaccines from tetanus toxoid conjugates of MUC1 glycopeptides with the Thomsen-Friedenreich antigen and a fluorine-substituted analogue. Angew. Chem. Int. Ed. Engl. 49, 8498-8503 (2010).

[258] Gaidzik, N., et al.: Synthetic antitumor vaccines containing MUC1 glycopeptides with two immunodominant domains-induction of a strong immune response against breast tumor tissues. Angew. Chem. Int. Ed. Engl. 50, 9977-9981 (2011).

[259] Cai, H., et al.: Variation of the glycosylation pattern in MUC1 glycopeptide BSA vaccines and its influence on the immune response. Angew. Chem. Int. Ed. Engl. 51, 1719-1723 (2012).

（吴国华译）

第七部分

计算机和输送系统

概 述

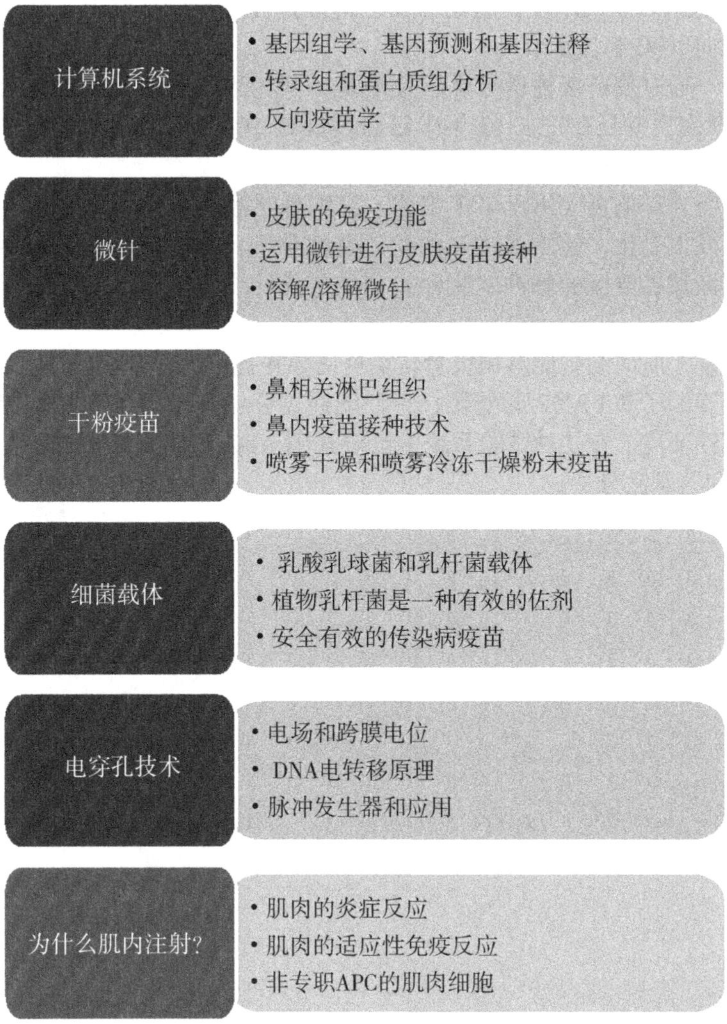

计算机系统
• 基因组学、基因预测和基因注释
• 转录组和蛋白质组分析
• 反向疫苗学
微针
• 皮肤的免疫功能
•运用微针进行皮肤疫苗接种
• 溶解/溶解微针
干粉疫苗
• 鼻相关淋巴组织
• 鼻内疫苗接种技术
• 喷雾干燥和喷雾冷冻干燥粉末疫苗
细菌载体
• 乳酸乳球菌和乳杆菌载体
• 植物乳杆菌是一种有效的佐剂
• 安全有效的传染病疫苗
电穿孔技术
• 电场和跨膜电位
• DNA电转移原理
• 脉冲发生器和应用
为什么肌内注射?
• 肌肉的炎症反应
• 肌肉的适应性免疫反应
• 非专职APC的肌肉细胞

除了适当的佐剂外，递送技术是疫苗开发的另一个关键因素。目前在该领域的研究内容将在以下章节中介绍。无痛和无针安全装置即将成为传统的多次注射的替代物。此外，新疫苗旨在诱导 CTL 或黏膜反应。与大多数常规疫苗不同，这种现代疫苗需要募集细胞效应机制，因此需要新的给药途径，并与新的佐剂结合（第八部分）。新的递送技术还应满足热稳定疫苗的需求，很有可能降低巨大的冷链储存成本。

疫苗设计 随着能够整合各种类型数据的系统生物学和生物信息学工具的发展，疫苗学的未来正在转向能够获得保护性免疫所涉及的各种因素的全景图。在实验确认免疫原性之前，表位的预测仍然是筛选病原体蛋白质编码序列的关键步骤。

微针 使用 MN 的皮内接种是将抗原递送到皮肤真皮层，而不使用皮下注射的最有吸引力的方法之一，皮下注射与发展中国家的感染传播和不当处置有关。

鼻腔递送 鼻黏膜很复杂，包括免疫系统的重要元件。因此，它是非侵入性疫苗递送的理想递送途径。已经证明喷雾干燥、冷冻干燥和喷雾冷冻干燥的粉末与常规液体制剂具有相同的免疫原性。

纳米技术和递送 纳米载体具有在大小方面与病原体相似的优点。因此，它们能够被免疫系统的抗原呈递细胞有效识别。ISCOMS 的特征是笼状结构，其中包含抗原。它们由脂质和 Quil A 组成，Quil A 是源自具有佐剂活性的植物皂树（*Quillaja saponaria*）的皂苷的活性成分。PADRE-PAMAM 树枝状聚合物纳米疫苗递送可以应对保护性疫苗所面临的一些主要挑战：提供高转染效率、适当靶向 APC 和佐剂效应。

细菌载体疫苗 通过疫苗接种诱导黏膜免疫是一项相当困难的任务。乳酸菌具有通常认可的安全（GRAS）状态，并且在过去 10 年中已经发展成为用于黏膜递送疫苗抗原的有效佐剂。乳酸乳球菌和乳酸杆菌已获得应用。

电穿孔技术 将外部电场应用于单个细胞、细胞悬浮液或生物组织会产生细胞跨膜电位的变化，导致膜结构发生变化，使膜透过其他非渗透性分子，这种现象称为电穿孔。电穿孔技术的基础是使用不同涂敷器电极的脉冲发生器，例如矩阵针或矩阵板，以向靶组织递送合适的电脉冲。

传统的肌内免疫 肌肉细胞能够积极参与免疫诱导并且作为非专职的 APC。

肌细胞表达细胞因子和 PAMP 的受体，它们能通过分泌细胞因子和趋化因子，以及表达黏附分子，响应炎症性环境。在几种试验系统中已观察到 APC 功能所必需的一些膜蛋白，如Ⅰ类和Ⅱ类 MHC 分子和共刺激分子。

第 42 章　后基因组时代疫苗设计考虑的因素

Christine Maritz-Olivier 和 Sabine Richards①

摘要

18 世纪 Edward Jenner 和 19 世纪 Louis Pasteur 奠定了接种疫苗的基础。在 20 世纪 30 年代和 40 年代，减毒活疫苗和灭活疫苗在该领域占主导地位。然后使用生化方法在培养的病原体中纯化抗原，引领亚单位疫苗时代。随着下一代测序技术的爆炸性增长以及第一代基因组的可用性，反向疫苗学领域以及相关的“组学”-变革逐渐成熟。这样就可以识别出以前没有开发其保护作用的有前景的抗原。通过将后者技术与免疫遗传学和免疫基因组学相结合，了解了感染/疫苗接种期间的免疫应答，并提供了保护性免疫的各种因素的全局图。在这个后基因组时代，疫苗开发正在从试错法转向以知识理论为基础的疫苗开发方法。

18 世纪 Edward Jenner 和 19 世纪 Louis Pasteur 奠定了接种疫苗的基础。在 20 世纪 30 年代和 40 年代，减毒活疫苗和灭活疫苗在该领域占主导地位，其中两个最典型的实例是鸡胚胎的流感病毒和脊髓灰质炎病毒，还有人和猴组织培养物中的其他病毒[1]。然后使用生化方法在培养的病原体中纯化抗原，引领亚单位疫苗时代。随着第一个细菌基因组的出现，Rappuoli 在 2000 年发明了反向疫苗学方法，随着下一代测序技术的发展和相关的“组学”革命，反向疫苗学已经成为热点。

血清群 B 脑膜炎球菌疫苗是根据反向疫苗学获得的首款疫苗，已经进行了最后的试验，可以在不久后证实反向疫苗学[2,3]。2007 年，Poland 及其同事开始了疫苗组学，这是旨在结合免疫遗传学和免疫基因组学的领域，提供对感染期间或疫苗接种期间免疫反应的深入了解。随着能够整合各种类型数据的系统生物学和生物信息学工具的发展，疫苗学的未来正在转向能够获得保护性免疫所涉及的各种因素的全局图。

尽管取得了这些进展，但仍需要了解关键的基础知识，以便从试错法转向基于知识的疫苗开发方法。包括彻底了解病原生物学、宿主免疫系统、宿主-病原体相互作用或疫苗接种过程中激活的免疫应答网络、反向疫苗学原理和该方法中使用的软件、免疫生物标记物的识别以及适当的疫苗配方和试验。这是本章的重点。

42.1　了解病原体：入侵、感染和生存的生物学

需要疫苗的病原体包括病毒、原核细菌、真核细胞内寄生虫（如引起疟疾的疟原虫）、细胞外寄生虫（如蠕虫）和体外寄生虫（如蜱和螨）。它们每一种都有其独特的生物学和对宿主免疫系统的影响，如果要成功开发疫苗，必须了解这一点。

---

① C. Maritz-Olivier，PhD（✉）（比勒陀利亚大学自然与农业科学学院遗传学系，南非比勒陀利亚林伍德路，E-mail：christine. maritz@ up. ac. za）

S. Richards，MSc（比勒陀利亚大学自然与农业科学学院遗传学系，南非比勒陀利亚林伍德路）

42.1.1 病原体计划：基因组学、基因预测和基因注释

能够在短时间内以合理的价格获得基因组 DNA 序列信息的能力无疑将塑造现代生物学。在疫苗学领域，基因组是了解病原体基因、宿主免疫基因以及感染和接种过程中参与宿主病原体相互作用的基因之间相互作用的关键。下一代测序已成功应用，可深入了解病原体和宿主遗传变异的检测、信息选择的压力、免疫逃逸机制、疫苗安全性、T 细胞和 B 细胞库的多样性、免疫调节和评估疫苗贮存质量[4]。

现在，使用密切相关生物基因组进行分析（泛-基因组分析）已成为常规方法。它能识别适用于联合疫苗的免疫原性和保守抗原，提供了抗遗传多样性菌株的保护作用。最好的实例是含有 4 种免疫原性和保守抗原的 Bexsero 疫苗，在超过 800 多种遗传多样性致病 MsnB 菌株中能够对抗其中 77%的致病性 MsnB 菌株[2]。对 8g 阳性致病性无乳链球菌基因组的比较研究揭示了核心基因组和非重要的基因组。有趣的是，发现具有保护作用的蛋白质并非全部来源于核心基因组。相比之下，发现只有 1 种核心基因组抗原和 3 种非重要的基因组抗原对一些菌株具有保护作用[5]。致病性和非致病性菌株之间的消减基因组学已成功用于识别来自致病性大肠杆菌的 ECOK I-3385 抗原[5]。

在这个后基因组时代，基因组序列组装和基因预测仍然是瓶颈，特别是对于缺乏基因序列片段的复杂真核病原体。后续基因与开放阅读框（ORF）的预测是基因组分析中最基本和最重要的步骤之一，对任何下游分析都有重大影响。识别 ORF 的一种方法是 BLAST[6]。另外，如果不存在其他基因证据，可以利用数学模型进行从头开始的基因预测[7]。有很多优化了原核生物 ORF 预测的程序，如 Glimmer[8]、GeneMark. hmm[9] 和 Prodigal[10]。通过评估原核生物的基因模型和检测异常提高基因预测准确性的方法也越来越成熟[11]。

在真核生物中，这并不是一个简单的过程，因为最近 ENCODE 方法的观察结果显示基因具有重叠的转录本以及两条 DNA 链转录[12]。此外，数学模型通常使用经典的模式生物例如果蝇[7]反应生物特异性基因组特征。这些预测的准确性取决于生物与这些真核生物的相关程度[7]。有些基因预测软件也可以使用 MAKER 等方法进行完善[13]。其他程序能够将外部信息（例如 EST 和 RNA-Seq 数据）纳入从头开始方法以改善预测。这类软件包括 AUGUSTUS[14] 和 GNOMON（http：//0-www. ncbi. nlm. nih. gov. innopac. up. ac. za/genome/guide/gnomon. shtml）。其他方法，例如 SAGE 文库标签-基因绘制图谱（由 BLAST 绘制标签基因组图谱，然后使用数据概率分布绘制标签基因图谱），需要获得对利什曼原虫基因组结构和转录的深入了解[15]。

之后对基因产物的功能注释也不是一个简单的过程。已经证明可成功使用 BLAST[16] 和基于 BLAST 的算法的同源搜索，并且许多在线数据库提供了用于注释的工具和信息，例如，PIR[17]、Pfam[18]、SWISS-PROT[19]、UniProt[20] 和 COG[21]。遗憾的是，这些方法中大多数依赖于确定特性的蛋白质信息的可用性。此外，并不总能从同源性推断功能相似性，成功进行功能预测需要额外的信息[18]。Gene Ontology（GO）的成功应用已经很广泛[22]，但它可能受到非计划、实验无法验证以及需要计算导出注释的限制[23]。人们逐渐接受的一种替代方案是使用蛋白质的相互作用来预测和识别功能关系。主要限制是缺乏蛋白质相互作用数据[24]，但预测的二级结构元件排列的几何关系可用于模拟和评估用于功能预测的假定结构相互作用[25]。可以将数据合并，进行推断和确认功能注释预测，例如蛋白质序列、基因表达和蛋白质相互作用数据[26]。

42.1.2 转录组和蛋白质组分析

转录组是一个高度动态的系统，代表一组确定的基因，这些基因在特定条件下由细胞在特定条件下表达。转录组分析为研究人员提供了有关罕见转录本、交替剪接、拷贝数变异、序列变异、调控机制和细胞功能相关生化途径的信息。RNA-Seq 是后基因组时代用于获得大转录组数据集的技术，据估计它检测的转录本比传统 DNA 微阵列多 25%[4]。到目前为止，已成功使用

RNA-Seq和常规转录组研究受试者对病原体或候选疫苗的免疫应答之间的差异[4]、miRNA在调节mRNA表达中的作用[4]、对病毒疫苗的表观遗传反应[4]、了解分子过程和生存机制（例如识别疫苗和药物靶标的心丝虫犬恶丝虫[27]）、确定其他表面表达的候选疫苗（例如脑膜炎奈瑟球菌[28]）、识别复杂病原体中表达的表面和分泌蛋白（例如顶丛寄生菌柔嫩艾美耳球虫[29]和现在可以用作候选疫苗的体外寄生牛蜱、微芽孢虫[30]）。

蛋白质组代表细胞表达的完整蛋白质补体，与核酸的分析相比，可以更直接地检测细胞反应，并可以深入了解转录后修饰。通常使用双向电泳评估蛋白质组，但是有一个困难，只能回收少量疏水蛋白用于下游分析。如今，高通量质谱法加速了膜蛋白分析[31]，并揭示尽管进行了广泛的基因组测序，但数据库中仍存在未知的质膜蛋白。蛋白质组分析对克服计算机定位预测软件的局限性至关重要。由于在疫苗接种过程中主要靶向表面暴露的蛋白质，因此表面蛋白组分析受到了相当多的关注。新方法包括使用“shaving”法分析膜蛋白，其中细胞用蛋白酶处理，通过LC-MS-MS分析所得肽[5]，并且采用高级质谱鉴定高或低反应相关HLA分子作为候选抗原的病毒肽[32]。

疟原虫Plasmodium vivax长期以来都被人们忽视。到目前为止，蛋白质组学为研究这类疟原虫提供了深入研究，识别了与先前识别的产物没有同源性的153种蛋白和29种新蛋白[33]，还提供了对克氏锥虫的病原生物学、基因和代谢途径的深入研究[34]，同时也为结核分枝杆菌中的表位疫苗和结构蛋白质组学的发展提供平台[35]。宿主蛋白质组数据的可用性和它们与其他“组学”数据（主要是转录数据）的整合对于未来疫苗的成功至关重要（本章第16.3节）。表42.1列出了与后基因组时代疫苗开发多学科领域相关的词汇。

表42.1　后基因组时代疫苗开发的多学科领域相关的术语

术语	定义/通用解释	参考文献
免疫遗传学	免疫系统的遗传分析 免疫系统中涉及的基因的研究	[36]
免疫基因组学	基因组学技术与免疫学相结合 参与免疫反应的相关基因及其表达谱的研究	[37]、[38]
系统生物学	系统生物学旨在如何定量地将生物系统的特性理解为其大分子组分的特征及其大分子间相互作用的功能 整个生物系统的研究，包括生物元素与其周围环境之间的所有相互作用	[39]
免疫蛋白组学	确定诱导免疫反应的相关蛋白质（免疫蛋白组） 参与免疫反应的所有蛋白的鉴定和功能研究	[40]
疫苗组学	该术语指用于疫苗免疫应答的免疫遗传学和免疫基因组学以及病原体和宿主对疫苗免疫应答的异源性潜在机制 结合免疫遗传学和免疫基因组学的综合方法，用于研究疫苗或病原体的免疫应答	[41]
反向疫苗学	基于基因组的疫苗研制方法 基于生物基因组信息的疫苗研制方法	[42]
免疫治疗	宿主免疫系统操纵过程中免疫细胞间复杂的相互作用网络 通过诱导、增强或抑制免疫应答治疗疾病	[43]

42.2　了解宿主免疫系统

有多种免疫细胞能够根据最开始遇到免疫细胞的方式和位置识别病原体。主要是先天免疫系统的细胞识别病原体，这些细胞没有抗原特异性受体，但是含有病原体识别受体（PRR），例如C型凝集素、Toll样受体（TLR）或NOD（核结合寡聚化结构域）蛋白质。它们存在于先天免疫细胞中，例如树突细胞（DC）、巨噬细胞、肥大细胞和嗜中性粒细胞以及内皮细胞和成纤维细胞（统称为抗原呈递细胞，APC）。结合后可以激活信号传导途径，随后合成并释放促炎细胞因子和趋化因子。然后，活化的抗原呈递细胞（主要是DC，也包括巨噬细胞和B细胞）处理源于病原体抗原，以递呈给MHC Ⅱ类的适应性免疫细胞。

由巨胞饮作用或外源蛋白的内吞作用介导，外源蛋白被递送到酸性内体，然后将其剪切成与MHC Ⅱ复合物的α亚基和β亚基结合并在细胞表面上表达的肽。活化的和成熟的APC迁移到淋巴结，它们介导抗原特异性激活和共刺激信号到幼稚T细胞（通过MHC和T细胞受体之间的相互作用）和B细胞，缩小先天性和适应性免疫之间的差距。T细胞能分化成$CD4^+$T辅助细胞（T_H）或$CD8^+$细胞毒性T淋巴细胞（CTL）。$CD4^+T_H$细胞可以被引导至细胞Th1，体液Th2或耐受性Treg反应中，这取决于与抗原呈递细胞（APC）的接触和特定细胞因子的诱导[43]。大多数情况下，MHC Ⅱ相关的肽诱导$CD4^+$T细胞和抗体反应[1]。效应T细胞的一部分分化成记忆细胞，记忆细胞在之后感染期诱导抗病原体反应。这种先天性和适应性免疫系统之间交互刺激是使用亚单位疫苗的主要困难。大多数简单的亚单位疫苗仅诱导$CD4^+$反应（激活B细胞和产生抗体），这些反应不足以预防慢性感染和细胞内病原体，还需要有效的激活细胞适应性$CD8^+$ T细胞反应（通过MHC Ⅰ或MHC Ⅱ）[1]。

为了深入了解人体免疫细胞中表达的大量基因，开发了免疫基因组计划[44]和IRIS[45]（计算机免疫应答）数据库。其中包括所有人和鼠的免疫细胞主要表达的基因概况，并为研究人员提供结合免疫学和计算生物学的工具。

该数据提供的内容包括免疫细胞的分化、在炎症和免疫中的作用、蛋白质的细胞定位和免疫系统的信号传导途径[45]。下一代测序技术可以对T细胞受体多样性进行大规模研究，从而提供TCR基因片段重排（主要是VDJ区域）作用，以及如何产生大量的抗原特异性T细胞。也扩展到B细胞的研究，鉴别了人免疫球蛋白重链可变区中的14个新等位基因的变异[4]，还更深入地了解了靶向树突细胞亚群的疫苗[46]。ChIP-Seq技术也成功用于深入了解B细胞和T细胞应答中涉及的转录因子[4]。

正在快速发展关于免疫细胞基因组中多态性作用的研究，了解对疫苗的异质性反应，为个性化疫苗铺平道路[47,48]。Jacobson及其同事和Tan及其同事进行了人类基因宿主多态性对疫苗反应的影响的经典研究，其中评估了单卵双生和双卵双生对MMR、乙型肝炎、口服脊髓灰质炎、破伤风类毒素和白喉疫苗的反应[47]。随后，发表了关于人白细胞抗原（HLA）基因的多态性（主要是SNP）和它们对MMR和风疹疫苗免疫力的研究（综述[47]）。其他疫苗的联系仍有待阐明，但目前的限制是，大规模全基因组连锁研究需要大量数据。已发表了关于风疹和许多佐剂的非HLA基因多态性和它们如何影响疫苗疗效的研究，如细胞因子和细胞因子受体中的SNP、Toll样受体、信号分子、维生素A和D受体、抗病毒效应物和先天性和适应性免疫相关基因[32,47]。

42.3 疫苗组学：了解感染和疫苗接种过程中诱导的宿主-病原相互作用和免疫类型

在感染后，病原体在介导保护性免疫的先天性和适应性免疫反应中激活大量免疫触发因子和长期抗病原体作用[49]。称其为“免疫应答网理论”，目的是整合环境、宿主和病原体或疫苗接种因子[47]。我们了解参与感染和疫苗接种的基因以及它们如何与保护性免疫相关联，将为改进疫苗开发提供必要的见解（图42.1）。后者由Bernstein及其同事明确指出：“未来的疫苗将不仅需要鉴别单个相关因子（或多个独立相关因子），还需要鉴别与免疫保护相关的多因子的特征

(即病原体、宿主) 和 [相互作用的特征) ”[49]，为实现通用疫苗的最终目标铺平道路。

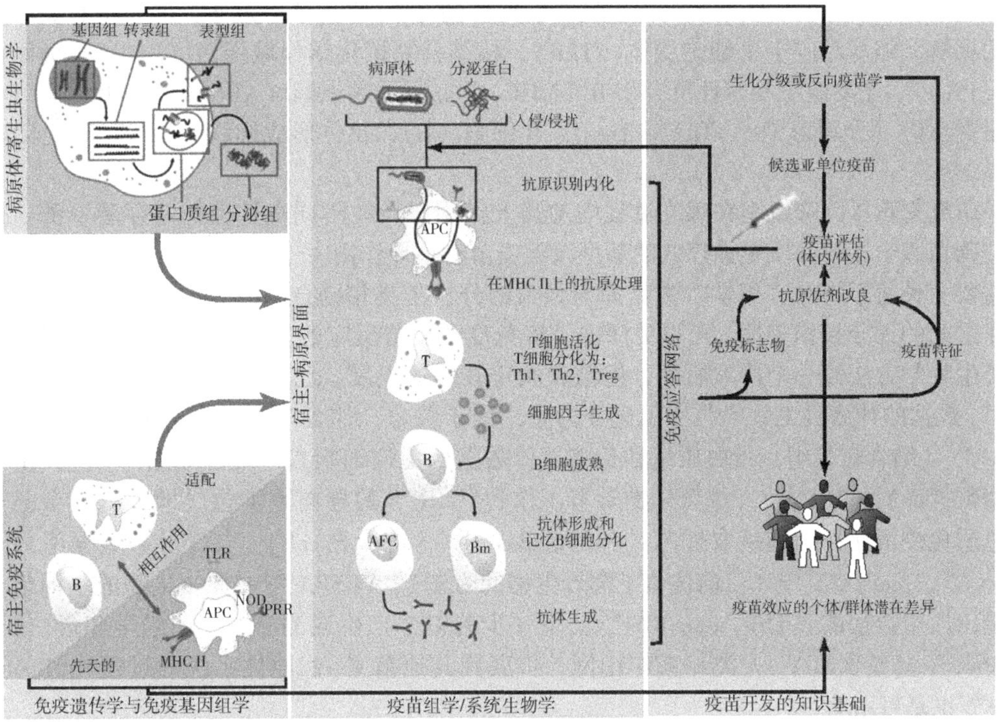

图 42.1　疫苗开发的考虑因素

2007 年 Poland 首次将疫苗组学定义为免疫遗传学和免疫基因组学的应用，免疫遗传学和免疫基因组学是研究与疫苗反应相关的遗传和表观遗传决定因素和途径。2011 年，延伸到包括病原体和宿主对疫苗接种反应的异质性机制，目的是提供对疫苗接种反应中观察到的变异性的理解[50]。

已成功使用综合方法深入了解未来疫苗。关于微生物生物学，Zhang 等已经发表了优秀的综述。[51]。例如包括同时分析风疹、天花和流感疫苗的高和低抗体应答者中的宿主和病原体转录组[32]，以及整合麻疹、腮腺炎、风疹、流感和天花疫苗的基因型/单倍型和表型[32]。将全基因组表达数据与流感病毒感染中的相互作用体（酵母双杂交）数据结合，识别调节干扰素产生和病毒复制的新病毒基因[50]。还将其用于了解 T 细胞对黄热病疫苗接种的反应（识别中和抗体反应的特征和 $CD8^{+}$T 细胞的两个额外特征）以及识别结核病患者中嗜中性粒细胞的两种不同的 IFN 反应[50]。

患者接受 MAGE-A3 肽疫苗治疗非小细胞肺癌，其整体转录谱表现的特征与治疗反应相关[43]。针对宿主 OAS 基因家族和西尼罗河病毒（WNV）的免疫遗传学研究表明，OASL 中的多态性与 WNV 感染的易感性有关，并且能够诱导 OAS1 活性的治疗剂/佐剂可以增强 WNV 疫苗的效率[50]。接种 Edwardsiella tarda 减毒活疫苗的斑马鱼的综合转录组研究揭示了抗原加工和急性期反应的途径，这提供了对斑马鱼免疫应答的潜在机制的深入了解和改进疫苗的方向。有趣的是，他们发现在疫苗接种过程中 MHC Ⅰ途径上调，MHC Ⅱ途径下调[52]。

在另一项研究中，金黄色葡萄球菌中的树突细胞基因表达谱在炎症过程和 T 辅助细胞极化过程以及易感动物和抗性动物之间 204 个不同的表达基因中显示出独特作用，还解释了金黄色葡萄球菌感染敏感性的差异[53]。全基因组数字基因表达（DGE）成功用于研究人畜共患病的病原体布鲁氏菌的宿主反应。

需要深入了解寄生虫/病原体生物学以及宿主免疫系统，才能充分理解感染过程和保护性免疫的发展。其可以通过结合基因组学、转录组和蛋白质组数据实现，并推动了反向疫苗学方法和亚单位疫苗的开发。综合系统生物学方法可以深入了解感染的复杂网络和保护性免疫的发展。将多态性与在疫苗应答中观察到的异质性和病原体的存活/毒力联系起来，能获得有关改进疫苗开发的见解。缩写对应于T细胞受体（TLR）、核结合寡聚化结构域蛋白（NOD）、病原体识别受体（PRR）、主要组织相容性复合物Ⅱ（MHC）、抗原呈递细胞（APC）、T辅助细胞1型和2型或耐受反应（分别为Th1、Th2或Treg）、B细胞（B）、抗体形成细胞（AFC）和记忆B细胞（Bm）。

该研究揭示了巨噬细胞在感染过程中的强大作用以及在不同毒性菌株的存活中诱导抗炎因子和抗凋亡因子，对开发更强疗效的新型减毒疫苗有指导作用[54]。

需要一种将不同的“组学”数据库与强大的分析工具相结合的综合系统生物学方法，提供数据库之间的评分和相关性，可以更深入了解免疫和疫苗反应的复杂特征[50]。这仍然是疫苗设计系统生物学方法的一个严重瓶颈，但该领域正在迅速发展。已经发表了一些综述，关于结合“组学”数据的计算机工具[51,55-58]。迄今为止，用于整合“组学”数据的软件包和数据库包括RefDIC[58]（结合转录组、蛋白质组和免疫遗传数据的免疫细胞参考基因数据库）、IIDB[59]（先天免疫数据库），WIBL[60]（综合生物学习工作台）、R[61]的导向聚类包（微阵列整合，全基因组染色质免疫沉淀和细胞摄取检测）和EchoBasE[62]（用于*E. coli*）。几个公共数据库，如integrOmics[63]、bioPIXIE[57]（目前优化了酵母生物的方法）、VESPA[64]（原核生物的基因组、转录组和蛋白质组数据的整合）、openBIS[65]（下一步的整合）生成测序、代谢组学和蛋白质组学、GPS-Prot[66]（整合HIV-人类相互作用网，扩展到其他宿主-病原体系统的过程）、Booly[67]和PARE[68]也是可用的。

最后，结构生物学已经提供了关于VRC01抗体如何中和90%的HIV-1株的见解，表明许多疫苗接种者和感染个体中抗自体HIV包膜序列的中和抗体具有多样性[4]。结构疫苗学将结构生物学与疫苗学结合起来，可用于改进疫苗设计。由于中和抗体需要识别特定的抗原结构，因此使用高分辨率结构分析识别这些区域很重要。保持中和结构的完整性，同时改变其他区域以使抗原更稳定、更有效以及消除可变区域，可以改进疫苗。

此外，有关某些抗原区域结构的知识将有助于提高我们对免疫原性和免疫优势的理解，从而有助于未来的疫苗设计[69]。有几项研究是按照结构疫苗学方法，例如，构建嵌合蛋白作为抗B组链球菌的疫苗[70]。这些微生物的一些蛋白质已经进化成不同分离物中的具有非相似特性的抗原区域，然而，具有相同的功能特性有助于逃避宿主免疫系统。因此，Nuccitelli等构建了一个嵌合蛋白，其中含有6种抗原变体的结构域，可提供抗6种试验分离株的保护作用[70]。

交叉反应性和自身免疫的挑战

抗相关致病菌株的交叉保护作用仍然是良好疫苗的特性。尽管对保守蛋白进行疫苗接种很有前景，但跨基因组保守的抗原不太可能引起有效的免疫应答，它们通常暴露于宿主免疫系统，处在较强的选择压力下[71]。如果抗原与宿主蛋白具有相似性，则很可能由于表位模拟导致免疫反应差，并且可以诱导自身免疫、靶向抗原和宿主蛋白[72-74]。不仅蛋白质可以诱导自身免疫，多糖也可以诱导。由于多糖对自身免疫的诱导，脑膜炎球菌中的疫苗不能用于脑膜炎奈瑟球菌[2]。

应使用BLAST等工具进行假定抗原和宿主蛋白之间的相似性搜索。可将预测的ORF与数据库中的已知基因和蛋白质进行比较。MHC配体非常短，仅有9个残基，短区域可能会引起自身免疫反应或低免疫原性[75]。使用相似性搜索也遵循消减法，可以识别致病菌株中存在的蛋白质，但不能识别非致病性菌株[76]。此外，还有表位保守工具，可以分析特定蛋白质如何保守或变异[77]。

42.4　反向疫苗学

反向疫苗学是一个过程，通过生物信息学工具分析基因组和转录组序列，以识别潜在的保护性分泌和表面显示抗原蛋白[2]。然后表达这些蛋白质，用于免疫适合的宿主以在进行生产和制备之前验证它们的免疫原性和保护能力（图 42.2）。与这种方法相关的一个主要问题是，疫苗学家无法改进选择候选物时的初始生物信息学，初始生物信息学能改善抗原性的预测，甚至更好地预测它们诱导保护性免疫的能力[2]。虽然反向疫苗学已应用于许多人和动物病原体（见［71］中的列表），但必须指出，其不适用于所有疫苗开发项目。

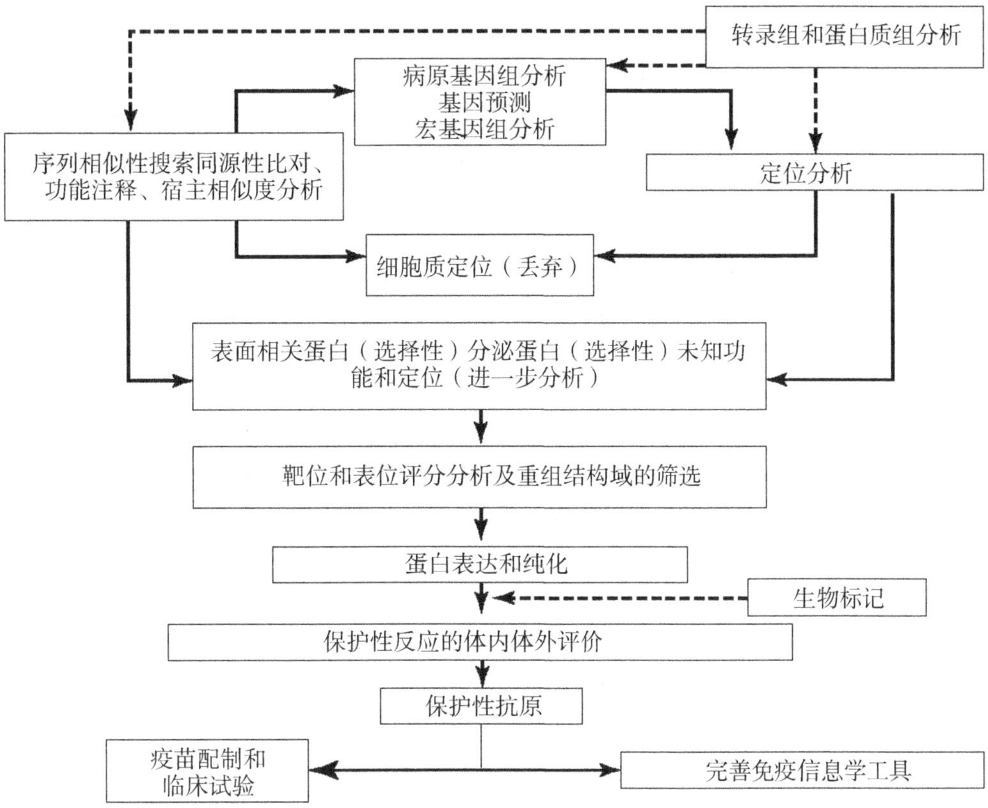

图 42.2　用反向疫苗学方法鉴定候选疫苗的策略（选自［78，79］）

42.4.1　亚细胞定位的预测

因为免疫因子可能不会进入细胞质或通过内膜蛋白，大多数保护性抗原需要病原体表面具有抗原决定簇[42,80,81]。迄今为止，已实现疫苗靶向表面的黏附素、纤维蛋白、GPI 锚定抗原、毒素、侵袭素和孔蛋白[82]。此外，有关抗原定位的内容为研究人员提供了蛋白质的生物学功能及其参与的生物途径[83]。

可免费获取大量蛋白质亚细胞定位的预测，选择适合您选择的生物体的程序具有挑战性。在许多情况下，需要采用组合方法识别分泌蛋白，如寄生线虫 *Strongyloides ratti*[84]。用于亚细胞定位预测的程序可以分为两组。第一种是使用氨基酸序列进行预测，而第二种是根据注释进行预测[85]。PSORT 是第一个基于蛋白质分选信号的计算机程序，于 1990 年发表[86]。随后进行了若干改进，例如结合机器学习技术，以提高各种生物体定位预测的准确性。PSORTb 3.0 是最新版本的 PSORT，于 2010 年发布，是最准确的细菌亚细胞定位预测程序之一，可用于所有原核生物[87]。

2006 年进行的一项研究比较了几种哺乳动物亚细胞的定位预测，发现没有一种方法能足够敏感的准确预测两个试验数据集中的蛋白质定位。此外，发现靶向分泌途径的蛋白质是最难预测的[88]。真核蛋白定位的新预测程序 iLoc-Euk 是最准确的程序，能将有些存在 1 个以上的亚细胞定位，甚至还会在这些之间移动的蛋白质考虑在内[83]。其他程序，如 SherLoc2，包括序列和注释的特征、系统发育特征以及从蛋白质序列获得的基因本体特征[85]。Cell-PLoc 包含 6 个预测程序包，EukmPLoc、Hum-mPLoc、Plant-PLoc、Gpos-PLoc、Gneg-PLoc 和 Virus-PLoc。这些程序结合不同的网站和方法，分别预测了真核生物、人类、植物、革兰氏阳性细菌、革兰氏阴性细菌和病毒中蛋白质的亚细胞定位[89]。

该过程从分析寄生虫/病原体的转录组和蛋白质组开始，在定位预测和序列相似性搜索后从中识别预测的表面相关蛋白和分泌蛋白。为了预防自身免疫，可使用中和抗体的同源性图谱、功能注释和宿主相似性搜索。可以对识别的靶标进行分级，用于预测表位和合适的重组结构域，并分析 B-细胞表位变异性和 T-细胞表位变异性。然后必须表达和纯化所选择的蛋白质，使用生物标志物信息评估免疫应答（如果可获得）。然后可将保护性抗原用于疫苗制剂和临床试验，并且所获得的数据可用于改进可用的免疫信息学工具。

感染细胞中病毒蛋白的亚细胞定位很有趣，因为这与宿主的健康和抗病毒药物设计直接相关[90]。Virus-mPLoc 是较新的预测程序之一，使用 Gene Ontology、功能域和进化信息。相比以前的病毒蛋白定位程序，主要改进之一是该程序可检测在多个位置出现的蛋白质[91]。另一个改进了功能的网络程序是 iLoc-Virus[90]。

大多数蛋白质是在氨基末端分选信号，分选到不同细胞器后在胞质中合成[92]。在反向疫苗学中信号肽的应用（参见 42.4 部分）很有意义，因为它们负责在真核生物中通过分泌途径转运蛋白质，因此其定位于或存在细胞表面之外[93,94]。2009 年进行的一项研究将 SignalP 3.0 确定为与其他 11 个相比最准确的预测软件，其次是信号肽的快速预测（RPSP）[95]。病毒需要宿主细胞机制进行复制和蛋白质合成，因此通常含有真核靶向信号和功能域，以利用宿主定位机制[96]。因此，可以使用适当的真核亚细胞定位预测软件。信号肽和跨膜区都具有疏水区域，因此当它们出现在蛋白质的氨基末端区域时，软件可能会错误理解[97]。为了解决这一问题，SignalP 4.0 版对数据库进行修整，以区分跨膜区和信号肽[97]。SignalP 3.0 和 4.0 版可用于真核生物、革兰氏阳性和革兰氏阴性细菌。其他程序例如 Phobius 网络服务器[98,99]或 SPOCTOPUS[100]，也可用于解决正确预测信号肽和跨膜区域的困难。

此外，考虑蛋白质跨膜螺旋的数量也很重要。首先，如果由于实验限制仅表达部分蛋白质，则应选择细胞外结构域，因为它更接近免疫系统。其次，表达完整的蛋白质时，应注意含有 1 个以上跨膜结构域的蛋白质更难表达[80]。因此，如果可以从许多可用蛋白质中选择抗原，与多跨膜蛋白质相比，应优先考虑单跨膜蛋白。

42.4.2 免疫原性和表位的预测

表位（可被宿主免疫系统识别并引发免疫应答的蛋白质序列）的预测仍是在确认免疫原性之前筛选病原体蛋白质编码序列的关键步骤。迄今为止，有几种根据序列和/或结构的计算方法（在［101-104］中综述）。

根据序列的方法假设类似的氨基酸序列生成类似的蛋白质结构和功能。检测 B 细胞和 T 细胞表位的能力还取决于蛋白质的序列模式和物理化学特性，如灵活性、可及性、亲水性和电荷[105]。在 Yu 及其同事进行的比较研究中发现，使用序列和蛋白质基序是预测 MHC 结合的最准确方式。然而，随着数据量的增加，利用机器学习技术的软件变得更加精确[106]。尽管使用同源建模等方法可以进行结构的预测，但这些方法相对较新、成本较高，还需要了解蛋白质的三维结构。因此，后一种方法尚未普遍使用[107]。最准确的预测包含几种技术，因为可以结合每种方法的优势[105]。

可以通过检测 MHC 结合物间接预测 T 细胞表位。由于 MHC Ⅰ和 MHC Ⅱ蛋白的不同构象，它们会结合不同的分子[108,109]。与 MHC Ⅰ相比，MHC Ⅱ可以结合可变的肽长度，导致预测 MHC Ⅱ结合的准确度比 MHC Ⅰ低得多[105]。这些预测并不总是准确的，因为有些蛋白质，即使与 MHC 结合，也不会引发免疫反应[108]。对于两种 MHC 类别，可以通过预测肽与 MHC 复合物的结合或通过在结合前预测肽的加工途径来识别配体[108]。最新的程序之一 POPISK 集中于预测由 T 细胞受体识别的 MHC Ⅰ-肽复合物序列。这种识别能引起 T 细胞活化和进一步的免疫应答[110]。NetCTLpan[111]是最准确的预测细胞毒性 T 淋巴细胞表位的程序之一。然而，众所周知，对于蛋白质是否引起免疫应答有很多不同的因素起作用。例如，与 MHC Ⅰ结合的亲和力是必要的，但不足以引起 T 细胞应答[110]。因此，为了能够充分利用表位预测方法，需要了解更多关于免疫系统中涉及的所有因子的复杂相互作用。

预测 B 细胞表位比预测 T 细胞更难以完成。其中一个原因是 MHC 蛋白在它们结合的分子中具有高度特异性，而大量短的线性肽具有结合抗体的潜力[108]。由于要区分两种类型的 B 细胞表位（连续和不连续）必须应用不同的预测方法。识别连续 B 细胞表位的方法类似于基于蛋白质性质的 T 细胞表位预测程序。相反地，不连续的 B 细胞表位预测需要了解抗原-抗体复合物的 3D 结构[107]。

因此，这类预测非常困难，并且当前用于此目的的软件显示出相对较差的性能。用于预测不连续 B 细胞表位的生物信息学工具始于获得残留溶剂可接受性（CEP 服务器）[112]。一些较新的程序使用贝叶斯分析中包含的大量生理化学和结构几何特性[113]、不同功能的共识评分（倾向、保守、能量、接触、表面平面性和二级结构组成）[114]，以及使用 B 因子和可接受表面积作为结构特征的逻辑回归算法[115]（除了考虑每个残基的空间环境之外）。识别不连续 B 细胞表位的另一种方法使用一种新概念，即抗原残基（根据距离的特征）和三维结构的空间特征[116]。

结合噬菌体的优势和深度测序的深度筛选，可以通过绘制多克隆抗血清特异性帮助我们理解对疾病的体液反应[117]。为了预测保护性细菌、病毒和肿瘤抗原，自交叉协方差方法以 VaxiJen 服务器的形式发布了与比对无关的方法[118]。包含支持向量机分类的方法显示出更好的结果，因此有改进反向疫苗学方法的潜力[81]。蛋白质微序列也可用于识别宿主感染期间表达的蛋白质和预测蛋白质抗原性[119,120]。

42.4.3 反向疫苗学平台

为了简化反向疫苗学方法，已经创建了几个平台。这使用户能够根据需要经常重复该过程，并改变严格条件，以获得用于进一步评估的较少的候选疫苗。NERVE（新的增强型反向疫苗学环境）是一种用于自动化反向疫苗学分析的服务器[121]，具有计算复杂的缺点[2]。使用了几种过滤器，可以预测蛋白质定位（PSORTb）[122]（需要非细胞质）和识别黏附素（SPAAN）[123]。此外，根据 HMMTOP[124]确定跨膜螺旋的数量（需要<2）。进行相似性搜索是为了防止与人类蛋白质的显著相似性，识别与已知抗原的同源性，并辅助功能性注释。

Vaxign 是一个网络平台，可搜索 70 多种致病基因组的抗原，识别基因组之间的蛋白质保守性，并排除非致病性生物体中存在的序列。它包括许多不同的程序，可用于预测蛋白质的亚细胞定位、跨膜螺旋的数量、粘附率、功能分析和免疫表位（MHC Ⅰ类和Ⅱ类）。此外，服务器可以将输入序列与宿主进行比较，以防止任何交叉反应和随后对宿主的伤害[73]。目前，该服务器仅能够处理微生物的高通量数据[73]。到目前为止，还没有可以研究真核生物的系统，但是 VaxiJen 对细菌、病毒和肿瘤抗原的预测准确率高达 70%~89%，对内寄生虫和真菌抗原的准确度为 78%~97%[125,126]。

42.5 疫苗候选物的验证

通过反向疫苗学方法，识别了数百种可诱导体液和/或细胞免疫的抗原。体外和体内测定需要相关标记，以评估疫苗效率，优化未来疫苗制剂和能够评估免疫原性的免疫生物标志物。

迄今为止的体外方法主要由 ELISA 和其变体组成，确定血清中存在的抗体的量。设定速度的平台是 MATS（脑膜炎球菌抗原分型系统），它是一种夹心 ELISA 系统，能够检测菌株表达的抗原量和疫苗中的抗原交叉反应，并将数据与来自许多国家大量菌株数据相关联[5]。ELISPOT 分析可以定量测量趋化因子和细胞因子的表达。也可以使用流式细胞仪或抑菌试验分析产生的抗血清[5]。许多常规技术仍在使用，例如用于检测抗体的凝集和沉淀测试、补体固定测试、抗毒素的中和测试，以及用于显现抗原-AB 复合物的荧光抗体测试。然而，这些技术并没有关于提供保护性免疫的复杂细胞环境的内容。对于体内评估，一种方法是在免疫小鼠之前汇集抗原以评估其免疫原性[2]。离体免疫系统模型可以提供昂贵的临床试验的替代方案，迫切需要免疫原性和疫苗应答检测的广泛的研究[32]。

42.5.1 免疫生物标志物

生物标志物是来自生物体的可检测产物，其可用作生物状态、病理过程或对治疗干预的药理学反应的指示物。它们有多种应用，包括改善现有药物和新药开发的药物预测和评估[127]。尽管许多疫苗可通过多种免疫反应机制发挥保护作用[128]，但免疫生物标志物有可能将特异性免疫应答与疫苗效率相关联[129]。

在癌症患者中使用生物标志物已经表明，多种标志物可用于指示疾病进展，选择正确的治疗[130]。然而，表示疾病状态和治疗选择的免疫生物标志物仍然有限[127]。例如，已知细胞因子干扰素-γ（IFN-γ）参与抗各种病原体的保护性免疫应答，包括结核分枝杆菌[131]和反刍兽艾利希体[132]。可以在疫苗接种之前和之后测试该生物标记物水平，以确定当前的疫苗抗原是否有治疗前景。关于宿主免疫系统和病原体相互作用的知识对于免疫生物标记物和有效疫苗的开发至关重要[128]。

42.5.2 蛋白质表达、佐剂和疫苗试验

为了评估亚单位疫苗的保护能力，需要重组表达和纯化重组蛋白。该过程有很多困难，其中最应关注的是宿主细胞内毒素的存在，为此已经开发了许多方法[71]。用于初步分析的一个重要方法是使用体外翻译系统（对于大规模疫苗生产来说很昂贵）。在 Cardoso 及其同事的一项研究中，他们使用体外转录和翻译系统表达了几乎整个疟原虫蛋白质组。然后使用 IFN-γELISpot 和流式微珠阵列细胞因子测定法分析这些蛋白质诱导 T 细胞免疫的能力，为高通量方法铺平道路，识别复杂病原体的 T 细胞靶点[133]。

灭活疫苗或亚单位疫苗不能有效激活先天免疫系统，在某些情况下需要增加应答，并克服免疫佐剂有限的困难，以改善其免疫原性反应[137]。最常用的 3 种佐剂是常用的佐剂（不溶性脂肪酸）[43,134]。首先，贮存型佐剂使用铝盐。目前它的作用方式涉及通过增加抗原摄取防止其降解，延长注射抗原的活性。弗氏不完全佐剂通过抗原呈递细胞，然后去稳定化，使抗原更容易加工，并且以 MontanideISA 50V2（矿物油的辅助剂）形式存在。此外，明矾可以激活 NLRP3。其次，颗粒佐剂，例如皂苷，有助于递送抗原促炎细胞因子。然而，强大的抗原呈递细胞能触发完整机制，并且诱导增强的免疫反应和免疫应答。

最后，免疫刺激佐剂通过共同刺激宿主免疫系统和抗原，增加细胞因子的产生，从而促进 T 辅助细胞的反应。例如，这些化合物含有复杂的微生物产物，树突状细胞和巨噬细胞很容易识别这些产物。这种佐剂的一个例子是壁酰二肽。当微粒或储存佐剂与免疫刺激剂结合时，也可

以产生非常有效的佐剂。这种混合物的一个例子是弗氏完全佐剂[135]。关于不同类型的佐剂及其在当前疫苗中的应用，已经发表了两篇有见地的评论[136,137]。

在疫苗配方中添加佐剂可能具有以下优点：产生足够的免疫应答以实现保护；延长免疫应答的持续时间，从而降低免疫次数；降低抗原剂量的要求；在注射多种化合物时防止抗原竞争；提高抗原的稳定性和反应的广度，并在某些情况下克服有限的免疫反应[137]。最常用的佐剂是明矾（不溶性铝盐）。其作用方式目前被认为是增加抗原呈递细胞对抗原的摄取，破坏抗原的稳定性，从而使处理和递呈更容易。此外，明矾可以激活 NLRP3 炎症小体，从而允许促炎细胞因子的分泌。然而，对由此产生的 TH2 反应的完整机制和诱导还不完全了解[137]。基于 web 的 Vaxjo 是一个数据库和分析系统，允许对疫苗佐剂及其在疫苗开发中的应用进行管理、存储和分析[136]。

疫苗试验必须事先计划，以提供反馈，并允许进一步假设的发展。这已经是药物开发试验的原则，但由于缺乏生物标记物，在疫苗试验中却缺失了很大一部分。含有鼠基因人类变体的基因敲除小鼠的可用性为研究疫苗反应和将全基因组关联研究与功能研究联系起来开辟了令人兴奋的新途径[50]。

综上所述，从对减毒活病原体进行反复试验到研制第一亚单位疫苗，疫苗的研制已经走过了漫长的道路。在后基因组时代，大量"组学"数据的可用性，加上整合这些数据的能力，使研究人员能够迅速向多学科、令人兴奋但又具有挑战性的疫苗设计未来迈进。本章提到的软件和数据库的词汇见表 42.2。

表 42.2　本章提及的软件和数据库词汇

名称	描述	参考文献
BLAST	Similarity searches for various applications	[16]
ORF identification		
Glimmer	Bacteria，archaea，viruses	[8]
GeneMark. hmm	Bacteria	[9]
Prodigal	Prokaryotes	[10]
AUGUSTUS	Eukaryotes	[14]
GNOMON	Eukaryotes	http：//0 - www. ncbi. nlm. nih.gov. innopac. up. ac. za/genome/guide/gnomon.shtml
MAKER	Eukaryotes，annotation pipeline	[13]
Functional annotation		
PIR	Protein information resource	[17]
Pfam	Protein families database	[18]
SWISS-PROT	Manually annotated and reviewed section of UniProt knowledge base	[19]
UniProt	Knowledge base	[20]
COG	Database	[21]
GO	Database	[22]
Subcellular localization		

（续表）

名称	描述	参考文献
PSORT	Bacteria，eukaryotes	[86]
PSORTb v3.0	Prokaryotes	[87]
iLoc-Euk	Eukaryotes	[83]
SherLoc2	Eukaryotes	[85]
Cell-PLoc	Eukaryotes，prokaryotes，viruses	[89]
Virus-mPloc	Viruses	[91]
iLoc-Virus	Viruses	[90]
Signal peptide		
SignalP 4.0	Signal peptide prediction	[97]
Phobius	Signal peptide prediction	[99]
SPOCTOPUS	Signal peptide prediction	[100]
Epitopes		
POPISK	MHC Ⅰ predictor	[110]
NetCTLpan	CTL predictor	[111]
CEP	Prediction of conformational epitopes	[112]
VaxiJen	Viral，tumour，bacterial protective antigens	[118]
Epitope conservancyDatabase of conserved epitopes [77] tool	Database of conserved epitopes	[77]
Reverse vaccinology platforms		
NERVE	Pipeline	[121]
Vaxign	Pipeline	[73]
Software packages and databases for integration of "omics" -data		
RefDIC	Reference genomics database of immune cells	[58]
IIDB	Innate immune database	[59]
WIBL	Workbench for integrative biological learning	[60]
Guided clustering	Combines experimental and clinical data	[61]
EchoBasE	Database for *E. coli*	[62]
integrOmics		[63]
bioPIXIE	Networkpredictions for *S. cerevisiae*	[57]
VESPA	Genomic annotation of prokaryotes	[64]
openBIS		[65]
GPS-Prot	Platform for integrating host - pathogen interaction data	[66]
Booly	Data integration platform	[67]

（续表）

名称	描述	参考文献
PARE	Protein abundance and mRNA expression	[68]
Other		
IRIS	Immune response in silico	[45]
Vaxjo	Curation, storage and analysis of vaccine adjuvants	[136]

参考文献

[1] Powell, M. F., Newman, M. J.: Vaccine Design-The Subunit and Adjuvant Approach, vol. 6, 1stedn. Plenum Press, New York (1995).

[2] Jones, D.: Reverse vaccinology on the cusp. Nat. Rev Drug Discov. 11, 175-176 (2012). doi: 10. 1038/nrd3679.

[3] Adu-Bobie, J., Arico, B., Giuliani, M. M.: Serruto, D., Chapter 9: The fi rst vaccine obtained through reverse vaccinology: the serogroup B Meningococcus Vaccine. In: Rappuoli, R., Serruto, D., Rappuoli, R. (eds.). Vaccine Design-Innovative and Novel Strategies, vol. 1, pp. 225-241. Caister Academic. Press, Norfolk (2011).

[4] Luciani, F., Bull, R. A., Lloyd, A. R.: Next generation deep sequencing and vaccine design: today and tomorrow. Trends Biotechnol. 30, 443-452 (2012).

[5] Seib, K. L., Zhao, X., Rappuoli, R.: Developing vac-cines in the era of genomics: a decade of reverse vaccinology. Clin. Microbiol. Infect. 18, 1-8 (2012).

[6] Cheng, H., Chan, W. S., Wang, D., Liu, S., Zhou, Y.: Small open reading frames: current prediction techniques and future prospect. Curr. Protein Pept Sci. 12, 503-507 (2011).

[7] Yandell, M., Ence, D.: A beginner's guide to eukary-otic genome annotation. Nat. Rev. 13, 329-342 (2012).

[8] Delcher, A. L., Bratke, K. A., Powers, E. C., Salzberg, S. L.: Identifying bacterial genes and endosymbiont DNA with Glimmer. Bioinformatics 23, 673-679 (2007). doi: 10. 1093/bioinformatics/btm009.

[9] Lukashin, A. V., Borodovsky, M.: GeneMark. hmm: new solutions for gene fi nding. Nucleic Acids Res. 26, 1107-1115 (1998). doi: 10. 1093/nar/26. 4. 1107.

[10] Hyatt, D., et al.: Prodigal: prokaryotic gene recogni-tion and translation initiation site identifi cation. BMC Bioinformatics 11, 119 (2010).

[11] Pinheiro, C. S., et al.: Computational vaccinology: an important strategy to discover new potential S. mansoni vaccine candidates. J. Biomed. Biotechnol. 2011, 503068 (2011). doi: 10. 1155/2011/503068.

[12] Ecker, J. R., et al.: Genomics: ENCODE explained. Nature 489, 52-55 (2012). doi: 10. 1038/489052a.

[13] Cantarel, B. L., et al.: MAKER: an easy-to-use anno-tation pipeline designed for emerging model organism genomes. Genome Res. 18, 188-196 (2008). doi: 10. 1101/gr. 6743907.

[14] Keller, O., Kollmar, M., Stanke, M., Waack, S.: A novel hybrid gene prediction method employing proteinmultiple sequence alignments. Bioinformatics (2011). doi: 10. 1093/bioinformatics/btr010.

[15] Smandi, S., et al.: Methodology optimizing SAGE library tag-to-gene mapping: application to Leishmania. BMC Res. Notes 5, 74 (2012). doi: 10. 1186/1756-0500-5-74.

[16] Altschul, S. F., Gish, W., Miller, W., Myers, E. W., Lipman, D. J.: Basic local alignment search tool. J. Mol. Biol. 215, 403-410 (1990).

[17] Barker, W. C., et al.: The protein information resource (PIR). Nucleic Acids Res. 28, 41-44 (2000).

[18] Punta, M., et al.: The Pfam protein families database. Nucleic Acids Res. 40, D290-D301 (2012). doi: 10. 1093/nar/gkr1065.

[19] Boeckmann, B., et al.: The SWISS-PROT protein knowledgebase and its supplement TrEMBL in. Nucleic Acids Res. 31, 365-370 (2003). doi: 10. 1093/ nar/gkg095 (2003).

[20] Magrane, M., Consortium, U.: UniProt Knowledgebase: a hub of integrated protein data. Database (2011). doi: 1 0. 1093/database/bar009 (2011).

[21] Tatusov, R. L., Galperin, M. Y., Natale, D. A., Koonin, E. V.: The COG database: a tool for genome-scale analysis of protein functions and evolution. Nucleic Acids Res. 28, 33-36 (2000). doi: 10. 1093/nar/28. 1. 33.

[22] Gene Ontology Consortium: The Gene Ontology (GO) database and informatics resource. Nucleic Acids Res. 32, D258-D261 (2004). doi: 10. 1093/nar/ gkh036.

[23] Yon Rhee, S., Wood, V., Dolinski, K., Draghici, S.: Use and misuse of the gene ontology annotations. Nat. Rev. Genet. 9, 509-515 (2008).

[24] Gomez, A., et al.: Gene ontology function predict ion in mollicutes using protein-protein association networks. BMC Syst. Biol. 5, 49 (2011). doi: 10. 1186/1752-0509-5-49.

[25] Zhang, Q. C., et al.: Structure-based prediction of pro-tein-protein interactions on a genome-wide scale. Nature 490, 556-560 (2012).

[26] Wass, M. N., Barton, G., Sternberg, M. J. E.: CombFunc: predicting protein function using heterogeneous data sources. Nucleic Acids Res. 40, W466-W470 (2012). doi: 10. 1093/nar/gks489.

[27] Fu, Y., et al.: Novel insights into the transcriptome of Dirofi laria immitis. PLoS One 7, e41639 (2012). doi: 10. 1371/journal. pone. 0041639.

[28] Hedman, A. K., Li, M. S., Langford, P. R., Kroll, J. S.: Transcriptional profi ling of serogroup B Neisseria meningitidis growing in human blood: an approach to vaccine antigen discovery. PLoS One 7, e39718 (2012). doi: 10. 1371/journal. pone. 0039718.

[29] Amiruddin, N., et al.: Characterisation of full-length cDNA sequences provides insights into the Eimeria tenella transcriptome. BMC Genomics 13, 21 (2012). doi: 10. 1186/1471-2164-13-21.

[30] Maritz - Olivier, C., van Zyl, W., Stutzer, C.: A system - atic, functional genomics and reverse vaccinology approach to the identifi cation of vaccine candidates in the cattle tick. Rhipicephalus microplus. Ticks Tick Borne Dis. 3, 179-189 (2012).

[31] Savas, J. N., Stein, B. D., Wu, C. C., Yates, J. R.: Mass spectrometry accelerates membrane protein analysis. Trends Biochem. Sci. 36, 388-396 (2011). doi: 10. 1016/j. tibs. 2011. 04. 005.

[32] Haralambieva, I. H., Poland, G. A.: Vaccinomics, pre-dictive vaccinology and the future of vaccine development. Future Microbiol. 5, 1757-1760 (2010). doi: 10. 2217/fmb. 10. 146.

[33] Acharya, P., et al.: Clinical proteomics of the neglected human malarial parasite Plasmodium vivax. PLoS One 6, e26623 (2011). doi: 10. 1371/journal. pone. 0026623.

[34] Minning, T. A., Weatherly, D. B., Atwood, J., Orlando, R., Tarleton, R. L.: The steady-state transcriptome of the four major life-cycle stages of Trypanosoma cruzi. BMC Genomics 10, 370 (2009). doi: 10. 1186/1471-2164-10-370.

[35] Jagusztyn-Krynicka, E. K., Roszczenko, P., Grabowska, A.: Impact of proteomics on anti-Mycobacterium t uberculosis (MTB) vaccine development. Pol. J. Microbiol. 58, 281-287 (2009).

[36] Lawrence, E.: Henderson's Dictionary of Biology. Pearson Education Limited, Harlow (2005).

[37] Lillehoj, H. S., Kim, C. H., Keeler, C. L., Zhang, S.: Immunogenomic approaches to study host immunity to enteric pathogens. Poult. Sci. 86, 1491-1500 (2007).

[38] Ohara, O.: From transcriptome analysis to immu-nogenomics: current status and future direction. FEBS Lett. 583, 1662-1667 (2009).

[39] Snoep, J. L., Bruggeman, F., Olivier, B. G., Westerhoff, H. V.: Towards building the silicon cell: a modular approach. Biosystems 83, 207-216 (2006).

[40] Tjalsma, H., Schaeps, R. M. J., Swinkels, D. W.: Immunoproteomics: from biomarker discovery to diagnostic applications. Proteomics Clin. Appl. 2, 167-180 (2008).

[41] Poland, G. A., Ovsyannikova, I. G., Jacobson, R. M., Smith, D. I.: Heterogeneity in vaccine immune response: the role of immunogenetics and the emerging fi eld of vaccinomics. Clin. Pharmacol. Ther. 82, 653-664 (2007). doi: 10. 1038/sj. clpt. 6100415.

[42] Rappuoli, R.: Reverse vaccinology. Curr. Opin. Microbiol. 3, 445-450 (2000).

[43] Buonaguro, L., Wang, E., Tornesello, M. L., Buonaguro, F. M., Marincola, F. M.: Systems biology applied to vaccine and immunotherapy development. BMC Syst. Biol. 5, 146-157 (2011).

[44] Heng, T. S., Painter, M. W.: The Immunological Genome Project: networks of gene expression in immune cells. Nat. Immunol. 9, 1091-1094 (2008). doi: 10. 1038/ni1008-1091.

[45] Abbas, A. R., et al.: Immune response in silico (IRIS): immune-specifi c genes identifi ed from a compendium of microarray expression data. Genes Immun. 6, 319 - 331 (2005). doi: 10. 1038/sj. gene. 6364173.

[46] Banchereau, J., et al.: Harnessing human dendritic cell subsets to design novel vaccines. Ann. N Y Acad. Sci. 1174, 24-32 (2009). doi: 10. 1111/j. 1749-6632. 2009. 04999. x.

[47] Ovsyannikova, I. G., Poland, G. A.: Vaccinomics: cur-rent fi ndings, challenges and novel approaches for vaccine development. AAPS J. 13, 438-444 (2011). doi: 10. 1208/s12248-011-9281-x.

[48] Poland, G. A., Kennedy, R. B., Ovsyannikova, I. G.: Vaccinomics and personalized vaccinology: is science leading us toward a new path of directed vaccine development and discovery? PLoS Pathog. 7, e1002344 (2011). doi: 10. 1371/journal. ppat. 1002344.

[49] Bernstein, A., Pulendran, B., Rappuoli, R.: Systems vaccinomics: the road ahead for vaccinology. OMICS 15, 529-531 (2011). doi: 10. 1089/omi. 2011. 0022.

[50] Kennedy, R. B., Poland, G. A.: The top fi ve "game changers" in vaccinology: toward rational and directed vaccine development. OMICS 15, 533-537 (2011). doi: 10. 1089/omi. 2011. 0012.

[51] Zhang, W., Li, F., Nie, L.: Integrating multiple 'omics' analysis for microbial biology: application and methodologies. Microbiology 156, 287-301 (2010). doi: 10. 1099/mic. 0. 034793-0.

[52] Yang, D., et al.: RNA-seq liver transcriptome analysis reveals an activated MHC Ⅰ pathway and an inhibited MHC Ⅱ pathway at the early stage of vaccine immunization in zebrafi sh. BMC Genomics 13, 319 (2012). doi: 10. 1186/1471-2164-13-319.

[53] Toufeer, M., et al.: Gene expression profi ling of dendritic cells reveals important mechanisms associated with predisposition to Staphylococcus infections. PLoS One 6, e22147 (2011). doi: 10. 1371/journal. pone. 0022147.

[54] Wang, F., et al.: Deep-sequencing analysis of the mouse transcriptome response to infection with Brucella melitensis strains of differing virulence. PLoS One 6, e28485 (2011). doi: 10. 1371/journal. pone. 0028485.

[55] Kaleta, C., de Figueiredo, L. F., Heiland, I., Klamt, S., Schuster, S.: Special issue: integration of OMICs datasets into metabolic pathway analysis. Biosystems 105, 107-108 (2011). doi: 10. 1016/j. biosystems. 2011. 05. 008.

[56] Joyce, A. R., Palsson, B. O.: The model organism as a system: integrating 'omics' data sets. Nat. Rev. Mol. Cell Biol. 7, 198-210 (2006). doi: 1 0. 1038/ nrm1857.

[57] Myers, C. L., Chiriac, C., Troyanskaya, O. G.: Discovering biological networks from diverse functional genomic data. Methods Mol. Biol. 563, 157-175 (2009). doi: 10. 1007/978-1-60761-175-29.

[58] Hijikata, A., et al.: Construction of an open-access database that integrates cross-reference information from the transcriptome and proteome of immune cells. Bioinformatics 23, 2934-2941 (2007). doi: 10. 1093/bioinformatics/btm430.

[59] Korb, M., et al.: The Innate Immune Database (IIDB). BMC Immunol. 9, 7 (2008). doi: 1 0. 1186/ 1471-2172-9-7.

[60] Lesk, V., Taubert, J., Rawlings, C., Dunbar, S., Muggleton, S.: WIBL: Workbench for Integrative Biological Learning. JIB 8, 156 (2011). doi: 10. 2390/ biecoll-jib-2011-156.

[61] Maneck, M., Schrader, A., Kube, D., Spang, R.: Genomic data integration using guided clustering. Bioinformatics 27, 2231-2238 (2011). doi: 10. 1093/ bioinformatics/btr363.

[62] Misra, R. V., Horler, R. S. P., Reindl, W., Goryanin, I. I., Thomas, H. G.: Echo BASE: an integrated post-genomic database for Escherichia coli. Nucleic Acids Res. (2005). doi: 10. 1093/nar/gki028.

[63] Le Cao, K. A., Gonzalez, I., Dejean, S.: IntegrOmics: an R package to unravel relationships between two omics datasets. Bioinformatics 25, 2855-2856 (2009). doi: 10. 1093/bioinformatics/btp515.

[64] Peterson, E. S., et al.: VESPA: software to facilitate genomic annotation of prokaryotic organisms through integration of proteomic and transcriptomic data. BMC Genomics 13, 131 (2012). doi: 10. 1186/1471-2164-13-131.

[65] Bauch, A., et al.: OpenBIS: a fl exible framework for managing and analyzing complex data in biology research. BMC Bioinformatics 12, 468 (2011). doi: 10. 1186/1471-2105-12-468.

[66] Fahey, M. E., et al.: GPS-Prot: a web-based visualiza-tion platform for integrating host-pathogen interaction data. BMC Bioinformatics 12, 298 (2011). doi: 10. 1186/1471-2105-12-298.

[67] Do, L. H., Esteves, F. F., Karten, H. J., Bier, E.: Booly: a new data integration platform. BMC Bioinformatics 11, 513 (2010). doi: 10. 1186/1471-2105-11-513.

[68] Yu, E. Z., Burba, A. E. C., Gerstein, M.: PARE: a tool for comparing protein abundance and mRNA expression data. BMC Bioinformatics 8, 309 (2007). doi: 10. 1186/1471-2105-8-309.

[69] Dormitzer, P. R., Ulmer, J. B., Rappuoli, R.: Structure-based antigen design: a strategy for next generation vaccines. Trends Biotechnol. 26, 659-667 (2008).

[70] Nuccitelli, A., et al.: Structure-based approach to rationally design a chimeric protein for an effective vaccine against Group B Streptococcus infections. Proc. Natl. Acad. Sci. 108, 10278-10283 (2011). doi: 10. 1073/pnas. 1106590108.

[71] Bagnoli, F., et al.: Designing the next generation of vaccines for global public health. OMICS 15, 545-566 (2011).

[72] Fujinami, R. S., Oldstone, M. B., Wroblewska, Z., Frankel, M. E., Koprowski, H.: Molecular mimicry in virus infection: crossreaction of measles virus phosphoprotein or of herpes simplex virus protein with human intermediate fi laments. Proc. Natl. Acad. Sci. 80, 2346-2350 (1983).

[73] He, Y., Xiang, Z., Mobley, H.: Vaxign: the fi rst web-based vaccine design program for reverse vaccinology and applications for vaccine development. J. Biomed. Biotechnol. (2010). doi: 10. 1155/2010/297505 (2010).

[74] Oldstone, M. B. A.: Molecular mimicry and immune-mediated diseases. FASEB J. 12, 1255-1265 (1998).

[75] Schatz, M. M., et al.: Characterizing the N-terminal processing motif of MHC class I ligands. J. Immunol. 180, 3210-3217 (2008).

[76] Moriel, D. G., et al.: Identifi cation of protective and broadly conserved vaccine antigens from the genome of extraintestinal pathogenic Escherichia coli. Proc. Natl. Acad. Sci. 107, 9072-9077 (2010). doi: 10. 1073/pnas. 0915077107.

[77] Bui, H. H., Li, W., Fusseder, N., Sette, A.: Development of an epitope conservancy analysis tool to facilitate the design of epitope-based diagnostics and vaccines. BMC Bioinformatics 8, 361 (2007). doi: 10. 1186/1471-2105-5-361.

[78] Bagnoli F., Norauis, N., Ferlenghi, I., Scarselli, M., Danati, C., Savina, S., Barocchi, M. A., Rappuoli, R. Chapter 2: deigning Vaccine in the Era of Genomics. In: Rappuoli, R., Serruto, D., Rappuoli, R. (eds.). Vaccine Design-Innovative and Novel Strategies, vol. 1, pp. 21-53. Caister Academic. Press, Norfolk (2011).

[79] Sollner, J., et al.: Concept and application of a compu-tational vaccinology workfl ow. Immunome Res. 6 (Suppl 2), S7 (2010). doi: 10. 1186/1745-75806-s2-s7.

[80] Pizza, M., et al.: Identifi cation of vaccine candidates against serogroup B. m eningococcus by whole-genome sequencing. Science 287, 1816-1820 (2000). doi: 10. 1126/science. 287. 5459. 1816.

[81] Bowman, B., et al.: Improving reverse vaccinology with a machine learning approach. Vaccine 29, 8156-8164 (2011).

[82] Vivona, S., et al.: Computer-aided biotechnology: from immuno-informatics to reverse vaccinology. Trends

Biotechnol. 26, 190–200 (2008). doi: 10. 1016/ j. tibtech. 2007. 12. 006.

[83] Chou, K. C., Wu, Z. C., Xiao, X.: iLoc–Euk: a multi–label classifi er for predicting the subcellular localization of singleplex and multiplex eukaryotic proteins. PLoS One 6, e18258 (2011). doi: 10. 1371/journal. pone. 0018258.

[84] Garg, G., Ranganathan, S.: In silico secretome analysis approach for next generation sequencing transcriptomic data. BMC Genomics 12, 514–524 (2011).

[85] Briesemeister, S., et al.: SherLoc2: a high–accuracy hybrid method for predicting subcellular localization ofproteins. J. Proteome Res. 8, 5363–5366 (2009). doi: 10. 1021/pr900665y.

[86] Nakai, K., Horton, P.: PSORT: a program for detect–ing sorting signals in proteins and predicting their subcellular localization. Trends Biochem. Sci. 24, 34–35 (1999).

[87] Yu, N. Y., et al.: PSORTb 3. 0: improved protein sub–cellular localization prediction with refi ned localization subcategories and predictive capabilities for all prokaryotes. Bioinformatics 26, 1608–1615 (2010). doi: 10. 1093/bioinformatics/btq249.

[88] Sprenger, J., Fink, J., Teasdale, R.: Evaluation and comparison of mammalian subcellular localization prediction methods. BMC Bioinformatics 7, S3 (2006). doi: 10. 1186/1471–2105–7–S5–S3.

[89] Chou, K. –C., Shen, H. –B.: Cell–PLoc: a package of web servers for predicting subcellular localization of proteins in various organisms. Nat. Protoc. 3, 153–162 (2008).

[90] Xiao, X., Wu, Z. –C., Chou, K. –C.: iLoc–Virus: a multi–label learning classifi er for identifying the subcellular localization of virus proteins with both single and multiple sites. J. Theor. Biol. 284, 42–51 (2011).

[91] Shen, H. –B., Chou, K. –C.: Virus–mPLoc: a fusion classifi er for viral protein subcellular location prediction by incorporating multiple sites. J. Biomol. Struct. Dyn. 28, 175–186 (2010).

[92] Bannai, H., Tamada, Y., Maruyama, O., Nakai, K., Miyano, S.: Extensive feature detection of N–terminal protein sorting signals. Bioinformatics 18, 298 – 305 (2002). doi: 10. 1093/bioinformatics/18. 2. 298.

[93] Lodish, H., Berk, A., Zipursky, S. L.: Molecular Cell Biology, 4th edn. W. H. Freeman, New York (2000).

[94] Nielsen, H., Engelbrecht, J., Brunak, S., von Heijne, G.: Identifi cation of prokaryotic and eukaryotic signal peptides and prediction of their cleavage sites. Protein Eng. 10, 1 (1997). doi: 10. 1093/protein/10. 1. 1.

[95] Choo, K. H., Tan, T. W., Ranganathan, S.: A comprehensive assessment of N–terminal signal peptides prediction methods. BMC Bioinformatics 10, S3 (2009). doi: 10. 1186/1471–2105–10–S15–S2.

[96] Scott, M. S., Oomen, R., Thomas, D. Y., Hallett, M. T.: Predicting the subcellular localization of viral proteins within a mammalian host cell. J. Virol. 3, 24 (2006). doi: 10. 1186/1743–422X–3–24.

[97] Petersen, T. N., Brunak, S., Von Heijne, G., Nielsen, H.: SignalP 4. 0: discriminating signal peptides from transmembrane regions. Nat. Methods 8, 784–786 (2011). doi: 10. 1038/nmeth. 1701.

[98] Käll, L., Krogh, A., Sonnhammer, E. L. L.: A combined transmembrane topology and signal peptide prediction method. J. Mol. Biol. 338, 1027–1036 (2004).

[99] Käll, L., Krogh, A., Sonnhammer, E.: Advantages of combined transmembrane topology and signal peptide prediction–the Phobius web server. Nucleic Acids Res. 35, W429–W432 (2007). doi: 10. 1093/nar/gkm256.

[100] Viklund, H. K., Bernsel, A., Skwark, M., Elofsson, A.: SPOCTOPUS: a combined predictor of signal peptides and membrane protein topology. Bioinformatics 24, 2928–2929 (2008). doi: 10. 1093/bioinformatics/btn550.

[101] Chen, P., Rayner, S., Hu, K. H.: Advances of bioinformatics tools applied in virus epitopes prediction. Virol. Sin. 26, 1–7 (2011). doi: 1 0. 1007/s12250–011–3159–4.

[102] Flower, D. R. Chapter 5: Vaccines: data driven predic–tion of binders, epitopes and immunogenicity. In: Flower, D. R. (ed.) Bioinformatics for Vaccinology, pages 167 – 216. Wiley – Blackwell, Chichester (2008).

[103] Iurescia, S., Fioretti, D., Fazio, V. M., Rinaldi, M.: Epitope–driven DNA vaccine design employing immunoinformatics against B – cell lymphoma: a biotech's challenge. Biotechnol. Adv. 30, 372 – 383

(2012).

[104] Sirskyj, D., Diaz-Mitoma, F., Golshani, A., Kumar, A., Azizi, A.: Innovative bioinformatic approaches for developing peptide-based vaccines against hypervariable viruses. Immunol. Cell Biol. 89, 81-89 (2011). doi: 10. 1038/icb. 2010. 65.

[105] Yang, X., Yu, X.: An introduction to epitope predic-tion methods and software. Rev. Med. Virol. 19, 77-96 (2009).

[106] Yu, K., Petrovsky, N., Schönbach, C., Koh, J., Brusic, V.: Methods for prediction of peptide binding to MHC molecules: a comparative study. Mol. Med. 8, 137-148 (2002).

[107] Davydov, Y. I., Tonevitsky, A. G.: Prediction of linear B-cell epitopes. Mol. Biol. 43, 150-158 (2009).

[108] Zhang, Q., et al.: Immune epitope database analysis resource (IEDB-AR). Nucleic Acids Res. 36, W513-W518 (2008). doi: 10. 1093/nar/gkn254.

[109] Van Bergen, J., et al.: Get into the groove! Targeting antigens to MHC class II. Immunol. Rev. 172, 87-96 (1999).

[110] Tung, C. W., Ziehm, M., Kamper, A., Kohlbacher, O., Ho, S. Y.: POPISK: T-cell reactivity prediction using support vector machines and string kernels. BMC Bioinformatics 12, 446 (2011). doi: 10. 1186/1471-2105-12-446.

[111] Stranzl, T., Larsen, M., Lundegaard, C., Nielsen, M.: NetCTLpan: pan-specifi c MHC class I pathway epitope predictions. Immunogenetics 62, 357-368 (2010). doi: 10. 1007/s00251-010-0441-4.

[112] Kulkarni-Kale, U., Bhosle, S., Kolaskar, A. S.: CEP: a conformational epitope prediction server. Nucleic Acids Res. 33, W168-W171 (2005). doi: 10. 1093/ nar/gki460.

[113] Rubinstein, N. D., Mayrose, I., Pupko, T.: A machine-learning approach for predicting B-cell epitopes. Mol. Immunol. 46, 840-847 (2009).

[114] Liang, S., Zheng, D., Zhang, C., Zacharias, M.: Prediction of antigenic epitopes on protein surfaces by consensus scoring. BMC Bioinformatics 10, 302 (2009). doi: 10. 1186/1471-2105-10-302.

[115] Liu, R., Hu, J.: Prediction of discontinuous B-cell epitopes using logistic regression and structural information. J. Proteomics Bioinformatics 4, 010-015 (2011).

[116] Zhang, W., et al.: Prediction of conformational B-cell epitopes from 3D structures by random forests with a distance-based feature. BMC Bioinformatics 12, 341 (2011). doi: 10. 1186/1471-2105-12-341.

[117] Ryvkin, A., et al.: Deep panning: steps towards probing the IgOme. PLoS One 7, e41469 (2012). doi: 10. 1371/journal. pone. 0041469.

[118] Doytchinova, I. A., Flower, D. R.: Identifying candidate subunit vaccines using an alignment-independent method based on principal amino acid properties. Vaccine 25, 856-866 (2007).

[119] Magnan, C. N., et al.: High-throughput prediction of protein antigenicity using protein microarray data. Bioinformatics 26, 2936-2943 (2010). doi: 10. 1093/ bioinformatics/btq551.

[120] Grandi, G.: Genomics and proteomics in reverse vac-cines. Methods Biochem. Anal. 49, 379-393 (2006).

[121] Vivona, S., Bernante, F., Filippini, F.: NERVE: new enhanced reverse vaccinology environment. BMC Biotechnol. 6, 35 (2006). doi: 1 0. 1186/1472-6750-6-35.

[122] Gardy, J. L., et al.: PSORT-B: improving protein subcellular localization prediction for gram-negative bacteria. Nucleic Acids Res. 31, 3613-3617 (2003). doi: 10. 1093/nar/gkg602.

[123] Sachdeva, G., Kumar, K., Jain, P., Ramachandran, S.: SPAAN: a software program for prediction of adhesins and adhesin-like proteins using neural networks. Bioinformatics 21, 483-491 (2005). doi: 10. 1093/bioinformatics/bti028.

[124] Tusnády, G. E., Simon, I.: Principles governing amino acid composition of integral membrane proteins: application to topology prediction. J. Mol. Biol. 283, 489-506 (1998).

[125] Doytchinova, I. A., Flower, D. R.: VaxiJen: a server for prediction of protective antigens, tumour antigens and subunit vaccines. BMC Bioinformatics 8, 4 (2007). doi: 10. 1186/1471-2105-8-4.

[126] Flower, D. R., Macdonald, I. K., Ramakrishnan, K., Davies, M. N., Doytchinova, I. A.: Computer

aided selec- tion of candidate vaccine antigens. Immunome Res. 6 (Suppl 2), S1 (2010). doi: 10. 1186/1745-7580-6-s2-s1.

[127] Plotkin, S. A.: Correlates of protection induced by vaccination. Clin. Vaccine Immunol. 17, 1055-1065 (2010).

[128] Thakur, A., Pedersen, L. E., Jungersen, G.: Immune markers and correlates of protection for vaccine induced immune responses. Vaccine 30, 4907-4920 (2012). doi: 10. 1016/j. vaccine. 2012. 05. 049.

[129] Whelan, M., Ball, G., Beattie, C., Dalgleish, A.: Biomarkers for development of cancer vaccines. Future Med. 3, 79-88 (2006).

[130] Mou, Z., He, Y., Wu, Y.: Immunoproteomics to iden-tify tumor-associated antigenseliciting humoral response. Cancer Lett. 278, 123-129 (2009). doi: 10. 1016/j. canlet. 2008. 09. 009.

[131] Walzl, G., Ronacher, K., Hanekom, W., Scriba, T. J., Zumla, A.: Immunological biomarkers of tuberculosis. Nat. Rev. Immunol. 11, 343-354 (2011). doi: 10. 1038/nri2960.

[132] Liebenberg, J., et al.: Identifi cation of Ehrlichia ruminantium proteins that activatecellular immune responses using a reverse vaccinology strategy. Vet. Immunol. Immunopathol. 145, 340-349 (2012). doi: 10. 1016/j. vetimm. 2011. 12. 003.

[133] Cardoso, F. C., Roddick, J. S., Groves, P., Doolan, D. L.: Evaluation of approaches to identify the targets of cellular immunity on a proteome-wide scale. PLoS One6, e27666 (2011). doi: 10. 1371/journal. pone. 0027666.

[134] Coffman, R. L., Sher, A., Seder, R. A.: Vaccine adju-vants: putting innate immunity to work. Immunity 33, 492-503 (2010).

[135] Tizard, I. A.: Veterinary Immunology-An Intro-duction, 8th edn. Philadelphia, Saunders (2008).

[136] Sayers, S., Ulysse, G., Xiang, Z., He, Y.: Vaxjo: a web-based vaccine adjuvant database and its application for analysis of vaccine adjuvants and their uses in vaccine development. J. Biomed. Biotechnol. 2012, 13 (2012). doi: 10. 1155/2012/831486.

[137] Skibinski, D. A. G., O'Hagan, D. T., Chapter 6: Adjuvants. In: Rappuoli, R., Serruto, D., Rappuoli, R. (eds.). Vaccine Design-Innovative and Novel Strategies, vol. 1, pp. 139-169. Caister Academic Press, Norfolk (2011).

(吴国华译)

第43章 使用微针进行疫苗递送

Ryan F. Donnelly, Sharifa Al-Zahrani,
Marija Zaric, Cian M. McCrudden,
Cristopher J. Scott 和 Adrien Kissenpfenning①

摘要

突破皮肤的角质层屏障将提高疫苗、基因载体、抗体、光敏剂甚至纳米颗粒的给药可能性，这些方法至少都需要先作用于皮肤细胞群。另外，皮内疫苗递送具有改善患者治疗效果的巨大潜力，特别对于发展中国家患者。多种微针疫苗递送方法已得到应用，在这里我们依次讨论每种方法，并且还描述了皮肤免疫生物学对微针介导的皮内疫苗接种的重要性。

43.1 疫苗接种

疫苗接种在控制传染病发病率方面的重要性不容小觑，据世界卫生组织报告，每年可挽救250多万儿童的生命。疫苗接种的目的是建立个体对特定疾病的免疫力。传统意义上疫苗为下面4种类型之一：灭活微生物疫苗、致弱微生物活疫苗、蛋白质亚单位疫苗和灭活毒性成分（类毒素）疫苗。正在开发许多创新性疫苗，例如重组载体和DNA疫苗。这些疫苗制剂类似于诱发疾病的微生物，刺激身体的免疫系统将该制剂识别为外来药物，将其破坏并“记住”它，以便免疫系统在之后的相遇中更容易处理这些微生物。

适当的疫苗接种途径是保证成功接种疫苗的关键因素。虽然肌内（M）和皮下（SC）途径用于大多数疫苗接种，但该方法并不是没有问题。这些方法需要训练有素的人员进行给药，并伴有疼痛和痛苦，这可能会导致患者依从性降低。此外，在发展中国家，皮下注射与患者中被针刺伤或重复使用受污染的针头而导致的交叉污染的高风险相关。在需要大规模的疫苗接种时，也可能出现生产和/或供应问题[1,2]。

大多数疫苗将被输送到皮下脂肪或肌肉中，很少有将疫苗输送到真皮[3]，局部或经皮应用于皮肤表面[4,5]，又称为表皮应用[5]，也很少见。上述方法都有效的原因是树突细胞（DC）能够摄取、加工并将抗原递呈给淋巴器官中的T淋巴细胞。尽管皮下脂肪和肌肉组织含有相对较少的DC，但真皮和表皮中有很多不同的DC亚基。因此，皮下注射的抗原递呈会绕过皮肤的免疫细胞，使疫苗接种效果较低。由于此原因，皮肤是疫苗接种的理想位置，在皮肤接种比肌内注射剂量低很多抗原剂量的疫苗能引起更强的免疫反应[6]。在一项临床试验中观察到皮肤免疫反应，即表皮流感疫苗接种诱导了流感特异性CD8 T细胞反应，而在经典的肌内注射途径中并未观察到[7]。能够节省剂量的方法对充分供应某些疫苗来说至关重要，特别是在大流行性疾病中[8]。

---

① R. F. Donnelly, PhD (✉) · S. Al-Zahrani, BSc

C. M. McCrudden, PhD · C. J. Scott, PhD（贝尔法斯特皇后大学药学院，英国贝尔法斯特，E-mail：r. donnelly@ qub. ac. uk）

M. Zaric, MB · A. Kissenpfenning, PhD（贝尔法斯特皇后大学感染和免疫中心，英国贝尔法斯特）

43.2　皮肤结构和功能

皮肤作为人体中最大和最复杂的器官之一（图 43.1），具有很多功能[9,10]。皮肤的屏障特性使皮肤具有防止物理、微生物或化学入侵等保护作用。皮肤由 3 层组成：表皮、真皮和皮下组织。表皮由活表皮和角质层组成。活表皮由组织学上 4 个不同的层组成：生发层、棘层、颗粒层和角质层。表皮的厚度不均匀，范围从眼睑 60 μm 到手掌 800 μm[11]。表皮层血管与下面的真皮毛细血管通过物质扩散来接收营养物。

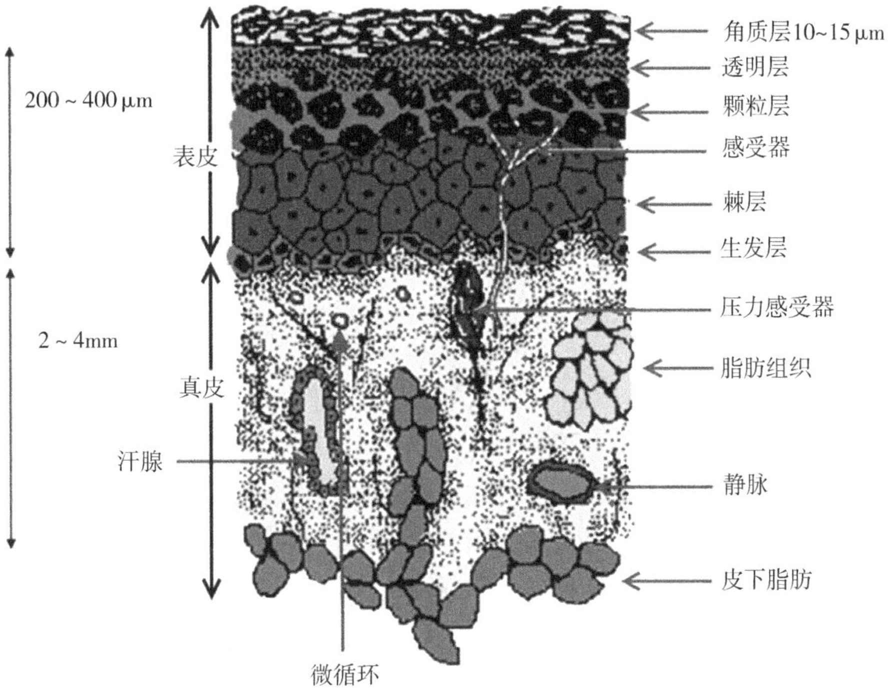

图 43.1　皮肤解剖学主要特征示意

真皮层（或真皮）位于皮下脂肪层上方，厚 3~5 mm[12]，由黏多糖基质组成，存在弹性蛋白和胶原蛋白，为皮肤提供弹性和结构[13,14]。真皮由神经末梢、淋巴管和血管网维持生理作用[15]。皮肤的血液供应为皮肤提供营养和氧气，带走废物。真皮下面是皮下脂肪层和皮下组织[13]。皮下组织主要由脂肪组织组成，作为绝缘层（因包含大量脂肪组织），也为真皮和表皮提供营养。皮下组织含有通向皮肤的主要血管和神经，也可能含有感觉器官[16]。皮下组织在真皮下方，在药物输送中不发挥主要作用[17]。

角质或称角质层，是表皮最外层，即皮肤。现在已经普遍认为角质层是大多数药物渗透的主要障碍。角质层是表皮细胞分化的最后阶段。该层的厚度通常为 10 μm，但是受许多因素影响，包括水合程度和皮肤位置。例如，手掌和脚掌角质层的平均厚度为 400~600 μm，而水合作用可以使厚度增加 4 倍。

角质层由 10~25 行死角质形成细胞组成，现在称为角质细胞，嵌入在板层体分泌的脂质中。角质细胞是扁平且细长的死细胞，没有细胞核和其他细胞器。细胞通过桥粒连接在一起，维持该层的黏结性。角质层的异质结构由干重 75%~80%蛋白质、5%~15%脂质和 5%~10%其他物质组成。

角质层中存在的大部分蛋白质是角蛋白，并且位于角质细胞内。角蛋白是 α-螺旋多肽家族。

单个分子聚集形成微丝（直径 7~10 nm，长数微米），由不溶性二硫桥稳定。这些微丝使角质细胞形成六边形的形状，并为角质层提供机械强度[12]。角质细胞的细胞周围有富含蛋白质的包膜，由前体蛋白形成，例如，外皮蛋白、兜甲蛋白和角质蛋白。转谷氨酰胺酶催化包膜蛋白之间 γ-谷氨酰交联的形成，使包膜具有抗性和高度不溶性。包膜蛋白将角质细胞与周围富含脂质的基质连接起来。

角质层中的主要脂质是神经酰胺、脂肪酸、胆固醇、胆固醇硫酸盐和甾醇/蜡酯。这些脂质在薄层的多个双层中排列（图 43.2），基本不存在磷脂，这是哺乳动物膜的独特特征。神经酰胺是角质层中最多的脂质，约占脂质总质量的一半，并且对角质层的脂质组织至关重要。

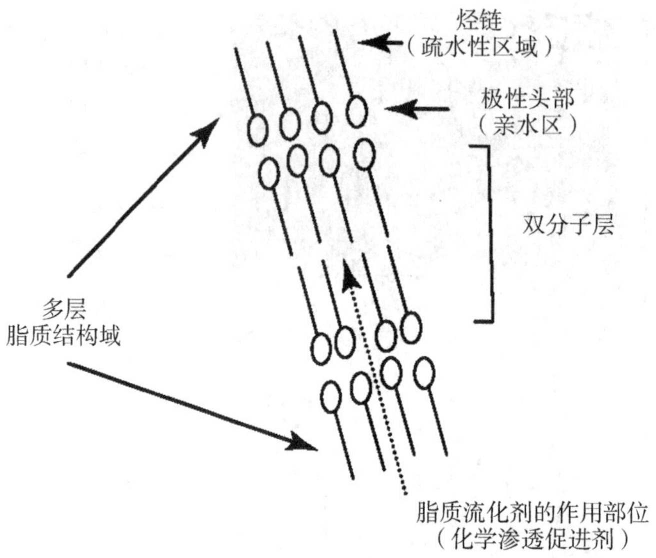

图 43.2　角质层中脂质的排列

角质层的砖和砂浆模型是描述角质层的常用模型。层砖层代表死角质化角质细胞的平行层，砂浆代表连续的间质脂质基质。要注意角质细胞实际上不是砖形的，而是细长且扁平的多边形，（厚度为 0.2~1.5 μm，直径为 34.0~46.0 μm）。“砂浆”是一种不均匀的基质。确切地说，脂质排列在层状相（水和脂质双层的交替层）中，有些脂质双层是凝胶或结晶状态。细胞外基质因内在和外在蛋白质（如酶）的存在进一步复杂化。细胞间隙中的多个脂质双层都具有角质层的屏障特性。这些双层能抑制水分流失，防止下面的组织干燥，并限制外部环境中的物质渗透。

43.3　皮肤的免疫功能

疫苗开发仍然是研究和制药领域的一个重要方向，除扩大新型疫苗的抗原谱外，改进接种途径以改善疫苗疗效仍然是一个挑战。在过去的 10 年中，人们越来越关注经过或进入皮肤的疫苗递送，主要是因为人们已逐渐认识到皮肤不同层的紧密半连续的免疫调节细胞网是疫苗接种的理想靶标。树突状细胞（DCs）、巨噬细胞和中性粒细胞是皮肤中的主要吞噬细胞（图 43.3），在正常皮肤中也存在许多获得性免疫系统的细胞，如 $CD8^+$T 细胞和 $CD4^+$T 细胞[3-7,18-44]。本书第 2 章中提供了树突状细胞的详细内容。

43.4　使用微针经皮肤进行疫苗接种

使用 MN 的皮内（ID）疫苗接种是在不使用皮下注射的情况下将抗原递送到皮肤真皮层的

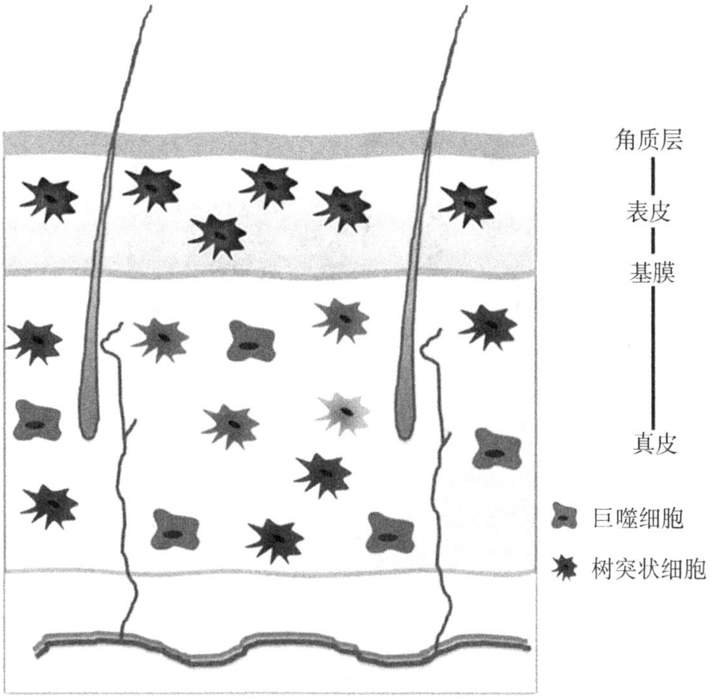

图 43.3 皮肤的免疫结构

最有吸引力的方法之一，它与发展中国家的感染传播和不当处置有关。MN 阵列（图 43.4）由多个连接到底部的高度为 25~2 000 μm 的微突出物组成[45]。当微针应用于皮肤表面时，可无痛地刺破角质层屏障，形成微米级的水性转运途径[46]。MN 产生的微孔易于运输各种小分子和大分子，例如免疫治疗剂，包括疫苗和蛋白质[47]。重要的是，MN 插入不会引起皮肤出血。

1998 年 Henry 等首先报道了 MN，从那以后 MN 一直是持续研究的课题[48]。MN 可以用各种材料制造，如金属、玻璃、硅和美国食品和药物管理局批准的聚合物[49]。Donnelly 等详细介绍了各种 MN 的开发和制造方法[50]。MN 有 4 种主要的作用方式，分别是刺入与贴剂、包被与刺入、刺入与释放、刺入与流入[51]（图 43.5）。

43.4.1 刺入与贴剂

刺入与贴剂的方法是使用固体 MN 刺入皮肤，然后将抗原应用于治疗区域，以便使抗原扩散进入皮肤，Bouwstra 小组已经开展了使用这种方法的大量研究，将微针刺入皮肤中以增加其渗透性，然后接种疫苗。Bouwstra 小组在一项报告中使用两种模型抗原：白喉类毒素（DT）和流感亚单位疫苗[52]研究了经皮肤免疫（TCI）处理后的小鼠的免疫反应。使用不锈钢 MN 阵列（16×300 μm MN）刺入小鼠皮肤，然后给予含有或不含有霍乱毒素（CT）的 DT 制剂。MN 刺入的皮肤应用 DT 引起血清 IgG 和中和毒素的抗体滴度显著高于未经处理的皮肤。CT 的存在使免疫应答增加至与皮下注射 AlPO4 吸附 DT（DT-明矾）相似的水平。与 DT 病例不同，MN 阵列预处理对单独的流感疫苗的免疫应答没有影响。CT 的加入很大程度地改善了免疫应答，与 MN 治疗无关。该研究的作者得出的结论是，在 MN 处理后，存在 CT 时，DT 的 TCI 引起的保护作用与注射 DT-明矾类似。

一项研究探讨了在应用 MN 阵列后，共同给予 DT 与各种佐剂对 TCI 小鼠免疫应答的调节作用[53]。共同给予小鼠 DT 和作为佐剂的脂多糖（LPS）、Quil-A、CpG 寡脱氧核苷酸（CpG）或 CT。MN 阵列预处理组显示出高水平的血清 IgG，共同给予佐剂后显著改善了这些水平。共同给

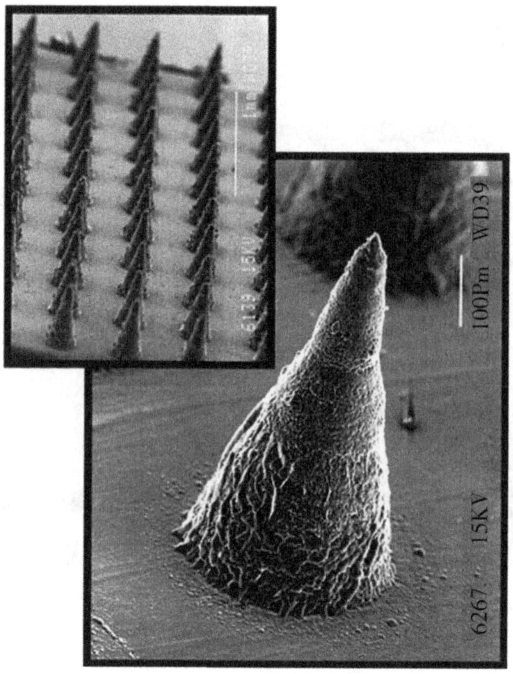

图 43.4　由结构性共聚物（甲基乙烯基醚-马来酸）的水性混合物制备的激光工程微模块化微针阵列

予 CT 与 DT 的治疗组的 IgG 水平与使用 DT-明矾皮下治疗组相似。MN 预处理后给予 N-三甲基壳聚糖和 DT 有利于增强免疫应答，但是发现携载 DT 的 N-三甲基壳聚糖纳米颗粒没有改善免疫应答[54]。

Bhowmik 等（2011）报道了透皮接种对黑素瘤发展的影响，他们使用 Dermaroller® MN 处理后，将一种新型微粒疫苗递送至皮肤[55]。将小鼠分成 4 组；第一组用 Dermaroller® 微针处理后，给予包封从 S-91 黑素瘤癌细胞获得的抗原的微粒。第二组皮下注射相同剂量的包封疫苗的微粒。第三组给予空白微粒，方式与第一组相同。最后一组皮下注射生理盐水。接种 8 周后，用活黑素瘤细胞刺激小鼠。在使用 Dermaroller® MN 和 SC 注射疫苗的治疗组中，在注射肿瘤 35 天后，没有检测到肿瘤生长。与对照组相比，透皮和皮下接种组的 IgG 抗体水平显著增加。然而，与 SC 组相比，在透皮接种组中观察到 IgG 抗体水平轻微增加。可能是由于表皮层中存在 Langerhans 细胞（LCs），在暴露于抗原时被激活。作者的结论是 MNs 技术可用于黑素瘤癌症的制剂开发，是预防黑素瘤癌症的新方法。

43.4.2　包被与刺入

包被与刺入的方法是将选择的抗原包被在固体 MN 上，可以一步递送。这可能是一个有很大吸引力的 ID 疫苗递送方法，因为较少量的抗原涂于微针能够诱导强烈的免疫应答。由于 MN 上的抗原包被呈固体形式[56]，因此应改善长期稳定性，确保最佳保质期[57]。

Georgia 学院的 Prausnitz 小组使用包被疫苗的 5 MNs 不锈钢碑形阵列进行了免疫研究。所使用的 MN 的长度约为 700 μm，使用激光切割不锈钢板制造。MN 介导的 ID 免疫包括编码丙型肝炎病毒和季节性流感的质粒：H1N1、H3N2、灭活病毒、流感病毒样颗粒和 BCG。优化了包被制剂后，使用海藻糖作为抗原稳定剂[58]包被 MN，刺入小鼠皮肤。结果显示，包被 MNs 诱导很强的免疫应答，具有与常规肌内注射抗致死性流感病毒相似的完全保护作用。该研究的结论是，由于包被制剂中包含稳定剂，因此实现了有效的疫苗接种，其中稳定剂起到保护抗原活性的作

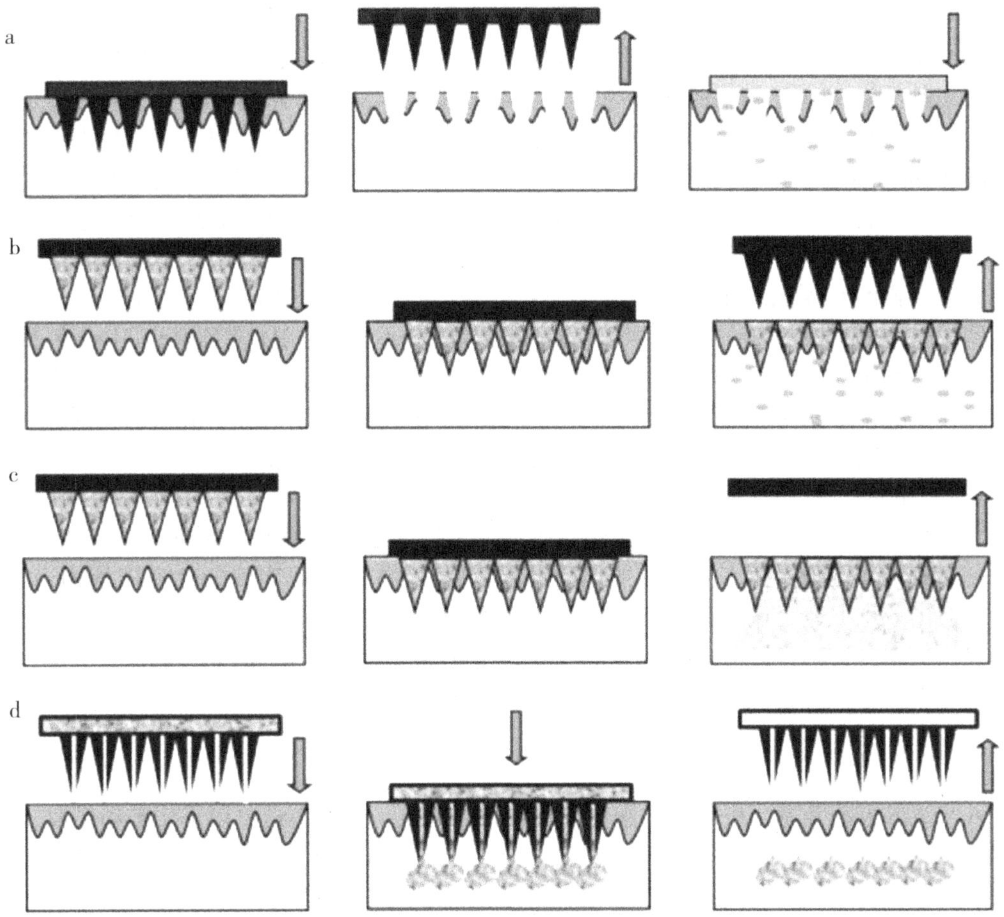

图 43.5　MN 应用于皮肤以增强透皮给药的方法示意图

(a)（穿入与贴剂）显示应用后再去除固体 MN，产生瞬时微孔，然后应用传统的透皮贴剂。(b)（包被与刺入）显示涂有药物的固体 MN 用于即时递送。在包被材料溶解后除去 MN。(c)（溶解微针）显示可溶性聚合物/碳水化合物 MN 含有药物，其随时间溶解于皮肤间质液中，从而递送药物。(d)（刺入与流入）显示使用空心 MN，将含有药物抗原的液体递送至皮肤中。

用。鉴于此，随后研究中的包被制剂含有海藻糖。

在 Koutsonanos 等的一项研究中，与 IM 注射观察到的结果相比，单次 MN 免疫接种灭活的 H3N2 流感病毒，血凝抑制（HI）滴度显著增加[2]。发现用灭活的流感病毒包被的固体金属 MN 在低抗原浓度或高抗原浓度下诱导相似水平的功能性抗体，在感染小鼠的肺部病毒的清除，提供保护和诱导短期反应，以及诱导记忆 B 免疫应答等方面的作用至少与传统的 IM 途径一样有效。IM 注射的血清 IgG 应答取决于剂量，但 MN 递送则不同，其在低抗原剂量或高抗原剂量下产生类似的应答。这一结果表明皮肤具有更高的产生免疫应答的能力。使用相同的体系评估 BCG 包被的 MN 疫苗贴剂的潜力[59]。向包被溶液中添加 15%海藻糖维持 BCG 的活力，提高活 BCG 疫苗的稳定性。将 BCG 包被的 10 微针贴剂（5×10^4CFU 的 BCG）施用于豚鼠皮肤。另一组豚鼠使用 26 号针和 1 mL 结核菌素注射器皮内注射 BCG（5×10^4CFU）。该研究的结果表明，与使用 26 号针相比，包被 BCG 疫苗的 MNs 在豚鼠的肺和脾中诱导较强的抗原特异性细胞免疫应答。发现 MN BCG 疫苗接种与皮下注射相比，对双功能 $CD4^+$T 细胞分泌 TNF-α 或分泌 IFN-γ 和 TNF-α 细胞因子的诱导作用相似。两种疫苗接种方法都诱导强烈的 IgG 反应。

该组评估了修饰的重组三聚体可溶性流感病毒血凝素（sHA GCN4pII）包被的 MNs 诱导保

护性免疫应答的能力[60]。对修饰和未修饰蛋白（sHA）的结果进行了比较。使用包被 sHA 三聚体的 MN 接种的小鼠诱导了抗流感病毒刺激的完全保护性免疫应答。与 SC 途径相比，sHA 和 sHA GCN4pII 包被的 MN 均改善了病毒的清除。包被 sHA GCN4pII 的 MN 在小鼠中诱导更强的 Th1 应答，表现在 $CD4^+$T 细胞的 IFN-γ^+与 IL-4^+分泌比例。该研究得出结论，包被稳定的 sHA 三聚体的 MN 与常规皮下给药一样，能够有效诱导保护性免疫应答，并具有相似的保护作用。

Mark Kendall 教授的研究率先开发了 Nanopatch™技术。Nanopatch™装置由硅制成，采用深反应蚀刻工艺。凸起是实心硅，涂有薄（~100 nm）金层，并含有 3 364个密集的凸起。这些阵列比标准针小两个数量级，也比典型的微针阵列小很多。Nanopatches™已用于小鼠抗西尼罗河病毒和基孔贡亚病毒的 ID 疫苗接种。灭活的整个基孔贡亚病毒疫苗和 DNA 递送的减毒西尼罗河病毒疫苗证明了 Nanopatch™免疫的效率[61]。Nanopatch™技术还可用作 IM 宫颈癌 Gardasil® 预防性疫苗的替代递送系统，并成功向小鼠耳朵递送高达 300ng 的疫苗。此外，就病毒中和反应而言，发现小鼠使用 Nanopatch™技术接种的血清样品结果与 IM 接种的对照组小鼠结果相同[62]。同样，当 Nanopatch™与流感疫苗 Fluvax®[63]一起使用时，也有类似的结果。

ZosanoPharma（前身为 ALZA 公司）的研究评估了 Macroflux®（另一种含有一系列微突出物的装置）的性能，用于皮内递送卵清蛋白抗原。结果显示，在所有抗原剂量下，使用 OVA 包被的 Macroflux® 引起的免疫应答与皮内给予相同剂量的抗原后观察到的免疫应答相当。此外，观察到与肌内或皮下途径递送等效剂量相比，采用 Macroflux® 递送 1 μg 和 5 μg 抗原，抗卵白蛋白水平增加 10 倍和 50 倍[64]。随后的机制研究显示免疫反应不受 MN 高度（225~600 μm）和密度（140/cm^2 和 657/cm^2，但依赖于抗原剂量）的影响[65]。

43.4.3 刺入与流入

刺入与流入方法是基于疫苗通过固体 MN 导管的扩散。抗原递送可以采用被动扩散，压力或电驱动流入[45,66]。然而，这种方法很容易受到与传统疫苗接种技术相同的一些问题的影响，包括需要冷链和需要经过培训的人员[62]。MN 形成的窄孔和皮肤的致密弹性也可能限制液体流入。Wang 等人发现在引入液体之前 MN 阵列的部分收缩抵消了部分皮肤限制；然而[67]，Frost 建议与透明质酸酶同时使用，它可以降解皮肤细胞外基质中的透明质酸，以降低皮肤抵抗力[68]。Martanto 等报道了影响空心 MN 流速的因素。由于空心 MN 可用于 TCI 或 ID 疫苗接种，因此是越来越受关注的方向[69]。

MicronJet® 是 NanoPass 开发的一种新型装置，专门用于皮内递送，由 4 个 MN 组成，每个长度为 450 μm。针是黏合到塑料适配器上的硅晶体，可以安装在任何标准的注射器上。Van Damme 等人进行了在健康成人中使用该装置的皮内注射免疫疫苗的安全性和有效性研究[70]。使用空心 MN 装置（MicronJet®）递送 α-RIX®（GSK Biologicals）流感疫苗，并评估其安全性和免疫原性。在 180 名健康成人受试者中进行的试验发现，MicronJet® 递送的低剂量流感疫苗诱导的免疫反应与在人类志愿者中 IM 注射每种菌株 15 μg HA 诱导的免疫反应类似。然而由于局部反应，受到一定限制，尽管这种反应很短暂。BD 技术也取得了类似的进展。

使用内径 76 μm，外径 178 μm 且暴露长度为 1 mm 的 34 G 不锈钢 MN 递送 3 种不同的流感疫苗。结果表明，这种递送模式能够诱导完全免疫应答，对于不同的抗原性质，其剂量比 IM 给药至少低 10 倍，并且疗效高达 100 倍[71]。在随后的研究中，相同的研究人员发现使用 MN 进行炭疽疫苗接种比肌内给药更利于节省剂量[72]。

43.4.4 溶解/可溶性微针

溶解 MN 是一种创新性的疫苗递送方法。这种 MN 是水溶性聚合物或碳水化合物材料，将药物包封在针状基质内。在刺入皮肤时，MN 完全溶解，递送其内容物。多方面原因使可溶性 MN 在疫苗递送方面很有前景。由于 MN 在刺入患者皮肤后会溶解，因此不会发生交叉感染。此外，该方法不需要进行任何特殊的后续处理，因为不会产生尖锐的生物危害废物。包含/包封的疫苗是固态的，减少了对冷链储存和运输的需要。此外，在大流行期间，可使用MN进行自我接种疫苗以及可以在发展中国家实施大规模免疫计划。由于这些能自我给药的MN没有常规疫苗接种技术的缺点，也没有与此处提到的MN方法相关的问题，因此刺入和贴剂MN在疫苗接种应用中的价值越来越受到关注。

Prausnitz小组首先采用可溶性MN贴剂用于疫苗接种。Sullivan等制造MN贴剂，并研究了使用简单贴剂系统接种流感疫苗的可能性。650 μm聚乙烯基吡咯烷酮MN贴剂，含有3 μg冷冻干燥灭活流感病毒疫苗，诱导了强大的抗体反应和细胞免疫应答，具有抗致死性流感病毒刺激的完全保护作用[73]。事实上，与IM疫苗接种相比，对于疫苗接种后肺病毒清除和细胞记忆反应，MN递送显示出更好的结果。

Kendall小组报道了一种Nanopatch™主模板的溶解MN阵列微模塑[74]。通过多次加工将羧甲基纤维素形成聚二甲基硅氧烷模具，制成复制版MN。包含卵清蛋白抗原和佐剂Quil-A的双层MN在小鼠中诱导免疫后改变抗体水平，虽然在MN中使用较低的抗原剂量，但免疫效果在第28天与IM卵清蛋白/ Quil-A免疫组类似，在第102天优于IM组。应用流感疫苗的研究也得到了类似的结果。

43.4.5　皮内基因递送

基因治疗定义为从个体的细胞和生物组织插入，改变或去除基因以治疗疾病。将功能基因插入非特定的基因组位置替换突变基因是最常用的基因疗法。然而，也有直接改变突变基因或修饰正常基因的其他方法。虽然该技术尚未充分发挥其潜力，但已经取得了一些成功。最常见的基因工程形式是将功能基因插入宿主细胞中，即通过分离和复制目的基因，产生含有所有遗传元件的、能进行正确表达的载体，然后将该载体插入宿主生物体中。皮肤内基因治疗的局部递送和表达为许多病理疾病提供新的治疗选择，包括遗传的皮肤病以及恶性肿瘤的非手术治疗。如果可以使用基因操控影响表皮细胞产生和分泌抗原分子，那么可以利用皮肤的有效免疫刺激特性提供对相关疾病的有效保护。

威尔士卡迪夫大学Birchall小组开发了使用各种微针阵列穿透角质层的表皮进行基因递送的方法，角质层是皮内递送基因载体的障碍。该小组最开始是为了确定经各向同性刻蚀/ BOSCH反应过程微加工的硅微针（150 μm高，底宽45~50 μm）是否能够在皮肤中产生足够大的微通道，使脂质/聚阳离子/质粒DNA（LPD）非病毒基因治疗载体通过[75]。使用扫描电子显微镜观察热分离的人表皮层在使用微针后产生的微孔。

分别通过光子相关光谱法和微电泳法确认粒子大小和粒子表面电荷之后，使用Franz型扩散池在体外建立荧光聚苯乙烯纳米球和LPD复合物经热分离的人表皮层的扩散。通过流式细胞术定量的体外细胞培养物用于确定人角质形成细胞（HaCaT细胞）中的基因表达。发现使用微针进行膜处理显著增强了100 nm直径的荧光聚苯乙烯纳米球的表皮扩散，荧光聚苯乙烯纳米球可用作LPD复合物的可量化的模型。还发现可将LPD复合物递送到膜微通道中或通过膜微通道递送。

在两种情况下，观察到颗粒与表皮之间有很大的相互作用。体外细胞培养显示，在适当的表面电荷下，LPD复合物介导了培养的人角质形成细胞中的报告基因有效表达。

该组的下一项研究中使用了包被铂的硅微针。它可用于产生微通道，直径约50 μm，并通过角质层和活表皮延伸[76]。在优化皮肤外植体培养技术和确认组织存活之后，发现可以使用微针递送β-半乳糖苷酶报告基因能介导基因表达。初步的研究证实了微针刺入皮肤后的局部递送，细胞内化作用和随后的pDNA基因表达。

根据这一结果，可以确定微制造的硅微针阵列是否能够有效地实现带电荷的大分子和质粒DNA（pDNA）的局部递送[77]。在用微针处理后，在人表皮膜上发现直径10~20 μm的微通道。

光学和荧光显微镜显示，已将生物分子β-半乳糖苷酶标记物和模拟带电荧光纳米颗粒的“非病毒基因载体”递送至微针处理组织的活表皮。

Franz 型扩散池法和模型合成膜生成的轨道刻蚀的渗透率表明，在约 2 小时内约有>50%的胶体颗粒悬浮液渗透通过膜孔。该小组的结果表明微针治疗可以实现向活表皮皮肤递送脂质/聚阳离子/ pDNA（LPD）基因载体和其他相关载体。初步基因表达研究证实，在微针破坏 SC 屏障后，分离的人皮肤中表达裸露的 pDNA。由于有限数量的微通道的基因表达呈阳性，因此还需进行进一步的研究以优化微针装置形态、应用方法和 pDNA 制备。

携带基因递送载体的大量溶液不能保留在皮肤表面上，因此在临床实践中应用有限。因此，该小组进行了使用持续释放的 pDNA 水凝胶制剂及其微针递送系统改善皮肤中质粒 DNA（pDNA）的递送的研究[78]。使用氢氧化钾湿法刻蚀硅再次制造微针。制备 Carbopol® 聚合物和热敏 PLGA-PEG-PLGA 三嵌段共聚物的水凝胶。使用分离的新鲜人体皮肤研究微针穿透（显微镜检测和皮肤水分流失检测）、凝胶在微通道中的停留、pDNA 扩散和报告基因（β-半乳糖苷酶）的表达。使用微针时，150~200 μm 深的通道增加了皮肤中的经皮水分流失。显示含有 pDNA 水凝胶，并逐渐释放 pDNA。微针辅助递送 pDNA 水凝胶到人皮肤后，在微通道附近的活表皮中发现 pCMV β 报道基因的表达。结论是靶向 pDNA 水凝胶可在活表皮中提供持续的基因表达。

最近，Prausnitz 小组也进行了微针介导的皮内基因传递的研究[79]。在该研究中，使用包被编码丙型肝炎病毒非结构蛋白 3/4A 质粒的微针接种，有效地激活特异性细胞毒性 T 淋巴细胞（CTL）。重要的是，微创微针在诱导 CTL 方面与更复杂或更具侵入性的递送技术一样有效，例如基因枪和皮下注射针。如上所述，Kendall 小组还进行了包被 Nanopatches™[80] 的基因递送研究。

虽然 Prausnitz 在 2005 年提出将微针与电穿孔联合使用[81]，但是 Hooper 等在 2007 年首次报道使用质粒 DNA 包被的微针阵列进行皮肤电穿孔的新方法，有效地提供了 4 种痘苗病毒基因（4pox）组成的实验性天花 DNA 疫苗[82]。电穿孔是使用高强度电场脉冲使细胞瞬时透化的过程。EasyVax® 递送系统由 80 个导电微针组成，每个微针高度约 1 mm，包被干燥的疫苗 DNA。

脉冲方案由 6 个 100V 脉冲，100ms 为脉冲持续时间和 125ms 为脉冲间隔组成。使用 4 个阵列（每个涂有 30 μg 不用的质粒）将 4pox DNA 疫苗递送至小鼠。将每个阵列用于不同的位置，即在每个部位分开使用（大腿内侧和外侧、右侧和左侧）。接种阴性对照质粒的小鼠使用一个阵列用于大腿内侧。天花 DNA 疫苗刺激 4 种免疫原产生了强大抗体反应，包括中和抗体滴度，产生的抗体滴度大于用于划痕的常规活病毒疫苗。此外，在接种疫苗的小鼠中发现了抗痘苗病毒株 IHD-J 致死性鼻内感染的完全保护作用。

该研究首次证明了微针介导的皮肤电穿孔能诱导保护性免疫反应，并表明此方向有进一步探索的可能。另外，Daugimont 等[83]接下来为了在皮肤表层中产生足以进行电穿孔的电场，研究了空心导电微针在无针皮内注射和电脉冲应用中的潜力。微针阵列与振动插入器一起破坏角质层，从而刺穿皮肤。实现了将蛋白质有效注射到皮肤中，诱导抗卵清蛋白模型抗原的免疫应答。然而，微针电极的双重功能对 DNA 电转移来说可能具有一些缺点。可能是由于数值计算所示的皮肤的电场分布和/或低剂量 DNA 的注射。作者得出的结论是，需要进一步研究这些因素，优化皮肤中的微创 DNA 电转移。

43.5 微针介导纳米颗粒作为靶向皮肤 DCs 的抗原呈递系统

在过去几年中，已提出将颗粒疫苗用于免疫。已将它们用于保护体内抗原稳定性，并以受控和持续的方式将其传递至作用部位[84]。

载药纳米颗粒（NPs）是胶体体系，直径为10~1 000 nm，将有治疗作用的药物包被、吸附或化学偶联到圆形基质上[85,86]。纳米颗粒广泛应用于将小分子药物，寡核苷酸和蛋白质抗原可控地递送至多种细胞中，包括树突状细胞[87]。在设计颗粒疫苗时所考虑的不同因素中，颗粒大小和颗粒物理化学性质对皮肤疫苗接种尤为重要。已经证明，直径<500 nm的聚合物纳米颗粒在各种APC细胞内具有高摄取率[88]。

几个研究小组已经证明纳米粒子具有与CFA或ALUM相当的佐剂效应，并且作为合成佐剂，可以激活DC以诱导抗包被的抗原的T细胞免疫应答[89-91]。重要的进展是，证明纳米颗粒作为佐剂促进了NLRP3炎症小体的激活[92]。

纳米颗粒由于其持续的药物释放而被广泛用于口服和注射给药研究[93,94]。纳米颗粒的这种性质也可以用于局部抗原给药，以在长时间内用抗原靶向皮肤DC。研究人员曾试图将纳米颗粒用于局部药物递送，他们发现皮肤表面上的纳米颗粒逐渐释放药物能增强药物的渗透，但没有发现将纳米颗粒递送到皮肤内的最好方式[95-97]。这表明，作为药物递送载体，纳米颗粒可以维持药物释放，但是，如果将其用作治疗皮肤病的药物，则必须将其递送到皮肤层，而不是停留在皮肤表面。

其他研究人员试图研究纳米颗粒在皮肤上的渗透，但发现只有少量的NP能够通过毛囊被动地渗透到皮肤中，而大多数NP因角质层限制无法渗透到皮肤中[98,99]。在使用全层皮肤的体外渗透实验中，发现NPs可以扩散到真皮和表皮中[100]。为了研究由微针产生的表皮上的微孔是否为NPs穿透皮肤的通道，已经进行体外研究并证明纳米颗粒可以穿过人体表皮膜，并进入皮肤层[101,102]。此外，Bal及其同事的研究表明，在体内进行皮内抗原递送至金属MNs预处理的皮肤时，相比于以可溶的形式递送，当它被包封到聚合物NP中时，皮肤DC能更有效地吸收抗原[103]。上述所有研究表明，在体内微针可能是皮内递送抗原包封的纳米颗粒的有效载体。

结论

树突状细胞作为免疫反应的关键调节因子，在现代疫苗的开发中发挥着关键作用[24,99]。皮肤包括这些细胞网，因此是免疫的重要靶标。然而，尚未成功解决的一个重要问题是DCs亚群的功能不同。随着人们对DCs亚群功能的了解，研究人员现在正试图通过利用特定DCs亚群的特定功能来改进皮内接种疫苗。将来，有可能将抗原与特定佐剂一起递送到特定的DC亚群，但是避免使用其他具有相反作用的佐剂。将抗原靶向特定的、功能明确的皮肤DC亚群是进一步开发保护性疫苗和治疗性疫苗的有前景的方法。

学术和专利文献中越来越多的证据表明，各种设计的MN能够成功实现皮内和透皮疫苗的传递，可以预见，在工业上一致努力可以开发出MN装置。此外，MN技术的新应用可能会成为最有前景的技术。MN阵列提取体液以确定疫苗接种疗效的能力特别令人感兴趣。随着技术的进步，MN阵列很可能成为未来的药物剂型和监测装置。然而，为了使微针技术取得进展，首先需要解决许多困难。

微针的递送和监测装置的最终商业成功将不仅取决于装置完成其预期功能的能力，还取决于医疗专业人员（例如医生、护士和药剂师）和患者的整体可接受性。因此，未来发展中还很有必要做更多努力以了解这些最终用户的态度。Birchall小组[104]在这方面的开创性研究提供了丰富的信息。参加该小组研究的大多数医疗保健专业人员和公众成员都了解使用微针的潜在优势，包括减少疼痛、组织损伤、传播感染和针刺伤的风险，还可以自我给药并且在儿童、害怕打针的人和/或糖尿病患者中都可使用。然而，提出了一些关于成功递送药物的有效方法的担忧（例如视觉剂量指示剂）、延迟起效、递送系统的成本、可能的意外使用、误用或滥用。医疗保健专业人员还担心皮肤厚度的个体差异，以及注射少量体积和感染风险相关的问题。本研究讨论了其他一些可能的问题（意外或错误）以及有关微针使用的疑问。总体而言，该小组报告称，

100%的公众参与者和74%的医疗保健专业人员对微针技术的未来持乐观态度[104]。在更多的人群中应用这些研究时，相关行业无疑将要采取必要行动解决一些问题，例如推出说明书和患者咨询策略，以确保安全有效地使用微针装置。显然，为了实现相比现有的广泛使用的传统输送系统的最大市场份额，营销策略也是至关重要的。

为了获得医疗保健专业人员、患者以及重要的监管机构（例如美国食品和药物管理局和英国的MHRA）的认可，很可能需要在整个微针“包装”内包含包被器辅助装置和“剂量指示器”，并且微针阵列为一次性使用，包被器/剂量指示器可重复使用。虽然已经在患者材料中公开了各种各样的包被器设计，但是仅描述了高冲击/速度插入或旋转装置的少数相对粗糙的设计。关于微针的插入深度，所用力度起重要作用。显然，患者无法“校准”他们的手，因此，可能会用不同的力度使用微针。除非有大量的研究显示，当手动插入微针时，微针药物递送的速率和程度呈正比，否则为了在整个群体中实现一致的给药剂量，需要提供包被器装置。此外，应确保微针装置已正确刺入患者皮肤中。当出现全球流行病暴发或出现生物恐怖主义事件时，必须要使用微针疫苗自我接种。此时确保微针装置已正确刺入患者皮肤中就更为重要。因此，在包被器装置或微针产品中，应有恰当的方法来确认微针已刺入皮肤。

从监管的角度来看，目前对长期使用微针装置所涉及的安全问题知之甚少。特别是，需要进行评估重复微孔化对恢复皮肤屏障功能影响的研究。然而，根据与使用皮下注射针相比使用微针在皮肤内产生微孔具有微创性，以及在患者一生中使用微针时不可能多次刺入完全相同的部位，可假设微针技术具有良好的安全性。实际上，已知无论微针刺入多长时间，皮肤屏障功能在微针移除的几个小时内完全恢复。对有些患者来说，可能会出现局部刺激或皮肤红斑（变红）的问题。由于皮肤是一个具有强免疫刺激的器官，因此了解重复使用微针是否会对药物、微针材料的赋形剂产生免疫反应以及这种效果是否会对患者造成严重影响很有意义。

关于使用微机械系统，微针必然会刺穿皮肤的保护性角质层屏障，因此感染一直是长期以来讨论的问题。然而，正如我们的结果显示[105]，微针产生的孔的微生物渗透是最小的。实际上也从未出现过任何关于微针导致皮肤或全身性感染的报道。这可能是由于皮肤的免疫成分或皮肤固有的非免疫并基于酶的防御作用。或者，由于微孔本质上是水性的，微生物可能更倾向于留在更疏水的角质层上。

关于是否需要在微针应用之前进行皮肤清洁还有待观察，这也是至关重要的问题。理想情况下不必进行，以避免给患者带来不必要的麻烦，在家庭环境中使用产品看起来更类似于自我给药的注射方式，而不是应用常规透皮贴剂。监管机构最终将根据现有证据的权重作出关键决策。根据其应用（例如药物/疫苗/活性药妆成分递送或微创监测），可以将微针装置分类为药物递送、消费产品或医疗用具。从递送的角度来看，将微针视为注射技术，而不是局部/透皮/皮内递送系统很重要，因为这将决定最终产品是否需要消毒，在无菌条件下制备或只携载低生物负荷。任何含有的微生物都可能需要鉴定和量化，热原含量也可能需要鉴定和量化。如果需要灭菌，那么所选择的方法将是至关重要的，因为最常用的方法（湿热、伽马或微波辐射、环氧乙烷）可能对微针本身和/或任何包含的活性成分（例如生物分子）产生不利影响。

参考文献

[1] Hegde, N. R., Kaveri, S. V., Bayry, J.: Recent advances in the administration of vaccines for infectious diseases: microneedles as painless delivery devices for mass vaccination. Drug Discov Today 16, 1061-1068 (2011).

[2] Koutsonanos, D., del Pilar Martin, M., Zarnitsyn, V., Sullivan, S., Compans, R., Skountzou, I., et al.: Transdermal infl uenza immunization with vaccine - coated microneedle arrays. PLoS One 4, e4773 (2009).

[3] Nicolas, J., Guy, B.: Intradermal, epidermal and transcutaneous vaccination: from immunology to clinical

practice. Expert Rev Vaccines 7, 1201-1214 (2008).

[4] Warger, T., Schild, H., Rechtsteiner, G.: Initiation of adaptive immune responses by transcutaneous immunization. Immunol Lett 109, 13-20 (2007).

[5] Stoitzner, P., Sparber, F., Tripp, C. H.: Langerhans cells as targets for immunotherapy against skin cancer. Immunol Cell Biol 88, 431-437 (2010).

[6] Kenney, R. T., Yu, J., Guebre-Xabier, M., Frech, S. A., Lambert, A., Heller, B. A., et al.: Induction of protective immunity against lethal anthrax challenge with a patch. J Infect Dis 190, 774-782 (2004).

[7] Combadiere, B., Vogt, A., Mahe, B., Costagliola, D., Hadam, S.: Preferential amplifi cation of CD8 effector-T cells after transcutaneous application of an inactivated infl uenza vaccine: a randomized phase I trial. PLoS One 5, e10818 (2010).

[8] Lambert, P. H., Laurent, P. E.: Intradermal vaccine delivery: will new delivery systems transform vaccine administration? Vaccine 26, 3197-3208 (2008).

[9] Wysocki, A. B.: Skin anatomy, physiology, and pathophysiology. Nurs Clin North Am 34, 777-797 (1999).

[10] Chuong, C. M., Nickoloff, B. J., Elias, P. M., Goldsmith, L. A., Macher, E., Maderson, P. A., et al.: What is the 'true' function of skin? Exp Dermatol 11, 159-187 (2002).

[11] Williams, A. C., Barry, B., Barry, B. W.: Skin absorption enhancers. Crit Rev Ther Drug Carrier Syst 9, 305-353 (1992).

[12] Wiechers, J. W.: The barrier function of the skin in relation to percutaneous absorption of drugs. Pharmaceutisch weekblad Scientifi c 11, 185-198 (1989).

[13] Tobin, D.: Biochemistry of human skin-our brain on the outside. Chem Soc Rev 35, 52-67 (2006).

[14] Asbill, C. S., El Kattan, A. F., Michniak, B.: Enhancement of transdermal drug delivery: chemical and physical approaches. Crit Rev Ther Drug Carrier Syst 17, 621-658 (2000).

[15] Menon, G.: New insights into skin structure: scratch-ing the surface. Adv Drug Deliv Rev 54, S3-S17 (2002).

[16] Siddiqui, O.: Physicochemical, physiological, and mathematical considerations in optimizing percutaneous absorption of drugs. Crit Rev Ther Drug Carrier Syst 6, 1-38 (1989).

[17] Scheuplein, R. J.: Permeability of the skin: a review of major concepts. Curr Probl Dermatol 7, 172-186 (1978).

[18] Steinman, R. M., Hawiger, D., Nussenzweig, M. C.: Tolerogenic dendritic cells. Annu Rev Immunol 21, 685-711 (2003).

[19] Banchereau, J., Briere, F., Caux, C., Davoust, J., Lebecque, S., Liu, Y. J., et al.: Immunobiology of dendritic cells. Annu Rev Immunol 18, 767-811 (2000).

[20] Kleijwegt, F. S., Jansen, D. T., Teeler, J., Joosten, A. M., Laban, S., Nikolic, T., Roep, B. O.: Tolerogenic dendritic cells impede priming of naïve CD8 (+) T cells and deplete memory CD8 (+) T cells. Eur J Immunol 43 (1), 85-92 (2012). doi: 10. 1002/ eji. 201242879.

[21] Ginhoux, F., Ng, L. G., Merad, M.: Understanding the murine cutaneous dendritic cell network to improve intradermal vaccination strategies. Curr Top Microbiol Immunol 351, 1-24 (2012).

[22] Teunissen, M. B., Haniffa, M., Collin, M. P.: Insight into the immunobiology of human skin and functional specialization of skin dendritic cell subsets to innovate intradermal vaccination design. Curr Top Microbiol Immunol 351, 25-76 (2012).

[23] Steinman, R. M., Hemmi, H.: Dendritic cells: translating innate to adaptive immunity. Curr Top Microbiol Immunol 311, 17-58 (2006).

[24] Merad, M., Ginhoux, F., Collin, M.: Origin, homeo-stasis and function of Langerhans cells and other langerin-expressing dendritic cells. Nat Rev Immunol 8, 935-947 (2008).

[25] Romani, N., Clausen, B. E., Stoitzner, P.: Langerhans cells and more: langerin-expressing dendritic cell subsets in the skin. Immunol Rev 234, 120-141 (2010).

[26] Takahara, K., Omatsu, Y., Yashima, Y., Maeda, Y., Tanaka, S., Iyoda, T., et al.: Identifi cation

and expression of mouse Langerin (CD207) in dendritic cells. Int Immunol 14, 433-444 (2002).

[27] Valladeau, J., Saeland, S.: Cutaneous dendritic cells. Semin Immunol 17, 273-283 (2005).

[28] Steinman, R. M., Nussenzweig, M. C.: Avoiding hor-ror autotoxicus: the importance of dendritic cells in peripheral T cell tolerance. Proc. Natl. Acad. Sci. U. S. A. 99, 351-358 (2002).

[29] Larregina, A. T., Falo Jr., L. D.: Changing paradigms in cutaneous immunology: adapting with dendritic cells. J Invest Dermatol 124, 1-12 (2005).

[30] Bursch, L. S., Wang, L., Igyarto, B., Kissenpfennig, A., Malissen, B., Kaplan, D. H., et al.: Identifi cation of a novel population of Langerin+dendritic cells. J Exp Med 204, 3147-3156 (2007).

[31] Ginhoux, F., Collin, M. P., Bogunovic, M., Abel, M., Leboeuf, M., Helft, J., et al.: Blood-derived dermal langerin+dendritic cells survey the skin in the steady state. J Exp Med 204, 3133-3146 (2007).

[32] Poulin, L. F., Henri, S., de Bovis, B., Devilard, E., Kissenpfennig, A., Malissen, B.: The dermis contains langerin+dendritic cells that develop and function independently of epidermal Langerhans cells. J Exp Med 204, 3119-3131 (2007).

[33] Romani, N., Koide, S., Crowley, M., Witmer-Pack, M., Livingstone, A. M., Fathman, C. G., et al.: Presentation of exogenous protein antigens by dendritic cells to T cell clones. Intact protein is presented best by immature, epidermal Langerhans cells. J Exp Med 169, 1169-1178 (1989).

[34] Stoitzner, P., Tripp, C. H., Eberhart, A., Price, K. M., Jung, J. Y., Bursch, L., et al.: Langerhans cells cross-present antigen derived from skin. Proc. Natl. Acad. Sci. U. S. A. 103, 7783-7788 (2006).

[35] Stoitzner, P., Green, L. K., Jung, J. Y., Price, K. M., Tripp, C. H., Malissen, B., et al.: Tumor immunotherapy by epicutaneous immunization requires langerhans cells. J Immunol 180, 1991 - 1998 (2008).

[36] Cunningham, A. L., Carbone, F., Geijtenbeek, T. B.: Langerhans cells and viral immunity. Eur J Immunol 38, 2377-2385 (2008).

[37] Kautz-Neu, K., Meyer, R. G., Clausen, B. E., von Stebut, E.: Leishmaniasis, contact hypersensitivity and graft-versus-host disease: understanding the role of dendritic cell subsets in balancing skin immunity and tolerance. Exp Dermatol 19, 760-771 (2010).

[38] Bennett, C. L., van Rijn, E., Jung, S., Inaba, K., Steinman, R. M., Kapsenberg, M. L., et al.: Inducible ablation of mouse Langerhans cells diminishes but fails to abrogate contact hypersensitivity. J Cell Biol 169, 569-576 (2005).

[39] Stoecklinger, A., Eticha, T. D., Mesdaghi, M., Kissenpfennig, A., Malissen, B., Thalhamer, J., et al.: Langerin+dermal dendritic cells are critical for $CD8^+$T cell activation and IgH gamma-1 class switching in response to gene gun vaccines. J Immunol 186, 1377-1383 (2011).

[40] Angel, C. E., Lala, A., Chen, C. J., Edgar, S. G., Ostrovsky, L. L., Dunbar, P. R.: $CD14^+$ antigen-presenting cells in human dermis are less mature than their $CD1a^+$ counterparts. Int Immunol 19, 1271-1279 (2007).

[41] Angel, C. E., Chen, C. J., Horlacher, O. C., Winkler, S., John, T., Browning, J., et al.: Distinctive localization of antigen-presenting cells in human lymph nodes. Blood 113, 1257-1267 (2009).

[42] Klechevsky, E., Morita, R., Liu, M., Cao, Y., Coquery, S., Thompson-Snipes, L., et al.: Functional specializations of human epidermal Langerhans cells and $CD14^+$ dermal dendritic cells. Immunity 29, 497-510 (2009).

[43] van der Aar, A. M., de Groot, R., Sanchez-Hernandez, M., Taanman, E. W., van Lier, R. A., Teunissen, M. B., et al.: Cutting edge: virus selectively primes human langerhans cells for CD70 expression promoting $CD8^+$T cell responses. J Immunol 187, 3488-3492 (2011).

[44] Garland, M., Migalska, K., Mahmood, T. M. T., Singh, T. R. R., Woolfson, A. D., Donnelly, R.: Microneedle arrays as medical devices for enhanced transdermal drug delivery. Expert Rev Med Devices 8, 459-482 (2011).

[45] Donnelly, R., Majithiya, R., Singh, T., Morrow, D., Garland, M., Demir, Y.: Design, optimization

and characterisation of polymeric microneedle arrays prepared by a novel laser-based micromoulding technique. Pharm Res 28, 41-57 (2011).

[46] Chen, X., Fernando, G. J. P., Crichton, M., Flaim, C., Yukiko, S., Corbett, H. J., et al.: Improving the reach of vaccines to low-resource regions, with a needle-free vaccine delivery device and long-term thermostabilization. J Controlled Release. 152, 349-355 (2011).

[47] Henry, S., McAllister, D. V., Allen, M. G., Prausnitz, M. R.: Microfabricated microneedles: a novel approach to transdermal drug delivery. J Pharm Sci 87, 922-925 (1998).

[48] Prausnitz, M., Mikszta, J., Cormier, M., Andrianov, A.: Microneedle-based vaccines. Curr Top Microbiol Immunol 333, 369-393 (2009).

[49] Donnelly, R., Raj Singh, T. R., Woolfson, A. D.: Microneedle-based drug delivery systems: microfabrication, drug delivery, and safety. Drug Deliv 17, 187-207 (2010).

[50] Zhou, C., Liu, Y., Wang, H., Zhang, P., Zhang, J.: Transdermal delivery of insulin using microneedle rollers in vivo. Int J Pharm 392, 127-133 (2010).

[51] Ding, Z., Verbaan, F. J., Bivas-Benita, M., Bungener, L., Huckriede, A., Kersten, G., et al.: Microneedle arrays for the transcutaneous immunization of diphtheria and infl uenza in BALB/c mice. J Controlled Release. 136, 71-78 (2009).

[52] Ding, Z., Van Riet, E., Romeijn, S., Kersten, G. F. A., Jiskoot, W., Bouwstra, J. A.: Immune modulation by adjuvants combined with diphtheria toxoid administered topically in BALB/c mice after microneedle array pretreatment. Pharm Res 26, 1635-1643 (2009).

[53] Bal, S. M., Slütter, B., van Riet, E., Kruithof, A. C., Ding, Z., Kersten, G. F. A., et al.: Effi cient induction of immune responses through intradermal vaccination with N-trimethyl chitosan containing antigen formulations. J Controlled Release. 142, 374-383 (2010).

[54] Bhowmik, T., D'Souza, B., Shashidharamurthy, R., Oettinger, C., Selvaraj, P., D'Souza, M.: A novel microparticulate vaccine for melanoma cancer using transdermal delivery. J Microencapsul 28, 294-300 (2011).

[55] Cleary, G.: Microneedles for drug delivery. Pharm Res 28, 1-6 (2011).

[56] Shah, U. U., Roberts, M., Orlu Gul, M., Tuleu, C., Beresford, M. W.: Needle-free and microneedle drug delivery in children: a case for disease-modifying antirheumatic drugs (DMARDs). Int J Pharm 416, 1-11 (2011).

[57] Kim, Y., Quan, F., Compans, R. W., Kang, S., Prausnitz, M. R.: Formulation and coating of microneedles with inactivated infl uenza virus to improve vaccine stability and immunogenicity. J Controlled Release. 142, 187-195 (2010).

[58] Hiraishi, Y., Nandakumar, S., Choi, S., Lee, J., Kim, Y., Prausnitz, M. R., et al.: Bacillus Calmette-Guérin vaccination using a microneedle patch. Vaccine 29, 2626-2636 (2011).

[59] Weldon, W., Martin, M., Zarnitsyn, V., Wang, B., Koutsonanos, D., Skountzou, I.: Microneedle vaccination with stabilized recombinant infl uenza virus hemagglutinin induces improved protective immu-nity. Clinical and vaccine immunol. 18, 647-654 (2011).

[60] Prow, T.: Nanopatch-targeted skin vaccination against West Nile Virus and Chikungunya virus in mice. Small 6, 1776-1784 (2010).

[61] Corbett, H., Chen, X., Frazer, I.: Skin vaccination against cervical cancer associated human papillomavirus with a novel micro-projection array in a mouse model. PLoS One 5, e13460 (2010).

[62] Fernando, G. J. P., Chen, X., Prow, T., Crichton, M., Fairmaid, E.: Potent immunity to low doses of infl uenza vaccine by probabilistic guided micro-targeted skin delivery in a mouse model. PLoS One 5, e10266 (2010).

[63] Matriano, J., Cormier, M., Johnson, J., Young, W., Buttery, M., Cormier, M., et al.: Macrofl ux microprojection array patch technology: a new and effi cient approach for intracutaneous immunization. Pharm Res 19, 63-70 (2002).

[64] Widera, G., Johnson, J., Kim, L., Libiran, L., Nyam, K., Daddona, P. E., et al.: Effect of delivery parameters on immunization to ovalbumin following intracutaneous administration by a coated mi-

croneedle array patch system. Vaccine 24, 1653-1664 (2006).

[65] Escobar-Chvez, J., Bonilla-Martinez, D., Villegas-Gonzúlez, M. A., Molina Trinidad, E., Casas Alancaster, N., et al.: Microneedles: a valuable physical enhancer to increase transdermal drug delivery. J Clin Pharmacol 51, 964-977 (2011).

[66] Amorij, J., Frijlink, H., Wilschut, J., Huckriede, A.: Needle-free infl uenza vaccination. Lancet Infect Dis 10, 699-711 (2010).

[67] Wang, P., Cornwell, M., Hill, J., Prausnitz, M.: Precise microinjection into skin using hollow microneedles. J Invest Dermatol 126, 1080-1087 (2006).

[68] Frost, G. I.: Recombinant human hyaluronidase (rHuPH20): an enabling platform for subcutaneous drug and fl uid administration. Expert Opin Drug Deliv 4, 427-440 (2007).

[69] Bal, S., Ding, Z., van Riet, E., Jiskoot, W., Bouwstra, J.: Advances in transcutaneous vaccine delivery: Do all ways lead to Rome? J Controlled Release. 148, 266-282 (2010).

[70] Van Damme, P., Oosterhuis-Kafeja, F., Van der Wielen, M., Almagor, Y., Sharon, O., Levin, Y.: Safety and effi cacy of a novel microneedle device for dose sparing intradermal infl uenza vaccination in healthy adults. Vaccine 27, 454-459 (2009).

[71] Alarcon, J., Hartley, A., Harvey, N., Mikszta, J.: Preclinical evaluation of microneedle technology for intradermal delivery of infl uenza vaccines. Clin Vaccine Immunol 14, 375-381 (2007).

[72] Mikszta, J., Dekker, J., Harvey, N., Dean, C., Brittingham, J., Huang, J., et al.: Microneedle-based intradermal delivery of the anthrax recombinant protective antigen vaccine. Infect Immun 74, 6806-6810 (2006).

[73] Sullivan, S., Koutsonanos, D., Del Pilar Martin, M., Lee, J., Zarnitsyn, V., Compans, R. W., et al.: Dissolving polymer microneedle patches for infl uenza vaccination. Nat Med 16, 915-920 (2010).

[74] Raphael, A., Prow, T., Crichton, M., Chen, X., Fernando, G. J. P., Prow, T.: Targeted, needle-free vaccinations in skin using multilayered, densely-packed dissolving microprojection arrays. Small 6, 1785-1793 (2010).

[75] Chabri, F., Bouris, K., Jones, T., Barrow, D., Hann, A., Allender, C., et al.: Microfabricated silicon microneedles for nonviral cutaneous gene delivery. Br J Dermatol 150, 869-877 (2004).

[76] Prow, T. W., Chen, X., Prow, N. A., Fernando, G. J., Tan, C. S., Raphael, A. P., et al.: Nanopatch-targeted skin vaccination against West Nile Virus and Chikungunya virus in mice. Small 16, 1776-1784 (2010).

[77] Birchall, J., Coulman, S., Pearton, M., Allender, C., Brain, K., Coulman, S., et al.: Cutaneous DNA delivery and gene expression in ex vivo human skin explants via wet-etch micro-fabricated micro-needles. J Drug Target 13, 415-421 (2005).

[78] Coulman, S. A., Barrow, D., Anstey, A., Gateley, C., Morrissey, A., Wilke, N., et al.: Minimally invasive cutaneous delivery of macromolecules and plasmid DNA via microneedles. Curr Drug Deliv 3, 65-75 (2006).

[79] Pearton, M., Allender, C., Brain, K., Anstey, A., Gateley, C., Wilke, N., et al.: Gene delivery to the epidermal cells of human skin explants using microfabricated microneedles and hydrogel formulations. Pharm Res 25, 407-416 (2008).

[80] Gill, H. S., Soderholm, J., Prausnitz, M. R., Sallberg, M., Sderholm, J., Sllberg, M.: Cutaneous vaccination using microneedles coated with hepatitis C DNA vaccine. Gene Ther 17, 811-814 (2010).

[81] Choi, S. O., Park, J. H., Gill, H. S., Choi, Y., Allen, M. G., M. R.: Prausnitz. Microneedles electrode array for electroporation of skin for gene therapy. Controlled Release Society 32nd Annual Meeting & Exposition Transactions. 318 (2005).

[82] Hooper, J., Golden, J., Ferro, A., King, A.: Smallpox DNA vaccine delivered by novel skin electroporation device protects mice against intranasal poxvirus challenge. Vaccine 25, 1814-1823 (2007).

[83] Daugimont, L., Baron, N., Vandermeulen, G., Pavselj, N., Miklavcic, D., Jullien, M., et al.: Hollow microneedle arrays for intradermal drug delivery and DNA electroporation. J Membr Biol 236, 117-125 (2010).

[84] Levine, M. M., Sztein, M. B.: Vaccine development strategies for improving immunization: the role of modern immunology. Nat Immunol 5, 460-464 (2004).

[85] Soppimath, K. S., Aminabhavi, T. M., Kulkarni, A. R., Rudzinski, W. E.: Biodegradable polymeric nanoparticles as drug delivery devices. J Control Release 70, 1-20 (2001).

[86] Delie, F., Blanco-Prieto, M. J.: Polymeric particu-lates to improve oral bioavailability of peptide drugs. Molecules 10, 65-80 (2005).

[87] McCarron, P. A., Donnelly, R. F., Marouf, W.: Celecoxib - loaded poly (D, L - lactide - co - glycolide) nanoparticles prepared using a novel and controllable combination of diffusion and emulsifi cation steps as part of the salting-out procedure. J Microencapsul 23, 480-498 (2006).

[88] Eniola, A. O., Hammer, D. A.: Artifi cial polymeric cells for targeted drug delivery. J Control Release 87, 15-22 (2003).

[89] Jaganathan, K. S., Vyas, S. P.: Strong systemic and mucosal immune responses to surface-modifi ed PLGA microspheres containing recombinant hepatitis B antigen administered intranasally. Vaccine 24, 201-4211 (2006).

[90] Gutierro, I., Hernandez, R. M., Igartua, M., Gascon, A. R., Pedraz, J. L.: Size dependent immune response after subcutaneous, oral and intranasal administration of BSA loaded nanospheres. Vaccine 21, 67-77 (2002).

[91] Lu, D., Garcia-Contreras, L., Xu, D., Kurtz, S. L., Liu, J., Braunstein, M., et al.: Poly (lactide-co - glycolide) microspheres in respirable sizes enhance an in vitro T cell response to recombinant Mycobacterium tuberculosis antigen 85B. Pharm Res 24, 1834-1843 (2007).

[92] Sharp, F. A., Ruane, D., Claass, B., Creagh, E., Harris, J., Malyala, P., et al.: Uptake of particulate vaccine adjuvants by dendritic cells activates the NALP3 infl ammasome. Proc. Natl. Acad. Sci. U. S. A. 106, 870-875 (2009).

[93] de Jalon, E. G., Blanco-Prieto, M. J., Ygartua, P., Santoyo, S.: PLGA microparticles: possible vehicles for topical drug delivery. Int J Pharm 226, 181-184 (2001).

[94] Jenning, V., Gysler, A., Schafer-Korting, M., Gohla, S. H.: Vitamin A loaded solid lipid nanoparticles for topical use: occlusive properties and drug targeting to the upper skin. Eur J Pharm Biopharm 49, 211-218 (2000).

[95] Alvarez-Roman, R., Naik, A., Kalia, Y. N., Guy, R. H., Fessi, H.: Enhancement of topical delivery from biodegradable nanoparticles. Pharm Res 21, 1818-1825 (2004).

[96] Alvarez-Roman, R., Naik, A., Kalia, Y. N., Guy, R. H., Fessi, H.: Skin penetration and distribution of polymeric nanoparticles. J Control Release 99, 53-62 (2004).

[97] Luengo, J., Weiss, B., Schneider, M., Ehlers, A., Stracke, F., Konig, K., et al.: Infl uence of nanoencapsulation on human skin transport of fl ufenamic acid. Skin Pharmacol Physiol 19, 190 - 197 (2006).

[98] Lademann, J., Richter, H., Teichmann, A., Otberg, N., Blume-Peytavi, U., Luengo, J., et al.: Nanoparticles-aneffi cient carrier for drug delivery into the hair follicles. Eur J Pharm Biopharm 66, 159-164 (2007).

[99] Toll, R., Jacobi, U., Richter, H., Lademann, J., Schaefer, H., Blume-Peytavi, U.: Penetration profi le of microspheres in follicular targeting of terminal hair follicles. J Invest Dermatol 123, 168 - 176 (2004).

[100] Coulman, S. A., Anstey, A., Gateley, C., Morrissey, A., McLoughlin, P., Allender, C., et al.: Microneedle mediated delivery of nanoparticles into human skin. Int J Pharm Jan. 366, 190-200 (2009).

[101] McAllister, D. V., Wang, P. M., Davis, S. P., Park, J. H., Canatella, P. J., Allen, M. G., et al.: Microfabricated needles for transdermal delivery of macromolecules and nanoparticles: fabrication methods and transport studies. Proc. Natl. Acad. Sci. U. S. A. 100, 13755-13760 (2003).

[102] Bal, S. M., Slutter, B., Jiskoot, W., Bouwstra, J. A.: Small is beautiful: N-trimethyl chitosan-ovalbumin conjugates for microneedle - based transcutaneous immunisation. Vaccine 29, 4025 - 4032 (2011).

[103] Ueno, H., Schmitt, N., Klechevsky, E., Pedroza-Gonzalez, A., Matsui, T., Zurawski, G., et al.: Harnessing human dendritic cell subsets for medicine. Immunol Rev 234, 199-212 (2010).

[104] Birchall, J. C., Clemo, R., Anstey, A., John, D. N.: Microneedles in clinical practice-an exploratory study into the opinions of healthcare professionals and the public. Pharm Res 28, 95-106 (2011).

[105] Donnelly, R. F., Singh, T. R., Tunney, M. M., Morrow, D. I., McCarron, P. A., O'Mahony, C., Woolfson, A. D.: Microneedle arrays allow lower microbial penetration than hypodermic needles in vitro. Pharm Res 26, 2513-2522 (2009).

(吴国华译)

第 44 章　鼻内干粉疫苗递送技术

Anthony J. Hickey，Herman Staats，Chad J. Roy，
Kenneth G. Powell，Vince Sullivan，Ginger Rothrock 和
Christie M. Sayes①

摘要

1 000 多年前在中国出现鼻内疫苗接种，即吸入可能含有活病毒的天花结痂。这种做法成为 18 世纪欧洲早期接种活病毒的基础。在过去的 10 年中，很多报道都集中在新的抗原、佐剂和递送系统上，但很少有将接种方法作为临床候选者来进行研究。本章概述了在递送系统中选取抗原的综合方法和筛选候选物理和生物化学佐剂所需的关键步骤。需要确定特殊制剂和设备的特性，评估由新型结构和佐剂实现的剂量节省，并开发快速筛选方法以鉴别有毒制剂。

44.1　引言

鼻黏膜十分复杂，包含许多免疫系统的重要元件。因此，它是非侵入性疫苗递送的理想途径。鼻腔的解剖结构如图 44.1 所示。鼻甲的纤毛表面将黏液递送到鼻腔，然后到喉咙、咽淋巴环（咽淋巴环）部位，特别是腺样体和扁桃体[1]。黏液纤毛清除沉积在鼻上皮上的环境微粒。正常的黏膜纤毛清除率从鼻咽部近端至远端需要 45~90 分钟[2]。鼻疫苗递送系统的目的是使疫苗在鼻黏膜上有足够的停留时间，从而引起所需免疫和保护性反应。

鼻疫苗递送建议靶向鼻相关淋巴组织（NALT），即最易接近的淋巴组织之一——黏膜相关淋巴组织（MALT）。该方法的原理是将选择的抗原递呈给 NALT 和咽淋巴环，并最终递呈给免疫学上重要的 M 细胞和树突细胞。这些细胞负责将抗原递呈给循环淋巴细胞产生免疫应答。

两种最常见（和有效）的鼻腔给药设备是多剂量泵或单剂量泵，实例分别是治疗骨质疏松症的降钙素和治疗过敏性鼻炎的皮质类固醇鼻内给药设备[3]。然而，美国食品和药物管理局批准的这些产品是溶液或悬浮液制剂，这些制剂并不是最理想的疫苗制剂。最近，已经提出使用粉末气雾剂进行疫苗接种，其优点是能增加抗原停留时间和抗原递呈机会[3,4]。

44.2　靶向疾病和抗原

由于肺炭疽的死亡率较高及最近关于使用致病微生物作为武器的担忧，科学家迫切需要研发易于使用的疫苗，能够以最小量存储并实现快速分发。炭疽病的致病微生物是炭疽芽孢杆菌

---

① A. J. Hickey，PhD（✉）· G. Rothrock，PhD · C. M. Sayes，PhD（三角研究国际气溶胶与纳米材料工程中心 美国北卡罗来纳州三角研究园区，E-mail：ahickey@ rti. org）

H. Staats，PhD（杜克大学医学中心病理学与免疫学系，美国北卡罗来纳州达勒姆）

C. J. Roy，PhD（新奥尔良杜兰大学医学院微生物学与免疫学系，美国洛杉矶）

K. G. Powell · V. Sullivan（BD 公司，美国北卡罗来纳州三角研究园区）

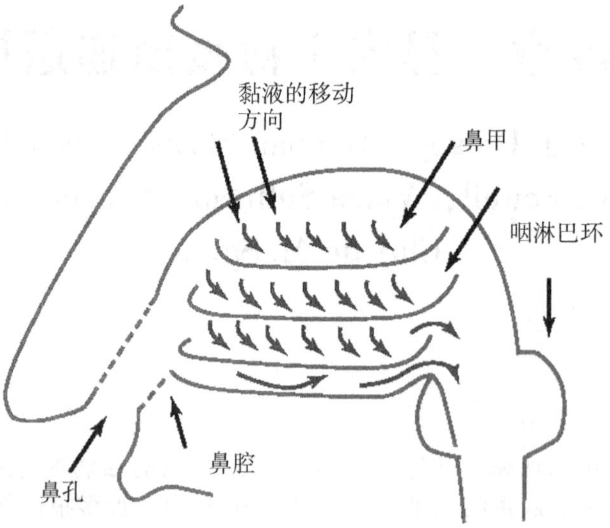

图 44.1　鼻腔示意图

(*B. anthracis*)。该病原体与动物感染密切相关，并且在接触受感染动物或动物产品后，可以导致人类皮肤和肺部疾病[5]。肺炭疽是一种特别危险的疾病，如果不及时进行治疗，可出现生命危险。炭疽杆菌生命周期的一个独特组成部分是形成芽孢，从而抵抗环境的剧烈变化。炭疽芽孢与其他微生物的芽孢一样，极易在空气中分散，然后沉积在整个呼吸道中。这些芽孢可以在极端环境压力生存几十年甚至几个世纪，造成公共卫生风险。炭疽杆菌芽孢易于递送和长期稳定的特性使得它们成为整个 20 世纪中叶国家生物武器发展高峰期以及最近关于生物恐怖主义关注的焦点。

在 2001 年 9 月 11 日的恐怖主义行动之后，随着信件中包含“武器化”炭疽粉末的事件的发生，人们对生物安全防御措施的需求也在逐渐增加[6,7]。急需药物（例如环丙沙星）可以用于治疗个体感染。特别是对于卫生保健人员和军事人员来说，能够以最小的储存量（特殊包装或温度）储存并能快速分发的疫苗成为研究和开发的重点[8]。在这种情况下，过去 10 年中，保护性疫苗和治疗药物的研发一直针对这一特定要求。

一般而言，将活的病原微生物作为疫苗系统呈现，即使病原微生物处于减毒状态，也会造成一定的健康风险。前面提到的中国和欧洲的天花治疗方法很有可能使接种者感染疾病，毋庸置疑这种治疗方法很危险。同样地，对于炭疽病来说，使用致病性的病原体也备受争议。目前的疫苗，如 BioThrax（Emergent Solutions）的生产使用了培养基，同样也引入了很多重要的抗原（致死因子［LF］、保护性抗原［PA］和水肿因子［EF］）[8]。针对炭疽免疫保护的新型免疫策略的重点是使用重组保护性抗原，在动物疾病模型中该抗原可以提供免疫保护作用[9]。事实上，刚刚宣布 rPA 疫苗 SparVaxTM（PharmAthene）正在进行 II 期临床试验。然而，有人建议使用佐剂以确保足够强的免疫反应。该建议可通过同时免疫佐剂分子来实现，佐剂分子种类繁多，但是效果最显著的是单磷酰脂质 A（MPL），经美国食品和药物管理局批准 MPL 佐剂 Cervarix 已应用于鼻内诺洛病毒疫苗的研究。

或者，使用小的靶向颗粒（纳米颗粒）的物理方法可以有效地将抗原有效地递送至 NALT。纳米颗粒具有特定的倾向，多年来作为载体和佐剂发挥着重要的作用[10-12]。值得注意的是，尽管存在黏膜给药途径，但一般认为 IgA 反应对炭疽的保护作用并不重要[9]。

然而，几个世纪前，Voltaire 发现“……中国人以不同的方式进行接种（接种），这些方式没有切口，而是通过鼻子给予天花疫苗，就像鼻烟一样。这种方式比较快乐（比皮肤递送），但同样的事情……”[13]

目前，我们不仅可以采用“友好的”（无针）和“同种方式”（一种保护性免疫反应）技术，也可以采用现有的存储稳定、易于使用且价格低廉的技术。稳定的可分散传递系统更适合干燥颗粒的传递。干燥颗粒悬浮液的喷雾疗效已经在许多疫苗应用中得到证实[14]。图 44.2 概述了研发安全有效的技术先进的鼻疫苗系统所需的组件。

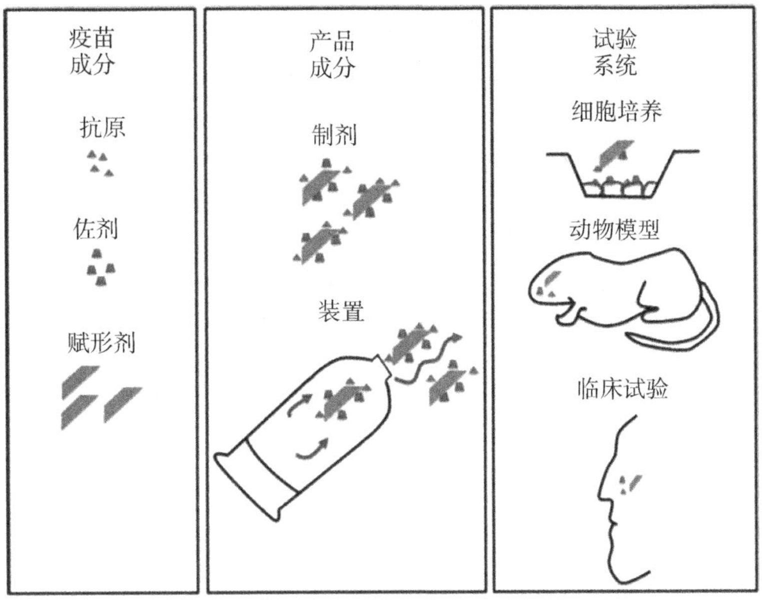

图 44.2 本章介绍的疫苗产品关键要素和评估方法示意图

44.3 鼻疫苗技术

44.3.1 制剂

为了开发具有全球销售潜力的疫苗制剂，干粉必须能长期储存且不需要冷藏，并且在不使用针头的情况下给药。喷雾干燥、冷冻干燥和喷雾冷冻干燥的粉末的免疫原性与常规液体制剂相当，其潜在优点是干粉可以避免液体疫苗制剂冷链问题。这些技术可以与常规和新颖的纳米颗粒制造方法匹配，例如沉淀、均质化和自上而下产生适合于黏膜递送的、空气动力学直径的、松散的纳米颗粒聚集体的干粉，例如先前已经证实的吸入型制剂。

44.3.2 装置

市场上或临床上只有少数的鼻腔粉末递送装置[15]，如图 44.3 所示。虽然在药物靶向沉积至鼻组织中，制剂和粉末形态起关键作用，但是使粉末解聚，并从装置中有效地递送的过程也很重要。

该装置的主要性能分为两类：影响临床疗效和反映人机作用的人为因素。首先，装置应有尽可能大的发射剂量分数。其次，发射的剂量必须能够到达目标组织。颗粒大小应该为 10~50 μm，小于 10 μm 的颗粒易递送至肺组织深处，而大于 50 μm 的颗粒可能难以到达免疫应答组织。在没有过多能量输入的情况下使颗粒的适当解聚是一个挑战。通常使用空气将颗粒解聚，并推进气道。然而，过量的气流使颗粒越过靶向位置，使得它们“通过”靶向位置，但不会在靶向位置沉降和引发免疫应答。从人为因素的角度来看，该装置比较直观且易于操作。临床医生目前使用该装置进行靶向给药，未经训练的人员也可能使用该装置进行生物防御疫苗的免疫。

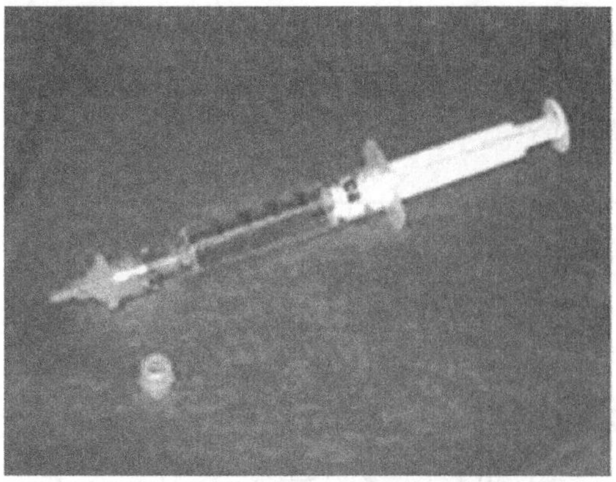

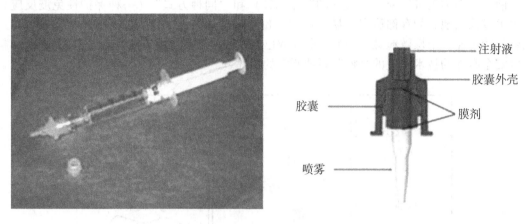

图 44.3 用于鼻腔疫苗给药的 BDSolovent® 装置[16]

该装置依赖于在注射气压下破裂的"破裂膜"，从而引起轻柔的弹出，并排出剂量。

最后，该装置适用于单剂量给药，不适用于多剂量给药，同一装置用于多人给药时会出现不必要的风险。此外，应避免该装置非预期用途的"重复使用"。

44.3.3 佐剂

铝盐（明矾）作为佐剂在近一个世纪以来取得了巨大成功，并且在促进保护性体液免疫方面特别有效。然而，对细胞介导的免疫力疾病的保护来说，明矾并不是最有效的。此外，经典的生物佐剂例如弗氏佐剂，在动物模型中显示更强的免疫保护作用，但其毒副作用限制了在人类中的广泛使用。因此，迫切需要研发能刺激细胞免疫的、没有毒副作用的、安全和无毒的、适用于黏膜递送重组和亚单位疫苗的佐剂。新型颗粒和生物佐剂克服了经典佐剂的缺陷。

最近的一项使用鼻腔递送干粉疫苗人体研究，使用含有诺瓦克病毒样颗粒（VLP）免疫原（即单磷酰脂质 A［MPL］）作为该疫苗佐剂，壳聚糖作为黏膜黏附剂[17]。人体对 Norwalk VLP 具有良好的耐受性，Norwalk VLP 也具有良好的免疫原性，表明将干粉疫苗制剂用于人类的可行性。聚合物纳米颗粒也具有双重功能，作为抗原递送系统，具有额外的佐剂效应[18]。我们假设使用商业装置递送含有重组保护性抗原（rPA）的干粉聚合物纳米颗粒气雾剂，靶向鼻黏膜，能够保护非人灵长类动物免受空气传播炭疽的感染。

在疫苗方面，佐剂可以在功能上定义为能够增强疫苗抗原免疫原性的物质。佐剂是有益的，它们能够提高在疫苗接种完成后机体获得保护性免疫的百分比[19]；能够增强免疫反应强度[19]，降低诱导保护性免疫所需的抗原剂量[20]，和/或减少诱导保护性免疫所需的免疫次数[21]。在一般情况下，佐剂可以增强抗原递送或增强树突细胞的活化状态，树突细胞是在接种疫苗后启动适应性免疫发展的主要抗原呈递细胞[22]。本综述的目的是，我们将简要讨论：①通过增加免疫系统抗原递送提高疫苗免疫原性的疫苗佐剂；②特异性激活先天免疫系统的佐剂。有关疫苗佐剂[23]和黏膜给药疫苗佐剂[24]现状的更多信息将在其他地方进行讨论。

佐剂可以通过增强抗原递送来增加经鼻递送疫苗的免疫原性。Kiyono 等最近的一项研究[25]报道了使用阳离子纳米凝胶作为鼻腔递送蛋白质抗原的佐剂。尽管阳离子纳米凝胶没有增强鼻树突细胞的活化状态，但由于改善了鼻腔中的抗原保留，更好地将抗原递送至鼻腔树突细胞，阳离子纳米凝胶显著增强了鼻疫苗的免疫原性[25]。正在开发将纳米乳剂佐剂体系作为鼻疫苗佐剂[26-28]，并将其描述为"……不含促炎成分，但是能穿透黏膜表面将抗原递送至树突细胞。"[28]

与使用液体疫苗制剂的鼻部免疫相比，喷雾冷冻干燥（SFD）液体疫苗制剂产生的干粉具有疫苗佐剂特性，能增强经鼻递送疫苗的免疫原性[29]。其他研究表明，疫苗制剂的鼻滞留时间增加与疫苗的免疫原性增加有关[25,30,31]。干粉疫苗制剂有可能通过增加鼻腔内抗原保留时间来增强疫苗的免疫原性，最终改善鼻树突细胞的抗原递送，从而更好地诱导抗原特异性适应性免疫应答。

佐剂还可以激活宿主先天免疫应答，从而增强鼻部免疫后抗原适应性免疫应答[23,24]。通常认为激活先天免疫系统的佐剂可以激活树突细胞[32]以增加其诱导抗原特异性适应性免疫应答的能力。能够激活先天免疫系统的多种黏膜佐剂包括促炎细胞因子[33,34]和 Toll 样受体配体（TLR 配体；单磷酰脂质 A、CpG 寡脱氧核苷酸、咪喹莫特）[33-35]。虽然霍乱毒素和相关毒素具有有效的黏膜佐剂活性，但毒素佐剂相关的不良事件[36,37]可能限制了它们在人类疫苗中的使用。在人类临床研究中，已将具有可接受的、安全性的细胞因子和 TLR 配体佐剂用作鼻疫苗佐剂，经额外的优化和安全性评估，表明了鼻内给予含有佐剂的疫苗的可行性[34,38]。

44.3.4　机制

许多毒性标志物也是促炎介质。这些物质诱导的反应有可能实现疫苗的保护性免疫作用。个别因素可能与具体现象有关（例如，细胞膜上超氧自由基的产生引起半胱天冬酶和加氧酶或细胞质渗出）。然而，这些非细胞的标志不能代表整个组织系统，重要的是，将每个细胞反应的强度和持续时间的综合反应视为从可逆“瞬时现象”转变为不可逆“毒性”反应的标志。例如，谷胱甘肽的氧化还原（即还原或氧化）状态所表示的氧化应激已被用于指示从正常瞬时现象到毒性细胞反应的这种转变[39-41]。

文献的结论指出，小颗粒（包括纳米颗粒）的两种主要的物理化学性质会影响它们的稳定性、流动性和毒性。这两种主要的物理化学性质如下：①表面修饰或与其他分子的结合；②与细胞的物理接触。前者影响颗粒聚集状态，决定单个粒子的悬浮性，并影响降解敏感性。后者基本上影响细胞摄取机制。小于 20 nm 的颗粒可以跨细胞膜转运，并积聚在多种亚细胞组分中。颗粒（或颗粒的聚集体）在 20~200 nm 可能会通过网格蛋白或小窝蛋白形成的主动转运机制被吸收到细胞中。大于 200 nm 的颗粒（或颗粒聚集体）更易通过巨胞饮作用内吞。最终，大多数颗粒被隔离在膜结合囊泡中，并在囊泡里开始酸化过程。酸化过程通常将颗粒分解成离子或分子，以浓缩剂量将它们释放到对最易受颗粒沉积影响的特定组织中。虽然并不清楚该研究领域中的所有机制，但大多数人认为，颗粒在细胞和组织的反应中，细胞摄取起着重要作用。

研究表明，在体外和体内系统中暴露的颗粒都可能诱导炎症相关的生物标志物的产生[40,42-45]。在工艺化颗粒被细胞内化后，例如吞噬细胞，可以触发炎症级联反应。炎症是细胞和组织对有害病原体和其他有毒物质的复杂生物反应。它是一种正向机制，可以去除这些有害的病原体，并产生修复酶。异常的炎症可导致许多疾病，如哮喘、动脉粥样硬化和类风湿性关节炎；因此，炎症通常受到身体的严格控制。图 44.4 表示颗粒诱导炎症级联反应的 4 个不同步骤。

44.3.5　动物模型

使用气雾剂递送新疫苗制剂的体内试验和评估需要仔细考虑特定实验动物和递送方式等因素，因为它涉及免疫原和相应的传染病病原体。特定实验动物选择的基础至少部分根据预测的血清学和/或黏膜免疫确定。由于缺少传染病模型可用性的文献记载，因此适用于这一类型的研究动物模型很少。此外，将动物模型中疫苗接种方式与最终的临床给药途径相匹配，进一步限制了选择范围。很多递送系统无法应用于临床前研究；尤其是吸入装置需要一定的呼吸顺应性以确保给药。因此，人们面临的挑战是提供用于动物实验的可替代性疫苗递送系统。

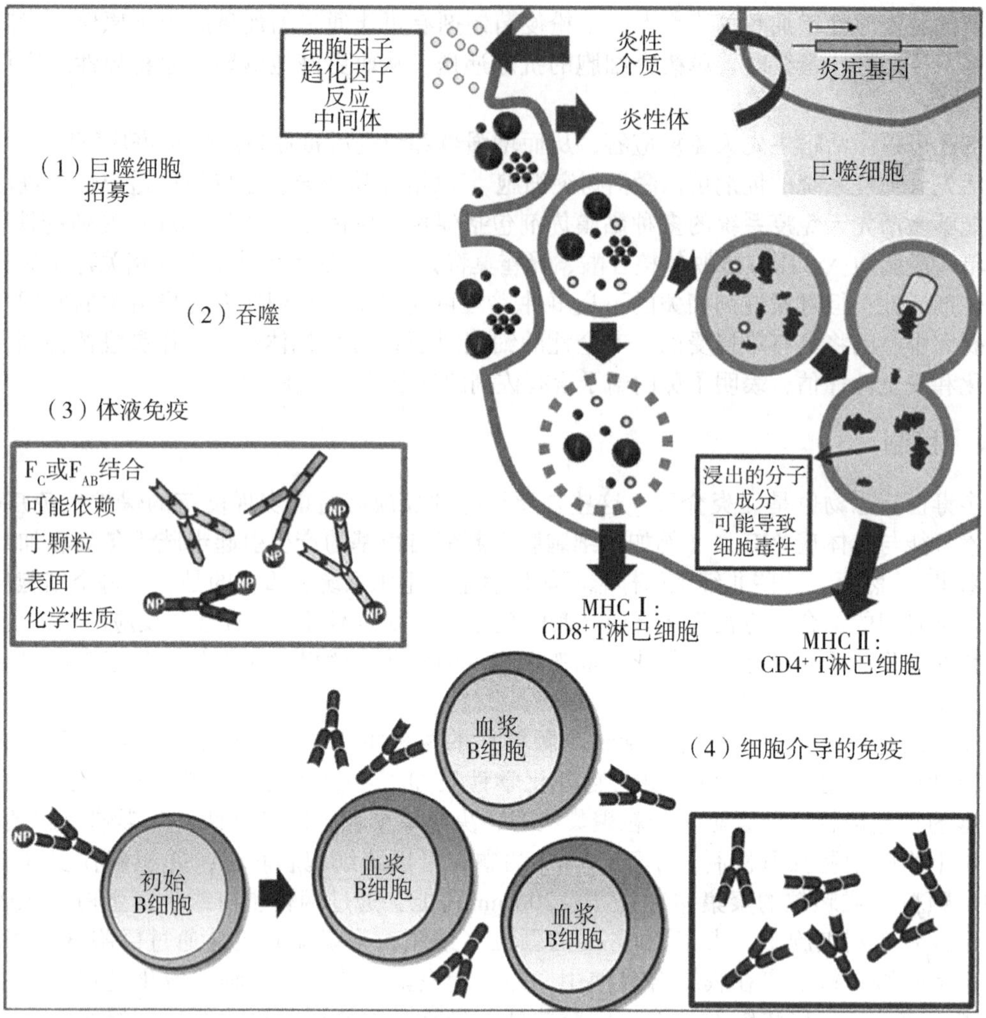

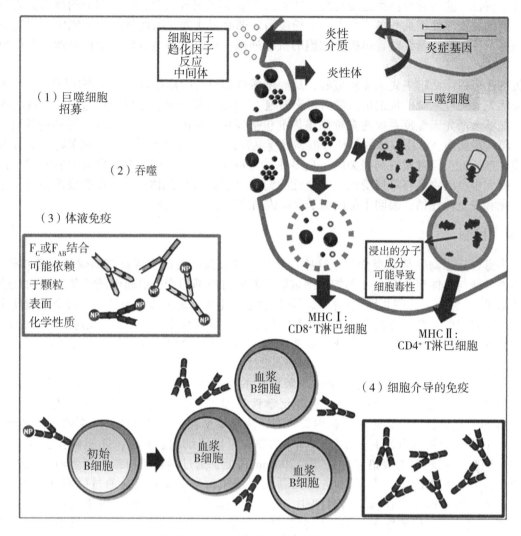

图 44.4 吸入颗粒后细胞毒性的一般机制

诱导细胞免疫或体液免疫的步骤包括:(1)吸入的颗粒诱导巨噬细胞聚集,释放细胞因子和一氧化氮介导炎症,引起局部组织损伤;(2)单个颗粒在巨噬细胞内吞途径中被吞噬和酸化,引起细胞毒素释放分子成分;(3)特定的纳米颗粒,与先天性体液的非特异性亚基结合;(4)B 细胞抗体结合部位识别颗粒,并分化成浆细胞,分泌颗粒特异性抗体,清除该特定物种中的颗粒。

对于具体的疫苗接种和免疫应答而言,很多时候是根据之前拥有相似目的研究中使用的类似免疫原,选择能产生所需要的免疫反应的实验动物。例如,选择兔作为炭疽的临床相关免疫性动物模型,是根据先前报道的动物模型,即培养滤液中疫苗产品(炭疽疫苗吸附,AVA)产生血清抗体[46]。兔吸入炭疽已有详细报道,是一个已熟知的感染模型,可用于检测特定疫苗制剂的效果[47]。在该特定实例中,通过动物模型和相应的免疫应答建立体外相关的免疫(α-PA IgG),从而预测疫苗制剂的临床效果。如果模型的基础(兔的吸入性炭疽)和相应的免疫标记物(α-PA IgG)尚未得到表征,则无法建立免疫相关性。

使用动物模型时,需要在选择实验动物之前对种间特异性生物反应进行严格评估。例如,用于研究超抗原毒素(Sag)的动物模型一般取决于动物物种中 MHC Ⅱ受体的丰富程度。MHC Ⅱ受体介导 Sag 毒素与循环 T 细胞直接结合,因此它是之后炎症反应中的关键途径。由于缺乏循环的 MHC Ⅱ受体,啮齿动物和类似啮齿动物的物种通常对纯化的葡萄球菌和链球菌肠毒素的刺激没有反应。虽然这些受体在进化程度更高的动物(如灵长类动物)中大量存在,但实验应谨慎选用较大的动物，与啮齿动物相比应仅在特定的时间间隔内使用。使用共刺激因子如 LPS 刺激或表达 MHC Ⅱ的转基因[48,49]。啮齿动物模型一直是 Sag 诱导的毒性疾病模型。Sag 诱导的毒性动物模型的具体实例表明，在设计研究之前对预测性反应进行关键评估，可能是确定疫苗产品疗效的挑战。

44.4　优点/缺点

在发生危险时，疫苗作为一种风险缓解策略，最有可能应对新出现的可能对公共健康造成危险的传染病，因此必须给卫生工作人员和军事人员提供一种快速分发和易于递送的方法。主要性能要求包括直径小于 100 nm（又名纳米颗粒）且含有在 1 周时间内释放的抗原，空气动力学直径约为 20 μm 的喷雾冷冻干燥的聚合物，并在不到 15 分钟内解聚成初级颗粒。因为黏膜纤毛清除率为 45~90 分钟，需要快速解聚使单个颗粒进入黏膜。在理想情况下，该产品可在室温下保持多年稳定。

结论

本综述的目的是为考虑进行鼻腔疫苗接种的人提供基础性概述。可向对此方法感兴趣的人更详细地介绍所遵循的基本原则及相关参考文献。数千年来，疫苗一直都作为一种保护性策略来抵抗疾病。鼻黏膜是多种给药途径之一。随着现代免疫系统的技术和知识的发展，如今我们能以理想的方式利用工程和免疫学原理开发这一重要给药途径的潜力。

我们已经探讨了鼻疫苗设计技术中需要考虑的重要因素，包括鼻黏膜的基本解剖学、生理学、细胞生物学和免疫学特征。前文介绍了鼻疫苗的重要组分，包括处方（抗原、佐剂和添加剂）、计量系统和装置。最后，我们总结了体外细胞培养和动物模型，这是评估鼻疫苗技术的基础。

疫苗在鼻黏膜部位沉淀及其疗效和安全性研究的深入，决定着经鼻传递疫苗的未来。在过去 10 年中，有关这一主题的文献显著增加，人们日益萌生对这种给药方式的兴趣。希望业界和监管机构最终能认可这种制剂的疗效和安全性，使其不仅能替代其他给药途径，还能成为一种常用的给药途径。

参考文献

[1] Hickey, A. J., Garmise, R. J.: Dry powder nasal vac-cines as an alternative to needle-based delivery. Crit. Rev. Ther. Drug. Carrier. Syst. 26, 1-27 (2009).

[2] Garmise, R. J., Staats, H. F., Hickey, A. J.: Novel dry powder preparations of whole inactivated infl uenza virus for nasal vaccination. AAPS PharmSciTech 8, E81 (2007).

[3] Hickey, A. J., Swift, D.: Aerosol measurement, prin-ciples, techniques and applications, 3rd edn, pp. 805-820. Wiley, Oxford (2011).

[4] Garmise, R. J., et al.: Formulation of a dry powder infl uenza vaccine for nasal delivery. AAPS PharmSciTech 7, E19 (2006).

[5] Alcomo, I. E.: Fundamentals of microbiology, pp. 263-266. Jones and Bartlett Publishers, Sudbury, Mass (2001).

[6] Krasner, R. I.: Biological weapons, the microbial chal-lenge, pp. 335-360. ASM Press, Washington, D. C (2002).

[7] Federal Bureau of Investigation, Famous Cases & Criminals: Amerithrax or Anthrax Investigation. www. fbi. gov/about-us/history/famous-cases/anthrax-amerithrax (2011).

[8] AHFS-DI, 'Anthrax Vaccine Adsorbed', in Bioterrorism Resource Manual, American Society of Health-System Pharmacists, Bethesda, MD, 360-377 (2002).

[9] Mantis, N. J., Morici, L. A., Roy, C. J: Mucosal vaccines for biodefense: critical factors in manufacture and delivery, pp. 181-195. Springer New York (2012).

[10] Csaba, N., Garcia-Fuentes, M., Alonso, M. J.: Nanoparticles for nasal vaccination. Adv. Drug Deliv. Rev. 61, 140-157 (2008).

[11] Koping-Hoggard, M., Sanchez, A., Alonso, M. J.: Nanoparticles as carriers for nasal vaccine delivery. Expert Rev. Vaccines 4, 185-196 (2005).

[12] Espueles, S., Gamazo, C., Blanco-Prieto, M. J., Irache, J. M.: Nanoparticles as adjuvant-vectors for vaccination, pp. 317-325. Informa Healthcare, New York (2007).

[13] Voltaire, F. M. A.: Philosophical letters. Dover, Mineola (2003).

[14] El-Kamary, S. S., et al.: Adjuvanted intranasal Norwalk virus-like particle vaccine elicits antibodies and antibody-secreting cells that express homing receptors for mucosal and peripheral lymphoid tissues. J. Infect. Dis. 202, 1649-1658 (2010).

[15] Administration, U. S. F. a. D.: Effi cacy Testing and Surrogate Markers of Immunity Workshop Vol. http: // www. fda. gov/downloads/BiologicsBloodVaccines/ NewsEvents/WorkshopsMeetingsConferences/TranscriptsMinutes/UCM054606. pdf. Center for Biologics Research and Evaluation (2002).

[16] Sewall, H.: The role of epithelium in experimental immunization. Science 62, 293-299 (1925).

[17] Wang, S. H., Kirwan, S. M., Abraham, S. N., Staats, H. F., Hickey, A. J.: Stable dry powder formulation for nasal delivery of anthrax vaccine. J. Pharm. Sci. 101, 31 - 47 (2011). doi: 10. 1002/jps. 22742.

[18] Jain, S., O'Hagan, D. T., Singh, M.: The long-term potential of biodegradable poly (lactide-co-glycolide) microparticles as the next-generation vaccine adjuvant. Expert Rev. Vaccines 10, 1731-1742 (2011).

[19] Jacques, P., et al.: The immunogenicity and reactoge-nicity profi le of a candidate hepatitis B vaccine in an adult vaccine non-responder population. Vaccine 20, 3644-3649 (2002).

[20] Keitel, W., et al.: Dose ranging of adjuvant and anti-gen in a cell culture H5N1 infl uenza vaccine: safety and immunogenicity of a phase 1/2 clinical trial. Vaccine 28, 840-848 (2010). doi: 10. 1016/j. vaccine. 2009. 10. 019.

[21] Nevens, F., et al.: Immunogenicity and safety of an experimental adjuvanted hepatitis B candidate vaccine in liver transplant patients. Liver Transpl. 12, 1489-1495 (2006).

[22] Soloff, A. C., Barratt-Boyes, S. M.: Enemy at the gates: dendritic cells and immunity to mucosal pathogens. Cell Res. 20, 872-885 (2010). doi: 10. 1038/ cr. 2010. 94.

[23] Levitz, S. M., Golenbock, D. T.: Beyond empiricism: informing vaccine development through innate immunity research. Cell 148, 1284-1292 (2012). doi: 10. 1016/j. cell. 2012. 02. 012.

[24] Lycke, N.: Recent progress in mucosal vaccine devel-opment: potential and limitations. Nat. Rev. Immunol. 12, 592-605 (2012). doi: 10. 1038/nri3251.

[25] Nochi, T., et al.: Nanogel antigenic protein-delivery system for adjuvant-free intranasal vaccines. Nat. Mater. 9, 572-578 (2010). doi: 10. 1038/nmat2784.

[26] Stanberry, L. R., et al.: Safety and immunogenicity of a novel nanoemulsion mucosal adjuvant W805EC combined with approved seasonal infl uenza antigens. Vaccine 30, 307-316 (2012). doi: 10. 1016/j. vaccine. 2011. 10. 094.

[27] Makidon, P. E., et al.: Nanoemulsion mucosal adju-vant uniquely activates cytokine production by nasal ciliated epithelium and induces dendritic cell traffi cking. Eur. J. Immunol. 42, 2073-2086 (2012). doi: 10. 1002/eji. 201142346.

[28] Bielinska, A. U., et al.: Mucosal immunization with a novel nanoemulsion-based recombinant anthrax protective antigen vaccine protects against Bacillus anthracis spore challenge. Infect. Immun. 75, 4020-4029 (2007). doi: 10. 1128/iai. 00070-07.

[29] Amorij, J. P., et al.: Pulmonary delivery of an inulin-stabilized infl uenza subunit vaccine prepared by

spray-freeze drying induces systemic, mucosal humoral as well as cell-mediated immune responses in BALB/c mice. Vaccine 25, 8707-8717 (2007).

[30] Gwinn, W. M., et al.: Effective induction of protective systemic immunity with nasally administered vaccinesadjuvanted with IL-1. Vaccine 28, 6901-6914 (2010). doi: 10. 1016/j. vaccine. 2010. 08. 006.

[31] Jaganathan, K. S., Vyas, S. P.: Strong systemic and mucosal immune responses to surface-modifi ed PLGA microspheres containing recombinant hepatitis B antigen administered intranasally. Vaccine 24, 4201-4211 (2006).

[32] Thompson, A. L., et al.: Maximal adjuvant activity of nasally delivered IL-1alpha requires adjuvant-responsive CD11c (+) cells and does not correlate with adjuvant-induced in vivo cytokine production. J. Immunol. 188, 2834-2846 (2012). doi: 10. 4049/ jimmunol. 1100254.

[33] Thompson, A. L., Staats, H. F.: Cytokines: the future of intranasal vaccine adjuvants. Clin. Dev. Immunol. 2011, 289597 (2011). doi: 10. 1155/2011/289597.

[34] Couch, R. B., et al.: Contrasting effects of type I inter-feron as a mucosal adjuvantfor infl uenza vaccine in mice and humans. Vaccine 27, 5344-5348 (2009). doi: 10. 1016/j. vaccine. 2009. 06. 084. S0264-410X (09) 00962-1 [pii].

[35] Duthie, M. S., Windish, H. P., Fox, C. B., Reed, S. G.: Use of defi ned TLR ligands as adjuvants within human vaccines. Immunol. Rev. 239, 178-196 (2011). doi: 10. 1111/j. 1600-065X. 2010. 00978. x.

[36] Lewis, D. J., et al.: Transient facial nerve paralysis (Bell's palsy) following intranasal delivery of a genetically detoxifi ed mutant of Escherichia coli heat labile toxin. PLoS One 4, e6999 (2009). doi: 10. 1371/journal. pone. 0006999.

[37] Mutsch, M., et al.: Use of the inactivated intranasal infl uenza vaccine and the risk of Bell's palsy in Switzerland. [see comment]. N. Engl. J. Med. 350, 896-903 (2004).

[38] Atmar, R. L., et al.: Norovirus vaccine against experi-mental human Norwalk Virus illness. N. Engl. J. Med. 365, 2178-2187 (2011). doi: 10. 1056/NEJMoa1101245.

[39] Hayes, J. D., Pulford, D. J.: The glutathione S-transferase supergene family: regulation of GST and the contribution of the isoenzymes to cancer chemoprotection and drug resistance. Crit. Rev. Biochem. Mol. Biol. 30, 445-600 (1995). doi: 10. 3109/10409239509083491.

[40] Sayes, C. M., Reed, K. L., Warheit, D. B.: Assessing toxicity of fi ne and nanoparticles: comparing in vitro measurements to in vivo pulmonary toxicity profi les. Toxicol. Sci. 97, 163-180 (2007). doi: 10. 1093/toxsci/ kfm018.

[41] Mittler, R.: Oxidative stress, antioxidants and stress tol-erance. Trends Plant Sci. 7, 405-410 (2002). doi: 1 0. 1016/ s1360-1385 (02) 02312-9. Pii s1360-1385 (02) 02312-9.

[42] Warheit, D., et al.: Comparative pulmonary toxicity assessment of single-wall carbon nanotubes in rats. Toxicol. Sci. 77, 117-125 (2004).

[43] Gurr, J. R., Wang, A. S. S., Chen, C. H., Jan, K. Y.: Ultrafi ne titanium dioxide particles in the absence of photoactivation can induce oxidative damage to human bronchial epithelial cells. Toxicology 213, 66-73 (2005). doi: 10. 1016/j. tox. 2005. 05. 007.

[44] Sayes, C. M., et al.: Correlating nanoscale titania structure with toxicity: a cytotoxicity and infl ammatory response study with human dermal fi broblasts and human lung epithelial cells. Toxicol. Sci. 92, 174-185 (2006). doi: 10. 1093/toxsci/kfj197.

[45] Zhu, S. Q., Oberdorster, E., Haasch, M. L.: Toxicity of an engineered nanoparticle (fullerene, C-60) in two aquatic species, Daphnia and fathead minnow. Mar. Environ. Res. 62, S5-S9 (2006). doi: 10. 1016/j. marenvres. 2006. 04. 059.

[46] Pitt, M. L., et al.: In vitro correlate of immunity in a rabbit model of inhalational anthrax. Vaccine 19, 4768-4773 (2001).

[47] Zaucha, G. M., Pitt, L. M., Estep, J., Ivins, B. E., Friedlander, A. M.: The pathology of experimental anthrax in rabbits exposed by inhalation and subcutaneous inoculation. Arch. Pathol. Lab. Med. 122, 982-992 (1998).

[48] Roy, C. J., et al.: Human leukocyte antigen-DQ8 trans-genic mice: a model to examine the toxicity of aerosolized staphylococcal enterotoxin B. Infect. Immun. 73, 2452-2460 (2005).

[49] LeClaire, R. D., et al.: Potentiation of inhaled staphy-lococcal enterotoxin B-induced toxicity by lipopolysaccharide in mice. Toxicol. Pathol. 24, 619-626 (1996).

(吴国华译)

第 45 章　纳米技术在疫苗递送中的应用

Martin J. D'Souza，Suprita A. Tawde，
Archana Akalkotkar，Lipika Chablani，
Marissa D'Souza 和 Maurizio Chiriva-Internati①

摘要

若干世纪以来，免疫一直是消除传染病的可靠疗法。有效的免疫疗法取决于若干因素，例如病原体的性质、疫苗递送系统、给药途径和宿主的免疫系统。随着纳米技术的进步，免疫疗法可用于治疗包括癌症和传染病在内的多种具有挑战性的疾病。随着多种增强免疫应答的疫苗佐剂不断发展，纳米技术的自身佐剂微粒正发挥着重要作用。

45.1　引言

纳米技术的进步带来了预防或治疗各种疾病的多种方法。这种技术对免疫疗法的影响增强了疫苗的传递和功效。

免疫疗法是一种特殊的消除疾病的方法，有助于免疫系统抵御外来抗原（传染病）或自身抗原（癌症）的攻击。数个世纪以来，免疫疗法已被广泛证明是一种预防疾病的经济实惠的措施。

随着不同挑战性疾病的出现，迫切需要开发用于这些疾病的疫苗，挽救全世界数百万人的生命。就目前的疫苗而言，仍需解决的问题有安全性、有效性、给药便捷性、制备时间以及成本。

随着免疫学和分子生物学对新型疫苗材料的探索，旨在触发机体对疫苗的记忆性免疫应答，这样可长时间保护宿主的免受疾病侵袭。疫苗的功效取决于是否能通过 Th1 和 Th2 免疫途径诱导记忆性 T 细胞和 B 细胞。传统疫苗材料包括完整的外来生物疫苗（活微生物、减毒微生物、杀灭的微生物或裂解物）、病原体的细胞碎片如细菌多糖和细菌毒素[1]。重组技术和 RT-PCR 的发展使更大规模地进行特异性抗原表达或合成肽成为可能，然后再用作疫苗。DNA 疫苗是最近开发的一种免疫治疗型疫苗，有些临床试验结果令人满意[2-4]。

疫苗有两种：预防性疫苗或治疗性疫苗。预防性疫苗可用于预防病毒、细菌或寄生虫等外来抗原引起的传染病，如流感、艾滋病毒、肺结核、疟疾、肺炎、脊髓灰质炎和天花。对于细胞自身改变引起的癌症，疫苗制剂需诱发针对自身细胞抗原的免疫应答，而不引起自身免疫应

---

① M. J. D'Souza，PhD（✉）· A. Akalkotkar

M. D'Souza（摩斯大学药学与健康科学学院药物科学系疫苗纳米技术实验室，美国佐治亚州亚特兰大，E-mail：dsouza_mj@ mercer. edu）

S. A. Tawde，PhD（爱克龙药业，美国伊利伊诺州费农山 60031）

L. Chablani，PhD（圣约翰费舍尔大学威格曼斯药学院药物科学系，美国纽约罗切斯特 14618）

M. Chiriva-Internati，PhD（得克萨斯理工大学健康科学中心医学院，美国德克萨斯州拉伯克市）

答，这颇具挑战。市场上用于预防癌症的商品化疫苗不多，例如治疗可导致宫颈癌的人乳头瘤病毒感染的 Gardasil® （默克）和 Cervarix® （葛兰素史克）。预防癌症的疫苗利用过度表达或发生突变的蛋白质、发生突变的致癌生长因子受体、热休克蛋白或其他肿瘤相关抗原预防肿瘤形成[5]。治疗性疫苗诱导机体抵御现有残存肿瘤细胞的免疫应答，与手术或化疗联合应用，旨在预防或延缓肿瘤复发[6]。市场上目前只有一种治疗前列腺癌的疫苗 Provenge® (Dendreon) 获得了 FDA 批准。研究人员正对黑色素瘤和结直肠癌进行研究[7]。据报道，癌症免疫疗法领域出现了持续生长，已开展了 DNA 或树突细胞（DC）或病毒载体疫苗的各种其他临床试验[8,9]。

疫苗的功效主要依赖于抗原的免疫原性。疫苗佐剂可以激活免疫细胞，可进一步增强抗原的免疫原性。研究人员正研究各种佐剂是否能诱导机体抵御多种抗原的体液、细胞和（或）黏膜免疫。研究发现，蛋白质佐剂主要诱发体液免疫应答。ISCOM、Montanide TM、Montanide ISA720、ISA 51 和病毒载体诱导机体细胞毒性 T 细胞应答。MF59 和 MPL® （单磷酰脂质）可增强 Th1 应答。病毒样颗粒、不可降解的纳米微粒和脂质体诱导细胞免疫。Douglas 等整合 Montanide ISA720 作为佐剂，可产生的针对恶性疟原虫 MSP1 的 T 细胞和 B 细胞应答等于或高于单独使用病毒或蛋白质佐剂所诱导的免疫应答。癌症疫苗商品 Cervarix® 中的单磷酰脂质 A（MPL）是用作靶向于 TLR-4 的佐剂，Gardasil® 含有佐剂明矾。复合物 AS04（明矾和单磷酰脂质 A 的组合物）也已用于抗乙型肝炎病毒的人疫苗。

已经批准能用于人体的佐剂包括明矾、复合物 AS04（明矾和单磷酰脂质 A 的组合物）、AS03 和 MF59。许多疫苗使用明矾作为佐剂，例如甲型肝炎病毒疫苗、乙型肝炎病毒疫苗、人乳头状瘤病毒疫苗、白喉疫苗、破伤风疫苗、B 型流感嗜血杆菌疫苗和肺炎球菌偶联疫苗。明矾不能诱导强的 Th1 免疫应答。已用动物模型进行过试验的佐剂包括细菌毒素，例如霍乱毒素、大肠杆菌的热不稳定肠毒素、霍乱毒素的无毒性 B 亚单位、Toll 样受体（TLR）9 激动剂和胞嘧啶磷酸鸟苷（CpG）二核苷酸[10]。治疗疟疾、癌症、流感、乙型肝炎、丙型肝炎、艾滋病和结核病等各种疾病的一些临床试验使用 Montanide、PLG、鞭毛蛋白、QS21、AS01、AS02、RC529、ISCOM、IC31、CpG、具有 MTP-PE 的 MF59、免疫刺激序列（ISS）和 1018 ISS 作为佐剂[11]。Heffernan 等发现，同时使用含有壳聚糖和 IL-12 的制剂可诱发机体产生针对模型蛋白疫苗的 Th1、IgG2a 和 IgG2b 抗体免疫应答。Denisov 等评估了各种佐剂（larifan、polyoxidonium、硫代硫酸钠、TNF-β 和 Ribi 佐剂系统）增强豚鼠对活布鲁氏菌病疫苗、布鲁氏菌流产株 82-PS（青霉素敏感）的免疫应答能力，结果发现联合应用 TNF-β 或聚氧鎓与 S82-PS 的保护作用最强。

陈等最近的研究结果推断细菌细胞内信号分子 3′ 5′-环二鸟苷酸（c-di-GMP）具有免疫刺激性，可用作疫苗佐剂。

Skountzou 等研究发现，大肠杆菌和沙门氏菌的细菌鞭毛蛋白与灭活的 A / PR / 8/34 (PR8) 病毒经鼻内联合给药可增强流感疫苗在小鼠体内的效力。这些蛋白质是黏膜疫苗佐剂的优良候选物，可改善机体对流感及其他传染病的抵抗作用。使用白细胞介素、IL-2、IL-12 和 GM-CSF 等细胞因子作为佐剂可以改善癌症疫苗的功效[12-15]。

提高疫苗功效的另一种方法是使用纳米技术。研究人员探索了纳米技术在传递小分子、蛋白质和多肽方面的不同应用。最近，各种制药方法已能够通过疫苗递送实现效力增强，易于传递，性能稳定。已用疫苗材料制成纳米载体，例如脂质体、聚合物纳米颗粒、ISCOM，树枝状大分子、胶束、病毒样颗粒和碳纳米管。

45.2 对微粒疫苗的需求

尽管目前市场上尚无商品化的微粒疫苗，但研究人员正对这此领域进行深入的研究，使微粒疫苗从试验室阶段进入临床试验阶段。纳米疫苗可刺激机体的体液免疫和细胞免疫，故治疗常见疾病的纳米疫苗颇具吸引力[16]。已证明纳米疫苗可产生黏膜免疫，抵抗经黏膜进入的病原体引起的传染病[17]。这些纳米载体的优点是其大小与病原体相似；可被免疫系统的抗原呈递细胞（APC）有效识别[18]。这些纳米载体汇入附近的淋巴结，激活机体的免疫细胞。这些免疫细胞汇入接收各种趋化因子信号的上皮看门细胞[19]。

与自然感染不同，单独使用疫苗无法产生高的抗体反应[20]。在单一实体中利用纳米粒子整合多种抗原作为疫苗可增强体液反应，产生细胞免疫[16,21-24]。Uddin、Lai 和 Yeboah 等分别使用微粒疫苗传递系统成功制备和测试了用于伤寒、黑色素瘤和肺结核的口服疫苗[25-27]。这些研究发现，微粒疫苗口服给药后产生的黏膜和血清抗体滴度（IgA 和 IgG）显著高于口服疫苗溶液。抗原递呈的持续时间也对增强免疫应答起重要作用[28]。抗原须以脉冲方式释放，以降低所需的加强剂数量。只有微粒保持完整，在胃部恶劣的酸性环境中才能不被降解，抗原才能持久。

修饰纳米微粒的外表面还可增加抗原呈递细胞的摄入。抗原可与免疫刺激或靶向部分偶联；可修饰表面的固有属性。表面电荷在微粒摄取中起重要作用，可影响免疫应答水平。结果表明，巨噬细胞和树突细胞（DCs）可以摄入阳离子微粒[29]。

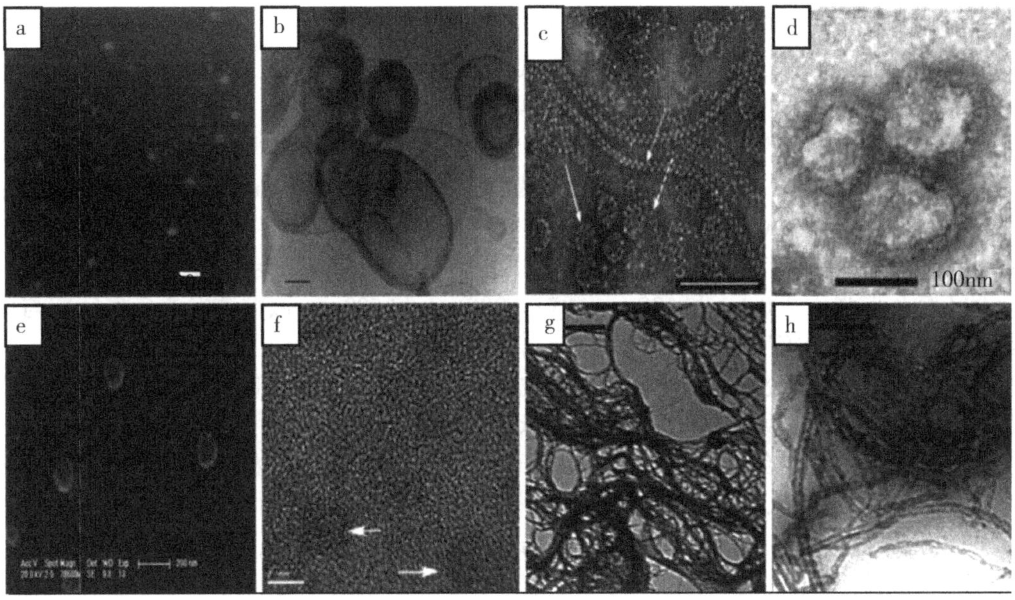

图 45.1　传递疫苗的不同纳米载体

（a）PEG-PLGA 纳米微粒的 TEM 图像（比例尺对应于 500 nm）（在获得 Elsevier 许可的情况下转载自 Bharali 等[34]）。（b）包埋 DNA 的阳离子脂质体的 Cryo-EM 图像（比例尺对应于 200 nm）（经 Elsevier 许可后，转载自 Perrie 等[35]）。（c）不同类型的 ISCOM 的 TEM 图像，例如典型的笼状（实心箭头）、螺旋（虚线箭头）和双螺旋（虚线箭头）（比例尺对应于 200 nm）（经 Elsevier 许可后，转载自 Sun 等[36]）。（d）H1N1 型流感病毒样颗粒的 EM 图像（比例对应于 100 nm）（转载自 Quan 等[37]，开放性文章）。（e）PEG-PEIPBLG 共聚物胶束的 SEM 图像（比例对应于 200 nm）（经 Elsevier 许可后，转载自 Tian 等[38]）。（f）PAMAM 树枝状大分子的 TEM 图像（经 ACS 许可后，转载自 Jackson 等[39]）。（G，H）单壁（g）和多壁（h）碳纳米管的 TEM 图像（比例尺分别对应于 1 μm 和 250 nm）（获得 Elsevier 许可后，转载自 Klumpp 等[40]）。

纳米微粒的另一个作用是整合各种免疫增强剂，以进一步增强免疫应答，其中包括有助于最大程度地降低疫苗不良影响的靶向配体。这些靶向配体包括 aleuria aurantia 凝集素（AAL）、 ulexeuropaeus 凝集素 1 凝集素（UEA-1）以及 Peyer 氏斑中 M 细胞靶向配体的小麦胚芽凝集素（WGA）。这有助于增加经小肠摄入的微粒，但不影响口服的耐受性[30,31]。这种制剂可用于包裹增加免疫应答的各种共刺激分子如白细胞介素或细胞因子。研究表明，树突细胞具有 IL-2 和 IL-12 受体；疫苗接种可提供外源性抗原，激活 MHC Ⅰ类（交叉递呈）和 MHC Ⅱ类途径[32,33]。

本综述旨在探讨这些纳米载体作为疫苗传递载体的功能，如图 45.1 所示。每种载体的简要说明如下。

45.3 聚合物纳米微粒

聚合物纳米微粒（图 45.1a）本身可用作佐剂，故研究人员对其用作疫苗传递载体进行了广泛研究。

聚合物纳米微粒可在口服给药时保护蛋白质和多肽，防止其在胃部降解，故疫苗是这些微粒的主要应用之一[41]。粒径小于 1 μm 的微粒可以促进抗原呈递细胞的靶向摄取和长期递呈，产生适应性免疫应答[42]。免疫应答还取决于抗原的溶解速率、表面形态、电荷和大小[43]。

口服给药更符合患者的要求，是首选的给药途径。肠道对这些微粒的摄取是决定口服疫苗功效的关键因素。一直以来，人们对使用纳米粒与微粒作为传递口服疫苗的载体争论不休。文献报道的诱导较强而持久免疫应答的理想粒径范围相互矛盾[44]。

Desai 等的此项研究表明，大鼠模型中对 100 nm 微粒的跨肠摄取能力大于 500 μm、1 μm 和 10 μm 微粒。这些微粒由聚乳酸—聚乙醇酸共聚物（50∶50）制备。传统纳米微粒因空间以及粘合的相互作用，易于包裹在黏液中。调整纳米微粒的大小可使这些相互作用过度，使这些微粒可经黏液扩散[45]。Primard 等报道，口服给药时，粒径大于 300 nm 的纳米微粒穿越肠黏液内层的效果较差。小肠 Peyer 小片的 M-细胞可吸收粒径为 200~250 nm 的微粒。

Gutierro 等的一项研究结果表明，模型蛋白质——牛血清白蛋白与 PLGA 整合的 1 000 nm 微粒诱发的 IgG 反应强于 200 nm 和 500 nm 微粒，200 nm 和 500 nm 微粒诱发的免疫反应与口服和皮下给药途径诱导的免疫反应相当[46]。

Wendorf 等的一项研究与此矛盾，该研究比较了粒径为 110 nm 和 800~900 nm 的聚丙交酯-共-乙交酯纳米微粒的功效，结果发现这些纳米微粒诱发的免疫应答相当[47]。van den Berg J. 等的研究评估了经皮肤途径免疫含 DNA 疫苗的阳离子纳米微粒。结果发现，静电相互作用使细胞外基质中的纳米微粒固定，这些阳离子纳米微粒阻断了疫苗诱导的小鼠和离体人皮肤中的抗原表达。PEG 化屏蔽了纳米微粒的表面电荷，可改善体内抗原表达[48]。聚乳酸是一种广泛用于疫苗传递的可生物降解聚合物。其疏水性和降解时产生的酸性微环境不利于抗原包封，使用受到限制。Jain 等的一项[49]研究使用 PEG 衍生物 PLA 嵌段共聚物包裹乙型肝炎表面抗原形成纳米微粒，用于乙型肝炎疫苗的黏膜免疫。结果发现，这些聚合物所诱导的分泌型免疫球蛋白 A 的黏膜免疫应答较好[49]，使用这些纳米微粒还可改变癌症的 T 细胞免疫应答[50]。已尝试了数种其他聚合物用于制备表 45.1 所列的疫苗纳米微粒。Primard 等进行了一项有趣的摄取研究，结果表明口服给药后聚（乳酸）纳米微粒跨肠黏膜至派尔氏结片，然后与底层 B 细胞和树突细胞相互作用[45]。聚合物纳米微粒具有这些优点，仍然是潜在的疫苗传递系统。

45.4 脂质体

虽然许多纳米载体可以用于疫苗传递，但是脂质体依然在药物和疫苗传递过程中发挥着重要的作用。Bangham 于 20 世纪 60 年代发现脂质体[51]，不久之后 Allison 发现其佐剂的功能[52]，从此以后脂质体许多研究都是围绕其佐剂的功效。由磷脂组成的具有纳米结构（图 45.1b）的

表 45.1 目前用于动物模型试验的各种疫苗聚合纳米颗粒研究综述

聚合物	大小/nm	电压/mV	疫苗制备	免疫反应	接种途径	制备方法	动物模型	参考文献
甲氧基聚乙二醇-聚(丙交酯-聚乙二醇)	150~200	NA	重组乙肝表面抗原(HBsAg)	抗-HBs 抗体	IP	双乳剂-溶剂蒸发法	BALB/cNCr 小鼠	[34]
聚甲基丙烯酸甲酯(PMMA)与亲水性天竺葵和聚乙二醇(乙二醇)链的内硬核	960±38	32.2±0.6	HIV-1 Tat	体液、细胞反应 Th1-型 T 细胞反应和 CTLs	IM	乳液聚合	BALB/c(H2kd)小鼠	[53]
聚(γ-谷氨酸)	200	NA	HIV-1 gp120	树突状细胞有效摄取，$CD8^{+}$ T 细胞反应，对 SHIV-IN 攻毒后的保护作用不明显	IN，SC	溶剂蒸发	猕猴	[54]
聚甲基丙烯酸甲酯(PMMA)内核与 Eudragit L100-55 亲水外壳	NA	NA	HIV-1 Tat	持久的细胞和体液反应	IM，SC，IN	乳液聚合	BALB/c 小鼠	[55]
聚丙烯硫化物	50	NA	卵清蛋白作为模型抗原	肺、脾组织细胞毒性 T 淋巴细胞反应及气道黏膜的体液反应	IN	NA	C57BL/6 小鼠	[56]
聚丙烯硫化物	NA	NA	卵清蛋白作为模型抗原	细胞毒性和辅助性 T 细胞应答	ID	乳液聚合	C57Bl/6 小鼠	[57]
N-三甲基壳聚糖和聚乳酸-乙醇酸	500	24.5±0.90	卵清蛋白作为模型抗原	当鼻腔给药时能产生体液免疫反应和黏膜免疫反应	IM，IN	乳化/溶剂萃取	Balb/c 小鼠	[58]
多糖壳聚糖	160~200	6~10	重组乙型肝炎表面抗原(rHBsAg)	抗-HBsAg IgG 水平	IM	温和离子凝胶技术 e	BALB/小鼠	[59]

IM 肌内；SC 皮下；IP 腹腔内；ID 皮内；IN 鼻内；NA 不适用。

脂质体，可以包裹亲水性和疏水性以及不同来源抗原。这些微粒不仅可作为载体保护抗原的生物活性部分，还具有一定的免疫原性，故是潜在的佐剂[35,60-62]。尽管随着隐形或聚乙二醇化脂质体的出现，循环纳米载体的半衰期已延长[64]，但传统的脂质体可被网状内皮系统摄取，从体内迅速清除，故未能用作疫苗微粒[63]。Doxil® 是多柔比星的聚乙二醇化脂质体，是一种利用该应用的上市产品，可治疗癌症。已上市的其他脂质体制剂包括 Ambisome®（Gilead）、Myocet®（Elan）和 Depocyt®（SkyePharma）。

研究人员已采用各种方法增强这些载体的免疫原性。Mohammed 等描述了使用阳离子脂质体改善其稳定性，产生了针对结核分枝杆菌的持续免疫作用[65]。研究人员对整合在脂质体中的佐剂是否能产生免疫刺激作用进行了研究；最近的研究发现，整合了单磷酰脂质 A 的二甲基二（十八烷基）溴化铵（DDA）和海藻糖 6，6′-二山嵛酸酯（TDB）脂质体可诱导体液免疫应答和细胞免疫免疫[66,67]。Altin 等进一步对脂质体和质膜囊泡（PMV）在作为载体靶向传递抗原的应用进行了综述[63]。有希望作为抗原载体的脂质体有各种其他形式，如病毒体、右菌体和蛋白体[68-70]。

这些纳米载体除具有免疫调节功能之外，其物理性质对其作为有效的疫苗传递载体也很重要。Xiang 等探讨了微粒疫苗开发中粒径的作用，并描述了各种粒径范围及其各自的摄取机制；较小的脂质体可以模拟病毒的摄取机制，而较大的脂质体遵循细菌途径[43]。如前所述，这些载体表面的电荷还决定了其作为微粒疫苗的功效；例如，阳离子脂质体的功效优于其他脂质体[71]。载体的物理性质的改变、免疫佐剂的使用以及其他未知性质增强了其作为微粒疫苗的用途[72]。鉴于脂质体产品在市场取得成功，希望不久后有脂质体疫苗上市。

45.5 免疫刺激复合体（ISCOM）

如图 45.1c 所示，ISCOM（免疫刺激复合物）是 40 nm 的微粒疫苗纳米载体，由胆固醇、磷脂和皂苷以及抗原组成。现在的 ISCOMATRIX™不含抗原，且与 ISCOM 的组成相同。该基质整合了可用作 ISCOMATRIX™疫苗的抗原，其免疫刺激活性与 ISCOM 相似。Quil A 是一种纯化自 Quillaja 皂苷的提取物，毒性较低，故这些复合物具有免疫刺激性。据报道，这些复合物可产生针对各种抗原如病毒、细菌、寄生虫或肿瘤的免疫应答[73,74]。

一些研究人员尝试改变或替换一些成分，如磷脂或 Quil A，以增强这些复合物的免疫刺激性[36]。数种 ISCOM™ 和 ISCOMATRIX™ 疫苗可诱导动物模型产生体液和细胞免疫应答（表 45.2）。这些系统可进入 MHC Ⅰ和 MHC Ⅱ 途径，作为先天免疫系统和适应性免疫系统的有效免疫调节剂。人经鼻内途径免疫流感 ISCOMATRIX™疫苗可诱导全身和黏膜免疫反应，故 ISCOMATRIX™佐剂可用作黏膜佐剂[75]。ISCOM 和 ISCOMATRIX™的癌症和传染病疫苗可诱导机体抗原特异性细胞毒性 T 淋巴细胞、T 辅助细胞和抗体[36]。改变 ISCOM 电荷可产生阳离子 ISCOM 衍生物（PLUSCOM），其阴离子抗原负载量增大，故 T 细胞应答强于针对模型蛋白抗原（卵清蛋白）的经典阴离子 ISCOM[84]。这些复合物可减少诱导免疫应答所需的抗原剂量[85]。表 45.2 列出了已进行了体内研究的针对各种感染的不同 ISCOM 和 ISCOMATRIX™。

表 45.2 使用动物模型研究的各种基于免疫刺激复合物（ISCOM）的疫苗概述

疫苗传递系统	疫苗制剂	免疫应答	给药途径	动物模型	参考文献
ISCOM	流感病毒 H3N2	体液免疫和细胞免疫	N A	短尾猴	[76]
ISCOM	用作人诺罗病毒 GII.4 HS66 株疫苗的佐剂	Th2 反应，细胞分泌 IgM、IgA 和 IgG 抗体显著升高	口服/IN	生猪	[77]

（续表）

疫苗传递系统	疫苗制剂	免疫应答	给药途径	动物模型	参考文献
ISCOM	禽流感 H5N1 亚型病毒	强烈抗体反应	IM	公鸡	[78]
ISCOM	流感病毒 A/PR8/34	黏膜以及全身抗体反应和细胞毒性 T 淋巴细胞反应	IN	BALB/c 小鼠	[79]
ISCOM	A 型流感病毒 H5N1 病毒颗粒	Th1 $CD4^+$细胞应答和强烈的抗体反应	IM	BALB/c 小鼠	[80]
ISCOMATRIX™	MEM 流感抗原	黏膜和血清抗体反应	IN	BALB/c 小鼠	[81]
ISCOMATRIX™ 和 ISCOM™	幽门螺杆菌	减少幽门螺杆菌定植	IN/SC	mice 小鼠	[82]
ISCOM	细胞内新芽孢虫的重组 NcSRS2 抗原	牛新芽孢虫特异性抗体和细胞反应	SC	BALB/c 小鼠	[83]

IM 肌内；SC 皮下；IN 鼻内；NA 不适用。

45.6　病毒样颗粒

如图 45.1d 所示，病毒样颗粒（VLPs）是一种可最有效传递疫苗的纳米载体[86]。顾名思义，这些微粒类似于 22~150 nm 大小范围的病毒，含各种病毒的自组装包囊或蛋白质。这些微粒缺乏遗传物质，故无传染性。根据 Noad 等报道，已有 30 多种病毒可以制备病毒样颗粒，使得这种方法成为了需求[87]。鉴于这种传递系统的各种优点，目前已有商品化的乙型肝炎病毒感染和人乳头状瘤病毒的病毒样颗粒疫苗[88,89]。这种微粒传递系统的各种临床试验正在进行中。最近，Buonaguro 等探讨了病毒样颗粒在微粒疫苗中的作用、对当前疫苗和临床试验的贡献，以及这些微粒诱导的免疫应答[90]。Grgacic 等详细综述了病毒样颗粒作为疫苗微粒在免疫应答中的作用[91]。从结构上看，根据衣壳蛋白周围有无脂质包膜，病毒样颗粒可分为有包膜的或无包膜的。

人乳头瘤病毒（HPV）的病毒样颗粒是一个很好的实例，这是一种以 L1 为主要衣壳蛋白组成的单衣壳无包膜病毒样微粒。这些病毒样颗粒可用酵母（Gardasil）以及杆状病毒（Cervarix）感染的昆虫细胞进行表达。Schiller 等综述了使用这些人乳头瘤病毒-病毒样颗粒进行的临床试验，并描述了其在治疗人乳头瘤病毒感染方面的功效[92]。该综述还强调指出，在临床试验期间病毒样颗粒疫苗十分安全。

已有针对甲型流感、乙型肝炎、丙型肝炎和若干种逆转录病毒的各种包膜病毒样颗粒。最近，Kang 等的研究表明，可使用病毒样颗粒包被的微针经透皮途径接种甲型流感疫苗，增强患者对这些纳米载体的依从性[93]。鉴于这些病毒样颗粒的广泛应用及其作为商品微粒疫苗取得的成功，其仍然是未来疫苗的潜在纳米载体[37,94-99]。

45.7　聚合物胶束

两亲性分子或亲水性区域保持适当浓度和温度时，自发形成的缔合胶体类[38,100]。这些胶束不通过共价键结合，容易解离。可根据其应用方法利用胶束的这种性质[17]。这些胶束在体外和体内都很稳定[101]。选择合适的亲水和疏水聚合物可控制聚合物胶束的物理和化学性质[102]。Morein 等在引起的肺炎的 PI-3 病毒 30 S 蛋白亚基胶束疫苗的攻毒试验中发现，该疫苗可诱导一定的抗体和保护性免疫[103]。Prabakaran 等进行了类似研究，使用大豆磷脂酰胆碱胶束抵御 H5N1 感染，[104]。研究发现血清 IgG、黏膜 IgA 和 HI 滴度高于游离抗原。胶束有希望用作疫苗抗原的载体。

45.8 树状聚合物

树枝状聚合物（图 45.1f）是高度分支化的单分散聚合物纳米微粒。树枝状聚合物由 3 种不同的组分组成：引发剂核心、分支和末端官能团。引发剂核心是树枝状聚合物的主要组分，分支是外部方向上的延伸。可根据电、亲水或亲脂性质对末端基团进行修饰[105]。这些聚合物类似于聚合物胶束，但与胶束不同的是这些聚合物经共价键连接，故化学键较强，不易于分离[106]。外表面很容易进行修饰，改变内腔使树枝状聚合物有望成为各种生物医学和工业应用的载体[107]。Baker 等最近的工作是将各种功能分子（包括传感单元、MRI 造影剂、触发装置和靶向分子）偶联到第 5 代树枝状聚合物（MW 25 000Da，直径 5 nm）的表面[108]。一种称为多抗原肽（MAP）系统的特定类别的树枝状聚合物已广泛用于疫苗。基于 MAP 的疟疾疫苗已进行了 I 期临床试验[109-111]。这些传递系统具有进入临床试验的潜力，预计不久就会上市。表 45.3 列出了目前正在研究的一些树枝状聚合物系统。

45.9 碳纳米管

最近，用于传递疫苗的纳米晶体、纳米线和纳米管等无机纳米材料日益受到关注。碳纳米管（图 45.1g、图 45.1h 所示）能够连接抗原，保持其构象，诱导抗原特异性抗体应答，故研究人员对其用作疫苗载体进行了研究。这些碳纳米管也可修饰成非免疫原性材料[115]。功能化的碳纳米管通过共价键形成用于传递抗原的纳米载体或通过非共价形成超分子组装体[40]。尽管碳纳米管仍是目前研究人员感兴趣的领域，但在进入临床试验之前仍需进行大量的工作。

45.10 挑战和未来方向

随着各种纳米载体用于传递疫苗，纳米技术不仅获得了当之无愧的风头，而且还吸引了监管机构的注意，在这些纳米载体上市之前仍需考虑一些挑战。

纳米颗粒疫苗的优点和缺点

纳米疫苗作为传递系统有其自身的优点和缺点。这些微粒由可生物降解的聚合物组成，安全性较高。纳米微粒疫苗具有不同的给药途径，例如肠胃外、口服、透皮、鼻、甚至肺部途径。经一种非侵入性的非肠胃外传递系统可无痛传递疫苗，优于常规疫苗[116]。如前所述，这些纳米微粒可有效激活免疫系统。纳米微粒可按所需的方式控制抗原的释放速率和时间[117]。

生产和储存这些疫苗的成本是一项基本问题。纳米疫苗的重现性是一个更大问题[70]。据报道，粒径大于 300 nm 的纳米微粒不能有效跨越肠黏膜内层，透过肠道中的派尔氏结摄取的微粒较少，对疫苗微粒的免疫应答较小[45]。纳米微粒的大小和电荷对疫苗制剂的功效起关键作用。应确保疫苗生产过程具有重现性，这需对微粒进行严格评估。另一个问题是需注意解决非热法灭菌[118]。小的纳米微粒在机体中被迅速清除，较大的聚集体在器官中蓄积可产生毒副作用。

表 45.3　树状大分子疫苗动物模型研究综述

所用聚合物	大小/nm	电压/mV	疫苗制备	免疫反应	接种途径	制备方法	动物模型	参考文献
聚醚胺（PETIM）	NA	NA	抗狂犬病的 PETIM-pDNA 复合物	病毒中和抗体反应	IM	NA	瑞士白化小鼠	[112]
聚（丙烯胺）	NA	21.3±0.33	编码 pRc/CMV-HBs [S] 的质粒 DNA	IgG 及其亚型-IgG1，IgG2a，IgG2b	IM	络合	Balb/c 母鼠	[113]
最小 B 细胞表位 J14 的聚丙烯酸酯核	20	-16	J 14B 细胞表位	IgG 亚型抗体反应-IgG1，IgG2b，and IgG3	SC	透析后“点击”反应	小鼠模型	[114]

IM 肌内；SC 皮下；NA 不适用。

"纳米"级使这些载体的前途充满希望，这也是对这些传递系统的关注点。研究人员指出，载体越小，功能越佳，且仍受身体网状内皮系统的保护。研究人员还设计了一些方法使这些载体具备隐形性质，以避免被吞噬细胞摄入。尽管所有这些性质使纳米载体成为潜在的传递系统，但也使其难以从体内清除，增加了"纳米毒性"。研究人员已进行了相对广泛的研究，以测定纳米大小分子而非纳米载体的毒性特征，对这种传递疫苗的纳米载体的毒性作用知之甚少。尽管业内仍对使用这些载体表示质疑，但正在进行各种研究以寻找答案，而监管机构也参与了这些争论。

利益声明如附件所示，Suprita A. Tawde、ArchanaAkalkotkar 和 LipikaChablani 是研究生，Marissa D'Souza 是一名夏季研究生，在佐治亚州亚特兰大默瑟大学药学与健康科学学院药学系疫苗纳米技术实验室 Martin J. D'Souza 教授的领导下工作。

参考文献

[1] Heffernan, M. J., Zaharoff, D. A., Fallon, J. K., Schlom, J., Greiner, J. W.: In vivo effi cacy of a chitosan/IL-12 adjuvant system for protein-based vaccines. Biomaterials 32, 926-932 (2011).

[2] Guimaraes-Walker, A., et al.: Lessons from IAVI-006, a phase I clinical trial to evaluate the safety and immunogenicity of the pTHr. HIVA DNA and MVA. HIVA vaccines in a prime-boost strategy to induce HIV-1 specifi c T-cell responses in healthy volunteers. Vaccine 26, 6671-6677 (2008).

[3] Beckett, C. G., et al.: Evaluation of a prototype dengue-1 DNA vaccine in a Phase 1 clinical trial. Vaccine 29 (5), 960-968 (2011).

[4] Martin, J. E., et al.: A SARS DNA vaccine induces neutralizing antibody and cellular immune responses in healthy adults in a Phase I clinical trial. Vaccine 26, 6338-6343 (2008).

[5] Finn, O. J., Forni, G.: Prophylactic cancer vaccines. Curr. Opin. Immunol. 14, 172-177 (2002).

[6] Arlen, P. M., Madan, R. A., Hodge, J. W., Schlom, J., Gulley, J. L.: Combining vaccines with conventional therapies for cancer. Update. Cancer. Ther. 2, 33-39 (2007).

[7] Tartaglia, J., et al.: Therapeutic vaccines against melanoma and colorectal cancer. Vaccine19, 2571-2575 (2001).

[8] Anderson, R. J., Schneider, J.: Plasmid DNA and viral vector-based vaccines for the treatment of cancer. Vaccine 25 (Suppl 2), B24-B34 (2007).

[9] Sheng, W. Y., Huang, L.: Cancer immunotherapy and nanomedicine. Pharm. Res. 28 (2), 200-214 (2011).

[10] Harandi, A. M., Medaglini, D., Shattock, R. J.: Vaccine adjuvants: a priority for vaccine research. Vaccine 28, 2363-2366 (2010).

[11] Mbow, M. L., De Gregorio, E., Valiante, N. M., Rappuoli, R.: New adjuvants for human vaccines. Curr. Opin. Immunol. 22, 411-416 (2010).

[12] Nguyen, C. L., et al.: Mechanisms of enhanced antigen-specifi c T cell response following vaccination with a novel peptide-based cancer vaccine and systemic interleukin-2 (IL-2). Vaccine 21, 2318-2328 (2003).

[13] Toubaji, A., et al.: The combination of GM-CSF and IL-2 as local adjuvant shows synergy in enhancing peptide vaccines and provides long term tumor protection. Vaccine 25, 5882-5891 (2007).

[14] Yin, W., et al.: A novel therapeutic vaccine of GM-CSF/TNFalpha surface-modifi ed RM-1 cells against the orthotopic prostatic cancer. Vaccine 28, 4937-4944 (2010).

[15] Germann, T., Rude, E., Schmitt, E.: The infl uence of IL12 on the development of Th1 and Th2 cells and its adjuvant effect for humoral immune responses. Res. Immunol. 146, 481-486 (1995).

[16] Fifi s, T., et al.: Size-dependent immunogenicity: therapeutic and protective properties of nanovaccines against tumors. J. Immunol. 173, 3148-3154 (2004).

[17] Chadwick, S., Kriegel, C, Amiji, M.: Nanotechnology solutions for mucosal immunization. Adv.

DrugDeliv. Rev. 62, 394–407 (2010).

[18] Randolph, G. J., Inaba, K., Robbiani, D. F., Steinman, R. M., Muller, W. A.: Differentiation of phagocytic monocytes into lymph node dendritic cells in vivo. Immunity 11, 753–761 (1999).

[19] Malik, B., Goyal, A. K., Mangal, S., Zakir, F., Vyas, S. P.: Implication of gut immunology in the design of oral vaccines. Curr. Mol. Med. 10, 47–70 (2010).

[20] Ada, G.: Vaccines and vaccination. N. Engl. J. Med. 345, 1042–1053 (2001).

[21] Fehr, T., Skrastina, D., Pumpens, P., Zinkernagel, R. M.: T cell – independent type I antibody response against B cell epitopes expressed repetitively on recombinant virus particles. Proc. Natl. Acad. Sci. U. S. A. 95, 9477–9481 (1998).

[22] Bachmann, M. F., et al.: The infl uence of antigen organization on B cell responsiveness. Science 262, 1448–1451 (1993).

[23] Bachmann, M. F., Zinkernagel, R. M.: Neutralizing antiviral B cell responses. Annu. Rev. Immunol. 15, 235–270 (1997).

[24] O'Hagan, D. T., Singh, M., Ulmer, J. B.: Microparticle–based technologies for vaccines. Methods 40, 10–19 (2006).

[25] Uddin, A. N., Bejugam, N. K., Gayakwad, S. G., Akther, P., D'Souza, M. J.: Oral delivery of gastroresistant microencapsulated typhoid vaccine. J. Drug Target. 17, 553–560 (2009).

[26] Yeboah, K. G., D'Souza, M. J.: Evaluation of albumin microspheres as oral delivery system for Mycobacterium tuberculosis vaccines. J. Microencapsul. 26, 166–179 (2009).

[27] Lai, Y. H., D'Souza, M. J.: Formulation and evalua–tion of an oral melanoma vaccine. J. Microencapsul. 24, 235–252 (2007).

[28] Storni, T., Ruedl, C., Renner, W. A., Bachmann, M. F.: Innate immunity together with duration of antigen persistence regulate effector T cell induction. J. Immunol. 171, 795–801 (2003).

[29] Thiele, L., Merkle, H. P., Walter, E.: Phagocytosis and phagosomal fate of surface–modifi ed microparticles in dendritic cells and macrophages. Pharm. Res. 20, 221–228 (2003).

[30] Akande, J., et al.: Targeted delivery of antigens to the gut–associated lymphoid tissues: 2. Ex vivo evaluation of lectin–labelled albumin microspheres for targeted delivery of antigens to the M–cells of the Peyer's patches. J. Microencapsul. 27, 325–336 (2010).

[31] Lai, Y. H., D'Souza, M. J.: Microparticle transport in the human intestinal M cell model. J. Drug Target. 16, 36–42 (2008).

[32] Pulendran, B., Banchereau, J., Maraskovsky, E., Maliszewski, C.: Modulating the immune response with dendritic cells and their growth factors. Trends Immunol. 22, 41–47 (2001).

[33] Banchereau, J., Steinman, R. M.: Dendritic cells and the control of immunity. Nature 392, 245–252 (1998).

[34] Bharali, D. J., Pradhan, V., Elkin, G., Qi, W., Hutson, A., Mousa, S. A., Thanavala, Y.: Novel nanoparticles for the delivery of recombinant hepatitis B vaccine. Nanomedicine 4, 311–317 (2008).

[35] Perrie, Y., Mohammed, A. R., Kirby, D. J., McNeil, S. E., Bramwell, V. W.: Vaccine adjuvant systems: enhancing the effi cacy of sub–unit protein antigens. Int. J. Pharm. 364, 272–280 (2008).

[36] Sun, H. X., Xie, Y., Ye, Y. P.: ISCOMs and ISCOMATRIX. Vaccine 27, 4388–4401 (2009).

[37] Quan, F. S., Vunnava, A., Compans, R. W., Kang, S. M.: Virus–like particle vaccine protects against 2009 H1N1 pandemic infl uenza virus in mice. PLoS One 5, e9161 (2010).

[38] Tian, H. Y., et al.: Biodegradable cationic PEG–PEI–PBLG hyperbranched block copolymer: synthesis and micelle characterization. Biomaterials 26, 4209–4217 (2005).

[39] Jackson, C. L., et al.: Visualization of dendrimer molecules by transmission electron microscopy (TEM): staining methods and cryo–TEM of vitrifi ed solutions. Macromolecules 31, 6259–6265 (1998).

[40] Klumpp, C., Kostarelos, K., Prato, M., Bianco, A.: Functionalized carbon nanotubes as emerging nanovectors for the delivery of therapeutics. Biochim. Biophys. Acta 1758, 404–412 (2006).

[41] des Rieux, A., Fievez, V., Garinot, M., Schneider, Y. J., Preat, V.: Nanoparticles as potential oral delivery systems of proteins and vaccines: a mechanistic approach. J. Control. Release 116, 1–27 (2006).

[42] Rice-Ficht, A. C., Arenas-Gamboa, A. M., Kahl-McDonagh, M. M., Ficht, T. A.: Polymeric particles in vaccine delivery. Curr. Opin. Microbiol. 13, 106-112 (2010).

[43] Xiang, S. D., et al.: Pathogen recognition and devel-opment of particulate vaccines: does size matter? Methods 40, 1-9 (2006).

[44] Desai, M. P., Labhasetwar, V., Amidon, G. L., Levy, R. J.: Gastrointestinal uptake of biodegradable microparticles: effect of particle size. Pharm. Res. 13, 1838-1845 (1996).

[45] Primard, C., et al.: Traffi c of poly (lactic acid) nanoparticulate vaccine vehicle from intestinal mucus to sub-epithelial immune competent cells. Biomaterials 31, 6060-6068 (2010).

[46] Gutierro, I., Hernandez, R. M., Igartua, M., Gascon, A. R., Pedraz, J. L.: Size dependent immune response after subcutaneous, oral and intranasal administration of BSA loaded nanospheres. Vaccine 21, 67-77 (2002). S0264410X02004358 [pii].

[47] Wendorf, J., et al.: A comparison of anionic nanopar-ticles and microparticles as vaccine delivery systems. Hum. Vaccin. 4, 44-49 (2008).

[48] van den Berg, J. H., et al.: Shielding the cationic charge of nanoparticle-formulated dermal DNA vaccines is essential for antigen expression and immunogenicity. J. Control. Release 141, 234-240 (2010).

[49] Jain, A. K., et al.: Synthesis, characterization and evaluation of novel triblock copolymer based nanoparticles for vaccine delivery against hepatitis B. J. Control. Release 136, 161-169 (2009).

[50] Demento, S., Steenblock, E. R., Fahmy, T. M.: Biomimetic approaches to modulating the T cell immune response with nano-and micro-particles. Conf. Proc. IEEE Eng. Med. Biol. Soc. 2009, 1161-1166 (2009).

[51] Bangham, A. D., Standish, M. M., Miller, N.: Cation permeability of phospholipid model membranes: effect of narcotics. Nature 208, 1295-1297 (1965).

[52] Allison, A. G., Gregoriadis, G.: Liposomes as immu-nological adjuvants. Nature 252, 252 (1974).

[53] Castaldello, A., Brocca-Cofano, E., Voltan, R., Triulzi, C., Altavilla, G., Laus, M., Sparnacci, K., Ballestri, M., Tondelli, L., Fortini, C., Gavioli, R., Ensoli, B., Caputo, A.: DNA prime and protein boost immunization with innovative polymeric cationic core-shell nanoparticles elicits broad immune responses and strongly enhance cellular responses of HIV-1 tat DNA vaccination. Vaccine 24, 5655-5669 (2006).

[54] Himeno, A., Akagi, T., Uto, T., Wang, X., Baba, M., Ibuki, K., Matsuyama, M., Horiike, M., Igarashi, T., Miura, T., Akashi, M.: Evaluation of the immune response and protective effects of rhesus macaques vaccinated with biodegradable nanoparticles carrying gp120 of human immunodefi ciency virus. Vaccine 28, 5377-5385 (2010).

[55] Caputo, A., Castaldello, A., Brocca-Cofano, E., Voltan, R., Bortolazzi, F., Altavilla, G., Sparnacci, K., Laus, M., Tondelli, L., Gavioli, R., Ensoli, B.: Induction of humoral and enhanced cellular immune responses by novel core-shell nanosphere-and microsphere-based vaccine formulations following systemic and mucosal administration. Vaccine 27, 3605-3615 (2009).

[56] Stano, A., van der Vlies, A. J., Martino, M. M., Swartz, M. A., Hubbell, J. A., Simeoni, E.: PPS nanoparticles as versatile delivery system to induce systemic and broad mucosal immunity after intranasal administration. Vaccine 29 (4), 804-812 (2011).

[57] Hirosue, S., Kourtis, I. C., van der Vlies, A. J., Hubbell, J. A., Swartz, M. A.: Antigen delivery to dendritic cells by poly (propylene sulfi de) nanoparticles with disulfi de conjugated peptides: cross-presentation and T cell activation. Vaccine 28, 7897-7906 (2010).

[58] Slutter, B., Bal, S., Keijzer, C., Mallants, R., Hagenaars, N., Que, I., Kaijzel, E., van Eden, W., Augustijns, P., Lowik, C., Bouwstra, J., Broere, F., Jiskoot, W.: Nasal vaccination with N-trimethyl chitosan and PLGA based nanoparticles: nanoparticle characteristics determine quality and strength of the antibody response in mice against the encapsulated antigen. Vaccine 28, 6282-6291 (2010).

[59] Prego, C., Paolicelli, P., Diaz, B., Vicente, S., Sanchez, A., Gonzalez-Fernandez, A., Alonso, M. J.: Chitosan-based nanoparticles for improving immunization against hepatitis B infection. Vaccine 28, 2607-2614 (2010).

[60] Gregoriadis, G.: Liposomes as immunoadjuvants and vaccine carriers: antigen entrapment. Immunomethods4, 210-216 (1994).

[61] Wang, D., et al.: Liposomal oral DNA vaccine (mycobacterium DNA) elicits immune response. Vaccine 28, 3134-3142 (2010).

[62] Karkada, M., Weir, G. M., Quinton, T., Fuentes-Ortega, A., Mansour, M.: A liposome-based platform, VacciMax, and its modifi ed water-free platform DepoVax enhance effi cacy of in vivo nucleic acid delivery. Vaccine 28, 6176-6182 (2010).

[63] Altin, J. G., Parish, C. R.: Liposomal vaccines-target-ing the delivery of antigen. Methods 40, 39-52 (2006).

[64] Immordino, M. L., Dosio, F., Cattel, L.: Stealth liposomes: review of the basic science, rationale, and clinical applications, existing and potential. Int. J. Nanomedicine 1, 297-315 (2006).

[65] Mohammed, A. R., Bramwell, V. W., Kirby, D. J., McNeil, S. E., Perrie, Y.: Increased potential of a cationic liposome-based delivery system: enhancing stability and sustained immunological activity in pre-clinical development. Eur. J. Pharm. Biopharm. 76 (3), 404-412 (2010).

[66] Nordly, P., Agger, E. M., Andersen, P., Nielsen, H. M., Foged, C.: Incorporation of the TLR4 agonist monophosphoryl lipid a into the bilayer of DDA/ TDB liposomes: physico-chemical characterization and induction of CD8 (+) T-cell responses in vivo. Pharm. Res. 28 (3), 553-562 (2011).

[67] Henriksen-Lacey, M., et al.: Liposomes based on dimethyldioctadecylammonium promote a depot effect and enhance immunogenicity of soluble antigen. J. Control. Release 142, 180-186 (2010).

[68] Gasparini, R., Lai, P.: Utility of virosomal adjuvated infl uenza vaccines: a review of the literature. J. Prev. Med. Hyg. 51, 1-6 (2010).

[69] Patel, G. B., Zhou, H., KuoLee, R., Chen, W.: Archaeosomes as adjuvants for combination vaccines. J. Liposome Res. 14, 191-202 (2004).

[70] Sharma, S., Mukkur, T. K., Benson, H. A., Chen, Y.: Pharmaceutical aspects of intranasal delivery of vaccines using particulate systems. J. Pharm. Sci. 98, 812-843 (2009).

[71] Henriksen-Lacey, M., et al.: Liposomal cationic charge and antigen adsorption are important properties for the effi cient deposition of antigen at the injection site and ability of the vaccine to induce a CMI response. J. Control. Release 145, 102-108 (2010).

[72] Zhong, Z., et al.: A novel liposomal vaccine improves humoral immunity and prevents tumor pulmonary metastasis in mice. Int. J. Pharm. 399, 156-162 (2010).

[73] Pearse, M. J., Drane, D.: ISCOMATRIX adjuvant: a potent inducer of humoral and cellular immune responses. Vaccine 22, 2391-2395 (2004).

[74] Pearse, M. J., Drane, D.: ISCOMATRIX adjuvant for antigen delivery. Adv. Drug Deliv. Rev. 57, 465-474 (2005).

[75] Drane, D., Pearse, M. J.: Immunopotentiators in modern vaccines, pp. 191-215. Elsevier Academic Press, Massachusetts, USA (2006).

[76] Rimmelzwaan, G. F., Baars, M., van Amerongen, G., van Beek, R., Osterhaus, A. D.: A single dose of an ISCOM infl uenza vaccine induces long-lasting protective immunity against homologous challenge infection but fails to protect Cynomolgus macaques against distant drift variants of infl uenza A (H3N2) viruses. Vaccine 20, 158-163 (2001).

[77] Souza, M., Costantini, V., Azevedo, M. S., Saif, L. J.: A human norovirus-like particle vaccine adjuvanted with ISCOM or mLT induces cytokine and antibody responses and protection to the homologous GII. 4 human norovirus in a gnotobiotic pig disease model. Vaccine 25, 8448-8459 (2007).

[78] Rimmelzwaan, G. F., Claas, E. C., van Amerongen, G., de Jong, J. C., Osterhaus, A. D.: ISCOM vaccine induced protection against a lethal challenge with a human H5N1 infl uenza virus. Vaccine 17, 1355-1358 (1999).

[79] Sjolander, S., Drane, D., Davis, R., Beezum, L., Pearse, M., Cox, J.: Intranasal immunisation with infl uenza-ISCOM induces strong mucosal as well as systemic antibody and cytotoxic T-lymphocyte responses. Vaccine 19, 4072-4080 (2001).

[80] Madhun, A. S., Haaheim, L. R., Nilsen, M. V., Cox, R. J.: Intramuscular Matrix-M-adjuvanted virosomal H5N1 vaccine induces high frequencies of multifunctional Th1 CD4^{+}cells and strong antibody responses in mice. Vaccine 27, 7367-7376 (2009).

[81] Sanders, M. T., Deliyannis, G., Pearse, M. J., McNamara, M. K., Brown, L. E.: Single dose intranasal immunization with ISCOMATRIX vaccines to elicit antibody-mediated clearance of infl uenza virus requires delivery to the lower respiratory tract. Vaccine 27, 2475-2482 (2009).

[82] Skene, C. D., Doidge, C., Sutton, P.: Evaluation of ISCOMATRIX and ISCOM vaccines for immunisation against Helicobacter pylori. Vaccine 26, 3880-3884 (2008).

[83] Pinitkiatisakul, S., Friedman, M., Wikman, M., Mattsson, J. G., Lovgren-Bengtsson, K., Stahl, S., Lunden, A.: Immunogenicity and protective effect against murine cerebral neosporosis of recombinant NcSRS2 in different iscom formulations. Vaccine 25, 3658-3668 (2007).

[84] McBurney, W. T., et al.: In vivo activity of cationic immune stimulating complexes (PLUSCOMs). Vaccine 26, 4549-4556 (2008).

[85] Boyle, J., et al.: The utility of ISCOMATRIX adju-vant for dose reduction of antigen for vaccines requiring antibody responses. Vaccine 25, 2541-2544 (2007).

[86] Plummer, E. M., Manchester, M: Viral nanoparticles and virus-like particles: platforms for contemporary vaccine design. WIREs Nanomed. Nanobiotechnol. 3, 174-196 (2011). doi: 10. 1002/wnan. 119.

[87] Noad, R., Roy, P.: Virus-like particles as immuno-gens. Trends Microbiol. 11, 438-444 (2003).

[88] Campo, M. S., Roden, R. B.: Papillomavirus prophy-lactic vaccines: established successes, new approaches. J. Virol. 84, 1214-1220 (2010).

[89] Ludwig, C., Wagner, R.: Virus-like particles-univer-sal molecular toolboxes. Curr. Opin. Biotechnol. 18, 537-545 (2007).

[90] Buonaguro, L., Tornesello, M. L., Buonaguro, F. M.: Virus-like particles as particulate vaccines. Curr. HIV Res. 8, 299-309 (2010).

[91] Grgacic, E. V., Anderson, D. A.: Virus-like particles: passport to immune recognition. Methods 40, 60-65 (2006).

[92] Schiller, J. T., Castellsague, X., Villa, L. L., Hildesheim, A.: An update of prophylactic human papillomavirus L1 virus-like particle vaccine clinical trial results. Vaccine26 (Suppl 10), K53-K61 (2008).

[93] Pearton, M., et al.: Infl uenza virus-like particles coated onto microneedles can elicit stimulatory effects on Langerhans cells in human skin. Vaccine 28, 6104-6113 (2010).

[94] Akahata, W., et al.: A virus-like particle vaccine for epidemic Chikungunya virus protects nonhuman primates against infection. Nat. Med. 16, 334-338 (2010).

[95] Quan, F. S., Huang, C., Compans, R. W., Kang, S. M.: Virus-like particle vaccine induces protective immunity against homologous and heterologous strains of infl uenza virus. J. Virol. 81, 3514-3524 (2007).

[96] Krammer, F., et al.: Infl uenza virus-like particles as an antigen-carrier platform for the ESAT-6 epitope of Mycobacterium tuberculosis. J. Virol. Methods 167, 17-22 (2010).

[97] Song, J. M., et al.: Protective immunity against H5N1 infl uenza virus by a single dose vaccination with virus-like particles. Virology 405, 165-175 (2010).

[98] Kang, S. M., et al.: Induction of long-term protective immune responses by infl uenza H5N1 virus-like particles. PLoS One 4, e4667 (2009).

[99] Muratori, C., Bona, R., Federico, M.: Lentivirus-based virus-like particles as a new protein delivery tool. Methods Mol. Biol. 614, 111-124 (2010).

[100] Torchilin, V. P.: Structure and design of polymeric surfactant-based drug delivery systems. J. Control. Release 73, 137-172 (2001).

[101] Torchilin, V. P.: Micellar nanocarriers: pharmaceuti-cal perspectives. Pharm. Res. 24, 1-16 (2007).

[102] O'Reilly, R. K.: Spherical polymer micelles: nano-sized reaction vessels? Philos. Transact. A Math. Phys. Eng. Sci. 365, 2863-2878 (2007).

[103] Morein, B., Sharp, M., Sundquist, B., Simons, K.: Protein subunit vaccines of parainfl uenza type 3 virus: immunogenic effect in lambs and mice. J. Gen. Virol. 64 (Pt 7), 1557-1569 (1983).

[104] Prabakaran, M., et al.: Reverse micelle-encapsulated recombinant baculovirus as an oral vaccine against H5N1 infection in mice. Antiviral Res. 86, 180-187 (2010).

[105] Bharali, D. J., Khalil, M., Gurbuz, M., Simone, T. M., Mousa, S. A.: Nanoparticles and cancer therapy: a concise review with emphasis on dendrimers. Int. J. Nanomedicine 4, 1-7 (2009).

[106] Patri, A. K., Majoros, I. J., Baker, J. R.: Dendritic polymer macromolecular carriers for drug delivery. Curr. Opin. Chem. Biol. 6, 466-471 (2002).

[107] Klajnert, B., Bryszewska, M.: Dendrimers: proper-ties and applications. Acta Biochim. Pol. 48, 199-208 (2001).

[108] Baker, J. R., Jr.: Dendrimer-based nanoparticles for cancer therapy. Hematology. Am. Soc. Hematol. Educ. Program. 1, 708-719 (2009).

[109] Moreno, C. A., et al.: Preclinical evaluation of a synthetic Plasmodium falciparum MAP malaria vaccine in Aotus monkeys and mice. Vaccine 18, 89-99 (1999).

[110] Nardin, E. H., et al.: A totally synthetic polyoxime malaria vaccine containing Plasmodium falciparum B cell and universal T cell epitopes elicits immune responses in volunteers of diverse HLA types. J. Immunol. 166, 481-489 (2001).

[111] Nardin, E. H., et al.: Synthetic malaria peptide vaccine elicits high levels of antibodies in vaccinees of defi ned HLA genotypes. J. Infect. Dis. 182, 1486-1496 (2000).

[112] Shampur, M., Padinjarenmattathil, U., Desai, A., Narayanaswamy, J.: Development and immunogenicity of a novel polyetherimine (PETIM) dendrimer based nanoformulated DNA rabies vaccine. Int. J. Infect. Dis. 14 (Suppl 1), E453 (2010). Elsevier Science.

[113] Dutta, T., Garg, M., Jain, N. K.: Poly (propyleneimine) dendrimer and dendrosome mediated genetic immunization against hepatitis B. Vaccine 26, 3389-3394 (2008).

[114] Skwarczynski, M., Zaman, M., Urbani, C. N., Lin, I. C., Jia, Z., Batzloff, M. R., Good, M. F., Monteiro, M. J., Toth, I.: Polyacrylate dendrimer nanoparticles: a self-adjuvanting vaccine delivery system. Angew. Chem. Int. Ed. Engl. 49, 5742-5745 (2010).

[115] in het Panhuis, M.: Vaccine delivery by carbon nano-tubes. Chem. Biol. 10, 897-898 (2003).

[116] Kendall, M.: Engineering of needle-free physical methods to target epidermal cells for DNA vaccination. Vaccine 24, 4651-4656 (2006). doi: 10. 1016/j. vaccine. 2005. 08. 066. S0264-410X (05) 00841-8 [pii].

[117] Kersten, G., Hirschberg, H.: Antigen delivery sys-tems. Expert Rev. Vaccines 3, 453-462 (2004). doi: 10. 1586/14760584. 3. 4. 453. ERV030423 [pii].

[118] Nandedkar, T. D.: Nanovaccines: recent develop-ments in vaccination. J. Biosci. 34, 995-1003 (2009).

（吴国华译）

第46章 疫苗递送系统的作用、挑战和最新进展

Aditya Pattani, Prem N. Gupta,
Rhonda M. Curran 和 R. Karl Malcolm①

摘要

绝大多数疫苗抗原是生物大分子，例如蛋白质和多糖，分子量常大于10 000。这些分子需以正确的构象递送到机体，才能诱导产生特异性免疫应答，有效地靶向免疫细胞。目前，大多数疫苗经皮内、皮下或肌内途径肠胃外给药，这在很大程度上取决于抗原是否处于吸附或非吸附状态。这些途径的主要缺点为使用针头引起的疼痛、针头污染的可能性、针头处置的实用性以及对保健工作者专业知识需求。现在特别注重开发旨在直接经口腔、鼻子、阴道和直肠黏膜表面给药的黏膜疫苗。

通常，当通过黏膜途径递送疫苗时，由于与黏膜保留和摄取相关的困难，简单的抗原溶液无免疫学功效。目前正积极研究各种制剂策略，包括微球、脂质体、纳米为和病毒样颗粒。除设计和选择抗原候选物之外，选择和制备抗原递送系统对实现疫苗接种的最终目标至关重要。本章概述了疫苗递送系统设计的作用和考虑因素，重点介绍胶体和纳米递送系统领域的挑战和最新进展。

46.1 引言

在目前分子疫苗时代，对产品安全的必要关注不可避免地影响临床疗效[1]。大多数候选抗原为生物大分子，故口服给药途径易致疫苗失活、丧失抗原天然构象和吸收不良。大多数疫苗使用针头注射通过肠胃外给药。抗原极易发生物理、化学及构象降解，导致疫苗失效[2]，这使这些问题更加复杂。目前的疫苗接种需训练有素的医务人员进行，并处置用过的针头和注射器[3]。诱导强烈的抗原特异性黏膜免疫应答是研发 HIV / AIDS 等疾病疫苗。实现这一目标可能需要在黏膜部位接种抗原，并采取特别措施克服与局部抗原降解[4,5]和保留（例如在鼻和阴道）相关的困难。全世界有许多正在研发的实用且价格实惠的疫苗递送策略，可维持或增强疫苗的递送和免疫效率。

46.2 疫苗递送系统的作用

已开发了多种先进的药物递送系统，以克服较传统的药物递送方法相关的各种问题和障碍。

---

① A. Pattani, PhD（KairavChemofarbe 工业有限公司 &NanoXpert 科技公司研发部，印度孟买）

P. N. Gupta, PhD（✉）（印度综合医学研究所制剂 & 药物递送研发部，印度查谟，E-mail：pngupta10@gmail. com）

R. M. Curran, PhD（阿尔斯特大学护理与健康研究所，英国北爱尔兰约旦城）

R. K. Malcolm, PhD（贝尔法斯特皇后大学药学院，英国贝尔法斯特）

表 46.1 概述了有效递送药物分子的主要制剂方法。

表 46.1　药物递送系统的作用

递送系统的作用	实例
防止化学和物理降解，延长保质期和消除冷链	冻干递送系统或喷雾干燥疫苗[6]
防止在黏膜部位发生化学或酶降解	微粒或纳米微粒[7]
调节免疫系统提高功效	脂质体[8,9]、纳米微粒[10,11]
通过持续释放提高功效	聚（丙交酯-共-乙交酯）微粒[12]
无须训练有素的医务人员	微针[13]
无须对针头或注射器进行特殊处理	溶解微针[14,15]
改善黏膜保留	阴道凝胶[16]、鼻凝胶[17]、阴道棒[18,19]

46.3　疫苗递送设计的重要因素和挑战

46.3.1　保护抗原天然构象

疫苗递送系统的制备过程或疫苗递送制剂的组成赋形剂可使疫苗发生物理或化学变化，导致活性丧失或改变[2]。必须使用各种免疫学和体内技术充分评估剂型的最终活性。最终目标应该是实现产品稳定性，以至于消除对冷链运输的需求。

46.3.2　疫苗递送速率应充分评估

许多递送系统旨在延长疫苗递送时间。众所周知，一些候选疫苗长期或重复给药可使机体产生耐受性[20,21]。应对此方面进行仔细评估。

46.3.3　佐剂组分常与抗原本身一样重要

佐剂有助于增强免疫应答，有时是必需的。应同时考虑佐剂处方、剂量和释放速率等因素，且可能需要考虑佐剂与伴随抗原的相关性。

46.3.4　剂型设计旨在消除对疫苗给药人员专业技能的需求

在理想情况下，没有经过培训的人员也可以进行快速且有效的大规模疫苗接种。这有助于在全球实现公平使用疫苗。

46.4　设计疫苗递送系统的理念

尽管亚单位疫苗的免疫原性经常受损，但临床结果表明亚单位疫苗的安全性优于传统疫苗。疫苗递送系统可克服此缺陷。

46.4.1　微粒抗原较可溶性抗原更有效

微粒疫苗相对于可溶性抗原的一个优点在于其促进抗原呈递细胞（APC）的摄取。微粒相关抗原可以模拟病原体的感染途径。例如微粒疫苗的粒径通常为几百纳米至几微米，与已进化的免疫系统所抵御的常见病原体相当，可促进抗原呈递细胞的有效摄取。已知吞噬体能够进行抗原交叉递呈，故经吞噬作用将微粒疫苗内化到吞噬体具有重要意义[22]。这使微粒疫苗在诱导细胞免疫应答方面颇具魅力，这与优先经 MHC Ⅱ类途径递呈且交叉递呈性不佳的可溶性抗原不同。微粒抗原的其他优点包括：①在抗原呈递细胞内递送相对的微粒相关抗原；②与可溶性抗原相比，延长抗原在细胞内[23]或细胞外[24]释放，延缓抗原递呈；③同时将抗原和免疫刺激成分递送到同一抗原呈递细胞[25]。

46.4.2 微粒递送系统具有内在佐剂作用

传统疫苗存在灭活不彻底、减毒的活微生物或减毒毒素等安全问题。某些病原体（如结核病和 HIV）无法制备传统疫苗[26]。如今，人们对表达蛋白质、多肽或抗原的 DNA 或 RNA 疫苗的兴趣日益增加。尽管这些疫苗比使用完整的微生物更安全，但它们的免疫原性不佳，需使用佐剂[27]。表 46.2 总结了已在进行临床试验的具有内在佐剂特性的微粒递送系统。佐剂可通过两种主要机制增强联合免疫抗原的特异性免疫应答[27]。①微粒递送系统通过抗原呈递细胞直接吞噬或形成延长暴露的抗原贮库，增加抗原呈递细胞对抗原的摄取；②微粒递送系统充当直接激活先天免疫细胞的免疫增强剂（例如细胞因子）。

表 46.2 已经在人体内试验的具有固有佐剂特性的递送系统

递送系统	组成（作用机制）	疾病（抗原）
脂质体	1 个或多个磷脂双层	流感（单价裂解）
脂质体 + 胞壁酰三肽-磷脂酰乙醇胺	脂质体包含合成的胞壁酰三肽-磷脂酰乙醇胺	HIV（gp120）
病毒体	脂质体包含了膜病毒融合蛋白。	HAV、流感、DT，TT
免疫刺激复合物	由皂甙 Quil-A、胆固醇和磷脂组成	流感（三价裂解）、HPV16（E6/E7）幽门螺杆菌
PLGA 微粒	由均聚物和共聚物或乳酸和乙醇酸制成的微粒。	TT、HIV
MF59®	用 Span85 和聚山梨酯 80 稳定的角鲨烯-水乳液	流感（三价裂解）、HBV、HSV-2（gB+gD）、HIV-1（gp120）、CMV（gB）
SBAS-2	包含 MPL® 和 QS21® 的的角鲨烯/水乳液	疟疾（RTS，S）、HIV-1（gp 120）
SBAS-4	含 MPL® 的明矾凝胶	HBV（HBsAg）、HSV
不完全弗氏佐剂	用甘露醇单油酸酯稳定的水/Drakeol 乳液	HIV-1、黑色素瘤（gp100）
Montanide ISA720	含可代谢油的乳液	疟疾（MSP1、MSP2）
Detox®	包含了 MPL® 和 QS21® 的 MPL® 和 CWS	疟疾（R32NS18）、黑色素瘤细胞裂解物

缩写：CMV 巨细胞病毒；CWS 草分枝杆菌的细胞壁骨架；DT 白喉类毒素；HAV 甲型肝炎病毒；HBV 乙型肝炎病毒；HPV 人乳头瘤病毒；HSV 单纯疱疹病毒；MPL 单磷酰脂质；MTP-PE 胞壁酰三肽二棕榈酰磷脂酰乙醇胺；PLGA 聚-（D，L）-丙交酯-共-乙醇酸；TT 破伤风毒素。

46.5　抗原递送系统的设计

目前各种试验和临床模型所测试的药物疫苗制剂通常为微粒状，通过抗原的聚集或交联[28]或在铝盐上吸附或沉淀抗原的方法制备。也可用化学方法将抗原连接到预先形成的特定载体[29]，或用化学或物理方法分布于微粒中或微粒表面，排列方式较规则（脂质体和病毒样颗粒），或不太规则（聚合物微球）[29,30]。通常情况下，任何类型的微粒都有助于专业抗原呈递细胞识别和摄入这些疫苗。在抗原呈递细胞摄取之前，制剂本身可影响所募集的吞噬细胞活性。如下所述，疫苗递送系统的设计也影响其他关键因素。

46.5.1　跨生物屏障的抗原转运

疫苗递送系统应改善抗原跨越肠黏膜和鼻黏膜等相关的生物屏障的能力。根据过去几十年的观察，已发现较小微粒一般可增强微粒跨越肠屏障转运抗原的能力[31]。虽然表46.3总结了影响微粒摄取的其他因素[7]，但黏膜疫苗载体的理想粒径范围为50~500 nm。

表46.3　影响聚合物微粒吸收的各种因素

粒子大小	用于评估的动物种属
微粒疏水性	动物年龄
微粒剂量（抗原剂量）	动物的进食状况
给药载体	黏膜层特性
聚合物组成	微粒中的靶向剂
添加剂效应	摄取程度的定量方法
微粒的表面电荷	

46.5.2　抗原的生物分布递送系统及其抗原呈递细胞靶向潜力

皮下、皮内或肌内给药后，聚合物微粒的大小可影响其在机体内的分布。经肌内或皮下途径给药时，20~100 nm粒径的微粒可穿透细胞外基质，直接进入淋巴管。随后由密集的免疫细胞群（如树突细胞）捕获，汇入淋巴结，产生有效的免疫应答[32]。除粒径的关键作用外，如果粒子表面具有阳离子，则巨噬细胞和树突细胞对纳米微粒的摄取可大大增强。[33]

46.5.3　抗原在递送系统中的稳定性

疫苗递送系统的优点是保护疫苗免受恶劣生理条件的影响。有机溶剂、极端温度或高能量输入可使抗原发生降解或聚集，故剂型开发条件至关重要。用于制造递送系统的材料或其降解产物也可增强蛋白质变性[34]。

46.5.4　抗原和共刺激分子的共递送

如果疫苗本身不能刺激活化T细胞，则在同一药物制剂中联合递送抗原和共刺激因子可改善淋巴细胞的激活。已用重组病毒载体和具有B7家族共刺激分子的脂质体证明了此概念[35]。

46.5.5 聚合物的免疫调节活性

一些聚合物可作为有效的蛋白质载体。开发基于脂质体、微球、纳米微粒或水溶性合成聚合物的疫苗递送系统，可定制符合特定抗原的物理、化学和免疫原性等方面的要求，这已受到大量关注[36]。有助于核内体酸化的抗原递送系统还可以促进其携带的抗原经 MHC Ⅱ类途径递呈。这种机制可能是聚丙交酯-聚-乙交酯纳米微粒和微米微粒发挥作用产生的。含碱性聚合物或赋形剂（例如壳聚糖、乙烯亚胺、胶原、阴离子脂质和表面活性剂）的药物制剂可通过 MHC Ⅱ类途径阻止抗原递呈。研究表明，壳聚糖可以促进 Th1 细胞因子应答和 MHC Ⅰ类抗原递呈[37]。

46.5.6 抗原剂量和结构影响免疫类型

启动免疫应答的抗原量和氨基酸序列也影响 CD4 T 细胞分化为不同的效应子集，抗原呈递细胞表面的高密度肽和低密度肽分别刺激 Th1 细胞应答或 Th2 细胞应答[38]。聚丙交酯-聚-乙交酯微球可损害抗原稳定性[39]，此时，不同表位的相对可及性可能已经改变，也可能影响免疫应答的 Th1／Th2 相位差。与 T 细胞受体相互作用较强的多肽常刺激 Th1 样反应，弱结合的多肽常刺激 Th2 样反应[38]。

46.6 先进的疫苗递送系统

46.6.1 脂质体

脂质体由 1 个或多个包裹水相的磷脂双层组成。抗原可包裹于水性隔室内，或与脂质体表面连接或包埋在脂质双层内，所有这些形式都可保护抗原免受周围环境的影响。改变大小、组成和物理化学特征可使脂质体成为抗原递送的通用平台。脂质体的免疫刺激性质源自：①延长抗原结合和释放的时间；②经抗原呈递细胞优先摄取。抗原的免疫利用率提高，可促进抗原呈递细胞的成熟和抗原递呈[40]。脂质体在溶液中的稳定性有限，常发生聚集而融合在一起，冻干在一定程度上可解决此问题。在过去几十年研究的多种类型的脂质体疫苗制剂中，阳离子脂质体的免疫原性较高。例如由二甲基二（十八烷基）溴化铵和具有免疫调节作用的糖脂海藻糖二山楂酸酯制成的脂质体可有效促进细胞介导的免疫应答和体液免疫应答，该类型脂质体目前正在进行结核病的 I 期临床评估[41]。

46.6.2 聚合物微粒或纳米微粒

20 世纪 90 年代早期，微粒首先用作包埋抗原的递送系统[42]。可生物降解的聚 D，L-丙交酯和聚 D，L-乳酸-聚-乙醇酸的医药应用历史悠久，故可能是研究最多的一种用于肠胃外和黏膜递送抗原的材料[43]。不同比例、不同分子量的聚合物制成的微粒经肠胃外给药后可诱导持久的免疫力，这些聚合物的水解时间很长。使用混合不同粒径大小的微粒也可诱导持久的免疫力，避免巨噬细胞摄取较大微粒，故分解较慢[44]。以此方式一次注射疫苗可产生持久的免疫力，无需进行加强免疫。除基于聚丙交酯-乙交酯（PLGA）的微粒产生令人鼓舞的免疫学性能之外，确保所包封蛋白质抗原的稳定性也是一个重要问题。实际上，一些蛋白质抗原在被捕获到 PLGA 中或从基质释放期间常发生聚集或降解。抗原在聚合物水解过程中暴露于有机溶剂和酸性微环境可导致抗原失活。采用优化的制造方法或添加稳定剂如 $Mg(OH)_2$ 其他蛋白质、表面活性剂或糖可部分解决这些问题[45,46]。

46.6.3 病毒体

病毒体是一种含有功能性病毒膜蛋白的脂质体，其粒径类似于病毒。例如，重构的空流感病毒包囊的病毒体具有免疫刺激性[47]。当病毒体携带异源性抗原时，可看作是具有内在佐剂活性的递送系统。迄今为止，在欧洲已经批准使用病原体作为甲型肝炎和流感疫苗递送系统，还制成了鼻内免疫制剂[48]。

46.6.4 免疫刺激复合物（ISCOMS）

ISCOMS 的特征是包含抗原的笼状结构。这种复合物由脂质和 Quil-A 组成，Quil-A 是源自皂树（Quillaja saponaria）的皂苷活性成分，具有佐剂活性[49]。疏水性抗原可直接嵌入或锚定在脂质胶体结构域中，亲水性抗原需进行修饰后才能被有效捕获[50]。ISCOM 在各种条件下具有良好的稳定性，这一特性非常有趣。已用人体临床试验评估了的 ISCOM 具有良好的耐受性，即使抗原剂量非常低，也能诱导强烈的体液应答和细胞毒性 T 淋巴细胞（CTL）应答[51]。尽管 ISCOM 的疫苗在临床试验中具有潜在而良好的表现，但仅获批用于兽医用途。

46.6.5 乳化液

乳化液作为疫苗佐剂的研究始于弗氏完全佐剂（矿物油、石蜡和已灭活的分枝杆菌的油包水乳液）[52]。尽管该乳化液能产生高滴度的抗体，但不良反应强，妨碍了其临床应用。无分枝杆菌成分的弗氏不完全佐剂毒性较小，仍在兽医中应用，但由于严重的不良事件，已禁用于人类疫苗。20 世纪 90 年代，开发了一种角鲨烯水包油乳液（MF59），可产生较高滴度的抗体，且其耐受性和安全性较好[53]。MF59 目前在欧洲获批用于流感疫苗。

46.6.6 黏膜黏合性聚合物或凝胶

一些黏膜黏合剂在注射给药时具有佐剂性质（例如海藻酸钠），已选择具有增强水性黏度性质的其他黏膜黏合剂（例如羧甲基纤维素钠）作为肠道外疫苗试验制剂的贮库剂。最近，用于经阴道免疫的疫苗黏膜黏附凝胶已受到很大关注。从制剂角度来看，经阴道宫颈注射含已溶解抗原的缓冲溶液以后，其诱导的抗原特异性免疫应答远非理想，这是因为可能在给药部位发生药物泄漏、抗原发生降解，以及月经周期的影响和抗原对黏膜相关淋巴组织的暴露不充分。为提高经阴道递送疫苗的效率，科学家已研发了各种黏膜黏附递送系统，包括基于羟乙基纤维素的流变性结构的凝胶载体[16]和冻干的固体剂型[18,54]。

最近，我们小组报道了脂质体凝胶制剂和新型冻干变体最新发展状况，其由 HIV-1 包膜糖蛋白、CN54gp140 构成，包裹在中性、带正电荷或带负电荷的脂质体中[19]。如图 46.1 所示，开发经阴道疫苗给药的冻干剂型的方案。在整合至羟乙基纤维素（HEC）水性凝胶之前评估 CN54gp140 脂质体的平均囊泡直径、多分散性、形态、zeta 电位和抗原包裹效率，然后

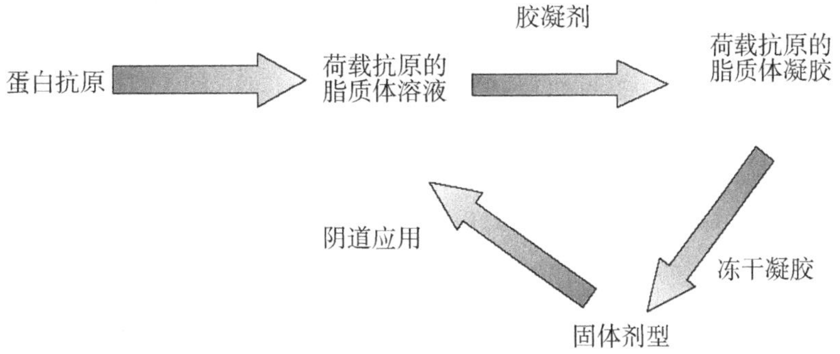

图 46.1 阴道疫苗冻干剂型研制方案

冻干成用于阴道的棒状固体剂型。评估冻干脂质体-HEC 棒在模拟阴道液中的水分含量和再分散性。这些杆状固体制剂可在阴道内给药后恢复凝胶形态，故评估了黏膜黏合性、机械性（可压缩性和硬度）和重整凝胶的流变性质。脂质体具有良好的抗原包裹效率，凝胶具有合适的黏膜黏附强度。冷冻干燥的脂质体-HEC 制剂是一种新的制剂策略，可制成稳定而实用的剂型。

结论

传统疫苗策略相关的安全问题限制了新型疫苗的研发进程，即免疫效力不高。疫苗递送系统不仅是简单的疫苗载体，还有佐剂和靶向剂的功能，是克服上述缺点不可或缺的方法。其他优点包括无针递送及稳定，促进了疫苗的广泛应用。疫苗递送系统的发展空间巨大，对未来全球免疫战略可以产生积极影响。

参考文献

[1] Bachmann, M. F., Jennings, G. T.: Vaccine delivery: a matter of size, geometry, kinetics and molecular patterns. Nat. Rev. Immunol. 10, 787-796 (2010).

[2] Pattani, A., et al.: Molecular investigations into vagi-nal immunization with HIV gp41 antigenic construct H4A in a quick release solid dosage form. Vaccine 30, 2778-2785 (2012).

[3] Donnelly, R. F., et al.: Design, optimization and char-acterisation of polymeric microneedle arrays prepared by a novel laser-based micromoulding technique. Pharm. Res. 28, 41-57 (2011).

[4] Duerr, A.: Update on mucosal HIV vaccine vectors. Curr. Opin. HIV AIDS 5, 397-403 (2010).

[5] Neutra, M. R., Kozlowski, P. A.: Mucosal vaccines: the promise and the challenge. Nat. Rev. Immunol. 6, 148-158 (2006).

[6] Wong, Y., et al.: Drying a tuberculosis vaccine without freezing. Proc. Natl. Acad. Sci. 104, 2591-2595 (2007).

[7] Vyas, S. P., Gupta, P. N.: Implication of nanoparticles/microparticles in mucosal vaccine delivery. Expert Rev. Vaccines 6, 401-418 (2007).

[8] Perrie, Y., Frederik, P. M., Gregoriadis, G.: Liposome-mediated dna vaccination: the effect of vesicle composition. Vaccine 19, 3301-3310 (2001).

[9] Yan, W., Chen, W., Huang, L.: Mechanism of adju-vant activity of cationic liposome: phosphorylation of a MAP kinase, ERK and induction of chemokines. Mol. Immunol. 44, 3672-3681 (2007).

[10] Arias, M. A., et al.: Carnauba wax nanoparticles enhance strong systemic and mucosal cellular and humoral immune responses to HIV-gp140 antigen. Vaccine 29, 1258-1269 (2011).

[11] Borges, O., et al.: Immune response by nasal delivery of hepatitis B surface antigen and co-delivery of a CPG ODN in alginate coated chitosan nanoparticles. Eur. J. Pharm. Biopharm. 69, 405-416 (2008).

[12] O'Hagan, D. T., Singh, M., Gupta, R. K.: Poly (lactide-co-g lycolide) microparticles for the development of single-dose controlled-release vaccines. Adv. Drug Deliv. Rev. 32, 225-246 (1998).

[13] Prausnitz, M. R., Mikszta, J. A., Cormier, M., Andrianov, A. K.: Microneedle-based vaccines. Curr. Top. Microbiol. Immunol. 333, 369-393 (2009).

[14] Sullivan, S. P., et al.: Dissolving polymer microneedle patches for infl uenza vaccination. Nat. Med. 16, 915-920 (2010).

[15] Donnelly, R. F.: Microneedle-mediated intradermal delivery. In: Donnelly, R. F., Thakur, R. R. S., Morrow, D. I. J., Woolfson, A. D. Microneedle-m ediated transdermal and intradermal drug delivery. Blackwell, Chichester, West Sussex (2011).

[16] Curran, R. M., et al.: Vaginal delivery of the recombi-nant HIV-1 clade-C trimeric gp140 envelope protein CN54gp140 within novel rheologically structured vehicles elicits specifi c immune responses. Vaccine 27, 6791-6798 (2009).

[17] Tiwari, S., et al.: Liposome in situ gelling system: novel carrier based vaccine adjuvant for intranasal delivery of recombinant protein vaccine. Procedia. Vaccinol. 1, 148-163 (2009).

[18] Donnelly, L., et al.: Intravaginal immunization using the recombinant HIV-1 clade-C trimeric envelope glycoprotein CN54gp140 formulated within lyophilized solid dosage forms. Vaccine 29, 4512-4520 (2011).

[19] Gupta, P. N., et al.: Development of liposome gel based formulations for intravaginal delivery of the recombinant HIV-1 envelope protein CN54gp140. Eur. J. Pharm. Sci. 46, 315-322 (2012).

[20] Black, C. A., et al.: Vaginal mucosa serves as an induc-tive site for tolerance. J. Immunol. 165, 5077-5083 (2000).

[21] Mestecky, J., Moldoveanu, Z., Elson, C. O.: Immune response versus mucosal tolerance to mucosally administered antigens. Vaccine 23, 1800-1803 (2005).

[22] Howland, S. W., Wittrup, K. D.: Antigen release kinet-ics in the phagosome are critical to cross-presentation effi ciency1. J. Immunol. 180, 1576 (2008).

[23] Shen, H., et al.: Enhanced and prolonged cross-presentation following endosomal escape of exogenous antigens encapsulated in biodegradable nanoparticles. Immunology 117, 78-88 (2006).

[24] Singh, M., et al.: Controlled release microparticles as a single dose hepatitis B vaccine: evaluation of immunogenicity in mice. Vaccine 15, 475-481 (1997).

[25] Singh, M., et al.: Cationic microparticles are an effective delivery system for immune stimulatory CPG DNA. Pharm. Res. 18, 1476-1479 (2001).

[26] O'Hagan, D. T., Rappuoli, R.: The safety of vaccines. Drug Discov. Today 9, 846-854 (2004).

[27] O'Hagan, D. T., Valiante, N. M.: Recent advances in the discovery and delivery of vaccine adjuvants. Nat. Rev. Drug Discov. 2, 727-735 (2003).

[28] Watanabe, M., Nagai, M., Funaishi, K., Endoh, M.: Effi cacy of chemically cross-linked antigens for acellular pertussis vaccine. Vaccine 19, 1199-1203 (2000).

[29] Jegerlehner, A., et al.: A molecular assembly system that renders antigens of choice highly repetitive for induction of protective B cell responses. Vaccine 20, 3104-3112 (2002).

[30] Kersten, G. F. A., Crommelin, D. J. A.: Liposomes and ISCOMS. Vaccine 21, 915-920 (2003).

[31] Gómez, S., et al.: Gantrez © AN nanoparticles as an adjuvant for oral immunotherapy with allergens. Vaccine 25, 5263-5271 (2007).

[32] Cubas, R., et al.: Virus-like particle (VLP) lymphatic traffi cking and immune response generation after immunization by different routes. J. Immunother. 32, 118-128 (2009).

[33] Foged, C., Brodin, B., Frokjaer, S., Sundblad, A.: Particle size and surface charge affect particle uptake by human dendritic cells in an in vitro model. Int. J. Pharm. 298, 315-322 (2005).

[34] Blanco, M. D., Alonso, M. J.: Development and character-ization of protein-loaded poly (lactide-co-g lycolide) nanospheres. Eur. J. Pharm. Biopharm. 43, 287-294 (1997).

[35] Zajac, P., et al.: Enhanced generation of cytotoxic T lymphocytes using recombinant vaccinia virus expressing human tumor-associated antigens and B7 co-stimulatory molecules. Cancer Res. 58, 4567-4571 (1998).

[36] Ríhová, B.: Immunomodulating activities of soluble synthetic polymer-bound drugs. Adv. Drug Deliv. Rev. 54, 653-674 (2002).

[37] Strong, P., Clark, H., Reid, K.: Intranasal application of chitin microparticles down-regulates symptoms of allergic hypersensitivity to dermatophagoides pteronyssinus and aspergillus fumigatus in murine models of allergy. Clin. Exp. Allergy 32, 1794-1800 (2002).

[38] Rogers, P. R., Croft, M.: Peptide dose, affi nity, and time of differentiation can contribute to the Th1/Th2 cytokine balance. J. Immunol. 163, 1205-1213 (1999).

[39] Johansen, P., Merkle, H. P., Gander, B.: Physico-chemical and antigenic properties of tetanus and diphtheria toxoids and steps towards improved stability. Biochim. Biophys. Acta 1425, 425-436 (1998).

[40] Hart, B. A., et al.: Liposome-mediated peptide loading of MHC-DR molecules in vivo. FEBS Lett. 409, 91-95 (1997).

[41] Van Dissel, J. T., et al.: Ag85B-ESAT-6 adjuvanted with IC31 © promotes strong and long-lived mycobacterium tuberculosis specifi c T cell responses in naïve human volunteers. Vaccine 28, 3571-3581 (2010).

[42] O'Hagan, D. T. E. A.: Biodegradable microparticles as controlled release antigen delivery systems. Immunology 73, 239-242 (1991).

[43] Putney, S. D., Burke, P. A.: Improving protein thera-peutics with sustained-release formulations. Nat. Biotechnol. 16, 153-157 (1998).

[44] Eldridge, J. H., et al.: Biodegradable microspheres as a vaccine delivery system. Mol. Immunol. 28, 287-294 (1991).

[45] Audran, R., Men, Y., Johansen, P., Gander, B., Corradin, G.: Enhanced immunogenicity of microencapsulated tetanus toxoid with stabilizing agents. Pharm. Res. 15, 1111-1116 (1998).

[46] Gupta, P. N., Khatri, K., Goyal, A. K., Mishra, N., Vyas, S. P.: M-cell targeted biodegradable plga nanoparticles for oral immunization against hepatitis B. J. Drug Target. 15, 701-713 (2007).

[47] Moser, C., Metcalfe, I. C., Viret, J. F.: Virosomal adju-vanted antigen delivery systems. Expert Rev. Vaccines 2, 189-196 (2003).

[48] Durrer, P., et al.: Mucosal antibody response induced with a nasal virosome-based infl uenza vaccine. Vaccine 21, 4328-4334 (2003).

[49] Barr, I. G., Sjölander, A., Cox, J. C.: ISCOMs and other saponin based adjuvants. Adv. Drug Deliv. Rev. 32, 247-271 (1998).

[50] Lenarczyk, A., et al.: ISCOM © based vaccines for cancer immunotherapy. Vaccine 22, 963-974 (2004).

[51] Ennis, F. A., et al.: Augmentation of human infl uenza A virus-specifi c cytotoxic t lymphocyte memory by infl uenza vaccine and adjuvanted carriers (ISCOMs). Virology 259, 256-261 (1999).

[52] Casals, J., Freund, J.: Sensitization and antibody formation in monkeys injected with tubercle bacilli in paraffin oil. J. Immunol. 36, 399-404 (1939).

[53] Ott, G., Barchfeld, G. L., Nest, G. V.: Enhancement of humoral response against human infl uenza vac-

cine with the simple submicron oil/water emulsion adjuvant MF59. Vaccine 13，1557-1562（1995）.

[54] Pattani，A.，et al.：Characterisation of protein stability in rod-insert vaginal rings. Int. J. Pharm. 430，89-97（2012）APC-Targeted（DNA）Vaccine Delivery Platforms：Nan.

（吴国华，赵志荀译）

第 47 章　纳米微粒辅助的 APC 靶向（DNA）疫苗递送平台

Pirouz Daftarian，Paolo Serafini，
Victor Perez 和 Vance Lemmon①

摘要

包括疫苗在内的所有药物给药后，只有一小部分药物可以到达其预期的靶组织或细胞。归巢疫苗在抗原呈递细胞（APC）发挥效力，几十年来一直是关注的焦点。这些疫苗具有适当的共刺激信号，只有特异性的抗原呈递细胞才能正确识别和处理抗原，随后刺激 T 细胞和 B 细胞产生特异性的免疫应答。随着现代免疫学和纳米技术领域的进步，新的疫苗递送平台在迈阿密大学米勒医学院正在兴起。抗原呈递细胞靶向疫苗才能诱发保护性免疫，降低生产成本，最大程度地降低疫苗脱靶效应引起的意外效应（减少免疫抑制机制和毒性），显著提高免疫效率。抗原呈递细胞体内靶向就无须对患者进行昂贵而复杂外周血单核细胞离体操作，这实际上是个体化免疫策略的最佳策略。本章阐述和比较了基于纳米微粒的靶向疫苗平台的新方法，评估了其分子作用机制、翻译能力以及效率和成本。

47.1　树突状细胞(DC)疫苗和免疫接种：体外负载至体内靶向

树突状细胞疫苗　树突状细胞是最强效的抗原呈递细胞，具有较高的抗原摄取和加工效率。这些细胞也可迁移至淋巴结，然后在自身主要组织相容性复合体以及适当的共刺激信号下递呈或交叉递呈抗原，刺激特异性 T 细胞和 B 细胞增殖。有趣的是，树突状细胞可在淋巴结中形成间隙连接，此过程可促进交叉启动[1,2]。树突状细胞疫苗来源于在白细胞介素-4 和粒细胞巨噬细胞集落刺激因子中培养的患者单核细胞。然后将抗原蛋白或肽加载至这些树突状细胞。再将最终产物负载抗原的树突状细胞注射回患者体内[2]。树突状细胞具有用于控制 MHC Ⅱ基因座的额外调节元件，产生高水平的 CIITA 转录（MHC Ⅱ分子表达的先决条件），从而导致更高水平的 MHC Ⅱ转录[3]。树突状细胞的抗原加工能力与其成熟度呈正相关，随着树突状细胞的成熟，MHC Ⅱ类分子的半衰期显著增加[3]。FDA 批准了第一种人前列腺癌的治疗性疫苗，该疫苗十分安全，能够在大部分疫苗接种的患者体内诱导肿瘤抗原特异性免疫应答，故使用负载抗原的自体树突状细胞作为疫苗给许多领域带来了希望，特别是癌症的治疗性疫苗。尽管基于树突状状细胞特异性前列腺癌疫苗是一项突破，但仍面临营销挑战，部分

---

① P. Daftarian，PhD（✉）（迈阿密大学眼科系，美国佛罗里达州迈阿密市，迈阿密大学生物化学与分子生物学系，美国佛罗里达州迈阿密市，E-mail：PDaftarian@ med. miami. edu）

P. Serafini，PhD（迈阿密大学米勒医学院微生物与免疫学系，美国佛罗里达州迈阿密市）

V. Perez，PhD（迈阿密大学眼科系，美国佛罗里达州迈阿密市，迈阿密大学米勒医学院微生物与免疫学系，美国佛罗里达州迈阿密市）

V. Lemmon，PhD（Miami Project to Cure Paralysis，University of Miami，Miami，FL，USA，迈阿密大学瘫痪治疗项目，美国佛罗里达州迈阿密市）

原因是其成本高达93美元，且操作复杂；在英国，只有3%的肿瘤医师表示他们可能会使用这种疫苗[1]。

科学家正在积极进行针对树突状细胞的其他尝试。尽管CD205在淋巴细胞（B细胞和T细胞）和粒细胞中的表达水平低，但在小鼠的成熟树突状细胞中表达水平极高[4]。有趣的是，早期使用抗dec205单克隆抗体抗原递送模型的研究工作表明，其诱导了耐受性和T细胞耗竭，而非肿瘤免疫[5]。如果使用佐剂（即抗CD40抗体），则产生强烈的体液和细胞免疫应答，能够治愈70%转移性黑素瘤攻毒的小鼠[6]。尽管在人体中DC205不限于树突状细胞，但这些试验表明，抗原呈递细胞靶向与足够的免疫佐剂是免疫疗法成功的关键。

其他靶标是单核细胞、巨噬细胞未成熟树突状细胞表达的甘露糖受体。但据报道，甘露糖受体在不同的上皮细胞亚群中也有表达[7]。抗原与甘露糖偶联可增强MHC Ⅰ类和MHC Ⅱ类抗原的递呈，促进体液和细胞介导的持续免疫反应[8]。在乳腺癌相关抗原Muc1偶联的氧化甘露聚糖的随机III期临床试验中，所有接受治疗的患者在5年内均无肿瘤发生，而27%用安慰剂治疗的患者出现乳腺癌复发[9]。

47.2 基因疫苗或DNA疫苗

启动免疫系统的裸质粒DNA疫苗最有希望替代传统疫苗。DNA疫苗给大规模生产带来诸多好处，这些优点在包括重组蛋白或全肿瘤细胞在内的其他形式疫苗中很难实现[10-13]。1990年，Jon Wolff报道DNA质粒免疫欧使小鼠产生免疫应答。1993年，Margaret Liu证明DNA疫苗能够诱导机体产生体液和细胞免疫应答。

DNA疫苗人体试验早在1996年就已开始进行。DNA或基因疫苗使用含预期抗原的DNA序列。哺乳动物的DNA质粒包括至少一种所需抗原的DNA序列和基于哺乳动物的启动子。将这种质粒注入宿主体内，然后使用细胞的翻译机制翻译DNA质粒编码的蛋白质。DNA疫苗能够利用抗原递呈的内源性和外源性途径。基因疫苗本质上优于主要外源性蛋白质疫苗。蛋白质疫苗面临的另一个问题是，现如今蛋白质苛刻的生产工艺很难使疫苗保留天然的蛋白质结构，而DNA疫苗使宿主表达天然构象的蛋白质[14]。

DNA疫苗可以诱导持续的免疫应答，这对免疫至关重要[15]。DNA疫苗可产生交叉中和抗体，这些应答对具有抗原变异和不同亚型的病原体至关重要。基因疫苗免疫类似于天然感染，不涉及病原体，没有任何临床表现。适当表位或抗原的鉴定是DNA疫苗设计面临的主要挑战。基因疫苗可包含多种候选抗原基因、删除免疫抑制元件、融合免疫增强基因或序列，极大改善了疫苗设计[16]，这对于非病毒诱发的癌症疫苗尤其重要。用编码各种候选抗原的质粒进行免疫机体可筛选出具有最佳保护性抗原[16]。

47.3 通过加强免疫改善DNA疫苗

DNA疫苗具有包括免疫原性差在内的自身缺点，并不完美。虽然许多临床试验证明这种具有成本效益的疫苗十分安全，但引发的免疫应答不足以清除已有的肿瘤或病原体，显示出其一定的局限性。例如哺乳动物细胞的细胞膜带负电荷，带负电荷的核酸不容易穿透。这种细胞膜是阻止DNA等大分子转运的强大屏障，使DNA疫苗效率不高。体内转染效率低，不能优先靶向专职抗原呈递细胞，以及DNA疫苗固有的佐剂活性是临床上和治疗性鼠模型中出现边缘效应的主要原因。一个公认的假设是，如果最大程度地向专业抗原呈递细胞（APC）递送DNA，且通过佐剂和（或）细胞因子的其他形式的免疫学辅助方法，则DNA疫苗的免疫原性可显著增加[17,18]。

DNA疫苗的递送属于非靶向，故其天然抗原的表达浓度低。尤其在抗原呈递细胞中，这是

DNA 疫苗的主要挑战。使用基因疫苗作为蛋白质免疫的加强免疫是克服该问题的一种方法[15]。科学的解释是，加强免疫需要适量的抗原、联合疫苗接种策略；常规方法使用蛋白质抗原，然后使用基因疫苗扩展 T 细胞克隆型，而 T 细胞克隆型识别宿主产生的内源性和外源性具有天然构象蛋白质。事实上，Nabel 和其同事的研究表明，蛋白质疫苗接种后的 DNA 疫苗的加强免疫增强了针对保守血凝素干表位的高强度抗体反应，且这些抗体在临床试验中能够中和多种菌株[15]。提高 DNA 疫苗接种效率有多种方法，包括体内电穿孔[14]或使用纳米载体[19]。体内电穿孔可产生局部 DNA 摄取增加及募集免疫相关组分的内源性表达和 T 细胞，是增强 DNA 疫苗接种的一种有效方法[16,20]。实际上，正在进行的各种体内电穿孔仪的临床试验表明，这些疫苗具有安全性和有效性[16]。

47.4 下一代疫苗：靶向递送

尽管 DNA 疫苗在临床前非治疗模型中取得了成功，证明其安全性，但在临床试验和严格的癌症小鼠模型中仅具有中度的治疗效果。缺乏有效的体内转染和 DNA 靶向抗原呈递细胞方法是失败的主要原因。如上所述，抗原呈递细胞（包括巨噬细胞、B 细胞和树突状细胞）对启动抵御外来病原体和肿瘤的保护性免疫应答至关重要[21]。高效转染是 DNA 疫苗充分释放其潜在治疗效果的主要途径。为达此目的，采用了 3 类不同的策略：①在抗原呈递细胞浓度达到一定值时用物理方法递送疫苗（即用真皮电穿孔靶向皮肤朗格汉斯抗原呈递细胞、经皮释放或使用无针喷射注射器）；②使用可植入微器械操纵时空免疫应答；③使用功能化的纳米微粒实现靶向抗原呈递细胞。

47.5 疫苗：使用物理方法递送至抗原呈递细胞

现已明确皮肤是不仅外部环境与内部器官分隔的物理屏障，还是机体与许多外部病原体发生首次相互作用的物理界面，是一种主动免疫器官[22,23]。皮肤的专职抗原呈递细胞包括表皮朗格汉斯细胞（CD207/langerin$^+$）、真皮郎罕细胞特异蛋白和真皮郎罕细胞特异蛋白+树突状细胞（DC）。人体皮肤中的朗格汉斯细胞可诱导细胞毒性 T 淋巴细胞，小鼠的这些细胞是肿瘤抗原交叉递呈，以及患黑色素瘤小鼠产生保护性免疫应答的关键[24]。真皮易于接近，且专业抗原呈递细胞的浓度高，是免疫的重要场所。将疫苗递送至该部位是促进专业抗原呈递细胞摄取 DNA 疫苗的物理方法。

体内真皮电穿孔已成为 DNA 免疫方法的金标准。疫苗通过微针介导的真皮电穿孔进行免疫，最大程度地促进 DNA 及危险信号进入细胞，促进朗格汉斯细胞的成熟，同时将天然形式的蛋白质呈递给免疫系统。尽管在幼鼠体内获得了理想结果，但在炎症期肿瘤携带者体内效果较差。实际上，真皮中除存在免疫原性抗原呈递细胞外，还有其他细胞群（包括致耐受性抗原呈递细胞），使用“物理”方法进行皮肤递送可能无法确保产生足够的保护性免疫。实际上，诱发强免疫应答的过程难以达到可重现性。强烈的免疫应答需在淋巴器官中产生大量活化的 B 细胞、T 细胞，特别是 CD4$^+$T 辅助细胞。由于该过程很难实现重复性，并且体内电穿孔无选择性地靶向所有细胞（例如肌肉细胞），故需对体内电穿孔的参数进行优化，并采取提高免疫系统 T 细胞臂的其他辅助策略。此外，使用普通注射器进行皮内注射需要丰富的专业知识，与此技术相关的困难是实验室间差异的主要原因。尽管如此，构建含有微针的微器械、纹身样疫苗免疫设备和无针注射器[25]使得靶向朗格汉斯细胞皮肤疫苗的使用更加便利。

47.6 通过植入式微器械调控时空免疫应答

Mooney 研究小组提出了一个简明的设想，通过分离树突状细胞前体，诱导其分化、加载抗原，然后促进其成熟的陈旧概念。作者不再尝试在体内选择性转染抗原呈递细胞，而是使用获得 FDA 批准的给予聚丙交酯-聚-乙交酯（PLG）的设备，重新描述了树突状细胞的成熟、脉冲和诱导成熟的概念，这与基于树突状细胞疫苗的概念相同[26]。这种基于合成聚合物的基质可在空间和时间上协调体内细胞因子、抗原和危险信号的释放[26]。作者基本上设想了一种将整个 GMP 试验室转移至可植入微器械的方法。

简而言之，通过气体发泡工艺可以将 PLG 制造成高度多孔的“海绵”。在发泡过程中将 GM-CSF、CpG 寡核苷酸和肿瘤抗原整合至支架中，其在植入体内后可进行局部的和持续的释放。结果表明，该装置释放 GM-CSF，吸引 $CD11b^+$ 树突状细胞前体，而 CpG-ODN 分子促进 $CD11c^+$PDCA1+浆细胞样树突状细胞的募集。当 GM-CSF 和 CpG-ODN 共同释放时，浆细胞样树突状细胞的数量显著增加，而骨髓树突状细胞募集未见差异。向支架中添加肿瘤裂解物不仅能促进抗原摄取，还最有可能通过释放内源性免疫原性激活剂（即 DNA、热休克蛋白）促进 IL-12 的原位分泌，从而产生 $CD8^+$ 树突状细胞[26]。

体内活化的和脉冲的树突状细胞亚群容易迁移至淋巴结导管，促进 9 日龄 B16 黑素瘤肿瘤的小鼠体生成强细胞毒性 T 淋巴细胞免疫，促进肿瘤消退。重要的是，这是一种极其严格的治疗方案，尚无其他免疫疗法产生类似的治疗效果。基于 PLG 的器械不仅对所需树突状细胞亚群进行靶向脉冲，还可能在树突状细胞募集期间调节微环境、抗原摄取和分化，与众不同地对体内免疫应答进行精细调节。尽管可以通过核酸疫苗对该方法进行改进，但是在保护 DNA 和（或）RNA 免受内源核酸酶降解方面具有一定困难。

47.7 通过功能化纳米微粒实现靶向抗原呈递细胞

纳米微粒可经组织扩散并穿透细胞膜，将负载的抗原运送到细胞内，故纳米微粒作为药物或疫苗靶向纳米载体颇具吸引力。靶向抗原呈递细胞递送 DNA 和 RNA 疫苗的纳米平台有两个主要组成部分：靶向分子和加载（包封）表面。有的纳米微粒并无靶向分子；其特异性依赖于纳米分子的大小、给药方式和未充分表征的化学特性。本章的其余内容重点阐述了利用动物模型和人体所保留的免疫系统和配体（受体）相互作用的研究基础，实现抗原呈递细胞靶向的纳米微粒。尽管还有其他重要的纳米平台，但均缺乏动物模型相关信息，难以应用，也难以将结果转化为临床。

随着对抗原呈递细胞不同亚群的受体特异性理解的加深，用于特异性递送疫苗的配体数量也在不断增长[30,31,34-39]。甘露糖受体、DC205、Fc 受体（I、II 和 III）、CD11c-CD18 整联蛋白和 MHC Ⅱ类分子是抗原呈递细胞常用的特异性靶向配体。靶向部分（肽、单克隆抗体或适体）修饰后的抗原、DNA 加载抗原或纳米载体可以归巢至抗原呈递细胞。用于该目的的靶向部分有 CD11b、CD169、甘露糖受体、Dec-205、CD11c、CD21／CD35、CX3CR1、趋化因子、IgG FC 受体和其他 Fc 受体。其他研究使用了 pH 依赖的靶向树突状细胞[27-32]。一些研究者甚至认为，这些存在于淋巴结中的细胞具有高吞噬性，如果使用大小（<500 nm）适宜的纳米微粒，则无需进行树突状细胞靶向，因此在不使用靶向配体的情况下靶向淋巴结[5,33-35]。

47.8 使用纳米载体改善 DNA 疫苗

树枝状聚合物的纳米载体与 DNA 或生物治疗剂非共价络合已成为广泛研究的重点[36,37]。为实现树突状细胞靶向，Cruz 和其同事描述了一种由聚 d，l-丙交酯-乙交酯包封的超顺磁性氧化铁组成的纳米微粒，其与树突状细胞特异性受体 DC-SIG 的抗体进行偶联。对这些纳米微粒进行荧光标记，可在 24 小时内追踪递送到内吞溶酶体腔中的抗原。这些平台含有超顺磁性氧化铁，需进行包封，使其具有生物相容性，故使用葡聚糖、聚丙交酯-乙交酯或其他聚合物进行了包封[39]。

脂质体是一种磷脂双层囊泡，一直用于递送药物和疫苗。用甘露糖包裹脂质体，可增强递送效果或提高树突状细胞和抗原呈递细胞的募集[40-42]。有趣的是，Geng 和其同事们发现，一种治疗高血压的药物阿米洛利能够在体内增强 DNA 向细胞内的递送过程，诱导编码抗原的强烈免疫反应[41]。

我们已报道了一个可与 DNA 络合，然后将其护送到宿主抗原呈递细胞的平台。该平台使用配体与包括专业抗原呈递细胞的 MHC Ⅱ类表达细胞在内的细胞进行结合（图 47.1）。下文将讨论此平台[37]。

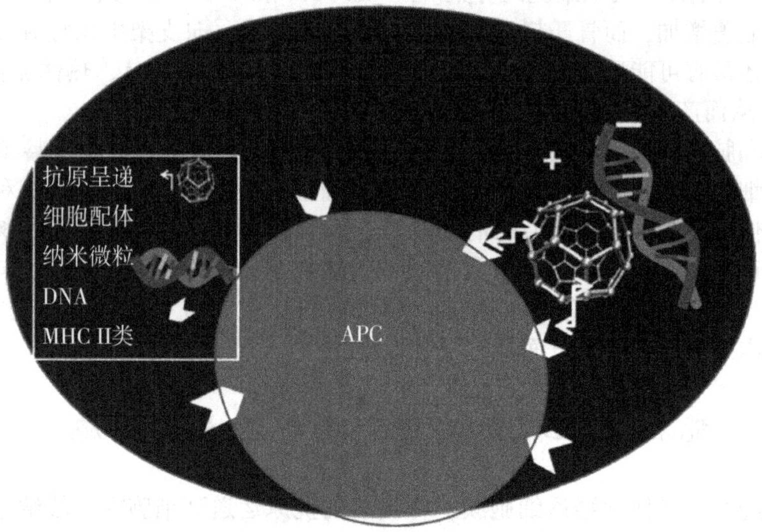

图 47.1 靶向抗原呈递细胞具有免疫增强作用的纳米微粒肽衍生的树枝状聚合物（PDD）

47.9 纳米微粒平台：PADRE 衍生化树状聚合物（PDD）

DNA 是一种具有负电荷的聚电解质，DNA 片段具有大量负电荷，阻碍遗传物质向细胞转运，阻碍遗传物质穿过细胞和核膜等疏水屏障。克服这一问题的一般方法是使用聚阳离子试剂络合 DNA，形成总电荷接近零的浓缩纳米微粒，后者可较容易地穿透膜屏障[43,44]。例如阳离子壳聚糖和树枝状聚合物已成为众多疫苗研究的重点[45,46]。树枝状聚合物有多种尺寸（通常达第 6 代），表面经醇、羧酸或胺基团封闭。

聚酰胺-胺（PAMAM）树枝状聚合物表面的氨基易于在生理 pH 下发生质子化，使其表面产生大量正电荷，促进其与含所需寡核苷酸序列的 DNA 质粒进行缩合，广泛用作递送基因。Tomalia 开发的 PAMAM 树枝状大分子已广泛用于 DNA 转染研究[37,47]。众所周知，含疏水性氨基酸的胺末大量合成和衍生化降低了树枝状聚合物的转染效率[48]，这是表面正电荷大量损失以及削弱了 DNA 的静电相互作用造成的[37,43,44,47]。PAMAM 树枝状聚合物是一种高度分支化的大分子，包括一个中央核心和一系列在结构和合成方面显著不同的层。树枝状大分子结构中的一层称为“代”，直接涉及大分子的逐步生长。如图 47.2 所示，第 5 代（G5）PAMAM 树枝状大分子是将 DNA（带负电）递送到细胞（也带负电荷）中的理想选择。

47.10 树枝状聚合物对抗原呈递细胞的识别能力

所有抗原呈递细胞均存在主要组织相容性复合物（MHC）基因组编码区域。MHC 是免疫系

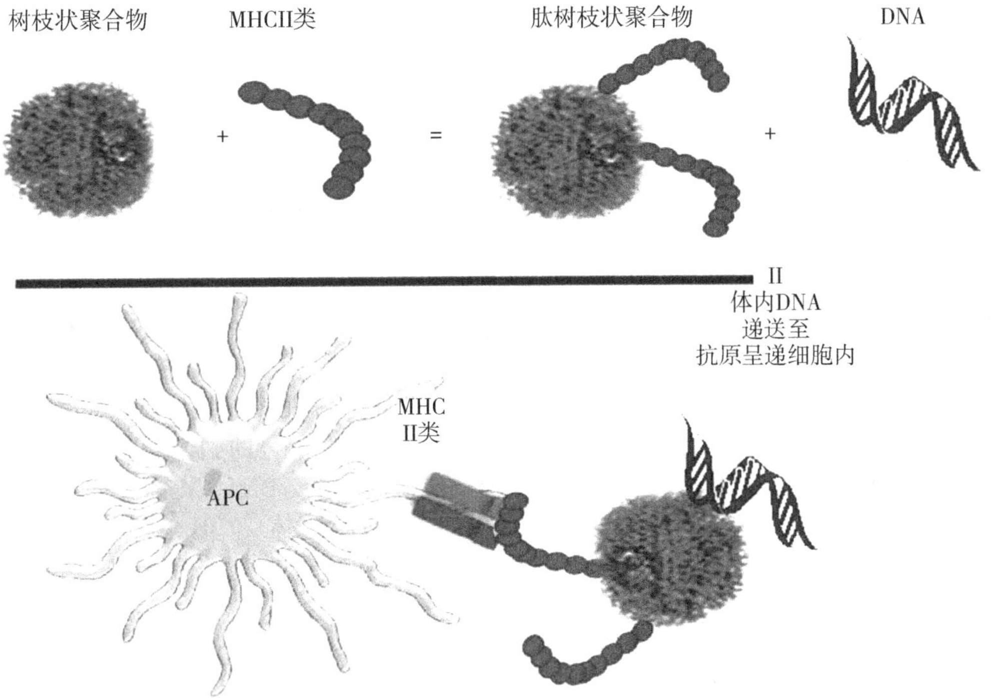

图 47.2 用于体内基因递送的肽-树枝状聚合物平台

靶向 MHC Ⅱ类的多肽与树枝状聚合物（PDD）共价连接。与 DNA 络合的 PDD 在体内有效地将疫苗靶向 MHC Ⅱ类+抗原呈递细胞。

统的重要组成部分，对病原体、肿瘤抗原以及自身免疫的免疫应答起重要作用。MHC 基因编码的蛋白质表达细胞表面，这些细胞包括表达自身抗原和非自身抗原的病原体、肿瘤细胞和各种 T 细胞，使这些细胞能够①有助于启动免疫应答；②辅助 T 细胞杀灭入侵的病原体、肿瘤细胞或感染病原体的细胞；③与已成熟的抗体抵抗非自身抗原。MHC Ⅱ类分子主要在树突状细胞、B 淋巴细胞、单核细胞和巨噬细胞上表达。辅助性 T 淋巴细胞识别 MHC Ⅱ类分子，然后刺激辅助性 T 淋巴细胞增殖，并放大所展示的特定免疫原多肽的免疫应答。

几乎所有的有核细胞都表达 MHC Ⅰ类分子，后者由细胞毒性 T 淋巴细胞（CTL）进行识别，然后细胞毒性 T 淋巴细胞破坏携带抗原的细胞。细胞毒性 T 淋巴细胞在肿瘤排斥和抵御病毒感染方面尤为重要。泛 D 结合肽 PADRE 能够与特定的 MHC 分子进行结合。例如，PADRE 对 6 种选定的 DRB1 亚型[49]和 PBMC 的抗原呈递细胞具有高亲和力[50]。事实上，PADRE 可与人和小鼠的 MHC Ⅱ类结合[51]。目前，已用 PADRE 结合抗原呈递细胞，然后激活 CD4 辅助 T 细胞，首次靶向于抗原呈递细胞（通过其与表面 MHC 结合）[37]。

其他以及我们已研究的通用 T 辅助表位，例如计算机生成的称为 PADRE 的泛 DR T 辅助表位可以作为疫苗的一种佐剂[52-60]。PADRE 是一种非天然表位，可与抗原呈递细胞（如 B 细胞、T 细胞、单核细胞、巨噬细胞、树突状细胞和朗格罕细胞）中的大多数 MHC Ⅱ类分子结合。我们设想，为使这种通用表位充当 CD4 辅助细胞，首先必须结合 MHC 类基因或 HLA-DR 等位基因，这样才可以在载体中作为细胞靶向的元件。MHC Ⅱ类主要由专业抗原呈递细胞表达，故其必须与抗原呈递细胞结合。然后我们设想将通用 T 辅助表位与纳米载体偶联，制备抗原呈递细胞靶向的纳米载体。

我们已对聚酰胺-胺（PAMAM）树枝状聚合物已进行了广泛研究，包括其合成、电荷、分子量和氨基数量等化学特征、表面官能化方法、毒性和生物相容性。我们开发了靶向抗原呈递细胞纳米载体，这种载体包括与泛 DR T 辅助表位结合的第五代树枝状聚合物或 PADRE（PADRE 树枝状聚合物，下文称为 PDD）、流感病毒血凝素（HA）分子的 T 辅助表位（PDD2），或使用随机对照肽的辅助表位。

将疫苗转化成人体临床研究以及在全球范围内分销疫苗的后勤可行性限制了一些具有科学性的平台的应用，使这些平台遭遇困难。例如，制备偶联 DNA 和佐剂的人源化单克隆抗体的纳米微粒可能在免疫途径调节、放大工艺和安全性上存在问题。决定纳米微粒产品成败的主要因素是化学、制造和质控（CMC）以及包括平台免疫原性在内的安全性。我们选择 CMC 相对简单以及安全性参数可接受的核心聚合物和细胞识别部分。

47.11 PDD 是一种抗原呈递细胞调理的体内免疫增强平台

这是一种可行的方法，我们尝试定制具有细胞靶向部分的纳米载体，这种载体除靶向抗原呈递细胞之外，还具有内在的佐剂活性。当与可能具有额外优势的纳米颗粒整合在一起时，DNA 可变成佐剂，且免于降解[61-67]。树枝状聚合物能保护 RNA / DNA，提高转染效率[68,69]；这些分子在体内效果有限。优化后的 PADRE-树枝状大分子纳米载体、PDD 平台应具有体内活性，可选择性地将 DNA 递送至抗原呈递细胞。PDD 的内在免疫增强特性与疫苗效应具有协同关系（表 47.1）。

该平台基于与 MHC Ⅱ类靶向肽偶联的第 5 代聚（酰氨基胺）（G5-PAMAM）树枝状大分子（表面负载 DNA），可选择性地将树枝状大分子递送至抗原呈递细胞，增强其免疫刺激效力。与该平台偶联的 DNA 可在体外有效地转染鼠和人抗原呈递细胞。当啮齿动物经皮下免疫 DNA-肽-树枝状大分子复合物时，可优先转染引流淋巴结中的树突状细胞（DC），促进高亲和力 T 细胞的生成，排斥已确定的肿瘤。

表 47.1 携带 DNA 的 APC 靶向疫苗平台具有哪些性质

传统疫苗的局限性	抗原呈递细胞靶向的纳米载体
尽管能诱导体液反应，但也诱导不良的 T 细胞反应 原因：抗原呈递细胞的疫苗数量不足	抗原呈递细胞靶向诱导体液反应和强烈的 T 细胞反应
许多抗原的免疫原性不佳 原因：T 细胞反应弱	激活细胞介导的反应，包括辅助性 T 细胞
抗原呈递细胞对疫苗的摄取不良 原因：蛋白质抗原、DNA 和 RNA 不能穿过细胞膜（即两者都载有负电荷，无转运机制）	通过胞吞作用诱导摄取
病原体和癌症抗原的免疫避免机制 原因：无强烈的 T 细胞反应和非靶向性抑制机制	激活细胞介导的反应，包括辅助性 T 细胞
成本高 原因：蛋白质的稳定性显著低于 DNA，制备成本高	可转运 DNA 疫苗，DNA 稳定，成本低

47.12 PPD 介导的 TRP2 疫苗接种排斥已有的黑色素瘤

如果 DNA 疫苗靶向 MHC Ⅱ类+抗原呈递细胞至关重要，则在治疗携带肿瘤的小鼠方面具有优势，而其预防性疗效并不能真正让人信服。肿瘤 B16 黑色素瘤治疗性小鼠易引发致耐受机制，故称为侵袭性模型，将抗原递送至专业抗原呈递细胞是治疗成功的关键[70]。使用严格的治疗性鼠肿瘤黑色素瘤模型 B16 测试 PPD 纳米颗粒（图 47.3）。B16 肿瘤相关抗原 TRP2 抗原性较弱，除非使用强效重组痘苗病毒[72]或联合疗法[71,73]进行抗原疫苗接种，TRP2 用于治疗已确诊的黑色素瘤时，疗效一般[71]。与先前的结果一致，在肿瘤注射后 2 天和 9 天使用未偶联多肽的对照树枝状大分子，不能诱导机体产生针对 TRP2 的免疫应答（图 47.3a）。

当相同的复合物与 PADRE 混合免疫小鼠后，其存活率与空载体接种的小鼠相比具有明显的统计学差异（$P=0.045$）（图 47.2）。在此情况下，PADRE 仅与树枝状聚合物混合，未发生偶联，故该效应可归因于 PADRE 的佐剂效应。当用 PPD 接种抵御 TRP2 时（图 47.3b、PPD-TRP2 组），与对照组相比，小鼠存活期和临床意义具有显著的统计学差异，50%的小鼠在 350 天的实验周期内内无肿瘤发生（PPD-TRP2 与 PPD-pcDNA3：$P<0.005$）。这些数据表明，PADRE 需与树枝状聚合物连接，每个树枝状聚合物平均需要 2~3 个肽，才能达到强效治疗效果。这些数据表明，专业抗原呈递细胞的特异性靶向作用和通用肽产生的佐剂效应可能是疫苗成功的关键。

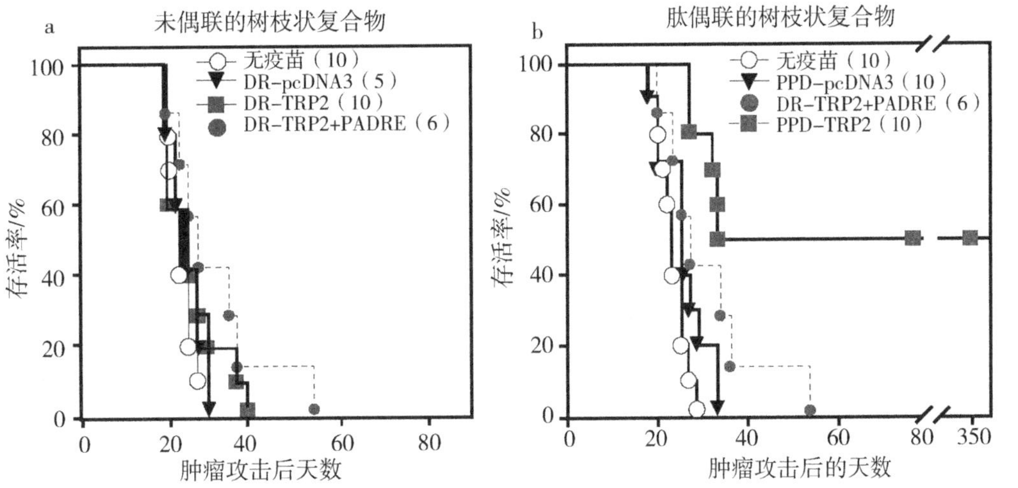

图 47.3 PPD 促进排斥已确立的 B16 黑色素瘤

4 只 C57Bl/6 荷瘤小鼠小鼠在肿瘤植入后第 2 天和第 9 天在解剖部位皮下注射作为疫苗的 cDNA3-TRP2（灰色方块）或 pcDNA3（用作对照，黑色三角形）（10 μg/部位）。使用未修饰的 DR（a）或使用 PPD（b）进行免疫。注射 PBS 的小鼠（白色圆圈）作为额外的对照或载有 pcDNA3-TRP2 的未经修饰的树枝状聚合物，然后与相同剂量的理论上连接 PPD（DR-TRP2+PADRE；灰色菱形）的 PADRE 肽混合。log-rank 检验 $P<0.001$，Holm-Sidak：PPD-TRP2 与 PPD-pcDNA3，$P=0.001$；PPD-TRP2 与 DR-TRP2+PADRE，$P=0.03$；DRTRP2+PADRE 与无疫苗，$P=0.04$（数字经许可使用）。

结论

幸运的是，近年来，已开发和报道了许多新型疫苗平台。靶向淋巴器官的载体可在免疫部位募集更多的抗原呈递细胞或树突状细胞，可诱导和增强树突状细胞向引流淋巴结迁移，也可携带共刺激分子或细胞因子。PDD 是一种靶向抗原呈递细胞的 DNA 纳米载体，使用具有内在免疫增强能力的细胞识别元件[17]。表 47.1 总结了此类疫苗与传统疫苗相比较存在的优势。自然界中许多肿瘤或病原体相关抗原的免疫原性较差，这属于免疫避免策略的一部分。靶向专业抗原呈递细胞并增强辅助 T 细胞作用的疫苗应可克服这一缺点。这些技术需要进一步完善，并制成单剂量制剂甚至口服制剂。

纳米载体疫苗平台还可与其他疫苗技术联合使用，包括蛋白质疫苗或通过体内电穿孔递送 DNA 疫苗[74]。递送 PADRE-PAMAM 树枝状大分子（PDD）纳米疫苗可解决诱导较强免疫反应疫苗所面临的一些重大挑战：转染效率高、正确靶向抗原呈递细胞、佐剂效应。该平台基于合成一种独特的纳米微粒，这两项技术已经完成单独试验，但效果有限（表 47.1）。通用肽 PADRE 可以结合或激活携带 15 种主要 HLA-DR 形式的细胞，并被广泛用作免疫佐剂[75-77]。

本报告确认了其佐剂效果（尽管有限）。与此形成鲜明对比的是，PAMAM 树枝状大分子与靶向 MHC Ⅱ类多肽的物理结合显著增加了该平台的免疫原性和 DNA 疫苗的功效，克服了目前 DNA 疫苗的局限性。鉴于疫苗剂量远低于药物或抗生素中使用的疫苗剂量，且由于靶向纳米疫苗使疫苗剂量进一步降低，预计许多纳米微粒的疫苗递送平台，尤其是 DNA 纳米疫苗递送平台，是可用于人体疾病治疗的候选免疫接种策略。

参考文献

[1] Payne, H., Bahl, A., Mason, M., Troup, J., De Bono, J.: Optimizing the care of patients with advanced prostate cancer in the UK: current challenges and future opportunities. BJU Int. 110, 658-667 (2012). doi: 10. 1111/j. 1464-410X. 2011. 10886. x.

[2] Lesterhuis, W. J., et al.: Route of administration modu-lates the induction of dendritic cell vaccine-induced antigen-specifi c T cells in advanced melanoma patients. Clin. Cancer Res. 17, 5725-5735 (2011). doi: 10. 1158/1078-0432. CCR-11-1261. 1078-0432. CCR-11-1261 [pii].

[3] Neefjes, J., Jongsma, M. L., Paul, P., Bakke, O.: Towards a systems understanding of MHC class I and MHC class II antigen presentation. Nat. Rev. Immunol. 11, 823-836 (2011). doi: 10. 1038/nri3084. nri3084 [pii].

[4] Witmer-Pack, M. D., Swiggard, W. J., Mirza, A., Inaba, K., Steinman, R. M.: Tissue distribution of the DEC-205 protein that is detected by the monoclonal antibody NLDC-145. II. Expression in situ in lymphoid and nonlymphoid tissues. Cell. Immunol. 163, 157-162 (1995). doi: 10. 1006/cimm. 1995. 1110. S0008-8749 (85) 71110-0 [pii].

[5] Bonifaz, L., et al.: Effi cient targeting of protein antigen to the dendritic cell receptor DEC-205 in the steady state leads to antigen presentation on major histocompatibility complex class I products and peripheral CD8+T cell tolerance. J. Exp. Med. 196, 1627-1638 (2002).

[6] Mahnke, K., et al.: Targeting of antigens to activated dendritic cells in vivo cures metastatic melanoma in mice. Cancer Res. 65, 7007-7012 (2005). doi: 10. 1158/0008-5472. CAN-05-0938. 65/15/7007 [pii].

[7] Irache, J. M., Salman, H. H., Gamazo, C., Espuelas, S.: Mannose-targeted systems fo the delivery of therapeutics. Expert Opin. Drug Deliv. 5, 703-724 (2008). doi: 10. 1517/17425247. 5. 6. 703.

[8] Tacken, P. J., de Vries, I. J., Torensma, R., Figdor, C. G.: Dendritic-cell immunotherapy: from ex vivo loading to in vivo targeting. Nat. Rev. Immunol. 7, 790-802 (2007). doi: 10. 1038/nri2173. nri2173 [pii].

[9] Apostolopoulos, V., et al.: Pilot phase III immuno-therapy study in early-stage breast cancer patients using oxidized mannan-MUC1 [ISRCTN71711835]. Breast Cancer Res. 8, R27 (2006). doi: 10. 1186/bcr1505. bcr1505 [pii].

[10] Widera, G., et al.: Increased DNA vaccine delivery and immunogenicity by electroporation in vivo. J. Immunol. 164, 4635-4640 (2000). ji_v164n9p4635 [pii].

[11] Cappelletti, M., et al.: Gene electro-transfer improves transduction by modifying the fate of intramuscular DNA. J. Gene Med. 5, 324-332 (2003). doi: 10. 1002/ jgm. 352.

[12] Mennuni, C., et al.: Effi cient induction of T-cell responses to carcinoembryonic antigen by a heterologous prime-boost regimen using DNA and adenovirus vectors carrying a codon usage optimized cDNA. Int. J. Cancer 117, 444-455 (2005). doi: 10. 1002/ijc. 21188.

[13] Rice, J., et al.: DNA fusion gene vaccination mobilizes effective anti-leukemic cytotoxic T lymphocytes from a tolerized repertoire. Eur. J. Immunol. 38, 2118-2130 (2008). doi: 10. 1002/eji. 200838213.

[14] Daftarian, P., et al.: In vivo electroporation and non-protein based screening assays to identify antibodies against native protein conformations. Hybridoma (Larchmt) 30, 409-418 (2011). doi: 10. 1089/ hyb.

2010. 0120.

[15] Ledgerwood, J. E., et al.: DNA priming and infl uenza vaccine immunogenicity: two phase 1 open label randomised clinical trials. Lancet Infect. Dis. 11, 916-924 (2011). doi: 10. 1016/S1473-3099 (11) 70240-7. S1473-3099 (11) 70240-7 [pii].

[16] Rice, J., Ottensmeier, C. H., Stevenson, F. K.: DNA vaccines: precision tools for activating effective immunity against cancer. Nat. Rev. Cancer 8, 108-120 (2008). doi: 10. 1038/nrc2326. nrc2326 [pii].

[17] Liu, M. A.: Immunologic basis of vaccine vectors. Immunity 33 (4), 504-515 (2010). doi: 10. 1016/j. immuni. 2010. 10. 004. S1074-7613 (10) 00364-X [pii].

[18] Leroux-Roels, G.: Unmet needs in modern vaccinol-ogy: adjuvants to improve theimmune response. Vaccine 28 (Suppl 3), C25-C36 (2010). doi: 10. 1016/j. vaccine. 2010. 07. 021. S0264-410X (10) 01004-2 [pii].

[19] Daftarian, P., et al.: Peptide-conjugated PAMAM den-drimer as a universal platform for antigen presenting cell targeting and effective DNA-based vaccinations. Cancer Res. (2011). doi: 10. 1158/0008-5472. CAN-11-1766. 0008-5472. CAN-11-1766 [pii].

[20] A hlen, G., et al.: In vivo electroporation enhances the immunogenicity of hepatitis C virus nonstructural 3/4A DNA by increased local DNA uptake, protein expression, infl ammation, and infi ltration of CD3+T cells. J. Immunol. 179, 4741-4753 (2007). 179/7/4741 [pii].

[21] Steinman, R. M., Mellman, I.: Immunotherapy: bewitched, bothered, and bewildered no more. Science 305, 197-200 (2004). 10. 1126/science. 1099688 305/5681/197 [pii].

[22] Nakajima, S., et al.: Langerhans cells are critical in epicutaneous sensitization with protein antigen via thymic stromal lymphopoietin receptor signaling. J. Allergy Clin. Immunol. 129, 1048-1055 e1046 (2012). doi: 10. 1016/j. jaci. 2012. 01. 063. S0091-6749 (12) 00191-1 [pii].

[23] Egawa, G., Kabashima, K.: Skin as a peripheral lymphoid organ: revisiting the concept of skin-associated lymphoid tissues. J. Invest. Dermatol. 131, 2178-2185 (2011). doi: 1 0. 1038/jid. 2011. 198. jid2011198 [pii].

[24] Stoitzner, P., et al.: Tumor immunotherapy by epicuta-neous immunization requires langerhans cells. J. Immunol. 180, 1991-1998 (2008). 180/3/1991 [pii].

[25] Kis, E. E., Winter, G., Myschik, J.: Devices for intradermal vaccination. Vaccine 30, 523-538 (2012). doi: 10. 1016/j. S0264-410X (11) 01787-7 [pii].

[26] Ali, O. A., Emerich, D., Dranoff, G., Mooney, D. J.: In situ regulation of DC subsets and T cells mediatestumor regression in mice. Sci. Transl. Med. 1, 8ra19 (2009). doi: 10. 1126/scitranslmed. 3000359. 1/8/8ra19 [pii].

[27] Karumuthil-Melethil, S., et al.: Dendritic cell-directed CTLA-4 engagement during pancreatic beta cell antigen presentation delays type 1 diabetes. J. Immunol. 184, 6695-6708 (2010). doi: 1 0. 4049/jimmunol. 0903130. jimmunol. 0903130 [pii].

[28] Kwon, Y. J., Standley, S. M., Goodwin, A. P., Gillies, E. R., Frechet, J. M.: Directed antigen presentation using polymeric microparticulate carriers degradable at lysosomal pH for controlled immune responses. Mol. Pharm. 2, 83-91 (2005). doi: 10. 1021/mp0498953.

[29] Malcherek, G., et al.: MHC class II-associated invariant chain peptide replacement by T cell epitopes: engineered invariant chain as a vehicle for directed and enhanced MHC class II antigen processing and presentation. Eur. J. Immunol. 28, 1524-1533 (1998).

[30] Matsuo, H., et al.: Engineered hepatitis B virus surface antigen L protein particles for in vivo active targeting of splenic dendritic cells. Int. J. Nanomedicine 7, 3341-3350 (2012). doi: 10. 2147/IJN. S32813. ijn-7-3341 [pii].

[31] Caminschi, I., Maraskovsky, E., Heath, W. R.: Targeting dendritic cells in vivo for cancer therapy. Front. Immunol. 3, 13 (2012). doi: 10. 3389/ fi mmu. 2012. 00013.

[32] Kwon, Y. J., James, E., Shastri, N., Frechet, J. M.: In vivo targeting of dendritic cells for activation of cellular immunity using vaccine carriers based on pH-responsive microparticles. Proc. Natl. Acad. Sci. U. S. A. 102, 18264-18268 (2005). doi: 10. 1073/ pnas. 0509541102. 0509541102 [pii].

[33] Faraasen, S., et al.: Ligand-specifi c targeting of microspheres to phagocytes by surface modifi cation with poly (L-lysine) -grafted poly (ethylene glycol) conjugate. Pharm. Res. 20, 237-246 (2003).

[34] Reddy, S. T., Rehor, A., Schmoekel, H. G., Hubbell, J. A., Swartz, M. A.: In vivo targeting of dendritic cells in lymph nodes with poly (propylene sulfi de) nanoparticles. J. Control. Release 112, 26-34 (2006).

[35] Bonifaz, L. C., et al.: In vivo targeting of antigens to maturing dendritic cells via the DEC-205 receptor improves T cell vaccination. J. Exp. Med. 199, 815-824 (2004).

[36] Crampton, H. L., Simanek, E. E.: Dendrimers as drug delivery vehicles: non-covalent interactions of bioactive compounds with dendrimers. Polym. Int. 56, 489-496 (2007). doi: 10. 1002/pi. 2230.

[37] Daftarian, P., et al.: Peptide-conjugated PAMAM den-drimer as a universal DNA vaccine platform to target antigen-presenting cells. Cancer Res. 71, 7452-7462 (2011). doi: 10. 1158/0008-5472. CAN-11-1766. 0008-5472. CAN-11-1766 [pii].

[38] Cruz, L. J., et al.: Multimodal imaging of nanovaccine carriers targeted to human dendritic cells. Mol. Pharm. 8, 520-531 (2011). doi: 10. 1021/mp100356k.

[39] Cho, N. H., et al.: A multifunctional core - shell nanoparticle for dendritic cell - based cancer immunotherapy. Nat. Nanotechnol. 6, 675-682 (2011).

[40] Vyas, S. P., Goyal, A. K., Khatri, K.: Mannosylated liposomes for targeted vaccines delivery. Methods Mol. Biol. 605, 177-188 (2010). doi: 10. 1007/978-1-60327-360-2_12.

[41] Geng, S., et al.: Amiloride enhances antigen specifi c CTL by facilitating HBV DNA vaccine entry into cells. PLoS One 7, e33015 (2012). doi: 10. 1371/journal. pone. 0033015. PONE-D-11-20345 [pii].

[42] Watson, D. S., Endsley, A. N., Huang, L.: Design con-siderations for liposomal vaccines: infl uence of f ormulation parameters on antibody and cell-mediated immune responses to liposome associated antigens. Vaccine 30, 2256-2272 (2012).

[43] Kesharwani, P., Gajbhiye, V., Tekade, R. K., Jain, N. K.: Evaluation of dendrimer safety and effi cacy through cell line studies. Curr. Drug Targets 12, 1478-1497 (2011). BSP/CDT/E-Pub/00269 [pii].

[44] Tekade, R. K., Dutta, T., Gajbhiye, V., Jain, N. K.: Exploring dendrimer towards dual drug delivery: pH responsive simultaneous drug-release kinetics. J. Microencapsul. 26, 287-296 (2009). doi: 1 0. 1080/02652040802312572. 902427985 [pii].

[45] van der Lubben, I. M., et al.: Chitosan microparticles for mucosal vaccination against diphtheria: oral and nasal effi cacy studies in mice. Vaccine 21, 1400-1408 (2003). S0264410X02006862 [pii].

[46] van der Lubben, I. M., Verhoef, J. C., Borchard, G., Junginger, H. E.: Chitosan for mucosal vaccination. Adv. Drug Deliv. Rev. 52, 139-144 (2001). S0169-409X (01) 00197-1 [pii].

[47] Wang, W., Kaifer, A. E.: Electrochemical switching and size selection in cucurbit [8] uril-mediated dendrimer self - assembly. Angew. Chem. Int. Ed. Engl. 45, 7042 - 7046 (2006). doi: 10. 1002/anie. 200602220.

[48] Kojima, C., Regino, C., Umeda, Y., Kobayashi, H., Kono, K.: Infl uence of dendrimer generation and polyethylene glycol length on the biodistribution of PEGylated dendrimers. Int. J. Pharm. 383, 293-296 (2010). doi: 10. 1016/j. ijpharm. 2009. 09. 015. S0378-5173 (09) 00628-0 [pii].

[49] Alexander, J., et al.: Development of high potency universal DR-restricted helper epitopes by modifi cation of high affi nity DR-blocking peptides. Immunity 1, 751-761 (1994).

[50] Neumann, F., et al.: Identifi cation of an antigenic peptide derived from the cancer-testis antigen NY-ESO-1 binding to a broad range of HLA-DR subtypes. Cancer Immunol. Immunother. 53, 589-599 (2004). doi: 10. 1007/s00262-003-0492-6.

[51] Kim, D., et al.: Role of IL-2 secreted by PADRE-specifi c $CD4^+$T cells in enhancing E7-specifi c $CD8^+$ T-cell immune responses. Gene Ther. 15, 677-687 (2008).

[52] Belot, F., Guerreiro, C., Baleux, F., Mulard, L. A.: Synthesis of two linear PADRE conjugates bearing a deca-or pentadecasaccharide B epitope as potential synthetic vaccines against Shigella fl exneri serotype 2a infection. Chemistry 11, 1625-1635 (2005).

[53] Alexander, J., et al.: Linear PADRE T helper epitope and carbohydrate B cell epitope conjugates induce

specifi c high titer IgG antibody responses. J. Immunol. 164, 1625-1633 (2000).

[54] Decroix, N., Pamonsinlapatham, P., Quan, C. P., Bouvet, J. -P.: Impairment by mucosal adjuvants and cross-reactivity with variant peptides of the mucosal immunity induced by injection of the fusion peptide PADRE-ELDKWA. Clin. Diagn. Lab. Immunol. 10, 1103-1108 (2003).

[55] Wei, J., et al.: Dendritic cells expressing a combined PADRE/MUC4-derived polyepitope DNA vaccine induce multiple cytotoxic T-cell responses. Cancer Biother. Radiopharm. 23, 121-128 (2008).

[56] Daftarian, P., et al.: Immunization with Th-CTL fusion peptide and cytosine-phosphate-guanine DNA in transgenic HLA-A2 mice induces recognition of HIV-infected T cells and clears vaccinia virus challenge. J. Immunol. 171, 4028-4039 (2003).

[57] Daftarian, P., et al.: Eradication of established HPV 16-expressing tumors by a single administration of a vaccine composed of a liposome-encapsulated CTL-T helper fusion peptide in a water-in-oil emulsion. Vaccine 24, 5235-5244 (2006).

[58] Daftarian, P., et al.: Novel conjugates of epitope fusion peptides with CpG-ODN display enhanced immunogenicity and HIV recognition. Vaccine 23, 3453-3468 (2005). doi: 1 0. 1016/j. vaccine. 2005. 01. 093. S0264-410X (05) 00134-9 [pii].

[59] Daftarian, P., et al.: Two distinct pathways of immuno-modulation improve potency of p53 immunization in rejecting established tumors. Cancer Res. 64, 5407-5414 (2004). doi: 10. 1158/0008-5472. CAN-04-0169. 64/15/5407 [pii].

[60] D aftarian, P. M., et al.: Rejection of large HPV-16 expressing tumors in aged mice by a single immunization of VacciMax encapsulated CTL/T helper peptides. J. Transl. Med. 5, 26 (2007). doi: 10. 1186/1479-5876-5-26. 1479-5876-5-26 [pii].

[61] Myhr, A. I., Myskja, B. K.: Precaution or integrated responsibility approach to nanovaccines in fish farming? A critical appraisal of the UNESCO precautionary principle. Nanoethics 5, 73-86 (2011). 10. 1007/s11569-011-0112-4 112 [pii].

[62] Clemente-Casares, X., Tsai, S., Yang, Y., Santamaria, P.: Peptide-MHC-based nanovaccines for the treatment of autoimmunity: a "one size fi ts all" approach? J. Mol. Med. (Berl) 89, 733-742 (2011). doi: 10. 1007/ s00109-011-0757-z.

[63] Skwarczynski, M., Toth, I.: Peptide - based subunit nanovaccines. Curr. Drug Deliv. 8, 282 - 289 (2011). BSP/CDD/E-Pub/00085 [pii].

[64] Yang, J., et al.: Preparation and antitumor effects of nanovaccines with MAGE-3 peptides in transplantedgastric cancer in mice. Chin. J. Cancer 29, 359-364 (2010). 1000-467X201004359 [pii].

[65] Nandedkar, T. D.: Nanovaccines: recent developments in vaccination. J. Biosci. 34, 995-1003 (2009).

[66] Salvador-Morales, C., Zhang, L., Langer, R., Farokhzad, O. C.: Immunocompatibility properties of lipid-polymer hybrid nanoparticles with heterogeneous surface functional groups. Biomaterials 30, 2231-2240 (2009).

[67] Danesh-Bahreini, M. A., et al.: Nanovaccine for leish-maniasis: preparation of chitosan nanoparticles containing Leishmania superoxide dismutase and evaluation of its immunogenicity in BALB/c mice. Int. J. Nanomedicine 6, 835-842 (2011). doi: 10. 2147/IJN. S16805. ijn-6-835 [pii].

[68] Pietersz, G. A., Tang, C. K., Apostolopoulos, V.: Structure and design of polycationic carriers for gene delivery. Mini Rev. Med. Chem. 6, 1285-1298 (2006).

[69] Tekade, R. K., Kumar, P. V., Jain, N. K.: Dendrimers in oncology: an expanding horizon. Chem. Rev. 109, 49-87 (2009).

[70] Hung, C. F., Yang, M., Wu, T. C.: Modifying profes-sional antigen-presenting cells to enhance DNA vaccine potency. Methods Mol. Med. 127, 199-220 (2006).

[71] Jerome, V., Graser, A., Muller, R., Kontermann, R. E., Konur, A.: Cytotoxic T lymphocytes responding to low dose TRP2 antigen are induced against B16 melanoma by liposome-encapsulated TRP2 peptide and CpG DNA adjuvant. J. Immunother. 29, 294-305 (2006).

[72] Bronte, V., et al.: Genetic vaccination with "self" tyrosinase - related protein 2 causes melanoma eradication but not vitiligo. Cancer Res. 60, 253-258 (2000).

[73] Cohen, A. D., et al.: Agonist anti-GITR antibody enhances vaccine-induced CD8 (+) T-cell responses and tumor immunity. Cancer Res. 66, 4904-4912 (2006).

[74] Frelin, L., et al.: Electroporation: a promising method for the nonviral delivery of DNA vaccines in humans? Drug News Perspect. 23, 647-653 (2010).

[75] Alexander, J., et al.: The optimization of helper T lymphocyte (HTL) function in vaccine development. Immunol. Res. 18, 79-92 (1998). doi: 10. 1007/BF02788751.

[76] Wu, C. Y., Monie, A., Pang, X., Hung, C. F., Wu, T. C.: Improving therapeutic HPV peptide-based vaccine potency by enhancing $CD4^+$ T help and dendritic cell activation. J. Biomed. Sci. 17, 88 (2010).

[77] Kim, D., et al.: Enhancement of $CD4^+$ T-cell help reverses the doxorubicin-induced suppression of antigen-specifi c immune responses in vaccinated mice. Gene Ther. 15, 1176-1183 (2008). gt200879 [pii] 10. 1038/gt. 2008. 79.

(吴国华译)

第 48 章　乳酸菌载体疫苗

Maria Gomes-Solecki①

摘要

目前根据新的抗原设计概念开发新型疫苗，这种抗原可促使先天免疫系统诱导适应性免疫，刺激应答所需 T 细胞的增殖。这种疫苗的传递途径和佐剂对黏膜病原体诱导的保护性免疫应答至关重要。然而，宿主通过对黏膜抗原的耐受反应维持黏膜微环境的平衡。因此，使用疫苗诱导黏膜免疫是一项相当困难的任务。但是，可使用潜在的黏膜免疫佐剂、载体和其他特殊的传递系统来达到这一效果。亟需开发有效的黏膜递送系统，从而避免抗原的降解，并促进抗原在胃肠道中的摄取，刺激机体的适应性免疫应答，避免使用可溶性抗原研究中出现的致耐受性免疫应答。

乳酸菌是公认安全的（GRAS），并且在过去 10 年中已经用于抗原黏膜免疫的有效佐剂。乳酸乳球菌和乳酸杆菌属已经得到了广泛的使用。本章综述了基于植物乳杆菌口服疫苗的平台技术发展，并以鼠疫进行了实例说明。

48.1　鼠疫和病原体

鼠疫耶尔森菌（Yersinia pestis）是一种肠杆菌科的、需氧型、无动力、革兰氏阴性杆菌，经跳蚤叮咬或气溶胶液传播给人类，能引起腺鼠疫或肺鼠疫[1,2]。

大多数人类瘟疫病主要表现为 3 种临床症状，即腺泡、败血症或肺炎。败血症、肺炎和脑膜炎是鼠疫最常见的并发症。美国每年平均报告 7 例病例。未经治疗的腺鼠疫病死亡率为 40%～60%，而未经治疗的继发性败血症和肺炎具有致命性[3,4]。大多数情况下，致死性病例包括患病后不能尽快就医或在就医时误诊的患者严重的感染性休克通常也会导致死亡[5,6]。

科学家已经完成了鼠疫杆菌全基因组测序。鼠疫菌基因与伪结核杆菌基因之间存着等位基因序列[7]。鼠疫杆菌可能是在 1 500～20 000年前由假单胞菌假单胞菌（*Y. pseudotuberculosis*）进化而来，经历了大规模的遗传变迁[8]。除 3 种毒力质粒（10 kb、70 kb 和 100 kb）外，鼠疫杆菌基因组的大小为 4. 63 Mbp[7,9,10]。编码纤溶酶原激活蛋白酶（Pla）[11]的 10kb 质粒和编码 F1 荚膜抗原的 100kb 质粒是鼠疫杆菌所特有的[11-13]。所有 3 种致病性耶尔森氏鼠疫杆菌均有 70 kb 质粒，该质粒编码的蛋白质是其重要毒力因子：III 型分泌系统、LcrV 和一系列耶尔森氏鼠疫杆菌外表面蛋白（Yops）[6,9,10,14]。

鼠疫杆菌的致病性源于其抑制哺乳动物宿主免疫系统的强大能力，并以大规模抑制宿主的生长发育。III 型分泌系统是多种细菌病原体所共有的细胞内靶向毒力因子[15]，该系统能够通过依赖于细菌-宿主细胞接触的机制产生分泌产物和毒力效应子因子。这些蛋白质诱导细胞的细胞毒作用，抑制炎症细胞的趋化性，诱导免疫调节细胞因子（如 IL-10）的分泌，防止形成保护性肉芽肿[16]。感染鼠疫杆菌的小鼠仅在死亡之前 IFN-γ 和 TNF-α 出现显著的升高。相反地，无

---

① M. Gomes-Solecki，DVM（田纳西大学健康科学中心微生物免疫学和生物化学系，美国田纳西州孟菲斯，E-mail：mgomesso@ uthsc. edu）

毒力鼠疫菌株感染后可诱导这些细胞因子迅速的合成[16]。

48.2 地方性兽疫流行周期与传播

鼠疫在非洲、亚洲、南美洲和北美的啮齿动物中具有地方流行性[17]。跳蚤通过吸食感染性血液或组织或食入感染性物质，造成鼠疫杆菌在其宿主之间传播。鼠疫杆菌可感染需索哺乳动物，以大鼠、松鼠、小鼠、草原土拨鼠、土拨鼠或沙鼠作为宿主，并以几种节肢动物为媒介进行传播[2,4,6,18]。这种人畜共患病感染在部分具有抗体的啮齿动物（地方性或维持宿主）体内以低水平呈长期周期性流行状态，宿主死亡率很低。当该病在人的发病率上升时，这些宿主或流行病中的偶然暴发或流行病（即传播死亡）将打断这种流行的长期性[19]，参见图 48.1。

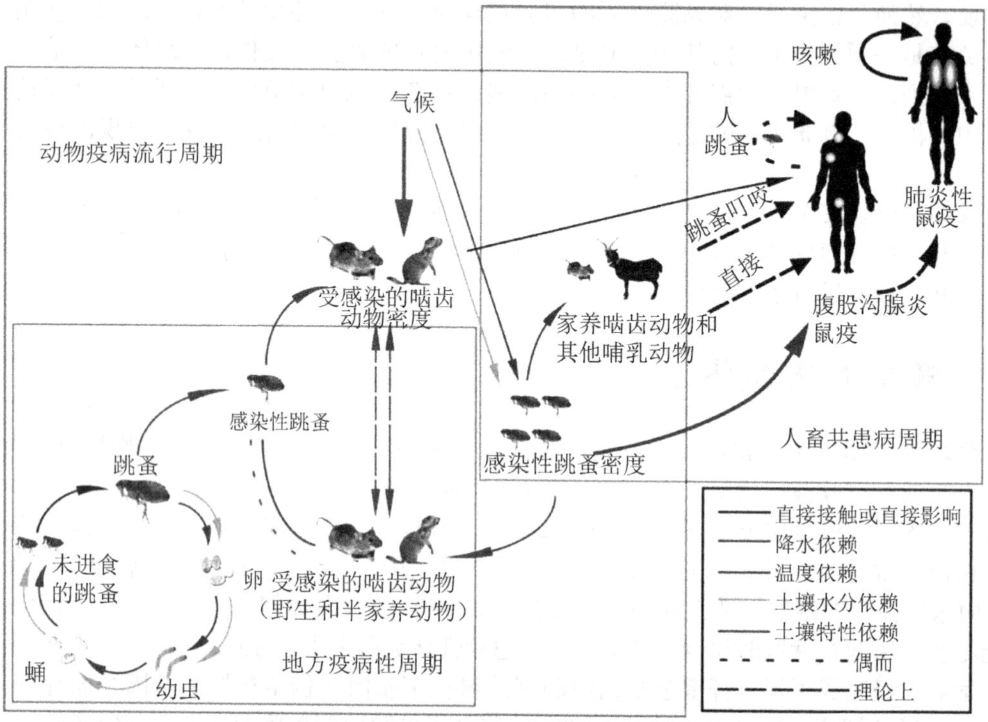

图 48.1　以小型哺乳动物为宿主、以跳蚤作为媒介的鼠疫传播周期示意图

箭头表示受气候影响的连接，颜色编码取决于该链路上的最大气流变量（即降水、温度和其他间接依赖性变量，如土壤特性和土壤湿度）。灰色矩形在某种程度上任意划定了流行性、地方性和人畜共患病的周期。请注意，尽管人体位于周期的末端，但可以提供鼠疫动态的唯一可用信息[19]。

动物跳蚤的非典型叮咬使人体发生这种人畜共患感染，有时是由于鼠疫导致动物死亡，此后跳蚤寻找新的血液来源。从跳蚤叮咬到出现症状的潜伏期为 2～10 天[20]。啮齿类动物种群及其跳蚤已足以维持鼠疫在自然界中流行，人是肺炎鼠疫短暂暴发的不良传播者，因此人类不能维持鼠疫在自然界中的流行。大多数受感染的跳蚤源自家养黑鼠 *Rattus rattus* 或棕色下水道鼠 *Rattus norvegicus*。尽管最常见和最有效的跳蚤媒介是印鼠客蚤（*Xenopsylla cheopis*），但许多其他种跳蚤也可传播鼠疫。东方大鼠跳蚤 *X. cheopis* 比其他跳蚤更易受含鼠疫菌血液阻断其消化道的前胃。被阻断的跳蚤无法清除中肠内受感染的血液，导致其反复叮咬，并将细菌传播到其下一个宿主的皮肤[17,21]。

鼠疫杆菌脾叮咬的感染部位扩散到局部淋巴结，大量繁殖后导致腹股沟炎，然后进入血液循环系统，大部分细菌在肝脏和脾脏中被清除。鼠疫杆菌在这些器官继续生长，扩散到其他器官，导致败血症，以败血症动物为食的跳蚤完成其感染周期。人的腺鼠疫可发展成肺部感染(继发性肺鼠疫)，导致气溶胶传播（原发性肺鼠疫)[2,6,22]。在这种情况下，病原体在肺组织中大量复制，出血性病变和免疫细胞浸润物破坏气道，促成快速传染和致命的感染病程，病原体进一步经气溶胶传播[23]。自然发生的肺鼠疫以及生物战可导致鼠疫杆菌的气溶胶传播。天然存在耐多种抗生素的鼠疫菌株，很容易进行生物改造[2]。鼠疫是一种 A 类生物恐怖物质，需要实施新型预防策略。

所有年龄和性别的人都易患此疾病。尽管世界卫生组织未提供鼠疫杆菌感染的年龄和性别分布[24]，但最近的病例见于儿童，男性略多[17]。鼠疫在疾病流行国家和地区具有季节性变化性。流行高发季对应兽疫流行高发季，伴随易感啮齿动物死亡。季节性与啮齿动物跳蚤的繁殖力、啮齿动物种群增加以及人类与感染动物接触密切相关[17]。

48.3　诊断和治疗

48.3.1　临床疾病

鼠疫是该疾病的典型形式。患者因跳蚤叮咬而接触病原体或开放性伤口暴露于受感染材料后 2~6 天内出现发烧、头痛、发冷和肿胀、极度压痛的淋巴结（腹股沟）症状。胃肠道症状，如恶心、呕吐和腹泻也很常见[25]。感染的初始部位很少出现皮肤损伤。发生肿胀之前有时出现淋巴结疼痛[26]，且可能出现在任何淋巴结，这取决于初始感染的部位。炎性淋巴腺肿通常见于腹股沟和股骨区域，但也见于其他部位[5]。腺鼠疫患者常发生菌血症或继发性鼠疫败血症[6,27]。

患者血培养阳性可确定为原发性鼠疫败血症，但没有明显的淋巴结病变。鼠疫败血症的临床症状类似于其他革兰阴性菌引起的败血症。患者出现发热、寒战、头痛、不适和胃肠道紊乱。有证据表明，鼠疫败血症患者腹痛的发生率高于腺鼠疫患者[25]。鼠疫败血症患者的死亡率相当高，这可能是因为常用于治疗其他败血症的抗生素对鼠疫杆菌无效[3,6,25,26]。

原发性肺鼠疫是一种罕见但致命的疾病，通过与感染者密切接触（60.96~152.4 cm）的呼吸道飞沫传播。它从发热性流感样疾病迅速发展成具有咳嗽和血痰的严重肺炎。原发性肺鼠疫的潜伏期为 1~3 天。一般而言，发展为继发性鼠疫肺炎的患者死亡率较高[6]。

48.3.2　实验室诊断

鼠疫的实验室诊断基于细菌学和（或）血清学[26]。用于诊断的样本包括鼠疫性脑膜炎患者的血液、腹膜抽出物、痰液、脑脊液和病变皮肤的刮屑（如果有）。尽管 Gram、Giemsa、Wright 或 Wayson 染色等染色技术可进行鼠疫诊断，但不能提供确凿证据[26]。鼠疫快速临床诊断的主要方法是 Gram 或 Wayson 染色后在显微镜下检查腹股沟淋巴结抽取物[17]。病原体的细菌培养可确诊鼠疫。鼠疫菌在大多数常规实验室培养基中培养 2 天后可出现菌落。菌落的外观不透明而光滑，边缘不规则，放大时具有“锤击金属”[6,26]。虽然血清学反应不是快速诊断技术，但常用于回溯性地确诊鼠疫病例。已经开发出需索鼠疫诊断的替代方法，包括用于检测 F1 的酶联免疫吸附试验定[17]、抗原捕获试验（即抗 F1 试纸测定），以及使用 F1 抗原结构基因的经典和实时 PCR 试验、纤溶酶原激活物、鼠毒素和 16S 核糖体 RNA[17,28]（表 48.1)。

表 48.1　检测临床标本中鼠疫耶尔森氏菌的诊断方法

方法	敏感性、特异性和检测时间
淋巴结抽取物、血液或痰标本的培养	如果患者未经治疗，该方法具有高度敏感性，2~3 天可完成检测
淋巴结抽取物或痰标本的革兰氏染色或威森染色	灵敏性、特异性中等，数分钟内即可完成检测
淋巴结抽取物或痰标本的免疫荧光抗体检测	灵敏性中等、高特异性，数分钟内完成检测
淋巴结抽取物中 F1 抗原的酶联免疫吸附测试验	高灵敏性、高特异性，数小时内完成检测
淋巴结抽取物中 F1 抗原的试纸条检测	高灵敏性、高特异性，数分钟内完成检测
淋巴结抽取物 F1 基因 PCR	灵敏性中等、高特异性，数小时内完成检测

48.3.3　经典疗法

所有疑似腺鼠疫的患者应在开始抗生素治疗后 2 天内进行隔离，以防患者发生继发性鼠疫肺炎时可能造成疾病的传播[6]。表 48.2 列出了用于治疗鼠疫菌感染的抗生素、方案，以及预防措施。

用链霉素治疗鼠疫已超过 45 年[29]，仍然是治疗首选药物。在没有链霉素的国家，庆大霉素是一种有效的替代品[30]。链霉素为溶菌药物，故应小心使用，以防发生内毒素休克。由于链霉素具有毒副作用，患者不应该在 10 天的疗程中都使用链霉素，而应逐渐转换为其他抗生素，常为四环素。四环素常也用于预防性治疗，建议用氯霉素治疗鼠疫性脑膜炎[31]。虽然鼠疫杆菌在体外对青霉素敏感，但在人体却无效[3]。在坦桑尼亚庆进行的大霉素和多西环素的随机比较研究表明，多西环素同样的效果，没有任何肾毒性[32]。强力霉素也是一种替代药物。据报道，只有 1 名患者成功接受了环丙沙星治疗[17,33]，但氟喹诺酮类和 β-内酰胺类均进行人体试验。抗生素耐药菌株十分罕见，出现频率未见增加[6]。

表 48.2　治疗鼠疫的抗菌剂

药物	临床经验
庆大霉素	链霉素禁止使用后，过去 10 年中在美国和坦桑尼亚庆大霉素治疗效果良好
链霉素	这是 1948 年后约 40 年常用的药物，大多数国家已在使用；在马达加斯加，链霉素仍在与甲氧苄氨嘧啶磺胺甲恶唑联合应用
强力霉素或四环素	首选口服药物，可替代庆大霉素；用于预防肺炎
甲氧苄氨嘧啶-磺胺甲恶唑	在马达加斯加，与链霉素联合使用；用于预防肺炎
氯霉素	具有骨髓毒性，虽然有效但很少使用
头孢菌素和其他 β-内酰胺类	不建议使用，但对实验动物感染有效
环丙沙星和其他氟喹诺酮类药物	不建议使用，但对实验动物感染有效

48.4　疫苗

目前的流行病学文献显示，全世界每年有 4 000例人感染鼠疫[34]。由于鼠疫杆菌可经气溶胶传播及其暴发性毒力，鼠疫杆菌归为 A 类生物恐怖主义病菌。与炭疽病不同，鼠疫菌不形成芽孢，也无法在体外存活。正由于这个原因，尚未出现雾化细菌的生物武器。此外，肺鼠疫流行病的传播能力严格受限于易感人群与垂死病人密切接触，尤其是在患者生命的最后一天[35]。人们夸大了使用这种病原体作为生物武器的危险[17,36]。在第二次世界大战期间，日军用飞机在中国城市上空释放了鼠疫感染的跳蚤[2]，值得怀疑的是，这种粗略的攻击方式是否会在如今使用。人们普遍认为，鼠疫杆菌很容易地进行遗传操作，产生具有特定功能的菌株，例如抵抗治疗疾病的抗生素。研究人员正努力开发预防鼠疫肺炎的新亚单位疫苗。肺鼠疫会经历细胞内阶段，需由细胞介导的免疫反应提供保护[37]，这是开发此类疫苗的主要挑战。

适当的药物分子经黏膜途径给药比全身给药具有许多重要优点，例如减少继发效应、易于给药、以及可能调节全身和黏膜免疫应答[38]。

最主要的技术难点是开发能够在胃肠道转运过程中存活的传递载体，同时保护治疗性或免疫原性分子免受低 pH、胆汁、蛋白水解酶、抗菌肽和肠道蠕动的影响[39]。

包括乳杆菌属、乳球菌属、明串珠菌属、片球菌属和链球菌属在内的乳酸菌（LAB）属于革兰阳性菌，无致病性，不形成芽孢。乳酸菌的生物合成能力有限，需以氨基酸、B 族维生素、嘌呤、嘧啶和糖作为碳源和能源。这些营养需求限制乳酸菌只能在所需化合物丰富的环境中生长。这些高度专业化的细菌存在于一系列微环境中，包括牛奶、植物表面、口腔、胃肠道和脊椎动物的阴道[40]。

数十个世纪以来，人类已在发酵食品中使用乳酸菌，其安全性非常高。这些内在优势使得乳酸菌成为人类新型的预防性和治疗性分子具有潜力的传递载体。已有许多报道关于基因工程病原菌或共生细菌类的口服疫苗[41-47]。乳酸菌可替代脂质体、微粒和减毒病原体等其他黏膜传递系统，颇具吸引力[40,48]。

在减毒疫苗研究过程中，通过改变致病因子，使其失去感染能力，同时保持其与宿主相互作用的能力，然后刺激机体产生保护性免疫应答[49]。分枝杆菌、沙门氏菌和博德特氏菌属衍生物等减毒致病菌是目前最常用的疫苗传递载体。这些载体能够特异性与黏膜表面分子相互作用，从而引起感染。活载体疫苗的主要缺点是病原体保持其致病力以及毒力返强的可能性。这些细菌具有毒力反强的可能性，故对人体使用并不完全安全，尤其是儿童和免疫抑制个体[48]。乳酸菌的疫苗可作为减毒疫苗，无安全问题。乳酸菌具有公认的安全状态（GRAS），不太可能造成伤害。

使用乳酸菌作为传递载体的兴趣源于大量免疫学研究，这些研究表明，需使用适当的传递系统避免抗原降解，促进抗原在胃肠道中摄取，刺激适应性免疫应答，而不产生可溶性抗原研究中常见的致耐受性免疫应答[40,50,51]。

乳酸菌可以细胞内 3 个不同的位置产生所需的抗原：①细胞内，尽管能使蛋白质避免细胞外部恶劣环境（例如胃液）的影响，但蛋白需要经过细胞裂解以后才能进行传递；②细胞外，可以将蛋白质释放到细胞外基质中，直接与环境（食品或消化道）相互作用；③锚定在细胞壁，结合了其他两个位置的优点（即避免蛋白酶水解，还具有细胞壁锚定蛋白与环境之间的相互作用）。鉴于此，一些研究比较了表达于乳酸菌 3 种不同位置抗原，评估了 3 种不同形式抗原的免疫原性[40,48]。这些研究表明，暴露于乳酸菌表面的细胞壁锚定抗原产生的免疫应答能力最好。最近大多数乳酸菌疫苗研究均使抗原暴露于乳酸菌表面，而不是细胞内或细胞外[48]。

树突状细胞（DC）在适应性免疫系统和先天免疫系统之间起着桥梁的作用[52-55]。树突状细胞遍布全身，尤其常见于黏膜表面。体内外环境之间只有一层上皮细胞，需不断进行微粒交换，肠道树突状细胞（DC）在维持肠内稳态以及调控入侵病原体的保护性免疫应答中起关键作用[56]。为避免激活自身反应性 T 细胞，限制不必要的反应（例如针对共生菌群的反应），树突状细胞可刺激 T 细胞产生耐受性[57]，参见图 48.2。

未成熟的树突状细胞富集在黏膜诱导位点的上皮底部，为捕获抗原做好准备。这些树突状

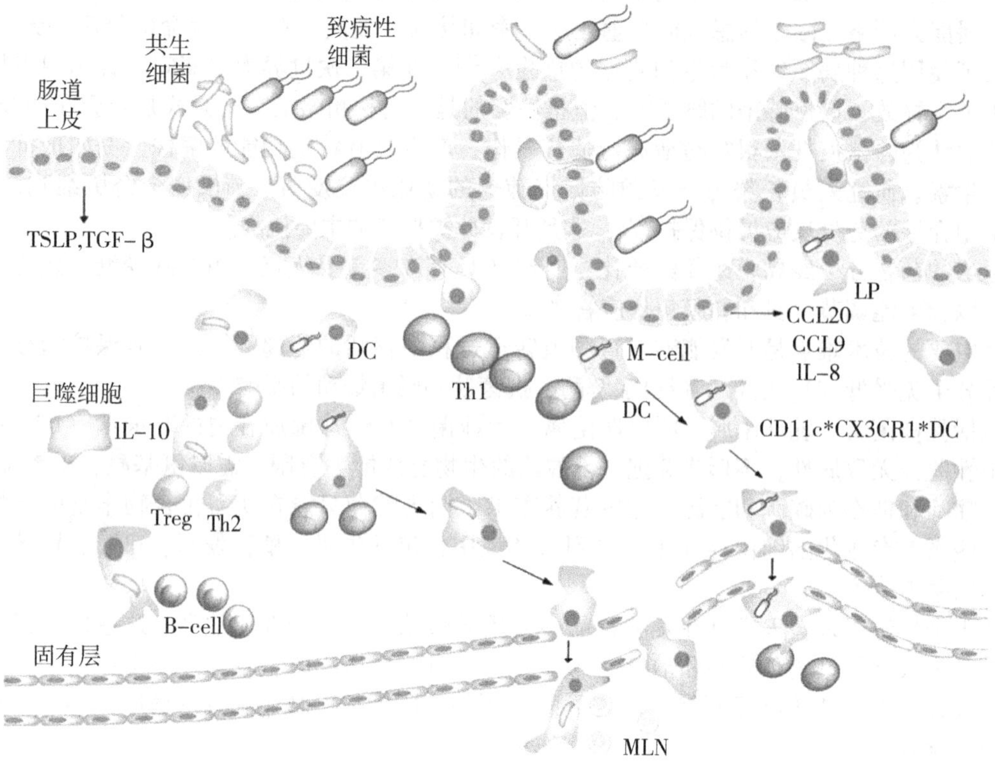

图 48.2　通过肠道树突状细胞（DCs）摄取细菌及其产物

细胞突起延伸至上皮细胞之间，直接摄取管腔内抗原[58]。通过上调 MHC 和共刺激分子表达，成熟的树突状细胞转化成高效的抗原呈递细胞[56]。抗原呈递细胞在 MHC Ⅱ类分子的共同作用下将抗原递呈至 $CD4^+T$ 细胞，而存在于 MHC Ⅰ类分子上的抗原表位可以刺激特异性 $CD8^+T$ 细胞[56]。当摄取抗原时，这些树突状细胞的共刺激分子表达量升高，表型发生改变，然后移动到 T 细胞的抗原作用位点进行抗原递呈。树突状细胞及其衍生的细胞因子在诱导抗原特异性效应子 Th 细胞应答中起关键作用。在此方面，靶向黏膜树突状细胞是诱导黏膜和系统免疫应答的有效策略[55,56]。

胃肠道中持续存在乳酸菌对基于乳酸菌疫苗的效力可能至关重要。持久性乳酸菌菌株比植物乳杆菌更能有效地诱发机体的特异性免疫应答，这表明持续促进了抗原的免疫原性[59]。研究表明，特定的乳杆菌属诱导炎症细胞因子，刺激树突状细胞的活化和成熟[60]。研究还表明，未成熟的树突状细胞可以有效捕获乳杆菌，这些细菌激活人树突状细胞，产生促炎细胞因子，如 IL12，增加 $CD4^+$和 $CD8^+$细胞的增殖，使 T 细胞免疫应答趋于参与有效清除微生物病原体的 Th1 途径[61,62]。参见图 48.3。在先前用于传递抗原的乳杆菌菌株中，我们研究了植物乳杆菌表达在肠道内表达抗原和在肠道内持续存在的能力[59]，以及其作为有效的佐剂的可能性[63]。

有证据表明，一些乳酸菌的肽聚糖层具有天然免疫佐剂的性能[64,65,66]，与细胞内蛋白或分泌蛋白相比，细胞壁的抗原更易接近机体免疫系统[67]。前导肽修饰的蛋白质更容易在细胞质膜上移动，脂质修饰对于蛋白质锚定于膜上和蛋白质功能都非常重要[68]。研究表明，成熟的伯氏疏螺旋体 OspA 蛋白的第一个氨基酸处的脂化是该蛋白质经 TLR2 诱导的免疫应答所必需的[69,70]。我们研究了 OspA 脂蛋白植物乳杆菌细胞定位对其翻译修饰的影响。结果发现，OspA 的前导肽将蛋白质靶向于乳酸杆菌的细胞包膜，植物乳杆菌细胞壁识别蛋白的 Cys 位点[17]，随后使蛋白质脂质化，再将蛋白质锚定于细胞膜表面。最终研究发现，该传递系统具有较强的免疫佐剂的功能[71]。

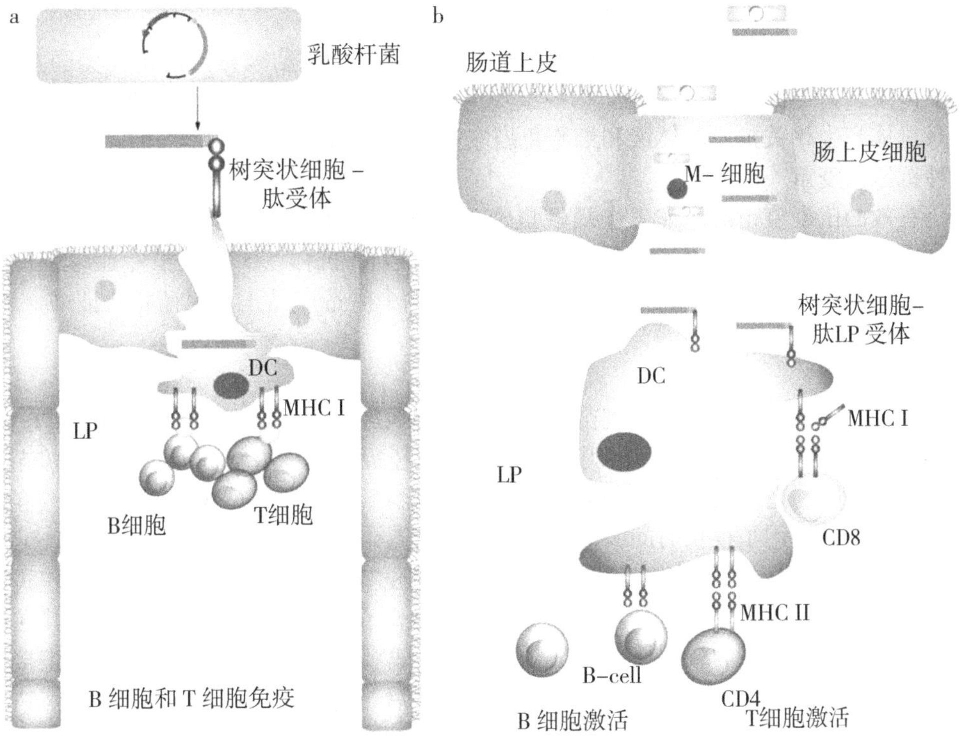

图 48.3 通过益生菌乳杆菌将免疫原性亚单位递送至肠道内膜细胞（树突状细胞）

（a）通过黏膜固有层（LP）中突出的树突状细胞直接摄取管腔内抗原。（b）乳酸杆菌经 M 细胞递送抗原并激活树突状细胞。

我们的基本原则是基于已知可使免疫应答趋向于保护性 Th1 反应的共生且完全安全的细菌（植物乳杆菌）设计一种口服传递系统。我们将 *B. burgdorferi* OspA 前导序列与重组蛋白融合，是重组蛋白表达于细菌细胞壁表面，产生隐形载体，使抗原免受胃肠攻击，再经 TLR2 进一步诱导免疫应答，增加了该系统的佐剂作用。我们的目标是建立可以抵御多种病原体的黏膜疫苗的技术平台。基于该系统的口服疫苗经树突状细胞摄取以后，使免疫应答偏向于 Th1，破坏肠道的天然免疫耐受性。

我们使用莱姆病小鼠模型证明，小鼠口服免疫表达 OspA 的植物乳杆菌后可产生了全身 IgG 和黏膜 IgA 体液免疫应答，保护其免受携带人伯氏疏螺旋体的蜱攻击[72]。这种疫苗可以刺激人树突状细胞产生细胞因子，可使 T 细胞向 Th1 型细胞应答发生极化。该疫苗不能刺激 T84 人上皮细胞系产生促炎趋化因子 IL8，这表明该疫苗在人胃肠道中不会诱导局部炎症反应[71]。

我们研发表达鼠疫菌 LcrV 蛋白的植物乳杆菌口服疫苗。低钙反应 V（LcrV）是一种鼠疫菌的分泌型毒力因子，重组蛋白肠外免疫小鼠后可保护小鼠免受鼠疫菌强毒的皮下感染[73]和气溶胶感染[74]。已证实 LcrV 是鼠疫耶尔森氏菌的一种候选疫苗。表达 LcrV 的乳杆菌口服免疫的小鼠后可产生 LcrV 全身特异性（血清学）IgG 和黏膜（肠、支气管肺泡和阴道）IgA 抗体应答。我们发现，表达 LcrV 的乳杆菌可以刺激人树突状细胞产生细胞因子，使 T 细胞免疫应答向 Th1 细胞免疫应答极化；疫苗接种 T 84 人上皮细胞系后不产生促炎趋化因子 IL8。这些结果再次确认了莱姆病疫苗所产生的保护性免疫应答反应。应该强调的是没有进行乳杆菌或 LcrV 疫苗的功毒试验，没有对疫苗抵御鼠疫杆菌的功效进行测定。除植物乳杆菌[75]外，用共生的非致病性细菌乳酸乳球菌表达 LcrV 也取得了一定成功[2,76,77]。

48.4.1 临床前开发、安全和功效

由于安全性和有效性问题，鼠疫菌全细菌灭活疫苗或鼠疫菌减毒活疫苗在美国并不受青睐，但理论上鼠疫菌减毒活疫苗可保护动物模型免受腺鼠疫和肺鼠疫的攻击。现代的鼠疫杆菌活疫苗应刺激机体产生相关抗原的体液和细胞免疫应答，在抵御鼠疫生化武器攻击方面强于仅基于1种或两种抗原的疫苗[2,78,79]。

安全有效的鼠疫疫苗的研究工作最近集中于开发重组亚单位疫苗，该疫苗刺激机体产生两种已充分表征的鼠疫杆菌抗原 F1 包囊和毒力蛋白 LcrV 的抗体[80,81,82]。虽然亚单位疫苗对一些非人类灵长类动物的功效还存在争议[83]，但目前有几个候选产品正在申请相关证书。

48.4.2 临床开发

尽管早期用于预防跳蚤鼠疫的疫苗已在疾病流行地区使用了半个多世纪，包括20世纪60年代和70年代年代部署到越南的1 100万美国军人，但经福尔马林灭活的全细胞鼠疫疫苗不能预防鼠疫性肺炎，美国制造商已于1998年停止该疫苗的生产。缺乏 *pgm* 基因的减毒活疫苗 EV76 已在欧洲和其他国家使用了很长时间，但尚无商品化疫苗[17]。

基于 rF1 和 rV 抗原的亚单位疫苗最具前景，已通过了 I 期和 II 期临床试验，正进入审批过程。由于伦理因素无法直接确定其功效，但人类对亚单位鼠疫疫苗的免疫应答与其在猕猴和小鼠体内免疫应答具有很好的相关性[84]。rV10 疫苗（截短的重组 LcrV 蛋白）目前正接受美国食品和药物管理局的临床研究前新药批准审评，以备以后进行 I 期试验[1]。目前，DynPort 疫苗公司开发的重组 F1V（rF1V）正进行 2b 期临床试验（www. clinicaltrials. gov）[2]。

48.5 优点与弱点

目前，由于黏膜疫苗的效力不高，故很少被使用。黏膜传播疾病感染后形成的天然免疫很难抵御异源病原体再次感染，这是一个常见话题。因此，疫苗需要诱导机体产生不同类型的免疫应答才能提供更完全的免疫保护[85]。

传统注射疫苗无法保护人体免受大量病原体（HIV、疟疾和结核病等）的感染，这使人们致力于研究和开发新的免疫方式（例如黏膜途径）。开发安全有效的黏膜佐剂仍是非复制性黏膜疫苗的重要部分。在病原体进入黏膜后就能诱导机体的免疫特异性应答。因此，有必要开发低免疫原性的非复制性抗原经黏膜免疫时无毒且有效的佐剂。黏膜给药的疫苗佐剂可实现黏膜免疫细胞靶向作用，产生特异性的免疫应答[85]。

通过对各种黏膜免疫方法的优点和局限性比较发现，目前尚无特别优势的方法，但这些方法的共同点问题是在疫苗研发阶段不同产品的抗原载量缺乏控制（相比于全身给药）。选择任何黏膜免疫途径以及合适的佐剂和制剂都会影响疫苗的设计、工艺和制造[85]。口服疫苗接种的主要优势可能是无针给药以及在资源贫乏国家的大量分销，这些国家缺乏药物和疫苗冷藏的复杂电力供应。

尽管仍有许多工作有待完成，但目前的研究仍在继续澄清某些概念，为探索黏膜疫苗的全部潜力提供所需的工具。本研究是用动物模型进行的，请牢记，由于解剖学、生理学和免疫原性的差异，这些模型通常不能推广到人类，这一点非常重要[85]。

参考文献

[1] Quenee, L. E., Schneewind, O.: Plague vaccines and the molecular basis of immunity against Yersinia pestis. Hum. Vaccin. 5, 817-823 (2009).

[2] Sun, W., Roland, K. L., Curtiss 3rd, R.: Developing live vaccines against plague. J. Infect. Dev. Ctries. 5, 614-627 (2011).

[3] Crook, L. D., Tempest, B.: Plague. A clinical review of 27 cases. Arch. Intern. Med. 152, 1253-1256 (1992).

[4] Craven, R. B., Maupin, G. O., Beard, M. L., Quan, T. J., Barnes, A. M.: Reported cases of human plague infections in the United States, 1970-1991. J. Med. Entomol. 30, 758-761 (1993).

[5] Butler, T.: The black death past and present. 1. Plague in the 1980s. Trans. R. Soc. Trop. Med. Hyg. 83, 458-460 (1989).

[6] Perry, R. D., Fetherston, J. D.: Yersinia pestis-etio-logic agent of plague. Clin. Microbiol. Rev. 10, 35-66 (1997).

[7] Parkhill, J., et al.: Genome sequence of Yersinia pestis, the causative agent of plague. Nature 413, 523-527 (2001). doi: 10. 1038/35097083.

[8] Achtman, M., et al.: Yersinia pestis, the cause of plague, is a recently emerged clone of Yersinia pseudotuberculosis. Proc. Natl. Acad. Sci. U. S. A. 96, 14043-14048 (1999).

[9] Cornelis, G. R.: Yersinia type III secretion: send in the effectors. J. Cell Biol. 158, 401-408 (2002). doi: 10. 1083/jcb. 200205077.

[10] Fields, K. A., Straley, S. C.: LcrV of Yersinia pestis enters infected eukaryotic cells by a virulence plasmid-independent mechanism. Infect. Immun. 67, 4801-4813 (1999).

[11] Galvan, E. M., Lasaro, M. A., Schifferli, D. M.: Capsular antigen fraction 1 and Pla modulate the susceptibility of Yersinia pestis to pulmonary antimicrobial peptides such as cathelicidin. Infect. Immun. 76, 1456-1464 (2008). doi: 10. 1128/IAI. 01197-07.

[12] Miller, J., et al.: Macromolecular organisation of recombinant Yersinia pestis F1 antigen and the effect of structure on immunogenicity. FEMS Immunol. Med. Microbiol. 21, 213-221 (1998).

[13] Li, B., Yang, R.: Interaction between Yersinia pestis and the host immune system. Infect. Immun. 76, 1804-1811 (2008). doi: 10. 1128/IAI. 01517-07.

[14] Hamad, M. A., Nilles, M. L.: Roles of YopN, LcrG and LcrV in controlling Yops secretion by Yersinia pestis. Adv. Exp. Med. Biol. 603, 225-234 (2007). doi: 10. 1007/978-0-387-72124-8_20.

[15] Hueck, C. J.: Type III protein secretion systems in bac-terial pathogens of animals and plants. Microbiol. Mol. Biol. Rev. 62, 379-433 (1998).

[16] Lathem, W. W., Crosby, S. D., Miller, V. L., Goldman, W. E.: Progression of primary pneumonic plague: a mouse model of infection, pathology, and bacterial transcriptional activity. Proc. Natl. Acad. Sci. U. S. A. 102, 17786-17791 (2005). doi: 10. 1073/pnas. 0506840102.

[17] Butler, T.: Plague into the 21st century. Clin. Infect. Dis. 49, 736-742 (2009). doi: 10. 1086/604718 18. Butler, T.: Plague and Other Yersinia Infections. Plenum Press, New York (1983).

[19] Ben-Ari, T., et al.: Plague and climate: scales matter. PLoS Pathog. 7, e1002160 (2011). doi: 10. 1371/journal. ppat. 1002160.

[20] Gage, K. L., et al.: Cases of cat-associated human plague in the Western US, 1977-1998. Clin Infect. Dis. 30, 893-900 (2000). doi: 10. 1086/313804.

[21] Jarrett, C. O., et al.: Transmission of Yersinia pestis from an infectious biofi lm in the fl ea vector. J. Infect. Dis. 190, 783-792 (2004). doi: 10. 1086/422695.

[22] Starnbach, M. N., Straley, S. C.: Yersinia: Strategies that Thwart Immune Defenses, pp. 71-92. Lippincott Williams and Wilkins, Philadelphia (2000).

[23] Meyer, K. F.: Pneumonic plague. Bacteriol. Rev. 25, 249-261 (1961).

[24] World Health Organization: Human plague in 2002 and 2004. Wkly. Epidemiol. Rec. 79, 301-308 (2004).

[25] Hull, H. F., Montes, J. M., Mann, J. M.: Septicemic plague in New Mexico. J. Infect. Dis. 155, 113-

118 (1987).

[26] Poland, J. D., Barnes, A. M.: Plague in CRC Handbook Series in Zoonoses. Section A. Bacterial, Rickettsial, Chlamydial and Mycotic Diseases, vol. I, pp. 515–559. CRC Press, Boca Raton (1979).

[27] Gage, K. L., Lance, S. E., Dennis, D. T., Montenieri, J. A.: Human plague in the United States: a review of cases from 1988–1992 with comments on the likelihood of increased plague activity. Border Epidemiol. Bull. 19, 1–10 (1992).

[28] Cavanaugh, D. C.: K F Meyer's work on plague. J. Infect. Dis. 129 (Suppl), S10–S12 (1974).

[29] Meyer, K. F.: Modern therapy of plague. J. Am. Med. Assoc. 144, 982–985 (1950).

[30] Boulanger, L. L., et al.: Gentamicin and tetracyclines for the treatment of human plague: review of 75 cases in new Mexico, 1985–1999. Clin. Infect. Dis. 38, 663–669 (2004). doi: 10. 1086/381545.

[31] Becker, T. M., et al.: Plague meningitis–a retrospective analysis of cases reported in the United States, 1970–1979. West. J. Med. 147, 554–557 (1987).

[32] Mwengee, W., et al.: Treatment of plague with gentamicin or doxycycline in a randomized clinical trial in Tanzania. Clin. Infect. Dis. 42, 614–621 (2006). doi: 10. 1086/500137.

[33] Kuberski, T., Robinson, L., Schurgin, A.: A case of plague successfully treated with ciprofl oxacin and sympathetic blockade for treatment of gangrene. Clin. Infect. Dis. 36, 521 – 523 (2003). doi: 10. 1086/367570.

[34] Stenseth, N. C., et al.: Plague: past, present, and future. PLoS Med. 5, e3 (2008). doi: 10. 1371/journal. pmed. 0050003.

[35] Kool, J. L.: Risk of person–to–person transmission of pneumonic plague. Clin. Infect. Dis. 40, 1166–1172 (2005). doi: 10. 1086/428617.

[36] Prentice, M. B., Rahalison, L.: Plague. Lancet 369, 1196–1207 (2007). doi: 10. 1016/S0140–6736 (07) 60566–2.

[37] Parent, M. A., et al.: Cell–mediated protection against pulmonary Yersinia pestis infection. Infect. Immun. 73, 7304–7310 (2005). doi: 10. 1128/IAI. 73. 11. 7304–7310. 2005.

[38] Bermudez–Humaran, L. G.: Lactococcus lactis as a live vector for mucosal delivery of therapeutic proteins. Hum. Vaccin. 5, 264–267 (2009).

[39] Amdekar, S., Dwivedi, D., Roy, P., Kushwah, S., Singh, V.: Probiotics: multifarious oral vaccine against infectious traumas. FEMS Immunol. Med. Microbiol. 58, 299–306 (2010). doi: 10. 1111/j. 1574-695X. 2009. 00630. x.

[40] Wells, J. M., Mercenier, A.: Mucosal delivery of thera–peutic and prophylactic molecules using lactic acid bacteria. Nat. Rev. Microbiol. 6, 349–362 (2008). doi: 10. 1038/nrmicro1840.

[41] Anderson, R., Dougan, G., Roberts, M.: Delivery of the Pertactin/P. 69 polypeptide of Bordetella pertussis using an attenuated Salmonella typhimurium vaccine strain: expression levels and immune response. Vaccine 14, 1384–1390 (1996).

[42] Ascon, M. A., Hone, D. M., Walters, N., Pascual, D. W.: Oral immunization with a Salmonella typhimurium vaccine vector expressing recombinant enterotoxigenic Escherichia coli K99 fi mbriae elicits elevated antibody titers for protective immunity. Infect. Immun. 66, 5470–5476 (1998).

[43] Peters, C., Peng, X., Douven, D., Pan, Z. K., Paterson, Y.: The induction of HIV Gag–specifi c $CD8^+$T cells in the spleen and gut–associated lymphoid tissue by parenteral or mucosal immunization with recombinant Listeria monocytogenes HIV Gag. J. Immunol. 170, 5176–5187 (2003).

[44] Lee, J. S., et al.: Mucosal immunization with surface–d isplayed severe acute respiratory syndrome coronavirus spike protein on Lactobacillus casei induces neutralizing antibodies in mice. J. Virol. 80, 4079–4087 (2006). doi: 1 0. 1128/JVI. 80. 8. 4079–4087. 2006.

[45] Wu, C. M., Chung, T. C.: Mice protected by oral immunization with Lactobacillus reuteri secreting fusion

protein of Escherichia coli enterotoxin subunit protein. FEMS Immunol. Med. Microbiol. 50, 354 – 365 (2007). doi: 10. 1111/j. 1574-695X. 2007. 00255. x.

[46] Kajikawa, A., Satoh, E., Leer, R. J., Yamamoto, S., Igimi, S.: Intragastric immunization with recombinant Lactobacillus casei expressing fl agellar antigen confers antibody-independent protective immunity against Salmonella enterica serovar Enteritidis. Vaccine 25, 3599-3605 (2007). doi: 10. 1016/j. vaccine. 2007. 01. 055.

[47] Daniel, C., et al.: Protection against Yersinia pseu-dotuberculosis infection conferred by a Lactococcus lactis mucosal delivery vector secreting LcrV. Vaccine 27, 1141-1144 (2009). doi: 10. 1016/j. vaccine. 2008. 12. 022.

[48] Bermudez-Humaran, L. G., Kharrat, P., Chatel, J. M., Langella, P.: Lactococci and lactobacilli as mucosal delivery vectors for therapeutic proteins and DNA vaccines. Microb. Cell Fact. 10 (Suppl 1), S4 (2011). doi: 10. 1186/1475-2859-10-S1-S4.

[49] Badgett, M. R., Auer, A., Carmichael, L. E., Parrish, C. R., Bull, J. J.: Evolutionary dynamics of viral attenuation. J. Virol. 76, 10524-10529 (2002).

[50] Lavelle, E. C., O'Hagan, D. T.: Delivery systems and adjuvants for oral vaccines. Expert Opin. Drug Deliv. 3, 747-762 (2006). doi: 10. 1517/17425247. 3. 6. 747.

[51] Neutra, M. R., Kozlowski, P. A.: Mucosal vaccines: the promise and the challenge. Nat. Rev. Immunol. 6, 148-158 (2006). doi: 10. 1038/nri1777.

[52] Steinman, R. M.: The dendritic cell system and its role in immunogenicity. Annu. Rev. Immunol. 9, 271-296 (1991). doi: 10. 1146/annurev. iy. 09. 040191. 001415.

[53] Banchereau, J., Steinman, R. M.: Dendritic cells and the control of immunity. Nature 392, 245 – 252 (1998). doi: 10. 1038/32588.

[54] Pulendran, B., Banchereau, J., Maraskovsky, E., Maliszewski, C.: Modulating the immune response with dendritic cells and their growth factors. Trends Immunol. 22, 41-47 (2001).

[55] Fujkuyama, Y., et al.: Novel vaccine development strategies for inducing mucosal immunity. Expert Rev. Vaccines 11, 367-379 (2012). doi: 10. 1586/ erv. 11. 196.

[56] Bedoui, S., et al.: Different bacterial pathogens, dif-ferent strategies, yet the aim is the same: evasion of intestinal dendritic cell recognition. J. Immunol. 184, 2237 – 2242 (2010). doi: 10. 4049/ jimmunol. 0902871.

[57] Artis, D.: Epithelial-cell recognition of commensal bacteria and maintenance of immune homeostasis in the gut. Nat. Rev. Immunol. 8, 411-420 (2008). doi: 10. 1038/nri2316.

[58] Rescigno, M., et al.: Dendritic cells express tight junction proteins and penetrate gut epithelial monolayers to sample bacteria. Nat. Immunol. 2, 361-367 (2001). doi: 10. 1038/86373.

[59] Grangette, C., et al.: Protection against tetanus toxin after intragastric administration of two recombinant lactic acid bacteria: impact of strain viability and in vivo persistence. Vaccine 20, 3304-3309 (2002).

[60] Christensen, H. R., Frokiaer, H., Pestka, J. J.: Lactobacilli differentially modulate expression of cytokines and maturation surface markers in murine dendritic cells. J. Immunol. 168, 171-178 (2002).

[61] Kalina, W. V., Mohamadzadeh, M.: Lactobacilli as natural enhancer of cellular immune response. Discov. Med. 5, 199-203 (2005).

[62] Mohamadzadeh, M., et al.: Lactobacilli activate human dendritic cells that skew T cells toward T helper 1 polarization. Proc. Natl. Acad. Sci. U. S. A. 102, 2880-2885 (2005). doi: 10. 1073/ pnas. 0500098102.

[63] Mohamadzadeh, M., Duong, T., Hoover, T., Klaenhammer, T. R.: Targeting mucosal dendritic cells with microbial antigens from probiotic lactic acid bacteria. Expert Rev. Vaccines 7, 163-174 (2008). doi: 10. 1586/14760584. 7. 2. 163.

[64] Perdigon, G., Alvarez, S., Pesce de Ruiz Holgado, A.: Immunoadjuvant activity of oral Lactobacillus ca-

sei: infl uence of dose on the secretory immune response and protective capacity in intestinal infections. J. Dairy Res. 58, 485-496 (1991).

[65] Pouwels, P. H., Leer, R. J., Boersma, W. J.: The poten-tial of Lactobacillus as a carrier for oral immunization: development and preliminary characterization of vector systems for targeted delivery of antigens. J. Biotechnol. 44, 183-192 (1996). doi: 10. 1016/0168-1656 (95) 00140-9.

[66] Maassen, C. B., et al.: Instruments for oral disease-intervention strategies: recombinant Lactobacillus casei expressing tetanus toxin fragment C for vaccination or myelin proteins for oral tolerance induction in multiple sclerosis. Vaccine 17, 2117-2128 (1999).

[67] Bermudez-Humaran, L. G., et al.: Controlled intra-or extracellular production of staphylococcal nuclease and ovine omega interferon in Lactococcus lactis. FEMS Microbiol. Lett. 224, 307-313 (2003).

[68] Navarre, W. W., Schneewind, O.: Surface proteins of gram-positive bacteria and mechanisms of their targeting to the cell wall envelope. Microbiol Mol Biol Rev 63, 174-229 (1999).

[69] Weis, J. J., Ma, Y., Erdile, L. F.: Biological activities of native and recombinant Borrelia burgdorferi outer surface protein A: dependence on lipid modifi cation. Infect. Immun. 62, 4632-4636 (1994).

[70] Sellati, T. J., et al.: Treponema pallidum and Borrelia burgdorferi lipoproteins and synthetic lipopeptides activate monocytic cells via a CD14-dependent pathway distinct from that used by lipopolysaccharide. J. Immunol. 160, 5455-5464 (1998).

[71] del Rio, B., Seegers, J. F., Gomes-Solecki, M.: Immune response to Lactobacillus plantarum expressing Borrelia burgdorferi OspA is modulated by the lipid modifi cation of the antigen. PLoS One 5, e11199 (2010). doi: 10. 1371/journal. pone. 0011199.

[72] del Rio, B., et al.: Oral immunization with recombi-nant lactobacillus plantarum induces a protective immune response in mice with Lyme disease. Clin. Vaccine. Immunol. 15, 1429-1435 (2008). doi: 10. 1128/CVI. 00169-08.

[73] Leary, S. E., et al.: Active immunization with recombi-nant V antigen from Yersinia pestis protects mice against plague. Infect. Immun. 63, 2854-2858 (1995).

[74] Alpar, H. O., Eyles, J. E., Williamson, E. D., Somavarapu, S.: Intranasal vaccination against plague, tetanus and diphtheria. Adv. Drug Deliv. Rev. 51, 173-201 (2001).

[75] del Rio, B., et al.: Platform technology to deliver prophylactic molecules orally: an example using the Class A select agent Yersinia pestis. Vaccine 28, 6714-6722 (2010). doi: 10. 1016/j. vaccine. 2010. 07. 084.

[76] Foligne, B., et al.: Prevention and treatment of colitis with Lactococcus lactis secreting the immunomodulatory Yersinia LcrV protein. Gastroenterology 133, 862-874 (2007). doi: 10. 1053/j. gastro. 2007. 06. 018.

[77] Ramirez, K., et al.: Neonatal mucosal immuniza-tion with a non-living, non-genetically modifi ed Lactococcus lactis vaccine carrier induces systemic and local Th1-type immunity and protects against lethal bacterial infection. Mucosal Immunol. 3, 159-171 (2010). doi: 10. 1038/mi. 2009. 131.

[78] Welkos, S., et al.: Determination of the virulence of the pigmentation-defi cient and pigmentation-/plasminogen activator-defi cient strains of Yersinia pestis in non-human primate and mouse models of pneumonic plague. Vaccine 20, 2206-2214 (2002).

[79] Smiley, S. T.: Current challenges in the development of vaccines for pneumonic plague. Expert Rev. Vaccines 7, 209-221 (2008). doi: 10. 1586/14760584. 7. 2. 209.

[80] Powell, B. S., et al.: Design and testing for a non-tagged F1-V fusion protein as vaccine antigen against bubonic and pneumonic plague. Biotechnol. Prog. 21, 1490-1510 (2005). doi: 10. 1021/bp050098r.

[81] Alvarez, M. L., et al.: Plant-made subunit vaccine against pneumonic and bubonic plague is orally immunogenic in mice. Vaccine 24, 2477-2490 (2006). doi: 10. 1016/j. vaccine. 2005. 12. 057.

[82] Cornelius, C. A., et al.: Immunization with recombi-nant V10 protects cynomolgus macaques from lethal

pneumonic plague. Infect. Immun. 76, 5588-5597 (2008). doi: 10. 1128/IAI. 00699-08.

[83] Smiley, S. T.: Immune defense against pneumonic plague. Immunol. Rev. 225, 256-271 (2008). doi: 10. 1111/j. 1600-065X. 2008. 00674. x.

[84] Williamson, E. D., et al.: Human immune response to a plague vaccine comprising recombinant F1 and V antigens. Infect. Immun. 73, 3598-3608 (2005). doi: 10. 1128/IAI. 73. 6. 3598-3608. 2005.

[85] Pavot, V., Rochereau, N., Genin, C., Verrier, B., Paul, S.: New insights in mucosal vaccine development. Vaccine 30, 142-154 (2012). doi: 10. 1016/j. vaccine. 2011. 11. 003.

（张强，吴国华译）

第 49 章　基于电穿孔的基因转移

Mattia Ronchetti，Michela Battista，
Claudio Bertacchini 和 RuggeroCadossi①

摘要

在特定脉冲条件下对混悬液或生物组织中的细胞施加外部电场，细胞膜的渗透性将瞬时增加，这种物理方法称为电穿孔，可用于将渗透性低或非渗透性分子导入细胞。电穿孔和化疗的组合称为电化学疗法。电化学疗法能够增强不易通过细胞膜的亲水性药物（例如博莱霉素）的局部细胞毒性。已证明这种联合疗法可有效地局部控制转移至皮肤的肿瘤结节，而与组织类型无关。用电脉冲的方式将遗传物质传递到靶组织或细胞中称为基于电穿孔的基因转移。采用不同的涂药器配置、电参数和传递的靶组织，可按目的优化体内电穿孔介导传递后的基因表达水平和动力学模式。本章讨论了非病毒基因传递的现有知识、DNA 电转移机制以及向骨骼肌和皮肤传递的临床应用，概述了目前可用于临床的基于电穿孔的基因转移的设备、组织电穿孔装置和电极。

49.1　细胞膜电穿孔原理

对单个细胞、细胞混悬液或生物组织施加外部电场可改变细胞跨膜电位。对细胞膜水平施加高于跨膜电位阈值（~1.5 V）的电场[1]可改变膜结构，使非渗透性分子透过膜[2]，这一现象称为电穿孔或电渗透。在确定的窗口期内调节施加的电场参数，可使膜暂时可渗透，恢复细胞的自然状态，保持其活力，此过程在历史上称为可逆电穿孔[3]，现已与化学治疗剂联合使用，则称为电化学疗法，也可作为传递核酸、基因电转移的方法[4]。

施加更高强度和（或）更长时间的电场可永久破坏细胞膜渗透性，导致细胞死亡。该过程通常称为不可逆电穿孔，是目前正在进行临床研究的一种新型软组织消融模式[3]。

49.2　电化学疗法

电化学疗法是对肿瘤组织进行局部可逆电穿孔，增强具有内在细胞毒性的非渗透性（即博来霉素）或低渗透性（即顺铂）药物的局部抗肿瘤活性[5]。电化学疗法的第一个临床研究可追溯到 20 世纪 90 年代早期[6]，据报道，可有效局部控制头颈部鳞状细胞癌结节。在第一次异质性临床试验之后，Sersa G[7]在 2006 年进行了一项前瞻性、多中心、国际性临床试验，这是一项 ESOPE（欧洲电化学疗法标准操作规程）研究，评估了电化学疗法治疗皮肤和皮下肿瘤结节的有效性和安全性，结果显示对所治疗的肿瘤结节的客观缓解率为 85%（完全缓解率为 73.7%）[8]。本研究根据电化学疗法标准操作规程的定义统一了治疗方案[9]。最近，对所有相关已发表文献的系统评价和正式荟萃分析进一步证实了电化学疗法的有效性，以及各种治疗条件的电化学疗法治疗肿瘤效果的潜在预测因子[10]。

---

① M. Ronchetti，BSc（✉）· M. Battista，PhD · C. Bertacchini，MSc · R. Cadossi，MD（IGEA S. P. A. 临床生物物理学实验室，意大利，卡尔皮，E-mail：m. ronchetti@ igeamedical. com）

49.3　基因电转移

基因疗法是将基因导入体内以治疗疾病，尤其是由遗传异常或缺陷引起的疾病，是一个很有前途的医学领域。安全、有效且可靠地将基因传递至靶组织是基因治疗的主要障碍。已证明 DNA 电转移是使用非病毒基因疗法传递裸 DNA 的一种有效且安全的方法。继 1982 年开创性的体外研究证明可用电脉冲将 DNA 导入活细胞[4,5]之后，已有多篇利用电穿孔将分子传递到真核细胞和各种组织的研究报告[11]。多年来，已开发出了电穿孔装置，并发明了电脉冲发生器和给药器，可在不同条件下向不同组织传递核酸。现代发生器可控制振幅、脉冲时长和数量，研究各种电参数对转染和表达水平的影响，并优化基因电转移方案[12]。

使用裸 DNA 消除了与使用病毒相关的限制问题，例如腺病毒中的编码序列长度等[13]，以及在逆转录病毒或慢病毒的情况下将 DNA 整合至宿主基因组中时形成插入突变的问题[14]。裸 DNA 安全且易于操作和制备，完全由双链 DNA 构成，无相关蛋白，而腺病毒蛋白可诱导免疫应答，无须再次进行病毒载体给药[15]；使用基因电转移传递的裸 DNA 可能多次给药。

使用非病毒基因治疗的其他物理方法包括直接注射[16]、质粒脂质体复合物[17]，是向细胞传递基因的最常用技术之一；将 DNA 转移到皮肤等浅表组织的生物弹道方法使用一种称为基因枪的装置，将质粒包被的金微粒经细胞膜推进至细胞质和细胞核，绕过了 DNA 可能受损的内体室。这种方法的局限性是金属颗粒到达整个组织的效率低，这是由于颗粒渗透性低造成的，金属颗粒到体内沉积，可能导致长期后果[18,19]。超声穿孔应用超声波增加细胞膜对包括质粒 DNA 在内的大分子的渗透性。事实上，注射 DNA 后向组织照射超声波可使基因表达增强[20]。在数秒钟内对小鼠快速静脉注射大量溶液可实现流体动力注射基因转移技术。由于注射的体积很大，液体蓄积在下腔静脉，注射的 DNA 主要被肝脏捕获[16,21]。由于注射的液体量非常大，这种方法的临床转化潜力似乎有限。

49.4　DNA 电转移的原则

DNA 电转移使用方便，安全而高效，故已迅速扩展并发展为一种传递非病毒基因的有效方法。数项实验表明，DNA 电转移时，细胞对外源 DNA 的摄入受不同因素的控制。高压短时电脉冲的主要作用是使膜可渗透。随后使用较低电压、较长持续时间的脉冲，DNA 在电泳力作用下向透化细胞膜移动，然后进入细胞质[22,23]。研究表明，当施加电场时，DNA 与膜直接相互作用，促进孔形成；故必须在电脉冲传递之前注射 DNA[22]。

使用组合脉冲研究了电脉冲在 DNA 电转移中的作用，已鉴定了 3 种用于 DNA 转移的脉冲方法：①仅施加短的高振幅脉冲（例如 6 个脉冲，100 μs 和 1.4 kV/cm）[24]，产生功效合理，低死亡率；②施加长的低幅度脉冲（例如 8 个脉冲，20 ms，200 kV/cm）[25]（较长脉冲的电泳效应更好，可提高转染率）；③短的高振幅脉冲，然后是长的低振幅脉冲[26]。该脉冲组合基于高振幅脉冲诱导渗透的概念，而随后的长时间低电压脉冲可驱动 DNA 穿过不稳定的膜。进一步的实验证明，高压和低压脉冲之间的时滞非常重要：高压和低压之间的时滞越短，转染效率越高[21]。

49.5　基因电转移至靶组织

自 Wolff 等的试验以来，已对基因向肌肉的转移进行了广泛研究[27]，结果表明，质粒 DNA 可被摄取，然后在小鼠骨骼肌细胞中获得表达。机体中有大量骨骼肌，易于接近，血管丰富，导入的基因在有丝分裂后可延长保留时间并表达，故是基因转移颇具吸引力的靶标[28,29]。转染效率的高度可变，所得蛋白质合成的不可预测性[25]为该方法的主要缺点。为克服这些限制，一些体内研究使用电穿孔将基因转移到骨骼肌。用相对大体积的等渗质粒溶液进行肌内注射 DNA，然后在注射部位周围植入电极施加电脉冲。非侵入性平板电极用于浅表肌肉，侵入性针电极用于治疗较深层组织。

为确保电场的分布均匀和电极之间的转染均一，促进质粒载体的表达，必须平行于肌肉纤维植入电极[28]。其他对基因表达效率至关重要的因素包括注射的质粒数量、启动子系统和转染部位[30]。已证明向肌肉组织的 DNA 电转移具有高效性，已发现转基因可稳定表达 1 年以上[25]，这克服了蛋白质表达水平的高度可变性。

此后，基因电转移已成功应用于其他组织，如角膜、睾丸、肺、肝、肾、膀胱肿瘤和皮肤[31]。最重要的是，皮肤保留了数个重要的特征，使其成为基因治疗颇具吸引力的靶组织：易于进行治疗，易于评估组织学和临床结果，而且涉及的包含抗原呈递细胞（朗格汉斯细胞、树突细胞）是免疫系统的一部分，因此皮肤成为研究 DNA 疫苗接种的适当靶点[31,32]。

对皮肤进行电穿孔时，必须使用适当的电极（非侵入性板和贴片电极或侵入式阵列、圆形或成对的针电极）施加电脉冲。已进行了数种脉冲组合试验；最佳的脉冲条件取决于转染的皮肤类型、电极类型和 DNA 类型[35]。皮肤的大小、厚度和年龄也必须考虑[33,34]。已评估了大量电学方案，最有效的基因转移方案是高压（HV）和低压（LV）脉冲的组合（1 HV 1 000 V/cm，100 μs+1 LV 100 V / cm，400 ms）[36]，数个短高压脉冲（6 HV 1，750 V / cm，100 μs）[37]和数个长抵押脉冲（8 LV 100 V / cm，150 ms）[38]。

已使用基因电转移实现数种临床相关质粒在皮肤的表达。这些质粒包括局部生长因子，如血管内皮生长因子（VEGF）、角质形成细胞生长因子（KGF）以及针对艾滋病、乙型肝炎、天花和疟疾疫苗的病毒靶标[31]。还研究了在进行电穿孔传递后，用于治疗黑素瘤的 IL-12 的表达[39,40]。

表 49.1　体内基因电转移至肌肉和皮肤组织的临床前研究总结

电极类型		动物模型	DNA 质粒	电压范围（V/cm）	作用时间	脉冲数范围	参考文献
	皮肤	猪，小鼠，大鼠	GFP，Luc，hepatitis B，	1 125~1 800	50~100 μs	2~18	[37，59-63]
			Hif-1α，KGF，VEGF，	HV			
			PSA，CEA，survivin	50~400	10~100 ms	6~10	[63-66]
				LV			
				1 000~10 125	50~100 μs	1~2	[35，63，67-69]
微创型				+	+	+	
				80~275	275~400ms	1~8	
				HV+LV			
	肌肉	小鼠，绵羊，大鼠，狗	IL-5，GFP，EPO，GHRH，JEV	100~250	50 ms	5~8	[89-93]

（续表）

电极类型		动物模型	DNA 质粒	电压范围（V/cm）	作用时间	脉冲数范围	参考文献
	皮肤	猪，小鼠，大鼠，家兔	NeoR，LacZ，GFP，OVA，	400-1 750（HV）	100~300 μs	1~8	[37，39，70-72]
			hepatitis B，IL-12，EPO	12~800	2~400 ms	1~12	[33，34，39，
				（LV）			71-77]
				700~1 000	700~1 000 μs	1	[35，36，41，72]
				+	+	+	[78-81]
				80~200	80~200 ms	1	
				（HV+LV）			
 非微创型	肌肉	小鼠，家兔，大鼠	GFP，Luc，IL-4，IL-1Ra，	1 000	100 μs	8	[84]
			VEGF，EPO	HV			
				50~600	100 μs~20 ms	2~8	[82，83，85-88]
				LV			
				100~2 000	100 μs	1	[36，82，84]
				+	+	+	
				80 V/cm	400 ms	1	
				HV+LV			

采用微创针电极和非微创电极。针电极已被用于针对皮肤和更深层次的组织，如肌肉较大的动物，而板电极已被用于治疗表面组织和啮齿动物的肌肉。

为简单起见，所采用的电压操作报告为总范围。

缩写：V/cm 伏/厘米，HV 高电压，LV 低电压。

基因转染肌肉可长期持续表达 1 年或 1 年以上，皮肤表达的持续时间为 3~4 周[41]。在不需要长期表达转染质粒且不必或不需要连续产生蛋白或抗原的情况下，皮肤是基因电转移的理想靶器官。

49.6 电穿孔装置

电穿孔技术基于使用不同给药器电极的脉冲发生器，例如，针式或板式阵列，向靶组织施加合适的电脉冲。完整的系统、脉冲发生器和给药器属于医疗器械。不同的系统生成脉冲的方法不同。当该技术用于人类时，无论是传递小分子还是基因转染，都必须符合安全性、可靠性和有效性的特定要求。目前，有市场上有数种用于临床前和实验室试验的电穿孔器；只有在欧盟获批人体临床应用的少数型号[42]具有 CE 标志。意大利 IGEA 有限公司开发和销售的 Cliniporator™就是这样的一款产品（图 49.1）。

可根据脉冲特性和产生脉冲的技术对电穿孔器进行分类。有几种技术可产生脉冲，产生的波形全都为正方形或指数形。长度短于 1 μs 的非常短的脉冲需特定电路，迄今为止仅用于电穿

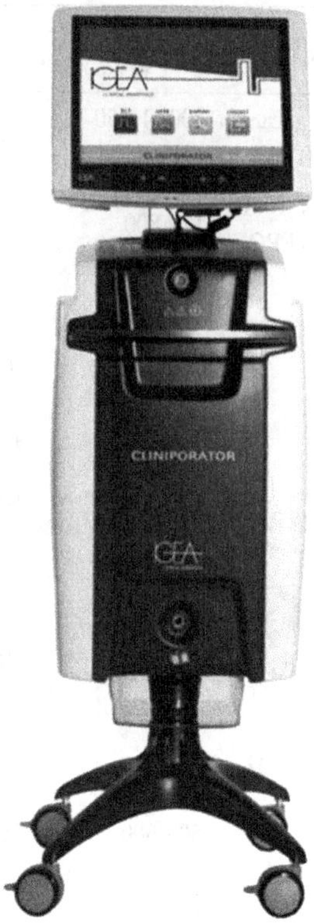

图 49.1　用于电穿孔的 Cliniporator™

孔研究。长度为 1~1 000 μs的短脉冲和长度为数毫秒的长脉冲可用不同方法产生：电容直接放电、脉冲变压器、高压模拟发电机或高电压方波生成[43,44]。用于基因电转移的电穿孔器的基本特征是产生高压（HV）脉冲和低压（LV）脉冲组合。这常需要两个不同的电路以确保适当的精度和良好地控制脉冲形状。

Cliniporator 等复杂的设备可测量负载上的实时电压和电流，迅速向用户反馈正在进行的治疗情况。对所施加的脉冲进行实时测量也是监控设备性能，并及早发现故障的理想方式。高压脉冲幅度在一百伏到几千伏之间[45]，法规和安全标准[42]要求高压电路的绝缘性非常好，以确保患者、操作员和设备的安全。另一个非常重要的安全问题决定于工作环境条件（例如手术室或临床医生办公室）的可变性，这些条件事先通常是未知的。传递的能量受组织或细胞阻抗的影响，可从点到点，也可从组织到组织。由于高电流可损伤细胞、组织甚至装置本身，故电流是控制的关键参数。

最后考虑设备的结构，在某些情况下用户界面直接控制电源部分，但在其他情况下，电源部分由独立的电子电路控制，用户界面简单地将治疗参数发送到电源部分。这种不同的方法可影响设备发生故障时的可靠性和安全性[43]。用于电穿孔的装置的设计必须严密控制传递给患者的能量。设计和实施应注重坚固快速的脉冲发生器，在发生故障时反应性高，使用冗余改善其安全性，用户界面应非常友好，最大限度地降低用户出现错误的风险[42,43,45]。

49.7　临床应用

49.7.1　基因治疗

基因转移的临床前研究涵盖了广泛的研究，包括遗传性肌肉疾病，如 Duchenne 营养不良的治疗[46]、基因疫苗接种[47]和分泌性治疗蛋白如促红细胞生成素（EPO）的系统递送[48]、VIII 因子等造血剂[49]、干扰素（IFN-α）等抗癌剂[50]和 metargidin 等抗血管生成因子[51]。表 49.2 总结了正在进行的以肌肉为基因转移的靶组织的临床试验[52]，表 49.3 总结了以真皮为靶组织的研究。

表 49.2 正在进行的肌内（IM）基因电转移临床试验

病理	受试者	主要结果	临床试验管理 ID
前列腺癌		使用或不使用电穿孔肌内注射 PSMA27/pDom 融合基因 DNA 疫苗治疗 HLA-A2+ 的前列腺癌患者 IV 期临床研究	UK-112
乳头瘤病毒感染（已完成）	24	评估成年女性患者术后或 2、3 级 CIN 消融治疗后辅助治疗时进行电穿孔肌内注射增加剂量 VGX-3100 的安全性和耐受性	NCTO0685412
慢性病型丙型肝炎感染	12	评估对接受首次治疗的低病毒载量慢性丙型肝炎患者进行电穿孔介导的 CHRONVAC-C® 肌内递送的安全性和耐受性	NCT00563 173
恶性 黑色素瘤（已完成）	25	评估对 IIB 期、IIC III 期或 IV 期黑色素瘤患者进行电穿孔介导的肌内注射小鼠酪氨酸酶质粒 DNA 疫苗的安全性和可行性	NCT00471133
HIV 感染 （已完成）	40	评估使用 TriGrid™ 电穿孔肌内注射 3 种剂量水平的基于 ADVAX DNA 的 HIV 疫苗的安全性	NCT00545987
健康成人	24	评估 MedPulser DDS 器件的耐受性	NCTO072 146
宫颈上皮内瘤样病变	348	以宫颈病变组织病理学消退至 CIN 1 或更低级别的参与者数量作为疗效的衡量标准	NCT01304524
恶性黑色素瘤	30	评估对早期肿瘤（即 III 期或 IV 期黑色素瘤）已扩散的黑色素瘤患者应用研究性免疫疗法 SCIB1 的安全性和耐受性	NCT01138410
白血病	184	血液系统恶性肿瘤患者肌内注射后肌内电穿孔进行 DNA 融合基因疫苗接种产生 WT1 免疫力的 II 期临床研究	NCT01334020
人乳头瘤病毒相关的头颈癌	21	评估对人乳头瘤病毒-16 相关的头颈癌患者使用肌内传递系统 TriGrid™ 进行 pNGVL4a-CRT / E7（Detox）DNA 疫苗联合环磷酰胺给药的安全性和可行性	NCT01493154
HIV-1 感染	12	评估 PENNVAX™-B（Gag、Pol、Env）+ 电穿孔对感染 HIV-1 成年人的安全性、耐受性和免疫原性	NCT01082692
慢性丙型肝炎	32	早期病毒动力学-第二阶段病毒下降的曲线斜率	NCT01335711

表 49.3 正在进行的皮内（ID）电转移给药临床试验

病理	受试者	主要结果	临床试验管理 ID
前列腺癌	18	评估前列腺癌复发患者联合应用皮内注射与电穿孔进行剂量递增的 pVAXrcPSAv531 DNA 疫苗给药的可行性和安全性	NCT00859729

（续表）

病理	受试者	主要结果	临床试验管理 ID
结肠直肠癌	20	评估联合应用 tetwtCEA DNA 与电穿孔进行 DNA 免疫接种方法的安全性和免疫原性	NCT01064375
H1 和 H5 流感病毒	100	评估用 9 种 H1 和 H5 HA 质粒多重组合制剂进行皮内电穿孔给药对健康成人的安全性和耐受性	NCT01405885
人类流感	50	评估用两种不同 H1 HA 质粒组成的流感 DNA 疫苗进行皮内电穿孔给药对健康老年人的安全性和耐受性	NCT01587131

49.7.2 DNA 疫苗接种

电穿孔介导的 DNA 疫苗接种通过电穿孔进行体内基因转移，很有前途。DNA 疫苗有许多优点。DNA 载体易于生产和操作，可快速测试、分离、易于储存和运输。DNA 疫苗可促进细胞和体液免疫反应[53]，必要时，可含有几种抗原表位[54]。与单独注射裸质粒相比，电穿孔的转染效率高，能够增强免疫应答；危险信号激活抗原呈递细胞（APC），施加电脉冲后发生的局部炎症以及向 DNA 给药部位募集免疫 B 细胞和 T 细胞进一步增强了免疫应答。对抗原呈递细胞进行直接转染可能对于 DNA 引发的 T 细胞启动以及电穿孔介导的 DNA 转移至皮肤的免疫应答增强至关重要[55,56]。联合应用腺病毒载体给药和 DNA 基因电转移的初次免疫—加强方案是另一种有前途的方法，已证明对抗原的免疫应答水平较高，携带 B 细胞淋巴瘤的犬的存活率增加[57]。

对健康志愿者的试验证实，人体可耐受 DNA 疫苗，不产生抗 DNA 抗体，向肌肉进行电穿孔介导的 DNA 传递后，pDNA 未整合至宿主染色体[58]（表 49.2 和表 49.3）。目前正在进行数项针对癌症的电转移 DNA 疫苗试验以及联合应用电穿孔抵御感染性病原（HIV、宫颈上皮内瘤变和丙型肝炎病毒）DNA 疫苗的 3 个 I 期临床研究[52]。

参考文献

[1] Kotnik, T., Bobanovic, F., Miklavcic, D.: Sensitivity of transmembrane voltage induced by applied electric fi elds-a theoretical analysis. Bioelectrochem. Bioenerg. 43, 285-291 (1997).

[2] Miklavcic, D., Semrov, D., Mekid, H., Mir, L. M.: A validated model of in vivo electric fi eld distribution for electrochemotherapy and for DNA electrotransfer for gene therapy. Biochim. Biophys. Acta 1523, 73-83 (2000).

[3] Rubinsky, B.: Irreversible electroporation in medi-cine. Technol. Cancer Res. Treat. 6, 255-260 (2007).

[4] Neumann, E., Schaefer-Ridder, M., Wang, Y., Hofschneider, P. H.: Gene transfer into mouse lyoma cells by electroporation in high electric fi elds. EMBO J. 1, 841-845 (1982).

[5] Silve, A., Mir, L. M.: Cell electropermeabilization and cellular uptake of small molecules: the electrochemotherapy concept. In: Kee, S. T., Gehl, J., Lee, E. W. (eds.) Clinical Aspects of Electroporation 1, pp. 69-82. Springer, New York (2011).

[6] Mir, L. M., et al.: Electrochemotherapy, a new antitu-mour treatment: first clinical trial. C. R. Acad. Sci. III 313, 613-618 (1991).

[7] Sersa, G.: The state of the art of electrochemotherapy before ESOPE study. Advantages and clinical use. Eur. J. Cancer. Suppl. 4, 52-59 (2006).

[8] Marty, M., et al.: Electrochemotherapy-an easy, highly effective and safe treatment of cutaneous and subcu-

taneous metastases. Results of ESOPE (European Standard Operating Procedures of Electrochemotherapy) study. Eur. J. Cancer. 4, 3-13 (2006).

[9] Mir, L. M., et al.: Standard operating procedures of the electrochemotherapy: instructions for the use of bleomycin or cisplatin administered either systemically or locally and electric pulses delivered by theCliniporator by means of invasive or noninvasive electrodes. Eur. J. Cancer. Suppl. 4, 14-25 (2006).

[10] Mali, B. Jarm, T., Snoj, M., Sersa, G., Miklavcic, D.: Antitumor effectiveness of electrochemotherapy: a systematic review and meta-analysis. Eur. J. Surg. Oncol. 39, 4-16 (2012). http://dx. doi. org/10. 1016/j. ejso. 2012. 08. 016.

[11] Mir, L. M., Moller, P. H., André, F., Gehl, J.: Electric pulse-mediated gene delivery to various animal tissues. Adv. Genet. 54, 83-114 (2005).

[12] Durieux, A. C., Bonnefoy, R., Manissolle, C., Freyssenet, D.: High-effi ciency gene electrotransfer into skeletal muscle: description and physiological applicability of a new pulse generator. Biochem. Biophys. Res. Commun. 296 (2), 443-450 (2002).

[13] Hacein-Bey-Abina, S., et al.: LMO2-associated clonal T cell proliferation in two patients after gene therapy for SCID-X1. Science 302 (5644), 415-419 (2003).

[14] Hacein-Bey-Abina, S., Le Deist, F., Carlier, F., Bouneaud, C., Hue, C., De Villartay, J. P., Thrasher, A. J., Wulffraat, N., Sorensen, R., Dupuis-Girod, S., Fischer, A., Davies, E. G., Kuis, W., Leiva, L., Cavazzana-Calvo, M.: Sustained correction of X-linked severe combined immunodefi ciency by ex vivo gene therapy. N. Engl. J. Med. 346 (16), 1185-1193 (2002).

[15] Couzin, J., Kaiser, J.: Gene therapy. As Gelsinger case ends, gene therapy suffers another blow. Science 307 (5712), 1028 (2005).

[16] Budker, V., et al.: Hypothesis: naked plasmid DNA is taken up by cells in vivo by a receptor-mediated process. J. Gene Med. 2, 76-88 (2000).

[17] Dauty, E., Remy, J. S., Blessing, T., Behr, J. P.: Dimerizable cationic detergents with a low cmc condense plasmid DNA into nanometric particles and transfect cells in culture. J. Am. Chem. Soc. 123, 9227-9234 (2001).

[18] Lin, M. T., Pulkkinen, L., Uitto, J., Yoon, K.: The gene gun: current application in cutaneous gene therapy. Int. J. Dermatol. 39, 161-170 (2000).

[19] Davidson, J. M., Krieg, T., Eming, S. A.: Particle-mediated gene therapy of wounds. Wound Repair Regen. 8, 452-459 (2000).

[20] Newman, C. M., Lawrie, A., Brisken, A. F., Cumberland, D. C.: Ultrasound gene therapy: on the road from concept to reality. Echocardiography 18, 339-347 (2001).

[21] Liu, F., Huang, L.: Improving plasmid DNA-mediated liver gene transfer by prolonging its retention in the hepatic vasculature. J. Gene Med. 3, 569-576 (2001).

[22] Sukharev, S. I., Klenchin, V. A., Serov, S. M., Chernomordik, L. V., Chizmadzhev, YuA.: Electroporation and electrophoretic DNA transfer into cells. The effect of DNA interaction with electropores. Biophys. J. 63 (5), 1320-1327 (1992).

[23] Klenchin, V. A., Sukharev, S. I., Serov, S. M., Chernomordik, L. V., Chizmadzhev, YuA.: Electrically induced DNA uptake by cells is a fast process involving DNA electrophoresis. Biophys. J. 60 (4), 804-811 (1991).

[24] Heller, R., Jaroszeski, M., Atkin, A., et al.: In vivo gene electroinjection and expression in rat liver. FEBS Lett. 389, 225-228 (1996).

[25] Mir, L. M., Bureau, M. F., Gehl, J., Rangara, R., Rouy, D., Caillaud, J. M., Delaere, P., Branellec, D., Schwartz, B., Scherman, D.: High-effi ciency gene transfer into skeletal muscle mediated by electric pulses. Proc. Natl. Acad. Sci. U. S. A. 96 (8), 4262-4267 (1999).

[26] Bureau, M. F., Gehl, J., Deleuze, V., Mir, L. M., Scherman, D.: Importance of association between permeabilization and electrophoretic forces for intramuscular DNA electrotransfer. Biochim. Biophys. Acta 1474, 353-359 (2000).

[27] Wolff, J. A., et al.: Direct gene transfer into mouse muscle in vivo. Science 247, 1465-1468 (1990).

[28] André, F., Mir, L. M.: DNA electrotransfer: its princi-ples and an updated review of its therapeutic applications. Gene Ther. 11 (1), S33-S42 (2004).

[29] Gehl, J.: Electroporation for drug and gene delivery in the clinic: doctors go electric. Methods Mol. Biol. 423, 351-359 (2008).

[30] Hojman, P., Gissel, H., Gehl, J.: Sensitive and precise regulation of haemoglobin after gene transfer of erythropoietin to muscle tissue using electroporation. Gene Ther. 14 (12), 950-959 (2007).

[31] Gothelf, A., Gehl, J.: Gene electrotransfer to skin; review of existing literature and clinical perspectives. Curr. Gene Ther. 10 (4), 287-299 (2010).

[32] Kutzler, M. A., Weiner, D. B.: DNA vaccines: ready for prime time? Nat. Rev. Genet. 9, 776-788 (2008).

[33] Zhang, L., Li, L., Hoffmann, G. A., Hoffman, R. M.: Depth-targeted effi cient gene delivery and expression in the skin by pulsed electric fi elds: an approach to gene therapy of skin aging and other diseases. Biochem. Biophys. Res. Commun. 220 (3), 633-636 (1996).

[34] Chesnoy, S., Huang, L.: Enhanced cutaneous gene delivery following intradermal injection of naked DNA in a high ionic strength solution. Mol. Ther. 5 (1), 57-62 (2002).

[35] Gothelf, A., Mahmood, F., Dagnaes-Hansen, F., Gehl, J.: Effi cacy of transgene expression in porcine skin as a function of electrode choice. Bioelechemistry 82 (2), 95-102 (2011).

[36] Andre, F. M., Gehl, J., Sersa, G., et al.: Effi ciency of High-and Low-Voltage Pulse Combinations for Gene Electrotransfer in Muscle, Liver, Tumor, and Skin. Hum. Gene Ther. 19 (11), 1261-1272 (2008).

[37] Drabick, J. J., Glasspool-Malone, J., King, A., Malone, R. W.: Cutaneous transfection and immune responses to intradermal nucleic acid vaccination are signifi cantly enhanced by in vivo electropermeabilization. Mol. Ther. 3 (2), 249-255 (2001).

[38] Heller, L. C., Jaroszeski, M. J., Coppola, D., Heller, R.: Comparison of electrically mediated and liposomecomplexed plasmid DNA delivery to the skin. Genet. Vaccines Ther. 6, 16 (2008).

[39] Heller, R., Schultz, J., Lucas, M. L., et al.: Intradermal delivery of interleukin-12 plasmid DNA by in vivo electroporation. DNA Cell Biol. 20 (1), 21-26 (2001).

[40] Daud, A. I., DeConti, R. C., Andrews, S., et al.: Phase I trial of interleukin-12 plasmid electroporation in patients with metastatic melanoma. J. Clin. Oncol. 26 (36), 5896-5903 (2008).

[41] Gothelf, A., Eriksen, J., Hojman, P., Gehl, J.: Duration and level of transgene expression after gene electrotransfer to skin in mice. Gene Ther. 17 (7), 839-845 (2010).

[42] European Standard EN60601-1: Medical Electrical Equipment-Part 1: General Requirements for Basic Safety and Essential Performance, 3rd edn. British Standards Institution, London (2007).

[43] Bertacchini, C., et al.: Design of an irreversible elec-troporation system for clinical use. Technol Cancer Res Treat 6 (4), 313-320 (2007).

[44] Rebersek, M., Miklavcic, D.: Advantages and disad-vantages of different concepts of electroporation pulse generation. ATKAFF 52, 12-19 (2011).

[45] Puc, M., Rebersek, S., Miklavcic, D.: Requirements for a clinical electrochemotherapy device-electroporator. Radiol. Oncol. 31, 368-373 (1997).

[46] Murakami, T., Nishi, T., Kimura, E., Goto, T., Maeda, Y., Ushio, Y., Uchino, M., Sunada, Y.: Full-length dystrophin cDNA transfer into skeletal muscle of adult mdx mice by electroporation. Muscle

Nerve 27 (2), 237-241 (2003).

[47] Rosati, M., et al.: Increased immune responses in rhe-sus macaques by DNA vaccination combined with electroporation. Vaccine 26 (40), 5223-5229 (2008).

[48] Payen, E., Bettan, M., Rouyer-Fessard, P., Beuzard, Y., Scherman, D.: Improvement of mouse beta-thalassemia by electrotransfer of erythropoietin cDNA. Exp. Hematol. 29 (3), 295-300 (2001).

[49] Long, Y. C., Jaichandran, S., Ho, L. P., Tien, S. L., Tan, S. Y., Kon, O. L.: FVIII gene delivery by muscle electroporation corrects murine hemophilia A. J. Gene Med. 7 (4), 494-505 (2005).

[50] Zhang, G. H., et al.: Gene expression and antitumor effect following im electroporation delivery of human interferon alpha 2 gene. Acta Pharmacol. Sin. 24 (9), 891-896 (2003).

[51] Trochon-Joseph, V., et al.: Evidence of antiangiogenic and antimetastatic activities of the recombinant disintegrin domain of metargidin. Cancer Res. 64, 2062-2069 (2004).

[52] ClinicalTrial. gov. US. National Institute of Health. www. clinicaltrials. gov (2012).

[53] Tuting, T., Storkus, W. J., Falo Jr., L. D.: DNA immunization targeting the skin: molecular control of adaptive immunity. J. Invest. Dermatol. 111 (2), 183-188 (1998).

[54] Medi, B. M., Hoselton, S., Marepalli, R. B., Singh, J.: Skin targeted DN vaccine delivery using electroporation in rabbits. I: effi cacy. Int. J. Pharm. 294 (1-2), 53-63 (2005).

[55] Hirao, L. A., Wu, L., Khan, A. S., Satishchandran, A., Draghia-Akli, R., Weiner, D. B.: Intradermal/subcutaneous immunization by electroporation improves plasmid vaccine delivery and potency in pigs and rhesus macaques. Vaccine 26 (3), 440-448 (2008).

[56] Liu, M. A.: DNA vaccines: a review. J. Intern. Med. 253 (4), 402-410 (2003).

[57] Peruzzi, D., et al.: A vaccine targeting telomerase enhances survival of dogs affected by B-cell lymphoma. Mol. Ther. 18, 1559-1567 (2010).

[58] Rune, K., Torunn, E. T., Dag, K., Jacob, M.: Clinical evaluation of pain and muscle damage induced by electroporation of skeletal muscle in humans abstract from American Society of Gene Therapy 7th annual meeting. June 2-6, 2004 Minneapolis, Minnesota, USA. Mol. Ther. 9 (Supp 1), S1-S435 (2004).

[59] Glasspool-Malone, J., Drabick, J. J., Somiari, S., Malone, R. W.: Effi cient nonviral cutaneous transfection. Mol. Ther. 2, 140-146 (2000).

[60] Byrnes, C. K., et al.: Electroporation enhances trans-fection effi ciency in murine cutaneous wounds. Wound Repair Regen. 12, 397-403 (2004).

[61] Marti, G., et al.: Electroporative transfection with KGF-1 DNA improves wound healing in a diabetic mouse model. Gene Ther. 11, 1780-1785 (2004).

[62] Lin, M. P., et al.: Delivery of plasmid DNA expression vector for keratinocyte growth factor-1 using electroporation to improve cutaneous wound healing in a septic rat model. Wound Repair Regen. 14, 618-624 (2006).

[63] Roos, A. K., et al.: Enhancement of cellular immune response to a prostate cancer DNA vaccine by intradermal electroporation. Mol. Ther. 13, 320-327 (2006).

[64] Kang, J. H., Toita, R., Niidome, T., Katayama, Y.: Effective delivery of DNA into tumor cells and tissues by electroporation of polymer-DNA complex. Cancer Lett. 265, 281-288 (2008).

[65] Liu, L., et al.: Age-dependent impairment of HIF-1alpha expression in diabetic mice: correction with electroporation-facilitated gene therapy increases wound healing, angiogenesis, and circulating angiogenic cells. J. Cell. Physiol. 217, 319-327 (2008).

[66] Ferraro, B., Cruz, Y. L., Coppola, D., Heller, R.: Intradermal delivery of plasmid VEGF (165) by electroporation promotes wound healing. Mol. Ther. 17, 651-657 (2009).

[67] Brave, A., et al.: Late administration of plasmid DNA by intradermal electroporation effi ciently boosts DNA-primed T and B cell responses to carcinoembryonic antigen. Vaccine 27, 3692-3696 (2009).

[68] Lladser, A., et al.: Intradermal DNA electroporation induces survivin-specifi c CTLs, suppresses angiogenesis and confers protection against mouse melanoma. Cancer Immunol. Immunother. 59, 81-92 (2009).

[69] Roos, A. K., Eriksson, F., Walters, D. C., Pisa, P., King, A. D.: Optimization of skin electroporation in mice to increase tolerability of DNA vaccine delivery to patients. Mol. Ther. 17, 1637-1642 (2009).

[70] Titomirov, A. V., Sukharev, S., Kistanova, E.: In vivo electroporation and stable transformation of skin cells of newborn mice by plasmid DNA. Biochim. Biophys. Acta 088, 131-134 (1991).

[71] Lucas, M. L., Jaroszeski, M. J., Gilbert, R., Heller, R.: In vivo electroporation using an exponentially enhanced pulse: a new waveform. DNA Cell Biol. 20, 183-188 (2001).

[72] Pavselj, N., Preat, V.: DNA electrotransfer into the skin using a combination of one high-and one lowvoltage pulse. J. Control. Release 106, 407-415 (2005).

[73] Maruyama, H., et al.: Skin-targeted gene transfer using in vivo electroporation. Gene Ther. 8, 1808-1812 (2001).

[74] Zhang, L., Nolan, E., Kreitschitz, S., Rabussay, D. P.: Enhanced delivery of naked DNA to the skin by noninvasive in vivo electroporation. Biochim. Biophys. Acta 1572, 1-9 (2002).

[75] Lee, P. Y., Chesnoy, S., Huang, L.: Electroporatic delivery of TGFbeta1 gene works synergistically with electric therapy to enhance diabetic wound healing in db/db mice. J. Invest. Dermatol. 123, 791 - 798 (2004).

[76] Thanaketpaisarn, O., Nishikawa, M., Yamashita, F., Hashida, M.: Tissue-specifi c characteristics of in vivo electric gene: transfer by tissue and intravenous injection of plasmid DNA. Pharm. Res. 22, 883-891 (2005).

[77] Heller, L. C., et al.: Optimization of cutaneous electri-cally mediated plasmid DNA delivery using novel electrode. Gene Ther. 14, 275-280 (2007).

[78] Vandermeulen, G., et al.: Optimisation of intradermal DNA electrotransfer for immunisation. J. Control. Release 124, 81-87 (2007).

[79] Vandermeulen, G., et al.: Skin-specifi c promoters for genetic immunisation by DNA electroporation. Vaccine 27, 4272-4277 (2009).

[80] Vandermeulen, G., et al.: Effect of tape stripping and adjuvants on immune response after intradermal DNA electroporation. Pharm. Res. 26, 1745-1751 (2009).

[81] Gothelf, A., Hojman, P., Gehl, J.: Therapeutic levels of erythropoietin (EPO) achieved after gene electrotransfer to skin in mice. Gene Ther. (2010). doi: 10. 1038/gt. 2010. 46.

[82] Ho, S. H., et al.: Protection against collagen-induced arthritis by electrotransfer of an expression plasmid for the interleukin 4. Biochem. Biophys. Res. Commun. 321, 759-766 (2004).

[83] Cukjati, D., Batiuskaite, D., André, F., Miklavcic, D., Mir, L. M.: Real time electroporation control for accurate and safe in vivo non-viral gene therapy. Bioelectrochemistry 70, 501-507 (2007).

[84] Hojman, P., et al.: Physiological effects of high - and low - voltage pulse combinations for gene electrotransfer in muscle. Hum. Gene Ther. 19, 1249-1260 (2008).

[85] Hojman, P., Zibert, J. R., Gissel, H., Eriksen, J., Gehl, J.: Gene expression profi les in skeletal muscle after gene electrotransfer. BMC Mol. Biol. 8, 56 (2007).

[86] Jeong, J. G., et al.: Electrotransfer of human IL-1Ra into skeletal muscles reduces the incidence of murine collagen-induced arthritis. J. Gene Med. 6, 1125-1133 (2004).

[87] Abruzzese, R. V., et al.: Ligand-dependent regulation of vascular endothelial growth factor and erythropoietin expression by a plasmid-based autoinducible gene-switch system. Mol. Ther. 2, 276-287 (2000).

[88] Bettan, M., et al.: High level protein secretion into blood circulation after electric pulse-mediated gene transfer into skeletal muscle. Mol. Ther. 2, 204-210 (2000).

[89] Aihara, H., Miyazaki, J.: Gene transfer into muscle by electroporation in vivo. Nat. Biotechnol. 16, 867-

870 (1998).

[90] Scheerlinck, J. P., et al.: In vivo electroporation improves immune responses to DNA vaccination in sheep. Vaccine 22, 1820-1825 (2004).

[91] Terada, Y., et al.: Effi cient and ligand-dependent regulated erythropoietin production by naked dna injection and in vivo electroporation. Am. J. Kidney Dis. 38, S50-S53 (2001).

[92] Tone, C. M., Cardoza, D. M., Carpenter, R. H., Draghia-Akli, R.: Long-term effects of plasmid-mediated growth hormone releasing hormone in dogs. Cancer Gene Ther. 11, 389-396 (2004).

[93] Wu, C. J., Lee, S. C., Huang, H. W., Tao, M. H.: In vivo electroporation of skeletal muscles increases the effi cacy of Japanese encephalitis virus DNA vaccine. Vaccine 22, 1457-1464 (2004).

（何继军译）

第 50 章 肌内免疫接种为什么有效？

Emanuela Bartoccioni①

摘要

抗原可持续停留在骨骼肌组织，故骨骼肌一直是免疫接种的部位。除免疫启动期间的被动功能外，最近的研究揭示，肌肉细胞（纤维、卫星细胞和基质细胞）对炎症和免疫应答具有积极作用。在这种情况下，抗原的化学性质（是否进入组织和细胞）以及佐剂的选择有助于确定抗原本身的结局和由此产生的免疫应答。特别是，这适用于使用核酸来局部合成蛋白质抗原的现代疫苗方法。基于这些新概念，根据最近的文献讨论是否能以治疗为目的调控肌肉组织水平的外周耐受机制。

50.1 引言

免疫包括诱导免疫记忆，利用 T 细胞和 B 细胞的特异性反应以及产生针对外来抗原的中和抗体预防随后发生病原体自然感染。通常，蛋白质抗原与佐剂混合，然后注射到肌肉中。触发先天免疫机制必需使用佐剂，这是决定适应性免疫应答结果的初始事件[1,2]。

免疫应答始终始于炎症反应，随后经单核细胞或巨噬细胞、树突状细胞（DC）和 B 细胞和 T 细胞的参与演变成抗原特异性反应。20 世纪 90 年代早期，人们建立了使用肌内注射编码蛋白质的非复制细菌质粒的新型基因治疗技术。这种新方法揭示，体内肌肉中以重组质粒表达的内源蛋白质刺激了免疫应答。事实上，肌肉组织可高水平表达蛋白质，同时重组 DNA 的直接肌内转移具有很多优势，例如对蛋白质抗原的特异性免疫应答、质粒的佐剂样作用、成本低和安全性高[3,4]。

50.2 骨骼肌

骨骼肌是人体最丰富的组织。骨骼肌由肌纤维组成，是一类大型合胞体多核细胞，是终末分化的永久细胞。成熟的肌纤维起源于单核卫星干细胞融合，可增殖以进行再生和修复受损组织：肌纤维受损后，卫星细胞成为活化的成肌细胞，增殖并融合成原始的多核细胞，然后分化为成熟的肌肉纤维[5,6]。这些活化的卫星细胞可在体外培养扩增，所得的成肌细胞可在肌管中融合，肌管是分化的成熟肌细胞的体外对应物。肌肉组织也有几种驻留巨噬细胞，在肌内免疫接种之前和之后，小鼠骨骼肌中存在的源自单核细胞和传统的树突状细胞最近得到了鉴定[7]。

50.3 肌肉炎症反应的触发

在先天免疫反应期间，炎症刺激是引发复杂细胞反应网络的触发因素，其物理、化学或生

① E. Bartoccioni，PhD（卡托利卡大学病理学研究所检验医学系，意大利罗马，E-mail：ebartoc@rm.unicatt.it）

物学性质决定了后续反应的质量。实际上，虽然任何性质的刺激都可引起先天免疫应答，但只有生物因子才能诱导其向适应性反应转化。微生物蛋白质、脂类、碳水化合物和核酸含有统称为PAMP（病原体相关分子模式）的共同结构模式，这些物质与细胞表面或响应细胞内体膜上存在的先天免疫受体（Toll 样受体，TLR）结合[8]。TLR 家族包括10多个成员，具有不同的配体特异性，细胞类型不同，表达也不同[9]。

虽然生物制剂中天然存在PAMP，但免疫过程需将具有相似功能的佐剂添加到纯化的蛋白质抗原中，以产生完全的适应性应答；这是激活树突状细胞所必需的，树突状细胞能将吸收摄入的抗原转运到引流淋巴结（LN）。这些佐剂可用于影响特定反应的大小和类型，以产生针对各种特定病原体最有效的免疫力。除传统的佐剂（如铝盐或MF59）外，现正使用天然以及合成配体定义 Toll 样受体，其中有许多物质现处于临床或临床前阶段的后期[1]

在过去几年中，日益增多的证据表明，肌肉细胞可对数种这样的分子发生响应，通过表达膜分子和细胞因子积极参与先天和适应性免疫反应的若干步骤，尤其是刺激 TLR2 和 TLR4 可诱导 CCL2 和 CXCL1 趋化因子的 mRNA。C2C12 小鼠成肌细胞[10]。类似地，刺激培养的人成肌细胞中的 TLR3 可激活转录因子 NF-κB，诱发 IL-8 分泌，而未甲基化的 CpG 二核苷酸与 TLR9 结合可触发 IL-12、TNFα 和 IFNα 释放，这是细胞免疫应答中的关键步骤[11]。主要 Toll 样受体的特征及其在肌肉中的表达总结如表50.1所示。

表50.1 几种 Toll 样受体（TLR）、天然配体、合成激动剂以及在肌肉中的亚细胞位置

	亚细胞位置	天然配体	合成激动剂	表达于肌肉细胞	参考文献
TLR2	质膜脂蛋白/脂肽	脂蛋白/脂肽肽多糖	BPPcysMPEG Pam3CSK4	是（不适用于成肌细胞）	[12-14]
TLR3	内体	dsRNA	Poly-IC, poly-ICLC, poly-IC_{12}-U	是	[15]
TLR4	质膜	脂多糖	膜磷脂	是	[14, 16]
TLR5	质膜	鞭毛蛋白	鞭毛蛋白抗原融合蛋白	是（不适用于成肌细胞）	[14]
TLR7/TLR8	内体	ssRNA	咪喹莫特、瑞喹莫德	?	[17]
TLR9	内体	细菌或病毒DNA	CpG 寡核苷酸	是	[14]

进行肌内免疫接种时，尽管针头导致的损伤本身是炎症的第一触发因素，但只有存在能够结合 Toll 样受体的外来生物分子才能启动适应性应答。进行 DNA 疫苗接种时，来自质粒骨架含有 CpG 二核苷酸等基序的细菌 DNA 可通过 TLR9 直接激活旁邻细胞。

直到最近，DNA 疫苗技术才因免疫原性不足受到影响；最近开发的新型基因传递技术，例如电转移，可在施加电场时增大传递和转基因表达，显著增强了免疫反应，这可能是电场导致肌肉组织产生损伤和炎症造成的[4,18]。

50.4 肌肉的适应性免疫应答：具体作用是什么？

适应性应答有两种，体液免疫和细胞介导的免疫，这两种免疫由免疫系统的不同组成部分产生，针对不同类型的微生物。

细胞外微生物（如进入细胞前的细菌和病毒）及其毒素（其产生的对宿主有毒分子）由 B 细胞分泌的抗体中和，细胞内微生物（如细胞内细菌和病毒）需更复杂的防御策略。事实上，细胞内微生物在宿主细胞和吞噬细胞中存活并增殖，循环抗体无法接近；必须用细胞介导的免疫（TH1 和 T CTL 细胞）破坏驻留在吞噬细胞中的微生物或杀死感染的细胞，才能消除感染。

用蛋白质抗原进行免疫接种主要诱导体液应答。注射的质粒 DNA 可诱导细胞外和（或）细

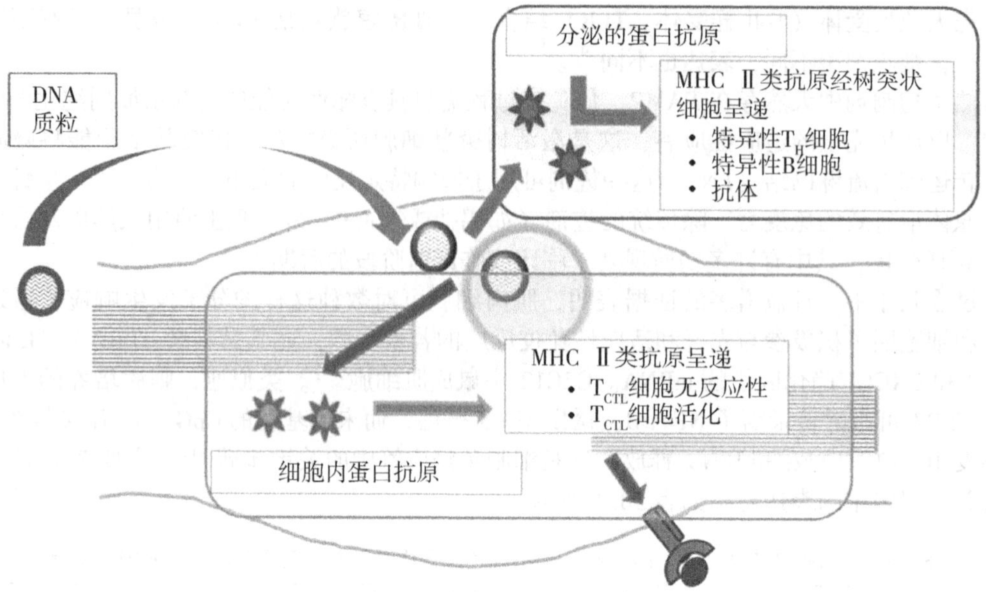

图 50.1　质粒 DNA 编码抗原在肌纤维中的表达

描绘了肌肉细胞中 DNA 驱动的抗原表达的不同潜在结果。分泌蛋白（即含有信号肽的蛋白质）释放在细胞外空间，成为可溶性抗原；可被树突状细胞摄入，然后经 MHC Ⅱ途径递呈，激活 $CD4^+T$ 细胞，使 B 细胞产生抗体。非分泌蛋白（红色途径）进入 MHC Ⅰ途径，基于由炎性微环境诱导的共刺激分子的同时表达，暴露这些蛋白表面可诱发 $CD8^+T$ 细胞无反应性或活化。注意，也可经 MHC Ⅰ 途径递呈分泌蛋白（未显示）

胞内抗原表达（图 50.1 和图 50.2），故 DNA 疫苗接种可刺激两种应答。注射的质粒 DNA 可进入（专业术语为“转染”）肌肉细胞无论哪种分泌细胞分泌蛋白质抗原，都可诱导正常的体液应答（类似于细胞外微生物）。特别是，转染的 APC 可通过两种不同途径将抗原递呈给 CTL 细胞；一种是以内源性蛋白的形式直接通过 MHCI 类分子递呈，另一种是通过所谓的交叉启动机制，即含有该蛋白的肌内细胞片段被 APC 摄取并通过 MHCI 类分子递呈给 T CTL 细胞（图 50.3）[19]。最近描述的另一种形式可能是由将旁邻细胞中 MHC Ⅰ-肽复合物直接转移至树突状细胞表面（“交叉敷料”）介导的[20]。捕获的抗原也经传统的 MHC Ⅱ类依赖性途径递呈给 T 辅助（T_H）淋巴细胞。迁移至引流淋巴结的树突状细胞是这些专业的抗原呈递细胞，以实现适应性免疫过程。

让我们仔细研究树突状细胞如何在适应性免疫中发挥作用。

50.5　用作专职抗原呈递细胞的树突状细胞

当外来抗原经 TLR 激活树突状细胞时，针对这些抗原适应性的免疫应答启动。启动的免疫应答类型取决于树突状细胞的类型，以及所接收的特定先天免疫信号。树突状细胞通过细胞因子和膜 B7 分子调节 $CD4^+T$ 辅助（T_H）细胞分化。这些分子是具有不同特征的家族。B7-1 和 B7-2 均激活或抑制 $CD4^+T$ 细胞，这取决于 $CD4^+T$ 细胞表达哪种配体：T 细胞上存在 CD28 可转导激活信号，CTLA-4 诱导 T 细胞抑制[21]。诱导型共刺激分子配体（ICOSL 也称为 B7-h、B7-H2）不仅刺激效应器或记忆 T 细胞应答，还刺激产生调节性 T（Treg）细胞[22,23]。

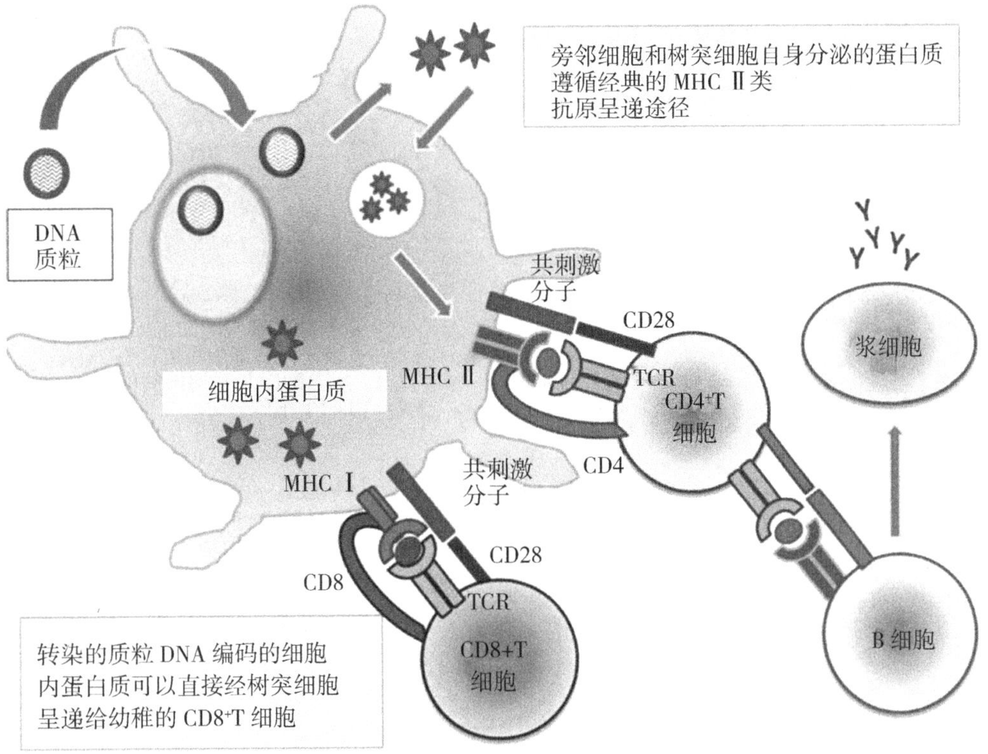

图 50.2　专业抗原呈递细胞在 DNA 疫苗接种中的核心作用

注射的 DNA 也靶向于抗原呈递细胞（树突状细胞），后者表达 DNA 编码的抗原。树突状细胞和周围细胞（肌肉纤维、基质细胞）分泌的蛋白质被摄入，然后经 MHC Ⅱ途径递呈给 $CD4^+$ T 细胞进行暴露。细胞内蛋白质抗原（“假感染”）经 MHC Ⅰ 途径递呈给幼稚 $CD8^+$ T 淋巴细胞。成功启动抗原特异性 $CD4^+$和 $CD8^+$ T 细胞需存在专业抗原呈递细胞表面表达的共刺激分子（B 7-1 和 B 7-2），且由 T 细胞受体共同受体（分别为 TCR 和 CD28）共同连接 MHC 分子。抗原呈递细胞和 T 淋巴细胞之间的这种复杂且高度专业化的多分子接触类比于神经元细胞之间的连接，故命名为“免疫突触”。还描绘了 B 细胞和 T 细胞之间的相互作用。B 细胞表达 MHC Ⅱ和共刺激分子，可将分泌的抗原（经 BCR 识别和捕获后）递呈给 T 细胞。与 T 辅助淋巴细胞接触可完全启动产生抗体的 B 细胞（T 细胞依赖性应答）。

已充分确定，T_H 细胞可响应于细胞内微生物，分化成产生记忆性 $CD8^+T$ 细胞所需的 T_{H1} 细胞，或响应于细胞外微生物分化成 T_{H2}细胞。这种不同的分化是树突状细胞产生的不同细胞因子组的一项功能，其反映从病原体接收的 TLR 介导的不同先天信号。

在稳态条件下未成熟的树突状细胞存在于组织中，TLR 激动剂刺激这些细胞成熟，然后促进其向引流淋巴结迁移。

未成熟和成熟的树突状细胞的特征在于膜分子不同，如 MHC Ⅱ类和 B7 家族分子和趋化因子受体。有两个主要的树突状细胞子集：骨髓（或传统或经典）树突状细胞和浆细胞样树突状细胞，这些细胞在小鼠和人体中并不完全重叠[24]。最近，人 CD141+树突状细胞已鉴定为小鼠 $CD8^+$树突状细胞的人体对应物；这些细胞能够捕获外源性抗原，然后经 MHC Ⅰ类分子（交叉递呈）递呈给 $CD8^+T$ CTL 细胞[25]。

最近一项有趣的研究表明，小鼠肌内免疫接种期间存在树突状细胞。结果表明，大量驻留的间质性树突状细胞能够捕获肌肉中的抗原，然后迁移至引流淋巴结，有效激活未成熟 T 细胞（图 50.4），在向无菌的铝佐剂化的抗原中添加脂多糖后，血液单核细胞产生的树突状细胞在很大程度上实现了这些能力[7]。

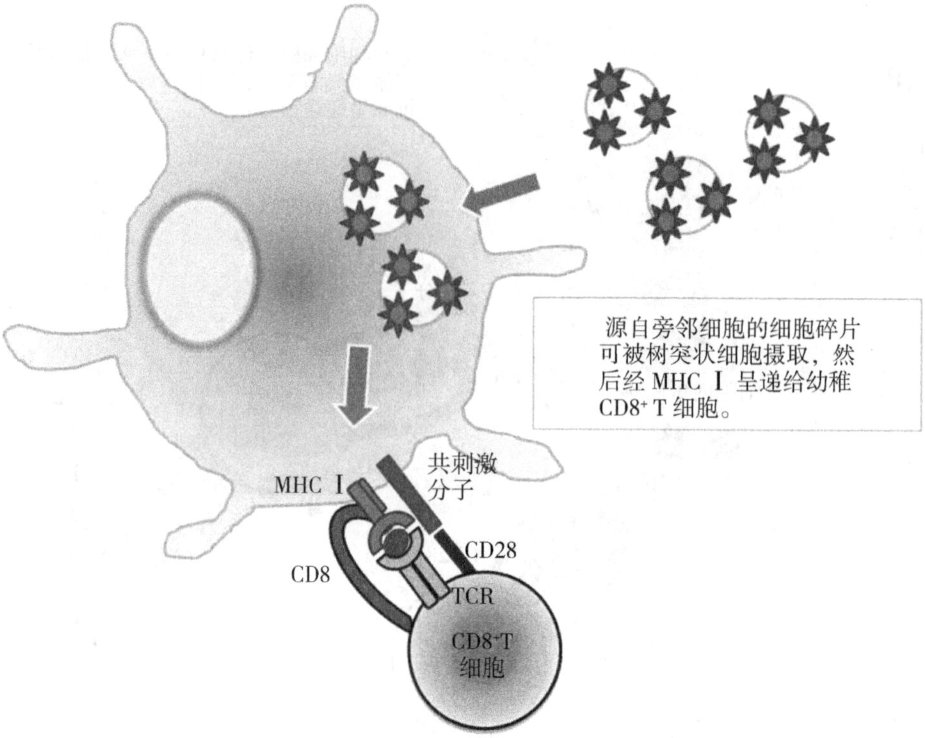

图 50.3　由树突状细胞交叉递呈

除利用内源性抗原（包括转染的 DNA 编码的抗原）以及作为细胞片段或微泡或从周围细胞摄取的外源性抗原外，树突状细胞还可启动 CD8+细胞。这种抗原递呈方式称为交叉递呈，涉及抗原的细胞内加工。

50.6　用作非专职抗原呈递细胞的肌肉细胞

即使转染的树突状细胞在 DNA 疫苗介导的免疫中起主导作用，也只是总诱导反应的一部分，局部非专职抗原呈递细胞在适应性免疫应答中具有重要作用。

肌肉细胞能积极参与免疫诱导，表现为非专职的抗原呈递细胞。肌细胞表达细胞因子和 PAMP 的受体，通过分泌细胞因子和趋化因子，表达黏附分子，使其能够对炎性环境做出应答[26]。特别是，抗原呈递细胞功能所必需的一些膜蛋白，如 MHC Ⅰ类和Ⅱ类分子以及共刺激分子，已在数个试验系统观察到[27-29]。重要的是，在质粒转染后，肌细胞表达抗原递呈所必需的几种分子，如 MHC Ⅰ类和如 BB-1 和 B7-1 等共刺激分子，将免疫原性 DNA 编码抗原递呈给 T CTL 细胞[30]。

以肌肉纤维作为抗原呈递细胞的思路面临许多重要问题。

首先，表达新抗原蛋白的肌细胞应像感染的“不同”细胞一样出现在免疫系统中，且被其诱导的相同细胞毒性特异性 T 细胞破坏；肌肉纤维是大型的合胞体有丝分裂后细胞，不太可能被淋巴细胞免疫杀死，这是因为有证据表明，DNA 转染的抗原在肌肉组织中广泛扩散，9 个月后仍然表达，分泌细胞的长期存活得到了确认[18]。

其次，虽然肌细胞似乎持久产生抗原，但伴随 DNA 转移的炎症活性消失，预计 DNA 转染所赋予的抗原呈递细胞功能实际上是短暂的（见上文）。但各种模型系统显示，记忆 T 细胞和 B 细胞都可以在没有抗原的情况下存活，即使在没有活性抗原递呈的情况下，也能确保疫苗接种的有效性[31-34]。

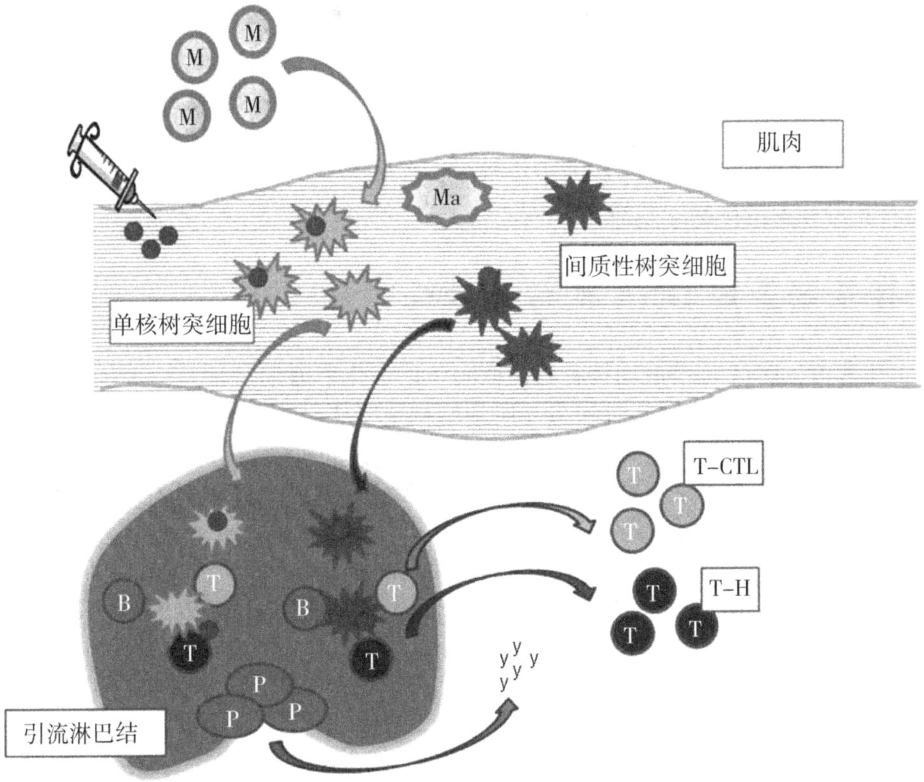

图 50.4 参与肌肉对所注射抗原应答的不同类型的树突状细胞肌肉含间质性或驻留树突状细胞（Int-DC），监测组织的免疫状态，并将注射的抗原转运至淋巴结（引流淋巴结），以激活特异性B细胞和T细胞应答。值得注意的是，驻留的树突状细胞的动员还需存在感染性刺激物（脂多糖是否为佐剂，参见文字）；另一类树突状细胞（Mo-DC）区别于炎症反应（Mo-DC）的情况下迁移至肌肉的单核细胞（M），可捕获抗原，然后将其运输到淋巴结，履行抗原呈递细胞的职责。这些不同类型的肌肉树突状细胞刚开始进行分子表征。描绘了分泌抗体（Y）的B细胞衍生的浆细胞（P）。巨噬细胞（Ma）可源自单核细胞或驻留

再次，抗原递呈肌细胞可以诱导耐受性，而不会诱导免疫力。

50.7 耐受性：疫苗的另一面

因此，似乎可以很容易地操纵骨骼肌，实现多种免疫相关分子的表达，将骨骼肌细胞作为靶细胞用于制备DNA疫苗接种将颇具前景。然而，免疫系统能够在该组织水平上诱导耐受性，这一点可以从针对肌肉的自身免疫性疾病的罕见性中看出。了解免疫耐受性的诱导和维持机制以及失能（对抗原无反应的状态），对解决疫苗接种的有效性至关重要。

一般来说，在非感染条件下，新抗原的表达可以诱导耐受性，而不会诱导免疫力。抗原在转基因动物中的表达可能就是这样。

最近研究表明，对肌肉自身抗原的免疫耐受性涉及自身反应外周 $CD8^+$T CTL 细胞的缺失[35]。在这个模型中，转基因小鼠在骨骼肌中表达了一种膜结合型卵清蛋白（OVA）新抗原（SM-OVA）。在这一类动物中没有自身免疫的迹象，但是用卵清蛋白免疫诱导了 $CD4^+$相关反应和抗卵清蛋白IgG抗体的产生，这表明小鼠卵清蛋白中存在卵清蛋白特异性 $CD4^+$细胞，并且对卵清蛋白没有固有的耐受性，除非在适当的炎症环境中出现，否则不需要考虑抗原。

相反地，当采用编码卵清蛋白水疱性口炎病毒（VSV-OVA）免疫引发卵清蛋白特异性 T CTL 细胞反应时，没有观察到细胞毒性；此外，采用过继转移的卵清蛋白特异性 $CD8^+$T CTL 细胞在 SM-OVA 小鼠外周有选择性缺失。这一过程与调节性 T 细胞无关，可能既涉及 APC 表面缺乏共刺激分子时的部分 T 细胞活化，也涉及非感染条件下缺乏细胞因子的存在。有趣的是，当抗原被肌纤维永久表达时，外周耐受性要高于免疫反应。

体内转染在多个方面模拟了转基因小鼠细胞内存在抗原，因此，肌肉诱导的耐受性代表了与 DNA 疫苗接种相关的生物学现象。

50.8 存在完美的 DNA 疫苗配方吗?

基于上述考虑，DNA 疫苗成功接种可能取决于以下几个因素。

（1）质粒 DNA 不仅应该转染肌细胞（肌细胞构成了大量对初始反应有用的组织），还应该转染树突状状细胞（树突状状细胞在炎性环境下启动复杂的机制），实现整体效应和记忆反应。

（2）质粒骨架中应含有未甲基化的 CpG 基序，结合细胞内 TLR9 作为佐剂发挥作用，使 T_H 对 T_{H1} 细胞的反应分化。

（3）还应使用佐剂，保证激活树突状状细胞所需的炎性触发。

（4）与免疫反应水平相关的表达抗原数量应足够高，表达时间应足以建立免疫记忆。建立记忆后，可能就不再需要抗原的存在（见上文）。

结论

关于肌肉免疫生物学的新知识清晰地表明，这种组织的作用远远不是简单的机械支持，而是积极地促进疫苗性免疫的建立。当涉及基于 DNA 的疫苗时尤是如此，因为肌肉纤维具有显著的蛋白质合成和分泌功能，并且还具有此前未知的与炎性细胞和细胞因子交互对话的功能。相应地，基于 DNA 的肌内注射疫苗作为规避传统疫苗策略（尤其是针对细胞内病原体）缺陷的一种途径，已经引起业内的极大期望，同时，正在进行该方向的第一次临床试验。

另一个会在很大程度上被忽视的可能性是利用肌内免疫特性操纵免疫耐受性。从转基因动物中获得的初步实验证据，加之肌肉导向自身免疫罕见性的临床经验，都支持了这一观点。

我们是否要使用肌肉诱导对 DNA 编码蛋白质的耐受性？如果使用，应该如何避免在这种情况下出现不必要的免疫反应？肌肉是对抗自身免疫的方法吗？

对于那些对未来既没有预测、也不相信免疫学教条理论的人来说，未来可能会带来许多惊喜。请拭目以待，保持肌肉健康。

参考文献

[1] Coffman，R. L.，Sher，A.，Seder，R. A.：Vaccine adju-vants：putting innate immunity to work. Immunity 33，492-503（2010）.

[2] Desmet，C. J.，Ishii，K. J.：Nucleic acid sensing at the interface between innate and adaptive immunity in vaccination. Nat. Rev. Immunol. 12，479-491（2012）.

[3] Rice，J.，Ottensmeier，C. H.，Stevenson，F. K.：DNA vaccines：precision tools for activating effective immunity against cancer. Nat. Rev. Cancer 8，108-120（2008）.

[4] Rochard，A.，Scherman，D.，Bigey，P.：Genetic immu-nization with plasmid DNA mediated by electrotransfer. Hum. Gene Ther. 22，789-798（2011）.

[5] Wiendl，H.，Hohlfeld，R.，Kieseier，B. C.：Immunobiology of muscle：advances in understanding an immunological microenvironment. Trends Immunol. 26，373-380（2005）.

[6] Bischoff, R., Franzini-Armstrong, C.: Satellite and stem cells in muscle regeneration. In: Engel, A., Franzini-Armstrong, C. (eds.) Myology, pp. 66-86. McGraw-Hill, New York (2004).
[7] Langlet, C., Tamoutounour, S., Henri, S., Luche, H., Ardouin, L., Grégoire, C., Malissen, B., Guilliams, M.: CD64 expression distinguishes monocyte-derived and conventional dendritic cells and reveals their distinct role during intramuscular immunization. J. Immunol. 188, 1751-1760 (2012).
[8] Janeway Jr., C. A.: Approaching the asymptote? Evolution and revolution in immunology. Cold Spring Harb. Symp. Quant. Biol. 54, 1-13 (1989).
[9] Zarember, K. A., Godowski, P. J.: Tissue expression of human Toll-like receptors and differential regulation of Toll-like receptor mRNAs in leukocytes in response to microbes, their products, and cytokines. J. Immunol. 168, 554-561 (2002).
[10] Gurunathan, S., Wu, C. Y., Freidag, B. L., Seder, R. A.: DNA vaccines: a key for inducing long-term cellular immunity. Curr. Opin. Immunol. 12, 442-447 (2000).
[11] Krieg, A. M.: CpG motifs in bacterial DNA and their immune effects. Annu. Rev. Immunol. 20, 709-760 (2002).
[12] Lombardi, V., Van Overtvelt, L., Horiot, S., Moussu, H., Chabre, H., Louise, A., Balazuc, A. M., Mascarell, L., Moingeon, P.: Toll-like receptor 2 agonist Pam3CSK4 enhances the induction of antigen-specifi c tolerance via the sublingual route. Clin. Exp. Allergy 38, 1819-1829 (2008).
[13] Prajeeth, C. K., Jirmo, A. C., Krishnaswamy, J. K., Ebensen, T., Guzman, C. A., Weiss, S., Constabel, H., Schmidt, R. E., Behrens, G. M.: The synthetic TLR2 agonist BPPcysMPEG leads to effi cient cross-priming against co-administered and linked antigens. Eur. J. Immunol. 40, 1272-1283 (2010).
[14] Boyd, J. H., Divangahi, M., Yahiaoui, L., Gvozdic, D., Qureshi, S., Petrof, B. J.: Toll-like receptors differentially regulate CC and CXC chemokines in skeletal muscle via NF-kappaB and calcineurin. Infect. Immun. 74, 6829-6838 (2006).
[15] Schreiner, B., Voss, J., Wischhusen, J., Dombrowski, Y., Steinle, A., Lochmuller, H., Dalakas, M., Melms, A., Wiendl, H.: Expression of toll-like receptors by human muscle cells in vitro and in vivo: TLR3 is highly expressed in infl ammatory and HIV myopathies, mediates IL-8 release and up-regulation of NKG2D-ligands. FASEB J. 20, 118-120 (2006).
[16] Casella, C. R., Mitchell, T. C.: Putting endotoxin to work for us: monophosphoryl lipid A as a safe and effective vaccine adjuvant. Cell. Mol. Life Sci. 65, 3231-3240 (2008).
[17] Gorden, K. B., Gorski, K. S., Gibson, S. J., Kedl, R. M., Kieper, W. C., Qiu, X., Tomai, M. A., Alkan, S. S., Vasilakos, J. P.: Synthetic TLR agonists reveal functional differences between human TLR7 and TLR8. J. Immunol. 174, 1259-1268 (2005).
[18] Mir, L. M., Bureau, M. F., Gehl, J., Rangara, R., Rouy, D., Caillaud, J. M., Delaere, P., Branellec, D., Schwartz, B., Scherman, D.: High-effi ciency gene transfer into skeletal muscle mediated by electric pulses. Proc. Natl. Acad. Sci. U. S. A. 96, 4262-4267 (1999).
[19] Joffre, O. P., Segura, E., Savina, A., Amigorena, S.: Cross-presentation by dendritic cells. Nat. Rev. Immunol. 12, 557-569 (2012).
[20] Wakim, L. M., Bevan, M. J.: Cross-dressed dendritic cells drive memory $CD8^{+}$T-cell activation after viral infection. Nature 471, 629-632 (2011).
[21] Krummel, M. F., Allison, J. P.: CD28 and CTLA-4 have opposing effects on the response of T cells to stimulation. J. Exp. Med. 182, 459-465 (1995).
[22] Sharpe, A. H., Freeman, G. J.: The B7-CD28 super-family. Nat. Rev. Immunol. 2, 116-126 (2002).
[23] I to, T., Yang, M., Wang, Y. H., Lande, R., Gregorio, J., Perng, O. A., Qin, X. F., Liu, Y. J., Gilliet, M.: Plasmacytoid dendritic cells prime IL-10-producing T regulatory cells by inducible costimulator ligand. J. Exp. Med. 204, 105-115 (2007).
[24] Palucka, K., Banchereau, J., Mellman, I.: Designing vaccines based on biology of human dendritic cell subsets. Immunity 33, 464-478 (2010).
[25] Haniffa, M., Shin, A., Bigley, V., McGovern, N., Teo, P., See, P., Wasan, P. S., Wang, X. N., Malinarich, F., Malleret, B., Larbi, A., Tan, P., Zhao, H., Poidinger, M., Pagan, S.,

Cookson, S., Dickinson, R., Dimmick, I., Jarrett, R. F., Renia, L., Tam, J., Song, C., Connolly, J., Chan, J. K., Gehring, A., Bertoletti, A., Collin, M., Ginhoux, F.: Human tissues contain CD141hi cross-presenting dendritic cells with functional homology to mouse CD103$^+$nonlymphoid dendritic cells. Immunity 37, 60-73 (2012).

[26] Marino, M., Scuderi, F., Provenzano, C., Bartoccioni, E.: Skeletal muscle cells: from local infl ammatory response to active immunity. Gene Ther. 18, 109-116 (2010).

[27] Goebels, N., Michaelis, D., Wekerle, H., Hohlfeld, R.: Human myoblasts as antigen presenting cells. J. Immunol. 149, 661-667 (1992).

[28] Behrens, L., Kerschensteiner, M., Misgeld, T., Goebels, N., Wekerle, H., Hohlfeld, R.: Human muscle cells express a functional costimulatory molecule distinct from B7. 1 (CD80) and B7. 2 (CD86) in vitro and in infl ammatory lesions. J. Immunol. 161, 5943-5951 (1998).

[29] Curnow, J., Corlett, L., Willcox, N., Vincent, A.: Presentation by myoblasts of an epitope from endogenous acetylcholine receptor indicates a potential role in the spreading of the immune response. J. Neuroimmunol. 115, 127-134 (2001).

[30] Shirota, H., Petrenko, L., Hong, C., Klinman, D. M.: Potential of transfected muscle cells to contribute to DNA vaccine immunogenicity. J. Immunol. 179, 329-336 (2007).

[31] Hou, S., Hyland, L., Ryan, K. W., Portner, A., Doherty, P. C.: Virus-specifi c CD8$^+$ T-cell memory determined by clonal burst size. Nature 369, 652-654 (1994).

[32] Lau, L. L., Jamieson, B. D., Somasundaram, T., Ahmed, R.: Cytotoxic T-cell memory without antigen. Nature 369, 648-652 (1994).

[33] Sallusto, F., Lanzavecchia, A., Araki, K., Ahmed, R.: From vaccines to memory and back. Immunity 33, 451-463 (2010).

[34] Zielinski, C. E., Corti, D., Mele, F., Pinto, D., Lanzavecchia, A., Sallusto, F.: Dissecting the human immunologic memory for pathogens. Immunol. Rev. 240, 40-51 (2011).

[35] Franck, E., Bonneau, C., Jean, L., Henry, J. P., Lacoume, Y., Salvetti, A., Boyer, O., Adriouch, S.: Immunological tolerance to muscle autoantigens involves peripheral deletion of autoreactive CD8$^+$T cells. PLoS One 7, e36444 (2012).

（何继军译）

第八部分

疫苗的专利申请、制造和注册

概　述

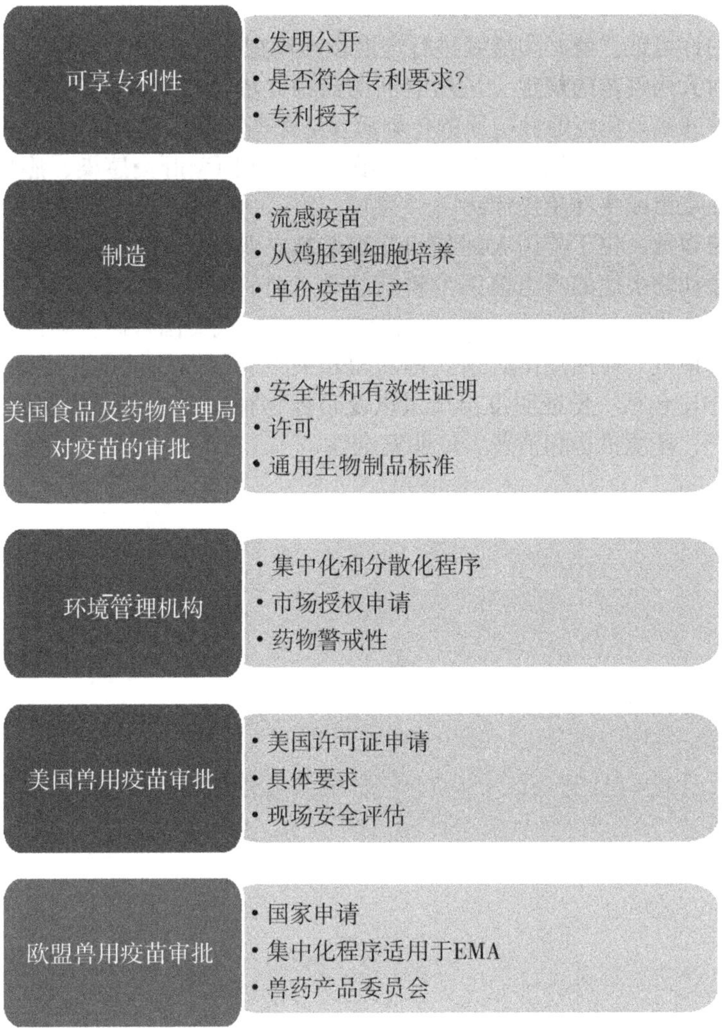

从可行性研究到产品发布是一条耗时耗力之路。一种新型疫苗的成功是药物非临床研究质量管理规范实验室、专利和市场营销部门精心策划、一步步实现的结果，而最后也是最重要的一点，是申请上市许可。概念验证要表明候选疫苗在原则上有效。根据国际药物临床试验质量管理规范指南进行成功的临床研究，是获得许可的基础。药品生产质量管理规范控制的制造过程（包括定期设备检查）是获得许可证的一部分。疫苗安全性通过药物安全检查在上市后进行控制。

可享专利性　美国和中国占据着第一（最早优先权）和第二（随后系列产品）的专利申请主导地位，声明疫苗的活性成分可预防传染病，欧洲和澳大利亚在疫苗领域第二专利申请中逐渐占据着主导地位。使用已知或新型疫苗组合物详述医学指征的权利要求，必须具有创新性，并充分披露。根据2000年《欧洲专利公约》第53（c）条规定的可享专利性排除的各种公共政策，《欧洲专利公约》对申请人申请确定的治疗、手术或诊断方法的疫苗作用做出了限制。

制造　开发和生产流感疫苗的基础是复杂的制造过程，从选择和开发最佳候选疫苗病毒开始，需要与监管机构和卫生保健官员进行各种动态互动。多数流感疫苗的生产以传统鸡胚技术为基础，30多年来，这种技术一直用于季节性疫苗的生产。相对基于鸡胚的制造方法，新型细胞培养技术具有多种优势，并且很可能成为对当前基于鸡胚技术的补充。

美国食品和药物管理局对人用疫苗的规定　疫苗许可证以安全性和有效性为基础，证明许可证持有人在指定和商定的范围内具有以一致的方式生产产品的能力。获得许可证后，只要生产商持有该产品的许可证，就必须继续进行定期设施检查等疫苗和生产活动监测。

欧洲药品局对人用疫苗的规定　虽然欧洲的药物立法没有正式规定预防措施，但疫苗通常被认为是含有1种或多种免疫原性抗原的传染病预防医药产品。含有1种或多种用于治疗疾病的免疫原性抗原药物，如慢性艾滋病病毒感染、慢性乙型或丙型肝炎感染、癌症或阿尔茨海默病，通常被称为治疗性疫苗或主动免疫疗法。

美国兽用疫苗审批　位于美国衣阿华州艾姆斯的农业部兽医生物制品中心（CVB）对用于诊断、预防和治疗动物疾病的本土兽医生物制品具有监管管辖权。在美国销售的所有兽医疫苗均须具有美国兽医生物制品许可证（在许可机构内生产）或美国兽医生物许可证。

欧洲兽用疫苗审批　兽用疫苗必须获得欧盟相关主管部门的批准。必须提交一份市场授权（MA）申请，其中应包括一份证明疫苗质量以及目标物种安全性和有效性的档案。需要注意的是，还必须向用户、环境机构和消费者证明安全性。

第 51 章　疫苗的可享专利性：实践角度

Stacey J. Farmer 和 Martin Grund①

摘要

本章对当前专利授予和保护机制的管理结构、可享专利性方法及与疫苗创新相关的显著法律趋势进行了一次非详尽调查。开发具有战略意义的全球专利组合，需要在经过深思熟虑的战略之间采取平衡的方法保障实际或预期市场地位，从而保持可靠的收入来源，并在全球专利体系和格局不断演变的情况下谨慎选择专利申请的发明。稳健的专利战略通常以核心技术为基础，着眼于从这些最初发明捕捉次要标的。对于疫苗的开发，首创专利申请可能包括提供新型结构或治疗作用的各种成分，如抗原（任何核酸序列、蛋白质，以及相应的表达技术）、佐剂和赋形剂、给药平台、给药机制以及单独或联合用药的目标患者群体。次要后续专利可以通过对现有技术教学的改善和/或进一步改进扩展这些基础技术的专有权。我们希望本章内容能够让读者认识并理解开发新型专利权作为一项重要商业工具的影响，并能够识别潜在机会，发现开发和/或利用疫苗技术的市场空白，并认识到现有专利权可能对计划或正在进行的研究措施构成障碍。

本章阐述了专利方案、趋势和具有里程碑意义的司法制度，这可能会影响基于疫苗的创新。我们的分析将主要针对欧洲和美国体系，即构成可申请专利主题的内容，通常用于保护面向疫苗发明的索赔策略。我们希望这项分析有助于开展获得知识产权的活动，帮助读者认识并理解现有专利权的影响（这些专利权可能对拟定或正在进行的研究措施构成障碍），有助于了解对现有权利授权可能需要的时间，并有助于发现疫苗技术可能存在的新机会或市场空白的潜在领域。

51.1　引言

生物技术产业依赖于强大的专利系统，至少可以为不断飙升的研发成本获得经费奖励和投资，特别是所有疫苗相关产品必须经历的漫长且昂贵的临床试验监管流程。此外，公司和组织可以依赖这些无形资产激发创新机会和合作关系，或者将其出售（无论是直接出售还是通过许可协议出售）。

毋庸置疑，当前的疫苗接种计划可以为易感个体提供免疫，避免疫苗可预防性疾病的威胁或发病，每年可有效防止全球数百万人的死亡。2012 年，世界知识产权组织（WIPO）公布了法国机构“法国创新科学与转移”（FIST）的一项大规模项目成果，该项目非常详细地剖析和分析了疫苗相关发明的全球专利历史进程。报告题为《选定传染疾病疫苗专利前景报告》（“报告”）[1]给出了一种颇具煽动性的全球专利和创新活动模式，涵盖了疫苗领域的各种技术，并对巴西、印度和中国等国家的此类活动进行了深入分析。报告显示，美国和中国在专利申请中占据着第一（最早优先权）和第二（随后系列产品）的主导地位，声明疫苗的活性成分可预防传染病，欧洲和澳大利亚在疫苗领域的次级专利申请中逐渐占据着主导地位[2]。这种活跃的融资活动模式表明，对于疫苗领域的企业来说，鉴于与研究、监管要求和技术商业化相关的高昂成

---

① S. J. Farmer, PhD（✉）· M. Grund, PhD（格伦德知识产权集团，德国慕尼黑，E-mail：farmer@grundipg. com）

本，奖励创新势在必行。这就是为什么协调可靠的专利保护方法（最好是在全球层面上）具有重大意义。

很多欧盟指令、相关欧共体条例、其他准则和解释性说明试图协调和规范生物技术活动，例如使用致弱和非致弱重组病毒载体活疫苗，在人类和动物两方面预防和治疗各种传染疾病。这些司法机构还负责管理疫苗从实验室推向市场的质量、非临床和临床方面的问题[3]。除基于病毒载体的疫苗之外，还存在其他类型的疫苗，包括细菌疫苗、基于DNA的疫苗[4]，包括质粒DNA疫苗、重组蛋白疫苗、亚单位和类毒素疫苗、联合疫苗、合成疫苗，以及免疫兽用药物产品的“联合”疫苗，其中任何一种特征都可以是灭活或致弱微生物的悬浮液或微生物的产品或衍生物。除这些疫苗形式之外，还可以加入抗原、抗体、佐剂和赋形剂等其他物质。除制造方法之外，所有这些生物物质、核酸序列、递送平台、表达用细胞系和培养物、治疗靶标和临床指征、相关肽和蛋白质（如抗体）以及对前述任何物质的改良，除受欧盟法律约束之外，也可以构成导致专利产权发明公开的一部分。

一项欧盟指令，特别是第98/44/EC指令，为保护整个欧盟的生物技术发明提供了统一的法律框架，并已在所有欧盟成员国以及《欧洲专利公约》的国家法律中成功实施，使之适用于向欧洲专利局提交的生物技术发明。目前在欧洲市场获得专利保护的主要方式（特别是对疫苗等生物技术的创新）是通过欧洲专利局授予欧洲专利（“EP专利”）。获得欧洲专利批准即视为一系列权利，在相关国家专利当局完成各种手续后，目前可在最多38个《欧洲专利公约》签约国（“《欧洲专利公约》成员国”）进行“验证”。

除意大利和西班牙之外，欧盟25个成员国仍在研究另一种专利实施方式，即提供统一的专利保护。这一新制度旨在为单一申请下的发明提供统一的专利文件，文件将根据欧洲专利局的现行程序进行实质性评估，并接受新“联合专利法院”的管辖，该法院将对与侵权和有效性相关的民事诉讼行使专属管辖权[5]。除了向个人和公司发明人提供行政和司法优势外，鉴于目前与欧洲专利验证程序相关的翻译成本显著降低，还可以实现显著的经济效益。虽然初始协议已经签署，但在单一专利制度产生法律效力之前，仍然必须根据特定协议获得一定数量国家的批准。

自30多年前首个《欧洲专利公约》（1973年《欧洲专利公约》）生效以来，专利业务在全球范围内一直蓬勃发展。《欧洲专利公约》行政理事会注意到，鉴于生物技术等技术的传播、各种国际条约的颁布以及近几年《欧洲专利公约》成员国数量激增，1973年《欧洲专利公约》不断更新，内容日益完善，于是发起了对《欧洲专利公约》的修订工作。修订目的在于使欧洲专利制度与时俱进，同时保持1973年《欧洲专利公约》中所载并经过证明的实质性和程序性专利法的基本原则。

促使人们认识到这种变化的一个主要因素是，欧洲专利法需要与各种立法措施相协调，如新制定和修订的欧盟指令，以及在全球舞台上对知识产权产生重大影响的其他国际协定，包括TRIPS协议（《与贸易有关的知识产权协议》，1994年乌拉圭回合圆桌会议协定制定的贸易相关知识产权）和欧洲专利局于2001年签署的专利法条约。另一个正在推进《欧洲专利公约》更新的情况是，《欧洲专利公约》签约国的数量越来越多（目前为38个），这就需要更简易灵活的欧洲专利授予前程序。

1973年《欧洲专利公约》修订版2000年《欧洲专利公约》是由签约国代表团、知识产权组织和参加2000年11月在欧洲专利局总部慕尼黑举行的外交会议的其他各方之间达成的。于2007年12月13日生效。《欧洲专利公约》修订的一项主要成果包括将《欧洲专利公约》程序的详细内容从条款（只能通过由所有签约国组成的外交会议进行修订）修订为实施条例（“规则”）（仅通过《欧洲专利公约》行政委员会的决议就可以进行快速修订）。在过去几年里，这种特殊变化允许《欧洲专利公约》随时调整组织操作方式，反映了在欧盟层面上制定的专利法律和政策的变化。其中最显著的变化是最近以生物技术为导向的规定变化，如人类胚胎干细胞的专利性，以及根据欧盟指令 98/44/EC[6] 对涉及遗传材料的生物技术发明的保护范围。

在本章中，我们将讨论《欧洲专利公约》条款的实质性和程序性，尤其是对疫苗创新专利的影响。图 51.1 为专利生命周期的顶层原理图。我们还会讨论建立专利组合时的生物技术行业常见专利申请策略。在进行这一分析的同时，我们会简要介绍美国体系下存在的有意义实际差异平行条款和判例法，特别关注了 2013 年 3 月 16 日生效的新专利法制度。该制度至少将对一项发明在美国专利商标局（“美国专利商标局”）的新颖性进行评估。

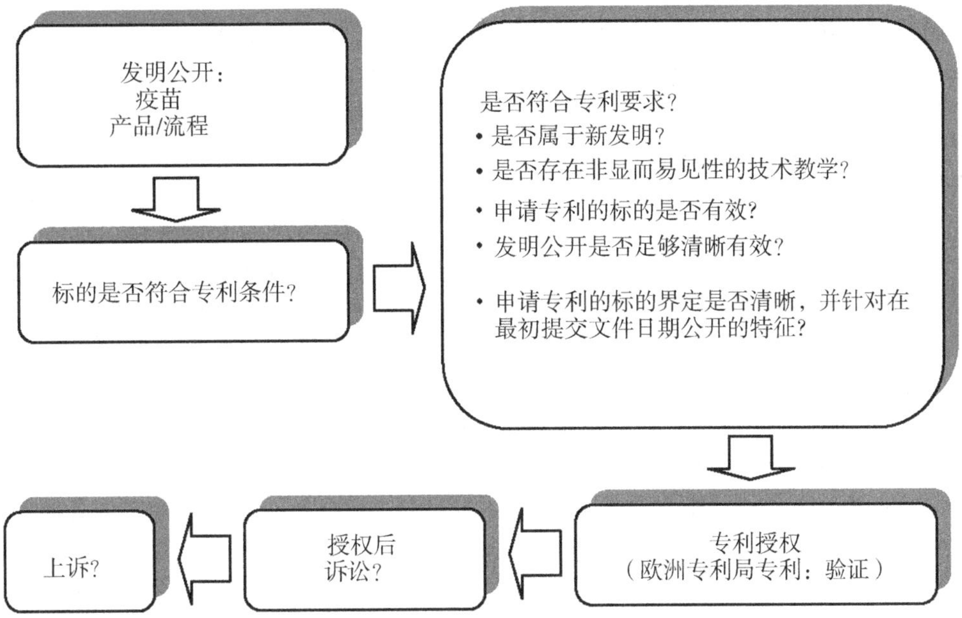

图 51.1　一项发明的申请周期

51.2　可享专利标的包括的例外与除外事项

在进行实质性的可享专利性分析之前，必须积极回答的一个门槛问题是：一项发明是否有资格获得专利保护。这项初步调查是在欧洲根据 2000 年《欧洲专利公约》第 52（1）条的授权进行的，公约规定：“所有技术领域的任何发明，只要是新颖的、涉及发明步骤，并易于在工业中应用的，都应授予欧洲专利。”这种说法与《与贸易有关的知识产权协议》第 27.1 节第一句的说法一致，因此清晰地表明，欧洲专利适用于所有产生技术教学的发明[7]。《欧洲专利公约》将生物技术发明定义为“涉及由生物材料组成，或含有生物材料，或利用生物材料生产加工或使用过程的发明”，应根据欧盟指令 98/44/EC 解释为补充性措施[8]。

然而，仍然有几个明显的例外和专利性除外事项会定期影响生物技术创新，分别见《欧洲专利公约》第 52 和 53 条。例如，欧洲专利公约第 52（2）条排除了包含发现和科学理论的发明性专利[9]。2000 年《欧洲专利公约》第 53（c）条反映了该组织的哲学立场，即在欧洲实践中，治疗方法（包括使用疫苗的方法）被视为“可享专利例外”[10]，因此禁止为任何理解为治疗、手术步骤或诊断特征等的方法授予欧洲专利，这些方法是在动物和人体活体上进行的，甚至可以公平地使用。因此，任何基于疫苗的索赔，如认为属于这种例外，都将被拒绝。有趣的是，与美国的操作方式不同，《欧洲专利公约》并没有因为这些发明缺乏（《欧洲专利公约》第 57 条规定的）工业适用性将这些方法排除在外[11]，因为这些很容易地就可以构成《欧洲专利公约》第 52（1）条所述的普通意义上的技术发明。考虑到公共健康和相关政策问题，《欧洲专利公约》将此类方法视为可享专利性的例外情况。例如，不应该仅仅因为医生为采取特定疗法而必须去协商获得专利许可，继而阻碍对患者的治疗。值得注意的是，2000年《欧洲专利公约》第53（c）条不适用于在任何此类例外方法中使用的产品，特别是物质或成分，以此作为对本条款所涵盖发明妥协的回报。

可申请专利标的，特别是生物技术创新标的，在很多司法管辖区一直处于司法调查的前线。最近，澳大利亚联邦法院判决，根据澳大利亚《专利法》[12]，从自然存在的细胞环境中分离出来的核酸（脱氧核糖核酸/DNA和核糖核酸/RNA）是人类干预的产物，不应作为“制造方式”申请专利。该判决的关键是Myriad基因检测公司争议专利[14]中要求保护的受试核酸成分，即包含乳腺癌基因BRCA1的全部或一部分BRCA1位点或突变BRCA1位点的分离多核苷酸被发现已人为地从细胞环境中清除，并因此“分离”出来。此外，使用这种分离核酸成分的公开体外结果显示，可用于要求确定和/或诊断个体对乳腺癌和/或卵巢癌的易感性。

在美国平行诉讼中，美国最高法院（“USSC”）于2013年6月13日就对应美国专利中描述和主张的Myriad“分离”基因[15]是否构成美国专利法[16]规定的可享专利标的问题作出了判决[17]。值得注意的是，该判决推翻了上诉法院——美国联邦巡回上诉法院（CAFC）的裁决，认为针对“分离”DNA分子成分提出的权利要求不是符合专利条件的自然产物，因为其代表的是自然界中存在的物质成分[18]。美国最高法院拒绝对这些分离出来但自然存在的DNA形式提供专利保护，其科学和司法基础在于，观察结果显示，Myriad没有创造或改变BRCA 1/BRCA2基因编码的遗传信息，也没有改变DNA本身的结构。尽管承认Myriad确实描述了一种“重要且有用的基因”，但简单地“将该基因从其周围的遗传物质中分离出来并不是一种发明行为”，而且“本身并不能满足第101项中的调查[19]。”美国最高法院判决的全部影响目前还不清楚，因为成千上万已发布的美国专利可能会失效——它们的权利要求包含了该判决含义内分离的、自然存在的核酸序列。重要的一点是，美国最高法院确实承认，cDNA序列等人工合成序列可能是符合专利资格的标的，因为它们是合成的、仅有外显子的创造物，在自然界中并不存在[20]。

然而，这一里程碑式的观点并没有直接回应美国联邦巡回上诉法院的审慎判决，仅仅是在“比较”或“分析”DNA序列的方法不具备专利资格，因为此类权利要求没有列举转化步骤，只有非专利性的抽象思维步骤[21]。因此，下级法院对这一方面的判决没有变化，因此对列举此类分析步骤权利要求的适用于标准也没有变化。

根据澳大利亚的判决，大家一定想知道欧洲专利侵权案件中是如何解释“分离”核酸的。英国最高法院在极具争议的孟山都案例[22]中，狭义地解释了“分离”一词，如一项针对“分离DNA序列”的权利要求所述，是指与细胞环境中的其他分子成分发生物理分离的片段。孟山都专利描述并要求保护的是对特定除草剂具有抗药性酶的编码基因，因此，该法院判决，进口含有这种基因的大豆没有侵犯孟山都公司的专利。法院认为，该标的不含分离基因序列，也没有根据孟山都公司专利要求的方式与某些抗体发生反应。欧洲法院同意这一观点，指出这种解释符合欧盟第98/44/ EC指令第9条的规定[23]。

虽然英国对孟山都公司作出的判决并没有在澳大利亚、美国或欧洲专利法中明文规定，但确实强调了在起草包含“分离”核酸的权利要求时，必须要特别注意。明智的做法可能是加入详尽的定义语言，至少要满足“制造方式”的要求或“物质构成”的条件，避免在特定法院中引发对该术语过于狭义的解释。此外，在疫苗范围内，对实际（预期）商品中出现的核酸成分或在疫苗生产过程中可分离的任何新中间物质，可能有必要加入针对核酸成分的权利要求。

我们不能夸大对分离核酸序列可享专利性的前述判决和未判决标的的影响，因为疫苗专利经常会描述并要求保护与分离核酸序列在某些方面相关的实施方案，如微生物序列或形成疫苗的其他生物成分，以及相关表达和递送载体。尽管2000年《欧洲专利公约》第29（2）条肯定地承认，人类基因序列为先有专利，任何公开序列都必须首先作为符合专利条件的标的通过审查，而不仅仅是简单的“发现”，否则《欧洲专利公约》第52（2）（a）条会将此排除在外[24]。相反地，这种序列必须从其自然出现的细胞环境中机械地“分离”（或采用“人工”方式获得美国专利)，而且是涉及所述发明概念的有用技术新教材，以足够清晰的方式充分公开并要求保护。

在美国的实际操作中，符合专利条件的标的受《美国法典》（U.S.C.）第35篇第101条的管辖，该条款违背了专利“效用”标准，即公开和要求保护的发明规定了具体的、实质性的可信效用[25]。在欧洲的实际操作中，这相当于要求所述标的具有“工业适用性”，并且发明为本领域带来了技术贡献[26]。疫苗成分（及其制造方法）通常不会引发效用问题，因为疫苗一般提供的是针对病毒或其他微生物的某种免疫效果，达到本领域普通技术人员所理解的可信效用。然而，我们还需要进一步研究，充分确定或至少合理确认，免疫原性治疗效果的公开效用可能会在“现实世界”中失败（《美国法典》第35篇第101条）。因此，在起草过程中，对公开和声明与防治某种未指明疾病或症状有关的标的，和/或使用无特征或部分特征蛋白质、治疗性蛋白质或抗体、编码完整开放阅读框（ORF）的DNA片段或受体靶标时，应当保持谨慎，企业最好能从结构上，或至少从功能上以直接与具体疗效相联系的方式进行定义。

美国联邦巡回法院已经根据《美国法典》第35篇第101条，就另一个似乎与疫苗创新高度相关标的的专利符合性问题进行审议，因为这些创新公开并声明了用药剂量和方法[27]。争议专利与婴儿传染病免疫接种程序有关，该程序旨在降低慢性免疫介导性疾病的风险。在3项有争议的专利中，其中两项要求对免疫接种程序信息与慢性病发病情况的筛选和比较方法提供保护，确定风险较低的程序，然后通过使用所述程序施用疫苗进行相应免疫。第三项有争议专利的权利要求列举了筛选和比较步骤，但不包含后续免疫步骤。

下级初审法院驳回了3项专利的可享专利性，理由是根据一些抽象观点，婴儿传染病免疫接种程序与特定慢性疾病的后续发病之间存在关系。然而，根据最高法院在《Bilski 判决》[28]，美国联邦上诉中级法院得出结论认为，“专利资格的排除应当严格适用。”[29]美国联邦上诉法院还认为，思维步骤的存在本身对《美国法典》第35篇第101条的合格性并没有决定性，即使严重怀疑权利要求的实质专利性，3项专利中的两项至少描述了符合专利资格的标的，因为它们在确定程序中列举了“免疫”的物理步骤，这足以满足《美国法典》第35篇第101条要求的具体、有形的应用。这些发明是否也符合其他实质专利性要求，需要根据美国专利法相关规定解决。然而，第三项专利被判决无效，因为其要求不符合专利要求标的，即《美国法典》第35条第101款中禁止为制定针对特定临床条件的最佳免疫接种程序而收集和比较免疫接种数据的纯粹思维抽象过程。

正如欧洲专利局扩大上诉委员会G 2/08号判决所述，美国关于用药制度的判例首先承认了“行政专利”的有效性，这似乎符合欧洲的操作模式。本案中有争议的专利要求使用已知烟酸或代谢为烟酸的化合物，通过特定口服给药方案治疗高脂血症。扩大委员会认为，“在已知使用药物治疗疾病的情况下，《欧洲专利公约》第54（5）条并不排除该药物通过治疗同一疾病而获得用于不同治疗的专利。”[30]但委员会还警告说，专利性“取决于是否遵守《欧洲专利公约》的其他规定，特别是新颖性和创造性步骤，”[31]这可能是所谓的行政权利要求的一个不可逾越的障碍。

这在欧洲专利局技术上诉委员会案件T 1760/08中得到了证实。委员会指出，专利说明书没有公开任何表明所述剂量的所述化合物和任何其他描述剂量相比，在提供更好的生物利用度方面给出了任何意料之外的改进[32]。因为其中一个所描述的工作实例表明，逐步增加所述化合物的给药剂量（即每天服用250 mg、500 mg或1 000 mg），效果会随着化合物用量的增加而更好。在权利要求中选择500 mg的剂量，只是从公开替代方案中作出的一个武断选择。因此，尽管剂量方案本身是符合专利资格的标的，但因为这种选择并没有产生优越的技术教学，该专利缺乏优于现有技术的创造性步骤，因此未能获得批准。

前述内容表明，专利申请起草人在准备和要求针对疫苗技术实施方案时，必须要非常注意。虽然具有免疫原性效果的疫苗成分似乎很容易满足效用阈值，但此类发明要获得足够广泛的保护，通常涉及到对疫苗本身之外的方面提供保护，如衍生微生物和相关表达载体的核酸序列，以及用于实现免疫保护效果的给药方案。每一项实施案例都必须首先成功满足各种全球专利系统法律中提出的效用要求，这可能是一项看似复杂的大工程。

51.3 新颖性原则和先有技术水平

在欧洲的操作中，《欧洲专利公约》第54（1）条规定，如果一项发明不构成先有技术的一部分，则将视为新发明。与世界上大多数其他专利法制度一样，《欧洲专利公约》采用绝对新颖性原则确定是否构成先有技术。如《欧洲专利公约》第54（2）条规定："先有技术应包括通过书面或口头说明、使用或以任何其他方式在欧洲专利申请提交日期之前向公众提供的所有内容。"[33]但是，在特定公开会议上或由于某种滥用而无意中公开的发明不属于此范围，但这种情况非常少[34]，对于在正式提交日期（或适用的优先权日期）[35]之前被视为"公开可用"的发明公开，则不存在公认宽限期。因此，潜在专利申请人必须积极进行保护，使其发明不会受到任何此类行为的影响，包括通过任何公开地、非保密地使用发明的行为，引起公众对该发明概念的注意，直至安全地提交至专利局。这种新颖性方法在全世界几乎所有的专利系统中都很常见，包括但不限于日本、中国、澳大利亚、加拿大、印度和巴西。

根据Leahy-Smith美国发明法案（AIA）实施的变革，美国专利系统至少在部分上采纳了上述"先申请制"原则，这是美国专利法的历史性改革，也是自1952年最后一次重大修订以来最全面的改革。新的第一发明人先申请制度规定，如果申请保护的发明在有效生效日期之前已经获得专利、在印刷出版物中描述、在公共场合使用、出售或以其他方式在世界任何地方可供公众使用，则该发明将不会被判定为新颖发明[36]。然而，新法律中保留了先有技术定义中的一种重要例外情况：如果这种公开行为是由发明人或联合发明人做出的，或是由直接或间接从发明人处获得公开标的的其他人做出的，且发生于申请保护发明生效日期之前1年或更短时间内，则美国专利商标局将不会认定公开内容（或其他公开可用的用途）属于先有技术的一部分[37]。此外，《美国发明法案》取消了Hilmer理论[38]，从而扭转了美国临时申请和外国优先权文件在先有技术地位方面的历史性不平等待遇。

新的"先申请"先有技术原则适用于在新的美国法律生效之日提交的新专利申请；[39]但是，如果申请人修改了原申请中缺乏支持的权利要求，那么对旧申请仍然有效的前"先发明"先有技术原则利益可能会造成不可逆转的丧失，转而支持新的结构。这种情况可能产生的后果是，根据旧法律排除在先有技术发明之外的先前内容，在定义先有技术状态的新方法后，可能会成为对应用的新颖性公开构成潜在破坏。

签署联合研究协议（这种安排在疫苗开发行业中显然很常见）的各方应意识到，如果标的和申请保护发明为共同拥有，或有义务在申请保护发明生效日期之前转让给同一个人，则另一方在申请或专利中所做出的任何公开不应视为先有技术[40]。这意味着申请保护发明必须是由联合研究协议规定范围内的活动做出的，该发明是由/代表在申请保护发明生效日期/之前生效的联合研究协议一方（或多方）做出的，同时，重要的是，申请文件公开了协议各方的名称。

回到欧洲对"先有技术"的定义，"先申请"原则中一个比较复杂的延伸在于"欧洲先有专利权利"的概念。这一概念的引入是为了防止双重专利。2000年《欧洲专利公约》第54（3）条[41]给出了定义："此外，提交的欧洲专利申请内容如申请日期在第（2）项所述日期之前，并且在该日期当天或之后公布，则应视为包含在先有技术中[42]。"因此，如具有较早优先权日期的欧洲（或欧洲-PCT）专利申请公开了相同的标的，并在欧洲申请的申请日之后有效发表，则会影响新颖性，但不得应用于评估发明步骤。因此，对先前申请中任何申请保护的"显而易见"同类发明，不应构成后续申请中先有技术的一部分。

然而，对于欧洲未决申请（包括欧洲-PCT）和2000年《欧洲专利公约》生效前授予的欧洲专利，不适用2000年《欧洲专利公约》第54（3）条的规定。相反地，"先有"欧洲权利只对与后一申请重复（或“冲突”）的指定国家赋予了这种先有技术效力，前提是根据1973年《欧洲专利公约》第54（4）条和第23a条有效支付指定费用。构成先有技术的文件不尽相同，所以这种情况的典型结果是对不同指定国家提出不同的权利要求。更复杂的是，如果早期申请中的指定国家在公布日期之后撤回申请，但已有效支付指定费用，则先有技术效果仍然适用于该欧洲专利公约的特定成员国。

说明1973年和2000年《欧洲专利公约》第54（3）条的实际效果：在之前的欧洲专利操作中，申请人只需支付欧洲申请EP-A的DEE、FR、GB的指定费用。发表后，EP-A对之后提交的欧洲专利申请只有在DEE、FR和GB联合指定和指定费用有效支付的情况下，才能具有先有技术效果（仅新颖性）。然而，根据2000年《欧洲专利公约》，申请人B应为后提交欧洲申请EP-B有效支付IT和SE的指定费用。在本示例中，EP-A在EP-B指定的所有《欧洲专利公约》签约国中，即使EP-B根本不指定DE、FR和GB的国家，对EP-B仍具有预先的技术效果（新颖性）。因此，根据2000年《欧洲专利公约》，“旧”欧洲权利的先有技术定义已经大大简化。

51.4　使用已知或新疫苗成分医学指征的权利要求必须具有创造性并充分公开

鉴于2000年《欧洲专利公约》第53（c）条规定的可享专利性排除的各种公共政策，《欧洲专利公约》对申请人申请确定为治疗、手术或诊断方法的疫苗功能做出了限制。基于疫苗的发明公开中常见的示例性实施方案包括与疾病诊断和/或其他检测手段、免疫调节、免疫刺激、治疗、预防治疗、监测和评估相关的方法，以及可能相当于外科手术步骤的疫苗施用技术。《欧洲专利公约》为此类医疗领域的创新提供了一种折中方案，但确实也存在一些例外，这在2000年《欧洲专利公约》第54（4）条和第54（5）条中有所体现：允许申请人在这些被排除的方法中分别为首次或进一步医疗用途要求已知物质或成分。

具体而言，2000年《欧洲专利公约》第54（4）条规定：“第（2）和（3）段不应排除先有技术中包含的任何物质或成分在2000年《欧洲专利公约》第53（c）条所述方法中使用的可享专利性，前提是先有技术中未使用该段所述的方法。”因此，该段为“第一医学指征权利要求”提供了必要的授权，内容可以宽泛地表述为考虑预期治疗目的，例如，“成分X作为药物”或“包含成分X的疫苗”。因此，这种格式对发明者来说非常有用，原因是他们可以在医学领域（如作为疫苗的治疗用途）首次描述一种新的或已知的“成分X”。如果先有技术公开合理地描述了相同的成分X可用作疫苗，这相当于以破坏新颖性的方式进行公开，除非专利申请中公开成分X是一项用于新的、具体的治疗环境中未知的先有技术，即第二次或进一步医学指征。

描述已知化合物的首次医疗用途之后，任何新发现已知物质或成分的第二次或进一步医疗用途都可以根据2000年《欧洲专利公约》第54（5）条申请专利，该条款规定：只要第4段中所指任何物质或成分在《欧洲专利公约》第53（c）条所指方法中的特殊用途不构成先有技术，则第2段和第3段也不得排除该物质或成分在第53（c）条中所指方法中的特殊用途的可享专利性。因此，本段旨在为之前已知的瑞士式“第二次医疗指征权利要求”提供明确的依据。因为第54（5）条具体规定该成分“用于”特定用途，针对已知疫苗新用途的权利要求应包含此类描述，避免清晰度不足和由此导致的新颖性异议。疫苗说明中的这种权利要求可以简单地表述为：“用于治疗/预防疾病Y的疫苗X”或“用于特定治疗用途的包含疫苗X的成分”。在这些示例性权利要求中，“疫苗成分X”已经作为药物为人所知，但并不用于权利要求中所述的特定的、与目的相关的用途。

即使专利说明书中仅有意义地公开了该物质的特定医疗用途，欧洲专利局也可以批准首次医疗用途的广泛权利要求[43]。但是，我们对有效支持某些类型疫苗权利要求所需的实验证据程度提出了警告，至少应满足《欧洲专利公约》第 83 条中规定的严格要求。该条款规定，一项发明应当“以足够清晰完整的方式公开，并可由本领域的技术人员实施。”对于分别根据第 54（4）条或第 54（5）条针对第一或第二医疗用途的权利要求，不仅需要使技术人员能够基于申请中的书面技术公开内容制造出申请保护的成分，而且考虑到技术领域的公知常识，申请中还必须提供直接和充分的证据，证明申请保护的疗效是为了满足《欧洲专利公约》第 83 条中的要求，因为所述效果代表权利要求的功能性技术特征[44]。

例如，列举“用于引发针对所述蛋白质（或化合物）的免疫应答治疗中来源于细菌 A 的蛋白质……或化合物”等的权利要求，需要对所述专利描述提供可信的证据或其他指导，证明所述蛋白质或化合物与体液或细胞免疫应答的诱导具有机械关联性。然而，功能上缺乏已知化合物或物质的权利要求，例如“用于预防疾病 Y 的疫苗治疗中的细菌 A”，必须满足要求较高的举证责任，充分证明所描述发明提供了超越单纯理论的有效公开，如感染细菌 A 后在预防疾病 Y 中具有可识别的疗效或可验证的免疫保护效果。这些原则足以确保阅读专利公开内容的技术人员可以实施本发明，并具有合理的成功预期，不会遇到任何不当问题，同时不需要为实施要求发明而进行实验研究。

显然，在现行的欧洲专利局的标准下，施用疫苗成分建立起有效的一般免疫保护作用，并与生理靶标联系起来，可能是一项艰巨的任务。例如，在欧洲专利局技术上诉委员会第 T 187/93 号判决中，该委员会拒绝了一项专利申请，因为该权利要求概括了所公开的关于特定单纯疱疹病毒 1 型或 2 型糖蛋白 D 蛋白对所有膜结合病毒蛋白（包括所有疱疹病毒膜蛋白和所有单纯疱疹病毒膜蛋白）的免疫保护作用的技术教学。我们认为，扩大申请公平技术教学是过度的且毫无根据，否则这种教学仅限于显示来自 1 型或 2 型单纯疱疹病毒糖蛋白 D 免疫保护作用的实验证据。因此，这项申请未通过《欧洲专利公约》第 83 条，因为不切实际的技术人员在申请中缺乏授权教学，所以不能落实广泛要求的发明。委员会还注意到，申请人关于享有广泛权利要求范围的论点与根据《欧洲专利公约》第 56 条成功支持发明步骤的论点不一致，该条款更狭隘地侧重于针对 1 型或 2 型单纯疱疹病毒膜糖蛋白 D 的方法，以及针对病毒株体内攻击的免疫受试者的诱导免疫保护[45]。

在欧洲专利局第 T 219/01 号判决中，一项专利公开并声称，作为一项重要的医疗用途，“一种包含 gp120 的艾滋病病毒疫苗”和“gp120 可用于激发对艾滋病病毒的保护性免疫反应”。不幸的是，尽管该专利确实提供了可信的技术数据证明黑猩猩接种了疫苗，但一项后发表的 AIDS-VAX 临床研究表明，这种疫苗不能为人类受试者提供完全的保护。根据反对者在授予专利后的反对程序中向欧洲专利局提交的证据，这项专利由于未能根据《欧洲专利公约》第 83 条提供充分允许的公开而被撤销，因为该权利要求的确包含了疫苗在人体内的使用，这在技术上不再具有合理性[46]。

如果专利说明书的描述仅仅提供了一种未确认化合物可能具有的医学用途的模糊说明，那么随后提交的更详细的发表后证据则不能用于弥补在公布日期披露该标的的根本性不足。《欧洲专利公约》第 54（5）条认可的权利要求格式要求达到所要求的治疗效果，即所要求的已知化合物的特定“用途”是权利要求的功能性技术特征。因此，根据《欧洲专利公约》第 83 条，除非技术人员在优先日期已经知道，否则申请必须公开产品是否适合所要求的治疗用途。在欧洲专利局第 T 609/02 号判决中，一项受攻击的授权专利权利要求描述了一种化合物的用途是“用于抑制类固醇激素应答或类固醇激素样化合物应答基因的过度表达”。然而，该申请未能提供任何数据，显示权利要求中实际考虑的类固醇激素反应性或类固醇样化合物，或如何防止这些基因的过度表达。这种观点的一个重要方面是，专利说明书中的简单陈述，即化合物 X 可用于治疗疾病 Y，只是不足以确保所要求保护药物的充分公开。相反地，对于新的医学用途，书面说明中有义务提供至少一些有意义的信息和/或证据，例如，实验测试表明，所述化合物对与所述临床状态相关的代谢机制具有直接影响[47]。

51.5 浅谈专利获批后授予程序

在获得欧洲专利许可之后，可以启动若干个授予后程序，这些程序可能会造成授予专利权的范围完全失效或至少部分权利受到限制。欧洲专利局的选择在生物技术领域很常见，即集中各部门的反对程序，联合第三方对欧洲专利局的专利有效性提起质疑，质疑必须在专利授权公布后9个月内提交。《欧洲专利公约》第100条规定的质疑范围规定，允许反对者以权利要求标的陈旧、不具有创造性或工业可应用性、不在可享专利性范围内、包含与申请相关的附加内容和/或未充分公开为由对专利提出质疑。在本章中，我们已经谈到了其中的许多问题。为了证明这一情况，反对者必须提交一份详细的简要说明，列出至少1个上述理由，并酌情提供科学出版物、专利文件、声明和其他专家证词形式的补充证据。专利所有人可以对通知作出回应，并根据需要提供支持证据反驳对方的主张。反对程序可能产生的结果包括专利完全撤销、要求修改限制性权利要求或完全不受影响。反对期结束后，只能在专利得到确认并由《欧洲专利公约》生效的成员国才能对欧洲专利提出质疑[49]。如果诉讼中受到不利影响的一方决定对反对方的决定提出上诉，欧洲专利局制定了详细的机制在二审时对该程序提供指导。

在2000年修订版《欧洲专利公约》中引入的中央限制程序，即第105条a–c款，为在所有已进行专利验证的《欧洲专利公约》成员国“重新开始”生效的专利所有者提供了单方面限制或完全撤销已授予欧洲专利的可能性[51]。这种集中程序可以让专利权人既有机会修改可能无效的欧洲专利实质内容，也有机会避免在每一个相关国家法院进行冗长而昂贵的诉讼。值得注意的是，这一程序完全由专利权人发起（尽管第三方可能提出意见）[52]；限制请求只针对权利要求的规定，而没有明确规定要修改说明书或附图（应当已经支持修改后的权利要求）；但可以根据请求进行此类修正。此外，欧洲专利局的限制程序并不凌驾于现有的国家程序（特别是撤销程序）之上，如果国家法院作出判决，则在欧洲专利局的最终裁定结果作出之前，这些程序可以暂停或中止。欧洲专利局根据2000年《欧洲专利公约》第105b条和《规则》实施的调查范围[53]，仅限于专利权人的修正案是否在事实上限制了权利要求（而不是纯粹的表面文章），修正案是否符合《欧洲专利公约》第84条的清晰简明要求，以及修正案是否符合《欧洲专利公约》第123（2）条和第（3）条对增加标的的严格要求（无论增加标的是否超出了申请内容）。

如果审查部门认为专利权人没有遵守上述要求，则可以自由裁量决定是否给予进一步的机会，因为专利权人将来有权提出额外的限制请求。有趣的是，限制程序并没有要求审查部门实质性地审查修改后的权利要求是否实际避免了先有技术，或者是否满足《欧洲专利公约》第52~57条规定的《欧洲专利公约》专利要求。限制请求获准后，专利权人除提交印刷费之外，还必须提供欧洲专利局3种官方语言的限制权利要求译文（如适用，应翻译成欧洲专利局每一个相关签约国的官方语言）。重要的是，在欧洲专利局拒绝限制或撤销请求的情况下，不得对该判决提起上诉。

对于美国的操作，《美国发明法案》最近引入和/或修改了多种专利授予后程序，这无疑将为处理许可后的美国专利开辟具有战略重要性的新选择。例如，《美国发明法案》建立了一个新的“授权后重审程序”（“PGR”）和“当事人间审查”[54]（“知识产权”），可以战略性地质疑并潜在摧毁竞争对手的美国专利。

更详细地说，授权后重审程序是一个类似于欧洲专利局程序的“反对式”机制，其中非专利权人第三方可以基于与《欧洲专利公约》第100条规定的理由类似的广泛理由，指控美国专利的一项或多项专利权利主张无效，而美国专利商标局之前的复审程序就是如此[55]。与欧洲的反对诉讼不同，美国专利可能会因缺乏明确性或确定性受到攻击，这是欧洲惯例（即《欧洲专利公约》第84条）所不允许的。然而，与欧洲专利局的反对意见相同，授权后重审程序的申请必须在专利授予日期或美国重新颁发专利的发布日期（视具体情况而定）后9个月内提交。

承认授权后重审程序诉讼的门槛要求是，如果反对者在起诉书中提供的信息没有遭到反驳，则表明“很有可能”，在起诉书中受到质疑的美国专利申请中，至少有一项是不可申请专利的[56]。这些要求似乎与欧洲专利程序相对应，欧洲专利程序要求“声明欧洲专利遭到反对的程度和反对的理由，以及支持这些理由的事实和证据的说明”[57]授权后重审程序诉讼是公开记录事项，由新成立的美国专利审判和上诉委员会（“PTAB”）进行，该委员会由一个3人行政专利法官小组组成，和欧洲的反对意见不同（在欧洲的反对意见中，反对意见庭由3名审查员组成，其中1名通常是进行授予诉讼的主审查员）。尽管授权后重申程序诉讼可以完全以书面形式进行，但各方均有权根据需要进行口头诉讼；《欧洲专利公约》中有相应的规定[58]。显然，与《欧洲专利公约》相比，美国专利审判和上诉委员会在允许发现范围方面更为有限，至少理论上不受限于双方根据《欧洲专利公约》第114（1）条提供的事实、证据和论据。此外，反对庭可以行使其酌情处理权，对甚至没有任何诉讼反对者提出的反对理由进行审查。授权后重审程序行动或者由美国专利审判和上诉委员会的判决终止，或者由双方达成协议终止，但这一决定最终由美国专利商标局（PTO）作出。

授予后的当事人间复审（“IPR”）旨在最终取代现有的当事人间复审程序，并允许第三方非专利所有人在自美国专利发布或重新发布之日起，①9个月之后（即在授权后重审程序期限到期之后）；②如授权后重审程序已经成立，则自其终止之日起计算；③请求人收到专利权人指控侵犯专利权的申诉后1年内对美国专利提出质疑。在美国专利审判和上诉委员会之前进行的当事人间复审程序在《欧洲专利公约》中没有相关的对应方。当事人间复审可以在美国专利的整个有效期内启动。与授权后重审程序不同，当事人间复审只能以新颖性和缺乏创造性为由，仅可使用专利和印刷出版物。

51.6 结论意见

上述内容对管理专利授予和辩护程序的现有结构以及某些可享专利性方法进行了非详尽概述，这些方法不仅适用于基于疫苗的创新，而且普遍适用于生物发明。强有力的专利地位通常始于建立在核心技术基础上，着眼于捕捉这些最初发明中经常出现的次要标的。对于疫苗创新，最初的专利申请可能侧重于细胞克隆、分离疫苗序列表达和治疗功能（可能会、也可能不会存放在官方认可机构），而次级后续专利可能会通过要求改进和/或进一步开发或相关技术教学（如纯化方法、药物输送方法和系统、药物组合物、给药方案和其他临床指征）扩展基础技术的排他性，其中任何一项都可能在需要监管部门批准的情况下，通过申请补充保护证书（SPC）延长欧盟的垄断期。从前述分析中可以明显看出，开发优化的全球专利组合需要一种平衡的方法，既要采取审慎策略保护期望市场地位，从而保持可靠的融资渠道，又要面对不断变化的专利系统和环境，仔细选择待申请专利的发明。

参考文献

[1] The Report can be viewed here：http：//www. wipo. int/ freepublications/en/patents/946/wipopub9463. pdf (Last visited 28 Feb 2013).

[2] See Report，Appendix 6 at Figures 1，2.

[3] EU/EC legislation regulating vaccine testing and authorization in Europe include (each website visited December 10，2012)：EU Directive 2001/83/ EC-on medicinal products for human use http：// ec. europa. eu/enterprise/pharmaceuticals/eudralex/vol-1/dir_2001_83_cons/dir2001_83_cons_20081230_en. pdf，EU Directive 2001/20/EC - Clinical trials directive h ttp：//europa. eu/eur - lex/pri/en/oj/dat/2001/ l _ 121/l _ 12120010501en00340044. pdf.

[4] Concept paper on guidance for DNA vaccines，EMEA/CHMP/308136/2007 Committee for the Medicinal

Products for Human Use (CHMP), 15 March 2012.

[5] For information about the Unitary Patent, including EU documents and current status, see: http://ec.europa.eu/internal_market/indprop/patent/index_en.htm (Last visited 1 Mar 2013).

[6] See Brüstle vs. Greenpeace, Case C-34/10 (18 October 2011) and Monsanto Technology LLC v Cefetra BV and Others, Case C-428/08 (6 July 2010), respectively.

[7] Basic Proposal for the Revision of the EPC, Preparatory Documents: CA/PL 6/99; CA/PL PV 9, points 24-27; CA/PL PV 14, points 143-156; CA/100/00, pages 37-40; CA/124/00, points 12-16; CA/125/00, points 45-73; MR/2/00, pages 43-44; MR/8/00; MR/15/00; MR/16/00; MR/24/00, pages 69-71).

[8] See Rule 26 EPC 2000, which specifi es additional patent-eligible inventions in the life sciences.

[9] See Article 52 (2) (a) EPC 2000.

[10] See Article 53 (c) EPC 2000.

[11] Article 57 EPC demands that inventions be industri-ally applicable, i. e., capable of being exploited in a commercial context. Since methods relating to treatment, diagnosis, and surgery can be readily adapted to a wide range of industries, they are indeed "industrially applicable" but are simply excepted from patentability for reasons relating to public policy.

[12] Cancer Voices Australia v Myriad Genetics Inc [2013] NSD643/2010, Federal Court of Australia (Sydney), decision dated 15 February 2013.

[13] See Part 3, Division 1, Section 18, Patents Act 1990, Act No. 83 of 1990 as amended, taking into account amendments up to Act No. 35 of 2012. This holding is also consistent with European practice; see Rule 27 (a) EPC.

[14] The patent in suit is Australian Pat. No. 686004, with a priority date of 12 August 1994, granted with 30 different claims. Only the validity of claims 1-3 was at issue in the proceedings.

[15] At issue were claims 1, 2, 5, 6, and 7 of U. S. Patent 5, 747, 282; claim 1 of U. S. Patent 5, 693, 473; and claims 1, 6, and 7 of U. S. Patent 5, 837, 492.

[16] 35 U. S. C. § 101 provides: "Whoever invents or dis-covers any new and useful process, machine, manufacture, or composition of matter, or any new and useful improvement thereof, may obtain a patent therefore, subject to the conditions and requirements of this title"

[17] The full text of the decision (Case No. 12-398) can be viewed here: http://www.supremecourt.gov/opinions/12pdf/12-398_1b7d.pdf. (Last viewed 29 July 2013).

[18] See Association for Molecular Pathology (AMP) and ACLU v. USPTO and Myriad Genetics 689 F. 3d, 1303 (Fed. Cir. 2012).

[19] See Slip Opinion, No. 12-398 at 2; 569 U. S.__ (2013) at 12.

[20] See Slip Opinion No. 12-398 at 3; 569 U. S.__ (2013) at 16-17.

[21] See Mayo Collaborative Servs. v. Prometheus Labs., Inc., 132 S. Ct. 1289 (Fed. Cir. 2012).

[22] See Monsanto Technology LLC v Cargill International SA&Anor [2007] EWHC 2257 (Pat) (10 October 2007).

[23] See Monsanto Technology LLC v Cefetra BV and Others, Case C-428/08 (6 July 2010).

[24] Rule 29 (2) EPC 2000 states: "An element isolated from the human body or otherwise produced by means of a technical process, including the sequence or partial sequence of a gene, may constitute a patentable invention, even if the structure of that element is identical to that of a natural element"

[25] 35 U. S. C. § 101 states: "Whoever invents or discovers any new and useful process, machine, manufacture, or composition of matter, or any new and useful improvement thereof, may obtain a patent therefore, subject to the conditions and requirements of this title".

[26] These requirements are set forth in Article 57 and Article 56 EPC, respectively.

[27] Classen Immunotherapies, Inc., v. Biogen Idec, 2006-1634-1649 (Fed. Cir. August 31, 2011).

[28] Bilski v. Kappos, 561 US__, 130 S. Ct. 3218, 177 L. Ed. 2d 792 (2010).

[29] Id. at 18-19.

[30] See EPO Enlarged Boards of Appeals decision G 2/08 at reason 6. 1.

[31] Id. at reason 5. 10. 9.

[32] See EPO Technical Boards of Appeal decision T 1760/08 at reason 3. 3.

[33] Both Article 54 (1) and Article 54 (2) EPC 2000 retain the same wording as their counterpart provisions in the EPC 1973 so no material changes to practice resulted when the EPC 2000 took effect.

[34] See Article 55 EPC.

[35] Unlike the US patent system, the EPC does not recog-nize a grace period prior to fi ling a patent application, whereby the inventor may disclose his invention without that disclosure forming part of the prior art. However, Article 55 EPC specifi es that certain "non-prejudicial disclosures" that protect an applicant from a novelty-destroying disclosure of the invention before the fi ling date caused either by "evident abuse" (e. g., violation of a confi dentiality agreement or theft of the invention) or where the invention was displayed at an EPO-certifi ed international exhibition. To qualify for Article 55 EPC protection, the disclosure must have occurred no earlier than 6 months prior to the application's fi ling (not priority) date.

[36] See new 35 U. S. C. § 102 (a) (1), which largely refl ects the prior art defi nitions stated in former 35 U. S. C. 102 (a) and 35 U. S. C. 102 (b). Furthermore, US patents, US Patent Application Publications, and PCT International Application Publications that were effectively fi led by a different applicant will be considered as part of the state of the art under new 35 U. S. C. § 102 (a) (2).

[37] See new 35 U. S. C. § 102 (b) (1), (2).

[38] In re Hilmer 359 F. 2 859 (C. C. P. A. 1966) held that a patent application's foreign fi ling date could be used as shield itself from later-fi led or published cited prior art at the US PTO, but could not be used as affi rmative prior art. This situation frequently disadvantaged foreign inventors who fi led a fi rst application outside the USA and then relied on Article 4 of the Paris Convention in fi ling a later US application. Under these circumstances, the foreign application could not be used as prior art against another US application because it was not "fi led in the United States." Therefore, the foreign priority application date received unequal treatment by the US PTO, since no offensive benefi t could be gained from the earlier disclosure.

[39] These changes under the AIA come into force on March 16, 2013.

[40] See new 35 U. S. C. § 102 (b) (2) (C).

[41] Article 54 (3) EPC 2000 differs from its predecessor by removing the language that the publication occur "under Article 93" and is regarded by the EPO as a "minor editorial amendment" (Preparatory documents: CA/PL 17/99; CA/PL PV 10, points 19-21; CA/PV 14, point 6; CA/100/00, pages 43-44; MR/6/00, pages 3-4; MR/2/00, pages 47-48; MR/24/00, page 71).

[42] Basic Proposal for the Revision of the EPC, Preparatory Documents: CA/PL 17/99; CA/PL PV 10, points 19-21; CA/PV 14, point 6; CA/100/00, pages 43-44: Article 54 (3) EPC 2000 partly simplifi es Article 54 (3) EPC 1973 practice since most European applications routinely designate all Contracting States. Therefore, the complicated prior art searches demanded by Article 54 (4) EPC 1973 appeared to confer an advantage to only a limited number of applicants.

[43] See Rule 42 (e) EPC 2000.

[44] An elegant discussion on this point can be found in EPO Technical Boards of Appeal decision T 1496/08, e. g., at reason 14.

[45] See EPO Technical Boards of Appeal decision T 187/93, e. g., at reasons 19-35.

[46] See EPO Technical Boards of Appeal decision T 219/01, reasons 4 and 5. 2, but compare: in decision T 716/08, the Board held that a therapeutic effect was found where an application disclosed that the mere presence of antibodies against the 48 kDa ISAV protein in the rabbit serum used for screening the lambda bacteriophage cDNA library was evidence that the 48 kDa ISAV protein was antigenic and, consequently, could also be a useful constituent of a subunit vaccine, see reasons at 17-24 and 47.

[47] See EPO Technical Boards of Appeal decision T 609/02, e. g., at reason 9.

[48] See Articles 99-105 EPC, which govern EPO opposition procedures.

[49] See Article 138 EPC for particular provisions that must be applied to these procedures.

[50] See Articles 106-111 EPC, which govern EPO appeals procedures.

[51] Article 68 EPC 2000: When the decision to revoke or limit the European patent is published per Article 105b

(3) EPC 2000, the effects of this decision apply ab initio or retroactively from the outset via Article 64 (rights conferred at patent grant) and Article 67 (rights conferred after publication of the European patent application).

[52] Observations may be submitted by third parties in an offi cial EPO language per Article 115 EPC 2000.

[53] For example, Rule 95 (2) EPC 2000 governs the scope of examination and admissibility of the limitation request, i. e., patentee's compliance with the substantive requirements for limitation of the European patent.

[54] These new procedures are found in the AIA at SEC. 6 in 35 U. S. C. § § 311–319 and 35 U. S. C. § § 321–329, respectively.

[55] U. S. C. 282 (b), any patentability ground specifi ed in 35 U. S. C. at Part II; 35 U. S. C. 282 (c), failure to comply with any requirement of 35 U. S. C. § § 112, 251 such as written description, defi niteness, and clarity.

[56] 35 U. S. C. § 324 (a), (b).

[57] Rule 76 (2) (c) EPC.

[58] Article 116 EPC.

[59] Rule 81 (1) EPC.

（何继军译）

第 52 章　细胞培养的流感疫苗的生产

Markus Hilleringmann，Björn Jobst 和 Barbara C. Baudner①

摘要

流感疫苗接种是目前降低或应对社区流感死亡率和发病率的主要手段。自 20 世纪 40 年代单价灭活疫苗配方早期开发以来，疫苗制造商采用了不同的策略生产各种流感疫苗（例如灭活全病毒疫苗、减毒活疫苗、去污剂或溶剂“裂解”疫苗、亚单位疫苗和佐剂疫苗）。实际上，可以分为 2 种主要的生产方式，一种是传统的鸡胚技术，另一种是最近基于细胞培养的方式。此外，有通过不同免疫途径的肌内、皮内和鼻内流感疫苗。

流感疫苗的开发和生产以复杂的制造过程为基础，从选择和开发最佳候选疫苗病毒开始，需要与监管机构和卫生保健官员进行各种动态互动。疫苗供应和使用规划以及其他相关保健资源的提供是全面应对季节性和大流行性流感的重要组成部分。如果要按时生产和交付疫苗，流感病毒在季节性流行病和偶发大流行期间的快速传播会将整个过程紧密地结合在一起。需要不断努力开发新型安全性流感疫苗，并改进用于毒株特异性功效测试的试剂，以更好地应对复杂流感的相关挑战。

52.1　引言

52.1.1　流感疾病

流感是由病毒引起的，通常会导致急性呼吸道疾病，主要影响鼻、喉咙、支气管，偶尔还会影响肺部（图 52.1），感染特征包括突发高热、肌肉疼痛、头痛和严重不适、干咳、喉咙痛和鼻炎。尽管可以通过突发高热和极度疲劳进行识别[1]，但在感染流感早期，很难将其与普通感冒区分开来。

病毒很容易通过感染者咳嗽或打喷嚏时产生的喷雾和小颗粒在人与人之间传播；因此，流感往往在每年秋季和冬季发生的季节性流行病中迅速传播[2]。

大多数人的发烧和其他症状会在 1 周内恢复，不需要就医。然而，高危人群（年幼、年老或慢性病患者）因流感住院和死亡的风险较高。在世界范围内，每年流行病都会导致 300 万～500 万例严重疾病和 25 万～50 万例死亡[3]。

52.1.2　流感病毒和流行病学

流感病毒分为 3 种：甲型、乙型（均为季节性）和丙型，甲型和乙型流感病例的发生频率

① M. Hilleringmann，PhD（✉）（Department of Applied Sciences and Mechatronics，FG Protein Biochemistry and Cellular Microbiology，University of Applied Sciences Munich，Munich，Germany，慕尼黑应用技术大学，FG 蛋白质生物化学与细胞微生物学系，应用科学与机电一体化系，德国慕尼黑，E-mail：markus. hilleringmann@hm. edu）

B. Jobst，PhD（诺华疫苗与诊断有限公司制造科学与技术（MS&T）部，德国马尔堡）

B. C. Baudner，PhD（诺华疫苗与诊断有限公司疫苗研究部，意大利锡耶纳）

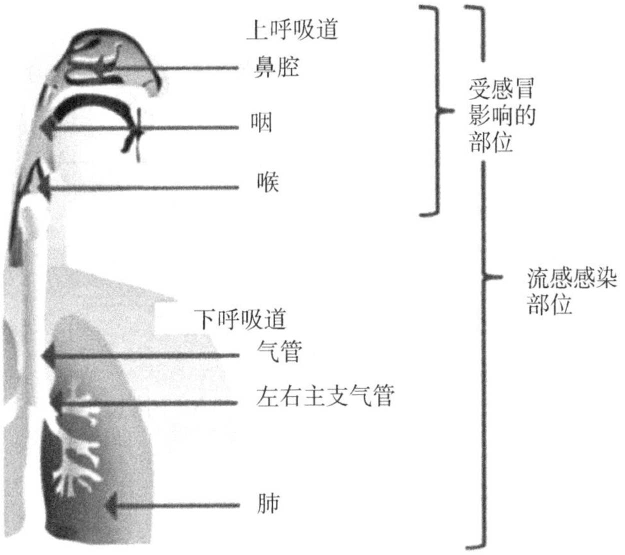

图 52.1　受感冒和流感影响的部位

远远高于丙型流感。甲型流感病毒根据病毒表面糖蛋白——血凝素（HA 或 H）和神经氨酸酶（NA 或 N）的不同种类和组合进一步分为亚型（图 52.2）[3,4]。在各种甲型流感病毒亚型中，目前正在人类之间传播的是甲型 H1N1 流感和甲型 H3N2 流感。流感病毒在人群中具有良好的适应性，并可以通过突变机制（抗原漂移）和远亲病毒株之间表面蛋白基因的交换（抗原转移），避开免疫系统[5]（图 52.2）。当发生重大变化时，就会出现人类大面积流行的风险[6]。

20 世纪暴发了 3 次流感，最严重的一次是西班牙流感。20 世纪的最后一次大流行发生在 1968 年，预计全世界有 100 万~300 万人感染了这种病毒。21 世纪第一次大流行发生在 2009 年，由 H1N1 病毒株引发（A/California/07/09）。

1997 年，一种新的禽流感病毒 H5N1 在中国首次感染人类。自出现以来，H5N1 病毒已从亚洲传播到欧洲和非洲，造成数百万只家禽和野生鸟类感染。世卫组织报告了数百例人类病例和死亡[7]。

52.1.3　流感药物

金刚烷（金刚烷胺和金刚乙胺）和神经氨酸酶抑制剂扎那米韦和奥司他韦（达菲）[8,9]等抗病毒药物已用于治疗流感；然而，一些流感病毒对抗病毒药物产生耐药性，限制了疗效。世卫组织在不断监测正在传播的流感病毒的抗病毒易感性[3]。

52.1.4　流感疫苗

世卫组织称，接种流感疫苗是预防流感病毒感染和严重后果的最有效方法。接种疫苗对因流感病毒感染而导致严重并发症风险较高的人，以及与高危人群共同生活或照顾他们的人尤其重要。

各种流感疫苗已经出现了很多年[10]（图 52.3）。季节性流感疫苗通常为三价疫苗，每剂含有 15 μg 两种甲型流感亚型（例如 H1N1 和 H3N2）和 15 μg 一种乙型流感毒株。主要灭活流感疫苗目前有 3 种：全病毒疫苗、裂解疫苗和亚单位疫苗[11,12]（图 52.3）

在健康成人中，流感疫苗可以预防 70%~90%的流感特异性疾病。在老年人中，疫苗可以减少高达 60%的严重疾病和并发症，减少 80%的死亡率[3]。

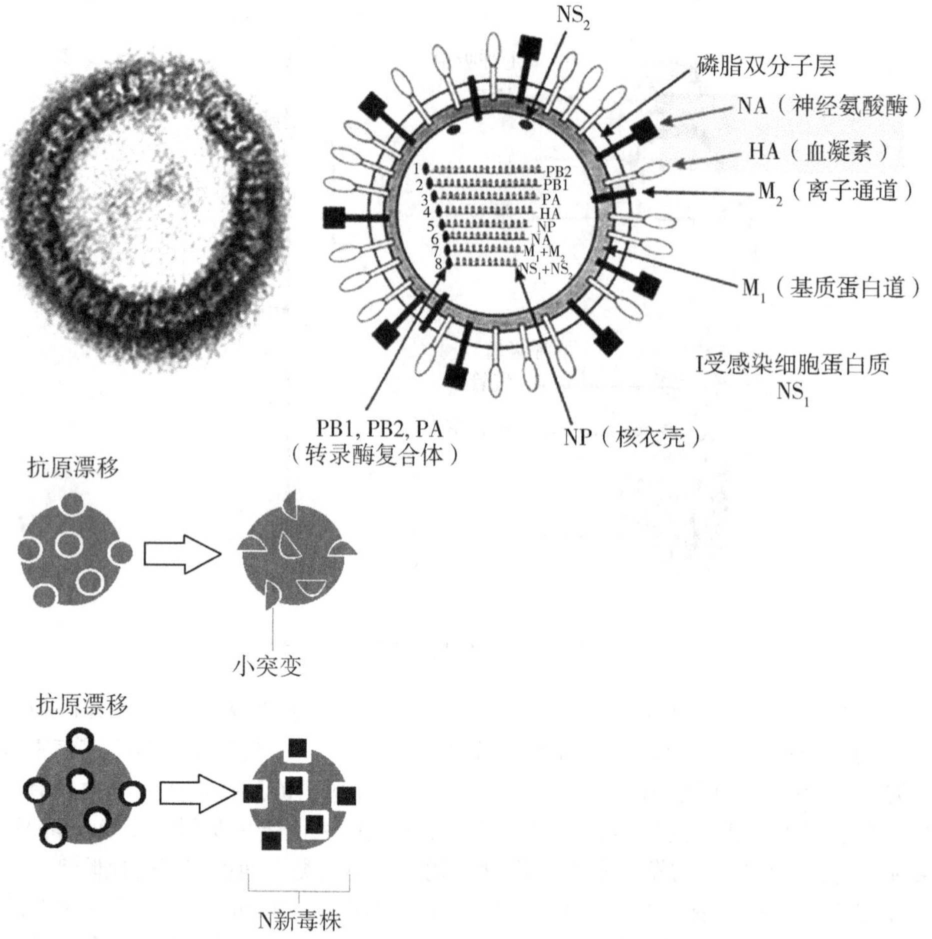

图 52.2　流感病毒；抗原漂移和抗原转移

抗原漂移——血凝素和神经氨酸酶蛋白在一种类型或亚型中的点突变逐渐改变。抗原漂移病毒导致先前菌株的抗体无法中和突变病毒。抗原漂移同时出现在甲型流感和乙型流感病毒中，并引起周期性流行病。抗原转移——甲型流感病毒在人类群体中的出现，该病毒将编码其 HA 蛋白或 NA 蛋白的基因与另一种流感病毒进行交换（这种病毒通常会感染动物而不是人类）。抗原转移是全球流行病暴发的罪魁祸首。

52.2　世界卫生组织（WHO）和流感监测

52.2.1　流感监测

由于基因突变，流感病毒进化频繁，新毒株会迅速取代旧毒株。因此，疫苗配制 1 年后，到下一年可能就会失效，所以需要对流感疫苗进行持续的全球监测和频繁研制。为此，世卫组织协调了一个国际监测系统，对流感病毒的流行病学进行监测（图 52.4）[13]。监测系统全年无间断运行，能够对分离于人体和动物的传播流感病毒进行详细分析，特别是鸟类（如 H5N1）和猪，并能够检测出人类可能易感的甲型流感（H3N2 和 H1N1）和乙型流感病毒的新进化抗原变异株[14]。

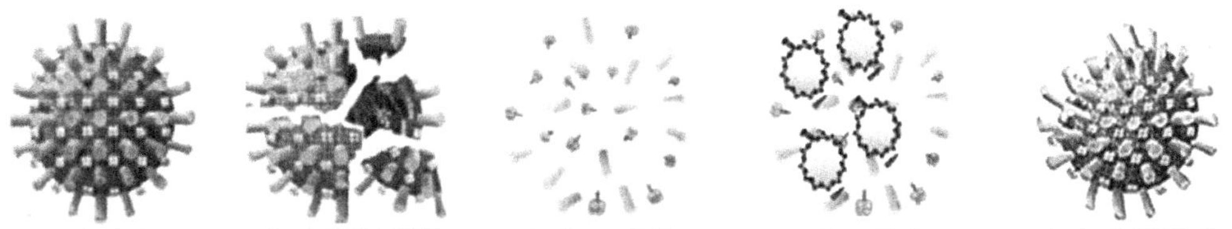

图 52.3　流感疫苗

（1）全病毒疫苗由被“杀死”或灭活的全病毒组成，因此不具传染性，但保留了毒株特异性抗原的特性。

（2）裂解病毒疫苗由去污剂处理破坏的灭活病毒颗粒组成。这些疫苗包含表面抗原和内部抗原。

（3）亚单位或表面抗原疫苗——主要由纯化的血凝素（HA）和神经氨酸酶（NA）组成，已经去除其他病毒成分。

（4）佐剂亚单位疫苗主要由纯化的表面抗原（HA 和 NA）组成，仅添加有佐剂。

（5）由弱的（非致病性）全病毒组成的致弱（冷适应）病毒活疫苗。其中疫苗中的活病毒只能在温度较低的鼻道中繁殖，需通过鼻内给药。

52.2.2　世卫组织的全球流感监测网络（GISN）

GISN 可以监测 1 年内在世界各地的人类中传播的流感病毒类型[13]。GISN 包括 5 个世卫组织合作中心（亚特兰大、北京、伦敦、墨尔本、东京）。

- 106 个国家建有 136 个国家流感中心
- 11 个 H5 参考实验室
- 4 个基本监管实验室 GISN 的主要技术职能是：

  监测人类流感疾病负荷

  监测季节性流感病毒的抗原漂移等变化（如抗病毒药物的耐药性）

  获取合适的病毒分离株，更新流感疫苗

  检测并获得感染人类的新流感病毒（特别是具有大流行潜力的病毒）分离株

52.2.3　流感疫苗建议

自 1973 年以来，世卫组织每年召开两次技术咨询会议，就疫苗中应包含的主要传播毒株类型提出建议——北半球在 2 月举办，南半球在 9 月举办。这些建议基于世卫组织全球流感监测网络（GISN）提供的信息，该网络现为世卫组织全球流感监测和反应系统（GISRS）（图 52.4）。最近，甲型 H5N1 流感、甲型 H9N2 流感和其他亚型流感病毒也已被世卫组织全球流感监测和反应系统纳入流感大流行准备工作的考虑范围[14]。

52.2.4　病毒种毒

世卫组织给出疫苗中应包含的主要流行毒株建议后，世卫组织合作中心就会开始生成并分析疫苗生产需要的病毒种株[14]。自 20 世纪 70 年代以来，这主要是通过甲型流感病毒株的基因重组实现的。用鸡胚同时感染为疫苗选择的野毒株和已知产量高的 A/PR8/34（或类似）供体毒株[15]。对高生长子代病毒进行分析，确认了病毒表面存在来源于野毒株的糖蛋白。第二项技术是反向遗传学，这是一项专利技术，可用于致弱高致病性病毒，并将致弱病毒的 HA 和 NA 与主干病毒重新结合[16]。开发完成后，候选重组病毒会根据抗原和遗传特性进行鉴定，然后根据要求发布给感兴趣的机构（图 52.4）。

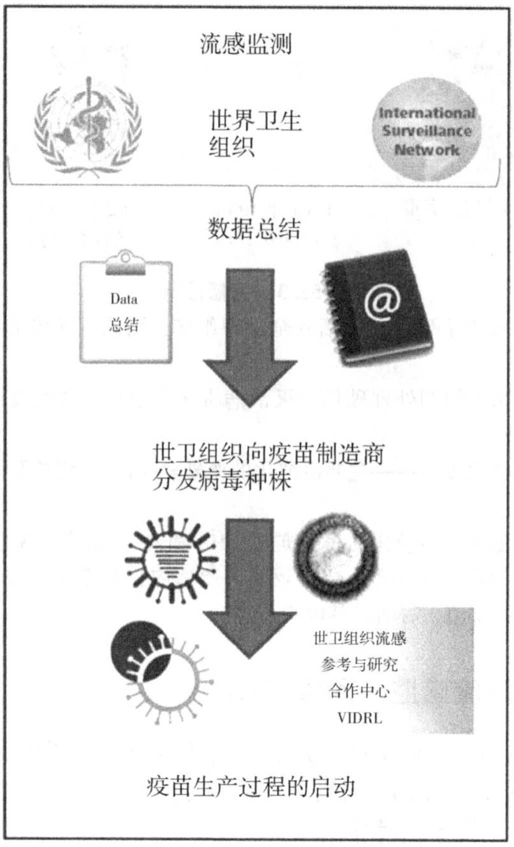

图 52.4　世卫组织全球流感监测和反应系统（GISRS）

52.2.5　疫苗功效/参考试剂

抗原标准和绵羊抗血清随后由基本监管实验室与疫苗制造商合作开发并标准化。可根据全球制造商的要求提供毒株特异性试剂。单次放射性免疫扩散（SRID）测试需要抗原标准和绵羊抗血清，以便量化流感疫苗原液中产生的抗原和投放疫苗[17]。

52.3　流感疫苗生产

流感疫苗的生产有几个独特方面，因此生产过程具有挑战性。

（a）流感疫苗成分必须与世卫组织的实际全球流行病学监测数据相匹配。因此，每次都会开发出最新的疫苗配方。

（b）每次修改都必须获得年度许可证。

（c）采纳流程和响应变化的时间非常紧迫，机会窗口也很短。生产过程是多个步骤中的一步，涉及到不同的机构。

（d）季节性流感疫苗和大流行性流感疫苗相互关联，对生产能力有重大影响。

因此，流感疫苗制造商正在不断优化现有工艺，开发新的制备季节性流感疫苗和候选大流行性疫苗的新方法。努力提高生产力，引入更多自动化流程，设计更灵活、更省时的制造方法。

多数流感疫苗的生产以传统鸡胚技术为基础，这种技术已经在季节性疫苗的生产商应用了30多年[18,19]。相对基于鸡胚的制造方法，新细胞培养技术具有多种优势，并且很可能会成为当前基于鸡胚技术的补充（图52.5）[20,21]。

不管使用哪种生产工艺，无论是基于鸡胚还是基于细胞培养，下游生产工艺都可以从收获的病毒液中生产全病毒疫苗、裂解疫苗或亚单位疫苗。所有制造过程在获得许可前都必须经过广泛的测试和验证，每批都受到严格控制。在本章中，我们将重点关注近期细胞培养流感疫苗的生产，介绍亚单位流感疫苗的流程。

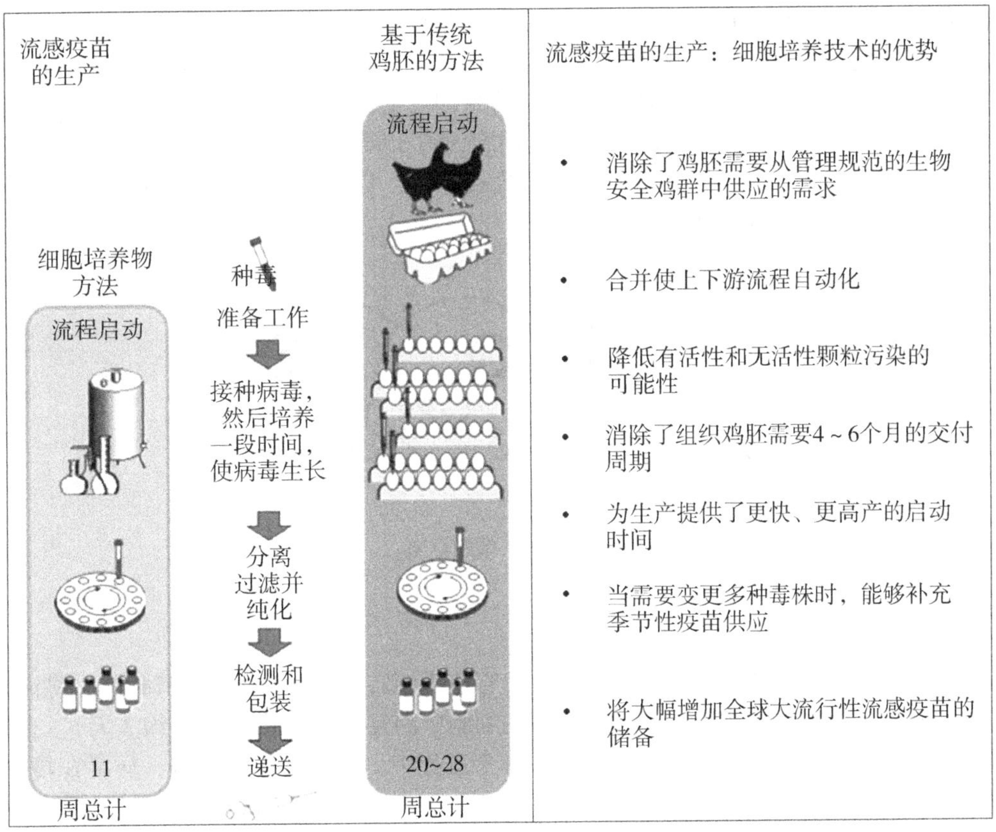

图 52.5　采用基于鸡胚的经典技术和新型细胞培养工艺生产疫苗：时间轴比较和基于细胞培养生产方法的优势

52.4　基于细胞培养的疫苗生产

世卫组织流感参考中心将疫苗种毒分发给疫苗生产商，评估其对疫苗生产的适用性。评估在细胞培养物（或鸡胚）中生长时的产量、抗原稳定性、灭活和纯化过程等因素。欧洲药品管理局（EMA）和美国食品药品管理局（FDA）将考虑所有参与公司在疫苗候选毒株制造商的经验，最终批准疫苗毒株。通常在世卫组织就疫苗组成作出初步决定后 1 个月各国举行会议进行批准。

不同细胞系可用于生产流感病毒，如来自非洲绿猴的肾细胞 VERO[22] 和 1958 年来自一只健康狗肾脏的 Madin-Darby 犬肾（MDCK）细胞[23]。昆虫细胞等重组流感蛋白的替代表达系统正在研究和评估之中。对 MDCK 细胞系进行优化，用于生产流感疫苗，并采用多种病毒变株进行测试。因为在悬浮液中生长，所以已经证明特别适合生产流感疫苗。这样不需要贴壁培养就能繁殖，大大简化了工业生产。

细胞培养流感疫苗生产产生的亚单位流感疫苗可分为以下 8 个步骤（图 52.6、图 52.7、图 52.8、图 52.9、图 52.10、图 52.11 和图 52.12），总结方案见图 52.13。

52.4.1 细胞繁殖

繁殖细胞系的准备工作从细胞系“种子”批次解冻开始。对于“首步”细胞系，繁殖应从种子细胞的小规模预培养繁殖开始。细胞储存在-196℃液氮中。在生产中，将所谓含有 1 mL 约 1 000万个细胞的冷冻管的最少量细胞解冻，并再从 10 L 到 100 L 然后到 1 000 L 体积的 3 个步骤中繁殖。在每个阶段，细胞在温度、氧气、酸碱度和营养供应方面都获得了最佳生长条件。计算机系统会持续监控发酵罐（不锈钢罐）中细胞的繁殖，该系统可自动检查所有数据，并准确记录各个步骤。细胞繁殖发生在所谓洁净室中的封闭发酵罐系统中，为员工、环境、患者和产品带来最大的安全性和纯净度。从液氮中取出细胞大约 3 周后，在 1 000 L 发酵罐中已经生长出足够的细胞。相比之下，在以鸡胚为基础的传统技术中，首次接种可能需要长达 6 个月的时间组织鸡胚供应、群设置等（图 52.6）。

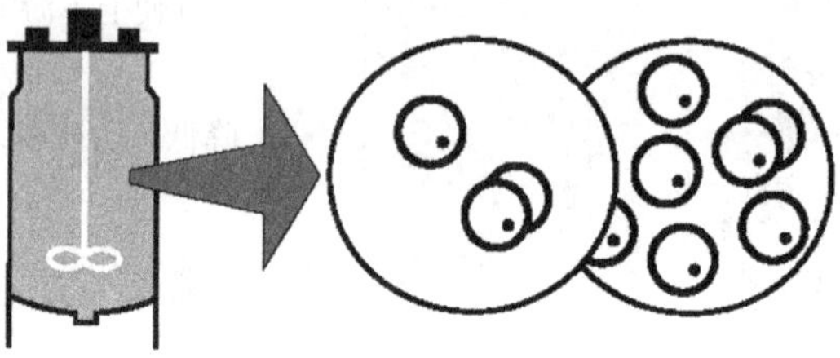

图 52.6　细胞繁殖

52.4.2 病毒生产

当达到一定的预定密度时，细胞将通过封闭管道系统转移到 2 500 L 发酵罐中。将流感病毒加入罐中，开始感染哺乳动物细胞。病毒在宿主细胞中的复制需要几天（大约 3 天），直到产生出充足的病毒量以满足下游的处理和生产。在这个过程中，细胞将死亡裂解，病毒在培养基/上清液中释放（图 52.7）。

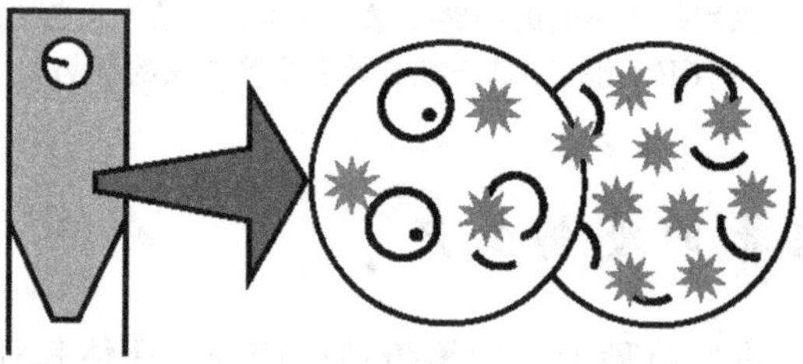

图 52.7　病毒繁殖

52.4.3 病毒纯化

一系列纯化过程的第一步是所谓的分离，病毒悬浮液通过分离从细胞残余物中分离出来。在随后的层析过程中，从培养基溶液中获得并分离出病毒。下一步将进行超滤，进一步浓缩中间产物。这一步发生在自动化和封闭系统中，而基于鸡胚的病毒收集主要是一个人工过程，需要提取受感染的细胞和液体，然后收集病毒（图 52.8）。

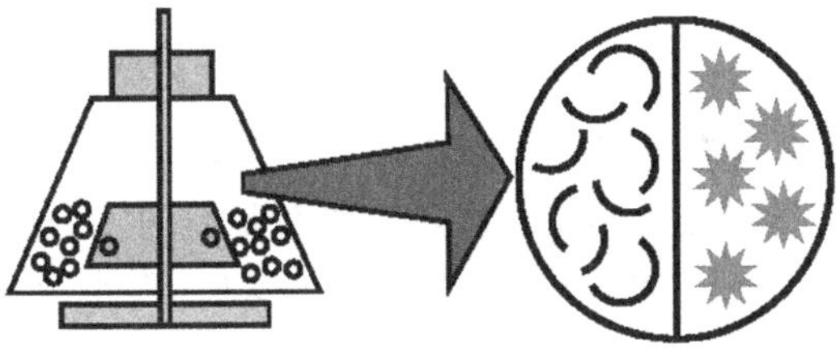

图 52.8　病毒纯化

52.4.4　病毒灭活、裂解和亚单位提取

纯化后，病毒通过化学反应过程灭活（不会引起感染）。大多数制造商为灭活病毒使用 β-丙内酯或甲醛。随后用去污剂分离/提取表面抗原（HA 和 NA），因为后续亚单位流感疫苗只需要特定表面蛋白部分（图 52.9）。

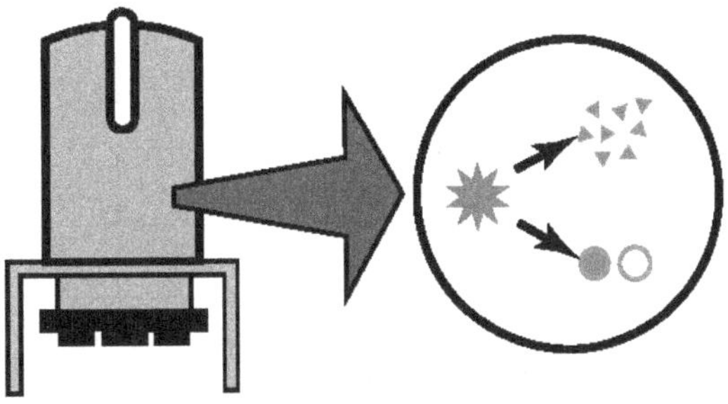

图 52.9　病毒灭活、裂解和亚单位提取

52.4.5　精制

病毒分裂后，通过超速离心去除不需要的病毒核心结构，这些结构在诱发中和抗体方面不起作用。这种纯化技术和基于鸡胚的疫苗过程基本类似，所得纯化亚单位疫苗在成分上与基于鸡胚的疫苗非常类似。在进一步纯化和浓缩步骤后，在最初感染产生的 2 500 L 抗原浓缩物中大约获得 10 L 病毒的抗原浓缩物。由于季节性流感疫苗含有 3 种（或最近 3 种）病毒株（H1N1、H3N2 和 1 种或两种 B 株），因此生产过程必须进行 3～4 次——每种病毒株 1 次（流感疫苗原液），以便生产 3 价或 4 价疫苗（图 52.10）。

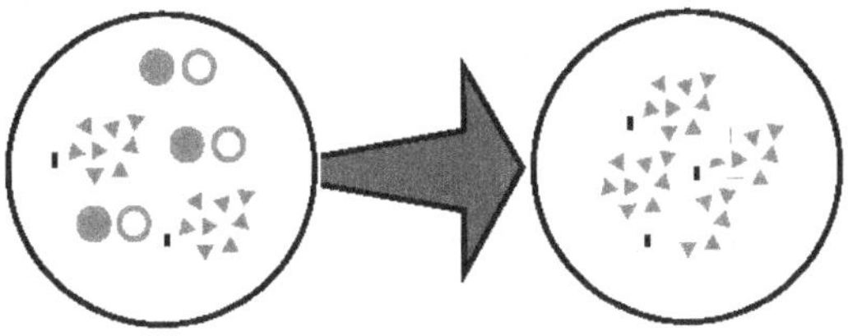

图 52.10　处理

52.4.6　疫苗特性和质量测试（原液释放）

随后，通过单个放射性免疫扩散（SRID）分析对 3 个毒株浓缩液的 HA 抗原含量进行分析（图 52.11）[24,25]。单个放射免疫扩散试验是免疫学中采用的一种技术，通过测量抗原样品周围

沉淀素复合物的直径确定抗原量，该复合物可标记抗原和悬浮在培养基（如琼脂凝胶）中抗体之间的边界。通过测量未知溶液和已知浓度标准抗原的沉淀盘直径可以用简单的方法估算出未知溶液的抗原浓度。

试剂需要对疫苗中的毒株具有特异性，生产和分发大约需要 3 个月的时间（见上文：疫苗功效/参考试剂）。这可能会延缓疫苗的发放，还需要对制造的每个原液产品和材料释放进行与工艺相关的杂质、无菌性和残留传染性病毒的额外测试以及稳定性研究（图 52.11）[26]。

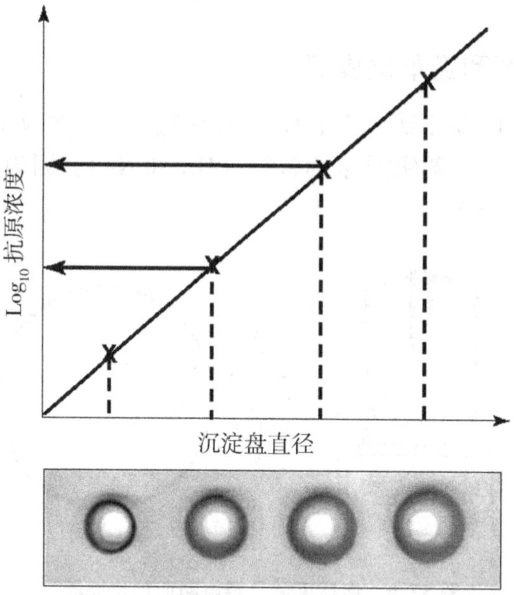

图 52.11　单个放射性免疫扩散（SRID）试验

52.4.7　配制（混合并填充）

使用浓度类似于体液的无菌磷酸盐缓冲盐溶液，将 3 种流感抗原原液稀释成标准剂量。然后将最终疫苗无菌过滤并添加到疫苗瓶、预装注射器或鼻腔喷雾容器中（图 52.12）。

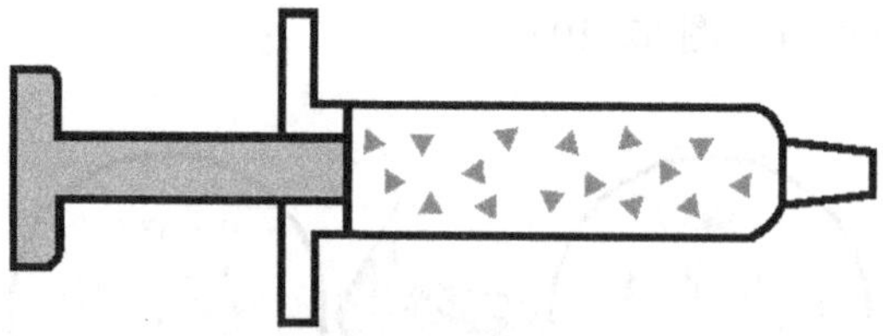

图 52.12　将已知浓度的标准抗原溶液混合并填充

52.4.8　最终质量控制和批号发布

每一个三价批次或“批号”（如单个原液）的疫苗均需经过严格的质量控制测试，确保在疫苗发布前符合监管机构的安全性、有效性和稳定性标准。此时，流感细胞培养亚单位疫苗的开发阶段已经完成。下面就是完成许可程序（图 52.13）。

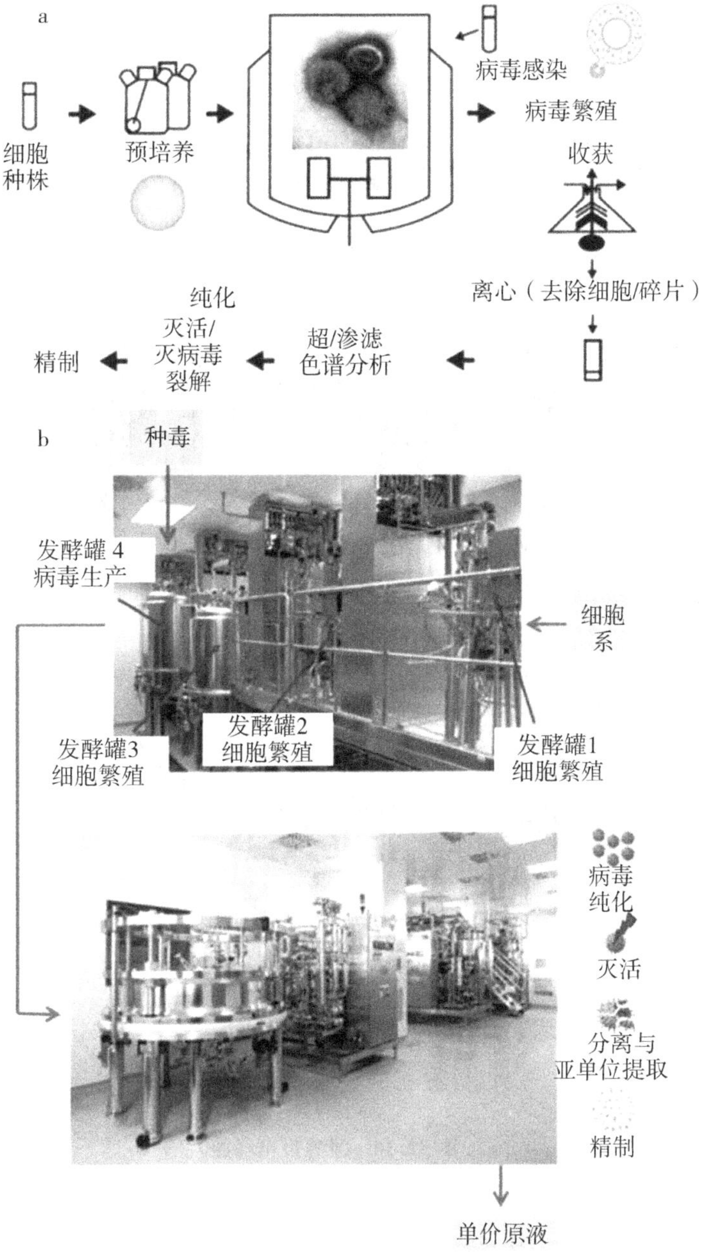

图 52.13　流感疫苗原液制备

从细胞解冻到最终抗原浓缩物，生产含有一种病毒株的单价原液大约需要 6 周。三价疫苗需要 3 轮原液生产。单价原液制备后，还需要 10 周的时间进行混合、灌装、包装和监管批准，才能装运最终疫苗产品［（a）：工作流程示意图；（b）：制造设备示例显示过程］。

52.5　流感疫苗释放

52.5.1　临床试验和监管机构审查

疫苗属于生物制品，由美国食品药品管理局（FDA）和欧洲药品管理局（EMA）监管。欧洲要求进行临床试验，证明每种新流感疫苗制剂的安全性和免疫原性。而在美国或南半球国家则不是如此。

疫苗批次也要仔细标记，防止分发后发现问题。只有在监管机构正式发布后，疫苗才能交付使用[26,27]（图 52.14）。

总体来说，从细胞提取到成品季节性流感疫苗包装并准备完毕进行交付大约需要 16 周。如上所述，毒株特异性试剂的可用性等外部因素会影响进程。图 52.15 示意描述了季节性流感疫苗接种活动的复杂过程，包括公共卫生机构/实验室和疫苗制造商的各种动态互动。

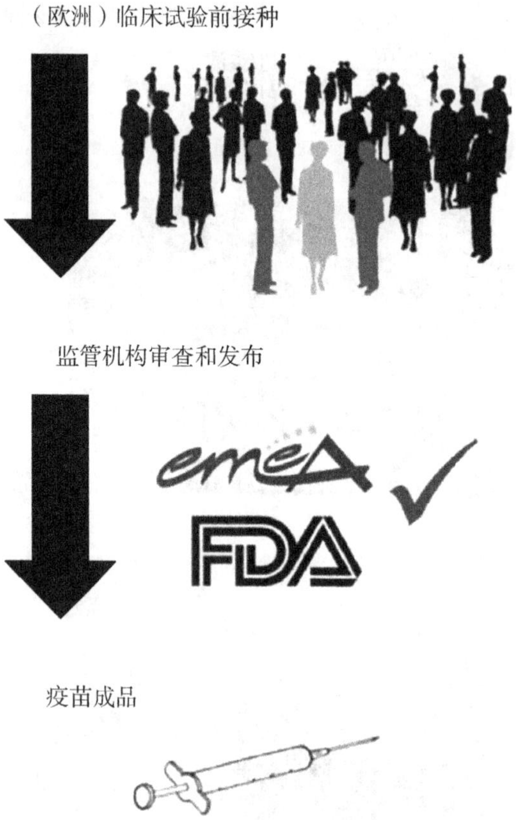

图 52.14　流感疫苗释放

52.6　流感疫苗的改进和未来趋势

为及时提供安全有效的流感疫苗，需要对高度复杂的生产过程精确协调，这个过程涉及到公共卫生机构、实验室和疫苗制造商的广泛专业知识。流感疫苗成分必须与世卫组织的实际全球流行病学监测数据相一致，这些数据反映的是流感病毒的持续演变。因此，几乎每年都会开发一种更新的疫苗配方。目前北半球和南半球季节性流感疫苗生产工艺调整在两个压缩生产周期内进行，即每批产品限时约 6 个月。这些时间限制为新技术和临床开发项目带来巨大的挑战。尤其是在流感大流行期间，疫苗生产面临着与时间的竞争。

尽管在 2009 年 H1N1 流感大流行期间，在大流行宣布后大约 3 个月就可获得第一批疫苗，但数量有限，只有少数人能够在大流行浪潮已经达到顶峰前获得接种。Dormitzer 等最近进行的详细审查显示，平台灵活度越高，疫苗生产速度越快，监管流程也更方便，这是非常理想的，制造商和政府以及研究机构都进行了一波创新激励[28]，开发出了流感疫苗生产新方法[29]。方法

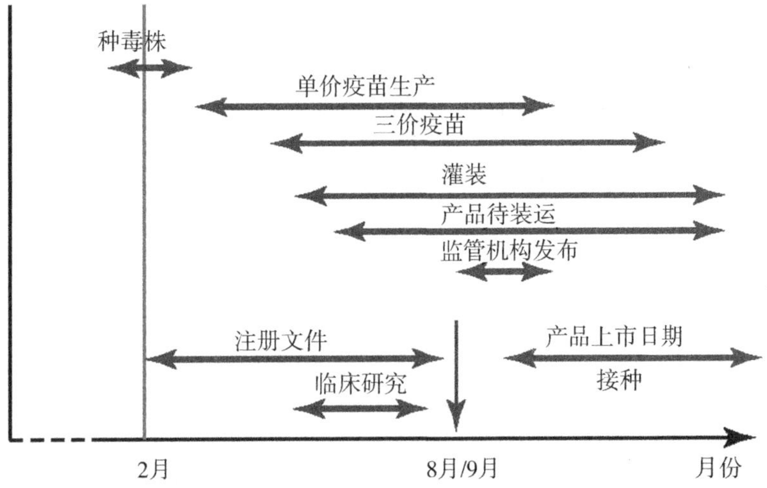

图 52.15 从流感疫苗种子毒株开始的北半球季节性疫苗接种活动周期

涵盖了从研究或早期开发项目到获得许可产品。然而，流感疫苗的新程序也可能提高疫苗生产和标准化的复杂性，提出必须应对的安全监管新挑战。

历史上，大多数工作都集中在对已获得许可的鸡胚疫苗进行改进及其生产上，重点是提高生产力和生产过程的自动化，与基于鸡胚的疫苗生产相比，基于细胞培养的流感疫苗生产具有很多优势应对未来流感相关挑战的潜力更大[20,21,30]。一般来说，毒株生产程序的改进和疫苗功效/参考试剂[31,32]的开发将加快流感疫苗的生产。新流感细胞培养疫苗生产介绍评估 HA 含量的分析策略可以进一步加速产品发布[33-35]。流感疫苗的其他新技术正在探索之中，包括重组策略(生成基于重组亚单位的流感疫苗[36-38]或基于重组病毒样颗粒（VLP）的流感疫苗)[39,40]以及生产“减毒活疫苗（LAIV）”[41-43]和“活载体流感疫苗”[44,45]的方法。

通过在疫苗制剂中添加佐剂，可以进一步提高流感疫苗的功效和生产能力[46]。佐剂是疫苗中包含的物质，用于提高疫苗接种人的抗体产量和免疫反应，或减少疫苗所需的抗原量（剂量)。后者是提高全球疫苗生产能力的最有效方式。几种佐剂已经证明，使用较低的抗原剂量，能够将抗体水平提高到具有保护作用的水平。重要的是，角鲨烯乳剂 MF59 在 2009 年 H1N1 全球大流行期间广泛使用，其安全性在各类人群中得到进一步验证，包括幼儿[47,48]和孕妇[49]。

为提高覆盖率，我们正在努力生产“高价”疫苗制剂，例如含有 4 种毒株的流感抗原的 4 价流感疫苗（QIV)[50,51]。最终目标将是生产一种“通用流感疫苗”，这种疫苗不需要随着流感病毒株的变化而改变。通用疫苗的方法之一是纳入适当表达的保守抗原，从而形成有效、持久的疫苗配方，广泛覆盖不同毒株[39,50-55]。

和传统肌内和皮下给药的新型给药方法（例如，口服、皮下、鼻内给药）不同，这种方法可以简化疫苗的应用，同时提高安全性和患者依从性[43,56-58]。

总之，技术和科学创新将进一步促进安全和创新的疫苗方案，从而显著扩大对流感的保护。全世界对有效流感疫苗需求的增加不仅是对疫苗公司技术能力的挑战，也是对其财务实力和战略眼光的挑战。重要的是，这些挑战必须与公共卫生官员有效落实疫苗、控制流感所做的工作相一致。

致谢

谨向 Heidi Trusheim 和 KarstenKattmann 对手稿的批判性阅读表示感谢。

参考文献

[1] Eccles, R.: Understanding the symptoms of the com-mon cold and infl uenza. Lancet Infect. Dis. 5, 718-25 (2005). doi: 10. 1016/S1473-3099 (05) 70270-X.

[2] Health topics infl uenza-WHO World Health Organization, Geneva, Switzerland. http: //www. who. int/topics/infl uenza/en/ (2013).

[3] Infl uenza (Seasonal) Fact sheet N°211-WHO World Health Organization, Geneva, Switzerland. http: //www. who. int/mediacentre/factsheets/fs211/en/ (2009).

[4] Seasonal infl uenza (fl u) /infl uenza (fl u) viruses-CDC Centers for Disease Control and Prevention, Atlanta, USA. http: //www. cdc. gov/fl u/about/viruses/index. htm.

[5] Webster, R. G., Laver, W. G.: Antigenic variation in infl uenza virus. Biology and chemistry. Prog. Med. Virol. 13, 271-338 (1971).

[6] Hay, A. J., Gregory, V., Douglas, A. R., Lin, Y. P.: The evolution of human infl uenza viruses. Philos. Trans. R. Soc. Lond. B Biol. Sci. 356, 1861-70 (2001). doi: 10. 1098/rstb. 2001. 0999.

[7] Gasparini, R., Amicizia, D., Lai, P. L., Panatto, D.: Afl unov (©): a prepandemic influenza vaccine. Expert Rev. Vaccines 11, 145-57 (2012). doi: 10. 1586/erv. 11. 170.

[8] Beigel, J., Bray, M.: Current and future antiviral therapy of severe seasonal and avian infl uenza. Antiviral Res. 78, 91-102 (2008). doi: 10. 1016/j. antiviral. 2008. 01. 003.

[9] Jefferson, T., et al.: Neuraminidase inhibitors for pre-venting and treating infl uenza in healthy adults and children. Cochrane Database Syst. Rev. 18, 1 (2012). doi: 10. 1002/14651858. CD008965. pub3.

[10] Oxford, J., Gilbert, A., Lambkin-Williams, R.: Infl uenza vaccines have a short but illustrious history of dedicated science enabling the rapid global production of A/Swine (H1N1) vaccine in the current pandemic. In: Del Giudice, G., Rappuoli, R. (eds.) Infl uenza Vaccines for the Future, pp. 115-147. Birkhauser Inc, Basel (2011).

[11] Verma, R., Khanna, P., Chawla, S.: Infl uenza vaccine: an effective preventive vaccine for developing countries. Hum. Vaccin. Immunother. 8, 675-8 (2012). doi: 10. 4161/hv. 19516.

[12] Ellebedy, A. H., Webby, R. J.: Infl uenza vaccines. Vaccine 27 (Suppl 4), D65-D68 (2009). doi: 10. 1016/j. vaccine. 2009. 08. 038.

[13] WHO Global Infl uenza Surveillance Network (GISN) Surveillance and Vaccine Development-WHO Collaborating Centre for Reference and Research on Infl uenza (VIDRL), North Melbourne, Australia. http: //www. infl uenzacentre. org/centre_GISN. htm.

[14] Gerdil, C.: The annual production cycle for infl uenza vaccine. Vaccine 21, 1776 - 1779 (2003). pii: S0264410X03000719.

[15] Burnet, F. M.: Growth of infl uenza virus in the allantoic cavity of the chick embryo. Aust. J. Exp. Biol. Med. Sci. 19, 291-295 (1941).

[16] Hoffmann, E., Neumann, G., Kawaoka, Y., Hobom, G., Webster, R. G.: A DNA transfection system for generation of infl uenza A virus from eight plasmids. Proc. Natl. Acad. Sci. U. S. A. 97, 6108-6113 (2000). doi: 10. 1073/pnas. 100133697.

[17] Expert Committee on Biological Standardization Geneva, 17 to 21 October 2011 Proposed Generic Protocol for the Calibration of Seasonal/Pandemic Influenza Antigen Working Reagents by WHO Essential Regulatory Laboratories (WHO/ BS/2011. 2183). http: //www. who. int/biologicals/ expert_committee/BS2011. 2183_Flu_vax_ERL_calibration_protocol. pdf.

[18] Hickling J., D'Hondt E. A review of production technologies for infl uenza virus vaccines, and their suitability for deployment in developing countries for infl uenza pandemic preparedness-WHO World Health Organization Initiative for Vaccine Research Geneva Switzerland Date: 20 December 2006. www. who. int/entity/vaccine_research/diseases/infl uenza/Flu_vacc_manuf_tech_report. pdf.

[19] Matthews, J. T.: Egg-based production of infl uenza vaccine: 30 years of commercial experience. The Bridge 36, 17-24 (2006).

[20] Rappuoli, R.: Cell-culture-based vaccine production: technological options. The Bridge 36, 25-30 (2006).

[21] Dormitzer, P. R.: Cell culture-derived influenza vaccines. In: Del Giudice, G., Rappuoli, R. (eds.) Infl uenza Vaccines for the Future, pp. 293-312. Birkhauser Inc, Basel (2011).

[22] Kistner, O., Barrett, P. N., Mundt, W., Reiter, M., Schober-Bendixen, S., Dorner, F.: Development of a mammalian cell (Vero) derived candidate infl uenza virus vaccine. Vaccine 16, 960-968 (1998). pii: S0264-410X (97) 00301-0.

[23] Palache, A. M., Brands, R., van Scharrenburg, G.: Immunogenicity and reactogenicity of infl uenza subunit vaccines produced in MDCK cells or fertilised chicken eggs. J. Infect. Dis. 176 (Suppl 1), S20-S23 (1997).

[24] Schild, G. C., Wood, J. M., Newman, R. W.: A single radial-immunodiffusion technique for the assay of infl uenza hemagglutinin antigen. WHO Bull. 52, 223-231 (1975).

[25] Wood, J. M., Schild, G. C., Newman, R. W., Seagroatt, V.: Application of an improved single radial-immunodiffusion technique for the assay of infl uenza hemagglutinin antigen content of whole virus and subunit vaccines. Dev. Biol. Stand. 39, 193-200 (1977).

[26] Recommendations for the production and control of infl uenza vaccine (inactivated) World Health Organization WHO Technical Report Series, No. 927, Annex 3 (2005), http://www.who.int/vaccine_research/diseases/infl uenza/ TRS_927_ANNEX_3_Infl uenza_2005.pdf.

[27] Wood, J. M., Levandowski, R. A.: The infl uenza vaccine licensing process. Vaccine 21, 1786-1788 (2003). pii: S0264410X03000732.

[28] Sambhara, S., Rappuoli, R.: Improving infl uenza vaccines. Expert Rev. Vaccines 11, 871-872 (2012). doi: 10.1586/erv.12.79.

[29] Dormitzer, P. R., Tsai, T. F., Del Giudice, G.: New tech-nologies for infl uenza vaccines. Hum. Vaccin. Immunother. 8, 45-58 (2012). doi: 10.4161/hv.8.1.18859.

[30] Montomoli, E., et al.: Cell culture-derived infl uenza vaccines from Vero cells: a new horizon for vaccine production. Expert Rev. Vaccines 11, 587-94 (2012). doi: 10.1586/erv.12.24.

[31] Strecker, T., et al.: Exploring synergies between aca-demia and vaccine manufacturers: a pilot study on how to rapidly produce vaccines to combat emerging pathogens. Clin. Chem. Lab. Med. 50, 1275-9 (2012). doi: 10.1515/cclm-2011-0650.

[32] WHO Writing Group, Ampofo W. K. et al.: Improving infl uenza vaccine virus selection: report of a WHO informal consultation held at WHO headquarters, Geneva, Switzerland, 14-16 June 2010. Infl uenza Other Respi. Viruses. 6, 142-152 (2012). doi: 10.1111/j.1750-2659.2011.00277.x.

[33] Kapteyn, J. C., et al.: HPLC-based quantifi cation of haemagglutinin in the production of egg and MDCK cell-derived infl uenza virus seasonal and pandemic vaccines. Vaccine 27, 1468-77 (2009). doi: 10.1016/j.vaccine.2008.11.113.

[34] Lorbetskie, B., et al.: Optimization and qualifi cation of a quantitative reversed-phase HPLC method for haemagglutinin in infl uenza preparations and its comparative evaluation with biochemical assays. Vaccine 29, 3377-89 (2011). doi: 10.1016/j.vaccine.2011.02.090.

[35] Williams, T. L., et al.: Quantifi cation of infl uenza virus hemagglutinins in complex mixtures using isotope dilution tandem mass spectrometry. Vaccine 26, 2510-20 (2008). doi: 10.1016/j.vaccine.2008.03.014.

[36] Cox, M. M.: Recombinant protein vaccines produced in insect cells. Vaccine 30, 1759-66 (2012). doi: 10.1016/j.vaccine.2012.01.016.

[37] Baxter, R., et al.: Evaluation of the safety, reactoge-nicity and immunogenicity of FluBlok © trivalent recombinant baculovirus-expressed hemagglutinin influenza vaccine administered intramuscularly to healthy adults 50-64 years of age. Vaccine 29, 2272-8 (2011). doi: 10.1016/j.vaccine.2011.01.039.

[38] Song, L., et al.: Efficacious recombinant infl uenza vaccines produced by high yield bacterial expression: a solution to global pandemic and seasonal needs. PLoS One 3, e2257 (2008). doi: 10. 1371/journal. pone. 0002257Infl uenza Cell-Culture Vaccine Production.

[39] Kang, S. M., Kim, M. C., Compans, R. W.: Virus-like particles as universal infl uenza vaccines. Expert Rev. Vaccines 11, 995-1007 (2012). doi: 10. 1586/erv. 12. 70.

[40] Haynes, J. R.: Influenza virus-like particle vaccines. Expert Rev. Vaccines 8, 435-45 (2009). doi: 1 0. 1586/erv. 09. 8.

[41] Gasparini, R., Amicizia, D., Lai, P. L., Panatto, D.: Live attenuated influenza vaccine-a review. J. Prev. Med. Hyg. 52, 95-101 (2011).

[42] Monto, A. S., et al.: Comparative effi cacy of inactivated and live attenuated influenza vaccines. N. Engl. J. Med. 361, 1260-7 (2009). doi: 10. 1056/NEJMoa0808652.

[43] Carter, N. J., Curran, M. P.: Live attenuated influenza vaccine (FluMist ©; Fluenz™): a review of its use in the prevention of seasonal influenza in children and adults. Drugs 71, 1591-622 (2011). doi: 1 0. 2165/11206860-000000000-00000.

[44] Kopecky-Bromberg, S. A., Palese, P.: Recombinant vectors as influenza vaccines. Curr. Top. Microbiol. Immunol. 333, 243-67 (2009). doi: 10. 1007/978-3-540-92165-3_13.

[45] Lambe, T.: Novel viral vectored vaccines for the prevention of infl uenza. Mol. Med. 18, 1153-60 (2012). doi: 10. 2119/molmed. 2012. 00147.

[46] O'Hagan, D. T., Tsai, T., Reed, S.: Emulsion-based adjuvants for improved influenza vaccines. In: Del Giudice, G., Rappuoli, R. (eds.) Infl uenza Vaccines for the Future, pp. 327-357. Birkhauser Inc, Basel (2011).

[47] Vesikari, T., Pellegrini, M., Karvonen, A., Groth, N., Borkowski, A., et al.: Enhanced immunogenicity of seasonal infl uenza vaccines in young children using MF59 adjuvant. Pediatr. Infect. Dis. J. 28, 563-571 (2009). doi: 10. 1097/INF. 0b013e31819d6394.

[48] Vesikari, T., Knuf, M., Wutzler, P., Karvonen, A., Kieninger-Baum, D., et al.: Oil-in-water emulsion adjuvant with infl uenza vaccine in young children. N. Engl. J. Med. 365, 1406-16 (2011). doi: 10. 1056/NEJMoa1010331.

[49] Heikkinen, T., Young, J., van Beek, E., Franke, H., Verstraeten, T., et al.: Safety of MF59-adjuvanted A/ H1N1 infl uenza vaccine in pregnancy: a comparative cohort study. Am. J. Obstet. Gynecol. 207, 177. e1-8 (2012). doi: 10. 1016/j. ajog. 2012. 07. 007.

[50] Ambrose, C. S., Levin, M. J.: The rationale for quadrivalentinfl uenza vaccines. Hum. Vaccin. Immunother. 8, 81-88 (2012). doi: 10. 4161/hv. 8. 1. 17623.

[51] Barr, I. G., Jelley, L. L.: The coming era of quadrivalent human influenza vaccines: who will benefi t? Drugs 72, 2177-85 (2012). doi: 10. 2165/11641110-000000000-00000.

[52] Shaw, A. R.: Universal infl uenza vaccine: the holy grail? Expert Rev. Vaccines 11, 923-927 (2012). doi: 10. 1586/erv. 12. 73.

[53] Du, L., Zhou, Y., Jiang, S.: Research and development of universal influenza vaccines. Microbes Infect. 12, 280-6 (2010). doi: 10. 1016/j. micinf. 2010. 01. 001.

[54] Kang, S. M., Song, J. M., Compans, R. W.: Novel vaccines against infl uenza viruses. Virus Res. 162, 31-38 (2011). doi: 10. 1016/j. virusres. 2011. 09. 037.

[55] Rudolph, W., Ben Yedidia, T.: A universal influenza vaccine: where are we in the pursuit of this "Holy Grail"? Hum. Vaccin. 7, 10-11 (2011). pii: 14925.

[56] Belshe, R. B., Newman, F. K., Cannon, J., Duane, C., Treanor, J., et al.: Serum antibody responses after intradermal vaccination against infl uenza. N. Engl. J. Med. 351, 2286-94 (2004). doi: 10. 1056/NEJMoa043555.

[57] Kenney, R. T., Frech, S. A., Muenz, L. R., Villar, C. P., Glenn, G. M.: Dose sparing with intradermal injection of influenza vaccine. N. Engl. J. Med. 351, 2295-301 (2004). doi: 10. 1056/NEJ-

Moa043540.
[58] Ansaldi, F., Durando, P., Icardi, G.: Intradermal influenza vaccine and new devices: a promising chance for vaccine improvement. Expert Opin. Biol. Ther. 11, 415-27 (2011). doi: 10. 1517/14712598. 2011. 557658.

(何继军译)

第53章 美国食品和药物管理局对疫苗的监管

Valerie Marshall①

摘要

疫苗开发是一个受监管要求指导的复杂过程，旨在确保安全有效产品获得许可。疫苗在其整个生命周期中都受到严格的监管，包括科学和临床评估。本章将重点介绍美国针对传染病指征的预防性疫苗开发和许可的监管程序。

53.1 总体要求

疫苗监管的法律框架主要来自《公共卫生服务法》(《美国法典》第42篇第262条第351节)[1]和《美国食品、药品和化妆品法》(FD&C法)(《美国法典》第21篇第321条)[2]。疫苗符合药品和生物制品标准，因为《食品、药品和化妆品法》将药品定义为“用于诊断、治疗、缓解、处理或预防疾病的物品”[2]。

《公共健康服务法》和《食品、药品和化妆品法》通过联邦规则汇编落实，其中包含联邦政府机构在《联邦公报》上公布的通用规则。专门适用于疫苗和其他生物制品许可的联邦法规为《联邦法规》第21篇第600~680节[3]。重要法规和立法适用于疫苗的开发、制造和许可见表53.1。

表53.1 适用于疫苗开发、制造和许可的美国立法和法规

法规名称	具体内容
《公共健康服务法》	(《美国法典》第42篇第262~263条)第351节
《食品、药品和化妆品法》	(《美国法典》第21篇第301~392条)
处方药使用者费用法	1992年、1997年、2002年、2007年、2012年
食品和药品管理局修正法案	2007年
食品药品监督管理局安全与创新法案	2012年

① V. Marshall，MPH（美国公共卫生署军官团，食品和药物管理局（FDA），生物制品评价与研究中心，疫苗研究和审查办公室，美国，马里兰州洛克维尔，E-mail：valerie. marshall@ fda. hhs. gov）

（续表）

法规名称	具体内容	
《联邦规则汇编》第 21 章	第 25 部分	环境影响因素
	第 50 部分	人体保护
	第 58 部分	非临床实验室研究良好实验室规范
	第 201 部分	标签
	第 202 部分	处方药广告
	第 210~211 部分	药品生产质量管理规范
	第 312 部分	临床试用新药（IND）申请
	第 601 部分	许可
	第 600 和第 610 部分	通用生物制品标准

疫苗许可证以安全性和有效性为基础，证明许可证持有人在指定和商定的范围内具有以一致的方式生产产品的能力。《公共健康服务法》（《美国法典》第 42 篇第 262 条）第 351 节规定，生物制品许可证申请可以在以下证明的基础上获得批准：“（a）申请主体的生物制品安全、纯净、有效；（b）制造、加工、包装或保存生物制品的设施符合旨在确保生物制品持续安全、纯净和有效的标准。”《联邦法规》第 21 章第 600 部分对安全性、纯度和功效给出了以下定义。

安全性指“在谨慎用药时，考虑到产品特性与接受者当时的状况，产品对直接或间接受影响人的相对无害性。”

纯度指“成品中相对没有对接受者有害或对产品有害的杂质。”

功效指“产品的特定能力或容量，通过适当的实验室测试或通过以预期方式使用该产品而获得充分控制的临床数据所示，以影响给定的结果。

考虑到新型疫苗产品的多样性，对这些标准的应用需要仔细考虑产品特性、制造方法、目标人群和指征。

美国食品和药物管理局定期发布指导文件，说明对法规的解读和/或机构当前对药物和生物产品的生产、临床前和临床评估特定方面的思考。与法规相反，指导文件没有约束力；因此，制造商为遵守律法规，可以选择其他方法替代相关指南中描述的方法。了解美国食品和药物管理局关于传染病指征预防性疫苗的产品测试和生产临床开发指导文件，见表 53.2 和表 53.3。

表 53.2 选择与产品测试和制造相关的美国食品和药物管理局指导文件

类型	内容
行业指南	用于生产传染性疾病指征病毒疫苗的细胞基质和其他生物材料的特性和质量标准，2010 年 2 月
行业指南	第 1 阶段药品生产管理规范，2008 年 7 月
行业指南	用于传染病适应证的质粒 DNA 疫苗的考量因素，2007 年 11 月
工业指南	预防和治疗性疫苗传染病指征发展毒性研究考量，2006 年 2 月
行业指南	分析程序和方法验证，2000 年 8 月
可预防疾病的联合疫苗评估行业指南	生产、测试和临床研究，1997 年 4 月

表 53.3　选择与临床研究相关的美国食品和药物管理局指导文件

类型	内容
行业指南草案	确定上市前后和批准后临床调查所需的安全数据收集范围，2012 年 2 月
行业非劣效临床试验指南草案	2010 年 3 月
行业指南	支持大流行性流感疫苗许可所需的临床数据，2007 年 5 月
行业指南	支持季节性灭活流感疫苗许可所需临床数据，2007 年 5 月
行业指南	参加预防疫苗临床试验的健康成人和青少年志愿者毒性分级表，2007 年 9 月

随着时间的推移，适用于疫苗开发的重要立法有所发展，部分原因是为了跟上制药和生物医学领域的技术和科学进步行业。1992 年首次颁布的《处方药使用者费用法案》（PDUFA）授权美国食品和药物管理局向制造商收取使用费，以加快药物和生物应用的审查，并根据绩效目标建立上市后药物安全活动[4]。该立法分别于 1997 年（PDUFA II）[5]、2002 年（PDUFA III）[6]、2007 年（PDUFA IV）[7]和 2012 年（PDUFA V）[8]重新获得授权。

2007 年《食品和药品管理局修正法案》（FDAAA）对药品和生物制品的监管进行了重大改革，并向美国食品和药物管理局提供了额外资金，建立了新管理局[7]。例如，FDAAA 在《食品、药品和化妆品法案》中增加了多项新条款，根据这些条款，如果满足某些条件，美国食品和药物管理局可以要求处方药和生物制品在批准时或获批上市后进行研究和临床试验，包括要求风险评估和减轻措施（REMS），以及与安全相关的标签变更[9]。

2012 年《食品和药物管理局安全和创新法案》（“FDASIA”）进一步修订了现有法律，包括重新授权和修改若干项药品和医疗器械条款，为生物仿制药和非专利药制定新的使用者费用法规，以及重新授权两项儿科药物开发鼓励计划[10]。除其他外，FDASIA 还为食品和药物管理局提供了有关药物短缺的新授权。

53.2　监管流程概述

美国食品和药物管理局生物制品评估和研究中心（CBER）是负责监管疫苗和其他生物制品的联邦机构。疫苗应用审查由生物制品评价与研究中心（CBER）疫苗研究和审查办公室、合规和生物制品质量办公室以及生物统计和流行病学办公室进行。生物制品评价与研究中心通过疫苗开发的 4 个主要阶段提供监管指导：新药预审（pre-IND）阶段、新药申请（IND）阶段、许可（生物制剂许可申请，BLA）阶段和上市后阶段（图 53.1），监管监督在每个阶段都有所增强。

生物制品评价与研究中心建立了管理评审流程，为所有监管提交的监管评审提供系统有效方法，确保批准生物制品的安全性和有效性。多学科审查小组由监管项目经理、临床/医疗官员、产品审查人员、统计人员、药理学/毒理学审查人员和其他具有各种背景的科学专家组成，根据美国食品和药物管理局规定的时间表审查疫苗应用和其他监管提交材料。

53.2.1　新药预审阶段（Pre-IND）

新药预审（pre-IND）的应用阶段主要包括实验室测试、生产工艺的开发和疫苗候选物鉴定方法的改进。在此阶段，候选疫苗在体内和体外临床前试验中进行评估，确定是否适合进入人体试验。

53.2.2　新药申请阶段

如有赞助商要对候选疫苗发起临床调查，必须提交一份新药申请报告。第 21 篇 312 节描述

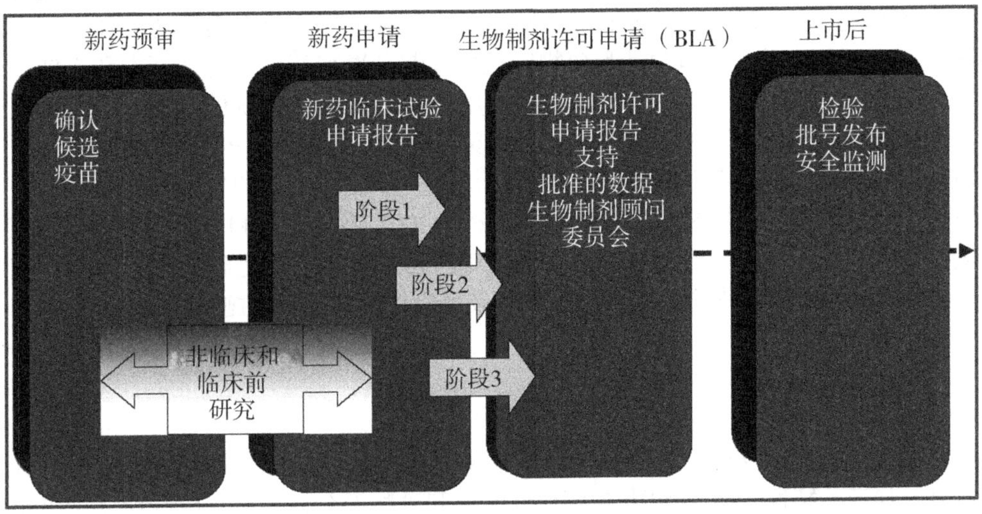

图 53.1　产品开发阶段和食品和药物管理局对疫苗开发的监督

了初始新药申请报告的内容以及新药申请报告条例下的临床试验管理要求[3]。简而言之，新药申请报告包括原材料描述、制造方法和质量检测、临床动物安全数据、拟定第一阶段临床方案以及临床研究者资格。疫苗的临床免疫原性、安全性和有效性在 21 CFR 312.21 中定义的研究性新药申请研究的不同阶段进行评估[11]。对新药申请报告的审查可以让美国食品和药物管理局监控临床试验受试者的安全性，并确保研究设计对疫苗的有效性和安全性进行彻底评估。产品开发过程中对新药申请的修正包括更明确的临床试验和产品质量测试，以及扩大生产的方法。

53.2.3　生物制剂许可申请（BLA）

新药申请阶段之后是许可阶段，新药申请研究完成，证明已经对用于特定用途的安全性和有效性数据进行了收集和分析。赞助商可以提交生物制剂许可申请（BLA），对产品进行生产和商业分销。有关提交生物制剂许可申请的规定见第 21 篇第 600~680 节[12]。生物制剂许可申请包含非临床和临床研究数据，证明疫苗满足安全性、纯度和功效的规定要求。生物制剂许可申请报告的其他强制性内容包括生产过程描述，证明生产过程的控制和一致性、稳定性数据，以及拟定包装说明书和标签。

美国食品和药物管理局可能要求制造商在生物制品评估和研究中心完成生物实验室评估之前，向疫苗和相关生物产品咨询委员会（VRBPAC）提交支持疫苗安全性和有效性的数据。生物制剂顾问委员会是一个由科学专家和临床医生组成的美国食品和药物管理局常务咨询委员会，负责审查数据是否能够支持目标人群的安全性和有效性，并可能对独特的复杂临床、制造和测试问题进行考量。生物制剂顾问委员会在决定发布疫苗许可证时，会强烈考虑委员会的建议。

53.2.4　上市后阶段

在获得许可证后，只要生产商持有该产品的许可证，就必须继续进行定期设施检查等疫苗和生产活动监测。对制造企业进行检查，确定许可产品是否按照许可申请中的说明和适用法规进行制造和测试。此外，所有获得许可的疫苗都必须按照第 21 篇 610 节的要求进行适当的批量发布测试[13]。

美国食品和药物管理局继续通过上市后监督对许可产品进行监督。制造商必须提供关于许可疫苗安全性的持续报告。疫苗不良事件报告系统（VAERS）接受来自医疗保健供应商、制造商和公众的任何可能与美国许可疫苗相关的不良事件报告。疾病控制和预防中心（CDC）及美国食品和药物管理局负责监测疫苗引起的不良反应或趋势。除疫苗不良事件报告系统之外，美国食品和药物管理局正在开发加强版上市后监测系统，对包括疫苗在内的药物和生物产品安全性进行跟踪。2008 年 5 月，美国食品和药物管理局发起了警戒行动，旨在开发和实施一个安全监测系统，作为对现有监测与药物和生物制品使用相关不良事件系统的补充。许可后快速免疫安全监测是警戒行动范围内的一个新型国家主动疫苗安全监测系统（PRISM）[14]。

疫苗获批后，疫苗制造商可进行上市后研究，进一步评估疫苗的安全性，这可能有助于确认在许可前研究中未检测到的罕见不良反应。此外，2007 年《食品和药品管理修正法案》第 9 条授予美国食品和药物管理局更大的权力，要求对药物和生物制品进行某些上市后研究和临床试验，以及在某些情况下与安全相关的标签变更及风险评估和减轻措施[9]。

53.3 临床前指南

临床前测试是将候选疫苗从实验室转移到临床的先决条件，包括将候选产品接种到人体前的所有测试，如产品特性、概念证明/免疫原性研究和动物安全性测试。除证明候选疫苗临床前安全性和生物活性外，临床前数据应支持拟定临床配方，还需要提供关于原材料的来源和质量以及制造过程的详细信息。临床前研究的具体要求具有产品特异性，取决于疫苗类型（即重组、活性、减毒）、生产工艺和作用机制。

临床前研究旨在确定候选疫苗的体外和体内特性，包括动物模型中的安全性和免疫原性评估。临床前动物研究是确认人类可能面临的风险、帮助规划后续临床研究方案的宝贵工具。美国食品和药物管理局对预防性疫苗的非临床安全性检测方法在世卫组织发布的指导文件《世卫组织疫苗非临床评估指南》中进行了概述[15]。本文件在个案方法的基础上提供了疫苗非临床安全性评估的基本原则和方法，同时可以测试要求的灵活性。

建议制造商在新药预审会议期间与生物制品评估和研究中心讨论临床前和早期临床检测的方法和要求。与生物制品评估和研究中心的对话将有助于阐明设计临床前和临床研究的方法，对开展临床试验提供支持。

53.4 临床指南

疫苗将在特定人群中进行广泛评估，生成可用作批准基础的安全性和有效性数据。

临床研究应遵守第 21 篇第 312 节的规定[3]。随机、双盲、对照试验是证明候选疫苗功效的黄金标准。

临床发展分为 3 个阶段，通常称为第一、第二和第三阶段研究。第一阶段研究旨在评估候选疫苗在少数健康受试者中的安全性和免疫原性。第二阶段研究可以招募数百名受试受试者，进一步评估疫苗的安全性和免疫反应，并提供对常见不良事件和实验室异常率的初步估计。第三阶段研究是大规模试验，提供支持许可所需的疫苗安全性和有效性的关键文件。第三阶段研究通常在代表产品目标人群的群体中进行。面向在第三阶段研究中获得的临床数据应由同一种产品生成，该产品通过特定工艺生产，所使用的产品规格面向美国市场销售。

在临床研究开始前应选择明确的终点，提高结果的科学性和统计可信度。临床疾病疗效终点可为评估疫苗提供最严格的科学依据。疗效的适当终点取决于传染病和候选疫苗的特点，通常与临床上显著的疾病发病率或死亡率有关[16]。疫苗是一项重要的公共卫生干预措施，可保护民众和社区免受与许多传染病有关的死亡率和发病率的影响。美国食品和药物管理局负责在整个复杂的疫苗开发和许可过程中提供监管监督。获得许可后，疫苗安全性通过批量发布测试、检查和产品监督，获得持续监控。美国食品和药物管理局通过其全面审查机制确保许可疫苗的安全性、有效性和可用性。

参考文献

[1] Public Health Service Act. July 1, 1944, Chap. 373, Title III, Sec. 351, 58 Stat. 702, currently codifi ed at 42 United States Code, Sec. 262.

[2] Federal Food, Drug and Cosmetic Act. 21 United States Code, Sec. 321 (1938).

[3] Code of Federal Regulations, Washington, DC, Offi ce of the Federal Register, National Archives & Records Administration. Title 21, Part 312 (2013).

[4] Prescription Drug User Fee Act, Public Law No. 102-571 (1992).

[5] Food and Drug Administration Modernization Act of 1997, Public Law No. 105-115 (1997).

[6] Public Health Security and Bioterrorism Preparedness and Response Act of 2002, Public Law No. 107-188 (2002).

[7] Food and Drug Administration Amendments Act of 2007, Public Law No. 110-85 (2007).

[8] Food and Drug Administration Safety and Innovation Act (FDASIA), Public Law No. 112-144 (2012).

[9] Gruber, M. F.: The review process for vaccines for pre-ventive and therapeutic infectious disease indications regulated by the US FDA: impact of the FDA Amendments Act 2007. Expert Rev. Vaccines 10 (7), 1011-1019 (2011).

[10] Regulatory Information. Food and Drug Administration Safety and Innovation Act (FDASIA). 13 July 2013. Food and Drug Administration. http: // www. fda. gov/RegulatoryInformation/Legislation/ FederalFoodDrugandCosmeticActFDCAct/ SignificantAmendmentstotheFDCAct/FDASIA/ucm20027187. htm. Accessed 25 July 2013.

[11] Code of Federal Regulations, Washington, DC, Offi ce of the Federal Register, National Archives & Records Administration. Title 21 CFR Part 312. 21 (2013).

[12] Code of Federal Regulations, Washington, DC, Offi ce of the Federal Register, National Archives & Records Administration. Title 21 CFR Parts 600 through 680 (2013).

[13] Code of Federal Regulations, Washington, DC, Offi ce of the Federal Register, National Archives & Records Administration. Title 21 CFR Part 610 (2013).

[14] Nguyen, M., Ball, R., Midthun, K., Lieu, T. A.: The Food and Drug Administration's Post-Licensure Rapid Immunization Safety Monitoring program: strengthening the federal vaccine safety enterprise. Pharmacoepidemiol. Drug Saf. 21 (Suppl 1), 291-297 (2012).

[15] World Health Organization, WHO Guidelines on Nonclinical Evaluation of Vaccines. http: //www. who. int/biologicals/publications/nonclinical_evaluation_vaccines_nov_2003. pdf. Accessed on 25 July 2013.

[16] Hudgens, M., Gilbert, P. G., Gulf, S. G.: Endpoints in vaccine trials. Stat. Methods Med. Res. 13 (2), 89-114 (2004).

(何继军译)

第 54 章　欧盟对疫苗的监管要求

Bettina Klug，Patrick Celis，
Robin Ruepp 和 James S. Robertson①

摘要

欧洲制药立法为疫苗的市场授权提供了全面框架；根据产品的性质，有 3 种不同应用途径。对于创新性疫苗（特别是重组疫苗），集中上市授权程序是强制性的。

本章节概述了欧洲预防和治疗性疫苗监管程序和市场许可申请的要求。重点介绍了疫苗质量、非临床和临床开发的最相关指南。

54.1　监管流程和定义

创新疫苗，特别是重组疫苗（重组蛋白疫苗和重组病毒载体疫苗），必须在欧盟通过集中程序进行评估和批准。在以下情况下，也可以集中批准其他新型疫苗。由申请人证明（符合第 726/2004 号法规（欧共体）第 3 条中概述的“可选范围”下的集中流程的资格）。对市场授权申请（MAA）的集中评估由欧洲药品管理局（EMA）科学委员会进行，评估时间长达 210 天，之后，人类用药委员会（CHMP）通过建议或拒绝授权的意见。如果是肯定意见，则由欧盟委员会（欧共体）给与市场授权，该授权在所有欧盟成员国有效（关于集中流程的更多信息见《申请人须知》2A[1]）。对于欧盟传统非重组疫苗的授权，开发商可以将市场授权申请提交给 1 个或多个国家药品主管部门进行审查（有关互认评审和分散流程的更多信息见《申请人须知》2A[2]）。图 54.1 描述了欧盟药品上市许可流程的不同途径。②

虽然欧洲医药法没有提供正式预防措施，但通常认为疫苗是含有 1 种或多种免疫原性抗原的传染病预防药品。含有 1 种或多种免疫原性抗原的治疗性药品，如慢性艾滋病病毒感染、慢性乙型或丙型肝炎感染、癌症或阿尔茨海默病，通常称为治疗性疫苗或主动免疫疗法。产品开发的科学原理与针对传染疾病的预防性疫苗相同。

基于病毒（或其他）载体或 DNA 质粒的传染病预防疫苗被明确排除在基因治疗药品（GT-MP）[3]定义之外，而包含肿瘤抗原的相同载体或质粒（所谓的“癌症疫苗”）在欧盟则被认为是基因治疗药品，因此属于先进治疗药品（ATMP）的定义和相应的监管框架。基于细胞的癌症免疫治疗产品（如美国许可的 sipuleucel-T）或针对传染病的基于细胞的疫苗（如载有人类病原体肽片段的树突状细胞）也是先进治疗药品（特别是体细胞疗法药品）。对于先进治疗药品，建立了一个专门的监管框架[4]，旨在通过先进治疗委员会（CAT）促进这些新产品的开发和许可。对于先进治疗药品的所有市场授权申请，先进治疗委员会在人类用药委员会采纳其最终意见之

---

① B. Klug，MD（✉）（保罗埃利希学院，德国兰根，E-mail：bettina. klug@ pei. de）

P. Celis，PhD · R. Ruepp，PhD（欧洲药品管理局（EMA），英国伦敦）

J. S. Robertson，PhD（国家生物标准与控制研究所，英国赫特福德郡波特斯巴）

② 本文中表达的观点仅代表作者本人的观点，不得理解或引用为代表或反映欧洲药品管理局或其委员会或工作组之一、Paul-Ehrlich 研究所或国家生物标准和控制研究所的立场。

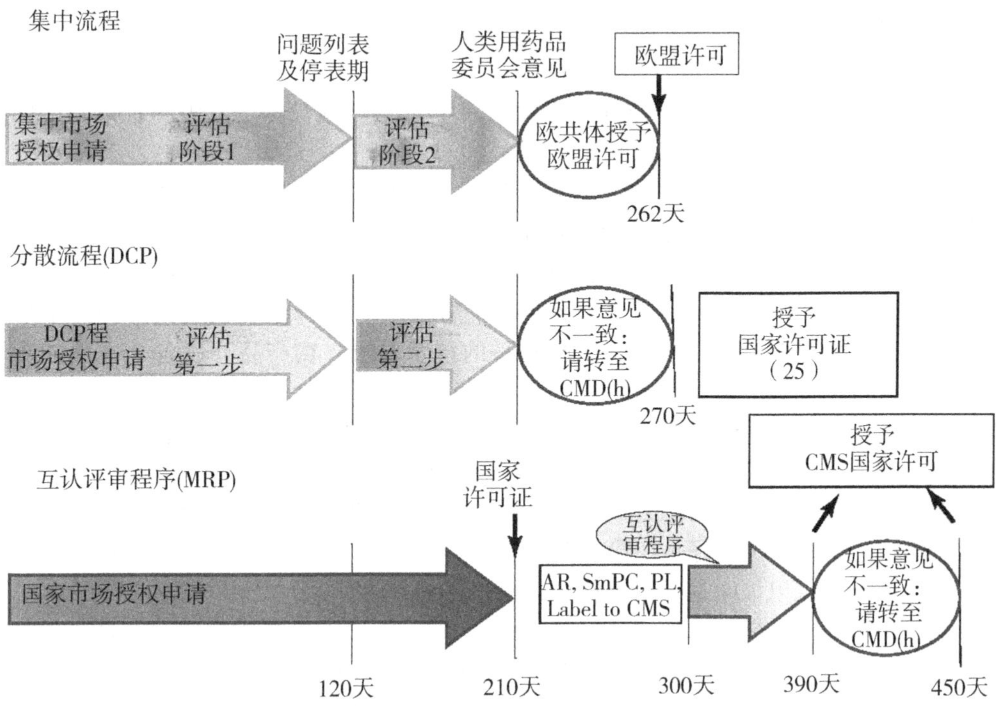

图 54.1　欧盟药品上市许可程序的 3 种途径说明

仅描述了主要流程步骤。该数字不包括转诊和复查程序。了解更多信息，请参阅申请人须知[1,2]。并非所有程序都对所有药物开放（例如，重组疫苗和细胞治疗药物必须遵循集中程序）。对于集中程序（基于 CHMP 的积极意见），欧盟委员会将授予在整个欧盟有效的市场授权（许可证）；在多个欧盟成员国（2~28 个）进行分散流程（DCP）和互认评审流程（MRP）后获得许可证。缩写：CHMP 人类用药品委员会，MAA 市场授权申请，EC 欧盟委员会、EU 欧盟欧洲联盟；CMD（h）互认评审和分散流程协调小组——人员、应收帐款评估报告，SmPC 产品特性总结，PL 包装说明书等。CMS 有关成员国。

前编写一份意见草案。

54.2　中小企业疫苗开发援助

新型预防性和治疗性疫苗通常由小型公司或作为学术周边产品开发。欧盟内部已经设立了多项激励措施，帮助开发商通过错综复杂的监管和科学要求。最显著的是在欧洲药品管理局设立了小微企业和中型企业办公室（中小企业办公室），同时也在一些国家药品管理局设立了该机构。中小企业办公室针对中小企业在上市前授权程序中遇到的主要财务和行政障碍提供了激励措施，具有中小企业地位的公司（由欧洲药品管理局指定）将从中受益。例如，可以从中小企业办公室的监管援助中受益，并获得科学咨询和其他科学服务的费用减免，而市场授权申请和相关检查费用则可以推迟缴纳。此外，先进治疗药品可以在实际提交市场授权申请之前申请质量/非临床数据认证。

另一个与新型疫苗和先进治疗药品开发商早期互动的重要工具是与欧洲药品管理局创新工作组（ITF）召开简要新闻发布会。创新工作组会议可让开发者和监管者就程序和科学问题进行非正式交流和对话。早期互动也可以通过与科学咨询工作小组（SAWP）（为科学咨询请求做准备）或罕见药品委员会（为罕见药品地位申请做准备）秘书处举行会议的形式进行[5]。

54.3 市场授权申请的结构和内容

《通用技术文件》规定了对市场授权申请结构和内容的要求，并为欧洲、日本和美国的申请提供了统一的结构和格式[6]。

在欧洲，2001/83/EC 指令附件 1 中规定了实施通用技术文件格式的法律规定，规定了本附件（包括疫苗）中规定的各类药品质量、非临床和临床文件的技术要求[7]。对于疫苗和先进治疗药品等具有特定特征的医药产品，需要对这些要求进行调整。提供量身定制的要求不仅能够解决这些产品的高度特殊性，而且可以采用灵活的方法应对生物技术领域技术和科学的快速发展。附件 1 中对市场授权申请内容提出了高水平的技术要求，在产品开发各个方面的具体规章和科学准则中进一步证实了这一点。

对于所有药品、疫苗和先进治疗药品，开发商必须向欧洲药品管理局儿科委员会（PDCO）提交儿科调查计划供审查和批准。儿科调查计划应描述计划对儿童进行的临床研究，包括研究拟议时间和适合儿童的配方调整。《儿科条例》（欧共体）第 1901/2006 号和第 1902/2006 号[8,9]要求，应尽早（最好是在第一阶段临床试验后不久）向欧洲药品管理局提交执行计划，并在提交市场授权申请时检查是否符合拟订执行计划。

市场授权申请还必须包括详细的风险管理计划（RMP），在计划中规定一套药物警戒活动和干预措施，用于识别、鉴定、预防或最小化与药品相关的风险，包括评估这些干预措施的有效性。新药物警戒法于 2012 年 7 月生效[10,11]，并引入了关于安全更新定期报告、风险管理计划、安全信号和授权后安全研究（PASS）等新规定。此外，欧盟层面的文献监测和产品安全审查工具也是该警戒法的一部分。欧洲药品管理局成立了药物警戒风险评估委员会（PRAC），其任务之一是对风险管理计划进行评估。新药物警戒法提供了进一步指导，已经以良好药物警戒操作（GVP）模块的形式发布[12]。良好药物警戒操作模块涵盖了通过欧洲药品管理局集中授权的药品以及国家级授权药品。全套 16 个最终模块计划于 2013 年年初推出。

市场授权申请还需要以环境风险评估（ERA）的形式，对医药产品造成的潜在环境风险进行评估[13]。

54.4 市场授权申请要求

抗原性质差异很大，包括复杂的病毒或微生物颗粒、合成多肽病毒、重组蛋白、病毒样颗粒、免疫调节抗体、基因治疗或基于细胞的产品。各种疫苗指南论述了疫苗的开发，特别是用于预防传染性疾病的抗原。

抗癌药品评价指南论述了治疗性癌症疫苗的开发[14]。

佐剂疫苗已经在欧盟获得许可，业内对为现有疫苗和新型疫苗开发的新型佐剂非常感兴趣。事实上，该领域相当复杂，新型佐剂的性质和作用方式也相当广泛。欧盟已经发布关于佐剂开发和监管批准的指南[15,16]。开发新型佐剂的一个重要方面是要表明其确实能增强对抗原的免疫反应，并具有相关临床益处。新型佐剂的安全性也是一个重要因素。

54.4.1 质量要求

市场授权申请的质量部分应说明药品（疫苗）的全面特征，从种株及细胞库的建立和特征说明，到疫苗最终配方制造过程的详细说明、制造过程中使用的所有原材料和成分说明，以及制造过程中和产品（疫苗）本身应用的所有质量控制测试的说明和验证。质量部分还应解决疫苗生产的一致性和疫苗的稳定性，并说明疫苗适当且有效的效价测定。

欧盟没有针对疫苗质量要求的通用指南；然而，这些要求与大多数生物药品的要求并无区别。指南适用于天花疫苗[17]、流感疫苗[18-20]、重组病毒载体疫苗[21]和DNA疫苗[22]等一些特定疫苗。流感疫苗和DNA疫苗的指南目前正在修订之中[23,24]。

54.4.2　非临床指南

关于预防传染病的疫苗，一般包括《临床前药理和临床前用药指南和毒理学试验指导说明》涵盖的方面[25]。对于预防传染病的疫苗，一般而言，《临床前药理和临床前用药指南和毒理学试验指导说明》所涵盖的内容，并不需要涵盖这些疫苗的所有传统非临床发展计划。例如，疫苗一般不需要药物动力学；然而，根据疫苗的不同，非临床免疫原性或挑战性研究可能具有相关性。含有佐剂的疫苗[15]、天花疫苗[17]和病毒重组活载体疫苗[21]可获得更具体的非临床指南。

与预防性疫苗相比，"治疗性疫苗"只有有限的非临床指南；然而，根据特定类型疫苗的特性，上述指南中的一些信息可能具有价值和适用性。关于基于细胞的医药产品[26]或抗癌药物[27]的具体指南可能对这类疫苗有用。

54.4.3　临床指南

《欧盟疫苗临床评估指南》[28]全面阐述了传染病预防性疫苗临床开发的要求。该指南还通过《病毒重组活载体疫苗质量、非临床和临床指南》的临床部分进行了补充[21]。

在任何新疫苗的开发中，应在临床开发计划中收集足够的免疫原性数据，包括免疫反应特征、适当剂量和进度研究、免疫持久性和加强剂量的考虑。此外，对于载体疫苗，应解决对载体原有免疫的测定和特征。

疫苗通常不需要药代动力学研究。然而，当采用新的递送系统或疫苗含有新型佐剂或赋形剂时，这种研究可能适用。

并非所有疫苗都需要和/或可行证明保护效力，因为这将受到目标传染病流行率和特征的影响。如果已经建立临床验证的相关保护，免疫原性研究可能被认为具有充分性，如乙型肝炎，破伤风或白喉疫苗。

疫苗功效反映了常规使用过程中的直接（疫苗诱导）和间接（人群相关）保护。因此，可以在授权后阶段评估疫苗功效，这也是非常可取的。

迄今为止，对治疗性疫苗临床发展的指导很有限。然而，治疗性疫苗免疫原性的评估应遵循与任何其他疫苗相同的原则[28]。为证明治疗性疫苗的有效性，请参考特定疾病的具体指南(如有)。

抗癌药物产品评价指南涉及治疗性抗癌疫苗的临床开发[14]。目标人群的选择以及结果评价的讨论都考虑到了疫苗的特殊性。传统细胞毒性化合物在有益生物活动或临床效果出现前就导致疾病恶化，与之相比，诱导有效的免疫和临床反应可能还需要更多的时间。

参考文献

[1] Notice to Applicants Volume 2A-procedures for marketing authorisation. Chapter 4: centralised procedure. http: //ec. europa. eu/health/fi les/eudralex/vol--2/a/chap4rev200604_en. pdf.

[2] Notice to Applicants Volume 2A-procedures for marketing authorisation. Chapter 2: mutual recognition. http: //ec. europa. eu/health/fi les/eudralex/vol-2/a/ vol2a_chap2_2007-02_en. pdf.

[3] Directive 2009/120/EC amending Directive 2001/83/EC of the European Parliament and of the Council on the Community code relating to medicinal products for human use as regards advanced therapy medicinal products. http: //ec. europa. eu/health/fi les/eudralex/ vol-1/dir_2009_120/dir_2009_120_en. pdf.

[4] Regulation (EC) No 1394/2007 of the European Parliament and of the Council of 13 November 2007 on advanced therapy medicinal products and amending Directive 2001/83/EC and Regulation (EC). http: // ec. europa. eu/health/files/eudralex/vol-1/ reg_2007_1394/reg_2007_1394_en. pdf.

[5] Klug, B., et al.: Regulatory structures for gene therapy medicinal products in the European Union. Methods Enzymol. 507, 337-354 (2012).

[6] Notice to Applicants Volume 2 B-presentation and format of the dossier-Common Technical Document (CTD). http: //ec. europa. eu/health/fi les/eudralex/vol--2/b/update_200805/ctd_05-2008_en. pdf.

[7] Annex 1 to directive 2001/83/EC-on medicinal products for human use. http: //ec. europa. eu/health/fi les/eudralex/ vol-1/dir_2001_83_cons2009/2001_83_cons2009_en. pdf.

[8] Regulation (EC) No 1901/2006 of the European Parliament and of the Council of 12 December 2006 on medicinal products for paediatric use and amending Regulation (EEC) No 1768/92, Directive 2001/20/ EC, Directive 2001/83/EC and Regulation (EC) No 726/2004. http: //ec. europa. eu/health/fi les/eudralex/ vol-1/reg_2006_1901/reg_2006_1901_en. pdf.

[9] Regulation (EC) No 1902/2006 of the European Parliament and of the Council of 20 December 2006 amending Regulation 1901/2006 on medicinal products for paediatric use. http: //ec. europa. eu/health/fi les/ eudralex/vol-1/reg_2006_1902/reg_2006_1902_en. pdf.

[10] Commission Regulation (EU) No 1235/2010 of the European Parliament and of the Council of 15 December 2010 amending, as regards pharmacovigilance of medicinalproducts for human use, Regulation (EC) No 726/2004 laying down Community procedures for the authorisation and supervision of medicinal products for human and veterinary use and establishing a European Medicines Agency, and Regulation (EC) No 1394/ 2007 on advanced therapy medicinal products. http: //ec. europa. eu/health/fi les/ eudralex/vol-1/reg_ 2010_1235/reg_2010_1235_en. pdf.

[11] Directive 2010/84/EU of the European Parliament and of the Council of 15 December 2010 amending, as regards pharmacovigilance, Directive 2001/83/EC on the Community code relating to medicinal products for human use. http: //ec. europa. eu/health/fi les/eudralex/ vol-1/dir_2010_84/dir_2010_84_en. pdf and corrigendum. http: //ec. europa. eu/health/fi les/eudralex/vol-1/ dir_2010_84_cor/dir_2010_84_cor_ en. pdf.

[12] Good pharmacovigilance practices. http: //www. ema. europa. eu/ema/index. jsp? curl=pages/regulation/ document_listing/document_listing_000345. jsp&mid=WC0b01ac058058f32c.

[13] Guideline on the environmental risk assessment of medicinal products for human use. EMEA/CHMP/ SWP/ 4447/00 corr. 1. http: //www. ema. europa. eu/ docs/en_GB/document_library/Scientific_guideline/2009/ 10/WC500003978. pdf.

[14] Guideline on the evaluation of anticancer medicinal products in man. CPMP/EWP/205/95 Rev. 4. http: // www. ema. europa. eu/docs/en_GB/document_library/ Scientifi c_guideline/2013/01/WC500137128. pdf.

[15] Guideline on adjuvants in vaccines for human use. EMEA/CHMP/VEG/134716/2004. http: //www. ema. europa. eu/docs/en_GB/document_library/Scientifi c_guideline/2009/09/WC500003809. pdf.

[16] Explanatory note on immunomodulators for the guideline on adjuvants in vaccines for human use. CHMP/ VWP/244894/2006. http: //www. ema. europa. eu/docs/en_GB/document_library/Scientifi c_guideline/ 2009/09/WC500003810. pdf.

[17] Note for guidance on the development of vaccinia virus-based vaccines against smallpox. CPMP/1100/02. http: // www. ema. europa. eu/docs/en _ GB/document _ library/ Scientifi c _ guideline/2009/09/ WC500003900. pdf.

[18] Note for guidance on the harmonisation of requirements for infl uenza vaccines. CPMP/BWP/214/96. h ttp: // www. ema. europa. eu/docs/en _ GB/document _ library/ Scientifi c _ guideline/2009/09/ WC500003945. pdf.

[19] Note for guidance on cell-culture-inactivated infl uenza vaccines. Annex to the note for guidance on the harmonisation of requirements for infl uenza vaccines. CPMP/BWP/2490/00. http: //www. ema. europa. eu/ docs/en_GB/document_library/Scientific_guideline/2009/09/WC500003877. pdf.

[20] Points to consider on the development of live attenuated infl uenza vaccines. EMEA/CPMP/BWP/1765/99. h ttp: // www. ema. europa. eu/docs/en _ GB/document _ library/ Scientifi c _ guideline/2009/09/ WC500003899. pdf.

[21] Guideline on quality, non-clinical and clinical aspects of live recombinant viral vectored vaccines. EMA/

CHMP/VWP/141697/2009. http: //www. ema. europa. eu/docs/en _ GB/document _ library/Scientifi c _ guideline/2010/08/WC500095721. pdf.

[22] Note for guidance on the quality, preclinical and clini - cal aspects of gene transfer medicinal products. CPMP/BWP/3088/99. http: //www. ema. europa. eu/ docs/en_GB/document_library/Scientific_guideline/ 2009/10/WC500003987. pdf.

[23] Concept paper on the revision of guidelines for infl uenza vaccines. EMA/CHMP/VWP/734330/2011. ht-tp: // www. ema. europa. eu/docs/en _ GB/document _ library/ Scientifi c _ guideline/2011/10/ WC500115612. pdf.

[24] Concept paper on guidance for DNA vaccines. EMEA/CHMP/308136/2007. http: //www. ema. europa. eu/ docs/en_GB/document_library/Scientifi c_guideline/2012/03/WC500124898. pdf.

[25] Note for guidance on pre - clinical safety evaluation of biotechnology - derived pharmaceuticals (ICH S6). CPMP/SWP/465/95. http: //www. ema. europa. eu/ docs/en_GB/document_library/Scientific_guideline/ 2009/10/WC500004004. pdf.

[26] Guideline on human cell - based medicinal products. EMEA/CHMP/410869/2006. http: //www. ema. europa. eu/docs/en_GB/document_library/Scientifi c_guideline/2009/09/WC500003894. pdf.

[27] Note for guidance on non-clinical evaluation for anti-cancer pharmaceuticals (ICH S9). EMEA/CHMP/ ICH/ 646107/2008. http: //www. ema. europa. eu/ema/ index. jsp? curl = pages/regulation/general/general _ content_000400. jsp&mid=WC0b01ac0580029570.

[28] Guideline on clinical evaluation of new vaccines. EMEA/CHMP/VWP/164653/2005. http: //www. ema. eu-ropa. eu/docs/en_GB/document_library/Scientifi c_ guideline/2009/09/WC500003870. pdf.

（郑海学译）

第 55 章 美国对兽医疫苗认证和许可的监管要求

Louise M. Henderson 和 Ada Mae Lewis①

摘要

位于爱荷华州艾姆斯的美国农业部兽医生物制品中心（CVB）对用于诊断、预防和治疗动物疾病的本土兽医生物制品具有监管管辖权。监管权限由 1913 年《病毒-血清毒素法（VSTA）[1]（1985 年修订）确立，该法案要求，美国境内所有可用的兽医生物制剂均须纯净、安全、有效（非廉价、危险、受污染或有害）。条例见《联邦规则汇编》第 9 篇第 5 章的第 101~122 部分。兽医服务备忘录（VS Memo）在线提供详细的要求和更具体的指导，该网站提供了相关指导文件的链接[3]。值得注意的是，美国农业部仅对专门用于预防和治疗动物疾病的兽医疫苗拥有监管权；食品和药品管理局（FDA）对用于其他目的的兽医疫苗（例如，针对某些激素或疼痛受体的疫苗）拥有监管管辖权。本章概述了美国农业部监管管辖范围内对疫苗的许可（美国生产的产品）和允许（美国境外生产，但在美国销售的产品）的法规和指南。

55.1 监管流程概述

位于美国爱荷华州艾姆斯的农业部兽医生物制品中心（CVB）对用于诊断、预防和治疗动物疾病的本土兽医生物制品具有监管管辖权。监管权限由 1913 年《病毒-血清毒素法（VSTA）[1]（1985 年修订）确立，该法案要求美国境内所有可用的兽医生物制剂都是纯净、安全和有效（非廉价、危险、受污染或有害）。条例见《联邦规则汇编》第 9 篇第 5 章的第 101~122 部分[2]。兽医服务备忘录（VS Memo）在线提供了详细的要求和更具体的指导，该网站提供了相关指导文件的链接[3]。值得注意的是，美国农业部仅对专门用于预防和治疗动物疾病的兽医疫苗拥有监管权；食品和药物（FDA）对用于其他目的的兽医疫苗（例如，针对某些激素或疼痛受体的疫苗）具有监管权限。本章将重点介绍美国农业部监管管辖范围内的疫苗。

兽医生物制品中心由两个职能部门构成：①政策、评估和许可部（PEL）负责制定监管政策、评估许可申请、评估支持许可申请的所有数据、在许可前评估拟用产品的风险、授予美国兽医生物制品机构许可证、授予美国兽医生物制品许可证（针对美国生产的产品）、授予美国兽医生物制品许可证（针对进口到美国的兽医生物制品）、在兽医生物制品中心实验室进行测试，以及生产用于测试许可生物制品的某些标准试剂；②检验与合规部负责制定和执行设施标准、记录保存、制造商人员资格审查和质量控制、检查制造生产和测试场所和一系列产品上市前的批准。

在美国销售的所有兽医疫苗均须具有美国兽医生物制品许可证（在许可机构内生产）或美国兽医生物许可证[4]。所有标签在使用前必须经过批准，确保对疫苗提供的声明与兽医生物制

---

① L. M. Henderson，PhD（✉）（亨德森咨询有限责任公司，兽医生物制剂顾问有限责任公司，美国艾奥瓦州艾姆斯，E-mail：lmhenderson@ consultantsforveterinarybiologics. com）

A. Lewis，PhD（刘易斯生物制品公司，兽医生物制剂顾问有限责任公司，美国艾奥瓦州艾姆斯）

品中心提供并批准的科学数据一致，且无虚假或误导性[5,6]。总体来说，无论由哪个国家生产，生产要求都非常相似[7,8]。如果要进口，许可证持有者必须安排专人在美国的许可设施中负责进口疫苗的接收和检疫，然后才能发布销售。只要符合进口国的要求，美国许可疫苗就可以从美国出口。此外，未经美国农业部许可的疫苗可以根据 1996 年《食品和药物管理局出口改革和加强法》出口；这些未经许可的产品必须符合进口公司的要求，符合生产质量管理规范要求或其他国际标准，并且标签上没有美国生物制品机构许可证[9,10]。

生物技术衍生疫苗[11]与传统衍生疫苗（活疫苗、改良活疫苗、灭活疫苗和亚单位疫苗）规定相同，尽管种株的特性包括在分子水平上对同一性和纯度的评价；唯一附加的相关规定是《国家环境政策法》[12]，该条例要求在释放到环境中前，应对具有复制能力的转基因活疫苗进行彻底的风险评估。

申请人不需提交完整档案；相反地，许可过程具有高度互动性。申请人将在政策、评估和许可部中指定一名职员（审查员），作为申请人的主要联系人。审查员负责在整个过程中与申请人沟通，在申请人开始对宿主动物进行必要的研究之前，就每次提交的资料提供反馈，并对拟定方案进行批准。在获得许可证或执照之前，兽医生物制品中心必须批准实验疫苗的生产和销售[13,14]。在对美国动物进行研究之前，兽医生物制品中心将审查之前提交的数据，确保在宿主动物中进行预期研究不会有不适当的风险；因此，初步安全性和稳定性数据必须为预期用途下的产品安全提供合理保证。申请人通过许可流程后，将指派一名生物制品专家（检查员）与公司合作，确保设施、记录保存系统和人员符合要求。这提供了一种在流程早期识别问题的方法，可以尽早缓解问题。

兽医生物制品中心本身不受生产质量管理规范的监管，但确实包括对同样问题的监督。由制造商编写并经兽医生物制品中心批准的疫苗生产大纲中规定了每种产品的详细要求，制造商必须遵守。这为疫苗的所有成分、种株、过程、条件、制造、储存和测试提供了监管监督。工艺必须控制在允许的时间、温度、酸碱度、测试结果等范围内，并由符合条件的人员进行监督。在每一系列疫苗（作为同质批次生产和测试的单一批次疫苗）发布之前，需要对测试结果进行 CVB 审查。这让我们相信，每一系列产品都是由合格人员根据获批制造设施中的既定要求生产的，并符合既定标准。每个系列的纯度、功效和安全性测试结果提交给兽医生物药品中心；如果可以接受，CVB-IC 将发布此系列产品（允许制造商出售该系列）[15]。兽医生物药品中心实验室可以在发布前对任何序列进行测试，如果在测定期间报告了不良事件，则可以在测定结束时对该序列进行检测。

55.2　总体要求

设施　为确保美国兽医疫苗生产符合要求，制造商必须满足非常具体的要求[16]。必须向兽医生物药品中心提交蓝图、计划和图例[17]。审查工厂文件并检查生产现场，确保符合所有标准[18]。对空气处理和工作流程进行评估，确保疫苗在最大限度降低污染风险的条件下生产。生产设施引入前，所有活性药剂必须经过批准。禁止使用某些外来动物疾病（FAD）制剂。成分必须满足特定要求，将生产过程中带入污染的风险降至最低。检查质量控制记录是否符合要求，检查设施是否符合规定[19]。

制造方法　兽医生物药品中心并没有提供具体的制造方法；相反地，由制造商提供详细的生产纲要，再由兽医生物药品中心审查和批准[20,21]。生产纲要详述了疫苗生产中使用的成分、方法、储存条件和测试要求。所有产品系列在发布前都必须经过纯度、安全性和功效测试[22]。进程内的程序必须经过充分验证（即设备必须经过校准，过程必须足以满足预期目标）[23]。生产大纲的修订必须获得批准后方可生产。生产大纲中详细说明的高度控制制造过程在需要检查的设备中应使用符合特定标准的成分，然后证实系列之间的变化是最小，产品系列具有纯净性、安全性和有效性。

55.3 具体要求

种株和细胞 美国监管体系的关键组成部分之一是建立和鉴定用于制造兽医疫苗的种株和/或细胞[24]。疫苗采用经批准的种株和/或细胞生产。虽然不是必需的，但几乎所有许可疫苗都是由获批种株和细胞制造的。种株是细菌或病毒病原体，细胞是为病毒和其他专性细胞内微生物的生长而建立的。种株和细胞属于单批次培养物，其纯度和性质已经制造商高度鉴定，纯度和特性已由兽医生物药品中心实验室验证。当使用原代细胞时，制造商必须测试每批原代细胞的纯度和特性。

对于基因工程疫苗，必须提交其他详细信息，说明种株的遗传和表型特征，包括用于生产重组种株的所有遗传材料来源、用于生产重组种株工艺，以及种株开发期间和之后的测试结果。信息必须以摘要信息格式（SIF）提供[25]。摘要信息格式专用于种株的预期用途。一类重组种株用于不可复制的产品。包括所有用于灭活（死）疫苗、亚单位疫苗、不能在目标宿主动物中复制的自杀载体和单克隆抗体（用于治疗、预防或诊断用途）。二类重组种株是没有外来核酸插入的活基因缺失微生物。三类重组种株用于活性引导疫苗，并提供充足的数据，以便在获得授权进行现场安全研究之前进行彻底的风险评估。

成分 在美国生产许可兽医疫苗所使用的所有成分必须符合质量标准，而且不能带入风险[26]。动物性成分必须证明来自可接受的来源，很小或没有牛海绵状脑病（BSE）的风险[27]。

无菌和安全性 所有系列疫苗在发布前必须证明纯度[28-30]。就灭活疫苗而言，第 9 条规定了所需进行的无菌测试[31]。对于活疫苗，制造商必须证明没有污染。所有系列还必须在少量动物身上进行安全性测试[32-38]。对于某些疫苗，实验室动物可以注射疫苗并观察不良反应。在其他情况下，必须给宿主动物注射该系列药物，并观察不良反应。

55.4 临床前研究

功效测定 所有系列都必须进行功效测试。功效测定旨在与功效相关联；因此，功效测定必须将每一系列疫苗中的免疫原性（抗原）含量与宿主-动物功效研究关联起来。许多疾病需要特定的功效测定，见 CFR 第 9 条[39]。对于改良活疫苗，病毒或细菌数量的计数可能是充分的；对于转基因活疫苗，插入抗原表达的稳定性也必须通过证明。

对于灭活疫苗，兽医生物药品中心允许进行宿主动物或实验室动物功效测定（疫苗接种血清学或接种挑战），但鼓励制造商在体外开发疫苗的定量免疫原性成分，并与宿主动物功效关联起来[40]。利用识别主要免疫原的单克隆抗体已广泛用于此类检测的开发。制造商必须开发和优化试验，并在试验中证明剂量反应。测定必须是敏感的、具体的、翔实的。我们鼓励在每次检测中使用纯化高度特征化的标准和类似产品的引用（如功效序列），并随着时间推移监测检测性能。制造商必须对效价测定进行监控，并证明稳定性和一致性。鼓励在药效研究开始之前提交功效分析验证，让兽医生物药品中心可以在获得宿主动物疗效之前对功效分析进行仔细评估，此时的功效分析将与功效相关。经过仔细规划并具备良好的特征标准后，制造商能够保留有效参考进行完善地检测，提供显示功效和信心的手段，同时减少未来的动物试验。兽医生物药品中心与欧盟和日本于 1996 年开始合作，参与协调兽医药品注册技术要求的国际合作，协调功效测定方法，减少动物试验[41]。

直肠研究 改良后的活病毒疫苗必须在通过宿主动物的 5 个连续直肠后保持致弱。这是为了确保致弱是重要分子突变的结果，在通过宿主动物的过程中不容易逆转，产生使用改良后活疫苗产生的毒性毒株[42]。

稳定性研究 使用根据标签建议储存疫苗的实时研究，必须证明疫苗的稳定性。建立后不需要证明每个系列的稳定性。

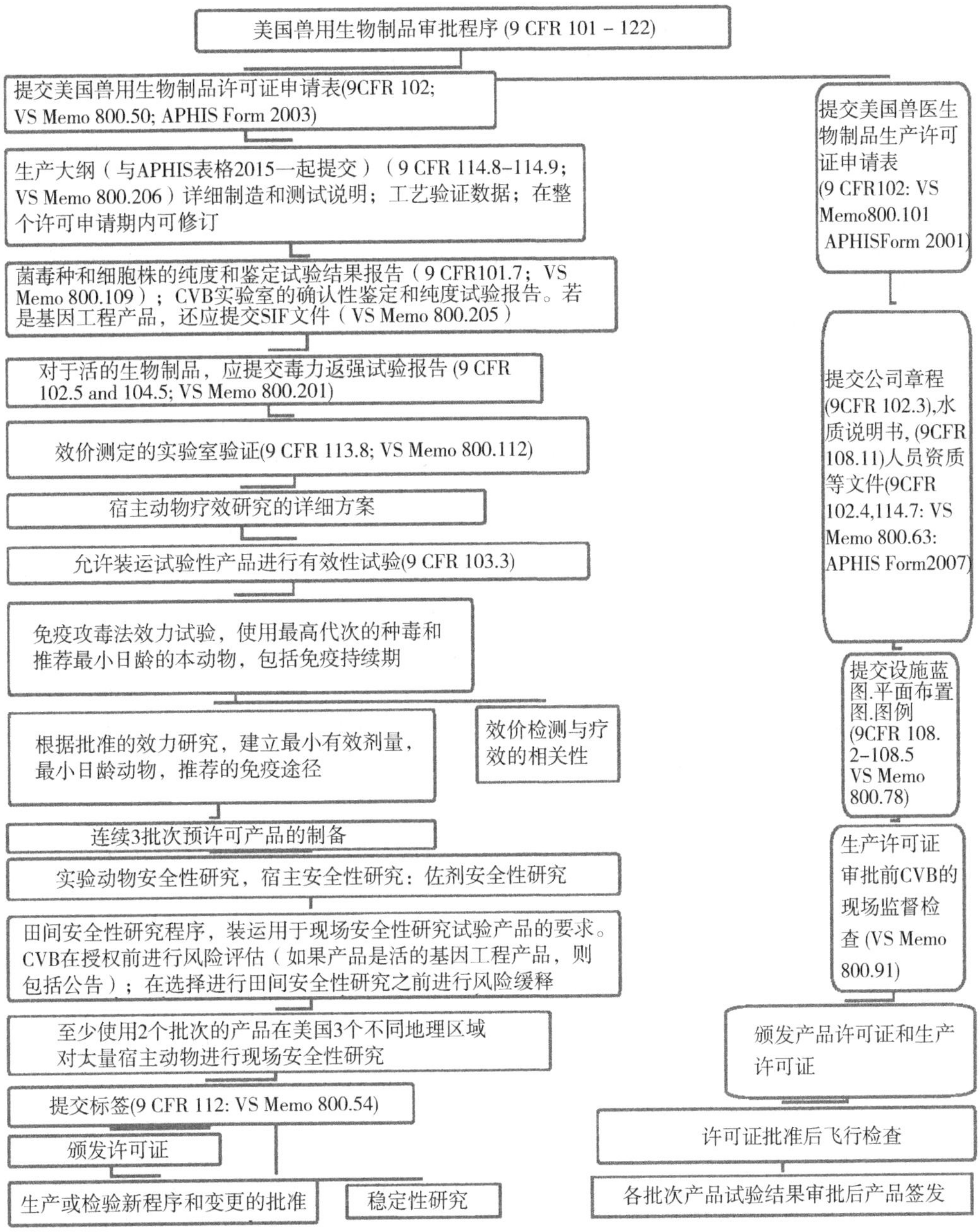

55.5 临床研究

包括与功效测定相关性的疗效研究　必须使用最小有效剂量的疫苗在宿主动物中证明每种疫苗的疗效（由制造商证明）。第 9 条中规定了某些疫苗的疗效研究要求，这可能会提供所需疗效研究的详细规范[43]。许多疫苗没有具体的疗效研究。必须证明每种疫苗在推荐的年龄最小的宿主中有效。在某些情况下，必须使用怀孕的动物；在其他情况下，必须使用新生或幼小动物。这些要求以疫苗的推荐使用方法为基础。研究必须是随机的、盲法的、科学严谨的。疗效研究要求使用随机分配的疫苗和接种安慰剂疫苗的对照动物（对照组）对疾病因子进行挑战。预计这一挑战足以在对照组中产生显著的临床症状。接种疫苗和对照组之间的差异必须具有统计学和临床意义[44]。如果没有已知的挑战模型，可以采用实地研究证明在自然条件下暴露的大量动物的疗效。疗效研究的结果将决定允许的权利要求。其他权利要求需要进行额外研究。具体功效研究的权利要求和研究期间的动物护理可以在网上查到[45,46]。兽医生物药品中心建议遵守《兽药注册国际协调局准则》[47]。

功效测定必须与宿主动物疗效研究相关，最好是在疗效研究开始时就进行疗效系列测试。每次测试系列产品时，通常会用疗效系列产品作参考；将被测序列与功效序列进行比较，必须显示出高于疗效序列的抗原含量。如检测到变质或过期，必须更换参照物。重要的是要注意，疫苗在到期时必须达到或高于系列疗效。对于改良活疫苗，要求连续释放时的细菌或病毒计数高于连续释放时的疗效。

免疫的持续时间研究 对于一些对接种疫苗诱导的免疫持续时间知之甚少的疾病，制造商必须确定接种疫苗的动物受到保护的时长。除非有足够的数据使血清学数据显示免疫力，否则通常需要额外的宿主动物免疫攻毒研究。

田间安全性研究 进行功效研究后，制造商必须使用至少两种系列产品证明美国动物的安全性，这些产品满足所有生产要求的纲要（包括疗效以上的效力）[48]。

这需要在美国至少 3 个地理位置不同的地区为大量动物（通常是数百只）接种疫苗。现场安全试验的目的是在不同畜牧环境或具有不同地方病条件的动物中进行。在接种疫苗后的几周内观察动物的不良反应。对于食用动物，还必须确定残留物清除率，为停用建议提供支持。

讨论 本章广泛概述了美国兽医疫苗的监管过程。更多详细信息可在兽医生物药品中心网站上查阅[49]。还有大量非常具体的需求在本次讨论中没有提到。法规往往是一般性的，兽医服务备忘录中有更具体的指导方针。提出要求的基础是确保美国境内或从美国销售的兽医疫苗的纯度、安全性、功效和疗效。随着疾病或新发动物疾病等新问题的暴发，我们会对要求进行调整。在完成疗效或功效测试之前，紧急情况下可以有条件地批准疫苗的许可证。当有科学依据或在出现危及动物或人类健康的紧急情况时，可给予例外规定。每年春季都会举办培训，我们鼓励制造商和国外监管官员参加，以了解更多关于该流程的信息，并与兽医生物药品中心人员会面[50]。

参考文献

[1] Virus-Serum-Toxin Act, US Code of Federal Regulations 21, Parts 151-159. http://www. aphis. usda. gov/animal_health/vet_biologics/vb_regs_and_guidance. shtml (2012).

[2] Code of Federal Regulations, Title 9 (9 CFR). http:// www. aphis. usda. gov/animal_health/vet_biologics/vb_regs_and_guidance. shtml (2012).

[3] CVB Web Site. http://www. aphis. usda. gov/animal_health/vet_biologics/vb_regs_and_guidance. shtml.

[4] 9 CFR 102, Licenses for biological products. http:// www. aphis. usda. gov/animal_health/vet_biologics/vb_cfr. shtml (2012).

[5] 9 CFR 112, Packaging and Labeling. http://www. aphis. usda. gov/animal_health/vet_biologics/vb_cfr. shtml (2012).

[6] Veterinary Services Memorandum 800. 54, Guidelines for Preparation and review ofLabeling Materials (February 17, 1986). http://www. aphis. usda. gov/animal_health/vet_biologics/vb_vs_memos. shtml (2012).

[7] Veterinary Services Memorandum 800. 50, Basic License Requirements and Guidelines for Submission of Materials in Support of Licensure (February 9, 2011). http://www. aphis. usda. gov/animal_health/ vet_biologics/vb_vs_memos. shtml (2012).

[8] Veterinary Services Memorandum 800. 101, U. S. Veterinary Biological Product Permits for Distribution and Sale (2012). http://www. aphis. usda. gov/animal_ health/vet_biologics/vb_vs_memos. shtml (2012).

[9] 21 CFR, Section 382. http://www. fda. gov/ (2012).

[10] Veterinary Services Memorandum 800. 94, FDA Export Reform and Enhancement Act of 1996, http://

www. aphis. usda. gov/animal_health/vet_biologics/ vb_vs_memos. shtml (2012).

[11] V eterinary Services Memorandum 800. 205, General Licensing Considerations: Biotechnology-derived Veterinary Biologics Categories I, II, III (May 28, 2003). http: //www. aphis. usda. gov/animal_health/vet_biologics/vb_vs_memos. shtml (2012).

[12] National Environmental Policy Act. http: //ceq. hss. doe. gov/ (1969).

[13] 9 CFR 103, Experimental production, distribution, and evaluation of biological products prior to licensure. http: //www. aphis. usda. gov/animal_health/vet_biologics/vb_regs_and_guidance. shtml (2012).

[14] Veterinary Services Memorandum 800. 67, Shipment of Experimental Veterinary Biological Products (November 16, 2011) h ttp: //www. aphis. usda. gov/ animal_ health/vet _biologics/vb _ vs _ memos. shtml (2012).

[15] Veterinary Services Memorandum 800. 53, Release of Biological Products (April 2, 2001). http: //www. aphis. usda. gov/animal_health/vet_biologics/vb_vs_memos. shtml (2012).

[16] 9 CFR 108, Facilities requirements for licensed estab-lishments. http: //www. aphis. usda. gov/animal_ health/ vet_biologics/vb_regs_and_guidance. shtml (2012).

[17] Veterinary Services Memorandum 800. 78, Preparation and Submission of Facilities Documents (November 11, 2010). http: //www. aphis. usda. gov/animal_health/ vet_biologics/vb_vs_memos. shtml (2012).

[18] Veterinary Services Memorandum 800. 91, Categories of Inspection for Licensed Veterinary Biologics Establishments (May 13, 1999). http: //www. aphis. usda. gov/animal_ health/vet _ biologics/vb _ vs _ memos. shtml (2012).

[19] 9 CFR 115, Inspections. http: //www. aphis. usda. gov/ animal_ health/vet _ biologics/vb _ regs _ and _ guidance. shtml (2012).

[20] 9 CFR 114, Production requirements for biological products. http: //www. aphis. usda. gov/animal_health/ vet_biologics/vb_regs_and_guidance. shtml (2012).

[21] Veterinary Services Memorandum 800. 206, General Licensing Considerations: Preparing Outlines of Production for Vaccines, Bacterins, Antigens, and Toxoids (April 13, 2012). http: //www. aphis. usda. gov/ animal_ health/vet_biologics/vb_vs_memos. shtml (2012).

[22] 9 CFR 113. 6, Animal and Plant Health Inspection Service Testing. http: //www. aphis. usda. gov/animal_ health/vet_biologics/vb_cfr. shtml (2012).

[23] 9 CFR 109, Sterilization and pasteurization at licensed establishments. http: //www. aphis. usda. gov/ animal_ health/vet_biologics/vb_regs_and_guidance. shtml (2012).

[24] 9 CFR 101. 7, Seed organisms. http: //www. aphis. usda. gov/animal_health/vet_biologics/vb_regs_and_ guidance. shtml (2012).

[25] Risk analysis for veterinary biologics. http: //www. aphis. usda. gov/animal_health/vet_biologics/vb_sifs. shtml (2011).

[26] 9 CFR 113. 50 through 113. 55, Ingredient require-ments. http: //www. aphis. usda. gov/animal_health/ vet_ biologics/vb_regs_and_guidance. shtml (2012).

[27] Veterinary Services Memorandum 800. 51, Additives in animal biological products (November 7, 2007). h ttp: //www. aphis. usda. gov/animal_health/vet_biologics/vb_vs_memos. shtml (2012).

[28] 9 CFR 113. 27, Detection of extraneous viable bacte-ria and fungi except in live vaccine. http: //www. aphis. usda. gov/animal_health/vet_biologics/vb_regs_and_guidance. shtml (2012).

[29] 9 CFR 113, 27, Detection of extraneous viable bacte-ria and fungi in live vaccines. http: //www. aphis. usda. gov/animal_health/vet_biologics/vb_regs_and_guidance. shtml (2012).

[30] 9 CFR 113. 28, Detection of mycoplasma contamina-tion. http: //www. aphis. usda. gov/animal_health/ vet_biologics/vb_regs_and_guidance. shtml (2012).

[31] 9 CFR sterility tests. http: //www. aphis. usda. gov/animal_health/vet_biologics/vb_regs_and_guidance. shtml (2012).

[32] 9 CFR 113. 33, Mouse safety tests. http: //www. aphis. usda. gov/animal_health/vet_biologics/vb_regs_ and_guidance. shtml (2012).

[33] 113. 38, Guinea pig safety test. http: //www. aphis. usda. gov/animal_health/vet_biologics/vb_regs_and_

guidance. shtml (2012).
[34] 9 CFR 113. 39, Cat safety tests. http: //www. aphis. usda. gov/animal_health/vet_biologics/vb_regs_and_guidance. shtml (2012).
[35] 9 CFR 113. 40, Dog safety tests. http: //www. aphis. usda. gov/animal_health/vet_biologics/vb_regs_and_guidance. shtml (2012).
[36] 9 CFR 113. 41, Calf safety test. http: //www. aphis. usda. gov/animal_health/vet_biologics/vb_regs_and_guidance. shtml (2012).
[37] 9 CFR 113. 44, Swine safety test. http: //www. aphis. usda. gov/animal_health/vet_biologics/vb_regs_and_guidance. shtml (2012).
[38] 9 CFR 113. 45, Sheep safety test. http: //www. aphis. usda. gov/animal_health/vet_biologics/vb_regs_and_guidance. shtml (2012).
[39] 9 CFR 113. 8, in vitro tests for serial release. http: // www. aphis. usda. gov/animal_health/vet_biologics/vb_regs_and_guidance. shtml (2012).
[40] Veterinary Services Memorandum 800. 112, guide-lines for validation of in vitro potency assays (August 29, 2011). http: //www. aphis. usda. gov/animal_health/ vet_biologics/vb_vs_memos. shtml (2012).
[41] International Cooperation on Harmonisation of Technical Requirements for Registration of Veterinary Medicinal Products. http: //www. vichsec. org/ (1996).
[42] Veterinary Services Memorandum 800. 201, general licensing considerations: backpassage studies (June 25, 2008). http: //www. aphis. usda. gov/animal_health/ vet_biologics/vb_vs_memos. shtml (2012).
[43] 9 CFR 113, Standard requirements. http: //www. aphis. usda. gov/animal_health/vet_biologics/vb_regs_and_guidance. shtml (2012).
[44] Veterinary services memorandum 800. 202, general licensing considerations: effi cacy studies (June 14, 2002). http: //www. aphis. usda. gov/animal_health/vet_biologics/vb_vs_memos. shtml (2012).
[45] Veterinary Services Memorandum 800. 200, general licensing considerations: study practices and documentation (June 14, 2002). http: //www. aphis. usda. gov/animal_health/vet_biologics/vb_vs_memos. shtml (2012).
[46] Veterinary Services Memorandum 800. 301, good clinical practices (July 26, 2001). http: //www. aphis. usda. gov/animal_health/vet_biologics/vb_vs_memos. shtml (2012).
[47] Veterinary Services Memorandum 800. 207, general licensing considerations: Target Animal Safety (TAS) studies prior to product licensure-VICH guideline 44 (July 6, 2010) http: //www. aphis. usda. gov/animal_health/vet_biologics/vb_vs_memos. shtml (2012).
[48] Veterinary Services Memorandum 800. 204, general licensing considerations: fi eld safety studies (March 16, 2007). http: //www. aphis. usda. gov/animal_health/ vet_biologics/vb_vs_memos. shtml (2012).
[49] Center for Veterinary Biologics. http: //www. aphis. usda. gov/animal_health/vet_biologics/vb_regs_and_guidance. shtml (2012).
[50] Institute for International Cooperation in Animal Biologics, Veterinary Biologics Training Program. http: // www. cfsph. iastate. edu/IICAB/ (2012).

(何继军译)

第 56 章　欧盟兽医疫苗的监管要求

Rhona Banks①

摘要

兽医疫苗必须获得欧盟相关主管部门的授权。必须提交一份市场授权（MA）申请，其中应包括一份证明疫苗质量以及目标物种安全性和有效性档案。特别是，必须证明疫苗对用户、环境和消费者具有安全性。

56.1　监管流程概述

兽医疫苗必须获得欧盟相关主管部门的授权。必须提交一份市场授权（MA）申请，其中应包括一份证明疫苗质量以及目标物种的安全性和有效性档案。特别是，还必须证明疫苗对用户、环境和消费者具有安全性。

所有疫苗必须按照 2001/82/EC 指令[1]和 2004/28/EC 指令[2]，以及 2009/09 指令[3]进行注册，并且具备可用的统一指令[4]，其中描述了支持产品权利要求所需的研究和信息。

申请市场授权有多种不同的途径，在《申请人须知》第 6A 卷中进行了概述[5]。

• 国家/适用于一个成员国。然后，该许可证可以成为其他成员国相互认可的基础。原产国将担任参考成员国。

• 相互认可——特定程序中可以包含若干个成员国，认可参考成员国的国家许可证。

• 分散流程——同时适用于多个成员国，选择 1 个成员国作为参考成员国。

• 集中流程——向欧洲药物管理局申请由兽医生物药品中心进行单一评估。如果兽医生物药品中心给出肯定意见，欧盟委员会将作出最终决定，并向所有欧盟成员国授权。210 天流程纲要见表 56.1。

表 56.1　集中程序概述（210 个评估日）。总结自《申请人须知》[6]

天	行动
xx	将档案提交给环境管理机构、起草人和联合起草人——推荐提交日期见欧洲药品管理局网站
xx	欧洲药品管理局向申请人发送档案接收确认书，并在接收后 10 个工作日内完成确认
xx	欧洲药品管理局将在验证程序结束时启动该流程。在流程启动后的 1 个月内，将档案所需部分发送给所有兽医生物药品中心成员
1	流程启动——由欧洲药品管理局制定并发送给申请人的时间表
70	提交给联合起草人、兽医生物药品中心成员和欧洲药品管理局秘书处的起草人评估报告

① R. Banks，MIBiol，PhD（Triveritas 有限公司生物制品监管事务部，英国布兰普顿，E-mail：rhona. banks @ triveritas. com）

（续表）

天	行动
85	将联合起草人对起草人评估报告的审查发送给起草人、兽医生物药品中心成员和欧洲药品管理局秘书处。欧洲药品管理局秘书处将这些报告发送给申请人（明确报告尚不代表兽医生物药品中心的立场）
100	起草人、联合起草人、其他兽生物医药品中心成员和欧洲药品管理局收到兽医生物药品中心成员的评论
115	起草人和联合起草人向兽医生物药品中心成员+欧洲药品管理局发送问题清单草案（包括总体结论和科学数据概述）
120	兽医生物药品中心批准问题清单、总体结论和科学数据审查，并由欧洲药品管理局发送给申请人。到第 120 天，兽医生物药品中心在必要时批准药品生产质量管理规范检查请求停表期。申请人有 3 个月的答复时间，但如果有正当理由，可以再申请 3 个月
121	提交回复（包括修订后的产品特性概述（SPC）、标签和英文包装说明书文本）。提交回复目标日期见欧洲药品管理局网站重启期——项目经理与起草人和联合起草人协商后，发送修订后的时间表（通常如下）
160	起草人和联合起草人向兽医生物药品中心成员、欧洲药品管理局和申请人发送联合反应评估报告（注意，评估报告并不代表兽医生物药品中心的立场）。如果适用，应进行药品生产质量管理规范检查
170	兽医生物药品中心成员向起草人和联合起草人、欧洲药品管理局和其他兽医生物药品中心成员提交意见的截止日期
180	兽医生物药品中心讨论意见草案（产品特性概述、标签和包装说明书），并决定是否需要申请人进行口头解释。如需要任何口头解释，应进入停表期，为申请人留出准备时间
181	重启期和口头解释（如需要）。项目经理向申请人发送最新的英文版产品特性概述和产品简介
210	批准兽医生物药品中心意见+兽医生物药品中心评估报告
211	向申请人发送兽医生物药品中心意见+兽医生物药品中心评估报告
211~237	产品特性概述和产品简介及管理程序的翻译

欧盟所有兽医疫苗的生产设施必须符合药品生产质量管理规范，同时疫苗必须接受欧盟机构的定期检查[7]。

采用 DNA 重组技术生产转基因生物（GMO）的医药产品必须通过集中流程[8]授权，同时此类疫苗还需遵守下文 2001/18/EC 指令[9]中规定的附加要求。

采用欧洲药品管理局认为的其他重大创新生物技术方法开发的医药产品，也可以通过集中流程获得授权。包括 DNA 病毒/核糖核酸和蛋白质在内的新型疫苗很可能有资格通过集中流程应用，因此这条路线是本章的主要重点内容。关于创新产品，可以在欧洲药管理品局网站上找到非常有用的问题和答案[10]。

这种监管体系听起来可能很复杂，但欧洲药品理管理局希望鼓励拥有创新产品的申请者尽早与之接触，因此将提供一些帮助。此外，可以就具体问题向科学咨询工作小组（SAWP）提出申请[11]。请注意，申请人应该陈述其对某个特定问题的观点，并对每个问题提供支持数据。科学咨询工作小组将考虑这些问题，并给出其基本原理的答案——所以，他们可能会同意申请人的立场，也可能不同意。

56.2 总体要求

市场授权持有者（MAH）必须在欧洲经济区成立，因此，这是申请市场授权前首先要解决的问题。市场授权申请应包括完整的档案，涵盖产品质量、安全和功效等所有相关方面。因此，产品整体评估是一次性完成的，而不是分阶段批准。

档案的第1部分必须包含产品特性概述，包括产品简要说明及介绍、所有权利要求、指征、使用方法和安全警告。档案应包括实验性试验、现场试验或参考书目，支持产品特性概述中的所有陈述内容。

欧洲药品管理局网站上有一个应使用的产品特性概述模板，还有一个产品特性概述免疫学指南[12]，给出了每个部分应包括的项目以及涵盖特定情况的一些标准措辞。

档案还必须包含药物警戒系统（DDPS）的详细说明，以及应遵守的欧洲药品局指南[13]。获得授权后，该产品将在药物警戒系统下接受上市后监控，并有规定的时间和方法报告进行定期安全更新。

欧盟会关注授权兽医药品的可获得性，特别是针对限制性使用/稀有物种（MUMS）。欧洲药品管理局已经制定了一个专门针对免疫限制性使用和稀有物种产品数据要求的指南[14]，其中概述了研究中可以认同的某些精简部分。有一份在各种动物物种中引发特定疾病的传染因子清单，目前的产品不能完全控制这些疾病。应向欧洲药品管理局提出申请，认可拟定产品具有限制性使用/稀有物种状态。兽医生物药品中心将考虑支持性文件，并授予限制性使用/稀有物种地位，还可能对急需产品给予财政激励。

委员会出台了条例[15]，旨在促进开发人用或兽用药物的小微型和中型企业的创新和新药开发。欧洲市场管理局的中小企业办公室可向中小企业提供援助，包括实际帮助和多项财政激励措施。欧洲药品管理局网站上的用户指南内容丰富，提供申请中小企业地位方法的详细信息[16]。

鼓励以电子方式提交档案，并在欧洲药品管理局网站上提供[17]问答文件。

56.3 具体要求

免疫兽药产品（IVMPs）的质量、安全性和疗效的总体要求和具体要求见2009/09指令[3]，其中还有效地给出了提交档案的格式。《欧洲药典》（Ph Eur）有适用于疫苗的通用和专用专著。应遵守通用专著《兽医用疫苗0062》[18]（注：该专著已于2013年初更新）。此外，还有关于安全性[19]和疫苗疗效[20]的专著。特别是特定类型的兽医疫苗有一些专门的专著，包括如何证明针对特定疾病的安全性和有效性的概述［例如，关于绵羊巴氏杆菌疫苗（灭活）的专著2072］[21]。请注意，兽医疫苗有许多专门的《欧洲药典》专著，因此在为欧洲开发新的免疫兽药产品时，必须要考虑这些内容。

新版欧洲药品局指南[22]已取代《申请人须知》第7卷中关于免疫兽药产品的几个通用和专用指南，最新免疫兽药产品指南见欧洲药品管理局网站[23]。新指南明确概述了质量、安全性和疗效方面的重要项目，这些项目在指令和《欧洲药典》专著中可能并不明显。然而，新指南中的信息并不如旧指南那么详细，所以申请人需要仔细考虑产品的开发方式。

含有或由转基因生物组成的疫苗应遵守第2001/18/EC指令的附加规定[9]。这建立了一个逐步批准的流程，因此在将任何转基因生物发布到环境中或投放市场之前，必须对人类健康和环境风险进行评估。对于现场试验等试验性发布，必须向拟发布转基因生物的成员国主管机构提交通知。通知必须包括技术档案，提供必要的信息进行环境风险评估。请注意，在大多数情况下，授予国家许可证并批准现场试验的监管部门与批准转基因生物试验性发布的主管部门不同。因此，需要与特定成员国的1个以上部门联系，以获得发布转基因生物和进行现场试验的许可。

提交转基因疫苗申请时，必须包含主管部门对有意发布到环境中的书面同意书副本。必须提供2001/28/EC指令附件3A中详述的转基因生物完整技术档案[9]和广泛的环境风险评估。这些信息应列入单独一卷档案中，便于向主管部门提供有关转基因生物的资料。欧洲药品管理局制定了一个特定标准流程，对在整个档案的正常评估过程中与国家转基因生物主管部门的磋商进行管理[24]。

56.4 临床前指南

欧洲药品管理局网站上有许多针对体外多药耐药监测系统的科学指南[25]，涵盖质量、安全和疗效问题。关于活疫苗菌株的毒性逆转试验[26]和验证活疫苗和灭活疫苗安全性的试验方法[27]，兽医注册国际协调局有两份指南（适用于欧盟、美国和日本）。通常预计，安全性实验室研究将按照药物非临床研究质量管理规范（GLP）的要求进行，功效实验室研究应按照药物临床试验质量管理规范的要求进行。在任何情况下，研究都应该按照方案进行适当规划，并完整记录在研究报告中。不论是否给出正面的研究结果，特定产品的所有数据均应该包括在档案中。

许多兽医疫苗可能含有1种以上活性成分（称为联合疫苗）。此外，还可以同时管理两个或多个单独授权的免疫兽药产品（视为一种联合药品）。目前正在编写一份关于此类疫苗评估方法的准则草案[28]。

可能影响所有类型兽医疫苗的一个问题是，母源抗体是否会干扰免疫反应的形成，从而降低疗效。有必要证明，在推荐接种疫苗时，母源抗体的水平已经下降，或者证明母源抗体不会干扰对致命疾病的疗效[29]。

对于用于食用动物的兽医疫苗，有必要考虑是否含有根据规例（欧共体）第470/2009号第1条[30]规定的最高残留限量（MRL）的药理活性物质。应该考虑免疫兽药产品中含有的任何佐剂或赋形剂，一份欧洲药品管理局指南[31]中列出了不属于规则范围的物质，还有一份欧洲药品管理局指南[32]列出了向可接受列表添加物质所需的数据。

56.5 临床指南

实验室试验通常应有现场试验的支持，现场试验使用的是根据市场授权申请中的生产工艺制备的疫苗批次。由于要使用商业批次，因此可在同一现场试验中进行安全性和有效性研究[3]。这些试验应当按照药物临床试验质量管理规范的要求进行，并且应当对结果进行预先确定的统计评估。在理想情况下，现场试验应在欧盟的两个不同地点进行（最好是在不同的国家，尽管在任何地方都没有明确规定）。一些《欧盟药典》专著还包括现场试验设计总结，例如新生仔猪大肠杆菌病疫苗（灭活）0962[33]。

最近还有一份关于评估鱼类疫苗安全性和有效性的研究设计指南[34]，其中包括实验室和现场试验的信息。

结论

欧盟对临时管理计划的监管要求相当复杂，但这些信息均为公开信息，在欧洲药品管理局网站上为集中申请流程提供了大量帮助。对于其他申请途径，药品管理局兽药主管网站也提供了有用的信息[35]。

在开发新型疫苗时，有必要向欧洲药品管理局申请科学建议，因为这可能会影响支持市场授权申请所需的研究。该建议对兽医生物药品中心不具约束力，但后者一般都会遵守，所以如果申请人遵守科学建议，成功概率会更高。

有一项相对较新的举措，即在档案第1部分中纳入效益风险评估的可能性。一些疫苗可能会对动物或人类产品的质量和安全构成一定的风险。有了这一举措，人们就能对这些疫苗的积极疗效进行科学评估[36]。效益风险评估一直是监管机构评估的一部分，但越来越多的人意识到，如果没有针对特定疾病的预防性或治疗性治疗，就存在潜在风险。

本章只能给出与欧盟兽医疫苗监管流程相关的法规、指令、指南、建议和一般经验的宽泛概念。各种兽医疫苗已经覆盖了许多不同的目标物种，希望能够通过新技术进一步推广。每种兽医疫苗通常都有独特的问题需要解决，因此有必要在早期阶段利用欧洲药品管理局的科学建议流程，并考虑免疫管理系统方法的相关性，与欧盟监管机构进行项目沟通。

参考文献

[1] Directive 2001/82/EC of the European Parliament and of the Council of 6 November 2001 on the Community Code relating to veterinary medicinal products. h ttp：// eur－lex. europa. eu/LexUriServ/LexUriServ. do? uri=OJ：L：2001：311：0001：0066：en：pdf（2001）.

[2] Directive 2004/28/EC of the European Parliament and of the Council of 31 March2004 amending Directive 2001/82/EC on the Community Code relating to veterinary medicinal products. http：//ec. europa. eu/ health/ files/eudralex/vol－5/dir_2004_28/ dir_2004_28_en. pdf（2004）.

[3] Commission Directive 2009/9/EC of 10 February 2009 amending Directive 2001/82/EC of the European Parliament and of the Council on the Community code relating to medicinal products for veterinary use. Offi cial Journal L 44，14/2/2009. p. 10－61. http：//ec. europa. eu/health/files/eudralex/vol－5/dir_2009_9/ dir_2009_9_en. pdf（2009）.

[4] Consolidated Directive 2001/82/EC as amended by Directive 2004/28/EC of the European Parliament and of the Council of 31 March 2004，Directive 2009/9/ EC of 10 February 2009，Regulation（EC）No 470/2009 of the European Parliament and of the Council of 6 May 2009，Directive 2009/53/EC of the European Parliament and of the Council of 18 June 2009 and Regulation（EC）No 596/2009 of the European Parliament and of the Council of 18 June 2009 http：//ec. europa. eu/health/fi les/eudralex/vol－5/ dir_2001_82_cons2009/ dir_2001_82_cons2009_en. pdf（2009）.

[5] EudraLex－Volume 6 notice to applicants and regula－tory guidelines for medicinal products for veterinary use. Volume 6A procedures for marketing authorisation. http：//ec. europa. eu/health/fi les/eudralex/vol－6/a/ vol6a_chap1_2007－01_en. pdf（2007）.

[6] EudraLex－Volume 6 notice to applicants and regula－tory guidelines for medicinal products for veterinaryuse. Volume 6A procedures for marketing authorisation chapter 4 centralised procedure. http：//ec. europa. eu/ health/files/eudralex/vol－6/a/vol6a_ chap4_2006_05_en. pdf（2006）.

[7] Commission Directive 91/412/EEC of 23 July 1991 laying down the principles and guidelines of good manufacturing practice for veterinary medicinal products http：//ec. europa. eu/health/fi les/eudralex/vol－5/ dir_1991_412/dir_1991_412_en. pdf（1991）.

[8] Regulation（EC）No 726/2004 of the European Parliament and of the Council of 31 March 2004 laying down Community procedures for the authorisation and supervision of medicinal products for human and veterinary use and establishing a European Medicines Agency. http：//eur－lex. europa. eu/LexUriServ/LexUriServ. do? uri=OJ：L：2004：136：000：0033：EN：PDF（2004）.

[9] Directive 2001/18/EC of the European Parliament and of the Council of 12 March2001 on the deliberate release into the environment of genetically modifi ed organisms and repealing Council Directive 90/220/ EEC. http：// ec. europa. eu/health/fi les/eudralex/vol－1/ dir_2001_18/dir_2001_18_en. pdf（2001）.

[10] European Medicines Agency：Veterinary presubmis－sion Q & A for innovative products. http：//www. emea. europa. eu/ema/index. jsp? curl = pages/regulation/general/general _ content _ 000171. jsp&mid = WC0b01ac058002d9ab（2012）.

[11] European Medicines Agency：EMA/CVMP/172329/2004－Rev. 3. Guidance for companies requesting scientifi c advice. h ttp：//www. emea. europa. eu/docs/en_GB/document_library/Regulatory_and_procedural_guideline/ 2009/10/WC500004147. pdf（2012）.

[12] EudraLex－Volume 6 notice to applicants and regula－tory guidelines for medicinal products for veterinaryuse. Volume 6C regulatory guidelines. Summary of the product characteristics SPC－immunologicals. h ttp：//ec.

europa. eu/health/fi les/eudralex/vol-6/c/spc_immunologicals_rev3_08-06-2007_en. pdf (2007).

[13] European Medicines Agency: EMA/1531641/2010. Pre-submission instruction on the detailed description of the pharmacovigilance system of a marketing authorisation holder; to be submitted with a marketing authorisation application for a veterinary medicinal product http: //www. ema. europa. eu/docs/en_GB/document_library/Regulatory_and_procedural_guideline/2012/03/WC500123503. pdf (2012).

[14] European Medicines Agency: EMA/CVMP/IWP/123243/2006-Rev. 2. Guideline on Data requirements for Immunological veterinary medicinal products intended for minor use or minor species/limited markets. http: //www. ema. europa. eu/docs/en_ GB/document _ library/Scientific _ guideline/2010/04/WC500089628. pdf (2010).

[15] Regulation (EC) No 2049/2005 of 15 December 2005 laying down, pursuant to Regulation (EC) No 726/2004 of the European Parliament and of the Council, rules regarding the payment of fees to, and the receipt of administrative assistance from, the European Medicines Agency by micro, small and medium-sized enterprises. http: //eur-lex. europa. eu/ LexUriServ/LexUriServ. do? uri = OJ: L: 2005: 329: 0004: 0007: EN: PDF (2005).

[16] European Medicines Agency: EMA/204919/2010 user guide for micro, small and medium-sized enterprises (SMEs) http: //www. ema. europa. eu/docs/ en_GB/document_library/Regulatory_and_procedural_guideline/2009/10/WC500004134. pdf (2010).

[17] European Medicines Agency: EMA/613295/2011-Rev. 1. Electronic submission of veterinary dossiers questions and answers. http: //www. emea. europa. eu/ docs/en_ GB/document _ library/Other/2012/05/WC500127591. pdf (2012).

[18] European Pharmacopoeia 7th Ed. Vaccines for veteri-nary use. Monograph 04/2013: 0062. European Pharmacopoeia, Strasbourg, France. (2013).

[19] European Pharmacopoeia 7th Ed. Evaluation of safety of veterinary vaccines and immunosera Monograph 04/2013: 50206. European Pharmacopoeia, Strasbourg, France. (2013).

[20] European Pharmacopoeia 7th Ed. Evaluation of effi cacy of veterinary vaccines and immunosera. Monograph 04/2008: 50207. (2013).

[21] European Pharmacopoeia 7th Ed. Pasteurella vaccine (inactivated) for sheep. Monograph 04/2013: 2072. (2013).

[22] European Medicines Agency: EMA/CVMP/IWP/206555/2010. Guideline on requirements for the production and control of immunological veterinary medicinal products. http: //www. ema. europa. eu/docs/ en_GB/document_library/Scientific_guideline/2012/06/WC500128997. pdf (2012).

[23] European Medicines Agency: Immunologicals g uidelines. h ttp: //www. ema. europa. eu/ema/index. jsp?curl = pages/regulation/general/general_content_000194. jsp&mid = WC0b01ac058002dd33 (2012).

[24] European Medicines Agency. SOP/V/4012. Evaluation of veterinary medicinal products containing or consisting of Genetically Modifi ed Organisms. http: //www. ema. europa. eu/docs/en _ GB/document _ library/Standard_Operating_Procedure_-_SOP/2009/09/WC500003079. pdf (2012).

[25] European Medicines Agency: Immunologicals guidelines. http: //www. emea. europa. eu/ema/index. jsp?curl = pages/regulation/general/general_content_000194. jsp&mid = WC0b01ac058002dd33 (2012).

[26] VICH Topic GL41. EMEA/CVMP/VICH/1052/2004. Guideline on target animal safety: examination of live veterinary vaccines in target animals for absence of reversion to virulence. http: //www. ema. europa. eu/docs/en_GB/document_library/Scientific_guideline/2009/10/WC500004552. pdf (2007).

[27] VICH Topic GL44 Step 7. EMEA/CVMP/ VICH/359665/2005. Guideline on target animal safety for veterinary live and inactivated vaccines. http: //www. ema. europa. eu/docs/en_GB/document_library/Scientifi c_guideline/2009/10/WC500004553. pdf (2008).

[28] European Medicines Agency: EMA/CVMP/ IWP/594618/2010 draft. Guideline on the requirements for combined vaccines and associations of immunological veterinary medicinal products (IVMPs). http: //www. ema. europa. eu/docs/en _ GB/ document _ library/Scientific _ guideline/2011/11/ WC500118227. pdf (2011).

[29] European Medicines Agency:. EMA/CVMP/IWP/439467/2007 Refl ection paper on the demonstration of a

possible impact of maternally derived antibodies on vaccine effi cacy in young animals. http：//www. ema. europa. eu/docs/en_GB/document_library/ Scientific_guideline/2010/03/WC500076626. pdf（2010）.

[30] Regulation（EC） No 470/2009 of the European Parliament and of the Council of 6 May 2009 laying down Community procedures for the establishment of residue limits of pharmacologically active substances in foodstuffs of animal origin, repealing Council Regulation（EEC） No 2377/90 and amending Directive 2001/82/EC ofthe European Parliament and of the Council and Regulation（EC） No 726/2004 of the European Parliament and of the Council. http：//ec. europa. eu/health/files/eudralex/vol-5/reg_2009-470/ reg_470_2009_en. pdf（2009）.

[31] European Medicines Agency：EMA/CVMP/519714/2009-Rev. 12 Substances considered as not falling within the scope of Regulation（EC） No 470/2009, with regards to residues of veterinary medicinal products in foodstuffs of animal origin. http：//www. ema. europa. eu/docs/en_GB/document_library/Regulatory_and_procedural_guideline/2009/10/WC500004958. pdf（2012）.

[32] European Medicines Agency：EMA/CVMP/516817/2009 Guideline on data to be provided in support of a request to include a substance in the list of substances considered as not falling within the scope of Regulation（EC） No 470/2009. http：//www. ema. europa. eu/docs/en_GB/document_library/Scientific_guideline/2010/11/WC500099149. pdf（2010）.

[33] European Pharmacopoeia 7th Ed. Neonatal piglet colibacillosis vaccine（inactivated）. Monograph 04/2013：0962.（2013）.

[34] European Medicines Agency：EMA/CVMP/IWP/314550/2010 Guideline on the design of studies to evaluate the safety and effi cacy of fi sh vaccines. http：//www. ema. europa. eu/docs/en_GB/document_library/Scientifi c_guideline/2011/11/WC500118226. pdf（2011）.

[35] Heads of Medicine Agency：Veterinary medicines. http：//www. hma. eu/veterinary. html（2012）.

[36] European Medicines Agency：EMEA/CVMP/248499/2007 recommendation on the evaluation of the benefi t-risk balance of veterinary medicinal products. http：//www. ema. europa. eu/docs/en_GB/document_library/Other/2009/10/WC500005264. pdf（2009）.

（何继军译）